Lecture Notes in Computer Science 4288

Commenced Publication in 1973
Founding and Former Series Editors:
Gerhard Goos, Juris Hartmanis, and Jan van Leeuwen

Tetsuo Asano (Ed.)

Algorithms and Computation

17th International Symposium, ISAAC 2006
Kolkata, India, December 18-20, 2006
Proceedings

 Springer

Volume Editor

Tetsuo Asano
JAIST, Japan Advanced Institute of Science and Technology
1-1, Asahidai, Nomi, Ishikawa 923-1292, Japan
E-mail: t-asano@jaist.ac.jp

Library of Congress Control Number: 2006937541

CR Subject Classification (1998): F.2, C.2, G.2-3, I.3.5, C.2.4, E.5

LNCS Sublibrary: SL 1 – Theoretical Computer Science and General Issues

ISSN   0302-9743
ISBN-10   3-540-49694-7 Springer Berlin Heidelberg New York
ISBN-13   978-3-540-49694-6 Springer Berlin Heidelberg New York

Springer is a part of Springer Science+Business Media

springer.com

© Springer-Verlag Berlin Heidelberg 2006
Printed in Germany

Typesetting: Camera-ready by author, data conversion by Scientific Publishing Services, Chennai, India
Printed on acid-free paper   SPIN: 11940128   06/3142   5 4 3 2 1 0

Preface

ISAAC 2006, the 17th International Symposium on Algorithms and Computation took place in Kolkata, India, December 18–20, 2006. It has been held in Tokyo (1990), Taipei (1991), Nagoya (1992), Hong Kong (1993), Beijing (1994), Cairns (1995), Osaka (1996), Singapore (1997), Taejon (1998), Chennai (1999), Taipei (2000), Christchurch (2001), Vancouver (2002), Kyoto (2003), Hong Kong (2004), and Hainan (2005).

The symposium provided a forum for researchers working in algorithms and the theory of computation from all over the world. In response to our call for papers, we received 255 submissions. The task of selecting the papers in this volume was carried out by our Program Committee and many other external reviewers. After a thorough review process and PC meeting, the committee selected 73 papers. We hope all accepted papers will eventually appear in scientific journals in a more polished form. Two special issues, one of *Algorithmica* and one of the *International Journal of Computational Geometry and Applications*, with selected papers from ISAAC 2006 are in preparation.

The best paper award was given for "Algorithmic Graph Minor Theory: Improved Grid Minor Bounds and Wagner's Contraction" by Erik Demaine, MohammadTaghi Hajiaghayi and Ken-ichi Kawarabayashi. The best student paper award was given for "Branching and Treewidth Based Exact Algorithms" by Serge Gaspers, Fedor Fomin and Saket Saurabh. Two eminent invited speakers, Kazuo Iwama, Kyoto University, Japan, and Tamal K. Dey, The Ohio State University, USA, also contributed to this volume.

I would like to thank the Conference Chair, Bhargab B. Bhattacharya and the Organizing Chair, Subhas C. Nandy, for their leadership, advice and help on crucial matters concerning the conference. I would like to thank the Program Committee and many external reviewers for their great efforts in the review process. I also thank the Advisory Committee members of ISAAC for their continuous encouragement.

Finally, I would like to acknowledge the EasyChair system, which is a free conference management system that is flexible, easy to use, and has many features to make it suitable for various conference models. Without the help of EasyChair, we could not have finished our review process within the deadline of notification.

December 2006

Tetsuo Asano
Program Chair
ISAAC 2006

Preface

Organization

Program Committee

Hee-Kap Ahn, Sejong Univ., Korea
Tetsuo Asano(Chair), JAIST, Japan
Mikhail Atallah, Purdue Univ., USA
Chanderjit Bajaj, Univ. Texas Austin, USA
Sergey Bereg, Univ. Texas Dallas, USA
Somenath Biswas, IIT Kanpur, India
Tamal K. Dey, The Ohio State Univ., USA
Benjamin Doerr, Max Planck Institute, Germany
Subir Ghosh, TIFR, India
Mordecai J. Golin, HKUST, Hong Kong
John Iacono, Polytechnic Univ., USA
Chuzo Iwamoto, Hiroshima Univ., Japan
Rolf Klein, Univ. Bonn, Germany
Sang-Ho Lee, Ewha Womens Univ., Korea
Kazuhisa Makino, Univ. Tokyo, Japan
Pat Morin, Carleton Univ., Canada
Stephan Näher, Univ. Trier, Germany
Subhas Chandra Nandy, ISI, Kolkata, India
Giri Narasimhan, Florida International Univ., USA
Ashwin Nayak, Univ. Waterloo, Canada
Kunsoo Park, Seoul National Univ., Korea
Md. Saidur Rahman, Bangladesh Univ. Eng.&Tech., Bangladesh
Desh Ranjan, New Mexico State Univ., USA
Peter Sanders, Univ. Karlsruhe, Germany
Sandeep Sen, IIT Kharagpur, India
Sung Yong Shin, KAIST, Korea
Hisao Tamaki, Meiji Univ., Japan
Akihisa Tamura, Keio Univ., Japan
Seinosuke Toda, Nihon Univ., Japan
Takeshi Tokuyama, Tohoku Univ., Japan
Ryuhei Uehara, JAIST, Japan
Gabriel Valiente, Tech. Univ. Catalonia, Spain
Alexander Wolff, Univ. Karlsruhe, Germany

Organizing Committee

Partha Bhowmik, Bengal Engineering and Science University
Arindam Biswas, Bengal Engineering and Science University

Nabendu Chaki, Calcutta University
Debesh Das, Jadavpur University
Sandip Das, Indian Statistical Institute
Parthasarathi Dasgupta, Indian Institute of Management
Rajat De Indian, Statistical Institute
Partha Pratim Goswami, Kalyani University
Arobindo Gupta, Indian Institute of Technology, Kharagpur
Susmita Sur-Kolay, Indian Statistical Institute
Mandar Mitra, Indian Statistical Institute
Pabitra Mitra, Indian Institute of Technology, Kharagpur
Subhamoy Moitra, Indian Statistical Institute
Dipti Prasad Mukherjee, Indian Statistical Institute
Krishnendu Mukhopadhyaya, Indian Statistical Institute
Subhas C. Nandy (Chair), Indian Statistical Institute
Sudeb K. Pal, Indian Institute of Technology, Kharagpur
Subhashis Pal, Indian Statistical Institute

Sponsors

1. Department of Science and Technology, Govt. of India
2. Council of Scientific and Industrial Research, Govt. of India
3. Reserve Bank of India
4. Department of Information Technology of the Govt. of West Bengal
5. Capgemini Consulting India Private Limited
6. Tata Consultancy Services
7. IBM India Software Laboratory
8. Cognizant Technology Solutions
9. Anshin Software

External Referees

Ashkan Aazami	Mahmoud Fouz	Shashank Mehta
V. Arvind	Satoshi Fujita	Atsuko Miyaji
Greg Plaxton	Xavier Goaoc	Elena Mumford
Adam Klivans	Robert Görke	Mridul Nandi
Surender Baswana	Masud Hasan	Martin Nöllenburg
Binay Bhattacharya	André Hernich	Enrico Pontelli
Ai Chen	Xiuzhen Huang	M. Sohel Rahman
Siu-Wing Cheng	Toshiya Itoh	Dana Ron
Joseph Cheriyan	Naoki Katoh	Kouichi Sakurai
Brian Cloteaux	Hartmut Klauck	Thomas Schank
Daniel Delling	Jochen Konemann	Anil Seth
Feodor F. Dragan	Martin Kutz	Gurdip Singh
Sandor Fekete	SN Maheshwari	Steve Tate

Antoine Vigneron
David Wood,
Binhai Zhu
Manindra Agarwal
Sang Won Bae
Vinay Siddahanavalli
Samrat Goswami
Michael Baur
Marina Blanton
Xiaomin Chen
Otfried Cheong
Taenam Cho
Sandip Das
Roman Dementiev
Khaled Elbassioni
Stephen Fenner
Tobias Friedrich
Stefan Funke
Mordecai Golin
Michel Habib
Herman Haverkort
Martin Holzer
H.-K. Hwang
Jesper Jansson
Bastian Katz
Christian Klein
Dariusz Kowalski
Stefan Langerman
Anil Maheshwari

Sascha Meinert
Takaaki Mizuki
Hyeon-Suk Na
Frank Neumann
Sudeb P. Pal
Mihai Prunescu
Bhaskaran Raman
Sasanka Roy
Eli Ben Sasson
Sandeep Sen
Chan-Su Shin
Masakazu Soshi
Mayur Thakur
Yusu Wang
Jian Xia
Martin Kutz
Lars Arge
Amitabha Bagchi
Inderjit Dhillon
Sugata Basu
Marc Benkert
Peter Brass
Jianer Chen
Sang Won Bae
Sunghee Choi
Sajal Das
Jon Derryberry
Will Evans
Henning Fernau

Keith Frikken
Marco Gaertler
Prosenjit Gupta
Sariel Har-Peled
Jing He
Seok Hee Hong
Giuseppe Italiano
Md. Abul Kashem
Akinori Kawachi
Christian Knauer
Amit Kumar
Hing Leung
Steffen Mecke
Damian Merrick
Mitsuo Motoki
Stefan Naeher
Tetsuro Nishino
Sangmin Park
Jaikumar Radhakrishnan
Edgar Ramos
Kunihiko Sadakane
Sanjeev Saxena
Seung-Hyun Seo
Akiyoshi Shioura
Andreas Spillner
Gerhard Trippen
Rephael Wenger
Yan Zhang

Message from the Conference Chair

It was our great pleasure to welcome you to the 17th Annual International Symposium on Algorithms and Computation (ISAAC 2006), which was held for the first time in Kolkata (formerly known as Calcutta), during December 18–20, 2006. This is the second ISAAC meeting organized in India; the first one was held in the city of Chennai in 1999. This symposium provided an excellent opportunity for sharing thoughts among the participants on the recent advances in algorithm design and their manifold applications to emerging areas. Thanks go to the members of the Advisory Committee of ISAAC for their concurrence to hold this symposium in Kolkata.

We would like to express our sincerest thanks to the invited speakers, Kazuo Iwama of the Kyoto University, Japan, and Tamal K. Dey of the Ohio State University, USA, who kindly agreed to speak on the frontier topics in algorithms and computation theory.

We are immensely grateful to Tetsuo Asano of the Japan Advanced Institute of Science and Technology, the Program Chair of the symposium, for compiling an outstanding technical program. On the advice of an excellent Program Committee of international experts, he followed stringent criteria for selecting only the very best technical papers out of a large number of submissions in order to preserve the high quality of the technical program of the symposium.

Our sincerest thanks are due to Sankar K. Pal, Director of the Indian Statistical Institute, for his support in co-sponsoring the symposium and for providing financial and infrastructural support. We also thank Anupam Basu of the Department of Computer Science and Engineering, Indian Institute of Technology, Kharagpur, for endorsing institutional cooperation. We also acknowledge, with thanks, the support we received from the Indian Association of Research in Computing Sciences (IARCS) for co-hosting the symposium. The financial support received from the Department of Science and Technology, Council of Scientific and Industrial Research, Reserve Bank of India, the Department of Information Technology of the Govt. of West Bengal, Capgemini Consulting India Private Limited, Tata Consultancy Services, IBM India Software Laboratory, Cognizant Technology Solutions, and Anshin Software for sponsoring various events, are also thankfully acknowledged.

We are also grateful to the local Organizing Committee for their excellent services that made the symposium a grand success.

We take this opportunity to extend our heartfelt thanks to all the participants, the authors, the reviewers, and the volunteers, who helped us immensely to make this symposium a success. We earnestly hope that the participants of the symposium enjoyed their stay in the wonderful and culturally vibrant city of Kolkata.

December 2006

Bhargab B. Bhattacharya
Indian Statistical Institute, Kolkata, India
Conference Chair
ISAAC 2006

Table of Contents

Session 2A: Approximation Algorithms

Session 2B: Graphs

Session 3A: Computational Geometry

Session 3B: Computational Complexity

Session 4A: Algorithms and Data Structures

Session 4B: Games and Networks

Session 5A: Combinatorial Optimization and Computational Biology

Session 5B: Graphs

Session 6A: Algorithms and Data Structures

Session 6B: Graphs

Session 7A: Approximation Algorithms

Session 7B: Graphs

Session 8A: Combinatorial Optimization and Quantum Computing

Stable Matching Problems

Kazuo Iwama*

School of Informatics, Kyoto University, Kyoto 606-8501, Japan
iwama@kuis.kyoto-u.ac.jp

Stable matching is one of the oldest problems studied from an algorithmic point of view, whose original version is defined as follows: An instance consists of N men, N women, and each person's preference list. A preference list is a totally ordered list including all members of the opposite sex depending on his/her preference. For a matching M between men and women, a pair of a man m and a woman w is called a *blocking pair* if both prefer each other to their current partners. A matching with no blocking pair is called *stable*.

The famous Gale-Shapley algorithm which finds a stable matching in linear time was presented in 1962 and Knuth also considered several related problems in 1976. On the other hand, it has a lot of real-world applications, including the National Resident Matching Program in the US which started in 1952. Thus the problem has a long history of theoretical research and application, but there still remain many interesting problems. For example, for its most general version, allowing both ties and incomplete lists, nothing was known until 1999 for its complexity. Very recently, Iwama, Miyazaki and Yamauchi found its approximation algorithm whose approximation factor is strictly less than two, improving the previous $2 - o(N)$.

This talk explains why this problem is so interesting and deep, from several different angles including the newest result above mentioned, different measures for goodness of matchings, different definitions for the stableness, related problems such as stable roommate problems and stable partition, and open problems.

* Supported in part by Scientific Research Grant, Ministry of Japan, 1609211 and 16300003.

T. Asano (Ed.): ISAAC 2006, LNCS 4288, p. 1, 2006.

Delaunay Meshing of Surfaces

Tamal K. Dey

Department of Computer Science and Engineering, The Ohio State University,
Columbus, Ohio 43210 USA

Abstract. Meshing of surfaces is an ubiquitous problem in many applications of science and engineering. Among different approaches available
for meshing surfaces, Delaunay meshing is often favored because of its
directional independence and good quality in general. As application
varies, so does the input form of the surface to be meshed. We present
algorithms to compute Delaunay meshes for various forms of input surfaces. Specifically, we consider surfaces input with (i) point cloud data,
(ii) implicit equations, and (iii) polyhedra. These algorithms come with
theoretical guarantees and some of them have been successfully implemented. In this talk we detail the algorithms, provide the mathematical
reasoning behind their designs, and show the results of some experiments.

T. Asano (Ed.): ISAAC 2006, LNCS 4288, p. 2, 2006.
© Springer-Verlag Berlin Heidelberg 2006

Algorithmic Graph Minor Theory: Improved Grid Minor Bounds and Wagner's Contraction

Erik D. Demaine[1], MohammadTaghi Hajiaghayi[1,2],
and Ken-ichi Kawarabayashi[3]

[1] MIT Computer Science and Artificial Intelligence Laboratory,
32 Vassar Street, Cambridge, MA 02139, USA
{edemaine, hajiagha}@mit.edu
[2] Department of Computer Science, Carnegie Mellon University,
Pittsburgh, PA 15213, USA
[3] Graduate School of Information Sciences, Tohoku University,
Aramaki aza Aoba 09, Aoba-ku Sendai, Miyagi 980-8579, Japan
k_keniti@dais.is.tohoku.ac.jp

Abstract. We explore the three main avenues of research still unsolved in the algorithmic graph-minor theory literature, which all stem from a key min-max relation between the treewidth of a graph and its largest grid minor. This min-max relation is a keystone of the Graph Minor Theory of Robertson and Seymour, which ultimately proves Wagner's Conjecture about the structure of minor-closed graph properties.

First, we obtain the only known polynomial min-max relation for graphs that do not exclude any fixed minor, namely, map graphs and power graphs. Second, we obtain explicit (and improved) bounds on the min-max relation for an important class of graphs excluding a minor, namely, $K_{3,k}$-minor-free graphs, using new techniques that do not rely on Graph Minor Theory. These two avenues lead to faster fixed-parameter algorithms for two families of graph problems, called minor-bidimensional and contraction-bidimensional parameters. Third, we disprove a variation of Wagner's Conjecture for the case of graph contractions in general graphs, and in a sense characterize which graphs satisfy the variation. This result demonstrates the limitations of a general theory of algorithms for the family of contraction-closed problems (which includes, for example, the celebrated dominating-set problem). If this conjecture had been true, we would have had an extremely powerful tool for proving the existence of efficient algorithms for any contraction-closed problem, like we do for minor-closed problems via Graph Minor Theory.

1 Introduction

Graph Minor Theory is a seminal body of work in graph theory, developed by Robertson and Seymour in a series of over 20 papers spanning the last 20 years. The original goal of this work, now achieved, was to prove Wagner's Conjecture [39], which can be stated as follows: every minor-closed graph property (preserved under taking of minors) is characterized by a finite set of forbidden minors. This theorem has a powerful algorithmic consequence: every minor-closed

T. Asano (Ed.): ISAAC 2006, LNCS 4288, pp. 3–15, 2006.
© Springer-Verlag Berlin Heidelberg 2006

graph property can be decided by a polynomial-time algorithm. A keystone in the proof of these theorems, and many other theorems, is a grid-minor theorem [37]: any graph of treewidth at least some $f(r)$ is guaranteed to have the $r \times r$ grid graph as a minor. Such grid-minor theorems have also played a key role for many algorithmic applications, in particular via the bidimensionality theory (e.g., [20, 13, 15, 11, 18, 17, 19]), including many approximation algorithms, PTASs, and fixed-parameter algorithms.

The grid-minor theorem of [37] has been extended, improved, and re-proved. The best bound known for general graphs is superexponential: every graph of treewidth more than 20^{2r^5} has an $r \times r$ grid minor [43]. This bound is usually not strong enough to derive efficient algorithms. Robertson et al. [43] conjecture that the bound on $f(r)$ can be improved to a polynomial $r^{\Theta(1)}$; the best known lower bound is $\Omega(r^2 \lg r)$. A tight linear upper bound was recently established for graphs excluding any fixed minor H: every H-minor-free graph of treewidth at least $c_H r$ has an $r \times r$ grid minor, for some constant c_H [18]. This bound leads to many powerful algorithmic results on H-minor-free graphs [18, 17, 19].

Three major problems remain in the literature with respect to these grid-minor theorems in particular, and algorithmic graph-minor theory in general. We address all three of these problems in this paper.

First, to what extent can we generalize algorithmic graph-minor results to graphs that do not exclude a fixed minor H? In particular, for what classes of graphs can the grid-minor theorem be improved from the general superexponential bound to a bound that would be useful for algorithms? To this end, we present polynomial grid-minor theorems for two classes of graphs that can have arbitrarily large cliques (and therefore exclude no fixed minors). One class, *map graphs*, is an important generalization of planar graphs introduced by Chen, Grigni, and Papadimitriou [10], characterized via a polynomial recognition algorithm by Thorup [45], and studied extensively in particular in the context of subexponential fixed-parameter algorithms and PTASs for specific domination problems [12, 9]. The other class, *power graphs*, e.g., fixed powers of H-minor-free graphs (or even map graphs), have been well-studied since the time of the Floyd-Warshall algorithm.

Second, even for H-minor-free graphs, how large is the constant c_H in the grid-minor theorem? In particular, how does it depend on H? This constant is particularly important because it is in the exponent of the running times of many algorithms. The current results (e.g., [18]) heavily depend on Graph Minor Theory, most of which lacks explicit bounds and is believed to have very large bounds. (To quote David Johnson [32], "for any instance $G = (V, E)$ that one could fit into the known universe, one would easily prefer $|V|^{70}$ to even constant time, if that constant had to be one of Robertson and Seymour's." He estimates one constant in an algorithm for testing for a fixed minor H to be roughly $2 \uparrow 2^{2^{2^{2 \uparrow (2 \uparrow \Theta(|V(H)|))}}}$, where $2 \uparrow n$ denotes a tower $2^{2^{2^{\cdot^{\cdot^{\cdot}}}}}$ involving n 2's.) For this reason, improving the constants, even for special classes of graphs, and presumably using different approaches from Graph Minors, is an important theoretical and practical challenge. To this end, we give explicit bounds for the case of $K_{3,k}$-minor-free graphs, an important class of apex-minor-free graphs (see, e.g., [4, 7, 26, 27]). Our bounds are not too small but are a vast improvement over previous bounds (in particular, much smaller than $2 \uparrow |V(H)|$); in addition, the proof techniques are interesting in their own right, being disjoint from most of Graph Minors. To the best of our knowledge, this is the only grid-minor theorem with an explicit bound other than for planar graphs [43] and bounded-genus graphs [13]. Our theorem also leads to several algorithms with explicit and improved bounds on their running time.

Third, to what extent can we generalize algorithmic graph-minor results to graph contractions? Many graph optimization problems are closed (only decrease) under edge contractions, but not under edge deletions (i.e., minors). Examples include dominating set, traveling salesman, or even diameter. Bidimensionality theory has been extended to such contraction-closed problems for the case of apex-minor-free graphs; see, e.g., [11, 13, 18, 17, 23]. The basis for this work is a modified grid-minor theorem which states that any apex-minor-free graph of treewidth at least $f(r)$ can be contracted into an "augmented" $r \times r$ grid (e.g., allowing partial triangulation of the faces). The ultimate goal of this line of research, mentioned explicitly in [16, 23], is to use this grid-contraction analog of the grid-minor theorem to develop a Graph Contraction Theory paralleling as much as possible of Graph Minor Theory. In particular, the most natural question is whether Wagner's Conjecture generalizes to contractions: is every contraction-closed graph property characterized by a finite set of excluded contractions? If this were true, it would generalize our algorithmic knowledge of minor-closed graph problems in a natural way to the vast array of contraction-closed graph problems. To this end, we unfortunately disprove this contraction version of Wagner's Conjecture, even for planar bounded-treewidth graphs. On the other hand, we prove that the conjecture holds for outerplanar graphs and triangulated planar graphs, which in some sense provides a tight characterization of graphs for which the conjecture holds.

Below we detail our results and techniques for each of these three problems.

1.1   Our Results and Techniques

Generalized grid-minor bounds. We establish polynomial relations between treewidth and grid minors for map graphs and for powers of graphs. We prove in Section 2 that any map graph of treewidth at least r^3 has an $\Omega(r) \times \Omega(r)$ grid minor. We prove in Section 3 that, for any graph class with a polynomial relation between treewidth and grid minors (such as H-minor-free graphs and map graphs), the family of kth powers of these graphs also has such a polynomial relation, where the polynomial degree is larger by just a constant, interestingly independent of k.

These results extend bidimensionality to map graphs and power graphs, improving the running times of a broad class of fixed-parameter algorithms for these graphs. See Section 4 for details on these algorithmic implications. Our results also build support for Robertson, Seymour, and Thomas's conjecture that all graphs have a polynomial relation between treewidth and grid minors [43].

Indeed, from our work, we refine the conjecture to state that all graphs of tree-width $\Omega(r^3)$ have an $\Omega(r) \times \Omega(r)$ grid minor, and that this bound is tight. The previous best treewidth-grid relations for map graphs and power graphs were given by the superexponential bound from [43].

The main technique behind these results is to use approximate min-max relations between treewidth and the size of a grid minor. In contrast, most previous work uses the seminal approximate min-max relation between treewidth and tangles or between branchwidth and tangles, proved by Robertson and Seymour [42]. We show that grids are powerful structures that are easy to work with. By bootstrapping, we use grids and their connections to treewidth even to prove relations between grids and treewidth.

Another example of the power of this technique is a result we obtain as a byproduct of our study of map graphs: every bounded-genus graph has tree-width within a constant factor of the treewidth of its dual. This is the first relation of this type for bounded-genus graphs. The result generalizes a conjecture of Seymour and Thomas [44] that, for planar graphs, the treewidth is within an additive 1 of the treewidth of the dual, which has apparently been proved in [35, 5] using a complicated approach. Such a primal-dual treewidth relation is useful, e.g., for bounding the change in treewidth when performing operations in the dual. Our proof crucially uses the connections between treewidth and grid minors, and this approach leads to a relatively clean argument. The tools we use come from bidimensionality theory and graph contractions, even though the result is not explicitly about either.

Explicit (improved) grid-minor bounds. We prove in Section 5 that the constant c_H in the linear grid-minor bound for H-minor-free graphs can be bounded by an explicit function of $|V(H)|$ when $H = K_{3,k}$ for any k: for an explicit constant c, every $K_{3,k}$-minor-free graph of treewidth at least $c^k r$ has an $r \times r$ grid minor. This bound makes explicit and substantially improves the constants in the exponents of the running time of many fixed-parameter algorithms from bidimensionality theory [13, 11, 18] for such graphs. $K_{3,k}$-minor-free graphs play an important role as part of the family of apex-minor-free graphs that is disjoint from the family of single-crossing-minor-free graphs (for which there exist a powerful decomposition in terms of planar graphs and bounded-treewidth graphs [41, 20]). Here the *family of $\mathcal{X}$-minor-free graphs* denotes the set of X-minor-free graphs for any fixed graph X in the class $\mathcal{X}$. $K_{3,k}$ is an *apex graph* in the sense that it has a vertex whose removal leaves a planar graph. For $k \geq 7$, $K_{3,k}$ is not a *single-crossing graph* in the sense of being a minor of a graph that can be drawn in the plane with at most one crossing: $K_{3,k}$ has genus at least $(k-2)/4$, but a single-crossing graph has genus at most 1 (because genus is closed under minors).

There are several structural theorems concerning $K_{3,k}$-minor-free graphs. According to Robertson and Seymour (personal communication—see [7]), $K_{3,k}$-minor-free graphs were the first step toward their core result of decomposing graphs excluding a fixed minor into graphs almost-embeddable into bounded-genus surfaces, because $K_{3,k}$-minor-free graphs can have arbitrarily large genus. Oporowski, Oxley, and Thomas [36] proved that any large 3-connected $K_{3,k}$-

minor-free graph has a large wheel as a minor. Böhme, Kawarabayashi, Maharry, and Mohar [3] proved that any large 7-connected graph has a $K_{3,k}$ minor, and that the connectivity 7 is best possible. Eppstein [26, 27] proved that a subgraph P has a linear bound on the number of times it can occur in $K_{3,k}$-minor-free graphs if and only if P is 3-connected.

Our explicit linear grid-minor bound is based on an approach of Diestel et al. [24] combined with arguments in [4, 3] to find a $K_{3,k}$ minor. Using similar techniques we also give explicit bounds on treewidth for a theorem decomposing a single-crossing-minor-free graph into planar graphs and bounded-treewidth graphs [41, 20], when the single-crossing graph is $K_{3,4}$ or K_6^- (K_6 minus one edge). Both proofs must avoid Graph Minor Theory to obtain the first explicit bounds of their kind.

Contraction version of Wagner's Conjecture. Wagner's Conjecture, proved in [39], is a powerful and very general tool for establishing the existence of polynomial-time algorithms; see, e.g., [28]. Combining this theorem with the $O(n^3)$-time algorithm for testing whether a graph has a fixed minor H [38], every minor-closed property has an $O(n^3)$-time decision algorithm which tests for the finite set of excluded minors. Although these results are existential, because the finite set of excluded minors is not known for many minor-closed properties, polynomial-time algorithms can often be constructed [14].

A natural goal is to try to generalize these results even further, to handle all contraction-closed properties, which include the decision versions of many important graph optimization problems such as dominating set and traveling salesman, as well as combinatorial properties such as diameter. Unfortunately, we show in Section 6 that the contraction version of Wagner's Conjecture is not true: there is a contraction-closed property that has no complete finite set of excluded contractions. Our counterexample has an infinite set of excluded contractions all of which are planar bounded-treewidth graphs. On the other hand, we show that the contraction version of Wagner's Conjecture holds for trees, triangulated planar graphs, and 2-connected outerplanar graphs: any contraction-closed property characterized by an infinite set of such graphs as contractions can be characterized by a finite set of such graphs as contractions. Thus we nearly characterize the set of graphs for which the contraction version of Wagner Conjecture's is true. The proof for outerplanar graphs is the most complicated, and uses Higman's theorem on well-quasi-ordering [31].

The reader is referred to the full version of this paper (available from the first author's website) for the proofs. See also [16] for relevant definitions.

2   Treewidth-Grid Relation for Map Graphs

In this section we prove a polynomial relation between the treewidth of a map graph and the size of the largest grid minor. The main idea is to relate the treewidth of the map graph, the treewidth of the radial graph, the treewidth of the dual graph, and the treewidth of the union graph.

Theorem 1. *If the treewidth of the map graph M is r^3, then it has an $\Omega(r) \times \Omega(r)$ grid as a minor.*

This theorem cannot be improved from $\Omega(r^3)$ to anything $o(r^2)$:

Proposition 1. *There are map graphs whose treewidth is $r^2 - 1$ and whose largest grid minor is $r \times r$.*

Robertson, Seymour, and Thomas [43] prove a stronger lower bound of $\Theta(r^2 \lg r)$ but only for the case of general graphs.

3   Treewidth-Grid Relation for Power Graphs

In this section we prove a polynomial relation between the treewidth of a power graph and the size of the largest grid minor. The technique here is quite different, analyzing how a radius-r neighborhood in the graph can be covered by radius-$(r/2)$ neighborhoods—a kind of "sphere packing" argument.

Theorem 2. *Suppose that, if graph G has treewidth at least cr^α for constants $c, \alpha > 0$, then G has an $r \times r$ grid minor. For any even (respectively, odd) integer $k \geq 1$, if G^k has treewidth at least $cr^{\alpha+4}$ (respectively, $cr^{\alpha+6}$), then it has an $r \times r$ grid minor.*

We have the following immediate consequence of Theorems 1 and 2 and the grid-minor theorem of [18] mentioned in the introduction:

Corollary 1. *For any H-minor-free graph G, and for any even (respectively, odd) integer $k \geq 1$, if G^k has treewidth at least r^5 (respectively, r^7), then it has an $\Omega(r) \times \Omega(r)$ grid minor. For any map graph G, and for any even (respectively, odd) integer $k \geq 1$, if G^k has treewidth at least r^7 (respectively, r^9), then it has an $\Omega(r) \times \Omega(r)$ grid minor.*

4   Treewidth-Grid Relations: Algorithmic and Combinatorial Applications

Our treewidth-grid relations have several useful consequences with respect to fixed-parameter algorithms, minor-bidimensionality, and parameter-treewidth bounds.

A *fixed-parameter algorithm* is an algorithm for computing a parameter $P(G)$ of a graph G whose running time is $h(P(G)) n^{O(1)}$ for some function h. A typical function h for many fixed-parameter algorithms is $h(k) = 2^{O(k)}$. A celebrated example of a fixed-parameter-tractable problem is vertex cover, asking whether an input graph has at most k vertices that are incident to all its edges, which admits a solution as fast as $O(kn + 1.285^k)$ [8]. For more results about fixed-parameter tractability and intractability, see the book of Downey and Fellows [25].

A major recent approach for obtaining efficient fixed-parameter algorithms is through "parameter-treewidth bounds", a notion at the heart of bidimensionality. A *parameter-treewidth bound* is an upper bound $f(k)$ on the treewidth of a graph with parameter value k. Typically, $f(k)$ is polynomial in k. Parameter-treewidth bounds have been established for many parameters; see, e.g., [1, 33, 29, 2, 6, 34, 30, 12, 20–22, 11, 15, 13]. Essentially all of these bounds can be obtained from the general theory of bidimensional parameters (see, e.g., [16]). Thus bidimensionality is the most powerful method so far for establishing parameter-treewidth bounds, encompassing all such previous results for H-minor-free graphs. However, all of these results are limited to graphs that exclude a fixed minor.

A parameter is *minor-bidimensional* if it is at least $g(r)$ in the $r \times r$ grid graph and if the parameter does not increase when taking minors. Examples of minor-bidimensional parameters include the number of vertices and the size of various structures, e.g., feedback vertex set, vertex cover, minimum maximal matching, face cover, and a series of vertex-removal parameters. Tight parameter-treewidth bounds have been established for all minor-bidimensional parameters in H-minor-free graphs for any fixed graph H [18, 11, 13].

Our results provide polynomial parameter-treewidth bounds for all minor-bidimensional parameters in map graphs and power graphs:

Theorem 3. *For any minor-bidimensional parameter P which is at least $g(r)$ in the $r \times r$ grid, every map graph G has treewidth $\mathrm{tw}(G) = O(g^{-1}(P(G)))^3$. More generally suppose that, if graph G has treewidth at least cr^α for constants $c, \alpha > 0$, then G has an $r \times r$ grid minor. Then, for any even (respectively, odd) integer $k \geq 1$, G^k has treewidth $\mathrm{tw}(G) = O(g^{-1}(P(G)))^{\alpha+4}$ (respectively, $\mathrm{tw}(G) = O(g^{-1}(P(G)))^{\alpha+6}$). In particular, for H-minor-free graphs G, and for any even (respectively, odd) integer $k \geq 1$, G^k has treewidth $\mathrm{tw}(G) = O(g^{-1}(P(G)))^5$ (respectively, $\mathrm{tw}(G) = O(g^{-1}(P(G)))^7$).*

This result naturally leads to a collection of fixed-parameter algorithms, using commonly available algorithms for graphs of bounded treewidth:

Corollary 2. *Consider a parameter P that can be computed on a graph G in $h(w) n^{O(1)}$ time given a tree decomposition of G of width at most w. If P is minor-bidimensional and at least $g(r)$ in the $r \times r$ grid, then there is an algorithm computing P on any map graph or power graph G with running time $[h(O(g^{-1}(k))^\beta) + 2^{O(g^{-1}(k))^\beta}] n^{O(1)}$, where β is the degree of $O(g^{-1}(P(G))$ in the polynomial treewidth bound from Theorem 3. In particular, if $h(w) = 2^{O(w)}$ and $g(k) = \Omega(k^2)$, then the running time is $2^{O(k^{\beta/2})} n^{O(1)}$.*

The proofs of these consequences follow directly from combining [11] with Theorems 1 and 2 below.

In contrast, the best previous results for this general family of problems in these graph families have running times $[h(2^{O(g^{-1}(k))^5}) + 2^{2^{O(g^{-1}(k))^5}}] n^{O(1)}$ [11, 14].

5   Improved Grid Minor Bounds for $K_{3,k}$

Recall that every graph excluding a fixed minor H having treewidth at least $c_H r$ has the $r \times r$ grid as a minor [18]. The main result of this section is an explicit bound on c_H when $H = K_{3,k}$ for any k:

Theorem 4. *Suppose G is a graph with no $K_{3,k}$-minor. If the treewidth is at least $20^{4k}r$, then G has an $r \times r$ grid minor.*

In [43], it was shown that if the treewidth is at least $f(r) \geq 20^{2^r}$, then G has an $r \times r$ grid as a minor. Our second theorems use this result to show the following. A *separation* of G is an ordered pair (A, B) of subgraphs of G such that $A \cup B = G$ and there are no edges between $A - B$ and $B - A$. Its *order* is $|A \cap B|$. Suppose G has a separation (A, B) of order k. Let A^+ be the graph obtained from A by adding edges joining every pair of vertices in $V(A) \cap V(B)$. Let B^+ be obtained from B similarly. We say that G is the *k-sum* of A^+ and B^+. If both A^+ and B^+ are minors of G other than G itself, we say that G is the *proper k-sum* of A^+ and B^+.

Using similar techniques as the theorem above, we prove the following two structural results decomposing $K_{3,4}$-minor-free and K_6^--minor-free graphs into proper k-sums:

Theorem 5. *Every $K_{3,4}$-minor-free graph can be obtained via proper 0-, 1-, 2-, and 3-sums starting from planar graphs and graphs of treewidth at most $20^{2^{15}}$.*

Theorem 6. *Every K_6^--minor-free graph can be obtained via proper 0-, 1-, 2-, and 3-sums starting from planar graphs and graphs of treewidth at most $20^{2^{15}}$.*

These theorems are explicit versions of the following decomposition result for general single-crossing-minor-free graphs (including $K_{3,4}$-minor-free and K_6^--minor-free graphs):

Theorem 7. *[41] For any fixed single-crossing graph H, there is a constant w_H such that every H-minor-free graph can be obtained via proper 0-, 1-, 2-, and 3-sums starting from planar graphs and graphs of treewidth at most w_H.*

This result heavily depends on Graph Minor Theory, so the treewidth bound w_H is huge—in fact, no explicit bound is known. Theorems 5 and 6 give reasonable bounds for the two instances of H we consider. Our proof of Theorem 5 uses a 15×15 grid minor together with the result in [40]. The latter result says roughly that, if there is a planar subgraph H in a non-planar graph G, then H has either a non-planar "jump" or "cross" in G such that the resulting graph is a minor of G. Our approach is to find a $K_{3,4}$-minor in a 13×13 grid minor plus some non-planar jump or cross. Similar techniques allow us to prove almost the same result for K_6^--free graphs in Theorem 6.

6   Contraction Version of Wagner's Conjecture

Motivated in particular by Kuratowski's Theorem characterizing planar graphs as graphs excluding $K_{3,3}$ and K_5 as minors, Wagner conjectured and Robertson and Seymour proved the following three results:

Theorem 8 (Wagner's Conjecture). [39] *For any infinite sequence G_0, G_1, $G_2, \ldots$ of graphs, there is a pair (i, j) such that $i < j$ and G_i is a minor of G_j.*

Corollary 3. [39] *Any minor-closed graph property[1] is characterized by a finite set of excluded minors.*

Corollary 4. [39, 38] *Every minor-closed graph property can be decided in polynomial time.*

The important question we consider is whether these theorems hold when the notion of "minor" is replaced by "contraction". The motivation for this variation is that many graph properties are closed under contractions but not under minors (i.e., deletions). Examples include the decision problems associated with dominating set, edge dominating set, connected dominating set, diameter, etc.

One positive result along these lines is about minor-closed properties:

Theorem 9. *Any minor-closed graph property is characterized by a finite set of excluded contractions.*

For example, we obtain the following contraction version of Kuratowski's Theorem, using the construction of the previous theorem and observing that all other induced supergraphs of $K_{3,3}$ have K_5 as a contraction.

Corollary 5. *Planar graphs are characterized by a finite set of excluded contractions.*

Another positive result is that Wagner's Conjecture extends to contractions in the special case of trees. This result follows from the normal Wagner's Conjecture because a tree T_1 is a minor of another tree T_2 if and only if T_1 is a contraction of T_2:

Proposition 2. *For any infinite sequence $G_0, G_1, G_2, \ldots$ of trees, there is a pair (i, j) such that $i < j$ and G_i is a contraction of G_j.*

Unfortunately, the contraction version of Wagner's Conjecture does not hold for general graphs:

Theorem 10. *There is an infinite sequence $G_0, G_1, G_2, \ldots$ of graphs such that, for every pair (i, j), $i \neq j$, G_i is not a contraction of G_j.*

Corollary 6. *There is a contraction-closed graph property that cannot be characterized by a finite set of excluded contractions.*

---

[1] A *property* is simply a set of graphs, representing the graphs having the property.

The graphs $G_i = K_{2,i+2}$ that form the counterexample of Theorem 10 and Corollary 6 are in some sense tight. Each G_i is a planar graph with faces of degree 4. If all faces are smaller, the contraction version of Wagner's Conjecture holds. A planar graph is *triangulated* if some planar embedding (or equivalently, every planar embedding) is triangulated, i.e., all faces have degree 3. Recall that the triangulated planar graphs are the maximal planar graphs, i.e., planar graphs in which no edges can be added while preserving planarity.

Theorem 11. *For any infinite sequence $G_0, G_1, G_2, \ldots$ of triangulated planar graphs, there is a pair (i, j) such that $i < j$ and G_i is a contraction of G_j.*

Another sense in which the counterexample graphs $G_i = K_{2,i+2}$ are tight is that they are 2-connected and are 2-outerplanar, i.e., removing the (four) vertices on the outside face leaves an outerplanar graph (with all vertices on the new outside face). However, the contraction version of Wagner's Conjecture holds for 2-connected (1-)outerplanar graphs:

Theorem 12. *For any infinite sequence $G_0, G_1, G_2, \ldots$ of 2-connected embedded outerplanar graphs, there is a pair (i, j) such that $i < j$ and G_i is a contraction of G_j.*

Corollary 7. *Every contraction-closed graph property of trees, triangulated planar graphs, and/or 2-connected outerplanar graphs is characterized by a finite set of excluded contractions.*

We can use this result to prove the existence of a polynomial-time algorithm to decide any fixed contraction-closed property for trees and 2-connected outerplanar graphs, using a dynamic program that tests for a fixed graph contraction in a bounded-treewidth graph.

7   Open Problems and Conjectures

One of the main open problems is to close the gap between the best current upper and lower bounds relating treewidth and grid minors. For map graphs, it would be interesting to determine whether our analysis is tight, in particular, whether we can construct an example for which the $O(r^3)$ bound is tight. Such a construction would be very interesting because it would improve the best previous lower bound of $\Omega(r^2 \lg r)$ for general graphs [43]. We make the following stronger claim about general graphs:

Conjecture 1. For some constant $c > 0$, every graph with treewidth at least cr^3 has an $r \times r$ grid minor. Furthermore, this bound is tight: some graphs have treewidth $\Omega(r^3)$ and no $r \times r$ grid minor.

This conjecture is consistent with the belief of Robertson, Seymour, and Thomas [43] that the treewidth of general graphs is polynomial in the size of the largest grid minor.

We also conjecture that the contraction version of Wagner's Conjecture holds for k-outerplanar graphs for any fixed k. If this is true, it is particularly interesting that the property holds for k-outerplanar graphs, which have bounded treewidth, but does not work in general for bounded-treewidth graphs (as we have shown in Theorem 10).

Acknowledgments

We thank László Lovász and Robin Thomas for helpful discussions.

References

1. J. Alber, H. L. Bodlaender, H. Fernau, T. Kloks, and R. Niedermeier. Fixed parameter algorithms for dominating set and related problems on planar graphs. *Algorithmica*, 33(4):461–493, 2002.
2. Jochen Alber, Henning Fernau, and Rolf Niedermeier. Parameterized complexity: exponential speed-up for planar graph problems. *Journal of Algorithms*, 52(1): 26–56, 2004.
3. Thomas Böhme, Ken-ichi Kawarabayashi, John Maharry, and Bojan Mohar. Linear connectivity forces large complete bipartite minors. Manuscript, October 2004. http://www.dais.is.tohoku.ac.jp/~k_keniti/KakNew8.ps.
4. Thomas Böhme, John Maharry, and Bojan Mohar. $K_{a,k}$ minors in graphs of bounded tree-width. *Journal of Combinatorial Theory, Series B*, 86(1):133–147, 2002.
5. Vincent Bouchitté, Frédéric Mazoit, and Ioan Todinca. Treewidth of planar graphs: connections with duality. In *Proceedings of the Euroconference on Combinatorics, Graph Theory and Applications (Barcelona, 2001)*, 2001.
6. Maw-Shang Chang, Ton Kloks, and Chuan-Min Lee. Maximum clique transversals. In *Proceedings of the 27th International Workshop on Graph-Theoretic Concepts in Computer Science (WG 2001)*, volume 2204 of *Lecture Notes in Computer Science*, pages 32–43, 2001.
7. Guantao Chen, Laura Sheppardson, Xingxing Yu, and Wenan Zang. Circumference of graphs with no $K_{3,t}$-minors. Submitted manuscript. http://www.math.gatech.edu/~yu/Papers/k3t2-10.pdf.
8. Jianer Chen, Iyad A. Kanj, and Weijia Jia. Vertex cover: further observations and further improvements. *Journal of Algorithms*, 41(2):280–301, 2001.
9. Zhi-Zhong Chen. Approximation algorithms for independent sets in map graphs. *Journal of Algorithms*, 41(1):20–40, 2001.
10. Zhi-Zhong Chen, Michelangelo Grigni, and Christos H. Papadimitriou. Map graphs. *Journal of the ACM*, 49(2):127–138, 2002.
11. Erik D. Demaine, Fedor V. Fomin, MohammadTaghi Hajiaghayi, and Dimitrios M. Thilikos. Bidimensional parameters and local treewidth. *SIAM Journal on Discrete Mathematics*, 18(3):501–511, December 2004.
12. Erik D. Demaine, Fedor V. Fomin, MohammadTaghi Hajiaghayi, and Dimitrios M. Thilikos. Fixed-parameter algorithms for (k, r)-center in planar graphs and map graphs. *ACM Transactions on Algorithms*, 1(1):33–47, 2005.
13. Erik D. Demaine, Fedor V. Fomin, MohammadTaghi Hajiaghayi, and Dimitrios M. Thilikos. Subexponential parameterized algorithms on graphs of bounded genus and H-minor-free graphs. *Journal of the ACM*, 52(6):866–893, 2005.

14. Erik D. Demaine and MohammadTaghi Hajiaghayi. Quickly deciding minor-closed parameters in general graphs. *European Journal of Combinatorics*. to appear.

15. Erik D. Demaine and MohammadTaghi Hajiaghayi. Equivalence of local treewidth and linear local treewidth and its algorithmic applications. In *Proceedings of the 15th ACM-SIAM Symposium on Discrete Algorithms (SODA'04)*, pages 833–842, January 2004.

16. Erik D. Demaine and MohammadTaghi Hajiaghayi. Fast algorithms for hard graph problems: Bidimensionality, minors, and local treewidth. In *Proceedings of the 12th International Symposium on Graph Drawing*, volume 3383 of *Lecture Notes in Computer Science*, pages 517–533, Harlem, NY, 2004.

17. Erik D. Demaine and MohammadTaghi Hajiaghayi. Bidimensionality: New connections between FPT algorithms and PTASs. In *Proceedings of the 16th Annual ACM-SIAM Symposium on Discrete Algorithms (SODA 2005)*, pages 590–601, Vancouver, January 2005.

18. Erik D. Demaine and MohammadTaghi Hajiaghayi. Graphs excluding a fixed minor have grids as large as treewidth, with combinatorial and algorithmic applications through bidimensionality. In *Proceedings of the 16th Annual ACM-SIAM Symposium on Discrete Algorithms (SODA 2005)*, pages 682–689, Vancouver, January 2005.

19. Erik D. Demaine, MohammadTaghi Hajiaghayi, and Ken-ichi Kawarabayashi. Algorithmic graph minor theory: Decomposition, approximation, and coloring. In *Proceedings of the 46th Annual IEEE Symposium on Foundations of Computer Science*, pages 637–646, Pittsburgh, PA, October 2005.

20. Erik D. Demaine, MohammadTaghi Hajiaghayi, Naomi Nishimura, Prabhakar Ragde, and Dimitrios M. Thilikos. Approximation algorithms for classes of graphs excluding single-crossing graphs as minors. *Journal of Computer and System Sciences*, 69(2):166–195, September 2004.

21. Erik D. Demaine, MohammadTaghi Hajiaghayi, and Dimitrios M. Thilikos. A 1.5-approximation for treewidth of graphs excluding a graph with one crossing. In *Proceedings of the 5th International Workshop on Approximation Algorithms for Combinatorial Optimization (Italy, APPROX 2002)*, volume 2462 of *Lecture Notes in Computer Science*, pages 67–80, 2002.

22. Erik D. Demaine, MohammadTaghi Hajiaghayi, and Dimitrios M. Thilikos. Exponential speedup of fixed-parameter algorithms for classes of graphs excluding single-crossing graphs as minors. *Algorithmica*, 41(4):245–267, February 2005.

23. Erik D. Demaine, MohammadTaghi Hajiaghayi, and Dimitrios M. Thilikos. The bidimensional theory of bounded-genus graphs. *SIAM Journal on Discrete Mathematics*, 20(2):357–371, 2006.

24. Reinhard Diestel, Tommy R. Jensen, Konstantin Yu. Gorbunov, and Carsten Thomassen. Highly connected sets and the excluded grid theorem. *Journal of Combinatorial Theory, Series B*, 75(1):61–73, 1999.

25. R. G. Downey and M. R. Fellows. *Parameterized complexity*. Springer-Verlag, 1999.

26. David Eppstein. Connectivity, graph minors, and subgraph multiplicity. *Journal of Graph Theory*, 17(3):409–416, 1993.

27. David Eppstein. Diameter and treewidth in minor-closed graph families. *Algorithmica*, 27(3-4):275–291, 2000.

28. Michael R. Fellows and Michael A. Langston. Nonconstructive tools for proving polynomial-time decidability. *Journal of the ACM*, 35(3):727–739, 1988.

29. Fedor V. Fomin and Dimitiros M. Thilikos. Dominating sets in planar graphs: Branch-width and exponential speed-up. In *Proceedings of the 14th Annual ACM-SIAM Symposium on Discrete Algorithms*, pages 168–177, 2003.

30. Gregory Gutin, Ton Kloks, Chuan Min Lee, and Anders Yao. Kernels in planar digraphs. *Journal of Computer and System Sciences*, 71(2):174–184, August 2005.

31. Graham Higman. Ordering by divisibility in abstract algebras. *Proceedings of the London Mathematical Society. Third Series*, 2:326–336, 1952.

32. David S. Johnson. The NP-completeness column: an ongoing guide (column 19). *Journal of Algorithms*, 8(3):438–448, 1987.

33. Iyad Kanj and Ljubomir Perković. Improved parameterized algorithms for planar dominating set. In *Proceedings of the 27th International Symposium on Mathematical Foundations of Computer Science (MFCS 2002)*, volume 2420 of *Lecture Notes in Computer Science*, pages 399–410, 2002.

34. Ton Kloks, C. M. Lee, and Jim Liu. New algorithms for k-face cover, k-feedback vertex set, and k-disjoint set on plane and planar graphs. In *Proceedings of the 28th International Workshop on Graph-Theoretic Concepts in Computer Science*, volume 2573 of *Lecture Notes in Computer Science*, pages 282–295, 2002.

35. Denis Lapoire. Treewidth and duality in planar hypergraphs. `http://www.labri.fr/Perso/~lapoire/papers/dual_planar_treewidth.ps`.

36. Bogdan Oporowski, James Oxley, and Robin Thomas. Typical subgraphs of 3- and 4-connected graphs. *J. Combin. Theory Ser. B*, 57(2):239–257, 1993.

37. Neil Robertson and P. D. Seymour. Graph minors. V. Excluding a planar graph. *Journal of Combinatorial Theory, Series B*, 41:92–114, 1986.

38. Neil Robertson and P. D. Seymour. Graph minors. XII. Distance on a surface. *Journal of Combinatorial Theory, Series B*, 64(2):240–272, 1995.

39. Neil Robertson and P. D. Seymour. Graph minors. XX. Wagner's conjecture. *Journal of Combinatorial Theory, Series B*, 92(2):325–357, 2004.

40. Neil Robertson, P. D. Seymour, and Robin Thomas. Non-planar extensions of planar graphs. Preprint, 2001. `http://www.math.gatech.edu/~thomas/ext.ps`.

41. Neil Robertson and Paul Seymour. Excluding a graph with one crossing. In *Graph structure theory (Seattle, 1991)*, pages 669–675. Amer. Math. Soc., 1993.

42. Neil Robertson and Paul D. Seymour. Graph minors. X. Obstructions to tree-decomposition. *Journal of Combinatorial Theory Series B*, 52:153–190, 1991.

43. Neil Robertson, Paul D. Seymour, and Robin Thomas. Quickly excluding a planar graph. *Journal of Combinatorial Theory, Series B*, 62(2):323–348, 1994.

44. Paul D. Seymour and Robin Thomas. Call routing and the ratcatcher. *Combinatorica*, 14(2):217–241, 1994.

45. Mikkel Thorup. Map graphs in polynomial time. In *Proceedings of the 39th Annual Symposium on Foundations of Computer Science*, pages 396–407, 1998.

Branching and Treewidth Based Exact Algorithms*

Fedor V. Fomin[1], Serge Gaspers[1], and Saket Saurabh[2]

[1] Department of Informatics, University of Bergen,
N-5020 Bergen, Norway
{fomin, serge}@ii.uib.no
[2] The Institute of Mathematical Sciences,
Chennai 600 113, India
saket@imsc.res.in

Abstract. Branch & Reduce and dynamic programming on graphs of bounded treewidth are among the most common and powerful techniques used in the design of exact (exponential time) algorithms for NP hard problems. In this paper we discuss the efficiency of *simple* algorithms based on combinations of these techniques. We give several examples of possible combinations of branching and programming which provide the fastest known algorithms for a number of NP hard problems: Minimum Maximal Matching and some variations, counting the number of maximum weighted independent sets. We also briefly discuss how similar techniques can be used to design parameterized algorithms. As an example, we give fastest known algorithm solving k-Weighted Vertex Cover problem.

1 Introduction

It is a common belief that exponential time algorithms are unavoidable when we want to find an exact solution of a NP hard problem. The last few years have seen an emerging interest in designing exponential time exact algorithms and we recommend recent surveys [4, 14] for an introduction to the topic.

One of the major techniques for constructing fast exponential time algorithms is the Branch & Reduce paradigm. Branch & Reduce algorithms (also called search tree algorithms, Davis-Putnam-style exponential-time backtracking algorithms etc.) recursively solve NP hard combinatorial problems using reduction rules and branching rules. Such an algorithm is applied to a problem instance by recursively calling itself on smaller instances of the problem.

Treewidth is one of the most basic parameters in graph algorithms. There is a well established theory on the design of polynomial (or even linear) time algorithms for many intractable problems when the input is restricted to graphs of bounded treewidth (see [1] for a comprehensive survey). What is more important for us here is that many problems on graphs with n vertices and treewidth at most ℓ can be solved in time $O(c^\ell n^{O(1)})$, where c is some problem dependent constant. This observation combined with upper bounds on treewidth was used to obtain fast exponential algorithms for NP hard problems on cubic, sparse and planar graphs [4, 5, 9]. For example, a maximum independent set of a graph given with a tree decomposition of width at most ℓ can be found in time

---

* Additional support by the Research Council of Norway.

T. Asano (Ed.): ISAAC 2006, LNCS 4288, pp. 16–25, 2006.

$O(2^{\ell}n)$ (see e.g. [1]). So, a quite natural approach to solve the independent set problem would be to branch on vertices of high degree and if a subproblem with all vertices of small degrees is obtained, then use dynamic programming. Unfortunately, such a simple approach still provides poor running time mainly because the best known upper bounds on treewidth of graphs with small maximum degree are too large to be useful.

In this paper we show two different approaches based on combinations of branching and treewidth techniques. Both approaches are based on a careful balancing of these two techniques. In the first approach the algorithm either performs fast branching, or if there is an obstacle for fast branching, this obstacle is used for the construction of a path decomposition of small width for the original graph. In the second approach the branching occurs until the algorithm reaches a subproblem with a small number of edges (and here the right choice of the size of the subproblems is crucial). We exemplify our approaches on the following problems.

- MINIMUM MAXIMAL MATCHING (MMM): Given a graph G, find a maximal matching of minimum size.

- #MAXIMUM WEIGHTED INDEPENDENT SET (#MWIS): Given a weighted graph G, count the number of independent sets in G of maximum weight.

For MMM, a number of exact algorithms can be found in the literature. Randerath and Schiermeyer [13] gave an algorithm of time complexity $O(1.44225^m)$ [1] for MMM, where m is the number of edges. Raman et al [12] improved the running time by giving an algorithm of time complexity $O(1.44225^n)$ for MMM, where n is the number of vertices. Here, using a combination of branching, dynamic programming over bounded treewidth and enumeration of minimal vertex covers we give an $O(1.4082^n)$ algorithm for MMM.

There was number of algorithms for #MWIS in the literature [2, 6, 7]. The current fastest algorithm is by Fürer and Kasiviswanathan [6] and runs in $O(1.2461^n)$. All mentioned algorithms are complicated and use many smart tricks (like splitting of a graph into its biconnected components and involved measure) and extensive case analysis.

In this paper we show how a combination of branching and dynamic programming can be used to obtain a simple algorithm solving #MWIS in time $O(1.2431^n)$. This is also the fastest known algorithm to find a maximum weighted independent set in a weighted graph G.

Finally we apply our technique to Parameterized Complexity. Here, we apply our technique to parameterized k-WEIGHTED VERTEX COVER.

- k-WEIGHTED VERTEX COVER (k-WVC): Given a graph $G = (V, E)$, a weight function $w : V :\to \mathbb{R}^+$ such that for every vertex v, $w(v) \geq 1$ and $k \in \mathbb{R}^+$, find a vertex cover of weight at most k. The weight of a vertex cover C is $w(C) = \sum_{v \in C} w(v)$.

For k-WEIGHTED VERTEX COVER, also known as REAL VERTEX COVER, Niedermeier and Rossmanith [11] gave two algorithms, one with running time $O(1.3954^k + kn)$ and polynomial space and the other one using time $O(1.3788^k + kn)$ and space $O(1.3630^k)$.

---

[1] We round the base of the exponent in all our algorithms which allows us to ignore polynomial terms and write $O(c^n n^{O(1)})$ as $O(c^n)$.

Their dedicated paper on k-WEIGHTED VERTEX COVER is based on branching, kernelization and the idea of memorization. Their analysis involves extensive case distinctions
when the maximum degree of the reduced graph becomes 3. Here, we give a very simple algorithm running in time $O(1.3570^k n)$ for this problem, improving the previous
$O(1.3788^k + kn)$ time algorithm of [11].

While the basic idea of our algorithms looks quite natural, the approach is generic
and the right application of our approach improves many known results.

2 Preliminaries

In this paper we consider simple undirected graphs. Let $G = (V, E)$ be a (weighted)
graph and let n denote the number of vertices and m the number of edges of G. We
denote by $\Delta(G)$ the maximum vertex degree in G. For a subset $V' \subseteq V$, $G[V']$ is the
graph induced by V', and $G - V' = G[V \setminus V']$. For a vertex $v \in V$ we denote the set of
its neighbors by $N(v)$ and its *closed neighborhood* by $N[v] = N(v) \cup \{v\}$. Similarly, for
a subset $D \subseteq V$, we define $N[D] = \cup_{v \in D} N[v]$. An *independent set* in G is a subset of
pair-wise non-adjacent vertices. A *matching* is a subset of edges having no endpoints in
common. A subset of vertices $S \subseteq V$ is a *vertex cover* in G if for every edge e of G at
least one endpoint of e is in S.

Major tools of our paper are tree and path decompositions of graphs. We refer to [1]
for definitions of tree decomposition, path decomposition, treewidth and pathwidth of
a graph. We denote by $\mathbf{tw}(G)$ and $\mathbf{pw}(G)$, treewidth and pathwidth of the graph G.

We need the following bounds on the pathwidth (treewidth) of sparse graphs. The
proof of Lemma 1 is simple and based on the result of [5] and by induction on the
number of vertices in a graph.

Lemma 1. *For any $\varepsilon > 0$, there exists an integer n_ε such that for every graph G with
$n > n_\varepsilon$ vertices and $m = \beta n$ edges, $1.5 \le \beta \le 2$, the treewidth of G is bounded by
$(m - n)/3 + \varepsilon n$.*

3 Minimum Maximal Matching

Given a graph $G = (V, E)$, any set of pairwise disjoint edges is called a *matching* of
G. The problem of finding a maximum matching is well studied in algorithms and
combinatorial optimization. One can find a matching of maximum size in polynomial
time but there are many versions of matching which are NP hard. Here, we give an exact
algorithm for one such version, that is MINIMUM MAXIMAL MATCHING (MMM).

We need the following proposition which is a combination of two classical results
due to Moon and Moser [10] and Johnson, Yannakakis and Papadimitriou in [8].

Proposition 1 ([8, 10]). *Every graph on n vertices contains at most $3^{n/3} = O(1.4423^n)$
maximal (with respect to inclusion) independent sets. Moreover, all these maximal independent sets can be enumerated with polynomial delay.*

Since for every $S \subseteq V$, S is a vertex cover of G if and only if $V \setminus S$ is an independent
set of G, Proposition 1 can be used for enumerating minimal vertex covers as well.

Our algorithm also uses the following characterization of a MMM.

Algorithm MMM(G)

Data: A graph G.

Result: A minimum maximal matching of G or a path decomposition of G.

  return findMMM($G, G, \emptyset$)

Function findMMM(G, H, C)

Input: A graph G, an induced subgraph H of G and a set of vertices $C \subseteq V(G) - V(H)$.

Output: A minimum maximal matching of G subject to H and C or a path decomposition of
    G.

  if $(\Delta(H) \geq 4)$ *or* $(\Delta(H) = 3$ *and* $|C| > 0.17385|V(G)|)$ **then**

      choose a vertex $v \in V(H)$ of maximum degree

      $M_1 \leftarrow$ findMMM($G, H - N[v], C \cup N(v)$)   (R1)

      $M_2 \leftarrow$ findMMM($G, H - \{v\}, C \cup \{v\}$)   (R2)

      return *the set of minimum size among* $\{M_1, M_2\}$

  else if $(\Delta(H) = 3$ *and* $|C| \leq 0.17385|V(G)|)$ *or* $(\Delta(H) \leq 2$ *and* $|C| \leq 0.31154|V(G)|)$ **then**

      output a path decomposition of G using Lemma 3

      The Algorithm stops.

  else

      $X \leftarrow E(G)$

      foreach *minimal vertex cover Q of H* **do**

          $M' \leftarrow$ a maximum matching of $G[C \cup Q]$

          Let $V[M']$ be the set of end points of M'

          $M'' \leftarrow$ a maximum matching of $G[C \cup V(H) \setminus V[M']]$

          if *$M' \cup M''$ is a maximal matching of G and* $|X| > |M' \cup M''|$ **then**

             $X \leftarrow M' \cup M''$

      return X

Fig. 1. Algorithm for Minimum Maximal Matching

Proposition 2 ([12]). *Let $G = (V, E)$ be a graph and M be a minimum maximal matching of G. Let*

$$V[M] = \{v \mid v \in V \text{ and } v \text{ is an end point of some edge of } M\}$$

be a subset of all endpoints of M. Let $S \subseteq V[M]$ be a vertex cover of G. Let M' be a maximum matching in $G[S]$ and M'' be a maximum matching in $G - V[M']$, where $V[M']$ is the set of the endpoints of edges of M'. Then $X = M' \cup M''$ is a minimum maximal matching of G.

Note that in Proposition 2, S does not need to be a *minimal* vertex cover.

    The proof of the next lemma is based on standard dynamic programming on graphs of bounded treewidth, and we omit it.

Lemma 2. *There exists an algorithm to compute a minimum maximal matching of a graph G on n vertices in time $O(3^p n)$ when a path decomposition of width at most p is given.*

The algorithm of Figure 1 outputs either a path decomposition of the input graph G of reasonable width or a minimum maximal matching of G. The parameter G of Function findMMM corresponds always to the original input graph, H is a subgraph of G and C is a vertex cover of $G - V(H)$. The algorithm consists of three parts.

Branch. The algorithm branches on a vertex v of maximum degree and returns the matching of minimum size found in the two subproblems created according to the following rules:
   (R1) Vertices $N(v)$ are added to the vertex cover C and $N[v]$ is deleted from H;
   (R2) Vertex v is added to the vertex cover C and v is deleted from H.
Compute path decomposition. The algorithm outputs a path decomposition using Lemma 3. Then the algorithm stops without backtracking. A minimum maximal matching can then be found using the pathwidth algorithm of Lemma 2.
Enumerate minimal vertex covers. The algorithm enumerates all minimal vertex covers of H. For every minimal vertex cover Q of H, $S = C \cup Q$ is a vertex cover of G and the characterization of Proposition 2 is used to find a minimum maximal matching of G.

The conditions under which these different parts are executed have been carefully chosen to optimize the overall running time of the algorithm, including the pathwidth algorithm of Lemma 2. Note that a path decomposition is computed at most once in an execution of the algorithm as findMMM stops right after outputting the path decomposition. Also note that the minimal vertex covers of H can only be enumerated in a leaf of the search tree corresponding to the recursive calls of the algorithm, as no recursive call is made in this part.

We define a *branch node* of the search tree of the algorithm to be a recursive call of the algorithm. Such a branch node is uniquely identified by the triple (G, H, C), that is the parameters of findMMM.

Theorem 1. *A minimum maximal matching of a graph on n vertices can be found in time $O(1.4082^n)$.*

Proof. The correctness of the algorithm is clear from the above discussions. Here we give the running time for the algorithm.

Time Analysis. In the rest of the proof we upper bound the running time of this algorithm. It is essential to provide a good bound on the width of the produced path decomposition of G. The following lemma gives us the desired bounds on the pathwidth. Its proof is easy and is based on the bound on the pathwidth given in Lemma 1.

Lemma 3. *Let $G = (V, E)$ be the input graph and (G, H, C) be a branch node of the search tree of our algorithm then the pathwidth of the graph is bounded by $pw(H) + |C|$. In particular,*
(a) If $\Delta(H) \leq 3$, then $pw(G) \leq (\frac{1}{6} + \varepsilon)|V(H)| + |C|$ for any $\varepsilon > 0$.
(b) If $\Delta(H) \leq 2$, then $pw(G) \leq |C| + 1$. A path decomposition of the corresponding width can be found in polynomial time.

Set $\alpha = 0.17385$ and $\beta = 0.31154$. First, consider the conditions under which a path decomposition may be computed. By combining the pathwidth bounds of Lemma 3 and the running time of the algorithm of Lemma 2, we obtain that MMM can be solved in time $O(max(3^{(1+5\alpha)/6}, 3^{\beta})^n)$ when the path decomposition part of the algorithm is executed.

Assume now that the path decomposition part is not executed. Then, the algorithm continues to branch when the maximum degree $\Delta(H)$ of the graph H is 3. And so, $|C| > \alpha n$ when $\Delta(H)$ first becomes 3. At this point, the set C has been obtained by branching on vertices of degree at least 4 and we investigate the number of subproblems obtained so far. Let L be the the set of nodes in the search tree of the algorithm that correspond to subproblems where $\Delta(H)$ first becomes 3. Note that we can express $|L|$ by a two parameter function $A(n, k)$ where $n = |V(G)|$ and $k = \alpha n$. This function can be upper bounded by a two parameter recurrence relation corresponding to the unique branching rule of the algorithm:

$$A(n, k) = A(n - 1, k - 1) + A(n - 5, k - 4).$$

When the algorithm branches on a vertex v of degree at least 4 the function is called on two subproblems. If v is not added to C ((**R1**)), then $|N[v]| \geq 5$ vertices are removed from H and $|C|$ increases by $|N(v)| \geq 4$. If v is added to C ((**R2**)), then both parameters decrease by 1.

Let r be the number of times the algorithm branches in the case (**R1**). Observe that $0 \leq r \leq k/4$. Let L_r be a subset of L such that the algorithm has branched exactly r times according to (**R1**) in the unique paths from the root to the nodes in L_r. Thus, $|L|$ is bounded by $\sum_{i=0}^{k/4} |L_i|$.

To bound the number of nodes in each L_i, let $l \in L_r$ and P_l be the unique path from l to the root in the search tree. Observe that on this path the algorithm has branched $k - 4i$ times according to (**R2**) and i times according to (**R1**). So, the length of the path P_l is $k - 3i$. By counting the number of sequences of length $k - 3i$ where the algorithm has branched exactly i times according to (**R1**), we get $|L_i| \leq \binom{k-3i}{i}$. Thus if the path decomposition is not computed, the time complexity $T(n)$ of the algorithm is

$$T(n) = O\left(\sum_{i=0}^{k/4} \binom{k-3i}{i} T'(n - 5i - (k - 4i)) \right) = O\left(\sum_{i=0}^{k/4} \binom{k-3i}{i} T'(n - i - k) \right) \quad (1)$$

where $T'(n')$ is the time complexity to solve a problem on a branch node (G, H, C) in L with $n' = |V_H|$. (Let us remind that in this case the algorithm branches on vertices of degree 3 and enumerates minimal vertex covers of H.) Let $p = (\beta - \alpha)n$. To bound $T'(n')$ we use similar arguments as before and use Proposition 1 to bound the running time of the enumerative step of the algorithm. Thus we obtain:

$$T'(n') = O\left(\sum_{i=0}^{p/3} \binom{p-2i}{i} 3^{\frac{n'-4i-(p-3i)}{3}} \right) = O\left(3^{(n'-p)/3} \sum_{i=0}^{p/3} \binom{p-2i}{i} 3^{-i/3} \right). \quad (2)$$

We bound $T(n')$ by $O(3^{(n'-p)/3} d^p)$ for some constant d, $1 < d < 2$ (the value of d is determined later). Substituting this in Equation (1), we get:

$$T(n) = O\left(\sum_{i=0}^{k/4} \binom{k-3i}{i} 3^{\frac{n-i-k-p}{3}} d^p \right) = O\left(3^{(1-\beta)n/3} d^p \sum_{i=0}^{k/4} \binom{k-3i}{i} 3^{-i/3} \right).$$

Further suppose that $\sum_{i=0}^{k/4} \binom{k-3i}{i} 3^{-i/3}$ sums to $O(c^k)$ for a constant c, $1 < c < 2$, then the overall time complexity of the algorithm is bounded by: $O((3^{(1-\beta)/3} d^{\beta-\alpha} c^{\alpha})^n)$.

Claim. $c < 1.3091$ and $d < 1.3697$.

Proof. The sum over binomial coefficients $\sum_{i=0}^{k/4} \binom{k-3i}{i} 3^{-i/3}$ is bounded by $(k/4)B$ where B is the maximum term in this sum. Let us assume that $B = \binom{k-3i}{i} 3^{-i/3}$ for some $i \in \{1, 2, \ldots, k/4\}$. To compute the constant c, let $r := c - 1$. We obtain $B = \binom{k-3i}{i} 3^{-i/3} \leq \frac{(1+r)^{k-3i}}{r^i} 3^{-i/3}$. Here we use the well known fact that for any $x > 0$ and $0 \leq k \leq n$, $\binom{n}{k} \leq \frac{(1+x)^n}{x^k}$. By choosing r to be the minimum positive root of $\frac{(1+r)^{-3}}{r} 3^{-1/3} - 1$, we arrive at $B < (1 + r)^k$ for $0.3090 < r < 0.3091$. Thus $c < 1.3091$. The value of d is computed in a similar way. $\square$

Finally, we get the following running time for Algorithm MMM by substituting the values for α, β, c and d:

$$O\left(max\left(3^{(1-\beta)/3} d^{\beta-\alpha} c^\alpha, 3^{(1+5\alpha)/6}, 3^\beta\right)^n\right) \leq O(1.4082^n) . \qquad \square$$

4   Counting Maximum Weighted Independent Sets

In this section we give an algorithm counting the number of maximum weighted independent sets in a graph, that is an algorithm for the #MWIS problem .

Most of the recent advances in counting maximum weighted independent sets are based on a reduction to counting satisfiable assignments of a 2-SAT formula. All these algorithms are based on the Branch & Reduce paradigm and involve detailed case distinctions. Here, we present a simple algorithm that combines Branch & Reduce and dynamic programming on graphs of bounded treewidth. It is well known (see for example [1]) that a maximum independent set can be found in time $O(2^\ell n)$ in a graph if a tree decomposition of width at most ℓ is given. This algorithm can be slightly modified to find the total number of maximum weighted independent sets in a graph with treewidth ℓ without increasing the running time of the algorithm.

Proposition 3. *Given a graph G with n vertices and a tree decomposition of G of width at most ℓ, all maximum weighted independent sets of G can be counted in time $O(2^\ell n)$.*

Our algorithm **#MWIS**, to count all maximum weighted independent sets of a graph, is depicted in Figure 2. The algorithm branches on a vertex v chosen by the following function which returns the vertex selected by the first applicable rule.

Pivot(G)

1. If $\exists v \in V$ such that $d(v) \geq 7$, then return v.
2. If $\exists v \in V$ such that $G - v$ is disconnected, then return v.
3. If $\exists u, v \in V$ such that $G - \{u, v\}$ is disconnected and $d(u) \leq d(v)$, then return v.
4. Return a vertex v of maximum degree such that $\sum_{u \in N(v)} d(u)$ is maximized.

When $|E| \leq 1.5185n$ the treewidth algorithm counts all maximum weighted independent sets. The procedure #MWISTW is a dynamic programming algorithm solving #MWIS of running time given in Proposition 3. When the algorithm branches on a vertex v, two subproblems are created according to the following rules:

```
Algorithm #MWIS (G = (V, E), w)
Input: A graph G = (V, E) and a weight function w : V → ℝ⁺.
Output: A couple (size, nb) where size is the maximum weight of an independent set of G
        and nb the number of different independent sets of this weight.
  if G is disconnected with connected components G₁, ···, Gₖ then
  |   s' ← ∑ᵢ₌₁ᵏ sᵢ,  n' ← ∏ᵢ₌₁ᵏ nᵢ where (sᵢ, nᵢ) ← #MWIS(Gᵢ, w)
  |_  return (s', n')
  if |E| ≤ 1.5185|V| then
  |   return #MWISTW(G)
  else
  |   v ← Pivot(G)
  |   (s₁, n₁) ← #MWIS(G − N[v], w)
  |   (s₂, n₂) ← #MWIS(G − {v}, w)
  |   s₁ ← s₁ + w(v)
  |   if s₁ > s₂ then
  |   |_  return (s₁, n₁)
  |   else if s₁ = s₂ then
  |   |_  return (s₁, n₁ + n₂)
  |   else
  |_  |_  return (s₂, n₂)
```

Fig. 2. An Algorithm to count all Maximum Weighted Independent Sets of a Graph

(R1) add v to the partially constructed independent set and delete $N[v]$ from the graph.
(R2) delete v from the graph.

The correctness of the algorithm is clear from the presentation. Now we discuss its time complexity in detail which is based on a careful counting of subproblems of different size. We also need the following lemma for the analysis of the time complexity.

Lemma 4 ([3,7]). *Let $G = (V, E)$ be a graph with n vertices, m edges, average degree $a > 0$ and v be a vertex chosen by Rule 4 of the function* **Pivot**. *Then the average degrees of $G - v$ and $G - N[v]$ are less than a.*

Theorem 2. *Let $G = (V, E)$ be a graph on n vertices. Algorithm* **#MWIS** *computes the number of* MAXIMUM WEIGHTED INDEPENDENT SETS *of G in time $O(1.2431^n)$.*

Proof. (1) If G has at most $1.5n$ edges, its treewidth is at most $(1/6 + \epsilon)n$ by Lemma 1. Since only the treewidth algorithm is executed in this case, the running time is $O(2^{(1/6+\epsilon)n}) = O(1.1225^n)$.

(2) Assume that G has at most $2n$ edges. The worst case of the algorithm is when it branches on a vertex chosen by Rule 4 of the function **Pivot**. In this case the algorithm branches on vertices of degree at least 4 and executes the treewidth algorithm when $m \leq 1.5185n$. Set $\beta = 1.5185$. Let xn be the number of times the algorithm branches. The constant x is at least 0 and satisfies the inequality $2n - 4xn \geq \beta(n - xn)$ which implies that $x \leq (2 - \beta)/(4 - \beta)$.

Let s be the number of times the algorithm branches according to (R1). Then, the algorithm branches $xn - s$ times according to (R2). When the treewidth algorithm is executed, the size of the remaining graph is at most $n - 5s - (xn - s) = n - xn - 4s$. By Lemma 1, $\mathbf{tw}(G)$ is bounded by $n(\beta - 1)/3$. Let $T_{tw}(n) = 2^{n(\beta-1)/3}$ be the bound on the running time of the treewidth algorithm when executed on a graph with n vertices and βn edges. The running time of the algorithm is then bounded by

$$T_1(n) = \sum_{s=0}^{xn} \binom{xn}{s} T_{tw}(n - xn - 4s) = 2^{\frac{\beta-1}{3}(1-x)n}(1 + 2^{-4\frac{\beta-1}{3}})^{xn} .$$

$T_1(n)$ is maximum when $x = (2 - \beta)/(4 - \beta)$. By replacing x by this value, we have that $T_1(n) \leq 1.20935^n$.

With an analysis similar to the one for the case when the graph has at most $2n$ edges, we can show that the Algorithm #MWIS takes $O(1.23724^n)$ time or $O(1.2431^n)$ time when the graph has at most $2.5n$ edges or $3n$ edges respectively.

(3) Now, if G has more than $3n$ edges then it contains vertices of degree at least 7 and hence the running time of the algorithm is smaller because the recurrence $T(n) = T(n - 8) + T(n - 1)$ solves to $O(1.2321^n)$.   □

5   Application to Parameterized Algorithms

Here we apply our technique to design a fixed parameter tractable algorithm for the parameterized version of WEIGHTED VERTEX COVER.

We need *kernelization* for our algorithm for weighted vertex cover. The main idea of *kernelization* is to replace a given instance (I, k) by a simpler instance (I', k') using some *data reduction rules* in polynomial time such that (I, k) is a yes-instance if and only if (I', k') is a yes-instance and $|I'|$ is bounded by a function of k alone. We state the kernelization proposition of [11] that we use in our algorithm.

Proposition 4 ([11]). *Let $G = (V, E)$ be a graph, $w : V :\to \mathbb{R}^+$ such that for every vertex v, $w(v) \geq 1$ and $k \in \mathbb{R}^+$. There is an algorithm that in time $O(kn + k^3)$ either concludes that G has no vertex cover of weight $\leq k$, or outputs a kernel of size $\leq 2k$.*

Our algorithm is very similar to the one presented for counting all maximum independent sets. First we apply Proposition 4 to obtain a kernel of size at most $2k$. Then, as long as $|E| > 3.2k$, the algorithm branches on a vertex v chosen by the function **Pivot** as in the algorithm presented in Figure 2. If $|E| \leq 3.2k$, then by Lemma 1, a tree decomposition of small width (tw) can be found in polynomial time and we can use a $O(2^{tw}n)$ dynamic programming algorithm to solve k-WEIGHTED VERTEX COVER.

Theorem 3. *k-WVC on a graph on n vertices can be solved in time $O(1.3570^k n)$.*

6   Conclusion

Branching and dynamic programming on graphs of bounded treewidth are very powerful techniques to design efficient exact algorithms. In this paper, we combined these two techniques in different ways and obtained improved exact algorithms for #MWIS and MMM. Other problems for which we obtain faster algorithms using this technique include variants of MMM that are MINIMUM EDGE DOMINATING SET and MATRIX DOMINATION SET. We also applied the technique to design fixed parameter tractable algorithms and obtained the best known algorithm for k-WVC which also shows the versatility of our technique. The most important aspects of this technique are that the resulting algorithms are very *elegant* and *simple* while at the same time the analysis of these algorithms is highly *non-trivial*. Our improvement in the runtime for #MWIS directly gives improved algorithms for #1-IN-k-SAT, #EXACT HITTING SET, #EXACT COVER and #WEIGHTED SET COVER. The definitions and reductions of these problems to #MWIS can be found in [2]. Other parameterized problems for which we obtain the fastest known algorithms using the techniques developed in this paper are the *weighted* and *unweighted* version of *parameterized minimum maximal matching* and *minimum edge dominating set*, which will appear in the longer version of this paper.

It would be interesting to find some other applications of the techniques presented here in the design of exact exponential time algorithms and fixed parameter tractable algorithms.

References

1. H. L. BODLAENDER, *A partial k-arboretum of graphs with bounded treewidth*, Theoretical Computer Science, 209 (1998), pp. 1–45.
2. V. DAHLLÖF AND P. JONSSON, *An algorithm for counting maximum weighted independent sets and its applications*, in 13th Annual ACM-SIAM Symposium on Discrete Algorithms (SODA 2002), ACM and SIAM, 2002, pp. 292–298.
3. V. DAHLLÖF, P. JONSSON, AND M. WAHLSTRÖM, *Counting models for 2SAT and 3SAT formulae*, Theoretical Computer Science, 332 (2005), pp. 265–291.
4. F. V. FOMIN, F. GRANDONI, AND D. KRATSCH, *Some new techniques in design and analysis of exact (exponential) algorithms*, Bulletin of the EATCS, 87 (2005), pp. 47–77.
5. F. V. FOMIN AND K. HØIE, *Pathwidth of cubic graphs and exact algorithms*, Information Processing Letters, 97 (2006), pp. 191–196.
6. M. FÜRER AND S. P. KASIVISWANATHAN, *Algorithms for counting 2-SAT solutions and colorings with applications*, Electronic Colloquium on Computational Complexity (ECCC), vol. 33, 2005.
7. M. K. GOLDBERG, D. BERQUE, AND T. SPENCER *A Low-Exponential Algorithm for Counting Vertex Covers*, Graph Theory, Combinatorics, Algorithms, and Applications, vol. 1, (1995), pp. 431–444.
8. D. S. JOHNSON, M. YANNAKAKIS, AND C. H. PAPADIMITRIOU, *On generating all maximal independent sets*, Information Processing Letters, 27 (1988), pp. 119–123.
9. J. KNEIS, D. MÖLLE, S. RICHTER, AND P. ROSSMANITH, *Algorithms based in treewidth of sparse graphs*, in Proceedings of the 31st International Workshop on Graph-Theoretic Concepts in Computer Science (WG 2005), vol. 3787 of LNCS, Springer, (2005), pp. 385–396.
10. J. W. MOON AND L. MOSER, *On cliques in graphs*, Israel Journal of Mathematics, 3 (1965), pp. 23–28.
11. R. NIEDERMEIER AND P. ROSSMANITH, *On efficient fixed-parameter algorithms for weighted vertex cover*, Journal of Algorithms, 47 (2003), pp. 63–77.
12. V. RAMAN, S. SAURABH, AND S. SIKDAR, *Efficient exact algorithms through enumerating maximal independent sets and other techniques*, Theory of Computing Systems, to appear.
13. B. RANDERATH AND I. SCHIERMEYER, *Exact algorithms for MINIMUM DOMINATING SET*, Technical Report zaik-469, Zentrum für Angewandte Informatik Köln, Germany, 2004.
14. G. WOEGINGER, *Exact algorithms for NP-hard problems: A survey*, in Combinatorial Optimization - Eureka, you shrink!, vol. 2570 of LNCS, Springer, (2003), pp. 185–207.

Deterministic Splitter Finding in a Stream with Constant Storage and Guarantees

Tobias Lenz

Freie Universität Berlin, Takustr. 9, 14195 Berlin, Germany
`tlenz@mi.fu-berlin.de`

Abstract. In this paper the well-known problem of finding the median of an ordered set is studied under a very restrictive streaming model with sequential read-only access to the data. Only a constant number of reference objects from the stream can be stored for comparison with subsequent stream elements. A first non-trivial bound of $\Omega\left(\sqrt{n}\right)$ distance to the extrema of the set is presented for a single pass over streams which do not reveal their total size n. For cases with known size, an algorithm is given which guarantees a distance of $\Omega\left(n^{1-\varepsilon}\right)$ to the extrema, which is an ε-approximation for the proven best bound possible.

1 Introduction

We study the problem of finding a good splitter—approximating the median—under very rigorous restrictions. The presented algorithms are deterministic, providing a guarantee on the quality of the output. Furthermore they should perform the task at hand in a single pass over the data storing only a constant number of elements.

Many applications require a good splitter. The crux of every divide-and-conquer algorithm is finding an element which partitions a data set of size n into two parts of roughly equal size. A simple example is quicksort which performs best splitting in each recursion at the median which is the element of rank $\lceil \frac{n}{2} \rceil$, where the rank of an element is its position in the sorted order of the set.

Sensor networks typically have very small and cheap components, so the amount of available memory on a single sensor is tiny. In a network with several sensors, e.g. measuring the temperature every few seconds, the median and other quantiles are significant quantities. No sensor has enough memory to store the whole data, so the information is broadcasted and has to be processed in real-time in the order of reception. This is modeled exactly by our approach with constant memory and no assumptions about the order of the stream elements.

Another class of application are databases. Fast estimation of quantiles is important for planning and executing queries, i.e. estimating the size of intermediate results or partitioning the data for parallel query processing. Obviously the estimation should be performed very fast and must happen under memory restrictions due to the size of todays databases compared to the size of main memory in today's computers. Modern servers process thousands of queries per

T. Asano (Ed.): ISAAC 2006, LNCS 4288, pp. 26–35, 2006.

second, so obviously each single process has only limited memory available. Constant memory turns out to be too little, compared to other algorithms allowing to store logarithmically many observations.

Considering the distribution of the data as unknown, a numerical approximation of the median is not appropriate. The value of such an approximation might be very close to the real median and simultaneously their positions in the sorted data might be arbitrarily far apart. Depending on the actual problem, a numerical approximation may be dissatisfying because in general no element exists with this value.

Blum et.al. [1] showed that finding the median is possible in linear time which is optimal. Their algorithm and all known modifications require random access to the data and either modify the original order of the elements or need $\Omega(n)$ addition memory. The past research for exact bounds on the number of comparisons needed and related information was collated by Paterson [9].

2   Counting Passes and the Streaming Model

Todays modern communication revived several models originated from storage on tape. Data which is streamed through a network comes in a strict sequential order forbidding random access and modification. Usually even if random access is possible, sequential access would be multiple times faster due to caching mechanisms. This justifies the idea of counting and minimizing the number of sequential passes over the data or even restricting this number to one.

Furthermore a data stream might not give away its total size beforehand. It might even be unknown throughout the network. Typically the total data will not fit into memory, e.g. on a sensor node, and might end without premonition. In some cases, a continuing stream has no designated end at all. An approximation algorithm for streams of unknown length must therefore maintain a good estimate at all times and cannot use the data size n in its decisions. We discuss streams both with known and unknown sizes in this paper.

The known optimal linear time median algorithms violate the above conditions and are therefore not applicable in this model.

3   Previous Work

Good splitters have high probability. A simple strategy for known n would be to pick a random element with probability $\frac{1}{n}$. This will split the data in a ratio not worse than 1 : 3 with a probability of 50%. Using the median over a constant sized random sample yields even better results. Vitter overcomes the problem of not knowing n with a special sampling technique called *reservoir sampling* [10].

With only five reference values Jain and Chlamtac [5] obtain good results but they approximate the value of the median by a special formula. This depends on the distribution and does not yield any guarantees concerning the rank of the returned element.

Granting more than constant memory allows a $(1+\varepsilon)$-approximation for finding the element with rank k. Manku et.al. [6] presented a single pass algorithm with $\mathcal{O}\left(\frac{1}{\varepsilon}\log^2 \varepsilon n\right)$ memory if n is known. This was improved to $\mathcal{O}\left(\frac{1}{\varepsilon}\log \varepsilon n\right)$ memory, not requiring the knowledge of n by Greenwald and Khanna [2]. In the former paper, the authors also showed a $(1+\varepsilon)$-approximation technique with probability δ using $\mathcal{O}\left(\frac{1}{\varepsilon}\log^2\left(\frac{1}{\varepsilon}\log\frac{1}{1-\delta}\right)\right)$ memory.

Munro and Paterson [7] studied the problem under several memory restrictions but for constant memory they only gave a multi-pass algorithm for the exact median without intermediate approximation results after each pass. The authors themselves call it "intolerable in practice". Later Munro and Raman [8] solved the problem of finding the median with minimal data movements in $\mathcal{O}\left(n^{1+\varepsilon}\right)$ steps with constant extra space but allowed random access to the read-only data.

Very recently Guha and McGregor [3] exploited the fact that the distribution of the elements in a stream is not completely adversarial in expectation, although the actual distribution might be unknown. They obtain an element with rank $\frac{n}{2} \pm \mathcal{O}\left(n^{\frac{1}{2}+\varepsilon}\right)$ using polylogarithmic space. Guha et al. [4] analyzed the entropy of streams even further and compared the random order model with several oracle models.

These results are with high probability but not guaranteed or require sophisticated structures and more than constant memory. This paper tackles the problem of finding a simple deterministic algorithm with constant memory which splits a stream with more than a constant number of elements in both parts.

4   The Model

The stream must contain comparable objects with a total order defined on them. For a weak or partial order, the median is not uniquely defined, but with a sensible definition or by embedding the data in a strict total order, the extension of our algorithms should be straightforward. Two cases are analyzed, one where the total number n of elements in the stream is known in advance and in the other case it is not known and cannot be guessed or computed from the stream.

The algorithm must not store more than a fixed number of references to objects in the stream, later called *markers*. These might be used in further comparisons. All elements from the stream which are not marked or the current ("top") element cannot be accessed. No arithmetic operations with the markers are allowed (like C-style pointer arithmetic) and the return value of the algorithm must be a marker pointing to an element.

Munro and Paterson [7] observed that there is no hope for more than a constant distance to the boundary in this setting with a purely comparison-based approach forbidding other operations. Therefore the number of markers is fixed and additionally a constant number of "statistics" is allowed, for example counting the number of elements smaller/larger than a marked element.

The measure for the quality of an algorithm is the minimum distance to the boundary (left boundary is 1, right boundary is the number of elements) of the index of the returned element in the sorted data $d(x)$. The median m has optimal distance $d(m) = \min\left(\lceil \frac{n}{2} \rceil, \lfloor \frac{n}{2} \rfloor\right) = \lfloor \frac{n}{2} \rfloor$ while a trivial algorithm with storage size s will return an element x of rank r with $d(x) = \min(r, n - r) \leq s$.

5   Obtaining Upper Bounds by Playing Games

We start with the following theorem, destroying all hope for very good approximations.

Theorem 1. *Every algorithm which finds an element in a stream of size n with distance to the median at most $\frac{n}{f(n)}$ has to store at least $\frac{f(n)}{4}$ stream elements.*

Proof: Assume an algorithm wants to achieve a distance to the median of at most $\frac{n}{f(n)}$. Stop after the first $\lceil \frac{n}{2} \rceil$ elements were read from the stream. We call the set containing these elements M. None of the elements in M can be ruled out being the median—just let the values of the following $\lfloor \frac{n}{2} \rfloor$ elements be smaller/larger properly. Covering the sorted order of M completely with intervals of the desired distance requires storing at least every $2\frac{n}{f(n)}$st element in the sorted order. This results in a total of $\lceil \frac{n}{2} \rceil \frac{f(n)}{2n} \geq \frac{n \cdot f(n)}{4n} = \frac{f(n)}{4}$ stored elements. □

This immediately implies the following result for constant storage with $f(n) \in \omega(1)$.

Corollary 1. *An algorithm which stores only a constant number of elements from a stream cannot achieve an approximation of the median's position in the sorted data which is sub-linear in the total number of elements n.*

For a rather tiny memory of one or two entries, the upper bound on d over all possible algorithms is modeled as an adversary game as follows. The adversary is the stream which selects the next element. The algorithm knows every element up to the current and might set a marker to the new element giving up an element currently marked. This game continues ad infinitum to simulate the asymptotic result on n. This might be considered as the online version of the problem.

Figure 1 shows that a single marker is arbitrarily bad: The adversary adds new minima. The algorithm might decide to keep its marker which remains close to the boundary (upper right). If the marker is moved to the new element, it will be stuck at the boundary by having the adversary switch the insertion position to the other end (lower right).

Corollary 2. *A single marker will always end at the boundary in the worst case.*

In this setting two markers are already sufficient to achieve a distance of $\lfloor \frac{n}{3} \rfloor$ by always keeping the median between the two markers and their distance as small as possible. This strength is due to the unrealistic assumption that the

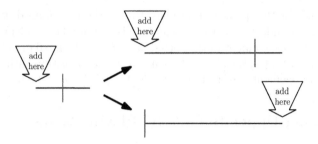

Fig. 1. The horizontal line represents the sorted order of the stream elements seen so far, the vertical line is the marker

algorithms always know the exact position of a new element in the sorted data. In reality nothing is known about how a new element partitions an existing interval between two markers, so we change the rules of the game slightly for the two marker case.

Assume that the adversary only specifies the interval between two markers or a marker and the boundary where the next element will lie in. The algorithm must select a marker which is set to that position or ignore the new element. Afterwards the adversary reveals the real position. A bound for this scheme with two markers m_1, m_2 is depicted in figure 2.

The adversary adds an element in the interval containing the median (top left). Not moving the marker does not increase d (top right) leading to an arbitrarily bad result. After one marker was set, the adversary reveals that it is next to the other marker (middle left). This is the only increase in $d(m_i)$ in the scheme. Now adding new extrema either keeps d (middle right) or sets a marker to the boundary (bottom left). Now an increasing number of extreme values can be inserted until both markers have the same distance to the boundary (bottom right) and the whole scheme restarts.

Lemma 1. *For two markers m_1, m_2, $\max(d(m_1), d(m_2)) \in \mathcal{O}(\sqrt{n})$ holds in the worst case.*

Proof: Assume $d(m_1) = d(m_2) = c$ (const) in the beginning. After one round $d(m_i) = c+1$ either for $i = 1$ or $i = 2$ and the total number of elements inserted is 2 (top left, middle left)$+(c + 1)$ (bottom left). In round j $d(m_i)$ increases by at most one while the total number of elements increases by $j + c + 2$. This gives $\sum_{j=1}^{k} (j + c + 2) = n$ for a total of n elements and k rounds. Obviously $k \in \Theta(\sqrt{n})$ and therefore $d(m_i) \leq k + c \Rightarrow d(m_i) \in \mathcal{O}(\sqrt{n})$. Please note that the case distinction is not complete because obviously bad choices are omitted. $\square$

This rule modification already fails for three markers and for an arbitrary fixed number it is not known whether the linear bound can be reached or not. The used adversary game seems to be a bad choice for a proof for bigger storage, because the algorithm might always make a good decision "by accident" even without the proper knowledge.

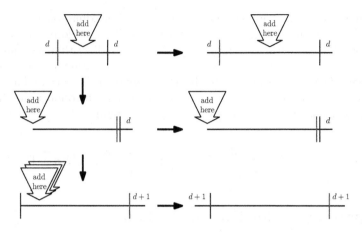

Fig. 2. Beating a two marker scheme

6   Dealing with Unknown Data Size

Every algorithm for streams with unknown total size is obviously also an algorithm for streams with known size n. The following algorithm achieves a distance of $\Omega\left(\sqrt{n}\right)$ to the boundary without using n. It uses two markers and is therefore optimal due to lemma 1. It is the best we can achieve without using n and this algorithm acts as a building block for the improved algorithm using n in the next section.

The general idea is to have roughly $\sqrt{n}$ rounds increasing the distance to the boundary by at least one in each round. We assume the existence of two stream operations: **next** provides the top element and removes it from the stream and **top** provides the top element without removing it.

Algorithm 1. Achieving a distance of at least $\sqrt{2(n+1)}-3$ to the boundary with only two markers m_1, m_2

$m_1 \leftarrow$ **next**
for $k = 2, 3, 4, \ldots$ **do**
    $m_2 \leftarrow$ **top**
    for *the upcoming k elements* **do**
        if next *is end of stream* **then   return** m_1
        if $(m_2 < $ **top** $< m_1)$ *or* $(m_2 > $ **top** $> m_1)$ **then** $m_2 \leftarrow$ **top**
    if *last k elements were all smaller or all larger than m_1* **then**
    $m_1 \leftarrow m_2$

Theorem 2. *Algorithm 1 returns an element with distance to the boundary in the sorted data at least $\sqrt{2(n+1)} - 3$ without knowing n beforehand using two markers.*

Proof: Starting with $k = 2$ the value of k increases by one in each round. The algorithm consumes k elements per round, with either at least one element smaller and one element larger than the current approximation m_1, hence increasing $d(m_1)$ by at least one, or ending with the element closest to m_1 in m_2. Since then all k elements were larger (symmetric: smaller) than m_1 and m_2 is the minimum (maximum) of these, setting m_1 to m_2 guarantees an increase of the distance of m_1 by at least one to the lower (upper) boundary and a distance of at least $k - 1$ to upper (lower) boundary. Therefore in round i the distance to the boundary is at least $i - 1$. We compute the number of complete rounds r for $n \geq 2$ as follows:

$$n - r \leq \sum_{k=1}^{r} k \leq n \Rightarrow n \leq \frac{r^2 + 3r}{2} \Rightarrow r \geq \sqrt{2n + \frac{9}{4}} - \frac{3}{2} > \sqrt{2(n+1)} - 2. \qquad \square$$

7   Improvements for Known Data Size

Knowing the number of elements, one wants to compute the median of, is a great deal of help. It allows to process specified fractions of the input as a block, which yields much better results than for an unknown size. The following algorithm uses the algorithm 1 as a subroutine to compute an approximation of the median over a block with $\frac{n}{t}$ elements. The next $\frac{n}{t}$ elements are used to verify the quality of the approximation, which is then refined in the next round. This leads to a total of $\frac{nt}{2n} = \frac{t}{2}$ rounds. In each round a lower and an upper filter are maintained where the lower filter has always "enough" elements above it and the upper filter has "enough" elements below it.

Algorithm 2. Achieving a distance of $\Omega\left(n^{\frac{2}{3}}\right)$ to the boundary with four markers, $t \in \Theta\left(n^{\frac{1}{3}}\right)$.

$low \leftarrow -\infty, high \leftarrow \infty$

during the whole algorithm: $l \leftarrow$ count elements smaller than low
during the whole algorithm: $h \leftarrow$ count elements larger than $high$
if *end of the stream is reached at any time during the algorithm* **then**
    if $l \geq h$ **then return** low **else return** $high$

for ∞ **do**
    $m_1 \leftarrow$ use algorithm 1 on $\frac{n}{t}$ elements $\in (low; high)$, skipping others
    $p = 0, q = 0$
    for *the next* $\frac{n}{t}$ *elements* **do**
        if top $\leq m_1$ **then** $p \leftarrow p + 1$
        if next $\geq m_1$ **then** $q \leftarrow q + 1$
    if $q > \frac{n}{2t}$ **then** $low \leftarrow m_1$ **else** $high \leftarrow m_1$

Lemma 2. *While processing* $3\frac{n}{t}$ *elements from the stream, the number of elements known to be smaller than* low *plus the number of elements known to be larger than* $high$ *increases by* $\Omega\left(\sqrt{\frac{n}{t}}\right)$.

Proof: Two cases are possible.

1. More than $2\frac{n}{t}$ elements are smaller than *low* or larger than *high* and are hence filtered out.
2. The algorithm 1 is applied to a block B of $\frac{n}{t}$ elements and returns an element m_1 with distance $\Omega\left(\sqrt{\frac{n}{t}}\right)$ to both boundaries or B by theorem 2. Counting over the next $\frac{n}{t}$ elements allows two symmetric cases, so we consider only the case that at least $\frac{n}{2t}$ elements are not smaller than m_1. The algorithm sets *low* to m_1, guaranteeing the claimed increase of *low*.   □

Lemma 3. *Algorithm 2 returns an element with distance* $\Omega\left(\frac{n}{t}\right) = \Omega\left(n^{\frac{2}{3}}\right)$ *to the boundary in the sorted data with* $t = n^{\frac{1}{3}}$ *and knowing n beforehand using four markers.*

Proof: The algorithm will only change *low*, if at least $\frac{n}{2t}$ larger elements are known. The symmetric holds for *high*, so we only have to care about pushing *low* and *high* as far as we can.

The n input elements can be seen as $\frac{t}{3}$ blocks with $3\frac{n}{t}$ elements each. Lemma 2 gives a total distance of $\frac{t}{3}\Omega\left(\sqrt{\frac{n}{t}}\right) = \Omega\left(t\sqrt{\frac{n}{t}}\right) = \Omega\left(n^{\frac{2}{3}}\right)$ to the boundary in the sorted data. At least half of this distance must be attained by *low* or *high* which is returned respectively.

The marker m_1 in this algorithm is the same as m_1 in the subroutine, so two markers are used by the subroutine plus two for *low* and *high*.   □

Please observe that t was chosen to maximize the minimum of the set size used for the estimation, $\frac{n}{t}$, and the least gain per round, $\sqrt{\frac{n}{t}}$ times the number of rounds $\frac{t}{2}$, omitting constants:

$$\frac{n}{t} = t\sqrt{\frac{n}{t}} = t^{\frac{1}{2}}n^{\frac{1}{2}} \Rightarrow n^{\frac{1}{2}} = t^{\frac{3}{2}} \Rightarrow t = n^{\frac{1}{3}}.$$

Theorem 3. *A distance to the boundary of* $\Omega\left(n^{1-\varepsilon}\right)$ *can be achieved with* $\mathcal{O}(\frac{1}{\varepsilon})$ *markers.*

Proof: Instead of using algorithm 1 as a subroutine, one can use algorithm 2 and then continue recursively needing two additional markers for the *low* and *high* values in each recursive step. Computing the optimal value for t by the formula above resolves to

$$\frac{n}{t} = t\left(\frac{n}{t}\right)^x \Rightarrow n^{1-x} = t^{2-x} \Rightarrow t = n^{\frac{1-x}{2-x}}.$$

Applying the proof of lemma 3 with this t yields an algorithm with distance

$$\Omega\left(\frac{n}{t}\right) = \Omega\left(n^{1-\frac{1-x}{2-x}}\right) = \Omega\left(n^{\frac{1}{2-x}}\right)$$

for a subroutine with a guarantee of $\Omega(n^x)$. In particular for a subroutine with guarantee $\Omega\left(n^{\frac{a}{a+1}}\right)$ the guarantee is raised to $\Omega\left(n^{\frac{1}{2-\frac{a}{a+1}}}\right) = \Omega\left(n^{\frac{a+1}{a+2}}\right)$. Starting with $\frac{1}{2}$, $\frac{a}{a+1}$ is reached after $a-1$ recursive levels, each requiring two markers,

$2a$ in total. For an $\Omega\left(n^{1-\varepsilon}\right)$ distance we pick $1-\varepsilon=\frac{a}{a+1}\Rightarrow a=\frac{1}{\varepsilon}-1$. This requires a total of $2a\in\mathcal{O}\!\left(\frac{1}{\varepsilon}\right)$ markers. $\qquad\square$

8   Multiple Passes

The algorithms from the prior sections give an approximation of the median in a single pass. It is fairly easy to apply these simple algorithms repeatedly to shrink the range of candidates for the value in question, i.e. the median or the element of rank k, in each iteration and finally obtain the exact value after several passes. In this section the number of required passes is analyzed. Problems of that kind typically require $\Theta(\log n)$ "passes" for random-access models.

The approximation m obtained after a single pass splits the original set into two sets, one with values smaller than m and the others larger. Only one of these two sets can contain the element in question, so dropping the other one reduces the number of elements. This is repeated until the remaining set completely fits into the constant storage and the desired element is deduced directly. Since it is not known whether the requested element is smaller or larger than the approximation m, every second pass is used to count the number of smaller/larger elements relative to m which reveals the rank of m and thereby determines which set to drop. By remembering the number of discarded elements, the new position in the reduced set can be computed.

The analysis treats the more involved case where n is known beforehand. The other case is contained in the proof by fixing $s=2$.

Theorem 4. *Using algorithm 2 with constant storage of size $s\geq 2$ repeatedly, the element of rank k in a set can be found after $\mathcal{O}\!\left(n^{\frac{1}{s}}\right)$ passes.*

Proof: Let the function $P(n)$ describe the number of passes necessary for n elements. Each pass removes $c_1 n^{\frac{1}{s}}$ elements for some fixed $c_1>0$ by theorem 3, so we have $P(n)\leq P(n-c_1 n^{1-\frac{1}{s}})+2$. Guess $P(n)\leq c_2 n^{\frac{1}{s}}$ for a fixed $c_2>0$.

$$P(n)\leq P(n-c_1 n^{1-\frac{1}{s}})+2\leq c_2\left(n-c_1 n^{1-\frac{1}{s}}\right)^{\frac{1}{s}}+2=c_2 n^{\frac{1}{s}}\left(1-\frac{c_1}{n^{\frac{1}{s}}}\right)^{\frac{1}{s}}+2\overset{!}{\leq}c_2 n^{\frac{1}{s}}$$

$$\Leftrightarrow\left(1-\frac{c_1}{n^{\frac{1}{s}}}\right)^{\frac{1}{s}}+\frac{2}{c_2 n^{\frac{1}{s}}}\overset{!}{\leq}1\Leftrightarrow 1-\frac{c_1}{n^{\frac{1}{s}}}\overset{!}{\leq}\left(1-\frac{2}{c_2 n^{\frac{1}{s}}}\right)^{s}$$

For $c_2=\frac{2s}{c_1}$ this is Bernoulli's inequality and since s and c_1 are fixed, c_2 is constant. $\qquad\square$

9   Conclusion

This is the first deterministic result for non-trivially approximating splitters in streams storing only a constant number of reference elements. The median problem itself occurs in many algorithms and hence is quite important in theory and practice. Especially for tiny devices like sensor nodes in a network, the model is appropriate and the results are novel. Although several solutions exist using randomization and polylogarithmic memory, it is nevertheless of interest whether the same results can be achieved in a (simple) deterministic way and/or with less memory.

For streams with unknown size, the case for storage size two is solved asymptotically optimal, for streams with known size an ε-approximation algorithm is presented. A multi-pass solution for finding the exact median or any element of a specified rank is presented. Several results are not only asymptotic, but are given explicitly with reasonable constants.

All the algorithms are fast and simple and have a tiny memory footprint in practice. They have worst-case guarantees but showed a much better average behavior tested on random streams with several distributions.

Whether a linear approximation is achievable with constant storage or not remains an interesting open problem.

Acknowledgment. Thanks to Britta Denner-Broser and Klaus Kriegel for their support.

References

1. M. Blum, R. W. Floyd, V. R. Pratt, R. L. Rivest, and R. E. Tarjan. Time bounds for selection. *J. Comput. Syst. Sci.*, 7(4):448–461, 1973.
2. M. Greenwald and S. Khanna. Space-efficient online computation of quantile summaries. In *SIGMOD '01: Proceedings of the 2001 ACM SIGMOD international conference on Management of data*, pages 58–66, New York, NY, USA, 2001. ACM Press.
3. S. Guha and A. McGregor. Approximating quantiles and the order of the stream. In *PODS*, 2006.
4. S. Guha, A. McGregor, and S. Venkatasubramanian. Streaming and sublinear approximation of entropy and information distances. In *SODA*, 2006.
5. R. Jain and I. Chlamtac. The p2 algorithm for dynamic calculation of quantiles and histograms without storing observations. *Commun. ACM*, 28(10):1076–1085, 1985.
6. G. S. Manku, S. Rajagopalan, and B. G. Lindsay. Approximate medians and other quantiles in one pass and with limited memory. *SIGMOD Rec.*, 27(2):426–435, 1998.
7. J. I. Munro and M. Paterson. Selection and sorting with limited storage. *Theoretical Computer Science*, 12:315–323, 1980.
8. J. I. Munro and V. Raman. Selection from read-only memory and sorting with minimum data movement. *Theoretical Computer Science*, 165(2):311–323, 1996.
9. M. Paterson. Progress in selection. In *SWAT '96: Proceedings of the 5th Scandinavian Workshop on Algorithm Theory*, pages 368–379, 1996.
10. J. S. Vitter. Random sampling with a reservoir. *ACM Trans. Math. Softw.*, 11(1):37–57, 1985.

Optimal Algorithms for Tower of Hanoi Problems with Relaxed Placement Rules⋆

Yefim Dinitz and Shay Solomon

Dept. of Computer Science
Ben-Gurion University of the Negev, Beer-Sheva 84105, Israel
{dinitz, shayso}@cs.bgu.ac.il

Abstract. We study generalizations of the Tower of Hanoi (ToH) puzzle with relaxed placement rules. In 1981, D. Wood suggested a variant, where a bigger disk may be placed *higher than* a smaller one if their size difference is less than k. In 1992, D. Poole suggested a natural disk-moving strategy, and computed the length of the shortest move sequence (algorithm) under its framework. However, other strategies were not considered, so the lower bound/optimality question remained open. In 1998, Beneditkis, Berend, and Safro were able to prove the optimality of Poole's algorithm for the first non-trivial case $k = 2$ only. We prove it be optimal in the general case. Besides, we prove a tight bound for the diameter of the configuration graph of the problem suggested by Wood. Further, we consider a generalized setting, where the disk sizes should not form a continuous interval of integers. To this end, we describe a finite family of potentially optimal algorithms and prove that for any set of disk sizes, the best one among those algorithms is optimal. Finally, a setting with the *ultimate* relaxed placement rule (suggested by D. Berend) is defined. We show that it is not more general, by finding a reduction to the second setting.

1   Introduction

The classic Tower of Hanoi (ToH) puzzle is well known. It consists of three pegs and disks of sizes $1, 2, \ldots, n$ arranged on one of the pegs as a "tower", in decreasing, bottom-to-top size. The goal of the puzzle is to transfer all disks to another peg, placed in the same order. The rules are to move a single disk from (the top of) one peg to (the top of) another one, at each step, subject to the divine rule: to never have a larger disk above a smaller one.

The goal of the corresponding mathematical problem, which we denote by $HT = HT_n$, is to find a sequence of moves ("algorithm") of a minimal length ("optimal"), solving the puzzle. We denote the pegs naturally as *source, target,* and *auxiliary*, while the size of a disk is referred as its name. The following algorithm γ_n is taught in introductory CS courses as a nice example of a recursive algorithm. It is known and easy to prove that it solves HT_n in $2^n - 1$ disk moves, and is the unique optimal algorithm for it.

⋆ Partially supported by the Lynn and William Frankel Center for Computer Science.

T. Asano (Ed.): ISAAC 2006, LNCS 4288, pp. 36–47, 2006.
© Springer-Verlag Berlin Heidelberg 2006

- If n is 1, move disk n from *source* to *target*.
- Otherwise:
  - recursively perform $\gamma_{n-1}(source, auxiliary)$;
  - move disk n from *source* to *target*;
  - recursively perform $\gamma_{n-1}(auxilary, target)$.

In the recent two decades, various ToH type problems were considered in the mathematical literature. Many algorithms were suggested, and extensive related analysis was performed. As usual, the most difficult, far not always achievable task is showing that a certain algorithm is optimal, by providing the matching *lower bound*. A distinguished example is the Frame-Stewart algorithm (of 1941), solving the generalization of the ToH problem to four or more pegs. It is simple, and an extensive research was conducted on its behavior, since then. However, its optimality still remains an open problem; the proof of its *approximate* optimality [6] was considered a breakthrough, in 1999. This paper contributes to the difficult sub-area of the ToH research—optimality proofs.

In 1981, D. Wood [7] suggested a generalization of HT, characterized by the *k-relaxed placement rule*, $k \geq 1$: *If disk j is placed higher than disk i on the same peg (not necessarily neighboring it), then their size difference $j - i$ is less than k.* In this paper, we refer it as the *bottleneck Tower of Hanoi problem* (following D. Poole [3]), and denote it $BTH_n = BTH_{n,k}$. Now, there are more than one legal way to place a given set of disks on the same peg, in general; we refer the decreasing bottom-to-top placement of all disks on the same peg as the *perfect* disk configuration. If k is 1, we arrive at the classic ToH problem.

In 1992, D. Poole [3] suggested a natural algorithm for BTH_n and declared its optimality. However, his (straightforward) proof is done under the fundamental assumption that before the last move of disk n to the (empty) target peg, *all* other $n-1$ disks are gathered on the spare peg. This situation is far not general, since before the last move of disk n, from some peg X to the target peg, any set of the disks $n-1, n-2, \ldots, n-k+1$ may be placed below disk n on peg X.

In 1998, S. Beneditkis, D. Berend, and I. Safro [1] gave a (far not trivial) proof of optimality of Poole's algorithm for the first non-trivial case $k = 2$ only.

We prove it for the general case, by different techniques. Besides, we prove a tight bound for the diameter of the configuration graph of BTH_n. In other words, we find the length of the *longest* one among all shortest sequence of moves, over all pairs of initial and final configurations, up to a constant factor. We also prove that the *average* length of shortest sequence of moves, over all pairs of initial and final configurations, is the same as the above diameter for all values of $n \leq k$ and $n > 3k$, up to a constant factor.

X. Chen et al. [2] considered independently a few ToH problems, including the bottleneck ToH problem. They also suggested a proof of optimality of Poole's algorithm; it is based on another technical approach, and is not less difficult than our proof.

Further, we consider the "subset" setting, generalizing $BTH_{n,k}$, where the disk sizes should not form a continuous interval of integers. Here, for different disks, there are different number of bigger disks allowed to be placed above them.

For this setting, we describe the finite family of potentially optimal algorithms, and prove that for any set of disk sizes, the best one among those algorithms is optimal. Poole's algorithm is the simplest one in this family; all of its other members do not obey the fundamental assumption made by Poole.

Finally, following a suggestion of D. Berend, we define the most general setting of relaxed placement rule, where the sets of bigger disks allowed to be placed above each disk may be arbitrary, obeying monotonicity only. We show that this "ultimate" setting is not more difficult than the "subset" setting. This is done in two ways: by generalizing the relevant proofs and by finding a reduction from the "ultimate" to the "subset" setting.

The preliminary version of this paper (with results on BTH_n only) was presented at the Workshop on the Tower of Hanoi and Related Problems (2005).

2   Definitions and Notation

A configuration of a disk set D is called *gathered*, if all disks in D are on the same peg. Such a configuration is called *perfect*, if D is an initial interval of naturals, and the order of disks (on a single peg) is decreasing. For any configuration C of D and any $D' \subseteq D$, the *restriction* $C|_{D'}$ is C with all disks not in D' removed.

A move of disk m from peg X to peg Y is denoted by the triplet (m, X, Y). For a disk set D, the configuration of $D \setminus \{m\}$ is the same before and after such a move; we refer it as the *configuration of $D \setminus \{m\}$ during (m, X, Y)*.

A *packet-move*, P, of D is a sequence of moves transferring D from one peg to another. W.r.t. P, the former peg is called *source*, the latter *target*, and the third peg *auxiliary*. The *length* $|P|$ of P is the number of moves in it. If both initial and final configurations of P are perfect, we call P a *perfect-to-perfect* (or *p.t.p.*, for short) packet-move of D.

For better mnemonics (following [2]), the entire set of disks $[1..m]$ is divided into $\lceil \frac{m}{k} \rceil$ *blocks* $B_i = B_i(m)$: $B_1 = [(m-k+1)..m]$, $B_2 = [(m-2k+1)..(m-k)]$, ..., $B_{\lceil \frac{m}{k} \rceil} = [1..(m - (\lceil \frac{m}{k} \rceil - 1) \cdot k)]$. Note that the set of disks in any block is allowed to be placed on the same peg in an arbitrary order. For any $m \geq 1$, let D_m denote $[1..m]$, and $Small(m)$ denote $D_m \setminus B_1(D_m)$. In the above notion, BTH_n concerns finding the shortest perfect-to-perfect packet-move of D_n.

We say that a packet-move P *contains* a packet-move P' if P' is a subsequence of P. Several packet-moves P_i, $1 \leq i \leq r$, contained in P, are called *disjoint* if the last move in P_i precedes the first move in P_{i+1}, for each $1 \leq i \leq r-1$.

For any sequence of moves S of D and any $D' \subseteq D$, the *restriction of S to D'*, denoted by $S|_{D'}$, is the result of omission from S all moves of disks not in D'. Note that any restriction of a legal sequence of moves is legal as well, and a restriction of a packet-move to D' is a packet-move of D'. Clearly, if D is partitioned into D' and D'', then $|P| = |P|_{D'}| + |P|_{D''}|$.

Consider a sequence of moves, S, containing two *consequent* moves of the same disk: (i, X, Y) and (i, Y, Z). Their replacement by the single move (i, X, Z), if $X \neq Z$, or the deletion of both, if $X = Z$, is called a *pruning* of S. We denote by $Prune(S)$ the result of all possible prunings, at S; it is easy to see that such a result is independent on the order of particular prunings and is legal, so $Prune(S)$ is well defined.

3 The Shortest "Somehow" Packet-Move

In this section, we consider general, not p.t.p. packet-moves.

In the attempts of S. Beneditkis, D. Berend, and I. Safro [1] to solve BTH_n, much like in [3], another related problem of "moving somehow", under the k-relaxed placement rule arose: *To move m disks* [1..m], *placed entirely on one peg, to another peg, in any order.* For solving it, the following *algorithm* $\beta_m = \beta_m(source, target)$ was presented:

– If m is at most k, move all disks from *source* to *target* one by one.
– Otherwise:
  • recursively perform $\beta_{m-k}(source, auxiliary)$;
  • move disks $[(m - k + 1)..m]$ from *source* to *target* one by one;
  • recursively perform $\beta_{m-k}(auxilary, target)$.

Notice that the sequence β_n is similar to γ_n, but deals with blocks, instead of single disks. When β_m is applied to the perfect disk configuration, it is legal, and results in the configuration, different from the perfect one by the *increasing* order of disks in $B_1(m)$. When β_m is applied to the latter configuration, it is legal and results in the perfect configuration.

For $k = 2$, it is proved in [1] that no sequence of moves shorter than β_m can transfer m disks from one peg to another. We generalize this result for general k. Let b_m denote the length of β_m. By definition of β_m, $b_m = m$ if $m \le k$, and $b_m = 2b_{m-k} + k$, otherwise. In [3], this recurrence relation is shown to imply the explicit formula $b_n = k \cdot (2^{\lfloor \frac{n}{k} \rfloor} - 1) + r \cdot 2^{\lfloor \frac{n}{k} \rfloor} = (k + r) \cdot 2^{\lfloor \frac{n}{k} \rfloor} - k$, where $r = n \bmod k$. As a consequence, $b_n - b_{n-1} = 2^{\lceil n/k \rceil - 1}$; in particular, the sequence $\{b_i\}$ is *strictly* monotonous.

Note that during a move (m, X, Y), all disks in $Small(m)$ are on the spare peg $Z \ne X, Y$. As a corollary, holds:

Fact 1. *If a sequence of moves S begins from a configuration, where disk m and Small(m) are on X, and finishes at a configuration, where they are on Y, $X \ne Y$, then it contains two disjoint packet-moves of Small(m): one (from X) before the first move of disk m in S and another (to Y) after its last move.*

Theorem 2. *Under the rules of BTH_m, the length of any packet-move of D_m is at least b_m.* (proof is omitted)

4 Optimal Solution to BTH_n

Let us describe *algorithm* α_n, presented first in [3] and independently afterwards in [1], solving BTH_n:

- perform $\beta_{n-1}(source, auxiliary)$;
- move disk n from *source* to *target*;
- perform $\beta_{n-1}(auxilary, target)$.

For $k = 2$, it is proved in [1] that α_n is an optimal solution to BTH_n. In this section, we generalize this result to the case of general k.

Let a_n denote the length of α_n. The explicit formula for a_n (established in [3]) is implied strait-forwardly by that for b_{n-1}: $a_n = 2(r + k) \cdot 2^{\lfloor \frac{n-1}{k} \rfloor} - 2k + 1$, where $r = (n - 1) \bmod k$. The following theorem implies that α_n is an optimal algorithm that solves the puzzle.

Theorem 3. *Any p.t.p. packet-move of D_n solving BTH_n is of length at least a_n.*

The rest of this section is devoted to the proof of this Theorem. Recall that $Small(m) = [1..(m - k)]$, $B_1(m) = [(m - k + 1)..m]$. In our study of a packet-move P of D_m, $m > k$, we usually consider separately $P|_{Small(m)}$ and $P|_{B_1(m)}$. The reason is that any move (m, X, Y) defines completely the placement of $Small(m)$—on the spare peg Z,—while the placement of disks in $B_1(m) \setminus \{m\}$ may be arbitrary, during such a move. For the analysis of $P|_{B_1(m)}$, we use the following statement:

Fact 4. *Any p.t.p. packet-move P of D_m contains at least two moves of any disk i, $i \neq m$: at least one before the first move of disk m and at least one after its last move. Hence, for any $D \subseteq D_{m-1}$, holds $|P|_D| \geq 2|D|$.*

4.1   Case 1: Disk n Never Moves to the Auxiliary Peg

We call a move of disk m, in a packet-move of D_m, *distinguished* if it is between *source* and *target* and disk $m - 1$ is at *auxiliary*, during that move.

Lemma 1. *Any packet-move P of D_m, which preserves the initial order between disks m and $m - 1$, and such that disk m moves only between the source and target pegs, contains a distinguished move of disk m.*

Proof. Define $P' := Prune(P|_{\{m-1,m\}})$. Note that P' preserves the initial order between disks m and $m-1$, since P preserves it. We will study P', since a move of m is distinguished in P' if and only if it is distinguished in P. By the definition of *Prune*, moves of disks m and $m - 1$ must interchange, in P'.

Claim. Under the conditions of the Lemma, the first move of disk $m - 1$ should be $(m - 1, source, auxiliary)$. (the proof is omitted)

Based on this Claim, we consider the possible scenarios of P'. Assume that disk $m - 1$ is initially above disk m. Then, necessarily, the first two moves are: $(m - 1, source, auxiliary)$, $(m, source, target)$; the latter one is distinguished. Assume that disk $m-1$ is initially below disk m. Then, necessarily, the first three moves are: $(m, source, target)$, $(m - 1, source, auxiliary)$, $(m, target, source)$; the latter one is distinguished. Hence, also P contains a distinguished move. $\square$

Proposition 1. *For any $m > k + 1$ and any packet-move P of D_m preserving the initial order between disks m and $m - 1$, s.t. disk m moves only between the source and target pegs, P contains four disjoint packet-moves of $Small(m - 1)$.*

Proof. Let us consider the first distinguished move of disk m during P; it exists by Lemma 1. Clearly, during such a move, all disks in $Small(m - 1)$ are on *auxiliary*, together with disk $m - 1$; by the k-relaxed placement rule, they are placed above disk $m - 1$. Hence, by Fact 1, the parts of P before and after that move contain two disjoint packet-moves of $Small(m - 1)$ each. Altogether, there are four disjoint packet-moves of $Small(m - 1)$, in P.   □

Corollary 1. *The length of any p.t.p. packet-move of D_n, such that disk n moves only between the source and target pegs, is at least a_n.*

Proof. Let P be a packet-move as in the Corollary. By Fact 4, $|P|_{B_1(n-1)}| \geq 2 \cdot |B_1(n-1)|$. If $n \leq k+1$, $|P| = |P|_{B_1(n-1)}| + |P|_{\{n\}}| \geq 2(n-1) + 1 = 2 \cdot b_{n-1} + 1 = a_n$. Otherwise, by Proposition 1 and Theorem 2, $|P| = |P|_{Small(n-1)}| + |P|_{B_1(n-1)}| + |P|_{\{n\}}| \geq 4 \cdot b_{n-k-1} + 2k + 1 = 2 \cdot b_{n-1} + 1 = a_n$.   □

4.2   Case 2: Disk n Moves to the Auxiliary Peg

Lemma 2. *If $m > k$ and a packet-move P of D_m contains a move of disk m to auxiliary, then $P|_{Small(m)}$ contains three disjoint packet-moves of $Small(m)$.* (the proof is omitted.)

Following is the central statement of this section.

Proposition 2. *For any $m, l \geq 0$, let P be a p.t.p. packet-move of D_m containing $2l + 1$ disjoint packet-moves of D_m. Then, $|P| \geq 2l \cdot b_m + 2 \cdot b_{m-1} + 1 = 2l \cdot b_m + a_m$, and this bound is tight.*

Proof. It can be easily proved that any packet-move as in the Proposition may be divided into t, $t \geq 2l+1$, packet-moves of D_m. If $t > 2l+1$, then, by Theorem 2 and the strict monotonicity of (b_i), $|P| \geq (2l + 2) \cdot b_m \geq 2l \cdot b_m + 2 \cdot b_{m-1} + 2$. We henceforth assume that P is divided into $2l + 1$ packet-moves of D_m.

We prove by a complete induction on m, for all l. *Basis*: $m \leq k$. (Note that in this basic case, the disks may be placed at pegs in an arbitrary order).

Lemma 3. *In the case $m \leq k$, the length of any p.t.p. packet-move P of D_m, divided into $2l + 1$ packet-moves of D_m, is at least $(2l + 2)m - 1$.*

Proof. We call a disk in D_m *expensive* w.r.t. some packet-move P', if it moves to the auxiliary peg during P'. Let us prove that there could be at most one disk, which is not expensive w.r.t. any one out of the $2l + 1$ packet-moves as in the Lemma. Assume to the contrary that there are two such disks, i and j. Then, after each packet-move, their order is reversed. Since there is an odd number of packet-moves, the order of i and j at the final configuration is inverse to that in the initial configuration,—a contradiction.

It follows that either all m disks or some $m - 1$ disks make at least $2l + 2$ moves each, while the remained disk, if any, makes at least $2l + 1$ moves. Altogether, at least $(2l + 2)m - 1$ moves are made.

As a corollary, since $b_r = r$, for any r, $1 \leq r \leq k$, holds $|P| \geq (2l+2)m - 1 = 2lm + 2(m-1) + 1 = 2l \cdot b_m + 2 \cdot b_{m-1} + 1$.

Induction step: $m > k$. We suppose that the claim holds for all lesser values of m and for all l, and prove it for m and all l.

Note that at the initial configuration of P, as well as at its final configuration, disk m is placed below disk $m - 1$. Since P is a composition of an odd number of disjoint packet-moves of D_m, there exists a packet-move among them, henceforth denoted by $\tilde{P}$, which preserves the order between disks m and $m - 1$. We bound $|P|_{Small(m)}|$ separately, for two complementary types of $\tilde{P}$.

Case 1: During $\tilde{P}$, disk m never moves to the auxiliary peg. By Proposition 1, $\tilde{P}$ contains four disjoint packet-moves of $Small(m-1)$. We notice that $\tilde{P}$ contains at least two moves of disk $m - k$: at least one before the first move of disk m, and at least one after its last move. By Theorem 2, $\tilde{P}|_{Small(m)}) \geq 4 \cdot b_{m-k-1} + 2 = 2 \cdot b_{m-1} - 2k + 2$. By Corollary 3, the other $2l$ packet-moves of D_m contain two disjoint packet-moves of $Small(m)$ each. Hence, their total length is at least $4l \cdot b_{m-k} = 2l(b_m - k)$. Therefore, $|P|_{Small(m)}| \geq 2l \cdot b_m - 2lk + 2 \cdot b_{m-1} - 2k + 2 = 2l \cdot b_m + 2 \cdot b_{m-1} - (2l+2)k + 2$. We denote this value by N.

Case 2: $\tilde{P}$ contains a move of disk m to the auxiliary peg. By Lemma 2, $\tilde{P}|_{Small(m)}$ contains three disjoint packet-moves of $Small(m)$. By Corollary 3, the other $2l$ packet-moves of D_m contain two disjoint packet-moves of $Small(m)$ each. Thus, $P|_{Small(m)}$ contains $4l + 3$ disjoint packet-moves of $Small(m)$. By the induction hypothesis, $|P|_{Small(m)}| \geq (4l+2) \cdot b_{m-k} + 2 \cdot b_{m-k-1} + 1 \geq 4l \cdot b_{m-k} + 4 \cdot b_{m-k-1} + 3 = 2l \cdot b_m + 2 \cdot b_{m-1} - (2l+2)k + 3 = N + 1$ (the second inequality holds since the sequence (b_i) is strictly monotonous).

By Lemma 3, $|P|_{B_1(m)}|$ is at least $(2l+2)k - 1$. So, $|P| = |P|_{Small(m)}| + |P|_{B_1(m)}| \geq N + |P|_{B_1(m)}| \geq 2l \cdot b_m + 2 \cdot b_{m-1} + 1$, as required.

The bound is tight, since the sequence composed from $2l$ β_m and one α_m, in this order, is a p.t.p. packet-move of length equal to this bound. $\square$

Now, Theorem 3 follows from Proposition 2 with $l = 0$ and $m = n$.

The difference of 1 between bounds at Cases 1 and 2, together with the tightness of the bound of Case 1, implies:

Corollary 2. 1. *No optimal algorithm for BTH_n contains a move of disk n to the auxiliary peg.*
  2. *Any optimal algorithm for BTH_n contains just a single move of disk n, from the source peg to the target peg.*
  3. *The only difference of an arbitrary optimal algorithm for BTH_n from α_n could be in choosing another optimal algorithms, for the two included optimal "somehow" packet-moves of D_{n-1}, instead of β_{n-1}.*

5   Diameter of the Configuration Graph of BTH_n

While the shortest perfect-to-perfect sequence of moves had been already found, we do not know what is such a sequence for transforming an arbitrary (legal) configuration to another given (legal) one, and what is its length. We study, what is the length of the *longest* one among all shortest sequences of moves, over all pairs of initial and final configurations. In other words, what is the diameter, denoted by $D(n, k)$, of the directed graph of all the configurations of the disks in $[1..n]$, under the k-relaxed placement rule?

The proof of the following theorem is omitted.

Theorem 5.

$$Diam(n, k) = \begin{cases} \Theta(n \cdot \log n) & \text{if } n \leq k \\ \Theta(k \cdot \log k + (n - k)^2) & \text{if } k < n \leq 2k \\ \Theta(k^2 \cdot 2^{\frac{n}{k}}) & \text{if } n > 2k \ . \end{cases}$$

Another question is what is the *average* length of shortest sequences of moves, over all pairs of initial and final configurations, denoted by $Avg(n, k)$. The following theorem states that it is the same as $D(n, k)$ for all values of $n \leq k$ and $n > 3k$, up to a constant factor (its proof is omitted).

Theorem 6.

$$Avg(n, k) = \begin{cases} \Theta(n \cdot \log n) & \text{if } n \leq k \\ \Theta(k^2 \cdot 2^{\frac{n}{k}}) & \text{if } n > 3k \ . \end{cases}$$

6   "Subset" Setting

In this section, the previous study is generalized to $BTH_D = BTH_{D,k}$, where the disk placement is still subject to the k-relaxed rule, but the disk set D is an *arbitrary* set, not necessarily a contiguous interval of naturals.

6.1   Preliminaries

Let us generalize some of the definitions given for BTH. For a non-empty disk set D, $|D| = n$, let us denote the disks of maximal and minimal size by $max(D)$ and $min(D)$, respectively, and by $s(D)$ the "stretch" $max(D) - min(D) + 1$. The ith biggest disk in D is denoted by $D(i)$. We define $D^- = D \setminus \{max(D)\}$. For a disk set D, its division into blocks is as follows: $B_1 = B_1(D) = D \cap [(max(D) - k + 1)..max(D)]$, and for any $i > 1$, s.t. $\bigcup_{j<i} B_j(D) \neq D$, $B_i = B_i(D) = B_1(D \setminus \bigcup_{j<i} B_j(D))$. The size of each block is between 1 and k, and the number of blocks $\#(D) = \#_k(D)$ is between $\lceil n/k \rceil$ and $\min\{n, s(D)/k\}$. We define $Small(D) = \bigcup_{j\geq 2} B_j(D)$. We generalize the optimal "moving somehow" algorithm β_m to the optimal algorithm $\beta_D = \beta_D(source, target)$:

- If $s(D) \leq k$, move all disks from *source* to *target* one by one.
- Otherwise:
  - recursively perform $\beta_{Small(D)}(source, auxiliary)$;
  - move disks in $B_1(D)$ from *source* to *target* one by one;
  - recursively perform $\beta_{Small(D)}(auxilary, target)$.

Denote by b_D the length of β_D. By definition of β_D, holds $b_D = n$, if $s(D) \leq k$, and $b_D = 2 \cdot b_{Small(D)} + |B_1(D)|$, otherwise. It can be proved that the value b_D is *strictly* inclusion-wise monotonous in D.

The proofs of the following statements are similar to those for the corresponding ones for BTH_n.

Theorem 7. *Any packet-move of D is of length at least b_D.*

Corollary 3. *For any packet-move P of D, $P|_{Small(D)}$ contains two disjoint packet-moves of $Small(D)$.*

Fact 8. *Any p.t.p. packet-move P of D contains at least two moves of any disk i, $i \neq max(D)$. Thus, for any $D' \subseteq D^-$, holds $|P|_{D'}| \geq 2|D'|$.*

6.2   The Set of Potentially Optimal Algorithms

Let us generalize algorithm α_n to α_D:

- perform $\beta_{D^-}(source, auxiliary)$;
- move disk $max(D)$ from $source$ to $target$;
- perform $\beta_{D^-}(auxilary, target)$.

We denote $\alpha_D^0 = \alpha_D$, and define α_D^i, $1 \leq i \leq \#(D) - 1 \leq n - 1$, as follows:

- perform $\beta_{Small(D)}(source, target)$.
- move disks in $B_1(D)$ from $source$ to $auxilary$ one by one;
- perform $\beta_{Small(D)}(target, source)$;
- move disks in $B_1(D)$ from $auxilary$ to $target$ one by one;
- recursively perform $\alpha_{Small(D)}^{i-1}(source, target)$.

We define $a_D^i = |\alpha_D^i|$. Clearly, $a_D^i = a_{Small(D)}^{i-1} + 2 \cdot b_{Small(D)} + 2 \cdot |B_1(D)|$. Let us denote $\bar{a}_D = min_{0 \leq i \leq \#(D)-1} a_D^i$.

It is easy to show, by induction on i, that each one of the algorithms α_D^i, $0 \leq i \leq \#(D) - 1$, solves BTH_D: the first and third items move out and return $Small(D)$ to $source$, the second and fourth items move $B_1(D)$ from $source$ to $target$, in the right order, and the fifth item does the same for $Small(D)$. The following Theorem shows that the length of the shortest sequence of moves solving BTH_D is $\bar{a}_D$.

Theorem 9. *The best one out of the $\#(D)$ algorithms α_D^i is optimal for BTH_D.*

The proof of this Theorem has the same structure as that of Theorem 3. It is based on the following statements, whose proofs are similar to those of the corresponding ones for BTH_n. The difficulty of proofs remains approximately the same, except for that of Proposition 4, which becomes more involved.

Proposition 3. *Let P be a packet-move of disks D', preserving the initial order between disks $max(D')$ and $max(D'^-)$, and such that disk $max(D')$ moves only between the source and target pegs. Then, $P|_{Small(D'^-)}$ contains four disjoint packet-moves of $Small(D'^-)$.*

Corollary 4. *If P is a p.t.p. packet-move of D, such that disk $max(D)$ moves only between the source and target pegs, then $|P| \geq a_D^0$.*

Lemma 4. *If a packet-move P of D' contains a move of disk $max(D')$ to the auxiliary peg, then $P|_{Small(D')}$ contains three disjoint packet-moves of $Small(D')$.*

Proposition 4. *Let P be a p.t.p. packet-move of D', containing $2l + 1$ disjoint packet-moves of D'. Then, $|P| \geq 2l \cdot b_{D'} + \bar{a}_{D'}$.*

Proof. The proof is made by induction on $\#(D')$, for all l. The basis of the induction is similar to that of Proposition 2. At the induction step, the case analysis is more involved. As in the proof of Proposition 2, we point our finger at a "guilty" packet-move, denoted by $\tilde{P}$, among the $2l + 1$ disjoint packet-moves contained in P. Due to space constraints, we show only the case satisfying the following conditions: 1. $\tilde{P}$ contains a move of disk $max(D')$ to the auxiliary peg; 2. $|B_1(D')| \geq 2$; 3. $|P|_{B_1(D')}| = (2l + 2)|B_1(D')| - 1$. When considering packet-moves forming P, we will use the names *source*, *target*, and *auxiliary* w.r.t. the currently considered packet-move.

It is easy to prove that there exists a disk $d \neq max(D')$ in $B_1(D')$ that does not move to *auxiliary* during P, and all disks in $B_1(D') - \{d\}$ move to *auxiliary* exactly once during P. In any packet-move other than $\tilde{P}$, d and $max(D')$ reverse their order. Since P is a composition of an odd number of disjoint packet-moves of D', d and $max(D')$ preserve their order in $\tilde{P}$. In the initial configuration of $\tilde{P}$, $max(D')$ must be placed higher than d, for otherwise they would reverse their order. The first move of $max(D')$ is to *auxiliary*. Before that move, all disks in $Small(D')$ have been moved to *target*. Before the move of disk d to *target*, all disks in $Small(D')$ have been moved to *auxiliary* above disk $max(D')$, otherwise disk $max(D')$ would not be able to move above disk d, on *target*. Before disk $max(D')$ moves from *auxiliary* to *target*, all disks in $Small(D')$ have been moved to *source*, and are due to be moved to *target* at some point. In total, $\tilde{P}|_{Small(D')} \geq 4 \cdot b_{Small(D')}$. Thus, $P|_{Small(D')} \geq (4l + 4) \cdot b_{Small(D')}$. We recall that $|P_{B_1(D')}| = (2l+2)|B_1(D')| - 1$. Altogether, $|P| = |P|_{Small(D')}| + |P_{B_1(D')}| \geq (4l + 4) \cdot b_{Small(D')} + (2l + 2)|B_1(D')| - 1 \geq$ (details are omitted) $2lb_{D'} + a_{D'}$.

6.3 Tightness and Other Related Issues

Recall that for the case $k = 1$, for any disk set D, $\alpha_D = \alpha_D^0$ is the shortest p.t.p. packet-move of D. It can be proven that it is the unique optimal algorithm.

Proposition 5. *For any $k \geq 2$, n, and $\lceil \frac{n}{k} \rceil \leq p \leq \lceil \frac{n}{2} \rceil$, there exists a set D, with $|D| = n$ and $\#(D) = p$, s.t. for any $0 \leq i, j \leq p - 1$, $|\alpha_D^i| = |\alpha_D^j|$.*
(the proof is omitted)

Theorem 10. *For any $k \geq 2$, n, $\lceil \frac{n-1}{k} \rceil + 1 \leq p \leq \lceil \frac{n}{2} \rceil$, and $0 \leq j \leq p - 2$, there exists a set D, with $|D| = n$ and $\#(D) = p$, s.t. for each $l \neq j$, $0 \leq l \leq p - 1$, holds $|\alpha_D^j| < |\alpha_D^l|$.*
(the proof is omitted)

Proposition 6. *For any $k \geq 2$, n, $\lceil (n-4)/k \rceil + 3 \leq p \leq n-1$, and $0 \leq j \leq p-1$, there exists a set D, with $|D| = n$ and $\#(D) = p$, s.t. for each $l \neq j$, $0 \leq l \leq p-2$, holds $|\alpha_D^j| > |\alpha_D^l|$.* (the proof is omitted)

We define the partial order "denser or equal" $\preceq_d$ between two integer sets, D_1 and D_2, of an equal cardinality, n, by $D_1 \preceq_d D_2 \Leftrightarrow \forall 1 \leq i \leq n-1 : D_1(i+1) - D_1(i) \leq D_2(i+1) - D_2(i)$.

Proposition 7.   *1. For any $D_1 \preceq_d D_2$, holds $\bar{a}_{D_1} \leq \bar{a}_{D_2}$.*
2. For any D, holds $a_{|D|} \leq \bar{a}_D \leq 2^{|D|} - 1$.
   (the proof is omitted)

In this paper, we considered a fixed value of k. The following statement shows that when k grows, BTH_D becomes more general.

Proposition 8. *For any two values $k_1 < k_2$ and any set D_1, there exists a set D_2, $|D_2| = |D_1|$, such that the instances BTH_{D_1,k_1} and BTH_{D_2,k_2} are equivalent in the following sense. The natural bijection $D_1(i) \leftrightarrow D_2(i)$ induces bijections between the families of legal configurations of BTH_{D_1,k_1} and BTH_{D_2,k_2} and thus between their families of legal sequences of moves.* (the proof is omitted)

7   Setting with the Ultimate Placement Rule

In this paper, we study shortest p.t.p. packet-moves under different types of relaxed placement rules. In this section, we consider the problem BTH_f, defined by the *ultimate* in a sense, relaxed placement rule, suggested by D. Berend. There are n disks $1, 2, \ldots, n$, and an (arbitrary) monotonous non-decreasing function $f : [1..n] \rightarrow [1..n]$, s.t. $f(i) \geq i$, for all $1 \leq i \leq n$. For each disk i, only disks of size at most $f(i)$ may be placed higher than i.

We consider the following restrictions natural: if some disk may be placed above i, then any smaller disk may also be placed above i, and if i may be placed above some disk, then i may also be placed above any larger disk. Besides, any disk smaller than i may be placed above i, since this is so at the initial, perfect configuration. Therefore, the free choice of f makes the above rule ultimate.

We state that BTH_f is equivalent to BTH_D. For this, let us first show that BTH_f is at least as general as BTH_D, by a reduction. For any instance of BTH_D, we set $n = |D|$ and consider the bijection $g_D : [1..n] \rightarrow D$, s.t. $g_D(i) = D(i)$. It induces naturally a function f as above, and thus an instance of BTH_f. It is easy to see that g_D and g_D^{-1} induce a bijection between the families of configurations at the two problem instances, which keeps the legality; thus, the two instances are equivalent.

Proving that BTH_D is as general as BTH_f becomes more complicated. One way is as follows. We define algorithms α_f^i and the value $\bar{a}_f$, similarly to α_D^i and $\bar{a}_D$. First, we observe that Theorem 11 below, analogous to Theorem 9, and certain analogues of Theorem 7, Proposition 5, Theorem 10, and Proposition 6 are valid, since the proofs can be generalized almost straightforwardly.

Theorem 11. *The best one out of the algorithms α_f^i is optimal for BTH_f.*

Besides, we have a reduction, which for any instance of BTH_f, constructs an instance of BTH_D, which is equivalent to it via the bijection g_D. The size of the instance constructed in this reduction is polynomial in n.

Hence, the theory of BTH_D can be extended to BTH_f.

Acknowledgment. Authors thank Daniel Berend for his suggestion to consider the question on the optimal algorithm for the bottleneck Tower of Hanoi problem, and for his constant willingness to help.

References

1. S. Beneditkis and I. Safro. Generalizations of the Tower of Hanoi Problem. Final Project Report, supervised by D. Berend, Dept. of Mathematics and Computer Science, Ben-Gurion University, 1998.
2. X. Chen, B. Tian, and L. Wang. Santa Claus' Towers of Hanoi. Manuscript, 2005.
3. D. Poole. The Bottleneck Towers of Hanoi Problem. *J. of Recreational Math.* **24** (1992), no. 3, 203-207.
4. P.K. Stockmayer. Variations on the Four-Post Tower of Hanoi Puzzle. *CONGRESSUS NUMERANTIUM* **102** (1994), 3-12.
5. P.K. Stockmayer. The Tower of Hanoi: A Bibliography. (1998). Available via http://www.cs.wm.edu/~pkstoc/h_Papers.html .
6. M. Szegedy, In How Many Steps the k Peg Version of the Towers of Hanoi Game Can Be Solved?, *Symposium on Theoretical Aspects of Computer Science* **1563** (1999), 356.
7. D. Wood. The Towers of Brahma and Hanoi revisited. *J. of Recreational Math.* **14** (1981-1982), no. 1, 17-24.

Flexible Word Design and Graph Labeling[*]

Ming-Yang Kao, Manan Sanghi, and Robert Schweller

Department of Electrical Engineering and Computer Science
Northwestern University
Evanston, IL 60208, USA
{kao, manan, schwellerr}@cs.northwestern.edu

Abstract. Motivated by emerging applications for DNA code word design, we consider a generalization of the code word design problem in which an input graph is given which must be labeled with equal length binary strings of minimal length such that the Hamming distance is small between words of adjacent nodes and large between words of non-adjacent nodes. For general graphs we provide algorithms that bound the word length with respect to either the maximum degree of any vertex or the number of edges in either the input graph or its complement. We further provide multiple types of recursive, deterministic algorithms for trees and forests, and provide an improvement for forests that makes use of randomization.

1   Introduction

This work can be viewed either as a generalization of codeword design or a special restricted case of the more general graph labeling problem. The problem of graph labeling takes as input a graph and assigns a binary string to each vertex such that either adjacency or distance between two vertices can be quickly determined by simply comparing the two labels. The goal is then to make the labels as short as possible (see [14] for a survey). Early work in the field [8, 9] considered the graph labeling problem with the restriction that the adjacency between two nodes must be determined solely by Hamming distance. Specifically, the labels for any two adjacent nodes must be below a given threshold, while the nodes between non-adjacent nodes must be above it.

We return to this restricted type of graph labeling motivated by growing applications in DNA computing and DNA self-assembly which require the design of DNA codes that exhibit non-specific hybridization. A basic requirement for building useful DNA self-assembly systems and DNA computing systems is the design of sets of appropriate DNA strings (code words). Early applications have simply required building a set of n equal length code words such that there is no possibility of hybridization between the words or Watson Crick complement of words [1, 4, 6, 7, 18, 23]. Using hamming distance as an approximation to how well a word and the Watson Crick complement of a second word will bind,

---

[*] Supported in part by NSF Grant EIA-0112934.

T. Asano (Ed.): ISAAC 2006, LNCS 4288, pp. 48–60, 2006.
© Springer-Verlag Berlin Heidelberg 2006

Table 1. Summary of our results

	Word Length	
	Lower Bound	Upper Bound
General Graphs (Matching Algorithm)	$\Omega(\gamma + n)$	$O(\gamma\hat{D} + \hat{D}n)$ Theorem 3
General Graphs (StarDestroyer)	Theorem 1	$O(\sqrt{\gamma^2\hat{m}n + \gamma\hat{m}n^2})$ Theorem 4
Forests (Randomized)	$\Omega(\gamma + \log n)$ Theorem 2	$O(\gamma D \log(\max\{f, \frac{n}{\gamma D}\}))$ Theorem 8

G : input graph (V, E)	m : number of edges in G
$\overline{G}$: complement of G	$\hat{m}$: smaller of the number of edges in G or $\overline{G}$
D : highest degree of any vertex in G	n : number of vertices in G
$\hat{D}$: smaller of the highest degree of any vertex in G or $\overline{G}$	f : maximum number of leaves in any tree in the input forest
γ : Hamming distance separation	

such a requirement can be achieved in part by designing a set of n words such that the Hamming distance between any pair in the set is large. There has been extensive work done in designing sets of words with this and other non-interaction constraints [5, 6, 10–13, 15–18, 21].

While the Hamming distance constraint is important for applications requiring that no pair of words in a code hybridize, new applications are emerging for which hybridization between different words in a code word set is desirable and necessary. That is, there is growing need for the efficient design of DNA codes such that the hybridization between any two words in the code is determined by an input matrix specifying which strands should bond and which should not. Aggarwal et al. [2, 3] have shown that a tile self assembly system that uses a set of glues that bind to one another according to a given input matrix, rather than only binding to themselves, greatly reduces the number of distinct tile types required to assemble certain shapes. Efficient algorithms for designing sets of DNA strands whose pairwise hybridization is determined by an input matrix may permit implementation of such tile efficient self-assembly systems.

Further, Tsaftaris et al. [19, 20] have recently proposed a technique for applying DNA computing to digital signal processing. Their scheme involves designing a set of equal length DNA strands, indexed from 1 to n, such that the melting temperature of the duplex formed by a word and the Watson Crick complement of another word is proportional to the difference of the indices of the words. Thus, this is again an example in which it is desirable to design a set of DNA words such that different words have varying levels of distinctness from one another.

Given an input graph G, we consider the problem of constructing a labeling such that the Hamming distance between labels of adjacent vertices is small, the Hamming distance between non-adjacent vertices is large, and there is a separation of at least γ between the small Hamming distance and the large Hamming distance. Breuer et al.[8, 9] first studied this problem for the special case of $\gamma = 1$ and achieved labels of size $O(Dn)$ for general graphs, where D is the degree of the node with the highest degree. By combining graph decompositions with codes similar in spirit to Hadamard codes from coding theory, we get a labeling of length $O(\gamma \hat{D} + \hat{D}n)$ where $\hat{D}$ is the smaller of the degree of the maximum degree node in G and its complement. We then explore more sophisticated graph decompositions to achieve new bounds that are a function of the number of edges in G. We also consider the class of trees and forests and provide various recursive algorithms that achieve poly-logarithmic length labels for degree bounded trees. Our forest algorithms also make use of probabilistic bounds from traditional word design to use randomization to reduce label length. Our results are summarized in Table 1.

Paper Layout: In Section 2, we introduce basic notation and tools and formulate the problem. In Section 3, we describe techniques for obtaining a restricted type of labeling for special graphs. In Section 4, we describe how to combine special graph labelings to obtain algorithms for labeling general graphs. In Section 5, we present recursive algorithms for labeling forests. In Section 6, we conclude with a discussion of future research directions. In the interest of space proofs and some algorithmic details are omitted in this version.

2   Preliminaries

Let $S = s_1 s_2 \ldots s_\ell$ denote a length ℓ bit string with each $s_i \in \{0, 1\}$. For a bit s, the *complement* of s, denoted by s^c, is 0 if $s = 1$ and 1 if $s = 0$. For a bit string $S = s_1 s_2 \ldots s_\ell$, the *complement* of S, denoted by S^c, is the string $s_1^c, s_2^c \ldots s_\ell^c$. For two bit strings S and T, we denote the concatenation of S and T by $S \cdot T$.

For a graph $G = (V, E)$, a length ℓ labeling of G is a mapping $\sigma : V \to \{0, 1\}^\ell$. Let $\deg(G)$ denote the maximum degree of any vertex in G and let $\bar{G} = (V, \bar{E})$ denote the complement graph of G. A γ-*labeling* of a graph G is a labeling σ such that there exist integers α and β, $\beta - \alpha \geq \gamma$, such that for any $u, v \in V$ the Hamming distance $\mathrm{H}(\sigma(u), \sigma(v)) \leq \alpha$ if $(u, v) \in E$, and $\mathrm{H}(\sigma(u), \sigma(v)) \geq \beta$ if $(u, v) \notin E$. We are interested in designing γ-labelings for graphs which minimize the length of each label, length(σ).

Problem 1 (Flexible Word Design Problem).
INPUT: Graph G; integer γ
OUTPUT: A γ-labeling σ of G. Minimize $\ell = $ length(σ).

Throughout this paper, for ease of exposition, we will assume the Hamming distance separator γ is a power of 2. For general graphs in Sections 3 and 4 we also assume the number of vertices n in the input graph is a power of 2. These assumptions can be trivially removed for these cases.

Theorem 1. *The required worst case label length for general n node graphs is $\Omega(\gamma + n)$.*

Theorem 2. *The required worst case label length for n node forests is $\Omega(\gamma + \log n)$.*

An important tool that we use repeatedly in our constructions is a variant of the Hadamard codes [22] from coding theory. The key property of this code is that it yields short words such that every pair of strings in the code has the exactly the same Hamming distance between them.

Hadamard Codes. We define two types of Hadamard codes using the Hadamard matrices. The size 2×2 Hadamard matrix is defined to be:

$$H_2 = \begin{bmatrix} 1 & 1 \\ 0 & 1 \end{bmatrix}$$

For n a power of 2 the size $n \times n$ Hadamard matrix H_n is recursively defined as:

$$H_n = \begin{bmatrix} H_{\frac{n}{2}} & H_{\frac{n}{2}} \\ H_{\frac{n}{2}}^c & H_{\frac{n}{2}} \end{bmatrix}$$

From the Hadamard matrices we define two codes. For n a power of 2 and γ a multiple of $\frac{n}{2}$, define the *balanced Hadamard* code $HR^B(n, \gamma)$ to be the set of words obtained by taking each row of H_n concatenated $\frac{2\gamma}{n}$ times. Similarly, define the *Hadamard* code $HR(n, \gamma)$ by taking each row of H_n, removing the last bit, and concatenating $\frac{2\gamma}{n}$ copies. The Hadamard and balanced Hadamard codes have the following properties.

Lemma 1. *Consider the codes $\mathrm{HR}(n, \gamma)$ and $\mathrm{HR}^B(n, \gamma)$ for n and γ powers of 2 and $2\gamma \geq n$. The following properties hold.*

1. *For any $S \in \mathrm{HR}(n, \gamma)$, length$(S) = 2\gamma - \frac{2\gamma}{n}$.*
2. *For any non-equal $S_i, S_j \in \mathrm{HR}(n, \gamma)$ (or any $S_i, S_j \in \mathrm{HR}^B(n, \gamma)$), $H(S_i, S_j) = \gamma$.*
3. *For any non-equal $S_i, S_j \in \mathrm{HR}(n, \gamma)$, $H(S_i, S_j^c) = \gamma - \frac{2\gamma}{n}$.*
4. *For any $S \in \mathrm{HR}^B(n, \gamma)$, length$(S) = 2\gamma$.*
5. *Let $F^B(n, \gamma) = \mathrm{HR}^B(n, \gamma) \setminus \{A_1, \ldots A_r\} \cup \{A_1^c, \ldots, A_r^c\}$ for an arbitrary subset $\{A_1, \ldots A_r\}$ of $\mathrm{HR}^B(n, \gamma)$. Then, properties 1 and 4 still hold for $F^B(n, \gamma)$.*
6. *The codes $\mathrm{HR}(n, \gamma)$ and $\mathrm{HR}^B(n, \gamma)$ can be computed in time $O(n \cdot \gamma)$.*

3   Exact Labelings for Special Graphs

In constructing a γ-labeling for general graphs, we make use of a more restrictive type of labeling called an *exact* labeling, as well as an *inverted* type of labeling. Such labelings can be combined for a collection of graphs to obtain labelings for larger graphs. We consider two special types of graphs in this section, matchings and star graphs, and show how to obtain short exact labelings for each. These results are then applied in Section 4 by algorithms that decompose arbitrary graphs into these special subgraphs efficiently, produce exact labelings for the subgraphs, and then combine the labelings to get a labeling for the original graph.

Definition 1 (Exact Labeling). *A γ-labeling σ of a graph $G = (V, E)$ is said to be* exact *if there exist integers α and β, $\beta - \alpha \geq \gamma$, such that for any two nodes $u, v \in V$ it is the case that $\mathrm{H}(\sigma(u), \sigma(v)) = \alpha$ if $(u, v) \in E$, and $\mathrm{H}(\sigma(u), \sigma(v)) = \beta$ if $(u, v) \notin E$. A labeling that only satisfies $\mathrm{H}(\sigma(u), \sigma(v)) = \alpha$ if $(u, v) \in E$, but $\mathrm{H}(\sigma(u), \sigma(v)) \geq \beta$ if $(u, v) \notin E$ is called a* lower exact labeling.

Definition 2 (Inverse Exact Labeling). *A labeling σ of a graph $G = (V, E)$ is said to be an* inverse exact labeling *for value γ if there exist integers α and β, $\beta - \alpha \geq \gamma$, such that for any two nodes $u, v \in V$ it is the case that $\mathrm{H}(\sigma(u), \sigma(v)) = \alpha$ if $(u, v) \notin E$, and $\mathrm{H}(\sigma(u), \sigma(v)) = \beta$ if $(u, v) \in E$.*

Thus, the difference between an exact γ-labeling and a γ-labeling is that an exact labeling requires the Hamming distance between adjacent vertices to be exactly α, rather than at most α, and the distance between non-adjacent nodes to be exactly β, rather than at least β. An inverse exact labeling is like an exact labeling except that it yields a large Hamming distance between adjacent nodes, rather than a small Hamming distance.

We are interested in exact γ-labelings because the exact γ-labelings for a collection of graphs can be concatenated to obtain a γ-labeling for their union. We define an *edge decomposition* of graph $G = (V, E)$ into $G_1, \ldots, G_r$ where $G_i = (V_i, E_i)$ such that $V_i = V$ for all i and $E = \bigcup_i E_i$.

Lemma 2. *Consider a graph G with edge decomposition $G_1, \ldots G_r$. For each G_i let σ_i be a labeling of G_i with length $\mathrm{length}(\sigma_i) = \ell_i$. Consider the labeling $\sigma(v) = \sigma_1(v) \cdot \sigma_2(v) \cdots \sigma_r(v)$ defined by taking the concatenation of each of the labelings σ for each vertex in G. Then the following hold.*

1. *If each σ_i is an exact γ_i-labeling of G_i with thresholds α_i and β_i, then for $\gamma = \min\{\gamma_i\}$ the labeling $\sigma(v)$ is a γ-labeling of G with thresholds $\alpha = \sum \beta_i - \gamma$ and $\beta = \sum \beta_i$.*
2. *If each σ_i is an inverse exact γ_i-labeling of G_i with thresholds α_i and β_i, then for $\gamma = \min\{\gamma_i\}$ the labeling $\sigma(v)$ is a γ-labeling of the complement graph $\overline{G}$ with thresholds $\alpha = \sum_{i=1}^{r} \alpha_i$ and $\beta = \sum_{i=1}^{r} \alpha_i + \gamma$.*

We now discuss how to obtain exact and inverse exact labelings for special classes of graphs. For the classes of graphs we consider, it is surprising that we are able to achieve the same asymptotic label lengths for exact labelings as for inverse exact labelings. In Section 4 we discuss algorithms that decompose general graphs into these classes of graphs, obtain exact or inverse exact labelings, and then combine them to obtain a γ-labeling from Lemma 2.

3.1   Matchings

A graph $G = (V, E)$ is said to be a matching if each connected component contains at most two nodes. To obtain an exact labeling for a matching we use Algorithm 1 MatchingExact. To obtain an exact inverse matching, there exists an algorithm InverseMatchingExact (details omitted).

Algorithm 1. MatchingExact(G, γ)

1. Let $\gamma' = max(\gamma, \frac{n}{2})$. Generate HR($n, \gamma'$).
2. Assign a distinct string from HR(n, γ') to each clique of G. That is, apply the labeling σ such that for each $v \in V$, $\sigma(v) \in$ HR(n, γ') and $\sigma(v) = \sigma(u)$ iff $(v, u) \in E$.
3. **Output** σ.

Lemma 3. *Algorithm 1 MatchingExact(G, γ) obtains an exact γ-labeling with $\alpha = 0$, $\beta = max(\gamma, \frac{n}{2})$, and length $O(\gamma + n)$, in run time $O(\gamma n + n^2)$.*

Lemma 4. *Algorithm InverseMatchingExact(G, γ) obtains an exact inverse labeling with $\alpha = \max(\gamma, \frac{n}{2})$, $\beta = 2 \cdot max(\gamma, \frac{n}{2})$, and length $O(\gamma + n)$, in run time $O(\gamma n + n^2)$.*

3.2   Star-Graphs

A graph is a *star graph* if there exists a vertex c such that all edges in the graph are incident to c. For such a graph, let A be the set of all vertices that are not adjacent to c and let B be the set of vertices that are adjacent to c. Algorithm 2 StarExact obtains an exact γ-labeling for a star graph G. (In fact, it achieves an exact 2γ-labeling). Figure 1 provides an example of the labeling assigned by StarExact. To obtain an exact inverse γ-labeling there exists an algorithm InverseStarExact (details omitted).

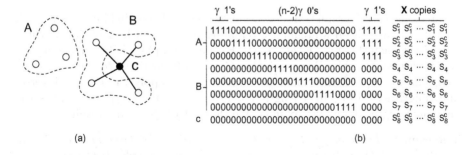

Fig. 1. (a) A star graph and (b) its corresponding exact labeling

Algorithm 2. StarExact(G, γ)

1. Let $\gamma' = max(\gamma, \frac{n}{2})$. Let $x = min(\gamma, \frac{n}{2})$. Arbitrarily index the n vertices of V as $v_1, v_2, \ldots, v_n$ with $c = v_n$.
2. Set the first $(n-1)\gamma$ bits of $\sigma(c)$ to be 0's.
3. **For** each vertex $v_i \neq c$, set the first $(n-1)\gamma$ bits to be all 0's except for the ith size γ word which is set to all 1's.
4. Append γ 1's to $\sigma(a)$ for each $a \in A$ and γ 0's to $\sigma(b)$ and $\sigma(c)$ for each $b \in B$.
5. **For** each $v_i \in A$ or $v_i = c$ append x copies of S_i^c to $\sigma(v_i)$ where S_i is the ith string in HR(γ', n).
6. **For** each $v_i \in B$ append x copies of $S_i \in$ HR(γ', n) to $\sigma(v_i)$.
7. **Output** σ.

Lemma 5. *Algorithm 2 StarExact(G, γ) obtains an exact γ-labeling for a Star graph G with $\alpha = \frac{\gamma n}{2}$, $\beta = 2\gamma + \frac{\gamma n}{2}$ and length $O(\gamma n)$, in run time $O(\gamma n^2)$.*

Lemma 6. *Algorithm InverseStarExact(G, γ) obtains an exact inverseγ-labeling for a Star graph G with $\alpha = \frac{\gamma n}{2}$, $\beta = 2\gamma + \frac{\gamma n}{2}$ and length $O(\gamma n)$, in run time $O(\gamma n^2)$.*

4   Labeling General Graphs

To obtain a γ-labeling for a general graph, we decompose either the graph or its complement into a collection of star and matching subgraphs. We then apply Lemmas 3 and 5 or Lemmas 4 and 6 to obtain exact or exact inverse labelings for these subgraphs, and then apply Lemma 2 to obtain a γ-labeling for the original graph. We first consider an algorithm that decomposes a general graph G into a collection of matchings.

4.1   Matching Decomposition

Lemma 7. *An edge decomposition of a graph $G = (V, E)$ into maximal matchings contains $\Theta(D)$ graphs where D is the maximum degree of any vertex in G.*

By breaking a given graph G into $\Theta(D)$ matchings and applying Lemmas 2 and 3, we have the algorithm MatchingDecomposition(G, γ) which yields a γ-labeling σ of G with length(σ) $= O(D \cdot (\gamma + n))$. For dense graphs whose vertices are all of high degree, MatchingDecomposition(G, γ) can be modified to decompose the complement graph $\overline{G}$ into maximal matchings and apply the routine Inverse-MatchingExact to obtain a length bound of $O(\overline{D} \cdot (\gamma + n))$ where $\overline{D}$ is the maximum degree of any vertex in $\overline{G}$. We thus get the following result.

Theorem 3. *For any graph G and γ, there exists a γ-labeling σ of G with length(σ) $= O(\hat{D}\gamma + \hat{D}n)$ that can be computed in time complexity $O(\gamma \hat{D}n + \hat{D}n^2)$ where $\hat{D}$ is the smaller between the degree of the maximum degree vertex in G and the maximum degree vertex in $\overline{G}$.*

4.2 Hybrid Decomposition (Star Destroyer)

The next algorithm for obtaining a γ-labeling adds the star decomposition to the basic matching algorithm. From Theorem 3, the matching algorithm may perform poorly even if there are just a few very high and very low degree vertices in the graph. The StarDestroyer(G, γ) algorithm thus repeatedly applies the star decomposition until all nodes have degree at most $\sqrt{m\gamma n}/\sqrt{\gamma + n}$, and then applies a final matching decomposition. With a few additional modifications we achieve the following.

Theorem 4. *For any graph G and γ, Algorithm StarDestroyer(G, γ) yields a γ-labeling σ of G with* length$(\sigma) = O(\sqrt{\gamma^2 \hat{m} n + \gamma \hat{m} n^2})$ *in time complexity $O(\sqrt{\gamma^2 \hat{m} n^3 + \gamma \hat{m} n^4})$ where $\hat{m} = min\{|E|, |\bar{E}|\}$.*

5 Trees and Forests

In this section we consider input graphs that are trees or forests and show that we are able to obtain substantially smaller labelings than what is possible for general graphs. For a collection of trees with a special type of γ-labeling, we show how to combine the collection into a single special γ-labeled tree. Thus, using recursive separators for trees we provide a recursive algorithm for tree labeling that achieves a length of $O(\gamma D \log n)$ where D is the largest degree node in the tree.

We then show how to improve this bound with a more sophisticated algorithm that assigns labels efficiently to paths as a base case, and recurses on the number of leaves in the tree rather than the number of nodes to achieve a length of $O(\gamma D \log(\max\{f, \frac{n}{\gamma D}\}))$ where f is the number of leaves in the tree. Note that this second bound is always at least as good as the first, and for trees with few leaves but high γ, is better. For example, consider the class of graphs consisting of $\log n'$ length $\frac{n'}{\log n'}$ paths, each connected on one end to a single node v. The number of nodes in this graph is $n = n' + 1$, the highest degree node has degree $D = \log n'$, and the number of leaves is $f = \log n'$. For $\gamma = \frac{n}{\log n'}$, the first bound yields $\ell = O(n \log n)$ while the second yields $\ell = O(n \log \log n)$.

5.1 Combining Trees

To make our recursive algorithms work, we need a way to take labelings from different trees and efficiently create a labeling for the tree resulting from combining the smaller trees into one. To do this, we will make use of a special type of γ-labeling.

Definition 3 (Lower Bounded Labeling). *A γ-labeling σ is said to be a lower bounded γ-labeling with respect to α_a, α_b, and β, $\alpha_a \leq \alpha_b < \beta$, $\beta - \alpha_b \geq \gamma$ if for any two nodes v and u, $\alpha_a \leq H(\sigma(v), \sigma(u)) \leq \alpha_b$ if v and u are adjacent, and $H(\sigma(v), \sigma(u)) \geq \beta$ if they are not adjacent.*

Given a collection of lower bounded labelings for trees, we can combine the labelings into a new lower bounded labeling with the same parameters according to Lemma 8. For the rest of this section, we will be dealing with a parameter D' which will be an upper bound on the maximum degree value of the input graph such that $D' + 1$ is a power of 2 greater than 2.

Algorithm 3. CombineTrees$(T = (V, E), v, \{\sigma_i\}_{i=1}^t)$

Input:

1. A degree t vertex v in tree T with $t \leq D'$.
2. An $\alpha_a = \gamma$, $\alpha_b = \frac{\gamma(D'-1)}{2}$, $\beta = \frac{\gamma(D'+1)}{2}$ lower bounded γ-labeling σ_i for each subtree of v.

Output: An $\alpha_a = \gamma$, $\alpha_b = \frac{\gamma(D'-1)}{2}$, $\beta = \frac{\gamma(D'+1)}{2}$ lower bounded γ-labeling of T.

1. **For** each labeling σ_i, append 0's such that length$(\sigma_i) = max_{i=1...t}\{$length$(\sigma_i)\}$.
2. **For** each of the child trees $T_1, \ldots T_t$ of v, **do**
   (a) Let v_i be the vertex in T_i adjacent to v and let $v_{i,j}$ denote the value of the jth character of $\sigma_i(v_i)$. For each $u \in T_i, u \neq v_i$, invert the jth character of $\sigma_i(u)$ if $v_{i,j} = 1$.
   (b) Set $\sigma_i(v_i)$ to all 0's.
3. Let $\sigma(v)$ be $max_{i=1...t}\{$length$(\sigma_i)\}$ 0's concatenated with $S_{t+1}^c \in HR(D' + 1, \frac{\gamma(D'+1)}{2})$. Let $\sigma(u) = \sigma_i(u)$ for each $u \in T_i$.
4. **For** $i = 1$ to t
   (a) **For** each $u \in T_i$, $\sigma(u) \leftarrow \sigma(u) \cdot S_i$ for $S_i \in HR(D' + 1, \frac{\gamma(D'+1)}{2})$.
5. **Output** σ.

Lemma 8 (Combining Trees). *Consider a vertex v in a tree T of degree t. Suppose for each of the t subtrees of v we have a corresponding length at most ℓ lower bounded γ-labeling σ_i with $\beta = \frac{\gamma(D'+1)}{2}$, $\alpha_b = \beta - \gamma$, and $\alpha_a = \gamma$ for some $D' \geq \max\{t, 2\}$, $D' + 1$ a power of 2. Then, Algorithm 3 Combine-Trees$(T, v, \{\sigma_i\}_{i=1}^t)$ computes a lower bounded γ-labeling with the same α_a, α_b, and β values and length $\ell' \leq \ell + \gamma D'$.*

5.2   Node Based Recursion

Define a *node separator* for a graph to be a node such that its removal leaves the largest sized connected component with at most $\lceil \frac{n}{2} \rceil$ vertices. Given Lemma 8 and the well known fact that every tree has a node separator, we are able to label a tree by first finding a separator, then recursively labeling the separated subtrees using lower bounded labeling parameters $\alpha_a = \gamma$, $\alpha_b = \frac{\gamma(D'-1)}{2}$, and $\beta = \frac{\gamma(D'+1)}{2}$ for $D' = O(D)$. Since it is trivial to obtain a lower bounded labeling satisfying such α_a, α_b, and β for a constant sized base case tree, we can obtain a $O(\gamma D \log n)$ bound on length of labelings for trees.

We can then extend this to a t tree forest by creating t length $\frac{\gamma(D'+1)}{2} \log t$ length strings such that each pair of strings has Hamming distance at least

$\frac{\gamma(D'+1)}{2}$ and appending a distinct string to the nodes of each forest. This yields the following result.

Theorem 5 (Node Recursive Forest). *Given an integer γ and a forest F with maximum degree D, a γ-labeling of F with length $O(\gamma D \log n)$ can be constructed in time $O(n\gamma D \log^2 n)$.*

5.3 Leaf Based Recursion

Instead of performing recursion by halving the number of nodes in the graph, we can instead halve the number of leaves in the graph and use an efficient path labeling algorithm to solve the base case. We first describe the efficient path labeling scheme.

Path Labeling. As a base case for our recursive algorithm, according to Lemma 8, we want to be able to produce a short lower bounded γ-labeling for a path graph with $\beta = \frac{\gamma(D'+1)}{2}$, $\alpha_b = \frac{\gamma(D'-1)}{2}$, and $\alpha_a \geq \gamma$ for any given D'. When called from the tree algorithm, D' will be on the order of the maximum degree of any node in the input tree. The Path algorithm will achieve $\alpha_a = \frac{\beta}{2} \geq \gamma$ to satisfy the desired constraints. The reason for this choice of α_a is that it is a power of 2, which is necessary for our algorithm. The basic structure of the Path algorithm is that it uses recursion based on node separators and Lemma 8 until the path is sufficiently short. Then, a labeling based on the Hadamard code is used. Recursive Algorithm 4 Path achieves the following result.

Algorithm 4. Path$(P = \langle v_1, \ldots v_n \rangle, \gamma, D')$

---

1. **If** $n \leq 2\gamma(D'+1) - 1$ **then**
   - (a) Compute $S_1, \ldots S_{\gamma(D'+1)} \in \mathrm{HR}(\gamma(D'+1), \frac{\gamma(D'+1)}{4})$.
   - (b) **For** $i = 1$ to $\gamma(D'+1) - 1$ **do**
     - i. $\sigma(v_{2i-1}) \leftarrow S_i.S_i$
     - ii. $\sigma(v_{2i}) \leftarrow S_i.S_{i+1}$
   - (c) $\sigma(v_{2\gamma(D'+1)-1}) \leftarrow S_{\gamma(D'+1)}.S_{\gamma(D'+1)}$
   - (d) **Output** σ.
2. **Else**
   - (a) Let $P_1 = \langle v_1, \ldots, v_{\frac{n}{2}-1} \rangle$, $P_2 = \langle v_{\frac{n}{2}+1} \ldots, v_n \rangle$.
   - (b) $\sigma_1 \leftarrow$ Path(P_1, γ, D'), $\sigma_2 \leftarrow$ Path(P_2, γ, D').
   - (c) **Output** CombineTrees$(P, v_{\frac{n}{2}}, \{\sigma_1, \sigma_2\})$.

Lemma 9. *For $D' \geq 3$, $D'+1$ a power of 2, and path P, Algorithm 4 Path(P,γ,D') generates a lower bounded γ-labeling σ of P with $\alpha_a = \frac{\gamma(D'+1)}{4}$, $\alpha_b = \frac{\gamma(D'-1)}{2}$, $\beta = \frac{\gamma(D'+1)}{2}$, and length$(\sigma) = O(\max\{\gamma D' \log(\frac{n}{\gamma D'}), \gamma D'\})$ in time $O(n \cdot (\max\{\gamma D' \log^2(\frac{n}{\gamma D'}), \gamma D'\}))$.*

Leaf Recursive Tree Algorithm. The leaf recursive tree algorithm recursively reduces the number of leaves in the tree until the input is a simple path, for which

Algorithm Path can be used. For a tree T with f leaves, a *leaf* separator is a node such that its removal reduces the largest number of leaves in any of the remaining connected components to at most $\lfloor \frac{f}{2} \rfloor + 1$. We start by observing that every tree must have a leaf separator.

Lemma 10. *Every tree has a leaf separator.*

Note that a leaf separator always reduces the number of leaves in a tree unless there are only 2 leaves, in which case the tree is a path which can be handled according to Lemma 9. Having removed a leaf separator and recursively solved for the sub trees, we can then apply Lemma 8 to finish the labeling. The details of the algorithm are given as follows. Here, the input parameter D' is the smallest integer such that $D' + 1$ is a power of 2 and D' is at least the degree of the highest degree node in the tree, or 3 in the case of an input path.

Algorithm 5. Tree($T = (V, E), \gamma, D'$)

---

1. **If** $\mathrm{Deg}(T) \leq 2$, **then output** Path(T, γ, D').
2. **Else**
   (a) Find a leaf separator v for T.
   (b) **For** each of the child trees $T_1, \ldots T_t$ of v, $\sigma_i \leftarrow$ Tree(T_i, γ, D').
   (c) **Output** CombineTrees($T, v, \{\sigma_i\}_{i=1}^t$).

Theorem 6 (Trees). *Consider a tree T with f leaves and integer $D' = 2^j - 1 \geq \max\{\deg(T), 3\}$. Then, Algorithm 5 Tree(T, γ, D') computes a length $O(\gamma D' \log(\max\{f, \frac{n}{\gamma D'}\}))$ γ-labeling of T in time complexity $O(n\gamma D' \log^2(\max\{f, \frac{n}{\gamma D'}\}))$.*

To extend this result to a forest of trees $T_1 \cdots T_t$, we can use Tree(T_i, γ, D') for each individual tree. We can then append a distinct string from a set of t strings to each tree such that the distance between any two strings is at least $\beta = \frac{\gamma(D'+1)}{2}$. Deterministically we can achieve such a set of strings trivially using additional length $O(\gamma D \log t)$ where $D = \deg(T)$. Alternately, we can use elements of HR(t, β) for an additional length of $O(t + \gamma D)$. These approaches yield the the following theorem.

Theorem 7 (Leaf Recursive Forest). *There exists a deterministic algorithm that produces a length $O(\min\{t, \gamma D \log t\} + \gamma D \log(\max\{f, \frac{n}{\gamma D}\}))$ γ-labeling for an input forest F in time complexity $O(n\min\{t, \gamma D \log t\} \gamma D \log^2(\max\{f, \frac{n}{\gamma D}\}))$, where $D = \deg(F)$, f is the largest number of leaves in any of the trees in F, and t is the number of trees in F.*

Alternately, we can use randomization to append shorter strings to each tree and avoid an increase in complexity. Kao et al.[15] showed that with high probability, a set of n uniformly generated random binary strings has Hamming distance at least x between any two words with high probability for words of length at least $10(x + \log n)$. Thus, we can produce a γ-labeling for a forest by first finding a labeling for each tree, making the length of the labels equal, and finally picking a random string of length $10(\beta + \log n)$ for each tree and appending the string to each of the nodes in the tree. We thus get the following result.

Theorem 8 (Randomized Forest). *There exists a randomized algorithm that produces a length $O(\gamma D \log(\max\{f, \frac{n}{\gamma D}\}))$ γ-labeling for an input forest F with probability at least $1 - \frac{1}{n+2^\gamma}$, in time complexity $O(n \cdot (\gamma D \log(\max\{f, \frac{n}{\gamma D}\})))$, where $D = \deg(F)$, and f is the largest number of leaves in any of the trees in F.*

6   Future Directions

There are a number of potential research directions stemming from this work. A few of these are as follows. First, can our technique for labeling general graphs by decomposing the graph into exact labelings be extended? We considered two different types of decompositions, stars and matchings. Are there other types of decompositions that can yield better bounds? Second, our lower bounds are straightforward and stem primarily from lower bounds for labeling for adjacency in general, rather than our much more restricted problem. It is likely that much higher bounds exist for flexible word design. Third, an important class of graphs that permits short labels for general graph labeling is the class of planar graphs. It would be interesting to know whether or not a flexible word labeling that is sublinear in the number of vertices exists as well. Fourth, we have initiated the use of randomization in designing labels. Randomization is used extensively in the design of standard DNA code word sets, and it would be interesting to know if more sophisticated randomized algorithms can be applied to achieve better flexible word labelings. Finally, although not included in this draft, we have also considered generalizations of flexible word design to both distance labelings and weighted graphs. These generalizations present many open problems and may have direct applications to applying DNA computing to digital signal processing.

References

[1] L. M. Adleman. Molecular computation of solutions to combinatorial problems. *Science*, 266:1021–1024, 1994.

[2] G. Aggarwal, Q. Cheng, M. H. Goldwasser, M.-Y. Kao, P. M. de Espanes, and R. T. Schweller. Complexities for generalized models of self-assembly. *SIAM Journal on Computing*, 34:1493–1515, 2005.

[3] G. Aggarwal, M. H. Goldwasser, M.-Y. Kao, and R. T. Schweller. Complexities for generalized models of self-assembly. In *Proceedings of the fifteenth annual ACM-SIAM symposium on Discrete algorithms*, pages 880–889, 2004.

[4] A. Ben-Dor, R. Karp, B. Schwikowski, and Z. Yakhini. Universal DNA Tag Systems: A Combinatorial Design Scheme. In *Proceedings of the 4th Annual International Conference on Computational Molecular Biology*, pages 65–75, 2000.

[5] A. Brenneman and A. E. Condon. Strand Design for Bio-Molecular Computation. *Theoretical Computer Science*, 287(1):39–58, 2001.

[6] S. Brenner. Methods for sorting polynucleotides using oligonucleotide tags, Feb. 1997. U.S. Patent Number 5,604,097.

[7] S. Brenner and R. A. Lerner. Encoded combinatorial chemistry. In *Proceedings of the National Academy of Sciences of the U.S.A.*, volume 89, pages 5381–5383, June 1992.

[8] M. Breuer. Coding vertexes of a graph. *IEEE transactions on Information Theory*, 8:148–153, 1966.

[9] M. Breuer and J. Folkman. An unexpected result on coding vertices of a graph. *Journal of Mathematical Analysis and Applications*, 20:583–600, 1967.

[10] R. Deaton, M. Garzon, R. C. Murphy, J. A. Rose, D. R. Franceschetti, and J. S. E. Stevens. Genetic search of reliable encodings for DNA-based computation. In *Proceedings of the 2nd International Meeting on DNA Based Computers*, 1996.

[11] A. G. Frutos, Q. Liu, A. J. Thiel, A. M. W. Sanner, A. E. Condon, L. M. Smith, and R. M. Corn. Demonstration of a word design strategy for DNA computing on surfaces. *Nucleic Acids Research*, 25(23):4748–4757, Dec. 1997.

[12] P. Gaborit and O. D. King. Linear constructions for DNA codes. *Theoretical Computer Science*, 334:99–113, 2005.

[13] M. Garzon, R. Deaton, P. Neathery, D. R. Franceschetti, and R. C. Murphy. A new metric for DNA computing. In *Proceedings of the 2nd Genetic Programming Conference*, pages 472–478. Morgan Kaufman, 1997.

[14] C. Gavoille and D. Peleg. Compact and localized distributed data structures. Technical Report RR-1261-01, Laboratoire Bordelais de Recherce en Informatique, 2001.

[15] M. Y. Kao, M. Sanghi, and R. Schweller. Randomized fast design of short dna words. In *Lecture Notes in Computer Science 3580: Proceedings of the 32nd International Colloquium on Automata, Languages, and Programming*, pages 1275–1286, 2005.

[16] O. D. King. Bounds for DNA Codes with Constant GC-content. *Electronic Journal of Combinatorics*, 10(1):#R33 13pp, 2003.

[17] A. Marathe, A. Condon, and R. M. Corn. On Combinatorial DNA Word Design. *Journal of Computational Biology*, 8(3):201–219, 2001.

[18] D. D. Shoemaker, D. A. Lashkari, D. Morris, M. Mittman, and R. W. Davis. Quantitative phenotypic analysis of yeast deletion mutants using a highly parallel molecular bar-coding strategy. *Nature Genetics*, 14(4):450–456, Dec. 1996.

[19] S. A. Tsaftaris. DNA Computing from a Signal Processing Viewpoint. *IEEE Signal Processing Magazine*, 21:100–106, September 2004.

[20] S. A. Tsaftaris. How can DNA-Computing be Applied in Digital Signal Processing? *IEEE Signal Processing Magazine*, 21:57–61, November 2004.

[21] D. C. Tulpan and H. H. Hoos. Hybrid Randomised Neighbourhoods Improve Stochastic Local Search for DNA Code Design. In Y. Xiang and B. Chaib-draa, editors, *Lecture Notes in Computer Science 2671: Proceedings of the 16th Conference of the Canadian Society for Computational Studies of Intelligence*, pages 418–433. Springer-Verlag, New York, NY, 2003.

[22] J. van Lint. *Introduction to Coding Theory*. Springer, third edition, 1998.

[23] E. Winfree, F. Liu, L. Wenzler, and N. Seeman. Design and self-assembly of two-dimensional DNA crystals. *Nature*, 394:539–544, August 1998.

Frequency Allocation Problems for Linear Cellular Networks[*]

Joseph Wun-Tat Chan[1], Francis Y.L. Chin[2], Deshi Ye[3],
Yong Zhang[2,4], and Hong Zhu[4]

[1] Department of Computer Science, King's College London, London, UK
jchan@dcs.kcl.ac.uk
[2] Department of Computer Science, The University of Hong Kong, Hong Kong
{chin, yzhang}@cs.hku.hk
[3] College of Computer Science, Zhejiang University, China
yedeshi@zju.edu.cn
[4] Department of Computer Science and Engineering, Fudan University, China
hzhu@fudan.edu.cn

Abstract. We study the online frequency allocation problem for wire-less linear (highway) cellular networks, where the geographical coverage area is divided into cells aligned in a line. Calls arrive over time and are served by assigning frequencies to them, and no two calls emanating from the same cell or neighboring cells are assigned the same frequency. The objective is to minimize the span of frequencies used.

In this paper we consider the problem with or without the assumption that calls have infinite duration. If there is the assumption, we propose an algorithm with absolute competitive ratio of $3/2$ and asymptotic competitive ratio of 1.382. The lower bounds are also given: the absolute one is $3/2$ and the asymptotic one is $4/3$. Thus, our algorithm with absolute ratio of $3/2$ is best possible. We also prove that the Greedy algorithm is $3/2$-competitive in both the absolute and asymptotic cases. For the problem without the assumption, i.e. calls may terminate at arbitrary time, we give the lower bounds for the competitive ratios: the absolute one is $5/3$ and the asymptotic one is $14/9$. We propose an optimal online algorithm with both competitive ratio of $5/3$, which is better than the Greedy algorithm, with both competitive ratios 2.

1 Introduction

Reducing channel interference and using frequencies effectively are fundamental problems in wireless networks based on Frequency Division Multiplexing (FDM) technology. In FDM networks, service areas are usually divided into cellular regions or *hexagonal cells* [7], each containing one base station. Base stations can allocate radio frequencies to serve the phone calls in their cells. The allocation strategy is to choose different frequencies for calls in the same cell or in the neighboring cells, so as to avoid interference.

---

[*] This research is supported by an Hong Kong RGC Grant.

T. Asano (Ed.): ISAAC 2006, LNCS 4288, pp. 61–70, 2006.

We consider the problems of *online frequency allocation* in *linear (or highway)* cellular networks where the cells are aligned in a line as shown in Fig. 1. Linear cellular networks can be used to cover the traffic on highways or long strips of busy metropolitan areas. There are many studies on using frequencies effectively so as to minimize interference and to reduce call blocking in linear networks [1, 2, 6, 8]. In this paper we study the performances of different strategies, which are to minimize the span of frequencies used to serve all calls without interference.

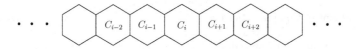

Fig. 1. Linear cellular network

A formal definition of our problem is described as follows. Given a linear cellular network, in which a sequence σ of calls arrive over time, where $\sigma = (C_{t_1}, C_{t_2}, \ldots, C_{t_k}, \ldots,)$ and C_{t_k} represents the cell from which the k-th call emanates. Each call C_{t_k} must be assigned upon its arrival, without information about future calls $\{C_{t_i} | i > k\}$, a frequency from the integer set $Z^+ = \{1, 2, \ldots\}$ of available frequencies, that is different from those of other calls in the same cell or neighboring cells. Let $\mathcal{A}(C_{t_k}) \in Z^+$ denote the integer frequency assigned to the k-th call. Then $\mathcal{A}(C_{t_k}) \neq \mathcal{A}(C_{t_i})$, where $i < k$ and C_{t_i} is adjacent to C_{t_k} or the same as C_{t_k}. The integer frequency once assigned to a call cannot be changed during the survival of this call. The *online frequency allocation problem for linear cellular network* (FAL for short) is to minimize the maximum assigned frequency, i.e., $\max\{\mathcal{A}(C_{t_k}) | k = 1, 2, \ldots, n\}$. If all the information of C_{t_k} is known in advance, we call this problem *off-line frequency allocation problem*. In this paper, we focus on the online version of FAL.

Two models of online frequency allocation problems will be investigated. The first model is that all calls have infinite duration [4]. We call this model *frequency allocation without deletion*. The second model is that each call may terminate at arbitrary time, i.e., each call is characterized by two parameters: arrival time and termination time. However, the termination time is not known even when the call arrives online. We call this model *frequency allocation with deletion*.

Performance Measures. We use competitive analysis [3] to measure the performance of online algorithms. For any sequence σ of calls, let σ_t denote the subsequence of calls served up to and at time t. Let $\mathcal{A}(\sigma_t)$ denote the cost of an online algorithm $\mathcal{A}$, i.e., the span of frequencies used by $\mathcal{A}$ at time t, and $\mathcal{O}(\sigma_t)$ the cost of the optimal off-line algorithm, which has the knowledge of the whole sequence σ in advance.

Let $\mathcal{A}(\sigma) = \max_t \mathcal{A}(\sigma_t)$ and $\mathcal{O}(\sigma) = \max_t \mathcal{O}(\sigma_t)$. The *(absolute) competitive ratio* of $\mathcal{A}$ is defined as $R_{\mathcal{A}} = \sup_\sigma \mathcal{A}(\sigma)/\mathcal{O}(\sigma)$. Meanwhile, when the number of calls emanating from the cells is large, the *asymptotic competitive ratio* of $\mathcal{A}$ is defined as

$$R_{\mathcal{A}}^{\infty} = \limsup_{n \to \infty} \max_{\sigma} \left\{ \frac{\mathcal{A}(\sigma)}{\mathcal{O}(\sigma)} \mid \mathcal{O}(\sigma) = n \right\}.$$

Clearly, for any online algorithm $\mathcal{A}$, we have $R_{\mathcal{A}}^{\infty} \leq R_{\mathcal{A}}$.

Related and Our Contributions. To our best knowledge, this is the first study on the online frequency allocation problem in linear cellular networks with the objective to minimize the span of frequencies used. It is easy to check that the off-line version can be solved in polynomial time. However, in many practical scenarios, the information of calls is not completely known until they arrive. The online problem is more suitable to model the mobile telephone networks problem.

A simple strategy for the online FAL problem is by fixed allocation assignment [7], in which cells are partitioned into independent sets with no neighboring cells in the same set. Each set is assigned a separate set of frequencies. The fixed allocation assignment algorithm gives an easy upper bound of 2 for the online FAL problem.

Another intuitive approach is by the *greedy algorithm* (Greedy) which assigns the minimum available frequency to a new call such that the call does not interfere with calls of the same or neighboring cells. We show that, Greedy is 3/2-competitive in the *without deletion* model, and 2-competitive in the *with deletion* model.

In this paper, new algorithms are proposed for both models. In the *without deletion* model, we present the algorithm HYBRID, which combines the idea of Greedy and fixed allocation strategy, and yields the absolute and asymptotic competitive ratios of 3/2 and 1.382, respectively. Contrasting with the lower bounds shown, 3/2 for the absolute case and 4/3 for the asymptotic case, HYBRID is also best possible in the absolute case and better than Greedy in the asymptotic case.

In the *with deletion* model, we propose the algorithm BORROW with both absolute and asymptotic competitive ratios 5/3. We also prove the lower bounds, which is 5/3 for the absolute case and 14/9 for the asymptotic case. Thus, BORROW is best possible in the absolute case, and also better than Greedy in the asymptotic case.

The rest of this paper is organized as follows. In Section 2, we analyze the performance of Greedy. Section 3 and Section 4 study respectively the *without deletion* and the *with deletion* models, in which upper and lower bounds are presented. Owing to the space limitation, the proofs of Lemma 1 and part of the proofs for Theorems 2 and 5 are omitted but they will be given in the full paper.

2   The Greedy Algorithm

We first consider the *without deletion* model and prove an upper bound of 3/2 for the Greedy. The proved ratio applies to both absolute and asymptotic cases. Greedy is in fact optimal in the absolute case because no online algorithm can achieve an absolute competitive ratio less than 3/2 (Theorem 6) in that case. Then, we show that Greedy is 2-competitive in both absolute and asymptotic cases of the *with deletion* model, and that is the best that Greedy can do.

Fig. 2. A line cellular network with cells A, B, C, D, E and F

Theorem 1. *In the without deletion model, the competitive ratio of Greedy for frequency allocation problem in linear cellular network is 3/2.*

Proof. Consider the network in Fig. 2. Suppose Greedy assigns the highest frequency to a call from cell C, and no more calls arrive after that. Let B and D be the left and right neighboring cells of C. Let f_X denote the set of frequencies used in cell X at the time when the highest frequency is assigned.

By the definition of Greedy, when the highest frequency, say h, is assigned to a call in C, the frequencies from 1 to $h-1$ must have been assigned to calls of C or its neighboring cells B and D. Thus, the span of frequencies used by Greedy is $h = |f_B \cup f_C \cup f_D|$.

Without loss of generality, assume the highest frequency among B and D appears in B. Since f_C and f_D cannot have common frequencies, those frequencies in $f_D - f_B$ must all appear in A. Therefore, $|f_A \cup f_B| \geq |f_B \cup f_D|$.

It is clear that the optimal span of frequencies used, say s^*, must be at least the maximum number of calls (frequencies used) from any two adjacent cells. Thus, we have $s^* \geq \max\{|f_A \cup f_B|, |f_B \cup f_C|, |f_C \cup f_D|\} \geq \max\{|f_B \cup f_D|, |f_B \cup f_C|, |f_C \cup f_D|\}$. Therefore, the competitive ratio of Greedy is at most

$$\frac{|f_B \cup f_C \cup f_D|}{\max\{|f_B \cup f_D|, |f_B \cup f_C|, |f_C \cup f_D|\}} \leq 3/2.$$

$\square$

Theorem 2. *In the with deletion model, the upper and lower bounds of the competitive ratio of Greedy are both 2 for the online frequency allocation problem in linear cellular networks.*

Proof. The upper bound proof is simple. Consider the network in Fig. 2. When the highest frequency, say h, appears in cell C, h is at most the total number of calls from C and its neighboring cells, B and D. The span of frequencies used by the optimal algorithm is at least the maximum among the numbers of calls from B and C and those from C and D. Thus, the upper bound of 2 follows. The lower bound proof is omitted in this paper. $\square$

3   FAL Without Deletion

We propose a generic online algorithm HYBRID for FAL in the *without deletion* model. HYBRID consists of two integer parameters, $\alpha \geq 1$ and $\beta \geq 0$. We prove that HYBRID is 3/2-competitive in the absolute case for any $\alpha \geq 1$ and $\beta \geq 0$. Moreover, with a proper ratio between the values of α and β, the asymptotic competitive ratio of HYBRID is at most 1.382, which is better than Greedy in the asymptotic case.

Conceptually, we divide the frequencies into groups, each of which consists of $\Delta = 3\alpha + \beta$ frequencies. A frequency f is said to be in group i if $\Delta i < f \leq \Delta(i+1)$. HYBRID partitions the set of all frequencies $\{1, 2, \ldots\}$ into 3 disjoint subsets. The first subset F_0 consists of $\alpha + \beta$ frequencies from each group, while each of the remaining 2 subsets F_1 and F_2 has α frequencies. The details of HYBRID are as follows:

Preprocessing Step: The cells of the linear cellular network are partitioned into two sets S_1 and S_2, e.g., cells $C_{2k+1} \in S_1$ and cells $C_{2k} \in S_2$, so that the cells in these two sets are interleaving each other. As mentioned above, the frequencies $\{1, 2, \ldots\}$ are partitioned into 3 disjoint subsets F_0, F_1 and F_2. Precisely, the frequencies of group i for each $i \geq 0$ are distributed to the three subsets as follows.

$$F_0 \leftarrow \{i\Delta + 3j + 1 \mid j = 0, 1, \ldots, \alpha - 1\} \cup \{i\Delta + 3\alpha + j \mid j = 1, \ldots, \beta\}$$
$$F_1 \leftarrow \{i\Delta + 3j + 2 \mid j = 0, 1, \ldots, \alpha - 1\}$$
$$F_2 \leftarrow \{i\Delta + 3j + 3 \mid j = 0, 1, \ldots, \alpha - 1\}$$

Frequency Assignment Step: Suppose a new call emanates from a cell C, which belongs to S_i, we assign a frequency x to the call either from F_i or F_0 according to the following scheme:

$\min\{x \mid x \in F_0 \cup F_i,$ s.t. x is not assigned to cell C or any of its neighboring cells$\}$

3.1 Asymptotic Competitive Ratio

We show that the asymptotic competitive ratio of HYBRID is $(5 - \sqrt{5})/2 \approx 1.382$ when $\alpha/\beta = (\sqrt{5}+1)/2$ and no online algorithm has an asymptotic competitive ratio less than $4/3$.

Lemma 1 lower bounds the number of frequencies required by the optimal offline algorithm, i.e., the total number of calls emanating from any two neighboring cells, which helps lead to a bound for the competitive ratio.

Lemma 1. *For a linear cellular network, if a cell A assigns a frequency from group k, then for $\alpha/\beta \geq (1 + \sqrt{5})/2$, the total number of calls from cell A and one of its neighbor is at least $(2\alpha + \beta)k$.*

Theorem 3. *In the without deletion model, the asymptotic competitive ratio of HYBRID for FAL approaches $(5 - \sqrt{5})/2 \approx 1.382$ when $\alpha/\beta \to (\sqrt{5}+1)/2$.*

Proof. If the highest frequency used by HYBRID, say h, is of group k, we have $h \leq (3\alpha+\beta)(k+1)$. Suppose the frequency h is assigned in a cell C. By Lemma 1, C and one of its neighbors together have at least $(2\alpha + \beta)k$ calls when $\alpha/\beta \geq (1 + \sqrt{5})/2$, in which the optimal algorithm has to settle with at least the same amount of frequencies. Therefore, the asymptotic competitive ratio of HYBRID is almost $\lim_{k\to\infty} \frac{(3\alpha+\beta)(k+1)}{(2\alpha+\beta)k} = (5 - \sqrt{5})/2$ when $\alpha/\beta \to (\sqrt{5}+1)/2$. $\qquad\square$

Next, we give a lower bound on the asymptotic competitive ratio for FAL in the without deletion model.

Theorem 4. *No online algorithm for FAL in the without deletion model has an asymptotic competitive ratio less than 4/3.*

Proof. Consider the network in Fig. 2 with cells A, B, C, and D in a row. The adversary initiates n calls from each of cells A and D. For any online algorithm S, S assigns n frequencies to each of A and D. Suppose in each of the two sets of frequencies, xn $(0 \leq x \leq 1)$ of the frequencies do not appear in the other set. Thus, the number of distinct frequencies (span of frequencies used) over the $2n$ frequencies assigned is $(2 - x)n$. If $x \leq 2/3$, the adversary stops and we have $R_S^\infty \geq 2 - x \geq 4/3$.

On the other hand, consider the case where $x > 2/3$. The adversary makes n new calls in each of B and C. S must use at least xn new frequencies in each of B and C. By now, S has used at least $(2+x)n$ distinct frequencies. However, the optimal algorithm can satisfy all these calls by $2n$ distinct frequencies. Therefore, $R_S^\infty \geq (2 + x)/2 \geq 4/3$. □

3.2   Absolute Competitive Ratio

We show that the absolute competitive ratio of HYBRID is $3/2$ for all $\alpha \geq 1$ and $\beta \geq 0$. We also give a matching lower bound proof for the problem, which shows that HYBRID, as well as Greedy, are both optimal.

Theorem 5. *In the without deletion model, the absolute competitive ratio of HYBRID algorithm for FAL is at most 3/2.*

Proof. We can prove that HYBRID is $3/2$-competitive for all $\alpha \geq 1$ and $\beta \geq 0$. For simplicity, we only prove the competitive ratio for the case $\alpha = 1$ and $\beta = 0$. The general proof will be given in the full paper.

Suppose the highest frequency used by HYBRID, say h, is of group k and assigned by a cell C of S_2 (which is worse than the case of S_1, which uses frequencies from F_1 that has smaller frequency values). We have h either $3k + 1$ from F_0 or $3k + 3$ from F_2. Consider the former case. C must have assigned k frequencies from F_2 before assigning h. Let $3i + 1$ for $i \leq k$ be the highest frequency from F_0 assigned to a neighboring cell of C, say B. B has at least $i + 1$ frequencies, where one is from F_0 and i from F_1. On the other hand, C has at least $k - i$ frequencies from F_0. Altogether, B and C consist of at least $k + i + 1 + k - i$, i.e., $2k + 1$, distinct frequencies/calls. The optimal algorithm must use at least the same amount of distinct frequencies. Thus the competitive ratio of HYBRID is at most $(3k + 1)/(2k + 1) \leq 3/2$.

For the latter case, following the same argument, C has at least $2k - i + 1$ distinct frequencies and B, the neighbor of C which has the highest frequency from F_0, has at least $i+1$ distinct frequencies. Then, the optimal algorithm must use at least $2k + 2$ distinct frequencies. The competitive ratio of HYBRID is at most $(3k + 3)/(2k + 2) = 3/2$. □

Next, we give the lower bound of absolute competitive ratio for FAL in the without deletion model.

Theorem 6. *No online algorithm for FAL in the without deletion model has an absolute competitive ratio less than 3/2.*

Proof. The proof is simple. Consider the network in Fig. 2 with cells A, B, C, and D in a row. The adversary begins with one call from each of A and D. For any online algorithm, if it assigns two different frequencies to these two calls, the adversary stops. The competitive ratio of the online algorithm is 2. Otherwise, the same frequency is assigned to both calls. One new call arrives at each of B and C. The online algorithm must use two new frequencies for the two calls. Thus, at least three different frequencies are used, while the optimal algorithm can use only two. Therefore, the absolute competitive ratio is at least 3/2.   □

4   FAL with Deletion

In this section we study the online frequency allocation problem in the linear cellular network in which the calls may terminate in arbitrary time. We call this the *with deletion* model. It is noted that the *without deletion* model considered above is a special case of the *with deletion* model. We present a new online algorithm BORROW with competitive ratio at most 5/3. A matching lower bound for problem is given for the absolute case which shows that our algorithm is best possible. For the asymptotic case, we show that no online algorithm has the competitive ratio less than 14/9, which leaves only a small gap between the upper and lower bounds.

4.1   Online Algorithm with Borrowing

The main idea of our algorithm is to reuse ("borrow") an existing frequency even if the frequency is not the smallest possible (i.e., Greedy). Consider Fig. 1. When a call emanates in cell C_i, we try to borrow existing frequencies from C_{i-2} or C_{i+2}, which does not create interference. If none can be borrowed from C_{i-2} or C_{i+2}, the call is satisfied by Greedy. In case there are more than one frequencies that can be borrowed, we select the frequency according to the following priority.

1. The frequency appears in both C_{i-2} and C_{i+2}. If there are more than one of these, pick one arbitrarily.
2. The frequency appears in either C_{i-2} or C_{i+2} which currently has more frequencies that do not appear in C_i. If there are more than one of these, pick one arbitrarily.
3. Pick one arbitrarily.

Theorem 7. *In the with deletion model, the competitive ratio of* BORROW *is at most 5/3 for FAL.*

Proof. Consider the network in Fig. 2 with cells A, B, C, D and E in a row. Suppose the highest frequency, say h, is assigned to a call from D. Note that without loss of generality, frequency h is assigned by the greedy approach. Hence, at the time when frequency h is assigned, all frequencies from 1 to h must appear in either C, D or E. We also consider another time instance, which is the latest time before frequency h is assigned, that either C or E assigns a frequency, say h', that does not exist in C or E. Without loss of generality, we assume that it is the cell C to assign the frequency h'. There are only two cases, either C assigns the frequency h' by the greedy approach or frequency h' is borrowed from A. For these two cases, we analyze the competitive ratio of BORROW.

By the greedy approach. Suppose when frequency h is assigned by D, the number of frequencies being used in C, D, E are $y+r_1$, x and $y+r_2$, respectively, where y is the number of common frequencies among cells C and E. Since frequency h is assigned by the greedy approach, we have $h = x + y + r_1 + r_2$, which is the number of distinct frequencies used in the three cells. In fact, for any algorithm to satisfies all calls from these cells, one has to use at least $x + y + \max\{r_1, r_2\}$ distinct frequencies.

Suppose when frequency h' is assigned by C, the number of frequencies being used in C and E are $y' + r_1'$ and $y' + r_2'$, respectively, where y' is the number of common frequencies among C and E. Note that as there are r_2' frequencies in E that C did not borrow, the r_2' frequencies must be used in B. Hence, the number of frequencies, and thus the number of calls, from cells B and C is at least $y + r_1 + r_2$. By the definition of frequency h', at the time frequency h' is assigned, the number of distinct frequencies among C and E, i.e., $y' + r_1' + r_2'$, must be at least $y + r_1 + r_2$. Any algorithm to satisfy the calls from B and C has to use at least $y' + r_1' + r_2' \geq y + r_1 + r_2$ frequencies.

As a result the competitive ratio of BORROW is at most

$$\frac{x + y + r_1 + r_2}{\max\{x + y + \max\{r_1, r_2\}, y + r_1 + r_2\}} \leq \frac{3}{2}.$$

By borrowing. Similar to the previous case, suppose when frequency h is assigned by D, the number of frequencies being used in C, D, E are $y + r_1$, x and $y + r_2$, respectively, where y is the number of common frequencies among C and E. For any algorithm to satisfies all calls from these cells, one has to use at least $x + y + \max\{r_1, r_2\}$ distinct frequencies. Suppose when frequency h' is assigned by C, the number of frequencies being used in C and E are $y' + r_1'$ and $y' + r_2'$, respectively, where y' is the number of common frequencies among C and E.

In this case, frequency h' assigned by C is borrowed from A but not E. There are two subcases by the algorithm: either all the r_2' frequencies in E which could be assigned to C are already in B (i.e., E has no candidate for C to borrow) or the number of frequencies in A is at least that in E. For the former subcase, we have the number of frequencies in B and C at least $y' + r_1' + r_2'$, and hence the analysis follows the previous case which yields a competitive ratio at most $3/2$. For the latter subcase, we have the number of frequencies in A but not in C at least that of frequencies in E but not in C, which is r'_2. In addition, what those frequencies that A could have but not in C are one frequency that is borrowed to C, and also the frequencies that are neither in C nor E and with frequency value less than h. There are at most x of them. That implies that $r'_2 \leq x$. Therefore, the competitive ratio of our algorithm is at most

$$\frac{x + y + r_1 + r_2}{\max\{x + y + \max\{r_1, r_2\}, y' + r'_1\}}$$

with the constraints that $r'_2 \leq x$ and $y + r_1 + r_2 \leq y' + r'_1 + r'_2$. By the constraints, we have $y' + r'_1 \geq y + r_1 + r_2 - x$. Together with the fact that $\max\{r_1, r_2\} \geq (r_1 + r_2)/2$, we can prove that the ratio is at most $5/3$. □

4.2 Lower Bound

Theorem 8. *There is no online algorithm for FAL, in the with deletion model, with an absolute competitive ratio less than $5/3$ or an asymptotic competitive ratio less than $14/9$.*

Proof. For the absolute competitive ratio, we give an adversary that any online algorithm will use at least five distinct frequency (with span of at least five), while the optimal algorithm uses only three.

Consider the network Fig. 2 with cells A, B, C, D, E and F in a row. The adversary has three calls emanate from each of A, C and F. In order for an algorithm to use less than five distinct frequency, either the sets of frequency in A and C differ by one frequency or the two sets are the same. In the following, we analyze these two cases to show that no online algorithm has an absolute competitive ratio less than $5/3$.

– If the sets of frequencies in A and C differ by one frequency, without loss of generality, we can assume that the set of frequencies in A is $\{1, 2, 3\}$ and that in C is $\{1, 2, 4\}$. In that case, the adversary terminates frequency 1 in A and frequency 2 in C, and make a call from B in which the fifth distinct frequency, say 5, has to be used. It is easy to see that the optimal can make use of three distinct frequencies only, and hence the competitive ratio is at least $5/3$.

– If the sets of frequencies in A and C is the same, without loss of generality, we can assume that both sets of frequency are $\{1, 2, 3\}$. Moreover, if less than five distinct frequencies are used, the sets of frequency in F must be in the form $\{1, 2, 3, 4\} - \{i\}$ for a fixed i with $1 \leq i \leq 4$. The aim of the adversary is to make a call in B such that frequency i must be assigned to serve the call. This can be done by terminating all calls in A except one and all calls in C except one, such that the remaining calls in A and C use a different frequency and none of the two frequencies are frequency i. Note that this can always be done since originally there are three calls in each of A and C. After frequency i is assigned by B, all calls in A and C are terminated and two new calls are made from B and three new calls are made from D. Since frequency i is used in B but not in F, the three frequencies assigned by D

cannot be the same to both of those in B and F. Then, applying the same argument as in the previous case, we can show that the online algorithm must use at least five distinct frequencies while the optimal algorithm can use only three. Hence, the competitive ratio is at least $5/3$.

For the asymptotic competitive ratio, the adversary makes n calls from each of the A, C and F. Let f_X denote the set of frequencies in cell X. For any online algorithm, let γ be the minimum between the numbers of common frequencies in A and F, and C and F, i.e., $\gamma = \min\{|f_A \cap f_F|, |f_C \cap f_F|\}$. The online algorithm uses at least $2n - \gamma$ distinct frequencies.

The adversary then terminates some calls in A and C such that $f_A \cap f_C = \emptyset$ and $f_A \cup f_C \subseteq f_F$ and $|f_A| = |f_C| = \gamma/2$. After that, $n - \gamma/2$ new calls are made from B, in which at least $\gamma/2$ of the frequencies assigned will not be in F. Then, all calls from A and C are terminated, $\gamma/2$ and n new calls are made from B and D, respectively. Since at least $\gamma/2$ of the frequencies in B are not in F and vice versa, D has at least $\gamma/4$ frequencies either not in B or F and vice versa, and without loss of generality assume that it is F. The adversary terminates some calls in D and F such that $f_D \cap f_F = \emptyset$ and $|f_D| = |f_F| = n/2 + \gamma/8$. Then, $n/2 - \gamma/8$ new calls are made from E in which the frequencies assigned must be different from those currently in D and F. The online algorithm must use at least $3n/2 + \gamma/8$ distinct different to satisfy all the calls in D, E and F. Including the case where γ is defined, the online algorithm uses at least $\max\{2n - \gamma, 3n/2 + \gamma/8\} \geq 14n/9$ distinct frequencies. As the optimal algorithm can use only n frequencies to satisfy all calls, the competitive ratio of the online algorithm is at least $14/9$. $\qquad\square$

References

1. S. Anand, A. Sridharam, and K. N. Sivarajan. Performance analysis of channelized cellular systems with dynamic channel allocation. *IEEE Transactions on vehicular technology*, 52(4):847-859, 2003.
2. M. Bassiouni and C. Fang. Dynamic channel allocation for linear macrocellular topology. *Wireless Personal Communications*, 19:121-138, 2001.
3. A. Borodin and R. El-Yaniv. *Online Computation and Competitive Analysis*. Cambridge University Press, 1998.
4. I. Caragiannis, C. Kaklamanis, and E. Papaioannou. Efficient on-line frequency allocation and call control in cellular networks. *Theory Comput. Syst.*, 35(5): 521–543, 2002. A preliminary version of the paper appeared in SPAA 2000.
5. A. Iera, S. Marano and A. Molinaro. Call-Level and Burst-Level Properties of Effective Management of Multimedia Sercies in UMTS. *Proceedings of IEEE INFOCOM*, 1363-1370, 1996.
6. H. Jiang and S. S. Rappaport. Hybrid channel borrowing and directed retry in highway cellular communications. *IEEE 46th Vehicular Technology Conference, 'Mobile Technology for the Human Race'*, 2:716-720, 1996.
7. V. H. MacDonald. The cellular concept. *The Bell System Techn. J.* 58:15-41, 1979
8. K.L. Yeung and T.P. Yum. Cell group decoupling analysis of a dynamic channel assignment strategy in linear microcellular radio systems. *IEEE Transactions on Communications*, 43:1289-1292, 1995.

Finite-State Online Algorithms and Their Automated Competitive Analysis

Takashi Horiyama, Kazuo Iwama, and Jun Kawahara

Graduate School of Informatics
Kyoto University Kyoto 606-8501, Japan
{horiyama, iwama, jkawahara}@kuis.kyoto-u.ac.jp

Abstract. In this paper we study the Revocable Online Knapsack Problem (ROKP) which is an extension of the Online Knapsack Problem [8]. We prove an optimal upper bound of $1/t$ for the competitive ratio of ROKP, where t is a real root of $4x^3 + 5x^2 - x - 4 = 0$ ($t \approx 0.76850$ and $1/t \approx 1.3012$). To prove this result, we made a full use of computer programs as follows: For the base algorithm that is designed in a conventional manner, we first construct an equivalent finite state diagram with about 300 states. Then for each state, we generate a finite set of inequalities such that the competitive ratio at that state is at most $1/t$ if the set of inequalities do not have a real solution. The latter can be checked by Mathematica. The number of inequalities generated was approximately 600 in total, and our computation time was 30 minutes using Athlon XP 2600+.

1 Introduction

When designing online algorithms (combinatorial optimization algorithms, more in general), we often have heuristics which must be useful to improve the performance of existing algorithms. In many cases, however, we have to give them up just because their performance analysis seems too difficult. We are especially unhappy if the difficulty is mainly due to the complexity of its case analysis, namely, too many cases are involved. Such a difficulty is rather common for some type of problems, including packing problems (see e.g., [10,11,13]).

In this paper, we study the Revocable Online Knapsack Problem (ROKP), which is an extension of the Online (Removable) Knapsack Problem (OKP) [8]. In OKP, we have a knapsack (or a bin) of size 1.0 and receive input items $u_1, u_2, \ldots, u_i, \ldots$ ($|u_i| \leq 1.0$) sequentially. For each u_i, we have to decide whether or not we take (and put into the bin) u_i. At the same time, we can discard zero or more items currently existing in the bin. The goal is to maximize the content of the bin at the end of the input. ROKP is a semi-online version of OKP: Other than the main bin, we can use an extra bin and can use it as a buffer for online decisions by allowing items to be moved between those two bins. Note that relaxation of the online rule has been quite popular, typically by allowing revoke of online decisions previously made or postponement of the decisions themselves. This type of relaxation includes, for example, allowing (restricted) repacking or

T. Asano (Ed.): ISAAC 2006, LNCS 4288, pp. 71–80, 2006.

lookahead in the online bin-packing [7,6]. Our present extension is in the same category.

Since the extension from OKP to ROKP gives us more freedom, we can naturally expect a better competitive ratio than $\frac{\sqrt{5}+1}{2}$ (≈ 1.6180) for OKP [8]. In fact it is not so hard to see that if we can increase the number k of extra bins then the competitive ratio approaches to 1.0 without limit. Unfortunately exact analysis for a small k, like $k = 1$ in the case of ROKP, seems to involve very complicated case analysis. To use a computer program is a natural attempt to cope with this kind of difficulty. The main purpose of this paper is to show that our online algorithm can be transformed into a *finite-state* diagram and we can make a full use of this fact for automated performance analysis.

Our Contribution. Our algorithm for ROKP achieves a competitive ratio of $1/t \approx 1.3012$ (t is a real root of $4x^3 + 5x^2 - x - 4 = 0$). We also prove a matching lower bound (by hand). Our algorithm can also been seen as an approximation algorithm for the knapsack (subset-sum) problem. It runs in time $O(n)$ and space $O(1)$ and its approximation ratio (the same as above) is currently best under the obviously strongest time/space restriction. (PTAS/FPTAS-type algorithms need at least a linear space.) Note that the constant factors hidden under the Big-O notation are not so large, either.

The basic idea of our scheme for automated competitive analysis is as follows: (i) We first design an online algorithm, called a *base algorithm*, in an ordinary way. Then we generate a state diagram G which can be seen as a *finite automaton* equivalent to the base algorithm. (ii) With each state of G, we associate a set of inequalities such that whether or not the competitive ratio at that state is within the target bound is equivalent to whether or not the set of inequalities are simultaneously satisfiable. (iii) We use Mathematica to solve the latter, i.e., the satisfiability of the inequalities. In this approach, we have two major difficulties:

(1) The number of states must be finite. First of all, input items can be arbitrarily small and hence we cannot bound the number of such items in the bins. Fortunately, we can prove that this class of items are not important for the competitive analysis, or we can assume that such items do not come. Thus it is enough to consider finitely many items in the bins. A more serious problem is that each input item takes a real value between 0 and 1. So, even if the bins contain a finite number of items, the state space is not finite. Our solution is to introduce a finite number of inequalities to each state which are implicit in the base algorithm. Now it is enough to consider a finite number of cases since the total number of inequalities is finite.

(2) Our target competitive ratio is not rational. Real numbers can be treated by using e.g., Mathematica but computation time badly slows down. Fortunately, there are few states which need really tight analysis; for others we can use approximated fractions.

In this project, we first proved an upper bound of 1.3334 and a lower bound of 1.2808. The proof of this upper bound required analysis with essential 13 cases. Although this proof was done by hand, improving this base algorithm required the explosion on the number of cases, and thus we adopted automated competitive analysis. We tried to prove an upper bound of 1.2808 but the system generated states whose competitive ratio is larger than this value. This gave us an important hint to prove the better lower bound of 1.3012 and the same upper bound was achieved by further modifying the base algorithm. The number of states generated is about 300 and the number of generated inequalities is about 600. The total computation time is less than 30 minutes using Athlon XP 2600+.

Related Work. Automated theorem proving is one of the oldest topics in computer science and has a number of successful applications [1,4,5,9,15] for design and analysis of algorithms. A recent beautiful example is given by Seiden against online bin-packing [14]. He proved that the asymptotic performance ratio of his algorithm, called Harmonic++, is at most 1.58889. The performance analysis is reduced to solving (finding an optimal solution of) a specific instance of a certain type of integer linear programming. The instance is then solved by using the standard branch and bound technique together with some edge-pruning techniques. Our approach in this paper is more direct, i.e., by describing an algorithm in terms of a finite-state diagram in which each state includes enough information to calculate its competitive ratio.

2   Problem Definitions and Lower Bounds

2.1   Revocable Online Knapsack Problem

An instance of the Online (Removable) Knapsack Problem (OKP) [8] is a sequence $\sigma = u_1, \ldots, u_n$ of *items*, where $0 < |u_i| \le 1$ is the *size* of item u_i. The online player has a *knapsack* (also called a *bin*) of size one and for each u_i, has to decide (i) whether u_i is put into the bin and (ii) which (zero or more) items among the ones currently in the bin are discarded. Our goal is to make the bin as full as possible.

The Revocable OKP (ROKP) is a semi-online version of the OKP. Other than the (main) bin we wish to fill as much as possible, the player can use an extra bin of size one and can use it for delaying discarding decisions. Since two bins are same and complete rearrangement of the items in the two bins is allowed, we do not have to distinguish the main and extra bins, namely our rule of the game can be simply stated as follows: We have two bins and let $B_1(t)$ and $B_2(t)$ be the set of items held by the first and the second bins, respectively, before step t. Initially $B_1(1) = B_2(1) = \phi$. In each step $t(\ge 1)$, we have to decide $B_1(t+1)$ and $B_2(t+1)$ so as to satisfy the condition that: (i) $|B_1(t+1)| \le 1$ and $|B_2(t+1)| \le 1$ ($|B_i(t+1)|$ denotes the total size of the items in the bin) and (ii) $B_1(t+1) \cup B_2(t+1) \subseteq B_1(t) \cup B_2(t) \cup \{u_t\}$.

Let X be an algorithm for ROKP. $|X(\sigma)|$ is the cost achieved by X for input $\sigma = u_1, \ldots, u_n$, and is defined by $|X(\sigma)| = \max(|B_1(n+1)|, |B_2(n+2)|)$. $|OPT(\sigma)|$ denotes the cost achieved by the off-line optimal algorithm. If $|X(\sigma)| \ge |OPT(\sigma)|/r$ for any input σ, then we say that the competitive ratio of algorithm X, denoted by $CR(X)$, is r.

2.2 Lower Bounds of Competitive Ratio

As mentioned in the first section, the extension from OKP to ROKP substantially increases the number of cases for analysing both lower and upper bounds. We first present a (tight) lower bound for the competitive ratio of ROKP.

Theorem 1. *Let X be any online algorithm for ROKP. Then $CR(X) > 1/t - \varepsilon$ for any $\varepsilon > 0$, where t is a real root of $4x^3 + 5x^2 - x - 4 = 0$ ($t \approx 0.76850$ and $1/t \approx 1.3012$).*

Proof. We use nine items $u, v, w, x, y, z, \bar{u}, \bar{v}$ and $\bar{y}$, where their sizes are defined as follows: $|u| = (t^2 + t)/2$ (≈ 0.67955), $|v| = 1 - t^2$ (≈ 0.40940), $|w| = t$ (≈ 0.76850), $|x| = t^2 + \varepsilon''$ (≈ 0.59056), $|y| = (-t^2 - t + 2)/2 + \varepsilon'$ (≈ 0.32045), $|z| = t^2 + t - 1$ (≈ 0.35910), $|\bar{u}| = 1 - |u|$ (≈ 0.32045), $|\bar{v}| = 1 - |v|$ (≈ 0.59060), and $|\bar{y}| = 1 - |y|$ (≈ 0.67955) (ε' and ε'' are small positive constants).

Now, the adversary first gives three items u, v, and w. The online player X has to discard at least one item, resulting in the following three cases: (Case 1) discarding u, (Case 2) discarding v, and (Case 3) discarding w.

In Case 1, the adversary gives $\bar{u}$ as the fourth item and stops the input. X can get at most $|w|$ as its cost since $|\bar{u}| < |v| < |\bar{u}| + |v| < |w| < 1$ and $|\bar{u}| + |w| > 1$. The optimal cost is $|u| + |\bar{u}| = 1$, and thus the CR is $\frac{|u| + |\bar{u}|}{|w|} = \frac{1}{t}$. Case 2 is similar; the adversary gives $\bar{v}$ and stops the input, and the CR is $\frac{|v| + |\bar{v}|}{|w|} = \frac{1}{t}$. In Case 3, the next item by the adversary is x. We have again three cases: (Case 3-1) discarding u, (Case 3-2) discarding v, and (Case 3-3) discarding x. In Cases 3-1 and 3-2, we can prove that the CR exceeds $\frac{1}{t}$ similarly to Cases 1 and 2. In Case 3-3, the adversary furthermore gives y and z. Now we have four items u, v, y and z, where at least one item should be discarded. It can be proved that the CR is at least $\frac{1}{t}$ for any of these four cases similarly. Thus the CR is at least $\frac{1}{t}$ for all the cases.   □

3   Base Algorithm

Recall that our primary goal is to design a finite state algorithm, denoted by $\mathcal{A}_{FS}$, for ROKP and prove its performance automatically. To do so, we need the base algorithm, denoted by $\mathcal{A}$, from which $\mathcal{A}_{FS}$ is generated as shown later.

Recall that t is a real root of equation $4x^3 + 5x^2 - x - 4 = 0$ ($t \approx 0.76850$). We need the following classification on the sizes of the items. Let $c_0, c_1, \dots, c_n$ be constant values satisfying $c_0 = 0 < c_1 < c_2 < \dots < c_n = 1$. An item u is said to be in class C_i if $c_{i-1} < |u| \leq c_i$ ($i = 1, 2, \dots, n$). In algorithm $\mathcal{A}$, we use seven classes: $c_1 = 1 - t$ (≈ 0.23150), $c_2 = 1 - t^2$ (≈ 0.40940), $c_3 = 2 - 2t$ (≈ 0.46299) $c_4 = 2t - 1$ (≈ 0.53701), $c_5 = t^2$ (≈ 0.59060), and $c_6 = t$ (≈ 0.76850). Classes $C_1, C_2, \dots, C_7$ are called XS, SS, MS, MM, ML, LL and XL, respectively. Items in class SS are denoted as $\ell_{SS1}, \ell_{SS2}, \dots$, or simply denoted as ℓ_{SS} if it causes no confusion. Similarly for other classes. Our algorithm $\mathcal{A}$ uses two bins B_1 and B_2, whose single round for an input item u is described as follows:

if ($|B_1| > t$ **or** $|B_2| > t$) discard u. $\cdots$ (1)
else if ($u \in XL$) $\cdots$ (2)
    put u into B_1, and discard all the other items.
else if (there exists items $\ell_1, \cdots, \ell_m$ in the two bins satisfying $t < |\ell_1| + \cdots + |\ell_m| + |u| \le 1$) $\cdots$ (3)
    put $\ell_1, \cdots, \ell_m$ and u into B_1, and discard all the other items.
else if ($u \in XS$) put u into B_1. $\cdots$ (4)
else if (it is possible to take u without discarding any item
    by suitably rearranging the items) put u into B_1 or B_2 $\cdots$ (5)
else decide what items should be discarded according to $B = \{$items in the bins$\} \cup \{u\}$. $\cdots$ (6)
    if (B includes at least two items in SS and at least one item in LL)
        compare the total sizes of the two ℓ_{SS}'s and the size of ℓ_{LL},
        and discard the larger item(s) (i.e., two SS's or one LL).
    else if (B includes at least two items in the same class)
        discard the largest item in the class.
    else if (B includes $\ell_{MS}, \ell_{ML}, \ell_{LL}$) discard ℓ_{ML}.
    else (B includes $\ell_{MM}, \ell_{ML}, \ell_{LL}$) discard ℓ_{ML}.

Here is an overview of algorithm $\mathcal{A}$: In (1), the target competitive ratio has already been achieved and the game ended already. In (2) and (3), the ratio will be achieved after this round. In (4) and (5), it is obviously safe to take the item u since we do not need to discard anything. (Notice that there is always a space for an XS item in (4) since neither bin has size t or more.) (6) is the main portion of $\mathcal{A}$, where we have to discard something. The basic idea is this: Suppose that we now have two items, x_1 and x_2 ($x_1 \le x_2$), in LL. Since two LL's do not fit a single bin, it is useless to hold both. So, we should discard one of them and we do so by discarding the *larger* one, i.e., x_2 for the following intuitive reason. Suppose that the offline solution includes x_2. Of course the online player that has discarded x_2 cannot use it but can use x_1 instead. Since $|x_2|/|x_1| \le t/t^2 = 1/t$, this is not harmful. The same idea applies to the case of two SS's and one LL.

4 Finite State Algorithm

4.1 State Diagrams

Now our goal is to construct a finite state algorithm $\mathcal{A}_{FS}$ from $\mathcal{A}$. For this purpose, there are two difficulties. The first one is the existence of XS items. Arbitrarily small items are undesirable for us to bound the number of states. Fortunately, we can prove the following lemma, by which without loss of generality, we can assume that each item has at lease size $1 - t$ (≈ 0.23150), i.e., the number of items in each bin is obviously finite.

Lemma 1. *Suppose that $\mathcal{A}$ achieves a competitive ratio of r for an input sequence which does not include XS items. Then $\mathcal{A}$ achieves the same competitive ratio for a general input.*

Proof. Let σ be an input sequence including XS items and σ' be the sequence obtained by deleting all the XS items in σ. Suppose that the competitive ratio of $\mathcal{A}$ is at most $1/t$ for σ' and consider the behavior of $\mathcal{A}$ for σ. Since the size of an XS item is at most $1 - t$, it must enter the bins unless the target ratio $1/t$ is already achieved. Thus if an XS item comes, the competitive ratio at that moment is always improved. Furthermore, one can see easily that if $\mathcal{A}$ has to discard XS items to take some (not XS) item, then the ratio $1/t$ must be achieved after that. Thus, we can assume that XS items always enter the bins and are never discarded, i.e., they always contribute to improving the competitive ratio. Thus $\mathcal{A}$ also achieves the ratio $1/t$ for σ. □

The second difficulty is that the size of each item takes a real value and our target competitive ratio is also real. As will be seen in a moment, we will escape this difficulty by introducing a finite number of inequalities. A directed graph G is called a *state diagram*, where each vertex is called a *state* and each arrow a *transition*. Each state and each transition have the following labels. (We sometimes call such a label itself a state or a transition.)

(1) A state consists of a set U and a set R. U consists of item classes with subscripts like $\{SS_1, SS_2, SS_3, LL_1\}$, which means we now have three SS items and one LL item in the bins. R consists of inequalities like $R = \{SS_1 \leq SS_2, SS_2 \leq SS_3, SS_1 + SS_2 < t, SS_1 + LL_1 > 1\}$. Recall that the size of SS and LL items, ℓ_{SS} and ℓ_{LL}, satisfies approximately that $0.23 < |\ell_{SS}| \leq 0.41$ and $0.59 < |\ell_{LL}| \leq 0.76$. Therefore $0.82 < |\ell_{SS}| + |\ell_{LL}| \leq 1.17$, namely there are two cases depending on whether ℓ_{SS} and ℓ_{LL} fit in a single bin or not. The last inequality in R, $SS_1 + LL_1 > 1$, means they do not.

(2) A transition from a state (U_1, R_1) to a state (U_2, R_2) consists of I, O and T. I is a class in *Class*, i.e., it means the class the current item u belongs to. $O \subseteq U_1 \cup \{u\}$, which shows the items that should be discarded. T is a set of inequalities similar to R above, but each inequality must include u. For example suppose that $I = LL$, $O = \{LL_1\}$, and $T = \{SS_1 + u > 1, u \leq LL_1\}$. This means the current item is an LL item which satisfies the inequalities in T against the items in U_1 of the state (U_1, R_1) (hence U_1 should include SS_1 and LL_1), and the item LL_1 in U_1 is discarded.

$\mathcal{A}_{FS}$ will be given in Sec. 5. In the rest of this section, we discuss some basic properties of $\mathcal{A}_{FS}$.

4.2   Feasibility of State Diagrams

We first introduce an important notion, *feasibility*, regarding a state diagram. A state diagram G is said to be *feasible* if the following two conditions are met:

(1) For each state (U, R) and each $C \in Class$, there is at least one transition from (U, R) whose I is equal to C. Suppose that there are two transitions $t_1 = (I_1, O_1, T_1)$ and $t_2 = (I_2, O_2, T_2)$ from (U, R) such that $I_1 = I_2$. Then for any item $u \in I_1 (= I_2)$, there must be exactly one set, either T_1 or T_2, of inequalities which are all satisfied. Namely, for the current state and input, the state transition must be determined uniquely.

(2) For each transition (I, O, T) from (U_1, R_1) to (U_2, R_2), the disjunction of T, R_1 and R_2 should be consistent in the following sense: Suppose for example that $U_1 = \{SS_1, SS_2, MM_1\}$, $I = LL$, and $T = \{SS_1 + SS_2 > u, \; SS_2 + u \leq 1\}$ and $O = \{SS_1\}$. (Namely, the current bins include two SS's and one MM. The current input u is in LL and u satisfies the two inequalities in T. The algorithm discards SS_1.) Then in the next state, the item u, which is put in the bins, becomes LL_1. Therefore, U_2 should be $\{SS_2, MM_1, LL_1\}$ and R_2 should be $R_1 \cup T - \{\text{all the inequalities including discarded items}\}$, namely $\{SS_2 + LL_1 \leq 1\}$ Furthermore, we change the subscripts of items so that they always start from one. Namely, the new state should be $U_2 = \{SS_1, MM_1, LL_1\}$ and $R_2 = \{SS_1 + LL_1 \leq 1\}$. We call this rule of subscript change the *item mapping*.

4.3 Execution Sequences

Now suppose that we are given a state diagram G and an item sequence $u_1, u_2, \ldots,$ u_n. Then we define a sequence $(S_0, A_0), (S_1, A_1), \ldots, (S_n, A_n)$, called an *execution sequence*, as follows:

(1) For each i, $S_i = (U_i, R_i)$ is a state. Suppose that $U_i = \{SS_1, SS_2, LL_1\}$. A_i is an assignment of a specific value to each item in U_i, say $\{SS_1 = 0.32, SS_2 = 0.33, LL_1 = 0.61\}$.

(2) S_0 is the initial state, i.e., $U_0 = R_0 = \phi$. A_0 is also ϕ.

(3) From (S_i, A_i) and $1 - t \leq u_i \leq 1.0$, (S_{i+1}, A_{i+1}) is determined (if any) as follows: (i) There is a transition (I_i, O_i, T_i) from S_i to S_{i+1} in G such that $u_i \in I_i$ and all the inequalities in T_i are satisfied under the assignment A_i and the value u_i. For example, suppose that U_i and A_i are as given in (1) above, $u_i = 0.39$ and $T_i = \{SS_1 + SS_2 + u > 1.0\}$. Then this inequality in T is met for $SS_1 = 0.32, SS_2 = 0.33$ and $u = 0.39$. (ii) A_{i+1} represents the change of assignment accordingly (for this, the item mapping rule should be considered, details are omitted.)

Now we can use a state diagram as an "algorithm", which is due to the following lemma (the proof is straightforward from the previous definitions and may be omitted.)

Lemma 2. *Suppose that G is a feasible transition diagram. Then for any sequence $u_1, u_2, \ldots, u_n$ of items such that $1 - t \leq u_i \leq 1.0$, its execution sequence is determined uniquely.*

4.4 Calculation of Competitive Ratio

Thus a feasible transition diagram has enough information as an algorithm. However, it is not enough to calculate its competitive ratio automatically. We thus add a bit more information to each state. In our new state diagram, called a *state diagram with history* or an *SDH*, each state consists of U, H and R. Here, U is exactly the same as before, i.e., a set of items in the bins. H is a set of items which have been discarded so far, which looks like $\{SS_1^H, SS_2^H, MS_1^H, MS_{2+}^H\}$. This means that two SS items have been discarded and at least two MS items (denoted by MS_{2+}^H) have been discarded. Note that at most two MS items fit a single bin and therefore we do not care about whether the number of them is two or more. R is a set of inequalities like $\{SS_1 + LL_1 > 1, SS_H + MS_1 < t\}$. Here SS_H means an SS item already discarded. Note that we prepare only one SS_H even if two or more SS items have been discarded. (Since there may be unlimitedly many such items, it is not possible to enumerate all of them.) As will be explained later, this inequality $SS_H + MS_1 < t$ is to hold for *any* SS item ever discarded.

Now we define the *feasibility* of an SDH. Its definition is the same as before excepting the set R_2 of inequalities in the next state. Suppose that $R_1 = \{SS_1 + LL_1 > 1, SS_H + LL_1 > 1, SS_H + MS_1 < t\}$, and SS_1 is discarded in this transition. Then when checking R_2, besides the same condition as before (given in (2) of Sec. 4.3), we make the following modification for inequality $SS_1 + LL_1 > 1$ (i.e., each inequality including the discarded item): $SS_1 + LL_1 > 1$ is changed to $SS_H + LL_1 > 1$ for R_2 if there is no SS in H or there is already $SS_H + LL_1 > 1$ in R_1 (this is the case in the above R_1). Otherwise, no inequalities including SS_1 or SS_H should be in R_2. There is no contradiction in this notation (details are omitted).

An *execution sequence* for an SDH is also defined as before. Only one difference is in assignments for discarded items. We have the following lemma similar to Lemma 2 (proof is omitted).

Lemma 3. *Suppose that G is a feasible SDH. Then for any sequence $u_1, \ldots, u_n$ of items such that $1 - t \leq u_i \leq 1.0$, its execution sequence is determined uniquely.*

Now we are ready to calculate the competitive ratio of an SDH. Suppose that a state $S = (U, H, R)$ has $U = \{SS_1, SS_2, LL_1\}$ and $H = \{SS_1^H, MS_1^H, MS_{2+}^H\}$. Then we prepare two sets $ALG(S)$ and $OPT(S)$. $ALG(S)$ includes all the nonempty subsets of the items in U. Namely, for the above example, $ALG(S) = \{\{SS_1\}, \{SS_2\}, \{LL_1\}, \{SS_1, SS_2\}, \ldots, \{SS_1, SS_2, LL_1\}\}$. $OPT(S)$ includes all the nonempty subsets of the items in $U \cup H$, namely, $OPT(S) = \{\{SS_1\}, \ldots, \{SS_1^H\}, \ldots, \{SS_1, MS_{2+}^H\}, \ldots, \{SS_1, SS_2, LL_1, SS_1^H, MS_1^H, MS_2^H\}\}$. Now we consider the following proposition P:

Proposition P. For any $\beta \in OPT(S)$, there exists $\alpha \in ALG(S)$ such that there is no assignment of values into items in $\alpha \cup \beta$ which satisfies $|\alpha|/|\beta| \leq t$, R, $|\alpha| \leq 1.0$, $|\beta| \leq 1.0$ and all the range restrictions.

Now we show, by the following lemma, that we can actually prove the competitive ratio by using P:

Lemma 4. *Suppose that G is a feasible SDH. Then if proposition P is true for every state of G, then G's competitive ratio is at most $1/t$.*

Proof. Suppose that G's competitive ratio is more than $1/t$. Then there is an input sequence $u_1, u_2, \ldots, u_n$ such that after u_n, there is no combination of items in the bins which achieves the desired competitive ratio. Now consider the execution sequence for $u_1, u_2, \ldots, u_n$ which is guaranteed to exist by Lemma 3.

After u_n we reach an state S and an assignment which determines all the values of the items in the bins and the items discarded so far at that state. Since the competitive ratio is more than $1/t$, we can select some β in $OPT(S)$ for any α in $ALG(S)$ such that $|\beta|/|\alpha| > t, |\alpha| \leq 1.0, |\beta| \leq 1.0$. By the definition of execution sequences, the values of the items must satisfy R and all the range restrictions. However this contradicts the fact that P is true for this state S. $\qquad\square$

Thus we have to check proposition P for every state of G to prove its competitive ratio: For each state S we enumerate $ALG(S)$ and $OPT(S)$, both of which are finite. Then for each $\alpha \in ALG(S)$ and each $\beta \in OPT(S)$, we generate the following set $EQ(\alpha, \beta)$ of inequalities and equalities: $R \cup \{|\alpha| \leq t|\beta|, |\alpha| \leq 1.0, |\beta| \leq 1.0\} \cup \{$All range restrictions$\} \cup \{4t^3 + 5t^2 - t - 4 = 0\}$ for t. For each S in G, if there exists α that $EQ(\alpha, \beta)$ has no solutions for all $\beta \in OPT(S)$, we can conclude that P is true. (This can be checked by Mathematica.)

5   Construction and Verification $\mathcal{A}_{FS}$

We now construct $\mathcal{A}_{FS}$ which should be a feasible SDH. The basic idea is to construct $\mathcal{A}_{FS}$ by simulating its base algorithm $\mathcal{A}$. We start with the initial state of $\mathcal{A}_{FS}$ which is (ϕ, ϕ, ϕ) and construct transitions and new states step by step. Suppose that we are now in state $S = (U, H, R)$. Then we consider a transition $t = (I, O, T)$ from S by an item class $c \in Class$. In order to determine the state S' to which t goes, we need the information of O and T of t (its I is c). Once I, O and T are determined, then the state S' is determined automatically by the feasibility condition. So, the obtained SDH is automatically feasible.

The only remaining problem is how to determine T and O. To do so, we developed a "symbolic simulator" of $\mathcal{A}$ which computes O and T from U, R and I. The idea is that there are only a small number of possible inequalities which can be in R and T, namely, they look like the following:

$$C_1 + C_2 > 1, \quad C_1 + C_2 \leq 1, \quad C_1 + C_2 < t, \quad C_1 + C_2 \geq t,$$

where C_1 and C_2 are item classes or u. So, the simulator exhaustively checks each (or sometimes two simultaneously) of those inequalities. Although details are omitted, this simulator is relatively a small program.

As a result, we obtained $\mathcal{A}_{FS}$ as an SDH with some 300 states, which is shown in http://www.lab2.kuis.kyoto-u.ac.jp/~jkawahara/rokp/. Thus by Lemma 4, we obtain

Theorem 2. $\mathcal{A}_{FS}$ *is a correct algorithm for ROKP without XS items, whose competitive ratio is at most $1/t$.*

It is not hard to modify $\mathcal{A}_{FS}$ into $\mathcal{A}'_{FS}$ so that it can accept XS items also. Note that we do not have a formal proof that $\mathcal{A}'_{FS}$ and $\mathcal{A}$ are equivalent since we do not have a formal relation between $\mathcal{A}$ and its simulator described in the previous section, but we do believe that they are equivalent. Thus we have the following final theorem.

Theorem 3. $\mathcal{A}'_{FS}$ *is a correct algorithm for ROKP, whose competitive ratio is at most* $1/t$.

6 Possibilities and Limits of the Approach

An obvious question against this work is the generality of this approach, namely, what kind of other problems can we use the same technique? There are two key issues: One is to describe the algorithm as a finite state diagram. This is reactively easier since the power of inequalities is quite high as seen in this paper. For example, the work-function algorithm for the 3-server problem can be described as a finite state diagram. The other is much harder, i.e., how to install a mechanism of performance analysis into the state diagram. In this paper we were able to do this by adding "history data" into each state. This does not seem always possible of course, but there might be alternatives, for example, checking all the paths of the state diagram up to some (finite) length.

References

1. K. Appel and W. Haken, Every planar map is four colorable. Part 1, II. Dischargin, *Illinois Journal of Mathematics*, vol.21, pp.429–597, 1977.
2. Y. Bartal and E. Koutsoupias, On the Competitive Ratio of the Work Function Algorithm for the k-Server Problem. *Proc. STACS*, pp.605–613, 2000.
3. W. W. Bein, M. Chrobak, L. L. Larmore, The 3-server problem in the plane, Theoretical Computer Science, vol.289/1, pp.335–354, 2002.
4. U. Feige and M. X. Goemans, Approximating the value of two prover proof systems, with applications to MAX-2SAT and MAX-DICUT, *Proc. ISTCS*, pp.182–189, 1995.
5. M. X. Goemans and D. P. Williamson, Improved approximation algorithms for maximum cut and satisfiability problems using semidefinite programming, *J. ACM*, vol.42, no.6, pp.1115–1145, 1995.
6. E. F. Grove, Online bin packing with lookahead, *Proc. SODA* pp.430–436, 1995.
7. Z. Ivkovic and E. L. Lloyd, Fully dynamic algorithms for bin packing: Being (Mostly) myopic helps, *Proc. ESA*, LNCS, pp.224–235, Springer, 1993.
8. K. Iwama and S. Taketomi, Removable on-line knapsack problems, *Proc. ICALP*, LNCS 2380, pp.293–305, 2002.
9. H. Karloff and U. Zwick, A 7/8-approximation algorithm for MAX 3SAT?, *Proc. FOCS*, pp.406–415, 1997.
10. O. Kullmann, New methods for 3-SAT decision and worst-case analysis, Theoretical Computer Science, vol.223/1-2, pp.1–72, 1999.
11. C. C. Lee and D. T. Lee, A simple on-line bin-packing algorithm, J. ACM, vol.32, no.3, pp.562–572, 1985.
12. M. Manasse, L. A. McGeoch, D. D. Sleator, Competitive algorithms for server problems, J. Algorithms, vol.11 pp.208–230, 1990.
13. M. B. Richey, Improved bounds for harmonic-based bin packing algorithms, Discr. Appli. Math., vol.34, pp.203–227, 1991.
14. S. S. Seiden, On the online bin packing problem, J. ACM, vol.49, no.5, pp.640–671, 2002.
15. L. Trevisan, G. B. Sorkin, M. Sudan, D. P. Williamson, Gadgets, Approximation, and Linear Programming, SIAM J. Comput., vol.29, no.6, pp.2074–2097, 2000.

Offline Sorting Buffers on Line

Rohit Khandekar[1] and Vinayaka Pandit[2]

[1] University of Waterloo, ON, Canada
rkhandekar@gmail.com
[2] IBM India Research Lab, New Delhi
pvinayak@in.ibm.com

Abstract. We consider the *offline sorting buffers* problem. Input to this problem is a sequence of requests, each specified by a point in a metric space. There is a "server" that moves from point to point to serve these requests. To serve a request, the server needs to visit the point corresponding to that request. The objective is to minimize the total distance travelled by the server in the metric space. In order to achieve this, the server is allowed to serve the requests in any order that requires to "buffer" at most k requests at any time. Thus a valid reordering can serve a request only after serving all but k previous requests.

In this paper, we consider this problem on a line metric which is motivated by its application to a widely studied disc scheduling problem. On a line metric with N uniformly spaced points, our algorithm yields the first *constant-factor approximation* and runs in quasi-polynomial time $O(m \cdot N \cdot k^{O(\log N)})$ where m is the total number of requests. Our approach is based on a dynamic program that keeps track of the number of pending requests in each of $O(\log N)$ line segments that are geometrically increasing in length.

1 Introduction

The sorting buffers problem arises in scenarios where a stream of requests needs to be served. Each request has a "type" and for any pair of types t_1 and t_2, the cost of serving a request of type t_2 immediately after serving a request of type t_1 is known. The input stream can be reordered while serving in order to minimize the cost of type-changes between successive requests served. However, a "sorting buffer" has to be used to store the requests that have arrived but not yet served and often in practice, the size of such a sorting buffer, denoted by k, is small. Thus a legal reordering must satisfy the following property: any request can be served only after serving all but k of the previous requests. The objective in the sorting buffers problem is to compute the minimum cost output sequence which respects this sequencing constraint.

Consider, as an example, a workshop dedicated to coloring cars. A sequence of requests to color cars with specific colors is received. If the painting schedule paints a car with a certain color followed by a car with a different color, then, a significant set-up cost is incurred in changing colors. Assume that the workshop has space to hold at most k cars in waiting. A natural objective is to rearrange

T. Asano (Ed.): ISAAC 2006, LNCS 4288, pp. 81–89, 2006.
© Springer-Verlag Berlin Heidelberg 2006

the sequence of requests such that it can be served with a buffer of size k and the total set-up cost over all the requests is minimized.

Consider, as another example, the classical disc scheduling problem. A sequence of requests each of which is a block of data to be written on a particular track is given. To write a block on a track, the disc-head has to be moved to that track. As discussed in [3], the set of tracks can be modeled by uniformly spaced points on a straight line. The cost of moving from a track to another is then the distance between those tracks on the straight line. We are given a buffer that can hold at most k blocks at a time, and the goal is to find a write-sequence subject to the buffer constraint such that the total head movement is minimized.

Usually, the type-change costs satisfy metric properties and hence we formulate the sorting buffers problem on a metric space. Let (V, d) be a metric space on N points. The input to the Sorting Buffers Problem (SBP) consists of a sequence of m requests, the ith request being labeled with a point $p_i \in V$. There is a server, initially located at a point $p_0 \in V$. To serve ith request, the server has to visit p_i. There is a sorting buffer which can hold up to k requests at a time. In a *legal schedule*, the ith request can be served only after serving at least $i - k$ requests of the first $i - 1$ requests. More formally, the output is given by a permutation π of $\{1, \ldots, m\}$ where the ith request in the output sequence is the $\pi(i)$th request in the input sequence. Observe that a schedule π is legal if and only if it satisfies $\pi(i) \leq i + k$ for all i. The cost of the schedule is the total distance that the server has to travel, i.e., $C_\pi = \sum_{i=1}^{m} d(p_{\pi(i-1)}, p_{\pi(i)})$ where $\pi(0) = p_0$ corresponds to the starting point. The goal in SBP is to find a legal schedule π that minimizes C_π. In the online version of SBP, the ith request is revealed only after serving at least $i - k$ among the first $i - 1$ requests. In the offline version, on the other hand, the entire input sequence is known in advance.

The car coloring problem described above can be thought of as the SBP on a uniform metric where all the pair-wise distances are identical while the disc scheduling problem corresponds to the SBP on a line metric where all the points lie on a straight line and the distances are given along that line.

1.1   Previous Work

On a general metric, the SBP is known to be NP-hard due to a simple reduction from the Hamiltonian Path problem. However, for the uniform or line metrics, it is not known if the problem remains NP-hard. In fact, no non-trivial lower bound is known on the approximation (resp. competitive) ratio of offline (resp. online) algorithms, deterministic or randomized. In [3], it is shown that the popular heuristics like shortest time first, first-in-first-out (FIFO) have $\Omega(k)$ competitive ratio on a line metric. In [5], it is shown that the popular heuristics like FIFO, LRU, and Most-Common-First (MCF) have a competitive ratio of $\Omega(\sqrt{k})$ on a uniform metric.

The offline version of the sorting buffers problem on any metric can be solved optimally using dynamic programming in $O(m^{k+1})$ time where m is the number of requests in the sequence. This follows from the observation that the algorithm can pick k requests to hold in the buffer from first i requests in $\binom{i}{k}$ ways when the $(i+1)$th request arrives.

The SBP on a uniform metric has been studied before. Räcke et al. [5] presented a deterministic online algorithm, called *Bounded Waste* that has $O(\log^2 k)$ competitive ratio. Englert and Westermann [2] considered a generalization of the uniform metric in which moving to a point p from any other point in the space has a cost c_p. They proposed an algorithm called Maximum Adjusted Penalty (MAP) and showed that it gives an $O(\log k)$ approximation, thus improving the competitive ratio of the SBP on uniform metric. Kohrt and Pruhs [4] also considered the uniform metric but with different optimization measure. Their objective was to maximize the reduction in the cost from that of the schedule without a buffer. They presented a 20-approximation algorithm for this variant and this ratio was improved to 9 by Bar-Yehuda and Laserson [1].

For SBP on line metric, Khandekar and Pandit [3] gave a polynomial time randomized online algorithm with $O(\log^2 N)$ competitive ratio. In fact, their approach works on a class of "line-like" metrics. Their approach is based on probabilistic embedding of the line metric into the so-called hierarchical well-separated trees (HSTs) and an $O(\log N)$-competitive algorithm for the SBP on a binary tree metric. No better approximations were known for the offline problem.

1.2 Our Results

The first step in understanding the structure of the SBP is to develop offline algorithms with better performance than the known online algorithms. We provide such an algorithm. Following is our main theorem.

Theorem 1. *There is a constant factor approximation algorithm for the offline SBP on a line metric on N uniformly spaced points that runs in quasi-polynomial time: $O(m \cdot N \cdot k^{O(\log N)})$ where k is the buffer-size and m is the number of input requests.*

This is the first constant factor approximation algorithm for this problem on any non-trivial metric space. The approximation factor we prove here is 15. However we remark that this factor is not optimal and most likely can be improved even using our techniques. Our algorithm is based on dynamic programming. We show that there is a near-optimum schedule with some "nice" properties and give a dynamic program to compute the best schedule with those nice properties. In Section 2.1, we give an intuitive explanation of our techniques and the Sections 2.2 and 2.3 present the details of our algorithm.

2 Algorithm

2.1 Outline of Our Approach

We start by describing an exact algorithm for the offline SBP on a general metric on N points. As we will be interested in a line metric as in the disc scheduling problem, we use the term "head" for the server and "tracks" for the points. Since the first k requests can be buffered without loss of generality, we fetch and store them in the buffer. At a given step in the algorithm, we define a *configuration* (t, C) to be the pair of current head location t and an N-dimensional vector C that specifies the number of requests pending at each track. Since there are N choices for t and a total of k requests pending, the number of distinct configurations is $O(N \cdot k^N)$. We construct a dynamic program that keeps track of the current configuration and computes the optimal solution in time $O(m \cdot N \cdot k^N)$ where m is the total number of requests. The dynamic program proceeds in m levels. For each level i and each configuration (t, C), we compute the least cost of serving i requests from the first $i + k$ requests and ending up in the configuration (t, C). Let us denote this cost by $\texttt{DP}[i, t, C]$. This cost can be computed using the relation

$$\texttt{DP}[i, t, C] = \min_{(t', C')} \left(\texttt{DP}[i - 1, t', C'] + d(t', t) \right)$$

where the minimum is taken over all configurations (t', C') such that while moving the head from t' to t, a request at either t' or t in C' can be served and a new request can be fetched to arrive at the configuration (t, C). Note that it is easy to make suitable modifications to keep track of the order of the output sequence.

Note that the high complexity of the above dynamic program is due to the fact that we keep track of the number of pending requests at *each* of the N tracks. We now describe our intuition behind obtaining much smaller dynamic program for a line metric on N uniformly spaced points. Our dynamic program keeps track of the number of pending requests only in $O(\log N)$ segments of the line which are geometrically increasing in lengths. The key observation is as follows: if the optimum algorithm moves the head from a track t to t' (thereby paying the cost $|t - t'|$), a constant factor approximation algorithm can safely move an additional $O(|t - t'|)$ distance and clear all the nearby requests surrounding t and t'. We show that instead of keeping track of the number of pending requests at each track, it is enough to do so for the ranges of length $2^0, 2^1, 2^2, 2^3, \ldots$ surrounding the current head location t. For each track t, we partition the disc into $O(\log N)$ ranges of geometrically increasing lengths on both sides of t. The configuration (t, C) now refers to the current head location t and an $O(\log N)$-dimensional vector C that specifies number of requests pending in each of these $O(\log N)$ ranges. Thus the new dynamic program will have size $O(m \cdot N \cdot k^{O(\log N)})$.

To be able to implement the dynamic program, we ensure the property that the new configuration around t' should be easily computable from the previous configuration around t. More precisely, we ensure that the partitions for t and t' satisfy the following property: outside an interval of length $R = O(|t - t'|)$ containing t and t', the ranges in the partition for t coincide with those in the partition for t' (see Figure 1). Note however that inside this interval, the two partitions may not agree. Thus when the optimum algorithm moves the head from t to t', our algorithm starts the head from t, clears all the pending requests in this interval and rests the head at t' and updates the configuration from the

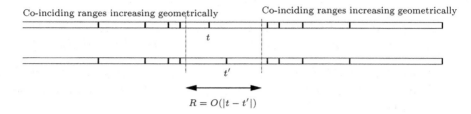

Co-inciding ranges increasing geometrically   Co-inciding ranges increasing geometrically

t

t'

$R = O(|t - t'|)$

Fig. 1. Division of the line into ranges for tracks t and t'

previous configuration. Since the length of the interval is $O(|t-t'|)$, our algorithm spends at most a constant factor more than the optimum.

2.2   Partitioning Scheme

Now we define a partitioning scheme and its properties that are used in our algorithm. Let us assume, without loss of generality, that the total number of tracks $N = 2^n$ is a power of two and that the tracks are numbered from 0 to $2^n - 1$ left-to-right. In the following, we do not distinguish between a track and its number. For tracks t and t', the quantity $|t - t'|$ denotes the distance between these tracks which is the cost paid in moving the head from t to t'. We say that a track t is to the right (resp. left) of a track t' if $t > t'$ (resp. $t < t'$).

Definition 1 (landmarks). *For a track t and an integer $p \in [1, n]$, we define pth landmark of t as $\ell_p(t) = (q + 1)2^p$ where q is the unique integer such that $(q - 1)2^p \leq t < q2^p$. We also define $(-p)$th landmark as $\ell_{-p}(t) = (q - 2)2^p$. We also define $\ell_0(t) = t$.*

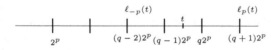

$\ell_{-p}(t)$ $\ell_p(t)$

t

2^p $(q - 2)2^p$ $(q - 1)2^p$ $q2^p$ $(q + 1)2^p$

Fig. 2. The pth and $(-p)$th landmarks of a track t

It is easy to see that $\ell_{-n}(t) < \cdots < \ell_{-1}(t) < \ell_0(t) < \ell_1(t) < \cdots < \ell_n(t)$. In fact the following lemma claims something stronger and follows easily from the above definition.

Lemma 1. *Let $p \in [1, n - 1]$ and $(q - 1)2^p \leq t < q2^p$ for an integer q.*

  - *If q is even, then $\ell_{p+1}(t) - \ell_p(t) = 2^p$ and $\ell_{-p}(t) - \ell_{-p-1}(t) = 2^{p+1}$.*
  - *If q is odd, then $\ell_{p+1}(t) - \ell_p(t) = 2^{p+1}$ and $\ell_{-p}(t) - \ell_{-p-1}(t) = 2^p$.*

In the following definition, we use the notation $[a, b) = \{t \text{ integer} \mid a \leq t < b\}$.

Definition 2 (ranges). *For a track t, we define a "range" to be a contiguous subset of tracks as follows.*

- $[\ell_{-1}(t), \ell_0(t) = t)$ and $[\ell_0(t) = t, \ell_1(t))$ are ranges.
- for $p \in [1, n-1]$, if $\ell_{p+1}(t) - \ell_p(t) = 2^{p+1}$ and $\ell_p(t) - \ell_{p-1}(t) = 2^{p-1}$ then $[\ell_p(t), \ell_p(t) + 2^p)$ and $[\ell_p(t) + 2^p, \ell_{p+1}(t))$ are ranges, **else** $[\ell_p(t), \ell_{p+1}(t))$ is a range.
- for $p \in [1, n-1]$, if $\ell_{-p}(t) - \ell_{-p-1}(t) = 2^{p+1}$ and $\ell_{-p+1}(t) - \ell_{-p}(t) = 2^{p-1}$ **then** $[\ell_{-p-1}(t), \ell_{-p-1}(t) + 2^p)$ and $[\ell_{-p-1}(t) + 2^p, \ell_{-p}(t))$ are ranges, **else** $[\ell_{-p-1}(t), \ell_{-p}(t))$ is a range.

The above ranges are disjoint and form a partition of the tracks which we denote by $\pi(t)$.

Note that in the above definition, when the difference $\ell_{p+1}(t) - \ell_p(t)$ and $\ell_{-p}(t) - \ell_{-p-1}(t)$ equals 4 times $\ell_p(t) - \ell_{p-1}(t)$ and $\ell_{-p+1}(t) - \ell_{-p}(t)$ respectively, we divide the intervals $[\ell_p(t), \ell_{p+1}(t))$ and $[\ell_{-p-1}(t), \ell_{-p}(t))$ into two ranges of length 2^p each. For example, in Figure 3, the region between $\ell_{p+2}(t)$ and $\ell_{p+3}(t)$ is divided into two disjoint ranges of equal size.

The following lemma proves a useful relation between the partitions $\pi(t)$ and $\pi(t')$ for a pair of tracks t and t': the ranges in the two partitions coincide outside the interval of length $R = O(|t - t'|)$ around t and t'. As explained in Section 2.1, such a property is important for carrying the information about the current configuration across the head movement from t to t'.

Lemma 2. *Let t and t' be two tracks such that $2^{p-1} \leq t' - t < 2^p$. The ranges in $\pi(t)$ and $\pi(t')$ are identical outside the interval $R = [\ell_{-p}(t), \ell_p(t'))$.*

Proof. First consider the case when $(q-1)2^p \leq t < t' < q2^p$ for an integer q, i.e., t and t' lie in the same "aligned" interval of length 2^p. Then clearly they also lie in the same aligned interval of length 2^r for any $r \geq p$. Thus, by definition, $\ell_r(t) = \ell_r(t')$ for $r \geq p$ and $r \leq -p$. Thus it is easy to see from the definition of ranges that the ranges in $\pi(t)$ and $\pi(t')$ outside the interval $[\ell_{-p}(t), \ell_p(t'))$ are identical.

Consider now the case when t and t' do not lie in the same aligned interval of length 2^p. Since $|t - t'| < 2^p$, they must lie in the adjacent aligned intervals of length 2^p, i.e., for some integer q, we have $(q-1)2^p \leq t < q2^p \leq t' < (q+1)2^p$ (See Figure 3). Let $q = 2^u v$ where $u \geq 0$ is an integer and v is an odd integer.

The following key claim states that depending upon how r compares with the the highest power of two that divides the "separator" $q2^p$ of t and t', either the rth landmarks of t and t' coincide with each other or the $(r+1)$th landmark of t coincides with the rth landmark of t'.

Claim. 1. $\ell_r(t) = \ell_r(t')$ for $r \geq p + u + 1$ and $r \leq -p - u - 1$.
2. $\ell_{r+1}(t) = \ell_r(t')$ for $p \leq r < p + u$,
3. $\ell_{-r}(t) = \ell_{-r-1}(t')$ for $p \leq r < p + u$,
4. $\ell_{p+u}(t') = \ell_{p+u}(t) + 2^{p+u}$ and $\ell_{p+u+1}(t) - \ell_{p+u}(t) = 2^{p+u+1}$,
5. $\ell_{-p-u}(t) = \ell_{-p-u-1}(t') + 2^{p+u}$ and $\ell_{-p-u}(t') - \ell_{-p-u-1}(t') = 2^{p+u+1}$,

Proof. The equation 1 follows from the fact that since 2^{p+u} is the highest power of two that divides $q2^p$, both t and t' lie in the same aligned interval of length 2^r for $r \geq p + u + 1$.

The equations 2, 3, 4, and 5 follow from the definition of the landmarks and the fact that t and t' lie in the different but adjacent aligned intervals of length 2^r for $p \leq r < p + u$ (see Figure 3).

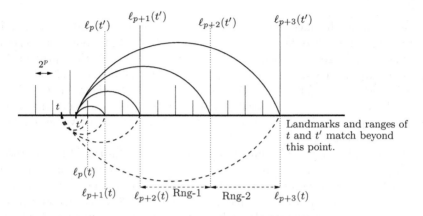

Fig. 3. Landmarks and ranges for tracks t and t' when $q = 4, u = 2$

Claim 2.2 implies that all but one landmarks of t and t' coincide with each other. For the landmarks of t and t' that coincide with each other, it follows from the definition of the ranges that the corresponding ranges in $\pi(t)$ and $\pi(t')$ are identical.

The landmarks of t, t' that do not coincide are $\ell_{p+u}(t') = \ell_{p+u}(t) + 2^{p+u}$ and $\ell_{-p-u}(t) = \ell_{-p-u-1}(t') + 2^{p+u}$. But, note that the intervals $[\ell_{p+u}(t), \ell_{p+u+1}(t))$ and $[\ell_{-p-u-1}(t'), \ell_{-p-u}(t'))$ are divided into two ranges each: $[\ell_{p+u}(t), \ell_{p+u}(t) + 2^{p+u})$, $[\ell_{p+u}(t) + 2^{p+u}, \ell_{p+u+1}(t))$ and $[\ell_{-p-u-1}(t'), \ell_{-p-u-1}(t') + 2^{p+u})$, $[\ell_{-p-u-1}(t') + 2^{p+u}, \ell_{-p-u}(t'))$. These ranges match with $[\ell_{p+u-1}(t'), \ell_{p+u}(t'))$, $[\ell_{p+u}(t'), \ell_{p+u+1}(t'))$ and $[\ell_{-p-u-1}(t), \ell_{-p-u}(t))$, $[\ell_{-p-u}(t), \ell_{-p-u+1}(t))$ respectively. This follows again from the Claim 2.2 and the carefully chosen definition of ranges. Thus the proof of Lemma 2 is complete.

For tracks t and t', where $t < t'$, let $R(t, t') = R(t', t)$ be the interval $[\ell_{-p}(t), \ell_p(t'))$ if $2^{p-1} \leq t' - t < 2^p$. Note that the length of the interval $R(t, t')$ is at most $|\ell_{-p}(t) - \ell_p(t')| \leq 4 \cdot 2^p \leq 8 \cdot |t - t'|$. Thus the total movement in starting from t, serving all the requests in $R(t, t')$, and ending at t' is at most $15 \cdot |t - t'|$.

2.3   The Dynamic Program

Our dynamic program to get a constant approximation for the offline SBP on a line metric is based on the intuition given in Section 2.1 and uses the partition scheme given in Section 2.2. Recall that according to the intuition, when the optimum makes a move from t to t', we want our algorithm to clear all the requests in $R(t, t')$. This motivates the following definition.

Definition 3. *A feasible schedule for serving all the requests is said to be "locally greedy" if there is a sequence of tracks $t_1, \ldots, t_l$, called "landmarks", which are visited in that order and while moving between any consecutive pair of tracks t_i and t_{i+1}, the schedule also serves all the current pending requests in the interval $R(t_i, t_{i+1})$.*

Since the total distance travelled in a locally greedy schedule corresponding to the optimum schedule is at most 15 times that of the optimum schedule, the best locally greedy schedule is a 15-approximation to the optimum. Our dynamic program computes the best locally greedy schedule. For a locally greedy schedule, let a configuration be defined as a pair (t, C) where t is the location of the head and C is an $O(\log N)$-dimensional vector specifying the number of requests pending in each range in the partition $\pi(t)$. Clearly the number of distinct configurations is $O(N \cdot k^{O(\log N)})$.

The dynamic program is similar to the one given in Section 2.1 and proceeds in m levels. For each level i and each configuration (t, C), we compute the least cost of serving i requests from the first $i + k$ requests and ending up in the configuration (t, C) in a locally greedy schedule. Let $\mathrm{DP}[i, t, C]$ denote this cost. This cost now can be computed as follows. Consider a configuration (t', C') after serving $i - r$ requests for some $r > 0$ such that while moving from a landmark t' to the next landmark t,

1. the locally greedy schedule serves exactly r requests from the interval $R(t', t)$,
2. it travels a distance of D, and
3. after fetching r new requests, it ends up in the configuration (t, C).

In such a case,
$$\mathrm{DP}[i - r, t', C'] + D$$
is an upper bound on $\mathrm{DP}[i, t, C]$. Taking the minimum over all such upper bounds, one obtains the value of $\mathrm{DP}[i, t, C]$.

Recall that the locally greedy schedule clears all the pending requests in the interval $R(t', t)$ while moving from t' and t and also that the ranges in $\pi(t)$ and $\pi(t')$ coincide outside the interval $R(t', t)$. Thus it is feasible to determine if after serving r requests in $R(t', t)$ and fetching r new requests, the schedule ends up in the configuration (t, C).

The dynamic program, at the end, outputs $\min_t \mathrm{DP}[m, t, \mathbf{0}]$ as the minimum cost of serving all the requests by a locally greedy schedule. It is also easy to modify the dynamic program to compute the minimum cost locally greedy schedule along with its cost.

3   Conclusions

Prior to this work, any offline algorithms with better approximation factors than the corresponding online algorithms were not known for the sorting buffers problem on any non-trivial metric. We give the first constant factor approximation for the sorting buffers problem on the line metric improving the previously known

$O(\log^2 N)$ competitive ratio. As the running time of our algorithm is quasi-polynomial, we suggest that there may be a polynomial time constant factor approximation algorithm as well. Proving any hardness results for the sorting buffers problem on the uniform or line metrics; or poly-logarithmic approximation results for general metrics remain as interesting open questions.

References

1. R. Bar-Yehuda and J. Laserson. 9-approximation algorithm for the sorting buffers problem. In *3rd Workshop on Approximation and Online Algorithms*, 2005.
2. M. Englert and M. Westermann. Reordering buffer management for non-uniform cost models. In *Proceedings of the 32nd International Colloquium on Algorithms, Langauages, and Programming*, pages 627–638, 2005.
3. R. Khandekar and V. Pandit. Online sorting buffers on line. In *Proceedings of the Symposium on Theoretical Aspects of Computer Science*, pages 616–625, 2006.
4. J. Kohrt and K. Pruhs. A constant approximation algorithm for sorting buffers. In *LATIN 04*, pages 193–202, 2004.
5. H. Räcke, C. Sohler, and M. Westermann. Online scheduling for sorting buffers. In *Proceedings of the European Symposium on Algorithms*, pages 820–832, 2002.

Approximating Tree Edit Distance Through String Edit Distance

Tatsuya Akutsu[1,*], Daiji Fukagawa[2,**], and Atsuhiro Takasu[2,**]

[1] Bioinformatics Center, Institute for Chemical Research, Kyoto University
Gokasho, Uji, Kyoto 611-0011, Japan
[2] National Institute of Informatics
Chiyoda-ku, Tokyo 101-8430, Japan
takutsu@kuicr.kyoto-u.ac.jp, {daiji, takasu}@nii.ac.jp

Abstract. This paper presents an $O(n^2)$ time algorithm for approximating the unit cost edit distance for ordered and rooted trees of bounded degree within a factor of $O(n^{3/4})$, where n is the maximum size of two input trees, and the algorithm is based on transformation of an ordered and rooted tree into a string.

1 Introduction

Recently, comparison of tree-structured data is becoming important in several diverse areas such as computational biology, XML databases and image analysis [3,9,16]. Though various measures have been proposed [3], the *edit distance* between rooted and ordered trees is widely-used [11,14,15,17]. This tree edit distance is a generalization of the edit distance for two strings [2,10,12,13], which is also widely-used for measuring the similarity between two strings. In this paper, we use *tree edit distance* and *string edit distance* to denote the distance between rooted and ordered trees and the distance between strings, respectively.

It is well-known that the string edit distance can be computed in $O(n^2)$ time, where n is the maximum length of input strings. Recently, extensive studies have been done on efficient (quasi linear time) approximation and low distortion embedding of string edit distances [2,10,12,13].

For the tree edit distance problem, Tai [14] first developed a polynomial time algorithm, from which several improvements followed [4,6,11,17]. Among these, a recent algorithm by Demaine et al. [6] is the fastest in the worst case and works in $O(n^3)$ time where n is the maximum size of input trees. They also proved an $\Omega(n^3)$ lower bound for the class of decomposition strategy algorithms.

Garofalakis and Kumar developed an algorithm for efficient embedding of trees [8], which can also be used for approximating tree edit distance. *However, the distance considered there is not the same as the tree edit distance: move*

---

* Supported in part by Grants-in-Aid "Systems Genomics" and #16300092 from MEXT, Japan.
** Supported in part by Grant-in-Aid "Cyber Infrastructure for the information-explosion Era" from MEXT, Japan.

T. Asano (Ed.): ISAAC 2006, LNCS 4288, pp. 90–99, 2006.

operations are allowed in their distance. Several practical algorithms have been
developed for efficient computation of lower bounds of tree edit distances [9,16],
but these algorithms do not guarantee any upper bounds. Therefore, it is required to develop algorithms for efficient approximation and/or low distortion
embedding for trees in terms of the original definition.

In order to approximate the tree edit distance, we studied a relation between
the tree edit distance and the string edit distance for the *Euler strings* [1]. It
was shown that the tree edit distance is at least half and at most $2h + 1$ of
the edit distance for the Euler strings, where h is the minimum height of two
trees. This result gives good approximation if the heights of input trees are
low. However, it does not guarantee any upper bounds of tree edit distances if
the heights of input trees are $O(n)$. In this paper, we improve this result by
modifying the Euler string. Though the modification is slight, a novel idea is
introduced and much more involved analysis is performed. We show that the
unit cost edit distance between trees is at least $1/6$ and at most $O(n^{3/4})$ of the
unit cost edit distance between the modified Euler strings, where we assume
that the maximum degree of trees is bounded by a constant. This result leads
to the first $O(n^{3-\epsilon})$ time algorithm for computing the tree edit distance with a
guaranteed approximation ratio (for bounded degree trees). Though this result
is not practical, it would stimulate further developments. It should be noted
that the current best approximation ratio within near linear time algorithms for
string edit distance is around $O(n^{1/3})$ [2] even though extensive studies have
been done in recent years. Though we consider the *unit cost* edit distances in
this paper, the result can be extended for more general distances to some extent.

2   String Edit Distance and Tree Edit Distance

Here we briefly review the string edit distance and the tree edit distance. We
consider strings over a finite or infinite alphabet Σ_S. For string s and integer i,
$s[i]$ denotes the i-th character of s, $s[i \ldots j]$ denotes $s[i] \ldots s[j]$, and $|s|$ denotes
the length of s. We may use $s[i]$ to denote both the character itself and the
position. An *edit operation on a string* s is either a *deletion*, an *insertion*, or a
substitution of a character of s. The *edit distance between two strings* s_1 and s_2
is defined as the minimum number of operations to transform s_1 into s_2. We use
$ED_S(s_1, s_2)$ to denote the edit distance between s_1 and s_2.

An *alignment* between two strings s_1 and s_2 is obtained by inserting *gap
symbols* (denoted by '-' where '-' $\notin \Sigma_S$) into or at either end of s_1 and s_2 such
that the resulting strings s_1' and s_2' are of the same length l, where it is not
allowed for each $i = 1, \ldots, l$ that both $s_1'[i]$ and $s_2'[i]$ are gap symbols. The *cost*
of alignment is given by $cost(s_1', s_2') = \sum_{i=1}^{l} f(s_1'[i], s_2'[i])$, where $f(x, y) = 0$ if
$x = y \neq$ '-', otherwise $f(x, y) = 1$. Then, an optimal alignment is an alignment
with the minimum cost. It is easy to see that the cost of an optimal alignment
is equal to the edit distance.

Next, we define the edit distance between trees (see [3] for details). Let T be
a rooted ordered tree, where "ordered" means that a left-to-right order among

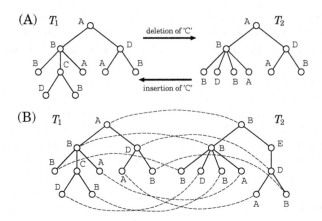

Fig. 1. (A) Insertion and deletion operations. (B) T_2 is obtained by deletion of a node with label 'C', insertion of a node with label 'E' and substitution for the root node. A mapping corresponding to this edit sequence is also shown by broken curves.

siblings is given in T. We assume that each node has a label from a finite alphabet Σ_T. $|T|$ denotes the size (the number of nodes) of T. An *edit operation on a tree* T is one of the following (see Fig. 1): [Deletion] Delete a non-root node v in T with parent u, making the children of v become the children of u. The children are inserted in the place of v as a subsequence in the left-to-right order of the children of u, [Insertion] Complement of delete. Insert a node v as a child of u in T making v the parent of a consecutive subsequence of the children of u, [Substitution] Change the label of a node v in T.

The *edit distance between two trees* T_1 and T_2 is defined as the minimum number of operations to transform T_1 into T_2, and is denoted by $ED_T(T_1, T_2)$. $M \subseteq V(T_1) \times V(T_2)$ is called an *ordered edit distance mapping* (or just a *mapping*) if the following conditions are satisfied for any two pairs (v_1, w_1), $(v_2, w_2) \in M$ [3]: (i) $v_1 = v_2$ iff. $w_1 = w_2$, (ii) v_1 is an ancestor of v_2 iff. w_1 is an ancestor of w_2, (iii) v_1 is to the left of v_2 iff. w_1 is to the left of w_2. Let $id(M)$ be the number of pairs having identical labels in M. It is well-known that the mapping M maximizing $id(M)$ corresponds to the edit distance, for which $ED_T(T_1, T_2) = |T_1| + |T_2| - |M| - id(M)$ holds.

3   Euler String

Our transformation from a tree to a string is based on the *Euler string* [11]. In this section, we review the Euler string (see Fig. 2) and our previous results [1].

For simplicity, we treat each tree T as an edge labeled tree: the label of each non-root node v in the original tree is assigned to the edge $\{u, v\}$ where u is the parent of v. It should be noted that information on the label on the root is lost in this case. But, it is not a problem because the roots are not deleted

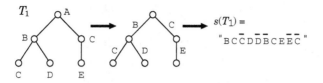

Fig. 2. Construction of an Euler string

or inserted. In what follows, we assume that the roots of two input trees have identical labels (otherwise, we just need to add 1 to the distance).

The depth-first search traversal of T (i.e., visiting children of each node according to their left-to-right order) defines an Euler tour (i.e., an Euler path beginning from the root and ending at the root where each edge $\{w, v\}$ is traversed twice in the opposite directions). We use $EE(T)$ to denote the set of directed edges in the Euler tour of T. Let $\Sigma_S = \{a, \bar{a} | a \in \Sigma_T\}$, where $\bar{a} \notin \Sigma_T$. Let $(e_1, e_2, \ldots, e_{2n-2})$ be the sequence of directed edges in the Euler path of a tree T with n nodes. From this, we create the Euler string $s(T)$ of length $2n - 2$. Let $e = \{u, v\}$ be an edge in T, where u is the parent of v. Suppose that $e_i = (u, v)$ and $e_j = (v, u)$ (clearly, $i < j$). We define $i_1(e)$ and $i_2(e)$ by $i_1(e) = i$ and $i_2(e) = j$, respectively. That is, $i_1(e)$ and $i_2(e)$ denote the first and second positions of e in the Euler tour, respectively. Then, we define $s(T)$ by letting $s(T)[i_1(e)] = L(e)$ and $s(T)[i_2(e)] = \overline{L(e)}$, where $L(e)$ is the label of e.

Proposition 1. [1,15] $s(T_1) = s(T_2)$ if and only if $ED_T(T_1, T_2) = 0$. Moreover, we can reconstruct T from $s(T)$ in linear time.

Lemma 1. [1] $ED_S(s(T_1), s(T_2)) \leq 2 \cdot ED_T(T_1, T_2)$.

Lemma 2. [1] $ED_T(T_1, T_2) \leq (2h + 1) \cdot ED_S(s(T_1), s(T_2))$, where h is the minimum height of two input trees.

It was shown in [1] that this bound is tight up to a constant factor. Fig. 3 gives an example such that $ED_S(s(T_1), s(T_2)) = 4$ and $ED_T(T_1, T_2) = \Theta(h)$.

4   Modified Euler String

As shown in the above, the approximation ratio of the tree edit distance through the edit distance between the Euler strings is not good if the minimum height of input trees is high. In order to improve the worst case approximation ratio, we modify labels of some edges in the input trees so that structures of small subtrees are reflected to the labels. For example, we consider trees shown in Fig. 3. Suppose that label "AC" is assigned to each edge just above each node having children with labels 'A' and 'C'. Similarly, suppose that labels "BD", "AD" and "BC" are assigned to appropriate edges. Then, $ED_S(s(T_1), s(T_2)) = \Theta(h)$ should hold. But, in a general case, changes of labels should be performed carefully in order to keep distance distortion not too large.

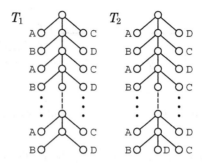

Fig. 3. Example for the case of $ED_T(T_1, T_2) = \Theta(h) \cdot ED_S(s(T_1), s(T_2))$ [1]. Nodes without labels have the same labels (e.g., label 'E').

For a node v in T_1 or T_2, $id(v)$ is an integer such that $id(v) = id(v')$ if and only if the tree induced by v and its descendants is isomorphic (including labels) to the tree induced by v' and its descendants. Since we only consider subtrees induced by some node v in T_1 or T_2 and its descendants, all $id(v)$ can be computed in $O(n)$ time and each $id(v)$ can be represented in a word (i.e., $O(\log n)$ bits) [7], where $n = \max(|T_1|, |T_2|)$.

We replace labels of some nodes in each input tree in the following way. Let $size(v)$ be the size (the number of nodes) of the subtree induced by v and its descendants. A subtree rooted at v is called *large* if $size(v) > \alpha$, where α is a parameter defined as $\alpha = n^{1/2}$. Otherwise, it is called *small*. We call w_i a *special node* if $size(w_i) \leq \alpha$ and $size(v) > \alpha$ where v is the parent of w_i.

Proposition 2. *For each node v in T, there exists at most one special node in the path from the root to v. Moreover, if $depth(v)$ (i.e., the length of the path from the root to v) $\geq \alpha$, there exists exactly one special node in the path.*

Next, we define edge labels (see Fig. 4), using which the modified Euler strings are constructed. Let v be a node in T_1. Let u be the parent of v and $w_1, \ldots, w_k$ be the children of v (Similarly, we define v', u', and $w'_1, \ldots$ for T_2). If none of w_i's are special, the original label (i.e., label in Σ_T) of v is assigned to edge $\{u, v\}$. Otherwise, let $w_{i_1}, \ldots, w_{i_h}$ be the special children of v. Then, let $id'(v, w_{i_1}, \ldots, w_{i_h})$ be an integer number such that $id'(v, w_{i_1}, \ldots, w_{i_h}) = id'(v', w'_{i_1}, \ldots, w'_{i_l})$ if and only if $h = l$, v and v' have identical labels, and $id(w_{i_j}) = id(w'_{i_j})$ holds for all $j = 1, \ldots, h$. As in the case of $id(v)$, the total time required for computing such numbers is $O(n)$. We assign $id'(v, w_{i_1}, \ldots, w_{i_h})$ to edge $\{u, v\}$ where we assume w.l.o.g. (without loss of generality) that $id'(\ldots) \notin \Sigma_T$. It should be noted that if v has at least one special children, information of the subtrees of the special children is reflected to the label of $\{u, v\}$. We call such edges *special edges*.

Using the above labeling of edges, we create a modified Euler string $ss(T)$ as in $s(T)$, where $ss(T)$ and $s(T)$ differ only on labels of special edges. It should be noted that $ss(T_1)$ and $ss(T_2)$ can be constructed in $O(n)$ time from T_1 and T_2.

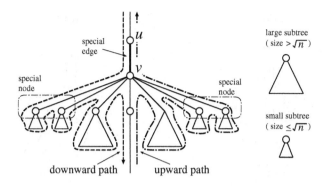

Fig. 4. Special nodes, special edges, and large subtrees

5 Analysis

In this section, we show the following main theorem using several propositions
and lemmas. In what follows, we may identify a directed edge (u, v), a node v
and the corresponding letter in $ss(T)$ if there is no confusion.

Theorem 1. $\frac{1}{O(n^{3/4})} \cdot ED_T(T_1, T_2) \leq ED_S(ss(T_1), ss(T_2)) \leq 6 \cdot ED_T(T_1, T_2).$

Since the righthand side inequality can be easily proven as in the proof of Lemma
1, we prove the lefthand side inequality here.

5.1 Construction of Tree Mapping from String Alignment

In this section, we show a procedure for obtaining a mapping between T_1 and
T_2 from an (not necessarily optimal) alignment AL_S between $ss(T_1)$ and $ss(T_2)$.
Before showing details of the procedure, we describe an outline. We first create a
mapping M_1 that is induced by corresponding downward paths, where downward
(and upward) paths are to be defined later. Next, we modify M_1 to M_2 so
that labeling information on special edges is reflected (i.e., mapping pairs for
small right subtrees rooted at special children are added to M_1). However, such
mappings (both M_1 and M_2) may contain pairs violating ancestor-descendant
relations. Thus, we delete inconsistent pairs from M_2 (the resulting mapping is
denoted as M_3). Finally, we add large subtrees included in upward paths, then
delete some inconsistent pairs from M_3, and get the desired mapping M_4.

[**Construction of M_1**] An edge (u, v) is called a *downward edge* if v is a child
of u. Otherwise, (u, v) is called an *upward edge*. For each downward edge e, $\bar{e}$
denotes the upward edge corresponding to e (i.e., $\bar{e} = (v, u)$ if $e = (u, v)$).

Let $\{(p_1^1, p_2^1), (p_1^2, p_2^2), \ldots\}$ be the set of maximal *substring* pairs (p_1^i from $ss(T_1)$
and p_2^i from $ss(T_2)$), each of which corresponds to a maximal consecutive region
(with length at least 2) in AL_S without insertions, deletions or substitutions.
Note that p_1^i and p_2^i correspond to isomorphic paths in T_1 and T_2. We divide each

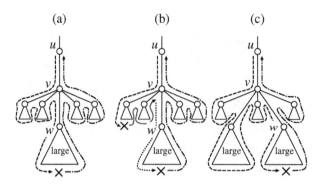

Fig. 5. Explanation of construction of M_2. In cases (a) and (c), mapping pairs for right subtrees are added, whereas these are not added in case (b) since (u, v) and (v, w) are included in different downward paths. In this figure, each cross means that there exists insertion, deletion or substitution at the point. It should be noted that the second and third paths in (b) are included in the same maximal substring.

region into two parts (see also Fig. 4 and Fig. 5): upward part and downward part, where one part may be empty. Let $p_1^i[k]$ be the first letter corresponding to a downward edge (u, v) such that a letter corresponding to (v, u) does not appear in p_1^i. Then, $(p_1^i[1 \ldots k-1], p_2^i[1 \ldots k-1])$ and $(p_1^i[k \ldots], p_2^i[k \ldots])$ are the *upward segment pair* and the *downward segment pair*, respectively. Two paths (one in T_1 and the other in T_2) corresponding to an upward segment pair are called *upward paths*. Two paths corresponding to a downward segment pair are called *downward paths*. A subtree that is fully included in a downward (resp. upward) path is called a *left subtree* (resp. *right subtree*). An edge (u, v) in an upward (resp. downward) path is called a *central edge* if (v, u) does not appear in the same path. Thus, central edges are the edges not appearing in any left or right subtrees.

Suppose that $ss(T_1)[i]$ corresponds to $ss(T_2)[i']$ in any downward segment pair, and downward edges (u, v) and (u', v') correspond to $ss(T_1)[i]$ and $ss(T_2)[i']$, respectively. Then, we let M_1 be the set of such (v, v')'s.

[Construction of M_2] Let $e = (u, v)$ and $e' = (u', v')$ be a pair of corresponding special edges in M_1. Suppose that for e (resp. e'), there exists an edge (v, w) such that (u, v) and (v, w) belong to the same downward path, and (v, u) and (w, v) belong to the same upward path. Then, we add matching pairs to M_1 that are induced by the corresponding subtrees of the special children of v and v' (see Fig. 5). Precisely, we only need to add matching pairs for small right subtrees since matching pairs for the left subtrees should already be included in M_1. We let the resulting mapping be M_2.

[Construction of M_3] Let (v, v') be the pair of nodes corresponding to the first letters of any downward segment pair $(p_1^i[k \ldots], p_2^i[k \ldots])$. Let P_v (resp. $P_{v'}$) be the set of nodes in the path from the root to v (resp. v'). We delete any $(u, u') \in M_2$ from M_2 if either $(u \in P_v$ and $u' \notin P_{v'})$ or $(u \notin P_v$ and $u' \in P_{v'})$ holds. We also delete small subtrees rooted at special children of u and u', where *deletion of a subtree* (resp. *a region or a node*) means that all mapping pairs containing a node in the subtree are deleted from the current mapping set. We execute this deletion procedure for all downward segment pairs (in any order). We let the resulting mapping be M_3. It should be noted that large left subtrees are never deleted: these will be consistent with other pairs in M_3 (and M_4).

[**Construction of M_4**] Finally, we add all large subtrees (i.e., subtrees with more than $n^{1/2}$ nodes) that are fully included in upward paths, and then delete inconsistent mapping pairs. It should be noted that deletion is required for regions where corresponding central edges in upward paths are different from those in downward paths (see Fig. 6 (B-1)). It should also be noted that large subtrees are not deleted. In this deletion phase, we consider the following two cases (see Fig. 6). (**A**) The number of large subtrees appearing in an upward path is at most δ, where δ is a constant (e.g., $\delta = 10$). (**B**) The number of large subtrees appearing in an upward path is greater than δ, where we assume w.l.o.g. that the central edges of an upward path are shared by the central edges in a downward path (otherwise, we can cut the upward path into multiple upward paths without affecting the order of the approximation ratio).

For case (A), we only show here the operation for the case where one large subtree is contained in an upward path. Let $z \ldots \overline{z}$ (resp. $z' \ldots \overline{z'}$) be the sequence of directed edges corresponding to the large subtree of T_1 (resp. T_2). Suppose that $x \in EE(T_1)$ (resp. $y' \in EE(T_2)$) is a parent of z (resp. z'). Suppose that $x' \in EE(T_2)$ corresponds to x in AL_S, $y \in EE(T_1)$ corresponds to y' in AL_S, and x' is an ancestor of y'. Then, we delete mapping pairs for the edges between x and y along with their small right subtrees and the edges between x' and y' along with their small right subtrees. We can do similarly if x' is not an ancestor of y', where details of the procedure and analysis are omitted in this paper.

For case (B), upward paths (excluding large subtrees) should have *periodicity*, where the length of a period is at most d. As in the case of (A), suppose that x and $\overline{x}$ correspond to x' and $\overline{y'}$ in AL_S, respectively. Then, the subtree consisting of the nodes in the path between x' and y' and the nodes in their right small subtrees is called a *block*. A subtree isomorphic to the block is also called a block. It can be seen that blocks appear repeatedly in the vertical direction both in T_1 and T_2. If a large subtree(s) on the right-hand side is attached to a block, it is called a *relevant block*. We consider the following two cases. (**B-1**) The size of a block is greater than β, where $\beta = dn^{1/4}$. In this case, the number of the central edges in an upward path is $O(n^{3/4})$. We delete the top and bottom blocks and the central edges from T_1 and T_2. (**B-2**) The size of a block is at most β. In this case, the (total) number of relevant blocks is $O(n^{1/2})$ since there are at most $O(n^{1/2})$ large subtrees. We delete all relevant blocks.

5.2   Analysis of Lower Bound of ED_S

Now we analyze the lower bound of $ED_S(ss(T_1), ss(T_2))$. For that purpose, we estimate the cost of M_4, assuming that the cost of AL_S is $d = ED_S(ss(T_1), ss(T_2))$. Before that, it is straight-forward to see the following.

Proposition 3. M_4 *is a valid mapping between T_1 and T_2.*

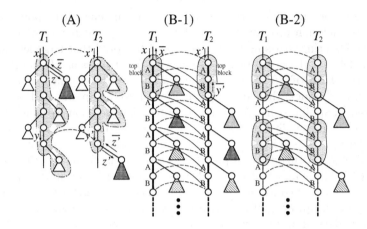

Fig. 6. Construction of M_4. Shaded triangles denote large subtrees. (A) Gray regions are deleted. (B-1) Top (shown in gray color) and bottom blocks and central edges (shown by bold lines) are deleted. (B-2) Relevant blocks shown in gray color are deleted.

In what follows, we estimate the number of mapping pairs deleted or ignored in the construction.

Proposition 4. *The number of nodes not appearing in any downward or upward path is $O(d)$.*

Proposition 5. *The number of downward (resp. upward) paths is $O(d)$.*

Proposition 6. *The total number of nodes in downward paths not appearing in M_1 is $O(d)$.*

Due to the above propositions, we only need to consider hereafter upward paths and deleted mapping pairs.

Lemma 3. *The number of nodes in the small subtrees in the upward paths that are not included in M_2 is $O(dn^{1/2})$.*

Proof. The number of special edges that are not taken into account for M_2 is $O(d)$. Since we assume that the maximum degree is bounded by a constant, the number of nodes in the small right subtrees rooted at special children of a special edge is $O(n^{1/2})$. There also may exist small subtrees that are fully included in the upward paths but are not attached to special edges. But, the number of such subtrees is $O(d)$. □

Next, we estimate the numbers of deleted mapping pairs in constructing M_3 and M_4 respectively, where the proofs are omitted in this version.

Lemma 4. *The number of deleted mapping pairs in the construction of M_3 is $O(d^2 n^{1/2})$.*

Lemma 5. *The number of deleted mapping pairs in the construction of M_4 is $O(d^2 n^{1/2} + dn^{3/4})$.*

Finally, assuming that $d = O(n^{1/4})$ (otherwise the cost of M_4 can be $O(n)$), we can see that the cost of M_4 is $O(d) + O(dn^{1/2}) + O(d^2 n^{1/2}) + O(d^2 n^{1/2} + dn^{3/4}) = O(dn^{3/4})$, which completes the proof of Theorem 1. Since the string edit distance can be computed in $O(n^2)$ time and construction of M_4 can also be done in $O(n^2)$ time, we have:

Corollary 1. *The unit cost edit distance for trees of bounded degree can be approximated within a factor of $O(n^{3/4})$ in $O(n^2)$ time.*

Acknowledgements

We would like to thank Tetsuji Kuboyama in the University of Tokyo for suggestions of several references and for helpful discussions.

References

1. Akutsu, T.: A relation between edit distance for ordered trees and edit distance for Euler strings. Information Processing Letters **100** (2006) 105–109
2. Batu, T., Ergun, F., Sahinalp, C.: Oblivious string embeddings and edit distance approximations. Proc. 17th ACM-SIAM Symp. Discrete Algorithms (2006) 792–801
3. Bille, P.: A survey on tree edit distance and related problem. Theoretical Computer Science **337** (2005) 217–239
4. Chen, W.: New algorithm for ordered tree-to-tree correction problem. Journal of Algorithms **40** (2001) 135–158
5. Cormode, G., Muthukrishnan, S.: The string edit distance matching problem with moves. Proc. 13th ACM-SIAM Symp. Discrete Algorithms (2002) 667–676
6. Demaine, E., Mozes, S., Rossman, B., Weimann, O.: An $O(n^3)$-time algorithm for tree edit distance. Preprint cs.DS/0604037 (2006)
7. Fukagawa, D., Akutsu, T.: Fast algorithms for comparison of similar unordered trees. Lecture Notes in Computer Science **3341** (2004) 452–463
8. Garofalakis, M., Kumar, A.: XML stream processing using tree-edit distance embedding. ACM Trans. Database Systems **30** (2005) 279–332
9. Guha, S., Jagadish, H.V., Koudas, N., Srivastava, D., Yu, T.: Approximate XML joins. Proc. ACM SIGMOD (2002) 287–298
10. Khot, S., Naor, A.: Nonembeddability theorems via Fourier analysis. Proc. 46th IEEE Symp. Foundations on Computer Science (2005) 101–110
11. Klein, P.N.: Computing the edit-distance between unrooted ordered trees. Proc. 6th European Symp. Algorithms (1998) 91–102
12. Krauthgamer, R., Rabani, R.: Improved lower bounds for embeddings into L_1. Proc. 17th ACM-SIAM Symp. Discrete Algorithms (2006) 1010–1017
13. Ostrovsky, R., Rabani, Y.: Low distortion embeddings for edit distance. Proc. 37th ACM Symp. Theory of Computing (2005) 218–224
14. Tai, K-C.: The tree-to-tree correction problem. J. ACM **26** (1979) 422–433
15. Valiente, G.: Algorithms on Trees and Graphs. Springer (2002)
16. Yang, R., Kalnis, P., Tang, A.K.H.: Similarity evaluation on tree-structured data. Proc. ACM SIGMOD (2005) 754–765
17. Zhang, K., Shasha, D.: Simple fast algorithms for the editing distance between trees and related problems. SIAM J. Computing **18** (1989) 1245–1262

A 6-Approximation Algorithm for Computing Smallest Common AoN-Supertree with Application to the Reconstruction of Glycan Trees

Kiyoko F. Aoki-Kinoshita[1], Minoru Kanehisa[2], Ming-Yang Kao[3,*],
Xiang-Yang Li[4,**], and Weizhao Wang[4]

[1] Dept. of Bioinformatics, Fac. of Engineering, Soka University
kkiyoko@t.soka.ac.jp
[2] Bioinformatics Center, Institute for Chemical Research, Kyoto University, and Human
Genome Center, Institute of Medical Science, University of Tokyo
kanehisa@kuicr.kyoto-u.ac.jp
[3] Dept. of Electrical Engineering and Computer Science, Northwestern University
kao@cs.northwestern.edu
[4] Dept. of Computer Science, Illinois Institute of Technology
xli@cs.iit.edu, wangwei4@iit.edu

Abstract. A node-labeled rooted tree T (with root r) is an all-or-nothing subtree
(called *AoN-subtree*) of a node-labeled rooted tree T' if (1) T is a subtree of the
tree rooted at some node u (with the same label as r) of T', (2) for each internal
node v of T, *all* the neighbors of v in T' are the neighbors of v in T. Tree T'
is then called an *AoN-supertree* of T. Given a set $\mathcal{T} = \{T_1, T_2, \cdots, T_n\}$ of n
node-labeled rooted trees, smallest common AoN-supertree problem seeks the
smallest possible *node-labeled rooted* tree (denoted as **LCST**) such that every
tree T_i in $\mathcal{T}$ is an *AoN-subtree* of **LCST**. It generalizes the smallest superstring
problem and it has applications in glycobiology. We present a polynomial-time
greedy algorithm with approximation ratio 6.

1 Introduction

In smallest AoN-supertree problem we are given a set $\mathcal{T} = \{T_1, T_2, \cdots, T_n\}$ of n
node-labeled rooted trees and we seek the smallest possible node-labeled rooted tree
LCST such that every tree T_i in $\mathcal{T}$ is an all-or-nothing subtree (called *AoN-subtree*) of
LCST. Here a tree T_i is an AoN-subtree of another tree T if (1) T_i is a subtree of T, and
(2) for each node v of tree T_i, either all children nodes of v in T are also children of v in
T_i, or none of the children nodes of v in T is a child node of v in T_i. The widely studied
shortest superstring problem (*e.g.*, [1,2,3,4,5,6,7]), which is known to be NP-hard and
even MAX-SNP hard [5], is a special case of smallest supertree problem where each
string can be viewed as a unary rooted tree. The best known approximation ratio for
shortest superstring problem is $2\frac{1}{2}$ [6]. The simple greedy algorithm was also proven to

---

* Supported in part by NSF Grant IIS-0121491.
** Partially supported by NSF CCR-0311174.

T. Asano (Ed.): ISAAC 2006, LNCS 4288, pp. 100–110, 2006.

be effective [5, 4], with the best proven approximation ratio $3\frac{1}{2}$ [4]. Here, we present a polynomial-time 6-approximation algorithm for smallest supertree problem.

The superstring problem has application in data compression and in DNA sequencing, while the supertree problem also has vast applications in glycobiology. In the field of glycobiology, for the study of glycans, or carbohydrate sugar chains (called glycome informatics), much work pertains to analyzing the database of known glycan structures themselves. Glycans are considered the third major class of biomolecules next to DNA and proteins. However, they are not studied as much as DNA or proteins due to their complex tree structure; they are branched structures. In recent years, databases of glycans [8] have taken off, and the application of theoretical computer science and data mining techniques have produced glycan tree alignment [9, 10], score matrices [11] and probabilistic models [12] for the analysis of glycans. In this work, we look at one of the current biggest challenges in this field, which is the characterization of glycan tree structures from mass spectrometry data. The retrieval of what glycan structures these data represent still remains a major difficulty. In this work, we will assess this problem theoretically in application to any glycan structure. By doing so, it would be straightforward to apply algorithms to quickly annotate any mass spectrometry data with accurate glycan structures, thus enabling the rapid population of glycan databases and resulting biological analysis.

2   Preliminaries and Problem Definition

In the remainder of this paper, unless explicitly stated otherwise, a tree is rooted. The relative positions of the children could be significant or non-significant. The tree is called an *ordered* tree if the relative positions of the children of each node is significant, that is, there is the first child, the second child, the third child, *etc.*, for each internal node. Otherwise it is called a *non-ordered* tree. The *size* of a tree T, denoted as $|T|$, is the number of nodes in T. The *distance* between nodes u and v in a tree T is the number of edges on the unique path between u and v in T. Given a node u in a tree T rooted at node r, the *level* of u is the distance between u and the root r. The *height* of a tree T is the maximum level over all nodes in the tree. A node w is an *ancestor* of a node u if it is on the path between u and r; the node u is then called a *descendant* of w. If all leaf nodes are on the same level, the tree is called *regular*. Given a rooted tree T, we use $r(T)$ to denote the root node of T.

In this paper, we consider the trees composed of nodes with *labels* that are not necessary to be unique. We assume that the labels of nodes are selected from a *totally ordered set*. Each node has x a unique ID. Given a tree T and a node u of T, a tree T' rooted at u is an **AoN-subtree** (representing *All-or-Nothing subtree*) of T if for each node v that is a descendant of u, either all children of v in tree T are in T' or none of the children of v in T is in T'. Note that the definition of the AoN-subtree is different from the traditional subtree definition. For example, consider a tree T in Figure 1 (a) and tree T_1 in Figure 1 (b). Tree T_1 is an AoN-subtree of T. Tree T_2 in Figure 1 is not an AoN-subtree of T since tree T_2 only contains one of the two children of node v_4. Given two trees T_1 and T_2, if T is an AoN-subtree of both T_1 and T_2, then T is the *common AoN-subtree* of T_1 and T_2. If T has the maximum number of nodes among all common AoN-subtrees,

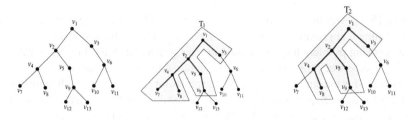

(a) Tree T (b) T_1 is an AoN-subtree of T (c) T_2 is not an AoN-subtree of T

Fig. 1. Illustration of AoN-subtree Notation

then T is the *maximum common AoN-subtree*. Given a tree T and an internal node u of T, let $T(u)$ be the tree composed of node u and all descendants of u in T. Obviously, $T(u)$ is an AoN-subtree of T.

If tree T' is an AoN-subtree of T, then T is an *AoN-supertree* of T'. In this paper, we assume that there is a set $\mathcal{T}$ of n rooted trees $\{T_1, T_2, \cdots, T_n\}$, where $r_i = r(T_i)$ is the root of the tree T_i. Here trees T_i could be *ordered* or *non-ordered*. If tree T is an AoN-supertree for every tree T_i for $1 \leq i \leq n$, then T is called a *common AoN-supertree* of $T_1, T_2, \cdots, T_n$. If T has the smallest number of nodes among all common AoN-supertrees, then T is *smallest common AoN-supertree* and is denoted as **LCST**$(\mathcal{T})$. In smallest AoN-supertree problem we are given a set $\mathcal{T}$ of n node-labeled rooted trees and we seek smallest common AoN-supertree **LCST**$(\mathcal{T})$.

3   Find the Maximum Overlap AoN-Subtree

Our algorithm for finding smallest common AoN-supertree is based on greedy merging of two trees that have the largest overlap. Given two trees T_1 and T_2, with root r_1 and r_2 respectively, if an internal node u of T_1 satisfying that (1) $u = r_2$ and (2) $T_1(u)$ is an AoN-subtree of T_2, then $T_1(u)$ is an *overlap AoN-subtree* of tree T_2 over T_1, denoted by $T_1(u) = T_1 \cap T_2$. Note that if tree T is an overlap AoN-subtree of T_2 over T_1, it is not necessary that T is an overlap AoN-subtree of T_1 over T_2. If T has the largest number of nodes among all overlap AoN-subtrees of T_2 over T_1, then T is the *largest overlap AoN-subtree*. Let $\mathcal{L}(T_1, T_2)$ be the largest overlap AoN-subtree of T_2 over T_1 and note that $\mathcal{L}(T_1, T_2)$ is not necessarily symmetric. If we remove $\mathcal{L}(T_1, T_2)$ from T_2, then the remaining forest is denoted as $T_2 - T_1$.

Here, we assume that the tree is non-ordered. If the tree is ordered, then find the largest overlap AoN-subtree is trivial. Without loss of generality, we assume that the labels of the tree are integers from $[1, m]$. We abuse the notations little bit here by using u to also denote the label of a node u with ID u if it is clear from context. Given two trees T_1 and T_2, we define a *total order-relation* $\prec$ of two trees as follows.

1. If $r(T_1) < r(T_2)$, we say $T_1 \prec T_2$. If $r(T_2) < r(T_1)$, we say $T_2 \prec T_1$.
2. If $r(T_1) = r(T_2)$, we further let that $\{u_1, u_2, \cdots, u_p\}$ be all the children of $r(T_1)$ in T_1 and $\{v_1, v_2, \cdots, v_q\}$ be all the children of $r(T_2)$ in T_2. W.l.o.g., we also assume that the children are sorted in an order such that $T_1(u_i) \succeq T_1(u_j)$ for

any $1 \leq i < j \leq p$ and $T_2(v_i) \succeq T_2(v_j)$ for any $1 \leq i < j \leq q$. Let k the smallest index such that either $T_1(u_k) \prec T_2(v_k)$ or $T_2(v_k) \prec T_1(u_k)$. We have three subcases: a) If $T_1(u_k) \prec T_2(v_k)$, we say $T_1 \prec T_2$; b) If $T_1(u_k) \prec T_2(v_k)$, we say $T_1 \prec T_2$; c) Such k does not exist. If $p < q$, then $T_1 \prec T_2$; if $p > q$ then $T_2 \prec T_1$; if $p = q$, then $T_1 = T_2$.

Notice that here $T_1 \preceq T_2$ if $T_1 \prec T_2$ or $T_1 = T_2$; $T_1 \succeq T_2$ if $T_1 \succ T_2$ or $T_1 = T_2$; $T_1 \succ T_2$ if $T_2 \prec T_1$. More formally, Algorithm 1 summarizes how to decide the order-relation between two non-ordered trees.

Algorithm 1. Decide the relationship of two trees.

Input: Two trees T_1 and T_2.

Output: The relationship between T_1 and T_2.

1: Label all internal nodes in T_1 WHITE and all leaf nodes BLACK.
2: **repeat**
3:    Pick any internal node in T_1 such that all children nodes are marked BLACK, say u. Sort all children nodes of $T_1(u)$ in the order as $\{u_1, u_2, \cdots, u_p\}$ such that $T_1(u_i) \succeq T_1(u_j)$ for any $1 \leq i < j \leq p$.
4:    Mark u BLACK.
5: **until** all internal nodes in T_1 are BLACK.
6: Mark all internal nodes in T_2 WHITE and all leaf nodes BLACK.
7: **repeat**
8:    Pick any internal node in T_2 such that all children nodes are with marked BLACK, say u. Sort all children nodes of $T_2(u)$ in the order as $\{v_1, v_2, \cdots, v_p\}$ such that $T_2(v_i) \succeq T_2(v_j)$ for any $1 \leq i < j \leq p$.
9:    Mark u BLACK.
10: **until** all internal nodes in T_2 are BLACK.
11: **If** $r(T_1) < r(T_2)$ **then** return $T_1 \prec T_2$. **end if**
12: **If** $r(T_1) > r(T_2)$ **then** return $T_1 \succ T_2$; **end if**
13: Assume $\{u_1, u_2, \cdots, u_p\}$ are children nodes of $r(T_1)$ and $\{v_1, v_2, \cdots, v_p\}$ are children nodes of $r(T_2)$.
14: **for** $i = 1$ to $\min(p, q)$ **do**
15:    If $T_1(u_i) \prec T_2(v_i)$ return $T_1 \prec T_2$; if $T_1(u_i) \succ T_2(v_i)$ return $T_1 \succ T_2$.
16: If $p < q$ return $T_1 \prec T_2$; if $p > q$ return $T_1 \succ T_2$; if $p = q$ return $T_1 = T_2$.

In Algorithm 1, we first compute a lexicographic ordering of a tree and the compute the order-relation of two trees. Note for any two siblings of a common parent, we can compare the order of them by a breadth first search. Thus, the worst case happens when the tree is a complete binary tree and all nodes have the same label, which takes time $O(n^2)$. Thus, for a tree T of n nodes, we have

Lemma 1. *Algorithm 1 computes the ordering of a tree T in time $O(n^2)$.*

We present a recursive method (Algorithm 2) that decides whether one tree is an AoN-subtree of another. Given two trees T_1 and T_2, we then show how to find the largest overlap tree of T_2 over T_1. First, we order the trees T_1 and T_2, and then find the internal node u such that $T_1(u)$ is an AoN-subtree of T_2 and $|T_1(u)|$ is maximum. From

Lemma 1, the ordering of trees T_1 and T_2 need $O(|T_1|^2 + |T_2|^2)$. Notice that for any internal node u of T_1, checking whether $T_1(u)$ is an AoN-subtree of T_2 takes time $O(|T_1(u)|)$. Thus, the total time needed is $\sum_{u \in T_1} |T_1(u)| \leq |T_1|^2$. Thus, we have

Lemma 2. *Finding largest overlap tree has time complexity* $O(|T_1|^2 + |T_2|^2)$.

We expect a better algorithm to find the largest overlap AoN-subtree based on the fact that there exists efficient linear time algorithm that can find a largest common substring of a set of strings. However, designing such efficient algorithm is not the scope of this paper. We leave it as a future work.

Algorithm 2. Decide whether a tree T_2 is an AoN-subtree of T_1.

1: Flag ← FALSE;
2: For each internal node in T_1, order its p children from left to right as $u_1, u_2, \cdots, u_p$ such that for any pair of children u_i and u_j, $T_1(u_i) \preceq T_1(u_j)$ for $i < j$. Similarly, we also order the children of each internal node in T_2 similarly. Assume that the children of $r(T_2)$ from left to right is $\{v_1, v_2, \cdots, v_q\}$.
3: **for** each internal node u of T_1 such that $u = r(T_2)$ and Flag==FALSE **do**
4:     Assume that the set of "sorted" children nodes of u is $\{u_1, u_2, \cdots, u_p\}$.
5:     Flag ← TRUE if (1) $p = q$, and (2) tree $T_2(u_i)$ is an AoN-subtree of $T_1(v_i)$ for evey u_i with $1 \leq i \leq p$.
6: Return Flag;

---

4   Approximate Smallest Common AoN-Supertree

We then consider how to find smallest common AoN-supertree given a set $\mathcal{T}$ of n regular trees $\{T_1, T_2, \cdots, T_n\}$. Here, we assume that no tree T_i is an AoN-subtree of another tree T_j. It is known that the problem of computing smallest common superstring, given n strings, is NP-Hard and even MAX-SNP hard [5]. Notice that computing the smallest common superstring is a special case of computing smallest common AoN-supertree when all trees are restricted to a rooted unary tree. Thus, we have

Theorem 1. *Computing smallest common AoN-supertree is NP-Hard.*

4.1   Understanding the Structure of LCST

Notice that if a tree T is a common AoN-supertree of $\mathcal{T} = \{T_1, T_2, \cdots, T_n\}$, then for each tree T_i, we can find an internal node u of T such that there is an AoN-subtree $T(u)$ of T root at u that matches T_i. When multiple such internal nodes u exist, we choose any node with the lowest level, denoted by $r_i(T)$. For notational simplicity, we also denote the AoN-subtree of T that equals to T_i rooted at $r_i(T)$ as T_i if it is clear from the context. If $r_i(T)$ is an ancestor of $r_j(T)$, then we also say that T_i is an *ancestor* of T_j. Similarly, if $r_i(T)$ is a descendant of $r_j(T)$, then we also say that T_i is a *descendant* of T_j. If T_i is an ancestor of T_j and there does not exist a tree T_k such that T_k is an ancestor of T_j and T_i is ancestor of T_k, then T_i is the *parent* of T_j and T_j is a *child* of T_i. Lemma 3 and 4 (whose proofs are omitted due to space limit) showed that the notation of child and parent is well defined in smallest common AoN-supertree **LCST**$(\mathcal{T})$.

Lemma 3. *If T_i is T_j's parent in tree $\textbf{\textit{LCST}}(\mathcal{T})$, then either $r_j(\textbf{\textit{LCST}}(\mathcal{T}))$ is a node in T_i or a child of some leaf node of T_i.*

Lemma 4. *There is a unique tree T_i such that $r_i(\textbf{\textit{LCST}}(\mathcal{T}))$ is the root of tree $\textbf{\textit{LCST}}(\mathcal{T})$.*

Given a tree set $\mathcal{T}$ and a common AoN-supertree T, if any node in tree T is in a tree T_i for some index i, then we call this common AoN-supertree *condensed common AoN-supertree*. If a common AoN-supertree T is not a condensed common AoN-supertree, then recursively apply the following process will generate a condensed common AoN-supertree. First, we pick any node $u \in T$ that is not in any tree T_i. Remove u, and let all children of u in T become the children of u's parent. Notice that this will not violate the all-or-nothing property of the AoN-supertree. Thus, we will only consider condensed common AoN-supertrees when we approximate smallest common AoN-supertree. Notice Lemma 3 and Lemma 4 implies the following lemma.

Lemma 5. *The optimum tree $\textbf{\textit{LCST}}(\mathcal{T})$ is a condensed common AoN-supertree.*

Notice that if T is a common AoN-supertree of $\mathcal{T}$, then for any tree T_i, its parent is unique. Together with Lemma 4, we have the following lemma.

Lemma 6. *Given $\mathcal{T}$ and a condensed common AoN-supertree T, for any node $r_i(T)$, either $r_i(T)$ is the root of T or there is a unique j where $r_j(T)$ is the parent of $r_i(T)$.*

If we treat each tree T_i as a node, then Lemma 6 reveals that we can construct a unique *virtual overlap tree* $\mathbb{VT}(T)$ as follows. Each vertex of the virtual overlap tree corresponds to a tree T_i. If $r_i(T)$ is the root of tree T, then T_i is the root. Otherwise, T_i's unique parent in T, denoted by $\mathcal{P}(T_i)$, becomes its parent in $\mathbb{VT}(T)$ and all children in T becomes its children in $\mathbb{VT}(T)$. When T_i is T_j's parent, from Lemma 3, the root of T_j is either in T_i or a child of a leaf node of T_i. If T_p and T_q are both children of T_i, then T_p and T_q are *siblings*. Following lemma reveals a property of the siblings.

Lemma 7. *If T_p and T_q are siblings, then T_p and T_q do not share any common nodes.*

Thus, given a virtual overlap tree $\mathbb{VT}(T)$, the size of the condensed common AoN-supertree is $|T| = |T_i| + \sum_{T_j \in \mathcal{T} - T_i} |T_j - \mathcal{P}(T_j)| = \sum_{T_j \in \mathcal{T}} |T_j| - \sum_{T_j \in \mathcal{T} - T_i} |\mathcal{P}(T_j) \cap T_j|$, where T_i is the root in $\mathbb{VT}(T)$. Algorithm 3 will reduce the size of a condensed tree T.

Theorem 2. *Algorithm 3 always maintains tree T as a condensed AoN-supertree. Moreover, for any tree T_j that is not a root, if T_j does not have a maximum overlap with its parent, then Algorithm 3 decreases the size of T by at least 1.*

PROOF. It is not difficult to observe that T^{temp} is a condensed common AoN-supertree. Thus, we focus on the second part. Without loss of generality, we assume that T_j, which is not the root, does not have the maximum overlap with its parent T_i. Notice that for each tree T_k who is a child of T_i and r_k is a descendant of u (including T_j), there is an overlap of T_k over T_i. It is not difficult to observe that the sum of the overlap is smaller than the size of $T_i(u)$. On the other hand, the maximum overlap of T_j over T_i is exactly $T_i(u)$. Thus, the overall overlap is increased by 1 at least. In order words, the size of T^{temp} is decreased at least by 1. □

Algorithm 3. Find the largest overlap AoN-subtree.

Input: A tree set $\mathcal{T}$ and a condensed common super tree T.
Output: A new tree T.

1: **for** each tree T_j in $\mathbb{VT}(T)$ that is not a root **do**
2:     **if** the overlap of T_i on T_j in tree T is not equal $\mathcal{L}(T_i, T_j)$ where T_i is $\mathcal{P}(T_j)$ **then**
3:        Find the the node $u \in T_i$ such that $T_i(u)$ is $\mathcal{L}(T_i, T_j)$.
4:        For each tree T_p who is a sibling of T_j and r_p is an descendant of u, we construct a new tree T_p^{new} that equals $T(r_p)$. Similarly, we also construct the tree T_j^{new}.
5:        For each tree T_p^{new}, remove $T_p^{new} - T_i$ from T. Tree T becomes T^{temp}.
6:        We overlap tree T_j^{new} at node u. For each tree T_p^{new}, we let root of T_p^{new} be the child of any leaf node in T^{temp}. Update tree T as T^{temp}.

Theorem 2 shows that for any condensed tree T, we can decrease its size by applying Algorithm 3 as long as some tree does not have a maximum overlap with its parent. Therefore, we can focus on the tree in which each non-root AoN-subtree always has a maximum overlaps with its parent. We denote this kind of common AoN-supertree as *maximum condensed common AoN-supertree* (MCCST). We then have

Theorem 3. *Smallest common AoN-supertree **LCST**$(\mathcal{T})$ is indeed a MCCST.*

4.2 Compute Good MCCST

With understanding of structures of LCST in Section 4.1, we are now ready to present our algorithm with constant approximation ratio. Notice that our focus now is to choose maximum condensed common supertree (MCCST) with smaller size among all MCC-STs. Given a tree set $\mathcal{T}$, the naive way is to first compute $\mathcal{L}(T_i, T_j)$ for each pair of T_i and T_j in $\mathcal{T}$. After that, for each T_j, we choose the T_i such that $\mathcal{L}(T_i, T_j)$ is maximum as its parent, which we call *treelization*. However, this solution does not guarantee a valid virtual overlap tree due to two reasons.

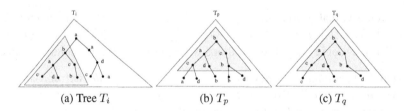

(a) Tree T_i (b) T_p (c) T_q

Fig. 2. Illustration of confliction

– First, it is possible that T_p and T_q choose the same tree T_i as their parent, and it is not possible for T_p and T_q to have maximum overlap with T_i simultaneously. In this case, we call tree T_p *conflicts* with T_q regarding T_i. One such example is shown in Figure 2. The maximum overlap of tree T_p and T_q over T_i are shown in Figure 2 (b) and (c) respectively. It is not difficult to observe that T_p and T_q can not maximally overlap with T_i simultaneously.

– Second, it is possible that T_j chooses T_i as its parent and T_i chooses a tree T_k who is a descendant of T_j as its parent. It thus creates a cycle in the virtual overlap graph, which we called *cycled tree*.

If we ignore the second problem, then any treelization avoids the first problem in the virtual overlap graph is a special forest with possibly several disconnected components such that each component is a tree whose root may have one backward edge toward its descendant. We call the forest a *cycled* forest.

In order to find the cycled forest with minimum size, we model it as a linear programming problem. Here, $x_{i,j} = 1$ if tree T_j chooses T_i as its parent; otherwise $x_{i,j} = 0$. Notice that for each tree T_j, it has exactly one parent, thus $\sum_{i:i\neq j} x_{i,j} = 1$ for each tree T_j. On the other hand, if $x_{i,j} = 1$, then in order to avoid the first problem, any tree T_k conflicting with T_j with respect to T_i should satisfy that $x_{i,k} = 0$. The objective is to minimize $\sum_{i\neq j} x_{i,j} \cdot (|T_j - \mathcal{L}(T_i, T_j)|)$, *i.e.*, the total size of the trees with cycles. Following is the Integer Programming we aim to solve, which is denoted as **IP1**.

$$\sum x_{i,j} \cdot (|T_j - \mathcal{L}(T_i, T_j)|). \tag{1}$$

$$\text{Subject to} \begin{cases} \sum_{i\neq j} x_{i,j} = 1 \ \forall \ T_j \\ x_{i,j} + x_{i,k} \leq 1 \ \forall \ i, j, k \text{ such that } T_j \text{ conflicts } T_k \text{ regarding } T_i \\ x_{i,j} = \{0, 1\} \ \forall \ i \neq j \end{cases} \tag{2}$$

Algorithm 4. Greedy Method To Compute A Cycled Forest.

Input: A tree set $\mathcal{T}$ and a condensed common AoN-supertree T.
Output: A cycled forest.

1: Compute $\mathcal{L}(T_i, T_j)$ for each pair of trees and sort them in a descending order. Initialize the tree set $S = \{\mathcal{L}(T_i, T_j) \mid \forall i \neq j\}$ and $\mathbb{TC} = \emptyset$.
2: **while** S is not empty **do**
3:    Choose the tree in S with the maximum size, say $\mathcal{L}(T_i, T_j)$.
4:    Add $\mathcal{L}(T_i, T_j)$ in $\mathbb{TC}$ and remove $\mathcal{L}(T_p, T_j)$ from S for any $p \neq i$ in S. Remove $\mathcal{L}(T_i, T_q)$ from S if T_q conflicts with T_j regarding T_i.
5: Set $x_{i,j} = 1$ if $\mathcal{L}(T_i, T_j)$ is in $\mathbb{TC}$, and $x_{i,j} = 0$ otherwise.

Algorithm 4 greedily selects the $\mathcal{L}(T_i, T_j)$ and it finds one solution to **IP1**.

Theorem 4. *Algorithm 4 computes a solution to the Integer Programming (1).*

PROOF. It is not difficult to verify that the solution does satisfy the constraints. Thus, we focus on the proof that it minimizes $\sum x_{i,j} \cdot (|T_j - \mathcal{L}(T_i, T_j)|)$. Since, $\sum_{i:i\neq j} x_{i,j} = 1$ for each T_j, $\sum x_{i,j} \cdot (|T_j - \mathcal{L}(T_i, T_j)|) = \sum_{T_i \in \mathcal{T}} |T_i| - \sum_{i\neq j} x_{i,j} \cdot |\mathcal{L}(T_i, T_j)|$. Thus, we only need to show that $\sum_{i\neq j} x_{i,j} \cdot |\mathcal{L}(T_i, T_j)|$ is maximized.

Without loss of generality, we assume that Algorithm 4 chooses $\mathcal{L}(T_{i_1}, T_{j_1})$, $\mathcal{L}(T_{i_2}, T_{j_2}), \cdots, \mathcal{L}(T_{i_n}, T_{j_n})$ in that order. We also assume that the solution to IP1 is x^{opt}, and all $\mathcal{L}(T_i, T_j)$ such that $x_{i,j}^{opt} = 1$ are ranked in a descending order $\mathcal{L}(T_{p_1}, T_{q_1})$, $\mathcal{L}(T_{p_2}, T_{q_2}), \cdots, \mathcal{L}(T_{p_n}, T_{q_n})$. Obviously, $\mathcal{L}(T_{i_1}, T_{j_1}) \geq \mathcal{L}(T_{p_1}, T_{q_1})$. Let k be the smallest index such that $i_k \neq p_k$ or $j_k \neq q_k$. If such k does not exist, then Algorithm 4 does compute a solution to IP1. Otherwise, such k must exist. Since greedy method always chooses the maximum $\mathcal{L}(T_i, T_j)$ that does not violate the constraint (2) of the Integer Programming **IP1**, $\mathcal{L}(T_{i_k}, T_{j_k}) \geq \mathcal{L}(T_{p_k}, T_{q_k})$. Without loss of generality, we assume that $x_{a,j_k}^{opt} = 1$ and $T_{b_1}, T_{b_2}, \cdots, T_{b_y}$ are the trees such that $x_{i_k, b_\ell}^{opt} = 1$ and T_{j_k} conflicts with T_{b_ℓ} regarding T_{i_k}. Let r_{b_i} be the root of tree $\mathcal{L}(T_{i_k}, T_{b_i})$, then r_{b_i} must be the descendant of root of tree $\mathcal{L}(T_{i_k}, T_{j_k})$. Since T_{b_i} does not conflict with T_{b_j} for any pair of i, j, then r_{b_i} is neither an ancestor nor a descendant of r_{b_j}. Now we modify the solution x^{opt} as follows. First, let $x_{i_k, j_k}^{opt} = 1$ and $x_{i_k, b_\ell}^{opt} = 0$ for $1 \leq \ell \leq y$. Then, set $x_{a, b_\ell}^{opt} = 1$ if it did not violate Constraint (2) and $x_{z, b_\ell}^{opt} = 0$ for any z that does not violate Constraint (2). The modified solution x^{opt} must satisfy Constraint (2). After this modification, the only difference between original solution and modified solution is the threes that overlap T_{i_k} and T_a. Let δ_1 be the increase of the overlap by replacing T_{j_k} with trees $T_{b_1}, T_{b_2}, \cdots, T_{b_y}$, and δ_2 be the decrease of the overlap by replacing $T_{b_1}, T_{b_2}, \cdots, T_{b_y}$ with T_{j_k}. Since, $\mathcal{L}(a, T_{T_{j_k}})$ is an AoN-subtree of $\mathcal{L}(T_{i_k}, T_{T_{j_k}})$, then $\delta_1 \geq \delta_2$. Thus, $\sum_{i \neq j} x_{i,j} \cdot |\mathcal{L}(T_i, T_j)|$ does not increase. This is a contradiction, which proves that Algorithm 4 computes a solution to the Integer Programming (1). $\Box$

Since $\sum x_{i,j}^{opt} \cdot (|T_j - \mathcal{L}(T_i, T_j)|) \leq |\mathbf{LCST}(\mathcal{T})|$, we found a cycled forest that is smaller than the size of **LCST**. Notice that cycled forest is not a valid tree because it violates the tree property. Following we will show that simple modification based on the cycled tree that was found by Algorithm 4 does output a valid common AoN-supertree. In the meanwhile, we also will show that the increase of the size is at most a constant time of the size of the original cycled forest.

Algorithm 5. Modify the cycled forest.

Input: Cycled Forest $\mathbb{CF}$.

Output: A valid virtual overlap tree.

1: Rank all cycled tree in cycled forest $\mathbb{CF}$ in arbitrary order, say $\mathbb{CF}_1, \mathbb{CF}_2, \ldots, \mathbb{CF}_k$.
2: For a cycled tree $\mathbb{CF}_i$, find the unique cycle C_i in tree $\mathbb{CF}_i$. Let r_i be any node in C_i and $\mathcal{P}(r_i)$ be its parent, then we remove the edge between r_i and $\mathcal{P}(r_i)$.
3: Concatenate the tree $\mathbb{CF}_i$ to $\mathbb{CF}_i$ without conflict for $i = 2, \cdots, k$, i.e., let r_i be a child of some node in $\mathbb{CF}_{i-1}$.
4: Output the final tree as a valid virtual overlap tree.

Borrowing some ideas from the construction of shortest common super-string (see [5] for more details), we have the following lemma

Lemma 8. *For any two cycles C_i and C_j in two different cycle tree $\mathbb{CF}_i$ and $\mathbb{CF}_j$, let s_i and s_j be any node in C_i and C_j respectively, then $\mathcal{L}(s_i, s_j) \leq |C_i| + |C_j|$.*

Theorem 5. *Algorithm 5 finds a common AoN-supertree of $\mathcal{T}$ with size $\leq 6 \cdot |\mathbf{LCST}(\mathcal{T})|$.*

PROOF. Let $\mathbb{CF}_1, \mathbb{CF}_2, \ldots, \mathbb{CF}_k$ be all the cycled trees computed by Algorithm 4. Then, $\sum_{i=1}^{k} |\mathbb{CF}_i| \leq |\mathbf{LCST}(\mathcal{T})|$. Let C_i be the cycle in cycled tree $\mathbb{CF}_i$, and s_i be the node whose corresponding tree has the largest size in cycle C_i. Lemma 8 shows that $\mathcal{L}(s_i, s_j) \leq |C_i| + |C_j|$ for any pair of cycles C_i and C_j. Unlike the string case, it is possible that two or more trees overlap with the same tree. However, if nodes $s_{j_1}, s_{j_2}, \ldots, s_{j_k}$ overlap with the tree s_i in **LCST**$(\mathcal{T})$, then $\sum_{\ell=1}^{k} |\mathcal{L}(s_i, s_{j_\ell})| \leq \sum_{\ell=1}^{k} |C_{j_\ell}| + |C_i| + |\mathbb{CF}_i|$. Thus, $\sum_{\ell=1}^{k} |\mathcal{L}(s_i, s_{j_\ell})| \leq \sum_{\ell=1}^{k} |C_{j_\ell}| + |C_i| + |\mathbb{CF}_i|$. For each s_i, let $\mathcal{P}(s_i)$ be its nearest ancestor in the virtual overlap graph of **LCST**$(\mathcal{T})$, then $\sum_{s_i} |\mathcal{P}(s_i) \cap s_i| \leq \sum_{s_i} |\mathcal{L}(\mathcal{P}(s_i), s_i)| \leq 2 \sum_{s_i} |C_i| + |\mathbb{CF}_i| \leq 4 \sum_{s_i} |\mathbb{CF}_i|$. Thus, $|\textbf{LCST}(\mathcal{T})| \geq \sum_{s_i} |s_i| - \sum_{s_i} |\mathcal{P}(s_i) \cap s_i| \geq \sum_{s_i} |s_i| - 4 \sum_{s_i} |\mathbb{CF}_i|$. Recall the virtual overlap tree computed by Algorithm 5 has the size at most $\sum_{s_i} |s_i| + \sum_{s_i} |\mathbb{CF}_i|$. Thus, $\sum_{s_i} |s_i| + \sum_{s_i} |\mathbb{CF}_i| \leq |\textbf{LCST}(\mathcal{T})| + 5 \sum_{s_i} |\mathbb{CF}_i| \leq 6|\textbf{LCST}(\mathcal{T})|$.   □

Theorem 6. *The time complexity of our approach is $O(n \cdot m^2)$, where n is the number of trees and m is the number of total nodes in these n trees.*

PROOF. Note that $m = \sum_{T_i \in \mathcal{T}} |T_i|$, and the time complexity to find the maximum overlap of T_j over T_i is $O(|T_i|^2 + |T_j|^2)$. Thus, finding the maximum overlap between each pair of trees is of time $O(n \cdot m^2)$. Algorithm 4 takes time $O(n^2 + n \log n)$ and Algorithm 5 only takes time $O(n)$. Thus, the overall time complexity is $O(n \cdot m^2)$.   □

5   Conclusion

In this paper, we gave a 6-approximation algorithm for smallest common AoN-supertree problem. It has applications in glycobiology. There are several interesting problems left for future research. It is known that the simple greedy algorithm will have an approximation ratio 3.5 (conjectured to be 2). It remains to be proved whether a similar technique as of [4] can be used to reduce the approximation ratio of our method to 5.5. Further, it remains an open problem what is the lower bound on the approximation ratio of the greedy method when all trees of the tree set $\mathcal{T}$ are k-nary trees. Secondly, currently the best approximation ratio for superstring problem is 2.5 [6] (not using the greedy method). Since superstring is a special case of the AoN-supertree problem, it remains an open question whether we can get similar approximation ratio for AoN-supertree problem. The last but not least important problem is to improve the time-complexity of finding the maximum overlapping subtree of two trees. Is there a linear time algorithm that can find the maximum overlap AoN-subtree of two trees?

References

1. Turner, J.S.: Approximation algorithms for the shortest common superstring problem. Information and Computation (1989) 1–20
2. Teng, S., Yao, F.: Approximating shortest superstrings. In: Annual Symposium on Foundations of Computer Science. (1993)
3. Weinard, M., Schnitger, G.: On the greedy superstring conjecture. In: FST TCS 2003: Foundations of Software Technology and Theoretical Computer Science. (2003) 387–398
4. Kaplan, H., Shafrir, N.: The greedy algorithm for shortest superstrings. Inf. Process. Lett. **93** (2005) 13–17
5. Blum, A., Jiang, T., Li, M., Tromp, J., Yannakakis, M.: Linear approximation of shortest superstrings. Journal of the ACM **41** (1994) 630–647

6. Sweedyk, Z.: A $2\frac{1}{2}$-approximation algorithm for shortest superstring. SIAM Journal of Computing **29** (1999) 954–986
7. Armen, C., Stein, C.: A 2 2/3-approximation algorithm for the shortest superstring problem. In: Combinatorial Pattern Matching. (1996) 87–101
8. Hashimoto, K., Goto, S., Kawano, S., Aoki-Kinoshita, K., Ueda, N., Hamajima, M., Kawasaki, T., Kanehisa, M.: Kegg as a glycome informatics resource. Glycobiology (2005).
9. Aoki, K., Yamaguchi, A., Ueda, N., Akutsu, T., Mamitsuka, H., Goto, S., Kanehisa, M.: Kcam (kegg carbohydrate matcher): A software tool for analyzing the structures of carbohydrate sugar chains. Nucleic Acids Research (2004) W267–W272
10. Aoki, K., Yamaguchi, A., Okuno, Y., Akutsu, T., Ueda, N., Kanehisa, M., Mamitsuka, H.: Efficient tree-matching methods for accurate carbohydrate database queries. In: Proceedings of the Fourteenth International Conference on Genome Informatics (Genome Informatics, 14). (2003) 134–143 Universal Academy Press.
11. Aoki, K.F., Mamitsuka, H., Akutsu, T., Kanehisa, M.: A score matrix to reveal the hidden links in glycans. Bioinformatics **8** (2005) 1457–1463
12. Ueda, N., Aoki-Kinoshita, K.F., Yamaguchi, A., Akutsu, T., Mamitsuka, H.: A probabilistic model for mining labeled ordered trees: capturing patterns in carbohydrate sugar chains. IEEE Transactions on Knowledge and Data Engineering **17** (2005) 1051–1064

Improved Approximation for
Single-Sink Buy-at-Bulk[*]

Fabrizio Grandoni[1] and Giuseppe F. Italiano[2]

[1] Dipartimento di Informatica, Università di Roma "La Sapienza", Via Salaria 113,
00198 Roma, Italy
grandoni@di.uniroma1.it
[2] Dipartimento di Informatica, Sistemi e Produzione, Università di Roma "Tor
Vergata", Via del Politecnico 1, 00133 Roma, Italy
italiano@disp.uniroma2.it

Abstract. In the single-sink buy-at-bulk network design problem we are given a subset of source nodes in a weighted undirected graph: each source node wishes to send a given amount of flow to a sink node. Moreover, a set of cable types is given, each characterized by a cost per unit length and by a capacity: the ratio cost/capacity decreases from small to large cables by economies of scale. The problem is to install cables on edges at minimum cost, such that the flow from each source to the sink can be routed simultaneously. The approximation ratio of this NP-hard problem was gradually reduced from $O(\log^2 n)$ to 65.49 by a long series of papers. In this paper, we design a better 24.92 approximation algorithm for this problem.

1 Introduction

Consider the problem of connecting different sites with an optical network. We know the distance and the traffic demand between each pair of sites. We are allowed to install optical cables, chosen from a limited set of available cable types: each cable type is characterized by a capacity and by a cost per unit length. By *economies of scale*, the cost per unit capacity decreases from small to large cables. The same kind of problem arises in several other applications, where optical cables are replaced by, e.g., pipes, trucks, and so on.

The essence of the mentioned problems is captured by *Multi-Sink Buy-at-Bulk* (MSBB) network design. In the MSBB we are given an n-node undirected graph $G = (V, E)$, with nonnegative edge lengths $c(e)$, $e \in E$. We distinguish a subset $P = \{(s_1, r_1), (s_2, r_2), \ldots, (s_p, r_p)\}$ of source-sink pairs: source s_i wishes to send $d(s_i)$ units of flow (the *demand* of s_i) to sink r_i. In order to support such flow, we are allowed to install *cables* on edges. There are k different *cable types*

---

[*] This work has been partially supported by the Sixth Framework Programme of the EU under Contract Number 507613 (Network of Excellence "EuroNGI: Designing and Engineering of the Next Generation Internet") and by MIUR, under Project ALGO-NEXT.

T. Asano (Ed.): ISAAC 2006, LNCS 4288, pp. 111–120, 2006.

$1, 2, \ldots, k$. Cables of type i have capacity μ_i and cost σ_i per unit length (that is, the cost of installing one such cable on edge e is $c(e)\sigma_i$). The cables satisfy the economies of scale principle: the cost $\delta_i = \sigma_i/\mu_i$ per unit capacity and unit length decreases from small to large cables. The aim is to find a minimum-cost installation of cables such that the flow between each source-sink pair can be routed simultaneously. In this paper we are concerned with the *Single-Sink* version of this problem (SSBB), where all the sources s_i send their flow to a unique sink r. The problem remains NP-hard also in this case (e.g., by reduction from the Steiner tree problem). The SSBB problem has been extensively studied in the literature. Meyerson, Munagala, and Plotkin [18] gave a $O(\log n)$ approximation. Garg et al. [9] described a $O(k)$ approximation, where k is the number of cable types. The first constant approximation is due to Guha, Meyerson, and Munagala [10]: the approximation ratio of their algorithm is roughly 2000. This approximation was reduced to 216 by Talwar [19], and later to 76.8 by Gupta, Kumar, and Roughgarden [12,15]. Using the same approach as Gupta et al., finally Jothi and Raghavachari [16] reduced the approximation factor to 65.49.

The contribution of this paper is a better approximation bound of 24.92 for the SSBB problem. Our improved bound is surprisingly obtained by a simple variant of the algorithm of Gupta et al. [12,15], combined with a more careful analysis. The algorithm by Gupta et al. works in two phases. First there is a *preprocessing phase*, where costs are rounded up and capacities are rounded down to the closest power of two. Part of the new cable types obtained in this way are "redundant" according to the new costs and capacities, and thus they can be safely discarded. Let $i(1)$, $i(2)$, $\ldots$, $i(k')$ be the remaining cable types, in increasing order of capacity. The second phase consists of a sequence of suitably defined *aggregation rounds*. In each round the demand is aggregated into a smaller, randomly selected subset of nodes, until all the demand is routed to the sink. In round t, only cables of type $i(t)$ and $i(t+1)$ are used. The initial rounding of this algorithm is responsible for a factor 4 in the approximation. Thus, it seems natural to wonder whether it is possible to improve substantially the approximation factor by replacing the first phase with a more sophisticated choice of the cable types to be used in the second phase (while preserving the same basic aggregation scheme). In this paper we present a simple, non-trivial cable selection rule which, in combination with a more careful analysis, reduces to 24.92 the approximation ratio for SSBB.

Related Work. Awerbuch and Azar [1] gave a $O(\log^2 n)$ approximation for MSBB, based on the tree embedding techniques by Bartal [2]. The improved tree embeddings in [3,5,8] lead to a $O(\log n)$ approximation. To the best of our knowledge, no constant approximation for MSBB is currently known. A problem closely related to MSBB is *Multi-sink Rent-Or-Buy* (MROB) network design [4,10,11,13,14,15,17]. As in MSBB, there is a set of source-sink pairs that wish to communicate. Now, instead of installing cables, we can either *buy* or *rent* edges: if we buy one edge, we pay a fixed cost c_{buy} per unit length, and we are then free to route an unbounded amount of flow on the bought edge. If we rent it, we pay a cost c_{rent} per unit length and unit flow along the edge. The current best approximation for MROB is 6.828 in the multi-sink case [4] and 3.55 in the single-sink case [15]. Another related problem (from the point of view of the techniques used to solve it) is *Virtual Private Network Design* (VPND) [6,7,15]. Here we have a set of terminals which wish to send flow to each other, but the traffic matrix is not know a priori: only upper bounds are given on the total amount of (unsplittable) flow that each terminal can send and receive. The aim is to find a minimum cost capacity reservation which supports every feasible traffic matrix. The current best approximation for VPND is 3.55 [7].

Preliminaries. For the sake of simplicity, in this extended abstract we assume that capacities, costs, and demands are non-negative integers. The same results can be extended to the case of real values. Let $1, 2, \ldots, k$ be the set of cable types, in increasing order of capacity: $\mu_1 \leq \mu_2, \ldots, \leq \mu_k$. Recall that $\delta_1 \geq \delta_2, \ldots, \geq \delta_k$ by economies of scale. Note that we can assume $\sigma_1 < \sigma_2, \ldots, < \sigma_k$. In fact, if $\sigma_i \geq \sigma_j$, for some $i < j$, we can eliminate the cable type i (without modifying the optimum). Following [15], and without loss of generality, we assume each node $v \in V$ has a demand $d(v)$, which is either zero or one. This can be achieved by duplicating nodes. The algorithm presented can be easily adapted so as to run in polynomial time even when the (original) demands are not polynomially bounded. The algorithm by Gupta et al. [15] is designed for capacities which are powers of two. Jothi and Raghavachari [16] designed a somehow complicated generalization of the algorithm in [15], in order to handle capacities which are powers of $(1 + \epsilon)$. Here we describe a simpler and more natural generalization of the algorithm in [15], which works for any value of the capacities. Our generalization is based on the following simple assumption: the sum of the demands $\sum_{v \in V} d(v)$ is a multiple of each capacity μ_i. This property can be enforced by adding *dummy demands* in the sink. By OPT we denote either the optimum solution or its actual value, where the meaning will be clear from the context. $OPT(s)$ is the cost paid by OPT to install cables of type s.

The remainder of this paper is organized as follows. In Section 2 we describe a generalization of the algorithm by Gupta et al., and analyze it under a generic cable selection paradigm. In Section 3 we introduce within this framework a more sophisticated cable selection rule, and prove that this yields the claimed 24.92 approximation bound for SSBB.

2   The Algorithm

One of the key steps in the approach of Gupta et al. [12,15] is aggregating demands over a tree in multiples of a given quantity. More precisely, consider a tree T and a given integer $U > 0$. Suppose each node v of T has integer weight $x(v) \in [0, U)$, and the sum of the weights is a multiple of U. They need to compute a flow moving weights along the tree such that: (1) The amount of flow along each edge is at most U, (2) The new weight $x'(v)$ at each node is either 0 or U, and (3) The expected weight at each node is preserved, that is: $Pr[x'(v) = U] = x(v)/U$. Gupta et al. give a randomized *aggregation algorithm* for this problem, which we describe next from a slightly different perspective. Replace each edge of T with two oppositely directed edges. Compute an Euler tour C' in the resulting directed graph T'. The same node v may appear several times in C': in that case assign the weight $x(v)$ to one of the occurrences of v, and zero to the others. Then replace each node with a path of length equal to its weight minus one (if the weight of a node is zero, remove the node and add one edge between its two neighbors). Now select a random subset of nodes in the resulting cycle $C = (w_0, w_1, \ldots, w_{|C|-1})$, such that the distance (number of hops) between any two consecutive selected nodes is U. This is possible since the total weight, which is equal to the total number of nodes of the cycle C, is a multiple of U by assumption. Eventually each node sends one unit of flow to the closest selected node in, say, clockwise direction. In particular, each selected node receives exactly $(U - 1)$ units of flow. The flow along C naturally induces a flow in the original graph. It is worth to mention that the duplication of nodes is not really necessary, but it is introduced here for the sake of simplicity.

We are now ready to describe our SSBB algorithm. We initially select a subset of cable types $i(1), i(2), \ldots, i(k')$ in increasing order of capacity, where we require that $i(1) = 1$ and $i(k') = k$ (that is, the first and last cable types are always selected). The selection rule will be described in Section 3. Note that there is no initial rounding. Then there is a sequence of rounds. In each round the demand is aggregated in a smaller and smaller randomly selected subset of nodes, until it is eventually routed to the sink. For ease of presentation, we distinguish the *initial* and *final* rounds from the remaining *intermediate* rounds. Let D_0 be the nodes with unit input demand. In the *initial round* we compute a ρ_{st}-approximate Steiner tree T_0 over $\{r\} \cup D_0$, and we apply the aggregation algorithm to T_0 with capacity $U = \mu_1$ and weights $x(v) = d(v)$ for each node v of T_0 (this is possible since by assumption the sum of the demands is a multiple of μ_1). The corresponding flow is supported by installing cables of type 1 (at most one on each edge of T_0). At the end of the round the demand at each node is either zero or μ_1. Now consider an *intermediate round* t, $t \in \{1, 2, \ldots, k' - 1\}$. By induction on the number of rounds, the initial demand $d_t(v)$ of node v is either zero or $\mu_{i(t)}$, while its final demand $d_{t+1}(v)$ is either zero or $\mu_{i(t+1)}$. The round consists of three steps. Initially the demand is collected at a random subset of aggregation points. Then a Steiner tree is computed on the aggregation points, and the demand is aggregated along such tree with the aggregation algorithm. Eventually the aggregated demand is redistributed back to the source nodes. Only cables of type $i(t)$ and $i(t + 1)$ are used in this process. We now describe the steps in more details. Let D_t denote the set of nodes with $d_t(v) = \mu_{i(t)}$.

Collection step: Each node in D_t is marked with probability $\sigma_{i(t)}/\sigma_{i(t+1)}$. Let D'_t be the set of marked nodes. Each node sends its demand to the closest node in $\{r\} \cup D'_t$ along a shortest path, using cables of type $i(t)$. Let $d'_t(w)$ be the new demand collected at each $w \in \{r\} \cup D'_t$.

Aggregation step: Compute a ρ_{st}-approximate Steiner tree T_t on $\{r\} \cup D'_t$. Apply the aggregation algorithm to T_t with $U = \mu_{i(t+1)}$ and weight $x(w) = d'_t(w)$ (mod $\mu_{i(t+1)}$) for each terminal node w (this is possible since the sum of the

$d'_t(w)$, and hence of the $x(w)$, is a multiple of $\mu_{i(t+1)}$). The corresponding flow is supported by installing cables of type $i(t+1)$ (at most one for each edge of T_t). Let $d''_t(w)$ be the new demand aggregated at each node w.

Redistribution step: For each node $w \in \{r\} \cup D'_t$, consider the subset of nodes $D_t(w) \subseteq D_t$ that sent their demand to w during the collection step (including w itself, if $w \neq r$). Uniformly select a random subset $\widetilde{D}_t(w)$ of $D_t(w)$ of cardinality $d''_t(w)/\mu_{i(t+1)}$. Send $\mu_{i(t+1)}$ units of flow back from w to each node in $\widetilde{D}_t(v)$ along shortest paths, installing cables of type $i(t+1)$.

Note that no demand is routed to the sink during the initial and intermediate rounds. The algorithm ends with a *final round*, where all the demands are sent to the sink along shortest paths, using cables of type $i(k') = k$. A generalization of the analysis given in [12,15] yields the following result, whose proof is omitted here for lack of space.

Lemma 1. *The* SSBB *algorithm above computes a solution of cost APX* $\leq \sum_{s=1}^{k} apx(s) OPT(s)$ *where*

$$apx(s) := 1 + \rho_{st} \frac{\sigma_{i(1)}}{\sigma_s} + \sum_{t=1}^{k'-1} \left(\left(2 + 2\frac{\delta_{i(t+1)}}{\delta_{i(t)}} \right) \left(1 - \frac{\sigma_{i(t)}}{\sigma_{i(t+1)}} \right) + \rho_{st} \right) \min \left\{ \frac{\sigma_{i(t+1)}}{\sigma_s}, \frac{\delta_{i(t)}}{\delta_s} \right\}. \tag{1}$$

3   An Improved Cable-Selection Rule

Let $i(1)$, $i(2)$, ..., $i(k')$ be the cable types, in increasing order of capacity, left after the first phase of the algorithm by Gupta et al. Such cables have the property that the σ's double and the δ's halve from one cable to the next one:

$$\forall t \in \{1, 2, \ldots, k'-1\}, \quad \sigma_{i(t+1)} \geq 2\sigma_{i(t)} \quad \text{and} \quad \delta_{i(t+1)} \leq \delta_{i(t)}/2.$$

Recall that all the remaining cables are used in the second phase. Thus, by Lemma 1, for every s,

$$apx(s) \leq 1 + (2 + 1 + \rho_{st}) \left(\sum_{i \geq 0} \frac{1}{2^i} + \sum_{j \geq 0} \frac{1}{2^j} \right) = 1 + (3 + \rho_{st})4.$$

Unfortunately, the initial rounding introduces an extra factor 4 in the approximation, thus leading to an overall $4(1 + (3 + \rho_{st})4) < 76.8$ approximation.

It is then natural to wonder whether it is possible to keep $apx(s)$ small, while avoiding rounding, by means of a more sophisticated cable selection rule. An intuitive approach could be selecting cables (in the original problem) in the following way: for a given selected cable type $i(t)$, starting from $i(1) = 1$, $i(t+1)$ is the smallest cable type such that $\sigma_{i(t+1)} \geq 2\sigma_{i(t)}$ and $\delta_{i(t+1)} \leq \delta_{i(t)}/2$. This way, we maintain the good *scaling* properties of σ's and δ's of selected cables. In particular, for any selected cable type $s = i(t')$, $apx(s) \leq 1 + (3 + \rho_{st})4$. Unluckily, this approach does not work for discarded cable types s, $i(t') < s < i(t'+1)$: in fact, in this case the *intermediate* term $(\min\{\sigma_{i(t'+1)}/\sigma_s, \delta_{i(t')}/\delta_s\})$ can be arbitrarily large. What can we do then? There is a surprisingly simple approach to tackle this problem. The idea is to slightly relax the scaling properties of the σ's: instead of requiring that $\sigma_{i(t+1)} \geq 2\sigma_{i(t)}$, we only require that $\sigma_{i(t+1)+1} \geq 2\sigma_{i(t)}$. More precisely, we use the following cable selection rule:

Improved cable selection rule: *Let* $i(1) = 1$. *Given* $i(t)$, $1 < i(t) < k$, $i(t+1)$ *is the smallest index such that*

$$\sigma_{i(t+1)+1} \geq 2\sigma_{i(t)} \quad and \quad \delta_{i(t+1)} \leq \delta_{i(t)}/2.$$

If such index does not exist, $i(t+1) = i(k') = k$.

Observe that the δ's halve at each selected cable (excluding possibly the last one), and the σ's double every other selected cable:

$$\forall t \in \{1, 2, \ldots, k'-2\}, \quad \delta_{i(t+1)} \leq \delta_{i(t)}/2 \quad and \quad \sigma_{i(t+2)} \geq \sigma_{i(t+1)+1} \geq 2\sigma_{i(t)}. \quad (2)$$

With this cable-selection policy we obtain $apx(s) \leq 1 + (3 + \rho_{st})7 < 32.85$ for every cable type s, including discarded ones. This is also a feasible bound on the overall approximation ratio since we avoided the initial rounding. This analysis can be refined by exploiting the telescopic sum hidden in Equation (1). This refinement improves to $4(1 + (3 + 2\rho_{st}) + (3 + \rho_{st})2) < 64.8$ the approximation bound of the algorithm by Gupta et al., and yields a better $16 + 7\rho_{st} < 26.85$ approximation bound if we use our approach.

Theorem 1. *The algorithm of Section 2, combined with the improved cable selection rule, yields a* $16 + 7\rho_{st} < 26.85$ *approximation bound for* SSBB.

Proof. Let us restrict to the case $s \leq i(k'-1)$. The case $i(k'-1) < s \leq i(k')$ is analogous, and thus it is omitted from this extended abstract. We distinguish between selected and discarded cables.

a) Discarded cable s, $i(t') < s < i(t'+1) \leq i(k'-1)$. By Lemma 1 and Equation (2), and observing that $\sigma_s \geq \sigma_{i(t')+1} \geq 2\sigma_{i(t'-1)}$ (for $t' > 1$), $apx(s)$ is bounded above by

$$1 + \rho_{st}\frac{\sigma_{i(1)}}{\sigma_s} + \sum_{t=1}^{t'-1}\left(3 + \rho_{st} - 3\frac{\sigma_{i(t)}}{\sigma_{i(t+1)}}\right)\frac{\sigma_{i(t+1)}}{\sigma_s}$$

$$+ \sum_{t=t'}^{k'-1}\left(2 + 2\frac{\delta_{i(t+1)}}{\delta_{i(t)}} + \rho_{st}\right)\min\left\{\frac{\sigma_{i(t+1)}}{\sigma_s}, \frac{\delta_{i(t)}}{\delta_s}\right\} \leq$$

$$1 + (3 + \rho_{st})\frac{\sigma_{i(t')}}{\sigma_s} + \rho_{st}\sum_{t=1}^{t'-2}\frac{\sigma_{i(t+1)}}{\sigma_s} + \sum_{t=t'}^{k'-1}\left(2 + 2\frac{\delta_{i(t+1)}}{\delta_{i(t)}} + \rho_{st}\right)\min\left\{\frac{\sigma_{i(t+1)}}{\sigma_s}, \frac{\delta_{i(t)}}{\delta_s}\right\} \leq$$

$$1 + (3 + \rho_{st}) + \rho_{st}\sum_{i=1}^{t'-2}\frac{1}{2^{\lceil i/2 \rceil}} + \sum_{t=t'}^{k'-1}\left(2 + 2\frac{\delta_{i(t+1)}}{\delta_{i(t)}} + \rho_{st}\right)\min\left\{\frac{\sigma_{i(t+1)}}{\sigma_s}, \frac{\delta_{i(t)}}{\delta_s}\right\} \leq$$

$$4 + 3\rho_{st} + \sum_{t=t'}^{k'-1}\left(2 + 2\frac{\delta_{i(t+1)}}{\delta_{i(t)}} + \rho_{st}\right)\min\left\{\frac{\sigma_{i(t+1)}}{\sigma_s}, \frac{\delta_{i(t)}}{\delta_s}\right\}.$$

From Equation (2), and observing that $\delta_s \geq \delta_{i(t'+1)}$, we get that $apx(s)$ is upper bounded by

$$4+3\,\rho_{st} + (3+\rho_{st})\min\left\{\frac{\sigma_{i(t'+1)}}{\sigma_s}, \frac{\delta_{i(t')}}{\delta_s}\right\} + (3+\rho_{st})\sum_{t=t'+1}^{k'-2}\frac{\delta_{i(t)}}{\delta_s} + (4+\rho_{st})\frac{\delta_{i(k'-1)}}{\delta_s} \leq$$

$$4+3\rho_{st} + (3+\rho_{st})\min\left\{\frac{\sigma_{i(t'+1)}}{\sigma_s}, \frac{\delta_{i(t')}}{\delta_s}\right\} + (3+\rho_{st})\sum_{j=0}^{k'-t'-3}\frac{1}{2^j} + (4+\rho_{st})\frac{1}{2^{k'-t'-2}} \leq$$

$$4+3\,\rho_{st} + (3+\rho_{st})\min\left\{\frac{\sigma_{i(t'+1)}}{\sigma_s}, \frac{\delta_{i(t')}}{\delta_s}\right\} + (3+\rho_{st})\left(\sum_{j=0}^{k'-t'-3}\frac{1}{2^j} + \frac{1}{2^{k'-t'-3}}\right)$$

Thus we get

$$apx(s) \leq 10 + 5\,\rho_{st} + (3+\rho_{st})\min\left\{\frac{\sigma_{i(t'+1)}}{\sigma_s}, \frac{\delta_{i(t')}}{\delta_s}\right\}. \tag{3}$$

We next show that

$$\min\left\{\frac{\sigma_{i(t'+1)}}{\sigma_s}, \frac{\delta_{i(t')}}{\delta_s}\right\} \leq 2. \tag{4}$$

Let $j(t')$ be the smallest index such that $\delta_{j(t')}/\delta_{i(t')} \leq 1/2$. Consider the case $s < j(t')$. By the definition of $j(t')$, $\delta_{j(t')-1}/\delta_{i(t')} > 1/2$. Therefore

$$\min\left\{\frac{\sigma_{i(t'+1)}}{\sigma_s}, \frac{\delta_{i(t')}}{\delta_s}\right\} \leq \frac{\delta_{i(t')}}{\delta_s} \leq \frac{\delta_{i(t')}}{\delta_{j(t')-1}} \leq 2.$$

Consider now the case $s \geq j(t') > i(t')$. Observe that $\sigma_{i(t'+1)}/\sigma_{i(t')} < 2$. In fact otherwise we would have

$$\sigma_{i(t'+1)-1+1}/\sigma_{i(t')} \geq 2 \quad \text{and} \quad \delta_{i(t'+1)-1}/\delta_{i(t')} \leq \delta_{j(t')}/\delta_{i(t')} \leq 1/2.$$

Thus cable $i(t'+1)-1$ should be selected, which contradicts the fact that $i(t'+1)$ is the first cable selected after $i(t')$. As a consequence

$$\min\left\{\frac{\sigma_{i(t'+1)}}{\sigma_s}, \frac{\delta_{i(t')}}{\delta_s}\right\} \leq \frac{\sigma_{i(t'+1)}}{\sigma_s} \leq \frac{\sigma_{i(t'+1)}}{\sigma_{i(t')}} \leq 2.$$

From (3) and (4),

$$apx(s) \leq 10 + 5\,\rho_{st} + (3+\rho_{st})\,2 = 16 + 7\rho_{st}. \tag{5}$$

b) Selected cable s, s = i(t') ≤ i(k' − 1). By basically the same arguments as for the case of discarded cables,

$$apx(s) \leq 1 + (3+\rho_{st})\frac{\sigma_{i(t')}}{\sigma_{i(t')}} + \rho_{st}\sum_{t=1}^{t'-2}\frac{\sigma_{i(t+1)}}{\sigma_{i(t')}} + (3+\rho_{st})\sum_{t=t'}^{k'-2}\frac{\delta_{i(t)}}{\delta_{i(t')}} + (4+\rho_{st})\frac{\delta_{i(k'-1)}}{\delta_{i(t')}}$$

$$\leq 1 + (3+\rho_{st}) + \rho_{st}\sum_{i=1}^{t'-2}\frac{1}{2^{\lfloor i/2 \rfloor}} + (3+\rho_{st})\left(\sum_{j=0}^{k'-t'-2}\frac{1}{2^j} + \frac{1}{2^{k'-t'-2}}\right)$$

$$\leq 1 + (3+\rho_{st}) + 3\rho_{st} + (3+\rho_{st})\,2$$

$$= 10 + 6\,\rho_{st}. \tag{6}$$

By (5) and (6)

$$APX \leq \sum_{s=1}^{k} apx(s) \, OPT(s) \leq (16 + 7\rho_{st})OPT.$$

$\square$

Remark 1. *In order to prove (4) the "relaxed" condition $\sigma_{i(t+1)+1} \geq 2\sigma_{i(t)}$ is crucial. The "naive" condition $\sigma_{i(t+1)} \geq 2\sigma_{i(t)}$ would not work properly.*

3.1 Adapting the Scaling Factors

The approximation can be further reduced to 24.92 by using better scaling factors. Let $\alpha > 1$ and $\beta > 1$ be two real parameters to be fixed later. Consider the following generalization of the improved cable selection rule:

Generalized cable selection rule: *Let $i(1) = 1$. Given $i(t)$, $1 < i(t) < k$, index $i(t + 1)$ is the smallest index such that*

$$\sigma_{i(t+1)+1} \geq \alpha \, \sigma_{i(t)} \quad and \quad \delta_{i(t+1)} \leq \delta_{i(t)}/\beta.$$

If such index does not exist, $i(t + 1) = i(k') = k$.

A proper choice of α and β leads to the following slightly refined approximation.

Theorem 2. *There is a 24.92 approximation algorithm for* SSBB.

Proof. Consider the algorithm of Section 2, with the generalized cable selection rule. For the sake of simplicity, let us assume $\beta \leq 3.77$, from which

$$(4 + \rho_{st}) \leq \left(2 + \frac{2}{\beta} + \rho_{st}\right) \frac{\beta}{\beta - 1}.$$

Consider first the case $i(t') < s < i(t' + 1) \leq i(k' - 1)$. By basically the same arguments as in the proof of Theorem 1, either $\delta_{i(t')}/\delta_s < \beta$ or $\sigma_{i(t'+1)}/\sigma_s < \alpha$. In the first case

$$apx(s) \leq 1 + \left(2 + \frac{2}{\beta} + \rho_{st}\right) \frac{\sigma_{i(t')}}{\sigma_s} + \rho_{st} \sum_{t=1}^{t'-2} \frac{\sigma_{i(t+1)}}{\sigma_s} + \left(2 + \frac{2}{\beta} + \rho_{st}\right) \left(\beta + \sum_{t=t'+1}^{k'-2} \frac{\delta_{i(t)}}{\delta_s}\right)$$

$$+ (4 + \rho_{st}) \frac{\delta_{i(k'-1)}}{\delta_s}$$

$$\leq 1 + \left(2 + \frac{2}{\beta} + \rho_{st}\right) + \rho_{st} \sum_{i=1}^{t'-2} \frac{1}{\alpha^{\lceil i/2 \rceil}} + \left(2 + \frac{2}{\beta} + \rho_{st}\right) \beta$$

$$+ \left(2 + \frac{2}{\beta} + \rho_{st}\right) \sum_{j=0}^{k'-t'-3} \frac{1}{\beta^j} + \frac{4 + \rho_{st}}{\beta^{k'-t'-2}}$$

$$\leq 1 + \left(2 + \frac{2}{\beta} + \rho_{st}\right) + \frac{2\rho_{st}}{\alpha - 1} + \left(2 + \frac{2}{\beta} + \rho_{st}\right) \beta + \left(2 + \frac{2}{\beta} + \rho_{st}\right) \frac{\beta}{\beta - 1}.$$

$$(7)$$

In the second case,

$$apx(s) \leq 1 + \left(2 + \frac{2}{\beta} + \rho_{st}\right)\frac{\sigma_{i(t'+1)}}{\sigma_s} + \rho_{st}\sum_{t=1}^{t'-1}\frac{\sigma_{i(t+1)}}{\sigma_s} + \left(2 + \frac{2}{\beta} + \rho_{st}\right)\sum_{t=t'+1}^{k'-2}\frac{\delta_{i(t)}}{\delta_s}$$

$$+ (4 + \rho_{st})\frac{\delta_{i(k'-1)}}{\delta_s}$$

$$\leq 1 + \left(2 + \frac{2}{\beta} + \rho_{st}\right)\alpha + \rho_{st}\sum_{i=1}^{t'-1}\frac{1}{\alpha^{\lfloor i/2 \rfloor}} + \left(2 + \frac{2}{\beta} + \rho_{st}\right)\frac{\beta}{\beta - 1}$$

$$\leq 1 + \left(2 + \frac{2}{\beta} + \rho_{st}\right)\alpha + \rho_{st}\frac{\alpha + 1}{\alpha - 1} + \left(2 + \frac{2}{\beta} + \rho_{st}\right)\frac{\beta}{\beta - 1}. \tag{8}$$

For any selected cable type $s = i(t') \leq i(k' - 1)$,

$$apx(s) \leq 1 + \left(2 + \frac{2}{\beta} + \rho_{st}\right)\frac{\sigma_{i(t')}}{\sigma_{i(t')}} + \rho_{st}\sum_{t=1}^{t'-2}\frac{\sigma_{i(t+1)}}{\sigma_s} + \left(2 + \frac{2}{\beta} + \rho_{st}\right)\sum_{t=t'}^{k'-2}\frac{\delta_{i(t)}}{\delta_s}$$

$$+ (4 + \rho_{st})\frac{\delta_{i(k'-1)}}{\delta_s}$$

$$\leq 1 + \left(2 + \frac{2}{\beta} + \rho_{st}\right) + \rho_{st}\sum_{i=1}^{t'-2}\frac{1}{\alpha^{\lfloor i/2 \rfloor}} + \left(2 + \frac{2}{\beta} + \rho_{st}\right)\sum_{j=0}^{k'-t'-2}\frac{1}{\beta^j} + \frac{4 + \rho_{st}}{\beta^{k'-t'-1}}$$

$$\leq 1 + \left(2 + \frac{2}{\beta} + \rho_{st}\right) + \rho_{st}\frac{\alpha + 1}{\alpha - 1} + \left(2 + \frac{2}{\beta} + \rho_{st}\right)\frac{\beta}{\beta - 1}. \tag{9}$$

In the case $i(k' - 1) < s \leq i(k') = k$ one obtains similarly:

$$apx(s) \leq 1 + \left(2 + \frac{2}{\beta} + \rho_{st}\right) + \frac{2\rho_{st}}{\alpha - 1} + (4 + \rho_{st})\beta, \tag{10}$$

$$apx(s) \leq 1 + (4 + \rho_{st})\alpha + \rho_{st}\frac{\alpha + 1}{\alpha - 1}, \tag{11}$$

$$apx(s) \leq 1 + (4 + \rho_{st}) + \rho_{st}\frac{\alpha + 1}{\alpha - 1}. \tag{12}$$

For a given choice of α and β, the approximation ratio is the maximum over (7)-(12). In particular, for $\alpha = \beta = 2$ we obtain the result of Theorem 1. The claim follows by imposing $\alpha = 3.1207$ and $\beta = 2.4764$. □

Acknowledgments. A special thank to Jochen Könemann for carefully reading a preliminary version of this paper and for helpful discussions.

References

1. B. Awerbuch and Y. Azar. Buy-at-bulk network design. In *IEEE Symposium on Foundations of Computer Science (FOCS)*, pages 542–547, 1997.
2. Y. Bartal. Probabilistic approximation of metric spaces and its algorithmic applications. In *IEEE Symposium on Foundations of Computer Science (FOCS)*, pages 184–193, 1996.
3. Y. Bartal. On approximating arbitrary metrics by tree metrics. In *IEEE Symposium on Foundations of Computer Science (FOCS)*, pages 161–168, 1998.

4. L. Becchetti, J. Konemann, S. Leonardi, and M. Pal. Sharing the cost more efficiently: improved approximation for multicommodity rent-or-buy. In *ACM-SIAM Symposium on Discrete Algorithms (SODA)*, pages 375–384, 2005.
5. M. Charikar, A. Chekuri, A. Goel, and S. Guha. Approximating a finite metric by a small number of tree metrics. In *IEEE Symposium on Foundations of Computer Science (FOCS)*, pages 379–388, 1998.
6. F. Eisenbrand and F. Grandoni. An improved approximation algorithm for virtual private network design. In *ACM-SIAM Symposium on Discrete Algorithms (SODA)*, pages 928–932, 2005.
7. F. Eisenbrand, F. Grandoni, G. Oriolo, and M. Skutella. New approaches for virtual private network design. In *International Colloquium on Automata, Languages and Programming (ICALP)*, pages 1152–1162, 2005.
8. J. Fakcharoenphol, S. Rao, and K. Talwar. A tight bound on approximating arbitrary metrics by tree metrics. In *ACM Symposium on the Theory of Computing (STOC)*, pages 448–455, 2003.
9. N. Garg, R. Khandekar, G. Konjevod, R. Ravi, F. Salman, and A. Sinha. On the integrality gap of a natural formulation of the single-sink buy-at-bulk network design problem. In *International Conference on Integer Programming and Combinatorial Optimization (IPCO)*, pages 170–184, 2001.
10. S. Guha, A. Meyerson, and K. Munagala. A constant factor approximation for the single sink edge installation problem. In *ACM Symposium on the Theory of Computing (STOC)*, pages 383–388, 2001.
11. A. Gupta, J. Kleinberg, A. Kumar, R. Rastogi, and B. Yener. Provisioning a virtual private network: a network design problem for multicommodity flow. In *ACM Symposium on the Theory of Computing (STOC)*, pages 389–398, 2001.
12. A. Gupta, A. Kumar, M. Pal, and T. Roughgarden. Approximation via cost-sharing: simpler and better approximation algorithms for network design. Manuscript.
13. A. Gupta, A. Kumar, M. Pal, and T. Roughgarden. Approximation via cost-sharing: a simple approximation algorithm for the multicommodity rent-or-buy problem. In *IEEE Symposium on Foundations of Computer Science (FOCS)*, pages 606–617, 2003.
14. A. Gupta, A. Kumar, and T. Roughgarden. A constant-factor approximation algorithm for the multicommodity. In *IEEE Symposium on Foundations of Computer Science (FOCS)*, pages 333–344, 2002.
15. A. Gupta, A. Kumar, and T. Roughgarden. Simpler and better approximation algorithms for network design. In *ACM Symposium on the Theory of Computing (STOC)*, pages 365–372, 2003.
16. R. Jothi and B. Raghavachari. Improved approximation algorithms for the single-sink buy-at-bulk network design problems. In *Scandinavian Workshop on Algorithm Theory (SWAT)*, pages 336–348, 2004.
17. A. Kumar and C. Swamy. Primal-dual algorithms for the connected facility location problem. In *International Workshop on Approximation Algorithms for Combinatorial Optimization*, pages 256–269, 2002.
18. A. Meyerson, K. Munagala, and S. Plotkin. Cost-distance: two metric network design. In *IEEE Symposium on Foundations of Computer Science (FOCS)*, pages 624–630, 2000.
19. K. Talwar. The single-sink buy-at-bulk LP has constant integrality gap. In *International Conference on Integer Programming and Combinatorial Optimization (IPCO)*, pages 475–486, 2002.

Approximability of
Partitioning Graphs with Supply and Demand
Extended Abstract

Takehiro Ito[1], Erik D. Demaine[2], Xiao Zhou[1], and Takao Nishizeki[1]

[1] Graduate School of Information Sciences, Tohoku University,
Aoba-yama 6-6-05, Sendai, 980-8579, Japan
[2] MIT Computer Science and Artificial Intelligence Laboratory,
32 Vassar St., Cambridge, MA 02139, USA
`take@nishizeki.ecei.tohoku.ac.jp`, `edemaine@mit.edu`,
`{zhou, nishi}@ecei.tohoku.ac.jp`

Abstract. Suppose that each vertex of a graph G is either a supply vertex or a demand vertex and is assigned a positive real number, called the supply or the demand. Each demand vertex can receive "power" from at most one supply vertex through edges in G. One thus wishes to partition G into connected components so that each component C either has no supply vertex or has exactly one supply vertex whose supply is at least the sum of demands in C, and wishes to maximize the fulfillment, that is, the sum of demands in all components with supply vertices. This maximization problem is known to be NP-hard even for trees having exactly one supply vertex and strongly NP-hard for general graphs. In this paper, we focus on the approximability of the problem. We first show that the problem is MAXSNP-hard and hence there is no polynomial-time approximation scheme (PTAS) for general graphs unless P = NP. We then present a fully polynomial-time approximation scheme (FPTAS) for series-parallel graphs having exactly one supply vertex. The FPTAS can be easily extended for partial k-trees, that is, graphs with bounded treewidth.

1 Introduction

Consider a graph G such that each vertex is either a supply vertex or a demand vertex. Each vertex v is assigned a positive real number; the number is called the *supply of v* if v is a supply vertex; otherwise, it is called the *demand of v*. Each demand vertex can receive "power" from at most one supply vertex through edges in G. One thus wishes to partition G into connected components by deleting edges from G so that each component C has exactly one supply vertex whose supply is at least the sum of demands of all demand vertices in C. However, such a partition does not always exist. So we wish to partition G into connected components so that each component C either has no supply vertex or has exactly one supply vertex whose supply is at least the sum of demands of all demand vertices in C, and wish to maximize the "fulfillment," that is, the sum

T. Asano (Ed.): ISAAC 2006, LNCS 4288, pp. 121–130, 2006.

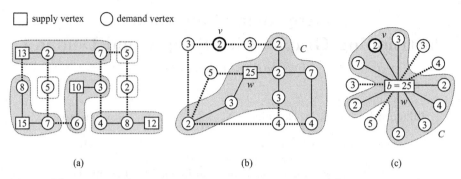

Fig. 1. (a) Partition of a graph with maximum fulfillment, (b) partition of a series-parallel graph G having exactly one supply vertex, and (c) a star S with a supply vertex at the center

of demands of the demand vertices in all components with supply vertices. We call this problem the *maximum partition problem* [4]. The maximum partition problem has some applications to the power supply problem for power delivery networks [4,6]. Figure 1(a) illustrates a solution of the maximum partition problem for a graph, whose fulfillment is $(2 + 7) + (8 + 7) + (3 + 6) + (4 + 8) = 45$. In Fig. 1(a) each supply vertex is drawn as a rectangle and each demand vertex as a circle, the supply or demand is written inside, the deleted edges are drawn by thick dotted lines, and each connected component with a supply vertex is shaded.

Given a set A of integers and an upper bound (integer) b, the *maximum subset sum problem* asks to find a subset C of A such that the sum of integers in C is no greater than the bound b and is maximum among all such subsets C. The maximum subset sum problem can be reduced in linear time to the maximum partition problem for a particular tree, called a star, with exactly one supply vertex at the center, as illustrated in Fig. 1(c) [4]. Since the maximum subset sum problem is NP-hard, the maximum partition problem is also NP-hard even for stars. Thus it is very unlikely that the maximum partition problem can be exactly solved in polynomial time even for trees. However, there is a fully polynomial-time approximation scheme (FPTAS) for the maximum partition problem on trees [4]. One may thus expect that the FPTAS for trees can be extended to a larger class of graphs, for example series-parallel graphs and partial k-trees, that is, graphs with bounded treewidth [1,2].

In this paper, we study the approximability of the maximum partition problem. We first show that the maximum partition problem is MAXSNP-hard, and hence there is no polynomial-time approximation scheme (PTAS) for the problem on general graphs unless P = NP. We then present an FPTAS for series-parallel graphs having exactly one supply vertex. The FPTAS for series-parallel graphs can be extended to partial k-trees. (The details are omitted from this extended abstract.) Figure 1(b) depicts a series-parallel graph together with a connected component C found by our FPTAS. One might think that it would be straight-foward to extend the FPTAS for the maximum subset sum problem in [3] to an

FPTAS for the maximum partition problem with a single supply vertex. However, this is not the case since we must take a graph structure into account. For example, the vertex v of demand 2 in Fig. 1(b) cannot be supplied power even though the supply vertex w has marginal power $25 - (2+3+2+2+3+7+4) = 2$, while the vertex v in Fig. 1(c) can be supplied power from the supply vertex w in the star having the same supply and demands as in Fig. 1(b). Indeed, we not only extend the "scaling and rounding" technique but also employ many new ideas to derive our FPTAS.

The rest of the paper is organized as follows. In Section 2 we show that the maximum partition problem is MAXSNP-hard. In Section 3 we present a pseudo-polynomial-time algorithm for series-parallel graphs. In Section 4 we present an FPTAS based on the algorithm in Section 3.

2   MAXSNP-Hardness

Assume in this section that a graph G has one or more supply vertices. (See Figs. 1(a) and 2(b).) The main result of this section is the following theorem.

Theorem 1. *The maximum partition problem is MAXSNP-hard for bipartite graphs.*

Proof. As in [7,8], we use the concept of "L-reduction" which is a special kind of reduction that preserves approximability. Suppose that both A and B are maximization problems. Then we say that A *can be L-reduced to* B if there exist two polynomial-time algorithms R and S and two positive constants α and β which satisfy the following two conditions (1) and (2) for each instance I_A of A:

(1) the algorithm R returns an instance $I_B = R(I_A)$ of B such that $OPT_B(I_B) \leq \alpha \cdot OPT_A(I_A)$, where $OPT_A(I_A)$ and $OPT_B(I_B)$ denote the maximum solution values of I_A and I_B, respectively; and

(2) for each feasible solution of I_B with value c_B, the algorithm S returns a feasible solution of I_A with value c_A such that $OPT_A(I_A) - c_A \leq \beta \cdot \big(OPT_B(I_B) - c_B\big)$.

Note that, by condition (2) of the L-reduction, S must return the optimal solution of I_A for the optimal solution of I_B.

We show that a MAXSNP-hard problem, called "3-occurrence MAX3SAT" [7,8], can be L-reduced to the maximum partition problem for bipartite graphs. However, due to the page limitation, we only show in this extended abstract that condition (1) of the L-reduction holds.

We now show that condition (1) of the L-reduction holds. An instance Φ of 3-occurrence MAX3SAT consists of a collection of m clauses $C_1, C_2, \cdots, C_m$ on n variables $x_1, x_2, \cdots, x_n$ such that each clause has exactly three literals and each variable appears at most three times in the clauses. The problem *3-occurrence MAX3SAT* is to find a truth assignment for the variables which satisfies the maximum number of clauses. Then it suffices to show that, from each instance

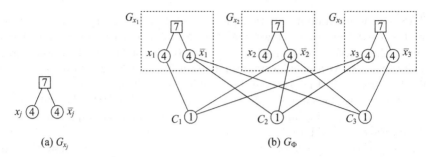

(a) G_{x_j} (b) G_Φ

Fig. 2. (a) Variable gadget G_{x_j}, and (b) the bipartite graph G_Φ corresponding to an instance Φ with three clauses $C_1 = (x_1 \lor \bar{x}_2 \lor x_3)$, $C_2 = (\bar{x}_1 \lor \bar{x}_2 \lor x_3)$ and $C_3 = (\bar{x}_1 \lor \bar{x}_2 \lor \bar{x}_3)$

Φ of 3-occurrence MAX3SAT, one can construct in polynomial time a bipartite graph G_Φ as an instance of the maximum partition problem such that

$$OPT_{MPP}(G_\Phi) \leq 26 \cdot OPT_{SAT}(\Phi), \tag{1}$$

where $OPT_{MPP}(G_\Phi)$ and $OPT_{SAT}(\Phi)$ are the maximum solution values of G_Φ and Φ, respectively: condition (1) of the L-reduction holds for $\alpha = 26$.

We first make a *variable gadget* G_{x_j} for each variable x_j, $1 \leq j \leq n$; G_{x_j} is a binary tree with three vertices as illustrated in Fig. 2(a); the root is a supply vertex of supply 7, and two leaves x_j and $\bar{x}_j$ are demand vertices of demands 4. The graph G_Φ is constructed as follows. For each variable x_j, $1 \leq j \leq n$, put the variable gadget G_{x_j} to the graph, and for each clause C_i, $1 \leq i \leq m$, put a demand vertex C_i of demand 1 to the graph. Finally, for each clause C_i, $1 \leq i \leq m$, join a demand vertex x_j (or $\bar{x}_j$) in G_{x_j} with the demand vertex C_i if and only if the literal x_j (or $\bar{x}_j$) is in C_i, as illustrated in Fig. 2(b). Clearly, G_Φ can be constructed in polynomial time, and is a bipartite graph. It should be noted that the degree of each demand vertex in the variable gadget for x_j is at most four since x_j appears at most three times in the clauses. Therefore, each supply vertex in the variable gadget G_{x_j} has enough "power" to supply all demand vertices C_i whose corresponding clauses have x_j or $\bar{x}_j$. Then one can verify Eq. (1), whose proof is omitted from this extended abstract.   □

3   Pseudo-polynomial-time Algorithm

Since the maximum partition problem is strongly NP-hard [5], there is no pseudo-polynomial-time algorithm for general graphs unless P = NP. However, Ito *et al.* presented a pseudo-polynomial-time algorithm for the maximum partition problem on series-parallel graphs [5]. In this section we present another pseudo-polynomial-time algorithm, which is suited to an FPTAS presented in Section 4. More precisely, we have the following theorem.

Theorem 2. *The maximum partition problem for a series-parallel graph G with a single supply vertex can be solved in time $O(F^2 n)$ if the demands and the supply* *are integers, where n is the number of vertices and F is the sum of all demands in G.*

3.1 Terminology and Definitions

Suppose that there is exactly one supply vertex w in a graph $G = (V, E)$, as illustrated in Figs. 1(b) and (c). Let sup(w) be the supply of w. For each demand vertex v, we denote by dem(v) the demand of v. Let dem$(w) = 0$ although w is a supply vertex. Then, instead of finding a partition of G, we shall find a set $C \subseteq V$, called a *supplied set* for G, such that

   (a) $w \in C$;
   (b) $\sum_{v \in C}$ dem$(v) \leq$ sup(w); and
   (c) C induces a connected subgraph of G.

The *fulfillment* $f(C)$ of a supplied set C is $\sum_{v \in C}$ dem(v). A supplied set C is called the *maximum supplied set* for G if $f(C)$ is maximum among all supplied sets for G. Then the *maximum partition problem* is to find a maximum supplied set for a given graph G. The *maximum fulfillment* $f(G)$ of a graph G is the fulfillment $f(C)$ of the maximum supplied set C for G. For the series-parallel graph G in Fig. 1(b), the supplied set C shaded in the figure has the maximum fulfillment, and hence $f(G) = f(C) = 23$, while $f(S) = 25$ for the star S in Fig. 1(c).

A *(two-terminal) series-parallel graph* G is defined recursively as a graph obtained from two series-parallel graphs by the so-called series or parallel connection [9]. The terminals of G are denoted by $v_s(G)$ and $v_t(G)$. Since we deal with the maximum partition problem, we may assume without loss of generality that G is a simple graph.

A series-parallel graph G can be represented by a "binary decomposition tree" T [9]. Every leaf of T represents a subgraph of G induced by a single edge. Each node u of T corresponds to a subgraph $G_u = (V_u, E_u)$ of G induced by all edges represented by the leaves that are descendants of u in T. G_u is a series-parallel graph for each node u of T, and $G = G_r$ for the root r of T. Since a binary decomposition tree of a given series-parallel graph G can be found in linear time [9], we may assume that a series-parallel graph G and its binary decomposition tree T are given.

3.2 Algorithm

In this subsection we give an algorithm to solve the maximum partition problem in time $O(F^2 n)$ as a proof of Theorem 2.

Let G be a series-parallel graph, let u, u' and u'' be nodes of a binary decomposition tree T of G, and let $G_u = (V_u, E_u)$, $G_{u'} = (V_{u'}, E_{u'})$ and $G_{u''} = (V_{u''}, E_{u''})$ be the subgraphs of G for nodes u, u' and u'', respectively, as illustrated in Fig. 3(a). Every supplied set C for G naturally induces subsets of V_u, $V_{u'}$ and $V_{u''}$. The supplied set C for G in Fig. 3(a) induces a single subset C_{st} of V_u in Fig. 3(b) such that $C_{st} = C \cap V_u$ and $v_s(G_u), v_t(G_u) \in C_{st}$. On the other

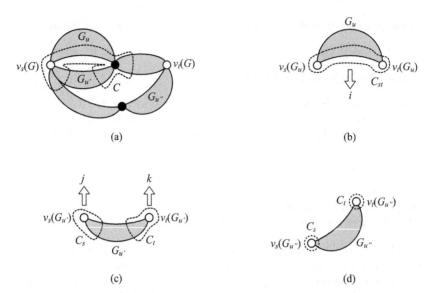

Fig. 3. (a) A supplied set C for a series-parallel graph G, (b) a connected set C_{st} for G_u, (c) a separated pair (C_s, C_t) of sets for $G_{u'}$, and (d) a separated pair (C_s, C_t) of isolated sets for $G_{u''}$

hand, C induces a pair of subsets C_s and C_t of $V_{u'}$ in Fig. 3(c) such that $C_s \cup C_t = C \cap V_{u'}$, $C_s \cap C_t = \emptyset$, $v_s(G_{u'}) \in C_s$ and $v_t(G_{u'}) \in C_t$. A set C_{st}, C_s or C_t is not always a supplied set for G_u or $G_{u'}$, because it may not contain the supply vertex w. C_{st} is a "connected set" for G_u, that is, C_{st} induces a connected subgraph of G_u, while the pair (C_s, C_t) is a "separated pair of sets" for $G_{u'}$, that is, C_s and C_t induce vertex-disjoint connected subgraphs of $G_{u'}$. The set C contains no terminals of $G_{u''}$ in Fig. 3(a). In such a case, we regard that $\text{dem}(v_s(G_{u''})) = \text{dem}(v_t(G_{u''})) = 0$ and C induces a separated pair of singleton sets (C_s, C_t) such that $C_s = \{v_s(G_{u''})\}$ and $C_t = \{v_t(G_{u''})\}$, as illustrated in Fig. 3(d).

If a set C_{st}, C_s or C_t contains the supply vertex w, then the set may have the "marginal" power, the amount of which is no greater than $\sup(w)$. If a set does not contain w, then the set may have the "deficient" power, the amount of which is no greater than $\sup(w)$. Thus we later introduce five functions g, h_1, h_2, h_3 and h_4; for a series-parallel graph G_u and a real number x, the value $g(G_u, x)$ represents the maximum marginal power or the minimum deficient power of connected sets for G_u; for a series-parallel graph G_u and a real number x, the value $h_i(G_u, x)$, $1 \le i \le 4$, represents the maximum marginal power or the minimum deficient power of separated pairs of sets for G_u. Our idea is to compute $g(G_u, x)$ and $h_i(G_u, x)$, $1 \le i \le 4$, from the leaves of T to the root r of T by means of dynamic programming.

We now formally define the notion of connected sets and separated pair of sets for a series-parallel graph G. Let $G_u = (V_u, E_u)$ be a subgraph of G for a node u of a binary decomposition tree T of G, and let $v_s = v_s(G_u)$ and $v_t = v_t(G_u)$.

We call a set $C \subseteq V_u$ a *connected set for G_u* if C satisfies the following three conditions (see Fig. 3(b)):

(a) $v_s, v_t \in C$;
(b) C induces a connected subgraph of G_u; and
(c) $\sum_{v \in C} \text{dem}(v) \leq \sup(w)$ if $w \in C$.

A pair of sets $C_s, C_t \subseteq V_u$ is called a *separated pair (of sets) for G_u* if C_s and C_t satisfy the following four conditions (see Fig. 3(c)):

(a) $C_s \cap C_t = \emptyset$, $v_s \in C_s$ and $v_t \in C_t$;
(b) C_s and C_t induce connected subgraphs of G_u;
(c) $\sum_{v \in C_s} \text{dem}(v) \leq \sup(w)$ if $w \in C_s$; and
(d) $\sum_{v \in C_t} \text{dem}(v) \leq \sup(w)$ if $w \in C_t$.

We then classify connected sets and separated pairs further into smaller classes. Let $\mathbb{R}_w = \{x \in \mathbb{R} : |x| \leq \sup(w)\}$, where $\mathbb{R}$ denotes the set of all real numbers. For each real number $i \in \mathbb{R}_w$, we call a connected set C for G_u an *i-connected set* if C satisfies the following two conditions:

(a) if $i > 0$, then $w \in C$ and $i + \sum_{x \in C} \text{dem}(x) \leq \sup(w)$; and
(b) if $i \leq 0$, then $w \notin C$ and $\sum_{x \in C} \text{dem}(x) \leq |i| = -i$.

An i-connected set C for G_u with $i > 0$ is a supplied set for G_u, and hence corresponds to some supplied set C_r for the whole graph $G = G_r$ such that $w \in C \subseteq C_r$, where r is the root of T; an amount i of the remaining power of w can be delivered outside G_u through v_s or v_t; and hence the "margin" of C is i. On the other hand, an i-connected set C for G_u with $i \leq 0$ is not a supplied set for G_u, but may correspond to a supplied set C_r for $G = G_r$ such that $w \notin C \subset C_r$ and $w \in C_r$; an amount $|i|$ of power must be delivered to C through v_s or v_t, and hence the "deficiency" of C is $|i|$. For an i-connected set C for G_u, let

$$f(C, i) = \sum_{x \in C} \text{dem}(x).$$

Then $f(C, i) = f(C)$ if $0 < i \leq \sup(w)$. On the other hand, if $-\sup(w) \leq i \leq 0$, then $f(C, i)$ represents the fulfillment of C when an amount $|i|$ of power is delivered to C from w in the outside of G_u.

Let $\sigma \notin \mathbb{R}_w$ be a symbol. For each pair of j and k in $\mathbb{R}_w \cup \{\sigma\}$, we call a separated pair (C_s, C_t) for G_u a *(j, k)-separated pair* if (C_s, C_t) satisfies the following seven conditions:

(a) if $j > 0$, then $w \in C_s$ and $j + \sum_{x \in C_s} \text{dem}(x) \leq \sup(w)$;
(b) if $j \leq 0$, then $w \notin C_s$ and $\sum_{x \in C_s} \text{dem}(x) \leq -j$;
(c) if $j = \sigma$, then $C_s = \{v_s\}$;
(d) if $k > 0$, then $w \in C_t$ and $k + \sum_{x \in C_t} \text{dem}(x) \leq \sup(w)$;
(e) if $k \leq 0$, then $w \notin C_t$ and $\sum_{x \in C_t} \text{dem}(x) \leq -k$;
(f) if $k = \sigma$, then $C_t = \{v_t\}$; and
(g) if $j + k \leq 0$, then $j \leq 0$ and $k \leq 0$.

Since G has only one supply vertex w, there is no (j, k)-separated pair (C_s, C_t) for G such that $j > 0$ and $k > 0$. A (j, k)-separated pair (C_s, C_t) for G_u with $j > 0$ corresponds to a supplied set C_r for the whole graph G such that $w \in C_s \subseteq C_r$; an amount j of the remaining power of w can be delivered outside C_s through v_s, and hence the margin of C_s is j. A (j, k)-separated pair (C_s, C_t) for G_u with $j \leq 0$ may correspond to a supplied set C_r for G such that $C_s \subset C_r$ and either $w \in C_t$ or $w \in C_r - C_s \cup C_t$; an amount $|j|$ of power must be delivered to C_s through v_s, and hence the deficiency of C_s is $|j|$. A (j, k)-separated pair (C_s, C_t) for G_u with $j = \sigma$ corresponds to a supplied set C_r for G such that $v_s \notin C_r$, that is, v_s is never supplied power. (See Figs. 3(a) and (d).) A (j, k)-separated pair (C_s, C_t) for G_u with $k > 0$, $k \leq 0$ or $k = \sigma$ corresponds to a supplied set C_r for G similarly as above. For a (j, k)-separated pair (C_s, C_t) for G_u, let

$$f(C_s, C_t, j, k) = \begin{cases} \sum_{x \in C_s \cup C_t} \mathrm{dem}(x) & \text{if } j, k \in \mathbb{R}_w; \\ \sum_{x \in C_s} \mathrm{dem}(x) & \text{if } j \in \mathbb{R}_w \text{ and } k = \sigma; \text{ and} \\ \sum_{x \in C_t} \mathrm{dem}(x) & \text{if } j = \sigma \text{ and } k \in \mathbb{R}_w. \end{cases}$$

Let

$$f(C_s, C_t, \sigma, \sigma) = \max\{f(C_u) \mid C_u \text{ is a supplied set for } G_u$$
$$\text{such that } v_s, v_t \notin C_u\};$$

let $f(C_s, C_t, \sigma, \sigma) = 0$ if G_u has no supplied set C_u such that $v_s, v_t \notin C_u$.

We now formally define a function g as follows: for a series-parallel graph G_u and a real number $x \in \mathbb{R}$,

$$g(G_u, x) = \max\{i \in \mathbb{R}_w \mid G_u \text{ has an } i\text{-connected set } C \text{ such that } f(C, i) \geq x\}.$$

If G_u has no i-connected set C with $f(C, i) \geq x$ for any number $i \in \mathbb{R}_w$, then let $g(G_u, x) = -\infty$. We then formally define a function h_1 as follows: for a series-parallel graph G_u and a real number $x \in \mathbb{R}$,

$$h_1(G_u, x) = \max\{j + k \mid G_u \text{ has a } (j, k)\text{-separated pair } (C_s, C_t) \text{ such that}$$
$$j, k \in \mathbb{R}_w, |j + k| \leq \sup(w), \text{ and } f(C_s, C_t, j, k) \geq x\}.$$

If G_u has no (j, k)-separated pair (C_s, C_t) with $f(C_s, C_t, j, k) \geq x$ for any pair of numbers j and k in $\mathbb{R}_w$, then let $h_1(G_u, x) = -\infty$. It should be noted that a (j, k)-separated pair (C_s, C_t) for G_u with $j, k \in \mathbb{R}_w$ corresponds to a supplied set C_r for G such that $C_s \cup C_t \subseteq C_r$, and hence we can simply take the summation of j and k as the marginal power or the deficient power of $C_s \cup C_t$. We next formally define a function h_2 as follows: for a series-parallel graph G_u and a real number $x \in \mathbb{R}$,

$$h_2(G_u, x) = \max\{j \in \mathbb{R}_w \mid G_u \text{ has a } (j, \sigma)\text{-separated pair } (C_s, \{v_t\})$$
$$\text{such that } f(C_s, \{v_t\}, j, \sigma) \geq x\}.$$

If G_u has no (j, σ)-separated pair $(C_s, \{v_t\})$ with $f(C_s, \{v_t\}, j, \sigma) \geq x$ for any number $j \in \mathbb{R}_w$, then let $h_2(G_u, x) = -\infty$. We then formally define a function h_3 as follows: for a series-parallel graph G_u and a real number $x \in \mathbb{R}$,

$$h_3(G_u, x) = \max\{k \in \mathbb{R}_w \mid G_u \text{ has a } (\sigma, k)\text{-separated pair } (\{v_s\}, C_t)$$
$$\text{such that } f(\{v_s\}, C_t, \sigma, k) \geq x\}.$$

If G_u has no (σ, k)-separated pair $(\{v_s\}, C_t)$ with $f(\{v_s\}, C_t, \sigma, k) \geq x$ for any number $k \in \mathbb{R}_w$, then let $h_3(G_u, x) = -\infty$. We finally define a function h_4 as follows: for a series-parallel graph G_u and a real number $x \in \mathbb{R}$,

$$h_4(G_u, x) = \begin{cases} 0 & \text{if } G_u \text{ has a } (\sigma, \sigma)\text{-separated pair } (\{v_s\}, \{v_t\}) \\ & \text{such that } f(\{v_s\}, \{v_t\}, \sigma, \sigma) \geq x; \\ -\infty & \text{otherwise.} \end{cases}$$

Our algorithm computes $g(G_u, x)$ and $h_i(G_u, x)$, $1 \leq i \leq 4$, for each node u of a binary decomposition tree T of a given series-parallel graph G from the leaves to the root r of T by means of a dynamic programming. Since $G = G_r$, one can compute the maximum fulfillment $f(G)$ of G from $g(G_r, x)$ and $h_i(G_r, x)$, $1 \leq i \leq 4$. However, due to the page limitation, we omit the details of our algorithm.

We now show that our algorithm takes time $O(F^2 n)$. Since all demands and the supply of vertices in a given series-parallel graph G are integers, $f(C_u)$ is an integer for any supplied set C_u for G_u. Similarly, $f(C, i)$ and $f(C_s, C_t, j, k)$ are integers for any i-connected set C and any (j, k)-separated pair (C_s, C_t), respectively. Then one can easily observe that it suffices to compute values $g(G_u, x)$ and $h_i(G_u, x)$, $1 \leq i \leq 4$, only for all integers x such that $0 \leq x \leq F$, because $f(G) \leq F = \sum_{v \in V} \text{dem}(v)$. We compute $g(G_u, x)$ and $h_i(G_u, x)$, $1 \leq i \leq 4$, for each internal node u of T from the counterparts of the two children of u in T. This is called *combining operation*, and can be done in time $O(F^2)$. Since T has at most $2n - 4$ internal nodes, the combining operation is executed no more than $2n$ times and hence one can compute $g(G, x)$ and $h_i(G, x)$, $1 \leq i \leq 4$, in time $O(F^2 n)$. Our algorithm thus takes time $O(F^2 n)$.

4   FPTAS

Assume in this section that the supply and all demands are positive real numbers which are not always integers. Since the maximum partition problem is MAXSNP-hard, there is no PTAS for the problem on general graphs unless P = NP. However, using the pseudo-polynomial-time algorithm in Section 3, we can obtain an FPTAS for series-parallel graphs having exactly one supply vertex, and have the following theorem.

Theorem 3. *There is a fully polynomial-time approximation scheme for the maximum partition problem on a series-parallel graph having exactly one supply vertex.*

We give an algorithm to find a supplied set C for a series-parallel graph G with $f(C) \geq (1 - \varepsilon)f(G)$ in time polynomial in n and $1/\varepsilon$ for any real number ε,

$0 < \varepsilon < 1$. Thus our approximate maximum fulfillent $\bar{f}(G)$ of G is $f(C)$, and hence the error is bounded by $\varepsilon f(G)$, that is,

$$f(G) - \bar{f}(G) = f(G) - f(C) \leq \varepsilon f(G). \qquad (2)$$

We now outline our algorithm and the analysis. We extend the ordinary "scaling and rounding" technique for the knapsack problem [3] and the maximum partition problem on trees [4] and apply it to the maximum partition problem for a series-parallel graph with a single supply vertex. For some scaling factor t, we consider the set $\{\cdots, -2t, -t, 0, t, 2t, \cdots\}$ as the range of functions g and h_i, $1 \leq i \leq 4$, and find the approximate solution $\bar{f}(G)$ by using the pseudo-polynomial-time algorithm in Section 3. Then we have

$$f(G) - \bar{f}(G) < 4nt. \qquad (3)$$

Intuitively, Eq. (3) holds because the combining operation is executed no more than $2n$ times and each combining operation adds at most $2t$ to the error $f(G) - \bar{f}(G)$. Let m_d be the maximum demand, that is, $m_d = \max\{\text{dem}(v) \mid v \in V_d\}$. Taking $t = \varepsilon m_d/(4n)$ and claiming $f(G) \geq m_d$, we have Eq. (2). One can observe that the algorithm takes time

$$O\left(\left(\left\lfloor \frac{F}{t} \right\rfloor + 1\right)^2 n\right) = O\left(\frac{n^5}{\varepsilon^2}\right),$$

because $F \leq nm_d$ and hence we have $F/t \leq 4n^2/\varepsilon$.

Acknowledgments

We thank MohammadTaghi Hajiaghayi for fruitful discussions.

References

1. S. Arnborg, J. Lagergren and D. Seese, Easy problems for tree-decomposable graphs, J. Algorithms, Vol. 12, pp. 308–340, 1991.
2. H. L. Bodlaender, Polynomial algorithms for graph isomorphism and chromatic index on partial k-trees, J. Algorithms, Vol. 11, pp. 631–643, 1990.
3. O. H. Ibarra and C. E. Kim, Fast approximation algorithms for the knapsack and sum of subset problems, J. ACM, Vol. 22, pp. 463–468, 1975.
4. T. Ito, X. Zhou and T. Nishizeki, Partitioning trees of supply and demand, International J. of Foundations of Computer Science, Vol. 16, pp. 803–827, 2005.
5. T. Ito, X. Zhou and T. Nishizeki, Partitioning graphs of supply and demand, Proc. of the 2005 IEEE Int'l Symposium on Circuits and Syst., pp. 160–163, 2005.
6. A. B. Morton and I. M. Y. Mareels, An efficient brute-force solution to the network reconfiguration problem, IEEE Trans. on Power Delivery, Vol. 15, pp. 996–1000, 2000.
7. C. H. Papadimitriou, Computational Complexity, Addison-Wesley, 1994.
8. C. H. Papadimitriou and M. Yannakakis, Optimization, approximation, and complexity classes, J. Computer and System Sciences, Vol. 43, pp. 425–440, 1991.
9. K. Takamizawa, T. Nishizeki and N. Saito, Linear-time computability of combinatorial problems on series-parallel graphs, J. ACM, Vol. 29, pp. 623–641, 1982.

Convex Grid Drawings of Plane Graphs with Rectangular Contours

Akira Kamada[1], Kazuyuki Miura[2], and Takao Nishizeki[1]

[1] Graduate School of Information Sciences, Tohoku University,
Sendai 980-8579, Japan
[2] Faculty of Symbiotic Systems Science, Fukushima University,
Fukushima 960-1296, Japan
kamada@nishizeki.ecei.tohoku.ac.jp, miura@sss.fukushima-u.ac.jp,
nishi@ecei.tohoku.ac.jp

Abstract. In a convex drawing of a plane graph, all edges are drawn as straight-line segments without any edge-intersection and all facial cycles are drawn as convex polygons. In a convex grid drawing, all vertices are put on grid points. A plane graph G has a convex drawing if and only if G is internally triconnected, and an internally triconnected plane graph G has a convex grid drawing on an $n \times n$ grid if G is triconnected or the triconnected component decomposition tree $T(G)$ of G has two or three leaves, where n is the number of vertices in G. In this paper, we show that an internally triconnected plane graph G has a convex grid drawing on a $2n \times n^2$ grid if $T(G)$ has exactly four leaves. We also present an algorithm to find such a drawing in linear time. Our convex grid drawing has a rectangular contour, while most of the known algorithms produce grid drawings having triangular contours.

1 Introduction

Recently automatic aesthetic drawing of graphs has created intense interest due to their broad applications, and as a consequence, a number of drawing methods have come out [11]. The most typical drawing of a plane graph is a *straight line drawing* in which all edges are drawn as straight line segments without any edge-intersection. A straight line drawing is called a *convex drawing* if every facial cycle is drawn as a convex polygon. One can find a convex drawing of a plane graph G in linear time if G has one [3,4,11].

A straight line drawing of a plane graph is called a *grid drawing* if all vertices are put on grid points of integer coordinates. This paper deals with a *convex grid drawing* of a plane graph. Throughout the paper we assume for simplicity that every vertex of a plane graph G has degree three or more, because the two edges incident to a vertex of degree two are often drawn on a straight line. Then G has a convex drawing if and only if G is "internally triconnected" [9]. One may thus assume without loss of generality that G is internally triconnected. If either G is triconnected [2] or the "triconnected component decomposition tree" $T(G)$ of G has two or three leaves [8], then G has a convex grid drawing on an $(n-1) \times (n-1)$

T. Asano (Ed.): ISAAC 2006, LNCS 4288, pp. 131–140, 2006.

grid and such a drawing can be found in linear time, where n is the number of vertices of G. However, it has not been known whether G has a convex grid drawing of polynomial size if $T(G)$ has four or more leaves. Figure 1(a) depicts an internally triconnected plane graph G, Fig. 2(b) the triconnected components of G, and Fig. 2(c) the triconnected component decomposition tree $T(G)$ of G, which has four leaves l_1, l_2, l_3 and l_4.

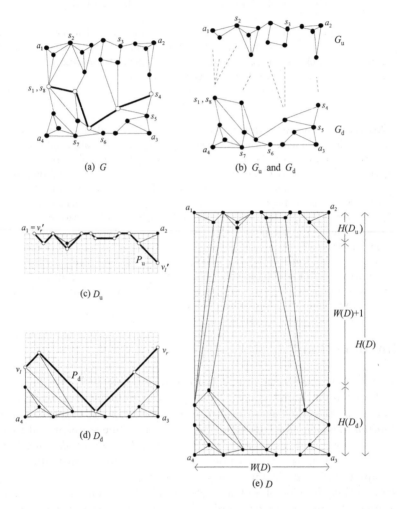

Fig. 1. (a) A plane graph G, (b) subgraphs G_u and G_d, (c) a drawing D_u of G_u, (d) a drawing D_d of G_d, and (e) a convex grid drawing D of G

In this paper, we show that an internally triconnected plane graph G has a convex grid drawing on a $2n \times n^2$ grid if $T(G)$ has exactly four leaves, and present an algorithm to find such a drawing in linear time. The algorithm is outlined as follows: we first divide a plane graph G into an upper subgraph G_u

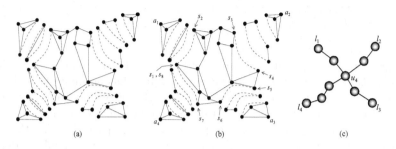

Fig. 2. (a) Split components of the graph G in Fig. 1(a), (b) triconnected components of G, and (c) a decomposition tree $T(G)$

and a lower subgraph G_d as illustrated in Fig. 1(b) for the graph in Fig. 1(a); we then construct "inner" convex grid drawings of G_u and G_d by a so-called shift method as illustrated in Figs. 1(c) and (d); we finally extend these two drawings to a convex grid drawing of G as illustrated in Fig. 1(e). This is the first algorithm that finds a convex grid drawing of such a plane graph G in a grid of polynomial size. Our convex grid drawing has a rectangular contour, while most of the previously known algorithms produce a grid drawing having a triangular contour [1,2,5,6,8,13].

2   Preliminaries

We denote by $W(D)$ the width of a minimum integer grid enclosing a grid drawing D of a graph, and by $H(D)$ the height of D. A plane graph G divides the plane into connected regions, called *faces*. The infinite face is called an *outer face*, and the others are called *inner faces*. The boundary of a face is called a *facial cycle*. We denote by $F_o(G)$ the outer facial cycle of G. A vertex on $F_o(G)$ is called an *outer vertex*, while a vertex not on $F_o(G)$ is called an *inner vertex*. In a convex drawing of a plane graph G, all facial cycles must be drawn as convex polygons. The convex polygonal drawing of $F_o(G)$ is called an *outer polygon*. We call a vertex of a polygon an *apex* in order to avoid the confusion with a vertex of a graph.

We call a pair $\{u, v\}$ of vertices in a biconnected graph G a *separation pair* if its removal from G results in a disconnected graph, that is, $G - \{u, v\}$ is not connected. A biconnected graph G is *triconnected* if G has no separation pair. A biconnected plane graph G is *internally triconnected* if, for any separation pair $\{u, v\}$ of G, both u and v are outer vertices and each connected component of $G - \{u, v\}$ contains an outer vertex. In other words, G is internally triconnected if and only if it can be extended to a triconnected graph by adding a vertex in the outer face and joining it to all outer vertices.

Let $G = (V, E)$ be a biconnected graph, and let $\{u, v\}$ be a separation pair of G. Then, G has two subgraphs $G'_1 = (V_1, E'_1)$ and $G'_2 = (V_2, E'_2)$ satisfying the following two conditions.

(a) $V = V_1 \bigcup V_2$, $V_1 \bigcap V_2 = \{u, v\}$; and
(b) $E = E_1' \bigcup E_2'$, $E_1' \bigcap E_2' = \emptyset$, $|E_1'| \geq 2$, $|E_2'| \geq 2$.

For a separation pair $\{u, v\}$ of G, $G_1 = (V_1, E_1' + (u, v))$ and $G_2 = (V_2, E_2' + (u, v))$ are called the *split graphs* of G with respect to $\{u, v\}$. The new edges (u, v) added to G_1 and G_2 are called the *virtual edges*. Even if G has no multiple edges, G_1 and G_2 may have. Dividing a graph G into two split graphs G_1 and G_2 is called *splitting*. Reassembling the two split graphs G_1 and G_2 into G is called *merging*. Merging is the inverse of splitting. Suppose that a graph G is split, the split graphs are split, and so on, until no more splits are possible, as illustrated in Fig. 2(a) for the graph in Fig. 1(a) where virtual edges are drawn by dotted lines. The graphs constructed in this way are called the *split components* of G. The split components are of three types: triconnected graphs; triple bonds (i.e. a set of three multiple edges); and triangles (i.e. a cycle of length three). The *triconnected components* of G are obtained from the split components of G by merging triple bonds into a bond and triangles into a ring, as far as possible, where a *bond* is a set of multiple edges and a *ring* is a cycle. Thus the triconnected components of G are of three types: (a) triconnected graphs; (b) bonds; and (c) rings. Two triangles in Fig. 2(a) are merged into a single ring, and hence the graph in Fig. 1(a) has ten triconnected components as illustrated in Fig. 2(b).

Let $T(G)$ be a tree such that each node corresponds to a triconnected component H_i of G and there is an edge (H_i, H_j), $i \neq j$, in $T(G)$ if and only if H_i and H_j are triconnected components with respect to the same separation pair, as illustrated in Fig. 2(c). We call $T(G)$ a *triconnected component decomposition tree* or simply a *decomposition tree* of G [7]. We denote by $\ell(G)$ the number of leaves of $T(G)$. Then $\ell(G) = 4$ for the graph G in Fig. 1(a). (See Fig. 2(c).) If G is triconnected, then $T(G)$ consists of a single isolated node and hence $\ell(G) = 1$.

The following two lemmas are known.

Lemma 1. *[9] Let G be a biconnected plane graph in which every vertex has degree three or more. Then the following three statements are equivalent to each other:*

(a) G has a convex drawing;
(b) G is internally triconnected; and
(c) both vertices of every separation pair are outer vertices, and a node of the decomposition tree $T(G)$ of G has degree two if it is a bond.

Lemma 2. *[9] If a plane graph G has a convex drawing D, then the number of apices of the outer polygon of D is no less than $\max\{3, \ell(G)\}$, and there is a convex drawing of G whose outer polygon has exactly $\max\{3, \ell(G)\}$ apices.*

Since G is an internally triconnected simple graph and every vertex of G has degree three or more, by Lemma 1 every leaf of $T(G)$ is a triconnected graph.

Lemmas 1 and 2 imply that if $T(G)$ has exactly four leaves then the outer polygon must have four or more apices. Our algorithm obtains a convex grid drawing of G whose outer polygon has exactly four apices and is a rectangle in particular, as illustrated in Fig. 1(e).

In Section 3, we will present an algorithm to draw the upper subgraph G_u and the lower subgraph G_d. The algorithm uses the following "canonical decomposition." Let $G = (V, E)$ be an internally triconnected plane graph, and let $V = \{v_1, v_2, \cdots, v_n\}$. Let v_1, v_2 and v_n be three arbitrary outer vertices appearing counterclockwise on $F_o(G)$ in this order. We may assume that v_1 and v_2 are consecutive on $F_o(G)$; otherwise, add a virtual edge (v_1, v_2) to the original graph, and let G be the resulting graph. Let $\Pi = (U_1, U_2, \cdots, U_m)$ be an ordered partition of V into nonempty subsets $U_1, U_2, \cdots, U_m$. We denote by G_k, $1 \leq k \leq m$, the subgraph of G induced by $U_1 \bigcup U_2 \bigcup \cdots \bigcup U_k$, and denote by $\overline{G_k}$, $0 \leq k \leq m - 1$, the subgraph of G induced by $U_{k+1} \bigcup U_{k+2} \bigcup \cdots \bigcup U_m$. We say that Π is a *canonical decomposition* of G (with respect to vertices v_1, v_2 and v_n) if the following three conditions (cd1)–(cd3) hold:

(cd1) $U_m = \{v_n\}$, and U_1 consists of all the vertices on the inner facial cycle containing edge (v_1, v_2).

(cd2) For each index k, $1 \leq k \leq m$, G_k is internally triconnected.

(cd3) For each index k, $2 \leq k \leq m$, all the vertices in U_k are outer vertices of G_k, and
    (a) if $|U_k| = 1$, then the vertex in U_k has two or more neighbors in G_{k-1} and has one or more neighbors in $\overline{G_k}$ when $k < m$; and
    (b) if $|U_k| \geq 2$, then each vertex in U_k has exactly two neighbors in G_k, and has one or more neighbors in $\overline{G_k}$.

A canonical decomposition $\Pi = (U_1, U_2, \cdots, U_{11})$ with respect to vertices v_1, v_2 and v_n of the graph in Fig. 3(a) is illustrated in Fig. 3(b).

3 Pentagonal Drawing

Let G be a plane graph having a canonical decomposition $\Pi = (U_1, U_2, \cdots, U_m)$ with respect to vertices v_1, v_2 and v_n, as illustrated in Figs. 3(a) and (b). In this section, we present a linear-time algorithm, called the *pentagonal drawing algorithm*, to find a convex grid drawing of G with a pentagonal outer polygon, as illustrated in Fig. 3(d). The algorithm is based on the so-called shift methods given by Chrobak and Kant [2] and de Fraysseix *et al.* [5], and will be used by our convex grid drawing algorithm in Section 4 to draw G_u and G_d.

Let v_l be an arbitrary vertex on the path going from v_1 to v_n clockwise on $F_o(G)$, and let $v_r (\neq v_l)$ be an arbitrary vertex on the path going from v_2 to v_n counterclockwise on $F_o(G)$, as illustrated in Fig. 3(a). Let V_l be the set of all vertices on the path going from v_1 to v_l clockwise on $F_o(G)$, and let V_r be the set of all vertices on the path going from v_2 to v_r counterclockwise on $F_o(G)$. Our

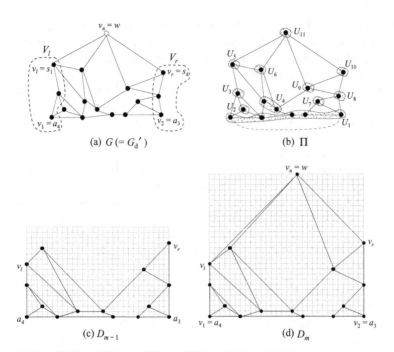

Fig. 3. (a) An internally triconnected plane graph $G(=G'_d)$, (b) a canonical decomposition Π of G, (c) a drawing D_{m-1} of G_{m-1}, and (d) a pentagonal drawing D_m of G

pentagonal drawing algorithm obtains a convex grid drawing of G whose outer polygon is a pentagon with apices v_1, v_2, v_r, v_n and v_l, as illustrated in Fig. 3(d).

We first obtain a drawing D_1 of the subgraph G_1 of G induced by all vertices of U_1. Let $F_o(G_1) = w_1, w_2, \cdots, w_t$, $w_1 = v_1$, and $w_t = v_2$. We draw G_1 as illustrated in Fig. 4, depending on whether (v_1, v_2) is a real edge or not, $w_2 \in V_l$ or not, and $w_{t-1} \in V_r$ or not.

We then extend a drawing D_{k-1} of G_{k-1} to a drawing D_k of G_k for each index k, $2 \le k \le m$. Let $F_o(G_{k-1}) = w_1, w_2, \cdots, w_t$, $w_1 = v_1$, $w_t = v_2$, and $U_k = \{u_1, u_2, \cdots, u_r\}$. Let w_f be the vertex with the maximum index f among all the vertices w_i, $1 \le i \le t$, on $F_o(G_{k-1})$ that are contained in V_l. Let w_g be the vertex with the minimum index g among all the vertices w_i that are contained in V_r. Of course, $1 \le f < g \le t$. We denote by $\angle w_i$ the interior angle of apex w_i of the outer polygon of D_{k-1}. We call w_i a *convex apex* of the

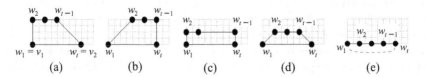

Fig. 4. Drawings D_1 of G_1

polygon if $\angle w_i < \pi$. Assume that a drawing D_{k-1} of G_{k-1} satisfies the following six conditions (sh1)–(sh6). Indeed D_1 satisfies them.

(sh1) w_1 is on the grid point $(0,0)$, and w_t is on the grid point $(2|V(G_{k-1})| - 2, 0)$.

(sh2) $x(w_1) = x(w_2) = \cdots = x(w_f)$, $x(w_f) < x(w_{f+1}) < \cdots < x(w_g)$, $x(w_g) = x(w_{g+1}) = \cdots = x(w_t)$, where $x(w_i)$ is the x-coordinate of w_i.

(sh3) Every edge (w_i, w_{i+1}), $f \le i \le g - 1$, has slope $-1, 0$, or 1.

(sh4) The Manhattan distance between any two grid points w_i and w_j, $f \le i < j \le g$, is an even number.

(sh5) Every inner face of G_{k-1} is drawn as a convex polygon.

(sh6) Vertex w_i, $f + 1 \le i \le g - 1$, has one or more neighbors in $\overline{G_{k-1}}$ if w_i is a convex apex.

We extend D_{k-1} to D_k, $2 \le k \le m$, so that D_k satisfies conditions (sh1)–(sh6). Let w_p be the leftmost neighbor of u_1, that is, w_p is the neighbor of u_1 in G_k having the smallest index p, and let w_q be the rightmost neighbor of u_r. Before installing U_k to D_{k-1}, we first shift $w_1, w_2, \cdots, w_p$ of G_{k-1} and some inner vertices of G_k to the left by $|U_k|$, and then shift $w_q, w_{q+1}, \cdots, w_t$ of G_{k-1} and some inner vertices of G_k to the right by $|U_k|$. After the operation, we shift all vertices of G_{k-1} to the right by $|U_k|$ so that w_1 is on the grid point $(0,0)$.

Clearly $W(D_1) = 2|V(G_1)| - 2$ and $H(D_1) \le 4$. One can observe that $W(D_k) = 2|V(G_k)| - 2$ and $H(D_k) \le H(D_{k-1}) + W(D_k)$ for each k, $2 \le k \le m$. We thus have the following lemma.

Lemma 3. *For a plane graph G having a canonical decomposition $\Pi = (U_1, U_2, \cdots, U_m)$ with respect to v_1, v_2 and v_n, the pentagonal drawing algorithm obtains a convex grid drawing of G on a $W \times H$ grid with $W = 2n - 2$ and $H \le n^2 - n - 2$ in linear time. Furthermore, $W(D_{m-1}) = 2(|V(G_{m-1})|) - 2$ and $H(D_{m-1}) \le |V(G_{m-1})|^2 - |V(G_{m-1})| - 2$.*

4 Convex Grid Drawing Algorithm

In this section we present a linear algorithm to find a convex grid drawing D of an internally triconnected plane graph G whose decomposition tree $T(G)$ has exactly four leaves. Such a graph G does not have a canonical decomposition, and hence none of the algorithms in [1], [2], [6], [8] and Section 3 can find a convex grid drawing of G.

Division. We first explain how to divide G into G_u and G_d. (See Figs. 1(a) and (b).) One may assume that the four leaves l_1, l_2, l_3 and l_4 of $T(G)$ appear clockwise in $T(G)$ in this order. Clearly, either exactly one node u_4 of $T(G)$ has degree four and each of the other non-leaf nodes has degree two as illustrated in Fig. 2(c), or two nodes have degree three and each of the other non-leaf nodes has degree two. In this extended abstract, we consider only the former case. Since each vertex of G is assumed to have degree three or more, all the four leaves of $T(G)$ are triconnected graphs. Moreover, according to Lemma 1, every bond has degree two in $T(G)$. Therefore, node u_4 is either a triconnected graph or a ring. We assume in this extended abstract that u_4 is a triconnected graph as in Fig. 2.

As the four apices of the rectangular contour of G, we choose four outer vertices a_i, $1 \leq i \leq 4$, of G; let a_i be an arbitrary outer vertex in the component l_i that is not a vertex of the separation pair of the component. The four vertices a_1, a_2, a_3 and a_4 appear clockwise on $F_o(G)$ in this order as illustrated in Fig. 1(a).

We then choose eight vertices $s_1, s_2, \cdots, s_8$ from the outer vertices of the component u_4. Among these outer vertices, let s_1 be the vertex that one encounters first when one traverses $F_o(G)$ counterclockwise from the vertex a_1, and let s_2 be the vertex that one encounters first when one traverses $F_o(G)$ clockwise from a_1, as illustrated in Fig. 1(a). Similarly, we choose s_3 and s_4 for a_2, s_5 and s_6 for a_3, and s_7 and s_8 for a_4.

We then show how to divide G into G_u and G_d. Split G for separation pairs $\{s_1, s_2\}$ and $\{s_3, s_4\}$ as far as possible, and let G' be the resulting split graph containing vertices a_3 and a_4. Then, G' is internally triconnected, and $T(G')$ has exactly two leaves. Consider all the inner faces of G' that contain one or more vertices on the path going from s_2 to s_3 clockwise on $F_o(G')$. Let G'' be the subgraph of G' induced by the vertices on these faces. Then $F_o(G'')$ is a simple cycle. Clearly, $F_o(G'')$ contains vertices s_1 and s_4. Let P be the path going from s_1 to s_4 counterclockwise on $F_o(G'')$. (P is drawn by thick lines in Fig. 1(a).)

Let G_d be the subgraph of G induced by all the vertices on or below P, and let G_u be the subgraph of G obtained by deleting all vertices in G_d as illustrated in Fig. 1(b). Let n_d be the number of vertices of G_d, and let n_u be the number of vertices of G_u. Then $n_d + n_u = n$.

Drawing G_d. We now explain how to draw G_d. Let G'_d be a graph obtained from G by contracting all the vertices of G_u to a single vertex w, as illustrated in Fig. 3(a) for the graph G in Fig. 1(a)D One can prove that the plane graph G'_d is internally triconnected.

The decomposition tree $T(G'_d)$ of G'_d has exactly two leaves, and a_3 and a_4 are contained in the triconnected graphs corresponding to the leaves and are not vertices of the separation pairs. Every vertex of G'_d other than w has degree three or more, and w has degree two or more in G'_d. Therefore, G'_d has a canonical decomposition $\Pi = (U_1, U_2, \cdots, U_m)$ with respect to a_4, a_3 and w, as illustrated in Fig. 3(b), where $U_m = \{w\}$. Let v_l be the vertex preceding w clockwise on the outer face $F_o(G'_d)$, and let v_r be the vertex succeeding w, as illustrated in Fig. 3(a). We obtain a pentagonal drawing D_m of G'_d by the algorithm in Section 3, as illustrated in Fig. 3(d). The drawing D_{m-1} of G_{m-1} induced by $U_1 \bigcup U_2 \bigcup \cdots \bigcup U_{m-1}$ is our drawing D_d of $G_d(= G_{m-1})$. (See Figs. 1(d) and 3(c).) By Lemma 3, we have $W(D_d) = 2n_d - 2$ and $H(D_d) \leq n_d^2 - n_d - 2$.

Drawing G_u. We now explain how to draw G_u. Let G'_u be a graph obtained from G by contracting all the vertices of G_d to a single vertex w'. Similarly to

G'_d, G'_u has a canonical decomposition $\Pi = (U_1, U_2, \cdots, U_m)$ with respect to a_2, a_1 and w'. Let v'_r be the vertex succeeding w' clockwise on the outer face $F_o(G'_u)$, and let v'_l be the vertex preceding w'. We then obtain a drawing D_{m-1} of $G_u(= G_{m-1})$ by the algorithm in Section 3, as illustrated in Fig. 1(c). By Lemma 3, we have $W(D_u) = 2n_u - 2$ and $H(D_u) \leq n_u^2 - n_u - 2$.

Drawing of G. If $W(D_d) \neq W(D_u)$, then we widen the narrow one of D_d and D_u by the shift method in Section 3. We may thus assume that $W(D_d) = W(D_u) = \max\{2n_d - 2, 2n_u - 2\}$. Since we combine the two drawings D_d and D_u of the same width to a drawing D of G, we have

$$W(D) = \max\{2n_d - 2, 2n_u - 2\} < 2n.$$

We arrange D_d and D_u so that $y(a_3) = y(a_4) = 0$ and $y(a_1) = y(a_2) = H(D_d) + H(D_u) + W(D) + 1$, as illustrated in Fig. 1(e).

Noting that $n_d + n_u = n$ and $n_d, n_u \geq 5$, we have

$$\begin{aligned}
H(D) &= H(D_d) + H(D_u) + W(D) + 1 \\
&< (n_d^2 - n_d - 2) + (n_u^2 - n_u - 2) + 2n + 1 \\
&< n^2.
\end{aligned}$$

We finally draw, by straight line segments, all the edges of G that are contained in neither G_u nor G_d. This completes the grid drawing D of G. (see Fig. 1(e).)

Validity of drawing algorithm. In this section, we show that the drawing D obtained above is a convex grid drawing of G. Since both D_d and D_u satisfy condition (sh5), every inner facial cycle of G_d and G_u is drawn as a convex polygon in D. Therefore, it suffices to show that the straight line drawings of the edges not contained in G_u and G_d do not introduce any edge-intersection and that all the faces newly created by these edges are convex polygons.

Since D_d satisfies condition (sh3), the absolute value of the slope of every edge on the path P_d going from v_l to v_r clockwise on $F_o(G_d)$ is at most 1. The path P_d is drawn by thick lines in Fig. 1(d). Similarly, the absolute value of the slope of every edge on the path P_u going from v'_r to v'_l counterclockwise on $F_o(G_u)$ is at most 1. Since $H(D) = H(D_d) + H(D_u) + W(D) + 1$, the absolute value of the slope of every straight line segment that connects a vertex in G_u and a vertex in G_d is larger than 1. Therefore, all the outer vertices of G_d on P_d are visible from all the outer vertices of G_u on P_u. Furthermore, G is a plane graph. Thus the addition of all the edges not contained in G_u and G_d does not introduce any edge-intersection.

Since D_d satisfies condition (sh6), every convex apex of the outer polygon of G_d on P_d has one or more neighbors in G_u. Similarly, every convex apex of the outer polygon of G_u on P_u has one or more neighbors in G_d. Therefore, every interior angle of a newly formed face is smaller than 180°. Thus all the inner faces of G not contained in G_u and G_d are convex polygons in D.

Thus, D is a convex grid drawing of G. Clearly the algorithm takes linear time. We thus have the following main theorem.

Theorem 1. *Assume that G is an internally triconnected plane graph, every vertex of G has degree three or more, and the triconnected component decomposition tree $T(G)$ has exactly four leaves. Then our algorithm finds a convex grid drawing of G on a $2n \times n^2$ grid in linear time.*

We finally remark that the grid size is improved to $2n \times 4n$ for the case where either the node u_4 of degree four in $T(G)$ is a ring or $T(G)$ has two nodes of degree three.

References

1. N. Bonichon, S. Felsner and M. Mosbah, *Convex drawings of 3-connected plane graphs -Extended Abstract-*, Proc. of GD 2004, LNCS 3383, pp. 60–70, 2005.
2. M. Chrobak and G. Kant, *Convex grid drawings of 3-connected planar graphs*, International Journal of Computational Geometry and Applications, 7, pp. 211–223, 1997.
3. N. Chiba, K. Onoguchi and T. Nishizeki, *Drawing planar graphs nicely*, Acta Inform., 22, pp. 187–201, 1985.
4. N. Chiba, T. Yamanouchi and T. Nishizeki, *Linear algorithms for convex drawings of planar graphs*, in Progress in Graph Theory, J. A. Bondy and U. S. R. Murty (Eds.), Academic Press, pp. 153–173, 1984.
5. H. de Fraysseix, J. Pach and R. Pollack, *How to draw a planar graph on a grid*, Combinatorica, 10, pp. 41–51, 1990.
6. S. Felsner, *Convex drawings of plane graphs and the order of dimension of 3-polytopes*, Order, 18, pp. 19–37, 2001.
7. J. E. Hopcroft and R. E. Tarjan, *Dividing a graph into triconnected components*, SIAM J. Compt.2, 3, pp. 135–138, 1973.
8. K. Miura, M. Azuma and T. Nishizeki, *Canonical decomposition, realizer, Schnyder labeling and orderly spanning trees of plane graphs*, International Journal of Foundations of Computer Science, 16, 1, pp. 117–141, 2005.
9. K. Miura, M. Azuma and T. Nishizeki, *Convex drawings of plane graphs of minimum outer apices*, Proc. of GD 2005, LNCS 3843, pp. 297–308, 2005, also to appear in International Journal of Foundations of Computer Science.
10. K. Miura, S. Nakano and T. Nishizeki, *Convex grid drawings of four-connected plane graphs*, Proc. of ISAAC 2000, LNCS 1969, pp. 254–265, 2000, also to appear in International Journal of Foundations of Computer Science.
11. T. Nishizeki and M. S. Rahman, *Planar Graph Drawing*, World Scientific, Singapore, 2004.
12. S. Nakano, M. S. Rahman and T. Nishizeki, *A linear time algorithm for four partitioning four-connected planar graphs*, Information Processing Letters, 62, pp. 315–322, 1997.
13. W. Schnyder and W. Trotter, *Convex drawings of planar graphs*, Abstracts of the AMS 13, 5, 92T-05-135, 1992.

Algorithms on Graphs with Small Dominating Targets

Divesh Aggarwal, Chandan K. Dubey, and Shashank K. Mehta*

Indian Institute of Technology, Kanpur - 208016, India
{divesh, cdubey, skmehta}@iitk.ac.in

Abstract. A dominating target of a graph $G = (V, E)$ is a set of vertices T s.t. for all $W \subseteq V$, if $T \subseteq W$ and induced subgraph on W is connected, then W is a dominating set of G. The size of the smallest dominating target is called dominating target number of the graph, $dt(G)$. We provide polynomial time algorithms for *minimum connected dominating set*, *Steiner set*, and *Steiner connected dominating set* in dominating-pair graphs (i.e., $dt(G) = 2$). We also give approximation algorithm for *minimum connected dominating set* with performance ratio 2 on graphs with small dominating targets. This is a significant improvement on *appx* $\leq d(opt + 2)$ given by Fomin et.al. [2004] on graphs with small d-octopus.

Classification: Dominating target, d-octopus, Dominating set, Dominating-pair graph, Steiner tree.

1 Introduction

Let $G = (V, E)$ be a simple (no loops, no multiple edges) undirected graph. For a subset $Y \subseteq V$, $G(Y)$ will denote the induced subgraph of G on vertex set Y i.e. $G(Y) = (Y, \{(x, y) \in E : x, y \in Y\})$. Since we will only deal with *induced* subgraphs in this paper, some times only Y may be used to denote $G(Y)$. For a vertex $x \in V$, *open neighborhood* denoted by $N(x)$ is given by $\{y \in V : (x, y) \in E\}$. The *closed neighborhood* is defined by $N[x] = N(x) \cup \{x\}$. Similarly, the closed and the open neighborhoods of a set $S \subset V$ are defined by $N[S] = \cup_{x \in S} N[x]$ and $N(S) = N[S] - S$ respectively. A vertex set S_1 is said to *dominate* another set S_2 if $S_2 \subseteq N[S_1]$. If $N[S_1] = V$, then S_1 is said to dominate G.

We address four closely related domination and connectivity problems on undirected graphs; minimum connected dominating set (MCDS), Steiner connected dominating set (SCDS), Steiner set (SS), and Steiner tree (ST), each is known to be NP-complete [1978]. Steiner set problem finds application in VLSI routing [1995], wire length estimation [1998a], and network routing [1990]. Minimum con- nected dominating set and Steiner connected dominating set problems have recently received attention due to their applications in wireless routing in ad hoc networks [2003a].

---

* Partly supported by Ministry of Human Resource Development, Government of India under grant no. MHRD/CD/2003/0320.

Many interesting graph classes such as permutation graphs, interval graphs, AT-free graphs [1997a, 1972, 1962, 1999] have a pair of vertices with a property that any path connecting them is a dominating set for the graph. This pair is called a *dominating pair* of the graph. The concept of *Dominating target* was introduced by Kloks et. al. [2001] as a generalization of the dominating pair. Any vertex set T in a graph $G = (V, E)$ is said to be a dominating target of G if the following property is satisfied: for every $W \subseteq V$, if $G(W)$ is connected and $T \subseteq W$, then W dominates V. The cardinality of the smallest dominating target is called the *dominating target number* of the graph G and it is denoted by $dt(G)$. The family of graphs with $dt(G) = 2$ are known as dominating-pair (DP) graphs and their dominating target is referred as dominating-pair. Minimum connected dominating set and Steiner set problems are polynomially solvable on the family of AT-free graphs [1993], which is a subclass of DP. We will present here efficient algorithms for MCDS, SS, and SCDS on dominating-pair graphs.

A relevant parameter to the current work is d-octopus, considered by by Fomin et. al. [2004]. A d-octopus of a graph is a subgraph $T = (W, F)$ of G s.t. W is a dominating set of G, and T is the union of d (not necessarily disjoint) shortest paths of G that have one endpoint in common. It is conjectured that $dt(G) \leq d$, where the graph has a d-octopus, [2004]. Let *opt* be the optimal solution of MCDS problem and *appx* be its approximation due to the algorithm by Fomin et.al., then $appx \leq d(opt + 2)$. The complexity of this algorithm is $O(|V|^{3d+3})$. We will present an $O(|V|^{dt(G)+1})$ approximation algorithm for MCDS with performance ratio 2, which is an improvement both in terms of complexity (assuming the conjecture) and approximation factor (for an introduction on approximation algorithms see [2003, 1992]).

2   Problem Definitions

In this paper we discuss the problem of computing following.

Minimum Connected Dominating Set (MCDS) Given a graph $G = (V, E)$, vertex set C is a *connected dominating set* (CDS) if $V = N[C]$ and $G(C)$ is connected. MCDS is a smallest cardinality CDS.

Steiner Connected Dominating Set (SCDS) Given a graph $G = (V, E)$ and a set $R \subseteq V$ of required vertices, vertex set C is a *connected $|R|$-dominating set* (R-CDS) if $R \subseteq N[C]$ and $G(C)$ is connected. SCDS of R is a smallest cardinality R-CDS.

Steiner Set (SS) Given a graph $G = (V, E)$ and a set $R \subseteq V$ of required vertices, vertex set S is an *R-connecting set* (R-CS) if $G(S \cup R)$ is connected. SS of R is a smallest cardinality R-CS.

Steiner Tree (ST) Given an edged-weighted graph $G = (V, E, w)$ (w is the edge-weight function) and a set $R \subseteq V$ of required vertices, a tree T is an *R-spanning tree* (R-SPN) if it contains all R-vertices. ST of R is a minimum weight (sum of the weights of the edges) R-SPN.

Note that Steiner set problem is equivalent to Steiner tree problem when the edge weights are taken to be 1; and MCDS is an instance of SCDS when R is the entire V.

3   Exact Algorithms on Dominating Pair Graphs

3.1   Minimum Connected Dominating Set

Let (u, v) be a dominating pair of the graph $G = (V, E)$ and $X = N[u]$ and $Y = N[v]$. For each $x \in X$ define $A_x = \{a : (a, x) \in E$ and $\{a, x\}$ dominates $X\}$. Define B_y in a similar way for each $y \in Y$. Now let Γ be as follows. Here $x \in X$, $y \in Y$, and $\alpha \ldots \beta$ denote a shortest path between α and β.

$$\Gamma = \{P | P = u \ldots v, \text{ or } u \ldots by, \text{ for } b \in B_y \text{ or}$$
$$xa \ldots v, \text{ for } a \in A_x \text{ or } xa \ldots by, \text{ for } a \in A_y \text{ and } b \in B_y\}$$

Balakrishnan et. al. [1993] have given $O(|V|^3)$ algorithms to compute MCDS and SS in AT-free graphs. They claim that the smallest cardinality path in Γ is a MCDS of the graph. Although the authors address the problem of MCDS in AT-free graphs, they do not use any property of this class other than the existence of a dominating pair. Contrary to our expectation, the algorithm does not work on all dominating pair (DP) graphs. In the graph of Figure 1 $\{x_1, x_2, x_5, x_6\}$ is an MCDS but no MCDS of size 4 is computable by their algorithm (no CDS of size 4 is in Γ).

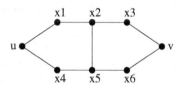

Fig. 1. A DP graph where Balakrishnan et.al. algorithm fails

Theorem 1. *Let $G = (V, E)$ be a dominating pair graph and $\{u, v\}$ any dominating pair with distance greater than 4. Then the shortest paths in Γ are MCDS of G.*

Proof. We show that if S is an MCDS then it can be transformed into another MCDS S' which belongs to Γ.

Case 1. $u \in S, v \in S$. In this case S must be a shortest path connecting u and v, which is already in Γ.

Case 2. $u \in S, v \notin S$ or $u \notin S, v \in S$. We consider the first situation only. There must exist a $y \in S \cap N(v)$. As S is connected, let P be a path from u to y contained in S. If $|S| - |P| \geq 1$ then $S' = P \cup \{v\}$ is the required MCDS in Γ.

So, assume that $S = P$. Let b be the vertex in P connected to y. If $b \in B_y$ then we are done. Else there must exist a $y' \in Y$ not dominated by $\{b, y\}$. As S is a MCDS, there must exist a $b' \in P$ s.t. $(b', y') \in E$. Then $S' = S \cup \{y', v\} - \{b, y\}$ is the required path in Γ.

Case 3. $u \notin S$, $v \notin S$. Therefore there exist S-vertices x and y such that $x \in X$ and $y \in Y$. Since S is connected there exists a path from x to y in S, say P. $P \cup \{u, v\}$ is a path connecting u and v so it must dominate entire graph. Therefore P must dominate $V - X - Y$. Further, the condition $d(u, v) > 4$ ensures that vertices that dominate any part of X are mutually exclusive from the vertices dominating any part of Y. We consider three cases.

$|S| - |P| \geq 2$ Here $S' = P \cup \{u, v\}$ is obviously in Γ.

$|S| - |P| = 1$ Let $S - P = \{p\}$. Now p must dominate either parts of X or parts of Y but not both. Without loss of generality assume that p dominates parts of X. So P must be dominating $V - X$. Thus $S' = S \cup \{u\} - \{p\}$, which is obviously connected, dominates entire V and $|S'| = |S|$. From Case 2 we know that there is a path $Q \in \Gamma$ such that it dominates V and $|Q| = |S'| = |S|$.

$|S| = |P|$ If the vertex a adjacent to x in P is in A_x and the vertex b adjacent to y in P is in B_y, then P is in Γ.

Next assume that vertex a adjacent to x in P is not in A_x or b adjacent to y in P is not in A_y. Without loss of generality assume the former. Then there must exist $x' \in X$ which is not dominated by $\{a, x\}$. Since both a and x dominate parts of X, they do not dominate any part of Y. Thus $P - \{x, a\}$ dominates Y. Let $S' = P \cup \{u, x'\} - \{x, a\}$. Clearly $S' \cup \{v\}$ is connected so it must dominate V. But $P - \{x, a\}$ dominates V so S' also dominates entire V. From Case 2 we know that there is a path $Q \in \Gamma$ such that it dominates V and $|Q| = |S'|$. But by construction $|S'| = |S|$ so $|Q| = |S|$. $\quad\square$

If $d(u, v) > 4$ then compute Γ and output the smallest path. In case $d(u, v) \leq 4$, then either a shortest path connecting u to v will be an MCDS or there exists an MCDS of size at most 4. This leads to an $O(|V|^5)$ algorithm to calculate an MCDS in DP graphs.

3.2   Steiner Set

Let $G = (V, E)$ be a graph and R a subset of its vertices. Define an edge-weighted graph $G_w(V, E, w)$ where $w(e) = 1$ if both vertices of the edge e are in $V - R$; $1/2$ if one vertex is in $V - R$; 0 if neither is in $V - R$. Define a function L over the paths of G as follows. Let P be a path of G and $length(P)$ denotes its length in G_w, then $L(P) = length(P) + 1$ if both end vertices of P are in $V - R$; $length(P) + 1/2$ is one end vertex of P is in $V - R$; $length(P)$ if neither end-vertex is in $V - R$. Observe that $L(P)$ is the number of $V - R$-vertices in P.

In describing the algorithm to compute Steiner set for a required set R in a dominating-pair graph, we will first assume that R is an independent set (no two R-vertices are adjacent). The general case will be shown to reduce into this case in linear time.

Theorem 2. *Let $G = (V, E)$ be a dominating-pair graph and R be an independent set of vertices in it. Then there exists a pair of vertices $u, v \in V$ such that for every minimum-L path P between u and v, $P - R$ is a Steiner set of R in G.*

Proof. Let S be a Steiner set for R in G. First we will assume that $|S| > 3$. The case of $|S| \leq 3$ will be handled by simple search. Let u', v' be a dominating pair of G. Let $P_1 = u'...u''u''' \equiv P_1'u''u'''$ be a G-shortest path from u' to the connected set $S \cup R$. Similarly let $P_2 = v'...v''v''' \equiv P_2'v''v'''$ be a G-shortest path from v' to $S \cup R$. Then u''', v''' are in $S \cup R$; $P_1 - \{u'''\}$ and $P_2 - \{v'''\}$ are outside $S \cup R$; and no vertex of P_1' or of P_2' dominates any R vertex. Observe that every path X connecting u'' and v'' dominates entire R because $P_1'.X.P_2'$ dominates entire graph. Let $u'''x_1x_2...x_{k-1}x_kv'''$ be a shortest path in $G(S \cup R)$. From the above observation $u''u'''x_1...x_kv'''v''$ dominates all the R vertices. For the convenience we will also label u''' and v''' with x_0 and x_{k+1} respectively.

Suppose there is an S-vertex s not in $\{x_i\}_{i \in [k+1]}$. Since a Steiner set is minimum, it must be dominating some R vertex which is not dominated by any x_i. Thus it must be dominated by u'' or v''. Let S' be the set of S-vertices outside $\{x_i\}_{i \in [k+1]}$. Define $S_1 = \{s \in S' : N[s] \cap R \cap N[u''] \neq \emptyset\}$ and $S_2 = \{s \in S' : N[s] \cap R \cap N[v''] \neq \emptyset\}$. From the above observation $S_1 \cup S_2 = S'$. We will show that $S_1 \cap S_2 = \emptyset$. Assume otherwise. Let $s \in S'$ such that $r_1 \in N[u''] \cap R \cap N[s]$ and $r_2 \in N[v''] \cap R \cap N[s]$. So $u''r_1sr_2v''$ is a path. From the earlier observation it dominates entire R. Thus $\{u'', s, v''\}$ is a Steiner set, but it contradicts an earlier assumption that SS has more than 3 vertices.

All paths connecting u'' to v'' dominate all R-vertices and minimum-L paths among them have L value at most $S - |S'| + 2$ because $L(u''x_0x_1...x_{k+1}v'') = |S| - |S'| + 2$. Using the path $P_3'' = u''x_0x_1...x_{k+1}v''$ we will find a pair of vertices u, v such that all paths connecting these vertices dominate R and among them minimum-L paths have $|S|$ non-R-vertices. We achieve this in two steps First we modify the u''-end of P_3'' and find u. Then work on the other end.

Case 1. $S_1 = \emptyset$. Starting from x_0, let x_{i_0} be the first S-vertex on the path $x_0, x_1, ..x_{k+1}$.

Claim. Either $N[u''] \cap R \subseteq N[i_0] \cap R$ or there is an index $j > i_0$ such that $u''rx_j...x_{k+1}v''$ is a path which dominates all R-vertices and $L(u''rx_jx_{j+1}...x_{k+1}v'')$ $\leq L(x_0x_1...x_{k+1}v'')$, where r is an R vertex.

Proof of the claim suppose u'' dominates an R vertex r which is not dominated by x_{i_0}. At least one S vertex must dominate it so let it be x_j. Consider the path $u'...u''rx_j...x_{k+1}v''...v'$. It dominates the graph so the subpath $u''rx_j...x_{k+1}v''$ must dominate all R-vertices. Further the number of non-R-vertices in this path cannot exceed that of $x_0...x_{k+1}v''$ because while the former has only one new vertex, it does not have x_{i_0}, an S vertex, which is present in the latter. **endproof**

Let $u = x_{i_0}$ if $N[u''] \cap R \subseteq N[x_{i_0}] \cap R$ else define $u = u''$. Let P_3' be the path $x_{i_0}x_{i_0+1}...x_{k+1}v''$ in the former case and $u''rx_jx_{j+1}...x_{k+1}v''$ in the latter case. Observe that in either case P_3' dominates all R-vertices (in former case there is at most one R-vertex between u'' and x_{i_0} and R-vertices do not dominate other R-vertices) and the number of non-R-vertices on it are no more than those in $x_0...x_{k+1}v''$, which is $|S| - |S_2| + 1$.

In addition, every path connecting u to v'' must dominate all R-vertices as the following reasoning shows. The case of $u = u''$ is already established. In case $u = x_{i_0}$, pad the path at the left with $P_1'u''x_0...x_{i_0-1}$ and to the right with P_2'. This path dominates the graph. P_2' does not dominate any R-vertex and $P_1'u''x_0...x_{i_0-1}$ does not dominate any R-vertex which is not already dominated by x_{i_0}. Since one path between u and v'', namely P_3', has L value $|S| - |S_2| + 1$, the minimum-L paths between these vertices have at most $|S| - |S_2| + 1$ non-R-vertices.

Case 2. $S_1 \neq \emptyset$. Then $P_3' = u''x_0...x_{k+1}v''$ has at most $|S| - |S_2| + 1$ non-R-vertices. Define $u = u''$. All path between u and v'' dominate entire R, because $u = u''$. The minimum L paths among them cannot have more than $|S| - |S_2| + 1$ non-R vertices since $L(P_3') = |S| - |S_2| + 1$.

Together these cases imply that there exists a vertex u such that all path between u and v'' dominate entire R and the minimum-L path among them have L value at most $|S| - |S_2| + 1$.

This completes the computation of u. To determine v we repeat the argument from the other end. Let x_{j_0} be the first S vertex on the path $x_{k+1}x_k...$ starting from x_{k+1}. Then $v = v''$ if S_2 is non-empty or if $N[v''] \cap R$ is not contained in $N[x_{j_0}] \cap R$. Otherwise $v = x_{j_0}$. Repeating the argument given above we see that all paths between u and v dominate all R-vertices and there is at least one path between these vertices with at most $|S|$ non-R-vertices. Therefore we conclude that all minimum-L path between u and v have at most $|S|$ non-R-vertices. $\square$

The algorithm to compute the Steiner set is as follows.

Data: A DP graph $G = (V, E)$ and a set $R \subseteq V$.
Result: A Steiner set for R.
1 For each set of at most 3 vertices check if it forms an R-connecting set. If any such set is found, then output the smallest of these sets;
2 Otherwise compute all-pair shortest paths on G_w. Compute the set Γ as the collection of those G_w-shortest paths that dominate R. Select a path P from Γ with minimum L-value. Output $P - R$.

Algorithm 1. Steiner set algorithm for independent set R in DP graphs

The time complexity of the first step is $O(|V|^3.(|E| + |V|))$. The cost of the second step is $O(|V|^3 + |V|^2.|E|)$ Hence the overall complexity is $|V|^3(|E| + |V|)$.

This completes the discussion for independent R case. The general case is easily reduced to this case. Let $G = (V, E)$ be a dominating pair graph and R be the required set of vertices. Shrink each connected components of $G(R)$ into a vertex. Then the resulting graph G' is also a dominating pair graph (if u, v is a dominating pair of G and u and v merge into u' and v' respectively after shrinking, then u', v' is a dominating pair of G'). Also the new required vertex set R' is an independent set in G' and each Steiner set for R' in G' is a Steiner set of R in G and its converse is also true.

3.3 Steiner Connected Dominating Set

Definition 1. *Let G be a graph and R be a subset of its vertices. A subset of vertices D_R is called R-dominating target if every connected subgraph of G containing D_R dominates R. In addition, if each vertex of D_R has some R vertex in its closed neighborhood, then we call it an essential-R-dominating-target.*

Lemma 1. *For any R there exists essential R-dominating target with cardinality at most $dt(G)$.*

Proof. We present a constructive proof. Let $D = \{d_i : i \in I\}$ be a dominating target of G of size $dt(G)$. Let r_0 be any vertex in R and $\mathfrak{p}_i$ be a path from r_0 to d_i for each $d_i \in D$. Let d_i' is the first vertex from d_i on $\mathfrak{p}_i$ such that $N[d_i'] \cap R \neq \emptyset$. Let $\mathfrak{p}_i'$ is the sub-path of $\mathfrak{p}_i$ from d_i to the vertex prior to d_i'. Now we show that $D_R = \{d_i' : i \in I\}$ is an essential R dominating target. By construction, each vertex of D_R has at least one R vertex in its neighborhood. Now consider arbitrary connected set C containing D_R. Append the paths $\mathfrak{p}_i'$ to C. The resulting graph is connected and contains all vertices of D so it dominates entire G. But $\mathfrak{p}_i'$ do not dominate any R-vertices so C must be dominating all the R-vertices. □

If G is a dominating pair graph, then an essential R dominating target D_R exists with at most 2 vertices. If it is a singleton, then SCDS problem becomes trivial because this vertex dominates the entire R. So in the remainder of this section we assume that $D_R = \{u, v\}$ and denote the distance $d(u, v)$ by d_0. D_R being an essential R-dominating target, $N[u] \cap R \neq \emptyset$ and $N[v] \cap R \neq \emptyset$.

Lemma 2. *Let S be a connected set of vertices in G, i.e., the induced graph on S is connected. Then S is a connected dominating set of R iff S dominates $N_2[u] \cap R$ and $N_2[v] \cap R$, here $N_2[.]$ denotes 2-distance closed neighborhood.*

Proof. "Only if" part is trivial since $N_2[u] \cap R$ and $N_2[v] \cap R$ are subsets of R.

As $\{u, v\}$ is an essential dominating target, $N[u] \cap R$ and $N[v] \cap R$ are non-empty. Let $r_1 \in N[u] \cap R$ and $r_2 \in N[v] \cap R$. So there must be some $x \in N_2[u] \cap S$ and $y \in N_2[v] \cap S$ s.t. r_1 and r_2 are adjacent to x and y respectively. Let $S_1 = \{r_1, u\}$ and $S_2 = \{r_2, v\}$. Then $S' = S \cup S_1 \cup S_2$ is connected and contains u and v. By the definition of R-dominating target, S' dominates all R-vertices. Thus S must dominate $R - (N_2[u] \cup N_2[v])$. Combining this with the given fact that S dominates $N_2[u] \cap R$ and $N_2[v] \cap R$, we conclude that S dominates entire R. □

Lemma 3. *Let $d(u, v) \geq 5$ and S be a connected set of vertices in G containing u. If S also contains a vertex x such that $d(x, v) \leq 2$, then S dominates $N_2[u] \cap R$.*

Proof. Let Q be a shortest path from x to v. Define $S' = S \cup Q$. By construction S' is connected and contains $\{u, v\}$ therefore it dominates R. In particular, it dominates $N_2[u] \cap R$. Vertices of $Q - \{x\}$ are contained in $N[v]$ and $d(u, v)$ is at least 5, so vertices of $Q - \{x\}$ do not dominate $N_2[u] \cap R$. Therefore S must dominate $N_2[u] \cap R$. □

Lemma 4. *Let $d(u, v) \geq 5$ and S be a connected R-dominating set. Let y be a cut vertex of $G(S)$ and $G(S - \{y\})$ has a component C such that $C \cup \{y\}$ contains all the S vertices within 3-neighborhood of v. If P is a path in G connecting y and u, then $S' = C \cup P$ is also a connected R-dominating-set.*

Proof. From Lemma 2 it is sufficient to show that S' is connected and it dominates $N_2[v] \cap R$ and $N_2[u] \cap R$. Firstly, $C \cup \{y\}$ is connected so S' is also connected. Next, S is an R-dominating-set and $S \cap N_3[v]$ is contained in $C \cup \{y\}$ so $C \cup \{y\}$ dominates $N_2[v] \cap R$. Finally, $N[v] \cap R$ is non-empty and S is an R-dominating set so S contains a vertex x such that $d(x, v) \leq 2$. All S-vertices within 3-neighborhood of v are in $C \cup \{y\}$ so $x \in S'$. Further, u also belongs to S' since it is in P. Using Lemma 3 we deduce that S' dominates $N_2[u] \cap R$. This completes the proof. □

Let S be a SCDS for R. We partition it into *levels* as follows. $x \in S$ is defined to be in level i if $d(u, x) = i$. Observe that there is at least one S-vertex at level 2 and at least one S-vertex at level $d_0 - 2$. Further, if $x \in S$ is the only vertex at level i where $2 < i < d_0 - 2$, then x is a cut vertex of $G(S)$.

Lemma 5. *Let $d_0 \geq 9$. Then there exists an SCDS for R which has a unique vertex x_0 with $d(u, x_0) = d_1$ for some $d_1 \in \{3, 4\}$ and a unique vertex y_0 with $d(v, y_0) = d_2$ for some $d_2 \in \{3, 4\}$.*

We omit the proof to save the space.

Theorem 3. *Suppose G has an essential R dominating target $\{u, v\}$ with $d(u, v) \geq 9$. Then every minimum vertex set, S, among the sets satisfying the following conditions is a SCDS of R.*

(a) $G(S)$ is connected.
(b) $\exists x_0 \in S$ with $d(u, x_0) = 3$ or 4 such that x_0 is a cut vertex of $G(S)$ and a component of $G(S - \{x_0\})$, C_u, is such that $C_u \cup \{x_0\}$ dominates $N_2[u] \cap R$.
(c) $\exists y_0 \in S$ with $d(v, y_0) = 3$ or 4 such that y_0 is a cut vertex of $G(S)$ and a component of $G(S - \{y_0\})$, C_v, is such that $C_v \cup \{y_0\}$ dominates $N_2[v] \cap R$.
(d) $S - C_u - C_v$ is a shortest path between x_0 and y_0.

Proof. From Lemma 2 every set satisfying the conditions is a connected R-dominating set. Therefore if a SCDS belongs to this collection of sets, then every smallest set satisfying the conditions must be a SCDS.

From Lemma 5 there exists a SCDS, S, of R with cut vertices x_0 at distance 3 or 4 from u such that $C_u = \{x \in S : d(u, x) < d(u, x_0)\}$ is a component of $G(S - \{x_0\})$. S being an SCDS, $\{x_0\} \cup C_u$ must dominate $N_2[u] \cap R$. Similarly y_0 at a distance 3 or 4 from v in S such that condition (c) is also satisfied. If we replace $S - C_u - C_v$ by a G-shortest path between x_0 and y_0 then also the set will be a CDS, from Lemma 2. Therefore minimality of S requires that $S - C_u - C_v$ is a shortest path connecting x_0 and y_0. Therefore S is one of the CDS that satisfy the conditions. Therefore the smallest sets that satisfy the conditions must be SCDS. □

Corollary 1. *If S is an SCDS, then $|C_u| \leq d(u, x_0)$ and $|C_v| \leq d(v, y_0)$.*

Proof. If C_u is replaced by a shortest path P between u and x_0 in S, then from Lemma 4 the resulting set is also R-CDS. Besides, the optimality of S requires that $|S| \leq |S| - |C_u| + |P| = |S| - |C_u| - d(u, x_0)$.   □

Algorithm 2 computes SCDS of any vertex set R in a DP graph with essential dominating pair $\{u, v\}$ with $d(u, v) \geq 9$.

Data: A DP graph $G = (V, E)$, a subset of vertices R, essential
  R-dominating-pair $\{u, v\}$ with $d(u, v) \geq 9$
Result: A Steiner connected dominating set of R
1 Compute all pair shortest paths;
2 **for** *all $x \in V$ s.t. $d(u, x) = 3$ or 4* **do**
3     $A_x = \{P_{ux}\} \cup \{A : G(A)$ is connected,
        $x \in A, |A| \leq d(u, x), N_2[u] \cap R \subset N[A]\}$;
        /* P_{ux} is a shortest path between u and x */
4     $A_x =$ smallest cardinality set in $\mathcal{A}_x$;
5 **end**
6 **for** *all $y \in V$ s.t. $d(v, y) = 3$ or 4* **do**
7     $A_y = \{P_{vy}\} \cup \{A : G(A)$ is connected,
        $y \in A, |A| \leq d(v, y), N_2[v] \cap R \subset N[A]\}$;
        /* P_{vy} is a shortest path between v and y */
8     $A_y =$ smallest cardinality set in $\mathcal{A}_y$;
9 **end**
10 $S = \{A_x \cup A_y \cup P_{xy} : d(u, x) = 3$ or $4, d(v, y) = 3$ or $4, P_{x,y}$ a shortest path
   between x and $y\}$;
11 **return** *the smallest set in S*;
  Algorithm 2. SCDS algorithm for DP graphs

The correctness of the Algorithm 2 is immediate from Theorem 3. Step 1 costs $O(|V|(|V| + |E|))$. Steps 2 and 6 each costs $O(|V|^4.|R|)$ Cost of the tenth step is $O(|V|^2)$. The total complexity of the algorithm is $O(|V|^4.|R|)$.

For the case with $d_0 \leq 8$ either the SCDS is a shortest path connecting u and v or it contains at most d_0 vertices. Therefore a simple way to handle this case is to test every set of up to d_0 cardinality for connectivity and R domination and select the smallest. If no such set exists, then the shortest path is the solution. This approach costs $O(|V|^8.|R|)$. The cost of computing an essential R-dominating-target is $O(|V| + |E|)$. Adding all the costs we have following theorem.

Theorem 4. *In a dominating-pair graph the Steiner connected dominating set for any subset R can be computed in $O(|V|^8.|R|)$ time. If the distance between the R-dominating pair vertices is greater than 8, then complexity improves to $O(|V|^4.|R|)$.*

4   Approximation Algorithms

Following result is by Fomin et.al.

Theorem 5 ([2004]). *Let $T = (W, F)$ be a d-octopus of a graph $G = (V, E)$, then*

- *T can be computed in $O(|V|^{3d+3})$.*
- *If $\gamma(G)$ is a minimum connected dominating set, then $|W| \le d.(\gamma(G) + 2)$.*

It is conjectured that $dt(G) \le d$ for a graph having a d octopus [2004]. We will present a $appx \le 2\gamma(G)$ algorithm with complexity $O(|V||E| + |V|^{dt(G)+1})$. Following theorem is stated without proof.

Theorem 6. *Let $G = (V, E, w)$ be an edge-weighted (non-negative weights) graph and $R \subseteq V$ be an arbitrary set of required vertices. Then a Steiner tree of R can be calculated in $O(|V|(|V| + |E|) + (|V| - |R|)^{|R|-2}|R|^2)$.*

Corollary 2. *Let $G = (V, E)$ be a graph and $R \subseteq V$ be an arbitrary set of required vertices. Then a Steiner set for R can be computed in $O(|V|(|V| + |E|) + (|V| - |R|)^{|R|-2}|R|^2)$.*

For convenience we define $f(k) = |V|(|V| + |E|) + |V|^k(k + 2)^2$.

4.1   Computation of a Minimum Dominating Target

Let $G = (V, E)$ be a graph. Then $T \subset V$ is a dominating target iff for all $W \subseteq V$ if $T \subseteq W$ and $G(W)$ is connected, then $N[W] = V$. The problem of computing a minimum dominating target is known to be NP-complete, [1981]. Here we generalize the algorithm given in [1993] to compute a dominating pair in AT-free graphs, to one that computes a dominating target in general graphs.

Lemma 6. *A set $S \subseteq V$ is a dominating target of G if and only if for every vertex $v \in V$, S doesn't lie in a single component of $G(V - N[v])$.*

First compute all neighborhood deleted components of the graph, which costs $O(|V|^{2.83})$ [2003b]. Starting with $t = 1$. Select each set of size t and check if it is completely contained in any of the pre-computed components. If any set is found which is not contained in any component, then it is a dominating target, otherwise increment t and repeat till one dominating target is found. This computation costs $O(dt(G) \cdot |V|^{dt(G)+1})$ time.

4.2   Minimum Connected Dominating Set

Theorem 7. *Let $G = (V, E)$ be a connected graph with dominating target number $dt(G)$. If the cardinality of MCDS is $opt(G)$, then in $O(|V|.|E| + |V|^{dt(G)+1})$ time a connected dominating set of G can be computed with cardinality no greater than $opt(G) + dt(G)$.*

Proof. Let D be a minimum dominating target of the graph. It can be computed in $O(|V|^{dt(G)+1})$ as described in section 2.3. Let T be a Steiner tree for the required set D. Hence from the definition of dominating targets, T is a connected dominating set for G. This can be calculated by algorithm of Theorem 6 in $O(f(dt(G) - 2))$.

Let M be any MCDS of G. In particular, it dominates D so $M \cup D$ is a connected set containing D. As T is the minimum connected set containing D, $|T| \leq |M \cup D| \leq |M| + |D| = |M| + dt(G)$. □

It is easy to see that $dt(G) \leq opt(G)$. So $appx \leq 2.opt(G)$.

4.3   Steiner Connected Dominating Set

Theorem 8. *Let $G = (V, E)$ be a connected graph with dominating target number $dt(G)$ and $R \subseteq V$. Let the Steiner connected dominating set (SCDS) of R have cardinality $opt(G, R)$. Then a connected R-dominating set (an approximation to SCDS for R), can be computed in $O(|V|.|E| + |V|^{dt(G)+1})$ time with cardinality no greater than $opt(G, R) + 2dt(G)$.*

Proof. As described in the proof of Lemma 1, compute an essential R-dominating-target D_R in $O(|V|^{dt(G)+1})$ time.

Compute Steiner tree of D_R, T using algorithm of Theorem 6. T is a connected set containing D_R so it dominates R. As $|D_R| \leq dt(G)$, the cost of the computation is bounded by $f(dt(G) - 2)$. Next we show that $|T| \leq opt(G, R) + 2.dt(G)$.

Let S be an SCDS of R in G. D_R is an essential dominating target for R so each member of D_R is adjacent to some R vertex. For each $d \in D_R$ let r_d denote any one vertex from R which adjacent to d. Let R_D denote the set $\{r_d : D \in D_R\}$. Since S dominates R, $S \cup R_D$ is connected. Further, by construction $S \cup R_D \cup D_R$ is connected also connected. By the definition of Steiner trees T is the smallest connected set containing D_R. So $|T| \leq |S \cup R_D \cup D_R| \leq |S| + |R_D| + |D_R| \leq opt(G, R) + 2.dt(G)$. The last inequality is due to the fact that $|R_D| \leq |D_R| \leq dt(G)$. □

$opt(G, R)$ = size of the smallest connected R-dominating set $\geq$ size of the smallest R-dominating target = D_R. Therefore from the last two lines of the above proof $appx \leq 3.opt(G, R)$.

4.4   Steiner Set

Corollary 3. *Let $G = (V, E)$ be a connected graph with dominating target number $dt(G)$ and $R \subseteq V$. Let $opt(G, R)$ denote the cardinality of a Steiner set of R, then an R-connecting set (Steiner set approximation) can be computed in $O(|V|.|E| + |V|^{dt(G)+1})$ time with cardinality not exceeding $opt(G, R) + 2dt(G)$.*

Proof (sketch). Reduce G to G' by shrinking each connected component, R_i, of R to a vertex r_i. Set R' is independent in G'. Observe that if S is an R-connecting set in G, then $S \cup R'$ is the union of R' and a connected R'-dominating set in G'. Conversely if C is a connected R' dominating set in G', then $C - R'$ is a connecting set of R' is G' which is also a connecting set of R in G. Therefore we can compute a Steiner set of R by first computing SCDS of R' in G'. The claim follows from the theorem. □

Future Work: It remains to decide whether MCDS, SS, and SCDS are NP-hard on graphs with bounded dominating targets.

References

[1996] Sudipto Guha and Samir Khuller: Approximation Algorithms for Connected Dominating Sets. Proceedings of ESA (1996) 179-193

[1993] Hari Balakrishnan and Anand Rajaraman and C. Pandu Rangan: Connected Domination and Steiner Set on Asteroidal Triple-Free Graphs. Proceedings of WADS (1993) 131-141

[2000] Gabriel Robins and Alexander Zelikovsky: Improved Steiner Tree Approximation in Graphs. Proceedings of SODA (2000) LNCS 770-779

[1998] Sudipto Guha and Samir Khuller: Improved Methods for Approximating Node Weighted Steiner Trees and Connected Dominating Sets. Proceedings of FSTTCS year (1998) LNCS 54-65

[1997] Hans Jürgen Prömel and Angelika Steger: RNC-Approximation Algorithms for the Steiner Problem. Proceedings of STACS (1997) LNCS 559-570

[1996a] Andrea E. F. Clementi and Luca Trevisan: Improved Non-Approximability Results for Vertex Cover with Density Constraints. Proceedings of COCOON (1996) **1090** LNCS 333-342

[1978] M. R. Garey and D. S. Johnson: Computers and Intractability. Freeman, San Francisco (1978)

[2003] G. Ausiello and P. Crescenzi and G. Gambosi and V. Kann and A. Marchetti-Spaccamela and M. Protasi: Complexity and Approximation. Springer, Heidelberg (2003)

[1995] A. B. Kahng and G. Robins: On Optimal Interconnections for VLSI. Kluwer Publsihers (1995)

[1999] Andreas Brandstädt and V. Bang Lee and Jeremy P. Spinrad: Graph Classes: A Survey. SIAM Monographs on Discrete Mathematics and Applications (1999)

[1998a] A. Caldwell and A. Kahng and S. Mantik and I. Markov and A. Zelikovsky: On Wirelength Estimations for Row-Biased Placement. Proceedings of International Symposium on Physical Design (1998) 4-11

[2001] T. Kloks and D. Kratsch H. Müller: On the Structure of Graphs with Bounded Asteroidal Number. Graphs and Combinatorics **17** (2001) 295-306

[1990] B. Korte and H. J. Prömel and A. Steger: Steiner Trees in VLSI Layouts. J. Paths flows and VLSI layout (1990)

[2003a] Xiuzhen Cheng and Xiao Huang and Deying Li and Weili Wu and Ding-Zhu Du: A Polynomial-Time Approximation Scheme for the Minimum-Connected Dominating Set in Ad Hoc Wireless Networks. Journal of Networks **42,4** (2003) 202-208

[1997a] Derek G. Corneil and Stephan Olariu and Lorna Stewart: Asteroidal Triple-Free Graphs. SIAM J. Discrete Math **10,3** (1997) 399-430

[1972] Shimon Even and Amir Pnueli and Abraham Lempel: Permutation Graphs and Transitive Graphs. J. ACM **19,3** (1972) 400-410

[1962] C. G. Lekkerkerker and J. Ch. Boland: Representation of a Finite Graphs by a Set of Intervals on the Real Line. J. Fund. Math **51** (1962) 245-264

[1992] Rajeev Motwani: Lecture Notes on Approximation Algorithms - Volume I. Dept. of Comp. Sc., Stanford University (1992)

[2004] Fedor V. Fomin and Dieter Kratsch and Haiko Müller: Algorithms for Graphs with Small Octopus. Journal of Discrete Applied Mathematics **134** (2004) 105-128

[2003b] Dieter Kratsch and Jeremy Spinrad: Between O(nm) and $O(n^\alpha)$. Prodeedings of SODA (2003) LNCS 709-716

[1981] Michel Habib: Substitution des Structures Combinatoires, Theorie et Algorithmes. These D'etat, Paris VI (1981)

Efficient Algorithms for Weighted Rank-Maximal Matchings and Related Problems

Telikepalli Kavitha and Chintan D. Shah

Indian Institute of Science, Bangalore, India
{kavitha, chintan}@csa.iisc.ernet.in

Abstract. We consider the problem of designing efficient algorithms for computing certain matchings in a bipartite graph $G = (\mathcal{A} \cup \mathcal{P}, \mathcal{E})$, with a partition of the edge set as $\mathcal{E} = \mathcal{E}_1 \dot{\cup} \mathcal{E}_2 \ldots \dot{\cup} \mathcal{E}_r$. A *matching* is a set of (a, p) pairs, $a \in \mathcal{A}, p \in \mathcal{P}$ such that each a and each p appears in at most one pair. We first consider the *popular matching* problem; an $O(m\sqrt{n})$ algorithm to solve the popular matching problem was given in [3], where n is the number of vertices and m is the number of edges in the graph. Here we present an $O(n^\omega)$ randomized algorithm for this problem, where $\omega < 2.376$ is the exponent of matrix multiplication. We next consider the *rank-maximal matching* problem; an $O(\min(mn, Cm\sqrt{n}))$ algorithm was given in [7] for this problem. Here we give an $O(Cn^\omega)$ randomized algorithm, where C is the largest rank of an edge used in such a matching. We also consider a generalization of this problem, called the weighted rank-maximal matching problem, where vertices in $\mathcal{A}$ have positive weights.

1 Introduction

In this paper we consider some problems in computing *optimal* matchings in bipartite graphs with one-sided preference lists. The input instance here is a bipartite graph $G = (\mathcal{A} \cup \mathcal{P}, \mathcal{E})$ and a partition $\mathcal{E} = \mathcal{E}_1 \dot{\cup} \mathcal{E}_2 \ldots \dot{\cup} \mathcal{E}_r$ of the edge set. We call the vertices in $\mathcal{A}$ *applicants*, the vertices in $\mathcal{P}$ *posts*, and the edges in $\mathcal{E}_i$ the edges of rank i. Each applicant can be considered to be ranking a subset of posts, i.e., its neighbors, in an order of preference (possibly, involving ties) and an edge $(a, p) \in \mathcal{E}_i$ implies that post p is an ith rank post for applicant a.

A matching M is a set of edges no two of which share an endpoint. A matching in G is an allocation of posts to applicants. The bipartite matching problem with a graded edge set is well-studied in the economics literature, see for example [1,11,13]. It models some important real-world problems, including the allocation of graduates to training positions [6], and families to government-owned housing [12]. In a matching M, a vertex $u \in \mathcal{A} \cup \mathcal{P}$ is either *unmatched*, or *matched* to some vertex, denoted by $M(u)$. One would like to fix some notion of optimality of matchings so that we can determine a "best" allocation, according to this notion of optimality. Various notions of optimality of matchings in this setting can be considered. For example, a matching is *Pareto-optimal* [2,1,11]

T. Asano (Ed.): ISAAC 2006, LNCS 4288, pp. 153–162, 2006.
© Springer-Verlag Berlin Heidelberg 2006

if no applicant can improve its allocation (say by exchanging posts with another applicant) without requiring some other applicant to be worse off. Here we consider two stronger notions of optimality, given by *popular* matchings and *rank-maximal* matchings.

If $(a, p) \in \mathcal{E}_i$ and $(a, p') \in \mathcal{E}_j$ with $i < j$, we say that a prefers p to p'. If $i = j$, we say that a is indifferent between p and p'. We say that an applicant a *prefers* matching M' to M if (i) a is matched in M' and unmatched in M, or (ii) a is matched in both M' and M, and a prefers $M'(a)$ to $M(a)$.

Definition 1. *M' is* more popular *than M if the number of applicants preferring M' to M exceeds the number of applicants preferring M to M'. A matching M is* popular *if and only if there is no matching M' that is more popular than M.*

Popular matchings were studied in [3]. It turns out that a popular matching may not always exist in a given graph (see [3] for an example). The popular matching problem is to determine if a given instance admits a popular matching, and to find such a matching, if one exists. An $O(m\sqrt{n})$ algorithm was given for this problem in [3]. Thus this algorithm takes time $\Theta(n^{5/2})$ when $m = \Theta(n^2)$. It is to be expected that the input graphs would be quite dense, since each applicant should typically rank a large subset of posts in some order of preference. In this paper we give a randomized algorithm with running time $O(n^\omega)$ for computing a popular matching, where $\omega < 2.376$ is the exponent of matrix multiplication. Thus our algorithm is faster than the previous algorithm whenever $m > n^{1.87}$.

Another definition of optimality is rank-maximality. A matching is *rank-maximal* [7] if it allocates the maximum number of applicants to their first rank posts, and then subject to this, the maximum number to their second rank posts, and so on. This problem was studied in [7] and an $O(\min(mn, Cm\sqrt{n}))$ algorithm was presented to compute such a matching, where $C \leq r$ is the maximum rank of an edge used in such a matching. Here we present an $O(Cn^\omega)$ randomized algorithm for this problem. Again, when the graph is dense and C is low, our algorithm is faster than the previous algorithm.

We also consider a generalization of this problem. Let us assume that some applicants are more important than some other applicants. We wish to give the more important applicants priority over the rest. A way to formalize this is to assume that there is a weight function $w : \mathcal{A} \to \mathbb{R}^+$, where every applicant a is given a positive weight $w(a)$. This model was used in [10], where the weighted popular matching problem is solved. Here we define a *weighted rank-maximal* matching as a matching which matches the maximum *weight* of applicants to their first rank posts, and then subject to this, the maximum *weight* to their second rank posts, and so on. Another way of defining a weighted rank-maximal matching is in terms of its *signature*.

Definition 2. *The* signature *$\rho(M)$ of a matching M is defined to be the r-tuple $(x_1, ..., x_r)$ where for each $1 \leq i \leq r$, x_i is the sum of the weights of applicants who are matched in M with one of their ith rank posts.*

The total order $\prec$ on signatures is given by: $(x_1, ..., x_r) \prec (y_1, ..., y_r)$ if $x_i = y_i$ for $1 \leq i < j$ and $x_j < y_j$, for some j. The weighted rank-maximal matching problem now is to compute a matching with the maximum signature. Let k be the number of distinct weights the applicants are given. We reduce the problem of computing a weighted rank-maximal matching in $G = (\mathcal{A} \cup \mathcal{P}, \mathcal{E}_1 \dot{\cup} \cdots \dot{\cup} \mathcal{E}_r)$ to computing a rank-maximal matching in a graph $G' = (\mathcal{A} \dot{\cup} \mathcal{P}, \mathcal{E}_{1,1} \dot{\cup} \cdots \dot{\cup} \mathcal{E}_{r,k})$ where the partition of the edge set is refined into rk different ranks. Thus we get an $O(\min(mn, Cm\sqrt{n}))$ deterministic algorithm and an $O(Cn^\omega)$ randomized algorithm for this problem, where $C \leq rk$ is the maximum rank of an edge used in a rank-maximal matching in G'.

2   Popular Matchings

We first present the algorithmic characterization of popular matchings given in [3]. For exposition purposes, we create a unique strictly-least-preferred post $l(a)$ for each applicant a. In this way, we can assume that every applicant is matched, since any unmatched applicant a can be paired with $l(a)$. From now on then, matchings are $\mathcal{A}$-perfect. Also, without loss of generality, we assume that preference lists contain no gaps, i.e., if a is incident to an edge of rank i, then a is incident to an edge of rank $i - 1$, for all $i > 1$.

Let $G_1 = (\mathcal{A} \cup \mathcal{P}, \mathcal{E}_1)$ be the graph containing only rank 1 edges. Then [3, Lemma 3.1] shows that a matching M is popular in G only if $M \cap \mathcal{E}_1$ is a maximum matching of G_1. Maximum matchings have the following important properties, which we use throughout the rest of the paper. $M \cap \mathcal{E}_1$ defines a partition of $\mathcal{A} \cup \mathcal{P}$ into three disjoint sets: a vertex $u \in \mathcal{A} \cup \mathcal{P}$ is *even* (resp. *odd*) if there is an even (resp. odd) length alternating path in G_1 (w.r.t. $M \cap \mathcal{E}_1$) from an unmatched vertex to u. Similarly, a vertex u is *unreachable* if there is no alternating path from an unmatched vertex to u. Denote by $\mathcal{O}, \mathcal{U}$, and $\mathcal{N}$ the sets of odd, unreachable, and even vertices respectively.

Lemma 1 (Gallai-Edmonds Decomposition). *Let $\mathcal{O}, \mathcal{U}$, and $\mathcal{N}$ be the sets of vertices defined by G_1 and $M \cap \mathcal{E}_1$ above. Then $\mathcal{O}, \mathcal{U}$, and $\mathcal{N}$ are pairwise disjoint, and independent of the maximum matching $M \cap \mathcal{E}_1$. In any maximum matching of G_1, every vertex in $\mathcal{O}$ is matched with a vertex in $\mathcal{N}$ and every vertex in $\mathcal{U}$ is matched with another vertex in $\mathcal{U}$. The size of a maximum matching is $|\mathcal{O}| + |\mathcal{U}|/2$. No maximum matching of G_1 contains an edge between a vertex in $\mathcal{O}$ and a vertex in $\mathcal{O} \cup \mathcal{U}$. Also, G_1 contains no edge between a vertex in $\mathcal{N}$ and a vertex in $\mathcal{N} \cup \mathcal{U}$.*

Using this vertex partition, we make the following definitions: for each applicant a, $f(a)$ is the set of odd/unreachable posts amongst a's most-preferred posts. Also, $s(a)$ is the set of a's most-preferred posts amongst all even posts. We refer to posts in $\cup_{a \in \mathcal{A}} f(a)$ as *f-posts* and posts in $\cup_{a \in \mathcal{A}} s(a)$ as *s-posts*. Note that f-posts and s-posts are disjoint. Also note that there may be posts in $\mathcal{P}$ that are neither f-posts nor s-posts. The next lemma characterizes the set of all popular matchings.

Lemma 2 ([3]). *A matching M is popular in G iff (i) $M \cap \mathcal{E}_1$ is a maximum matching of $G_1 = (\mathcal{A} \cup \mathcal{P}, \mathcal{E}_1)$, and (ii) for each applicant a, $M(a) \in f(a) \cup s(a)$.*

We present in Fig. 1 the algorithm from [3], based on Lemma 2, for solving the popular matching problem in a given graph $G = (\mathcal{A} \cup \mathcal{P}, \mathcal{E})$.

---

1. Construct a maximum matching M of $G_1 = (\mathcal{A} \cup \mathcal{P}, \mathcal{E}_1)$.
2. Construct the graph $G' = (\mathcal{A} \cup \mathcal{P}, \mathcal{E}')$, where $\mathcal{E}' = \{(a, p) : a \in \mathcal{A}$ and $p \in f(a) \cup s(a)\}$.
3. Remove any edge in G' between a vertex in $\mathcal{O}$ and a vertex in $\mathcal{O} \cup \mathcal{U}$.
4. Augment M in G' until it is a maximum matching of G'.
5. Return M if it is $\mathcal{A}$-perfect, otherwise return *"no popular matching"*.

Fig. 1. An $O(m\sqrt{n})$-time algorithm for the popular matching problem (from [3])

It is easy to see that the running time of this algorithm is $O(m\sqrt{n})$ by using the $O(m\sqrt{n})$ Hopcroft-Karp algorithm [5] for computing maximum matchings. A maximum matching M in G_1 is computed by the Hopcroft-Karp algorithm and this identifies the odd, unreachable, and even vertices and so the graph G' is constructed in $O(m\sqrt{n})$ time. M is repeatedly augmented (by the Hopcroft-Karp algorithm) to obtain the final matching.

2.1 Our Improvement

Lovász [8] showed that it is possible to test whether a given graph has a perfect matching in randomized time $O(n^\omega)$, where $\omega < 2.376$ is the exponent of matrix multiplication. Mucha and Sankowski [9] showed that it is possible to actually compute a maximum matching in randomized time $O(n^\omega)$. In our algorithm we need to determine the Gallai-Edmonds decomposition of G to construct G' and for that we could either use this maximum matching algorithm or use the $O(n^\omega)$ Gallai-Edmonds decomposition algorithm of Cheriyan [4]. We also use the $O(n^\omega)$ perfect matching algorithm from [9], where a simple algorithm, based on the LUP factorization algorithm of Hopcroft and Bunch, is given for computing perfect matchings in bipartite graphs. We first give a characterization of popular matchings, which our algorithm uses. Its proof will be given in the full version of the paper.

Lemma 3. *M is a popular matching in G iff it is an $\mathcal{A}$-perfect matching in G' (the pruned graph after Step 3 in Fig. 1) that matches all the f-posts.*

It follows from Lemma 3 that we seek an $\mathcal{A}$-perfect matching in G' that matches all the f-posts. If G' admits an $\mathcal{A}$-perfect matching, then obviously $|\mathcal{P}| \geq |\mathcal{A}|$. If $|\mathcal{P}| > |\mathcal{A}|$, then an $\mathcal{A}$-perfect matching does not match all vertices in $\mathcal{P}$, and we need to ensure that all f-posts are matched. Note that if G admits a popular matching, then we know that there *is* an $\mathcal{A}$-perfect matching in G' that matches all the f-posts. In fact, a popular matching is such a matching. The $\mathcal{A}$-perfect matching returned by the algorithm in Fig. 1 matches all the f-posts since the matching M obtained in Step 1 necessarily matches all the f-posts (since M is maximum in G_1) and we obtain the final matching by *augmenting* M, so all f-posts continue to remain matched. In our algorithm below, we present a simple way of accomplishing that all f-posts are matched by an $\mathcal{A}$-perfect matching in G'.

1. Compute the Gallai-Edmonds decomposition of G and using it, construct the graph $G' = (\mathcal{A} \cup \mathcal{P}, \mathcal{E}')$, where $\mathcal{E}' = \{(a, p) : a \in \mathcal{A} \text{ and } p \in f(a) \cup s(a)\}$. Remove all edges in G' between vertices in $\mathcal{O}$ and vertices in $\mathcal{O} \cup \mathcal{U}$.
2. Modify G' as follows:
   - add $|\mathcal{P}| - |\mathcal{A}|$ new vertices $x_1, \ldots, x_{|\mathcal{P}|-|\mathcal{A}|}$ to $\mathcal{A}$;
   - make every s-post adjacent to all the vertices $x_1, \ldots, x_{|\mathcal{P}|-|\mathcal{A}|}$.
3. Test if G' admits a perfect matching. If not, then return *"no popular matching"*.
4. Else compute a perfect matching M in G'. Delete all edges (x_i, y) from M, where x_i is a dummy vertex. Return M.

Remark. Note that if G admits a popular matching, then the graph G' after Step 2 does admit a perfect matching. Any popular matching of G along with unmatched vertices (each of which is an s-post) paired with the new vertices, leads to a perfect matching in the new graph G'.

Lemma 4. *The matching M returned by our algorithm is a popular matching.*

Proof. The matching M is an $\mathcal{A}$-perfect matching in G'. We need to show that all f-posts are still matched by M after deleting the edges incident on the new vertices x_i in Step 4; then by Lemma 3, we can conclude that M is a popular matching. But this is obvious: the new vertices were adjacent only to s-posts. Hence none of the edges deleted in Step 4 is incident on an f-post. Thus M is an $\mathcal{A}$-perfect matching that matches all f-posts, thus it is popular (by Lemma 3). $\qquad\square$

Hence we can conclude the following theorem, since the Gallai-Edmonds decomposition and testing if a perfect matching exists and if so, computing a perfect matching take randomized $O(n^\omega)$ time.

Theorem 1. *The popular matching problem can be solved in randomized $O(n^\omega)$ time on an n-vertex bipartite graph $G = (\mathcal{A} \cup \mathcal{P}, \mathcal{E})$.*

3   Rank-Maximal Matchings

In this section we first present the algorithm from [7] to compute a rank-maximal matching. This algorithm computes a rank-maximal matching by reducing it to a maximum matching computation in a subgraph G' of the given graph $G = (\mathcal{A} \cup \mathcal{P}, \mathcal{E}_1 \cup \ldots \cup \mathcal{E}_r)$. However, note that not every maximum matching of G'

is a rank-maximal matching. The pruning of G to G' uses the properties of *odd, unreachable, even* vertices given in Lemma 1.

A rank-maximal matching is a matching M in $G = (\mathcal{A} \cup \mathcal{P}, \mathcal{E}_1 \cup \ldots \cup \mathcal{E}_r)$ that matches the maximum number of applicants to their rank one posts, subject to this, the maximum number of applicants to their rank two posts, and so on. So this gives us the property that $M_i = M \cap \mathcal{E}_{\leq i}$ is a rank-maximal matching in $G_i = (\mathcal{A} \cup \mathcal{P}, \mathcal{E}_1 \cup \ldots \cup \mathcal{E}_i)$. The algorithm in [7] iteratively computes matchings M_i for $i = 1, \ldots, r$ such that M_i is a rank-maximal matching in G_i. We present below the algorithm from [7] to compute a rank-maximal matching.

3.1   The Rank-Maximal Matching Algorithm from [7]

Let $G_i = (\mathcal{A} \cup \mathcal{P}, \mathcal{E}_1 \cup \ldots \cup \mathcal{E}_i)$. Initialize $G'_1 = G_1$, and M_1 to any maximum matching in G'_1. For $i = 1$ to $r - 1$ do the following steps, and output M_r.

1. Partition the vertices of $\mathcal{A} \cup \mathcal{P}$ into three disjoint sets: $\mathcal{N}_i$, $\mathcal{O}_i$, and $\mathcal{U}_i$, which are the even, odd and unreachable vertices in G'_i. Delete all edges incident to a vertex in $\mathcal{O}_i \cup \mathcal{U}_i$ from $\mathcal{E}_j, \forall j > i$. Delete all $\mathcal{O}_i$ - $\mathcal{O}_i$ and $\mathcal{O}_i$ - $\mathcal{U}_i$ edges from G'_i. Add the edges in $\mathcal{E}_{i+1}$ to G'_i. Call the resulting graph G'_{i+1}.
2. Determine a maximum matching M_{i+1} in G'_{i+1} by augmenting M_i.

We will not present a formal proof of correctness of the above algorithm and refer the reader to [7]. The proof of correctness relies on the following facts:

Fact 1: for all $1 \leq i \leq r$, every rank-maximal matching in G_i is a maximum matching in G'_i (note that the converse is not true).

Fact 2: for all $1 < i \leq r$, we have that the matching M_i contains as many rank $\leq i - 1$ edges as M_{i-1} (this is due to the edges that we delete and the fact that M_i is obtained from M_{i-1} by augmentation).

These facts put together yield that the matching M_i is a rank-maximal matching in G_i for all i. Suppose that M_{i-1} is a rank-maximal matching in G_{i-1}. Let M_{i-1} contain s_k number of edges of rank k for $1 \leq k \leq i - 1$. Then $M_i \cap \mathcal{E}_{\leq i-1}$ is also a rank-maximal matching in G_{i-1} because if M_i contains r_k number of edges of rank k, then for every $k \leq i - 1$, we have $\sum_{j=1}^{k} r_j = \sum_{j=1}^{k} s_j$ (by Fact 2 above), which means that $r_k = s_k$ for all k. Also, the cardinality of M_i is the same as the cardinality of a rank-maximal matching of G_1 since by Fact 1, a rank-maximal matching is a maximum cardinality in G'_i and so is M_i. Thus M_i contains the same number of rank j edges as a rank-maximal matching for all $j \leq i$. Thus M_i is a rank-maximal matching in G_i.

Running time of this algorithm. We will now analyze the running time of the algorithm. The expensive step of iteration i is the augmentation of M_i to M_{i+1}. Using the algorithm of Hopcroft and Karp [5], this takes $O(\min(\sqrt{n}, |M_{i+1}| - |M_i| + 1) \cdot m)$ time. The number of iterations is r and hence the overall running time is $O(\min(r\sqrt{n}, n) \cdot m)$. It is easy to replace r by C, the maximum rank of edge used in a rank-maximal matching. At the beginning of each iteration, say iteration i, first check whether M_i is already a maximum matching in H_r, where H_r denotes the graph consisting of all edges, of all ranks, that are still present at the beginning of phase i. This takes time $O(m)$. If M_i is a maximum matching in H_r, then stop; otherwise, continue as described above. In this way, only C iterations are executed. Thus this is an $O(\min(Cm\sqrt{n}, mn))$ algorithm for computing a rank-maximal matching.

3.2 Our Improvement

Now we present a randomized $O(Cn^\omega)$ algorithm, where $\omega < 2.376$ is the exponent of matrix multiplication, for computing a rank-maximal matching in $G = (\mathcal{A} \cup \mathcal{P}, \mathcal{E}_1 \cup \cdots \cup \mathcal{E}_r)$. Our algorithm is based on the same approach as the algorithm in [7]. However it does not compute M_r by successive augmentations. We first present an $O(rn^\omega)$ algorithm and the same idea as in [7] can be used to replace r with C. The graph $G'_1 = G_1 = (\mathcal{A} \cup \mathcal{P}, \mathcal{E}_1)$.

1. For $i = 1$ to r do the following steps.

   – Partition the vertex set $\mathcal{A} \cup \mathcal{P} = \mathcal{N}_i \,\dot\cup\, \mathcal{O}_i \,\dot\cup\, \mathcal{U}_i$. Delete all edges incident to a vertex in $\mathcal{O}_i \cup \mathcal{U}_i$ from $\mathcal{E}_j, \forall j > i$. Delete all $\mathcal{O}_i$-$\mathcal{O}_i$ and $\mathcal{O}_i$-$\mathcal{U}_i$ edges from G'_i. If $i < r$, then add the edges of $\mathcal{E}_{i+1}$ to G'_i, call this graph G'_{i+1}.

2. Compute the cardinality k of a maximum matching in the graph G'_r. Modify G'_r as follows:

   – Add $|\mathcal{P}| - k$ vertices $x_1, \ldots, x_{|\mathcal{P}|-k}$ to $\mathcal{A}$ and add $|\mathcal{A}| - k$ vertices $y_1, \ldots, y_{|\mathcal{A}|-k}$ to $\mathcal{P}$. Make each vertex in $\cap_{i=1}^r \mathcal{N}_i \cap \mathcal{A}$ adjacent to all of $y_1, \ldots, y_{|\mathcal{A}|-k}$. Make each vertex in $\cap_{i=1}^r \mathcal{N}_i \cap \mathcal{P}$ adjacent to all of $x_1, \ldots, x_{|\mathcal{P}|-k}$.

3. Compute a perfect matching M in G'_r. Delete all edges in M incident on the vertices $\{x_1, \ldots, x_{|\mathcal{P}|-k}, y_1, \ldots, y_{|\mathcal{A}|-k}\}$. Return M.

   The proofs of Lemma 5 and 6 will be given in the full version of the paper.

Lemma 5. *The graph G'_r (in Step 3 of the above algorithm) always admits a perfect matching.*

Lemma 6. *The matching M returned by the algorithm above is a rank-maximal matching in G.*

This shows that a rank-maximal matching in G can be computed in randomized $O(rn^\omega)$ time. We can replace r by C, the maximum rank used by an edge in a rank-maximal matching, using the same idea as was used in Section 3.1: in Step 1 for every i, check if a maximum matching in G'_i is still maximum when the existing edges of the sets $\mathcal{E}_{>i}$ are added to G'_i. If so, then we can stop the algorithm. Otherwise, we proceed as usual. Hence we can conclude the following theorem now.

Theorem 2. *A rank-maximal matching in a $G = (\mathcal{A} \cup \mathcal{P}, \mathcal{E})$ can be computed in randomized $O(Cn^\omega)$ time, where n is the number of vertices in G and C is the maximum rank used by an edge in the rank-maximal matching.*

4   Weighted Rank-Maximal Matchings

In this section we consider the problem of computing a rank-maximal matching where each applicant a has a weight $w(a)$ assigned to it. Let k be the number of distinct weights that the applicants are given; let these weights be $w_1 > w_2 > \cdots > w_k$. So the set $\mathcal{A}$ can be partitioned into sets $C_j, j = 1, \ldots, k$, where set C_j consists of all applicants of weight w_j.

Let us consider a simple case first. Consider the problem of computing a weighted rank-maximal matching in the graph $G_1 = (\mathcal{A} \cup \mathcal{P}, \mathcal{E}_1)$. It is easy to show the following lemma.

Lemma 7. *A rank-maximal matching in* $(\mathcal{A} \cup \mathcal{P}, \mathcal{E}_{1,1} \cup \cdots \cup \mathcal{E}_{1,k})$ *is a maximum weight matching in* $G_1 = (\mathcal{A} \cup \mathcal{P}, \mathcal{E}_1)$.

We will show that the above relationship holds for all graphs $G_i = (\mathcal{A} \cup \mathcal{P}, \mathcal{E}_1 \cup \cdots \cup \mathcal{E}_i)$. In other words, a weighted rank-maximal matching in G_i is the same as a rank-maximal matching in G_i', which is the same graph without weights on applicants and instead of i ranks, the edge set consists of ik different ranks, that is, we look at the graph $G_i' = (\mathcal{A} \cup \mathcal{P}, \mathcal{E}_{1,1} \cup \cdots \cup \mathcal{E}_{i,k})$ where edges of $\mathcal{E}_{x,y}$ have rank $(x-1)k + y$. Thus edges of $\mathcal{E}_{x,y}$ have a better rank than edges of $\mathcal{E}_{x',y'}$ if $x < x'$, or $x = x'$ and $y < y'$. So this yields the simple algorithm below to compute a weighted-rank maximal matching in G.

1. Partition the edge set $\mathcal{E}$ as $\mathcal{E} = \dot{\cup}_{i,j}\mathcal{E}_{i,j}$ where i ranges over the r ranks and j ranges over the k weights. The set $\mathcal{E}_{i,j}$ consists of edges of rank i incident upon applicants of weight w_j.
2. Ignore the applicant weights and compute a rank-maximal matching in $G = (\mathcal{A} \cup \mathcal{P}, \cup_{i,j}\mathcal{E}_{i,j})$ where the edges of $\mathcal{E}_{i,j}$ have rank $k(i-1) + j$.

Now we come to the proof of correctness of the above algorithm. This is shown by the following lemma.

Lemma 8. *For all* $1 \le i \le r$, *a rank-maximal matching in* $(\mathcal{A} \cup \mathcal{P}, \mathcal{E}_{1,1} \cup \cdots \cup \mathcal{E}_{i,k})$, *where edges in* $\mathcal{E}_{x,y}$ *have rank* $(x-1)k + y$, *is the same as a weighted rank-maximal matching in* $G_i = (\mathcal{A} \cup \mathcal{P}, \mathcal{E}_1 \cup \cdots \cup \mathcal{E}_i)$.

Proof. We prove this by induction. Lemma 7 proves the case $i = 1$. So let us assume by induction hypothesis that a rank-maximal matching in $(\mathcal{A} \cup \mathcal{P}, \mathcal{E}_{1,1} \cup \cdots \cup \mathcal{E}_{i-1,k})$ is a weighted rank-maximal matching in $G_{i-1} = (\mathcal{A} \cup \mathcal{P}, \mathcal{E}_1 \cup \cdots \cup \mathcal{E}_{i-1})$. Denote the edges $\mathcal{E}_{1,1} \cup \cdots \cup \mathcal{E}_{i-1,k}$, with edges in $\mathcal{E}_{x,y}$ having rank $(x-1)k + y$, as $\mathcal{F}$. Let M be a rank-maximal matching in $G_i' = (\mathcal{A} \cup \mathcal{P}, \mathcal{F} \cup \mathcal{E}_{i,1} \cdots \cup \mathcal{E}_{i,k})$. Now we make the following claim.

Claim. Any rank-maximal matching M in $G_i' = (\mathcal{A} \cup \mathcal{P}, \mathcal{F} \cup \mathcal{E}_{i,1} \cdots \cup \mathcal{E}_{i,k})$ is also a rank-maximal matching in $H_i = (\mathcal{A} \cup \mathcal{P}, \mathcal{F} \cup \mathcal{E}_i)$.

We will prove the claim later, but first let us assume the claim and complete the proof of the lemma. The claim implies that the number of rank i edges used in M is the largest number of rank i edges that any matching that is rank-maximal in $G'_{i-1} = (\mathcal{A} \cup \mathcal{P}, \mathcal{F})$ can use. Since M is a rank-maximal matching in G'_i, the matching $M \cap \mathcal{F}$ is a rank-maximal matching in G'_{i-1}, so it is a weighted rank-maximal matching in G_{i-1}, by the induction hypothesis.

So M has the following properties: (i) M uses the maximum number of rank i edges possible by any rank-maximal matching in G'_{i-1} and (ii) under the constraint that M has to be rank-maximal in G'_{i-1}, M contains the maximum number of rank i edges incident on weight w_1 applicants, subject to this constraint, it contains the maximum number of rank i edges incident on weight w_2 applicants, and so on.

Property (i) follows from the claim, and property (ii) follows from the definition that M is rank-maximal in $\mathcal{F} \cup \mathcal{E}_{i,1} \cup \cdots \cup \mathcal{E}_{i,k}$. Let us assume that $(c_1, \ldots, c_{i-1})$ is the signature (refer Definition 2) of a weighted rank-maximal matching in G_{i-1}. Then M has prefix signature $(c_1, \ldots, c_{i-1})$ over $\mathcal{E}_1 \cup \cdots \cup \mathcal{E}_{i-1}$ and under this constraint, it has the largest possible weight of rank i edges, by properties (i) and (ii). Thus M is a weighted rank-maximal matching in G_i. □

Proof of the Claim. We need to show that M is a rank-maximal matching in $H_i = (\mathcal{A} \cup \mathcal{P}, \mathcal{F} \cup \mathcal{E}_i)$. We know that M is a rank-maximal matching in $G'_i = (\mathcal{A} \cup \mathcal{P}, \mathcal{F} \cup \mathcal{E}_{i,1} \cup \cdots \cup \mathcal{E}_{i,k})$. The rank-maximal matching algorithm from [7] (refer Section 3.1) on input G'_i obtains M as follows: it first executes $(i-1)k$ rounds corresponding to edges in $\mathcal{F}$. This yields a rank-maximal matching (call it M_0) in $(\mathcal{A} \cup \mathcal{P}, \mathcal{F})$. Then the following k rounds for $j = 1, \ldots, k$ are executed:

1. add edges of the existing set $\mathcal{E}_{i,j}$ to the current graph. Call this graph X_j.
   (recall that the original set $\mathcal{E}_{i,j}$ gets pruned during the first $(i-1)k$ rounds)
2. augment M_{j-1} in this graph, call the new matching M_j.
3. partition $\mathcal{A} \cup \mathcal{P}$ into sets $\mathcal{O}_j, \mathcal{U}_j, \mathcal{N}_j$ (odd, unreachable, even).
   (i) delete all $\mathcal{O}_j$-$\mathcal{O}_j$ and $\mathcal{O}_j$-$\mathcal{U}_j$ edges from X_j.
   (ii) delete all edges incident on a vertex in $\mathcal{O}_j \cup \mathcal{U}_j$ from $\mathcal{E}_{i,j+1}, \ldots, \mathcal{E}_{i,k}$. Note that all these edges are of the form (a, p), where $p \in \mathcal{O}_j \cup \mathcal{U}_j$ and $w(a) \in \{w_{j+1}, \ldots, w_k\}$, since there are *no* edges from $\mathcal{E}_{i,j+1}, \ldots, \mathcal{E}_{i,k}$ incident on an applicant with weight w_j.

The final matching M_k is our matching M. This is a maximum matching in X_k. We need to show that if we add to the graph X_k all the rank i edges that we deleted in Steps 3(i) and 3(ii) during the above k rounds, then M is still a maximum cardinality matching in this new graph (call this graph Y). That proves that M is a rank-maximal matching in $(\mathcal{A} \cup \mathcal{P}, \mathcal{F} \cup \mathcal{E}_i)$, since subject to the constraint that it is rank-maximal in $\mathcal{F}$, the matching M contains the maximum number of rank i edges.

We want to show that there is no augmenting path in Y with respect to M. The main observation here is that none of the edges that was deleted in Steps 3(i) and 3(ii) during the above k rounds is incident on a post outside $\cup_{\ell=1}^{k}(\mathcal{O}_\ell \cup \mathcal{U}_\ell)$. So when we build the Hungarian tree, which is the tree of all alternating paths, rooted at an unmatched applicant, we never see a post outside $\cup_{\ell=1}^{k}(\mathcal{O}_\ell \cup \mathcal{U}_\ell)$ in this tree. Any post in $\cup_{\ell=1}^{k}(\mathcal{O}_\ell \cup \mathcal{U}_\ell)$ is already matched in M. Thus we never see a free post in this Hungarian tree, thus there is no augmenting path with respect to M in Y.

The observation sketched above can be formalized (the details will be given in the full version). Hence there is no augmenting path with respect to M. This finishes the proof of the claim.   □

We conclude this section with the following theorem, which follows from Lemma 8, that a weighted rank-maximal matching in $G = (\mathcal{A} \cup \mathcal{P}, \mathcal{E}_1 \cup \cdots \cup \mathcal{E}_i)$ can be computed as a rank-maximal matching in the graph $(\mathcal{A} \cup \mathcal{P}, \mathcal{E}_{1,1} \cup \cdots \cup \mathcal{E}_{r,k})$.

Theorem 3. *A weighted rank-maximal matching in a graph $G = (\mathcal{A} \cup \mathcal{P}, \mathcal{E}_1 \cup \cdots \cup \mathcal{E}_r)$ with m edges and n vertices, where each vertex in $\mathcal{A}$ has one of the weights $w_1 > \cdots > w_k$ assigned to it, can be computed in deterministic $O(\min(C\sqrt{n}m, (n + C)m))$ time and randomized $O(Cn^\omega)$ time, where $C \le rk$ is the largest rank of an edge that is used in any rank-maximal matching of $(\mathcal{A} \cup \mathcal{P}, \mathcal{E}_{1,1} \cup \cdots \cup \mathcal{E}_{r,k})$.*

References

1. A. Abdulkadiroğlu and T. Sönmez. *Random serial dictatorship and the core from random endowments in house allocation problems.* Econometrica, 66(3):689–701, 1998.
2. D.J. Abraham, K. Cechlárová, D.F. Manlove, K. Mehlhorn. Pareto-optimality in house allocation problems. In *Proc. of 15th ISAAC*, pages 3-15, 2004.
3. D.J. Abraham, R.W. Irving, T. Kavitha, and K. Mehlhorn. Popular matchings. In *Proc. of 16th SODA*, pages 424-432, 2005.
4. Joesph Cheriyan. Randomized $\tilde{O}(M(|V|))$ algorithms for problems in matching theory. SIAM Journal on Computing, 26(6):1635–1655, 1997.
5. J.E. Hopcroft and R.M. Karp. *A $n^{5/2}$ Algorithm for Maximum Matchings in Bipartite Graphs.* SIAM Journal on Computing, 2:225–231, 1973.
6. A. Hylland and R. Zeckhauser. The efficient allocation of individuals to positions. *Journal of Political Economy*, 87(2):293–314, 1979.
7. R.W. Irving, T. Kavitha, K. Mehlhorn, D. Michail, and K. Paluch. Rank-maximal matchings. In *Proc. of 15th SODA*, pages 68–75, 2004.
8. L. Lovász. On determinants, matchings and random algorithms. In *Fundamentals of Computation Theory*, pages 565–574, 1979.
9. Marcin Mucha and Piotr Sankowski. Maximum Matchings via Gaussian Elimination In *Proc. of 45th FOCS*, pages 248–255, 2004.
10. J. Mestre. Weighted popular matchings. In *Proc. of 33rd ICALP*, LNCS 4051, pages 715–726, 2006.
11. A.E. Roth and A. Postlewaite. *Weak versus strong domination in a market with indivisible goods.* Journal of Mathematical Economics, 4:131–137, 1977.
12. Y. Yuan. Residence exchange wanted: a stable residence exchange problem. *European Journal of Operational Research*, 90:536–546, 1996.
13. L. Zhou. On a conjecture by Gale about one-sided matching problems. *Journal of Economic Theory*, 52(1):123–135, 1990.

On Estimating Path Aggregates over Streaming Graphs

Sumit Ganguly and Barna Saha

Indian Institute of Technology, Kanpur
{sganguly, barna}@cse.iitk.ac.in

Abstract. We consider the updatable streaming graph model, where edges of a graph arrive or depart in arbitrary sequence and are processed in an online fashion using sub-linear space and time. We study the problem of estimating aggregate path metrics P_k defined as the number of pairs of vertices that have a simple path between them of length k. For a streaming undirected graph with n vertices, m edges and r components, we present an $\tilde{O}(m(m-r)^{-1/4})$ space[1] algorithm for estimating P_2 and an $\Omega(\sqrt{m})$ space lower bound. We show that estimating P_2 over directed streaming graphs, and estimating P_k over streaming graphs (whether directed or undirected), for any $k \geq 3$ requires $\Omega(n^2)$ space. We also present a space lower bound of $\Omega(n^2)$ for the problems of (a) deterministically testing the connectivity, and, (b) estimating the size of transitive closure, of undirected streaming graphs that allow both edge-insertions and deletions.

1 Introduction

The data streaming model has gained popularity as a computational model for a variety of *monitoring* applications, where, data is generated rapidly and continuously, and must be analyzed very efficiently and in an online fashion using space that is significantly sub-linear in the data size. An emerging class of monitoring applications is concerned with *massive dynamic graphs*. For example, consider the dynamic web graph, where nodes are web-pages and edges model hyperlinks from one page to another. The edges in the web-graph are generated in a streaming fashion by web-crawlers [8]. Significant changes in the size, connectivity and path properties of web-communities of interest can be glimpsed by computing over these stream of edges. Another example is the citations graph [7], where, nodes are published articles and directed edges denote a citation of one article by another. Consider the query: find the top-k *second-level* frequent citations, where the second-level citation number of an article A is the number of (distinct) articles C that cite an article B that cite A.

---

[1] $f(m)$ is said to be $\tilde{O}(g(m))$ if $f(m) = O(\frac{1}{\epsilon^{O(1)}}(\log m)(\log n)(\log \frac{1}{\delta})^{O(1)} g(n))$. Similarly, $f(m)$ is said to be $\tilde{\Omega}(g(m))$ if $g(m)$ is $\tilde{O}(f(m))$.

T. Asano (Ed.): ISAAC 2006, LNCS 4288, pp. 163–172, 2006.

Graph Streaming Models. In the *updatable* edge-streaming model of graphs, the stream is viewed as a sequence of tuples of the form $(u, v, +)$ or $(u, v, -)$, corresponding, respectively, to the insertion or the deletion of the edge (u, v). In the updatable model, once an edge (u, v) is inserted, it remains *current* in the graph until a tuple of the form $(u, v, -)$ appears in the stream to delete the edge. The current state of the graph $G = (V, E)$ is defined by the set of current edges E; the set of vertices V are those vertices that are incident to any of the current edges. Multi-graphs are modelled by allowing an edge to be inserted multiple times. Edges may be inserted and deleted in arbitrary order; however, an edge may be deleted at most as many times as it is inserted. The *insert-only* streaming model [1,3,7] only allows tuples of the form $(u, v, +)$ to appear in the stream. Graph streaming models that allow use of external memory and extra passes over stored data have been proposed—these include the semi-streaming graph model [2] and the W-stream model [1]. In this paper, we do not consider computational models over streaming graphs that allow multiple passes.

Path Aggregates. The path aggregate P_k is defined as the number of pairs of vertices (u, v) such that there is a simple path of length k from u to v. In this work, we consider the problem of estimating the path aggregate P_k, for $k \geq 2$ over updatable streaming graphs. The continuous monitoring of path aggregates enables online detection of changing path properties of a dynamic graph. For example, an article can be said to be frequently cited at level l, provided, the number of its level l-citations exceeds P_l/s, for a parameter s. The problem also has applications in database query size estimation. For example, let $R(A, B)$ be a binary relation over attributes A and B, over the same domain. Then, the P_2 over the binary relation R viewed as a graph represents the number of distinct pairs in the self-join (the *distinct self join*) of its relations.

Prior work in estimating path aggregates. [5] presents the JDSKETCH algorithm for estimating the *Join-Distinct size* of two data streams, $R = R(A, B)$ and $S(B, C)$ defined as $JD(R, S) = |\pi_{A,C}(R \bowtie S)|$. If $R = S$, then, $JD(R, R) = P_2$ and therefore, the JDSKETCH algorithm can be used to estimate P_2. The space requirement of the JDSKETCH algorithm is $\tilde{O}(m^2/P_2)$ [5]. In particular, for complete bi-partite graphs, chain graphs, etc., the JDSKETCH requires $\Omega(m)$ space.

Contributions. We present the RS algorithm for estimating P_2 for undirected streaming graphs and multi-graphs to within accuracy factors of $1 \pm \epsilon$ and confidence $1 - \delta$, where, $0 < \epsilon, \delta < 1$. For a graph with n vertices, m edges and r-components, the algorithm requires $O(\frac{1}{\epsilon^2} \frac{m}{(m-r)^{-1/4}} (\log n)(\log \frac{1}{\delta}))$ bits. For graphs with $\frac{m}{2}$ or less components, the space complexity of the algorithm is $\tilde{O}(m^{3/4})$ bits. We present a lower bound of $O(\sqrt{m})$ bits for estimating P_2 for undirected and connected streaming graphs. For directed streaming graphs, we show that the estimating P_k, for any $k \geq 2$, to within any approximation factor, requires $\Omega(m)$ bits of space. We also show that estimating P_k, for $k \geq 3$, for undirected streaming graphs to within a factor of $1 \pm \frac{3}{4}$, requires $\Omega(n^2)$ bits of space. Finally, we present a space lower bound of $\Omega(n^2)$ for the problems of (a) deterministically testing the connectivity, and, (b) estimating the size of transitive closure, of undirected streaming graphs that allow both edge-insertions and deletions.

Organization. Section 2 presents the *RS* algorithm for estimating P_2 and Section 3 presents lower bound results.

2   Estimating P_2

In this section, we present the *RS* algorithm for estimating P_2 for *undirected* graphs and multi-graphs. We first consider insert-only streaming graphs and prove Theorem 1 and then generalize it to updatable edge streaming graphs.

Theorem 1. *For $0 \leq \epsilon < \frac{1}{6}$ and $0 < \delta < 1$, there exists an algorithm that takes as input an insert-only streaming graph with r components, m edges and n vertices and returns an estimate $\hat{P}_2$ satisfying $\Pr\{|\hat{P}_2 - P_2| \leq \epsilon P_2\} \geq 1 - \delta$ using $O(\epsilon^{-2}m(m-r)^{-1/4}(\log \frac{1}{\delta})(\log n))$ bits.*

2.1   Random Subgraph *RS* of Graph Streams

Given a graph $G = (V, E)$, the random subgraph *RS* is obtained by sampling the *vertices* of V uniformly and independently with probability p, and storing the adjacency list of each sampled vertex . We now design an *adaptive RS* structure for streaming graphs, given a *sampling probability function* $p(m)$ (for e.g., $p(m) = \frac{1}{\sqrt{m}}$) and space function $s = s(m) = 8mp(m)$.

Data Structure. The *current level counter* l_{curr} is initialized to 1 and takes increasing values between 1 and $\log|F|$. The *current sampling probability*, denoted by p_{curr}, is given by $p_{\text{curr}} = 2^{-l_{\text{curr}}+1}$. The current upper limit on the number of edges of the graph is given by m_{curr} that is initialized to $O(1)$. We maintain the invariant that $m_{\text{curr}} = \max(4m, O(1))$. The value of m_{curr} is doubled periodically as necessary. The counter s_{curr} denotes the *current space* provided to the portion of the data structure that stores the adjacency list of the sampled vertices. The invariant $s_{\text{curr}} = s(m_{\text{curr}})$ is maintained. Let S denotes the actual space (in words) used to store the adjacency lists of the sampled vertices and is initialized to 0. The set V_l stores the current set of sampled vertices. For every vertex in V_l, its adjacency list is also stored. The value of m is tracked by the data structure. This can be done exactly for simple graphs; for multi-graphs, an ϵ-approximation to the number m of distinct edges m can be tracked using space $O(\frac{1}{\epsilon^2}(\log n)(\log \frac{1}{\delta}))$ using a standard technique for counting the number of distinct items in a data stream [4,6].

Let $e = \{u, v\}$ be an incoming streaming edge. The set of vertices that are adjacent to a given vertex $u \in V$ is denoted by $adj(u)$. If $u \in V_l$, then we add v to $adj(u)$. If $u \notin V_l$, then, we insert u into V_l with probability p_{curr} and initialize $adj(u)$ as $\{v\}$. If u is not sampled, then, no further action is taken. The procedure is repeated for v independently and the space incurred S is incremented suitably. After processing an incoming edge, we check whether $S < s_{\mathrm{curr}}$, that is, whether there is room for further insertions. If not, then, we perform a sub-sampling operation, if $m < \frac{m_{\mathrm{curr}}}{2}$, or, increase available space, if $m \geq \frac{m_{\mathrm{curr}}}{2}$. In the former case, we *sub-sample*, that is, the sampling probability p_{curr} is halved and for every $u \in V_l$, we retain u and its adjacency list with probability $1/2$ (and, otherwise, u and and its adjacency list are dropped). In the latter case, if $m \geq \frac{m_{\mathrm{curr}}}{2}$ and $S = s_{\mathrm{curr}}$, then, we increase the available space from s_{curr} to $s_{\mathrm{curr}} = s(2m_{\mathrm{curr}})$ and update $m_{curr} = 2m_{curr}$.

Analysis. It is quite straightforward to see that the algorithm maintains the following invariants: $s_{\mathrm{curr}} = s(m_{\mathrm{curr}})$ and $m_{\mathrm{curr}} \leq \max(O(1), 4m)$. The first invariant holds at initialization and at all subsequent space increases. Therefore, space used (in words) is $S = O(s_{\mathrm{curr}}) = O(s(m_{\mathrm{curr}})) = O(s(4m)) = O(s(m))$, since, $s(m)$ is a sub-linear function, and, therefore, $s(4m) \leq 4s(m)$.

For $u \in V$, define an indicator variable y_u that is 1 iff $u \in V_l$ and is 0 otherwise. The space used by the data structure (in words of size $\log n$ bits) is $S = \sum_{u \in V} \deg(u) y_u$. Thus, $\mathsf{E}[S] = \sum_{u \in V} \deg(u) \mathsf{Pr}\{y_u = 1\} = (2m)p_{\mathrm{curr}}$. By Markov's inequality, $\mathsf{Pr}\{S \leq 4\mathsf{E}[S]\} = \mathsf{Pr}\{S \leq 8mp_{\mathrm{curr}}\} \geq \frac{3}{4}$. Therefore, $\mathsf{Pr}\{p_{\mathrm{curr}} \geq \frac{S}{8m}\} = \mathsf{Pr}\{S \leq 8mp_{\mathrm{curr}}\} \geq \frac{3}{4}$. In view of this calculation, we keep $s_2 = O(\log \frac{1}{\delta})$ independent copies of the data structure. Suppose we call the current state of the data structure as *concise* if $p_{\mathrm{curr}} \geq \frac{S}{8m}$. At the time of inference, we consider only the *concise* copies, obtain estimates of P_2 from the concise copies and return the median of these estimates. By Chernoff's bounds, the number of concise copies is $O(\log \frac{1}{\delta})$ with probability $1 - \frac{\delta}{2}$. The space requirement is $O(m \cdot p(m)(\log \frac{1}{\delta})(\log n))$. The above data structure can be *extended to updatable streaming graphs* using a combination of existing data structures [9].

Estimator. An estimate $\hat{P}_2$ is obtained from a concise copy of the RS structure with sampling probability $p = p(m)$ as follows. Let EP_2 denote the number of unordered vertex pairs u and v that are both sampled and have a common neighbor.

$$\hat{P}_2 = \frac{1}{p^2} EP_2 = \frac{1}{p^2} |\{\{u, v\} \mid u, v \in V_l \text{ and } adj(u) \cap adj(v) \neq \phi\}|$$

Finally, we return the median of $t = O(\log \frac{1}{\delta})$ independent estimates.

2.2   Analysis: Graph Based Properties of P_2

For an undirected simple graph $G = (V, E)$ and a vertex $u \in V$, let $\deg(u)$ denote the degree of u in G and let $\deg_2(u)$ denote the number of vertices in $V - \{u\}$ that can be reached from u in two hops.

Lemma 2. *In any graph $G = (V, E)$, $\deg_2(u) \leq (4P_2)^{3/4}$.*

Proof. Let r denote $\deg(u)$ and let T be the set of vertices, not including u, that can be reached from u in two hops. Let $s = |T|$. The vertices adjacent to u contribute $A = \binom{r}{2}$ to P_2. Let B denote the contribution to P_2 by vertex pairs in T. For each fixed value of s, B is minimized if each vertex of $adj(u)$ has either $\lceil \frac{s}{r} \rceil$ or $\lfloor \frac{s}{r} \rfloor$ neighbors in T and no two vertices of $adj(u)$ has any common neighbor (except u). Therefore, $B \geq r\binom{s/r}{2}$. Since, each vertex pair may be counted at most twice, that is once in both A and B, $P_2 \geq \frac{1}{2}(A + B) \geq \frac{1}{2}\binom{r}{2} + \frac{r}{2}\binom{s/r}{2}$. The expression in the *RHS* attains a minimum at $r \approx \frac{s^{2/3}}{2^{1/3}}$ and the corresponding minimum value of P_2 is greater than $\frac{s^{4/3}}{4}$. Thus, $s = \deg_2(u) \leq (4P_2)^{3/4}$. $\square$

Lemma 3 presents a lower bound on the value of P_2 for simple undirected graph.

Lemma 3. *For a connected graph $G = (V, E)$ such that $|E| = m$, $P_2 \geq m - \sqrt{m}$. For a graph with r components, $P_2 \geq m - \sqrt{mr}$.*

Proof. We first show, by induction, that for a connected graph $G = (V, E)$ with m edges, $P_2 \geq m - \sqrt{m}$. Base Case: A connected graph G with one edge, that is, $m = 1$, $0 = P_2 \geq 1 - \sqrt{1} = 0$.

Induction Case. Suppose that the statement of the theorem holds true for graphs with number of edges between 1 and $m - 1$. Consider a connected graph G with m edges. Let x be a lowest degree vertex among all vertices in the connected graph G that are not cut-vertices and let $\deg(x) = q$. (Note that in any graph G, the end vertices of any longest path are not cut-vertices; hence, we can always find x.). Let $y_1, y_2, \ldots, y_q$ denote the neighbors of x. Let $z_1, z_2, \ldots, z_s$ be the set of neighboring vertices of $y_1, \ldots, y_q$, not including x.

Suppose $s \geq q$. Since x is not a cut-vertex of G, deleting x from G leaves G connected. In the resulting graph, G', there are $m - q$ edges, and therefore, by the induction hypothesis, $P_2(G') \geq m - q - \sqrt{m - q}$. In G, x is connected by a path of length 2 to $z_1, z_2, \ldots, z_s$ respectively. Therefore, $P_2 \geq m - q - \sqrt{m - q} + s \geq m - \sqrt{m}$, since, $s \geq q$.

Suppose $s < q$. We first claim that none of $y_1, y_2, \ldots, y_q$ are cut-vertices. To prove this, suppose that y_j is a cut-vertex. Then, by removing y_j from G, $G - \{y_j\}$ has two or more components. Thus, in $G - \{y_j\}$, there is a z_k that is in a different component than x and z_k is adjacent to y_j. The component in $G - \{y_j\}$ that contains x also contains $y_1, \ldots, y_{j-1}, y_{j+1}, \ldots, y_q$. Therefore, there is no edge between y_i and z_k, for, $1 \leq i \leq q, i \neq j$ or between x and z_k. Thus, among the y_i's, z_k is attached only to y_j. Continuing this argument, we can show that if $y_{j_1}, y_{j_2}, \ldots, y_{j_p}$ are cut-vertices in G, then, there exist vertices $z_{k_1}, z_{k_2}, \ldots, z_{k_p}$ distinct from each other such that z_{k_r} is attached to y_{j_r} only and to none of the other y_i's or to x.

Not all of the y_i's can be cut vertices, since, this implies that the number of z_k's is at least q, which contradicts the assumption that $s < q$. Therefore, there exists at least one of the y_i's that is not a cut-vertex, say y_a. Suppose further that there is at least one cut-vertex y_j. Let y_j be attached to z_k such that z_k and x lie in different components in the graph $G - \{y_j\}$. Consider the degree of y_a. It is attached to x and is not attached to z_k. Therefore, $\deg(y_a) \leq 1 + (s-1) = s$. Since, $s < q$, $\deg(y_a) < q = \deg(x)$. By assumption, x is the vertex with the smallest degree among all vertices that are not cut-vertices in G. Since, y_a is not a cut-vertex, and $\deg(y_a) < \deg(x)$, this is a contradiction. Thus, the only conclusion possible is that none of the y_i's are cut-vertices, proving the claim.

Further, since, none of the vertices y_i are cut-vertices, their degree is at least $\deg(x) = q$. Therefore, other than x, each y_i is connected to at least $q-1$ of the z_i's. Since $s < q$, this implies that $s = q-1$, and each of $y_1, y_2, \ldots, y_q$ is attached to each of x and $z_1, z_2, \ldots, z_{q-1}$. The subgraph of the y_i's in one partition and the z_j's and x in the other partition (y_i's and z_j's are disjoint, otherwise $s \geq q$, since G is a simple graph) is the complete bi-partite subgraph $K_{q,q}$. If there are no other edges in the graph, then, we can calculate m and P_2 for $K_{q,q}$ as follows.

$$m = q^2, \; P_2 = q(q-1), \; \text{and} \; P_2 = m - \sqrt{m}$$

which satisfies the statement of the lemma.

Suppose there are edges in addition to the $K_{q,q}$ subgraph formed above. Note that since, $s = q-1$, if there is any edge in the graph G other than the $K_{q,q}$ subgraph, then, there must be an edge attaching some z_k to some vertex u (since, vertices x and $y_1, \ldots, y_q$ are saturated with respect to degree). The vertex u is neither x nor one of $y_1, \ldots, y_q$. We now remove the vertex y_1 from G. The reduced graph G' is still connected since y_1 was not a cut-vertex and has $m - \deg(y_1) = m - q$ edges. Therefore, by the induction hypothesis, $P_2(G') \geq m - q - (m-q)^{1/2}$. In G, y_1 is at distance 2 from each of $y_2, \ldots, y_q$. In addition, y_1, by virtue of the edges (y_1, z_k) and (z_k, u), has a path of length 2 to u. Therefore, $\deg_2(y_1) \geq q - 1 + 1 = q$. Thus, $P_2 \geq (m-q) - (m-q)^{1/2} + q \geq m - \sqrt{m}$.

We can now prove Lemma 3. Let m_c denote the number of edges of component number c, $1 \leq c \leq r$. Since, each component is connected, therefore, $P_2 \geq \sum_{c=1}^{r}(m_c - \sqrt{m_c}) \geq r\left(\frac{m}{r} - \sqrt{\frac{m}{r}}\right) = m - \sqrt{rm}$. $\qquad \square$

2.3   Analysis: Space Usage of the Estimator

For $u \in V$, define an indicator random variable x_u such that $x_u = 1$ iff $u \in V_l$.

Lemma 4. $\mathsf{E}[EP_2] = p^2 P_2$ and $\mathsf{E}[\hat{P}_2] = P_2$.

Proof. $EP_2 = \sum_{\{u,v\} \in P_2} x_u x_v$. So, $\mathsf{E}[EP_2] = p^2 P_2$ and $\mathsf{E}[\hat{P}_2] = \mathsf{E}\left[\frac{EP_2}{p^2}\right] = P_2$. $\qquad \square$

Lemma 5. $\mathsf{Var}[\hat{P}_2] = \frac{P_2}{p^2} + \frac{1}{2p}\sum_{u \in V} \deg_2^2(u)$.

Proof. Since, $EP_2 = \sum_{\{u,v\} \in P_2} x_u x_v$,

$$EP_2^2 = \Big(\sum_{\{u,v\} \in P_2} x_u x_v\Big)^2 = \sum_{\{u,v\} \in P_2} x_u x_v$$

$$+ \sum_{\substack{\{u,v\} \in P_2 \\ \{u,v'\} \in P_2 \\ v \neq v'}} x_u x_v x_{v'} + \sum_{\substack{\{u,v\} \in P_2 \\ \{u',v'\} \in P_2 \\ \{u,v\} \cap \{u',v'\} = \phi}} x_u x_v x_{u'} x_{v'}$$

Taking expectations,

$$\mathsf{E}[EP_2^2] \le p^2 P_2 + \sum_{\substack{u \in V \\ v' \in adj(u) \\ v \ne v'}} \sum_{\substack{v \in adj(u)}} p^3 + \sum_{\substack{\{u,v\} \in P_2 \\ \{u',v'\} \in P_2 \\ \{u,v\} \cap \{u',v'\} = \phi}} p^4$$

$$\le p^2 P_2 + p^3 \sum_{u \in V} \binom{\deg_2(u)}{2} + (p^2 P_2)^2 \; .$$

Using Lemma 4,

$$\mathsf{Var}[EP_2] = \mathsf{E}[EP_2^2] - (\mathsf{E}[EP_2])^2 \le p^2 P_2 + p^3 \sum_{u \in V} \binom{\deg_2(u)}{2} \; .$$

So, $\mathsf{Var}[\hat{P}_2] = \mathsf{Var}[\frac{EP_2}{p^2}] = \frac{1}{p^4}\mathsf{Var}[EP_2] < \frac{P_2}{p^2} + \frac{1}{2p}\sum_{u \in V} \deg_2^2(u)$.   □

Lemma 6. $\mathsf{Pr}\{|\hat{P}_2 - P_2| > \epsilon P_2\} < \frac{2}{9}$, if $p \ge \max(\frac{3}{\epsilon\sqrt{P_2}}, \frac{6}{\epsilon^2 P_2^2}\sum_{u \in V}\deg_2^2(u))$.

Proof. By Chebychev's inequality, $\mathsf{Pr}\{|\hat{P}_2 - \mathsf{E}[\hat{P}_2]| > \epsilon P_2\} \le \frac{\mathsf{Var}[\hat{P}_2]}{25\epsilon^2 P_2^2} < \frac{1}{p^2\epsilon^2 P_2} + \frac{\sum_{u \in V}\deg_2^2(u)}{\epsilon^2 P_2^2 p} + \frac{2}{25} < \frac{1}{9} + \frac{1}{9}$.   □

Lemma 7. *Let G have r components, m edges and n vertices. Then, $\mathsf{Pr}\{|\hat{P}_2 - P_2| \le 6\epsilon P_2\} \ge 1 - \delta$. The space requirement is $O(\frac{m}{\epsilon^2(m-r)^{1/4}}(\log\frac{1}{\delta})(\log n))$ bits, with probability $1 - \delta$.*

Proof. The space requirement is $O(mp)$, where, by Lemma 6, $mp = O(\max(\frac{m}{\epsilon^2\sqrt{P_2}}, \frac{m}{\epsilon^2 P_2^2}\sum_{u \in V}\deg_2^2(u)))$. By Lemma 3, $P_2 \ge m - \sqrt{rm}$. Therefore, $\frac{m}{\epsilon\sqrt{P_2}} = \frac{m}{\epsilon(m-\sqrt{rm})} = \frac{m^{1/2}}{(m-r)^{1/2}}$. Further, since, $\sum_{u \in V}\deg_2(u) = 2P_2$, we have, by Lemma 2,

$$\sum_{u \in V} \deg_2^2(u) \le (\max_{w \in V}\deg_2(w)) \sum_{u \in V} \deg_2(u) \le (4P_2)^{3/4}(2P_2) \le 8P_2^{7/4}.$$

By Lemma 3, $P_2 \ge m - \sqrt{mr} = \sqrt{m}(\sqrt{m} - \sqrt{r}) = \sqrt{m}\frac{m-r}{\sqrt{m}+\sqrt{r}} \ge \frac{m-r}{2}$, since, $r \le m$. Using this, it follows that $\frac{m}{\epsilon^2 P_2^2}\sum_u \deg_2^2(u) \le \frac{8m}{\epsilon^2 P_2^{1/4}} \le \frac{16m}{(m-r)^{1/4}}$. To boost the confidence to $1 - \delta$, we keep $O(\log\frac{1}{\delta})$ independent copies and return the median from the concise copies. The space required is therefore $O(\frac{m}{\epsilon^2(m-r)^{1/4}}(\log\frac{1}{\delta})(\log n))$ bits.   □

3   Lower Bounds

In this section, we present space lower bounds.

Lemma 8. *An algorithm that estimates P_2 for undirected and connected streaming graphs in the insert-only model to within a factor of $1 \pm \frac{1}{8}$ with probability $\frac{2}{3}$ requires $\Omega(n + \sqrt{m})$ bits.*

Proof. We reduce a special case of the two-party set disjointness problem in which parties A and B are each given a subset of $\{0, 1, \ldots, n-1\}$ of size at least $\frac{n}{3}$ with the promise that the subsets are either disjoint or have exactly one element in common. The parties have to determine whether the sets are disjoint. This problem has communication complexity $\Omega(n)$ bits. Suppose there is an algorithm $\mathcal{A}$ satisfying the premises of the lemma. A and B each construct in their local memory a complete graph whose nodes correspond to the items in the subset given to it. A inserts the edges corresponding to its complete graph into the data structure for $\mathcal{A}$ and sends it to B. B inserts the edges of its complete graph into the data structure of $\mathcal{A}$ and estimates P_2. If the sets are disjoint, then, $P_2 \leq \frac{5n^2}{16}$, and otherwise, $P_2 \geq \frac{7n^2}{16}$, allowing $\mathcal{A}$ to distinguish between the two cases. Hence, $\mathcal{A}$ requires $\Omega(n)$ bits. In the constructed graph, $m = \Theta(n^2)$, hence, the space complexity is $\Omega(\sqrt{m})$.

In the above construction, the graph is either connected (when the subsets intersect) or has two components (disjoint case). An additional tree-structure ensures that the graph is always connected. For $i \in \{0, 1, \ldots, n-1\}$, B inserts new vertices v_{2i} and v_{3i}, with edges between v_i and v_{2i} and between v_{2i} and v_{3i}. The nodes $\{v_{3i} : 0 \leq i \leq n-1\}$ are then made the leaf nodes of a complete binary tree (as much as possible) by adding new vertices. The resulting graph is connected. The contribution to P_2 by the new vertices is as follows. $\deg_2(v_{2i}) = 1 + \deg(v_i)$, for $0 \leq i \leq n-1$, and the contribution to P_2 by the remaining tree vertices is at most $n-1$ (vertex pairs at the same level) $+ n - 3$ (vertex pairs where one vertex is a grandparent of the other) $= 2n - 4$. Thus, total P_2 of the new graph is $n + 2n - 4 + 2$ oldP_2, where, oldP_2 is the P_2 of the graph prior to the addition of the tree structure. The rest of the argument proceeds as before. □

Lemma 9. *Deterministically estimating P_2 over streaming graphs to within factor of $(1 \pm \frac{1}{4})$ requires $\Omega(m)$ space.*

The proof of this lemma may be found in [9].

Estimating P_k over directed streaming graphs. We show that for *directed* streaming graphs, estimating P_k for $k \geq 2$, to any multiplicative factor requires $\Omega(m)$ space. The reductions use the standard bit vector index problem: Party A is given a bit-vector v of size r and party B is given an index i, $1 \leq i \leq r$. B has to determine whether $v[i] = 1$. The communication allowed is one-way from A to B. This problem has communication complexity of $\Omega(r)$ [7,3].

Lemma 10. *Estimating P_k for directed streaming graphs to within any multiplicative accuracy factor requires $\Omega(m)$ bits.*

Proof. We will reduce a special case of the bit-vector index problem, where, it is given that exactly $\frac{r}{2}$ bits of v have value 1. The communication complexity of this problem is also $\Omega(r)$. Let $r = 2n$ and let $\mathcal{A}$ be an algorithm for estimating P_2. For every $v[i] = 1$ in the bit-vector v, party A inserts a directed edge $(r+1, i)$ to the summary structure of algorithm $\mathcal{A}$. A then sends the summary structure to B. Given index j, B adds the set of directed edges $\{(j, k) \mid r+2 \leq k \leq r+n+2\}$, to the summary structure that it received from A. If $v[j] = 1$, then $P_2 = n$, else $P_2 = 0$, proving the claim for P_2. The extension for P_k is analogous. □

Lemma 11. *For $k \geq 3$, estimating P_k to within factor of $1 \pm \frac{1}{3}$ with probability $\frac{3}{4}$ over undirected streaming graphs with n vertices requires $\Omega(n^2)$ bits.*

Proof. We reduce the bit-vector index problem to the problem of estimating P_3. Let $r = \frac{n(n-1)}{2}$ and let $v[1 \dots r]$ be the given vector of 0's and 1's. Let $\mathcal{B}$ be an algorithm for estimating P_2 with the specified accuracy and confidence. Each index $1 \leq r \leq \frac{n(n-1)}{2}$ is written uniquely as a pair of distinct numbers, (u, w), each lying between 0 and $n - 1$. This mapping is used to create a graph $G = (V, E)$, where, $V = \{1, 2, \dots, 9n\}$. For every index $j = (u, w)$ such that $v[j] = 1$, we add an edge $(u, w) \in E$. Next, for the given index $i = (c, d)$, we add $8n$ new vertices to the graph, and attach $4n$ of them to c and $4n$ of them to d. These edges are given as input stream to $\mathcal{B}$. We now use $\mathcal{B}$ to estimate P_3. $v[b] = 1$ iff there is an edge between c and d in G. In this case, $P_3 \geq 16n^2$, and, otherwise, $P_3 \leq 8n^2 + \binom{n}{2}$. Therefore the space requirement by P_3 is $\Omega(r) = \Omega(n^2)$. The proof can be easily extended to P_k, $k > 3$. □

Theorem 12. *A deterministic algorithm for testing connectivity of an undirected graph in the updatable streaming graph model requires $\Omega(n^2)$ space.*

Proof. Let $G = (V, E)$ be a connected graph and let $G' = (V, E')$ be the edge-complement graph on the same set of vertices. Consider the family of graphs for which G and G' are both connected. For this family of graphs, checking for edge-membership can proceed as below. (u, v) is an edge in G iff there is a sequence of edges $e_1, \dots, e_{k-1}, e_k = (u, v)$ in G, such that after the deletion of $e_1, e_2, \dots, e_{k-1}$ in sequence, the graph remains connected, but gets disconnected after $e_k = (u, v)$ is deleted thereafter. The sequence of edges $e_1, \dots, e_{k-1}$ can be thought of as a certificate of membership of (u, v) in G. Analogously, if (u, v) is not in G, then, it is in G', and therefore, there exists a certificate for membership of (u, v) in G'. This certificate serves as a certificate that (u, v) is not in G. Hence checking edge membership reduces to connectivity testing problem.

Given an algorithm that maintains a summary structure for testing connectivity of a streaming graph, we use it to maintain a pair of summaries corresponding to G and its complement G'. This is easily done by letting $E = \phi$ and $E' = K_n$, where K_n is the clique of n vertices. Corresponding to each edge update, we propagate the update to the summary structure for G and propagate the complement of the update to the summary structure for G'.

We now obtain a lower bound on the number of graph-complement pairs (G, G') over n vertices such that both G and G' are connected. Consider the complete graph K_n on n vertices, for $n > 2$. Choose a spanning tree C of K_n that is a chain. Consider the remaining graph defined by the set of edges in

$K_n - C$. This graph remains connected for $n \geq 4$. Let D be a spanning tree of the graph defined by edges in $K_n - C$. Place the set of edges in C in G and the set of edges in D in G'. The number of remaining edges is $\binom{n}{2} - 2(n-1)$. Each of these edges can be placed either in G or in G' in $2^{\binom{n}{2}-2(n-1)}$ ways. Each of these ways gives a different (G, G') pair. By construction, G and G' contain C and D respectively, and are therefore connected. Therefore, the number of graph-complement pairs (G, G') over n vertices such that both G and G' are connected is at least $2^{\binom{n}{2}-2(n-1)}$.

The algorithm that tests for edge-membership must have a different memory pattern for each of the graph-complement pairs (G, G'). Otherwise, given two distinct pairs (G, G') and (H, H'), there are edge pairs (e, e') that distinguish them. Mapping them to the same pattern causes the algorithm to make at least one error when presented with the certificates of the edges e and e', respectively. Hence checking edge-membership requires space $\Omega(\log(2^{\binom{n}{2}-2(n-1)})) = \Omega(n^2)$ bits . Since edge-membership can be reduced to connectivity testing, the statement of the lemma follows.   □

Corollary 13. *Deterministic algorithms for the following problems require $\Omega(n^2)$ space in the updatable graph streaming model: (a) estimating the size of the transitive closure of an undirected graph to within a factor of $1 \pm \frac{1}{5}$, and (b) estimating the diameter of an undirected graph to within any approximation factor.*

The proof of the corollary may be found in [9]. Note that testing connectivity and maintaining the size of transitive closure is easily solved using $O(n \log n)$ space in the insert-only streaming model [7,3].

References

1. C. Demetrescu, I. Finocchi, and A. Ribichini. "Trading off space for passes in graph streaming problems". In *Proceedings of ACM SODA*, 2006.
2. J. Feigenbaum, S. Kannan, A. McGregor, S. Suri, and J. Zhang. On graph problems in a semi-streaming model. In *Proceedings of ICALP*, pages 531–543, 2004.
3. J. Feigenbaum, S. Kannan, A. McGregor, S. Suri, and J. Zhang. Graph distances in the streaming model: the value of space. In *Proceedings of ACM SODA*, 2005.
4. Philippe Flajolet and G.N. Martin. "Probabilistic Counting Algorithms for Database Applications". *J. Comp. Sys. and Sc.*, 31(2):182–209, 1985.
5. S. Ganguly, M.N. Garofalakis, A. Kumar, and R. Rastogi. "Join-distinct aggregate estimation over update streams". In *Proceedings of ACM PODS*, 2005.
6. P. B. Gibbons and S. Tirthapura. "Estimating simple functions on the union of data streams". In *Proceedings of ACM SPAA*, 2001.
7. M. Henzinger, P. Raghavan, and S. Rajagopalan. Computing on data streams. Technical Note 1998-011, Digital Systems Research, Palo Alto, CA, May 1998.
8. S. Muthukrishnan. "*Data Streams: Algorithms and Applications*". Foundations and Trends in Theoretical Computer Science, *Vol. 1, Issue 2*, 2005.
9. Barna Saha. "Space Complexity of Estimating Aggregate Path Metrics over Massive Graph Streams and Related Metrics". Master's thesis, IIT Kanpur, Computer Science, 2006.

Diamond Triangulations Contain Spanners of Bounded Degree*

Prosenjit Bose, Michiel Smid, and Daming Xu

School of Computer Science, Carleton University, Ottawa, ON, Canada K1S 5B6
{jit, michiel, dxu5}@scs.carleton.ca

Abstract. Given a triangulation G, whose vertex set V is a set of n points in the plane, and given a real number γ with $0 < \gamma < \pi$, we design an $O(n)$-time algorithm that constructs a connected spanning subgraph G' of G whose maximum degree is at most $14 + \lceil 2\pi/\gamma \rceil$. If G is the Delaunay triangulation of V, and $\gamma = 2\pi/3$, we show that G' is a t-spanner of V (for some constant t) with maximum degree at most 17, thereby improving the previously best known degree bound of 23. If G is the graph consisting of all Delaunay edges of length at most 1, and $\gamma = \pi/3$, we show that G' is a t-spanner (for some constant t) of the unit-disk graph of V, whose maximum degree is at most 20, thereby improving the previously best known degree bound of 25. Finally, if G is a triangulation satisfying the diamond property, then for a specific range of values of γ dependent on the angle of the diamonds, we show that G' is a t-spanner of V (for some constant t) whose maximum degree is bounded by a constant dependent on γ.

1 Introduction

Let V be a set of n points in the plane and let $t \geq 1$ be a real number. An undirected graph G with vertex set V is called a *t-spanner* of V, if for any two vertices u and v of V, G contains a path between u and v, whose length is at most $t|uv|$, where $|uv|$ denotes the Euclidean distance between u and v.

The problem of constructing a t-spanner with $O(n)$ edges for any given point set has been studied intensively; see the book by Narasimhan and Smid [11].

In this paper, we focus on spanners that are *plane*, i.e. the interiors of any two (straight-line) edges of the spanner are disjoint. Chew [5] and Dobkin *et al.* [7] were the first to show the existence of plane spanners. Dobkin *et al.* proved that the *Delaunay triangulation* of V is a t-spanner of V, for some constant $t = ((1 + \sqrt{5})/2)\pi$. Keil and Gutwin [8] improved the analysis, and showed that the Delaunay triangulation is a t-spanner for $t = 4\pi\sqrt{3}/9$. A more general result appears in Bose *et al.* [2]: For every two vertices u and v of V, the Delaunay triangulation contains a path between u and v of length at most $\frac{4\pi\sqrt{3}}{9} \cdot |uv|$, all of whose edges have length at most $|uv|$.

---

* This research was supported by the Natural Science and Engineering Research Council of Canada.

T. Asano (Ed.): ISAAC 2006, LNCS 4288, pp. 173–182, 2006.
© Springer-Verlag Berlin Heidelberg 2006

Das and Joseph [6] generalized these results to triangulations that satisfy the so-called *diamond property*: Let G be a triangulation of V, and let α be a real number with $0 < \alpha < \frac{\pi}{2}$. Let e be an edge of G, and consider the two isosceles triangles Δ_1 and Δ_2 with base e and base angle α. We say that the edge e satisfies the α-*diamond property*, if at least one of Δ_1 and Δ_2 does not contain any point of V in its interior. We say that the triangulation G satisfies the α-*diamond property*, if every edge e of G satisfies this property.

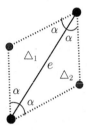

Fig. 1. An illustration of the diamond property. At least one of the triangles Δ_1 and Δ_2 does not contain any point of V.

Das and Joseph [6] showed that any triangulation satisfying the α-diamond property is a t-spanner, for some real number t that only depends on the value of α. (In fact, Das and Joseph considered plane graphs that, additionally, satisfy the so-called *good polygon property*.) The analysis was refined by Lee [9], who showed that $t \leq \frac{8(\pi-\alpha)^2}{\alpha^2 \sin^2(\alpha/4)}$. It is clear that the Delaunay triangulation satisfies the α-diamond property, for $\alpha = \pi/4$. Das and Joseph proved that both the greedy triangulation and the minimum weight triangulation satisfy the α-diamond property, for some constant α.

None of the results mentioned above lead to plane spanners in which the degree of every vertex is bounded by a constant. Bose *et al.* [1] were the first to show the existence of a plane t-spanner (for some constant t), whose maximum vertex degree is bounded by a constant. To be more precise, they showed that the Delaunay triangulation of any set V of n points in the plane contains a spanning subgraph, which is a t-spanner for V, where $t = \frac{4\pi(\pi+1)\sqrt{3}}{9}$, and whose maximum degree is at most 27. This result was improved by Li and Wang [10]: For any real number γ with $0 < \gamma \leq \pi/2$, the Delaunay triangulation contains a spanning subgraph, which is a t-spanner, where $t = \max\{\frac{\pi}{2}, 1 + \pi \sin \frac{\gamma}{2}\} \cdot \frac{4\pi\sqrt{3}}{9}$, and whose maximum degree is at most $19 + \lceil 2\pi/\gamma \rceil$. For $\gamma = \pi/2$, the degree bound is 23. In this paper, we further improve the degree bound:

Theorem 1. *Let V be a set of n points in the plane, and let γ be a real number with $0 < \gamma \leq 2\pi/3$. Assume that we are given the Delaunay triangulation G of V. Then, in $O(n)$ time, we can compute a spanning subgraph G' of G, such that G' is a t-spanner of V, where*

$$t = \begin{cases} \frac{4\pi\sqrt{3}}{9} \cdot \max\left\{\frac{\pi}{2}, 1 + \pi \sin\frac{\gamma}{2}\right\} & \text{if } \gamma < \pi/2, \\ \frac{4\pi\sqrt{3}}{9}\left(\pi + 9 \cdot \max\left\{\frac{\pi}{2}, 1 + \pi \sin\frac{\gamma}{2}\right\}\right) & \text{if } \pi/2 \leq \gamma \leq 2\pi/3, \end{cases}$$

and the maximum degree of G' is at most $14 + \lceil 2\pi/\gamma \rceil$. Thus, for $\gamma = 2\pi/3$, the degree bound is 17.

We obtain this result by designing a linear algorithm that, when given an arbitrary triangulation G of the point set V, computes a spanning subgraph G' of G, that satisfies the degree bound in Theorem 1.

We also extend this result to the *unit-disk graph*, which is a graph where every two distinct points u and v in the vertex set are connected by an edge if and only if $|uv| \leq 1$. A t-spanner of the unit-disk graph is a subgraph of the unit-disk graph with the property that for every edge (u, v) of the unit-disk graph, there exists a path between u and v whose length is at most $t|uv|$. Let G be the graph consisting of all edges in the Delaunay triangulation of V, whose length is at most one. It follows from the result of Bose *et al.* [2] which was mentioned above, that G is a $\frac{4\pi\sqrt{3}}{9}$-spanner of the unit-disk graph of V. This construction for the Delaunay triangulation was modified by Li and Wang [10] to obtain a plane t-spanner (for some constant t) of the unit-disk graph whose maximum degree is at most 25. By modifying our algorithm, we obtain the following result:

Theorem 2. *Let V be a set of n points in the plane, and let γ be a real number with $0 < \gamma \leq \pi/3$. Assume that we are given the Delaunay triangulation G of V. Then, in $O(n)$ time, we can compute a plane graph G', such that G' ($G' \subseteq G$ does not necessarily hold) is a t-spanner of the unit-disk graph of V, where*

$$t = \frac{4\pi\sqrt{3}}{9} \cdot \max\left\{\frac{\pi}{2}, 1 + \pi \sin\frac{\gamma}{2}\right\},$$

and the maximum degree of G' is at most $14 + \lceil 2\pi/\gamma \rceil$. Thus, for $\gamma = \pi/3$, the degree bound is 20.

Bounded degree plane spanners of the unit-disk graph have applications in topology control of wireless ad hoc networks, see [10]. In such a network, a vertex u can only communicate with those vertices that are within the communication range of u. If we assume that this range is equal to one for each vertex, then the unit-disk graph models a wireless ad hoc network. Many algorithms for routing messages in such a network require that the underlying network topology is plane. Moreover, if the maximum degree is small, then the throughput of the network can be improved significantly.

In the final part of the paper, we show the following result:

Theorem 3. *Let V be a set of n points in the plane, let α be a real number with $0 < \alpha \leq \pi/2$, and let G be a triangulation of V that satisfies the α-diamond property. Then, in $O(n)$ time, we can compute a spanning subgraph G' of G, such that G' is a t-spanner of V, where*

$$t = \left(1 + \frac{2(\pi - \alpha)}{\alpha \sin \frac{\alpha}{4}} \cdot \max\left\{1, 2\sin\frac{\alpha}{2}\right\}\right) \frac{8(\pi - \alpha)^2}{\alpha^2 \sin^2 \frac{\alpha}{4}},$$

and the maximum degree of G' is at most $14 + \lceil 2\pi/\alpha \rceil$.

Thus, by combining Theorem 3 with the results of Das and Joseph [6], it follows that both the greedy triangulation and the minimum weight triangulation contain a t-spanner (for some constant t) whose maximum degree is bounded by a constant.

Full details of all proofs are available in [4].

2 Computing a Bounded-Degree Spanning Subgraph of a Triangulation

Let V be a set of n points in the plane, and let $G = (V, E)$ be a planar graph. We define a numbering of the elements of V in the following way: We pick a vertex of minimum degree and assign it label 1. Then delete this vertex together with its incident edges, increment label by 1 and recursively define a numbering of the remaining $n - 1$ vertices. The resulting numbering $(v_1, v_2, \ldots, v_n)$ of the vertex set V is called a *low-degree numbering*.

The following algorithm computes a bounded-degree spanning subgraph of any given triangulation; for an illustration, refer to Figure 2. In this algorithm, $N(v)$ denotes the set of *neighbors* of the vertex v in G, i.e., $N(v) = \{w \in V : (v, w) \in E\}$.

Algorithm *BDegSubgraph*(G, γ)
Input: A triangulation $G = (V, E)$ whose vertex set V is a set of n points in
       the plane, and a real number γ with $0 < \gamma < \pi$.
Output: A spanning subgraph $G' = (V, E')$ of degree at most $14 + \lceil\frac{2\pi}{\gamma}\rceil$.
1.   compute a low-degree numbering $(v_1, v_2, \cdots, v_n)$ of $G = (V, E)$;
2.   label each vertex of V as "unprocessed";
3. $E' = \emptyset$;
4.   **for** $i = n$ **downto** 1
5.   **do if** v_i has "unprocessed" neighbors
6.   **then** compute the closest "unprocessed" neighbor x of v_i;
7.   divide the plane into cones $C_1, \ldots, C_{\lceil 2\pi/\gamma\rceil}$ with apex v_i
                    and angle at most γ such that the segment $v_i x$ is on the
                    boundary between C_1 and C_2;
8.   add the edge (v_i, x) to E'
9.   **else** go to line 18
10.   **for** each cone $C \notin \{C_1, C_2\}$
11.   **do** compute the closest "unprocessed" vertex $w \in V \cap C \cap N(v_i)$;
12.   **if** w exists
13.   **then** add the edge (v_i, w) to E'

14.   let $w_0, w_1, \cdots, w_{d-1}$ be the vertices in $N(v_i)$, ordered in clockwise
             order around v_i;
15.   **for** $k = 0$ **to** $d - 1$
16.   **if** w_k and $w_{(k+1) \bmod d}$ are both "unprocessed"
17.   **then** add the edge $(w_k, w_{(k+1) \bmod d})$ to E';
18.   label v_i as "processed";
19. **return** the graph $G' = (V, E')$

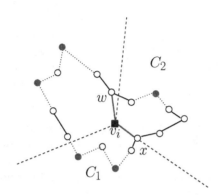

Fig. 2. An illustration of algorithm BDEGSUBGRAPH(G, γ), for $\gamma = 2\pi/3$, when processing vertex v_i. The figure shows v_i and all vertices in $N(v_i)$. Solid vertices represent the "processed" vertices, whereas hollow vertices represent the "unprocessed" vertices. When processing v_i, the algorithm adds the solid edges to the graph G'.

Lemma 1. *Algorithm* BDEGSUBGRAPH(G, γ) *returns a connected spanning subgraph G' of G, whose maximum degree of G' is at most $14 + \lceil \frac{2\pi}{\gamma} \rceil$.*

Proof. Let u be an arbitrary element of V. Since the algorithm processes the vertices in the reverse order of the low-degree numbering, we know that, before u is processed, it has at most 5 processed neighbors $u_1, u_2, \cdots, u_k$ where $k \leq 5$ due to the low-degree numbering. In the worst case, each u_j can increase the degree of u by 3. (This is depicted in Figure 2, where the degree of the unprocessed vertex w is increased by 3 when v_i is processed.) During the processing of u, the degree of u is increased by at most $\lceil \frac{2\pi}{\gamma} \rceil - 1$. After u has been processed, the degree of u does not change. Therefore, the degree of u is at most $15 + \lceil \frac{2\pi}{\gamma} \rceil - 1$. The connectivity of G' follows from results in the next section. □

3 Bounded-Degree Spanners of the Delaunay Triangulation

In this section, we assume that $G = (V, E)$ is the Delaunay triangulation of the point set V. Let $G' = (V, E')$ be the output of algorithm BDEGSUBGRAPH(G, γ). In this section, we will prove that G' is a spanner of V. Our analysis uses the following two lemmas.

Lemma 2. *[3] Given the following four conditions: (a) $0 < \gamma < \pi/2$ and u, v, and v' are three points of V such that (u,v) and (u,v') are Delaunay edges with $\angle vuv' \leq \gamma$; (b) $v' = s_1, s_2, \ldots, s_{k-1}, s_k = v$ are the Delaunay neighbors of u between v' and v sorted in angular order around u; (c) $|uv'| \leq |us_i|$ for all i with $1 \leq i \leq k$; (d) P_{uv} is the path $u, v', s_2, s_3, \ldots, s_{k-1}, v$ in G between u and v. Then the length of P_{uv} is at most $t_\gamma |uv|$ where*

$$t_\gamma = \max \left\{ \frac{\pi}{2}, 1 + \pi \sin \frac{\gamma}{2} \right\}.$$

Lemma 3. *[1] Let u, x, and y be three points of V, such that (u,x) and (u,y) are Delaunay edges and $\angle xuy < \pi/2$. Let $x = s_1, s_2, \ldots, s_{k-1}, s_k = y$ be the Delaunay neighbors of u between x and y, sorted in angular order around u. Let Q_{xy} be the path $x, s_2, s_3, \ldots, s_{k-1}, y$ in G between x and y. Then, the length of Q_{xy} is at most $\frac{\pi}{2} (|ux| + |uy|)$.*

We can now complete the proof of Theorem 1. We fix an angle γ with $0 < \gamma \leq 2\pi/3$.

Recall that G is a $\frac{4\pi\sqrt{3}}{9}$-spanner of V; see Keil and Gutwin [8]. Also, G' is a spanning subgraph of G. Therefore, it suffices to show that for each edge (u,v) of $G \setminus G'$, the graph G' contains a path between u and v, whose length is at most a constant factor times $|uv|$.

Throughout the rest of the proof, we fix an edge (u,v) of $G \setminus G'$. We assume that u is processed before v. Let C be the cone with apex u and having angle at most γ that contains v and that is constructed when vertex u is processed. Let v' be the closest Delaunay-neighbor of u that is contained in C and that is unprocessed at the moment when u is processed. We assume that (u, v') is clockwise to the right of (u, v).

Case 1: $\angle vuv' < \frac{\pi}{2}$.
Consider the path P_{uv} of Lemma 2 between u and v; refer to Figure 3(a). By Lemma 2, the length of this path is at most $t_\gamma |uv|$.

If, at the moment when u is processed, all Delaunay-neighbors of u between v and v' are unprocessed, then it follows from algorithm BDegSubgraph that P_{uv} is a path in G'. Assume that at least one Delaunay-neighbor u' of u between v and v' has already been processed; refer to Figure 3(b). Let u' be the first such Delaunay-neighbor in clockwise order from (u,v). Let s be the Delaunay-neighbor of u such that s is between v and u' and $\triangle_{usu'}$ is a Delaunay-triangle. Then u' is processed before s. Also, u' is processed before u. Thus, during the processing of u', the algorithm adds the edge (u,s) to G'. Let Q be the subpath of P_{uv} that starts at s and ends at v. Let Q' be the path obtained by concatenating the edge (u,s) and the subpath Q. It follows from algorithm BDegSubgraph that Q' is a path in G' between u and v. By the triangle inequality, the length of Q' is at most the length of P_{uv}.

Case 2: $\angle vuv' \geq \frac{\pi}{2}$ and there is at least one Delaunay-neighbor w of u such that $\angle vuw < \frac{\pi}{2}$ and $\angle wuv' < \frac{\pi}{2}$.

This case is depicted in Figure 4. The analysis for this case appears in the full paper.

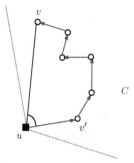
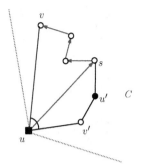

(a) All Delaunay-neighbors of u are unpro- (b) At least one Delaunay-neighbor of u
cessed   has been processed

Fig. 3. Case 1 in the proof of Theorem 1

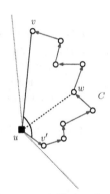
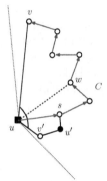

(a) All Delaunay-neighbors of u are unpro-(b) At least one Delaunay-neighbor of u has
cessed   been processed

Fig. 4. Case 2 in the proof of Theorem 1

Case 3: $\angle vuv' \geq \frac{\pi}{2}$ and there is no Delaunay-neighbor w of u such that $\angle vuw < \frac{\pi}{2}$ and $\angle wuv' < \frac{\pi}{2}$.

In this case, there exist two Delaunay-edges (u, x) and (u, y) such that $\angle yuv' \geq \frac{\pi}{2}$, $\angle vux \geq \frac{\pi}{2}$, and (x, y) is a Delaunay-edge. (Refer to Figure 5(a).)

Let P_{ux} be the path that starts at u, follows the edge (u, v'), and then follows the Delaunay-neighbors of u from v' to x. Let Q_{yv} be the path that starts at y, and follows the Delaunay-neighbors of u from y to v. Let P be the concatenation of P_{ux}, the edge (x, y), and Q_{yv}. Then the length of P is equal to the sum of the lengths of P_{ux}, Q_{yv}, and $|xy|$. Since $\angle vux \geq \frac{\pi}{2}$, we have $\angle xuv' < \frac{\pi}{2}$. Therefore, it follows from Lemma 2 that the length of P_{ux} is at most $t_\gamma |ux|$. Since $\angle yuv' \geq \frac{\pi}{2}$, we have $\angle vuy < \frac{\pi}{2}$. Thus, by Lemma 3, the length of Q_{vy} is at most $\frac{\pi}{2}(|uv|+|uy|)$. Thus, the length of P is at most $t_\gamma |ux| + |xy| + \frac{\pi}{2}(|uv| + |uy|)$. We will prove an upper bound on this quantity in terms of $|uv|$.

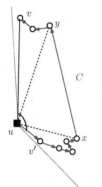
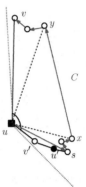

(a) All Delaunay-neighbors of u are unpro-   (b) At least one Delaunay-neighbor of u
cessed   has been processed

Fig. 5. Case 3 in the proof of Theorem 1

If x is inside $\triangle_{uyv'}$, then, by convexity, $|xy| + |v'x| \leq |uy| + |uv'|$. By considering $\triangle_{yuv'}$, and recalling that $\angle yuv' \geq \frac{\pi}{2}$, we have $|uy| + |uv'| \leq 2|v'y|$. If x is outside $\triangle_{uv'y}$, let C' be the circle through u, v', and y. Since $\triangle_{uv'y}$ is not a triangle in the Delaunay triangulation of $\{u, v', x, y\}$, x is contained in C'. Since $\angle yuv' \leq \gamma \leq \frac{2\pi}{3}$, it can be shown that $|xy| + |v'x| \leq 2|v'y|$. Thus, for any location of x, we have $|xy| + |v'x| \leq 2|v'y|$.

Since $|ux| \leq |uv'| + |v'x|$, the total length of the path P_{ux} and edge (x, y) is at most

$$
\begin{aligned}
t_\gamma |ux| + |xy| &\leq t_\gamma(|ux| + |xy|) \\
&\leq t_\gamma(|uv'| + |v'x| + |xy|) \\
&\leq t_\gamma(|uv'| + 2|v'y|) \\
&= t_\gamma |uv'| + 2t_\gamma |v'y|.
\end{aligned}
$$

If y is inside $\triangle_{uvv'}$, then, by convexity, $|vy| + |v'y| \leq |uv| + |uv'| \leq 2|uv|$. If y is outside $\triangle_{uvv'}$, let C'' be the circle through u, v, and v'. Since $\triangle_{uvv'}$ is not a triangle in the Delaunay triangulation of $\{u, v', y, v\}$, y is contained in C''. Since $\angle vuv' \leq \gamma \leq \frac{2\pi}{3}$, it can be shown that $|vy| + |v'y| \leq 2|vv'| \leq 2(|uv| + |uv'|) \leq 4|uv|$. Thus, for any location of y, we have $|vy| + |v'y| \leq 4|uv|$.

Since $|uy| \leq |uv| + |vy|$, the length of the path P is at most

$$
\begin{aligned}
t_\gamma |uv'| + 2t_\gamma |v'y| + \frac{\pi}{2}(|uv| + |uy|) &\leq t_\gamma |uv| + 2t_\gamma |v'y| + \frac{\pi}{2}(|uv| + |uv| + |vy|) \\
&\leq (\pi + t_\gamma)|uv| + 2t_\gamma(|v'y| + |vy|) \\
&\leq (\pi + 9t_\gamma)|uv|.
\end{aligned}
$$

If, at the moment when u is processed, all Delaunay-neighbors of u between v and v' are unprocessed (see Figure 5(a)), then P is a path in G' between u and v. Otherwise, at least one Delaunay-neighbor u' of u between v and v' had already been processed (see Figure 5(b)). In this case, by a similar argument as in Case 1, G' contains a path between u and v which is obtained by shortcutting P.

Since $t_\gamma \geq \frac{\pi}{2}$, we have $\pi + 9t_\gamma \geq 3\pi + 5t_\gamma$. Therefore, for the case when $\pi/2 \leq \gamma \leq 2\pi/3$, Cases 1–3 prove that G' is a t-spanner of V, where $t = \pi + 9t_\gamma$.

4 Bounded-Degree Spanners of the Unit-Disk Graph

Let V be a set of n points in the plane, and let $G = (V, E)$ be the graph with vertex set V, and whose edge set E is the set of all edges in the Delaunay triangulation of V whose length is at most one. In general, G is not a triangulation, even though it is a plane graph. Nevertheless, we can run algorithm BDEGSUB-GRAPH(G, γ), and the degree bound in Lemma 1 is still valid. Observe that in line 17 of this algorithm, edges may be added to the graph G' which are not in G. Using an analysis that is almost identical to the one in Li and Wang [10], it can be shown that, when $0 < \gamma \leq \pi/3$, (i) the graph G' is plane, and (ii) G' is a spanning subgraph of the unit-disk graph of V.

Consider an arbitrary edge (u, v) of $G \setminus G'$. Then, using the same proof technique as in Theorem 1, it can be shown that, again when $0 < \gamma \leq \pi/3$, G' contains a path between u and v whose length is at most $t_\gamma |uv|$. Since Bose et al. [2] have shown that G is a $\frac{4\pi\sqrt{3}}{9}$-spanner of the unit-disk graph of V, Theorem 2 follows.

5 Bounded-Degree Spanners of a Diamond Triangulation

Let V be a set of n points in the plane, let α be a real number with $0 < \alpha < \frac{\pi}{2}$, and let $G = (V, E)$ be a triangulation that satisfies the α-diamond property; see Section 1 for the definition of this property. In this section, we complete the proof of Theorem 3, by showing that the output $G' = (V, E')$ of algorithm BDEGSUBGRAPH(G, α) is a t-spanner, for some value t that only depends on α. The key of the proof is to use the following lemma, which generalizes Lemma 2. This lemma can be proved using results by Das and Joseph [6] and Lee [9].

Lemma 4. *Let u, v, and v' be three points of V, such that $\angle vuv' \leq \alpha$. Let $v' = s_1, s_2, \cdots, s_k = v$ be the neighbors of u in G between v' and v, sorted in angular order around u. Assume that $|uv'| \leq |us_i|$ for all i with $1 \leq i \leq k$. Let P_{uv} be the path $u, v', s_2, s_3, \cdots, s_{k-1}, v$. Then, P_{uv} is a path in G between u and v, whose length is at most $t'_\alpha |uv|$, where*

$$t'_\alpha = 1 + \frac{2(\pi - \alpha)}{\alpha \sin \frac{\alpha}{4}} \cdot \max\left\{1, 2\sin\frac{\alpha}{2}\right\}.$$

We are now able to complete the proof of Theorem 3: Let (u, v) be an arbitrary edge of $G \setminus G'$. Using the same proof technique as in Theorem 1, and using Lemma 4, it can be shown that G' contains a path between u and v whose length is at most $t'_\alpha |uv|$. Das and Joseph [6] and Lee [9] have shown that G is a t'-spanner of V, for $t' = \frac{8(\pi - \alpha)^2}{\alpha^2 \sin^2 \frac{\alpha}{4}}$.

References

1. P. Bose, J. Gudmundsson, and M. Smid. Constructing plane spanners of bounded degree and low weight. *Algorithmica*, 42:249–264, 2005.
2. P. Bose, A. Maheshwari, G. Narasimhan, M. Smid, and N. Zeh. Approximating geometric bottleneck shortest paths. *Computational Geometry: Theory and Applications*, 29:233–249, 2004.
3. P. Bose and P. Morin. Online routing in triangulations. *SIAM Journal on Computing*, 33:937–951, 2004.
4. P. Bose, M. Smid, and D. Xu. Diamond triangulations contain spanners of bounded degree. *Carleton University, Computer Science Technical Report*, 2006.
5. L. P. Chew. There are planar graphs almost as good as the complete graph. *Journal of Computer and System Sciences*, 39:205–219, 1989.
6. G. Das and D. Joseph. Which triangulations approximate the complete graph? In *Proceedings of the International Symposium on Optimal Algorithms*, volume 401 of *Lecture Notes in Computer Science*, pages 168–192, Berlin, 1989. Springer-Verlag.
7. D. P. Dobkin, S. J. Friedman, and K. J. Supowit. Delaunay graphs are almost as good as complete graphs. *Discrete & Computational Geometry*, 5:399–407, 1990.
8. J. M. Keil and C. A. Gutwin. Classes of graphs which approximate the complete Euclidean graph. *Discrete & Computational Geometry*, 7:13–28, 1992.
9. A. W. Lee. Diamonds are a plane graph's best friend. Master's thesis, School of Computer Science, Carleton University, Ottawa, 2004.
10. X.-Y. Li and Y. Wang. Efficient construction of low weighted bounded degree planar spanner. *International Journal of Computational Geometry & Applications*, 14:69–84, 2004.
11. G. Narasimhan and M. Smid. *Geometric Spanner Networks*. Cambridge University Press, Cambridge, UK, 2007.

Optimal Construction
of the City Voronoi Diagram
Extended Abstract

Sang Won Bae[1], Jae-Hoon Kim[2], and Kyung-Yong Chwa[1]

[1] Div. of Computer Science, Dept. of EECS,
Korea Advanced Institute of Science and Technology, Daejeon, Korea
{swbae, kychwa}@jupiter.kaist.ac.kr
[2] Div. of Computer Engineering, Pusan University of Foreign Studies, Busan, Korea
jhoon@pufs.ac.kr

Abstract. We address proximity problems in the presence of roads on the L_1 plane. More specifically, we present the first optimal algorithm for constructing the city Voronoi diagram. We apply the continuous Dijkstra paradigm to obtain an optimal algorithm for building a shortest path map for a given source, and then it extends to that for the city Voronoi diagram. Moreover, the algorithm applies to other generalized situations including metric spaces induced by roads and obstacles together.

1 Introduction

We are interested in a city scene. In a modern city that is well planned and developed, such as Manhattan, lots of buildings stand evenly and densely in rows and columns so that people are forced to move among those buildings, which act as obstacles, and thus every movement is supposed to be vertical or horizontal. However, transportation networks like bus networks, roads, or taxi systems, are normally equipped with such a modern city as well, and they serve faster movement for citizens to reach their destinations.

In our sense, a *transportation network* models such a real transportation network. A transportation network is defined as a plane graph and each edge is called a *road*. We assume that along each road of the network, one moves at a certain fixed speed that is faster than out of the network, and he/she can access or leave the road at any point on it. In this situation, shortest (travel time) paths using the given network are of considerable interest, and so are Voronoi diagrams when we are given a set of equally attractive facilities. We thus address these proximity problems in the presence of roads, in particular, on the L_1 plane.

Shortest paths considering roads under the L_1 metric in fact induce a special metric, called a *city metric*, and the Voronoi diagram under such a city metric is called the *city Voronoi diagram* [2]. The first result about the city Voronoi diagram was Abellanas et al. [1], and was extended by Aichholzer et al. [2] whose setting is exactly the same as ours. The authors present an $O(n^2 \log n + m \log m)$ time and $O(n + m)$ space algorithm for constructing the city Voronoi

T. Asano (Ed.): ISAAC 2006, LNCS 4288, pp. 183–192, 2006.
© Springer-Verlag Berlin Heidelberg 2006

diagram when a transportation network of n axis-parallel roads and m sites are given. Afterwards, there has been some improved results but they seem not very successful; $O(n(n + m) + m \log(n + m))$ time and $O(n + m)$ space [4]; $O((n+m) \log^5(n+m) \log \log(n+m))$ time and $O((n+m) \log^5(n+m))$ space [6]. Some other variations also have been studied. Bae and Chwa [4, 3] considers more general transportation networks consisting of roads that may have arbitrary directions and speeds under the Euclidean metric and even convex distances. Ostrovsky-Berman [9] discusses a discrete version of transportation networks. All these previous results solve the shortest path problem also (e.g. for building a shortest path map) due to its strong interconnection with the city Voronoi diagram problem.

This paper, to our extent, presents the first optimal algorithm that builds a city Voronoi diagram and also a shortest path map (hereafter, we may call it SPM) in the presence of a transportation network which in particular consists of axis-parallel roads of equal speed. Indeed, we first obtain an optimal algorithm for building a SPM for a single source by applying the continuous Dijkstra paradigm, and then extend to that for the Voronoi diagram. The following is our main result.

Theorem 1. *A shortest path map under the city metric induced by n axis-parallel roads of equal speed can be constructed in $O(n \log n)$ time and $O(n)$ space. Also, the city Voronoi diagram for m sites can be computed in $O((n + m) \log(n + m))$ time and $O(n + m)$ space. These bounds are worst-case optimal.*

The resulting map or diagram has information of shortest paths to the given source or to the nearest site, and of their lengths. With this information, we are able to get the length of the shortest path from a query point in logarithmic time, or the path itself in additional time proportional to the complexity of the path, with the aid of a proper point location structure.

Our algorithm can be easily extended to more general situations. We are able to allow roads to have a constant number of speeds while the algorithm still runs within the same runtime and space. Another interesting extension is to composite (geodesic) metric spaces by roads together with obstacles; only by a little modification on our algorithm for processing obstacle vertices, as described in [7]. Subsequently, we can construct the SPM and the Voronoi diagram in the same time and space bounds, letting n be the number of all the endpoints introduced by the roads and the obstacles. Moreover, as other minor extensions, we can take polygonal regions with additive weights as input sources into account, and also one-way roads as in [3].

For other underlying metrics than the L_1 metric, such as the Euclidean metric and convex distances, the SPM and the Voronoi diagram can be constructed in the same manner. However, the details are not so trivial that another discussion is seemingly required.

As mentioned above, the city Voronoi diagram can be computed by using the algorithm for the SPM. Thus, we shall focus on the algorithm for constructing a SPM for a given source s in the body of this paper. In Section 2, we introduce some preliminaries related to our work and, in Section 3, we describe our algorithm. In Section 4, we analyze the correctness and the complexity of the algorithm and hence prove Theorem 1 at last.

2   Preliminaries

A *transportation network* on the L_1 plane is represented as a planar straight-line graph $G(V, E)$ such that each edge $e \in E$ has its supporting *speed* $\nu(e) > 1$. An edge in E is often called a *road* and a vertex in V a *node*. A transportation network together with its underlying metric induces a new metric, called a *transportation distance*, which is defined as the shortest travel time between two points using G [4, 3].

In this paper, we are given a transportation network G under the L_1 metric consisting of n axis-parallel roads of equal speed ν. This setting induces so-called the *city metric* [2]. We let d_1 be the L_1 distance on the plane and d the city metric (or the transportation distance, interchangeably) induced by G and d_1.

A *needle* is a generalized Voronoi site proposed by Bae and Chwa [4] for easy explanation of geometry induced by transportation networks. A needle can be viewed as a generalized type of a line segment with additive weight; a needle, indeed, is a line segment with a weight function that is linear over all points along the segment.

More specifically, a needle p can be represented by a 4-tuple $(p_1(p), p_2(p), t_1(p), t_2(p))$ with $t_2(p) \geq t_1(p) \geq 0$, where $p_1(p), p_2(p)$ are two endpoints and $t_1(p), t_2(p)$ are additive weights of the two endpoints, respectively. We let $s(p)$ be the segment $\overline{p_1(p)p_2(p)}$. The L_1 distance from any point x to a needle p is measured as $d_1(x, p) = \min_{y \in s(p)} \{d_1(x, y) + w_p(y)\}$, where $w_p(y)$ is the weight assigned to y on $s(p)$, given as $w_p(y) = t_1(p) + (t_2(p) - t_1(p)) \times d_1(y, p_1(p))/d_1(p_2(p), p_1(p))$, for all $y \in s(p)$.

It was shown that a SPM in our setting may have at most linear size [2] and further can be represented as a Voronoi diagram of $O(n)$ needles under the L_1 metric, which can be computed in optimal time and space [3].

We apply the continuous Dijkstra paradigm to obtain a shortest path tree (SPT, in short) rooted at a given source s. The framework of our algorithm is not so different from that of Mitchell [8] but most of specific details are quite distinguishable. The continuous Dijkstra method simulates the wavefront propagation from s, where the wavefront can be defined as the set $\{p \in \mathbb{R}^2 \mid d(p) = \delta\}$ for any positive δ, where we denote $d(s, p)$ just by $d(p)$. This can be done by tracking effects of the *pseudo-wavefront* that is easy to maintain but sufficient to express the *true* wavefront.

In our case, the pseudo-wavefront is represented by a set of straight line segments, called *wavelets*. A wavelet ω is an arc of a "circle" centered at *root* $r(\omega)$.[1] Note that the root $r(\omega)$ of a wavelet ω will be a needle along a certain road in our situation. Each wavelet ω has a left and a right *track*, denoted by $\alpha(\omega)$ and $\beta(\omega)$, respectively, and expands with its endpoints sliding along its tracks.

---

[1] Here, a "circle" means the set of all the equidistant points from a given center, generally being a set of points or a needle.

Each track is either a portion of a vertical or horizontal line or a portion of a bisecting curve between two certain roots. A bisector $B(r, r')$ between two roots r and r' under the L_1 metric is piecewise linear and can be computed explicitly in constant time, even when the two roots are needles in general [3].

On the L_1 plane, wavelets are basically inclined at angle either $45°$ or $135°$. However, a road e inclined at angle θ makes wavelets inclined at angles $\theta \pm \tan^{-1} 1/\nu(e)$, where $\nu(e)$ is the speed of e [1, 2]. We shall denote the set of such angles by A_e. Since we deal only with axis-parallel roads of speed ν, either $A_e = \{\tan^{-1} 1/\nu, -\tan^{-1} 1/\nu\}$ or $A_e = \{\pi/2 + \tan^{-1} 1/\nu, \pi/2 - \tan^{-1} 1/\nu\}$. We thus have only 6 angles for the inclinations of wavelets.

After running the continuous Dijkstra method, we obtain a SPT, a vertex-labeled tree rooted at the given source s such that every path to s through its vertices in the tree leads us to a shortest path with respect to the city metric d. Let $V' := V \cup \{s\} \cup \{q \in F(p) \mid p \in V \cup \{s\}\}$, where $F(p)$ is the set of at most four points q such that for each axis-parallel ray γ starting at p, q is the first meeting point along γ on any road in E. We take V' as the set of vertices of our SPT. We observe that V' still has a linear number of vertices and guarantees that we detect and handle combinatorial and geometric changes of wavelets by the following lemma based on the lemma for primitive paths [3] and other previous results [2]. Furthermore, we are able to determine a SPT uniquely with the vertices V'.

Lemma 2. *For any point $t \in \mathbb{R}^2$, there exists an L_1 shortest path $\pi = (s = v_0, \cdots, v_k = t)$ connecting s and t using the transportation network G such that $v_i \in V'$ for $i = 0, 1, \cdots, k - 1$.*

3   The Algorithm for the SPM

Our algorithm works with two steps: we compute the SPT rooted at s by applying the continuous Dijkstra method and then construct a shortest path map from the SPT.

Each vertex $v \in V'$ has a label $\ell(v)$ and initially $\ell(v) = \infty$. After completing the continuous Dijkstra method, v will have a finite label $\ell(v) < \infty$ and its predecessor $pre(v)$ in the resulting SPT; indeed, at the end of the first step, $\ell(v)$ will become equal to $d(v)$, which is the length of the corresponding shortest path from s to v.

For that purpose, we maintain the set of wavelets and keep track of their effects by tracking *events* at every step. An event is associated with each wavelet and has its corresponding *event point* and *event distance*. We predict and handle two sorts of events during the algorithm, namely, closure events and vertex events: *Closure events* occur when two tracks of a wavelet meet at a point, namely the *closure point*, and the associated wavelet degenerates to the closure point. *Vertex events* occur when a wavelet, either its interior or its endpoint along a track, hits a vertex in V'.

We initially set the current event distance δ to zero. As the current event distance δ increases, we keep track of *active* wavelets that represent the circle with radius δ centered at s (or, the true wavefront) with respect to our city metric d. This can be done by maintaining the following data structures: The *event queue* $\mathcal{Q}$ is a priority queue containing active wavelets indexed by event distance. A wavelet ω is stored in $\mathcal{Q}$ with its left/right tracks $\alpha(\omega)$ and $\beta(\omega)$, its root $r(\omega)$, its left/right neighbors $L(\omega)$ and $R(\omega)$, and its corresponding event. We call a wavelet *active* when it is stored in $\mathcal{Q}$. The *segment dragging structure* efficiently answers segment dragging queries that ask which vertex in V' is first encountered by a given wavelet as it propagates. $SPM(\delta)$-*subdivision* is a (partial) subdivision of the plane, which consists of polygonal *cells*. Each cell in $SPM(\delta)$-subdivision is either a triangle or a quadrilateral, and represents the locus of a wavelet until event distance δ from when it has been created or modified. In particular, we allow $SPM(\delta)$ cells to overlap since two wavelets can collide with each other. We, however, fix up such unpleasant situations in a conservative way so that we will have no overlap among $SPM(\delta)$ cells at the end of the algorithm. A cell is called *open* if it consists of an active wavelet in its boundary or, otherwise, *closed*. Every open cell contains exactly one active wavelet in the pseudo-wavefront at distance δ.

We instantiate, modify, or terminate a wavelet when a certain event occurs. Performing such operations for an active wavelet is accompanied with updates for data structures such as the event queue $\mathcal{Q}$ and $SPM(\delta)$-subdivision. Every time we instantiate or modify a wavelet, we also do the following procedure; (1) we determine the corresponding event point and event distance, (2) insert the wavelet into $\mathcal{Q}$ or modify it in $\mathcal{Q}$, and (3) create a new corresponding cell in $SPM(\delta)$-subdivision when instantiating a new wavelet, or make the corresponding $SPM(\delta)$ cell closed and create a new one when modifying an existing wavelet (so that we maintain all the $SPM(\delta)$ cells to be triangles or quadrilaterals). When we terminate a wavelet, we remove it from $\mathcal{Q}$ and set the corresponding cell in $SPM(\delta)$-subdivision to be closed by making up its boundary appropriately.

Determining the event point and the event distance for a wavelet is performed by computing the distance to its closure point (for a closure event), if any, and the distance to the first vertex $v \in V'$ that is encountered as the wavelet propagates (for a vertex event). The second one can be computed via a segment dragging query. Also note that even under non-degeneracy a vertex event and a closure event can occur at the same place and time. We break this type of ties by the "closure-first" rule.

While such events completely check when a wavelet disappears or when a wavelet collides with a vertex, what remains difficult is detecting collisions among wavelets. We detect such collisions also by vertex events. When a vertex event occurs at $v \in V'$ due to wavelet ω with root $r = r(\omega)$, the label $\ell(v)$ is set to the current event distance δ if $\ell(v) = \infty$ yet. Otherwise, if $\ell(v) < \infty$ (in fact, $\ell(v) < \delta$), this means that another wavelet ω' has already hit the vertex v and hence two wavelets have been colliding with each other. This collision is also represented by an overlap, which has been swept over twice, in $SPM(\delta)$-subdivision. In order to fix up this kind of errors, we shall use subroutines *Clip-and-Merge* and *Trace-*

Bisector when such collisions among wavelets (overlaps in $SPM(\delta)$-subdivision, equivalently) are detected by a vertex event. (Thus, we handle collisions among wavelets in a conservative way.) These subroutines are summarized as follows: *Clip-and-Merge* works with two phases: Find a point q on the bisector between two sets of roots corresponding to overlapped cells by walking along path π from v to $p_1(r)$, checking which overlapped cells we are in. Then, run *Trace-Bisector* (as described below) with q as input to trace out the merge curve γ. *Trace-Bisector* traces γ with q as a starting point as is done in merging two Voronoi diagrams, making up corresponding $SPM(\delta)$ cells appropriately. These subroutines are indeed small modifications of two subroutines in Mitchell [8]. Thus, for more details about them, we refer to Mitchell [8].

Before going inside the main loop of the algorithm, we need some preprocessing: (1) Computing V' can be transformed to finding closest segments in a vertical or horizontal direction from a set of points. This can be done by a usual plane sweep method. All the effort on this task is at most $O(n \log n)$ time and $O(n)$ space. (2) For segment dragging queries, we can just make use of the same structures as in Mitchell [7]. It is easy to see that, for wavelets, both the number of inclinations and the number of possible directions of track rays are bounded by a constant. Thus, we have only a constant number of configurations of wavelets so that we can find the answer for a segment dragging query during our algorithm with $O(n \log n)$ preprocessing time, $O(\log n)$ query time, and $O(n)$ space.

3.1   Computing the SPT

First, we set δ to be 0 and $\ell(v)$ to be ∞ for all $v \in V'$ except for s, and $\ell(s)$ to be 0. We instantiate four zero-length wavelets along 4 vertical or horizontal tracks from s that are inclined at $45°$ or $135°$ and have root s (or, equivalently, needle $(s, s, 0, 0)$).

While $\mathcal{Q}$ is not empty, we repeat the following procedure: We extract the upcoming event from the front of $\mathcal{Q}$. Let p be the event point, $r := r(\omega)$ the root of the wavelet ω that caused the event, and δ the event distance. According to its type, we process the event as follows.

Closure Event. First we check whether or not a portion of the bisector $B(r(L(\omega)), r(R(\omega)))$ has been swept over doubly by $L(\omega)$ and $R(\omega)$. This can be done by just looking locally. If so, we run *Trace-Bisector* with p as input. Otherwise, we terminate ω at p and modify $L(\omega)$ and $R(\omega)$ to have the ray starting at p along $B(r(L(\omega)), r(R(\omega)))$ as the right track and the left track, respectively.

Vertex Event. If ω causes a vertex event on p, p is a vertex $v \in V'$ incident to at most 4 roads. Let $E(v)$ denote the set of roads incident to v. If $\ell(v) < \infty$, v has already been swept over by another wavelet, so we call the subroutine *Clip-and-Merge*. Otherwise, we first label v with $\ell(v) = \delta = d_1(v, r)$, which is the length of the path from s through $p_1(r)$ to v, and set $pre(v)$ to be $p_1(r)$. We also store the set R_v of roots r' of wavelets coming out from v such that $p_1(r') = v$. Consider a needle r_e for each $e \in E(v)$ such that $p_1(r_e) = v$, $p_2(r_e) = q_e$, $t_1(r_e) = \ell(v)$, and $t_2(r_e) = \ell(v) + d_1(v, q_e)/\nu$, where q_e is the other node of e than v. This needle r_e plays a role as the root of wavelets propagating from v along e. We create new wavelets or modify existing ones according to each case.

1. *(v is hit by the interior of ω.)* In this case, no portion of any road in $E(v)$ has been swept over by ω and hence $|E(v)| \leq 2$.

   (a) If $|E(v)| = 1$, let e be the only road in $E(v)$. If the inclination θ of ω is in A_e, we take three rays as tracks; $\alpha(\omega)$, $\beta(\omega)$, and a ray along e away from v. With these rays, we split ω into two wavelets with inclination θ and root r.

   Otherwise, if $\theta \notin A_e$, we take five tracks; $\alpha(\omega)$, $\beta(\omega)$, a ray along e away from v, and two rays away from v obtained from the bisector $B(r, r_e)$. Note that $B(r, r_e)$ must go through v and we can get two rays at v by using two local directions of the bisector at v. With these five rays, we instantiate four wavelets and then terminate ω. Two of them taking e as a track are inclined at an appropriate angle in A_e and have root r_e. The other two are inclined at θ and have root r. Finally, we set up the neighbor information among new wavelets.

   (b) If $|E(v)| = 2$, then say $E(v) = \{e_1, e_2\}$. If $\theta \in A_{e_1}$, we take 6 tracks; $\alpha(\omega)$, $\beta(\omega)$, a ray along e_2, a ray along e_1, and two rays away from v obtained from the bisector $B(r, r_{e_2})$. We instantiate 5 wavelets in a similar way as Case 1(a), and terminate ω. The case of $\theta \in A_{e_2}$ is analogous. Also, we note that there is no chance of $\theta \in A_{e_1} \cap A_{e_2}$ since $\nu > 1$ and hence $A_{e_1} \cap A_{e_2} = \varnothing$.

   If $\theta \notin A_{e_1} \cup A_{e_2}$, we take 7 tracks; $\alpha(\omega)$, $\beta(\omega)$, two rays along e_1 and e_2, respectively, and three from the bisectors among $\{r, r_{e_1}, r_{e_2}\}$. Then, we instantiate 6 wavelets with these 7 tracks, and terminate ω.

2. *(v is hit by the left track $\alpha(\omega)$ of ω.)* There are two cases; either $\alpha(\omega)$ goes along a road e or not. In either case, $\beta(L(\omega)) = \alpha(\omega)$ and we thus process the neighbor wavelet $L(\omega)$, too.

   (a) If $\alpha(\omega)$ goes along a road e, $r = r(\omega) = r(L(\omega))$. We gather a set T of tracks; $\alpha(L(\omega))$, $\beta(\omega)$, axis-parallel rays from v that have not been swept over yet by ω or $L(\omega)$, rays from v obtained from the bisectors among $\{r\} \cup \{r_{e'} \mid e' \neq e \in E(v)\}$. We also note that $|T|$ is a constant; $|T|$ is at most 5 if $|E(v)| = 1$, at most 6 if $|E(v)| = 2$, at most 7 of $|E(v)| = 3$, and at most 9 if $|E(v)| = 4$.

   With T, we modify ω and $L(\omega)$, and instantiate $|T| - 3$ number of new wavelets to have two neighboring tracks in T. If a new wavelet ω' has a track along a road $e_i \in E(v)$, ω' is inclined at an appropriate angle in A_{e_i} and has root $r(\omega') = r_{e_i}$. Otherwise, if neither two tracks of ω' are along a road in $E(v)$, ω' is inclined at 45° or 135° and has root $r(\omega') = (v, v, \ell(v), \ell(v))$. We finally set the neighbor information accordingly. (In fact, this procedure is a generalization of those for Cases 1(a) and 1(b).)

   (b) If $\alpha(\omega)$ is not along any road, then $\alpha(\omega)$ is axis-parallel and $|E(v)| \leq 3$ since we consider the L_1 metric as the underlying metric and deal with only axis-parallel roads. This case can be handled in a similar way as in Case 2(a), though possibly $r \neq r(L(\omega))$.

3. *(v is hit by the right track $\beta(\omega)$ of ω.)* This case is analogous to Case 2.

3.2   Building the SPM from the SPT

The SPT consists of labeled vertices and directed links among the vertices. From this information, we gather the set of needles (or roots), $\mathcal{N} = \bigcup_{v \in V'} R_v$. Note that the Voronoi diagram $\mathcal{V}(\mathcal{N})$ under the L_1 metric coincides with a SPM for the source s. This has been already argued in earlier results [2, 4]. Consequently, we can build a SPM from the resulting SPT in $O(n \log n)$ time and $O(n)$ space.

4   Correctness and Complexity of the Algorithm

In this paper, we skip the proof of the correctness of the algorithm, which has been shown in the previous paper of the authors [5].

Theorem 3. *The algorithm correctly constructs a SPT. In other words, if the event distance is $\delta > 0$ during the algorithm, for all vertices v with $d(v) < \delta$, v has been correctly labeled with $\ell(v) = d(v) = d(s, v)$.*

We now discuss the complexity of the algorithm. In doing so, we investigate nearest neighbor graphs for vertices under the city metric. The *nearest neighbor graph* of a set P of points under a metric is built as follows: For each point $p \in P$, there is a directed edge to point $q \in P$ such that q is the nearest neighbor of p with respect to the given metric among the points P.

    Although Mitchell [8] proved the efficiency of his algorithm in a different way, we observe that the maximum in-degree of any nearest neighbor graph of vertices among obstacles under the Euclidean metric is at most 6. Since at least one of 7 or more wavelets must be clipped before all of them hit the same vertex, any point in the plane can be swept over at most 6 times by the pseudo-wavefront in the presence of obstacles. A similar argument can be naturally applied also to our situation. It is already known that the maximum in-degree of any directed nearest neighbor graph in the presence of n roads is $\Theta(\nu \min(\nu, n))$ [5]. This bound, however, is for arbitrary finite sets of points so that a more careful analysis for vertices in V', which constitute a particular structure, reduces the bound significantly.

    Now, we give an upper bound on the maximum in-degree, say Δ, of the nearest neighbor graph for $V' \cup \{p\}$ for any point p in the plane. By the construction of V', it is revealed in Lemma 5 that Δ is surprisingly bounded by a constant, while it may increase to $\Theta(\nu \min(\nu, n))$ if we consider arbitrary finite sets of points. This observation will be very helpful in showing the optimality of the algorithm through the following sequence of lemmas and corollaries. This process starts with the following simple fact.

Fact 4. *Let P be a finite set of points in the plane. The nearest neighbor graph for P under the L_1 metric has maximum in-degree of at most four.*

Proof (Sketch). Pick a point $p \in P$. Let N_p be the set of points in P whose nearest neighbor is p and D_q be the L_1 disk centered at q whose radius is $d_1(q, p)$ for each $q \in N_p$. Then, D_q touches p and contains no other points $q' \in N_p$ in its interior. Subdividing the plane into four unbounded regions by cutting the plane along two lines which pass through p and have slopes of 1 and -1, we observe that no two points in N_p lie in the same region and hence $|N_p| \leq 4$.   □

Lemma 5. *For any point p in the plane, the in-degree of p in the nearest neighbor graph for $V' \cup \{p\}$ is at most a constant.*

Proof. We first observe a good property of nearest neighbors graph among $V' \cup \{p\}$: Consider the shortest path π from any vertex $v \in V'$ to p. By the construction of V', if π goes through a road that is not incident to v, then it must pass through another vertex $v' \in V'$, which implies p cannot be the nearest neighbor of v since v' is closer to v along π than p is. Thus, any vertex $v \in N_p$ approaches p either by an L_1 path or by a path using only one road incident to v if any. Also, D_v is either just an L_1 disk or a union of at most four needle shapes in the sense of Bae and Chwa [4, 3].

Next, we consider a subset D'_v of D_v defined as follows: D'_v is the L_1 disk centered at v with radius $d_1(v, p)$ if the path from v to p is not using any road. Otherwise, if the path is using one road e incident to v, D'_v is the disk centered at v as if e is the only road incident to v. See Figure 1. Note that D'_v still contains p, and that since D_v does not contain any other vertices $v' \in N_p$ in its interior, neither does D'_v.

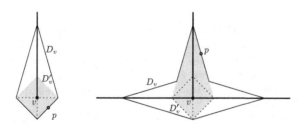

Fig. 1. D_v (regions bounded by solid thin segments) and D'_v (gray regions)

Now, the problem is switched to how many such D'_v's can be there with each containing no other vertices $v' \in N_p$ than v. By Fact 4, we know that there are at most 4 D'_v's whose shapes are L_1 disks. For vertical needle shapes, we divide the plane into four unbounded regions by two lines which pass through p and have slopes ν and $-\nu$ in a similar way as we did in the proof of Fact 4. Then, we can easily see that no two vertices in N_p that make vertical needle shapes cannot lie in the same region. An analogous discussion works for horizontal needle shapes. Thus, we conclude that (roughly) the in-degree of p is at most 12.[2]   □

---

[2] This is only a rough bound. One may make it tight by a more careful discussion.

Almost the same proof shows a bit more extended fact, which helps to show Lemma 7.

Corollary 6. *Let P be a set of points and $N_p \subseteq P$ be the set of points q such that the nearest neighbor of q is p over all points in P. Then, $|N_p| \leq 12$ if the following condition holds for every $q \in N_p$; the shortest path from q to p is either (1) the L_1 path or (2) using a single road incident to q if any.*

Lemma 7. *Any point can be swept over at most a constant number of times by the pseudo-wavefront during the algorithm.*

Lemma 7 implies the efficiency of the algorithm as follows.

Lemma 8. *The algorithm processes at most $O(n)$ events.*

Lemma 9. *The total number of $SPM(\delta)$ cells created is at most $O(n)$.*

Lemma 10. *The overall time spent by the subroutines* Clip-and-Merge *and* Trace-Bisector *during the algorithm is only $O(n)$.*

Each event, before occurring, may involve a segment dragging query among the vertices in V', which costs $O(\log n)$ query time, $O(n \log n)$ preprocessing time, and $O(n)$ space. The city Voronoi diagram for m sites can be obtained just by initializing the event queue to consist of the initial wavelets for each input site and regarding each site as a vertex. Furthermore, the $\Omega(n \log n)$ lower bound for the problem can be easily obtained by reduction from the sorting problem. Finally, we conclude the main theorem stated at the beginning.

References

[1] M. Abellanas, F. Hurtado, C. Icking, R. Klein, E. Langetepe, L. Ma, B. Palop, and V. Sacristán. Proximity problems for time metrics induced by the L_1 metric and isothetic networks. *IX Encuetros en Geometria Computacional*, 2001.

[2] O. Aichholzer, F. Aurenhammer, and B. Palop. Quickest paths, straight skeletons, and the city Voronoi diagram. In *Proc. 18th Annu. ACM Sympos. Comput. Geom.*, pages 151–159, 2002.

[3] S. W. Bae and K.-Y. Chwa. Shortest paths and Voronoi diagrams with transportation networks under general distances. In *Proc. 16th Annu. Internat. Sympos. Algorithms Comput.*, volume 3827 of *LNCS*, pages 1007–1018, 2005.

[4] S. W. Bae and K.-Y. Chwa. Voronoi diagrams for a transportation network on the Euclidean plane. *Internat. J. Comp. Geom. Appl.*, 16(2–3):117–144, 2006.

[5] S. W. Bae, J.-H. Kim, and K.-Y. Chwa. L_1 shortest paths with isothetic roads. Technical Report CS-TR-2005-241, KAIST, 2005.

[6] R. Görke and A. Wolff. Computing the city Voronoi diagram faster. In *Proc. 21st Euro. Workshop on Comput. Geom.*, pages 155–158, 2005.

[7] J. S. B. Mitchell. L_1 shortest paths among polygonal obstacles in the plane. *Algorithmica*, 8:55–88, 1992.

[8] J. S. B. Mitchell. Shortest paths among obstacles in the plane. *Internat. J. Comput. Geom. Appl.*, 6(3):309–331, 1996.

[9] Y. Ostrovsky-Berman. Computing transportation Voronoi diagrams in optimal time. In *Proc. 21st Euro. Workshop on Comput. Geom.*, pages 159–162, 2005.

Relations Between Two Common Types of Rectangular Tilings

Yusu Wang

Dept. of Comp. Sci. and Engineering, Ohio State University, Columbus, OH 43016
yusu@cse.ohio-state.edu

Abstract. Partitioning a multi-dimensional data set (array) into rectangular regions subject to some constraints (error measures) is an important problem arising from applications in parallel computing, databases, VLSI design, and so on. In this paper, we consider two most common types of partitioning used in practice: the Arbitrary partitioning and $(p \times p)$ partitioning, and study their relationships under three widely used error metrics: Max-Sum, Sum-SVar, and Sum-SLift.

1 Introduction

Partitioning a multi-dimensional data set into rectangular regions (*tiles*) subject to some constraints is an important problem arising from various applications in parallel computing, databases, load balancing, and VLSI designs. For example, in VLSI chip design, it is essential to lower the total power consumption for a given chip, which can be considered as a two-dimensional array where each cell stores the square of required voltage at that position and the total power consumption is proportional to the sum of all cell values. Unfortunately, supplying each position with a different voltage requires plotting large amount of lines as well as much overhead in voltage shifting devices. One practical solution to this problem is to partition the input chip into a few "voltage islands" [9,18]. Within each island, all cells are supplied with the same voltage, which is the highest voltage needed by any cell in it. Obviously, this increases the total power consumption. Given a limit in the power consumption, the goal is then to partition the input chip into smallest number of voltage islands without exceeding the power limit.

In general, given a way to evaluate a partitioning, called its *error* (e.g, the increase in power consumption in the previous example), the partitioning problem asks for the smallest partitioning whose error is smaller than some error threshold δ. In this paper, we study the relations between two most common types of partitioning under different error metrics.

1.1 Preliminaries and Problem Definition

We follow the definitions and terminologies of previous papers, especially from [13]. Let A be an $n \times m$ array, and $A(c)$ or $A[i][j]$ the value of element at cell $c \in A$ or at position (i, j), for $1 \leq i \leq n$ and $1 \leq j \leq m$. A *tile* of A is a

T. Asano (Ed.): ISAAC 2006, LNCS 4288, pp. 193–202, 2006.
© Springer-Verlag Berlin Heidelberg 2006

rectangular subarray of A. Note that a tile can also be viewed as an array itself. A *partitioning* $\mathcal{R}$ of array A is a set of disjoint tiles $\mathcal{R} = \{R_1, \ldots, R_k\}$ that cover A; $k = |\mathcal{R}|$ is the *size* of this partitioning.

There are three common types of partitioning schemes: (i) *Arbitrary*; (ii) *Hierarchical*, where the partitioning can be obtained by recursively cutting current subarray into two new ones by a horizontal or vertical line (like a quadtree); and (iii) *(p × p)*, a partitioning of size p^2 resulted from p horizontal and p vertical cutting lines[1].

The *rectangle tiling problem* asks for the smallest partitioning $\mathcal{R}^*$ of a given array A so that some *error measure* is below a given threshold; $\kappa_x(A, \delta) = |\mathcal{R}^*|$ denote the size of such an optimal partitioning, where $x =$ 'a', 'h', or 'p', corresponding to Arbitrary, Hierarchical, or $(p \times p)$ partitionings. The error measure is defined both for each tile in a partitioning and for the entire partitioning. This paper considers the following three error metrics: (1) Max-Sum: where the error for a tile R is defined as $\mathcal{E}(R) = \sum_{c \in R} R(c)$, and the error for a partitioning $\mathcal{R}$, $\mathcal{E}(\mathcal{R})$, is the maximum error of any of tile in $\mathcal{R}$. (2) Sum-SVar: $\mathcal{E}(R) = \sum_{c \in R}(R(c) - \mu(R))^2$ where $\mu(R)$ is the mean of all elements in R, and $\mathcal{E}(\mathcal{R}) = \sum_{R \in \mathcal{R}} \mathcal{E}(R)$. (3) Sum-SLift: $\mathcal{E}(R) = \sum_{c \in R}(\rho(R) - R(c))$, where $\rho(R)$ is the maximum value of elements in R, and $\mathcal{E}(\mathcal{R}) = \sum R \in \mathcal{R}\mathcal{E}(R)$. The Max-Sum and Sum-SVar metrics are wildly used (e.g., construct the so-called V-optimal histograms or equi-depth high dimensional histograms for summarizing database contents [15]). The Sum-SLift metric is the metric used in the VLSI chip design example at the beginning.

A $(p \times p)$-partitioning of A can also be regarded as a $p \times p$ array M, called a *reduced array*, such that each cell of M is a tile (subarray) of A. Abusing the notation slightly, we use M to refer to both the $(p \times p)$-partitioning and the corresponding reduced array. Any partitioning $\mathcal{S}$ over M induces a partitioning $I(\mathcal{S})$ over A. The *induced error* of $\mathcal{S}$ is simply $\widehat{\mathcal{E}}(\mathcal{S}) = \mathcal{E}(I(\mathcal{S}))$, and the *optimal induced partitioning* of M w.r.t. δ is the smallest partitioning of M with induce error at most δ; let $\kappa_x(M|A, \delta)$ denote the size of this optimal induced partitioning.

1.2 Related Work

The rectangular tiling problem and its variants have a rich history [1,4,5,11,12]. But the more formal and theoretical studies of exact and approximation algorithms for them are mostly recent [3,7,8,10,13,14,16,17]. For one dimensional arrays, the problem is well solved and there are efficient algorithm for most error metrics [8]. For dimension higher than one, however, the tiling problem is NP-hard for both Arbitrary and $(p \times p)$ partitioning under most common error metrics [6,7,13]. In fact, even approximating them within some constant factor is NP-hard [7,13].

The optimal Hierarchical partitioning, on the other hand, can be computed exactly by a dynamic programming approach [13]. However, under Sum-SVar and

---

[1] Our results hold for $(p \times q)$ partitioning. In this paper, for simplicity, we consider $(p \times p)$ partitioning only.

Sum-SLift metrics, the algorithm requires $O(n^5\kappa^2)$ time and $O(n^4\kappa)$ space for an $n \times n$ input array A, where κ is the size of the optimal solution. If approximation is allowed, it takes $\tilde{O}(n^{3.2}\kappa^2)$ time to compute a partitioning of at most 9κ tiles within the same error bound δ (i.e, an 9-approximation of $\kappa_h(A, \delta)$) by a neat divide and rounding technique. The same technique can be applied recursively to reduce the time complexity to $O(n^{2+\varepsilon}\kappa^2)$ but at a cost of increasing the approximation factor to $O(1/\varepsilon^2)$.

For $(p \times p)$-partitioning, Muthukrishnan and Suel [14] present a simple randomized algorithm to compute, for an $n \times n$ array A and an error threshold δ, a *double-sided approximation* in time (i) $\tilde{O}(n^2 + \kappa^3 \log n)$ under Max-Sum error metric and (ii) $O(n^2 + (n + \kappa^2)\kappa \log n)$ under Sum-SVar and Sum-SLift metrics, where the optimal $(p \times p)$-partitioning with error at most δ has size $\kappa \times \kappa$. A double-sided (α, β)-approximation of $\kappa_p(A, \delta)$ means that the algorithm outputs a partitioning $\mathcal{R}$ of size $p \times p$ such that $\mathcal{E}(\mathcal{R}) \leq \alpha\delta$ and $p \leq \beta\kappa$.

For Arbitrary partitioning, it follows from results for rectilinear BSP [2] that any Arbitrary partitioning $\mathcal{R}_1$ of a two-dimensional array A can be refined into a Hierarchical partitioning $\mathcal{R}_2$ of size at most $2|\mathcal{R}_1| - 1$. As long as the error metric is *super-additive*, which roughly means that refining a partitioning also decreases its total error, we have $\mathcal{E}(\mathcal{R}_2) \leq \mathcal{E}(\mathcal{R}_1)$. Hence $\kappa_a(A, \delta) \leq \kappa_h(A, \delta) \leq 2\kappa_a(A, \delta)$. This framework provides so far the best approximation algorithms for Arbitrary partitioning under metrics Sum-SVar and Sum-SLift. For metric Max-Sum, the problem is better understood [3,10].

Given that there is no simple and practical algorithm for approximating Arbitrary partitioning other than under Max-Sum metric, Wu *et al.* used the following heuristics for the chip design problem [18]: first, construct a $(p \times p)$-partitioning M of input array A with $\mathcal{E}(M) \leq \delta$. Next, compute the optimal hierarchical partitioning for the reduced array M using dynamic programming. This Two-step algorithm is very simple, and greatly reduced the time and space requirement in practice, as M is generally very small (less than 100×100, while A is usually $100K \times 100K$ in chip designs). However, although each step has a guarantee, it is not clear whether the optimal solution over the reduced array M indeed approximates the optimal solution of the original array A.

1.3   Our Results

Given that $\kappa_a(A, \delta) \leq \kappa_h(A, \delta) \leq 2\kappa_a(A, \delta)$, in this paper, we focus on relations between the Arbitrary partitioning and $(p \times p)$-partitioning under three common error metrics: Max-Sum, Sum-SVar, and Sum-SLift. First, observe that $\kappa_a(A, \delta) \leq \kappa_p(A, \delta) \leq \kappa_a^2(A, \delta)$ under any error metric. It is easy to construct an example showing that this bound is also asymptotically tight in worst case.

The main results of this paper focus on the following question: given an array A, let $\kappa = \kappa_a(A, \delta)$ and a $(p \times p)$ partitioning M with $\mathcal{E}(M) \leq \delta$, what is the relation between $\kappa = \kappa_a(A, \delta)$ and $\kappa_a(M|A, \delta)$? In Section 2, we show that for error metric Max-Sum, $\kappa_a(M|A, \delta)$ 4-approximates κ, i.e, $\kappa \leq \kappa_a(M|A, \delta) \leq 4\kappa$. (The results can in fact be extended for higher dimensional arrays.) This implies that performing a $(p \times p)$ partitioning does not destroy the optimal structure for $\kappa_a(A, \delta)$ much. For metrics Sum-SVar and Sum-SLift, however, there are examples where M is an $(\Omega(\kappa) \times \Omega(\kappa))$ partitioning, but $\kappa_a(M|A, \delta) \geq c\kappa^2$ for some constant $c > 0$.

On the other hand, although $\kappa_a(M|A, \delta)$ does not approximate $\kappa_a(A, \delta)$ within a constant factor under metric Sum-SVar, it turns out that if we also relax the error threshold δ, then one can achieve a double-sided approximation for A from M. More specifically, we show in Section 3 that M produces an $(2, 7)$-approximation of $\kappa_a(A, \delta)$, i.e, $\kappa_a(A, 2\delta) \leq \kappa_a(M|A, 2\delta) \leq 7\kappa_a(A, \delta)$. Unfortunately, such result does not hold for Sum-SLift metric, which we prove by a counter-example.

We remark that the above results imply that the two-step algorithm [18] approximates $\kappa = \kappa_a(A, \delta)$ in a double-sided manner under metric Sum-SVar. The running time is near-quadratic when κ is small (which is usually the case in practice). Although the algorithm has double-sided approximation, it is simple to implement and more efficient for small κ, say, when $\kappa = o(n^{0.64})$, than previous best known algorithms [13]. Unfortunately, under metric Sum-SLift, the two-step algorithm can generate $\Omega(\kappa^2)$ tiles in worst case, no matter how much extra error we allow.

2   Can $(p \times p)$-Partitioning Approximate Arbitrary Partitioning?

Given an $n \times n$ array A, if we first perform a $(p \times p)$ partitioning M of A while keeping the error below δ, can we still recover the optimal structure for $\kappa_a(A, \delta)$, either exactly or approximately, from this reduced array M? In other words, how does $\kappa_a(A, \delta)$ and $\kappa_a(M|A, \delta)$ relate? In what follows, we sometimes omit 'a' from κ_a when it is clear that it refers to Arbitrary partitioning.

2.1   Error Metric Max-Sum

Given an array A, let $W = \sum_{c \in A} A(c)$ be the sum of all elements in A. Easy to see that $\kappa(A, \delta) \geq \lceil \frac{W}{\delta} \rceil$ under Max-Sum metric. The following lemma upper bounds $\kappa(A, \delta)$ by $O(\lceil \frac{W}{\delta} \rceil)$. The proof is straightforward (similar to the one for Theorem 3 from [3]) and omitted.

Lemma 1. *Under error metric* Max-Sum, *for any d-dimensional array A, we can compute a partitioning $\mathcal{R}$ of A such that $\mathcal{E}(\mathcal{R}) \leq \delta$ and $\lceil \frac{W}{\delta} \rceil \leq \kappa(A, \delta) \leq |\mathcal{R}| \leq 2d \lceil \frac{W}{\delta} \rceil$.*

Given a reduced array M of A, for any cell $X \in M$, set $M(X) = \sum_{c \in A_X} A(c)$, where A_X is the set of cells from A covered by X. Using this value assignment, the total weight of M is the same as that of A, i.e, $\sum_{X \in M} M(X) = W$; and for any partitioning $\mathcal{S}$ of M, we have that the induced error $\widehat{\mathcal{E}}(\mathcal{S})$ for A is the same as $\mathcal{E}(\mathcal{S})$ for M. This, together with Lemma 1, implies the following result:

Lemma 2. *For any $(p \times p)$ partitioning M of a d-dimensional array A such that $\mathcal{E}(M) \leq \delta$, we have that: $\kappa(A, \delta) \leq \kappa(M|A, \delta) \leq 2d\,\kappa(A, \delta)$.*

2.2 Error Metric Sum-SVar and Sum-SLift

The nice relationship under error metric Max-Sum does not exist for metrics Sum-SVar and Sum-SLift. Below we describe one counter example for metric Sum-SVar. The same example also works for metric Sum-SLift.

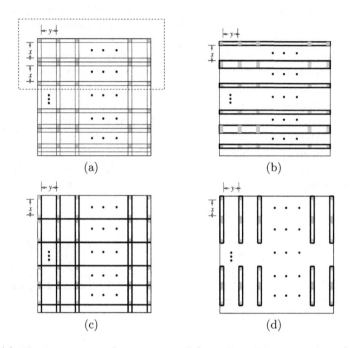

Fig. 1. (a) The input array A is a unit grid (not all grid lines are shown): dark cells have value 1, otherwise 0. The dashed region contains $2x + 4$ rows. It is repeated (vertically) r times, and the number of rows between two consecutive such regions is much larger than x^2. There are altogether c columns containing dark cells, and y all-white columns between two consecutive such columns. (b) The optimal partitioning of A w.r.t. $\delta = r^2x/(x + 2)$ has k tiles (thick segments). In (c), one $(\Theta(k) \times \Theta(k))$-partitioning M of A such that $\mathcal{E}(M) = \delta$. In (d), thick segments bound those tiles containing dark cells from the optimal induced partitioning of M with error at most δ.

Consider the array A in Figure 1 (a) with $c = r/4$ and $y = cx/2(c-1)$. It can be shown that if the error threshold $\delta = r^2x/(x + 2)$, then the optimal partitioning for A is shown in (b), where $k = \kappa(A, \delta) = 3c/2$. On the other hand, consider the $(\Theta(k) \times \Theta(k))$ partitioning M as shown in Figure 1 (c). Straightforward calculation shows that $\mathcal{E}(M) = \delta$. Furthermore, call a tile *dark* if it contains any dark cell. We claim that the minimum number of dark tiles for any induced partitioning of M with error at most δ is shown in Figure 1 (d) (details removed due to lack of space). As there are $\Theta(k^2)$ number of them, we conclude that:

Lemma 3. *Under metrics Sum-SVar and Sum-SLift, there exists an $(\Omega(k) \times \Omega(k))$ partitioning M of A, so that $\mathcal{E}(M) \leq \delta$, but $\kappa(M|A, \delta) = \Theta(\kappa^2(A, \delta))$.*

3   Double-Sided Approximations Under Sum-SVar

Previous section shows that under metric Max-Sum, one can perform a $(p \times p)$-partitioning safely: the optimal partitioning for the resulting reduced array is a constant-factor approximation of the original optimal partitioning. However, under metrics Sum-SVar and Sum-SLift, the optimal induced partitioning may have size $\Omega(k^2)$ while the optimal partitioning of the original array is of size $O(k)$. The next natural question is: if we also relax the error requirement, is there a partitioning from the reduced array of size $O(k)$? More precisely, given a reduced array M of A with $\mathcal{E}(M) \leq \delta$, is there a partitioning $\mathcal{R}$ of M such that $|\mathcal{R}| = O(\kappa(A, \delta))$ and the induced error in A $\widehat{\mathcal{E}}(\mathcal{R}) \leq c\delta$ for some constant c.

It turns out that for error metric Sum-SLift, the same example from previous section shows that there is no such constant. In fact, no matter how much extra error we allow, the optimal induced partitioning may have size $\Omega(k^2)$. This is not true for error metric Sum-SVar (details omitted in this extended abstract). In what follows, we show that for any reduced array M of A with $\mathcal{E}(M) \leq \delta$, one can compute a *double-sided* (α, β)-*approximation* for $\kappa(A, \mathcal{E})$ with $\alpha = 2$ and $\beta = 7$: i.e, $\kappa(A, 2\delta) \leq \kappa(M|A, 2\delta) \leq 7\kappa(A, \delta)$.

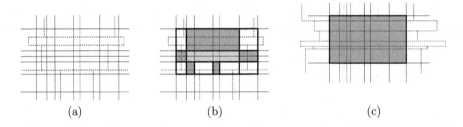

(a)   (b)   (c)

Fig. 2. (a) Solid thin lines form a $(p \times p)$-partitioning M. Dashed lines are from $\mathcal{X}$. We modify $\mathcal{X}$ into $\mathcal{P}$ in (b) (bounded by thick solid segments) (c) The shaded tile R is from $\mathcal{P}_3$. Its corresponding $\mathcal{R}_s$ has seven cells of M from the same row, while $\mathcal{R}_o$ has six tiles of same width, each of which is the intersection of R and some tile from $\mathcal{X}$.

First, given any partitioning $\mathcal{X}$ of A, we modify it into a partitioning $\mathcal{P}$ of M as follows (Figure 2): For any cell in M, if it contains the corner of some tile from $\mathcal{X}$, we add it as a tile into the set $\mathcal{P}_1$ (empty tiles in Figure 2 (b)). Next, for every tile $R \in \mathcal{X}$, add the largest tile of M completely contained in the interior of R, if it exists, into the set $\mathcal{P}_2$ (the light-colored one in Figure 2 (b)). Now the only cells of M left uncovered are those that intersect boundaries of tiles from $\mathcal{X}$. Each boundary edge is possibly broken into several pieces, some already covered by tiles from $\mathcal{P}_1$ and $\mathcal{P}_2$. For each maximal uncovered piece, we add the corresponding tile of M intersecting it into $\mathcal{P}_3$ (dark colored in Figure 2 (b)); a tile R from $\mathcal{P}_3$ is thus either a $1 \times |R|$ or a $|R| \times 1$ subarray of M. Tiles from $\mathcal{P}_1$, $\mathcal{P}_2$, and $\mathcal{P}_3$ are disjoint, and $\mathcal{P} = \mathcal{P}_1 \cup \mathcal{P}_2 \cup \mathcal{P}_3$. In the remaining of this section, we first show that $\mathcal{E}(\mathcal{P}) \leq \mathcal{E}(\mathcal{X}) + \mathcal{E}(M)$. We then bound the size of $\mathcal{P}$

by $7|\mathcal{X}|$. Hence if $\mathcal{X}$ is the optimal partitioning of A w.r.t. δ, $\mathcal{P}$ then provides a $(2,7)$-approximation of it.

3.1   Upper Bound $\mathcal{E}(\mathcal{P})$

Given an $s \times t$ array R, let $\mu = \frac{1}{st} \sum_{i,j} R[i][j]$ be the average of all elements in R, $\alpha[i] = \frac{1}{t} \sum_{j=1}^{t} R[i][j]$ the average of elements in the i'th column, for $1 \leq i \leq s$, and $\beta[j]$ be the mean of elements in the j'th row, for $1 \leq j \leq t$.

Lemma 4. *Let C_i and R_j denote the i'th column and j'th row of R respectively. Then we have that $\mathcal{E}(R) \leq \sum_{i=1}^{s} \mathcal{E}(C_i) + \sum_{j=1}^{t} \mathcal{E}(R_j)$.*

Proof. Easy to verify that the claim is the same as

$$\sum_{i=1}^{s} \sum_{j=1}^{t} (R[i][j] - \mu)^2 \leq \sum_{i=1}^{s} \sum_{j=1}^{t} (R[i][j] - \alpha[i])^2 + \sum_{j=1}^{t} \sum_{i=1}^{s} (R[i][j] - \beta[j])^2.$$

First, observe that if we change every element $R[j][j]$ by some arbitrary constant c, then the left and right terms of the above inequality remains the same. As such, we can now assume that $\mu = 0$, which in turn implies that $\sum_{i=1}^{s} \alpha[i] = \sum_{j=1}^{t} \beta[j] = 0$. Furthermore,

$$\sum_{i=1}^{s} \sum_{j=1}^{t} (R[i][j] - \alpha[i])^2 + \sum_{j=1}^{t} \sum_{i=1}^{s} (R[i][j] - \beta[j])^2 - \sum_{i=1}^{s} \sum_{j=1}^{t} R^2[i][j]$$

$$= \sum_{i=1}^{s} \sum_{j=1}^{t} (R^2[i][j] - 2R[i][j](\alpha[i] + \beta[j]) + \alpha^2[i] + \beta^2[j])$$

$$= \sum_{i=1}^{s} \sum_{j=1}^{t} (R[i][j] - \alpha[i] - \beta[j])^2 - 2 \sum_{i=1}^{s} \sum_{j=1}^{t} \alpha[i]\beta[j]$$

$$= \sum_{i=1}^{s} \sum_{j=1}^{t} (R[i][j] - \alpha[i] - \beta[j])^2 \;\; \geq \;\; 0.$$

The last line follows from the fact that

$$\sum_{i=1}^{s} \sum_{j=1}^{t} \alpha[i]\beta[j] = \sum_{i=1}^{s} (\alpha[i] \sum_{j=1}^{t} \beta[j]) = 0.$$

The lemma then follows.

To bound $\mathcal{E}(\mathcal{P}) = \mathcal{E}(\mathcal{P}_1) + \mathcal{E}(\mathcal{P}_2) + \mathcal{E}(\mathcal{P}_3)$, we need the following observation, called the *super-additive property* in [13], which holds for all three metrics we consider.

Observation 1 ([13]). *Given any two disjoint tiles H and G, let T be the tile obtained by merging H and G, then $\mathcal{E}(T) \geq \mathcal{E}(H) + \mathcal{E}(G)$.*

The above property implies that given any two sets of tiles $\mathcal{R}_1$ and $\mathcal{R}_2$, if $\mathcal{R}_2$ is (a subset of) some refinement of $\mathcal{R}_1$, i.e., for any $R \in \mathcal{R}_2$, there exists some $R' \in \mathcal{R}_1$ such that $R \subseteq R'$, then we have $\mathcal{E}(\mathcal{R}_2) \leq \mathcal{E}(\mathcal{R}_1)$. Now consider any tile $R \in \mathcal{P}_3$: R is either a $|R| \times 1$ or a $1 \times |R|$ subarray of M. Assume that it is the first case (Figure 2 (c)). Then it intersects a set of boundary edges from $\mathcal{X}$, all of which cut through R horizontally, as it contains no vertex of $\mathcal{X}$. Let $\mathcal{R}_s = \mathcal{R}_s(R)$ denote the set of cells from M covered by R, and $\mathcal{R}_o = \mathcal{R}_o(R)$ the set of intersections between R and tiles from $\mathcal{X}$ (Figure 2 (c)). By Observation 1, we have $\mathcal{E}(\mathcal{R}_s) \geq \sum_{j=1}^{k} \mathcal{E}(R_j)$ and $\mathcal{E}(\mathcal{R}_o) \geq \sum_{i=1}^{s} \mathcal{E}(C_i)$ (the case where R intersects vertical boundaries of $\mathcal{X}$ is symmetric). It then follows from Lemma 4 that

$$\mathcal{E}(R) \ \leq \ \sum_{i=1}^{m} \mathcal{E}(C_i) + \sum_{j=1}^{k} \mathcal{E}(R_j) \ \leq \ \mathcal{E}(\mathcal{R}_s) + \mathcal{E}(\mathcal{R}_o). \tag{1}$$

Lemma 5. *Given a reduced array M of A and some partitioning $\mathcal{X}$ of A, let $\mathcal{P}$ be the corresponding modified partitioning over M as described earlier. We have that $\widehat{\mathcal{E}}(\mathcal{P}) \leq \mathcal{E}(\mathcal{X}) + \mathcal{E}(M)$.*

Proof. Recall that $\mathcal{P} = \mathcal{P}_1 \bigcup \mathcal{P}_2 \bigcup \mathcal{P}_3$. Let M_i be the set of cells from M covered by $\mathcal{P}_i$, for $i = 1, 2$, and 3; M_i's are disjoint, and their union form the set of cells in M. For $\mathcal{E}(\mathcal{P}_1)$, observe that $\mathcal{E}(\mathcal{P}_1) = \sum_{C \in M_1} \mathcal{E}(C)$ since each tile in $\mathcal{P}_1$ is a single cell in M (thus in M_1). To bound $\mathcal{E}(\mathcal{P}_2)$ and $\mathcal{E}(\mathcal{P}_3)$, let $\mathcal{X}_3$ be the intersections of tiles from $\mathcal{X}$ with tiles from $\mathcal{P}_3$; $\mathcal{X}_3$ is the collection of $\mathcal{R}_o$'s (as introduced above) for all $R \in \mathcal{P}_3$. Furthermore, tiles from $\mathcal{P}_2 \bigcup \mathcal{X}_3$ provide a subset of some refinement of $\mathcal{X}$. As such, by Observation 1, we have that $\mathcal{E}(\mathcal{P}_2) + \mathcal{E}(\mathcal{X}_3) \leq \mathcal{E}(\mathcal{X})$. It then follows from Eqn (1) that

$$\mathcal{E}(\mathcal{P}_3) \leq \sum_{R \in \mathcal{P}_3} \mathcal{E}(\mathcal{R}_s) + \sum_{R \in \mathcal{P}_3} \mathcal{E}(\mathcal{R}_o) \leq \sum_{C \in M_3} \mathcal{E}(C) + \mathcal{E}(\mathcal{X}_3)$$
$$\Rightarrow \ \mathcal{E}(\mathcal{P}) = \mathcal{E}(\mathcal{P}_1) + \mathcal{E}(\mathcal{P}_2) + \mathcal{E}(\mathcal{P}_3)$$
$$\leq \sum_{C \in M_1 \cup M_3} \mathcal{E}(C) + \mathcal{E}(\mathcal{P}_2) + \mathcal{E}(\mathcal{X}_3) \ \leq \ \mathcal{E}(\mathcal{X}) + \mathcal{E}(M).$$

3.2 Upper Bound $\kappa(M|A, 2\delta)$

We now bound the size of $\mathcal{P}$. First, easy to see that $|\mathcal{P}_1|$ is bounded by the number of vertices in $\mathcal{X}$, a trivial bound of which is $4|\mathcal{X}|$ as each tile may produce four vertices. However, since each vertex is shared by at least two tiles in $\mathcal{X}$, we can improve it to $|\mathcal{P}_1| \leq 2|\mathcal{X}|$. Second, by definition, each tile in $\mathcal{X}$ can produce at most one tile in $\mathcal{P}_2$. Hence $|\mathcal{P}_2| \leq |\mathcal{X}|$. Finally, we bound the size of $\mathcal{P}_3$ as follows: recall that tiles from $\mathcal{P}_3$ are disjoint subarrays of some row or column of M. We classify these tiles into *row-tiles* and *column-tiles*, respectively, depending on whether they cover horizontal or vertical boundary edges of tiles from $\mathcal{X}$. Consider any row of M and count the number of row-tiles within this row. Note that the neighbors of such tiles within the same row are necessary tiles from $\mathcal{P}_1$. In particular, if there are s vertex-covering tiles in this row, there are at most $s - 1$ row-tiles. Hence the total number of row-tiles is at most $|\mathcal{P}_1| - 1$. Similarly, the total number of column-tiles is at most $|\mathcal{P}_1| - 1$, implying that $|\mathcal{P}_3| < 4|\mathcal{X}|$. Overall, $|\mathcal{P}| = |\mathcal{P}_1| + |\mathcal{P}_2| + |\mathcal{P}_3| \leq 7|\mathcal{X}|$. Now if $\mathcal{X}$ is the optimal partitioning of A w.r.t. error δ and if $\mathcal{E}(M) \leq \delta$, then there exists a partitioning $\mathcal{P}$ of M such that $\widehat{\mathcal{E}}(\mathcal{P}) \leq 2\delta$ and $|\mathcal{P}| \leq 7 \cdot \kappa(A, \delta)$. Putting everything together, we conclude with our main theorem:

Theorem 2. *Given any $(p \times p)$-partitioning M of A such that $\mathcal{E}(M) \leq \delta$ under Sum-SVar metric, the optimal induced partitioning of M with error at most 2δ provides an $(2, 7)$-approximation of $\kappa_a(A, \delta)$, that is, $\kappa_a(A, 2\delta) \leq \kappa_a(M|A, 2\delta) \leq 7\kappa_a(A, \delta)$.*

Corollary 1. *Given an $n \times n$ array A and an error threshold δ, under Sum-SVar metric, the Two-step algorithm from [18] can compute a $(2, 7)$-approximation of $\kappa_h(A, \delta)$, thus a $(2, 14)$-approximation of $k = \kappa_a(A, \delta)$ in $O(n^2 + k^7)$ time and $O(n^2 + k^5)$ space.*

We remark that we can remove the double-sided approximation by computing a $(p \times p)$ partitioning w.r.t. error $\delta/2$ in the first step, at the cost of increasing the running time to $O(n^2 + \kappa_a^7(A, \delta/2))$. This gives rise to a 14-approximation of $\kappa_a(A, \delta)$. Furthermore, the n^2 term in time and space comes from storing the input array as well as weights for some subarrays (details omitted here). So the constant hidden is very small. In comparison, the algorithm from [13] 9-approximates $\kappa_h(A, \delta)$, thus 18-approximates $\kappa_a(A, \delta)$ in $O(n^{3.2}\kappa^2)$ time. The two-step algorithm outperforms it in efficiency when $\kappa = o(n^{0.64})$ and is much simper.

4   Conclusion and Discussion

Our results on relation between Arbitrary and $(p \times p)$ partitioning imply that there is no theoretical guarantee for the Two-step algorithm under Sum-SLift error metric, although it performs well in practice [18]. How to develop simple and efficient approximation algorithms under Sum-SLift metric is one important future research direction. We also remark that it is natural to extend the array partitioning problem to dimensions higher than two. However, while approximation algorithms for multi-dimensional arrays exist for $(p \times p)$ partitioning [14], for Arbitrary partitioning, such algorithm is only available under Max-Sum metric [3]. One main reason is because in three dimensions and higher, the nice result on the size of rectilinear BSP trees no longer holds. It is thus an interesting open problem to develop different frameworks (other than using BSP) for approximating Arbitrary partitioning under these metrics that can be extended to higher dimensions.

Acknowledgment. The author wish to thank an anonymous reviewer for the current much simpler proof of Lemma 4 and thank Sariel Har-Peled for helpful discussions.

References

1. S. Anily and A. Federgruen. Structured partitioning problems. *Operations Research*, pages 130–149, 1991.
2. P. Berman, B. Dasgupta, and S. Muthukrishnan. Exact size of binary space partitionings and improved rectangle tiling algorithms. *SIAM J. Discrete Math*, 15(2):252–267, 2002.
3. P. Berman, B. DasGupta, S. Muthukrishnan, and S. Ramaswami. Efficient approximation algorithms for tiling and packing problems with rectangles. *J. Algorithms*, 41(2):443–470, 2001.
4. S. Bokhari. Partitioning problems in paralle, pipelined, and distributed computing. *IEEE Transactions on Computers*, 37:38–57, 1988.
5. G. Fox, M. Johnson, G. Lyzenga, S. Otto, J. Salmon, and D. Walker. *Solving problems on concurrent processors, Volumn 1*. Prentice-Hall, Englewood Cliffs, New Jersey, 1998.
6. M. Grigni and F. Manne. On the complexity of the generalized block distribution. In *IRREGULAR'96*, pages 319–326, 1996. Lecture notes in computer science 1117, Springer.
7. S. Khanna, S. Muthukrishnan, and M. Paterson. On approximating rectangle tiling and packing. In *SODA '98: Proceedings of the ninth annual ACM-SIAM symposium on Discrete algorithms*, pages 384–393, 1998.
8. S. Khanna, S. Muthukrishnan, and S. Skiena. Efficient array partitioning. In *ICALP*, pages 616–626, 1997.
9. D. E. Lackey, P. S. Zuchowski, T. R. Bednar, D. W. Stout, S. W. Gould, and J. M. Cohn. Managing power and performance for system-on-chip designs using voltage islands. In *Proceedings of the 2002 IEEE/ACM international conference on Computer-aided design table of contents*, pages 195–202, 2002.
10. K. Lorys and K. E. Paluch. New approximation algorithm for RTILE problem. *Theor. Comput. Sci.*, 2-3(303):517–537, 2003.
11. F. Manne. *Load Balancing in Parallel Sparse Matrix Computations*. PhD thesis, Dept. of Informatics, Univ. of Bergen, Norway, 1993.
12. F. Manne and T. Sorevik. Partitioning an array onto a mesh of processors. In *Workshop on Applied Parallel Computing in Industrial Problems*, 1996.
13. S. Muthukrishnan, V. Poosala, and T. Suel. On rectangular partitionings in two dimensions: Algorithms, complexity, and applications. In *ICDT '99: Proceeding of the 7th International Conference on Database Theory*, pages 236–256, 1999.
14. S. Muthukrishnan and T. Suel. Approximation algorithms for array partitioning problems. *Journal of Algorithms*, 54:85–104, 2005.
15. V. Poosala. *Histogram-based estimation techniques in databases*. PhD thesis, Univ. of Wisconsin-Madison, 1997.
16. J. P. Sharp. Tiling multi-dimensional arrays. In *International Symposium on Fundamentals of Computation Theory*, pages 500–511, 1999.
17. A. Smith and S. Suri. Rectangular tiling in multi-dimensional arrays. In *ACM/SIAM Symposium on Discrete Algorithms (SODA)*, pages 786–794, 1999.
18. H. Wu, I. Liu, M. D. F. Wong, and Y. Wang. Post-placement voltage island generation under performance requirement. In *IEEE/ACM International Conference on Computer-Aided Design (ICCAD)*, pages 309–316, 2005.

Quality Tetrahedral Mesh Generation for Macromolecules

Ho-Lun Cheng and Xinwei Shi

National University of Singapore

Abstract. This paper presents an algorithm to generate quality tetrahedral meshes for the volumes bounded by the molecular skin model defined by Edelsbrunner. The algorithm applies the Delaunay refinement to the tetrahedral meshes bounded by quality surface meshes. In particular, we iteratively insert the circumcenters of bad shape tetrahedra with a priority parameterized by its distance from the surface. We achieve a bounded radius-edge ratio for the tetrahedral mesh after the refinement. Finally, we apply the sliver exudation algorithm to remove 'slivers'. The algorithm terminates with guarantees on the tetrahedral quality and an accurate approximation of the original surface boundary.

1 Introduction

This paper studies the quality mesh generation for macromolecules, in particular, the quality Delaunay mesh generation for the volumes bounded by molecular skin models. This section introduces the motivation of this work and reviews previous work on Delauany refinement for quality tetrahedral mesh generation.

Motivation. Electrostatics potential is one of the fundamental energy terms to model a molecular system. It defines the potential energy at a particular location near a molecule created by the system of molecular charges. The study of the electrostatic potential within a molecule or the interactions among different molecules is necessary to investigate the protein folding and protein-protein interactions. Thus, modeling and computation of the electrostatic of molecules have become a central topic in molecular modeling studies [1].

The Poisson-Boltzmann equation(PBE) is one of the most popular approach to model the electrostatic of large molecules. Using the solution of PBE to predict the electrostatic property of molecules achieved good agreement with experimental results [1]. Since the PBE is a non-linear partial differential equation, an analytical solution is not available and numerical methods are necessary, for example, using the finite element methods. The accuracy and stability of the solution with numerical methods depend on the quality of the elements used to decompose the molecular volume. Moreover, the solution of PBE is sensitive to the boundary of the molecular models [1]. As a result, quality volumetric meshes of molecules that conforms to their boundary are desirable for the computing the molecular electrostatic by solving the PBE.

T. Asano (Ed.): ISAAC 2006, LNCS 4288, pp. 203–212, 2006.
© Springer-Verlag Berlin Heidelberg 2006

Therefore, it is still challenging to construct quality tetrahedral meshes conforming to the boundary of the molecular models. Most of the previous works use regular grids[15]. The mesh elements generated by these methods have a biased alignment to the axis. Moreover, since the boundary of a molecule is smooth and curved, the resolution of the grids cannot be infinitely fine to conform the boundary. On the other hand, Delaunay meshes have no such problems and support efficient construction algorithms with quality guarantees [16]. In this paper, we use Delaunay refinement methods to generate quality tetrahedral meshes for the macromolecules.

Related Work. Delaunay refinement methods generate quality tetrahedral meshes by placing new mesh vertices carefully in the domain and maintaining the Delaunay triangulation until all mesh elements satisfy the quality constraints. Shewchuk [16] used the Delaunay refinement to generate tetrahedral meshes for 3D domains bounded by piecewise linear complexes(PLC). The algorithm eliminates poor quality tetrahedra by iteratively inserting their circumcenters. Consequently, the resulting meshes achieve an upper bound on the radius-edge ratio, which is the ratio of the circumradius of a tetrahedron with its shortest edge. However, the algorithm requires the PLC domains have no acute input angles. This problem was addressed recently by Cheng et. al [9], in which the algorithm generates Delaunay meshes for polyhedra with small input angles with guarantees on the radius-edge ratio, except for the tetrahedra near the small input angles. Bounded radius-edge ratio eliminates all kinds of bad shape tetrahedra except slivers, which are tetrahedra with bounded radius-edge ratio and extreme small dihedral angles. Cheng et. al [7] introduced sliver exudation algorithm by assigning weights to the mesh vertices of slivers such that the resulting weighted Delaunay triangulation is sliver free. Subsequently, Edelbrunner et. al [14] perturb the vertices to clean up the slivers in the mesh. However, both sliver removal algorithms cannot handle the boundaries because they are only applied to periodic sets. Recently, Cheng and Dey [6,10] combined Delaunay refinement with sliver exudation to construct quality tetrahedra meshes for polyhedra. The algorithm guarantees the quality for all tetrahedra except the ones near the corners with small input angles.

It is natural to apply the Delaunay refinement to generate quality meshes for the domains bounded by smooth surfaces. However, this attempt needs to overcome the following obstacles. First, there are no general piecewise linear representations for smooth surfaces defined by implicit or parametric equations. Although a number of surface polygonization and triangulation algorithms have been proposed, none of them can guarantee the output surface meshes without acute angles [2,3,8]. As a result,the Delaunay refinement algorithms for polyhedra cannot be applied to mesh the volumes of smooth surfaces. Second, sliver removal algorithms devote further study for the domains bounded by smooth surfaces because Cheng and Dey's algorithm [6,10] cannot handle the slivers near acute input angles. Third, the analysis of the mesh size for PLC domains using local feature size is not suitable for analyzing the size of volumetric meshes of smooth surfaces. Instead, the curvature and the local feature size of the smooth surface are important for the analysis of the mesh density.

We present an algorithm for generating quality tetrahedral meshes of the volumes bounded by molecular skin surfaces. The algorithm overcomes the challenges arising from applying the Delaunay refinement methods to the domains with smooth surface boundaries, and generates quality tetrahedral meshes with bounded radius-edge ratio and free of slivers. In addition, the boundary of the final tetrahedral mesh is also an accurate approximation of the input molecular skin surface.

Outline. The remainder of this paper is organized as follows. We introduce some basic concepts and definitions used in our algorithm in Section 2. Our meshing algorithm is described in Section 3. Experiments results are illustrated in Section 4 and the paper is concluded in Section 5.

2 Background

In this section, we review the necessary backgrounds to understand our meshing algorithm for the macromolecules. We first introduce the molecular skin model and the recent result on skin meshing. Then, we introduce the distance function defined by the skin surface. Finally, the measure for the tetrahedral quality is introduced.

Molecular Skin Model. The skin surface F_B specified by a finite set of spheres B is a closed C^1-continuous surface in $\mathbb{R}^3$. To model a molecule with the skin surface, we consider each atom as a sphere $b_i = (z_i, w_i) \in B$. That is, the position z_i is the center of an atom, and its radius w_i is $\sqrt{2}$ times the summation of the atom's van der Waals radius with the radius of the probe sphere, which is usually chosen as 1.4 Angstrom to represent the water as solvent. Then, the skin surface F_B gives a model of the molecule.

Adaptive Surface Mesh. Denote the maximum principle curvature at $x \in F_B$ as $\kappa(x)$. The reciprocal $1/\kappa(x)$ is called the *local length scale* at x, denoted as $\varrho(x)$. The local length scale varies slowly on the skin surface and satisfying the 1-Lipschitz condition, that is, $|\varrho(x) - \varrho(y)| \leq \|x - y\|$, in which $x, y \in F_B$. This property facilitates adaptive homeomorphic triangulations for the skin surface. We use the recent result from [5], in which the algorithm generates quality skin surface mesh adaptive to the maximum curvature. Specifically, the surface mesh is the restricted Delaunay triangulation $R_B(P)$, of an ε-sampling P of the skin surface. The surface mesh has the following properties [5,4].

Lemma 1 (Small Circumradius Lemma). *The cricumradius R_{abc} of a triangle $abc \in R_B(P)$ has an upper bound, $R_{abc} < \frac{\varepsilon}{1-\varepsilon} \varrho_{abc}$,*[1]

This lemma says that the triangles in the surface mesh have a small circumradius compared to local length scale at their vertices. The smallest circumsphere of

---

[1] In which ϱ_{abc} is the local length scale of abc defined as $\varrho_{abc} = \min\{\varrho(a), \varrho(b), \varrho(c)\}$.

a surface triangle abc is also called the *protecting ball* of abc. The union of the protecting balls of all the surface triangles forms the *protecting region* that detects the intrusion of the newly inserting mesh vertices. We also have the following property of the skin surface mesh [5,4].

Lemma 2 (Dihedral Angle Lemma). *For two triangles $abc, abd \in R_B(P)$ with shared edge ab, the dihedral angle at edge ab has a lower bound of $\pi - 2 \arcsin(\frac{2\varepsilon}{1-\varepsilon})$.*

These nice properties of the input skin surface mesh facilitate our Delaunay refinement algorithm for quality tetrahedral mesh generation. Together with a specific priority for the order of the new mesh vertices insertion, the termination of the algorithm and the mesh quality are guaranteed. Next, we introduce the distance function to define the priority.

Distance Function. Given a skin surface F_B, the distance function to F_B is defined over $\mathbb{R}^3$ by assigning each point its distance to the surface, $d(x) = \inf_{p \in F_B} \|x - p\|, \forall x \in \mathbb{R}^3$.

We approximate the function using the ε-sampling P of skin the skin surface, that is, $d'(x) = \min_{p \in P} \|x - p\|, \forall x \in \mathbb{R}^3$.

The approximation has been used by Dey [11] to reconstruct the smooth surface from a point cloud. We use the function value to parameterize the priority for new mesh vertices insertion during the Delaunay refinement. The new mesh vertices are the circumcenters of the tetrahedra with poor quality. Next, we introduce the measures for the quality of a tetrahedron.

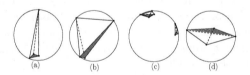

Fig. 1. A classification of the bad shape tetrahedra

Tetrahedral Quality. Let r and l be the circumradius and the lengthes of the shortest edge a tetrahedron respectively. The bad shape tetrahedra can be characterized by using the radius-edge ratio $\frac{r}{l}$. We call a tetrahedron τ *skinny* if its radius-edge ratio is larger than a constant $c > 1$, that is, $\frac{r}{l} \geq c$. Otherwise, we call the tetrahedron τ has *ratio property* c. A tetrahedral mesh K has the ratio property for a constant c means that every tetrahedron $\tau \in K$ has ratio property c. Figure 1 (a), (b) and (c) illustrate the examples of skinny tetrahedra.

The *sliver* is a special kind of bad shape tetrahedra with bounded radius-edge ratio. See Figure 1 (d) for an example. The vertices of a sliver are almost on a big circle of its circumsphere and form dihedral angles approaching to 0. We call a tetrahedron *sliver* if it has the ratio property c and its minimum dihedral angle is smaller than a constant ζ.

The goal of our meshing algorithm is to generate tetrahedral meshes for the skin volumes with bounded ratio property and free of slivers. We introduce our algorithm in the next section.

3   Algorithm

The algorithm is divided into three stages, namely, *sculpture, Delaunay refinement with priority and sliver removal by pumping vertices.* The procedure of sculpture builds a coarse tetrahedral mesh T_0 for the volume bounded by the skin F_B, denote as V_B. Since we have the restricted Delaunay triangulation of F_B, $R_B(P)$, defined by point set P as the surface mesh, the tetrahedral mesh T_0 can be constructed by taking the subset of the Delaunay triangulation of P that includes all the tetrahedra whose circumcenters lie inside the skin volume V_B. Note that the boundary of T_0 is exactly the input surface mesh, namely, the restricted Delaunay triangulation $R_B(P)$. Starting with this coarse tetrahedral mesh T_0, we improve the mesh quality by running Delaunay refinement and sliver removal sequentially. Next, we introduce the prioritized Delaunay refinement methods and analyze the quality guarantees when the refinement terminates.

3.1   Prioritized Delaunay Refinement

Delaunay refinement methods improve the mesh quality by inserting the circumcenters of the poor shape tetrahedra incrementally. After each new mesh vertex is inserted, the Delaunay triangulation is maintained and this process is repeated until all the tetrahedra satisfies the quality constraints. The new mesh vertices can be inserted in a random way or in a certain order. Shewchuk [16] inserted new mesh vertices with a priority parameterized by the radius-edge ratio. That is, the circumcenter of the tetrahedron with largest radius-edge ratio is always inserted. This priority decreases the number of inserting points in some cases. Edelsbrunner and Guoy [12] defined the sink as the circumcenters that are contained inside their own tetrahedra. A circumcenter is inserted as a new mesh vertex only when it is a sink. The priority facilitates parallel implementation of the Delaunay refinement.

We introduce a new priority parameterized by the distance function value of the circumcenters. That is, the circumcenter t of a skinny tetrahedron τ that has the largest distance $d'(t)$ to the surface is inserted in each iteration of the Delaunay refinement. The reason beyond this priority is that new mesh vertices are restricted to be as far as possible from such that the circumcenters close to the media axis of the surface are inserted with high priority to improve the mesh quality as much as possible. Once the circumcenters near the surface are necessary to be inserted, the mesh quality satisfies the quality constraints because the input surface mesh has guaranteed quality.

The Delaunay refinement process adapts the incremental algorithm for Delaunay triangulation computation. Starting from the initial coarse tetrahedral mesh T_0, the circumcenter t_i of a skinny tetrahedron τ_i is inserted and forms four new tetrahedra with the faces of the tetrahedron τ_i. The Delaunay property is restored by flipping algorithm and we get the Delaunay tetrahedral mesh T_i with mesh vertices $P_i = P_{i-1} \bigcup \{t_i\}$, for $P_0 = P$. The procedure is described in the following pseudo code.

Algorithm 1. PDeloneRefine()

1: Test the radius-edge ratio for all the tetrahedra in T_0 and push the skinny tetrahedron to a queue Q prioritized by their distance function values;
2: **while** $Q \neq \varnothing$ **do**
3:    $\tau = \text{ExtractMax}(Q)$;
4:    **if** the τ is a tetrahedron in T_{i-1} with its circumcenter t_i that falls outside of the protecting region **then**
5:       Compute the Delaunay triangulation of $P_i = P_{i-1} \cup \{t_i\}$ by the incremental method;
6:       Update Q by adding the new skinny tetrahedra;
7:    **end if**
8: **end while**

We analyze the behavior of the Algorithm 1 to validate the termination and quality guarantees of the prioritized refinement procedure. First, we prove that all the inserted circumcenters lies inside the skin volume V_B and outside the protecting region. This property ensures the input surface mesh is stable during the refinement process. Then, we prove the refinement process terminates with a upper bound of the radius-edge ratio c depending on the constants ε and γ that specify the surface mesh quality.

Lemma 3 (Circumcenters Lemma). *Let t be the circumcenter of a skinny tetrahedron $\tau \in T_i$. The circumcenter t is contained inside the underlying space of T_i, namely, $|T_i|$.*

Proof. We prove the lemma using a deductive method.

In the case of $i = 0$, the claim is true on the base of our sculpture procedure. Since T_0 consists of all the tetrahedra whose circumcenters lies inside the volume of the skin surface V_B. The difference between V_B and $|T_0|$ is the space between the skin surface and the surface triangles when the local shape of the surface is convex. According to the dihedral angle Lemma 2, a tetrahedron with its circumcenter inside V_B and outside $|T_0|$ must be a sliver. And a sliver is never a skinny tetrahedron. Thus, the claim is true when $i = 0$.

We assume the claim is true when $i = k$ and prove all the circumcenters of tetrahedra in T_{k+1} are inside $|T_{k+1}|$. To get a contradiction, let a circumcenter t locates outside $|T_{k+1}|$. The circumcenter t must be a Voronoi vertex of a mesh vertex q that either lie on the boundary or the interior of $|T_{k+1}|$. In the case of the mesh vertex q lie inside $|T_{k+1}|$, one of the Voronoi edge of q must penetrate the surface mesh and we denote the intersection with a surface triangle abc as u. As a result, the point u must be inside the protecting sphere of abc and its distance to a, b and c is smaller than the distance to q since q is always outside the protecting region. It contradicts with the definition of the Voronoi cell for the point q. In the case of the mesh vertex q is on the boundary, then t must be the circumcenter of a tetrahedron with its four vertices on the boundary. It is impossible because all the new tetrahedra we created during the refinement must connect to the newly inserted internal nodes. Thus, all the circumcenters of tetrahedra in T_{k+1} are inside $|T_{k+1}|$ and the claim follows.

The Lemma 3 implies that the boundary of the tetrahedral mesh T_n conforms to the input surface mesh. Next, we prove the algorithm terminates with bounded radius-edge ratio for all the tetrahedral in T_n.

Theorem 1. *The algorithm terminates with quality tetrahedra mesh having ratio property for the constant $c \leq \frac{2\varepsilon}{\gamma(1-\varepsilon)}$.*

Proof. We first prove the algorithm terminates. Since the algorithm only inserts the circumcenters of tetrahedra with radius-edge ratio larger than 1, the edge length of newly created edges during the Delaunay refinement never shrinks. In the other words, the inter-vertices distances are bounded from below. Moreover, the algorithm only inserts points in the domain and never deletes any points. As a result, the algorithm must terminate because the volume of V_B is finite.

Then, we prove the bound for the radius edge ratio. To get a contradiction, we assume a tetrahedron τ has radius-edge ratio $\frac{R}{l} > \frac{2\varepsilon}{\gamma(1-\varepsilon)}$ when the algorithm terminates. Then, the circumcenter t of τ must be inside a protecting sphere of a surface triangle abc. Otherwise, Algorithm 1 will insert the t as a mesh vertex. The situation is illustrated in Figure 2. Let one of the vertices of τ be q, then we have $R = \|t - q\| < \|t - a\| < 2R_{abc}$, in which R_{abc} is the radius of the protecting sphere of triangle abc. Combining with Lemma 1, we have $R < \frac{2\varepsilon}{(1-\varepsilon)}\varrho_{abc}$. On the other hand, the length of the shortest edge of the tetrahedron τ must be longer than the length of any edges on the boundary incident to a. That is, $l > \|a - b\| \geq \gamma\varrho_{ab}$. According to the assumption,

$$R > l\frac{2\varepsilon}{\gamma(1-\varepsilon)} > \gamma\varrho_{ab}\frac{2\varepsilon}{\gamma(1-\varepsilon)} = \frac{2\varepsilon}{(1-\varepsilon)}\varrho_{ab}. \tag{1}$$

Since $\varrho_{abc} \leq \varrho_{ab}$, Equation (1) and Equation (2) contradict to each other. As a result, all the tetrahedra have ratio property for the constant $c \leq \frac{2\varepsilon}{\gamma(1-\varepsilon)}$.

With feasible values for $0 < \gamma < 0.218$ and $0 < \varepsilon < 0.279$, a conservative lower bound for c would be 3.5 [5]. Our experiments show that we can achieve a much better bound of 1.5 on the radius-edge ratio in practice. However, slivers still frequently exist inside the tetrahedral mesh after the Delaunay refinement terminated with a bounded radius-edge ratio. Next, we adapt the sliver exudation algorithm to remove the slivers.

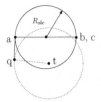

Fig. 2. A tetrahedron τ with its circumcenter t inside a protecting sphere defined by the surface triangle abc. q is one of the vertices of τ.

3.2   Sliver Removal by Pumping Vertex

We adapts the sliver exudation algorithm [7] to remove the slivers. The algorithm achieves sliver free tetrahedral mesh by assigning feasible weight to the vertices of Delaunay mesh with bounded radius-edge ratio so that the resulting weighted Delaunay triangulation contains no slivers. However, the sliver exudation algorithm only applies to the periodic point set, which is an infinite set without boundary. In order to apply the algorithm to our bounded domains, we only pump the non-boundary mesh vertices incident to a sliver. In addition, we further restrict the assignment of the weights not to challenge the boundary. That is, the weight for x should be small enough so that the boundary triangles will stay in the weighted Delaunay triangulation. For the slivers with four mesh vertices on the surface, we remove them from the tetrahedral mesh directly without influencing the boundary approximation.

4   Experimental Results

We implemented the algorithm on the PC platform with C++ on the base of the skin surface meshing software built by the authors. The construction of the Delaunay triangulation and weighted Delaunay triangulation partially reuse the prior software on alpha shapes. One point worth noting here is the computation of the distance function value for each circumcenter of a skinny tetrahedra. We utilize the Delaunay triangulation of the input surface mesh vertices for this purpose. That is, we locate the tetrahedron contains the circumcenter first and search its nearest neighbor locally. We tested our implementation to generate quality tetrahedral meshes for some molecular skin models. The experimental results show that the prioritized Delaunay refinement performs excellently and it achieves an upper bound of 1.5 on the radius-edge ratio. At the same time, the dual Voronoi diagram the Delaunay triangulation decompose the volumes into well shaped convex polyhedra. Such a decomposition may be useful for the numerical computations using control volume methods. Moreover, our implementation of the sliver exudation algorithm eliminates most of the slivers. Table 1 gives the statistics of the tetrahedral mesh quality.

Table 1 illustrates the statistics of the mesh quality for the molecule Crambin. The input surface mesh includes 27,341 mesh vertices and 50,222 triangular faces.

Table 1. The distribution of the radius-edge ratios and minimal dihedral angles of the tetrahedral mesh of molecule Crambin

R/l	0-1	1-1.5	1.5-2	2-3	≥ 3	$\zeta(°)$	0-5	5-10	10-20	20-30	≥ 30
T_0	2,675	4,451	3,327	25,570	47,726	T_0	6,964	12,254	10,787	7,335	46,409
T_n	117,854	138,609	2	0	0	T_n	1,292	3,309	14,009	35,877	201,978
$\hat{T}_n$	117,654	13,8258	12	0	0	$\hat{T}_n$	13	212	14,479	39,566	201,654

The minimum angle in the surface mesh is 20.1°. The Delaunay refinement takes around 8 minutes on Pentium 4 PC to insert 26,709 vertices inside the volume and improve the radius-edge ratio to 1.5. Totally 1,292 slivers exist in the final tetrahedral mesh, in which 300 slivers have four vertices on the surface. After we performed the sliver removal, only 13 slivers are left. The distribution of the radius-edge ratio and minimal dihedral angle in the coarse tetrahedral mesh before prioritized Delaunay refinement T_0, the mesh after the Delaunay refinement T_n and the final mesh after the sliver removal $\hat{T}_n$ is presented in the table 1.

5 Discussion

In this paper, we present an algorithm for generating quality tetrahedral mesh for the volumes of the macromolecules. The algorithm improves the mesh quality of a coarse mesh using Delaunay refinement prioritized by the distance function value followed by a sliver removal process. The prioritized Delaunay refinement process terminates with guarantees on the upper bound of the radius-edge ratio of the tetrahedral mesh. We give a proven upper bound 3.5 on the ratio. In addition, our experimental results show that an upper bound of 1.5 on the radius-edge ratio can be achieved, which is much better than the theoretical bound and implies that there are no any skinny tetrahedra in the mesh after the prioritized Delaunay refinement. The following sliver removal process removes the slivers effectively and improves the minimal dihedral angle in the mesh. The boundary of the final tetrahedral mesh also approximates the original surface accurately.

We also note that the statistics results of our algorithm are different from the experiments performed on the smooth surface by Edelsbrunner and Guoy [13]. Firstly, the radius-edge ratio distribution is slightly different. We achieved a better bound for the radius-edge ratio after the refinement. For example, the upper bound of the radius-edge ratio for all the tetrahedra in our meshes is 1.57, comparing with 2.14 in [13]. This may be explained from two aspects. On one hand, the prioritized Delaunay refinement may work better than sink insertion for the volumes bounded by smooth surfaces. On the other hand, the surface meshes we used have better quality than the surface meshes used in [13]. Secondly, the number of tetrahedra with $\zeta < 5°$ in our experimental results is more than that of [13]. This may due to our algorithm only pumps the vertices of the slivers.

References

1. Nathan A. Baker and J. Andrew McCammon. *Structural Bioinformatics*. Wiley-Liss, Inc., 2003.
2. Jules Bloomenthal. *An Implicit Surface Polygonizer*. Academic Press, Boston, 1994.
3. J. D. Boissonnat and S. Oudot. Provably good surface sampling and approximation. In *Proceedings of the Eurographics/ACM SIGGRAPH symposium on Geometry processing*, pages 9–18. Eurographics Association, 2003.
4. H. Cheng, T. K. Dey, H. Edelsbrunner, and J.Sullivan. Dynamic Skin Triangulation. *Discrete Comput. Geom.*, 25:525–568, 2001.
5. Ho-Lun Cheng and Xinwei Shi. Quality mesh generation for molecular skin surfaces using restricted union of balls. In *Proceedings of IEEE Visualization*, pages 399–405, 2005.
6. Siu-Wing Cheng and Tamal K. Dey. Quality meshing with weighted delaunay refinement. In *Proceedings of the thirteenth annual ACM-SIAM symposium on Discrete algorithms*, pages 137–146. Society for Industrial and Applied Mathematics, 2002.
7. Siu-Wing Cheng, Tamal K. Dey, Herbert Edelsbrunner, Michael A. Facello, and Shang-Hua Teng. Sliver exudation. In *Proceedings of the fifteenth annual symposium on Computational geometry*, pages 1–13. ACM Press, 1999.
8. Siu-Wing Cheng, Tamal K. Dey, Edgar Ramos, and Tathagata Ray. Sampling and meshing a surface with guaranteed topology and geometry. In *Proceedings of ACM Annual Symposium on Computational Geometry*, 2004.
9. Siu-Wing Cheng, Tamal K. Dey, Edgar A. Ramos, and Tathagata Ray. Quality meshing for polyhedra with small angles. In *Proceedings of the twentieth annual symposium on Computational geometry*, pages 290–299. ACM Press, 2004.
10. Siu-Wing Cheng, Tamal K. Dey, and Tathagata Ray. Weighted delaunay refinement for polyhedra with small angles. *Proceedings of 14th International Meshing Roundtable*, ,, pages 323–342, 2005.
11. Tamal K. Dey, Joachim Giesen, Edgar A. Ramos, and Bardia Sadri. Critical points of the distance to an epsilon-sampling of a surface and flow-complex-based surface reconstruction. In *SCG '05: Proceedings of the twenty-first annual symposium on Computational geometry*, pages 218–227, New York, NY, USA, 2005. ACM Press.
12. Herbert Edelsbrunner and Damrong Guoy. Sink-insertion for mesh improvement. In *SCG '01: Proceedings of the seventeenth annual symposium on Computational geometry*, pages 115–123, 2001.
13. Herbert Edelsbrunner and Damrong Guoy. An experimental study of sliver exudation. *Eng. Comput. (Lond.)*, 18(3):229–240, 2002.
14. Herbert Edelsbrunner, Xiang-Yang Li, Gary Miller, Andreas Stathopoulos, Dafna Talmor, Shang-Hua Teng, Alper Ungor, and Noel Walkington. Smoothing and cleaning up slivers. In *Proceedings of the thirty-second annual ACM symposium on Theory of computing*, pages 273–277. ACM Press, 2000.
15. Walter Rocchia, Sundaram Sridharan, Anthony Nicholls, Emil Alexov, Alessandro Chiabrera, and Barry Honig. Rapid grid-based construction of the molecular surface and the use of induced surface charge to calculate reaction field energies: Applications to the molecular systems and geometric objects. *Journal of Computational Chemistry*, 23(1):128 – 137, June 2001.
16. Jonathan Richard Shewchuk. *Delaunay refinement mesh generation*. PhD thesis, 1997.

On Approximating the TSP with Intersecting Neighborhoods

Khaled Elbassioni[1], Aleksei V. Fishkin[2], and René Sitters[1]

[1] Max-Planck-Institut für Informatik, Saarbrücken, Germany
{elbassio, sitters}@mpi-sb.mpg.de
[2] University of Liverpool, UK
avf@csc.liv.ac.uk

Abstract. In the TSP with neighborhoods problem we are given a set of n regions (neighborhoods) in the plane, and seek to find a minimum length TSP tour that goes through all the regions. We give two approximation algorithms for the case when the regions are allowed to intersect: We give the first $O(1)$-factor approximation algorithm for intersecting convex fat objects of comparable diameters where we are allowed to hit each object only at a finite set of specified points. The proof follows from two packing lemmas that are of independent interest. For the problem in its most general form (but without the specified points restriction) we give a simple $O(\log n)$-approximation algorithm.

1 Introduction

In the *TSP with neighborhoods problem* we are given a set of n subsets of the Euclidean plane and we have to find a tour of minimum length that visits at least one point from each subset. This generalizes both the classical Euclidean TSP and the group Steiner tree problems [6] with applications in VLSI-design, and other routing-related applications (see e.g. [10,13]). Although the problem has been extensively studied in the last decade after Arkin and Hassin [1] introduced it in 1994, still large discrepancies remain between known inapproximability and approximation ratios for various cases. Safra and Schwartz [14] showed that the problem in the general case is NP-hard to approximate within any constant factor, and is APX-hard if each set forms a connected region in the plane. For connected polygonal regions, Mata and Mitchell [9] gave an $O(\log n)$-approximation in $O(N^5)$-time based on "guillotine rectangular subdivisions", where N is the total number of vertices of the polygons. Gudmundsson and Levcpoulos [7] reduced the running time to $O(N^2 \log N)$.

In another interesting *discrete* variant of the problem, sometimes called the *group-TSP problem*, we are given n connected subsets of the plane, often refred to as regions or objects, and one set of points P. The TSP-tour must hit each region in one or more of the points of P. Typically, this restriction makes the problem harder. For the most general version in which the subsets are unrestricted, and the metric is not necessarily Euclidean, the gap is almost closed: Garg et al. [6] gave a randomized $O(\log N \log \log N \log k \log n)$-approximation algorithm for the

T. Asano (Ed.): ISAAC 2006, LNCS 4288, pp. 213–222, 2006.

group Steiner tree problem, the variant in which we are given a graph with N vertices and a set of n groups with at most k vertices per group, and seek a minimum cost Steiner tree connecting to at least one point from each group. Slavík [15] showed that the problem can be approximated within $O(k)$. On the negative side, Halperin and Krauthgamer [8] gave an inapproximability threshold of $\Omega(\log^{2-\epsilon} n)$ for any fixed $\epsilon > 0$.

With this being the situation for the general case, recent research has considered the cases where the given subsets are connected regions in the plane. We speak of the *continues* case if we can hit any point of the connected region, and we speak of the *discrete* case if we are only allowed to hit the region in one of the specified points. Previous results also distinguish between *fat* regions, such as disks, and *non-fat* regions, such as line-segments, and between instances with *disjoint* regions and *intersecting* regions. Non-fatness and intersections seem to make the problem much harder and, in fact, no constant factor approximation algorithm is known for the general case of intersecting non-fat regions.

For the continuous case when the regions are translates of disjoint convex polygons, and for disjoint unit disks, Arkin and Hassin [1] presented constant-factor approximations. Dumitrescu and Mitchell [4] gave an $O(1)$-approximation algorithm for intersecting unit disks. For disjoint varying-sized convex fat regions, de Berg et al. [3] presented an $O(\alpha^3)$-approximation algorithm, where α is a measure of fatness of the regions. A much simpler algorithm was given in [5] with an improved approximation factor of $O(\alpha)$, where also an $O(1)$-approximation algorithm was given for the discrete case with intersecting unit disks. Very recently, Mitchell [12] gave a PTAS for the continuous case with disjoint fat neighborhoods, even under a weaker notion of fatness than the one used in this paper.

Perhaps the two most natural extensions for which no constant factor algorithm is known, are that of non-fat disjoint objects, and that of fat intersecting objects. In Section 2 we consider the discrete version of the latter problem and give an $O(\alpha^3)$-approximation algorithm for intersecting convex α-fat objects of comparable size. The proof follows from two lemmas given in Section 2.2. Lemma 1) is interesting on its own and gives a relation between the length of the optimal TSP tour inside a square and the distribution of the points. Additionally, we give in Section 3 a simple alternative $O(\log n)$-algorithm for the general problem of connected regions, which does not require the regions to be simple polygons.

There are several definitions of fatness in the literature and the following is commonly used for the problem we consider [3,5,16].

Definition 1. *An object $O \subseteq \mathbb{R}^2$ is said to be α-fat if for any disk D which does not fully contain O and whose center lies in O, the area of the intersection of O and D is at least $1/\alpha$ times the area of D.*

Notice for example that the plane $\mathbb{R}^2$ has fatness 1, a halfspace has fatness 2 and a disk has fatness 4.

2   Intersecting Convex Fat Objects - The Discrete Case

We denote the given set of objects by $\mathcal{O} = \{O_1, \ldots, O_n\}$. In this section, we assume that each object can be hit only at specified points, i.e., we are given a set of points P and the required TSP tour must hit O_i at some point in $S_i \equiv P \cap O_i$. We consider the case when $O_1, \ldots, O_n$ are intersecting convex α-fat objects of the same (or comparable) diameter δ. We assume $P = S_1 \cup \ldots \cup S_n$. We first present the algorithm. In subsection 2.2, we derive a packing lemma that will be used in analyzing the performance of the algorithm. We briefly comment on the analysis of the approximation ratio in subsection 2.3.

2.1   The Algorithm

A subset $P' \subseteq P$ is called a *hitting pointset* for $\mathcal{O}$ if $P' \cap S_i \neq \emptyset$ for $i = 1, \ldots, n$ and a *minimal hitting pointset* if for every $x \in P'$ there exists an $i \in [n]$ such that $(P' \setminus \{x\}) \cap S_i = \emptyset$. A minimal hitting set can be found by the natural greedy algorithm: Set $P' = P$, and keep deleting points from P' as long as it is still a hitting set. An axis-aligned square B is called *a covering box* for the set of objects $\mathcal{O}$ if B contains a hitting pointset for $\mathcal{O}$, and a *minimum covering box* if it is smallest size amongst all such covering boxes. Since a minimum covering box is determined by at most three points of P on its boundary, there are only $O(|P|^3)$ such candidates. By enumerating over all such boxes, and verifying if they contain a hitting set, one can compute a minimum covering box.

Consider the following algorithm for the Group TSP problem on sets $S_1, \ldots, S_n$ (which is essentially the same as the one used in [5] for unit disks):

Algorithm $\mathcal{B}$:
(1) Compute a minimum covering box B of $\mathcal{O}$.
(2) Find a minimal hitting pointset $P' \subseteq P$ for $\mathcal{O}$ inside B.
(3) Compute a $(1 + \epsilon)$-approximate TSP tour on P'.

The last step can be done efficiently for any $\epsilon > 0$ using techniques from [2] and [11].

Theorem 1. *Algorithm $\mathcal{B}$ is an $O(\alpha^3)$-approximation algorithm for the Group TSP problem, with convex and α-fat neighborhoods of the same diameter.*

To analyze the performance of the algorithm, we need to show that, even though a collection of convex fat objects, with exactly one point in each, might be intersecting, they still exhibit a *packing property* that admits a "short tour" visiting all the points. This is the content of the packing lemmas, stated and proved in the next section.

2.2   Two Packing Lemmas

We give two lemmas which are of independent interest. The first relates the length of a TSP tour through a set of points in the plane to the distribution of the points. Call a circular sector with head angle $\theta \leq \pi$, and radius γ a (γ, θ)-sector.

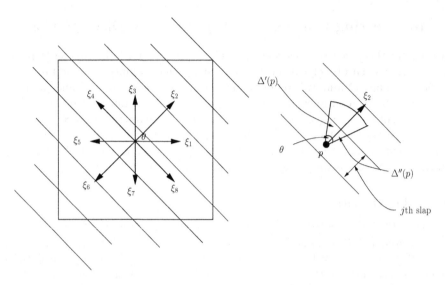

Fig. 1. (a) Partitioning the covering box. (b) A $(\beta L, \theta)$-sector $S(p) \in \mathcal{S}_{i,j}$.

Lemma 1. *Let $P = \{p_1, \ldots, p_n\} \subseteq \mathbb{R}^2$ be a a set of points with covering box of size L, and $\beta > 0$ and $0 < \theta < \pi/2$ be two constants. Then there exists an absolute constant $c = c(\beta, \theta)$ such that the following holds:*

If for every point $p \in P$ there is a $(\beta L, \theta)$-sector centered at p which contains no other point from P, then the optimum TSP tour on P has length $|\text{OPT}| \leq cL$.

Proof. Let P be a point set satisfying the conditions of the lemma, i.e. for every point $p \in P$ there is a $(\beta L, \theta)$-sector $S(p)$, centered at p, which contains no other point from P. We begin by partitioning the set of sectors $\mathcal{S} = \{S(p) : p \in P\}$ into $k = hg$ groups, depending on their orientations and locations, where $h = \lceil 2\pi/\theta \rceil$ and $g = \lceil \sqrt{2}/(\beta \cos \theta) \rceil$. The precise partitioning is done as follows. Fix h directions, $\xi_1, \ldots, \xi_h$, where ξ_i, for $i \in [h]$, makes an angle of $(i-1)\theta$ with the horizontal direction. For each direction ξ_i, we partition the covering box, into at most g parallel slabs $\rho_{i,j}$ ($j = 1 \ldots g$), along the direction $\xi_i^\perp$ orthogonal to ξ_i. See Figure 1-(a) for an example. For $i \in [h]$ and $j \in [g]$, let $\mathcal{S}_{i,j} \subseteq \mathcal{S}$ be the set of sectors $S(p)$ with the following two properties (see Figure 1-(b)): (i) the line through p with direction ξ_i intersects the circular arc part of $S(p)$, and (ii) p lies in the jth slab with respect to the direction of $\xi_i^\perp$.

Since $h = \lceil 2\pi/\theta \rceil$ we can find for each $p \in P$ a direction ξ_i such that (i) is satisfied. Clearly (ii) is satisfied for some value j given the direction ξ_i for p. Hence, $\bigcup_{i \in [h], \, j \in [g]} \mathcal{S}_{i,j} = \mathcal{S}$.

Claim 1. *Fix $i \in [h]$ and $j \in [g]$. Then there exists a path T on the set of points $\{p \in P : S(p) \in \mathcal{S}_{i,j}\}$ of length at most $(4\sqrt{2}/\sin(\theta/2) + \sqrt{2})L$.*

Proof. By performing the appropriate rotation, we may assume without loss of generality that ξ_i is the vertical direction, and thus the slab $\rho_{i,j}$ determined by the pair (i, j) is horizontal. Note that, since the diameter of the covering box is at most $\sqrt{2}L$, the width of such a slab is at most $\sqrt{2}L/g \leq L\beta\cos\theta$. In particular, if we consider any point $p \in P$ such that $S(p) \in \mathcal{S}_{i,j}$, then the circular arc of $S(p)$ lies completely outside $\rho_{i,j}$ (see Figure 1-(b)), and thus the boundary of the intersection of $S(p)$ and $\rho_{i,j}$ is a triangle $\Delta(p)$, with head angle θ. A line passing through p parallel to the direction ξ_i divides this triangle into two, one on the left $\Delta'(p)$ and one on the right $\Delta''(p)$ of the line (see Figure 1-(b)). Clearly, the angle with head p in one of these triangles is at least $\theta/2$. Now we partition $\mathcal{S}_{i,j}$ further into two groups of sectors: $\mathcal{S}'_{i,j}$ is the set of sectors $S(p)$ whose left triangle $\Delta'(p)$ makes an angle of at least $\theta/2$ with the vertical direction, and $\mathcal{S}''_{i,j} = \mathcal{S} \setminus \mathcal{S}'_{i,j}$.

We claim that there is a path λ connecting all the points in $P' = \{p \in P : S(p) \in \mathcal{S}'_{i,j}\}$, with total length

$$|\lambda| \leq \sqrt{2}L\left(\frac{1 + \cos(\theta/2)}{\sin(\theta/2)}\right) \leq 2\sqrt{2}L/\sin(\theta/2). \tag{1}$$

To see this, we may assume without loss of generality that each triangle $\Delta'(p)$, for $p \in P'$ makes an angle of exactly $\theta/2$ with the vertical direction. The path λ is obtained by traversing the boundary of these triangles from left to right as shown in Figure 2-(a). By projecting the sides of each such triangle on the big dotted triangle Δ_0 containing all of them (see Figure 2-(a)), we observe that the sum of all these lengths is at most the sum of the two non-horizontal sides of Δ_0, which in turn implies 1. Applying the same for $\mathcal{S}''_{i,j}$ and connecting both paths by a segment of length at most $\sqrt{2}L$ implies Claim 1. □

Now construct an Eulerian graph by taking two copies of the minimum covering box, together with the hg paths of the claim above, but extended to start and end at the covering box, which adds at most L for each slab. The total length is at most

$$8L + \left(4\sqrt{2}/\sin(\theta/2) + \sqrt{2} + 1\right) Lhg,$$

where $h = \lceil 2\pi/\theta \rceil$ and $g = \lceil \sqrt{2}/(\beta\cos\theta) \rceil$. □

Notice that the upper bound in the previous lemma does not depend on the number of points. An infinite set of points could still satisfy the condition of the lemma. The next lemma is the analogue for the TSP with intersecting neighborhoods.

Lemma 2. *Let B be a box of size L containing a set of points $P = \{p_1, \ldots, p_n\} \subseteq \mathbb{R}^2$. Assume that there is a collection of n convex α-fat objects $\mathcal{O} = \{O_1, \ldots, O_n\}$, each of diameter δ, such that (i) each point $p \in P$ is contained in exactly one object $O(p) \in \mathcal{O}$ (ii) each object O contains exactly one point $p(O) \in P$. Then there exists a tour T on P with length $O(L^2\alpha^2/\delta)$.*

Proof. Consider an object O with its unique point $p = p(O) \in P$. We will prove that there is $(\beta L, \theta)$-sector with center p that lies completely inside O, with $\theta = 2\pi/(3\alpha)$ and $\beta = \delta/(4L)$.

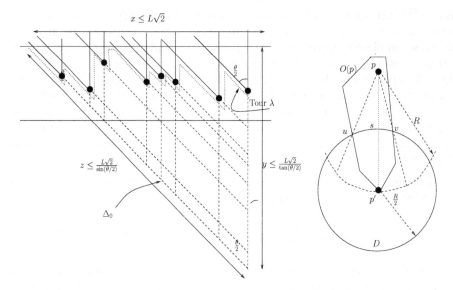

Fig. 2. (a) Bounding the tour length in the proof of Lemma 1. (b) Illustration for the proof of Lemma 2.

Let p' be a point in O at maximum distance, say R, from p (See Figure 2(b)). Obviously, $R \geq \delta/2$. Let s be the point in the middle of line segment pp' and consider a disk D with center p' and radius $R/2$. Let u and v be points in O on the circumference of D such that the angle $\langle upv \rangle$ is maximum. Finally, denote by S the (R, ϕ)-sector passing through u and v, with center p, radius R, and head angle upv, where $\phi = \langle upv \rangle$.

Denote by $A(U)$ the area of a given region U of the plane. By definition of fatness we have $A(O \cap D) \geq \pi(R/2)^2/\alpha$. Further, $A(S \cap D) \leq 3/4 \cdot A(S) = 3/4 \cdot R^2\phi/2$. It follows from the definition of u, v, and p', and the convexity of O that $O \cap D \subseteq S \cap D$. Therefore, $\pi(R/2)^2/\alpha \leq A(O \cap D) \leq A(S \cap D) \leq 3R^2\phi/8$. Hence, $\pi/\alpha \leq 3\phi/2$, implying $\phi \geq 2\pi/(3\alpha)$. The sector with center p, radius $R/2$ and head angle upv is contained in O.

Now we apply Lemma 1 with $\theta = 2\pi/(3\alpha)$ and $\beta = \delta/(4L)$. Note that if $\alpha \to \infty$, then $\cos(\theta/2) \to 1$ and $\sin(\theta/2) = O(1/\alpha)$. We conclude that the length of the optimum tour is $O(L^2\alpha^2/\delta)$. $\qquad \square$

2.3   Analysis of the Approximation Ratio

The analysis is similar to the one given in [5] modulo Lemma 2 stated above. We only give a sketch here and leave the details for the full version.

Take a *maximal independent set* of objects $\mathcal{O}' \subseteq \mathcal{O}$, i.e., a maximal collection of $k = |\mathcal{O}'|$ pairwise disjoint objects. If we pick an arbitrary point in each of these objects and consider an optimal TSP tour on this set, then its length is at most OPT$+O(k)$, where δ is hidden in the constant. Since the independent set is maximal we can partition $\mathcal{O}$ in k clusters and by Lemma 2 we can connect all objects in one cluster by a tour of length $O(\alpha^2)$. The lemma below is taken from [5] and will also be used in the proof of Theorem 3, part (ii). It sates that OPT= $\Omega(|\mathcal{O}'|/\alpha)$. Hence, Theorem 1 follows.

Lemma 3 ([5]). *The length of the shortest path connecting k disjoint α-fat objects in $\mathbb{R}^2$ is at least $(k/\alpha - 1)\pi\delta/4$, where δ is the diameter of the smallest object.*

3   General Objects - The Continuous Case

De Berg et al. [3] prove that the TSP with connected neighborhoods problem is APX-hard. The constant was raised to $2 - \varepsilon$ by Safra and Schwartz [14]. Both their reduction use curved objects of varying sizes. We can show that the problem is even APX-hard for the very restricted case where all objects are line-segments of nearly the same length. The proof is omitted in this abstract.

Theorem 2. *The TSP with neighborhood problem is APX-hard, even if all objects are line segments of approximately the same length, i.e, the lengths differ by an arbitrarily small constant factor.*

If we do not restrict the shape of the objects then no better approximation algorithm than $O(\log n)$ is known. Gudmundsson and Levcpoulos [7] used a guillotine subdivision of the plane to obtain an algorithm which runs in time $N^2 \log N$, where N is the total number of vertices of the polygonal objects. Here, we give a very simple approximation algorithm, with the same approximation guarantee, which does not require the objects to be simple polygons.

As before, we denote the objects by $O_1, \ldots, O_n$ and their respective diameters by $\delta_1, \ldots, \delta_n$.

Lemma 4. *Let $\mathcal{O}$ be a set of connected objects with diameter at least δ and a minimum covering box of size at most L. Then there exists a TSP tour T connecting all the objects in $\mathcal{O}$ of length $|T| \leq 4 \left[\frac{L}{\delta}\sqrt{2} + 1\right] L$.*

Proof. Consider covering box B of size at most L. A grid G of granularity $\delta/\sqrt{2}$ in B covers all the objects in $\mathcal{O}$, and the total length of all the lines in G is at most $2(L\sqrt{2}/\delta + 1)L$. By doubling each line of G we can build a TSP tour with length as stated in the lemma.   □

For two objects $O, O' \in \mathcal{O}$, define their distance $d(O, O') = \min\{\|x - y\| : x \in O, \; y \in O'\}$, and for an object $O \in \mathcal{O}$ and $r \in \mathbb{R}_+$, define the *r-neighborhood* of O to be $N(O, r) = \{O' \in \mathcal{O} : d(O, O') \leq r\}$. Finally, we fix a constant c and for $i = 1, \ldots, n$, define the neighborhood of O_i as $N(O_i) = N(O_i, c\delta_i)$.

Algorithm $\mathcal{A}$:

(1) For $i = 1, 2, \ldots$, let Q_i be the *smallest* object in $\mathcal{O}$ not belonging to $N(Q_1) \cup N(Q_2) \cup \ldots \cup N(Q_{i-1})$. Let k be the largest value of i for which such Q_i exists.

(2) Let B be a minimum covering box for $\mathcal{O}$, and B_i be a minimum covering box for $\{O \cap B : O \in N(Q_i)\}$, for $i = 1, \ldots, k$.

(3) Pick arbitrary points $p_i \in Q_i \cap B_i$, for $i = 1, \ldots, k$. Construct a $(1 + \epsilon)$-approximate TSP tour T_0 on $\{p_1, \ldots p_k\}$.

(4) For $i = 1, \ldots, k$, let T_i be the TSP tour guaranteed by Lemma 4 on the set of objects $N(Q_i)$ with the covering box B_i.

(5) Combine the tours $T_0, T_1, \ldots, T_k$ into a single TSP tour T.

(6) Output the minimum of T and the tour implied by Lemma 4 on the set $\mathcal{O}$.

Step (5) above can be done by adding two copies of a line segment connecting p_i to the closest point of T_i, for $i = 1, \ldots, k$. This yields an Eulerian tour that can be shortcut to a TSP tour T.

Theorem 3. *(i) Algorithm $\mathcal{A}$ gives an $O(\log n)$-approximate solution for the TSP with connected neighborhoods.*

(ii) If all the neighborhoods have the same (or comparable) diameters, then $\mathcal{A}$ is an $O(1)$-approximation algorithm.

Proof. Let OPT and OPT$'$ be respectively optimal TSP tours on $\mathcal{O}$ and $\mathcal{O}' = \{Q_1, \ldots, Q_k\}$. Note that $\delta_1 \leq \delta_2 \leq \ldots \leq \delta_k$. Let $L, L_1, \ldots, L_k$ be respectively the sizes of the minimum covering box $B, B_1, \ldots, B_k$.

(i) We first establish the following claim.

Claim 2. *If $k \geq 1$ then $|\text{OPT}'| \geq c \sum_{i=1}^{k-1} \delta_i / \log k$.*

Proof. Fix an orientation of OPT$'$ and define W_i as the arc of this directed tour (that connects exactly 2 objects in $\mathcal{O}'$) starting from Q_i. For any $i = 1, \ldots, k$, let $Q_{h(i)}$ be an object with smallest diameter among the two objects on the arc W_i. Then for $i = 1, \ldots, k$, $|W_i| \geq c\delta_{h(i)}$. (This follows from the fact that if W_i connects $Q_{h(i)}$ and Q_j then $Q_j \notin N(Q_{h(i)})$.) Consequently, $|\text{OPT}'| = \sum_{i=1}^{k} |W_i| \geq c \sum_{i=1}^{k} \delta_{h(i)}$. Let $\mathcal{O}'' = \mathcal{O}' \setminus \{Q_{h(i)} : i = 1, \ldots, k\}$. Note that $|\mathcal{O}''| \leq |\mathcal{O}'|/2$. Moreover, if OPT$''$ is an optimum TSP tour connecting the objects in $\mathcal{O}''$ then $|\text{OPT}''| \leq |\text{OPT}'|$. Thus applying the above argument to $\mathcal{O}''$, we obtain $|\text{OPT}''| \geq c \sum_{i: Q_i \in \mathcal{O}''} \delta_{h'(i)}$, where $h'(i)$ is the index of the object with smallest diameter among the two consecutive objects on the ith part of optimal path OPT$''$. Next we let $\mathcal{O}''' = \mathcal{O}'' \setminus \{Q_{h'(i)} : i \text{ s.t. } Q_i \in \mathcal{O}''\}$ and note again that $|\mathcal{O}'''| \leq |\mathcal{O}''|/2$. This process can continue for at most $\log k$ steps leading to at most $\log k$ lower bounds on $|\text{OPT}|$. Adding together these inequalities and noting that $|\text{OPT}'| \geq |\text{OPT}''| \geq |\text{OPT}'''| \geq \ldots$, we arrive at the bound stated in the claim. $\qquad\square$

Claim 3. $|T_0| \leq (1 + \epsilon) \left[4|\text{OPT}| + 2 \sum_{i=1}^{k-1} \delta_i \right]$.

Proof. Let OPT$''$ be an optimum tour on $\{Q_1, \ldots, Q_{k-1}\}$ and q_i be a point in $Q_i \cap \text{OPT}''$, for $i = 1, \ldots, k-1$. Then the union of OPT$''$ with two copies of each of the segments $p_i q_i$, for $i = 1, \ldots, k-1$, and $p_1 p_k$, forms a connected Eulerian graph that can be shortcut to a TSP tour T' on $\{p_1, \ldots, p_k\}$ of length

$$|T'| \leq |\text{OPT}''| + 2 \sum_{i=1}^{k-1} |p_i q_i| + 2|p_1 p_k| \leq |\text{OPT}''| + 2 \sum_{i=1}^{k-1} \delta_i + 2|p_1 p_k|$$

$$\leq (1 + 2\sqrt{2})|\text{OPT}| + 2 \sum_{i=1}^{k-1} \delta_i,$$

since $|\text{OPT}''| \leq |\text{OPT}|$, $p_i, q_i \in Q_i$, $p_1, p_k \in B$, and thus $|p_i q_i| \leq \delta_i$ and $|p_1 p_k| \leq \sqrt{2}L \leq \sqrt{2}|\text{OPT}|$. □

From the definition of $Q_1, \ldots, Q_k$, it follows that the minimum diameter of objects in the set $N(Q_i)$ is δ_i. It also follows from the definition of $N(Q_i)$ that $L_i \leq (2c+1)\delta_i$. Thus Lemma 4 gives, for $i = 1, \ldots, k$,

$$|T_i| \leq 4 \left[\frac{L_i}{\delta_i} \sqrt{2} + 1 \right] L_i \leq 4[(2c+1)\sqrt{2} + 1]L_i = O(L_i). \tag{2}$$

The length of the tour T returned by the algorithm can be bounded as follows:

$$|T| \leq |T_0| + \sum_{i=1}^{k} |T_i| + 2 \sum_{i=1}^{k} L_i \leq O(1)|\text{OPT}| + O(\sum_{i=1}^{k-1} \delta_i) = O(\log k)|\text{OPT}|,$$

using Claims 2 and 3, $L_i \leq (2c+1)\delta_i$, and $L_k \leq L \leq |\text{OPT}|$.

(ii) Let δ be the diameter of the smallest object, and assume that the diameters of all other objects are bounded by $\rho\delta$, for some constant $\rho \geq 1$. We require that the constant c used in the definition of the neighborhood is at least $\sqrt{2}(1+\rho)$. In case the objects have comparable diameters, we can strengthen Claim 2 as follows.

Claim 4. $|\text{OPT}'| \geq \left[\frac{k}{4} - 1 \right] \frac{\pi\delta}{4} \sqrt{2}$.

Proof. Let D_i, for $i = 1, \ldots, k$, be a disk of diameter $\delta_i \sqrt{2}$, enclosing object Q_i. We observe that, since $c \geq 1+\rho$, all the disks are disjoint. (Otherwise, there exist two indices $i < j$, such that $D_i \cap D_j \neq \emptyset$, implying that $d(Q_i, Q_j) \leq \sqrt{2}(\delta_i + \delta_j) \leq \sqrt{2}(1+\rho)\delta_i \leq c\delta_i$, and contradicting the fact that $Q_j \not\subseteq N(Q_i)$.) Thus we can apply Lemma 3 to the disks $D_1, \ldots, D_k$, (using $\alpha = 4$ for disks) to conclude that any TSP tour connecting these disks, and hence connecting the objects inside them, must have length bounded as stated in the claim. □

To bound the tour T returned by the algorithm, we observe that $|T_0| \leq (1+\epsilon)(|\text{OPT}| + 2\sum_{i=1}^{k} \delta_i)$, and combine this with 2 and Claim 4 to get

$$|T| \leq |T_0| + \sum_{i=1}^{k} |T_i| + 2 \sum_{i=1}^{k} L_i = O(\rho)|\text{OPT}|,$$

assuming that $L \geq \delta$ (otherwise, a tour of length at most $4(\sqrt{2}+1)|\text{OPT}|$ is guaranteed by Lemma 4 and Step (6) of the algorithm). □

References

1. E. M. Arkin and R. Hassin, *Approximation algorithms for the geometric covering salesman problem.*, Discrete Applied Mathematics **55** (1994), no. 3, 197–218.
2. S. Arora, *Nearly linear time approximation schemes for euclidean TSP and other geometric problems*, J. ACM **45** (1998), no. 5, 1–30.
3. M. de Berg, J. Gudmundsson, M.J. Katz, C. Levcopoulos, M.H. Overmars, and A. F. van der Stappen, *TSP with Neighborhoods of varying size*, J. of Algorithms **57** (2005), 22–36.
4. A. Dumitrescu and J.S.B. Mitchell, *Approximation algorithms for TSP with neighborhoods in the plane*, J. Algorithms **48** (2003), no. 1, 135–159.
5. K. Elbassioni, A.V. Fishkin, N. Mustafa, and R. Sitters, *Approximation algorithms for Euclidean group TSP.*, Proc. 15th Internat. Coll. on Automata, Languages and Programming, Lecture Notes in Computer Science, vol. 3580, Springer, 2005, pp. 1115–1126.
6. N. Garg, G. Konjevod, and R. Ravi, *A polylogarithmic approximation algorithm for the group Steiner tree problem*, J. Algorithms **37** (2000), no. 1, 66–84.
7. J. Gudmundsson and C. Levcopoulos, *A fast approximation algorithm for TSP with neighborhoods.*, Nordic J. Computing **6** (1999), no. 4, 469–488.
8. E. Halperin and R. Krauthgamer, *Polylogarithmic inapproximability*, Proc. 35th. Annual ACM Symposium on Theory of Computing, 2003, pp. 585–594.
9. C.S. Mata and J.S.B. Mitchell, *Approximation algorithms for geometric tour and network design problems (extended abstract).*, Proc. 11th. Annual ACM Symposium on Computational Geometry, 1995, pp. 360–369.
10. J.S.B. Mitchel, *Handbook of computational geometry*, ch. Geometric shortest paths and network optimization, pp. 633–701, Elsevier, North-Holland, Amsterdam, 2000.
11. J.S.B. Mitchell, *Guillotine subdivions approximate polygonal subdivisons: A simple polynomial-time approximation scheme for geometric TSP, k-MST and related problems*, SIAM J. Computing **28** (1999), no. 4, 1298–1309.
12. _____, *A PTAS for TSP with neighborhoods among fat regions in the plane*, Proc. 18th. Annual ACM-SIAM Symposium on Discrete Algorithms, 2007, To appear.
13. G. Reich and P. Widmayer, *Beyond Steiner's problem: a VLSI oriented generalization*, Proc. 15th. Int. Workshop on Graph-theoretic Concepts in Computer Science, Springer, 1990, pp. 196–210.
14. S. Safra and O. Schwartz, *On the complexity of approximating TSP with Neighborhoods and related problems.*, Proc. 11th. Annual European Symposium on Algorithms, Lecture Notes in Computer Science, vol. 2832, Springer, 2003, pp. 446–458.
15. P. Slavik, *The errand scheduling problem*, Tech. report, SUNY, Buffalo, USA, 1997.
16. A. F. van der Stappen, *Motion planning amidst fat obstacles*, Ph.d. dissertation, Utrecht University, Utrecht, the Netherlands, 1994.

Negation-Limited Complexity of Parity and Inverters

Kazuo Iwama[1], Hiroki Morizumi[1], and Jun Tarui[2]

[1] Graduate School of Informatics, Kyoto University, Kyoto 606-8501, Japan
{iwama, morizumi}@kuis.kyoto-u.ac.jp
[2] Department of Information and Communication Engineering,
University of Electro-Communications, Chofu, Tokyo 182-8585, Japan
tarui@ice.uec.ac.jp

Abstract. We give improved lower bounds for the size of negation-limited circuits computing Parity and for the size of negation-limited inverters. An inverter is a circuit with inputs $x_1, \ldots, x_n$ and outputs $\neg x_1, \ldots, \neg x_n$. We show that (1) For $n = 2^r - 1$, circuits computing Parity with $r - 1$ NOT gates have size at least $6n - \log_2(n + 1) - O(1)$ and (2) For $n = 2^r - 1$, inverters with r NOT gates have size at least $8n - \log_2(n + 1) - O(1)$. We derive our bounds above by considering the minimum size of a circuit with at most r NOT gates that computes Parity for *sorted* inputs $x_1 \geq \cdots \geq x_n$. For an *arbitrary* r, we completely determine the minimum size. For odd n, it is $2n - r - 2$ for $\lceil \log_2(n + 1) \rceil - 1 \leq r \leq n/2$, and it is $\lfloor 3/2\, n \rfloor - 1$ for $r \geq n/2$. We also determine the minimum size of an *inverter for sorted inputs* with at most r NOT gates. It is $4n - 3r$ for $\lceil \log_2(n + 1) \rceil \leq r \leq n$. In particular, the negation-limited inverter for sorted inputs due to Fischer, which is a core component in all the known constructions of negation-limited inverters, is shown to have the minimum possible size. Our fairly simple lower bound proofs use gate elimination arguments.

1 Introduction and Summary

Although exponential lower bounds are known [4], [6] for the monotone circuit size, at present we cannot prove a superlinear lower bound for the size of circuits computing an explicit Boolean function; the largest known lower bound [9], [7] is $5n - o(n)$. It is natural to ask: What happens if we allow a limited number of NOT gates? The hope is that by the study of *negation-limited* complexity of Boolean functions under various scenarios ([3], [14], [13], [2], [1], [11]), we obtain a better understanding of the power of NOT gates.

An *inverter* for n Boolean inputs $x_1, \ldots, x_n$ is a circuit whose outputs are the negations of the inputs, i.e., $\neg x_1, \ldots, \neg x_n$. We denote this n-input n-output function by Inv_n. Beals, Nishino, and Tanaka [3] have shown that one can construct a size-$O(n \log n)$ depth-$O(\log n)$ inverter with $\lceil \log_2(n + 1) \rceil$ NOT gates.

Following previous works, which we will explain below, we consider the circuit complexity of Parity$_n$ and Inv$_n$ with a *tightly limited* number of NOT gates: We assume that $n = 2^r - 1$ and we consider computations of Parity$_n$ and Inv$_n$

T. Asano (Ed.): ISAAC 2006, LNCS 4288, pp. 223–232, 2006.
© Springer-Verlag Berlin Heidelberg 2006

with $r - 1$ and r NOT gates respectively. For Parity$_n$ and Inv$_n$, $n = 2^r - 1$ is the maximum n such that computations are possible with $r - 1$ and r NOT gates respectively. An r-*circuit* is a circuit with at most r NOT gates. For a Boolean function f, let size$_r(f)$ and size$_{\text{mono}}(f)$ respectively denote the minimum size of r-circuits and monotone circuits computing f. The Boolean function Parity$_n(x_1, \ldots, x_n)$ is 1 iff $\sum x_i \equiv 1 \pmod 2$, and the Boolean function Majority$_n(x_1, \ldots, x_n)$ is 1 iff $\sum x_i \geq n/2$. We give the following lower bounds.

Theorem 1. *For* $n = 2^r - 1$,

$$\text{size}_{r-1}(\text{Parity}_n) \geq 2n - \log_2(n+1) - 1 + \text{size}_{\text{mono}}(\text{Majority}_n)$$
$$\geq 6n - \log_2(n+1) - O(1).$$

Theorem 2. *For* $n = 2^r - 1$,

$$\text{size}_r(\text{Inv}_n) \geq 4n - \log_2(n+1) + \text{size}_{\text{mono}}(\text{Majority}_n)$$
$$\geq 8n - \log_2(n+1) - O(1).$$

Now we explain the previously known lower bounds shown in Table 1, and how we obtain our improvements focusing on Parity$_n$.

Let C be a circuit computing Parity$_n$ with a tightly limited number of NOT gates as in Theorem 1. Then, the first NOT gate N, i.e., a unique NOT gate that is closest to the inputs, must compute $\neg$Majority$_n$, and the subcircuit C' at the immediate predecessor of N is a monotone circuit computing Majority$_n$. Long [8] has shown that such a monotone circuit has size at least $4n - O(1)$:

Proposition 1. *([8])* size$_{\text{mono}}$(Majority$_n$) $\geq 4n - O(1)$.

We want to show that in addition to those gates in the subcircuit C', the circuit C must contain a certain number of gates; i.e., we want to show as good a lower bound as possible for the number of gates in $C - C'$. Tanaka, Nishino, and Beals [14] showed that there are at least $3\log_2(n+1)$ additional gates; Sung [12] and Sung and Tanaka [13] showed that there are at least about $1.33n$ additional gates; we show that there are at least about $2n$ additional gates. We show this in the following way.

We argue that a part of $C - C'$ must be computing what we call a sorted parity function, and we show that a circuit computing a sorted parity function has size at least about $2n$ when the number of NOT gates is tightly limited. A Boolean

Table 1. The lower bounds of previous works and this paper

	Parity	Inverter
Tanaka-Nishino-Beals [14]	$4n + 3\log_2(n+1) - O(1)$	
Beals-Nishino-Tanaka [3]		$5n + 3\log_2(n+1) - O(1)$
Sung [12]/Sung-Tanaka [13]	$5.33n + \log_2(n+1)/3 - O(1)$	$7.33n + \log_2(n+1)/3 - O(1)$
this paper	$6n - \log_2(n+1) - O(1)$	$8n - \log_2(n+1) - O(1)$

function $f : \{0,1\}^n \rightarrow \{0,1\}$ is a *sorted parity* function if for all sorted inputs $x_1 \geq x_2 \geq \cdots \geq x_n$, $f(x_1,\ldots,x_n) = \text{Parity}(x_1,\ldots,x_n)$. A function f is a sorted ¬parity function if for all sorted inputs $x_1 \geq x_2 \geq \cdots \geq x_n$, $f(x_1,\ldots,x_n) = \neg\text{Parity}(x_1,\ldots,x_n)$.

In fact, we completely determine the minimum size of a circuit with at most r NOT gates computing Sorted Parity$_n$ and Sorted ¬Parity$_n$, where a parameter r is an arbitrary nonnegative integer. From about $2n$, the minimum size decreases by 1 with each additional NOT gate. This decrease stops at about $1.5n$: one cannot make a circuit smaller using more NOT gates.

We also consider the minimum size of an *inverter for sorted inputs*, i.e., a circuit with Boolean inputs $x_1,\ldots,x_n$ that outputs $\neg x_1,\ldots,\neg x_n$ for all the sorted inputs $x_1 \geq \cdots \geq x_n$. The negation-limited inverter for sorted inputs due to Fischer [5] (shown in Figure 3 in the last page) is a core component in all the known constructions of negations-limited inverters [3], [14], [5]. We again completely determine the minimum size of an inverter for sorted inputs with at most r NOT gates for any r. In particular, we show that Fischer's inverter for sorted inputs has the minimum possible size.

We think that our complete determination of size$_r$(Sorted Parity$_n$) and size$_r$(Sorted Inv$_n$) are interesting in their own. For the trade-off of size versus the number of NOT gates, an *asymptotically* tight result has been shown by Amano, Maruoka, and Tarui [2]. They showed that for $0 \leq r \leq \log_2 \log_2 n$, the minimum size of a circuit computing Merge with r NOT gates is $\Theta(n \log n / 2^r)$; thus they showed a smooth trade-off between the monotone case of $\Theta(n \log n)$ and the general case of $\Theta(n)$. But as far as we know, our result for Sorted Parity and inverters for sorted inputs is the first one that establishes an *exact* trade-off.

Our fairly simple lower bound proofs use gate elimination arguments in a somewhat novel way. The following are precise statements of our results.

Theorem 3. *The size and the number of AND/OR/NOT gates in smallest circuits with at most r NOT gates that compute Sorted Parity$_n$ and Sorted ¬Parity$_n$ are as shown in Table 2 and Table 3. In particular, for $n = 2^s - 1$, a smallest circuit with $s - 1$ NOT gates computing Sorted Parity$_n$ has size $2n - s - 1 = 2n - \log_2(n+1) - 1$.*

Theorem 4. *For $\lceil \log_2(n+1) \rceil \leq r \leq n$, a smallest inverter for sorted inputs with at most r NOT gates has size $4n - 3r$ consisting of $2n - 2r$ AND gates, $2n - 2r$ OR gates, and r NOT gates. In particular, for $n = 2^r - 1$, a smallest inverter for sorted inputs with r NOT gates has size $4n - 3r = 4n - 3\log_2(n+1)$.*

Table 2. the size and the number of AND/OR/NOT gates in a smallest circuit with $\leq r$ NOTs computing Sorted Parity

	size	AND	OR	NOT
$\lfloor n/2 \rfloor \leq r$	$\lfloor 3/2\, n \rfloor - 1$	$\lfloor n/2 \rfloor$	$\lceil n/2 \rceil - 1$	$\lfloor n/2 \rfloor$
$\lceil \log_2(n+1) \rceil - 1 \leq r \leq \lfloor n/2 \rfloor$, n odd	$2n - r - 2$	$n - r - 1$	$n - r - 1$	r
$\lceil \log_2(n+1) \rceil - 1 \leq r \leq \lfloor n/2 \rfloor$, n even	$2n - r - 1$	$n - r$	$n - r - 1$	r
$r < \lceil \log_2(n+1) \rceil - 1$	not computable			

Table 3. the size and the number of AND/OR/NOT gates in a smallest circuit with $\leq r$ NOTs computing Sorted ¬Parity

	size	AND	OR	NOT
$\lfloor n/2 \rfloor \leq r$	$\lceil 3/2\,n \rceil - 1$	$\lceil n/2 \rceil - 1$	$\lfloor n/2 \rfloor$	$\lceil n/2 \rceil$
$\lceil \log_2(n+2) \rceil - 1 \leq r \leq \lfloor n/2 \rfloor,\ n$ odd	$2n - r$	$n - r$	$n - r$	r
$\lceil \log_2(n+2) \rceil - 1 \leq r \leq \lfloor n/2 \rfloor,\ n$ even	$2n - r - 1$	$n - r - 1$	$n - r$	r
$r < \lceil \log_2(n+2) \rceil - 1$	not computable			

2   Lower Bounds for Parity and Inverters

2.1   Preliminaries

Markov [10], [5] precisely determined the minimum number of NOT gates necessary to compute a Boolean function. We state a special case of Markov's result relevant to our work, and include a proof sketch.

Proposition 2. *(Markov [10], [5]) The maximum n such that Inv_n is computable by an r-circuit is $n = 2^r - 1$. The maximum n such that $Parity_n$ is computable by an r-circuit is $n = 2^{r+1} - 1$.*

Proof Sketch. We only show the upper bounds for n; we show that if Inv_n is computable by some r-circuit, then $n \leq 2^r - 1$; a similar argument yields the upper bound for $Parity_n$. Proceed by induction on r, and for the sake of contradiction assume that C is an r-circuit computing Inv_n for $n = 2^r$. Let N be a NOT gate in C such that if G is the immediate predecessor of N, i.e., there is a wire from G to N, then the subcircuit at G is monotone. An arbitrary monotone function $f : \{0,1\}^m \to \{0,1\}$ has a minterm or a maxterm of size $\leq \lceil m/2 \rceil$, i.e., there are $\lceil m/2 \rceil$ inputs x_i's such that fixing x_i's to be 1 (or 0) fix f to be 1 (or 0). Fixing such a term for G fixes N, and yields an $(r-1)$-circuit computing $Inv_{n/2}$; a contradiction.   □

Let C be an r-circuit computing Inv_n for $n = 2^r - 1$. Let G_1 be the immediate predecessor of a NOT gate N_1 such that the subcircuit at G_1 is monotone. By a similar analysis we can see that all the maxterms and the minterms of G have size $(n+1)/2$, i.e., G computes $Majority_n$. By an inductive analysis we can see that the immediate predecessors $G_1, \ldots, G_r$ of r NOT gates are such that for each $x \in \{0,1\}^n$, $G_1(x) \cdots G_r(x)$ is the binary representation of $|\{i : x_i = 1\}|$. An $(r-1)$-circuit computing $Parity_n$ for $n = 2^r - 1$ has a similar property: $G_1(x) \cdots G_{r-1}(x)$ are the $r-1$ significant bits of the binary representation. What we have just stated about G_i's is due to Beals, Nishino, and Tanaka [3].

We will use the following result by Sung and Tanaka[13]. We include a proof sketch in the appendix.

Lemma 1. *([13]) For $n = 2^r - 1$, $size_r(Inv_n) \geq size_{r-1}(Parity_n) + 2n + 1$.*

2.2 Crossing Wires

We introduce the notion of *crossing wire* and show simple lemmas. The lemmas are not strictly necessary for our proofs of the theorems, but their statements and proofs should be helpful for understanding our framework, and we think that the lemmas may be useful for further investigations of negation-limited circuits. A similar notion has been introduced in [12] as a *boundary gate*. We focus on wires as opposed to gates.

Fix a circuit C. A gate g in C is *black* if there is a path from some input to g going through a NOT gate, including the case where g itself is a NOT gate. Otherwise, g is *white*; inputs $x_1, \ldots, x_n$ are white.

Say that a wire going from g to h is a *crossing* wire if g is white and h is black. The white gates and inputs constitute the *monotone part* of C, and the black gates constitute the *nonmonotone part*.

Lemma 2. *Distinct crossing wires go into distinct gates.*

Proof. Let w_1 from g_1 to h_1 and w_2 from g_2 to h_2 be distinct crossing wires. By definition, g_1 and g_2 are white. If $h_1 = h_2$, this single gate is white; this contradicts the assumption that w_1 and w_2 are crossing wires. □

Lemma 3. *Let C be a circuit computing a nonmonotone Boolean function f. Suppose that there are $a_0, \ldots, a_k \in \{0,1\}^n$ such that $a_0 < \cdots < a_k$ and $f(a_i) \neq f(a_{i+1})$ for $0 \leq i < k$. Then, the number of crossing wires in C are at least k.*

Proof. The output gate T of a nonmonotone circuit C is black. Hence any path in C from an input x_i to T contains a crossing wire. If the values on all crossing wires remain the same, then the output remains the same. The value of a crossing wire changes only monotonically. The lemma follows. □

We note that the two lemmas above immediately yield an n lower bound for the size of nonmonotone area of circuits computing Parity_n and Inv_n.

2.3 Proofs of Theorems 1 and 2

We prove Theorems 1 and 2 using the lower bound for Sorted Parity_n in Theorem 3, which will be proved in Section 3.

Proof of Theorem 1. Let C be an $(r-1)$-circuit that computes Parity_n for $n = 2^r - 1$. As explained after Proposition 2, there is a NOT gate N in C such that the subcircuit C' at its immediate predecessor is a monotone circuit computing $\mathrm{Majority}_n$. All the gates in C' are white, and by Proposition 1 the number of them is at least $\mathrm{size}_{\mathrm{mono}}(\mathrm{Majority}_n) \geq 4n - O(1)$.

We can convert the nonmonotone, black part of C into a circuit computing Sorted Parity for new inputs $y_1, \ldots, y_n$ as follows. Consider the chain $\langle a_0 = 0^n, a_1 = 10^{n-1}, \ldots, a_n = 1^n \rangle$, and the computation of C on $a_0, \ldots, a_n$. When the input changes from a_{i-1} to a_i ($1 \leq i \leq n$), some crossing wires change the value from 0 to 1. Let W_i be the set of such crossing wires. Note that each W_i is nonempty, and by Lemma 2 the sets W_i's are mutually disjoint.

Connect a new input y_i to all the gates g in C such that some crossing wire w in W_i goes into g. Let D be the circuit thus obtained. Clearly, D computes Sorted Parity for $y_1 \geq \cdots \geq y_n$, and the number of gates in D is a lower bound for the number of black gates in C. By the lower bound for Sorted Parity$_n$ in Theorem 3, the size of D is at least $2n - (r - 1) - 2 = 2n - \log_2(n + 1) - 1$.

Adding up the lower bounds for the number of white gates in C and the number of black gates in C yields the theorem.   □

Theorem 2 immediately follows from Theorem 1 and Lemma 1. We note that instead of using Lemma 1, we can argue similarly as above using the lower bound in Theorem 4, and obtain a lower bound that is smaller by $2 \log_2 n$ than the bound in Theorem 2.

3   Sorted Input Case: The Minimum Size Determined

The upper bounds of Theorem 3 and Theorem 4 can be shown by straightforward constructions as we will explain in section 3.2. We prove the lower bounds of Theorem 3 and Theorem 4 in section 3.1.

3.1   Lower Bounds

We use well-known gate elimination arguments: We fix x_i, one at a time, to be $0/1$ and eliminate some gates. A gate g is eliminated if its value is fixed or else the value of one wire coming into g is fixed. In the latter case, the other input wire of g replaces all the out-going wires of g, and g is eliminated. A lower bound for the total number of eliminations is a lower bound for the number of gates in a circuit.

Proof of the lower bound of Theorem 3.   Assume that n is odd and let C be a circuit computing Sorted Parity$_n$ for $x_1 \geq \cdots \geq x_n$ at the top output gate T. Starting from $(0, 0, \ldots, 0)$, consider flipping and fixing $x_i = 1$ for $i = 1, \ldots, n-1$, in this order one at a time: Fix $x_i = 1$ after $x_1, \ldots, x_{i-1}$ have been fixed and remain to be 1. Each time we flip and fix $x_i = 1$, the value of T changes flipping from 0 to 1 or 1 to 0. There must be a path p from x_i to T such that all the gates on p flip the values when we fix $x_i = 1$. Call such a path a *propagating* path with respect to x_i.

Consider fixing $x_i = 1$. Let p be a propagating path for x_i. Consider the gates on p from x_i towards T. If all the gates on p (including T) are ORs, fixing $x_i = 1$ will fix $T = 1$; this is a contradiction. Thus there is either an AND or a NOT in p. Let g be the first non-OR gate in p. All the OR gates, if any, before g are *fixed* to be 1 once we fix $x_i = 1$. Thus one input wire of g is fixed to be 1.

(1) If g is AND, g is eliminated.

(2) If g is NOT, g is fixed to be 0 and is eliminated. In this case, there must be at least one AND/OR gate in p beyond g: If all the gates beyond g are NOTs, all their values are fixed; this is a contradiction. Hence at least one AND/OR gate (the first AND/OR beyond g) gets eliminated.

Now assume that the circuit C contains s NOT gates. From (1) and (2) we see that there are at least $n-1$ AND/OR gates; thus there are at least $n-1+s$ gates. This bound becomes meaningful when s is large. In particular, combined with the bounds we derive below it will be easy to see that a smallest circuit for Sorted Parity$_n$ does not contain more than $\lfloor n/2 \rfloor$ NOT gates. By (1), at least $n-1$ AND/NOT gates are eliminated; thus the circuit contains at least $n-1-s$ ANDs.

Starting from $(1,1,\ldots,1)$, consider flipping $x_i = 0$ for $i = n, n-1, \ldots, 2$ in this order one at a time. Dual arguments yield the same lower bound for the number of ORs.

Consider the case where n is even. In this case the circuit obtained after fixing $x_i = 1$ for $i = 1, \ldots, n-1$ must contain one NOT gate; thus at most $s-1$ NOT gates are eliminated, and hence the lower bound for the number of ANDs increases by 1. A similar increase occurs for odd n and Sorted ¬Parity$_n$.   □

For Theorem 4 we want to show a lower bound about twice as large by showing that the number of AND/OR gates eliminated is twice as large.

In the lower bound proof of Theorem 3 above, the eliminations of gates are always due to the fact that the value of a gate has been *determined* by having fixed some inputs. In the lower bound proof of Theorem 4, we also eliminate a gate when its value is not necessarily determined for an arbitrary input, but its value *must stay constant* for *sorted* inputs. With this additional argument we proceed similarly as in the lower bound proof of Theorem 3.

Proof of the lower bound of Theorem 4. Let C be an inverter for n sorted inputs $x_1 \geq \cdots \geq x_n$. Starting from $(0,0,\ldots,0)$, consider flipping and fixing $x_i = 1$ for $i = 1, \ldots, n$: Fix $x_i = 1$ after $x_1, \ldots, x_{i-1}$ have been fixed and remain to be 1. Each time we flip and fix $x_i = 1$, the output $\overline{x_i}$ changes flipping from 1 to 0. There must be a path p from x_i to $\overline{x_i}$ such that all the gates on p flip the values when we fix $x_i = 1$. Call such a path a *propagating* path for x_i.

Consider fixing $x_i = 1$. Let p be a propagating path for x_i. Consider the gates on p from x_i towards $\overline{x_i}$. If all the gates on p are ORs, fixing $x_i = 1$ will fix $\overline{x_i} = 1$; this is a contradiction. Thus there is either an AND or a NOT in p.

Let g be the first non-OR gate in p. The gate g gets eliminated after fixing $x_i = 1$. Note that if g is an AND, the value of g is 1 after fixing $x_i = 1$ since all the gates, if any, before g are ORs.

Let h be the last non-OR gate in p. All the gates, if any, beyond h are ORs. After fixing $x_i = 1$, the values of all the gates between h and the output $\overline{x_i}$, including h and $\overline{x_i}$, are 0.

We claim that we can fix h to be 0 and thus eliminate h from the circuit in the following sense. We have fixed $x_1, \ldots, x_i$ to be 1; $x_{i+1}, \ldots, x_n$ are 0 at present. We will further flip and fix $x_{i+1}, \ldots, x_n$ to be 1 one at a time; but in this process the value of gate h must remain to be 0 since if the gate h has value 1, the output $\overline{x_i}$ gets flipped back from 0 to 1 contradicting to the fact that x_i has been fixed and remains to be 1. Since the gate h will always be 0, we can fix h to be 0 and eliminate h; the resulting circuit behaves in the same way. We note that if we set $x_{i+1}, \ldots, x_n$ to be a *non*-sorted 0/1 sequence, it is possible that the gate h evaluates to 1 even if $x_1, \ldots, x_i$ are all 1.

It is possible that the gate g and h are the same NOT gate, i.e., $g = h$. But they can *not* be the same AND gate since after fixing $x_i = 1$, h is 0 and g is 1 if g is an AND. Thus unless both g and h are NOTs, $g \neq h$. Therefore if the circuit C contains s NOT gates, we can eliminate a total of at least $2n - 2s$ AND gates, and hence C contains at least $2n - 2s$ AND gates.

The dual argument about starting from $(1, 1, \ldots, 1)$ and fixing $x_i = 0$ for $i = n, n-1, \ldots, 1$ yields the same lower bound for the number of ORs.   □

3.2   Upper Bounds

Proof of the upper bound of Theorem 3.   We can construct a smallest circuit computing Sorted Parity$_n$ with at most r NOT gates for odd n as follows. Constructions for even n and for Sorted ¬Parity will be explained in the end.

CASE 1: $r = \lceil \log_2(n+1) \rceil - 1$ and $n = 2^{r+1} - 1$: See Figure 1.

CASE 2: $r = \lceil \log_2(n+1) \rceil - 1$ and $2^r \leq n < 2^{r+1} - 1$: See Figure 2.

In cases 1 and 2 it is easy to see that y_i's are sorted if x_i's are sorted, and that the circuit consists of $n - r - 1$ ANDs, $n - r - 1$ ORs, and r NOTs.

CASE 3: $r > \lceil \log_2(n+1) \rceil - 1$: Construct a circuit of the following form:

$$(x_1 \wedge \overline{x_2}) \vee \cdots \vee (x_{2s-1} \wedge \overline{x_{2s}}) \vee \text{Sorted Parity}_{n-2s}(x_{2s+1}, \ldots, x_n),$$

where Sorted Parity$_{n-2s}$ is computed by a circuit in case 1 or case 2: Let s be the maximum integer satisfying $2^{(r-s)+1} - 1 \geq n - 2s \geq 1$. Use s NOT gates for s pairs $(x_1, x_2), \ldots, (x_{2s-1}, x_{2s})$, and use $\lceil \log_2(n - 2s + 1) \rceil - 1$ NOT gates for $x_{2s+1}, \ldots, x_n$ as in cases 1 and 2. As for the size, the analysis for cases 1 and

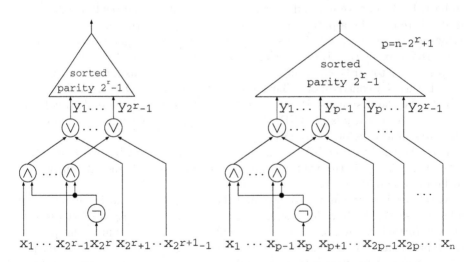

Fig. 1. Sorted Parity for $n = 2^{r+1} - 1$ with r NOTs

Fig. 2. Sorted Parity for $2^r \leq n < 2^{r+1} - 1$ with r NOTs

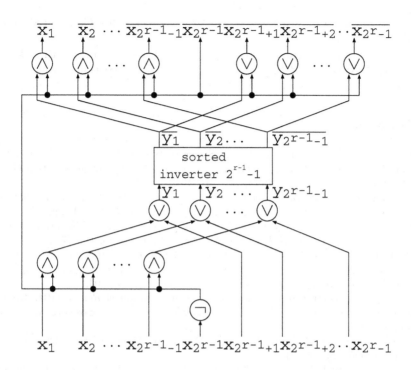

Fig. 3. Fischer's inverter for $n = 2^r - 1$ sorted inputs $x_1 \geq \cdots \geq x_n$ with r NOTs

2 applies for the subcircuit for $x_{2s+1}, \ldots, x_n$, and we are using s ANDs, s ORs, and s NOTs additionally.

For even n, construct a circuit as SortedParity$(x_1, \ldots, x_{n-1}) \wedge \overline{x_n}$. For Sorted $\neg$Parity$_n$, construct a circuit as $\overline{x_1} \vee$ Sorted Parity$(x_2, \ldots, x_n)$. $\qquad \square$

Proof of the upper bound of Theorem 4. Construct a circuit as follows.
CASE 1: $r = \lceil \log_2(n+1) \rceil$, $n = 2^r - 1$: Figure 3 shows the circuit due to Fischer.
CASE 2: $r = \lceil \log_2(n+1) \rceil$, $2^{r-1} \leq n < 2^r - 1$: Use $\overline{x_p}$ instead of $\overline{x_{2^r-1}}$ similarly as in case 2 of Sorted Parity.
CASE 3: $r > \lceil \log_2(n+1) \rceil$: Similarly as in case 3 of Sorted Parity, apply s NOTs directly to inputs $x_1, \ldots, x_s$ to obtain outputs $\overline{x_1}, \ldots, \overline{x_s}$, and use $\lceil \log_2(n - s + 1) \rceil$ NOT gates for $x_{s+1}, \ldots, x_n$ to obtain $\overline{x_{s+1}}, \ldots, \overline{x_n}$.

It is easy to see that the circuit thus constructed has size $4n - 3r$ consisting of $2n - 2r$ ANDs, $2n - 2r$ ORs, and r NOTs. $\qquad \square$

4   Open Problems

In a recent paper, Sato, Amano, and Maruoka [11] consider the problem of inverting k-tonic 0/1 sequences, where a k-tonic sequence is a natural generalization of a bitonic sequence. They consider restricting the number of NOT gates to be $O(\log n)$, and show that for constant k, one can construct a k-tonic inverter of size $O(n)$ and depth $O(\log^2 n)$ using $O(\log n)$ NOT gates. Can we reduce the depth to $O(\log n)$?

Which functions are computable by circuits with size $O(n)$, depth $O(\log n)$, and $O(\log n)$ NOT gates? In particular, under what restrictions on inputs can we construct inverters with these parameters?

References

1. K. AMANO AND A. MARUOKA, A Superpolynomial Lower Bound for a Circuit Computing the Clique Function with at most $(1/6) \log \log n$ Negation Gates, *SIAM J. Comput.* (2005) **35(1)**, pp. 201–216.
2. K. AMANO, A. MARUOKA AND J. TARUI, On the Negation-Limited Circuit Complexity of Merging, *Discrete Applied Mathematics* (2003) **126(1)**, pp. 3–8.
3. R. BEALS, T. NISHINO AND K. TANAKA, On the Complexity of Negation-Limited Boolean Networks, *SIAM J. Comput.* (1998) **27(5)**, pp. 1334–1347.
4. R. BOPPANA AND M. SIPSER, The Complexity of Finite Functions, *Handbook of Theoretical Computer Science, Volume A: Algorithms and Complexity*, J. v. Leeuwen editor, Elsevier/MIT Press (1990), pp. 757–804.
5. M. FISCHER, Lectures on Network Complexity, *Technical Report* 1104, CS Department, Yale University, http://cs-www.cs.yale.edu/homes/fischer, 1974, revised 1996.
6. D. HARNIK AND R. RAZ, Higher Lower Bounds on Monotone Size, *Proc. of 32nd STOC* (2000), pp. 378–387.
7. K. IWAMA AND H. MORIZUMI, An Explicit Lower Bound of $5n - o(n)$ for Boolean Circuits, *Proc. of 27th MFCS* (2002), LNCS vol. 2420, pp. 353–364.
8. D. LONG, The Monotone Circuit Complexity of Threshold Functions, Unpublished manuscript, University of Oxford, (1986).
9. O. LACHISH AND R. RAZ, Explicit Lower Bound of $4.5n - o(n)$ for Boolean Circuits, *Proc. of 33rd STOC* (2001), pp. 399–408.
10. A. A. MARKOV, On the Inversion Complexity of a System of Functions, *J. ACM* (1958) **5(4)**, pp. 331–334.
11. T. SATO, K. AMANO, AND A. MARUOKA, On the Negation-Limited Circuit Complexity of Sorting and Inverting k-tonic Sequences, *Proc. of 12th COCOON*, LNCS vol. 4112, (2006), pp. 104–115.
12. S. SUNG, On Negation-Limited Circuit Complexity, *Ph.D. thesis*, Japan Advanced Institute of Science and Technology, 1998.
13. S. SUNG AND K. TANAKA, Lower Bounds on Negation-Limited Inverters, *Proc. of 2nd DMTCS: Discrete Mathematics and Theoretical Computer Science Conference* (1999), pp. 360–368.
14. K. TANAKA, T. NISHINO AND R. BEALS, Negation-Limited Circuit Complexity of Symmetric Functions, *Inf. Process. Lett.* (1996) **59(5)**, pp. 273–279.

Appendix: Proof Sketch of Lemma 1. Let C be an inverter with r NOTs for $n = 2^r - 1$ inputs $x_1, \ldots, x_n$. The subcircuit D at the immediate predecessor of the last NOT gate N in C is an $(r-1)$-circuit for Parity$_n$. For $1 \le i \le n$, let e_i be the n-bit vector such that the i-th bit is 1 and the other bits are 0's. Consider a propagating path p_i with respect to $(0, 0, \ldots, 0)$ and e_i. Let g_i be the last OR gate in p_i from N towards $\overline{x_i}$; such g_i must exist, and $g_i \ne g_j$ for $i \ne j$; thus there are at least n ORs in $C - D$. We can argue similarly for ANDs. $\square$

The Complexity of Quasigroup Isomorphism and the Minimum Generating Set Problem

V. Arvind[1] and Jacobo Torán[2]

[1] The Institute of Mathematical Sciences, Chennai 600 113, India
arvind@imsc.res.in
[2] Theoretische Informatik, Universität Ulm, D-89069 Ulm, Germany
jacobo.toran@uni-ulm.de

Abstract. Motivated by Papadimitriou and Yannakakis' paper on limited nondeterminism [19], we study two questions arising from their work: *Quasigroup Isomorphism* and the *Minimum generating set problem* for groups and quasigroups.

1 Introduction

The Isomorphism problem for combinatorial and algebraic structures has motivated important notions like Arthur-Merlin games, lowness, and interactive proof systems, and has inspired advances in areas like counting classes and derandomization. In this paper, we study Quasigroup Isomorphism, QGROUP-ISO, and related problems. A finite *quasigroup* is an algebraic structure $(G, \cdot)$, where the finite set G is closed under the binary operation $\cdot$ that has unique left and right inverses. Quasigroups are more general than groups: they are nonassociative and need not have an identity element. The input for QGROUP-ISO is a pair (G_1, G_2) of quasigroups of order n given by multiplication tables (of size $n^2 \times \log n$), and the problem is to test if they are isomorphic. An isomorphism between G_1 and G_2 is a bijection $\varphi : G_1 \longrightarrow G_2$ such that for $i, j \in G_1$, $\varphi(ij) = \varphi(i)\varphi(j)$.[1] The complexity of this problem has been studied for close to three decades. Quasigroups of order n have generating sets of size $\log n$ computable in time polynomial in n. This gives an easy $n^{\log n + O(1)}$ time isomorphism test: compute a $\log n$-size generating set S for G_1. Map S bijectively to $S' \subset G_2$ (for each of the $\binom{n}{|S|} \times |S|!$ ordered subsets S' of G_2). For each of these maps check, using the multiplication table of G_1, if it extends to an isomorphism from G_1 and G_2. This algorithm is due to Tarjan [18]. Tarjan's algorithm can be seen as a polynomial-time nondeterministic procedure for QGROUP-ISO that uses only $\log^2 n$ nondeterministic bits (to guess the injective map from S into G_2). Papadimitriou and Yannakakis observe this in their paper [19] on limited nondeterminism, classifying QGROUP-ISO in NP($\log^2 n$) (details in Section 2.1).

We study QGROUP-ISO in terms of bounded nondeterminism, parameterized complexity and restricted queries to NP. Papadimitriou and Yannakakis

---

[1] For convenience we represent operations of both quasigroups by concatenation.

T. Asano (Ed.): ISAAC 2006, LNCS 4288, pp. 233–242, 2006.
© Springer-Verlag Berlin Heidelberg 2006

ask [19] whether QGROUP-ISO is $NP(\log^2 n)$-complete. We give negative evidence by showing a suitable complexity upper bound on Quasigroup Nonisomorphism (QGROUP-NONISO). As QGROUP-ISO is polynomial-time reducible to Graph Isomorphism which is in AM, clearly QGROUP-NONISO $\in$ AM. However, this upper bound is not entirely satisfactory because QGROUP-ISO $\in$ $NP(\log^2 n)$ unlike Graph Isomorphism. For QGROUP-NONISO we give an AM protocol with constant error probability, in which Arthur uses $O(\log^c n)$ random bits and Merlin uses $O(\log^c n)$ nondeterministic bits for a constant c. In fact, we first give such a IP protocol for QGROUP-NONISO. Then we do an IP to AM conversion via a *scaled down* Goldwasser-Sipser simulation [12] of IP by AM protocols that preserves polylog message size and random bits. That the Goldwasser-Sipser simulation scales down in this manner is interesting in its own right.[2] Let $NTIME[s(n), t(n)]$ denote the class of languages accepted by nondeterministic Turing machines in time $O(t(n))$ with $O(s(n))$ nondeterministic bits. We show, using the above protocol, that QGROUP-ISO cannot be $NP(\log^2 n)$ complete unless the coNP-complete problem $\overline{\text{CLIQUE}}$ is in the non-uniform class $NTIME[n^{O(1)}, 2^{O(\sqrt{n} \log n)}]/\text{poly}$. I.e. for inputs of length n, $\overline{\text{CLIQUE}}$ has polynomial-size proofs that can be verified in time $2^{O(\sqrt{n} \log n)}$ with a polynomial-size advice. We also consider a parameterized version of QGROUP-ISO and show it is unlikely to be hard for the parameterized complexity classes $W[P]$ or even $W[1]$ unless $\overline{\text{CLIQUE}}$ is in $NTIME[n^{O(1)}, 2^{o(n)}]/\text{poly}$.

In Section 3 we consider the *minimum* generating set problem MIN-GEN: given a quasigroup G as a table, find a minimum cardinality generating set for G. We show that for general groups MIN-GEN is in $DSPACE(\log^2 n)$, and for nilpotent groups it is in P. Finally, in Section 4 we give a sufficient condition for QGROUP-ISO to be in P.

Most concepts used in the paper are defined where needed. Standard complexity theory definitions can be found in a textbook like [4].

2   Quasigroup Isomorphism

Let G be a quasigroup with n elements. By definition, for $a, b \in G$ there are unique elements $x, y \in G$ with $ax = b$ and $ya = b$. It follows that quasigroups have the cancellation property: $ba = ca$ implies $b = c$, and $ab = ac$ implies $b = c$. Suppose $H \subset G$ is a proper *sub-quasigroup*, and $x \in G \setminus H$. Then $Hx \cap H = \emptyset$ as H is a quasigroup. Moreover, because of cancellation $|Hx| = |H|$. Thus, $|H| \leq |G|/2$. It follows easily that G has a generating set of size at most $\log n$. Moreover, a $\log n$ size generator set is easy to find in polynomial time by a greedy algorithm. As observed in the introduction, QGROUP-ISO is in $NP(\log^2 n)$.

For each constant $k > 0$, let $IP_k(r(n), s(n))$ denote the class of languages accepted by k-round IP protocols, with error probability $1/4$, in which the verifier uses $O(r(n))$ random bits in each round and the size of each message in the protocol is bounded by $O(s(n))$. Similarly, denote by $AM_k(r(n), s(n))$ the class

---

[2] This AM protocol is conceptually simpler than the one given by the authors for Group Isomorphism in [3].

of languages accepted by k-round AM protocols, with error probability $1/4$, in which Arthur uses $O(r(n))$ random bits in each round and Merlin uses $O(s(n))$ nondeterministic bits. Formally, a language L is in $\text{AM}_2(r(n), s(n))$ if there is a set $B \in \text{P}$ such that for all x, $|x| = n$,

$$x \in A \Rightarrow \text{Prob}_{w \in_R \{0,1\}^{r'(n)}}[\exists y, |y| = s'(n) : \langle x, y, w \rangle \in B] \geq 3/4,$$

$$x \notin A \Rightarrow \text{Prob}_{w \in_R \{0,1\}^{r'(n)}}[\forall y, |y| = s'(n) : \langle x, y, w \rangle \in B] \leq 1/4,$$

where r' and s' are functions in $O(r(n))$ and $O(s(n))$ respectively. Notice that $\text{AM}_2(n^{O(1)}, n^{O(1)})$ is the usual 2-round AM class. The first goal of this section is to show that QGROUP-NONISO is in $\text{AM}_2(\log^c n, \log^c n)$ for some constant c and with constant success probability.

Let G be an n element quasigroup. A k-sequence $(g_1, \ldots, g_k)$ of G is a *generating k-sequence* if $\langle g_1, \ldots, g_k \rangle = G$ (i.e. the distinct elements in the sequence form a generating set for G). An element can appear more than once in the sequence. We describe an easy polynomial-time algorithm that on input an n-element quasigroup G, a generating set X for G, and a 1-1 mapping $\phi : X \rightarrow \{1, \ldots, |X|\}$ outputs a table for a quasigroup G' isomorphic to G.

Proposition 1. *Let G be an n element quasigroup with a k element generating set X. Let $\phi : X \rightarrow [k]$ be a bijection. Then there is a polynomial-time algorithm that computes the table for an n element quasigroup G' isomorphic to G with an isomorphism extending ϕ.*

Proof. First, we extend ϕ to a bijection from G to $[n]$ in a canonical way (for example, using the lexicographic ordering of G). Now, it is easy to see that we can define the quasigroup G' as follows: For two elements $\phi(x), \phi(y)$ in G',

$$\phi(x) \cdot \phi(y) = \phi(\phi^{-1}(\phi(x)) \cdot \phi^{-1}((\phi(y)))) = \phi(x \cdot y).$$

■

We next describe a procedure $\mathcal{B}$ that on input a quasigroup table G with n elements outputs with high probability an $O(\log n)$ size generating sequence G.

Theorem 1. *There is a randomized polynomial-time algorithm that takes as input a quasigroup table G and a parameter $\epsilon > 0$ and with probability $1 - \epsilon$ outputs a $c \cdot \log n$ size generating sequence for G. Otherwise, with probability at most ϵ the algorithm halts with "fail" as output. Here, the constant c depends on ϵ.*

Proof. The algorithm starts with $S = \emptyset$ and proceeds in stages by including new randomly picked elements from G into S at every stage. Thus, it computes a sequence of subsets $S_1 = \emptyset \subseteq S_2 \subseteq \ldots \subseteq S_m$, where m will be appropriately fixed in the analysis. Let H_i denote the sub-quasigroup generated by S_i for each $i > 0$.

We analyze the probability that S_m generates the entire quasigroup G.

Claim. For $m = 4 \log n + 1$ the probability that S_m generates G is at least $1/3$.

The proof is an application of Markov's inequality. Define indicator random variables $Z_i, 1 \le i \le 4 \log n$.

$$Z_i = \begin{cases} 1 \text{ if } H_i = H_{i+1} \text{ and } H_i \ne G; \\ 0 \text{ otherwise.} \end{cases}$$

Let $Z = \sum_{i=1}^{4 \log n} Z_i$. We bound the expectation of each Z_i. If $H_i = G$ then clearly $E[Z_i] = 0$. Suppose $H_i \ne G$. Then $|H_i| \le |G|/2$. Therefore, a random $x \in G$ lies in H_i with probability at most $1/2$. Thus, $\text{Prob}[Z_i = 1] \le \text{Prob}[x \in H_i] \le 1/2$. Putting it together, $\mu = E[Z] \le 2 \log n$. By Markov's inequality, $\text{Prob}[Z > 3 \log n] \le \text{Prob}[Z > 3\mu/2] \le 2/3$. It follows that with probability $1/3$, S_m generates G which proves the claim.

Finally, note that we can boost the success probability to $1 - \epsilon$ by $O(\log(1/\epsilon))$ repetitions. ∎

Combining Proposition 1 and Theorem 1 we get a randomized polynomial-time algorithm that takes as input a quasigroup G with $|G| = n$, and using $O(\log^2 n)$ random bits samples from quasigroups isomorphic to G with failure probability bounded by a given constant ϵ: the algorithm uses Theorem 1 to pick a generating sequence $X \subset G$ of size $c \log n$ (using $c \log^2 n$ random bits). Then, using Proposition 1 the algorithm deterministically generates a quasigroup table G' isomorphic to G. Denote the distribution of quasigroup tables thus obtained by $\mathcal{D}(G, \epsilon)$. The next proposition follows directly from Theorem 1.

Proposition 2. *Let G_1 and G_2 be two n element quasigroups and $\epsilon > 0$ be any constant.*

1. *If $G_1 \cong G_2$ then $\mathcal{D}(G_1, \epsilon)$ and $\mathcal{D}(G_2, \epsilon)$ are identical distributions.*
2. *$G_1 \not\cong G_2$ then $\mathcal{D}(G_1, \epsilon)$ and $\mathcal{D}(G_2, \epsilon)$ have disjoint support (i.e. their statistical difference is 1).*

Let G be an n element quasigroup G and let ϵ be fixed for the following discussion. For any other quasigroup G' let $p_G(G')$ denote the probability of G' w.r.t. the distribution $\mathcal{D}(G, \epsilon)$. Define the set $C(G) = \{G' \mid p_G(G') > 0\}$. Notice that the size $C(G)$ is bounded by $n^{c \log n}$, where c is given by Theorem 1.

Let G_1 and G_2 be two n element quasigroups. By Proposition 2 $C(G_1)$ and $C(G_2)$ are identical or disjoint. Let X denote $C(G_1) \cup C(G_2)$. Although $|X| \le 2n^{c \log n}$, elements in X have polynomial in n length. Assume that elements of X are encoded as integers so we can use Chinese remaindering fingerprints to bound the message lengths in our IP protocol. Since $|X| = n^{O(\log n)}$, for a randomly picked $\log^3 n$ bit prime p, with probability greater than $1 - 2^{-\log^2 n}$ we have $x \ (mod \ p) \ne y \ (mod \ p)$ for every pair $x, y \in X$ such that $x \ne y$. Call such a p a *good* prime. Let $X_p = \{x \ (mod \ p) \mid x \in X\}$. Then $|X_p| = |X|$ for a good prime p. Elements of X_p are bit strings of length $t = \log^3 n$.

In the protocol we are about to describe, the verifier picks a random $\log^3 n$ bit prime number p by sampling random $\log^3 n$ bit numbers and using the AKS primality test [1]. With a sample of size $O(\log^3 n)$, the verifier has constant success probability of picking a random $\log^3 n$ bit prime. Thus, $O(\log^6 n)$ random bits suffice for this purpose. We now describe the limited resource IP protocol for Quasigroup Nonisomorphism.

Input: A pair of quasigroup tables (G_1, G_2).
Verifier: Randomly sample $\log^3 n$ bit positive integers until a prime number p is found. If after $5 \log^3 n$ trials no prime number has been found, then reject the input. Pick $a \in \{1, 2\}$ randomly and, using Theorem 1, pick a random generating sequence of length $c \log n$ for G_a. Using Proposition 1, compute from G_a the quasigroup table G'. Now compute the $\log^3 n$ bit integer $z = G' \pmod{p}$. Send $\langle z, p \rangle$ to Prover.
Prover: Sends back an integer $b \in \{1, 2\}$.
Arthur: Accept the pair (G_1, G_2) as nonisomorphic iff $a = b$.

This is clearly an $\mathrm{IP}_2(\log^6 n, \log^3 n)$ protocol. The analysis given before the protocol shows that the IP protocol accepts nonisomorphic pairs with probability $1/2 + \epsilon$ and rejects isomorphic pairs with probability $1/2 + \epsilon$, for a constant $\epsilon > 0$. We bound the error probability, by bounding the probability of each bad event causing an incorrect output. First, the verifier fails to pick a random $\log^3 n$ bit prime p with probability bounded by $1/n$. Next, the probability that p is not a good prime is bounded by $n^{-\log n}$. Denote the distribution of pairs $\langle z, p \rangle$ generated by the randomized verifier by $\mathcal{D}'(G, \epsilon)$. The following claim is a direct consequence of Proposition 2.

Claim. Let G_1 and G_2 be two n element quasigroups and $\epsilon > 0$ be any constant. If $G_1 \cong G_2$ then $\mathcal{D}'(G_1, \epsilon)$ and $\mathcal{D}'(G_2, \epsilon)$ are identical distributions. If $G_1 \not\cong G_2$ then the statistical difference between $\mathcal{D}'(G_1, \epsilon)$ and $\mathcal{D}'(G_2, \epsilon)$ is at least $1 - (\epsilon + n^{-\log n})$.

Proof. The first part is obvious. To see the second part it suffices to notice that a pair $\langle z, p \rangle$ can have nonzero probability w.r.t. both distributions $\mathcal{D}'(G_1, \epsilon)$ and $\mathcal{D}'(G_2, \epsilon)$ only if p is not a good prime or the algorithm $\mathcal{B}'$ outputs "fail". The probability of that event is bounded by $n^{-\log n} + \epsilon$. ∎

Thus, if $G_1 \not\cong G_2$, the (all powerful) honest prover will correctly compute the index a with probability at least $1 - (1/n + \epsilon + n^{-\log n})$. If $G_1 \cong G_2$ then a cheating prover can find a with probability at most $1/2$.

Theorem 2. *Quasigroup Nonisomorphism is in* $\mathrm{IP}_2(\log^6 n, \log^3 n)$.[3]

Goldwasser and Sipser [12] showed that for every k, k-round IP protocols can be simulated by $k + 2$ round AM protocols. They considered IP protocols that use polynomially many random bits and polynomial size messages. Careful examination of their proof shows that their AM simulation appropriately scales down to polylog random bits used by the verifier and polylog size messages. We

---

[3] Based on this protocol we can obtain a statistical zero knowledge protocol for QGROUP-ISO that uses polylog random bits with polylog message size. We leave details to the full version of the paper.

summarize this observation as a theorem (details will appear in a full version). Likewise, by examining the proof of the result that $AM_k = AM_2$ for any constant $k \geq 2$ [5], we have a similar version for polylog message size and random bits.

Theorem 3. *For* $a, b, k > 0$

- $IP_k(\log^a n, \log^b n) \subseteq AM_{k+2}(k^3 \log^{4a+b} n, k^2 \log^{4a+b} n)$.
- $AM_k(\log^a n, \log^b n) \subseteq AM_2(\log^{kb+a} n, \log^b n)$.

Combining Theorems 2 and 3 we get the following.

Theorem 4. *Quasigroup Nonisomorphism is in* $AM_2(\log^{135} n, \log^{27} n)$.

Since the actual constants are not important for the following discussion, let $c = 135$ and note that QGROUP-NONISO is in $AM_2(\log^c n, \log^c n)$.

2.1   Limited Nondeterminism

Complexity subclasses of NP with bounded nondeterminism are studied in different contexts: [11] is a nice survey. Kintala and Fischer in [16] introduced NP subclasses with polylogarithmic nondeterminism. Let $NP(\log^k n) \subseteq NP$ denote the subclass in which only $O(\log^k n)$ nondeterministic bits are allowed for the accepting NP machine on inputs of size n. As noted in the introduction, QGROUP-ISO is in $NP(\log^2 n)$. The class $NP(\log^2 n)$ contains complete problems [8,19] under polynomial time reductions, e.g. the tournament dominating set problem. Papadimitriou and Yannakakis ask in [19] whether QGROUP-ISO is $NP(\log^2 n)$-complete. Using Theorem 4 we give strong negative evidence. We consider the following version of the general clique problem: CLIQUE $= \{G \mid G$ has n vertices and a clique of size $n/2\}$.

Theorem 5. *If* QGROUP-ISO *is many-one complete for* $NP(\log^2 n)$ *under polynomial time reductions then* CLIQUE *is in* $coNTIME[n^{O(1)}, 2^{O(\sqrt{n} \log n)}]/poly$. *I.e. for inputs of length* n, $\overline{CLIQUE}$ *has polynomial-size proofs which can be verified in* $2^{O(\sqrt{n} \log n)}$ *time with the help of a polynomial-size advice.*

Proof. The problem log-CLIQUE $= \{(G, k) \mid G$ has n vertices, $k \leq \log n$ and G has a clique of size $k\}$ is clearly in $NP(\log^2 n)$. If QGROUP-ISO is $NP(\log^2 n)$ complete, then log-CLIQUE is many-one reducible to QGROUP-ISO. Hence, by Theorem 4, log-$\overline{CLIQUE} \in AM(\log^c n, \log^c n)$ for some constant $c > 0$.

We now apply an idea of Feige and Kilian [10]: Let G be an n-node graph, as an instance of CLIQUE. For simplicity suppose $n/2$ is a perfect square. In time $n^{O(\sqrt{n})}$ convert G to a pair (G', ℓ') such that G' is a graph with at most $\binom{n}{\sqrt{n/2}}$ nodes. Here, each node of G' corresponds to a $\sqrt{n/2}$ size subset of $V(G)$, and two nodes of G' are adjacent iff their corresponding sets of nodes in G are disjoint and together they form a clique of size $2\sqrt{n/2}$ in G. Clearly G' has a $\sqrt{n/2}$ size clique iff $G \in$ CLIQUE. Set ℓ' to $\sqrt{n/2}$. The number of vertices in G' is $N = \binom{n}{\sqrt{n/2}}$. As $\ell' \leq \log N$, (G', ℓ') is an instance of log-CLIQUE. The AM protocol on input (G', ℓ') uses $O(\log^c N)$ random bits and $O(\log^c N)$ nondeterministic bits. In the AM protocol, the final deterministic computation by Arthur after the communication rounds is of time polynomial in N. Consequently, $\overline{\text{CLIQUE}}$ has an $\text{AM}(n^{O(1)}, n^{O(1)})$ protocol with error probability $1/4$, where the final deterministic computation done by Arthur, after all the communication rounds, takes time $2^{O(\sqrt{n} \log n)}$.

A parallel repetition of the protocol $n^{O(1)}$ times, with Arthur deciding by majority vote yields an $\text{AM}(n^{O(1)}, n^{O(1)})$ protocol for $\overline{\text{CLIQUE}}$, where the final deterministic computation done by Arthur is still of time $2^{O(\sqrt{n} \log n)}$ and the error probability is $2^{-n^{O(1)}}$. For each n, we derandomize the protocol by fixing the random choices to a polynomial-size advice string. Hence, $\overline{\text{CLIQUE}}$ is in $\text{NTIME}[n^{O(1)}, 2^{O(\sqrt{n} \log n)}]/\text{poly}$. ∎

Consider the parameterized complexity [7] problem k-QGROUP-ISO: let G_1 and G_2 be quasigroups with n elements each given by multiplication tables and generating sets S_1 and S_2 of size k each. The problem is to test if the groups are isomorphic. The following theorem is analogous to Theorem 5.

Theorem 6. *If k-QGROUP-ISO is hard for $W[1]$ w.r.t. parameterized reductions then $\overline{\text{CLIQUE}}$ is in $\text{NTIME}[n^{O(1)}, 2^{o(n)}]/\text{poly}$.*

3 The Minimum Generating Set Problem

The complexity of MIN-GEN — finding a *minimum size* generating set for a quasigroup G — is first examined in [19]. Since G has generating sets of size $\log n$, the problem has an easy $n^{O(\log n)}$ time algorithm. Note that MIN-GEN is related to QGROUP-ISO: if (G_1, G_2) is a QGROUP-ISO instance and k bounds G_1's minimum generating set, then Tarjan's isomorphism test takes $n^{O(k)}$ time. The decision version of MIN-GEN is MIN-GEN $= \{(G, k) \mid$ the quasigroup G has a generating set of size $k\}$. The complexity of MIN-GEN is left open for groups and quasigroups in [19]. However, if T is the multiplication table of an *arbitrary* binary operation then checking if T has a $\log n$-size generating set is $\text{NP}(\log^2 n)$ complete [19]. We first note that MIN-GEN for *groups* is in $\text{DSPACE}(\log^2 n)$.

Proposition 3. MIN-GEN *for groups is in* $\text{DSPACE}(\log^2 n)$.

Proof. Let (G, k) be a instance of MIN-GEN. Let X denote its Cayley graph with vertex set G and directed edges labeled by elements of G such that (u, v) is an edge labeled by g if $ug = v$. The $\text{DSPACE}(\log^2 n)$ machine cycles through k-element subsets $S \subset G$. W.l.o.g. $k \leq \log n$, thus S takes $O(\log^2 n)$ space. For each S, consider the graph X_S to have only edges labeled by $g \in S$. Clearly, S generates G iff there is a directed path in X_S from 1 to v for each $v \in G$. This test is clearly an NL predicate and hence in $\text{DSPACE}(\log^2 n)$. The overall $\text{DSPACE}(\log^2 n)$ machine accepts if some S passes the test. ∎

Theorem 7. *The MIN-GEN problem for nilpotent groups given by Cayley table is in deterministic polynomial time.*

Proof. Let G be the input group with n elements and $n = p_1^{e_1} p_2^{e_2} \cdots p_k^{e_k}$ be the prime factorization of n. Recall that G is *nilpotent* iff $G = S_{p_1} \times \cdots \times S_{p_k}$, where each S_{p_i} is the unique (hence normal) p_i-Sylow subgroup of G. Let $n_i = np_i^{-e_i}$. Then $S_{p_i} = \{g^{n_i} \mid g \in G\}$. This gives an easy (well-known) nilpotence test for G. If $\langle g_1, \cdots, g_k \rangle = G$ then $\langle g_1^{n_i}, \cdots, g_k^{n_i} \rangle = S_{p_i}$. Thus, if G has a generating set of size k then so does S_{p_i} for each i. Conversely, suppose $\langle g_{i1}, g_{i2}, \cdots, g_{ik} \rangle = S_{p_i}$ for each i. Letting $g_j = \prod_{i=1}^{k} g_{ij}$, $1 \leq j \leq k$, note that $\langle g_1, \cdots, g_k \rangle = G$, because $\langle g_1^{n_i}, \cdots, g_k^{n_i} \rangle = S_{p_i}$. Thus, it suffices to solve MIN-GEN for each S_{p_i}. I.e., we have polynomial-time reduced MIN-GEN for nilpotent groups to MIN-GEN for p-groups (groups of p-power order). Henceforth, let G be a p-group. The *Frattini subgroup* $\Phi(G)$ of a finite group G is the intersection of all *maximal proper* subgroups of G. Clearly, $\Phi(G) \lhd G$. We claim that G has a size k generating set iff $G/\Phi(G)$ has a size k generating set. To see this, suppose $\{x_1\Phi(G), x_2\Phi(G), \ldots, x_k\Phi(G)\}$ generates $G/\Phi(G)$. Let $H = \langle x_1, \cdots, x_k \rangle$. Then $H\Phi(G) = G$. By [13, Theorem 1.1, Chapter 5], since $\Phi(G)$ consists of *nongenerators*, it follows that $H = G$. The forward implication of the claim is trivial.

Now, using the above claim we can polynomial-time reduce MIN-GEN for nilpotent groups to MIN-GEN for elementary abelian groups which, in turn, has an easy polynomial-time algorithm. By [13, Theorem 1.3, Chapter 5], $G/\Phi(G)$ is an elementary abelian p-group as G is a p-group. We now show that $\Phi(G)$ is computable from G in polynomial time when G is nilpotent. We recall some simple properties of the Frattini subgroup for nilpotent groups [20, Chapter 5.2]. The commutator subgroup G' of G is generated by $\{xyx^{-1}y^{-1} \mid x, y \in G\}$. We know $G' \lhd G$ and G/G' is abelian. Now, G is nilpotent iff $G' \subseteq \Phi(G)$ [20, Theorem 5.2.16]. Furthermore, $N \lhd G$ and $N \subseteq \Phi(G)$ implies $\Phi(G/N) = \Phi(G)/N$. In particular, $\Phi(G/G') = \Phi(G)/G'$. Therefore, it suffices to compute $\Phi(G/G')$ to find $\Phi(G)$. Now, G' can be computed in polynomial time from G. Hence, we have the table for G/G'. As G/G' is abelian, in polynomial time we can easily decompose G/G' as a product of cyclic groups Thus, $G/G' = H_1/G' \times H_2/G' \times \cdots H_t/G'$. Let H_j be generated by $y_jG' \in G/G'$, where y_jG' has order, say, p^{m_j}, $1 \leq j \leq t$. The Frattini subgroup of a product group is the product of the Frattini subgroups of the constituent groups, i.e. $\Phi(G/G') = \Phi(H_1/G') \times \Phi(H_2/G') \times \cdots \Phi(H_t/G')$. However, H_j/G' is cyclic of p-power order. Hence it has a unique maximal proper subgroup, namely, $\langle y_j^p G' \rangle$, of index p. Therefore, $\Phi(H_j/G') = \langle y_j^p G' \rangle$ for each j. Hence, $\Phi(G/G') = \langle y_1^p G', \cdots, y_t^p G' \rangle$. It follows that $\Phi(G) = \langle y_1^p, \cdots, y_t^p, G' \rangle$. Finally, since $G/\Phi(G)$ is an elementary abelian p-group (isomorphic to some $\mathbb{Z}_p^k$), we can easily compute a minimum generating set $\{x_1\Phi(G), \cdots, x_k\Phi(G)\}$ for it which has to be of size k. By the above claim, $\{x_1, \cdots, x_k\}$ is a minimum generating set for G. ∎

4 Quasigroup Isomorphism and Parallel Queries to NP

For a function $f : \mathbb{N} \longrightarrow \mathbb{N}$, let $\mathrm{FP}_{||}^{\mathrm{NP}}[f]$ denote functions computable in polynomial time with $O(f(n))$ *parallel* queries to NP and let $\mathrm{FP}^{\mathrm{NP}}[f]$ denote functions computable in polynomial time with $O(f(n))$ many *adaptive* queries to NP. Considerable research has examined the hypothesis $\mathrm{FP}_{||}^{\mathrm{NP}}[\mathrm{poly}] = \mathrm{FP}^{\mathrm{NP}}[\log]$. When the range of the functions is $\{0,1\}$ these classes coincide [14]. In general, this hypothesis implies $\mathrm{NP} = \mathrm{RP}$ [6,21]. An important open question is whether the same assumption yields $\mathrm{NP} = \mathrm{P}$. Jenner and Torán in [15] make a detailed investigation. In [2], it is shown that $\mathrm{FP}_{||}^{\mathrm{NP}}[\mathrm{poly}] = \mathrm{FP}^{\mathrm{NP}}[\log]$ implies Graph Isomorphism is in P. In this section we show that the *weaker* assumption $\mathrm{FP}_{||}^{\mathrm{NP}}[\log^2] \subseteq \mathrm{FP}^{\mathrm{NP}}[\log]$ implies QGROUP-ISO is in P. In $\mathrm{FP}_{||}^{\mathrm{NP}}[\log^2]$, the base machine can make only $O(\log^2 n)$ parallel queries to NP.

Definition 1. [9] *A promise problem is a pair of sets (Q,R). A set L is called a* solution *of the promise problem (Q,R) if for all $x \in Q$, $x \in L \Leftrightarrow x \in R$.*

A promise problem of interest is $(1\mathrm{SAT}, \mathrm{SAT})$, where 1SAT contains formulas with at most one satisfying assignment. Any solution of $(1\mathrm{SAT}, \mathrm{SAT})$ agrees with SAT in formulas with a unique satisfying assignment as well as the unsatisfiable formulas. By [21], $\mathrm{FP}_{||}^{\mathrm{NP}}[\mathrm{poly}] = \mathrm{FP}^{\mathrm{NP}}[\log]$ implies $(1\mathrm{SAT}, \mathrm{SAT})$ has a solution in P. An easy consequence of this shown in [2] is the following.

Theorem 8. [2] *Suppose $L \in \mathrm{NP}$ is accepted by a deterministic polynomial-time oracle machine M with access an NP oracle A such that M makes only queries to A having at most one nondeterministic solution.[4] Then $\mathrm{FP}_{||}^{\mathrm{NP}} = \mathrm{FP}^{\mathrm{NP}}[\log]$ implies that $L \in \mathrm{P}$.*

Our result based on a weaker hypothesis follows.

Theorem 9. $\mathrm{FP}_{||}^{\mathrm{NP}}[\log^2] \subseteq \mathrm{FP}^{\mathrm{NP}}[\log]$ *implies that* QGROUP-ISO *is in P.*

Proof. Suppose $L \in \mathrm{NP}(\log^2 n)$ as witnessed by NP machine M. Analogous to SAT, consider the promise problem $(1L, L)$ where $1L$ is the set of all $x \in \Sigma^*$ such that either $x \in \bar{L}$ or M has exactly one accepting computation path (of length $\log^2 n$ bits). Notice that, following [21], the hypothesis $\mathrm{FP}_{||}^{\mathrm{NP}}[\log^2] \subseteq \mathrm{FP}^{\mathrm{NP}}[\log]$ implies for every $L \in \mathrm{NP}(\log^2)$ that any solution to the promise problem $(1L, L)$ is in P. Thus, in order to derive QGROUP-ISO $\in \mathrm{P}$ from the hypothesis it suffices to show that there is set $L \in \mathrm{NP}(\log^2 n)$ such that QGROUP-ISO $\in \mathrm{P}^A$ for any solution A to the promise problem $(1L, L)$.

This claim is based on [2, Theorem 13] where it is shown that Graph Isomorphism is in P^B, where B is any solution to the promise problem $(1\mathrm{SAT}, \mathrm{SAT})$. The crucial reason why we can replace SAT by a set $L \in \mathrm{NP}(\log^2 n)$ for the case of QGROUP-ISO is because of the following property of n element quasigroups G: the automorphism group $Aut(G)$ is a subgroup of S_n but has a *base* B of size $\log n$. I.e. $Aut(G)$'s elements are completely determined by their action on B. This follows from the fact that G has $\log n$ size generating sets. ∎

---

[4] Here, nondeterministic solution means a nondeterministic computation path for A's NP machine.

References

1. M. AGRAWAL, N. KAYAL AND N. SAXENA, PRIMES is in P, Annals of Mathematics, (2) 160 (2004), 781-793.
2. V. ARVIND, PIYUSH P KURUR, Graph Isomorphism is in SPP, *Information and Computation*, Volume 204, Issue 5, pp. 835-852, May 2006.
3. V. ARVIND, J. TORÁN, Solvable Group Isomorphism is (almost) in NP ∩ coNP, in *Proc. 19th IEEE Computational Complexity Conference Conference*, 91–103, 2004.
4. J. L. BALCÁZAR, J. DÍAZ, J. GABARRÓ, *Structural Complexity I*, EATCS Monographs on Theoretical Computer Science, Springer-Verlag, 1989.
5. L. BABAI, Trading group theory for randomness, *Proc. 17th ACM Symposium on Theory of Computing*, 421–429, 1985.
6. R. Beigel. NP-hard sets are P-superterse unless R = NP, January 04 1988.
7. R.G. DOWNEY AND M.R. FELLOWS *Parameterized Complexity,* Springer Verlag 1992.
8. J. DÍAZ, J. TORÁN, Classes of bounded nondeterminism, *Math. Systems Theory*, 23, 1990, 21–32.
9. S. Even, A. L. Selman, and Y. Yacobi. The complexity of promise problems with applications to public-key cryptography. *Information and Control*, 61(2):159–173, May 1984.
10. U. FEIGE, J. KILIAN, On Limited versus Polynomial Nondeterminism, *Chicago Journal of Theoretical Computer Science*, March (1997).
11. J. GOLDSMITH, M. LEVY AND M. MUNDHENK, Limited nondeterminism in *SIGACT news*, June 1996.
12. SHAFI GOLDWASSER AND MICHAEL SIPSER, Private coins versus public coins in interactive proof systems, In *Silvio Micali, editor, Advances in Computing Research*, volume 5, pp. 73–90. JAC Press, Inc., 1989.
13. D. GORENSTEIN, *Finite Groups*, Harper and Row Publishers, New York, 1968.
14. L. HEMACHANDRA, The strong exponential hierarchy collapses, in *Proc. 19th ACM Symposium on Theory of Computing* , 1987, 110–122.
15. B. Jenner and J. Torán. Computing functions with parallel queries to NP. *Theoretical Computer Science*, 141(1–2):175–193, 1995.
16. C. KINTALA AND P. FISCHER, Refining nondeterminism in relativized polynomial time computations, *SIAM J. on Computing*, 9, 1980, 46–53.
17. J. KÖBLER, U. SCHÖNING, AND J. TORÁN, *Graph Isomorphism: its Structural Complexity*, Birkhäuser, Boston, 1992.
18. G.L. MILLER, On the $n^{\log n}$ isomorphism technique, in *Proc. 10th ACM Symposium on the Theory of Computing*, 1978, 51–58.
19. C. PAPADIMITRIOU, M. YANNAKAKIS On limited nondeterminism and the complexity of the VC dimension. In *Journal of Computer and System Sciences*, 53(2): 161-170, 1996.
20. D.J.S. ROBINSON, *A Course in the Theory of Groups*, Graduate Texts in Mathematics, Springer Verlag, 1996.
21. A. L. Selman. A taxonomy of complexity classes of functions. *Journal of Computer and System Sciences*, 48(2):357–381, 1994.
22. M. SIPSER,A complexity theoretic approach to randomness. In *Proc. 15th ACM Symp. Theory of Computer Science* 1983, 330–335.

Inverse HAMILTONIAN CYCLE and Inverse 3-D MATCHING Are coNP-Complete

Michael Krüger and Harald Hempel

Institut für Informatik, Friedrich-Schiller-Universität Jena
{krueger, hempel}@minet.uni-jena.de

Abstract. In this paper we show that the inverse problems of HAMIL-TONIAN CYCLE and 3-D MATCHING are coNP complete. This completes the study of inverse problems of the six natural NP-complete problems from [2] and answers an open question from [1]. We classify the inverse complexity of the natural verifier for HAMILTONIAN CYCLE and 3-D MATCHING by showing coNP-completeness of the corresponding inverse problems.

Keywords: computational complexity, coNP-completeness, inverse NP-problems, HAMILTONIAN CYCLE, 3-DIMENSIONAL MATCHING.

1 Introduction

The influential book by Garey and Johnson [2] lists six natural NP-complete languages: 3SAT, VERTEX COVER (VC), CLIQUE, HAMILTONIAN CYCLE (HC), 3-D MATCHING (3DM) and PARTITION. When it comes to studying the complexity of inverse NP problems it seems desirable to start by investigating the inverse problems of the above six examples. The inverse problems of 3SAT, VC, CLIQUE, and PARTITION have been shown to be coNP-complete in [4, 1].

In this paper we show that the inverse problems for the remaining two problems HC and 3DM, are coNP-complete. This settles an open question from [1] and contributes to the growing knowledge about the complexity of inverse NP problems [4, 1, 3]. In particular we show that inverting the natural verifiers of HC and 3DM, is complete for the class coNP.

The complexity class NP is often referred to as the class of problems having polynomial-time verifiers. A polynomial-time verifier V is a polynomial-time computable function mapping from $\Sigma^* \times \Sigma^*$ to $\{0, 1\}$ such that there exists a polynomial p such that for all $x, \pi \in \Sigma^*$, $V(x, \pi) = 1$ implies $|\pi| \leq p(|x|)$. The language $L(V)$ associated with a verifier V is defined as $L(V) = \{x \in \Sigma^* : (\exists \pi \in \Sigma^*)[V(x, \pi) = 1]\}$. It is well-known that NP $= \{L(V) : V$ is a polynomial-time verifier$\}$. The inverse problem for a verifier V is given a set $\Pi \subseteq \Sigma^*$ (represented as a list) to decide if there exists a string x such that $\Pi = \{\pi \in \Sigma^* : V(x, \pi) = 1\}$. It appears that inverting verifiers has an Σ_2^p upper bound but this bound only holds for so called fair verifiers [1]. However, even though there do exist verifiers such that their inverse problems are

T. Asano (Ed.): ISAAC 2006, LNCS 4288, pp. 243–252, 2006.
© Springer-Verlag Berlin Heidelberg 2006

Σ_2^p complete, the inversion of natural verifiers for some NP-complete languages is complete for the class coNP [4, 1, 3]. It is known that different verifiers for one and the same NP problem may have inverse problems of different complexity [1]. However for many NP-complete languages L there seems to exist a verifier that deserves to be called "the natural verifier" for L and we will focus on the inverse problems relative to those natural verifiers. We mention in passing that a different definition of the inverse problem for a verifier has been studied and a very general coNP-hardness result been proven in [3]. The difference between the two concepts is the representation of the input as a list (see [1] and above) or as a boolean circuit (see [3]).

Our paper is organized as follows. In Chap. 2 we define the basic concepts and some useful graph modules. The two main theorems of this paper are stated in Sect. 3. Due to space restrictions we only give a proof for the coNP-completeness of the inverse problem of HC. We mention that the proof idea can be modified to also work for the inverse problem of HC in directed graphs and thus coNP-completeness holds in the directed case as well.

2   Preliminaries

We assume the reader to be familiar with the basic definitions and notations from graph theory [6] and complexity theory [5]. Let $\Sigma = \{0, 1\}$ be our alphabet.

2.1   Inverse Problems

The class NP is also called the class of problems with short proofs. Essential for a definition that follows that informal description is the notion of a verifier.

Definition 1.   *1. A (polynomial-time) verifier V is a polynomial-time computable function $V : \Sigma^* \times \Sigma^* \to \{0, 1\}$ such that there exists a polynomial p satisfying that for all $x, \pi \in \Sigma^*$, $(x, \pi) \in V \implies |\pi| \le p(|x|)$.*
2. For a (polynomial-time) verifier V let $V(x)$ denote the set of proofs for a string $x \in \Sigma^*$, that is, for all $x \in \Sigma^*$, $V(x) = \{\pi \in \Sigma^* : V(x, \pi) = 1\}$.
3. The language $L(V)$ accepted by a polynomial-time verifier V is defined as $L(V) = \{x \in \Sigma^* : V(x) \ne \emptyset\}$.

It is well known that NP is the class of languages that can be accepted by (polynomial-time) verifiers. Inverse problems are defined relative to a verifier V.

Definition 2 ([1]). *The inverse problem V^{-1} for a verifier V is defined as*

$$V^{-1} = \{\Pi \subseteq \Sigma^* : (\exists x \in L(V))[V(x) = \Pi]\}.$$

The inverse problem of a language $A \in$ NP can clearly only be defined relative to a verifier accepting A. We will study the inverse problems of the natural verifiers for 3SAT and HC.

The concept of a candidate function is a useful tool when studying the complexity of inverse problems.

Definition 3 ([1]). *Let V be a verifier. A polynomial-time computable mapping $c : \mathcal{P}(\Sigma^*) \to \Sigma^*$ is called a candidate function for V if and only if for all $\Pi \subseteq \Sigma^*$: if there exists a $z \in \Sigma^*$ such that $V(z) = \Pi$ then $V(c(\Pi)) = \Pi$.*

It is not clear if all verifiers do have candidate functions. However, many natural verifiers for NP-complete languages, such as 3SAT or VC, have candidate functions. Note that if a verifier V has a candidate function c then we have an obvious coNP upper bound for the complexity of V^{-1}, namely given Π, compute $c(\Pi)$, and then check if for all π such that $|\pi| \le p(|c(\Pi)|)$ (where p is the polynomial that bounds the length of witnesses with respect to the verifier V) we have $\pi \in \Pi \iff V(c(\Pi), \pi) = 1$.

Observation 1. *If V is a verifier with a candidate function, then $V^{-1} \in$ coNP.*

2.2   HAMILTONIAN CYCLE

A cycle in a graph $G = (V, E)$ is assumed to be a set $C \subseteq E$ such that there exist pairwise different vertices $x_1, x_2, x_3, \ldots x_{k-1}, x_k$, $k \ge 3$, such that $C = \{\{x_1, x_2\}, \{x_2, x_3\}, \ldots, \{x_{k-1}, x_k\}, \{x_k, x_1\}\}$. A Hamiltonian cycle in a (simple and undirected) graph is a cycle in G that contains every vertex of G.

The NP-complete problem HC is defined as the set of all (simple and undirected) graphs that contain a Hamiltonian cycle.

Definition 4. *The verifier V_{HC} is defined as $V_{HC}(G, C) = 1$ if G is a simple, undirected graph and C is a Hamiltonian cycle in G and $V_{HC}(G, C) = 0$ otherwise.*

Clearly, V_{HC} is a verifier for HC and since it appears to be the most natural verifier for HC we will choose Invs-HC as a more intuitive notation for V_{HC}^{-1}.

Following a more general concept from [1] we will call a collection of sets of edges $\Pi = \{C_1, C_2, \ldots, C_k\}$ well-formed if and only if $C_1, C_2, \ldots, C_k$ are cycles over a common vertex set V such that $||C_1|| = ||C_2|| = \cdots = ||C_k|| = ||V||$. It is not hard to see that non well-formed sets Π can not be in Invs-HC. Obviously, testing if an instance Π is well-formed can be done in polynomial time.

Looping back to the notion of a candidate function (see Definition 3) it is not hard to see that the verifier V_{HC} has a candidate function c_{HC}. Given a well-formed collection of sets of edges $\Pi = \{C_1, C_2, \ldots, C_k\}$ let $c_{HC}(\Pi)$ be the graph induced by $C_1, C_2, \ldots, C_k$, that is $c_{HC}(\Pi) = (V, E)$ such that $V = \{v : (\exists u)[\{u, v\} \in C_1]$ and $E = C_1 \cup C_2 \cup \cdots \cup C_k$. The following corollary is an immediate consequence of Observation 1.

Corollary 1. *The problem Invs-HC is in coNP.*

2.3   3-DIMENSIONAL MATCHING

The NP-complete problem 3-DIMENSIONAL MATCHING(3DM) is defined as follows.

3-DIMENSIONAL MATCHING

Input: A 4-tuple (S, X, Y, Z) of sets such that S is a subset of $X \times Y \times Z$ and X, Y and Z have the same number of elements.

Question: Does S contain a 3D-Matching, i.e., a subset $M \subseteq S$ such that $|M| = |X|$ and no two elements of M agree in any coordinate.

As many other NP-complete problems 3DM has a natural verifier.

Definition 5. *The verifier V_{3DM} is defined via $V_{3DM}((S, X, Y, Z), M) = 1$ if S is a subset of $X \times Y \times Z$, X, Y and Z are sets, having the same number of elements and M is a 3D-Matching for S. Otherwise $V_{3DM}((S, X, Y, Z), M) = 0$.*

It is easy to see that V_{3DM} is a verifier for 3DM. Since we feel that V_{3DM} is the most natural verifier for 3DM we let Invs-3DM denote the language V_{3DM}^{-1}.

2.4   3-SATISFIABILITY

One of the standard NP-complete problems is 3-SATISFIABILITY (3SAT), the set of all satisfiable boolean formulas in 3-conjunctive normal form (3-CNF), that is any clause has at most three literals. 3SAT will play an important role in the proofs of our main theorems.

A natural verifier for 3SAT is the following: $V_{3SAT}(F, \alpha) = 1$ if F is a boolean formula in 3-CNF and α is a satisfying assignment for the variables of F, and $V_{3SAT}(F, \alpha) = 0$ otherwise. The inverse problem V_{3SAT}^{-1} also has been denoted by Invs-3SAT or 3SAT^{-1} [1, 4]. Throughout this paper we will use Invs-3SAT to denote V_{3SAT}^{-1}.

As it has been the case for V_{HC} there are easy to check properties that a proof set for V_{3SAT} has to have in order to be in Invs-3SAT. Since any assignment for an n-variable boolean formula is represented by a string from $\{0, 1\}^n$, the notion of well-formed proof sets Π with respect to V_{3SAT} only requires that all strings from Π have the same length.

Theorem 2 ([4]). *Invs-3SAT is coNP-complete.*

A concept that will be useful for our purposes as well was defined in [4].

Definition 6 ([4]). *Let Π be a set of boolean assignments for $x_1, x_2, \ldots, x_n$.*

1. *An assignment α for $x_1, \ldots, x_n$ is said to be $\{x_i, x_j, x_k\}$-compatible with Π, $1 \leq i < j < k \leq n$, if and only if there exists an assignment $\beta \in \Pi$, such that α and β assign the same truth values to x_i, x_j, x_k.*
2. *An assignment for $x_1, \ldots, x_n$ is called 3-compatible with Π if and only if it is $\{x_i, x_j, x_k\}$-compatible with Π for each triplet x_i, x_j, x_k of variables, $1 \leq i < j < k \leq n$.*

The notion of 3-compatibility leads to a useful characterization of Invs-3SAT.

Theorem 3 ([4]). *A well-formed set of proofs Π for V_{3SAT} is in Invs-3SAT if and only if it is closed under 3-compatibility, i. e., if and only if for each assignment α it holds, that if α is 3-compatible with Π then $\alpha \in \Pi$.*

2.5   Some Helpful Graph Modules for Hamiltonian Cycles

For the proof of our main result we need two simple but very useful graph modules, that will help us to direct a Hamiltonian cycle through a given graph. The first module is the parity-module, introduced in [5].

The parity module "connects" two edges e and f of a graph G and forces each Hamiltonian cycle to either enter and leave the parity module through the endpoints of the original edge e or enter and leave the parity module through the endpoints of the original edge f, but not both, i.e., each Hamiltonian cycle either "uses" e or f but not both. In all forthcoming figures we will use the symbol shown in Fig. 1 to express that a pair of edges is connected by a parity-module.

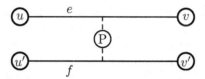

Fig. 1. Symbolic depiction of two parity-connected edges $e = \{u, v\}$ and $f = \{u', v'\}$

It is not hard to extend this idea to connecting f with several edges $e_1, e_2, \ldots, e_k$ via parity-modules. We obtain a module that relates the edges $e_1, e_2, \ldots, e_k$ of a graph in such a way that each Hamiltonian cycle of the modified graph uses all of the edges $e_1, e_2, \ldots, e_k$ or none of them.

Lemma 1. *Let G be an undirected graph with an edge f, that is used by each Hamiltonian cycle of G and let $e_1, e_2, \ldots, e_k$ be pairwise different edges of G. The graph G is modified to G' by first inserting an edge f' that connects the endpoints of f and then inserting parity modules between f' and each of $e_1, e_2, \ldots, e_k$.*

Each Hamiltonian cycle of G' uses either all of $e_i, 1 \leq i \leq n$ or none of them.

The proof is omitted. We will call edges that are connected in the sense of the above Lemma 1 *all-connected* and symbolize this connection as done in Fig. 2. The symbolization does not include the connection to the edge f. In the forthcoming proof it will always be clear which edge will play the role of the edge f.

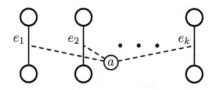

Fig. 2. Symbolization for all-connected edges $e_1, e_2, \ldots, e_k$

3   Main Results

We now state the main results of this paper. However, due to space restrictions we will only prove the coNP-completeness of Invs-HC by giving a reduction from Invs-3SAT in the remainder of this section.

Theorem 4. *Invs-3DM is* coNP*-complete.*

Theorem 5. *Invs-HC is* coNP*-complete.*

Proof of Theorem 5: Due to Corollary 1 it suffices to give a $\leq_m^p$-reduction from the coNP-complete problem Invs-3SAT (Theorem 2) to Invs-HC. If a proof set Π_{3SAT} is not well-formed (with respect to V_{3SAT}) then it is not in Invs-3SAT and hence we will map it to a fixed non-member of Invs-HC. Given a well-formed proof set Π_{3SAT} for V_{3SAT}, we will construct a graph $G_{\Pi_{3SAT}}$ that will contain a Hamiltonian cycle for each assignment (via an one-to-one correspondence) that is 3-compatible with Π_{3SAT}. The proof set Π_{HC} can then be easily extracted from $G_{\Pi_{3SAT}}$, and it will contain exactly those Hamiltonian cycles from $G_{\Pi_{3SAT}}$ that correspond to the assignments in Π_{3SAT}. Recall that by Theorem 3 for any proof set Π_{3SAT} it holds that Π_{3SAT} is in Invs-3SAT if and only if Π_{3SAT} is closed under 3-compatibility. Our construction will ensure that the latter is the case if and only if the candidate graph of Π_{HC} contains exactly the Hamiltonian cycles from Π_{HC}.

Now let Π_{3SAT} be well-formed. Hence, there exist $n, m \in \mathbb{N}$ such that $\Pi_{3SAT} = \{\alpha_1, \alpha_2, \ldots, \alpha_m\}$ and for all i, $1 \leq i \leq m$, $\alpha_i \in \{0,1\}^n$. The strings α_i will be interpreted as assignments to n boolean variables $y_1, y_2, \ldots, y_n$ in the canonical way, and we will write $\alpha_i(y_j)$ to denote the jth bit of α_i. Recall that by Theorem 3 $\Pi_{3SAT} \in$ Invs-3SAT if and only if Π_{3SAT} is closed under 3-compatibility, that is, if and only if Π_{3SAT} contains all assignments, that are 3-compatible with Π_{3SAT}.

We will now construct a graph $G_{\Pi_{3SAT}}$ in stages that will contain exactly one Hamiltonian cycle for each assignment that is 3-compatible with Π_{3SAT}.

Construction Step 1. *The construction starts with a simple cycle C_l, $l = n + 2\binom{n}{3} + 1$. Fix n consecutive edges $e_1, e_2, \ldots, e_n$ in C_l and for each i, $1 \leq i \leq n$, add one new edge f_i to C_l that connects the endpoints of e_i (and so produce a chain of n double edges). Let $G'_{\Pi_{3SAT}}$ be the graph constructed so far.*

Even though $G'_{\Pi_{3SAT}}$ is not a simple graph the upcoming stages of the construction will ensure that the final graph $G_{\Pi_{3SAT}}$ is simple.

The chain of "double" edges e_i, f_i, $1 \leq i \leq n$ will be called n-chain in the following. For each i, we associate the edges e_i, f_i with the variable y_i. Note that the n-chain induces exactly 2^n Hamiltonian cycles in $G'_{\Pi_{3SAT}}$, one for each possible assignment of the variables $y_1, y_2, \ldots, y_n$. The edges e_i (f_i), $1 \leq i \leq n$, will also be called 0-edges (1-edges) referring to the desired correspondence between Hamiltonian cycles and assignments. The usage of a 0-edge e_i (1-edge f_i) in a Hamiltonian cycle will represent assigning the boolean value 0 (1) to y_i. Hence any Hamiltonian cycle traversing the n-chain canonically corresponds to an assignment for $y_1, y_2, \ldots, y_n$ and vice versa.

The remaining part of the construction of $G_{\Pi_{3SAT}}$ consists of the insertion of subgraphs into $G'_{\Pi_{3SAT}}$ in order to restrict the set of Hamiltonian cycles to those that correspond to assignments that are 3-compatible with Π_{3SAT}. Recall that an assignment β is called 3-compatible with a set of assignments Π (over the same variable set) if for each three-element set of variables $\{y_{i_1}, y_{i_2}, y_{i_3}\}$ the assignment β is $\{y_{i_1}, y_{i_2}, y_{i_3}\}$-compatible with Π.

We will now define gadgets H_i, $1 \leq i \leq \binom{n}{3}$, one gadget for each three-element set of variables $\{y_{i_1}, y_{i_2}, y_{i_3}\}$, that will eventually be subgraphs of the to be constructed graph $G_{\Pi_{3SAT}}$. The structure of a subgraph H_i associated with the three variables $y_{i_1}, y_{i_2}, y_{i_3}$ and its connections to the remaining graph, in particular the n-chain, will ensure that every Hamiltonian cycle in $G_{\Pi_{3SAT}}$ corresponds to an assignment that is $\{y_{i_1}, y_{i_2}, y_{i_3}\}$-compatible with Π_{3SAT}.

The gadgets H_i, $1 \leq i \leq \binom{n}{3}$, will all have the same structure and so without loss of generality we will only describe the construction of the gadget for the three variables y_1, y_2, y_3, call it H_1. The gadget H_1 will also be constructed in stages.

First we define the define set

$$\Pi^1_{3SAT} = \{a_1 a_2 a_3 \in \{0,1\}^3 : (\exists \alpha \in \Pi_{3SAT})(\forall i : 1 \leq i \leq 3)[\alpha(y_i) = a_i]\}.$$

So, Π^1_{3SAT} is the set of pairwise different partial $\{y_1, y_2, y_3\}$-assignments in Π_{3SAT}, i.e., assignments from Π_{3SAT} restricted to the variables y_1, y_2, and y_3. In other words Π^1_{3SAT} consists of those possible triples of values $(\beta(y_1), \beta(y_2), \beta(y_3))$ for an assignment β that is $\{y_1, y_2, y_3\}$-compatible with Π_{3SAT}. Let k_1 denote the number of elements in Π^1_{3SAT} and note that $k_1 \leq 8$.

Construction Step 2a. *The construction of the gadget H_1 starts with a path of four edges. After "doubling" the first three edges, i.e., inserting new edges connecting their endpoints and so building a graph consisting of a chain of three double edges followed by a (single) edge, we obtain a graph K'. Connect k_1 copies of K', call them $K'_1, K'_2, \ldots, K'_{k_1}$ in a path-like manner by identifying the start and end vertices of the original path of four edges of consecutive copies of K' (see also Fig. 3).*

Each chain of (three) consecutive double edge will be called a 3-chain and every 3-chain will correspond to an element $a_1 a_2 a_3$ from Π^1_{3SAT}. This correspondence will play an important role in the upcoming Construction Step 2b. Within each 3-chain of H_1 we will associate the first double edge with the variable y_1, the second with y_2 and the third with y_3, where one edge participating in a double edge will be called 0-edge while the other will be called 1-edge.

The next construction step for H_1 deals with the issue that H_1 is supposed to handle $\{y_1, y_2, y_3\}$-compatibility. Informally put, the traversal of a Hamiltonian cycle through the n-chain, i.e. the usage of 0- and 1-edges in the n-chain, will effect the traversal of that Hamiltonian cycle through the gadget H_1. So in the upcoming step we describe how that yet to be completely defined gadget H_1 is connected to the n-chain.

Construction Step 2b. *Let e' be a 0-edge in H_1. Suppose e' is part of a 3-chain that is associated with the partial assignment $a_1a_2a_3 \in \Pi^1_{3SAT}$, $a_1, a_2, a_3 \in \{0,1\}$, and let e' be associated with the variable y_i, $1 \leq i \leq 3$, within that 3-chain. Connect e' with f_i (the 1-edge in the n-chain that is associated with y_i) via a parity module if and only if $a_i = 1$ and connect e' with e_i (the 0-edge in the n-chain that is associated with y_i) via a parity module if and only if $a_i = 0$.*

Suppose that C is a Hamiltonian cycle in the to be constructed graph $G_{\Pi_{3SAT}}$ that by its traversal through the n-chain defines an assignment β. Consider a 3-chain K in H_1 that is associated with a partial assignment $a_1a_2a_3$ and the 0-edge (edge) in that 3-chain associated with y_1. Observe that by the above insertion of the parity modules we have that C does not use the 0-edge in K that is associated with y_1 if and only if $a_1 = \beta(y_1)$.

It follows that a Hamiltonian cycle C corresponds to an assignment (the assignment defined by the Hamiltonian cycle's traversal of the n-chain) that is $\{y_1, y_2, y_3\}$-compatible with Π_{3SAT} if and only if there exists one 3-chain K in H_1 such that C does not use any of the three 0-edges in K.

The next step introduces three auxiliary vertices in H_1 that will force any Hamiltonian cycle in $G_{\Pi_{3SAT}}$ to avoid all three 0-edges in at least one 3-chain K and so in light of the above comment force any Hamiltonian cycle in $G_{\Pi_{3SAT}}$ to correspond to an assignment that is (y_1, y_2, y_3)-compatible with Π_{3SAT}.

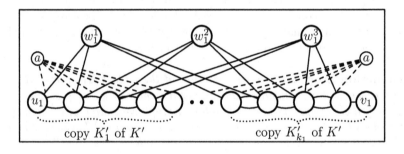

Fig. 3. The structure of the gadget H_1

Construction Step 2c. *Add three new vertices w^1_1, w^1_2, w^1_3 to the gadget H_1, that will be associated with the variables y_1, y_2, and y_3, respectively. For each i, $1 \leq i \leq 3$, and each 3-chain K in H_1 add edges from the endpoints of its y_i-double-edge to w^1_i. Furthermore, for each 3-chain K all-connect (see Lemma 1) the six edges between K and the vertices w^1_1, w^1_2, w^1_3 [1] (see Fig. 3).*

In order to see that the introduction of w^1_1, w^1_2, w^1_3 has the desired effect note that the three new vertices have to be visited by each Hamiltonian cycle. Each edge leading from a 3-chain to one of the new vertices is all-connected to the

---

[1] Note that all-connection requires the existence of an auxiliary edge, that is used by each Hamiltonian cycle. However the 3-chain K contains an edge that has not been doubled in Construction Step 2a and so it can be used for exactly this purpose.

other five edges that connect this 3-chain with one of w_1^1, w_2^1, w_3^1. It follows that w_1^1, w_2^1, w_3^1 will be visited by any Hamiltonian cycle via edges coming from one and the same 3-chain.

Now it easy to see that each Hamiltonian cycle avoids all three 0-edges in at least one 3-chain, namely in the 3-chain, from which the Hamiltonian cycle visits w_1^1, w_2^1, w_3^1. If a Hamiltonian cycle C would use one of these 0-edges there would be a small cycle in C consisting of the 0-edge and those two edges between the 3-chain and w_1^1, w_2^1, w_3^1, that start at the endpoints of the 0-edge, a contradiction.

Observe that with respect to a potential Hamiltonian cycle C in the graph $G_{\Pi_{3SAT}}$ and C's traversal of the n-chain the parity-connections between the n-chain and the gadget H_1 uniquely define C's way through the series of 3-chains inside H_1 and hence also the way the vertices w_1^1, w_2^1, w_3^1 are traversed. The latter is determined by the 3-chain in which C avoids all three 0-edges, in other words by the 3-chain that witnesses the $\{y_1, y_2, y_3\}$-compatibility of the assignment defined by C's traversal through the n-chain.

This completes the construction of H_1. It is obvious how the gadgets for other triples of variables have to be constructed. The overall structure of the gadgets $H_2, \ldots, H_{\binom{n}{3}}$ is identical to the structure of H_1 as shown in Fig. 3, except for the names of the vertices and the number of copies of K'. The connections of a gadget to the n-chain (not shown in Fig. 3) also differ between the gadgets.

This concludes the construction of the gadgets H_i and we will return to the construction of $G_{\Pi_{3SAT}}$. Let for all i, u_i and v_i be the "first" and "last" vertices of H_i (see Fig. 3).

Construction Step 3. *Insert the gadgets $H_1, \ldots H_{\binom{n}{3}}$ into the graph $G'_{\Pi_{3SAT}}$ as shown in Fig. 4. In particular, recall that the graph $G'_{\Pi_{3SAT}}$ contains a simple path of $2\binom{n}{2} + 1$ edges. Replace any second edge of that simple path by one gadget (replacing an edge $\{u'', v''\}$ by H_i means removing the edge $\{u'', v''\}$ from $G'_{\Pi_{3SAT}}$ and identifying the vertices u'' and v'' with the vertices u_i and v_i—the "first" and "last" vertices of H_i—respectively). Note that by inserting a gadget H_i any connections from H_i to the n-chain, i.e., the parity modules spoken of in Construction Step 2b, are also inserted in $G_{\Pi_{3SAT}}$.*

This completes the construction of the graph $G_{\Pi_{3SAT}}$. As mentioned at the beginning of the proof, $G_{\Pi_{3SAT}}$ has the property that it contains exactly one Hamiltonian cycle for each assignment that is 3-compatible with Π_{3SAT}.

Lemma 2. *Let Π_{3SAT} be a well-formed set of proofs for V_{3SAT}. There is a bijective mapping between the set of assignments that are 3-compatible with Π_{3SAT} and the set of Hamiltonian cycles in $G_{\Pi_{3SAT}}$.*

Note that the size of the constructed graph $G_{\Pi_{3SAT}}$ is polynomial in the length n of the strings in Π_{3SAT} and thus in $|\Pi_{3SAT}|$. Also, the construction of $G_{\Pi_{3SAT}}$ can be done in time polynomial in $|\Pi_{3SAT}|$. Furthermore, it follows from the construction of the graph $G_{\Pi_{3SAT}}$ that given a well-formed set of proofs Π_{3SAT} for V_{3SAT} and an assignment $\alpha \in \Pi_{3SAT}$ the Hamiltonian cycle that is associ-

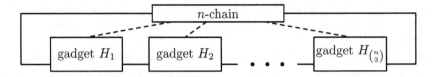

Fig. 4. The overall structure of $G_{\Pi_{3SAT}}$. The dashed lines represent the "connections" between the gadgets H_i and the n-chain that are realized by several parity-modules.

ated with α via the bijective mapping spoken of in the above Lemma 2 can be constructed in time polynomial in Π_{3SAT}.

We will now turn to formally define the function f that reduces Invs-3SAT to Invs-HC. As already mentioned at the beginning of this proof f maps non well-formed proof sets Π_{3SAT} to a fixed non-member of Invs-HC. For well formed proof-sets Π_{3SAT} we define $f(\Pi_{3SAT})$ to be the set of those Hamiltonian cycles in $G_{\Pi_{3SAT}}$ that correspond to the assignments from Π_{3SAT} via the mapping spoken of in Lemma 2. Note that any assignment from Π_{3SAT} is 3-compatible with Π_{3SAT} and hence there does indeed exist a Hamiltonian cycle in $G_{\Pi_{3SAT}}$ for every assignment from Π_{3SAT} (Lemma 2). Since checking whether a proof set is well-formed, constructing the graph $G_{\Pi_{3SAT}}$, and also extracting Hamiltonian cycles corresponding to given assignments can all be done in polynomial time it follows that f is computable in polynomial time.

It remains to show that for all Π_{3SAT} we have $\Pi_{3SAT} \in$ Invs-3SAT $\longleftrightarrow$ $f(\Pi_{3SAT}) \in$ Invs-HC. Due to space restrictions that part of the proof is omitted.

$\Box$

Acknowledgement. We thank Tobias Berg for all the inspiring and helpful discussions and we thank the anonymous reviewers for their useful comments.

References

[1] H. Chen. Inverse NP problems. *Mathematical foundations of computer science (2003), 338–347, Lecture Notes in Comput. Sci., 2747, Springer, Berlin, 2003*

[2] M. Garey and D.Johnson. *Computers and Intractability: A Guide to the Theory of NP-completeness*,W.H.Freeman Company, 1979

[3] E. and L.Hemaspaandra, H. Hempel. All superlinear inverse schemes are coNP-hard. *Theoretical Computer Science 345 (2005), no. 2-3, 345–358.*

[4] D. Kavvadias and M. Sideri. The inverse satisfiability problem. *SIAM Journal on Computing 28(1) (1998), no. 1, pp. 152–163*

[5] C. H. Papadimitriou. *Computational Complexity*, Addison-Wesley, 1994

[6] D.B. West. *Introduction to Graph Theory*, Prentice Hall, 2001

Parameterized Problems on Coincidence Graphs

Sylvain Guillemot

LIRMM - 161 rue Ada - 34392 Montpellier Cedex 5
sguillem@lirmm.fr

Abstract. A (k, r)-tuple is a word of length r on an alphabet of size k. A graph is (k, r)-representable if we can assign a (k, r)-tuple to each vertex such that two vertices are connected iff the associated tuples agree on some component. We study the complexity of several graph problems on (k, r)-representable graphs, as a function of the parameters k, r; the problems under study are MAXIMUM INDEPENDENT SET, MINIMUM DOMINATING SET and MAXIMUM CLIQUE. In this framework, there are two classes of interest: the graphs representable with tuples of logarithmic length (i.e. graphs (k, r)-representable with $r = O(k \log n)$), and the graphs representable with tuples of polynomial length (i.e. graphs (k, r)-representable with $r = poly(n)$). In both cases, we show that the problems are computationally hard, though we obtain stronger hardness results in the second case. Our hardness results also allow us to derive optimality results for MULTIDIMENSIONAL MATCHING and DISJOINT r-SUBSETS.

1 Introduction

A (k, r)-tuple is a word of length r on an alphabet of size k. A graph is (k, r)-representable if we can assign a (k, r)-tuple to each vertex such that two vertices are connected iff the associated tuples agree on some component. In this paper, we consider the complexity of several problems on (k, r)-representable graphs: MAXIMUM INDEPENDENT SET, MINIMUM DOMINATING SET, MAXIMUM CLIQUE. The focus is on the parameterized complexity of these problems with respect to the parameters k, r. To address this question, we use the toolkit developed by Downey and Fellows (fpt algorithms, classes $W[t]$ and $M[t]$).

Our results are as follows:

- in Section 3, we consider the MAXIMUM INDEPENDENT SET (MIS) problem. We show that the problem can be solved in $O(n^k)$ time in the general case, and in $O(2^{O(pr)} rn)$ time if we seek an independent of size $\geq p$. Moreover, we show that the problem is M[1]-hard when $r = O(k \log n)$, and W[1]-complete when $r = poly(n)$.
- in Section 4, we consider the MINIMUM DOMINATING SET (MDS) problem. We show that the problem can be solved in $O(n^k)$ time in the general case, and in $O((pr)^{2pr} rn)$ time if we seek a dominating set of size $\leq p$. We also show that the problem is M[2]-hard when $r = O(k \log n)$, and W[2]-complete when $r = poly(n)$.

T. Asano (Ed.): ISAAC 2006, LNCS 4288, pp. 253–266, 2006.

- in Section 5, we consider the MAXIMUM CLIQUE (MC) problem. We show that even when $r = O(k \log n)$, there exists a value of k for which the problem is NP-hard (under randomized reductions).
- in Section 6, we show that our hardness results imply optimality results for various FPT algorithms, under complexity-theoretic assumptions. Namely, we show that MIS and MDS on (k, r)-representable graphs cannot be solved in $2^{o(kr)} n^c$ time, or in $2^{o(pr)} n^c$ time, under likely assumptions (FPT $\neq$ M[1] in the first case, FPT $\neq$ M[2] in the second case); we also show that the problems MULTIDIMENSIONAL MATCHING and DISJOINT r-SUBSETS cannot be solved in $2^{o(pr)} n^c$ time if FPT $\neq$ M[1].

2   Definitions

We use the notation $[k]$ to denote either the set of integers $\{1, ... k\}$ or the set of integers $\{0, ..., k-1\}$, depending on the context.

2.1   Graphs

Let $G = (V, E)$ be a graph. We recall that a *k-coloring* of G is a function $\gamma : V \to [k]$ s.t. $\gamma(x) \neq \gamma(y)$ whenever $\{x, y\} \in E$. We say that $\gamma : V \to [k]$ is a *k-cocoloring* of G iff γ is a k-coloring of $\bar{G}$. A *k-colored graph* is a triple $G = (V, E, \gamma)$ where γ is a k-coloring of (V, E). A *k-cocolored graph* is a triple $G = (V, E, \gamma)$ where γ is a k-cocoloring of (V, E).

A (k, r)-*tuple* is an element of $[k]^r$. If u is a (k, r)-tuple, for every $i \in \{0, ..., r-1\}$ the i-th component of u is denoted $u[i]$. A (k, r)-*family* is a family of (k, r)-tuples: note that this is a family and not a set, meaning that a given tuple may have several occurences. Given two (k, r)-tuples u, v, we say that u, v are *coincident* (denoted by $u \parallel v$) iff there exists i s.t. $u[i] = v[i]$. We say that u, v are *non coincident* (denoted by $u \nparallel v$) otherwise. The *coincidence graph* of a (k, r)-family $\mathcal{F}$ is the graph of the coincidence relation on $\mathcal{F}$, i.e. this is the graph $G = (V, E)$, where $V := \mathcal{F}$, and where we have $\{u, v\} \in E$ whenever $u \parallel v$. We denote by $G[\mathcal{F}]$ the coincidence graph of $\mathcal{F}$.

We say that G is (k, r)-*representable* iff there exists a (k, r)-family $\mathcal{F}$ s.t. $G = G[\mathcal{F}]$, or equivalently s.t. there exists a bijection $\phi : V \to \mathcal{F}$ s.t. $\{x, y\} \in E \Leftrightarrow \phi(x) \parallel \phi(y)$. Note that in this case, for every $i \in [r]$, the function $\gamma_i : V \to [k]$ defined by $\gamma_i(x) = \phi(x)[i]$ is a k-cocoloring of G. We say that G is (k, r)-*corepresentable* iff there exists a (k, r)-family $\mathcal{F}$ s.t. $\bar{G} = G[\mathcal{F}]$, or equivalently s.t. there exists a bijection $\phi : V \to \mathcal{F}$ st $\{x, y\} \in E \Leftrightarrow \phi(x) \nparallel \phi(y)$. Note that in this case, for every $i \in [r]$, the function γ_i is a k-coloring of G.

If G is a graph, we denote by $\alpha(G)$ the size of an independent set of maximum cardinality, and by $\beta(G)$ the size of a dominating set of minimum cardinality. We will also use these notations for families, e.g. $\alpha(\mathcal{F})$ and $\beta(\mathcal{F})$.

2.2 Parameterized Complexity

Let A be an alphabet. A *parameterized language* is a set $L \subseteq A^* \times \mathbb{N}$. Given two parameterized languages L, L', we say that L *fpt-reduces* to L' iff there exists a computable function $f : \mathbb{N} \to \mathbb{N}$, a function $g : \mathbb{N} \to \mathbb{N}$, a constant $c \in \mathbb{N}$, and an algorithm A s.t. from every instance $I = (x, k)$ of L, A computes an instance $I' = (x', k')$ of L' in time $f(k)n^c$, s.t. $k' \leq g(k)$ and $I \in L \Leftrightarrow I' \in L'$. We write $L \leq_{fpt} L'$ to mean that L fpt-reduces to L'.

We define the following parameterized languages. 1 is the trivial language with two elements. MINIATURIZED 3-SATISFIABILITY (MINI3SAT) asks: given a parameter k, an integer n encoded in unary, and a 3CNF Φ with $k \log n$ variables, decide if Φ is satisfiable. MINIATURIZED SATISFIABILITY (MINISAT) asks: given a parameter k, an integer n encoded in unary, and a CNF Φ with $k \log n$ variables s.t. $|\Phi| \leq n$, decide if Φ is satisfiable. INDEPENDENT SET (IS) asks: given a parameter k, and a graph G, decide if $\alpha(G) \geq k$. DOMINATING SET (DS) asks: given a parameter k, and a graph G, decide if $\beta(G) < k$.

A *class* is a set $\mathcal{C}$ of parameterized languages, closed by fpt-reduction: if $L \in \mathcal{C}$ and $L' \leq_{fpt} L$, then $L' \in \mathcal{C}$. Given a parameterized language L, we denote by L^{fpt} the class generated by L, this is the set of parameterized languages L' s.t. $L' \leq_{fpt} L$. Given a class $\mathcal{C}$, we say that a parameterized language L is $\mathcal{C}$-complete iff $\mathcal{C} = L^{fpt}$ and that L is $\mathcal{C}$-hard iff $\mathcal{C} \subseteq L^{fpt}$. We define the classes FPT, M[1], W[1], M[2] and W[2] as follows: FPT $= 1^{fpt}$, M[1] $=$ MINI3SAT^{fpt}, W[1] $=$ IS^{fpt}, M[2] $=$ MINISAT^{fpt} and W[2] $=$ DS^{fpt} (this is consistent with the usual definitions of these classes). We have the inclusions FPT $\subseteq$ M[1] $\subseteq$ W[1] $\subseteq$ M[2] $\subseteq$ W[2], and these inclusions are conjectured to be proper.

3   Complexity of Maximum Independent Set

In this section, we determine the complexity of the MAXIMUM INDEPENDENT SET problem on (k, r)-representable graphs (given a representation). Formally, we define the problem MAXIMUM INDEPENDENT SET ON FAMILIES (MISF): given a parameter k, a (k, r)-family $\mathcal{F}$, and an integer p, decide if $\alpha(\mathcal{F}) \geq p$.

In Section 3.1, we show that the problem can be solved in $O(n^k)$ time for any r by exhaustive search, and in $O(2^{2kr}rn)$ time by dynamic programming. We also show that the problem of deciding if a (k, r)-family has an independent of size $\geq p$ can be solved in $O(2^{2pr}rn)$ time via a randomized algorithm. These results give an alternative to known algorithms for the MULTIDIMENSIONAL MATCHING problem (since this problem is equivalent to MISF).

In Section 3.2, we obtain hardness results for MISF. Since the problem can be solved in $O(n^k)$ time, the question arises whether it could be fixed-parameter tractable. We settle this question negatively. Let MISF-1 denote the restriction of MISF to $(k, O(k \log n))$-families. Let MISF-2 denote the restriction of MISF to $(k, poly(n))$-families. We prove that MISF-1 is M[1]-hard, and that MISF-2 is W[1]-complete.

3.1   Solving MISF

First, observe that:

Proposition 1. *There is an algorithm which, given a (k, r)-family $\mathcal{F}$, computes $\alpha(\mathcal{F})$ in $O(n^k)$ time.*

Proof. This follows from the fact that for a (k, r)-family $\mathcal{F}$, we have $\alpha(\mathcal{F}) \leq k$. Indeed, if we have a set of $k+1$ elements $S = \{v_1, ..., v_{k+1}\} \subseteq \mathcal{F}$, then there exists i, j s.t. $v_i[0] = v_j[0]$, and thus S is not an independent. □

We now give an algorithm solving the problem in $O(2^{2kr}rn)$ time, which is faster than the previous algorithm when $r \leq \log_2 n/2$:

Proposition 2. *There is an algorithm which, given a (k, r)-family, computes $\alpha(\mathcal{F})$ in $O(2^{2kr}rn)$ time.*

Proof. Before describing the algorithm, we introduce a few definitions. A *filter* is a tuple $X = (x_1, ..., x_r)$, where $x_i \subseteq [k]$ for every i. Let $\emptyset = (\emptyset, ..., \emptyset)$ denote the *empty filter*, and let $X_0 = ([k], ..., [k])$ denote the *full filter*. We say that X is *thin* iff $|x_i| = 1$ for every i.

Consider three filters X, Y, Z. We denote the union of X, Y by $X \sqcup Y$, and we denote their intersection by $X \sqcap Y$ (obtained by taking unions, resp. intersections, componentwise). We say that Y *dominates* X (denoted $X \sqsubseteq Y$) if $x_i \subseteq y_i$ for every i. We say that (X, Y) is a *partition* of Z iff (i) $X \neq \emptyset, Y \neq \emptyset$, (ii) $X \sqcap Y = \emptyset$ and (iii) $X \sqcup Y = Z$.

Let $Allowed_{\mathcal{F}}(X)$ denote the set of elements $t \in \mathcal{F}$ s.t. $t[i] \in x_i$ for every i. Let $IS_{\mathcal{F}}(X)$ denote the family of independent sets of $Allowed_{\mathcal{F}}(X)$; let $is_{\mathcal{F}}(X)$ denote the size of a largest set in $IS_{\mathcal{F}}(X)$. Obviously, $\alpha(\mathcal{F}) = is_{\mathcal{F}}(X_0)$.

Consider the following algorithm, which computes $is_{\mathcal{F}}(X)$ by dynamic programming:

1. if $|Allowed_{\mathcal{F}}(X)| = 0$, return 0;
2. else if X is thin, return 1;
3. else: compute $is_{\mathcal{F}}(X) := \max\{is_{\mathcal{F}}(X_1) + is_{\mathcal{F}}(X_2) : (X_1, X_2) \text{ partition of } X\}$.

The termination of the algorithm is ensured by the fact that if (X_1, X_2) is a partition of X, we have $X_1 \sqsubset X$ and $X_2 \sqsubset X$. The correctness follows from the observation that if X is not thin and $I \in IS_{\mathcal{F}}(X)$, then there is a partition (X_1, X_2) of X and a partition I_1, I_2 of I s.t. $I_i \in IS_{\mathcal{F}}(X_i)$ (and conversely).

Obviously, this algorithm has running time $O(2^{2kr}rn)$, since for a given tuple X, computing $|Allowed_{\mathcal{F}}(X)|$ requires time $O(rn)$. □

Observe that the running time can be shown to be $O(3^{kr}rn)$ by a more precise analysis. Note also that the complexity can be lowered to $O(2^{kr}k^r rn)$ if in Step 3 we consider only the partitions (X_1, X_2) of X where X_1 is thin.

Using the above algorithm, we can show that:

Proposition 3. *There is an algorithm which, given a (k, r)-family $\mathcal{F}$ and an integer p, decides whether $\alpha(\mathcal{F}) \geq p$ in RPTIME($O(2^{O(pr)}rn)$).*

Proof. Let $\mathcal{F}$ be a (k,r)-family s.t. we want to decide whether $\alpha(\mathcal{F}) \geq p$. Consider a tuple of functions $F = (f_1, ..., f_r)$ where each $f_i : [k] \rightarrow [p]$. For every $t \in \mathcal{F}$, let $\gamma_F(t) = (f_1(t[1]), ..., f_r(t[r]))$, and let $\mathcal{F}_F = \{\gamma_F(t) : t \in \mathcal{F}\}$. We claim that:

1. if $\alpha(\mathcal{F}) \geq p$, then $Pr_F[\alpha(\mathcal{F}_F) = p] \geq \frac{1}{e^{pr}}$.
2. if $\alpha(\mathcal{F}) < p$, then $Pr_F[\alpha(\mathcal{F}_F) = p] = 0$.

Now, consider the following algorithm. Repeat e^{pr} times the following procedure: choose a random tuple $F = (f_1, ..., f_r)$, and test if $\alpha(\mathcal{F}_F) = p$ using the algorithm described in Proposition 2: return "yes" if this is the case, continue otherwise; return "no" if no such F is found. Then this algorithm has running time $2^{O(pr)}n$, and by the above observation:

1. if $\alpha(\mathcal{F}) \geq p$, then the algorithm returns "yes" with probability $\geq 1/2$;
2. if $\alpha(\mathcal{F}) < p$, then the algorithm always returns "no".

$\square$

Note that the above algorithm could be derandomized using the following notion:

Definition 1. *A (p,r)-perfect hash family on a base set S is a family $\mathcal{F}$ of tuples $(f_1, ..., f_r)$, where each $f_i : S \rightarrow [p]$, satisfying the following:*
for every $x_1, ..., x_r \subseteq S$, $|x_i| = p$, there exists a tuple $(f_1, ..., f_r) \in \mathcal{F}$ s.t. f_i is injective on x_i.

Note that this is a generalization of the notion of p-perfect hash family [2] (which corresponds to the particular case $r = 1$). Observe that:

Lemma 1. *Let S be a base set of size k. Fix N, and let $\mathcal{F}$ be a random family of N tuples $(f_1, ..., f_r)$, where each $f_i : S \rightarrow [p]$. Then: if $N \geq (1 + \epsilon)e^{pr}pr \log k$, then $\mathcal{F}$ is a (p,r)-perfect hash family with high probability.*

Proof. Consider a tuple of sets $X = (x_1, ..., x_r)$, where each x_i is a p-subset of S. Say that a tuple of functions $F = (f_1, ..., f_r)$ *validates* X iff every f_i is injective on x_i. Say that X is *good* iff there exists $F \in \mathcal{F}$ which validates X, say that X is *bad* otherwise. Let N_b denote the number of bad tuples. For a given $F \in \mathcal{F}$, and for a given tuple X, we compute the probability that X is validated by F:

$$Pr[X \text{ validated by } F] = \prod_{i=1}^{r} Pr[f_i \text{ injective on } x_i] \geq \frac{1}{e^{pr}}$$

Now, we compute the expected number of bad tuples:

$$E[N_b] \leq \sum_X Pr[X \text{ good}] \leq k^{pr}(1 - \frac{1}{e^{pr}})^N \leq e^{pr \log k - Ne^{-pr}}$$

which is $o(1)$ as soon as $N \geq (1 + \epsilon)e^{pr}pr \log k$. We obtain that $\mathcal{F}$ is a (p,r)-perfect hash family with probability $1 - o(1)$. $\square$

Since the above result gives a probabilistic construction of a (p, r)-perfect hash family, this yields a randomized algorithm for deciding whether $\alpha(\mathcal{F}) \geq p$ in $\text{ZPTIME}(2^{O(pr)} rn \log k)$ (improving the RPTIME algorithm of Proposition 3, at the expense of a $\log k$ factor). We conjecture that it is possible to give a deterministic algorithm with the same time complexity; note that this would require the explicit construction of a (p, r)-perfect hash family on $[k]$, having size $2^{O(pr)} \log k$. This algorithm could then be combined with the kernelization algorithm for MULTIDIMENSIONAL MATCHING described in [10], lowering the running time to $O(n + 2^{O(pr)})$.

3.2   Hardness Results

First, we show M[1]-hardness of MISF-1. To this end, we introduce the problem MINIBIS. A (k, p)-*partitioned graph* is a tuple $G = (V, E, \mathcal{P})$, where $\mathcal{P}$ is a partition of V in k classes $V_1, ..., V_k$, each of size p. A K-*balanced independent* of G is a set $I \subseteq V$ s.t. (i) I is an independent of G, (ii) $|I \cap V_i| = |V_i|/K$ for every i. Let MINIBIS-(K, Δ) denote the following problem: given a parameter k, and a $(k, O(\log n))$-partitioned graph G s.t. G is Δ-regular, decide if G has a K-balanced independent. We prove that:

Proposition 4. MINIBIS-$(3, 4)$ *is* M[1]-*hard.*

Proof. Let 3SAT-E2,2 denote the problem 3SAT, restricted to instances in which each variable has exactly 2 positive occurences and 2 negative occurences. Let MINI3SAT-E2,2 denote its miniaturized version. Then:

- MINI3SAT-E2,2 is M[1]-hard: indeed, the reduction from 3SAT to 3SAT-E2,2 runs in linear time, and thus induces an fpt-reduction between the miniaturized problems;
- MINI3SAT-E2,2 reduces to MINIBIS-$(3, 4)$ via the usual reduction from SAT to IS. Indeed, let Φ be an instance of MINI3SAT-E2,2, with $p = O(k \log n)$ clauses and q variables.
  The reduction constructs a graph $G = (V, E, \mathcal{P})$ where V is the set of literals of Φ, two elements $u, v \in V$ are connected if: (i) either they belong to the same clause of Φ, or (ii) they are complementary literals, i.e. $u = x, v = \neg x$. The partition $\mathcal{P} = \{V_1, ..., V_k\}$ is defined as follows. Let $\mathcal{C}$ be the set of clauses of Φ, then partition $\mathcal{C}$ in k sets $\mathcal{C}_1, ..., \mathcal{C}_k$ of size $O(\log n)$, and for each i, let V_i be the set of literals appearing in clauses of $\mathcal{C}_i$.
  Then G is a $(k, O(\log n))$-partitioned graph. Moreover, since each clause of Φ has size 3, and since each variable of Φ has two positive and two negative occurences, it follows that every vertex of G has degree 4, thus G is 4-regular. Hence, G is an instance of MINIBIS-$(3, 4)$; besides, its construction takes polynomial time.
  The correctness of the reduction follows by showing that: Φ is satisfiable iff G has a 3-balanced independent (proof omitted).

$\square$

We are now ready to prove that:

Proposition 5. MISF-1 *is* M[1]-*hard.*

Proof. We reduce from MINIBIS-(K, Δ). Let $I = (G, k)$ be a an instance of MINIBIS-(K, Δ). Then $G = (V, E, \mathcal{P})$ is a Δ-regular (k, p)-partitioned graph with $p = O(\log n)$. Thus $\mathcal{P}$ is a partition of V in k classes $V_1, ..., V_k$, each of size p. Let $\Delta' = \frac{\Delta}{2}$. We construct the following instance $I' = (\mathcal{F}, k)$ of MISF-1. We set $l := k + 1$, $r := \Delta' kp + 1$, and $\mathcal{F}$ is the (l, r)-family constructed as follows.

Since G is Δ-regular, we have $|E| = \Delta' kp$. Let $e_1, ..., e_{|E|}$ be an enumeration of E. For every $i \in [k]$, let $\mathcal{S}_i$ denote the family of independent sets $I \subseteq V_i$ s.t. $|I| = p/K$. Then, for every $u \in \mathcal{S}_i$ we define the tuple $s_{i,u}$ as follows. Let $j \in [r]$, we set:

$$\begin{cases} s_{i,u}[0] & = i \\ s_{i,u}[j] & = k + 1 \text{ if } j \geq 1, e_j \cap u \neq \emptyset \\ s_{i,u}[j] & = i \text{ if } j \geq 1, e_j \cap u = \emptyset \end{cases}$$

Let $\mathcal{F} = \{s_{i,u} : i \in [k], u \in \mathcal{S}_i\}$. Then the construction takes polynomial time. The validity of the reduction follows by proving that: G has a K-balanced independent iff $\alpha(\mathcal{F}) \geq k$. $\quad\square$

Now, we show W[1]-completeness of MISF-2. This relies on the following Lemma, on the representability of k-cocolored graphs:

Lemma 2. *Let* $G = (V, E, \gamma)$ *be a* k-*cocolored graph. Then* G *is* $(k + 1, O(n^2))$-*representable.*

Proof. Let $V_1, ..., V_k$ be the color classes of G, and let $e_1, ..., e_m$ be an enumeration of E. We will construct a $(k + 1, m + 1)$-family $\mathcal{F}$ as follows. For every $v \in V$ we create a tuple s_v. Suppose that $v \in V_i$, and consider $j \in [m + 1]$, then we set:

$$\begin{cases} s_v[0] & = i \\ s_v[j] & = i \text{ if } j \geq 1, v \notin e_j \\ s_v[j] & = k + 1 \text{ if } j \geq 1, v \in e_j \end{cases}$$

We take $\mathcal{F} = \{s_v : v \in V\}$. It follows that $G = G[\mathcal{F}]$, by proving that for every $u, v \in V$, we have: $\{u, v\} \in E$ iff $s_u \parallel s_v$. $\quad\square$

We consider the problem PARTITIONED INDEPENDENT SET (PIS) which asks: given a parameter k, and a k-cocolored graph G, decide if $\alpha(G) = k$. Using a similar argument to [13], we show that:

Proposition 6. PIS *is* W[1]-*complete.*

Proof. Clearly, PIS is in W[1], since it fpt-reduces to the IS problem.

To prove that PIS is W[1]-hard, we reduce from IS. Let $I = (G, k)$ be an instance of IS, where $G = (V, E)$ is a graph of which we seek an independent of size k. Consider the k-cocolored graph $G' = (V', E', \gamma')$, where:

$$V' = [k] \times V$$
$$E' = \{\{(i, x), (j, y)\} : i = j \vee x = y \vee \{x, y\} \in E\}$$
$$\gamma'(i, x) = i$$

It is straightforward to verify that G' is k-cocolored. Moreover, we have: $\alpha(G') = k$ iff $\alpha(G) \geq k$ (proof omitted). □

We are now ready to prove that:

Proposition 7. MISF-2 *is* W[1]-*complete*.

Proof. Membership is obvious. Hardness follows from Proposition 6 and Lemma 2. □

4   Complexity of Minimum Dominating Set

In this section, we investigate the complexity of the MINIMUM DOMINATING SET problem on (k, r)-representable graphs (given a representation). Formally, we define the problem MINIMUM DOMINATING SET ON FAMILIES (MDSF): given a parameter k, a (k, r)-family $\mathcal{F}$, and an integer p, decide if $\beta(\mathcal{F}) \leq p$?

In Section 4.1, we show that the problem can be solved in $O(n^k)$ time by exhaustive search. We also show that the problem of deciding if a (k, r)-family has a dominating set of size $\leq p$ can be solved in $O((pr)^{2pr} n)$ time.

In Section 4.2, we obtain hardness results for MDSF, ruling out the possibility of having a FPT algorithm. More precisely, we show the following. Let MDSF-1 denote the restriction of MDSF to $(k, O(k \log n))$-families. Let MDSF-2 denote the restriction of MDSF to $(k, poly(n))$-families. We prove that MDSF-1 is M[2]-hard, and that MDSF-2 is W[2]-complete.

4.1   Solving MDSF

First, observe that:

Proposition 8. *There is an algorithm which, given a (k, r)-family $\mathcal{F}$, computes $\beta(\mathcal{F})$ in $O(n^k)$ time.*

Proof. This follows from the fact that for a (k, r)-family $\mathcal{F}$, we have $\beta(\mathcal{F}) \leq k$. Indeed, for every $i \in [k]$, let $\mathcal{F}_i = \{s \in \mathcal{F} : s[0] = i\}$. Then for every $i \in [k]$ s.t. $\mathcal{F}_i \neq \emptyset$, choose an element $v_i \in \mathcal{F}_i$. Clearly, $D = \{v_i : i \in [k], \mathcal{F}_i \neq \emptyset\}$ is a dominating set of G. □

Moreover, if we have a bound p on the size of the dominating set sought, then we can solve the problem efficiently:

Proposition 9. *There is an algorithm which, given a (k,r)-family $\mathcal{F}$ and an integer p, decides whether $\beta(\mathcal{F}) \leq p$ in $O((pr)^{2pr} n)$ time.*

Proof. Say that a *partial (k,r)-tuple* is a tuple $u \in ([k] \cup \{\bot\})^r$. Say that u is *full* iff $u[i] \neq \bot$ for every i; in the following, we identify the (k,r)-tuples and the full partial (k,r)-tuples. If u is a partial (k,r)-tuple, if $i \in [r]$ and $v \in [k]$, let $u[i \to v]$ denote the partial (k,r)-tuple u' s.t.: $u'[j] = u[j]$ $(j \neq i)$, $u'[i] = v$. Denote by ϵ the partial (k,r)-tuple s.t. $\epsilon[i] = \bot$ for every i. Let u, v be two partial (k,r)-tuples, say that u *extends* v iff $u(i) = v(i)$ whenever $v(i) \neq \bot$.

We say that a partial (k,r)-tuple u is *realizable* by $\mathcal{F}$ iff there exists $v \in \mathcal{F}$ which extends u. Given a set D of partial (k,r)-tuples, we say that a (k,r)-tuple x is *dominated by* D iff there exists $y \in D, i \in [r]$ s.t. $x[i] = y[i]$.

Say that a *state* is a tuple $\mathcal{S} = (\mathcal{R}, U)$, where $U = (u_1, ..., u_p)$ is a list of partial (k,r)-tuples, and $\mathcal{R} = \{x \in \mathcal{F} : x \text{ not dominated by } U\}$. Say that $\mathcal{S}$ is *final* if u_i is full for every i. Define the *initial state* $\mathcal{S}_0 = (\mathcal{F}, U_0)$ with $U_0 = (\epsilon, ..., \epsilon)$. Say that $\mathcal{S}$ is *accepting* if there exists $v_1, ..., v_p \in \mathcal{F}$ s.t.:

- for every i, v_i extends u_i;
- $\{v_1, ..., v_p\}$ is a dominating set of $\mathcal{F}$.

Then obviously: $\beta(\mathcal{F}) \leq p$ iff $\mathcal{S}_0$ is accepting.

Let $\mathcal{S} = (\mathcal{R}, U)$ be a state. We will define a function $IsAccepting(\mathcal{S})$ which determines if $\mathcal{S}$ is accepting. We rely on the following observation. For every $x \in [k], i \in [r]$, denote $nocc_{\mathcal{S}}(x, i) = |\{s \in \mathcal{R} : s[i] = x\}|$. Then:

Lemma 3. *Suppose that $\mathcal{S}$ is accepting. Then there exists $v \in [k], i \in [r]$ s.t. $nocc_{\mathcal{S}}(v, i) \geq |\mathcal{R}|/pr$.*

Proof. Since $\mathcal{S}$ is accepting, there exists $v_1, ..., v_p \in \mathcal{F}$ as prescribed by the definition. Now, since $\mathcal{R} \subseteq \mathcal{F}$, $D = \{v_1, ..., v_p\}$ is a dominating set of $\mathcal{R}$. Thus, for every $x \in \mathcal{R}$, there exists $\phi(x) = (i, j)$ s.t. $x[i] = v_j[i]$. Thus, we have a function $\phi : \mathcal{R} \to [r] \times [p]$; let $x \in [r] \times [p]$ s.t. $|\phi^{-1}(x)|$ is maximal. Then $x = (i, j)$, let $v = v_j[i]$, thus we have $nocc_{\mathcal{S}}(v, i) \geq |\mathcal{R}|/pr$ as claimed. $\square$

This observation allows us to define the function $IsAccepting(\mathcal{S})$ as follows:

The correctness of the algorithm follows from the following observations. Let $\mathcal{S}$ be a state, define $m(\mathcal{S}) = |\{(i, j) \in [p] \times [r], u_i[j] = \bot\}|$. When calling $IsAccepting(\mathcal{S})$, we have $\mathcal{S} = (\mathcal{R}, U)$ where:

1. $\mathcal{R} = \{x \in \mathcal{F} : x \text{ not dominated by } U\}$;
2. every u_i is realizable by $\mathcal{F}$;
3. for every call $IsAccepting(\mathcal{S}')$ issued, we have $m(\mathcal{S}') < m(\mathcal{S})$.

These observations can be proved by induction on $pr - m(\mathcal{S})$. Clearly, they hold for $\mathcal{S} = \mathcal{S}_0$, and when issuing a new call $IsAccepting(\mathcal{S}')$ at Line 14, using induction hypothesis, we check that Point 1 is verified (because of Line 9) and that Point 2 is verified (because of Lines 11). Point 3 is clear since $m(\mathcal{S}') = m(\mathcal{S}) - 1$.

IsAccepting(S)

1: **if** S is final **then**
2:      return true if $\mathcal{R} = \emptyset$, false otherwise
3: **else**
4:      choose $i \in [r]$
5:      for every v, compute $t[v] = nocc_S(v, i)$
6:      choose v s.t. $t[v] \geq |\mathcal{R}|/pr$
7:      choose $j \in [p]$ s.t. $u_j[i] = \perp$
8:      let $S' = (\mathcal{R}', U')$ be the state defined as follows
9:      $\mathcal{R}' = \{x \in \mathcal{R} : x[j] \neq v\}$
10:     $U' = (u'_1, ..., u'_p)$ where $u'_l = u_l$ $(l \neq j)$ and $u'_j = u_j[i \rightarrow v]$
11:     **if** u'_j is not realizable by $\mathcal{F}$ **then**
12:         return false
13:     **else**
14:         return *IsAccepting(S')*
15:     **end if**
16: **end if**

Now, the correctness of the algorithm follows by proving by induction on $m(S)$ that: for every state S, S is accepting iff *IsAccepting(S)* returns true.

Finally, we justify the running time of the algorithm. Observe that the algorithm builds a search tree, where the nodes are the states, and the leaves are the final states. Moreover, observe that:

- the search tree has height pr, since $m(S)$ decreases by one at each step.
- each internal node has degree $\leq (pr)^2$: indeed, when processing an internal node, the algorithm makes choices at lines 4, 6, 7. Note that Line 4 induces r possible choices, Line 6 induces $\leq pr$ possible choices, and Line 7 induces $\leq p$ possible choices. Thus, the degree of the node is $\leq (pr)^2$.
- each node is processed in $O(n)$ time.

Thus, the algorithm has running time $O((pr)^{2pr}n)$ as claimed. □

We conjecture that MDSF can be solved in $2^{O(pr)}n$ time, but we have been unable to devise an algorithm with this complexity.

4.2   Hardness Results

First, we show:

Proposition 10. MDSF-1 *is* M[2]-*hard.*

Proof. We give an fpt-reduction from MINISAT. Let $I = (\Phi, k)$ be an instance of MINISAT. Then Φ is a CNF with $p := O(k \log n)$ variables and $O(n)$ clauses. Let $\mathcal{V}$ be its variable set, and consider a partition of $\mathcal{V}$ in sets $\mathcal{V}_1, ..., \mathcal{V}_k$, each of size $O(\log n)$. Let $L = \{x, \neg x : x \in \mathcal{V}\}$, and for every i, let $L_i = \{x, \neg x : x \in \mathcal{V}_i\}$. Let $l_1, ..., l_{2p}$ be an enumeration of L. For every i, we define $\mathcal{S}_i$ as the sets $X \subseteq L_i$ s.t. for every $x \in L_i$, $|X \cap \{x, \neg x\}| = 1$.

We construct an instance $I' = (\mathcal{F}, k)$ of MDSF-1 as follows. We set $l := k+2$ and $r := 2p+1$. We define $\mathcal{F}$ to be a (l, r)-family, with the following tuples:

- for every $i \in [k]$, we define $s_i := (i, ..., i)$;
- for every $i \in [k]$, for every $X \in \mathcal{S}_i$, we define $t_{i,X}$ as follows:

$$\begin{cases} t_{i,X}[0] & = i \\ t_{i,X}[j] & = i \text{ if } j \geq 1, l_j \notin X \\ t_{i,X}[j] & = k+1 \text{ if } j \geq 1, l_j \in X \end{cases}$$

- for every clause C of Φ, we define u_C as follows.

$$\begin{cases} u_C[0] & = k+1 \\ u_C[i] & = k+1 \text{ if } i \geq 1, l_i \in C \\ u_C[i] & = k+2 \text{ if } i \geq 1, l_i \notin C \end{cases}$$

We set $\mathcal{F}_1 = \{s_i : i \in [k]\}$, $\mathcal{F}_2 = \{t_{i,X} : i \in [k], X \in \mathcal{S}_i\}$, $\mathcal{F}_3 = \{u_C : C \text{ clause of } \Phi\}$. We set $\mathcal{F} = \mathcal{F}_1 \cup \mathcal{F}_2 \cup \mathcal{F}_3$.

We can show that: Φ is satisfiable iff $\beta(\mathcal{F}) \leq k$ (proof omitted). □

Now, we prove W[2]-completeness of MDSF-2. Let PARTITIONED DOMINATING SET (PDS) denote the following problem: given a parameter k, and a k-cocolored graph $G = (V, E)$, decide if $\beta(G) < k$. We show that:

Proposition 11. PDS *is* W[2]-*complete.*

Proof. Obviously, PDS is in W[2], since it fpt-reduces to the DS problem.

To prove that PDS is W[2]-hard, we reduce from DS. Let $I = (G, k)$ be an instance of DS, where $G = (V, E)$ is a graph of which we seek a dominating set of size $< k$. Consider the k-cocolored graph $G' = (V', E', \gamma')$, where:

$$V' = [k] \times V$$
$$E' = \{\{(i, x), (j, y)\} : i = j \vee x = y \vee \{x, y\} \in E\}$$
$$\gamma'(i, x) = i$$

Obviously, G' is k-cocolored. We can show that: $\beta(G) < k$ iff $\beta(G') < k$ (proof omitted). □

We are now ready to prove that:

Proposition 12. MDSF-2 *is* W[2]-*complete.*

Proof. Membership is obvious. Hardness follows from Proposition 11 and Lemma 2. □

5   Complexity of Maximum Clique

In this section, we consider the complexity of MAXIMUM CLIQUE on (k, r)-representable graphs (given a representation). The formal definition of the problem is MAXIMUM CLIQUE ON FAMILIES (MCF): given a parameter k, a (k, r)-family $\mathcal{F}$, and an integer p, decide if $\omega(\mathcal{F}) \geq p$.

Let MCF-1 denote the restriction of MCF to $(k, O(k \log n))$-families. We prove that MCF-1 is NP-hard for some value of k (Proposition 13). This result relies on the following lemma, concerning the (co)representability of graphs of bounded degree. Its proof uses a simple variant of Welsh-Powell's greedy algorithm ([14]).

Lemma 4. *Let $G = (V, E)$ be a graph of maximum degree $\leq \Delta$. Let $n = |V|$. Let $f \geq 3$ and $g > 6f$, and let $h = \frac{g}{3f}$. Then G is $(f\Delta, g\Delta \log n)$-corepresentable, and such a representation can be found in $O(|G|)$ time with probability $\geq 1 - \frac{1}{n^{h-2}}$.*

Proof. We use a probabilistic argument. Let $k = f\Delta$ and $r = g\Delta \log n$. Consider the following probabilistic algorithm. Enumerate the vertices of G in some fixed order, and assign a tuple $\gamma(x) \in [k]^r$ to each vertex $x \in V$ in the following way. At each step, choose a vertex $u \in V$, let $v_1, ..., v_d$ be its neighbors which have been already examined, and let $c_1, ..., c_d$ be the tuples assigned to them (i.e. $c_i = \gamma(v_i)$ for every $i = 1..d$). We then define $c := \gamma(u)$ as follows. For every $j \in [r]$, let $C_j = [k] \backslash \{c_i[j] : i = 1..d\}$: since $d \leq \Delta$, we have $|C_j| \geq (f - 1)\Delta$; let $c[j]$ be an element chosen randomly (and uniformly) in C_j.

Let $\mathcal{F} = \{\gamma(x) : x \in V\}$ be the (k, r)-family obtained at the end of the algorithm. We claim that $G[\mathcal{F}] = \bar{G}$ with high probability. We need to check that $\mathcal{F}$ satisfies the two following conditions with high probability: (i) for every $\{x, y\} \in E$ then $\gamma(x) \nparallel \gamma(y)$; (ii) for every $\{x, y\} \notin E$ then $\gamma(x) \parallel \gamma(y)$. Condition (i) is verified by construction. We need to show that Condition (ii) holds with high probability. Indeed:

- for $\{x, y\} \notin E$, say that $\{x, y\}$ is *bad* if $\gamma(x) \nparallel \gamma(y)$. Then we can show that:

$$Pr[\{x, y\} \text{ bad}] \leq \frac{1}{n^h}$$

- Let N_b denote the number of bad pairs obtained for a given execution of the algorithm. We compute the expected number of bad pairs:

$$E[N_b] \leq \frac{n^2}{2} \frac{1}{n^h} \leq \frac{1}{n^{h-2}}$$

Say that an execution is *good* if it produces no bad pairs, and *bad* otherwise. Thus, by Markov inequality, we obtain that the probability that an execution is bad is $\leq \frac{1}{n^{h-2}}$, and thus an execution is good with probability $\geq 1 - \frac{1}{n^{h-2}}$ as claimed.

$\square$

Note that the result still holds for graphs of bounded degeneracy. But in our case, we only need the result for graphs of bounded degree. This allows us to prove that:

Proposition 13. *The problem* MCF-1 *is* NP-*hard for some k (under randomized reductions).*

Proof. Let 3-MIS denote the MIS problem restricted to cubic graphs. Then 3-MIS is NP-hard [11]. Now, Lemma 4 gives a randomized polynomial-time reduction from 3-MIS to MCF-1 (with $k = 9$). This proves that the problem MCF-1 is NP-hard for $k = 9$ (under randomized reductions).   □

Note that Lemma 4 implies that for every problem Π which is NP-hard on bounded-degree graphs, the complementary $co\Pi = \{\bar{G} : G \in \Pi\}$ is NP-hard on $(k, O(k \log n))$-representable graphs for some k (under randomized reductions), and thus is not in ZPP unless NP $\subseteq$ ZPP.

6   Optimality Results

In the last years, several techniques have yielded numerous optimality results for parameterized problems [5,8,6,4,7]. In particular, M[1]-hardness results have been used by several authors to prove optimality of FPT algorithms. Two notorious examples are VERTEX COVER and PLANAR DOMINATING SET. The problem of finding a Vertex Cover of size k can be solved in $2^{O(k)}n$ time (see [12,3] for the first algorithms), and cannot be solved in $2^{o(k)}n^c$ time unless FPT $=$ M[1] ([4]). The problem of finding a Dominating Set of size k in a planar graph can be solved in $2^{O(\sqrt{k})}n$ time ([1]), and cannot be solved in $2^{o(\sqrt{k})}n^c$ time unless FPT $=$ M[1] ([4]).

We observe that our M[1]-hardness results for MISF-1 and MDSF-1 yield interesting optimality results for these problems, as well as for the problems MULTIDIMENSIONAL MATCHING and DISJOINT r-SUBSETS. The problem MULTIDIMENSIONAL MATCHING asks: given sets $V_1, ..., V_r$, and a family of tuples $\mathcal{F} \subseteq V_1 \times ... \times V_r$, can we find a matching $M \subseteq \mathcal{F}$ of size p? The problem DISJOINT r-SUBSETS asks: given a set X, and a family of $\mathcal{F}$ of r-subsets of X, can we find p elements of $\mathcal{F}$ that are pairwise disjoint? These problems can be solved in $O((pr)!2^{O(pr)}n^c)$ time using perfect hashing [9]. Moreover, MULTIDIMENSIONAL MATCHING can be solved in $O(n + 2^{O(pr)})$ time (using reduction to a problem kernel of size p^r, [10]). We show that these problems cannot be solved in $2^{o(pr)}n^c$ time under complexity-theoretic assumptions.

Our optimality results for these problems rely on the following observation:

Lemma 5 ([4]). *A parameterized problem is in* FPT *if it is solvable in time* $O(2^{O(s(n)t(k))}p(n))$ *for some unbounded and nondecreasing function* $s(n)$ $= o(\log n)$.

Using this Lemma, we show:

Proposition 14. *We have the following optimality results:*

1. MISF *cannot be solved in* $2^{o(kr)}n^c$ *or* $2^{o(pr)}n^c$ *time, unless* FPT $=$ M[1];
2. MDSF *cannot be solved in* $2^{o(kr)}n^c$ *or* $2^{o(pr)}n^c$ *time, unless* FPT $=$ M[2];
3. MULTIDIMENSIONAL MATCHING *cannot be solved in* $2^{o(pr)}n^c$ *time unless* FPT $=$ M[1];
4. DISJOINT r-SUBSETS *cannot be solved in* $2^{o(pr)}n^c$ *time unless* FPT $=$ M[1].

Proof. 1. follows from Proposition 5 and Lemma 5.

2. follows from Proposition 10 and Lemma 5.

3. is a direct consequence of 1. since the problems MISF and MULTIDIMEN-SIONAL MATCHING are equivalent.

4. is shown by reducing MISF to DISJOINT r-SUBSETS. If $\mathcal{F}$ is a (k, r)-family given as instance of MISF, then we construct an instance $\mathcal{F}'$ of DISJOINT r-SUBSETS as follows. The base set is $B = [k] \times [r]$, and the family of sets is $\mathcal{F}' = \{t_x : x \in V\}$. To each $x \in V$, if $s_x = (a_1, ..., a_r)$, then we associate the set $t_x = \{(1, a_1), ..., (r, a_r)\}$. $\qquad\square$

References

1. J. Alber, H. Bodlaender, H. Fernau, and R. Niedermeier. Fixed parameter algorithms for planar dominating set and related problems. In *Proceedings of SWAT'00*, volume 1851 of *Lecture Notes in Computer Science*, pages 97–110. Springer-Verlag, 2000.

2. N. Alon, R. Yuster, and U. Zwick. Color-coding. *Journal of the Association for Computing Machinery*, 42(4):844–856, 1995.

3. J.F. Buss and J. Goldsmith. Nondeterminism within P. *SIAM Journal on Computing*, 22:560–572, 1993.

4. L. Cai and D. Juedes. On the existence of subexponential parameterized algorithms. *Journal of Computer and System Sciences*, 67(4):789–807, 2003.

5. J. Chen, B. Chor, M. Fellows, X. Huang, D. Juedes, I.A. Kanj, and G. Xia. Tight lower bounds for certain parameterized NP-hard problems. *Information and Computation*, 201(2):216–231, 2005.

6. F. Dehne, M. Fellows, and F. Rosamond. An FPT algorithm for Set Splitting. In *Proceedings of WG'03*, volume 2880 of *Lecture Notes in Computer Science*, pages 180–191. Springer-Verlag, 2003.

7. Frank K. H. A. Dehne, Michael R. Fellows, Michael A. Langston, Frances A. Rosamond, and Kim Stevens. An $O(2^{O(k)}n^3)$ FPT Algorithm for the Undirected Feedback Vertex Set Problem. In *Proceedings of COCOON'05*, volume 3595 of *Lecture Notes in Computer Science*, pages 859–869. Springer-Verlag, 2005.

8. R. Downey, V. Estivill, M. Fellows, E. Prieto, and F. Rosamond. Cutting up is hard to do: the parameterized complexity of k-cut and related problems. In *Proceedings of CATS'03*, volume 78 of *ENTCS*, 2003.

9. R.G. Downey and M.R. Fellows. *Parameterized Complexity*. Springer Verlag, 1999.

10. M. Fellows, C. Knauer, N. Nishimura, P. Ragde, F. Rosamonds, U. Stege, D. Thilikos, and S. Whitesides. Faster fixed-parameter tractable algorithms for matching and packing problems. In *Proceedings of ESA'04*, volume 3221 of *Lecture Notes in Computer Science*, pages 311–322. Springer-Verlag, 2004.

11. M. R. Garey and D. S. Johnson. *Computers and Intractability: A Guide to the Theory of NP-Completeness*. Freeman, New-York, 1979.

12. K. Mehlhorn. *Data Structures and Algorithms*. Springer Verlag, 1990.

13. K. Pietrzak. On the parameterized complexity of the fixed alphabet shortest common supersequence and longest common subsequence problems. *Journal of Computer and System Sciences*, 67(4):757–771, 2003.

14. D.J.A. Welsh and M.B. Powell. An upper bound for the chromatic number of a graph and its application to timetabling problems. *The Computer Journal*, 10:85–86, 1967.

On 2-Query Codeword Testing with Near-Perfect Completeness

Venkatesan Guruswami[*]

Department of Computer Science and Engineering
University of Washington
Seattle, WA 98195

Abstract. A codeword tester is a highly query-efficient spot checking procedure for ascertaining, with good confidence, proximity of a given string to its closest codeword. We consider the problem of binary codeword testing using only two queries. It is known that three queries suffice for non-trivial codeword testing with perfect completeness (where codewords must be accepted with probability 1). It is known that two queries are not enough for testing with perfect completeness, whereas two queries suffice if one relaxes the requirement of perfect completeness (this is akin to the polynomial-time decidability of 2SAT and the APX-hardness of Max 2SAT, respectively).

In this work, motivated by the parallel with 2-query PCPs and the approximability of near-satisfiable instances of Max 2SAT, we investigate 2-query testing with completeness close to 1, say $1 - \varepsilon$ for $\varepsilon \to 0$. Our result is that, for codes of constant relative distance, such testers must also have soundness $1 - O(\varepsilon)$ (and this is tight up to constant factors in the $O(\varepsilon)$ term). This is to be contrasted with 2-query PCPs, where assuming the Unique Games Conjecture, one can have completeness $1 - \varepsilon$ and soundness $1 - O(\sqrt{\varepsilon})$. Hence the ratio $(1 - s)/(1 - c)$ can be super-constant for 2-query PCPs while it is bounded by a constant for 2-query LTCs. Our result also shows a similar limitation of 2-query PCPs of proximity, a notion introduced in [1].

1 Introduction

A *locally testable code* (LTC) is an error-correcting code that has a testing procedure which can ascertain whether a string is a codeword or far from every codeword, by querying very few (eg. a small constant number of) locations of the string. LTCs have been the subject of much research over the years and there has been heightened activity and progress on them recently (eg., see the survey [6] and the references therein). The construction of good LTCs has gone hand-in-hand with constructions of Probabilistically Checkable Proof (PCP) systems. A PCP system specifies a proof format that allows a verifier to ascertain membership of a string in an NP language by probing very few locations into the proof/witness. The key parameters in both LTCs and PCPs are the number

---

[*] Research supported in part by NSF CCF-0343672 and a Sloan Research Fellowship.

of queries q, completeness c which is the probability of accepting correct codewords (proofs in the case of PCPs), and soundness s which is the probability of accepting strings far-off from codewords (false proofs in the case of PCPs).

The most interesting and non-trivial setting for the LTC case (resp. PCP case) is when length of the codewords (resp. proofs) is polynomially long in the message length (resp. size of the input). Additionally, for LTCs, the distance of the code should be a constant fraction of the codeword length. For this setting, which is our focus, there is a great deal of parallel between what is achievable with respect to the three parameters q, c, s in the LTC and PCP worlds. Qualitatively, PCP constructions, or in other words negative results for approximation algorithms, have parallel positive results for LTCs. Conversely, results showing limitations of PCPs (i.e., approximation algorithms) have parallel negative results for LTCs. We describe below some of the manifestations of this phenomenon (these are also discussed briefly in [2]):

- When $q = 3$, there is an absolute constant $\alpha_0 < 1$, for which there are known constructions of binary PCPs and LTCs that have $c = 1$ (referred to as *perfect completeness*) and $s = \alpha_0$. Therefore, three queries suffice for non-trivial codeword testing and proof checking, even with perfect completeness. (The analogy w.r.t approximation algorithms is the hardness of approximating Max 3SAT, even on satisfiable instances.)

- When $q = 2$, there is an absolute constant $\alpha_1 < 1$ for which there are known constructions of PCPs and LTCs over ternary alphabets that have perfect completeness and soundness $s = \alpha_1$. Thus, two queries suffice for LTCs[1] and PCPs with perfect completeness over alphabets of size at least 3. (The analogy in terms of approximability is the hardness of approximating Max 2SAT over domain size d, even on satisfiable instances, for all $d \geq 3$.[2])

- For two queries and binary alphabet, there are no non-trivial PCPs and LTCs with perfect completeness and soundness $s < 1$. Thus, two queries do **not** suffice for binary LTCs and PCPs if we insist on perfect completeness. (The algorithmic analogy for this is the polynomial-time decidability of 2SAT.)

- For two queries and binary alphabet, there are absolute constants $0 < \alpha_2 < \alpha_3 < 1$ (one can take $\alpha_3 = 11/12$) for which non-trivial LTCs and PCPs are known with $c = \alpha_3$ and $s = \alpha_2$. (The analogy in terms of approximability is the APX-hardness of Max 2SAT.)

Despite these striking similarities, we stress that there is no generic translation of LTCs into PCPs with similar parameters, or vice versa (cf. the discussion/warning in [7]).

By the above discussion if either alphabet size or number of queries is bigger than 2, LTCs with perfect completeness exist, while imperfect completeness is necessary and sufficient for 2-query binary LTCs. This work investigates the potential of 2-query binary LTCs when the completeness approaches 1 (this

---

[1] The codes must necessarily be non-linear for testing with two queries [2].

[2] A 2SAT clause on variables x, y over general domains is of the form $(x \neq a) \vee (y \neq b)$ for some values a, b from the domain.

setting is typically called *near-perfect completeness*). Below we describe some of the motivation and context for considering this question which at first sight might seem rather esoteric.

To describe the context, consider the Max 2SAT problem when the instance is promised to be *near-satisfiable*, i.e., have an assignment that satisfies most (say, a fraction $(1 - \varepsilon)$ for small $\varepsilon > 0$) of the clauses. What fraction of clauses can one then satisfy in polynomial time? Establishing a hardness result for this problem was in fact one of the original motivations for the formulation of the Unique Games Conjecture (UGC) by Khot [9]. In fact, assuming the UGC, and using a deep theorem by Bourgain [3], Khot showed the following: For every $t > 1/2$, it is NP-hard to decide, given a 2CNF formula, whether it is $(1 - \varepsilon)$-satisfiable or at most $(1 - O(\varepsilon^t))$-satisfiable. Recently, a proof of the Majority is Stablest conjecture [10] improved this result to $t = 1/2$. In particular, assuming the UGC, for all $\varepsilon > 0$, there is a 2-query PCP for NP languages that has completeness $1 - \varepsilon$ and soundness $1 - O(\sqrt{\varepsilon})$.

Given the above-mentioned parallel between the PCP world and the LTC world, the original motivation for this work was to see if one could *unconditionally* establish that 2-query binary codeword testing is possible with completeness $1 - \varepsilon$ and soundness $1 - O(\sqrt{\varepsilon})$, or weaker still, soundness $1 - \varepsilon h(\varepsilon)$ where $h(\varepsilon) \to \infty$ as $\varepsilon \to 0$. This would imply that the ratio $(1 - s)/(1 - c)$ of the rejection probabilities for strings far away from the code and strings that are codewords can be super-constant. (We remark that it is quite easy to get completeness $1 - \varepsilon$ and soundness $1 - B\varepsilon$ for some absolute constant $B > 1$.[3]) Such a result could then perhaps be viewed, using the "empirically" observed parallel between PCPs and LTCs, as a comforting (but definitely non-rigorous) evidence towards the UGC, or at least towards the super-constant hardness of Min 2CNF deletion (one of the first corollaries of the UGC). Unfortunately, as described below, such a result does not exist.

Zwick [11] gave a polynomial time approximation algorithm for near-satisfiable instances of Max 2SAT. His algorithm finds an assignment that satisfies a fraction $1 - O(\sqrt[3]{\varepsilon})$ of the clauses in a 2SAT formula if the optimal assignment satisfies a fraction $1 - \varepsilon$ of the clauses. Recently, Charikar *et al* [4] gave a better polynomial time algorithm that matches the UGC based hardness result and satisfies a fraction $1 - O(\sqrt{\varepsilon})$ of the clauses given a $(1 - \varepsilon)$-satisfiable instance. This also hints at a potential *negative* result for 2-query codeword testing that says that if completeness is $1 - \varepsilon$ the soundness must be at least $1 - f(\varepsilon)$ for some function f where $f(\varepsilon) \to 0$ as $\varepsilon \to 0$.

Indeed, the result presented here is a (stronger) negative result in this spirit. In fact, we prove that we can take $f(\varepsilon) = B\varepsilon$ for a constant B that depends on the distance of the code. (This negative result is of course tight up to the

---

[3] Indeed, one can start with a 2-query tester T with completeness c and soundness s for say $c = 11/12$ and $s < 11/12$. We can define a new tester T' that simply accepts with probability $1 - \varepsilon$ without making any queries, and with the remaining probability runs T. It is clear that the completeness and soundness of T' are both $1 - \Theta(\varepsilon)$.

constant factor in $O(\varepsilon)$ term.) In particular, our result precludes the existence of a 2-query locally testable code of fixed relative distance with completeness $1 - \varepsilon$ and soundness $1 - \varepsilon^a$ for $a < 1$ and arbitrarily small $\varepsilon > 0$. This shows that one cannot get a result for 2-query codeword testing similar to what is known (assuming the UGC) for 2-query PCPs. Note that our result for LTCs is unconditional, and we appeal to the UGC only for contrasting with the PCP world.

Our proof is an easy adaptation of the proof of impossibility of 2-query codeword testing with perfect completeness from [2]. Our contribution is therefore not technical but rather conceptual, as it points out that under the UGC, at least with the usual notion of soundness, the parallel between PCPs and LTCs is subtle and ceases to hold for 2-queries and near-perfect completeness (which was essentially the only remaining qualitative case that wasn't understood).[4]

We also want to highlight the *"qualitative"* nature of the discrepancy between these worlds: the ratio of $(1 - s)/(1 - c)$ can be super-constant (in the input length) for 2-query PCPs (conditioned on the UGC), whereas it is bounded by a fixed constant for 2-query testing of any code of constant relative distance. Showing just a fixed constant gap between these quantities would not be very interesting. As an example of such a difference, for the case of perfect completeness and three queries, PCPs can achieve soundness $s \rightarrow 1/2$, and this cannot be achieved for LTCs whose relative distance is small.

PCPs of Proximity and Codeword Testing

As remarked earlier, there is no generic way known to translate PCPs into LTCs. However, a notion related to PCPs, called *PCPs of proximity* (PCPP), which was introduced in [1] (and independently in [5] where the notion was called Assignment Testers), can be used in a generic way to construct locally testable codes, as described in [1]. In a nutshell, the queries a verifier of a PCPP makes into the input also count towards its query complexity, while in a PCP the input can be read fully and only queries into the proof count. However, a PCPP verifier is required to do a weaker job; it needs to only check proximity of the input to a string in the language, i.e., it only needs to reject with good probability inputs that are δ-far from every member of the language, for some constant $\delta > 0$ which is called the *proximity parameter* (larger the δ, easier the verifier's job).

The high-level approach to use PCPPs to construct LTCs due to [1] is as follows. One starts with an *arbitrary* binary linear code C_0 of block length n and large distance, say $n/10$. Then define a new code C where the encoding $C(x) = C_0(x)^t \pi(C_0(x))$ where $C_0(x)^t$ denotes the encoding of x by C_0 repeated t times (for some larger t) and $\pi(z)$ is a PCP of proximity for the fact that $z \in C_0$ (which can be checked in polynomial time). Given input $\langle w_1, w_2, \ldots, w_t, \pi \rangle$, the tester

---

[4] With a relaxed notion of "soundness" that is motivated from the PCP application, long codes can be tested using two queries with completeness $1 - \varepsilon$ and "soundness" $1 - O(\sqrt{\varepsilon})$ [3], and this forms the basis for the inapproximability result for near-satisfiable instances of Max 2SAT in [9].

for C operates as follows. With probability $1/2$ it picks two w_k, w_ℓ and an index $i \in \{1, 2, \ldots, n\}$, and checks that the i'th symbol of w_k and w_ℓ are equal (this involves two queries). With probability $1/2$, it picks a random $i \in \{1, 2, \ldots, n\}$ and runs the PCPP verifier for input w_i and proof π. This translation of a PCPP construction into a LTC preserves the query complexity and completeness, and further preserves the quantity $1 - s$ up to a small constant factor. The distance to the code for which the codeword tester rejects with probability at least $1 - s$ is related to the proximity parameter of the PCPP construction.

In light of the above, our result on the limitation of 2-query LTCs also implies that a 2-query PCP of proximity cannot have $c = 1 - \varepsilon$ and $s = 1 - \varepsilon^a$ for $a < 1$. This is the case even if we want PCPPs for languages only in P, since the above-mentioned construction of LTCs from PCPPs only requires PCPPs for the class P. We simply state our result concerning limitations of 2-query PCPs of proximity below, and do not elaborate further on PCPPs. Instead, we point the interested reader to [1] where PCPPs and their connection to locally testable codes are discussed at length.

Theorem 1 (Limitations of PCPPs). *For every $a < 1$ and $\delta < 1/20$, the following holds. For all sufficiently small $\varepsilon > 0$, there exists a polynomial time decidable language L that does not admit a probabilistically checkable proof of proximity (PCPP) for proximity parameter δ with query complexity 2, completeness $(1 - \varepsilon)$, and soundness $(1 - \varepsilon^a)$.*

2 Preliminaries and Definitions

We consider binary (error-correcting) codes $C \subseteq \{0, 1\}^n$ in this work. The *block length* of C is n, and its *size* is its cardinality $|C|$. The elements of C are referred to as *codewords*. For $x, y \in \{0, 1\}^n$, we denote by $\Delta(x, y)$ their Hamming distance, i.e., the number of coordinates i where $x_i \neq y_i$. The *distance* of a code C, denoted $\mathrm{dist}(C)$, is the minimum Hamming distance between two distinct codewords in C, i.e., $\mathrm{dist}(C) = \min_{x \neq y \in C} \Delta(x, y)$. The *relative distance* of a code, denoted $\delta(C)$, is the normalized distance $\mathrm{dist}(C)/n$. The distance of a word w from the code C, denoted $\Delta(w, C)$, is $\min_{x \in C} \Delta(w, x)$.

In coding theory, one is typically interested in a family of codes of increasing block lengths whose relative distance is bounded away from 0. A code of distance $d = \delta n$ enables, in principle, detection of up to $d - 1$ errors, and correction of up to $(d - 1)/2$ errors. The latter task in particular involves determining if there is a codeword within distance $\delta n/2$ from a string w, i.e., whether $\Delta(w, C) < \delta n/2$. In the subject of *codeword testing*, which has witnessed intense activity due to its connections to property testing, PCPs, and coding theory, one is interested in spot checking procedures that attempt to ascertain this fact using very few queries. This brings us to the following definitions concerning local testing of binary codes.

Definition 1 (Codeword tester). *A $[q, \rho, c, s]$-codeword tester for a code $C \subseteq \{0, 1\}^n$ is a randomized oracle machine T that has oracle access to $w \in \{0, 1\}^n$*

(viewed as a function $w : \{1, 2, \ldots, n\} \rightarrow \{0, 1\}$*) and satisfies the following three conditions:*

- [QUERY COMPLEXITY]: *T makes at most q queries to w. (The queries may be made adaptively, that is the location of a query can depend on both T's randomness and the answers of the oracle to its previous queries.)*
- [COMPLETENESS]: *For any $w \in C$, given oracle access to w, T accepts with probability at least c.[5]*
- [SOUNDNESS]: *For any w for which $\Delta(w, C) > \rho\delta n$, given oracle access to w, T accepts with probability at most s.*

When $c = 1$, the tester is said to have perfect completeness. *A tester is said to be* non-adaptive *if its queries only depend on its random coin tosses and not on previous answers of the oracle.*

Definition 2 (Locally testable code). *A code C is said to be $[q, \rho, c, s]$-locally testable if there exists a $[q, \rho, c, s]$-codeword tester for C.*

3 2-Query Codeword Testing with Near-Perfect Completeness

3.1 Statement of Result

We now state our main result, which in particular implies that 2-query codeword testers for binary codes with completeness very close to 1, say $1 - \varepsilon$, must also have soundness close to 1, specifically soundness at least $1 - B\varepsilon$ for some constant B that depends on the relative distance of the code.

Theorem 2 (Main). *Fix $\rho < 1/2$ and $\delta > 0$, and let $M(\rho, \delta) = \left(1 + 2^{1/(\delta(1-2\rho))}\right)$. Let C be a binary code of relative distance δ with $|C| \geq M(\rho, \delta)$ and which is $[2, \rho, c, s]$-locally testable. Then $(1 - s) \leq 2(1 - c)M(\rho, \delta)$.*

Note that the case when $c = 1$, the above says that $s = 1$ as well, as was shown in [2]. The binary alphabet hypothesis is essential, as there are ternary codes that are $[2, 1/3, 1, s]$-locally testable for some $s < 1$.

The above lets us conclude the following, which can viewed as the main message from this work.

Corollary 1. *Fix an arbitrary choice of $\delta > 0$, $\rho < 1/2$, and $a < 1$. Then, for small enough ε, there is no non-trivial binary code with relative distance δ that is $[2, \rho, 1 - \varepsilon, 1 - \varepsilon^a]$-locally testable. Here by a non-trivial code we mean any code that has at least $M(\rho, \delta)$ codewords.*

---

[5] One can imagine a stronger condition requiring that T must accept with probability at least c all w's for which $\Delta(w, C)$ is small (as opposed to being 0). This notion was recently studied under the label "Tolerant testing" [8]. Since we prove a negative result in this work, our result is only stronger when working with the weaker condition where T needs to only accept exact codewords with probability at least c.

As mentioned earlier, the above is to be contrasted with the $(1-\varepsilon, 1-O(\sqrt{\varepsilon})$-hardness for approximating 2SAT, or equivalently a 2-query PCP that has completeness $1 - \varepsilon$ and soundness $1 - O(\sqrt{\varepsilon})$, which is known to exist for all $\varepsilon > 0$ assuming the Unique Games Conjecture (UGC) [9,10].

3.2 Proof

We now prove Theorem 2. The proof is a simple generalization of the proof for the perfect completeness case from [2], and proceeds by examining the digraph induced by the set of possible pairs of queries made by the codeword tester.

Let C be a code with $|C| \geq M(\rho,\delta)$, and let T be a (possibly adaptive) 2-query tester for C with completeness c and soundness s. We first modify T to a non-adaptive tester T' with completeness $c' \geq c$ and soundness $s' \leq (1+s)/2$, where each test made by T' is a 2SAT clause (the clause could just depend on one literal, or could even be a constant 0 or 1).

Every test made by T is a decision tree of depth at most 2, i.e., T reads one bit, say x, first, and depending on its value does its next move (for example, it could read y if $x = 0$, and base the decision on (x, y), and read z if $x = 1$ and base its decision on (x, z)). Note that the second query must lead to one positive and one negative outcome, otherwise it will be unnecessary. It follows that every test made by T can be described by a 2CNF with at most two clauses.[6] We can get a non-adaptive verifier T' which operates just like T, except that for each of T's tests ϕ, it picks at random one of the at most two 2SAT clauses in ϕ and just checks this clause. Since T' is performing only more lax tests compared to T, the completeness of T' is at least that of T. Moreover, the probability that T' rejects is at least $1/2$ times the probability that T rejects. Hence $1 - s' \geq (1-s)/2$, as desired.

Let n denote the block length of C and let x_i denote the bit in position i of the string to be tested. We will denote by ℓ_i a literal that stands for either x_i or $\bar{x}_i$. Define the following digraph H on vertices $\{x_i, \bar{x}_i \mid 1 \leq i \leq n\} \cup \{0,1\}$ corresponding to the operation of the tester T'. For each test of form $(\ell_i \vee \ell_j)$, we add directed edges $(\bar{\ell}_i \to \ell_j)$ and $(\bar{\ell}_j \to \ell_i)$. For tests of the form ℓ_i, we add edges $(1 \to \ell_i)$ and $(\bar{\ell}_i \to 0)$. Each of these edges has a weight equal to half the probability that T' makes the associated tests. If T' makes the trivial test that always accepts, we add a self-loop on 1 with weight equal to the probability that T' makes this test.

For each $x \in \{0,1\}^n$, denote by H_x the subgraph of H consisting of precisely those edges that are satisfied by x, where an edge $(\ell_i \to \ell_j)$ is said to be satisfied iff the implication $(\ell_i \Rightarrow \ell_j)$ is satisfied; the loop on 1, if it exists, is satisfied by every x. By definition of H, the probability that T' accepts x is precisely the total weight of edges in H_x.

Let $M = M(\rho,\delta)$, and let $c_1, c_2, \ldots, c_M$ be distinct codewords in C (recall that we are assuming $|C| \geq M$). Define the digraph $G = \bigcap_{i=1}^{M} H_{c_i}$, i.e., an edge of H is present in G if and only if it is satisfied by all the codewords $c_1, c_2, \ldots, c_M$. By the completeness of T', the total weight of edges in G is at least $1 - M(1 - c')$.

---

[6] Thanks to an anonymous reviewer for this observation.

Now we can use the exact same argument for the perfect completeness case from [2] and construct a word v with $\Delta(v, C) > \rho \delta n$ that passes all tests in G. In other words, T' accepts v with probability at least $1 - M(1 - c')$. But since $\Delta(v, C) > \rho \delta n$, by the soundness condition T' accepts v with probability at most s'. Therefore, $s' \geq 1 - M(1 - c')$. Recalling $1 - s' \geq (1 - s)/2$ and $c' \geq c$, we get $1 - s \leq 2M(1 - c)$, as desired.

We stress that the remainder of the proof proceeds exactly as in [2], and we include it only for completeness.

Let us consider the decomposition of G into strongly connected components (SCCs). (Recall that a SCC is a maximal set of vertices W such that any ordered pair $w, w' \in W$, there is a directed path from w to w'.) Call a SCC of G to be *large* if its cardinality is at least $\delta \alpha n$, where $\alpha \overset{\text{def}}{=} (1 - 2\rho)$. Let L be the set of large SCCs. Clearly $|L| \leq 1/(\delta \alpha)$. Also, for every SCC C in G, and each $j = 1, 2, \ldots, M$, the value of c_j on all variables in C must be the same (since all edges in G are satisfied by every c_j). Let A be the set of variables that belong to *large* SCCs. Since $M > 2^{1/(\delta \alpha)}$, there must exist two codewords, say c_1, c_2, which agree on all the variables in A. Let γ denote the common assignment of c_1, c_2 to the variables in A. Note that by construction, γ satisfies all the edges in G. Let L be the set of literals associated with A, together with the nodes corresponding to constants $0, 1$. Let us say a variable v is forced by the assignment γ to A if one of the following conditions hold:

(i) $v \in A$.
(ii) There exists $\ell' \in L$ with $\gamma(\ell') = 1$ such that there is a directed path in G from ℓ' to either v or $\bar{v}$ (or equivalently, there exist $\ell'' \in L$ with $\gamma(\ell'') = 0$ and a directed path in G from either v or $\bar{v}$ to ℓ'').
(iii) There exists a directed path from v to $\bar{v}$ (in this case v is forced to 0) or from $\bar{v}$ to v (in this case v is forced to 1).

The rationale behind this notion is that since all edges in G are satisfied by both c_1, c_2, and c_1, c_2 agree on their values for literals in L, they must also agree on every variable forced by A. Let σ be the partial assignment that is the closure of γ obtained by fixing all forced variables (and their negations) to the appropriate forced values. Let U denote the set of variables that are *not* forced. Since c_1, c_2 satisfy all edges in G, they agree outside U, and so their Hamming distance satisfies $\Delta(c_1, c_2) \leq |U|$. Since c_1, c_2 are distinct codewords, $\Delta(c_1, c_2) \geq \delta n$. It follows that $|U| \geq \delta n > 2\rho \delta n$.

Let H be the subgraph of G induced by the variables in U and their negations. We can define a transitive relation $\prec$ on SCCs of H where $S \prec S'$ iff there exists $w \in S$ and $w' \in S'$ with a directed path from w to w'. We note that for each $u \in U$, there will be two SCCs in G, one containing u and a disjoint one containing $\bar{u}$ (this follows since G, and therefore H, has a satisfying assignment and so no variable and its negation can be in the same SCC). Moreover these two SCCs are incomparable according to $\prec$, since otherwise there must be a directed path from u to $\bar{u}$ (or from $\bar{u}$ to u), and in such a case u will be a forced variable. Therefore, we can pick a topological ordering $S_1, S_2, \ldots, S_k$ of strongly connected components where each S_i contains either a variable in U or its negation, and for each $u \in U$, the S_i's will comprise of exactly one of the two SCCs that contain u and $\bar{u}$ respectively.

For $j = 0, 1, \ldots, k$, let $v^{(j)}$ be the assignment extending σ defined by:

$$v^{(j)}(S_i) = \begin{cases} 0 & \text{if } i \leq j \\ 1 & \text{if } i > j \end{cases}$$

Since each assignment $v^{(j)}$ respects the topological ordering of the S_i's, each $v^{(j)}$ satisfies all edges in G. Since the weight of edges in G is at least $1 - M(1 - c')$ and the soundness of T' is $s' < 1 - M(1 - c')$, each $v^{(j)}$ is within distance $\rho\delta n$ from some codeword of C, say $w^{(j)}$. Now,

$$\begin{aligned} 2\rho\delta n < |U| &= \Delta(v^{(0)}, v^{(k)}) \\ &\leq \Delta(v^{(0)}, w^{(0)}) + \Delta(w^{(0)}, w^{(k)}) + \Delta(w^{(k)}, v^{(k)}) \\ &\leq 2\rho\delta n + \Delta(w^{(0)}, w^{(k)}) \end{aligned}$$

which implies $w^{(0)} \neq w^{(k)}$, and hence $\Delta(w^{(k)}, w^{(0)}) \geq \delta n$. Since $\Delta(w^{(k)}, v^{(k)}) \leq \rho\delta n$, we have

$$\Delta(v^{(k)}, w^{(0)}) \geq \Delta(w^{(k)}, w^{(0)}) - \Delta(w^{(k)}, v^{(k)}) \geq (1 - \rho)\delta n . \tag{1}$$

Also, we have $\Delta(v^{(j)}, v^{(j+1)}) < (1 - 2\rho)\delta n$ for each $j = 0, 1, \ldots, k - 1$ since $|S_i| < \alpha\delta n$ for each $i = 1, 2, \ldots, k$ (recall our threshold $\alpha\delta n$ for considering a SCC to be *large* where $\alpha = 1 - 2\rho$). Together with $\Delta(v^{(0)}, w^{(0)}) \leq \rho\delta n$ and (1), we conclude that there must exist j, $1 \leq j < k$, such that

$$\rho\delta n < \Delta(v^{(j)}, w^{(0)}) < (1 - \rho)\delta n .$$

This implies that $\Delta(v^{(j)}, C) > \rho\delta n$, and yet $v^{(j)}$ satisfies all edges of G. The existence of such a word $v^{(j)}$ is what we needed to conclude that $s' \geq 1 - M(1 - c')$, and this completes the proof.

References

1. E. Ben-Sasson, O. Goldreich, P. Harsha, M. Sudan, and S. Vadhan. Robust PCPs of proximity, shorter PCPs and application to coding. In *Proceedings of the 36th Annual ACM Symposium on Theory of Computing (STOC)*, pages 1–10, 2004.
2. E. Ben-Sasson, O. Goldreich, and M. Sudan. Bounds on 2-query codeword testing. In *Proceedings of the 7th International Workshop on Randomization and Approximation Techniques in Computer Science (RANDOM)*, pages 216–227, 2003.
3. J. Bourgain. On the distribution of the Fourier spectrum of boolean functions. *Israel Journal of Mathematics*, 131:269–276, 2002.
4. M. Charikar, K. Makarychev, and Y. Makarychev. Note on Max2SAT. Technical Report TR06-064, Electronic Colloquium on Computational Complexity, 2006.
5. I. Dinur and O. Reingold. Assignment Testers: Towards a combinatorial proof of the PCP-Theorem. In *Proceedings of 45th Annual Symposium on Foundations of Computer Science (FOCS)*, pages 155–164, 2004.

6. O. Goldreich. Short locally testable codes and proofs (Survey). *ECCC Technical Report TR05-014*, 2005.
7. O. Goldreich and M. Sudan. Locally testable codes and PCPs of almost linear length. In *Proceedings of 43rd Symposium on Foundations of Computer Science (FOCS)*, pages 13–22, 2002.
8. V. Guruswami and A. Rudra. Tolerant locally testable codes. In *Proceedings of the 9th International Workshop on Randomization and Approximation Techniques in Computer Science (RANDOM)*, pages 306–317, 2005.
9. S. Khot. On the power of unique 2-prover 1-round games. In *Proceedings of the 34th ACM Symposium on Theory of Computing (STOC)*, pages 767–775, May 2002.
10. E. Mossel, R. O'Donnell, and K. Oleszkiewicz. Noise stability of functions with low influences: invariance and optimality. In *Proceedings of the 46th IEEE Symposium on Foundations of Computer Science (FOCS)*, pages 21–30, 2005.
11. U. Zwick. Finding almost satisfying assignments. In *Proceedings of the 30th ACM Symposium on Theory of Computing (STOC)*, pages 551–560, May 1998.

Poketree: A Dynamically Competitive Data Structure with Good Worst-Case Performance

Jussi Kujala and Tapio Elomaa

Institute of Software Systems
Tampere University of Technology
P. O. Box 553, FI-33101 Tampere, Finland
jussi.kujala@tut.fi, elomaa@cs.tut.fi

Abstract. We introduce a new $O(\lg \lg n)$-competitive binary search tree data structure called poketree that has the advantage of attaining, under worst-case analysis, $O(\lg n)$ cost per operation, including updates. Previous $O(\lg \lg n)$-competitive binary search tree data structures have not achieved $O(\lg n)$ worst-case cost per operation. A standard data structure such as red-black tree or deterministic skip list can be augmented with the dynamic links of a poketree to make it $O(\lg \lg n)$-competitive. Our approach also uses less memory per node than previous competitive data structures supporting updates.

1 Introduction

Among the most widely used data structures are different binary search trees (BSTs). They support a variety of operations, usually at least searching for a key as well as updating the set of stored items through insertions and deletions. A successful search for a key stored into the BST is called an *access*. BST data structures are mostly studied by analyzing the cost of serving an arbitrary sequence of operation requests. Moreover, the realistic *online* BST algorithms, which cannot see future requests, are often contrasted in competitive analysis against *offline* BST algorithms that have the unrealistic advantage of knowing future requests. Examples of BST algorithms include 2-3 trees, red-black trees [1], B-trees [2], splay trees [3], and an alternative to using the tree structure, skip lists [4, 5].

In an attempt to make some progress in resolving the *dynamic optimality conjecture* of Sleator and Tarjan [3], Demaine et al. [6] recently introduced Tango, an online BST data structure effectively designed to reverse engineer Wilber's [7] first lower bound on the cost of optimal offline BST. Broadly taken the dynamic optimality conjecture proposes that some (self-adjusting) BST data structure working online, possibly splay tree, would be competitive up to a constant factor with the best offline algorithm for any access sequence. Tango does not quite attain constant competitiveness, but Demaine et al. showed an asymptotically small competitive ratio of $O(\lg \lg n)$ for Tango in a static universe of n keys.

T. Asano (Ed.): ISAAC 2006, LNCS 4288, pp. 277–288, 2006.
© Springer-Verlag Berlin Heidelberg 2006

Table 1. Known asymptotic upper bounds on the performance of competitive BST algorithms. Memory is given in bits, where letter w is a shorthand for "words".

	Tango	MST	Poketree	Poketree(RB)	Poketree(skip)
Search, worst-case	$O(\lg \lg n \lg n)$	$O(\lg^2 n)$	$O(\lg n)$	$O(\lg n)$	$O(\lg n)$
Search, amortized	$O(\lg \lg n \lg n)$	$O(\lg n)$	$O(\lg n)$	$O(\lg n)$	$O(\lg n)$
insert/delete	N/A	$O(\lg^2 n)$	N/A	$O(\lg n)$	$O(\lg^2 n)$
memory per node	4w+2lg w+2	7w+lg w+1	6w+1	6w+2	4w+lg lg w

Prior to this result the best known competitive factor for an online BST data structure was the trivial $O(\lg n)$ achieved by any balanced BST.[1]

Tango only stores a static set of keys, it does not support update operations on stored data items. It has $O(\lg n \lg \lg n)$ worst-case access cost and cannot guarantee better performance under amortized analysis for all access sequences.[2] Wang, Derryberry, and Sleator [8] put forward a somewhat more practical BST data structure, in the sense that it supports update operations as well, *multisplay trees* (MST) — a hybrid of Tango and splay trees — which also attains the double logarithmic competitive ratio $O(\lg \lg n)$. The worst-case complexity of a MST is $O(\lg^2 n)$, but it achieves $O(\lg n)$ amortized access cost. The costs for update operations in a MST are similar. MSTs inherit some interesting properties from splay trees; it is, e.g., shown that the sequential access lemma [9] holds for MSTs.

In this paper we introduce poketree, which is a generic scheme for attaining $O(\lg \lg n)$-competitive BST data structures with different static data structures underlying it. As static data structures we use red-black trees [1] and deterministic skip lists [5]. By using different static data structures we can balance between the amount of augmented information in the nodes and efficiency of operations. In contrast to Tango and MST, poketree has $O(\lg n)$ cost per operation under worst-case analysis. The $O(\lg^2 n)$ update cost with skip lists can be, in practice, lowered to $O(\lg n)$ cost per update [5]. See Table 1 for a summary of the characteristic costs of Tango, MST, and poketree.

The contribution of this paper is a dynamically $O(\lg \lg n)$-optimal data structure that has the best combination of space and cost requirements with updates supported. Poketree is not a strict binary search tree, because nodes in it can have up to one additional pointer, but it supports the same set of operations. As a side result we give a lower bound to the cost of BST algorithms, proven in a similar manner to the bound of Demaine et al. [6], which has a small increase in the additive term from the previous best of $-n - 2r$ [8] to $-n/2 - r/2$.

The remainder of this paper is organized as follows. In Section 2 we explain the lower bound on the cost of offline BST algorithms that we use. It is slightly tighter than the one in [6] and formulated to support other than strict 2-trees. In Section 3 we demonstrate the idea behind poketree in case where the static

---

[1] By $\lg n$, for a positive integer n, we denote $\lceil \log_2 n \rceil$ and define $\lg 0 = \lg 1 = 1$.

[2] Subsequent manuscripts available at the Internet mention that Tango can be modified to support $O(\lg n)$ worst-case access.

structure is a perfectly balanced BST with no update operations supported. In Sections 4 and 5 we generalize to other static structures, now supporting update operations. The concluding remarks of this paper are presented in Section 6.

2   Cost Model and the Interleave Bound

Because we want to compare the costs of online and offline BST algorithms, we need a formal model of cost. Earlier work [3, 7, 6, 8] has mostly used a model which charges one unit of cost for each node accessed and for each rotation. However, a simpler cost model was used by Demaine et al. [6] in proving a lower bound. They only counted the number of nodes *touched* in serving an access. A node is touched during an access if the data structure reads or writes information on the node. These two models, and several others, differ at most by a constant factor because a BST can be transformed to any other one containing the same set of keys using a number of rotations that is twice the number of nodes in the tree [10]. We adopt the latter model and take care that no computation is too costly when compared to the number of nodes accessed.

Wilber [7] gave two lower bounds on the cost of dynamic BST algorithms; the first of these is also known as the *interleave bound*. It has been used to prove the $O(\lg \lg n)$-competitiveness of the recent online BST algorithms [6, 8]. We will also use it to show the $O(\lg \lg n)$-competitiveness of poketree. Let us describe the interleave bound briefly. Let P be a static BST on the items that are accessed. The BST P is called a *reference tree*. It may be a proper BST, 2-3 tree, or any 2-...-* tree. Using BSTs results in the tightest bound and, hence, they have been used previously. However, we do not necessarily need as tight results as possible, therefore, we give a slightly different version of the interleave bound.

We are given an access sequence $\sigma = \sigma_1, \ldots, \sigma_m$ that is served using P. For each item i in P define the *preferred child* of i as the child into whose subtree the most recent access to the subtree of i was directed. Let $IB(\sigma, P, i)$ be the number of switches of preferred child of i in serving σ. The interleave bound $IB(\sigma, P)$ is the sum of these over all the nodes of P: $\sum_{i \in P} IB(\sigma, P, i)$. Let r be the number of rotations in P while serving σ.

Theorem 1. *Any dynamic binary search tree algorithm serving an access sequence σ has a cost of at least $IB(\sigma, P)/2 + m - n/2 - r/2$.*

This bound is slightly tighter than previously known best lower bound $IB(\sigma, P)/2 + m - n - 2r$ by Wang, Derryberry, and Sleator [8]. The proof for the new bound is given is given in Appendix B.

3   Poketree — A Dynamic Data Structure

Let P_t be the state of the reference tree P at time t. In addition to P, P_t contains information on preferred children. *Preferred paths* follow preferred child pointers in P_t. Both Tango and MST keep each preferred path of a reference tree P_t on a separate tree; these trees make up a tree of trees. When the interleave bound on P increases by k, then exactly k subtrees are touched. Moreover, the algorithms provide an efficient way to update the structure to correspond to the new reference tree P_{t+1}. Thus the access cost is at most $k \lg(\#$ of items on a path$) \leq k \lg \lg n$ when the data structure for subtrees is chosen suitably.

We, rather, augment a standard balanced search structure with *dynamic links* to support a kind of binary search on preferred paths. The balanced search structure corresponds to the structure of the reference tree, which is used to provide a lower bound on the cost. In our approach we take advantage of the static links in searching for an item, whereas Tango and MST maintain the reference tree information but do not really use it in searching. This enables us to implement update operations using less space than MST.

We first describe how our method, called poketree, works on a perfectly balanced tree. In the following sections we relax these assumptions and generalize it to handle trees with less balance and to support update operations insert and delete. Interpreted most strictly, poketree is not really a BST, because items are not necessarily searched through a search tree, but using dynamic links. However, this is rather a philosophical than a practical point, since the same operations are supported in any case. The nodes of a poketree are augmented with a dynamic link to make it dynamically competitive. The idea is to follow a dynamic link whenever possible, and descend via a static link otherwise. Dynamic links allow to efficiently find a desired location on a preferred path and they can be efficiently updated to match the new reference tree P_{t+1}. We consider later the requirement of inserting (removing) an item to the head of a preferred path brought along by supporting insertion (deletion).

Let us fix a preferred path $a = a_1, \ldots, a_l$ and note that a_l is always a leaf in the static tree but a_1 is not necessarily its root. If the dynamic links would implement a binary search on the path a, then it would be possible to travel from a_1 to any a_i in $\lg l \leq \lg \lg n$ time. However, this idea does not work as such, because updating dynamic links would be difficult and, even worse, since the preferred path a is a path in a BST, it is not necessarily ordered, making it impossible to carry out binary search on it.

For now we just augment each node to contain the smallest and the largest item in the subtree rooted at the node (we later lift this requirement in Section 5). We still have to set the dynamic links to implement a kind of binary search including quick updates to the structure of the preferred paths in P. There are two types of nodes, of type SDD and S. Define a *static successor* of a node N to be the child in which the last search through N went. The dynamic link of a node of type SDD leads to the same node as following two dynamic links starting from its static successor. A node of type S, on the other hand, has its dynamic link pointing to its static successor. Note that the dynamic links always point lower in the tree, except in the special case of a leaf node.

The key trick is to choose the type of a node. The rule for this is as follows. A node is of type SDD if the length of the dynamic link of its successor equals the length of the dynamic link of the dynamic link of the successor. Otherwise, the

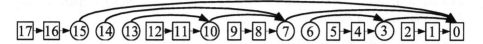

Fig. 1. An example of dynamic links. Static links are omitted, they point always to the next item in the list. The round nodes are of type *SDD* and the square ones of type *S*. The node with label 0 is a leaf.

node is of type *S* or a leaf node, which does not have a dynamic link. Observe that to satisfy this requirement, links have to be set up from a leaf to the root.

An example of the resulting structure is given in Figure 1. Intuitively, dynamic links let us recursively split a path into two parts and have one dynamic link arc over the first part and another one over the second part. The two *D*s in *SDD* stand for these arcs. Now note that if these links are set from the last item to the first one, the distance that a dynamic link traverses is a function of its distance to the end of the preferred path. Every possible path starting from a particular node is of the same length because we have assumed that a poketree is perfectly balanced. These facts together make it possible to update dynamic links quickly when a preferred path in *P* changes direction. We describe next the search operation and how to update the structure corresponding to P_t to correspond to P_{t+1}.

The search attempts at each node to use the dynamic link to potentially skip a number of static links. However, if following the dynamic link leads to a node that does not contain the searched item in the range of items in its subtree, we say that the dynamic link *fails*. If the dynamic link fails, then the search backtracks to the node where the dynamic link was used and resorts to using the appropriate static pointer. Using a dynamic link has a cost, because an additional node has to be visited. It is possible to augment the parent of a dynamic link with the range information of the dynamic link, but this costs more space.

After the searched item has been found, we may have to update those nodes in which a static link was used, because the preferred path may have changed. Assume that we know whether a node is of type *S* or *SDD*, for example by reading a bit that contains this information. After performing a search, in going from the bottom to the top, the dynamic link in each node can be (re)set because the the links in the subtree rooted at the preferred child have already been set up. Let us start by stating the most obvious result.

Theorem 2. *The worst-case cost for any search in a poketree is* $O(\lg n)$.

Proof. Clearly, in the worst case all dynamic links fail because otherwise some static link can be skipped with no cost. Thus, the worst-case cost is bounded to be a constant factor away from the number of static links traversed. This bound is of the order $O(\lg n)$ because the static structure is balanced and a constant amount of work is done during each static link.

The following theorem states that dynamic links are fast enough.

Theorem 3. *No more than $O(\lg \lg n)$ time is spent on a preferred path on the way to the accessed element.*

The proof is given in Appendix A.

Each change in the preferred paths, and subsequently in dynamic links, corresponds to one switch in dynamic pointers. Poketree does at most a constant amount of work for each switch in the dynamic links. Thus, by Theorems 3 and 1, it follows:

Theorem 4. *Poketree is $O(\lg \lg n)$-competitive among the class of all binary search tree algorithms.*

4   Insertions and Deletions: Poketree(RB)

In previous section we assumed that each root-to-leaf path has the same length. In reality, a data structure must support insertions and deletions and, thus, we cannot rely on idealized perfect balance to make things easy for us. Nevertheless, it is possible to retain the use of dynamic links while supporting these important operations. Some data structures are in a sense always perfectly balanced. For example, in a red-black tree [1, 11] every root-to-leaf path has the same number of black nodes and in balanced 2-3 tree implementations the nodes (each containing one or two items) are always in perfect balance.

In order to argue competitiveness, we need a cost model that can handle updates. During updates the structure of the reference tree changes to correspond to the static structure of the poketree. An insertion causes a search and insertion to the location in the reference tree where the item should be, and a deletion causes a search to both the deleted item and its successor in the key space, after which the item is deleted as usual. The actual cost charged from an offline BST algorithm is the number of preferred child pointers that switch, more precisely a constant factor of that number. Wang et al. [8] implement updates and use a similar model, which however is not as tight, because in a deletion they search for both predecessor and successor as well as rotate the item in the offline BST to a leaf.

We now describe how to support insertion and deletion by using a red-black tree as the static structure. A red-black tree can be viewed as a 2-3-4 tree, where a node of the 2-3-4 tree corresponds to a black node and its red children. We choose the reference tree P to be the 2-3-4 tree of the red-black tree. In the red-black tree the dynamic links point only from black nodes to black nodes. First, note that in the 2-3-4 tree view of a red-black tree, the tree increases height only from the top, and when splitting nodes the distance from nodes to leaves stays the same (in the 2-3-4 tree). Second, during updates nodes in the red-black tree may switch nodes in the 2-3-4 tree, because there might be repaints. Third, it is known that during an update to a red-black tree the amortized number of repaints and the worst-case number of rotations are both constants [12]. The resulting algorithm, poketree(RB), has the same *asymptotic* competitiveness properties as a perfectly balanced poketree, but now with updates supported. Note that the competitive ratio is not exactly the same, because red nodes are without dynamic links and, thus, there is a constant factor overhead in the competitive ratio. On the other hand, less unnecessary work is done if the access sequence does not conform to a structure that allows a BST algorithm to serve it in less than $\Theta(m \lg n)$ cost.

To support update operations, we must be able to decide the type of a new root — S or SDD — assuming that the types in the subtree rooted at the root are set up correctly. If we are given such a node, then it is possible to count the number of consecutive SDD nodes its static successor and dynamic link of the static successor have. If these are equal, then the node is of type SDD, otherwise it is of type S. This holds, because the rule for choosing the type of a node depends on the fact that the length of the dynamic link in its successor and in the dynamic link of the successor are the same, which is true if the same number of SDD nodes have been chosen in a row in those two locations. Leaf nodes (tails in a preferred path) make an exception, because they do not have a dynamic link.

In poketree each node carries information about the interval of keys in its subtree. In poketree(RB) these bounds are maintained in the form of strict lower and upper bounds, i.e., a lower bound cannot be the smallest key in the tree, but could be the predecessor to the smallest key. The reason is that these bounds can be efficiently handled during updates.

To insert a key to a poketree(RB) we need to search for its predecessor and successor (note that there is no need to actually know their values), actually insert it and set the lower and upper bound in the corresponding node, update the dynamic links, and finally fix any violations to red-black invariants while taking care in each repaint of the nodes that the poketree invariant holds in the tree at the level of the current operation. In general a repaint from black to red deletes the dynamic link and a repaint from red to black sets the dynamic link according to the type of the node which can either be obtained from some other node or as described above in the case of a new root node. For completeness we describe what to do during the fixup of the red-black tree as it might not be obvious. These cases correspond to ones in [11, pp. 284–286]. Unfortunately, the page limit does not allow for a more complete presentation.

Case 1: Swap memory locations of the A and C nodes and set static pointers and dynamic pointer in A. This ensures that dynamic links upper in the tree point to a correct node. Then set the type of D to be the type of C and remove dynamic link in C and set the dynamic link in D. This case represents a split of a node in the 2-3-4 tree view.

Case 2: Do nothing.

Case 3: Swap memory locations of B and C and set the static pointers. Thus B obtains the dynamic link that was in C and dynamic links upper in the tree point to a correct node.

Finally, if the root was repainted from red to black, obtain a new type using the procedure described bove and set the type info and dynamic link accordingly.

Note that we may need to update information about lower and upper bounds on the nodes. The total time of insertion is $O(\lg n)$, because the type-procedure needs to be called at most once at the root.

A deletion can be implemented using the same ideas as the insertion, but the details are slightly more complicated. In a deletion the following sequence of operations is executed: find the item to be deleted; find its successor by using dynamic links, bound information, and comparing the lower bound to the deleted item; set dynamic links; delete the item as usual; finally, call the RB-fixup to fix possible invariant violations. In fixup there are several cases to consider, but the general idea is that in the tree there might be an extra black with an associated memory address (because a node further up in the tree may point to it through a dynamic link), which floats upper in the tree, and the address might change, until it is assigned in another location. Things to consider during RB-fixup:

Case 1: Swap memory locations of B and D items and set the static pointers.

Case 2: (There is no dynamic link pointing to D). Delete dynamic information on D. Swap memory location and type of A with the extra black. This represents a merge of two nodes in the 2-3-4 tree.

Case 3: Swap memory locations of C and D items and set the static pointers.

Case 4: Set static pointers, delete dynamic information on D, set type of E to type of D and set the dynamic pointer, swap memory location and type of A with extra black, set memory location and type of B to those of the extra black, set static pointers and update the dynamic pointer on B and A accordingly.

If there is an extra black in the root, just delete it and set the type to be the type of a black child node before the update.

Note that the lower bound must be set on the nodes along the dynamic links on the preferred path from the successor node to the location where the successor previously was. This ensures that the dynamic links pointing from above of the successor to its subtree do not make a mistake if the successor is later accessed.

All operations done during updates have a cost of $O(\lg \lg n)$ per switch of preferred child pointer in the reference tree. We conclude that poketree(RB) is $O(\lg \lg n)$-competitive, even when updates to the tree are taken into account.

5   Reducing Memory Consumption: Poketree(Skip)

So far we have augmented each node in a poketree with a lower bound and an upper bound on key values in its subtree. This constitutes a problem, since each node consumes two words of precious memory. It is no surprise that it is possible to fare better, as a part of the information about lower and upper bounds seems to be redundant between nodes. We suggest a poketree based on a variant of a deterministic skip list by Munro, Papadakis, and Sedgewick [5].

In a skip list items form a list. A node of this list is an array that contains a key, a pointer to the next item in the list, and a varying number of links that point progressively further in the list. More precisely, each additional link on an item points about twice as far as the previous link. The number of additional links on an item is referred to as its *height*. Approximately $1/2^h$th part of the nodes have h additional links and links of same height are nearly uniformly distributed in the list. Thus, the search cost is logarithmic and the total space consumed is upper bound by $n + n + n \sum_{i=1}^{\infty} 1/2^i = 3n$. What makes this structure desirable for us is that if we search for an item and can see only a node of the list, then it is easy to check whether the searched item is between the current node and a node pointed by a link of some particular height. Hence, we can lower the memory overhead of storing bounds for keys in subtrees, but have to fetch one additional node to see its key.

A deterministic skip list corresponds to a perfectly balanced 2-3 tree [5]. Using this correspondence, it is possible to relate the performance of a skip list augmented with dynamic links to BST algorithms. More specifically, if there is a dynamic link for each additional static link, then it is possible to associate each dynamic link to a node in the 2-3 tree view. Items between an item and a particular static link on it correspond to a subtree rooted to a node in the 2-3 tree. The dynamic link associated with this static link corresponds to the dynamic link of that node in the 2-3 tree. We do not go into the details, because of lack of space.

Insertion and deletion can be implemented similarly as in a poketree(RB). Unfortunately, an insertion takes $O(\lg^2 n)$-time in a deterministic skip list, so we have a trade-off here. However, Munro et al. [5] argue that in practice the update operation can be implemented in $O(\lg n)$-time if memory for nodes is allocated in powers of two.

6   Conclusions

We have presented poketree algorithm, which is $O(\lg \lg n)$-competitive against the best dynamic offline BST algorithm and that is founded on same ideas as previous such algorithms, like Tango [6]. Our implementation supports update operations, like MST [8] does, and has better worst case performance.

Acknowledgments

This work was supported by Academy of Finland project "INTENTS: Intelligent Online Data Structures". Moreover, the work of J. Kujala is financially supported by Tampere Graduate School in Information Science and Engineering (TISE).

References

1. Bayer, R.: Symmetric binary B-trees: Data structure and maintenance algorithms. Acta Informatica **1** (1972) 290–306
2. Bayer, R., McCreight, E.M.: Organization and maintenance of large ordered indices. Acta Informatica **1** (1972) 173–189

3. Sleator, D.D., Tarjan, R.E.: Self-adjusting binary search trees. Journal of the ACM **32**(3) (1985) 652–686
4. Pugh, W.: Skip lists: A probabilistic alternative to balanced trees. Communications of the ACM **33**(6) (1990) 668–676
5. Munro, I., Papadakis, T., Sedgewick, R.: Deterministic skip lists. In: Proceedings of the 3rd Annual ACM-SIAM Symposium on Discrete Algorithms, SIAM (1992) 367–375
6. Demaine, E.D., Harmon, D., Iacono, J., Pătraşcu, M.: Dynamic optimality – almost. In: Proceedings of the 45th Annual IEEE Symposium on Foundations of Computer Science, IEEE Computer Society Press (2004) 484–490
7. Wilber, R.: Lower bounds for accessing binary search trees with rotations. SIAM Journal on Computing **18**(1) (1989) 56–67
8. Wang, C.C., Derryberry, J., Sleator, D.D.: $O(\log \log n)$-competitive dynamic binary search trees. In: Proceedings of the 17th Annual ACM-SIAM Symposium on Discrete Algorithms, ACM Press (2006) 374–383
9. Tarjan, R.E.: Sequential access in splay trees takes linear time. Combinatorica **5**(4) (1985) 367–378
10. Culik II, K., Wood, D.: A note on some tree similarity measures. Information Processing Letters **15**(1) (1982) 39–42
11. Cormen, T.H., Leiserson, C.E., Rivest, R.L., Stein, C.: Introduction to Algorithms. Second edn. McGraw-Hill (2001)
12. Tarjan, R.E.: Data Structures and Network Algorithms. SIAM (1983)

A   Proof of Theorem 3

The definition of dynamic links gives the following recursive rule for the length $L(m)$ of the dynamic link of the mth item counted from the tail of a preferred path.

$$L(m) = \begin{cases} 0 & m \leq 1 \\ 1 + 2L(m-1) & \text{if } L(m-1) = L(m-1-L(m-1)) \\ 1 & \text{otherwise.} \end{cases}$$

Let $b_1, \ldots, b_j$ be the items accessed on the preferred path when searching for b_j, or the place where the search deviates from the path, and the first item on the path is b_1. Associate to each b_i a the length of its dynamic link k_i. We will prove that the sequence $K = \langle k_1, \ldots, k_j \rangle$ is of the form where it first increases and then decreases and, moreover, that no number appears more than twice in a row. Together with the fact that numbers k_i are of form $L(\mathbb{N}) = \{0, 1, 3, 7, 15, 31, 63, \ldots, 2^i - 1, \ldots\}$ and a maximal k_i is at most a length of a root-to-leaf path l this implies that j can be at most $4 \lg l \leq 4 \lg \lg n$, if $l \leq \lg n$.

As a tool we use a more intuitive presentation of the sequence $L(m)$: $L(m)$ is the mth item in a sequence $\langle 0, S \rangle$, where S is a infinite sequence. Let $S_{1:i}$ be the prefix of S containing the first i numbers. The sequence S has been generated by repeatedly applying a rule to generate a longer prefix of S:

$$S_{1:2}^1 = \langle 1, 1 \rangle$$
$$S_{1:2(i+1)}^k = \langle S_{1:i}^{k-1}, (i+1), S_{1:i}^{k-1}, (i+1) \rangle.$$

Here the superscript k counts how many times this rule has been used. Equivalence of these two presentations can be verified by simple induction on k. In induction step assume that numbers in $S_{1:i}^k$ equal numbers given by the function L, i.e. $S_{1:i}^k = \langle L(2), \ldots, L(i+1) \rangle$. By inductive assumption $S^k = \langle S^{k-1}, L(i+1), S^{k-1}, L(i+1) \rangle$ and by definition the first i numbers in S^{k+1} equal S^k and $S_{i+1}^{k+1} = 2L(i+1)+1 = L(i+2)$. Now, $L(i+3) = 1$ because $L(i+2)$ must be larger than numbers in $\langle L(1), \ldots, L(i+1) \rangle$, and $L(i+3) = 1$ because $L(i+1) > 1$ and $L(i+2) = 1$. In fact, $\langle L(2), \ldots, L(i+2) \rangle = \langle L(i+3), \ldots, L(2i+3) \rangle$, because $L(i+2)$ is larger than numbers in $\langle L(2), \ldots, L(i+1) \rangle$ and thus it behaves like 0 in the definition of L until the number $L(i+2)$ itself is generated again, which is not until $L(2i+3)$. Thus $\langle L(i+3), \ldots, L(2i+4) \rangle = \langle S^k, 2i+1 \rangle$ and we can conclude that the correspondence between L and S holds.

Let us denote by k_M the maximal element in the sequence K. The prefix of K that is $\langle k_1, \ldots, k_{i+1}, k_{i+2}, \ldots, k_M \rangle$ is a non-decreasing sequence with at most two repetitions of the same value because of the following structure in a subsequence of S:

$$\langle k_M, \ldots, k_{i+1} \text{ or } k_{i+2}, \underbrace{1, 1, \ldots}_{k_i-1 \text{ items}}, k_i, \underbrace{\ldots}_{k_i-1 \text{ items}}, k_i, 2k_i + 1 \rangle.$$

Now the subscript in the item marked by $k' = (k_{i+1} \text{ or } k_{i+2})$ depends on which k_i corresponds to the actual item that is visited during k_i in K (we use k_i as both an item in the sequence K and a numerical value). Due to the rule generating S, k' must be either 0, which is impossible in our case, $2k_i + 1$, or a larger value.

On the other hand, $\langle k_M, \ldots, k_i, \ldots, k_j \rangle$ is a non-increasing sequence with at most two repetitions of same value, because we can again write a subsequence S as:

$$\langle \ldots, \underbrace{1, 1, \ldots}_{k_j \text{ is here}}, k_i, \underbrace{1, 1, \ldots}_{\text{or here}}, k_i, 2k_i + 1, \ldots, k_M, \ldots \rangle.$$

Here the parts indicated by underbraces are of length $k_i - 1$.

B   Proof of Theorem 1

A BST algorithm serving $\sigma = \sigma_1, \ldots, \sigma_m$ defines a sequence of trees $T_0, \ldots, T_m$ and touches items during each access, let these be connected subtrees $S_1, \ldots, S_m$. In the spirit of Demaine et al. [6], we will play with marbles. Our argument is similar, but not quite the same as theirs. More precisely, for each change of a preferred child, we will place a marble on an item. They are placed so that at any time there is at most one marble on an item. Furthermore, no more than two marbles per item in $S_j - \sigma_j$ are discarded during σ_j. Two can be discarded because after the first discard a new marble might be placed and then discarded. Thus, half of the number of the marbles discarded is a lower bound on the cost of the BST algorithm minus m. The number of marbles discarded can be at most that of marbles placed M_{placed} and is at least $M_{\text{placed}} - n$, because there is at most one marble on an item at any given time. So the total cost of any BST algorithm is at least $M_{\text{placed}}/2 - n/2 + m$. Note that in our argument the trees T_i are BSTs, but P can have nodes with several items. As such, this is not an improvement to the results of Demaine et al., because if for example a 2-3 tree P is given as a BST it gives a tighter bound, but we are able to get a smaller additive term of $-n/2$ to the bound.

Let us now describe a method of placing marbles. On an access σ_j we first discard marbles on $S_j - \sigma_j$. Then for each switch in preferred children a marble is placed. It is placed to the *least common ancestor* (LCA) in T_j of items in the subtree of formerly preferred child. Note that T_j is the tree after an access σ_j. After placing marbles, again discard marbles on $S_j - \sigma_j$.

Why are two marbles never on the same item? First, note that *distinct* subtrees rooted at items in P form continuous intervals in key values. Thus their LCAs must be distinct, because the LCA of a continuous interval belongs to that interval. This implies that marbles placed during the same time step do not mix; the previously preferred subtrees are distinct, so their LCAs are distinct too. Second, marbles placed at different time steps do not mix. To see why, assume that there is an item a that already has a marble when we try to place another on it. Preferred child pointer of a node v changes; it previously pointed to subtree P_a of P containing a. Let s_1 be the access during which the first marble was placed and s_2 the access trying to place the second marble. There are two separate cases depending on where the first marble has been placed: above of v in P or on v or below it. In the first case we must either have touched a during the access to s_2, because a must have been an ancestor of s_2, or a must have been touched while it was rotated from being an ancestor to s_2. In the second case the first marble has been placed on v or in P_a and a has been touched when an item in P_a was last accessed, which must be after or during s_1 because the preferred child pointer of v points to P_a and s_1 is in P_a. In any case, a cannot hold the first marble anymore and the second marble can be safely inserted to a.

Assume now that we may do rotations on BST P. What happens if we rotate an item a above of b? If before the rotation tree P was safe in the sense that an access to any item would not place two marbles on the same item, then by removing a marble from a certain item on P, we can guarantee that after the rotation the new tree P' is safe as well. To find this item, note that preferred child pointers of a and b point to two subtrees and it is safe to switch the pointers to these subtrees and place a marble to the LCA of those subtrees. After the rotation, if the preferred child pointers are set to their natural places, at most one of these two pointers points to a different tree (this can be verified by going through all four — eight, counting the mirror images — possible cases). If we remove the marble on the LCA of the items in this tree, then P' is safe. This implies a lower bound of $(\text{IB}(\sigma, P) - n - r)/2 + m$, where r is the number of rotations. This is the tightest known bound formulated using a reference tree P.

Efficient Algorithms for the Optimal-Ratio Region Detection Problems in Discrete Geometry with Applications[*]

Xiaodong Wu

Departments of Electrical & Computer Engineering and Radiation Oncology,
the University of Iowa, Iowa City, IA 52242, USA
xiaodong-wu@uiowa.edu

Abstract. In this paper, we study several interesting *optimal-ratio region detection (ORD)* problems in d-D ($d \geq 3$) discrete geometric spaces, which arise in high dimensional medical image segmentation. Given a d-D voxel grid of n cells, two classes of geometric regions that are enclosed by a single or two coupled smooth *heightfield surfaces* defined on the entire grid domain are considered. The objective functions are normalized by a function of the desired regions, which avoids a bias to produce an overly large or small region resulting from data noise. The normalization functions that we employ are used in real medical image segmentation. To our best knowledge, no previous results on these problems in high dimensions are known. We develop a unified algorithmic framework based on a careful characterization of the intrinsic geometric structures and a nontrivial graph transformation scheme, yielding efficient polynomial time algorithms for solving these ORD problems. Our main ideas include the following. We show that the optimal solution to the ORD problems can be obtained via the construction of a convex hull for a set of $O(n)$ unknown 2-D points using the hand probing technique. The probing oracles are implemented by computing a minimum s-t cut in a weighted directed graph. The ORD problems are then solved by $O(n)$ calls to the minimum s-t cut algorithm. For the class of regions bounded by a single heighfield surface, our further investigation shows that the $O(n)$ calls to the minimum s-t cut algorithm are on a monotone parametric flow network, which enables to detect the optimal-ratio region in the complexity of computing a single maximum flow.

1 Introduction

In this paper, we study several *optimal-ratio region detection* problems in discrete geometry, which aims to find a "best" well-shaped region in a given d-dimensional voxel grid $\Gamma = [1..N]^d$ of $n = N^d$ cells. Many applications such as

---

[*] This research was supported in part by an NIH-NIBIB research grant R01-EB004640, in part by a faculty start-up fund from the University of Iowa, and in part by a fund from the American Cancer Society through an Institutional Research Grant to the Holden Comprehensive Cancer Center, the University of Iowa, Iowa City, Iowa, USA.

data mining [8, 5], data visualization [1], and computer vision [3], require that the target region be of a "good" shape. We develop in this paper efficient algorithms for computing two interesting classes of optimal-ratio geometric regions, the *coupled-surfaces-bounded regions* and the *smooth lower-half regions*, which model important applications in medical image analysis [18, 2, 15, 20].

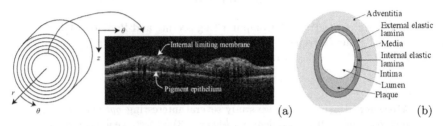

Fig. 1. (a) A retinal layer in a slice of a 3-D optical coherence tomography (OCT) image. Each 3-D OCT image is composed of a number of 2-D radial scans, as schematically shown on the left. (b) A schematic cross-sectional anatomy of a diseased artery.

A central problem in medical image analysis is image segmentation, which aims to define accurate boundaries for the objects of interest captured by image data. Accurate three and higher dimensional (e.g., 3-D + time) image segmentation promises to revolutionize the current medical imaging practice. Many medical anatomies express layered structures. Detecting regions bounded by a single or coupled surfaces from image data is highly demanded in medical practice. Figure 1(a) shows the retinal layer in a slice of a 3-D optical coherence tomography (OCT) image; the retinal layer is bounded by two coupled heightfield (terrain-like) surfaces, the internal limiting membrane and the pigment epithelium. Figure 1(b) illustrates a schematic cross-sectional anatomy of a diseased artery. For non-heightfield structures, a common segmentation approach [18, 15] for those objects is to perform a resampling of the original image to produce another new image $\mathcal{I}$ in the geometric **xyz**-space. This so-called "unfolding" operation is done in such a way that the sought object surfaces in $\mathcal{I}$ are terrain-like ones. Thus, each of the sought surfaces in $\mathcal{I}$ for the object boundaries contains exactly one voxel in every column of $\mathcal{I}$ that is parallel to the **z**-axis. Then, the region bounded by two coupled terrain-like surfaces is segmented from $\mathcal{I}$. Some geometric constraints on these surfaces should be satisfied by the segmentation. First, the two sought surfaces must be non-crossing and within a given range of distance apart. Since many anatomical structures are smooth, the segmented surfaces must be sufficiently "smooth". Generally speaking, the smoothness is related to the surface curvature and means that an object boundary cannot change abruptly. Motivated by this medical image segmentation and other applications, we formulate the following optimal-ratio region detection problems.

Let Γ be a given d-D ($d \geq 3$) voxel grid $[1..N]^d$ of $n = N^d$ cells. The domain $\mathcal{D}$ of Γ is the projection of Γ onto the first $d-1$ dimensions (i.e., $\mathcal{D} = [1..N]^{d-1}$). For each $\mathbf{x} = (x_1, x_2, \ldots, x_{d-1} \in \mathcal{D}$, the voxel subset $\{(x_1, x_2, \ldots, x_{d-1}, z \mid 1 \leq$

$z \leq N\}$ forms a column $Col(\mathbf{x})$, called the $\mathbf{x}$-*column* of Γ. We denote the voxel $(x_1, x_2, \ldots, x_{d-1}, z)$ of $Col(\mathbf{x})$ by $\mathbf{x}_z$. Two columns, $Col(\mathbf{x})$ and $Col(\mathbf{y})$ with $\mathbf{x} = (x_1, x_2, \ldots, x_{d-1})$, $\mathbf{y} = (y_1, y_2, \ldots, y_{d-1}) \in \mathcal{D}$, are *adjacent in the p-th dimension* if $\sum_{i=1}^{d-1} |x_i - y_i| = 1$ and $|x_p - y_p| = 1$ ($1 \leq p < d$). Each voxel $\mathbf{x}_z \in \Gamma$ is assigned an *on-surface cost* $b(\mathbf{x}_z)$ and an *in-region cost* $c(\mathbf{x}_z)$; both are an arbitrary real value. The on-surface cost of a voxel is closely related to the likelihood that it may appear on a desired surface, while the in-region cost measures the probability of a given voxel preserving the expected regional property (e.g., we assign the logarithm of the probability as the on-surface/in-region cost of each voxel). In image segmentation, both on-surface and in-region costs can be determined using low-level image features [18].

A *heightfield surface* in Γ is defined by a function $S : \mathcal{D} \rightarrow \{1, 2, \ldots, N\}$ such that it satisfies the following *smoothness constraint*: Given $d - 1$ smoothness parameters $\{\Delta_i \geq 0 \mid i = 1, 2, \ldots, d - 1\}$ with each for one of the first $d - 1$ dimensions, for any two voxels, $\mathbf{x}_z$ and $\mathbf{y}_{z'}$, on the surface (i.e., $z = S(\mathbf{x})$ and $z' = S(\mathbf{y})$), if their corresponding columns $Col(\mathbf{x})$ and $Col(\mathbf{y})$ are adjacent in the p-th dimension ($p = 1, 2, \ldots, d - 1$), then $|z - z'| \leq \Delta_p$. Intuitively, the smoothness constraint defines the maximum allowed change in the d-th coordinate of a feasible heightfield surface along each unit distance change in the first $d - 1$ dimensions. A *coupled-surfaces-bounded region* (briefly called a *csb-region*) is a region R in Γ enclosed by two interrelated heightfield surfaces S_1 and S_2 (i.e., $R = \{\mathbf{x}_z \mid \mathbf{x} \in \mathcal{D}, S_1(\mathbf{x}) \leq z \leq S_2(\mathbf{x})\}$), whose interrelation is specified by the *surface separation constraint*: Given two parameters $\delta^u \geq \delta^l \geq 0$, for each $\mathbf{x} \in \mathcal{D}$, $\delta^l \leq S_2(\mathbf{x}) - S_1(\mathbf{x}) \leq \delta^u$. The second class of regions is called the *smooth lower-half region*, which is a region R bounded by a heightfield surface S (i.e., $R = \{\mathbf{x}_z \mid \mathbf{x} \in \mathcal{D}, 1 \leq z \leq S(\mathbf{x})\}$). The *net-cost* $\beta(R)$ of a region R is defined as the total on-surface cost of the heightfield surface(s) enclosing R plus the total in-region cost of the voxels in R.

The **optimal-ratio region detection (ORD)** problem seeks a desired region R in Γ such that the *ratio cost* $\alpha(R)$ induced by R, with $\alpha(R) = \frac{\beta(R)}{g(R)}$, is maximized, where $g(R)$ is a non-negative real-valued function of R. The objective function $\alpha(R)$ incorporates both boundary and regional information of the desired region R. The normalization over $g(R)$ is to avoid a bias to produce an overly large or small region resulting from data noise. The normalization techniques [19, 17] is commonly used for image segmentation. Let $|R|$ denote the number of voxels in the region R. In this paper, we consider two normalization functions. One is $g(R) = \sqrt{|R|(n - |R|)}$, which is closely related to the *inter-class variance* in discriminant analysis [13]. We denote by **ORDI** this version of the ORD problem. The second normalization function that we consider is $g(R) = |R|$, which is equivalent to the *volume* of R in the discrete space, and the problem is denoted by **ORDV**.

Previous work on the optimal region detection problems mainly focuses on computing optimal well-shaped regions in low dimensions. Fukuda *et al.* [8] considered computing optimal rectangular, connected x-monotone, and rectilinear convex regions in a 2-D pixel grid Γ of n cells, and gave $O(n^{1.5})$-, $O(n)$-, and

$O(n^{1.5})$-time algorithms, respectively. Asano *et al.* [2] considered to detect a connected x-monotone region in a 2-D image while maximizing the interclass variance. Their algorithm employed the hand-probing technique in computational geometry and dynamic programming schemes. In studying the optimal pyramid problem [5], Chun *et al.* gave a linear time algorithm for computing an optimal region called *point-stabbed union of rectangles* in 2-D. The optimal region detection problems in higher dimensions are so under-explored that few known methods actually address them. Chen *et al.* [4] extended Chun *et al.*'s 2-D approach to searching for a minimum-weight stabbed union of orthogonal regions in d-D ($d \geq 3$). Wu *et al.* [20] considered several classes of less restricted regions in the d-D discrete geometric space. However, both approaches bias to find an overly small region in a given voxel grid Γ.

In this paper, we develop an interesting algorithmic framework for solving each of the ORD problems. We exploit a set of interesting geometric observations and show that each of the ORD problems is closely related to the construction of a convex hull for a set of $O(n)$ unknown 2-D points using the hand probing technique [7, 6]. The optimal solution actually defines a vertex of the constructed convex hull. Asano *et al.* [2] observed a similar property for the detection of a connected x-monotone region in 2-D. The implementation of the probing oracle, which recognizes either one hull vertex or one edge at each call, is essentially the main challenge here. By judiciously characterizing the intrinsic structures of the problems, we are able to implement such a probing oracle by computing a minimum s-t cut in a weighted directed graph. The ORD problems are then solvable by $O(n)$ calls to the minimum s-t cut algorithm. Interestingly, we are able to do much better for computing an optimal smooth lower-half region using either normalization criterion (called the ORDI-SLH and the ORDV-SLH problems, respectively). For the ORDI-SLH problem, we observe that the $O(n)$ calls to the minimum s-t cut algorithm are on a sequence of weighted directed graphs, which forms a *monotone parametric flow network* [12, 9]. Hence, the ORDI-SLH problem can be solved in the complexity of computing a single maximum flow using Gusfield and Martel's algorithm [12]. For the ORDV-SLH problem, we establish a connection between our convex hull model for the problem and the traditional Newton based approach for the fractional programming problem (see, e.g., Gondran and Minoux [11]). This connection enables us to apply Gallo *et al.*'s simple parametric minimum s-t cut algorithm [9], yielding an $O(n^2 \log n)$ time algorithm for solving the ORDV-SLH problem.

2   Our Algorithms for the ORDI Problems

This section presents our polynomial-time algorithms for the ORDI problems. We illustrate our algorithmic framework using the computation of an optimal coupled-surfaces-bounded region in Γ (called the ORDI-CSB problem) as an example. An improved $O(n^2 \log n)$ time algorithm for computing an optimal smooth lower half region in Γ (called the ORDI-SLH problem) is then obtained by exploiting the applicability of the parametric minimum s-t cut algorithm [12].

2.1 Convex Hull for the ORDI-CSB Problem

The objective function that we want to maximize is the ratio cost $\alpha(R)$ of a csb-region R with $\alpha(R) = \frac{\beta(R)}{\sqrt{|R|(n-|R|)}}$, where $\beta(R)$ is the net-cost of R. Note that $|R|$ denotes the total number of voxels in R and ranges from 0 to n. Observe that for each k $(0 \leq k \leq n)$, if we are able to compute an optimal csb-region R_k^* in Γ such that the size of R_k^* is k and the net-cost $\beta(R_k^*) = \sum_{i=1}^2 \sum_{\mathbf{x}_z \in S_i} b(\mathbf{x}_z) + \sum_{\mathbf{x}_z \in R_k^*} c(\mathbf{x}_z)$ is maximized, then we solve the problem. Unfortunately, that is not an easier problem at all. However, the view of the problem in such a way lays down a base for further exploiting the geometric structure of the problem.

For each $k = 0, 1, \ldots, n$, the pair $(k, \beta(R_k^*))$ defines a point in the 2-D $(\mathbf{x}, \mathbf{y})$-plane, thus forming a set P of points with $P = \{(k, \beta(R_k^*)) \mid k = 0, 1, \ldots, n\}$. Note that for some k's, the problem may not have a feasible solution; we then simply let the net-cost be $-\infty$. Actually, we may not need to compute all the points in P in order to find the optimal solution R^* for the ORDI-CSB problem, as stated in Lemma 1.

Lemma 1. *The point* $(|R^*|, \beta(R^*))$ *defined by an optimal csb-region* R^* *for the ORDI-CSB problem, must be a vertex of the upper chain* $UH(P)$ *of the convex hull* $CH(P)$ *of* P.

Thus, we only need to compute a subset of points in P that lie on the upper chain $UH(P)$ of convex hull $CH(P)$ (note that the convex hull model here is mainly used to get rid of the denominator $g(R)$ of the objective function). However, directly computing the hull vertices of $UH(P)$ appears to be quite involved. Inspired by the *hand probing* method [7, 6], which can be viewed as recognizing a convex polygon by " touching with lines", we use the following *probing oracle* to construct the upper chain $UH(P)$ even the points in P is unknown.

Given a slope θ, *report the tangent line with slope* θ *to* $CH(P)$ *and the tangent point as well.*

With this probing oracle, the convex polygonal chain $UH(P)$ can be constructed, as follows. Start with slopes $+\infty$ and $-\infty$ to find the two endpoints (leftmost and rightmost points) of $UH(P)$. Now suppose that we have computed two vertices u and v on the hull and there is no vertex of $UH(P)$ between u and v being computed so far. Let θ be the slope of the line through u and v. Then, perform a probing oracle with respect to θ. Consequently, we either find a new vertex on $UH(P)$ between u and v or know that $\overline{uv}$ is a hull edge of $UH(P)$ (i.e., no vertex of $UH(P)$ between u and v). Thus, performing a probing oracle results in either a new vertex or a new hull edge of $UH(P)$. Hence, the convex polygonal chain with m vertices can be computed with $O(m)$ probing oracles [7].

Based on Lemma 1, if we are able to implement such a probing oracle, the ORDI-CSB problem can be solved by performing $O(n)$ probing oracles since $|UH(P)| = O(n)$. In the next section, we address the main challenge of efficient implementation of the probing oracles.

2.2   Implementation of the Probing Oracle

Given a real-valued parameter θ, we define the *parametric net-cost* of a *csb*-region R in Γ as the net-cost of R minus $\theta|R|$ (i.e., $\beta(R) - \theta|R|$), denoted by $\beta_\theta(R)$. We show in this section that the probing oracle can be implemented via computing in Γ an optimal *csb*-region with a maximized parametric net-cost.

For a given parameter θ, let $R^*(\theta)$ be an optimal *csb*-region with a maximized parametric net-cost. Recall that R_k^* denotes the optimal *csb*-region with size of k such that $\beta(R_k^*) = \max_{R \subseteq \Gamma, |R| = k} \beta(R)$. Lemma 2 follows immediately from the relevant definitions.

Lemma 2. $\max_{R(\theta)} \beta_\theta(R(\theta)) = \max_k [\beta(R_k^*) - k\theta]$

Lemma 3. *There exists a tangent line to $UH(P)$ at the point $(j, \beta(R_j^*))$ with a slope θ if and only if $|R^*(\theta)| = j$ and $\beta(R^*(\theta)) = \beta(R_j^*)$.*

Consequently, for a given slope θ, we need to compute an optimal *csb*-region $R^*(\theta)$ in Γ. If the size of $R^*(\theta)$ is j, based on Lemma 3, the line l: $y = \theta x + (\beta(R^*(\theta)) - j \cdot \theta)$ is a tangent line to $UH(P)$ at the point $(j, \beta(R^*(\theta)))$ with slope θ. We thus let $R_j^* = R^*(\theta)$. Next, we develop an efficient algorithm for computing such an optimal *csb*-region $R^*(\theta)$ in Γ.

2.3   The Algorithm for Maximizing the Parametric Net-Cost

Our algorithm for computing an optimal *csb*-region $R^*(\theta)$ with a maximized parametric net-cost (namely, the **MPNC** problem) is inspired by our previous algorithm for solving the layered net surface problem [20]. Instead of seeking an optimal "partition" of a given voxel grid into multiple disjoint regions using heightfield surfaces as in [20], here we "select" an optimal region bounded by coupled heightfield surfaces in Γ. We characterize the self-closure structures of the MPNC problem, and then model it as a maximum-cost closed set problem [16, 14] based on a nontrivial graph transformation scheme.

The self-closure structures of the MPNC problem. Recall that a *csb*-region $R(\theta)$ in Γ is a region enclosed by two coupled heightfield surfaces, S_1 and S_2, satisfying the surface separation constraint. WOLG, we assume that S_2 is "on top" of S_1 (i.e., for any $\mathbf{x} \in \mathcal{D}$, $S_2(\mathbf{x}) > S_1(\mathbf{x})$).

Given a set of $d-1$ smoothness parameters $\{\Delta_i \geq 0 | i = 1, 2, \ldots, d-1\}$ with each for one of the first $d-1$ dimensions, consider every voxel $\mathbf{x}_z \in \Gamma$ (i.e., $\mathbf{x} \in \mathcal{D}$ and $1 \leq z \leq N$) and each column $Col(\mathbf{y})$ adjacent to $Col(\mathbf{x})$ in the p-th dimension for every $p = 1, 2, \ldots, d-1$. The *lowest neighbor* of $\mathbf{x}_z$ on $Col(\mathbf{y})$ is the voxel $\mathbf{y}_{z'}$ with $z' = \max\{1, z - \Delta_p\}$ (i.e., the voxel on $Col(\mathbf{y})$ with the smallest d-th coordinate that can possibly appear together with $\mathbf{x}_z$ on a same feasible heightfield surface in Γ). To help exploit the spatial relations between two bounding surfaces S_1 and S_2 of a feasible *csb*-region, we define below the *upstream* and *downstream* voxels of any voxel $\mathbf{x}_z \in \Gamma$ for the given surface separation constraint specified by two parameters δ^l and δ^u: the *upstream* (resp., *downstream*) of $\mathbf{x}_z$ is $\mathbf{x}_{z+\delta^l}$ (resp., $\mathbf{x}_{\max\{1,z-\delta^u\}}$) if $z + \delta^l \leq N$ (resp., $z - \delta^l \geq 1$). Intuitively, if $\mathbf{x}_z \in S_1$ (resp., $\mathbf{x}_z \in S_2$), then the upstream (resp., downstream) voxel of $\mathbf{x}_z$ is the voxel on $Col(\mathbf{x})$ with the smallest d-th coordinate that can be on S_2 (resp., S_1). We say that a voxel $\mathbf{x}_z$ is *below* a heightfield surface S if $S(\mathbf{x}) > z$, and denote by $LO(S)$ the subset of all voxels of Γ that are on or below S. The following self-closure structures shown in Observations 1 and 2 are crucial to our MPNC algorithm and suggests a connection between our target problem and the maximum-cost closed set problem [16, 14].

Observation 1. For any feasible heightfield surface S in Γ, if a voxel $\mathbf{x}_z$ is in $LO(S)$, then every lowest neighbor of $\mathbf{x}_z$ is also in $LO(S)$.

Observation 2. For any feasible *csb*-region R enclosed by S_1 and S_2, the upstream (resp., downstream) voxel of each voxel in $LO(S_1)$ (resp., $LO(S_2)$) is in $LO(S_2)$ (resp., $LO(S_1)$).

In our MPNC approach, instead of directly searching for an optimal *csb*-region $R^*(\theta)$ bounded by S_1 and S_2, we look for optimal $LO(S_1)$ and $LO(S_2)$ in Γ, such that $LO(S_1)$ and $LO(S_2)$ uniquely define S_1 and S_2, respectively.

Computing an optimal *csb*-region with a maximum parametric net-cost. For a given θ, we construct a vertex-weighted directed graph $G(\theta) = (V, E)$ from Γ, such that the maximum-cost closed set in $G(\theta)$ specifies an optimal *csb*-region $R^*(\theta)$ with a maximized parametric net-cost in Γ. The construction of $G(\theta)$ crucially relies on the self-closure structures shown in Section 2.3. $G(\theta)$ contains two vertex disjoint subgraphs $\{G_i = (V_i, E_i) | i = 1, 2\}$; each G_i is constructed in reflecting the single surface self-closure structure of the MPNC problem and is used for the search of S_i of $R^*(\theta)$. The separation constraints between S_1 and S_2 are enforced in $G(\theta)$ by a set of edges E_s, connecting the corresponding subgraphs G_1 and G_2, in such a way to reflect the inter-surface self-closure structure. Thus, $V = V_1 \cup V_2$ and $E = E_1 \cup E_2 \cup E_s$.

We first show the construction of each $G_i = (V_i, E_i)$ $(i = 1, 2)$. Every voxel $\mathbf{x}_z \in \Gamma$ corresponds to exactly one vertex $v_i(\mathbf{x}_z) \in V_i$. For each $\mathbf{x}$ in the domain $\mathcal{D}$ of Γ and $z = 2, 3, \ldots, N$, vertex $v_i(\mathbf{x}_z)$ has a directed edge to the vertex $v_i(\mathbf{x}_{z-1})$, forming a *chain* $Ch_i(\mathbf{x})$: $v_i(\mathbf{x}_N) \to v_i(\mathbf{x}_{N-1}) \to \ldots \to v_i(\mathbf{x}_1)$ in G_i for $Col(\mathbf{x})$. We next put directed edges between every two adjacent chains (i.e., their corresponding columns in Γ are adjacent) in G_i to enforce the surface smoothness constraint. Based on Observation 1, for every voxel $\mathbf{x}_z$ of each $Col(\mathbf{x})$ in Γ and its lowest neighbor $\mathbf{y}_{z'}$ on each adjacent $Col(\mathbf{y})$ of $Col(\mathbf{x})$, we put into E_i a directed edge from $v_i(\mathbf{x}_z) \in Ch_i(\mathbf{x})$ to $v_i(\mathbf{y}_{z'}) \in Ch_i(\mathbf{y})$. We then put directed edges into E_s between G_1 and G_2, to incorporate the surface separation constraint. For each vertex $v_1(\mathbf{x}_z)$ with $z \leq N - \delta^l$ on the chain $Ch_1(\mathbf{x})$ in G_1, a directed edge is put in E_s from $v_1(\mathbf{x}_z)$ to $v_2(\mathbf{x}_{z+\delta^l})$ on $Ch_2(\mathbf{x})$ in G_2. On the other hand, each vertex $v_2(\mathbf{x}_z)$ with $z > \delta^l$ on $Ch_2(\mathbf{x})$ of G_2 has a directed edge in E_s to vertex $v_1(\mathbf{x}_{z'})$ with $z' = \max\{1, z - \delta^u\}$ ($\mathbf{x}_{z'}$ in Γ is the downstream voxel of $\mathbf{x}_z$).

The following lemma establishes the connection between a closed set in $G(\theta)$ and a feasible *csb*-region in Γ.

Lemma 4. *(1) Any feasible csb-region in Γ defines a closed set $\mathcal{C} \neq \emptyset$ in $G(\theta)$. (2) Any closed set $\mathcal{C} \neq \emptyset$ in $G(\theta)$ specifies a feasible csb-region in Γ.*

Our goal is to compute a non-empty maximum-cost closed set in $G(\theta)$, which can specify an optimal *csb*-region in Γ. Thus, we need to further assign a cost $w(\cdot)$ to each vertex in $G(\theta)$. Using the following vertex-cost assignment scheme, we can show that the parametric net-cost $\beta_\theta(R)$ of the *csb*-region R defined by a closed set $\mathcal{C} \neq \emptyset$ is equal to the total vertex cost $w(\mathcal{C})$ of $\mathcal{C}$, and vice verse. For every $\mathbf{x} \in \mathcal{D}$,

$$
w(v_1(\mathbf{x}_z)) = \begin{cases} b(\mathbf{x}_z) & \text{if } z = 1, \\ [b(\mathbf{x}_z) - b(\mathbf{x}_{z-1})] + [\theta - c(\mathbf{x}_{z-1})] & \text{for } z = 2, 3, \ldots, N - \delta^l. \end{cases} \tag{1}
$$

$$
w(v_2(\mathbf{x}_z)) = \begin{cases} b(\mathbf{x}_z) + \sum_{z'=1}^{z} c(\mathbf{x}_{z'}) & \text{if } z = \delta^l + 1, \\ [b(\mathbf{x}_z) - b(\mathbf{x}_{z-1})] + [c(\mathbf{x}_z) - \theta] & \text{for } z = \delta^l + 2, \ldots, N. \end{cases} \tag{2}
$$

Based on Lemma 4, we have the following fact.

Lemma 5. *For a given θ, the region $R^*(\theta)$ specified by a maximum-cost non-empty closed set $\mathcal{C}$ in $G(\theta)$ is an optimal csb-region with a maximized parametric net-cost in Γ.*

As in [16,9], we obtain a maximum non-empty closed set $\mathcal{C}^*$ in $G(\theta)$ by computing a minimum s-t cut. We then define in Γ a feasible *csb*-region R bounded by two coupled heightfield surfaces S_1 and S_2 from $\mathcal{C}^*$, as follows. Recall that we search for each S_i in subgraph G_i ($i = 1, 2$). Let $\mathcal{C}_i = \mathcal{C}^* \cap V_i$. For each voxel $\mathbf{x} \in \mathcal{D}$, denote by $\mathcal{C}_i(\mathbf{x})$ the set of vertices of $\mathcal{C}_i$ on the chain $Ch_i(\mathbf{x})$ of G_i. Based on the construction of G_i, it is not hard to show that $\mathcal{C}_i(\mathbf{x}) \neq \emptyset$. Let $r_i(\mathbf{x})$ be the largest d-th coordinate of the vertices in $\mathcal{C}_i(\mathbf{x})$. Then, define the function S_i as $S_i(\mathbf{x}) = r_i(\mathbf{x})$ for every $\mathbf{x} \in \mathcal{D}$. By applying a similar argument as in [20], we can prove that each S_i is a heightfield surface in Γ and S_1 and S_2 satisfy the surface separation constraint. By using Goldberg and Tarjan's minimum s-t cut algorithm [10], we compute in $G(\theta)$ a maximum-cost closed set $\mathcal{C}^* \neq \emptyset$ in $O(n^2 \log n)$ time.

Lemma 6. *For a given θ, the MPNC problem can be solved in $O(n^2 \log n)$ time.*

In summary, it suffices to compute the upper chain $UH(P)$ of the convex hull $CH(P)$ to solve the ORDI-CSB problem by Lemma 1, where $P = \{(k, \beta(R_k^*)) | k = 0, 1, \ldots, n\}$. We can perform $O(n)$ probing oracles to obtain all vertices on $UH(P)$. Each probing oracle can be implemented in $O(n^2 \log n)$ time by Lemmas 3 and 6. Thus, the total running time is $O(n^3 \log n)$.

Theorem 1. *Given a d-D ($d \geq 3$) voxel grid Γ of n cell, an optimal csb-region R^* with a maximum ratio cost $\alpha(R^*) = \max_R \dfrac{\beta(R)}{\sqrt{|R|(n-|R|)}}$ can be computed in $O(n^3 \log n)$ time.*

2.4 Computing an Optimal-Ratio Smooth Lower-Half Region

This section presents our $O(n^2 \log n)$ time algorithm for solving the ORDI-SLH problem, which is achieved by exploiting the monotonicity of the parametric graph used for the search of the bounding heightfield surface.

Recall that a smooth lower-half region R in Γ is bounded by a heightfield surface S, that is, $R = LO(S)$. A key subroutine here is that, for a given parameter θ, computing an optimal smooth lower-half region $R^*(\theta)$ in Γ such that the parametric net-cost $\beta_\theta(R^*(\theta))$ of $R^*(\theta)$ is maximized. Note that the single surface self-closure structure as in Section 2.3 holds for the ORDI-SLH problem. Thus, a similar graph transformation scheme as in Section 2.3 is used to construct a vertex-weighted directed graph $G(\theta) = (V, E)$, yet here $G(\theta)$ has only one subgraph used for the search of the only bounding heightfield surface S of the smooth lower-half region $R(\theta)$ for a given θ. The cost of each vertex in $G(\theta)$ is assigned, as follows.

$$w(v(\mathbf{x}_z)) = \begin{cases} b(\mathbf{x}_z) + [c(\mathbf{x}_z) - \theta] & \text{if } z = 1, \\ [b(\mathbf{x}_z) - b(\mathbf{x}_{z-1})] + [c(\mathbf{x}_z) - \theta] & \text{for } z = 2, 3, \dots, N. \end{cases} \tag{3}$$

Note that computing a maximum-cost closed set $\mathcal{C}$ in $G(\theta)$ is equivalent to computing a minimum s-t cut in $G_{st}(\theta)$ [16, 9]. In $G_{st}(\theta)$, the source s has a directed edge to every vertex $v(\mathbf{x}_z)$ in $G(\theta)$ with a cost of $w(v(\mathbf{x}_z))$, the sink t has a directed edge with a cost of 0 from every vertex in $G(\theta)$, and the cost of all other edges is $+\infty$. Obviously, based on the vertex-cost assignment scheme (3), the cost of every edge from source s is a non-increasing function of θ and all other edges in $G_{st}(\theta)$ have a constant cost with respect to θ. Thus, $G_{st}(\theta)$ is a monotone parametric flow network [12, 9].

To compute an optimal-ratio smooth lower-half region in Γ, as in the ORDI-CSB algorithm, we need to compute an optimal smooth lower-half region with a maximum parametric net-cost for each of a sequence of parameters $\{\theta_1, \theta_2, \dots, \theta_m\}$ generated by the hand probing process (see Section 2.1), where $m = O(n)$. Due to the monotonicity of $G_{st}(\theta)$, we can apply Gusfield and Martel's parametric minimum cut algorithm [12] to compute all those $O(n)$ optimal smooth lower-half regions in Γ in the complexity of solving a single maximum flow problem.

Theorem 2. *Given a d-D (d $\geq$ 3) voxel grid Γ of n cell, the ORDI-SLH problem can be solved in $O(n^2 \log n)$ time.*

3 Our Algorithms for the ORDV Problems

In this section, we present our efficient algorithms for the ORDV problem, which seeks a maximum-ratio coupled-surfaces-bounded region (called the *ORDV-CSB* problem) or a maximum-ratio smooth lower-half region (called the *ORDV-SLH* problem) in a given d-D (d $\geq$ 3) voxel grid Γ. The algorithmic framework follows that for the ORDI problems. Our ORDV-SLH algorithm utilizes an observation that establishes a connection between our convex hull model for the ORDV-SLH problem and the traditional Newton-based approach for the fractional programming problem (see, e.g., Gondran and Minous [11]). This connection enables us to find an order to compute the vertices on the convex hull, such that all those vertices can be computed by using Gallo *et al.*'s simple parametric minimum *s-t* cut algorithm [9] in the complexity of solving a single maximum flow problem. The detailed algorithms are left to the full version of this paper.

Theorem 3. *Given a d-D (d $\geq$ 3) voxel grid Γ of n cell, an optimal csb-region R^* with a maximum ratio cost $\alpha(R^*) = \max_R \frac{\beta(R)}{|R|}$ can be computed in $O(n^3 \log n)$ time.*

Theorem 4. *Given a d-D (d $\geq$ 3) voxel grid Γ of n cell, the ORDV-SLH problem can be solved in $O(n^2 \log n)$ time.*

References

1. A. Amir, R. Kashi, N.S. Netanyalm, Analyzing Quantitative Databases: Image Is Everything, *Proc. 27th Int. Conf. on Very Large Data Bases*, Italy, 2001, pp. 89-98.
2. T. Asano, D.Z. Chen, N. Katoh, and T. Tokuyama, Efficient Algorithms for Optimization-Based Image Segmentation, *Int'l J. of Computational Geometry and Applications*, 11(2001), pp. 145-166.
3. I. Bloch, Apatial Relationship between Objects and Fuzzy Objects using Mathematical Morphology, in *Geometry, Morphology and Computational Imaging, 11th Dagsthul Workshop on Theoretical Foundations of Computer Vision*, April 2002.
4. D.Z. Chen, J. Chun, N. Katoh, and T. Tokuyama, Efficient Algorithms for Approximating a Multi-dimensional Voxel Terrain by a Unimodal Terrain, *Lecture Notes in Computer Science*, Vol. 3106, Springer Verlag, *Proc. of the 10th Int. Computing and Combinatorics Conf. (COCOON)*, Jeju Island, Korea, August 2004, pp. 238-248.
5. J. Chun, K. Sadakane, T. Tokuyama, Linear Time Algorithm for Approximating a Curve by a Single-Peaked Curve, *Lecture Notes in Computer Science*, Vol. 2906, Springer Verlag, *Proc of the 14th Int. Symp. on Algorithms and Computation (ISAAC)*, Kyoto, Japan, Dec 2003, pp. 6-15.
6. R. Cole and C.K. Yap, Shape from Probing, *J. of Algorithms*, 8(1987), pp. 19-38.
7. D. Dobkin, H. Edelsbrunner, and C.K. Yap, Probing Convex Polytopes, *Proc. 18th Annual ACM Symp. on Theory of Computing*, 1986, pp. 387-392.
8. T. Fukuda, Y. Morimoto, S. Morishita, and T. Tokuyama, Data Mining with Optimized Two-Dimensional Association Rules, *ACM Transaction on Database Systems* 26(2001), pp. 179-213.
9. G. Gallo, M.D. Grigoriadis, and R.E. Tarjan, A Fast parametric maximum flow algorithm and applications, *SIAM J. Comput.*, 18(1989), pp. 30-55.
10. A.V. Goldberg and R.E. Tarjan, A New Approach to the Maximum-flow Problem, *J. Assoc. Comput. Mach.*, 35(1988), pp. 921-940.
11. M. Gondran and M. Minous, *Graphs and Algorithms*, John Wiley, New York, 1984.
12. D. Gusfield and C. Martel, A Fast Algorithm for the Generalized Parametric Minimum Cut Problem and Applications, *Algorithmica*, 7(1992), pp. 499-519.
13. D.J. Hand, *Discrimination and Classification*, John Wiley & Sons, 1981.
14. D.S. Hochbaum, A New-old Algorithm for Minimum-cut and Maximum-flow in Closure Graphs, *Networks*, 37(4)(2001), pp. 171-193.

15. K. Li, X. Wu, D.Z. Chen, and M. Sonka, Optimal Surface Segmentation in Volumetric Images – A Graph-Theoretic Approach, *IEEE Trans. on Pattern Analysis and Machine Intelligence,* 28(2006), pp. 119 - 134.
16. J.C. Picard, Maximal Closure of a Graph and Applications to Combinatorial Problems, *Management Science,* 22(1976), 1268-1272.
17. J. Shi and J. Malik, Normalized Cuts and Image Segmentation, *IEEE Trans. on Pattern Analysis and Machine Intelligence,* 22(8)(2000), pp. 888-905.
18. M. Sonka, V. Hlavac, and R. Boyle, *Image Processing, Analysis, and Machine Vision,* 2nd edition, Brooks/Cole Publishing Company, Pacific Grove, CA, 1999.
19. J. Stahl and S. Wang, Convex Grouping Combining Boundary and Region Information, *IEEE Int. Conf. on Computer Vision,* Volume II, pp. 946-953, 2005.
20. X. Wu, D.Z. Chen, K. Li, and M. Sonka, The Layered Net Surface Problems in Discrete Geometry and Medical Image Segmentation, *Lecture Notes in Computer Science,* Vol. 3827, Springer Verlag, *Proc. of the 16th Int. Symposium on Algorithms and Computation (ISAAC),* Sanya, China, December 2005, pp. 17-27.

On Locating Disjoint Segments with Maximum Sum of Densities

Hsiao-Fei Liu[1] and Kun-Mao Chao[1,2,3,*]

[1] Department of Computer Science and Information Engineering
[2] Graduate Institute of Biomedical Electronics and Bioinformatics
[3] Graduate Institute of Networking and Multimedia
National Taiwan University, Taipei, Taiwan 106
kmchao@csie.ntu.edu.tw

Abstract. Given a sequence A of n real numbers and two positive integers l and k, where $k \leq \frac{n}{l}$, the problem is to locate k disjoint segments of A, each has length at least l, such that their sum of densities is maximized. The best previously known algorithm, due to Bergkvist and Damaschke [1], runs in $O(nl + k^2l^2)$ time. In this paper, we give an $O(n + k^2 l \log l)$-time algorithm.

1 Introduction

Given a sequence $A = (a_1, a_2, \ldots, a_n)$ of n real numbers and two positive integers l and k, where $k \leq \frac{n}{l}$, let $d(A[i,j])$ denote the *density* of segment $A[i,j]$, defined as $\frac{a_i + a_{i+1} + \ldots + a_j}{j-i+1}$. The problem is to find k disjoint segments $\{s_1, s_2, \ldots, s_k\}$ of A, each has length at least l, such that $\sum_{1 \leq i \leq k} d(s_i)$ is maximized.

For $k = 1$, this problem was well studied in computational biology [3, 6, 5, 7, 9]. A closely related problem in data mining, which basically deals with a binary sequence, was independently formulated and studied by Fukuda *et al.* [4]. For general k, Chen *et al.* [2] proposed an $O(nkl)$-time algorithm and an improved $O(nl + k^2l^2)$-time algorithm was given by Bergkvist and Damaschke [1]. In this paper, we propose an $O(n + k^2 l \log l)$-time algorithm.

Lin *et al.* [8] formulated a related problem: Given a sequence A of n real numbers and two positive integers l and k, where $k \leq \frac{n}{l}$, find a sequence $\Gamma = (\gamma_1, \gamma_2, \ldots, \gamma_k)$ of k disjoint segments of A such that for all i, γ_i is either a maximum-density segment of length between l and $2l - 1$ not overlapping any of the first $i - 1$ segments of Γ or NIL if all segments of length between l and $2l - 1$ overlap some of the first $i - 1$ segments of Γ. For this related problem, Lin *et al.* [8] proposed a heuristic $O(n \log k)$-time algorithm and an optimal $O(n + k \log k)$-time algorithm was given by Liu and Chao [10].

The rest of this paper is organized as follows. In Section 2, we introduce some preliminary knowledge. In Section 3, we show how to reduce the length of the input sequence. In Section 4, we give our main algorithm. Section 5 summarizes our results.

---

* Corresponding author.

T. Asano (Ed.): ISAAC 2006, LNCS 4288, pp. 300–307, 2006.
© Springer-Verlag Berlin Heidelberg 2006

2 Preliminaries

Let $PS[0, \ldots, n]$ be the prefix-sum array of $A[1, \ldots, n]$, i.e., $PS[i] = a_1 + a_2 + \ldots + a_i$ for $i > 0$ and $PS[0] = 0$. PS can be computed in linear time by set $PS[0]$ to 0 and $PS[i]$ to $PS[i - 1] + A[i]$ for $i = 1, 2, \ldots, n$. Since $d(A[i,j]) = (PS[j] - PS[i - 1])/(j - i + 1)$, the density of any segment can be computed in constant time after the prefix-sum array is constructed.

A simpler version of the following lemma was first presented in [6].

Lemma 1. *Let* $S' = \{s'_1, s'_2, \ldots, s'_k\}$ *be a set of* k *disjoint segments of length at least* l *such that* $\sum_{1 \leq i \leq k} d(s'_i)$ *is maximized. There exists a set* $S = \{s_1, s_2, \ldots, s_k\}$ *of* k *disjoint segments of length between* l *and* $2l - 1$ *such that* $\sum_{1 \leq i \leq k} d(s_i) = \sum_{1 \leq i \leq k} d(s'_i)$.

Proof. We prove this lemma by showing that each s'_i has a subsegment s_i of length between l and $2l - 1$ such that $d(s'_i) = d(s_i)$. If s'_i is of length between l and $2l - 1$, then let $s_i = s'_i$. Otherwise, let $s'_i = A[p, q]$. Suppose for the contradiction that $d(A[p, p + l - 1]) \neq d(A[p + l, q])$. Without loss of generality assume $d(A[p, p + l - 1]) < d(A[p + l, q])$. Since the length of s'_i is larger than $2l - 1$ and the length of $A[p, p + 1 - l]$ is l, the length of $A[p + 1, q]$ is at least l. It follows that $\{s'_1, \ldots, s'_{i-1}, A[p + 1, q], s'_{i+1}, \ldots, s'_k\}$ is a better solution than S', a contradiction. Thus, $d(A[p, p + l - 1])$ must be equal to $d(A[p + l, q])$. Since $d(A[p, p + l - 1]) = d(A[p + l, q]) = d(s'_i)$, we can let $s_i = A[p, p + l - 1]$.   □

Lemma 1 states that there exists a solution for the problem instance (A, k, l) composed of segments of length between l and $2l - 1$. It allows us to redefine the problem as follows: Given a sequence $A = (a_1, a_2, \ldots, a_n)$ of n real numbers and two positive integers l and k, where $k \leq \frac{n}{l}$, find k disjoint segments $\{s_1, s_2, \ldots, s_k\}$ of A, each has length between l and $2l - 1$, such that $\sum_{1 \leq i \leq k} d(s_i)$ is maximized. From now on, we shall adopt this problem definition.

3 Preprocessing

In this section, we show how to compress an input sequence A of length $n \geq 2kl$ into a sequence A' of length $O(kl)$ in $O(n + k \log k)$. First we have to find a sequence $\Gamma = (\gamma_1, \gamma_2, \ldots, \gamma_{2k})$ of $2k$ disjoint segments of length between l and $2l - 1$ such that for all i, γ_i is either a maximum-density segment of length between l and $2l - 1$ not overlapping any of the first $i - 1$ segments in Γ or NIL if all segments of length between l and $2l - 1$ overlap some of the first $i - 1$ segments of Γ. Let $\gamma_i = A[p_i, q_i]$ for all i. We extend each segment γ_i to get $\gamma'_i = A[p'_i, q'_i]$, where

$$p'_i = \begin{cases} p_i - 2l + 1 & \text{if } p_i \geq 2l, \\ 1 & \text{if } p_i < 2l, \end{cases} \text{ and } q'_i = \begin{cases} q_i + 2l - 1 & \text{if } q_i \leq n - 2l + 1, \\ n & \text{if } q_i > n - 2l + 1. \end{cases}$$

We say that $A[p, q]$ is a segment consisting of only elements in $\bigcup_{i=1}^{2k} \gamma'_i$ if and only if for each index $j \in [p, q]$, there exists a $\gamma'_i = A[p'_i, q'_i]$ such that

$j \in [p_i', q_i']$. A segment $A[p, q]$ consisting of only elements in $\bigcup_{i=1}^{2k} \gamma_i'$ is maximal if and only if $A[p, q]$ is not a subsegment of any other segment consisting of only elements in $\bigcup_{i=1}^{2k} \gamma_i'$. Note that any two different maximal segments consisting of only elements in $\bigcup_{i=1}^{2k} \gamma_i'$ must be disjoint according to our definition. Let $R = (r_1, r_2, \ldots, r_{|R|})$ be all of the maximal segments, in left-to-right order, consisting of only elements in $\bigcup_{i=1}^{2k} \gamma_i'$. We set A' to $r_1 \cdot (-\infty) \cdot r_2 \cdot (-\infty) \cdot r_3 \ldots (-\infty) \cdot r_{|R|}$, where the symbol "$\cdot$" means concatenation. Since $\sum_{1 \le i \le |R|} r_i \le \sum_{1 \le i \le k} \gamma_i' \le 6kl$, A' is of length $O(kl)$

The correctness follows from the next lemma which ensures that it is safe to delete elements not in any segments of R.

Lemma 2. *There exists a solution $S = \{s_1, s_2, \ldots, s_k\}$ for the problem instance (A, k, l) such that each segment in S is a subsegment of some segment of R.*

Proof. First we show that there exists a solution $S = \{s_1, s_2, \ldots, s_k\}$ for the problem instance (A, k, l) such that s_i overlaps some segment of Γ for $i = 1, \ldots, k$. Let $S' = \{s_1', s_2', \ldots, s_k'\}$ be a solution with fewest segments in it not overlapping any segments of Γ. Let s_i' be a segment not overlapping any segment of Γ. Since each segment in S' has length shorter than $2l$ and each segment of Γ has length at least l, each segment in S' can overlap at most two segments of Γ. It follows that at most $2(k-1)$ segments of Γ are overlapped with some segment in S'. Since Γ is composed of $2k$ segments, there exists some γ_j not overlapping any segment in S'. By the specification for Γ, we know $d(\gamma_j) \ge d(s_i')$. Thus, $(S'/\{s_i'\}) \cup \{\gamma_j\}$ is a solution with fewer segments in it not overlapping any segment of Γ, a contradiction.

It remains to prove that each segment in S is a subsegment of some segment of R. Let s_i be overlapped with γ_{j_i} for $i = 1, 2, \ldots, k$. Since each s_i is of length shorter than $2l$, s_i must be a subsegment of γ_{j_i}' for $i = 1, 2, \ldots, k$. Since each s_i is a subsegment of γ_{j_i}' and each γ_{j_i}' is a subsegment of some segment of R, each s_i is a subsegment of some segment of R. □

Now we start to analyze the time complexity of our preprocessing.

Lemma 3. *It takes $O(n + k \log k)$ time to compute A'.*

Proof. Liu and Chao [10] proposed an $O(n + k \log k)$-time algorithm for computing Γ. Let $\Gamma' = (\gamma_1', \ldots, \gamma_k')$. It is clear that Γ' can be computed in $O(kl)$ time and R can be computed in $O(n)$ time. Since $\sum_{1 \le i \le |R|} r_i \le \sum_{1 \le i \le k} \gamma_i' \le 6kl$, A' is of length $O(kl)$ and can be computed in $O(kl)$ time. The total complexity is therefor $O(n + kl + k \log k) = O(n + k \log k)$ time. □

The next theorem summarizes the work of this section.

Theorem 1. *Given a problem instance $(A[1 \ldots n], k, l)$, we can reduce it to a new problem instance (A', k, l) in $O(n + k \log k)$ time, where A' is of length $O(kl)$.*

Proof. Immediate from Lemmas 2 and 3. □

4   The Main Algorithm

In the following, we shall describe an $O(nk \log l)$-time algorithm for finding a solution for the problem instance $(A[1 \ldots n], k, l)$.

Definition 1. *Let $S_{i,j}$ be a solution for the problem instance $(A[1 \ldots i], j, l)$ such that the position of the leftmost element of the rightmost segment in $S_{i,j}$ is maximized. Define $D^j[i]$ to be the sum of densities of segments in $S_{i,j}$, $S^j[i]$ to be the rightmost segment in $S_{i,j}$, and $P^j[i]$ to be the position of the leftmost element of $S^j[i]$.*

Let $DC((p, q), P^j[p], P^j[q])$ be a procedure for computing $D^j[i]$, $S^j[i]$, and $P^j[i]$ for all $i \in (p, q)$, where $q - p \leq l$. A sketch of our main algorithm is given in Figure 1. For simplicity, we assume n is a multiple of l.

Algorithm MAIN$(A[1 \ldots n], k, l)$
1 **for** $j \leftarrow 1$ to k **do**
2   Compute $D^j[i]$, $S^j[i]$, and $P^j[i]$ for all $i \in \{jl, (j+1)l, (j+2)l, \ldots, n\}$.
3   Run $DC((tl, (t+1)l), P^j[tl], P^j[(t+1)l])$ for all $j \leq t \leq \frac{n}{l} - 1$.
4 **end for**
5 Compute the solution with the help of $\bigcup_{j=1}^{k}\{S^j[1 \ldots n]\}$ and $\bigcup_{j=1}^{k}\{P^j[1 \ldots n]\}$.

Fig. 1. A sketch of the main algorithm

We now start to explain the algorithm in detail. By the next lemma, a solution can be found in $O(k)$ time with the help of $\bigcup_{j=1}^{k}\{S^j[1 \ldots n]\}$ and $\bigcup_{j=1}^{k}\{P^j[1 \ldots n]\}$.

Lemma 4. *After $S^j[1 \ldots n]$ and $P^j[1 \ldots n]$ are known for $j = 1, \ldots, k$, a solution for the problem instance $(A[1 \ldots n], k, l)$ can be found in $O(k)$ time*

Proof. Suppose now S^j and P^j are known for $j = 1, \ldots, k$. We describe a procedure for finding a solution for the problem instance $(A[1 \ldots n], k, l)$ in $O(k)$ time as follows.

1. Initiate i with n and Y with $\{\}$.
2. For $j = k, k-1, \ldots, 1$ do
   (a) $Y \leftarrow Y \cup S^j[i]$.
   (b) $i \leftarrow P^j[i] - 1$.
3. Return Y.

Since each iteration takes constant time, the total time complexity is $O(k)$. The correctness is easy to verify by observing the loop invariant: there exists a solution such that Y is its last $|Y|$ segments in left-to-right order.   □

Thus, the challenge now lies on computing D^j, S^j, and P^j for all j in $[1, k]$ in $O(nk \log l)$ time. The computation consists of k iterations. In the jth iteration, D^j, S^j, and P^j are computed in $O(n \log l)$ time. In the following, we shall describe how to compute D^j, S^j, and P^j in $O(n \log l)$ time for each j by utilizing Lemma 5 and the Chung-Lu algorithm [3]. For technical reason, we define $D^j[i]$, $S^j[i]$, and $P^j[i]$ to be $-\infty$, NIL, and 0 respectively if $i < jl$.

Lemma 5. $P^j[1] \leq P^j[2] \leq \ldots \leq P^j[n-1] \leq P^j[n]$ for $j = 1, \ldots, k$.

Proof. Suppose not. Let $p < q$ and $P^j[p] > P^j[q]$. Let $S_{p,j} = \{s_1, s_2, \ldots, s_j\}$ be a solution for the problem instance $(A[1, p], j, l)$, in left-to-right order, such that the position of the leftmost element of s_j is $P^j[p]$. Let $S_{q,j} = \{s'_1, s'_2, \ldots, s'_j\}$ be a solution for the problem instance $(A[1, q], j, l)$, in left-to-right order, such that the position of the leftmost element of s'_j is $P^j[q]$. Let $s_j = A[l_1, r_1]$ and $s'_j = A[l_2, r_1]$. r_1 is less than r_2; otherwise $\sum_{1 \leq i \leq j} d(s'_i) \leq \sum_{1 \leq i \leq j} d(s_i)$ and $l_2 < l_1$, a contradiction. Thus, we can let $L = A[\bar{l}_2, \bar{l}_1 - 1]$ and $R = A[r_1 + 1, r_2]$. It is clear that $\sum_{1 \leq i \leq j-1} d(s_i) \geq \sum_{1 \leq i \leq j-1} d(s'_i)$, so $d(s_j) < d(s'_j)$; otherwise $\sum_{1 \leq i \leq j} d(s_i) \geq \sum_{1 \leq i \leq j} d(s_j)$ and $l_2 < l_1$, a contradiction. Since $d(s_j) < d(s'_j)$, we have $d(R) > d(L \cup s_j)$. Suppose for the contradiction that $d(L) < d(s_j)$. By $d(R) > d(L \cup s_j)$, we have $d(R) > d(L \cup s_j) > d(L)$. It follows that $d(s_j \cup R) > d(L)$, so $d(s_j \cup R) > d(L \cup s_j \cup R) = d(s'_j)$, a contradiction. Thus, we have $d(L) \geq d(s_j)$. By $d(R) > d(L \cup s_j)$ and $d(L) \geq d(s_j)$, we have $d(R) > d(L \cup s_j) \geq d(s_j)$. By $d(R) > d(L \cup s_j) \geq d(s_j)$ and $|L \cup s_j| > |L|$, we have

$$d(s'_j) - d(L \cup s_j) < d(s_j \cup R) - d(s_j). \tag{1}$$

Since $S_{p,j}$ is a solution for the problem instance $(A[1, p], j, l)$ and $\{s'_1, \ldots, s'_{j-1}, L \cup s_j\}$ is a set of j disjoint segments of $A[1, p]$ of length between l and $2l - 1$, we have

$$\sum_{1 \leq i \leq j-1} d(s'_i) + d(L \cup s_j) \leq \sum_{1 \leq i \leq j} d(s_i). \tag{2}$$

By (1) and (2), we have $\sum_{1 \leq i \leq j-1} d(s'_i) + d(s'_j) < \sum_{1 \leq i \leq j-1} d(s_i) + d(s_j \cup R)$. It follows that $S_{q,j}$ is not a solution for the problem instance $(A[1, q], j, l)$, a contradiction. □

The Chung-Lu Algorithm [3]. *Given a sequence A of n number pairs (v_i, w_i) with $w_i > 0$ and two positive numbers $l \leq u$, define the density and length of a segment $A[i, j]$ to be $(v_i + v_{i+1} + \ldots + v_j)/(w_i + w_{i+1} + \ldots + w_j)$ and $(w_i + w_{i+1} + \ldots + w_j)$ respectively. The Chung-Lu algorithm can find a maximum-density segment of A with length bwtueen l and u in an online manner in $O(n)$ time.*

First we describe how to compute $D^j[i]$, $S^j[i]$, and $P^j[i]$ for all i in $\{jl, (j+1)l, (j+2)l, \ldots, n\}$ in $O(n)$ time.

Lemma 6. *Computing $D^j[i]$, $S^j[i]$, and $P^j[i]$ for all i in $\{jl, (j+1)l, (j+2)l, \ldots, n\}$ can be done in $O(n)$ time.*

Proof. The procedure consists of $\frac{n}{l}$ iterations, and in the ith iteration $D^j[il]$, $S^j[il]$, and $P^j[il]$ are computed in $O(l)$ time. The first iteration can be completed in $O(1)$ time by setting $D^j[jl]$, $S^j[jl]$, and $P^j[jl]$ to $D^{j-1}[(j-1)l]+d(A[(j-1)l+1,jl])$, $A[(j-1)l+1,jl]$, and $(j-1)l+1$ respectively. Suppose now we are in the ith iteration, where $i > 1$. We first find the the maximum-density segments s_t of $A[t,il]$ with length between l and $2l-1$ for $t = il, il-1, \ldots, (i-1)l-2l+1$ in $O(l)$ time by taking $A[t]$ as a pair $(v = A[t], w = 1)$ and using the Chung-Lu algorithm to scan A from position il to position $(i-1)l-2l+1$. Let t' be the largest t such that $D^{j-1}[t-1]+d(s_t)$ is maximized. Then there are two cases to consider. Case 1: $P^j[i] \leq (i-1)l-2l$. In this case, it is clear that $D^j[jl]$, $S^j[jl]$, and $P^j[jl]$ are equal to $D^j[(i-1)l]$, $S^j[(i-1)l]$, and $P^j[(i-1)l]$ respectively. Case 2: $P^j[i] \geq (i-1)l-2l+1$. In this case, $D^j[il]$, $S^j[il]$, and $P^j[il]$ are set to $D^{j-1}[t'-1]+s_{t'}$, $s_{t'}$, and t' respectively. We distinguish between Case 1 and Case 2 by comparing $D^j[(i-1)l]$ with $D^{j-1}[t'-1] + s_{t'}$. If $D^j[(i-1)l] > D^{j-1}[t'-1] + s_{t'}$, then it is Case 1; otherwise it is Case 2. □

Now we begin to describe how to compute $D^j[i]$, $S^j[i]$, and $P^j[i]$ for all $i \in \bigcup_{j \leq t \leq \frac{n}{l}-1}(tl, (t+1)l)$ in $O(n \log l)$ time.

Lemma 7. *If the procedure* $DC((p,q), P^j[p], P^j[q])$ *can be implemented to run in* $O((m'+n')\log n')$ *time, where* $n' = q-p+1$ *and* $m' = P^j[q]-P^j[p]+1$, *then* $D^j[i]$, $S^j[i]$, *and* $P^j[i]$ *for all* $i \in \bigcup_{j \leq t \leq \frac{n}{l}-1}(tl, (t+1)l)$ *can be computed in* $O(n \log l)$ *time.*

Proof. Let $m_t = P^j[(t+1)l]-P^j[tl]+1$ for $t = j, (j+1), \ldots, \frac{n}{l}-1$. By Lemma 5, $\sum_{j \leq t \leq \frac{n}{l}-1} m_t = O(n)$. Since $(t+1)l - tl = l$, $D^j[i]$, $S^j[i]$, and $P^j[i]$ for all $i \in (tl, (t+1)l)$ can be computed in $O((m_t+l)\log l)$ time by calling $DC((tl, (t+1)l), P^j[tl], P^j[(t+1)l])$. The total complexity for computing $D^j[i]$, $S^j[i]$, and $P^j[i]$ for all $i \in \bigcup_{j \leq t \leq \frac{n}{l}-1}(tl, (t+1)l)$ is therefore $O(\sum_{j \leq t \leq \frac{n}{l}-1}(m_t+l)\log l) = O(n \log l)$. □

Lemma 8. *The procedure* $DC((p,q), P^j[p], P^j[q])$ *can be implemented to run in* $O((m'+n')\log n')$ *time, where* $n' = q-p+1$ *and* $m' = P^j[q]-P^j[p]+1$.

Proof. Let $c = \lfloor(p+q)/2\rfloor$. We first compute $D^j[c]$, $S^j[c]$, and $P^j[c]$ in $O(n'+m')$ time and then recursively call $DC((p,c), P^j[p], P^j[c])$ and $DC((c,q), P^j[c], P^j[q])$. Let $T(n', m')$ be the run time of $DC((p,q), P^j[p], P^j[q])$. Since $P^j[c]$ is between $P^j[p]$ and $P^j[q]$ by Lemma 5, we have $T(n', m') \leq T(\lfloor\frac{n'}{2}\rfloor-1, x)+T(\lceil\frac{n'}{2}\rceil-1, m'-x+1)$ for some integer x in $[1, m']$. It follows that $T(n', m') = O((m'+n')\log n')$. It remains to explain how to compute $D^j[c]$, $S^j[c]$, and $P^j[c]$ in $O(n'+m')$ time. Note that since $q-p \leq l$, we have $P^j[q] \leq p+1$. First we compute the maximum-density segments s_t of $A[t,c]$ of length between l and $2l-1$ with the position of the rightmost element not in $(P^j[q], p+1)$ for $t = P^j[p], P^j[p]+1, \ldots, P^j[q]$. It can be done in $O(n'+m')$ time by using the Chung-Lu algorithm to process $A[P^j[p], c]$ from right to left by taking $A[i]$ as a pair $(v = A[i], w = 1)$ for i not in $[p^j[q]+1, p+1]$ and the whole segment $A[p^j[q]+1, p+1]$ as a pair $(v = \sum_{P^j[q]+1 \leq i \leq p+1} A[i], w = p-P^j[q]+1)$. Let t' be the largest t such that

$D^{j-1}[t-1]+d(s_t)$ is maximized. Then there are two cases to consider. Case 1: the position of the rightmost element of $S^j[c]$ can be $\leq p$. In this case, $D^j[c]$, $S^j[c]$, and $P^j[c]$ are set to $D^j[p]$, $S^j[p]$, and $P^j[p]$ respectively. Case 2: the position of the rightmost element of $S^j[c]$ has to be $> p$. In this case, $D^j[c]$, $S^j[c]$, and $P^j[c]$ are set to $D^{j-1}[t'-1]+d(s_{t'})$, $s_{t'}$, and t' respectively. We can distinguish between Case 1 and Case 2 as follows: If $D^j[p] > D^{j-1}[t'-1]+d(s_{t'})$ or both $D^j[p] = D^{j-1}[t'-1]+d(s_{t'})$ and $t' = P^j[p]$, then it is Case 1; otherwise it is Case 2.   □

The next theorem summarizes the work of this section.

Theorem 2. *Given a problem instance $(A[1\ldots n], k, l)$, we can find a solution in $O(nk\log l)$ time.*

Proof. By Lemmas 6, 7, and 8, computing D^j, S^j, and P^j can be done in $O(n\log l)$ time in the jth iteration for $j = 1,\ldots,k$. Thus, computing D^j, S^j, and P^j for all j in $[1, k]$ can be done in $O(nk\log l)$ time. After S^j and P^j are found for $j = 1,\ldots,k$, a solution for the problem instance $(A[1\ldots n], n, k)$ can be found in $O(k)$ time by Lemma 4.   □

By Theorems 1 and 2, we get our main result.

Theorem 3. *Given a problem instance $(A[1\ldots n], k, l)$, we can find a solution in $O(n + k^2 l\log l)$ time.*

5   Concluding Remarks

We give an $O(n + k\log k)$-time preprocessing algorithm for reducing the length of the input sequence to $O(kl)$ and a main algorithm which finds a solution in $O(nk\log l)$ time without preprocessing. By combining the preprocessing algorithm with the main algorithm, we get an $O(n + k^2 l\log l)$-time algorithm.

Acknowledgments

We thank Prof. Kunihiko Sadakane for kindly informing us of the elegant results by Fukuda *et al.* We thank Hung-Lung Wang for verifying our proof. Hsiao-Fei Liu and Kun-Mao Chao were supported in part by NSC grants 94-2213-E-002-018 and 95-2221-E-002-126-MY3 from the National Science Council, Taiwan.

References

1. Anders Bergkvist and Peter Damaschke. Fast Algorithms for Finding Disjoint Subsequences with Extremal Densities. In *Proceedings of the 16th Annual International Symposium on Algorithms and Computation*, 714-723, 2005.
2. Yen Hung Chen, Hsueh-I Lu, Chuan Yi Tang. Disjoint Segments with Maximum Density. In *Proceedings of the 5th Annual International Conference on Computational Science*, 845-850, 2005.

3. Kai-Min Chung and Hsueh-I Lu. An Optimal Algorithm for the Maximum-Density Segment Problem. *SIAM Journal on Computing*, 34:373-387, 2004.
4. Takeshi Fukuda, Yasuhiko Morimoto, Shinichi Morishita and Takeshi Tokuyama. Mining Optimized Association Rules for Numeric Attributes. *Journal of Computer and System Sciences*, 58:1-12, 1999.
5. Michael Goldwasser, Ming-Yang Kao and Hsueh-I Lu. Linear-Time Algorithms for Computing Maximum-Density Sequence Segments with Bioinformatics Applications. *Journal of Computer and System Sciences*, 70:128-144, 2005.
6. Xiaoqiu Huang. An Algorithm for Identifying Regions of a DNA Sequence that Satisfy a Content Requirement. *Computer Applications in the Biosciences*, 10:219-225, 1994.
7. Sung Kwon Kim. Linear-Time Algorithm for Finding a Maximum-Density Segment of a Sequence. *Information Processing Letters*, 86:339-342, 2003.
8. Yaw-Ling Lin, Xiaoqiu Huang, Tao Jiang and Kun-Mao Chao. MAVG: Locating Non-Overlapping Maximum Average Segments in a Given Sequence. *Bioinformatics*, 19:151-152, 2003.
9. Yaw-Ling Lin, Tao Jiang and Kun-Mao Chao. Efficient Algorithms for Locating the Length-Constrained Heaviest Segments with Applications to Biomolecular Sequence Analysis. *Journal of Computer and System Sciences*, 65:570-586, 2002.
10. Hsiao-Fei Liu and Kun-Mao Chao. An Optimal Algorithm for Iteratively Locating Non-Overlapping Maximum Density Segments. *Information Processing Letters*, submitted.

Two-Tier Relaxed Heaps[*]

Amr Elmasry[1], Claus Jensen[2], and Jyrki Katajainen[2]

[1] Department of Computer Engineering and Systems
Alexandria University, Alexandria, Egypt
[2] Department of Computing, University of Copenhagen
Universitetsparken 1, 2100 Copenhagen East, Denmark

Abstract. We introduce an adaptation of run-relaxed heaps which provides efficient heap operations with respect to the number of element comparisons performed. Our data structure guarantees the worst-case cost of $O(1)$ for *find-min*, *insert*, and *decrease*; and the worst-case cost of $O(\lg n)$ with at most $\lg n + 3 \lg \lg n + O(1)$ element comparisons for *delete*, improving the bound of $3 \lg n + O(1)$ on the number of element comparisons known for run-relaxed heaps. Here, n denotes the number of elements stored prior to the operation in question, and $\lg n$ equals $\max \{1, \log_2 n\}$.

1 Introduction

In this paper we study (min-)heaps which support the following set of operations:

find-min(H). Return the location of a minimum element held in heap H.
insert(H, e). Insert element e into heap H and return the location of e in H.
delete(H, p). Remove the element at location p from heap H.
decrease(H, p, e). Replace the element at location p in heap H with element e, which must be no greater than the element earlier located at p.

Observe that *delete-min*(H), which removes the current minimum of heap H, can be accomplished by invoking *find-min* and thereafter *delete* with the location returned by *find-min*. In the heaps studied, the location abstraction is realized by storing elements in nodes and passing pointers to these nodes.

The research reported in this paper is a continuation of our earlier work aiming to reduce the number of element comparisons performed in heap operations. In [5] (conference version) and [7] (journal version), we described how the comparison complexity of heap operations can be improved using a multi-component data structure which is maintained by moving nodes from one component to another. In a technical report [6], we were able to add *decrease* having the worst-case cost of $O(1)$ to the operation repertoire. Unfortunately, the resulting data structure is complicated. In this paper we make the data structure simpler

---

[*] Partially supported by the Danish Natural Science Research Council under contracts 21-02-0501 (project Practical data structures and algorithms) and 272-05-0272 (project Generic programming—algorithms and tools).

T. Asano (Ed.): ISAAC 2006, LNCS 4288, pp. 308–317, 2006.

and more elegant by utilizing the connection between number systems and data structures (see, for example, [14]).

For the data structures considered our basic requirement is that the worst-case cost of *find-min*, *insert*, and *decrease* is $O(1)$. Given this constraint, our goal is to reduce the number of element comparisons involved in *delete*. Binary heaps [17] are to be excluded based on the fact that $\lg \lg n \pm O(1)$ element comparisons are necessary and sufficient for inserting an element into a heap of size n [11]. Also, pairing heaps [9] are excluded because they cannot guarantee *decrease* at a cost of $O(1)$ [8]. There exist several heaps that achieve a cost of $O(1)$ for *find-min*, *insert*, and *decrease*; and a cost of $O(\lg n)$ for *delete*. Fibonacci heaps [10] and thin heaps [13] achieve these bounds in the amortized sense. Run-relaxed heaps [4], fat heaps [12, 13], and the meldable heaps described in [1] achieve these bounds in the worst case.

For all of the aforementioned heaps guaranteeing a cost of $O(1)$ for *insert*, $2 \lg n - O(1)$ is a lower bound on the number of element comparisons performed by *delete*, and this is true even in the amortized sense (for binomial heaps [16], on which many of the above data structures are based, this is proved in [6, 7]). Run-relaxed heaps have a worst-case upper bound of $3 \lg n + O(1)$ on the number of element comparisons performed by *delete* (see Section 2). For fat heaps the corresponding bound is $4 \log_3 n + O(1) \approx 2.53 \lg n + O(1)$, and for meldable heaps the bound is higher.

In this paper we present a new adaptation of run-relaxed heaps. In Section 2, we give a brief review of the basic operations defined on run-relaxed heaps. In Section 3, we discuss the connection between number systems and data structures; among other things, we show that it is advantageous to use a zeroless representation of a run-relaxed heap which guarantees that any non-empty heap always contains at least one binomial tree of size one. In Section 4, we describe our data structure, called a *two-tier relaxed heap*, and prove that it guarantees the worst-case cost of $O(1)$ for *find-min*, *insert*, and *decrease*, and the worst-case cost of $O(\lg n)$ with at most $\lg n + 3 \lg \lg n + O(1)$ element comparisons for *delete*.

2 Run-Relaxed Heaps

Since we use run-relaxed heaps as the basic building blocks of the two-tier relaxed heaps, we recall the details of run-relaxed heaps in this section. However, we still assume that the reader is familiar with the original paper by Driscoll et al. [4], where the data structure was introduced.

A *binomial tree* [16] is a rooted, ordered tree defined recursively as follows: A binomial tree of rank 0 is a single node; for $r > 0$, a binomial tree of rank r consists of the root and its r binomial subtrees of ranks $0, 1, \ldots, r-1$ connected to the root in that order. We denote the root of the subtree of rank 0 the *smallest child* and the root of the subtree of rank $r - 1$ the *largest child*. The size of a binomial tree is always a power of two, and the *rank* of a tree of size 2^r is r.

Each node of a binomial tree stores an element drawn from a totally ordered set. Binomial trees are maintained *heap-ordered* meaning that the element stored at a node is no greater than the elements stored at the children of that node. Two heap-ordered trees of the same rank can be linked together by making the root that stores the non-smaller element the largest child of the other root. We refer to this as a *join*. A *split* is the inverse of a join, where the subtree rooted at the largest child of the root is unlinked from the given binomial tree. A join involves a single comparison, and both a join and a split have a cost of $O(1)$.

A *relaxed binomial tree* [4] is an almost heap-ordered binomial tree where some nodes are denoted *active*, indicating that the element stored at that node may be smaller than the element stored at the parent of that node. Nodes are made active by *decrease*, even though no heap-order violation is introduced, and remain active until the potential heap-order violation is explicitly removed. From the definition, it follows that a root cannot be active. A *singleton* is an active node whose immediate siblings are not active. A *run* is a maximal sequence of two or more active nodes that are consecutive siblings.

Let τ denote the number of trees in any collection of relaxed binomial trees, and let λ denote the number of active nodes in the *entire* collection of trees. A *run-relaxed heap* is a collection of relaxed binomial trees where $\tau \leq \lfloor \lg n \rfloor + 2$ and $\lambda \leq \lfloor \lg n \rfloor$, n denoting the number of elements stored.

To keep track of the active nodes, a *run-singleton structure* is maintained as described in [4]. All singletons are kept in a *singleton table*, which is a resizable array accessed by rank. In particular, this table must be implemented in such a way that growing and shrinking at the tail is possible at the worst-case cost of $O(1)$, which is achievable, for example, by doubling, halving, and incremental copying. Each entry of the singleton table corresponds to a rank; pointers to singletons having this rank are kept in a list. For each entry of the singleton table that has more than one singleton of the same rank a counterpart is kept in a *pair list*. The last active node of each run is kept in a *run list*. All lists are doubly linked, and each active node should have a pointer to its occurrence in a list (if any). The bookkeeping details are quite straightforward so we will not repeat them here, but refer to [4]. The fundamental operations supported are an addition of a new active node, a removal of a given active node, and a removal of at least one arbitrary active node if λ is larger than $\lfloor \lg n \rfloor$. The cost of each of these operations is $O(1)$ in the worst case.

As to the transformations needed for reducing the number of active nodes, we again refer to the original description given in [4]. The rationale behind the transformations is that, when there are more than $\lfloor \lg n \rfloor$ active nodes, there is at least one pair of singletons that root a subtree of the same rank, or there is a run of two or more neighbouring active nodes. In that case, it is possible to apply the transformations—a constant number of singleton transformations or run transformations—to reduce the number of active nodes by at least one. The cost of performing any of the transformations is $O(1)$ in the worst case. Hereafter one application of the transformations together with all necessary changes to the run-singleton structure is referred to as a λ-*reduction*.

To keep track of the trees in a run-relaxed heap, the roots are doubly linked together in a *root list*. Each tree is represented as a normal binomial tree [3], but to support the transformations used for reducing the number of active nodes each node stores an additional pointer. That is, a node contains sibling pointers, a child pointer, a parent pointer, a rank, and a pointer to its occurrence in the run-singleton structure. The occurrence pointer of every non-active node has the value null; for a node that is active and in a run, but not the last in the run, the pointer is set to point to a fixed sentinel. To support our two-tier relaxed heap, each node should store yet another pointer to its counterpart held at the upper store (see Section 4), and vice versa.

Let us now consider how the heap operations are implemented. A reader familiar with the original paper by Driscoll et al. [4] should be aware that we have made modifications to the implementation of the heap operations to adapt them for our purposes.

A minimum element is stored at one of the roots or at one of the active nodes. To facilitate a fast *find-min*, a pointer to the node storing a minimum element is maintained. When such a pointer is available, *find-min* can be accomplished at the worst-case cost of $O(1)$.

An insertion is performed in the same way as in a worst-case efficient binomial heap [6, 7]. To obtain the worst-case cost of $O(1)$ for *insert*, all the necessary joins cannot be performed at once. Instead, one join is done in connection with each insertion, and the execution of any remaining joins is delayed for forthcoming *insert* operations. A way of facilitating this is to maintain a logarithmic number of pointers to unfinished joins on a stack. In one *join step*, the pointer at the top of the stack is popped, the two roots are removed from the root list, the corresponding trees are joined, and the root of the resulting tree is put in the place of the two. If there exists another tree of the same rank as the resulting tree, a pointer indicating this pair is pushed onto the stack. In *insert* a join step is executed, if necessary, and a new node is added to the root list. If the given element is smaller than the current minimum, the pointer indicating the location of a minimum element is updated. If there exists another tree of rank 0, a pointer to this pair of trees is pushed onto the stack. When one join is done in connection with every *insert*, the ongoing joins are disjoint and there is always space for new elements (for a formal proof, see [2, p. 53 ff.]). To summarize, *insert* has the worst-case cost of $O(1)$ and requires at most two element comparisons. An alternative way of achieving these bounds is described in [4].

There exists at most two trees of any given rank in a relaxed binomial heap. In fact, a tighter analysis shows that the number of trees is bounded by $\lfloor \lg n \rfloor + 2$. Namely, it can be shown that *insert* (as well as *delete*) maintains the invariant that between any two ranks holding two trees there is a rank holding no tree (see [2, p. 53 ff.] or [12]).

In *delete* we rely on the same borrowing technique as in [4]: the root of a tree of the smallest rank is borrowed to fill in the hole created by the node being removed. To free a node that can be borrowed, a tree of the smallest rank is repeatedly split, if necessary, until the split results in a tree of rank 0. In one *split step*, if x denotes the root of a tree of the smallest rank and y its largest child, the tree rooted at x is split, and if y is active, it is made non-active and its occurrence is removed from the run-singleton structure.

Deletion has two cases depending on whether one of the roots or one of the internal nodes is to be removed. Let z denote the node being deleted, and assume that z is a root. If the tree rooted at z has rank 0, then z is removed and no other structural changes are done. Otherwise, the tree rooted at z is repeatedly split and, when the tree rooted at z has rank 0, z is removed. In each split step all active children of z are retained active, but they are temporarily removed from the run-singleton structure (since the structure of runs may change). Thereafter, the freed tree of rank 0 (the node borrowed) and the subtrees rooted at the children of z are repeatedly joined by processing the trees in increasing order of rank. Finally, the active nodes temporarily removed are added back to the run-singleton structure. The resulting tree replaces the tree rooted at z in the root list. It would be possible to handle the tree used for borrowing and the tree rooted at z symmetrically, with respect to the treatment of the active nodes, but when *delete* is embedded into our two-tier relaxed heap it would be too expensive to remove all active children of z in a single *delete*. To complete the operation, all roots and active nodes are scanned to update the pointer indicating the location of a minimum element. Singletons are found by scanning through all lists in the singleton table. Runs are found by accessing their last node via the run list and following the sibling pointers until a non-active node is reached.

The computational cost of deleting a root is dominated by the repeated splits, the repeated joins, and the scan over all minimum candidates. In each of these steps a logarithmic number of nodes is visited, so their total cost is $O(\lg n)$. Splits as well as updates to the run-singleton structure do not involve any element comparisons. In total, joins may involve at most $\lfloor \lg n \rfloor$ element comparisons. Even though a tree of the smallest rank is split, after the joins the number of trees is at most $\lfloor \lg n \rfloor + 2$. If the number of active nodes is larger than $\lfloor \lg(n-1) \rfloor$ (the size of the heap is now one smaller), a single λ-reduction is performed which involves $O(1)$ element comparisons. To find the minimum of $2\lfloor \lg n \rfloor + 2$ elements, at most $2\lfloor \lg n \rfloor + 1$ element comparisons are to be done. To summarize, this form of *delete* performs at most $3\lg n + O(1)$ element comparisons.

Assume now that the node z being deleted is an internal node, and let x be the node borrowed. Also in this case the tree rooted at z is repeatedly split, and after removing z the tree of rank 0 rooted at x and the subtrees of the children of z are repeatedly joined. The resulting tree is put in the place of the subtree rooted earlier at z. If z was active and contained the current minimum, the pointer to the location of a minimum element is updated. If x is the root of the resulting subtree, node x is made active. Finally, the number of active nodes is reduced, if necessary, by performing a λ-reduction once or twice (once because one new node may become active and possibly once more because of the decrement of n, since the difference between $\lfloor \lg n \rfloor$ and $\lfloor \lg(n-1) \rfloor$ can be one).

Similar to the case of deleting a root, the deletion of an internal node has the worst-case cost of $O(\lg n)$. If z did not contain the current minimum, only at most $\lg n + O(1)$ element comparisons are done; at most $\lfloor \lg n \rfloor$ due to joins and $O(1)$ due to λ-reductions. However, if z contained the current minimum, at most $2\lfloor \lg n \rfloor + 1$ additional element comparisons may be necessary. That is, the total number of element comparisons performed is bounded by $3 \lg n + O(1)$. To sum up, each *delete* has the worst-case cost of $O(\lg n)$ and requires at most $3 \lg n + O(1)$ element comparisons.

In *decrease*, after making the element replacement, the corresponding node is made active, an occurrence is inserted into the run-singleton structure, and a single λ-reduction is performed if the number of active nodes is larger than $\lfloor \lg n \rfloor$. Moreover, if the given element is smaller than the current minimum, the pointer indicating the location of a minimum element is corrected to point to the active node. All these actions have the worst-case cost of $O(1)$.

3   Number Systems and Data Structures

The way our two-tier framework works (see Section 4) suggests that the run-relaxed heaps are to be modified before they can be used. One of the main operations required for our framework is the ability to borrow a node from the structure at a cost of $O(1)$, such that this borrowing would only produce $O(1)$ new roots in the root list. Accordingly, we introduce an operation *borrow* which fulfils this requirement. We rely on the observation that, in a run-relaxed heap, there is a close connection between the sizes of the relaxed binomial trees of the heap and the number representation of the current size of the heap denoted by n. Three different number representations are relevant:

Binary representation:
$$n = \sum_{i=0}^{\lfloor \lg n \rfloor} d_i 2^i, \text{ where } d_i \in \{0, 1\} \text{ for all } i \in \{0, \dots, \lfloor \lg n \rfloor\}.$$

Redundant representation:
$$n = \sum_{i=0}^{\lfloor \lg n \rfloor} d_i 2^i, \text{ where } d_i \in \{0, 1, 2\} \text{ for all } i \in \{0, \dots, \lfloor \lg n \rfloor\}.$$

Zeroless representation:
$$n = \sum_{i=0}^{k} d_i 2^i, \text{ where } k \in \{-1, 0, \dots, \lfloor \lg n \rfloor\} \text{ and } d_i \in \{1, 2, 3, 4\} \text{ for all } i \in \{0, \dots, k\}.$$

For each such representation, the heap contains d_i relaxed binomial trees of size 2^i, appearing in the root list in increasing order of rank. Now *insert* can be realized elegantly by imitating increments in the underlying number system. The worst-case efficiency of *insert* is directly related to how far a carry has to be propagated. If the binary representation is used as in [3], *insert* has the worst-case cost of $\Theta(\lg n)$. Both the redundant and zeroless representations can reduce the worst-case cost of *insert* to $O(1)$; the zeroless representation can also support *borrow* at the worst-case cost of $O(1)$.

In Section 2, the redundant representation was used. For the zeroless representation, two crucial changes are made. First, the relaxed binomial trees of the same rank are maintained in sorted order according to the elements stored at the roots. The significant consequence of this ordering is that *delete* has to only consider one root per rank when finding a minimum element stored at the roots. Second, every carry (digit 4 which corresponds to four consecutive relaxed binomial trees of the same rank) and every borrow (digit 1) are kept on a stack in rank order. When a carry/borrow stack is available, increments and decrements can be performed as follows [2, p. 56]:

1) Fix the topmost carry or borrow if the stack is not empty.
2) Add or subtract one as desired.
3) If the least significant digit becomes 4 or 1, push a pointer to this carry or borrow onto the stack.

Let x be a digit in the used number system. To fix a carry, $x4$ is converted to $(x + 1)2$, after which the top of the stack is popped. Analogously, to fix a borrow, $x1$ is converted to $(x - 1)3$ and the top of the stack is popped. If a fix creates a new carry or borrow, an appropriate pointer is pushed onto the stack. In terms of relaxed binomial trees, fixing a carry means that a join is made, which produces a relaxed binomial tree whose rank is one higher; and fixing a borrow means that a split is made, which produces two relaxed binomial trees whose rank is one lower.

To summarize, *insert* is carried out by doing at most one join or one split depending on the contents of the carry/borrow stack, and injecting a new node as a relaxed binomial tree of rank 0 into the root list. Correspondingly, after a join or split, if any, *borrow* ejects one relaxed binomial tree from the root list. Due to the zeroless representation, we can be sure that the rank of every ejected relaxed binomial tree is 0, i.e. it is a single node. However, if the ejected node contains the current minimum and the heap is not empty, another node is borrowed, and the first is inserted back into the data structure. The correctness of *insert* and *borrow* is proved in [2, p. 56 ff.] (see also [12]) by showing that both increments and decrements maintain the representation *regular*, i.e. between any two digits equal to 4 there is a digit other than 3, and between any two digits equal to 1 there is a digit other than 2. (In [2] digits $\{-1, 0, 1, 2\}$ are used, but this is equivalent to our use of $\{1, 2, 3, 4\}$.) Clearly, *insert* and *borrow* can be accomplished at the worst-case cost of $O(1)$.

4   Two-Tier Relaxed Heaps

The two-tier relaxed heap is composed of two components, the *lower store* and the *upper store*. The lower store stores the actual elements of the heap. The reason for introducing the upper store is to avoid the scan over all minimum candidates when updating the pointer to the location of a minimum element; pointers to minimum candidates are kept in a heap instead. Actually, both components are realized as run-relaxed heaps modified to use the zeroless representation as discussed in Section 3.

Upper-store operations. The upper store is a modified run-relaxed heap storing pointers to *all* roots of the trees held in the lower store, pointers to all active nodes held in the lower store, and pointers to some earlier roots and active nodes. In addition to *find-min*, *insert*, *delete*, and *decrease*, which are realized as described earlier (however *delete* could also use *borrow*), it should be possible to mark nodes to be deleted and to unmark nodes if they reappear at the upper store before being deleted. Lazy deletions are necessary at the upper store when, at the lower store, a join is done or an active node is made non-active by a λ-reduction. In both situations, a normal upper-store deletion would be too expensive. The algorithms maintain the following invariant: for each marked node whose pointer refers to a node y in the lower store, in the same tree there is another node x such that the element stored at x is no greater than the element stored at y.

To provide worst-case efficient lazy deletions, we adopt the global-rebuilding technique from [15]. When the number of unmarked nodes becomes equal to $m_0/2$, where m_0 is the current size of the upper store, we start building a new upper store. The work is distributed over the forthcoming $m_0/4$ upper-store operations. In spite of the reorganization, both the old structure and the new structure are kept operational and used in parallel. All new nodes are inserted into the new structure, and all old nodes being deleted are removed from their respective structures. Since the old structure does not handle any insertions, it can be emptied using *borrow*. In connection with each of the next at most $m_0/4$ upper-store operations, four nodes are borrowed from the old structure; if a node is unmarked, it is inserted into the new structure; otherwise, it is released and in its counterpart in the lower store the pointer to the upper store is given the value null. When the old structure becomes empty, it is dismissed and thereafter the new structure is used alone. During the $m_0/4$ operations at most $m_0/4$ nodes can be deleted or marked to be deleted, and since there were $m_0/2$ unmarked nodes in the beginning, at least half of the nodes are unmarked in the new structure. Therefore, at any point in time, we are constructing at most one new structure. We emphasize that each node can only exist in one structure and whole nodes are moved from one structure to the other, so that pointers from the outside remain valid.

Given that the cost of each *borrow* and *insert* is $O(1)$, the reorganization only adds an additional cost of $O(1)$ to all upper-store operations. A *find-min* may need to consult both the old and the new upper stores, but its worst-case cost is still $O(1)$. The cost of marking and unmarking is clearly $O(1)$. If m denotes the total number of unmarked nodes currently stored, at any point during the rebuilding process, the total number of nodes stored is $\Theta(m)$, and all the time during this process $m_0 = \Theta(m)$. Therefore, since in both structures *delete* is handled normally, except that it may take part in reorganizations, it has the worst-case cost of $O(\lg m)$ and requires $3 \lg m + O(1)$ element comparisons.

Let n be the number of elements in the lower store. The number of trees in the lower store is at most $4(\lfloor \lg n \rfloor + 1)$, and the number of active nodes is at most $\lfloor \lg n \rfloor$. At all times at most a constant fraction of the nodes stored at the upper store can be marked to be deleted. Hence, the number of pointers is $O(\lg n)$. That is, at the upper store the worst-case cost of *delete* is $O(\lg \lg n)$, including at most $3 \lg \lg n + O(1)$ element comparisons.

Lower-store operations. The lower store is a modified run-relaxed heap storing all the elements. Minimum finding relies on the upper store; an overall minimum element is either in one of the roots or in one of the active nodes held in the lower store. The counterparts of the minimum candidates are stored at the upper-store, so communication between the lower store and the upper store is necessary each time a root or an active node is added or removed, but not when an active node is made into a root.

In addition to the modifications described in Section 3, *insert* requires three further modifications in places where communication between the lower store and upper store is necessary. First, in each join the counterpart of the root of the loser tree must be lazily deleted from the upper store. Second, in each split a counterpart of the largest child of the given root must be inserted into the upper store, if it is not there already. Third, after inserting a new node its counterpart must be added to the upper store. After these modifications, the worst-case cost of *insert* is still $O(1)$.

Deletion is done as described in Section 2, but now borrowing is done by invoking *borrow* instead of repeated splitting. As a consequence of *borrow*, a lazy deletion or an insertion may be necessary at the upper store. As a consequence of *delete*, a removal of a root or an active node will invoke *delete* at the upper store, and an insertion of a new root or an active node will invoke *insert* at the upper store. A λ-reduction may invoke one or two lazy deletions (a λ-reduction can make up to two active nodes non-active) and at most one insertion at the upper store. In total, lazy deletions and insertions have the worst-case cost of $O(1)$. Also *borrow* has the worst-case cost of $O(1)$. At most one real upper-store deletion will be necessary, which has the worst-case cost of $O(\lg \lg n)$ and includes $3 \lg \lg n + O(1)$ element comparisons. Therefore, *delete* has the worst-case cost of $O(\lg n)$ and performs at most $\lg n + 3 \lg \lg n + O(1)$ element comparisons.

In *decrease*, three modifications are necessary. First, each time a new active node is created, *insert* has to be invoked at the upper store. Second, each time an active node is removed by a λ-reduction, the counterpart must be lazily deleted from the upper store. Third, when the node whose value is to be decreased is a root or an active node, *decrease* has to be invoked at the upper store as well. If due to a λ-reduction an active node is made into a root, no change at the upper store is required. After these modifications, the worst-case cost of *decrease* is still $O(1)$.

The following theorem summarizes the main result of the paper.

Theorem 1. *Let n be the number of elements of the heap prior to each operation. A two-tier relaxed heap guarantees the worst-case cost of $O(1)$ for find-min, insert, and decrease; and the worst-case cost of $O(\lg n)$ including at most $\lg n + 3 \lg \lg n + O(1)$ element comparisons for delete.*

We conclude the paper with two remarks. 1) It is relatively easy to extend the data structure to support *meld* at the worst-case cost of $O(\min \{\lg m, \lg n\})$, where m and n are the number of elements of the two melded heaps. 2) It is an open problem whether it is possible or not to achieve a bound of $\lg n + O(1)$ element comparisons for *delete*, when fast *decrease* is to be supported. Note that the worst-case bound of $\lg n + O(1)$ is achievable [6, 7], when *decrease* is not supported.

References

[1] G.S. Brodal. Worst-case efficient priority queues. *Proceedings of the 7th ACM-SIAM Symposium on Discrete Algorithms*, ACM/SIAM (1996), 52–58.

[2] M.J. Clancy and D.E. Knuth. A programming and problem-solving seminar. Technical Report STAN-CS-77-606, Department of Computer Science, Stanford University (1977).

[3] T.H. Cormen, C.E. Leiserson, R.L. Rivest, and C. Stein. *Introduction to Algorithms*, 2nd Edition. The MIT Press (2001).

[4] J.R. Driscoll, H.N. Gabow, R. Shrairman, and R.E. Tarjan. Relaxed heaps: An alternative to Fibonacci heaps with applications to parallel computation. *Communications of the ACM* **31** (1988), 1343–1354.

[5] A. Elmasry. Layered heaps. *Proceedings of the 9th Scandinavian Workshop on Algorithm Theory, Lecture Notes in Computer Science* **3111**, Springer-Verlag (2004), 212–222.

[6] A. Elmasry, C. Jensen, and J. Katajainen. A framework for speeding up priority-queue operations. CPH STL Report 2004-3. Department of Computing, University of Copenhagen (2004). Available at http://cphstl.dk.

[7] A. Elmasry, C. Jensen, and J. Katajainen. Multipartite priority queues. Submitted for publication (2004).

[8] M.L. Fredman. On the efficiency of pairing heaps and related data structures. *Journal of the ACM* **46** (1999), 473–501.

[9] M.L. Fredman, R. Sedgewick, D.D. Sleator, and R.E. Tarjan. The pairing heap: A new form of self-adjusting heap. *Algorithmica* **1** (1986), 111–129.

[10] M.L. Fredman and R.E. Tarjan. Fibonacci heaps and their uses in improved network optimization algorithms. *Journal of the ACM* **34** (1987), 596–615.

[11] G.H. Gonnet and J.I. Munro. Heaps on heaps. *SIAM Journal on Computing* **15** (1986), 964–971.

[12] H. Kaplan, N. Shafrir, and R. E. Tarjan. Meldable heaps and Boolean union-find. *Proceedings of the 34th Annual ACM Symposium on Theory of Computing*, ACM (2002), 573–582.

[13] H. Kaplan and R.E. Tarjan. New heap data structures. Technical Report TR-597-99, Department of Computer Science, Princeton University (1999).

[14] C. Okasaki. *Purely Functional Data Structures*. Cambridge University Press (1998).

[15] M.H. Overmars and J. van Leeuwen. Worst-case optimal insertion and deletion methods for decomposable searching problems. *Information Processing Letters* **12** (1981), 168–173.

[16] J. Vuillemin. A data structure for manipulating priority queues. *Communications of the ACM* **21** (1978), 309–315.

[17] J.W.J. Williams. Algorithm 232: Heapsort. *Communications of the ACM* **7** (1964), 347–348.

The Interval Liar Game

Benjamin Doerr[1], Johannes Lengler[2], and David Steurer[3]

[1] Max–Planck–Institut für Informatik, Saarbrücken, Germany
[2] Mathematics Department, Saarland University, Saarbrücken
johnny@math.uni-sb.de
[3] Computer Science Department, Princeton University, Princeton
dsteurer@cs.princeton.edu

Abstract. We regard the problem of communication in the presence of faulty transmissions. In contrast to the classical works in this area, we assume some structure on the times when the faults occur. More realistic seems the "burst error model", in which all faults occur in some small time interval.

Like previous work, our problem can best be modelled as a two-player perfect information game, in which one player ("Paul") has to guess a number x from $\{1, \ldots, n\}$ using Yes/No-questions, which the second player ("Carole") has to answer truthfully apart from few lies. In our setting, all lies have to be in a consecutive set of k rounds.

We show that (for big n) Paul needs roughly $\log n + \log \log n + k$ rounds to determine the number, which is only k more than the case of just one single lie.

1 Introduction and Results

Communication in the presence of transmission faults is a well-studied subject. Pelc's [Pel02] great survey lists more than a hundred references on such problems.

1.1 Communication Model with Errors

The customary model is that there are two entities, "Sender" and "Receiver". Sender wants to send a message to Receiver. The message is represented by a number x from $[n] := \{1, \ldots, n\}$. If we have an error-free channel, it is clear that Sender needs to send $\log(n) := \log_2(n)$ bits (and Receiver only needs to listen).

In the model with errors, however, some of the bits sent by Sender are flipped. Of course, we need some restriction on the occurrence of errors, as otherwise no reliable communication is possible. Typically, we assume that such errors only occur a certain number of times, at a certain rate or according to a certain probability distribution.

To compete with the errors, we often assume a two-way communication, that is, Receiver may send out information to Sender. However, we typically think of the situation as not symmetric: Bits sent from Receiver to Sender are never flipped (no errors occur). This model is justified in many practical situations where one communication partner has much less energy available and thus his sendings are more vulnerable to errors.

T. Asano (Ed.): ISAAC 2006, LNCS 4288, pp. 318–327, 2006.

1.2   Liar Games

We often adopt a worst-case view. Hence we do not assume the errors to be random, but rather to be decided on by a malevolent adversary. In fact, we may think of that sender not really wanting to share his secret x, but rather trying to keep it by intentionally causing errors (lying). This leads to a so-called *liar game*. In the following, we adopt the language usually used in the analysis of such games. In particular, Sender/Lier will be called "Carole", an anagram of oracle, and Receiver, who is questioning Carole to reveal the secret, will be called "Paul" in honor of Paul Erdős, the great questioner.

The rules of the game are as follows: Carole decides on a number (secret) $x \in [n]$. There are q rounds. Each round, Paul asks a Yes/No-question, which Carole answers. In doing so, Carole may lie according to further specifications. Paul wins the game, if after q such rounds, he knows the number.

To make this a perfect information game (in-line with our worst-case view), let us assume that Carole does not have to decide on the number x beforehand, but rather tries to answer in a way that is consistent with some secret. For technical reasons, we shall also allow that she lies in a way that is inconsistent with any secret, which will be viewed as a win for Paul as well.

We remark that, depending on the parameters n, q, and on the lying restrictions either Paul or Carole has a winning strategy. So we say that Paul wins if he has a winning strategy.

Note that this set-up perfectly models the communication problem with errors. There is one more remark regarding Paul's questions. It seems that his communication effort is much higher, since each question can only be represented by a n bit string.

This could be justfied by the stronger battery Paul has compared to Carole, but there is a more natural explanation: If Paul and Carole agree on a communication protocol beforehand, then Paul does not need to transmit his questions. It suffices that he merely repeats the bit he just received and Carole can deduce the next question from this and the agreed-on protocol.

In the following, we rather use the language of games than that of communication protocols. With the above equivalence at hand, this is merely a question of taste and we follow the authors of previous work in this respect.

1.3   Previous Results

As said, liar games are an intensively studied subject. We now briefly state the main results relevant for our work and refer to the survey paper Pelc [Pel02] for a more complete coverage.

The first to notice the connection between erroneous communication and such games was Alfréd Rényi [Rén61, Rén76]. However, for a long time most of this community was not aware of Rényi's work and cited Ulam [Ula76] as inventor of liar games.

Pelc [Pel87] was the first to completely analyse the game with one lie. He showed that Paul wins for even n if $n \leq 2^q/(q+1)$, and for odd n if $n \leq (2^q - q + 1)/(q+1)$. There are numerous results for $k = 2, 3$, or 4 lies, which we will not discuss here.

Spencer [Spe92] solved the general problem for any fixed number k of lies. Here Paul wins if $n \leq 2^q / \binom{q}{\leq k} (1 + o(1))$, where $\binom{q}{\leq k} = \sum_{i=0}^{k} \binom{q}{i}$.

All results above concern the fully adaptive ('real game') setting with unrestricted questions and fixed numbers of lies. The problem has a quite different nature if only comparison questions ("Is $x \leq s$?" for some $s \in [n]$) are allowed [BK93], a constant fraction of lies is allowed in any initial segment of rounds [Pel89], or Paul's questions have to come in two batches, where Carole gives her answers only after having received the whole batch [CM99].

1.4   Our Contribution

Translating the above results back into the model of erroneous communication, the errors occur independently at arbitrary times. While this might be true for some types of errors, we feel that it is much more likely that the errors occur in bunchs. We think, e.g., of atmospheric disorders. Here, not only a single bit will be affected, but a whole sequence of bits sent.

In the game theoretic setting, we allow Carole to lie up to k times, but only in a way that all lies occur in k consecutive rounds. Note that, in these k rounds, Carole may lie, but of course she does not have to.

The additional interval restriction makes Carole's position much harder. Roughly speaking, Paul only needs k more questions than in the one-lie game. This shows that, in scenarios where it can be assumed, using our interval assumption is a valuable improvement. More precisely, we show the following.

Theorem 1. *Let $n, q \in \mathbb{N}$ and $k \in \mathbb{N}_{\geq 2}$.*

 (i) Paul wins if $q \geq \lceil \log n \rceil + k + \lceil \log \log 2n \rceil$ and $q \geq \lceil \log n \rceil + 2k$.
 (ii) Carole wins if $q < \log n + 2k$.
(iii) Carole wins if $q < \log n + k + \log \log 2n - 1$.

We assumed $k \geq 2$ as otherwise the game in consideration would revert to the searching game with just one lie.

Note that Theorem 1 gives almost matching lower and upper bounds on the number of questions Paul needs to reliably distinguish n integers. Specifically, for all choices of n and k, the upper and lower bound differ by at most 3.

2   Notation and Preliminaries

We describe a game position by a non-negative vector $P = (x_k, \ldots, x_0)$, where x_i is the number of integers for which (assuming it to be the correct answer) Carole is allowed to lie within the next i questions. Note that for the analysis, it does not matter which are the particular integers that Carole may lie for i times, it is only their number that matters.

In particular, x_k is the number of integers for which Carole has never lied, and x_0 is the number of integers for which Carole must not lie anymore. Note that $\sum_{i=0}^{k} x_i \leq n$, and this is strict if there are integers for which Carole would have lied at two times separated by at least k rounds. For the initial position, denoted P^0, we have $x_k = n$ and $x_0 = \ldots = x_{k-1} = 0$.

We continue formalizing the questions Paul is asking. Note first that a Yes/No-questions can always be expressed in the form "$x \in S$?" for some $S \subseteq [n]$. Since again for the analysis the particular integers are not so relevant, we describe the question via an integer vector $v = (v_k, \ldots, v_0)$, where v_i is the number of integers that (i) are in S and (ii) Carole may lie i times for. Consequently, we have $0 \leq v_i \leq x_i$ for all $i \in \{0, \ldots, k\}$. To ease the language, we identify questions with their corresponding vectors.

Depending on Carole's answer there are two possibilities for the next game position P', namely $P' = YES(P, v)$ and $P' = NO(P, v)$, where

$$YES(P, v) = (v_k, x_k - v_k, x_{k-1}, x_{k-2}, \ldots, x_1 + v_0)$$
$$NO(P, v) = YES(P, P - v) = (x_k - v_k, v_k, x_{k-1}, x_{k-2}, \ldots, x_1 + x_0 - v_0)$$

Note that neither $YES(P, v)$ nor $NO(P, v)$ depends on any v_i with $0 < i < k$. For the integers corresponding to these entries, Carole's answer does not affect the state of the game.

For a position $P = (x_k, \ldots, x_0)$, a question (i.e. an integer vector v with $0 \leq v \leq P$) is a *perfect bisection* if $v_0 = \frac{1}{2}x_0$ and $v_k = \frac{1}{2}v_k$.

Recall that $YES(P, v)$ and $NO(P, v)$ do not depend on $v_1, \ldots, v_{k-1}$, so if Paul can make a perfect bisection, then the successor state does not depend on Carole's answer.

We call a question a *quasi-perfect bisection* if $v_i \in \{\lfloor x_i/2 \rfloor, \lceil x_i/2 \rceil\}$ for $i = 0$ and $i = k$.

We conclude this section by explaining when some position is better than another:

Lemma 2. *Let $P = (x_k, \ldots, x_0)$ and $P' = (x'_k, \ldots, x'_0)$ be positions (= non-negative inegral vectors). Assume that P and P' have the following property:*

$$\sum_{i=j}^{k} x_i \leq \sum_{i=j}^{k} x'_i \qquad \text{for all } j = 0, \ldots, k. \tag{1}$$

Then for any q, we have the implication

$$\text{Paul can win } P' \text{ in } q \text{ rounds} \implies \text{Paul can win } P \text{ in } q \text{ rounds}$$

In this case, we call position P at least as good as P', and we call P' at most as good as P.

Proof. Though the statement is rather technical, the idea is simple: We can generate P' out of P by (*i*) allowing Carole some additional lies and (*ii*) adding some more numbers to the search space. Clearly, both operations will make the game harder for Paul, so if he has a winning strategy for P' in q rounds, then exactly the same strategy will also win P.

So we want to prove that we can indeed transform P into P' by operations (*i*) and (*ii*). We use an inductive argument. Firstly, we add some numbers of type x_0 to P until we get equality for $j = 0$, i.e., $\sum_{i=0}^{k} x_i = \sum_{i=0}^{k} x'_i$.

Now we have $x_0 = \sum_{i=0}^{k} x_i - \sum_{i=1}^{k} x_i \geq \sum_{i=0}^{k} x'_i - \sum_{i=1}^{k} x'_i = x'_0$, so $x_0 - x'_0 \geq 0$. We choose $x_0 - x'_0$ numbers in P at the x_0-position. For these numbers, we allow

Carole to lie in the next step. So we get a new position $P^1 = (x_k, \ldots, x_2, x_1 + x_0 - x'_0, x'_0)$, and we know that P is at least as good as P^1.

Now inductively we produce a sequence $P^0 := P, P^1, P^2, \ldots, P^k$ with the following properties:

- P^{i-1} is at least as good as P^i (in the sense of equation (1)).
- P^i is generated from P^{i-1} by operations of type *(i)* and *(ii)*.
- For $0 \le j < i$ we have $x^i_j = x'_j$, where x^i_j is the j-entry of P^i.
- $\sum_{j=0}^k x^i_j = \sum_{j=0}^k x'_j$ for $i > 0$.

Indeed, we have already constructed P^1. Out of P^{i-1}, by the same construction we get P^i, namely by allowing one additional lie for some numbers from x_i^{i-1}. (Formally by setting $P^i := (x_k^{i-1}, \ldots, x_{i+1}^{i-1}, x_i^{i-1} + x_{i-1}^{i-1} - x'_{i-1}, x'_{i-1}, x_{i-2}^{i-1}, \ldots, x_0^{i-1}))$. Note that P^{i-1} and P^i are identical except for the components $i - 1$ and i. It is easy to check that P^i has the desired properties.

Finally, we end up with P^k, which is automatically identical to P'.

Altogether, we have constructed P' out of P by the feasible operations *(i)* and *(ii)*. This proves the claim.

3 Upper Bounds and Strategies for Paul

In this section, we give a strategy for Paul. In this way, we derive upper bounds on the number of questions Paul needs in order to reveal the secret $x \in [n]$. We show (Corollary 6) that for n being a power of 2, Paul can win if

$$q \ge \max \left\{ k + \log n + \lceil \log \log n \rceil, \ 2k + \log n \right\}.$$

Our strategy is constructive, that is, immediately yields an efficiently executable protocol for the underlying communication problem.

Here is an outline of the strategy. Assume that n is a power of two. Clearly, some strategy working for a larger n will also work for a smaller one, hence this assumption is fine (apart from possible a minor loss in the resulting bounds). If all x_i are even, Paul can ask the question $v = \frac{1}{2}P$. He does so for the first $\log n$ rounds of the game (Main Game), resulting in a position with $x_k = 1$. Now the aim is to get rid of this one integer Carole has not lied for yet. To do so, we ask a "trigger question", roughly $(1, 0, \ldots, 0)$. Either we succeeded with our plan and simply repeat asking for half of the x_0-integers (Endgame I), or we end up with very few possible integers altogether (Endgame II), allowing an easy analysis.

Lemma 3 (Main Game). *If n is a power of 2, then with the first $m = \log n$ questions Paul can reach position*

$$P^m = (1, 1, 2, \ldots, 2^{k-2}, (m - k + 1)2^{k-1}).$$

Proof. In the first m rounds, Paul can always ask questions of the form $v = P/2$, where P is the current game position. The position after k such perfect bisections is

$$P^k = (2^{m-k}, 2^{m-k}, 2^{m-k+1}, \ldots, 2^{m-1}).$$

A simple inductive argument shows that the position after $k + \nu$ questions with $\nu \leq m - k$ is

$$P^{k+\nu} = (2^{m-k-\nu}, 2^{m-k-\nu}, 2^{m-k-\nu+1}, \ldots, 2^{m-\nu-2}, (\nu + 1) \cdot 2^{m-\nu-1}).$$

For $\nu = m - k$, we get the statement of the lemma.

After the first m questions, Paul asks a "trigger question" $v^{m+1} = (1, 0, \ldots, 0, 2^{k-2})$. If k is sufficiently small compared to n, Carole will not give up the relatively many possibilities encoded in x_0 and therefore answer "No". The following two lemmas deal with both possible successor positions, namely $YES(P^m, v^{m+1})$ and $NO(P^m, v^{m+1})$.

Lemma 4 (Endgame I). *From position*

$$NO(P^m, v^{m+1}) = (0, 2^0, 2^0, \ldots, 2^{k-3}, (m - k + 1)2^{k-1})$$

Paul wins the game (by reaching position $(0, \ldots, 0, 1)$), with at most $k - 1 + \lceil \log m \rceil$ questions.

Proof. With $k-2$ perfect bisections, Paul reaches the position with $x_k = \ldots = x_2 = 0$, $x_1 = 1$ and $x_0 = 2(m - k + 1) + \sum_{i=1}^{k-2} 2^{i-1}/2^{i-1} = 2m - k$.

In the next question, Paul asks for $m - \lceil k/2 \rceil$ integers corresponding to the last entry of the position. So the next position is no more than

$$(0, \ldots, 0, m - \lfloor k/2 \rfloor + 1) \leq (0, \ldots, 0, m).$$

From this position on, the game reverts to classical "Twenty Questions" problem for a universe of size m. So Paul can win with $\lceil \log m \rceil$ additional questions.

The total number of questions is at most

$$k - 2 + 1 + \lceil \log m \rceil \leq k - 1 + \lceil \log m \rceil.$$

Lemma 5 (Endgame II). *Paul can win with at most $2k - 1$ questions from position*

$$YES(P^m, v^{m+1}) = (1, 0, 2^0, 2^1, \ldots, 2^{k-3}, 2^{k-2})$$

Proof. With $k - 2$ quasi-perfect bisections, Paul reaches a position at least as good as

$$(1, 0, \ldots, 0, \sum_{i=0}^{k-2} 2^i/2^i) = (1, 0 \ldots, 0, k - 1).$$

Now Paul asks for the number corresponding to the first entry of the position, that is, the question $v = (1, 0, \ldots, 0)$. If the answer is "Yes", Paul wins instantly. Otherwise, the position is $(0, 1, 0, \ldots, k - 1)$. Playing the "Twenty Questions" game on the $k - 1$ integers corresponding to the last entry, we reach with $t \leq \lceil \log k \rceil$ additional questions a position with $x_0 = x_{k-1-t} = 1$ and all other entries naught. From this position, Paul can win in $k - t$ questions.

The total number of questions is at most

$$k - 2 + 1 + k = 2k - 1.$$

Corollary 6. *For* $\log n \in \mathbb{N}$, *Paul can win if*

$$q \geq \max \left\{ k + \log n + \lceil \log \log n \rceil, \; 2k + \log n \right\}.$$

Proof. By Lemma 3, we need $\log n$ questions for the main game. Then Paul asks one "trigger question". Depending on Carole's answer, Paul either plays Endgame I or Endgame II. In the first case, he needs $k + \lceil \log \log n \rceil - 1$ further questions to win the game (Lemma 4). In the latter case, Paul wins with $2k - 1$ questions (Lemma 5).

If n is not a power of two, we can replace the starting position $P = (n, 0, \ldots, 0)$ by $(2^{\lceil \log(n) \rceil}, 0, \ldots, 0)$, which is at most as good as P. By the Corollary, Paul can still win if

$$q \geq \max \left\{ k + \lceil \log n \rceil + \lceil \log \log n \rceil, \; 2k + \lceil \log n \rceil \right\},$$

which is the statement in Theorem 1 (i). □

4   Lower Bound

In this section, we prove lower bounds showing that our strategies given in the previous section are optimal up to a small constant number of questions. We start by defining the following *formal weight function*:

$$w_j(x_k, \ldots, x_0) = (j - k + 2)2^{k-1}x_k + \sum_{i=0}^{k-1} 2^i x_i.$$

The weight function is supposed to determine whether it is possible for Paul to find out the correct number in j rounds. It does not quite so, but it solves only a formal relaxation of the problem. (That's why it is called *formal* weight function.)

Note that the weight function is linear in its variables.

The following lemma summarises the important properties of such a formal weight function.

Lemma 7. *(i)* **Triangle equality:** *For all* $j \geq k + 1$ *and for all integral vectors* P *and* v,

$$w_j(P) = w_{j-1}(YES(P, v)) + w_{j-1}(NO(P, v)).$$

(Note: We do not require that the entries of P and v are positive.)
(ii) **Formal descent:** *For all* $j \geq k + 1$ *and for all integral* P, *there is a formal choice* v *for Paul, such that*

$$w_{j-1}(YES(P, v)) = w_{j-1}(NO(P, v)), \text{ if } w_j(P) \text{ is even.}$$

$$w_{j-1}(YES(P, v)) = w_{j-1}(NO(P, v)) + 1, \text{ if } w_j(P) \text{ is odd.}$$

By a formal choice, we mean an integral vector with possibly negative entries.
(iii) **Starting condition:** *For* $j = k$, *if P is a state with non-negative integral entires, we have* $w_k(P) \leq 2^k$ *if and only if Paul can win the situation P in k rounds.*

Proof. Let $P = (x_k, \ldots, x_0)$, $v = (v_k, \ldots v_0)$. Direct calculation proves the assertion:

$$w_{j-1}(YES(P, v)) + w_{j-1}(NO(P, v))$$

$$= \left((j - k + 1)2^{k-1}v_k + 2^{k-1}(x_k - v_k) + \sum_{i=1}^{k-2} 2^i x_{i+1} + (v_0 + x_1) \right)$$

$$+ \left((j - k + 1)2^{k-1}(x_k - v_k) + 2^{k-1}v_k + \sum_{i=1}^{k-2} 2^i x_{i+1} + (x_0 - v_0 + x_1) \right)$$

$$= (j - k + 1)2^{k-1}x_k + 2^{k-1}x_k + \sum_{i=1}^{k-2} 2^{i+1} x_{i+1} + 2x_1 + x_0$$

$$= (j - k + 1)2^{k-1}x_k + \sum_{i=2}^{k-1} 2^i x_i + 2x_1 + x_0$$

$$= (j - k + 1)2^{k-1}x_k + \sum_{i=0}^{k-1} 2^i x_i$$

$$= w_j(P)$$

This proves the triangle equality.

Obviously, if $P = (0, \ldots, 0, 2, 0, \ldots, 0)$, then Paul can choose $v = (0, \ldots, 0, 1, 0, \ldots, 0)$, and thus obtain $w_{j-1}(YES(P, v)) = w_{j-1}(NO(P, v))$. (Because by symmetry $YES(P, v) = NO(P, v)$.)

But w_j is linear in all entries, so it suffices to prove the claim for $P = (0, \ldots, 0, 1, 0, \ldots, 0)$, with the i-th entry $= 1$. Let $P' = (0, \ldots, 0, 1, 0, \ldots, 0)$, but with the $i - 1$-th entry $= 1$. Now put $a := w_j(P) - 2w_{j-1}(P')$. We must distinguish two cases:

- $w_j(P)$ is even: Then also a is even. Put $v := P + (0, \ldots, 0, \frac{a}{2})$. Then $YES(P, v) = P' + (0, \ldots, 0, \frac{a}{2})$, so $w_{j-1}(YES(P, v)) = w_{j-1}(P') + \frac{a}{2} = \frac{1}{2}w_j(P)$. On the other hand, by the triangle equality, $w_{j-1}(NO(P, v)) = w_j(P) - w_{j-1}(YES(P, v)) = \frac{1}{2}w_j(P) = w_{j-1}(YES(P, v))$.
- $w_j(P)$ is odd: Then also a is odd. Put $v := P + (0, \ldots, 0, \frac{a+1}{2})$. Then $YES(P, v) = P' + (0, \ldots, 0, \frac{a+1}{2})$, so $w_{j-1}(YES(P, v)) = w_{j-1}(P') + \frac{a+1}{2} = \frac{1}{2}(w_j(P) + 1)$. On the other hand, by the triangle equality, $w_{j-1}(NO(P, v)) = w_j(P) - w_{j-1}(YES(P, v)) = \frac{1}{2}(w_j(P) - 1) = w_{j-1}(YES(P, v)) - 1$.

For the starting condition, note that due to $j = k$, the weight function simplifies to $w_k(x_k, \ldots, x_0) = \sum_{i=0}^{k} 2^i x_i$.

Case 1: $x_k \geq 1$

In this case, there is a chip C_1 on the x_k-position.

First assume that the weight is $\geq 2^k$. Then there is some other chip C_2. Now Carol can take the following strategy: In the remaining k rounds, she always says that C_2 is the correct chip. Then after the k moves, C_2 is still in the game. But so is C_1, because it takes at least k moves to travel down all the way to the x_0-position and one more to be kicked out. Hence, there are two chips left and Paul cannot decide which one is correct.

Now assume that the weight is $\leq 2^k$. Then C_1 is the only chip, and Paul has already won.

Case 2: $x_k = 0$

First assume that the weight is $\leq 2^k$. Then Paul chooses the question $v := (0, \ldots, 0, \lfloor \frac{1}{2}x_0 \rfloor)$. ($\lfloor \ \rfloor$ means rounding down to the next integer.) The two possible consecutive states differ only at the x_0-position, and it is better for Carole to take $NO(P, v) = (0, x_k, \ldots, x_2, x_1 + \lceil \frac{1}{2}x_0 \rceil)$, having weight

$$w(NO(P,v)) = \left\lceil \frac{1}{2}x_0 \right\rceil + x_1 + \sum_{i=1}^{k-1} 2^i x_{i+1}$$

$$= \left\lceil \frac{1}{2}x_0 \right\rceil + \sum_{i=1}^{k} 2^{i-1} x_i$$

$$= \left\lceil \frac{w(P)}{2} \right\rceil \leq \left\lceil \frac{2^k}{2} \right\rceil = 2^{k-1}.$$

So Paul can assure that in the following state, the weight is $\leq 2^{k-1}$. By induction, after k rounds the weight is ≤ 1, implying that only one chip is left. Hence, Paul wins the game.

Now assume that the weight is $> 2^k$. Paul asks a question, and Carol choses the answer that leaves more chips on the x_0-position. The other positions are indifferent against Carols choice, and the consecutive state is $P_{new} = (0, x_k, \ldots, x_2, x_1 + \tilde{x}_0)$, with some $\tilde{x}_0 \geq \frac{1}{2}x_0$.

Then the weight of the new position is at least

$$w(P_{new}) \geq \tilde{x}_0 + x_1 + \sum_{i=1}^{k-1} 2^i x_{i+1}$$

$$\geq \frac{1}{2}x_0 + \sum_{i=1}^{k} 2^{i-1} x_i$$

$$= \frac{w(P)}{2} > \frac{2^k}{2} = 2^{k-1}.$$

So Carol can assure that in the following state, the weight is $> 2^{k-1}$. By induction, after k rounds the weight is > 1. But during those rounds, all chips must move all the way down to the x_0-position. So all chips have weight 1, implying that there is more than one chip left. Hence, Carol wins the game.

Corollary 8. *If P is a state in the liars game, and if $j \geq k$ with $w_j(P) > 2^j$, then Paul can not win the game within j moves.*

Hence, $\max\{j \geq k \mid w_j(P) \leq 2^j\}$ is a lower bound for the minimal number of questions that Paul needs.

Proof. Assume Paul had a strategy that would yield him victory in j moves. Then Carol does the following: In each round, she picks the answer with the higher weight function. By the triangle equality, the new weight will be at least half the old weight. Hence, we have the invariant that $w_i(P_i) > 2^i$, where P_i is the state when there are i questions left.

In particular, for $i = k$, we have $w_k(P_k) > 2^k$, and by our assumption, Paul can still win within k moves. This is a contradiction to the starting condition of our theorem.

We now show an almost tight lower bound for the case that $n \leq 2^{2^k}$. To do so, we need the following lemma.

Lemma 9. *For $n = 2$, Paul needs at least $2k + 1$ questions to win the game.*

Proof. For the first k questions Carole claims that $x = 1$, and for the next k questions she claims $x = 2$. Now Paul needs one additional questions to finally determine Carole's choice.

The above lower bound for $n = 2$ extends in the following way to arbitrary n.

Lemma 10. *Paul needs at least $\log n + 2k$ questions to win the game.*

Proof. From the start position $(n, 0, \ldots, 0)$, Paul needs at least $\log n - 1$ questions to reach a position $P = (x_k, \ldots, x_0)$ with $x_k = 2$, if Carole always chooses an answer that yields the largest entry in the first component of the successor position. Lemma 9 implies that Paul needs at least $2k + 1$ questions to win the game from position P.

Thus the total number of questions needed for Paul to win the game is at least $\log n + 2k$.

Theorem 1 (ii) is now a corollary of the lemma above.

Proof (Theorem 1 (ii)). If $n < 2^{q-2k}$ then Paul may ask less than $\log n + 2k$ questions and henceforth cannot win the game by Lemma 10.

References

[BK93] Ryan S. Borgstrom and S. Rao Kosaraju. Comparison-based search in the presence of errors. In *STOC*, pages 130–136, 1993.

[CM99] Ferdinando Cicalese and Daniele Mundici. Optimal binary search with two unreliable tests and minimum adaptiveness. In Jaroslav Nesetril, editor, *ESA*, volume 1643 of *Lecture Notes in Computer Science*, pages 257–266. Springer, 1999.

[Pel87] Andrzej Pelc. Solution of Ulam's problem on searching with a lie. *J. Comb. Theory, Ser. A*, 44(1):129–140, 1987.

[Pel89] Andrzej Pelc. Searching with known error probability. *Theor. Comput. Sci.*, 63(2): 185–202, 1989.

[Pel02] Andrzej Pelc. Searching games with errors - fifty years of coping with liars. *Theor. Comput. Sci.*, 270(1-2):71–109, 2002.

[Rén61] Alfréd Rényi. On a problem of information theory. *MTA Mat. Kut. Int. Kozl.*, 6B: 505–516, 1961.

[Rén76] Alfréd Rényi. Napl'o az információelméletről. Gondolat, Budapest, 1976. (English translation: A Diary on Information Theory, Wiley, New York, 1984.)

[Spe92] Joel Spencer. Ulam's searching game with a fixed number of lies. *Theor. Comput. Sci.*, 95(2):307–321, 1992.

[Ula76] Stanislaw M. Ulam. *Adventures of a Mathematician*. p. 281, Scribner, New York, 1976.

How Much Independent Should Individual Contacts Be to Form a Small–World?*

Extended Abstract

Gennaro Cordasco and Luisa Gargano

Dipartimento di Informatica ed Applicazioni, Università di Salerno, Italy
{cordasco, lg}@dia.unisa.it

Abstract. We study Small–World graphs in the perspective of their use in the development of efficient as well as easy to implement network infrastructures. Our analysis starts from the Small–World model proposed by Kleinberg: a grid network augmented with directed long–range random links. The choices of the long–range links are independent from one node to another. In this setting greedy routing and some of its variants have been analyzed and shown to produce paths of polylogarithmic expected length. We start from asking whether all the independence assumed in the Kleinberg's model among long–range contacts of different nodes is indeed necessary to assure the existence of short paths. In order to deal with the above question, we impose (stringent) restrictions on the choice of long–range links and we show that such restrictions do not increase the average path length of greedy routing and of its variations. Diminishing the randomness in the choice of random links has several benefits; in particular, it implies an increase in the clustering of the graph, thus increasing the resilience of the network.

1 Introduction

In this paper we consider Small-World (SW) networks based on Kleinberg's model [5]. We investigate the possibility of diminishing the amount of randomness that nodes need in the choice of their long–range links, while keeping the short routes of the original model. Our proposal has the advantage that the obtained networks show high clustering and, hence, they are particularly resilient to failures.

1.1 Small–World (SW) Networks

The study of many large–scale real world networks shows that such networks exhibit a set of properties that cannot be totally captured by the traditional models: regular graphs and random graphs. Indeed, many biological and social networks occupy a position which is intermediate between completely regular and random graphs. Such networks, commonly called *Small–World* networks, are characterized by the following main properties:

- they tend to be sparse;
- they tend to have short paths (as random graphs);
- they tend to be clustered (unlike sparse random graphs).

---

* This work was partially supported by the Italian FIRB project "WEB-MINDS", http://web-minds.consorzio-cini.it/

T. Asano (Ed.): ISAAC 2006, LNCS 4288, pp. 328–338, 2006.

The study of SW graph was pioneered by Milgram [10] in the 1960s. In his famous experiment, people were asked to send letters to unknown targets only through acquaintances. The result confirmed the believe that random pairs of individuals are connected by short chains of acquaintances.

The work by Milgram has been followed by several studies finalized to a better understanding and modeling of the SW phenomenon. In particular, Watts and Strogatz [16] noticed that, unlikely random graphs, real data networks tend to be clustered. Thus, Watts and Strogatz proposed thinking about SW networks as combining an underlying regular (high–diameter) graph with a few random links. They showed that several real networks fall into this category, e.g. the network of actors cooperations or the neural wiring diagram of the worm *C. elegans*.

Recently, Kleinberg [5] reconsidered an important algorithmic aspect of Milgram's experiment: not only short path exist but individuals are able to deliver messages to unknown targets using short routes. Kleinberg proposed a basic model that uses a two-dimensional grid as underlying interconnection, the grid is then augmented with random links: Each node has an undirected local link to each of its grid neighbors and one directed long-range random link; this last link connects to a node at lattice distance d with probability proportional to d^{-2}. Kleinberg proved that such graphs are indeed *navigable*, that is, a simple *greedy* algorithm finds routes between any source and target using only $O(\log^2 n)$ expected hops in a network on n nodes [5]. Notice that navigability is an interesting property for a graph. Such graphs, in fact, can be easily used in the development of efficient network infrastructures, such as for Peer–to–peer systems, where neither flooding nor complex routing protocols are to be used.

Subsequently, routing strategies that make use of an augmented topological awareness of the nodes have been investigated. In particular it was shown that if each node knows the long-range contacts of its closest $O(\log n)$ neighbors on the grid (*Indirect greedy routing* [8, 13]) then $O(\log^{1+1/s} n)$ expected hops suffice in an s–dimensional lattice, for any fixed s. Papers [11] and [12] consider the improvements obtainable over greedy routing in the case in which the topological awareness of a node is augmented by the knowledge of the long–range contacts of all neighbors of the node (*Neighbor–of–neighbors* routing).

Augmenting an overlay network with random links is also at the base of randomized Peer–to–peer networks. Two examples are randomized Chord and Symphony [12]. Such networks are both obtained by adding $O(\log n)$ random long range links to each node of a ring. They allow to obtain optimal routing [12].

1.2 Low Randomness Small–World Networks

In a SW network, the additional long–range random links represent the chance, that play a large role in creating short paths through the network as a whole. In this paper we consider the following question:

> *Do the long–range contacts really need to be completely random, or some "long–range clustering" could instead be envisaged in such a "navigable" network?*

In other words, we investigate the problem of whether all the independence assumed in the Kleinberg's model among different long–range contacts is indeed necessary to assure the existence of short paths. We show that, up to a certain extend, the answer to the above question is that the same (greedy) routing performances can be maintained if we limit the amount of randomness nodes use in the choice of long–range contacts.

Why to reduce randomization?
In the perspective of using SW graphs as a base for the development of network infrastructures, we notice that besides diameter and degree, a very important property for a such an infrastructure is the resilience to simultaneous node failures.

The resilience of a network grows with the clustering of the graph which provides the overlay network. High clustering provides several alternative pathways throw which flow can pass, thus avoiding the failed component [14]. Clustering is a very interesting concept that is found in many natural phenomena and, roughly speaking, determines how tightly neighbors of a given node link to each other [9]. In a graph the clustering is related to the level of randomization of the graph itself. In particular, it was shown in [16] that the smaller is the randomization the higher is the clustering of the considered system. It is also worth noticing that the use of randomization increases the difficulties in the implementation and testing of applications.

Therefore, it is worth investigating if, in analogy with real SW graphs [16], a SW interconnection can be obtained by using a limited amount of randomness. Such a small value would allow both SW requirements (small average path length and high clustering) to be obtained together with easy routing strategies.

1.3   Our Results

We start from Kleinberg's model (that is an s–dimensional lattice augmented with long–range links) and proceed in two steps. In a first step, we limit the choice of long–ranges of a node u to be done only among those other nodes that differ from u itself in exactly one coordinate; namely, a node $u = \langle u_1, \ldots, u_s \rangle$ has its long–range contacts chosen among nodes $v = \langle v_1, \ldots, v_s \rangle$ for which there exists i ($1 \leq i \leq s$) such that $v_j = u_j$ for each $j \neq i$. We show that all the routing properties immediately translate to this restricted model, some times with easier proofs.

In a second part, we enforce even more the restriction nodes have in establishing their long–range contacts. We introduce the notion of *communities*: keeping the restriction that long–range links can only connect two nodes differing in exactly one coordinate, nodes are partitioned into (random) groups and all nodes belonging to the same group are subject to additional restrictions in the choice of their long range contacts depending on the group they belong to.

We show that a logarithmic (in the number of nodes) number of different communities is sufficient to assure the SW property for the resulting graph. Namely, we analyze the routing performances of greedy, indirect and neighbor–of–neighbor routing strategies in dependence of the number of communities; in particular, if the number of communities is at least logarithmic all the routing strategies attain the same performances as in Kleinberg's original model.

The main results presented in this paper, together with those known for the original Kleinberg model, are summarized in Table 1.

Road Map: In Section 2 we give the basic definitions and present the notation used in the paper. In Section 3 we define Restricted–Small–World networks and analyze their routing properties. Section 4 is devoted to the study of SW networks with communities. Section 5 concludes the paper.

1.4 Related Work

The Small–World phenomenon was first demonstrated by Milgram's famous experiment [10]. Such an experiment showed that not only pairs of people were connected by short chains of acquaintances but also that people were able to route letters in a few hops by forwarding them to one of their acquaintances. Since then there has been quite a lot of work in the literature concerning the SW phenomenon and models for SW networks. A first model was proposed by Watts and Strogatz [16]. Following the notion that SW graph interpolate between completely regular graphs and purely random graphs, such model is based on randomly rewiring each edge of a regular graph with a given probability p. For suitable values of p, this gives rise to networks having the short path lengths and the high level of clustering observed in studied real networks (e.g. the network of actors cooperations or the neural wiring diagram of the worm *C. elegans*).

The model we consider in this paper (namely, an s-dimensional grid augmented with long–range links according to a given probability distribution – cfr. Definition 1) was introduced by Kleinberg in [5]. Kleinberg's proposal gave rise to a vast subsequent literature on routing in SW like networks. The $O(\log^2 n)$ expected number of hops shown by Kleinberg for greedy routing was proved tight in [1]. Moreover [13] and [8] prove that if the knowledge of a node is augmented with the knowledge of the long–range contacts of some of its neighbors then $O(\log^{1+1/s} n)$ can be achieved by greedy routing in a s–dimensional lattice, for any fixed s. A review of the vast literature on the subject can be found in [7].

Table 1. Performance of variants of greedy routing (see Section 2.1 for more details). Routing strategies: Greedy (Greedy routing), IR (Indirect greedy routing), NoN (Neighbor–of–neighbor greedy routing). Grid networks having n nodes, s-dimensions and q long-range contacts are considered: $\mathcal{K}(n, s, q)$ (Kleinberg–Small–World network $\mathcal{K}(n, s, q, p)$ with probability $p(d)$ proportional to d^{-s}, cfr. Definition 1), $\mathcal{R}(n, s, q)$ (Restricted–Small–World network, cfr. Definition 2), $\mathcal{R}_c(n, s, q)$ (Small–World network with communities, cfr. Definition 3).

Paper	Results	Avg #steps	Networks	Observations
[5]	Greedy	$O((\log^2 n)/q)$	$\mathcal{K}(n, s, q)$	
[1, 13]	Greedy	$\Omega((\log^2 n)/q)$	$\mathcal{K}(n, s, q)$	
[11, 12]	NoN	$O((\log^2 n)/(q \log q))$	$\mathcal{K}(n, s, q)$	$s = 1$
[13]	IR	$O(\log^{1+1/s} n)$	$\mathcal{K}(n, s, q)$	$s = O(1)$
[8]	IR	$O((\log^{1+1/s} n)/(q^{1/s}))$	$\mathcal{K}(n, s, q)$	$s = O(1)$
[8]	IR	$\Omega(\log^{1+1/s} n)$	$\mathcal{K}(n, s, q)$	$s = O(1)$
This paper	Greedy	$O((\log^2 n)/q)$	$\mathcal{R}(n, s, q), \mathcal{R}_c(n, s, q)$	
This paper	IR	$O((\log^{1+1/s} n)/(q^{1/s}))$	$\mathcal{R}(n, s, q), \mathcal{R}_c(n, s, q)$	
This paper	NoN	$O((\log^2 n)/(q \log q))$	$\mathcal{R}(n, s, q), \mathcal{R}_c(n, s, q)$	

SW like networks have been considered in the context of routing in peer–to–peer systems. In particular [12] analyzes the use of neighbor–of–neighbor greedy routing strategy, were the local knowledge of a node is augmented with the knowledge of the long–range links of all its contacts. They show that such routing reaches the optimal $O(\log n / \log \log n)$ expected number of hops in SW percolation networks and in the randomized version of the Chord network [15] (both having $O(\log n)$ degree). Paper [2] introduces, in order to speed-up bootstrap in the Chord ring, the concept of classes; such concept has some resemblance to that of communities.

2 Preliminary Notation and Definitions

In the following we denote by d the distance between two points $v = \langle v_1, \ldots, v_s \rangle$ and $u = \langle u_1, \ldots, u_s \rangle$ on an s-dimensional toroidal lattice having $n = m^s$ nodes. The metric distance is defined as $d(v, u) = \sum_{i=1}^{s} d_i$ where $d_i = (u_i - v_i) \bmod m$.

Definition 1. *Kleinberg–Small–World network* ($\mathcal{K}(n, s, q, p)$): *Consider n nodes lying on a toroidal s-dimensional grid $\{0, \ldots, m-1\}^s$ where each node maintains two types of connections:*

> **short–range contacts** *($2s$ connections): Each node has a direct connection to every other node within distance 1 on the grid;*
> **long–range contacts** *(q connections): Each node v establishes q directed links independently according to the probability distribution p (on the integers): each link has endpoint u with probability $p(d(u, v))$.*

All reported results on Kleinberg's model assume $p(d)$ proportional to d^{-s} with normalization factor $\sum_{u,v} d(u, v)^{-s}$. We notice that in Definition 1, we assume the interconnection be formed by a torus while the original Kleinberg's model was based on a grid. For sake of simplicity we will show our results on the torus, however they could be also obtained in case of a grid (e.g. with no wrap–around).

In the following we will denote by:

- $N(v)$ the neighborhood of v, that is, set of $2s$ neighbors of v on the s-dimensional torus;
- $L(v)$ the set of the q long-ranges contacts of node v;
- $N_r(v) = \{v \mid d(u, v) \le r\}$ the ball of radius r and center v.

We recall that,

$$|N_r(v)| = \frac{2^s}{s!} \cdot r^s + \nu(r), \tag{1}$$

where $\nu(x)$ is a positive polynomial of degree $s - 1$[3].

2.1 Routings Strategies

Consider a SW network $\mathcal{K}(n, s, q, p)$. We shortly review the routing strategies adopted in the Small–World related literature and that will be subsequently used in this paper. We denote by v be the node currently holding the message and by t the target key.

Greedy Routing uses only the local knowledge of v: a message is forwarded along the link (either short or long) that takes it closest to the target t.

Indirect greedy routing. and *neighbor–of–neighbor greedy routing* are obtained through an additional *topological awareness* given to the nodes: Each node v is in fact aware of the long-range contacts of some other nodes.

Indirect Greedy Routing (IR) assumes that node v is aware of the long-range contacts of its $|N_r(v)|$ closest neighbors, for some $r > 0$. Formally, IR entails the following decision:

1. Let $L_r(v) = \bigcup_{u \in N_r(v)} L(u)$ denote the set of all long-ranges contacts of nodes in $N_r(v)$;

2. Among the nodes in $L_r(v) \cup N(v)$, assume that z is the closest to the target (with respect to the distance $d()$): If $z \in L(v) \cup N(v)$ then route the message from v **to** z directly; otherwise, let $z \in L(u)$ ($u \in N_r(v)$ for some $u \neq v$) and route the message from v **to** z **via** u.

Neighbor–of–Neighbor Greedy Routing (NoN) assumes that each node knows its long-range contacts, and on top of that it holds the long-range contacts of its neighbors. Here we restrict ourself to consider only the long-range contacts of nodes in $L(v)$, this will be sufficient to get the improvements over greedy assured by NoN. Formally, a NoN greedy step entails the following decision:

1. Consider the set $L^2(v) = \cup_{u \in L(v) \cup \{v\}} L(u)$ of long–range contacts of the long–ranges of v;

2. Among the nodes in $L^2(v) \cup N(v)$, let z be the closest to the target (with respect to the distance $d()$): If $z \in L(v) \cup N(v)$ then route the message from v **to** z directly, otherwise let $z \in L(u)$ ($u \in L(v)$ for some $u \neq v$) and route the message from v **to** z **via** u.

We recall that in the *NoN routing*, u may not be the long-range neighbor of v which is the closest to the target; indeed the algorithm could be viewed as a greedy algorithm on the square of the graph induced by the long-range contacts.

3 Restricted–Small–World Networks

In a restricted–Small–World network we allow each node to make long–range connections only with nodes that differ from it in exactly one coordinate. A connection between two nodes u and v is created with probability proportional to $d^{-1}(u, v)$. In particular this probability is $p(d(u, v)) = \frac{1}{\lambda \cdot d(u,v)}$, where λ is the inverse normalized coefficient, $\lambda = s \cdot \sum_{j=1}^{m} \frac{1}{j} \approx \ln n$. Different connections are established by independent trials.

Definition 2. *A* **Restricted–Small–World** *network* $(\mathcal{R}(n, s, q))$*: is a network* $\mathcal{K}(n, s, q, p)$ *with probability distribution* p *s.t. for any* u *and* v*, the probability of having a long–range link from* u *to* v *is*

$$p(d(u, v)) = \begin{cases} \frac{1}{d(u,v)\ln n} & \text{if } v \text{ and } u \text{ differ in exactly one dimension} \\ 0 & \text{otherwise.} \end{cases}$$

It is easy to observe that any outgoing link goes along a generic dimension i with probability:

$$\sum_{j=1}^{m} p(j) \approx \frac{\ln n^{1/s}}{\ln n} = 1/s.$$

We will show that all the results obtained on Kleinberg's SW networks [5, 13, 8, 12] can be easily proved to remain valid in spite of the restrictions we impose on the long-range connections.

Greedy Routing

Theorem 1. *The average path length is $O\left(\frac{\log^2 n}{q}\right)$ for the greedy routing on $\mathcal{R}(n, s, q)$ when $1 \leq q \leq \log n$.*

Proof. Consider a generic node $v = \langle v_1, \ldots, v_s \rangle \in \{0, \ldots, m-1\}^s$ that holds a message destined for node $t = \langle t_1, \ldots, t_s \rangle$ at distance d. For each $i = 1, \ldots, s$, let d_i the distance between v and t on dimension i that is $d_i = (t_i - v_i) \bmod m$.

The routing proceeds by greedily diminishing the distance on dimension 1 first, until $d_1 = 0$, then the distance on dimension 2 is considered, and so on until the target is reached.

Consider now a fixed dimension i. Clearly the analysis is equivalent to that of greedy routing on a ring with long-range connections added according to Definition 1.

Let ϕ denote the event that the current node is able to diminish the remaining distance, from d_i to at most d_i/k in one hop. The expected number of nodes encountered before a successful event ϕ occurs is $O\left(\frac{k \log n}{q}\right)$. Since the maximum number of times the remaining distance could possibly be diminished is $\log_k d_i \leq \log_k m$, $d_i \leq m$, it follows that the average number of hops we need on each dimension is $O\left(\frac{k \log n}{q} \cdot \log_k m\right)$. By repeating for each dimension (that is, up to s times),

$$O\left(s \cdot \frac{k}{\log k} \frac{\log m \log n}{q}\right) = O\left(\frac{k}{\log k} \frac{\log^2 n}{q}\right). \tag{2}$$

By choosing $k = 2$, the result follows. $\square$

Indirect Greedy Routing

In the following we analyze *indirect greedy routing*. Step 2 in the definition of IR routing can be specialized to $\mathcal{R}(n, s, q)$ as follows:

2'. Among the nodes in $L_r(v) \cup N(v)$, assume that z is the closest to the target (with respect to the distance $d()$): If $z \in L(v) \cup N(v)$ then route the message from v **to** z directly, otherwise first route the message from v **to** $u \in N_r(v)$ (at most r hops), then use the u's long-range to z (1 hop), and finally go from z **to** a node w using a inverse path with respect to the path from v to u, in such a way that w differs from v in exactly one dimension (at most r hops). Formally, if $u = \langle v_1 + d_1, v_2 + d_2, \ldots, v_s + d_s \rangle$ then $w = \langle z_1 - d_1, z_2 - d_2, \ldots, z_s - d_s \rangle$.

Let $N = |N_r(v)|$. We can repeat the proof of Theorem 1 by noticing that Nq long-ranges contacts are available at each greedy step and choosing the parameter k in equation (2) so that $qN = k \ln n$. Hence, we get that the number of indirect steps to reach the destination is $O\left(\frac{k}{\log k} \frac{\log^2 n}{qN}\right) = O(\log_k n)$. Each step requires at most $2r + 1$ hops where, by (1), $r = O\left(s\left(\frac{k \ln n}{q}\right)^{1/s}\right)$. Hence, the average path length is

$$O\left(r \log_k n\right) = O\left(\frac{s \ln n}{\ln k}\left(\frac{k \ln n}{q}\right)^{1/s}\right).$$

Choosing $k = e^s$, we obtain the following result.

Corollary 1. *The average path length is* $O\left(\frac{\log^{1+1/s} n}{q^{1/s}}\right)$ *for the indirect routing on* $\mathcal{R}(n, s, q)$ *when each node is aware of the long-range contacts of its* $\frac{e^s \ln n}{q}$ *closest neighbors.*

Remark 1. Unlikely in [8, 13], where the same result holds only for value of the parameter s independent of n (multiplicative factors in s are discarded in the asymptotic notation), our results are expressed also for a non-constant number s of dimensions. In particular the results in [8, 13] are obtained using an awareness of $O(\log n/q)$ neighbors; this awareness allows to obtain an average path length $O\left(s \cdot \frac{\log^{1+1/s} n}{q^{1/s}}\right)$.

Corollary 2. *The average path length is* $O(\log n)$ *for the indirect routing on* $\mathcal{R}(n, s, q)$ *when* $s \geq \ln \ln n$ *and each node is aware of the long-range contacts of its* $\ln^2 n/q$ *closest neighbors.*

NoN Greedy Routing

In the case of neighbor–of–Neighbor greedy routing, we obtain the following result.

Theorem 2. *The average path length is* $O\left(\frac{\log^2 n}{q \log q}\right)$ *for NoN routing on* $\mathcal{R}(n, s, q)$ *when* $1 < q \leq \log n$.

4 Small–World Networks with Communities

In this section we impose more strict restrictions on the choice of long–range contacts by the nodes in the network. Namely, we assume that nodes are partitioned into c groups, called *communities*. Each node randomly choose one of the communities to belong to; each node in community i, for $0 \leq i < c$, can choose its long–range contacts only among a subset of nodes depending on the parameter i.

Definition 3. *A Small–World network with communities* ($\mathcal{R}_c(n, s, q)$): *is a network* $\mathcal{K}(n, s, q, p)$ *with probability distribution p such that for any u and v the probability of having a long–range from v to u is obtained as follows:*

i) Node v chooses the community its belongs to by uniformly at random selecting an integer c_v in the interval $[0, c)$.

ii) Node v chooses uniformly at random an integer σ in the set $[1, s]$.
Let $t = q \bmod s$, $T = \{\sigma, \sigma + 1 \bmod s \dots, \sigma + t - 1 \bmod s\}$, and, for $i = 1 \dots s$,

$$q_i = \begin{cases} \lceil q/s \rceil & \text{if } i \in T \\ \lfloor q/s \rfloor & \text{otherwise.} \end{cases} \tag{3}$$

For $i = 1, \dots, q$, the ith long–range link from v has endpoint u at distance $d(v, u)$ with probability

$$p(d(v, u)) = \begin{cases} \frac{1}{q} & \text{if } v \text{ and } u \text{ differ in exactly one dimension and } u \text{ is a feasible} \\ & \text{endpoint (i.e. } d(v, u) = \lfloor \gamma_i^{\ell + \frac{c_v}{c}} \rfloor \text{ for some } \ell = 0, \dots, q_i - 1) \\ 0 & \text{otherwise.} \end{cases}$$

where γ_i denotes a real number satisfying $\ln \gamma_i = (\ln m)/q_i$.

Observation 1. *For each node v there are exactly q feasible endpoints each with probability $\frac{1}{q}$. Hence a feasible endpoint of a node v is a long-range of v with probability $1 - \left(1 - \frac{1}{q}\right)^q \geq 1 - e^{-1} > 1/2$.*

4.1 Routing in Small–World Networks with Communities

In this section we show that by introducing communities we reduce the amount of randomness with no harm to the efficiency of the system.

The following preliminary result will be a tool in the analysis of the performances of the various routing strategies.

Lemma 1. *Fix a dimension i. Let $k > 1$ an integer and let Φ denote the probability that a node of a $\mathcal{R}_c(n, s, q)$ network (where $1 \leq q \leq \log n$) is able to diminish with one hop the distance on a fixed dimension i, from d_i to at most d_i/k, then $\Phi = \Omega\left(\frac{q}{k \log n}\right)$ if $c \geq \frac{2k \ln n}{q}$.*

Greedy Routing

Theorem 3. *The average path length is $O\left(\frac{\log^2 n}{q}\right)$ for the greedy routing on $\mathcal{R}_c(n, s, q)$ when $1 \leq q \leq \log n$ and $c \geq \frac{4 \ln n}{q}$.*

Proof. By following the same arguments as in proof of Theorem 1 we only need to show that $\Phi = \Omega\left(\frac{q}{\log n}\right)$, where Φ denotes the probability that the current node is able to halve the distance on a fixed dimension i. Using Lemma 1 with $k = 2$ we obtain the desired value. $\qquad \square$

Indirect Greedy Routing

Consider an indirect greedy routing step as described in Section 3. Let v be the node currently holding the message, consider $N_r(v)$ such that it contains a set $N_c \subseteq N_r(v)$ of nodes which belong to different communities with $|N_c| = O\left(\frac{k \ln n}{q}\right)$. Fix any dimension i for which the distance from v to the target on dimension i is $d_i > r$ (otherwise we route on this dimension using short–range links).

For each node in N_c, consider the event that one of its long–ranges allows to to diminish in one indirect step the distance on dimension i from d_i to at most d_i/k; since nodes in N_c belong to different communities, such events are independent. Then applying Lemma 1, the probability that at least one node in N_c is able to diminish, with one indirect step, the distance to the target on dimension i, from d_i to at most d_i/k is $\Omega(1)$.

If $c \geq \frac{2k \ln n}{q}$ and $|N_r(v)| \geq \frac{k \ln n}{q}$ we have $|N_c| = O\left(\frac{k \ln n}{q}\right)$ and we can reach the destination using $O(\log_k n)$ indirect greedy routing step. Therefore, since $r = O\left(s\left(\frac{k \ln n}{q}\right)^{1/s}\right)$ the average path length is as for Restricted–Small–World.

Corollary 3. *The average path length is* $O\left(\frac{\log^{1+1/s} n}{q^{1/s}}\right)$ *for indirect routing on* $\mathcal{R}_c(n,$ $s, q)$ *when* $1 \leq q \leq \log n$, $c \geq \frac{2e^s \ln n}{q}$ *and each node knows the long-range contacts of its* $\frac{e^s \ln n}{q}$ *closest neighbors.*

NoN Greedy Routing

By analyzing NoN Greedy Routing in SW network with communities, we can prove the following Theorem.

Theorem 4. *The average path length is* $O\left(\frac{\log^2 n}{q \log q}\right)$ *for the NON routing on* $\mathcal{R}_c(n, s, q)$ *when* $1 < q \leq \log n$ *and* $c > \log n$.

5 Conclusions

Our Theorems 3 and 4 answer in a positive way our initial question: *Do the long–range contacts really need to be completely random, or some "long–range clustering" could instead be envisaged in such a "navigable" network?* In a sense, we show that it is not necessary to use a completely eclectic network in order to obtain a SW. Indeed, such result can be obtained using a limited amount of heterogeneity, namely only a logarithmic number of communities.

A part of their theoretical interest, such networks can be used toward the design of efficient as well as easy to implement network infrastructures based on the SW approach. Diminishing the amount of randomness used for random links increases the clustering of the network. Hence, one can get interconnected networks which, in addition to convenient graph properties (such as low average path length and degree) and beside providing the efficient and easy routing algorithms (as offered by Kleinberg's model) offer an increased resilience (due to a higher clustering).

References

1. Lali Barrière, Pierre Fraigniaud, Evangelos Kranakis and Danny Krizanc. "Efficient Routing in Networks with Long Range Contacts". Proc. *DISC 01*, LNCS 2180, pp. 270–284, Springer, 2001.
2. Giovanni Chiola, Gennaro Cordasco, Luisa Gargano, Alberto Negro, and Vittorio Scarano. "Overlay networks with class". *Proc. I-SPAN 2005*, IEEE Press, Dec. 2005.
3. John H. Conway and Neil J.A. Sloane. "Low Dimensional Lattices VII: Coordination Sequences". *Proc. Royal Soc. A453*, pages 2369–2389, 1997.
4. Peter Sheridan Dodds, Roby Muhamad, and Duncan J. Watts. "An Experimental study of Search in Global Social Networks". In *Science Volume 301*, pages 827–829, August 2003.
5. Jon M. Kleinberg. "The Small–World Phenomenon: An Algorithm Perspective". In *Proc. of 32th ACM Symp. on Theory of computing (STOC '00)*, pages 163–170, May 2000.
6. Jon M. Kleinberg. "The Small–World Phenomena and the Dynamics of informations". In *Advances in Neural Information Processing Systems (NIPS)*, December 2001.
7. Jon M. Kleinberg. "Complex Networks and Decentralized Search Algorithms". In *International Congress of Mathematicians (ICM)*, 2006.
8. P. Fraigniaud, C. Gavoille, C. Paul. "Eclecticism Shrinks Even Small Worlds" *PODC 2004*.
9. D. Loguinov and A. Kumar and V. Rai and S. Ganesh Graph-Theoretic, "Analysis of Structured Peer-to-Peer Systems: Routing Distances and Fault Resilience". *In Proc. of the ACM SIGCOMM '03 Conference,* 2003.
10. Stanley Milgram. "The Small World Problem". In *Psychology Today*, pp. 60–67, May 1967.
11. Gurmeet Singh Manku, Mayank Bawa, and Prabhakar Raghavan. "Symphony: Distributed hashing in a Small World". In *Proc. of USENIX Symp.on Internet Technologies and Systems (USITS)*, March 2003.
12. Gurmeet Singh Manku, Moni Naor, and Udi Wieder. "Know thy Neighbor's Neighbor: The Power of Lookahead in Randomized P2P Networks". *Proceedings of 36th ACM Symp. on Theory of Computing (STOC '04)*, pages 54–63, June 2004.
13. Chip Martel and Van Nguyen. "Analyzing Kleinberg's (and other) Small–world models". *In the 23rd ACM Symposium on Principles of Distributed Computing, pages 179–188, 2004.*
14. David Newth and Jeff Ash. "Evolving cascading failure resilience in complex networks". In *Proc. of 8th Asia Pacific Symp. on Intelligent and Evolutionary Systems*, December 2004.
15. Ion Stoica, Robert Morris, David Liben-Nowell, David R. Karger, M. Frans Kaashoek, Frank Dabek, and Hari Balakrishnan. "Chord: A Scalable Peer-to-Peer Lookup Protocol for Internet Applications". In *IEEE/ACM Trans. Networking, Vol. 11, No. 1*, pp. 17–32, 2003.
16. Duncan J. Watts and Steven H. Strogatz. "Collective dynamics of 'smallworld' networks". *In Nature, volume 393, Macmillan Publishers Ltd*, pages 440–442, June 1998.

Faster Centralized Communication in Radio Networks

Ferdinando Cicalese[1,*], Fredrik Manne[2,**], and Qin Xin[2,**]

[1] AG Genominformatik, Technische Fakultät, Universität Bielefeld, Germany
nando@cebitec.uni-bielefeld.de
[2] Department of Informatics, The University of Bergen, Norway
{fredrikm, xin}@ii.uib.no

Abstract. We study the communication primitives of broadcasting (one-to-all communication) and gossiping (all-to-all communication) in known topology radio networks, i.e., where for each primitive the schedule of transmissions is precomputed based on full knowledge about the size and the topology of the network. We show that gossiping can be completed in $O(D + \frac{\Delta \log n}{\log \Delta - \log \log n})$ time units in any radio network of size n, diameter D and maximum degree $\Delta = \Omega(\log n)$. This is an almost optimal schedule in the sense that there exists a radio network topology, such as: a Δ-regular tree in which the radio gossiping cannot be completed in less than $\Omega(D + \frac{\Delta \log n}{\log \Delta})$ units of time. Moreover, we show a $D + O(\frac{\log^3 n}{\log \log n})$ schedule for the broadcast task. Both our transmission schemes significantly improve upon the currently best known schedules in Gąsieniec, Peleg and Xin [PODC'05], i.e., a $O(D + \Delta \log n)$ time schedule for gossiping and a $D + O(\log^3 n)$ time schedule for broadcast. Our broadcasting schedule also improves, for large D, a very recent $O(D + \log^2 n)$ time broadcasting schedule by Kowalski and Pelc.

Keywords: Centralized radio networks, broadcasting, gossiping.

1 Introduction

We consider the following model of a radio network: an undirected connected graph $G = (V, E)$, where V represents the set of nodes of the network and E contains unordered pairs of distinct nodes, such that $(v, w) \in E$ iff the transmissions of node v can directly reach node w and vice versa (the reachability of transmissions is assumed to be a symmetric relation). In this case, we say that the nodes v and w are *neighbours* in G. Note that in a radio network, a message transmitted by a node is always sent to all of its neighbors.

The *degree* of a node w is the number of its neighbours. We use Δ to denote the *maximum degree* of the network, i.e., the maximum degree of any node in the network. The *size of the network* is the number of nodes $n = |V|$.

Communication in the network is synchronous and consists of a sequence of communication steps. In each step, a node v either transmits or listens. If v transmits, then

---

* Supported by the Sofja Kovalevskaja Award 2004 of the Alexander von Humboldt Stiftung.
** Supported by the Research Council of Norway through the SPECTRUM project.

T. Asano (Ed.): ISAAC 2006, LNCS 4288, pp. 339–348, 2006.

the transmitted message reaches each of its neighbours by the end of this step. However, a node w adjacent to v successfully receives this message iff in this step w is listening and v is the only transmitting node among w's neighbors. If node w is adjacent to a transmitting node but it is not listening, or it is adjacent to more than one transmitting node, then a *collision* occurs and w does not retrieve any message in this step.

The two classical problems of information dissemination in computer networks are the *broadcasting* problem and the *gossiping* problem. The broadcasting problem requires distributing a particular message from a distinguished *source* node to all other nodes in the network. In the gossiping problem, each node v in the network initially holds a message m_v, and the aim is to distribute all messages to all nodes. For both problems, one generally considers as the efficiency criterion the minimization of the time needed to complete the task.

In the model considered here, the running time of a communication schedule is determined by the number of time steps required to complete the communication task. This means that we do not account for any internal computation within individual nodes. Moreover, no limit is placed on the length of a message which one node can transmit in one step. In particular, this assumption plays an important role in the case of the gossiping problem, where it is then assumed that in each step when a node transmits, it transmits all the messages it has collected by that time. (i.e., the ones received and its own one.)

Our schemes rely on the assumption that the communication algorithm can use complete information about the network topology. Such topology-based communication algorithms are useful whenever the underlying radio network has a fairly stable topology/infrastructure. As long as no changes occur in the network topology during the execution of the algorithm, the tasks of broadcasting and gossiping will be completed successfully. In this extended abstract we do not touch upon reliability issues. However, we remark that it is possible to increase the level of fault-tolerance in our algorithms, at the expense of some small extra time consumption. We defer this issue to the extended version of this paper.

Our results. We provide a new (efficiently computable) deterministic schedule that uses $O(D + \frac{\Delta \log n}{\log \Delta - \log \log n})$ time units to complete the gossiping task in any radio network of size n, diameter D and maximum degree $\Delta = \Omega(\log n)$. This significantly improves on the previously known best schedule, i.e., the $O(D + \Delta \log n)$ schedule of [10]. Remarkably, our new gossiping scheme constitutes an almost optimal schedule in the sense that there exists a radio network topology, specifically a Δ-regular tree, in which the radio gossiping cannot be completed in less than $\Omega(D + \frac{\Delta \log n}{\log \Delta})$ units of time.

For the broadcast task, we show a new (efficiently computable) radio schedule that works in time $D + O(\frac{\log^3 n}{\log \log n})$, improving the currently best published result for arbitrary topology radio networks, i.e., the $D + O(\log^3 n)$ time schedule proposed by Gąsieniec *et al.* in [10]. It is noticeable that for large D, our scheme also outperforms the very recent (asymptotically optimal) $O(D + \log^2 n)$ time broadcasting schedule by Kowalski and Pelc in [12]. This is because of the significantly larger coefficient of the D term hidden in the asymptotic notation. In fact, in our case the D term comes with coefficient 1.

Related work. The work on communication in known topology radio networks was initiated in the context of the broadcasting problem. In [3], Chlamtac and Weinstein prove that the broadcasting task can be completed in time $O(D \log^2 n)$ for every n-vertex radio network of diameter D. An $\Omega(\log^2 n)$ time lower bound was proved for the family of graphs of radius 2 by Alon et al [1]. In [5], Elkin and Kortsarz give an efficient deterministic construction of a broadcasting schedule of length $D + O(\log^4 n)$ together with a $D + O(\log^3 n)$ schedule for planar graphs. Recently, Gąsieniec, Peleg and Xin [10] showed that a $D + O(\log^3 n)$ schedule exists for the broadcast task, that works in *any* radio network. In the same paper, the authors also provide an optimal randomized broadcasting schedule of length $D + O(\log^2 n)$ and a new broadcasting schedule using fewer than $3D$ time slots on planar graphs. A $D + O(\log n)$-time broadcasting schedule for planar graphs has been showed in [13] by Manne, Wang and Xin. Very recently, a $O(D + \log^2 n)$ time deterministic broadcasting schedule for any radio network was proposed by Kowalski and Pelc in [12]. This is asymptotically optimal unless $NP \subseteq BPTIME(n^{O(\log \log n)})$ [12]. Nonetheless, for large D, our $D + O(\frac{\log^3 n}{\log \log n})$ time broadcasting scheme outperforms the one in [12], because of the larger coefficient of the D term hidden in the asymptotic notation describing the time evaluation of this latter scheme.

Efficient radio broadcasting algorithms for several special types of network topologies can be found in Diks et al. [4]. For general networks, however, it is known that the computation of an optimal (radio) broadcast schedule is NP-hard, even if the underlying graph is embedded in the plane [2, 15].

Radio gossiping in networks with known topology was first studied in the context of radio communication with messages of limited size, by Gąsieniec and Potapov in [8]. They also proposed several optimal or close to optimal $O(n)$-time gossiping procedures for various standard network topologies, including lines, rings, stars and free trees. For general topology radio network a $O(n \log^2 n)$ gossiping scheme is provided and it is proved that there exists a radio network topology in which the gossiping (with unit size messages) requires $\Omega(n \log n)$ time. In [14], Manne and Xin show the optimality of this bound by providing an $O(n \log n)$-time gossiping schedule with unit size messages in any radio network. The first work on radio gossiping in known topology networks with arbitrarily large messages is [9], where several optimal gossiping schedules are shown for a wide range of radio network topologies. For arbitrary topology radio networks, an $O(D + \Delta \log n)$ schedule was given by Gąsieniec, Peleg and Xin in [10]. To the best of our knowledge no better result is known to date for arbitrary topology.

2   Gossiping in General Graphs with Known Topology

The gossiping task can be performed in two consecutive phases. During the first phase we gather all individual messages in one (central) point of the graph. Then, during the second phase, the collection of individual messages is broadcast to all nodes in the network. We start this section with the presentation of a simple gathering procedure that works in time $O((D + \Delta)\frac{\log n}{\log \Delta - \log \log n})$ in free trees. Later we show how to choose a spanning breadth-first (BFS) tree in an arbitrary graph G in order to gather (along its branches) all messages in G also in time $O((D + \Delta)\frac{\log n}{\log \Delta - \log \log n})$, despite the additional edges in G which might potentially cause additional collisions. Finally, we show how the gathering process can be pipelined and sped up to run in $O(D + \frac{\Delta \log n}{\log \Delta - \log \log n})$ time.

A super-ranking procedure. Given an arbitrary tree, we choose its central node c as the root. Then, the nodes in the tree (rooted at c) are partitioned into consecutive layers $L_i = \{v \mid dist(c, v) = i\}$, for $i = 0, .., r$ where r is a radius of the tree. We denote the size of each layer L_i by $|L_i|$.

We use a non-standard approach for ranking the nodes in a rooted tree, which we call *super-ranking*. The super-ranking depends on an integer parameter $2 \le x \le \Delta$, that for our purposes will be optimized later. Specifically, for every leaf v we define $\text{rank}(v, x) = 1$. Then, for a non-leaf node, v with children $v_1, \ldots, v_k$, we define $\text{rank}(v, x)$ as follows. Let $\hat{r} = \max_{i=1,\ldots,k}\{\text{rank}(v_i, x)\}$. If at least x of the children of v have rank $\hat{r}$, then $\text{rank}(v, x) = \hat{r} + 1$ otherwise $\text{rank}(v, x) = \hat{r}$.

For each $x \ge 2$, we define $r_{max}^{[x]} = \max_{v \in T} \text{rank}(v, x)$. As an immediate consequence of the definition of $\text{rank}(\cdot, \cdot)$ we have the following.

Lemma 1. *Let T be a tree with n nodes of maximum degree Δ. Then, $r_{max}^{[x]} \le \lceil \log_x n \rceil$, for each $2 \le x \le \Delta$.*

Note that when $x = 2$ we obtain the standard ranking procedure, that has been employed in the context of radio communication in known topology networks in [6, 9, 10]. Previously this same ranking had been used to define the *Strahler number* of binary trees, introduced in hydrogeology [16] and extensively studied in computer science (cf. [17] and the references therein).

The schedule for gathering messages at the root is now defined in stages using the super-ranked tree under the assumption that the value of the parameter x has been fixed. For the sake of the analysis, we will optimize its value later. We partition the nodes of the tree into different *rank sets* that are meant to separate the stages in which nodes are transmitting, i.e., nodes from different rank sets transmit in different stages. For $y \ge 2$, let $r_{max}^{[y]}$ be the maximum rank for a node of T according to the super-ranking with parameter y. Recall that $r_{max}^{[y]} \le \lceil \log_y n \rceil$. Then, let $R_i(y) = \{v \mid \text{rank}(v, y) = i\}$, where $1 \le i \le r_{max}^{[y]}$.

We use the above rank sets to partition the node set as follows. In particular, we shall use the ranking of the nodes both for the parameter y set to a fixed parameter $x > 2$ and to 2.

Definition 1. *We partition the set of nodes as follows:*
The fast transmission set is given by $F_j^k = \{v \mid v \in L_k \cap R_j(2) \text{ and } parent(v) \in R_j(2)\}$. *Also define* $F_j = \bigcup_{k=1}^{D} F_j^k$ *and* $F = \bigcup_{j=1}^{r_{max}^{[2]}} F_j$.
The slow transmission set is given by $S_j^k = \{v \mid v \in L_k \cap R_j(2) \text{ and } parent(v) \in R_p(2), \text{ for some } p > j; \text{ and } \text{rank}(v, x) = \text{rank}(parent(v), x), x > 2\}$. *Also define* $S_j = \bigcup_{k=1}^{D} S_j^k$ *and* $S = \bigcup_{j=1}^{r_{max}^{[2]}} S_j$.
The super-slow transmission set is given by $SS_j^k = \{v \mid v \in L_k \cap R_j(x) \text{ and } parent(v) \in R_i(x), i > j\}$. *Accordingly, define* $SS_j = \bigcup_{k=1}^{D} SS_j^k$ *and* $SS = \bigcup_{j=1}^{r_{max}^{[x]}} SS_j$.

Lemma 2. *Fix positive integers* $i \leq r_{max}^{[x]}$, $j \leq r_{max}^{[2]}$ *and* $k \leq D$. *Then, during the ith stage, all nodes in* F_j^k *can transmit to their parents simultaneously without any collisions.*

Proof. Consider any two distinct nodes u and v in F_j^k, and suppose they interfere with each other. This is true if they have a neighbor in L_{k-1} in common. Obviously, u and v are on the same level and must therefore have the same parent y in the tree. Moreover, according to the definition of the fast transmission set F_j^k, $u, v, y \in R_j(2)$. However, according to the definition of the super-ranking procedure, if $\texttt{rank}(u, 2) = \texttt{rank}(v, 2) = j$ then $\texttt{rank}(y, 2)$ must be at least $j+1$. Hence the nodes u and v cannot both belong to F_j^k, which leads to a contradiction. ∎

Lemma 3. *Fix positive integers* $i \leq r_{max}^{[x]}$, $j \leq r_{max}^{[2]}$ *and* $k \leq D$. *Then, all messages from nodes in* $S_j^k \cap R_i(x)$ *can be gathered in their parents in at most* $x - 1$ *time units.*

Proof. By Definition 1 we have: For each node v in $S_j^k \cap R_i(x)$ we have that $parent(v)$ has at most $x - 1$ children in $S_j^k \cap R_i(x)$, for $i = 1, 2, ..., r_{max}^{[x]} \leq \lceil \log_x n \rceil$, $j = 1, 2, ..., r_{max}^{[2]} \leq \lceil \log n \rceil$ and $k = 1, .., D$. Now, using the above claim, the desired result is achieved by letting each parent of nodes in $S_j^k \cap R_i(x)$ collect messages from one child at a time. ∎

We shall use the following result from [10].

Proposition 1. *[10] There exists a gathering procedure* Γ *such that in any graph* G *of maximum degree* Δ_G *and diameter* D_G *the gossiping task, and in particular the gathering stage, can be completed in time* $O(D_G + \Delta_G \log n)$.

The following procedure moves messages from all nodes v with $\texttt{rank}(v, x) = i$ into their lowest ancestor u with $\texttt{rank}(u, x) \geq i + 1$, where $x > 2$, using the gathering procedure Γ from the previous proposition.

Procedure SUPER-GATHERING(i);

1. Move messages from nodes in $(F \cup S) \cap R_i(x)$ to SS_i;
   using the gathering procedure Γ *in Proposition 1.*
2. Move messages from nodes in SS_i to their parents;
   all parents collect their messages from their children in SS_i *one by one.*

Note that the subtrees induced by the nodes in $R_i(x)$ have maximum degree $\leq x$. Thus, by Proposition 1 and Lemma 3, we have that the time complexity of step 1 is $O(D + x \log n)$. The time complexity of step 2 is bounded by $O(\Delta)$, where Δ is the maximum degree of the tree. By Lemma 1, $r_{max}^{[x]} \leq \lceil \log_x n \rceil$. Thus, we have that the procedure SUPER-GATHERING completes the gathering stage in time $O((D+\Delta+x \log n) \log_x n)$. Since we can follow this with the trivial broadcasting stage following in time $O(D)$, we have proved the following.

Theorem 1. *In any tree of size* n, *diameter* D *and maximum degree* Δ, *the gossiping task can be completed in time* $O((D + \Delta + x \log n) \log_x n)$, *where* $2 < x \leq \Delta$. *In* *particular when* $\Delta = \Omega(\log n)$, *by choosing* $x = \frac{\Delta}{\log n}$, *we obtain the bound* $O((D + \Delta)\frac{\log n}{\log \Delta - \log \log n})$.

Gathering messages in arbitrary graphs. We start this section with the introduction of the novel concept of a *super-gathering spanning tree* (SGST). These trees play a crucial role in our gossiping-scheme for arbitrary graphs. We shall show an $O(n^3)$-time algorithm that constructs a SGST in an arbitrary graph G of size n and diameter D. In the concluding part of this section, we propose a new more efficient schedule that completes message gathering in time $O(D + \frac{\Delta \log n}{\log \Delta - \log \log n})$.

A *super-gathering spanning tree* (SGST) for a graph $G = (V, E)$ is any BFS spanning tree T_G of G, ranked according to the super-ranking above and satisfying[1]

(1) T_G is rooted at the central node c of G,
(2) T_G is ranked,
(3) all nodes in F_j^k of T_G are able to transmit their messages to their parents simultaneously without any collision, for all $1 \leq k \leq D$ and $1 \leq j \leq r_{max}^{[2]} \leq \lceil \log n \rceil$
(4) every node v in $S_j^k \cap R_i(x)$ of T_G has following property: $parent(v)$ has at most $x - 1$ neighbours in $S_j^k \cap R_i(x)$, for all $i = 1, 2, ..., r_{max}^{[x]} \leq \lceil \log_x n \rceil$, $j = 1, 2, ..., r_{max}^{[2]} \leq \lceil \log n \rceil$ and $k = 1, .., D$.

Any BFS spanning tree T_G of G satisfying only conditions (1),(2), and (3) above is called a *gathering spanning tree*, or simply *GST*. Figure 1 shows an example of a *GST*. We recall the following result from [10].

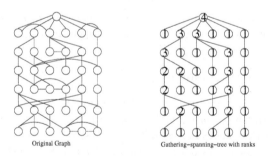

Original Graph   Gathering–spanning–tree with ranks

Fig. 1. Creating a gathering spanning tree

Theorem 2. *There exists an efficient* ($O(n^2 \log n)$ *time) construction of a GST on an arbitrary graph G. (see Theorem 2.5 in [10])*

The procedure SUPER-GATHERING-SPANNING-TREE constructs a super-gathering-spanning-tree SGST $\subseteq G$ on the basis of a GST$\subseteq G$ using a *pruning process*. The pruning process is performed layer by layer starting from the bottom (layer D) of the

---

[1] We use the definition 1 of the ranking partitions given above.

GST. For each layer we gradually fix the parents of all nodes which violate condition (4) above, i.e., each v in $S_j^k \cap R_i(x)$ of *GST*, such that $parent(v)$ has at least x neighbours in $S_j^k \cap R_i(x)$. In fact, for our gathering-scheme, v is a node which is potentially involved in collisions. In each layer, the pruning process starts with the nodes of highest rank in the current layer. We use $NB(v)$ to denote the set of neighbours of the node v in the original graph G. In Figure 2, we show the output of the SUPER-GATHERING-SPANNING-TREE procedure when it is run on the GST presented in Figure 1.

Procedure SUPER-GATHERING-SPANNING-TREE(*GST*);
(1) Fix $\text{rank}(w, 2)$ for every node $w \in V$;
(2) For $k := D$ down to 1 do
(3)   For $i := r_{max}^{[x]}$ down to 1 do
(4)   For $j := r_{max}^{[2]}$ down to 1 do
(5)   While $\exists v \in S_j^k \cap R_i(x)$ in *GST* such that $|S_j^k \cap R_i(x) \cap NB(parent(v))| \geq x$ do
(6) $\text{rank}(parent(v), x) = i + 1$; //$\text{rank}(v, x) = i$
(7) $UPDATE = \{u | u \in S_j^k \cap R_i(x) \cap NB(parent(v))\}$;
(8) $SS_{\text{rank}(v,x)}^k = SS_{\text{rank}(v,x)}^k \cup UPDATE$;
(9) $E_{GST} = E_{GST} - \{(u, parent(u)) | u \in UPDATE\}$;
(10) $E_{GST} = E_{GST} \cup \{(u, parent(v)) | u \in UPDATE\}$;
(11) $S_j^k = S_j^k - \{u | u \in UPDATE\}$;
(12)   re-set $\text{rank}(w, x)$ for each $w \in V$;
(13)   recompute the sets S and SS in *GST*

We now prove that Procedure SUPER-GATHERING-SPANNING-TREE constructs the SGST of an arbitrary graph $G = (V, E)$ in time $O(n^3)$. The following technical lemma is easily proved by induction.

Lemma 4. *After completing the pruning process at layer k in GST, the structure of edges in GST between layers $k - 1, .., D$ is fixed, i.e., each node v within layers $k, .., D$ in all sets $S_j^k \cap R_i(x)$, satisfy the following property: $parent(v)$ has at most $x - 1$ neighbours in $S_j^k \cap R_i(x)$, for $i = 1, .., r_{max}^{[x]} \leq \lceil \log_x n \rceil$ and $j = 1, .., r_{max}^{[2]} \leq \lceil \log n \rceil$.*

By the above lemma, Theorem 2 and the fact that procedure SUPER-GATHERING-SPANNING-TREE preserves the property of the *GST* it starts with, we get

Theorem 3. *For an arbitrary graph there exists an $O(n^3)$ time construction of a SGST.*

$O((D + \Delta) \frac{\log n}{\log \Delta - \log \log n})$**-time gossiping.** Using the ranks computed on the *SGST*, the nodes of the graph are partitioned into distinct *rank sets* $R_i = \{v | \text{rank}(v, x) = i\}$, where $1 \leq i \leq r_{max}^{[x]} \leq \lceil \log_x n \rceil$. This allows the gathering of all messages into the central node c, stage by stage, using the structure of the *SGST* as follows. During the ith stage, all messages from nodes in $(F \cup S) \cap R_i(x)$ are first moved to the nodes in SS_i. Later, we move all messages from nodes in SS_i to their parents in *SGST*. In order to avoid collisions between transmissions originating at neighbouring BFS layers we divide the sequence of transmission time slots into three separate (interleaved)

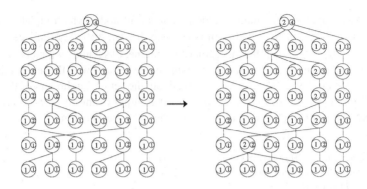

Fig. 2. From gathering-spanning-tree to super-gathering-spanning-tree

subsequences of time slots. Specifically, the nodes in layer L_j transmit in time slot t iff $t \equiv j \pmod 3$.

Lemma 5. *In stage i, the nodes in the set SS_i of the SGST transmit their messages to their parents in time $O(\Delta)$.*

Proof. By [9, Lemma 4], one can move all messages between two partitions of a bipartite graph with maximum degree Δ (in this case two consecutive BFS layers) in time Δ. The solution is based on the use of the *minimal covering set*. Note that during this process a (possibly) combined message m sent by a node $v \in SS_i$ may be delivered to the parent of another transmitting node $w \in SS_i$ rather than to $parent(v)$. But this is fine, since now the time of delivery of the message m to the root of the tree is controlled by the delivery mechanism of the node w. Obviously this flipping effect can be observed a number of times in various parts of the tree, though each change of the route does not change the delivering mechanism at all.

In order to avoid extra collisions caused by nodes at neighbouring BFS layers, we use the solution with three separate interleaved subsequences of time slots incurring a slowdown with a multiplicative factor of 3.   ∎

When the gathering stage is completed, the gossiping problem is reduced to the broadcasting problem. We distribute all messages to every node in the network by reversing the direction and the time of transmission of the gathering stage. In section 3 we prove that the broadcasting stage can be performed faster in graphs with large Δ, i.e., in time $D + O(\frac{\log^3 n}{\log \log n})$.

Theorem 4. *In any graph G with $\Delta = \Omega(\log n)$, the gossiping task can be completed in time $O((D + \Delta)\frac{\log n}{\log \Delta - \log \log n})$.*

Proof. During the ith stage, all messages from $(F \cup S) \cap R_i(x)$ are moved to SS_i. Because of property (4) of the SGST, Proposition 1 assures that this can be achieved in time $O(D + x \log n)$. By Lemma 5, all nodes in the set SS_i can transmit their messages to their parents in $SGST$ in time $O(\Delta)$. By Lemma 1, this process is repeated at most $\log_x n$ times. Thus, the gossiping time can be bounded by $O((D + \Delta + x \log n) \log_x n)$. The desired result follows directly by setting $x = \frac{\Delta}{\log n}$.   ∎

$O(D + \frac{\Delta \log n}{\log \Delta - \log \log n})$ **-time gossiping.** The result of Theorem 4 is obtained by a transmission process consisting of $\lceil \log_x n \rceil$ separate stages, each costing $O(D + \Delta + x \log n)$ units of time. We shall now show that the transmissions of different stages can be pipelined and a new gossiping schedule obtained of length $O(D + \frac{\Delta \log n}{\log \Delta - \log \log n})$.

The communication process will be split into consecutive blocks of 9 time units each. The first 3 units of each block are used for fast transmissions from the set F, the middle 3 units are reserved for slow transmissions from the set S and the remaining 3 are used for super-slow transmissions of nodes from the set SS. We use 3 units of time for each type of transmission in order to prevent collisions between neighbouring BFS layers, like we did in the last section. Recall that we can move all messages between two consecutive BFS layers in time Δ [9, Lemma 4]. Moreover, the same result in [9] together with property (4) of the GSTS, allows us to move all messages stored in $S_j^k \cap R_i(x)$ to their parents in $SGST$ within time $x - 1$.

We compute for each node $v \in S_j \cap R_i(x)$ at layer k the number of a step $1 \leq s(v) \leq x - 1$ in which node v can transmit without interruption from other nodes in $S_j \cap R_i(x)$ also in layer k. We also compute for each node $u \in SS_i$ at layer k the number of a step $1 \leq ss(u) \leq \Delta$ in which the node u can transmit without interruption from other nodes in SS_i also in layer k.

Let v be a node at layer k and with $\text{rank}(v, 2) = j$ and $\text{rank}(v, x) = i$, in SGST. Depending on if v belongs to the set F, to the set S or to the set SS, it will transmit in the time block $t(v)$ given by:

$$
t(v) = \begin{cases}
(D - k + 1) + (j - 1)(x - 1) + (i - 1)(\Delta + (x - 1)\log n) & \text{if } v \in F \\
(D - k + 1) + (j - 1)(x - 1) + s(v) + (i - 1)(\Delta + (x - 1)\log n) & \text{if } v \in S \\
(D - k + 1) + \log n(x - 1) + (i - 1)(\Delta + (x - 1)\log n) + ss(v) & \text{if } v \in SS
\end{cases}
$$

We observe that any node v in the $SGST$ requires at most D fast transmissions, $\log n$ slow transmissions and $\log_x n$ super-slow transmissions to deliver its message to the root of the $SGST$ if there is no collision during each transmission. Moreover, the above definition of $t(v)$ results in the the following lemma, whose proof is deferred to the full version of the paper.

Lemma 6. *A node v transmits its message as well as all messages collected from its descendants towards its parent in SGST successfully during the time block allocated to it by the transmission pattern.*

Since the number of time blocks used is $\leq D + (x \cdot \log n + \Delta) \cdot (\log_x n + 1)$, we have

Theorem 5. *In any graph G, the gossiping task can be completed in time $O(D + (x \cdot \log n + \Delta) \log_x n)$, where $2 \leq x \leq \Delta$. In particular when $\Delta = \Omega(\log n)$, by setting $x = \frac{\Delta}{\log n}$ the bound becomes $O(D + \frac{\Delta \log n}{\log \Delta - \log \log n})$.*

By employing the solution of the equation $\Delta = x \log x$ one can obtained an improved $O(D + \frac{\Delta \log n}{\log \Delta - \log \log n + \log \log \log^* n})$-time gossiping schedule. Moreover, a recursive procedure can be employed to attain the bound $O(D + \frac{\Delta \log n}{\log \Delta - \log \log n + \log^c \log \log^* n})$, where c is some constant.

3   Final Remarks: Broadcasting in Graphs with Known Topology

By exploiting the structure of the SGST it is possible to obtain a very efficient scheduling algorithm for completing the broadcasting task in a general known topology radio
network. The following theorem summarizes our findings. Due to the space constaints
we defer the details to the full version of the paper.

Theorem 6. *For any n node radio network of diameter D, a broadcasting schedule of
length $D + O(\frac{\log^3 n}{\log \log n})$ can be deterministically constructed in polynomial time.*

References

1. N. Alon, A. Bar-Noy, N. Linial and D. Peleg. A lower bound for radio broadcast. *J. Computer
    and System Sciences* 43, (1991), 290 - 298.
2. I. Chlamtac and S. Kutten. On broadcasting in radio networks-problem analysis and protocol
    design. *IEEE Trans. on Communications* 33, (1985), pp. 1240-1246.
3. I. Chlamtac and O. Weinstein. The wave expansion approach to broadcasting in multihop
    radio networks. *IEEE Trans. on Communications* 39, (1991), pp. 426-433.
4. K. Diks, E. Kranakis, D. Krizanc and A. Pelc. The impact of information on broadcasting
    time in linear radio networks. *Theoretical Computer Science*, 287, (2002), pp. 449-471.
5. M. Elkin and G. Kortsarz. Improved broadcast schedule for radio networks. *Proc. 16th ACM-
    SIAM Symp. on Discrete Algorithms*, 2005, pp. 222-231.
6. I. Gaber and Y. Mansour. Broadcast in radio networks. *Proc. 6th ACM-SIAM Symp. on Dis-
    crete Algorithms*, 1995, pp. 577-585.
7. L. Gąsieniec, E. Kranakis, A. Pelc and Q. Xin. Deterministic M2M multicast in radio net-
    works. *Proc. 31st ICALP*, 2004, LNCS 3142, pp. 670-682.
8. L. Gąsieniec and I. Potapov, Gossiping with unit messages in known radio networks. *Proc.
    2nd IFIP Int. Conference on Theoretical Computer Science*, 2002, pp. 193-205.
9. L. Gąsieniec, I. Potapov and Q. Xin. Efficient gossiping in known radio networks. *Proc. 11th
    SIROCCO*, 2004, LNCS 3104, pp. 173-184.
10. L. Gąsieniec, D. Peleg and Q. Xin. Faster communication in known topology radio networks.
    Proc. 24th Annual ACM SIGACT-SIGOPS PODC, 2005, pp. 129-137.
11. L. Gąsieniec, T. Radzik and Q. Xin. Faster deterministic gossiping in ad-hoc radio networks.
    Proc. 9th Scandinavian Workshop on Algorithm Theory, 2004, LNCS 3111, pp. 397-407.
12. D. Kowalski and A. Pelc. Optimal deterministic broadcasting in known topology radio net-
    works. *Distributed Computing*, (2006), to appear.
13. F. Manne, S. Wang and Q. Xin. Faster radio broadcast in planar graphs. *Manuscript, 2006.*
14. F. Manne and Q. Xin. Optimal gossiping with unit size messages in known radio networks.
    Proc. 3rd Workshop on Combinatorial and Algorithmic Aspects of Networking, to appear.
15. A. Sen and M.L. Huson. A new model for scheduling packet radio networks. *Proc. 15th Joint
    Conf. of IEEE Computer and Communication Societies*, 1996, pp. 1116-1124.
16. A.N. Strahler. Hypsometric (area-altitude) analysis of erosional topology. *Bull. Geol. Soc.
    Amer.* 63, (1952), pp. 117–1142.
17. X.G. Viennot. A Strahler bijection between Dyck paths and planar trees. *Discrete Mathemat-
    ics* 246, (2002), pp. 317–329.

On the Runtime and Robustness of Randomized Broadcasting[*]

Robert Elsässer and Thomas Sauerwald

University of Paderborn
Institute for Computer Science
33102 Paderborn, Germany
{elsa, sauerwal}@upb.de

Abstract. One of the most frequently studied problems in the context of information dissemination in communication networks is the broadcasting problem. In this paper, we study the following randomized broadcasting protocol. At some time t an information r is placed at one of the nodes of a graph. In the succeeding steps, each informed node chooses one neighbor, independently and uniformly at random, and informs this neighbor by sending a copy of r to it.

In this work, we develop tight bounds on the runtime of the algorithm described above, and analyze its robustness. First, it is shown that on Δ-regular graphs this algorithm requires at least $\log_{2-\frac{1}{\Delta}} N + \log_{(\frac{\Delta}{\Delta-1})^\Delta} N - o(\log N)$ rounds to inform all N nodes. For general graphs, we prove a slightly weaker lower bound and improve the upper bound of Feige et. al. [8] to $(1+o(1))N \ln N$ which implies that $K_{1,N-1}$ is the worst-case graph. Furthermore, we determine the worst-case-ratio between the runtime of a fastest deterministic algorithm and the randomized one.

This paper also contains an investigation of the robustness of this broadcasting algorithm against random node failures. We show that if the informed nodes are allowed to fail in some step with probability $1-p$, then the broadcasting time increases by a factor of at most $6/p$. Finally, the previous result is applied to state some asymptotically optimal upper bounds for the runtime of randomized broadcasting in Cartesian products of graphs and to determine the performance of agent based broadcasting [6] in graphs with good expansion properties.

1 Introduction

The study of information spreading in large networks has various fields of application in distributed computing. Consider for example the maintenance of replicated databases on name servers in a large network [5]. There are updates

---

[*] This work is partially supported by German Science Foundation (DFG) Research Training Group GK-693 of the Paderborn Institute for Scientific Computation (PaSCo) and by Integrated Project IST-15964 "Algorithmic Principles for Building Efficient Overlay Computers" (AEOLUS) of the European Union.

T. Asano (Ed.): ISAAC 2006, LNCS 4288, pp. 349–358, 2006.

injected at various nodes, and these updates must be propagated to all the nodes in the network. In each step, a processor and its neighbor check whether their copies of the database agree, and if not, they perform the necessary updates. In order to be able to let all copies of the database converge to the same content, efficient broadcasting algorithms have to be developed.

There is an enormous amount of experimental and theoretical study of broadcasting algorithms in various models and on different network topologies. Several (deterministic and randomized) algorithms have been developed and analyzed. In this paper we only concentrate on the efficiency of randomized broadcasting and mainly consider the runtime of the so called *push algorithm* [5] which is defined in the following way: In a graph $G = (V, E)$, where $N := |V|$, we place at some time t an information on one of the nodes. Then, in every succeeding round, each *informed* vertex sends the information to one of its neighbors selected independently and uniformly at random.

The advantage of randomized broadcasting is in its inherent robustness against several kinds of failures and dynamical changes compared to deterministic schemes that either need substantially more time [9] or can tolerate only a relatively small number of faults [15]. Most papers dealing with randomized broadcasting analyze the runtime of the push algorithm in different graph classes. Pittel [19] proved that with a certain probability an information is spread to all nodes by the push algorithm within $\log_2 N + \ln N + O(1)$ steps in a complete graph. Feige et al. [8] determined asymptotically optimal upper bounds for the runtime of this algorithm on several graph classes. They showed that in random graphs and hypercubes of size N, all nodes of the graph receive the information within $O(\log N)$ steps, with high probability[1] and also proved that in bounded degree graphs the number of broadcasting steps is bounded by $O(D + \log N)$, where D is the diameter of the graph, and provided simple bounds on the broadcasting time in general graphs. In [7], we could prove optimality also on Star graphs [1].

A model related to the push algorithm has been introduced in [5] and is called *pull model*. Here, any (informed or uninformed) node is allowed to call a randomly chosen neighbor, and the information is sent from the called to the calling node. Please note, these kind of transmission makes only sense if new or updated informations occur frequently in the network so that every node places a random call in each round anyway.

It was observed in complete graphs of size N that the push algorithm needs at least $\Omega(N \log N)$ transmissions to inform all nodes of the graph, w.h.p. However, in the case of the pull algorithm if a constant fraction of the nodes are informed, then within $O(\log \log N)$ additional steps every node of this graph becomes informed as well, w.h.p. [5, 14]. This implies that in such graphs at most $O(N \log \log N)$ transmissions are needed if the distribution of the information is stopped at the right time.

---

[1] When we write "with high probability" or "w.h.p." we mean with probability at least $1 - 1/N$.

In [6], we introduced the so called *agent based broadcasting*. In this model, N agents are distributed among the nodes and jump from one node to another via edges which are chosen uniformly at random in each time step. An information r placed initially on one node is carried by the agents to other vertices. If an agent visits an informed node, then the agent becomes informed, and any node visited by an informed agent becomes informed as well. Also in this environment, $O(\log N)$ steps are sufficient to distribute r among all nodes in Random graphs. We also could prove in this model that bounded degree graphs support optimal broadcasting. Furthermore, we point at several examples, on which the agent-based broadcasting algorithm and the push-algorithm considered in this paper behave differently.

We should also note that several broadcasting models have been analyzed in scenarios that allow nodes and/or edges to fail during the algorithm is executed (e.g. [13, 14, 17]). Most of these papers deal with the worst case asymptotic behavior of broadcasting algorithms when the failures are goverened by an adversary, however, in some papers the random failure scenario is also considered.

In this paper we study the push algorithm and examine its runtime and robustness in general graphs. Section 2 consists of several lower and upper bounds on the runtime. First we show that in any Δ-regular graph the push algorithm requires $\log_{2-\frac{1}{\Delta}} N + \log_{(\frac{\Delta}{\Delta-1})^\Delta} N - o(\log N)$ rounds to inform all nodes of the graph, w.h.p. Since this function is decreasing in Δ, this result implies that among regular graphs, this algorithm performs best in complete graphs (cf. [19]). We also analyze the runtime of the push algorithm in general graphs, and prove that $\log_{2-\frac{1}{\Delta}} N + \log_4 N - o(\log N)$ steps are necessary, w.h.p., to inform all nodes of an arbitrary graph. Moreover, we extend a result of Feige et.al.[8] by proving that the graph $K_{1,N-1}$ is the worst-case-graph and determine in several graphs the worst-case-ratio between the runtime of an optimal deterministic algorithm and the push algorithm.

In section 3 we also analyze its robustness against random failures, and show that if the informed nodes are allowed to fail in some step with probability $1 - p$, then the broadcasting time increases by a factor of at most $6/p$. An important implication of this result is that agent based broadcasting [6] has in many graphs the same asymptotic runtime as the broadcasting algorithm defined above. Finally, we derive asymptotically tight bounds for the runtime of randomized broadcasting in Cartesian products of graphs, and compare the runtime of deterministic and randomized broadcasting in general graphs. The last section contains our conclusions. Due to space limitations some proofs are omitted in this extended abstract.

2   Bounds on the Broadcasting Time

2.1   Notations and Definitions

Throughout this paper, let $G = (V, E)$ be an unweighted, simple and connected graph of size $N := |V(G)|$ and diameter $\text{diam}(G)$. We denote by δ and Δ the minimum and the maximum degree of G, respectively. As already mentioned, in this paper we mainly consider the following randomized broadcasting algorithm (known also as the push model [5]). At the beginning round $t = 0$ one arbitrary node owns an information which is to be sent to all other nodes in G. In the following rounds $t = 1, 2, \ldots$, each *informed* vertex contacts one neighbor selected independently and uniformly at random and sends the information to it.

In this paper we focus on the number of required rounds. Let $I(t)$ be the set of informed nodes at time t and $H(t) := V \setminus I(t)$. Let $\mathsf{RT}(G, p) = \min_{t \in \mathbb{N}} \{ \mathbf{Pr}\,[\,I(t) = V\,] \geq p \}$ denote the runtime of our randomized algorithm in G (guaranteed with probability p), i.e., the number of rounds needed by the push algorithm to inform all vertices of G with a given probability p. Additionally, let $\mathbb{E}\,[\,\mathsf{RT}(G)\,]$ denote the expected runtime. Clearly, in any graph G at least $\max\{\mathrm{diam}(G) + \log_2 N\}$ rounds are always required to inform all nodes.

2.2   Lower Bounds

There exists several techniques to prove lower bounds for deterministic broadcasting. In most cases these bounds make use of a bounded maximum degree which often leads to expressions using Fibonacci-Numbers, see e.g. [3], or rely on the special structure of some graphs [13].

Theorem 1. *Let $G = (V, E)$ be an arbitrary Δ-regular graph, where $\Delta \geq 2$. Then it holds that*

$$\mathsf{RT}(G, \Theta(1)) \geq \log_{2 - \frac{1}{\Delta}} N + \log_{(\frac{\Delta}{\Delta - 1})^\Delta} N - o(\log N).$$

Proof. In order to show the theorem we consider two cases. First, we assume that $2 \leq |I(t)| \leq N/4$. Since the set of informed nodes always forms a connected subgraph, any node $v \in I(t)$ has at least one informed neighbor. Therefore, an informed node contacts in step t some informed neighbor with probability at least $1/\Delta$, which implies that $\mathbb{E}\,[\,|I(t)| \mid |I(t-1)| = k\,] \leq (2 - 1/\Delta)k$. Then we obtain by using conditional expectations (see e.g. [18]).

$$\mathbb{E}\,[\,|I(t)|\,] = \sum_{k=0}^{N} \mathbf{Pr}\,[\,|I(t-1)| = k\,] \cdot \mathbb{E}\left[\,|I(t)| \;\middle|\; |I(t-1)| = k\,\right]$$

$$\leq \left(2 - \frac{1}{\Delta}\right) \cdot \mathbb{E}\,[\,|I(t-1)|\,] \leq \left(2 - \frac{1}{\Delta}\right)^t.$$

Now, if $c > 0$ is any constant, the Markov inequality leads to

$$\mathbf{Pr}\left[\,|I(\log_{2 - \frac{1}{\Delta}}(c\tfrac{N}{4}))| \geq N/4\,\right] \leq \frac{\mathbb{E}\left[\,|I(\log_{2 - \frac{1}{\Delta}}(c\tfrac{N}{4}))|\,\right]}{\frac{N}{4}} = \frac{4cN}{4N} = c.$$

In the second case, let t_0 be the first time step when $|I(t)| \geq N/4$. If v is a node in V, then the probability that in some round v will not be contacted is exactly $(\frac{\Delta-1}{\Delta})^\Delta$. As in the first case, we have $\mathbb{E}\left[|H(t)|\right] \geq (\frac{\Delta-1}{\Delta})^{(t-t_0)\Delta}N/2$, and applying the same methods as before, we conclude that if $|\bar{I}(t_0)| \leq N/2$ at time t_0, then after $\log_{(\frac{\Delta}{\Delta-1})^\Delta}(cN/2)$ additional rounds some node is uninformed with constant probability. $\qquad\square$

Using the Azuma-Hoeffdings Inequality [18], we could guarantee the lower bound with much higher probability, however, the analysis would be more complicated.

In [19] Pittel showed that on the complete graph $\log_2 N + \ln N + o(1)$ rounds are both sufficient and necessary with probability $1 - o(1)$, which matches asymptotically our lower bound for regular graphs if $\Delta = N - 1$. Since $\log_{2-\frac{1}{\Delta}} N + \log_{(\frac{\Delta}{\Delta-1})^\Delta} N$ is monotonously decreasing in $\Delta \in \mathbb{N}$, as can be shown by using Taylor series, the complete graphs have asymptotically the lowest runtime among all regular graphs.

It is worth mentioning that the assumption of regularity is crucial for the last proof. One can easily construct non-regular graphs and choose $N/2$ (connected) informed nodes such that after less than $\ln N$ rounds all nodes will be informed w.h.p.

Theorem 2. *Let $G = (V, E)$ be an arbitrary graph, where $N \geq 3$. Then,*

$$\mathsf{RT}(G, \Theta(1)) \geq \log_{2-\frac{1}{\Delta}} N + \log_4 N - o(\log N).$$

Proof. Again, we consider two cases. First, we assume that $\delta(G) \geq 2$. For a fixed node $v \in V$ denote by $\mathbb{E}[v]$ the expected number of nodes which contact v in one round. Note, that $\sum_{v \in V} \mathbb{E}[v] = N$. We call a node v *good* if $\mathbb{E}[v] < 1 + \gamma$ for some $\gamma = \Omega(1/\log\log N)$. If we denote by x the number of good nodes, then it holds that $x \geq N \cdot (1 - \frac{1}{1+\gamma})$.

Now, for any good node v we have $\mathbb{E}[v] = \sum_{w \in N(v)} \frac{1}{\deg(w)} \leq 1 + \gamma$, where $N(v)$ represents the set of neighbors of v, and $\deg(w)$ is the degree of w. Then, in some round node v is not contacted with probability

$$\prod_{w \in N(v)} \left(1 - \frac{1}{\deg(w)}\right) = \prod_{w \in N(v)} \left(1 - \frac{1}{\deg(w)}\right)^{\frac{\deg(w)}{\deg(w)}} \geq \left(\frac{1}{4}\right)^{1+\gamma},$$

since $(1 - 1/\deg(w))^{\deg(w)} \geq 1/4$ whenever $\deg(w) \geq 2$. Using the same techniques as in the proof of Theorem 1, we can show that, with probability at least $1/2$, more than $\log_{2-1/\Delta}(\frac{N}{4} \cdot (1 - \frac{1}{1+\gamma})) - o(\log N) = \log_{2-1/\Delta} N - o(\log N)$ rounds are needed to inform $\frac{N}{2} \cdot (1 - \frac{1}{1+\gamma})$ nodes. Therefore, with constant probability, there are still $\frac{N}{2} \cdot (1 - \frac{1}{1+\gamma})$ uninformed good nodes, which are not contacted in some further round t with probability $(\frac{1}{4})^{1+\gamma}$. The same arguments as before imply that, with probability at least $1/2$, we need more than $\log_{4^{1+\gamma}}(\frac{N}{4} \cdot (1 - \frac{1}{1+\gamma})) - o(\log N) = \frac{1}{1+\gamma}\log_4 N - o(\log N)$ additional rounds to inform all nodes in the graph.

The case $\delta = 1$ is similar, but is omitted due to space limitations. $\qquad\square$

2.3   Upper Bounds

Proposition 1. *For any Δ-regular graph $G = (V, E)$ it holds that $\mathbb{E}\left[\,\mathsf{RT}(G)\,\right] = O(N)$, and for any Δ there exists a Δ-regular graph G such that $\mathbb{E}\left[\,\mathsf{RT}(G)\,\right] = \Omega(N)$.*

Of course, a similar results holds if δ and Δ differ only by a constant factor.

In [8] it is shown that $\mathsf{RT}(G, 1 - 1/N) \leq 12N \ln N$. Moreover they gave $K_{1,N-1}$ as an example for a graph with a runtime of $\Omega(N \ln N)$. The following theorem tightens their upper bound and in particular implies that $K_{1,N-1}$ is the worst-case graph.

Theorem 3. *For any graph $G = (V, E)$ it holds that*

$$\mathsf{RT}(G, 1 - o(1)) \leq (1 + o(1))N \ln N.$$

It is easy to see, that the runtime in $K_{1,N-1}$ reduces to the Coupon-Collector-Problem [18] and $(1 + o(1))N \ln N$ rounds are both sufficient and necessary.

2.4   Price of Randomness

In this subsection, we compare the runtime of a fastest deterministic broadcasting algorithm with the runtime of the randomized broadcasting algorithm. Let $\mathsf{PR}(\mathcal{G}) = \sup_{G \in \mathcal{G}}(\mathsf{RT}(G, 1/2)/\mathsf{DT}(G))$ denote the *price of randomness* [14] for some graph class $\mathcal{G}$, where $\mathsf{DT}(G)$ is the runtime of an optimal determinsitic algorithm in G.

Theorem 4. *Let $\mathcal{R}$ be the set of regular graphs, $\mathcal{E}$ the set of regular edge-transitive graphs, and $\mathcal{G}$ the set of general graphs according to Section 2.1. Then, we have*

$$\mathsf{PR}(\mathcal{G}) = \Theta(N), \ \ \mathsf{PR}(\mathcal{R}) = \Theta\left(\frac{N}{\log N}\right) \ \text{and} \ \mathsf{PR}(\mathcal{E}) = O(\log N).$$

3   Robustness of Randomized Broadcasting and Applications

In this section we analyze the robustness of the push algorithm against random node failures. Then this result is applied to derive new bounds on the runtime of the push-algorithm in Cartesian product of graphs and also related to agent based broadcasting.

3.1   A Robustness Result

In this section we consider the robustness of the push algorithm against random failures. We assume here that in each round t, any informed node is allowed to fail with probability $1 - p$ for some $p \in (0, 1)$, independently of any failure in other rounds. However, there might exist failure dependencies between nodes within one round. We should note that our model is somehow a generalization of the probabilistic failure-model examined in [13], in which no dependencies between failures within the same round are allowed.

As described above, only informed vertices are allowed to fail. If an informed vertex fails in some round t, then it does not choose any communication partner to send the message. If it is functional, then it executes the push algorithm as described before. If some informed node is able to send a message, then the transmission will be completed.

It is worth mentioning, however, that this model can be extended to other random failure models as well. We should also note here that the results below can be generalized to the case when restricted dependencies are allowed between the time steps (e.g. if a node fails in some step $t + 1$ after being functional in step t, then it fails for $O(1)$ further rounds). Therefore, this model is well-suited to describe restricted asynchronicity in a network, in which even if some nodes are busy for a time period, the messages sent to these nodes do not get lost.

Denote by $\mathsf{RT}'(G, p) = \min_{t \in \mathbb{N}} \{\mathbf{Pr}\,[\,I(t) = V\,] \geq p\}$ the runtime of the push-algorithm in the previously described failure model.

Theorem 5. *For any graph G it holds that*

$$\mathsf{RT}'(G, 1 - O(1/N)) \leq \frac{6}{p} \cdot \mathsf{RT}(G, 1 - O(1/N)).$$

Proof. In this proof, we are going to show that any instance of the push algorithm in the failure model can be related to an instance of the push algorithm without failures. Then, we show that, with very high probability, there is no large difference between the runtimes of the corresponding instances.

For an instance T of the push algorithm (without failures) let $N_{T,j}(v)$ denote the neighbor of v chosen in step $i(v) + j$, where $i(v)$ denotes the time step in which v has been informed (according to instance T). Accordingly, let $(N_{T,j}(v))_{j=1}^{\infty}$ be the sequence of nodes chosen by v in steps $i(v)+1, \ldots, \infty$. Similarly, for any instance T' of the push algorithm in the failure model, let $N'_{T',j}(v)$ denote the neighbor of v chosen in step $i'(v) + X_{T',j}(v) + j$, where $i'(v)$ denotes the time step in which v has been informed according to T' and $X_{T',j}(v)$ is the number of failures of v before v has been functional j times, i.e., the number of failures within the first $X_{T',j}(v) + j$ steps after v becomes informed. Again, let $(N'_{T',j}(v))_{j=1}^{\infty}$ be the sequence of nodes chosen by v in the steps v is functional. Furthermore, let $\mathsf{RT}(T)$ be the *exact* runtime of the push algorithm for instance T, and let $\mathsf{RT}'(T')$ be the runtime of the push algorithm (in the failure model) for instance T'. Hereby, an instance T of the push algorithm is completely described by the set of sequences $\cup_{v \in V}(N_{T,j}(v))_{j=1}^{\infty}$ and the node informed at the beginning. However, an instance T' is only described by both sets of sequences $\cup_{v \in V}(N'_{T',j}(v))_{j=1}^{\infty}$, $\cup_{v \in V}(X_{T',j}(v))_{j=1}^{\infty}$, and the node informed at the beginning. In the following paragraphs, $N_{T,j}(v)$ is simply denoted by $N_j(v)$ for any j and v. Let now $T'(\cup_{v \in V}(N_j(v))_{j=1}^{\infty})$ denote the set of instances in the failure model, which contain the set of sequences $\cup_{v \in V} (N_j(v))_{j=1}^{\infty}$. Now we are going to show for any set of sequences $\cup_{v \in V} (N_j(v))_{j=1}^{\infty}$ that at least $(1 - O(1/N))|T'(\cup_{v \in V} (N_j(v))_{j=1}^{\infty})|$ instances of $T'(\cup_{v \in V} (N_j(v))_{j=1}^{\infty})$ will have a runtime which is less than $\frac{6}{p} \mathsf{RT}(\cup_{v \in V} (N_j(v))_{j=1}^{\infty})$.

To show this, we first consider the push algorithm without failures, and analyze for the instance $\cup_{v \in V} (N_j(v))_{j=1}^{\infty}$ some path used by the information to reach from the starting node u a node v. Let $P(u, v) := (u = u_1 \to u_2 \to \cdots \to u_n = v)$ be this path, and define $d_j := i(u_{j+1}) - i(u_j)$ as the time the information needs to reach u_{j+1} from u_j, i.e., the time difference between the time step u_i gets the information and the time step u_i sends the information directly to u_{i+1}. If $d(P(u, v)) := \sum_{j=1}^{n-1} d_j = i(v)$, then $\max_{v \in V} d(P(u, v)) = \mathsf{RT}(\cup_{v \in V} (N_j(v))_{j=1}^{\infty}) \geq \log_2 N$.

Now we consider some instance T' containing $\cup_{v \in V} (N_j(v))_{j=1}^{\infty}$. Obviously, the path $P(u, v) = (u = u_1 \to u_2 \to \cdots \to u_n = v)$ still exists in T', however, the time needed for the information to reach u_{i+1} from u_i is now $d'_j := i'(u_{j+1}) - i'(u_j) = X_{T', d_j}(u_j) + d_j$. If $d'(P(u, v)) := \sum_{j=1}^{n-1} (d_j + X_{T', d_j}(u_j))$, then $\mathsf{RT}'(T') \leq \max_{v \in V} d'(P(u, v))$. In order to estimate $d'(P(u, v))$ we define for any time step t the random variable X_i, which is 0 if the last node informed before step t on $P(u, v)$ fails in step t, and 1 otherwise. Since any (informed) node fails in some round with probability $1 - p$, independently of the other rounds, $X_t = 0$ with probability $1 - p$, and $X_t = 1$ with probability p, independently of any other X_j.

Now we show that $\sum_{t=1}^{6/p \cdot \max\{d(P(u,v)), \log_2 N\}} X_t \geq d(P(u, v))$ with probability $\geq 1 - 1/N^2$. Since all X_t are independent from each other, using the Chernoff bounds [4, 11] (cf. Appendix) where $\delta = 5/6$, we obtain

$$\mathbf{Pr} \left[\sum_{t=1}^{6/p \cdot d(P(u,v))} X_t \leq (1 - \frac{5}{6}) 6 d(P(u, v)) \right] \leq e^{\frac{-6d(P(u,v))25}{72}} \leq e^{-2d(P(u,v))} \leq \frac{1}{N^2},$$

whenever $d(P(u, v)) \geq \log_2 N$. If $d(P(u, v)) < \log_2 N$, then using the same technique we can show that $\mathbf{Pr} \left[\sum_{t=1}^{6/p \cdot \log_2 N} X_t \leq d(P(u, v)) \right] \leq 1/N^2$. This implies that $d'(P(u, v)) \leq 6/p \cdot \max\{d(P(u, v)), \log_2 N\}$ with probability $1 - O(1/N^2)$ for any node $v \in V$. By Markovs inequality, the last statement holds for any node $v \in V$ with probability at least $1 - O(1/N)$. Since $\max_{v \in V} d(P(u, v)) \geq \log_2 N$, the theorem follows. $\square$

3.2   Applications

In the following paragraphs, we use Theorem 5 to derive bounds on the runtime of agent based broadcasting [6] (cf. Section 1 for the description), which is denoted in the following paragraphs by $\mathsf{AT}(G, 1 - O(1/N))$ according to the definition of $\mathsf{RT}(G, 1 - O(1/N))$. Since any agent performs an ergodic random walk in the network, the distribution of the agents converges towards the stationary distribution [16]. However, the distribution of the agents in some round $t + 1$ depends on their distribution in round t. Therefore, we introduce so-called log-expanding graphs. Their good expansion properties guarantee that the distribution of the agents possess a certain "renewal"-property.

Definition 1. *A graph $G = (V, E)$ is a log-expanding graph, if $\forall v \in V(G) \, \exists \gamma = O(1) : |B_\gamma(v)| \geq 5 \ln N$, where $B_\gamma(v) := \{w \in V(G) \mid \text{dist}(w, v) \leq \gamma\}$.*

Among others, classical random graphs [2], hypercubes [12] and Star graphs [7] are log-expanding graphs. In [8] and [7] it has been shown that all these graphs support an optimal runtime of $O(\log N)$.

Theorem 6. *Let $G = (V, E)$ be a log-expanding graph with $\delta = \Theta(\Delta)$, and assume that N agents are initially distributed independently and according to the stationary distribution. Then,*

$$\mathsf{AT}(G, 1 - O(1/N)) \leq O(\mathsf{RT}(G, 1 - O(1/N))).$$

Proof. Note, that the stationary distribution of a random walk is given by the vector $\pi(v) = \frac{\deg(v)}{2|E|}$ for any node $v \in V(G)$. Therefore, on an arbitrary node lies at least one agent with a constant probability c. Now consider some node v together with $B_\gamma(v)$. Since $|B_\gamma(v)| \geq 5 \ln N$, there are $\Theta(|B_\gamma(v)|)$ agents in $B_\gamma(v)$ with probability $1 - O(1/N^5)$ by using a Chernoff-Bound. This guarantees that within the first $O(N^2)$ rounds, there are always $\Theta(|B_\gamma(v)|)$ agents in $B_\gamma(v)$, with probability $1 - O(1/N^3)$. Hence with probability $1 - O(1/N^2)$, for any fixed $t = O(N^2)$, a node v will be visited by at least one agent within the time interval $[t, t + O(1)]$ with a constant probability p. Thus, we can apply Theorem 5, and the theorem follows. □

Using Theorem 5 we also state new results on the runtime of randomized broadcasting in Cartesian products of graphs [10]. Denote by $G := G_1 \times G_2$ the product of two connected graphs, G_1 and G_2 of size N_1 and N_2, resp. It is easy to see that $\mathsf{DT}(G) \leq \mathsf{DT}(G_1) + \mathsf{DT}(G_2)$. For the randomized case, we can state the following.

Corollary 1. *For $p := \min\{\frac{\delta_1}{\delta_1 + \Delta_2}, \frac{\delta_2}{\Delta_1 + \delta_2}\}$ it holds that $\mathsf{RT}(G, 1 - 1/N) \leq \frac{1}{p} \cdot O\left(\mathsf{RT}(G_1, 1 - 1/N_1) + \mathsf{RT}(G_2, 1 - 1/N_2)\right)$ and for $p := \min\{\frac{\delta_1}{\delta_1 + \Delta_2}\}$ it holds that $\mathsf{RT}(G, 1 - o(1)) \leq \frac{1}{p} \cdot O\left(\mathsf{RT}(G_1, 1 - 1/N_1) \cdot N_2\right)$.*

As shown in [7], $\mathsf{RT}(K_{\sqrt{N}} \times C_{\sqrt{N}}, O(1)) = \Omega(\log N \cdot \sqrt{N})$, so that the second bound is tight.

4   Conclusion

In this paper, we developed tight lower and upper bounds on the runtime of the push algorithm, and analyzed its robustness against random node failures. We generalized the lower bound of Pittel [19] and improved the upper bound of Feige et.al. [8] for general graphs. One open problem is to close the gap between the lower bound for regular and the one for general graphs. Moreover, we determined in several graph classes the worst-case ratio between the runtime of an optimal deterministic algorithm and the push algorithm. We also investigated the robustness of randomized broadcasting against random node failures. After that, we related the robustness result to broadcasting on Cartesian product of graphs and to the agent-based broadcasting model, introduced in [6].

References

1. S.B. Akers, D. Harel, and B. Krishnamurthy. The star graph: An attractive alternative to the n-cube. In *Proc. of ICPP*, pages 393–400, 1987.
2. B. Bollobás. *Random Graphs*. Academic Press, 1985.
3. M. Capocelli, L. Gargano, and U. Vaccaro. Time bounds for broadcasting in bounded degree graphs. In *Proc. of WG'89*, pages 19–33, 1989.
4. H. Chernoff. A measure of asymptotic efficiency for tests of a hypothesis based on the sum of observations. *Ann. Math. Stat.*, 23:493–507, 1952.
5. A. Demers, D. Greene, C. Hauser, W. Irish, J. Larson, S. Shenker, H. Sturgis, D. Swinehart, and D. Terry. Epidemic algorithms for replicated database maintenance. In *Proc. of PODC'87*, pages 1–12, 1987.
6. R. Elsässer, U. Lorenz, and T. Sauerwald. Agent based information handling in large networks. In *Proc. of MFCS'04*, pages 686–698, 2004.
7. R. Elsässer and T. Sauerwald. On randomized broadcasting in star graphs. In *Proc. of WG'05*, pages 307–318, 2005.
8. U. Feige, D. Peleg, P. Raghavan, and E. Upfal. Randomized broadcast in networks. *Random Structures and Algorithm*, 1(4):447–460, 1990.
9. L. Gasieniec and A. Pelc. Adaptive broadcasting with faulty nodes. *Parallel Computing*, 22:903–912, 1996.
10. J. L. Gross and J. Yellen (eds.). *Handbook of Graph Theory*. CRC Press, 2004.
11. T. Hagerup and C. Rüb. A guided tour of Chernoff bounds. *Information Processing Letters*, 36(6):305–308, 1990.
12. L.H. Harper. Optimal assignment of numbers to vertices. *J. Soc. Ind. Appl. Math.*, 12:131–135, 1964.
13. J. Hromkovič, R. Klasing, A. Pelc, P. Ružička, and W. Unger. *Dissemination of Information in Communication Networks*. Springer, 2005.
14. R. Karp, C. Schindelhauer, S. Shenker, and B. Vöcking. Randomized rumor spreading. *Proc. of FOCS'00*, pages 565–574, 2000.
15. F. Leighton, B. Maggs, and R. Sitamaran. On the fault tolerance of some popular bounded-degree networks. In *Proc. of FOCS'92*, pages 542–552, 1992.
16. L. Lovász. Random walks on graphs: A survey. *Combinatorics, Paul Erdös is Eighty*, 2:1–46, 1993.
17. D. Malkhi, Y. Mansour, and M.K. Reiter. On diffusion updates in a byzantine environment. In *Proc. of 18th IEEE Symp. on Reliable Distributed Systems*, pages 134–143, 1999.
18. M. Mitzenmacher and E. Upfal. *Probability and Computing*. Cambdrige University Press, 2005.
19. B. Pittel. On spreading rumor. *SIAM Journal on Applied Mathematics*, 47(1):213–223, 1987.

Local Search in Evolutionary Algorithms: The Impact of the Local Search Frequency

Dirk Sudholt*

FB Informatik, LS2, Universität Dortmund, 44221 Dortmund, Germany
Dirk.Sudholt@udo.edu

Abstract. A popular approach in the design of evolutionary algorithms is to integrate local search into the random search process. These so-called memetic algorithms have demonstrated their efficiency in countless applications covering a wide area of practical problems. However, theory of memetic algorithms is still in its infancy and there is a strong need for a rigorous theoretical foundation to better understand these heuristics. Here, we attack one of the fundamental issues in the design of memetic algorithms from a theoretical perspective, namely the choice of the frequency with which local search is applied. Since no guidelines are known for the choice of this parameter, we care about its impact on memetic algorithm performance. We present worst-case problems where the local search frequency has an enormous impact on the performance of a simple memetic algorithm. A rigorous theoretical analysis shows that on these problems, with overwhelming probability, even a small factor of 2 decides about polynomial versus exponential optimization times.

1 Introduction

Solving optimization problems is a fundamental task in computer science. Theoretical computer science has developed powerful techniques to design problem-specific algorithms and to provide guarantees on the worst-case runtime and the quality of solutions. Nevertheless, these algorithms can be quite complicated and difficult to implement. Moreover, practitioners often have to deal with problems where they have only limited insight into the structure of the problem, thus making it impossible to design specific algorithms.

The advantage of randomized search heuristics like randomized local search, tabu search, simulated annealing, and evolutionary algorithms is that they are easy to design and easy to implement. Despite the lack of performance guarantees, they often yield good results in short time and they can be applied in scenarios where the optimization problem at hand is only known as a black box.

Therefore, practitioners often apply randomized search heuristics like, e. g., evolutionary algorithms to find good solutions during a random search process. Often the performance of evolutionary algorithms can be enhanced if (problem-specific) local search techniques are integrated. These hybrid algorithms are

---

* This work was supported by the Deutsche Forschungsgemeinschaft (DFG) as part of the Collaborative Research Center "Computational Intelligence" (SFB 531).

T. Asano (Ed.): ISAAC 2006, LNCS 4288, pp. 359–368, 2006.

known as memetic algorithms. Using problem-specific local search can provide a better guidance for the random search process while preserving the low costs of implementation. That way, the advantages of problem-specific algorithms and simple randomized search heuristics can be combined. It is therefore not surprising that practitioners have applied memetic algorithms to a wide range of applications, see Moscato [6] for a survey or Hart, Krasnogor, and Smith [3] for various applications.

However, from a theoretical point of view this situation is unsatisfactory because these algorithms are presently not considered in the theory of algorithms. Despite a broad activity in the area of memetic algorithms, theory on memetic algorithms is hanging behind and rigorous theoretical results are rare.

We present a brief survey of theoretical approaches concerning memetic algorithms. Hart [2] empirically investigates the role of the local search frequency and the local search depth, i. e., the maximal number of iterations in one local search call, on three artificial test functions. Lourenço, Martin, and Stützle [4] empirically analyze the runtime distribution of memetic algorithms on problems from combinatorial optimization. Merz [5] adapts the parameterization of memetic algorithms to the given problem by using problem-specific knowledge gained from empirical analysis of the problem structure. Sinha, Chen, and Goldberg [8] present macro-level theoretical results on the design of global-local search hybrids explaining how to balance global and local search. Finally, Sudholt [10] compares a simple memetic algorithm with two well-known randomized search heuristics and proves rigorously for an artificial function that the local search depth has a large impact on the behavior of the algorithm.

In the design of memetic algorithms it is essential to find a proper balance between evolutionary (global) search and local search. If the effect of local search is too weak, we fall back to standard evolutionary algorithms. If the effect of local search is too strong, the algorithm may get stuck in local optima of bad quality. Moreover, the algorithms is likely to rediscover the same local optimum over and over again, wasting computational effort. Lastly, when dealing with population-based algorithms, too much local search quickly leads to a loss of diversity within the population.

A common design strategy is to apply local search with a fixed frequency, say every τ generations for some $\tau \in \mathbb{N}$. At present, there are no guidelines available for the choice of this parameter. Hence, an interesting question is what impact the local search frequency has on the performance of the algorithm. We will define a simple memetic algorithm and prove that in the worst case (w. r. t. the problem instance) even small changes to the local search frequency can totally change the algorithm's behavior and decide about polynomial versus exponential optimization times, with overwhelming probability.

In Section 2 we define a simple memetic algorithm, the (1+1) Memetic Algorithm. In Section 3 we define so-called race functions where local search effects compete with global search effects. Section 4 proves rigorously that the local search frequency has a large impact on the (1+1) MA on race functions and that even a factor of 2 makes an enormous difference. We conclude in Section 5.

Due to space limitations, we restrict ourselves to sketches of proofs. An extended version of this work with full proofs can be found in [9].

2   Definitions

The (1+1) Memetic Algorithm ((1+1) MA) is a simple memetic algorithm with population size 1 that has already been investigated in [10]. It employs the following local search procedure. The algorithm is defined for the maximization of pseudo-boolean functions $f: \{0,1\}^n \to \mathbb{R}$ including problems from combinatorial optimization. $H(x,y)$ denotes the Hamming distance between x and y.

Procedure 1 (Local Search(y) with depth δ)
> $t := 1.$
> *While* $t \leq \delta$ *and* $\exists z\colon (H(z,y) = 1$ *and* $f(z) > f(y))$ *do*
> > $y := z.\ t := t + 1.$

If there is more than one Hamming neighbor with larger fitness, z may be chosen arbitrarily among them.

Algorithm 1 ((1+1) Memetic Algorithm ((1+1) MA))
1. **Initialization:** gen := 1. *Choose* x *uniformly at random. Local Search(x).*
2. **Mutation:** $y := x.$ *Flip each bit in* y *independently with probability* $1/n.$
3. **Local Search:** *If* gen mod $\tau = 0$ *then Local Search(y).*
4. **Selection:**   *If* $f(y) \geq f(x)$ *then* $x := y.$
5. **Loop:**   gen := gen +1. *Continue at line 2.*

We do not specify a termination condition as we are only interested in the number of f-evaluations until a global optimum is found. Note that an iteration of local search may require up to n f-evaluations.

Definition 1. *An event E occurs with overwhelming probability (w. o. p.) if $Prob(E) = 1 - 2^{-\Omega(n^\varepsilon)}$ for a constant $\varepsilon > 0$, n the search space dimension.*

We say that an algorithm $\mathcal{A}$ is efficient *on a function f iff $\mathcal{A}$ finds a global optimum on f in a polynomial number of f-evaluations w. o. p.*

We say that an algorithm $\mathcal{A}$ fails *on a function f iff $\mathcal{A}$ does not find a global optimum in an exponential number of f-evaluations w. o. p.*

When constructing the race functions, we will make use of so-called long K-paths. A long K-path is a sequence of Hamming neighbors where all points are different. The following definition is taken from [1].

Definition 2 (Long K-paths). *Let $K, N \in \mathbb{N}$ with $(N-1)/K \in \mathbb{N}$. The long K-path of dimension N is a sequence of strings from $\{0,1\}^N$ defined recursively. The long K-path of dimension 1 is $P_1^K := (0,1)$. Let $P_{N-K}^K = (v_1, \ldots, v_\ell)$ be the long K-path of dimension $N - K$. The long K-path of dimension N is the concatenation $S_0 \cdot B \cdot S_1$, where S_0, B, S_1 result from prepending K bits to strings from P_{N-K}^K:*

$$S_0 := (0^K v_1, 0^K v_2, \ldots, 0^K v_\ell),$$
$$B := (0^{K-1} 1 v_\ell, 0^{K-2} 1^2 v_\ell, \ldots, 01^{K-1} v_\ell),\ and$$
$$S_1 := (1^K v_\ell, 1^K v_{\ell-1}, \ldots, 1^K v_1).$$

S_0 and S_1 differ in the K leading bits and B represents a bridge between them. If $N = K^2 + 1$, the length of the path is $\Omega(2^K)$. Moreover, for all $0 < i < K$ the following statement holds. Let x be a point on the long K-path. If x has at least i successors on the path, then the ith successor has Hamming distance i of x and all other successors of x on the path have Hamming distances different from i (a proof is given in [1]). This implies that all successors on the path except the $K - 1$ next ones have Hamming distance at least K to x. The index of a point z on a long K-path is denoted by $i(z)$. If z is not on the path, $i(z) := -1$.

3   Race Functions: Where Local Search and Global Search Compete

Now we will define the aforementioned race functions where local search effects compete with global search effects. The idea behind the construction is quite intuitive. We will identify two non-overlapping blocks of the bit string of length N, referred to as x' and x'' if x is the current bit string. These partial bit strings span subspaces of the original search space. Then, subfunctions are defined on those subspaces such that the value of the original superior function is the (weighted) sum of the subfunctions' values for an important part of the search space.

The two subfunctions are defined as follows. The function on the left block x' is based on a connected subsequence (subpath) of a long K-path of adjustable length. The fitness is increasing on the subpath, thus it can be optimized efficiently by local search. The function on the right block x'' consists of a much shorter subpath, but only every third search point on the path has positive fitness. Hence, this subfunction contains a sequence of isolated peaks with increasing fitness and mutation can help to jump to the next peak by mutations flipping three specific bits.

To conclude, the function on the left block (or shortly, the left path) can be optimized efficiently by local search and the right path can only be optimized by mutations. The (1+1) MA on the superior function now optimizes the two subfunctions in parallel. If the local search frequency is high, we expect the algorithm to optimize the left path prior to the right path. Contrarily, if the local search frequency is low, then we expect the right path to be optimized prior to the left one.

By defining special fitness values for cases where some path end is reached, we obtain a function where it makes a large difference which path is optimized first. For example, we can define the end of the left path as being globally optimal. However, if the right path is optimized prior to the left one, the function turns into a so-called deceptive function giving hints to move away from all global optima and to get stuck in a local optimum. That way, the (1+1) MA typically optimizes this function efficiently if the local search frequency is high and it gets stuck in a local optima if the local search frequency is low. Another function can be defined analogously where it is optimal to reach the end of the right path.

Before giving formal definitions for these race functions, we present our main theorem that will be proved in Section 4 according to the ideas described above.

Theorem 1 (Main Theorem). *Let $\delta, \tau \in \mathbb{N}$ be defined such that $\delta = poly(n)$, $\delta \geq 22$, $\delta/\tau \geq 2/n$, $\tau = \omega(n^{2/3})$, and $\tau = O(n^3)$ hold. There exist functions $\mathrm{Race}_{\ell,r}^{\mathrm{left}}, \mathrm{Race}_{\ell,r}^{\mathrm{right}} : \{0,1\}^n \to \mathbb{R}$ such that*

- *the (1+1) MA with local search frequency $1/\tau$ is efficient on $\mathrm{Race}_{\ell,r}^{\mathrm{left}}$ while the (1+1) MA with local search frequency $1/(2\tau)$ fails on $\mathrm{Race}_{\ell,r}^{\mathrm{left}}$ and*
- *the (1+1) MA with local search frequency $1/\tau$ fails on $\mathrm{Race}_{\ell,r}^{\mathrm{right}}$ while the (1+1) MA with local search frequency $1/(2\tau)$ is efficient on $\mathrm{Race}_{\ell,r}^{\mathrm{right}}$.*

Definition 3. *Let $n = 4N$ and $N = K^2 + 1$ with $K/3 \in \mathbb{N}$. Let P_i be the ith point on the long K-path of dimension N.*

The (1+1) MA initializes uniformly at random. However, we want the optimization of the two paths to start with specific starting points. Therefore, we use a construction that is explained in detail in Section 3 of [10] (here, we use a slight transformation of the search space which is immaterial to the algorithm). In a nutshell, we append additional $2N$ bits denoted by x''' to the $2N$ bits used by the two blocks x' and x''. The following subfunction ZZO guides the algorithm to reach $x''' = 0^{2N}$ (i.e. a concatenation of $2N$ zeros) and then to reach specific starting points for x' and x'', namely $x'x'' = 0^N P_{n^5-1}$ ($n^5 - 1$ is the multiple of 3 closest to n^5 due to the choice of K and n). Afterwards, x''' is turned into $x''' = 1^{2N}$. Once all these bits are ones, the optimization of the two paths begins.

Definition 4. *Let $x = x'x''x'''$ with $x', x'' \in \{0,1\}^N$ and $x''' \in \{0,1\}^{2N}$. We define ZZO: $\{0,1\}^n \to \mathbb{R}$ as*

$$\mathrm{ZZO}(x) := \begin{cases} -H(x''', 0^{2N}) - 4N & \text{if } x'x'' \neq 0^N P_{n^5-1}, x''' \neq 0^{2N}, \\ -H(x'x'', 0^N P_{n^5-1}) - 2N & \text{if } x'x'' \neq 0^N P_{n^5-1}, x''' = 0^{2N}, \\ -H(x''', 1^{2N}) & \text{if } x'x'' = 0^N P_{n^5-1}. \end{cases}$$

Definition 5 (Race Functions). *Call a search point $x = x'x''x'''$ well-formed iff $i(x') \geq 0$, $i(x'') \geq 0$, $i(x'')/3 \in \mathbb{N}$, and $x''' = 1^{2N}$. Given $\ell, r \in \mathbb{N}$ we define*

$$\mathrm{Race}_{\ell,r}^{\mathrm{left}}(x) := \begin{cases} \mathrm{ZZO}(x) & \text{if } H(x''', 1^{2N}) \geq 3, \\ i(x') \cdot n + i(x'') & \text{if } x \text{ well-formed}, i(x') < \ell, i(x'') < r, \\ 2^N + H(x', P_\ell) & \text{if } x \text{ well-formed}, i(x') < \ell, i(x'') = r, \\ 2^{2N} & \text{if } x \text{ well-formed}, i(x') = \ell, \\ -\infty & \text{otherwise.} \end{cases}$$

$$\mathrm{Race}_{\ell,r}^{\mathrm{right}}(x) := \begin{cases} \mathrm{ZZO}(x) & \text{if } H(x''', 1^{2N}) \geq 3, \\ i(x') \cdot n + i(x'') & \text{if } x \text{ well-formed}, i(x') < \ell, i(x'') < r, \\ 2^N + H(x'', P_r) & \text{if } x \text{ well-formed}, i(x') = \ell, i(x'') < r, \\ 2^{2N} & \text{if } x \text{ well-formed}, i(x'') = r, \\ -\infty & \text{otherwise.} \end{cases}$$

In a typical run, after random initialization the function ZZO is optimized guiding the search towards the well-formed search point $x = x'x''x'''$ with $i(x') = 0$, $i(x'') = P_{n^5-1}$, and $x''' = 1^{2N}$. There is a gap between the ZZO-dependent search points and all well-formed search points since all points with one or two zero-bits in the x'''-part have fitness $-\infty$. However, this gap can easily be jumped over by mutation in expected time $O(n^3)$. Moreover, the probability that at least $K = \Theta(n^{1/2})$ bits flip in this jump is exponentially small. Thus, it is very likely that we reach points close to the desired starting points in polynomial time. For a proof of a result similar to the following corollary, we refer the reader to [10].

Corollary 1. *With overwhelming probability, the (1+1) MA on either* $\text{Race}_{\ell,r}^{\text{left}}$ *or* $\text{Race}_{\ell,r}^{\text{right}}$ *reaches some well-formed search point* x^* *with* $i(x^{*\prime}) < K$ *and* $|i(x^{*\prime\prime}) - (n^5 - 1)| < K$ *within the first* n^4 *generations.*

4   Analyzing the Impact of the Local Search Frequency

To prove our main theorem, we will investigate the progress of the algorithm on the two paths. The progress will be estimated by separating the effects of different operations and proving bounds for the cumulated progress for single types of operations.

For the rest of the section, we consider the (1+1) MA on $\text{Race}_{\ell,r}^{\text{left}}$ or $\text{Race}_{\ell,r}^{\text{right}}$ after some well-formed search point has been reached. In a generation with local search, the mutation only affects the algorithm if the outcome of local search is accepted in the selection step. Thus, we only have to take into account those mutations where the outcome of the following local search call is accepted.

Lemma 1. *Let* $x = x'x''x'''$ *be the current population,* x *well-formed, let* $y = y'y''y'''$ *be an offspring created by mutation, and let* $z = z'z''z'''$ *be the result of local search applied to* y*. Then* z *is accepted in the selection step only if* y *has Hamming distance at most 1 to a well-formed search point.*

Proof. Let $w = w'w''1^{2N}$ be a well-formed search point with minimal Hamming distance to y. We distinguish three cases according to $H(y''', x''')$, i.e., the number of zero-bits in y'''.

- If $H(y''', x''') \geq 2$, the function to be optimized during the local search process is ZZO since the fitness of all search points with one or two zero-bits in the x'''-part is $-\infty$ and the fitness is $\text{ZZO}(\cdot) > -\infty$ in case of three or more zero-bits. However, due to the gap between ZZO-dependent search points and well-formed search points, local search cannot reach a well-formed search point. Hence, the offspring z is rejected in the selection step.
- In case $y''' = 1^{2N}$ and $H(w'w'', y'y'') \geq 2$ we have fitness $-\infty$ for y and all Hamming neighbors of y. Hence, local search stops immediately in this case.
- Lastly, if $H(y''', x''') = 1$ and $H(w'w'', y'y'') \geq 1$ we have fitness $-\infty$ and the fitness cannot be increased by flipping single bits in $y'y''$. The Hamming neighbor obtained by flipping the unique zero-bit in y''' has fitness $-\infty$ and so do all Hamming neighbors with a larger number of zero-bits in the x''' part. Thus, local search stops immediately, here.   □

An important observation is that mutations followed by local search are in some sense more powerful than mutations without local search. Imagine a mutation yielding a non-well-formed search point with Hamming distance 1 to a well-formed one. Then local search reaches the well-formed search point within its first iteration and the outcome of local search may be accepted by the algorithm (note that Lemma 1 provides a necessary condition, not a sufficient one). Hence, mutations followed by local search are more likely to yield an accepted search point than mutations without local search and the first iteration of local search plays a crucial role, here. As a consequence, we may regard the first iteration of local search as being part of the mutation instead of local search.

Definition 6. *An* extended mutation *is either a mutation reaching a well-formed search point or a mutation followed by one iteration of local search in case the mutant is not well-formed.*

Using these insights, we now formally define the intuitive notion of progress. In a generation without local search, the progress by one mutation on, say, the left path is defined as $i(y') - i(x')$ if y is accepted and 0 otherwise. In a generation with local search let x be the current search point, y be the individual obtained by an extended mutation, and z be the result of local search. Then the progress by one extended mutation is defined as $i(y') - i(x')$ if z is accepted and 0 otherwise and the progress by local search is $i(z') - i(y')$ if z is accepted and 0 otherwise. The progress on the right path is defined analogously.

In the following lemmas, we will prove lower and upper bounds on the cumulated progress for specific operations, namely mutations in generations without local search, extended mutations, and the remaining iterations of local search after extended mutations. Full proofs can be found in [9].

Lemma 2. *Let $\Delta_{\text{mut}}^{\text{left}}$ ($\Delta_{\text{mut}}^{\text{right}}$) be the progress on the left (right) path in $T = \Omega(n^4), T = poly(n)$ mutations. Then with probability $1 - 2^{-\Omega(n^{1/2})}$ for $\varepsilon > 0$*

$$(1 - \varepsilon) \cdot T/(en) < \Delta_{\text{mut}}^{\text{left}} < (1 + \varepsilon) \cdot T/(en) \text{ and}$$
$$(1 - \varepsilon) \cdot T/(en^3) < \Delta_{\text{mut}}^{\text{right}} < (1 + \varepsilon) \cdot T/(en^3) \ .$$

Sketch of Proof. Here and in the following proofs, we consider a typical run of the algorithm. Events preventing a run from being typical are called errors and the total error probability is bounded by the sum of single error probabilities. If there is only a polynomial number of exponentially small single error probabilities, a run is typical with overwhelming probability.

In a typical mutation less than K bits flip simultaneously. Upper bounds on $\Delta_{\text{mut}}^{\text{left}}$ and $\Delta_{\text{mut}}^{\text{right}}$ are proved in the same way. We consider a well-formed search point x and the sequence of bits $b_1, \ldots, b_K$ that have to be flipped in that order to climb the next K steps on the path. Mutations of x flipping other bits are not accepted, thus we only consider mutations not flipping bits outside of $\{b_1, \ldots, b_K\}$ occurring with probability close to $1/e$. Suppose the mutation operator decides sequentially for each bit whether to flip it or not and that $b_1, \ldots, b_K$ are processed in the specified order. Then the progress in one mutation can be modelled by a (bounded) geometric distribution with parameter $(1-1/n)$ for $\Delta_{\text{mut}}^{\text{left}}$ and $(1-1/n^3)$ for $\Delta_{\text{mut}}^{\text{right}}$, resp. The upper bounds are then obtained by applying a variance-based tail inequality (Theorem 3.44 in Scheideler [7]).

A progressing step is a mutation flipping exactly the bits differing from the next well-formed point on the considered path which happens with probability $1/n \cdot (1-1/n)^{n-1} \geq 1/(en)$ for the left path and $1/n^3 \cdot (1-1/n)^{n-3} \geq 1/(en^3)$ for the right path. Changes on the left path dominate changes on the right path due to the larger weight in the definition of the race functions. Thus, we may have an accepted mutation stepping back on the right path if the same mutation yields progress on the left path. However, the probability for such an event is $O(1/n^4)$ and the regress on the right path can be bounded by the above-mentioned tail inequality. For both paths, Chernoff bounds show that the number of progressing steps is large enough to prove the claimed lower bounds w. o. p.   □

Lemma 3. *Let $\Delta_{\text{enh}}^{\text{right}}$ be the progress on the right path in $T = O(n^4)$ extended mutations of parents whose index on the right path is greater than 0. Let $\delta \geq 6$, then with probability $1 - 2^{-\Omega(n^{1/4})}$*

$$-4T^{3/4}n^{-3/2} - n^{1/2} < \Delta_{\text{enh}}^{\text{right}} < 4T^{3/4}n^{-3/2} + n^{1/2} \ .$$

Sketch of Proof. Since $\delta \geq 6$ and the left path has a larger weight, it is likely that local search climbs the left path yielding an accepted search point if the extended mutation reaches a well-formed search point. This typically holds regardless whether the index on the right path has increased or decreased. Hence, the process describing the progress by extended mutations on the right path resembles a martingale. It is easy to see that an upper bound on the progress also represents a lower bound on the negative progress. Hence, we only prove an upper bound by distinguishing two cases according to T.

Let $T < n^{5/2}/9$. An extended mutation yields progress $3i$ for some $0 < i < K/3$ iff it reaches the $(3i)$th-next path point directly or if it reaches a Hamming neighbor thereof. Exactly $3i$ Hamming neighbors have distance $3i - 1$, thus dominating the probability to have progress $3i$: it can be estimated by $(1 + o(1))/e \cdot 3i \cdot n^{-3i+1}$ which is $(1 + o(1))/e \cdot 3n^{-2}$ for $i = 1$. We now imagine a sequence of binary random variables where each variable is set to 1 independently with probability $3n^{-2}$. Then the probability to have a block of i ones is larger than the probability to have progress $3i$, hence the random process describing the number of ones in T binary random variables dominates the random process describing the progress on the right path divided by 3. Applying Chernoff bounds to the former process shows that the probability of having progress larger than $n^{1/2}$ is exponentially small.

Now let $T \geq n^{5/2}/9$. We exploit the fact that the process is almost a martingale. Progresses of at least 6 or at most -6 are unlikely and their effect can be bounded by rather crude estimates. Then only ± 3-steps remain, and apart from pathological steps with expected progress $o(1)$, the situation is completely symmetric due to the hypothesis on the parents' indices. The upper bound follows by the method of bounded martingale differences (Theorem 3.67 in [7]).   □

Lemma 4. *Let $\Delta_{\text{enh}}^{\text{left}}$ be the progress on the left path in $T \geq n^{1/2}, T = poly(n)$ extended mutations. Then with probability $1 - 2^{-\Omega(n^{1/2})}$ for $\varepsilon > 0$*

$$-(1 + \varepsilon) \cdot T/e < \Delta_{\text{enh}}^{\text{left}} < (1 + \varepsilon) \cdot T/e .$$

Lemma 4 can be proved using ideas from Lemma 3, thus we omit a proof.

The probability that an extended mutation leads to an offspring that is accepted after local search converges to $2/e$ as it is dominated by the probability that the mutation flips at most one bit. Afterwards, either $\delta - 1$ or δ iterations of local search yield progress on the left path. By Chernoff bounds, the following lemma can be proved.

Lemma 5. *Let $\Delta_{\text{ls}}^{\text{left}}$ be the progress on the left path in $T = poly(n)$ calls of local search. Then with probability $1 - 2^{-\Omega(T)}$ for $\varepsilon > 0$*

$$(\delta - 1) \cdot (1 - \varepsilon) \cdot 2T/e \leq \Delta_{\text{ls}}^{\text{left}} \leq \delta \cdot (1 + \varepsilon) \cdot 2T/e .$$

Proof of the Main Theorem. Let $\ell = (1 - \varepsilon)/e \cdot (n^3 + (2\delta - 4)n^4\tau^{-1})$ be the length of the left path and $r = n^5 - 1 + (1 + \varepsilon)/e \cdot n + 4n^{3/2}\tau^{-3/4} + n^{1/2} + 2K$ be the length of the right path for a small enough constant $\varepsilon > 0$. W. l. o. g. $r/3 \in \mathbb{N}_0$.

We investigate typical runs of the $(1+1)$ MA with local search frequency $1/\tau$ and $1/(2\tau)$ on $\text{Race}_{\ell,r}^{\text{left}}$ and $\text{Race}_{\ell,r}^{\text{right}}$, resp. The following statements hold w. o. p. By Corollary 1, the $(1+1)$ MA reaches some well-formed search point x_{first} with $i(x_{\text{first}}') < K$ and $|i(x_{\text{first}}'') - (n^5 - 1)| < K$ within the first n^4 steps.

We consider a period of n^4 generations of the $(1+1)$ MA with local search frequency $1/\tau$ after x_{first} has been reached. Let Δ^{left} be the total progress on the left path and Δ^{right} be the total progress on the right path in n^4 generations. Then we apply Lemmas 2, 4, and 5 w. r. t. $n^4 - n^4/\tau$ mutations, n^4/τ extended mutations or n^4/τ local search calls, respectively. Simple calculations yield $i(x_{\text{first}}') + \Delta^{\text{left}} \geq \ell$, thus the end of the left path is reached within the considered period.

Moreover, we show that the end of the right path is not reached within this period. First, we consider the very last search point with index $i(x_{\text{first}}'') + \Delta^{\text{right}}$ on the right path and show that the probability of $i(x_{\text{first}}'') + \Delta^{\text{right}} > r - K$ is exponentially small, i. e., the last considered search point is by at least K path points away from the end of the right path. This is done by applying Lemmas 2 and 3 where Lemma 3 can be applied since $i(x_{\text{first}}'') \geq n^5 - 1 - K$ and n^4 steps can only decrease the index by $n^4 \cdot K$ implying that the index on the right path cannot become 0. Observe that the probability to reach the end of a path cannot increase with decreasing number of generations. Hence, this bound also holds for all other search points reached within the period and the error probability increases by a factor of n^4.

Together, the $(1+1)$ MA with local search frequency reaches the end of the left path within $O(n^4)$ generations and $O(n^4 + n \cdot \delta/\tau) = poly(n)$ function evaluations. This implies that on $\text{Race}_{\ell,r}^{\text{left}}$, a global optimum is found and the $(1+1)$ MA is efficient. On $\text{Race}_{\ell,r}^{\text{right}}$, however, the Hamming distance to the end of the right path is at least K. As all search points closer to the right path now have worse fitness, the only way to reach a global optimum is a direct jump flipping at least K bits. The probability for such an event is at most $n^{-K} = 2^{-\Omega(n^{1/2} \log n)}$, thus the (1+1) MA fails on $\text{Race}^{\text{right}}_{\ell,\text{r}}$.

The argumentation for the (1+1) MA with local search frequency $1/(2\tau)$ is symmetric. Repeating the arguments from above, the end of the right path is reached within a period of $\sqrt{2}n^4$ generations while all search points traversed during this period are at least K away from the end of the left path. Thus, the (1+1) MA is efficient on $\text{Race}^{\text{right}}_{\ell,\text{r}}$ and it gets trapped on $\text{Race}^{\text{left}}_{\ell,\text{r}}$, which completes the proof. $\qquad \square$

5   Conclusions

We have presented a rigorous theoretical analysis of a simple memetic algorithm, the (1+1) MA, thus showing that these randomized search heuristics can be analyzed in terms of computational complexity. On worst-case instances we have shown that the choice of the local search frequency has an enormous impact on the performance of the (1+1) MA: with overwhelming probability, even altering the parameterization by a factor of 2 turns a polynomial runtime behavior into an exponential one and vice versa.

References

1. S. Droste, T. Jansen, and I. Wegener. On the analysis of the (1+1) evolutionary algorithm. *Theoretical Computer Science*, 276:51–81, 2002.
2. W. E. Hart. *Adaptive Global Optimization with Local Search*. PhD thesis, University of California, San Diego, CA, 1994.
3. W. E. Hart, N. Krasnogor, and J. E. Smith, editors. *Recent Advances in Memetic Algorithms*, vol. 166 of *Studies in Fuzziness and Soft Computing*. Springer, 2004.
4. H. R. Lourenço, O. Martin, and T. Stützle. Iterated local search. In *Handbook of Metaheuristics*, vol. 57 of *International Series in Operations Research & Management Science*, pages 321–353. Kluwer Academic Publishers, Norwell, MA, 2002.
5. P. Merz. Advanced fitness landscape analysis and the performance of memetic algorithms. *Evolutionary Computation*, 12(3):303–326, 2004.
6. P. Moscato. Memetic algorithms: a short introduction. In D. Corne, M. Dorigo, and F. Glover, editors, *New Ideas in Optimization*, pages 219–234. McGraw-Hill, 1999.
7. C. Scheideler. *Probabilistic Methods for Coordination Problems*. HNI-Verlagsschriftenreihe 78, University of Paderborn, 2000. Habilitation Thesis, available at http://www14.in.tum.de/personen/scheideler/index.html.en.
8. A. Sinha, Y. Chen, and D. E. Goldberg. Designing efficient genetic and evolutionary algorithm hybrids. In [3], pages 259–288.
9. D. Sudholt. Local search in memetic algorithms: the impact of the local search frequency. Technical Report CI-208/06, Collaborative Research Center 531, University of Dortmund, June 2006. Available at http://sfbci.cs.uni-dortmund.de.
10. D. Sudholt. On the analysis of the (1+1) memetic algorithm. In *Proceedings of the Genetic and Evolutionary Computation Conference (GECCO 2006)*, pages 493–500. ACM Press, New York, NY, 2006.

Non-cooperative Facility Location and Covering Games

Martin Hoefer[*]

Department of Computer & Information Science, Konstanz University, Germany
`hoefer@inf.uni-konstanz.de`

Abstract. We study a general class of non-cooperative games coming from combinatorial covering and facility location problems. A game for k players is based
on an integer programming formulation. Each player wants to satisfy a subset
of the constraints. Variables represent resources, which are available in costly
integer units and must be bought. The cost can be shared arbitrarily between
players. Once a unit is bought, it can be used by all players to satisfy their constraints. In general the cost of pure-strategy Nash equilibria in this game can be
prohibitively high, as both prices of anarchy and stability are in $\Theta(k)$. In addition,
deciding the existence of pure Nash equilibria is NP-hard. These results extend to
recently studied single-source connection games. Under certain conditions, however, cheap Nash equilibria exist: if the integrality gap of the underlying integer
program is 1 and in the case of single constraint players. In addition, we present
algorithms that compute cheap approximate Nash equilibria in polynomial time.

1 Introduction

Analyzing computational environments using game-theoretic models is a quickly evolving research direction in theoretical computer science. Motivated in large parts by the
Internet, the resulting dynamics of introducing selfish behavior of distributed agents into
a computational environment are studied. In this paper we follow this line of research
by considering a general class of non-cooperative games based on general integer covering problems. Problems concerning service installation or clustering, which play an
important role in large networks like the Internet, are modeled formally as some variant of covering or partition problems. Our games can serve as a basis to analyze these
problems in the presence of independent non-cooperative selfish agents.

The formulation of our games generalizes an approach by Anshelevich et al [2],
who proposed games in the setting of Steiner forest design. In particular, we consider
a covering optimization problem given as an integer linear program and turn this into
a non-cooperative game as follows. Each of the k non-cooperative players considers
a subset of the constraints and strives to satisfy them. Each variable represents a resource, and integer units of resources can be bought by the players. The cost of a unit
is given by the coefficient in the objective function. In particular, players pick as strategy a payment function that specifies how much they are willing to pay for the units
of each resource. A unit is considered *bought* if the cost is paid for by the amount the

---

[*] Supported by DFG Research Training Group 1042 "Explorative Analysis and Visualization
of Large Information Spaces" and in part by DFG grant Kr 2332/1-2 within Emmy Noether
research group on "Algorithmic Game Theory".

T. Asano (Ed.): ISAAC 2006, LNCS 4288, pp. 369–378, 2006.

players offer. Bought units can then be used by *all* players simultaneously to satisfy their constraints – no matter whether they contribute to the cost or not. A player strives to minimize the sum of her offers, but insists on satisfaction of her constraints. A variety of integer covering problems, most prominently variants of set cover and facility location, can be turned into a game with the help of this model. We study our games with respect to the existence and cost of stable outcomes of the game, which are exact and approximate Nash equilibria. At first, we characterize prices of anarchy [14] and stability [1]. They measure the social cost of the worst and best Nash equilibria in terms of the cost of a social optimum solution. Note that a social optimum solution is the optimum solution to the underlying integer program. As the cost of exact Nash equilibria can be as high as $\Theta(k)$, we then consider a two-parameter optimization problem to find (α, β)-approximate Nash equilibria. These are solutions in which each player can reduce her contribution by at most a factor of α by unilaterally switching to another strategy, and which represent a β-approximation to the socially optimum cost. We refer to α as the *stability ratio* and β as the *approximation ratio*.

Related Work. *Competitive location* is an active research area, in which game-theoretic models for spatial and graph-based facility location have been studied in the last decades [7, 18]. These models consider facility owners as players that selfishly decide where to place and open a facility. Clients are modeled as part of player utility, e.g. they are always assumed to connect to the closest facility. Recent examples of this kind of location games are also found in [21, 5]. According to our knowledge, however, none of these models consider the clients as players that need to create connections and facilities without central coordination.

Closer to our approach are cooperative games and mechanism design problems based on optimization. In [6] strategyproof cost sharing mechanisms have been presented for games based on set cover and facility location. For set cover games this work was extended in [20,15] by considering different social desiderata and games with items or sets being agents. Furthermore, in [11] lower bounds on budget-balance for cross-monotonic cost sharing schemes were investigated. Cooperative games based on integer covering/packing problems were studied in [4]. It was shown that the core of such games is non-empty if and only if the integrality gap is 1. In [8] similar results are shown for a class of facility location games and an appropriate integer programming formulation. Cooperative games and the mechanism design framework are used to model selfish service receivers who can either cooperate to an offered cost sharing or manipulate. Our game, however, is strategic and non-cooperative in nature and allows players a much richer set of actions. We investigate distributed uncoordinated covering scenarios rather than a coordinated environment with a mechanism choosing customers, providing service and charging costs. Our model is suited for a case in which players have to directly investment into specific resources. Nevertheless our model has some connections to the cooperative setting, which we will outline in the end of Sect. 2.1.

The non-cooperative model we consider stems from [2], who proposed a game based on the Steiner forest problem. They show that prices of anarchy and stability are in $\Theta(k)$ and give a polynomial time algorithm for $(4.65 + \epsilon, 2)$-approximate Nash equilibria. In our uncapacitated facility location (UFL) game we assume that each of the clients must be connected directly to a facility. We can introduce a source node s and connect all facilities f to it, furthermore direct all edges from clients to facilities. The costs for the new edges (f, s) are given by the opening costs $c(f)$ of the corresponding facilities. This creates a single-source connection game on a directed graph. If we allow indirect connections to facilities, the game can be turned into an undirected single-source connection game (SSC) considered in [2, 10]. For both UFL and SSC games results in [2] suggest that the price of anarchy is k and the price of stability is 1 if each player has a single client. Algorithms for $(3.1 + \epsilon, 1.55)$-approximate Nash equilibria in the SSC game were proposed in [10]. In a very recent paper [3] we considered our game model for the special case of vertex covering. Prices of anarchy and stability are in $\Theta(k)$ and there is an efficient algorithm computing $(2, 2)$-approximate Nash equilibria. For a lower bound it was shown that both factors are essentially tight. In addition, for games on bipartite graphs and games with single edge players the price of stability was shown to be 1. This paper extends and adjusts these results to a much larger class of games based on general covering and facility location problems.

Our results. We study our games with respect to the quality and existence of pure strategy exact and approximate Nash equilibria. We will not consider mixed equilibria, as our model requires concrete investments rather than a randomized action, which would be the result of a mixed strategy. Our contributions are as follows.

Section 2 introduces the facility location games. Even for the most simple variant, the metric UFL game, the price of anarchy is exactly k and the price of stability is at least $k - 2$. Furthermore, it is NP-hard to determine whether a game has a Nash equilibrium. For the metric UFL game there is an algorithm to compute $(3, 3)$-approximate Nash equilibria in polynomial time. There is a lower bound of 1.097 on the stability ratio. For the more general class of facility location problems considered in [8] the price of stability is 1 if the integrality gap of a special LP-relaxation is 1. The best Nash equilibrium can be derived from the optimum solution to the LP-dual. Furthermore, if every player has only a single client, the price of stability is 1. We translate the lower bounds from the UFL game to SSC games [2] showing that it is NP-hard to determine Nash equilibrium existence and the price of stability is at least $k-2$. In addition, there is a lower bound of 1.0719 for the stability ratio in the SSC game. This negatively resolves a recent conjecture that the price of stability is 1 for SSC games with more than two terminals per player [10].

In Section 3 we consider general covering games. Even for the case of vertex cover it has been shown in [3] that prices of anarchy and stability are k and at least $k - 1$, respectively, and it is NP-hard to decide the existence of exact Nash equilibria. We show that for covering games, in which the integrality gap of the ICP-formulation is 1, the price of stability is 1. The best Nash equilibrium can be derived from the optimum solution to the LP-dual in polynomial time. If each player holds a single item, the price of stability is 1. There is an algorithm to get $(\mathcal{F}, \mathcal{F})$-approximate Nash equilibria in set cover games in polynomial time, where $\mathcal{F}$ is the maximum frequency of any item in the sets. This generalizes results for vertex cover games on bipartite graphs and an algorithm for $(2, 2)$-approximate Nash equilibria for general vertex cover games [3]. Proofs omitted from this extended abstract will be given in the full version of the paper.

2   Facility Location Games

Consider the following non-cooperative game for the basic problem of uncapacitated facility location (UFL). Throughout the paper we denote a feasible solution by S and the social optimum solution by S^*.

A complete bipartite graph $G = (T \cup F, T \times F)$ with vertex sets F of n_f *facilities* and T of n_t *clients* or *terminals* is given. Each of the k non-cooperative players holds a set $T_i \subset T$ of terminals. Each facility $f \in F$ has nonnegative opening costs $c(f)$, and for each terminal t and each facility f there is a nonnegative connection cost $c(t, f)$. The goal of each player is to connect her terminals to opened facilities at the minimum cost. Consider an integer programming (IP) formulation of the UFL problem:

$$
\begin{aligned}
\text{Min} \quad & \sum_{f \in F} c(f) y_f + \sum_{t \in T} c(t, f) x_{tf} \\
\text{subject to} \quad & \sum_{f \in F} x_{tf} \geq 1 && \text{for all } t \in T \\
& y_f - x_{tf} \geq 0 && \text{for all } t \in T, f \in F \\
& y_f, x_{tf} \in \{0, 1\} && \text{for all } t \in T, f \in F.
\end{aligned}
\tag{1}
$$

Each player insists on satisfying the constraints corresponding to her terminals $t \in T_i$. She offers money to the connection and opening costs by picking as a *strategy* a pair of two *payment functions* $p_i^c : T \times F \to \mathbb{R}_0^+$ and $p_i^o : F \to \mathbb{R}_0^+$, which specify her contributions to the connection and opening costs, resp. These are her offers to the cost of raising the x_{tf} and y_f variables. If the total offers of all players exceed the cost coefficient in the objective function (e.g. for a facility $\sum_i p_i(f) \geq c(f)$), the variable is raised to 1. In this case the corresponding connection or facility is considered bought or opened, resp. This affects *all* constraints, as all players can use bought connections and opened facilities for free, no matter whether they contribute to the cost or not. A *payment scheme* is a vector of strategies specifying for each player a single strategy. An (α, β)-*approximate Nash equilibrium* is a payment scheme in which no player can reduce her payments by more than a factor of α by unilaterally switching to another strategy, and which purchases a β-approximation to the socially optimum solution S^*. We refer to α as the *stability ratio* and β as the *approximation ratio*. Using this concept a payment scheme purchasing S^* is an $(\alpha, 1)$-approximate Nash equilibrium, and an exact Nash equilibrium is $(1, \beta)$-approximate.

The following observations can be used to simplify a game. Suppose a terminal is not included in any of the terminal sets T_i. This terminal is not considered by any player and has no influence on the game. Hence, we will assume that $T = \bigcup_{i=1}^{k} T_i$.

Suppose a terminal t is owned by a player i and a set of players J, i.e. $t \in T_i \cap (\bigcap_{j \in J} T_j)$. Now consider an (approximate) Nash equilibrium for an adjusted game in which t is owned only by i. If t is added to T_j again, the covering requirement of player j increases. Contributions of j to resource units satisfying the constraint of t might have been superfluous previously, but become mandatory now as t is included in T_j. Thus j's incentive to deviate to another strategy does not increase. So if the payment scheme is an (α, β)-approximate Nash equilibrium for the adjusted game, it can yield only a smaller stability ratio for the original game. We will thus assume that all terminal sets

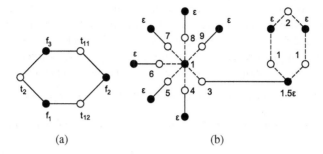

Fig. 1. (a) A metric UFL game without Nash equilibria – player 1 owns terminals labeled t_{11} and t_{12}, player 2 owns terminal t_2; (b) a metric UFL game with price of stability close to $k - 2$ for small ϵ – terminal labels indicate player ownership, facility labels specify opening costs. Black vertices are facilities, white vertices are terminals. All solid edges have cost 1, all dashed edges cost $\epsilon > 0$, all other edge costs are given by the shortest path metric.

T_i are mutually disjoint, as our results continue to hold if the sets T_i are allowed to overlap.

2.1 Metric UFL Games

In this section we present results on exact and approximate Nash equilibria for the metric UFL game. For lower bound constructions we only consider a subset of *basic* edges, for which we explicitly specify the connection cost. All other edge costs are given by the shortest path metric over basic edges.

Even in the metric UFL game the price of anarchy is exactly k. The lower bound is derived by an instance with two facilities, f_1 with cost k and f_2 with cost 1. Each player i has one terminal t_i, and all connection costs are $\epsilon > 0$. If each player pays a cost of 1 for f_1 and her connection cost, then no player has an incentive to switch and purchase f_2 completely. $\mathcal{S}^*$ is derived by opening only f_2 and connecting all terminals to it. This yields a lower bound on the price of anarchy arbitrarily close to k. For an upper bound suppose there is a Nash equilibrium with cost larger than $kc(\mathcal{S}^*)$. Then at least one player pays at least the cost $c(\mathcal{S}^*)$ and can thus deviate to purchase $\mathcal{S}^*$ completely by herself. This contradicts the assumption of a Nash equilibrium. The argumentation allows to show a price of anarchy of exactly k even for non-metric games. To derive a bound on the price of stability, we note that there are games without Nash equilibria.

Lemma 1. *There is a metric UFL game without Nash equilibria.*

Consider the game in Fig. 1(a). We assume that $c(f_1) = c(f_3) = 1$ and $c(f_2) = 1.5$. Player 1 either contributes to f_2 or to f_1 and f_3. If she purchases only $c(f_2)$, for it is best for player 2 to open one other facility, e.g. f_1. In this case it is better for player 1 to connect to f_1 and pay for opening f_3 as well. Then player 2 can drop f_1 and simply connect to f_3. This will create an incentive for player 1 to return to paying only for f_2. Although this is not a formal proof, it illustrates the cycling objectives inherent in the game. In a Nash equilibrium each terminal must be connected to an opened facility. Thus, formally all Nash equilibria can be considered by seven cases – depending on the different sets of opened facilities. It can be shown that for each set of opened facilities the costs cannot be purchased by a Nash equilibrium payment scheme. This game and the game outlined for the lower bound on the price of anarchy can be combined to a class of games that yields a price of stability of $k - 2$. The construction is shown in Fig. 1(b). In addition, deciding the existence of Nash equilibria is NP-hard.

Theorem 1. *The price of stability in the metric UFL game is at least $k - 2$.*

Theorem 2. *It is NP-hard to decide whether a metric UFL game has a Nash equilibrium.*

Both results extend easily to non-metric games. Thus, exact Nash equilibria can be quite costly and hard to compute. For some classes of games, however, there is a cheap Nash equilibrium. In particular, results in [2] can be used to show that UFL games with a single terminal per player allow for an iterative improvement procedure that improves both stability and approximation ratio. The price of stability is 1, and $(1 + \epsilon, 1.52)$-approximate Nash equilibria can be found using a recent approximation algorithm [16] to compute a starting solution. In addition, we show that there is another class of games with cheap equilibria, which can be computed efficiently.

Theorem 3. *For any metric UFL game, in which the underlying UFL problem has integrality gap 1, the price of stability is 1. An optimal Nash equilibrium can be computed in polynomial time.*

The payments are determined as follows. Reconsider the IP formulation (1) and its corresponding LP-relaxation obtained by allowing $y_f, x_{tf} \geq 0$. The integrality gap is assumed to be 1, so the optimum solution (x^*, y^*) to (1) is optimal for the relaxation. Using the optimum solution (γ^*, δ^*) to the dual of the LP-relaxation we assign $p_i^o(f) = y_f^* \left(\sum_{t \in T_i} \delta_{tf}^* \right)$ for player i and each facility f. In addition, we let player i contribute $p_i^c(t, f) = x_{tf}^*(\gamma_t^* - \delta_{tf}^*)$ for each $t \in T_i$ and $f \in F$. The argument that this gives a Nash equilibrium relies on LP duality and complementary slackness.

For general games we consider approximate Nash equilibria. This concept is motivated by the assumption that stability evolves if each player has no *significant* incentive to deviate. Formally, for (α, β)-approximate Nash equilibria the stability ratio $\alpha \geq 1$ specifies the violation of the Nash equilibrium inequality, and $\beta \geq 1$ is the approximation ratio of the social cost.

Theorem 4. *For the metric UFL game there is a primal-dual algorithm to derive $(3, 3)$-approximate Nash equilibria in polynomial time.*

Proof. In Algorithm 1 we denote a terminal by t, a facility by f, and the player owning t by i_t. The algorithm raises budgets for each terminal, which are offered for purchasing the connection and opening costs. Facilities are opened if the opening costs are covered by the total budget offered, and if they are located sufficiently far away from other opened facilities. For the approximation ratio of 3 we note that the algorithm is a primal-dual method for the UFL problem [17, 19].

For the analysis of the stability ratio consider a single player i and her payments. Note that the algorithm stops raising the budget of a terminal by the time it becomes directly or indirectly connected. We will first show that for the final budgets $\sum_{t \in T_i} B_t$ is a lower bound on the cost of any deviation for player i. For any terminal t we denote by $f(t)$ the facility t is connected to in the calculated solution. Verify that $c(t, f) \geq B_t$ for any terminal t and any opened facility $f \neq f(t)$. Hence, if a player has a deviation that improves upon B_t, it must open a new facility and connect some of her agents to it. By opening a new facility, however, the player is completely independent of the cost contributions of other players. Similar to [19] we can argue that the final budgets yield a feasible solution for the dual of the LP-relaxation. Hence, they form a 3-approximately budget balanced core solution for the cooperative game [13]. Now suppose there is a deviation for a player, which opens a new facility f and connects a subset of her terminals T_f to f thereby improving upon the budgets. Then the cost of $c(f) + \sum_{t \in T_f} < \sum_{t \in T_f} B_t$. This, however, would mean that the coalition formed by T_f in the coalitional game has a way to improve upon their budgets, which is a contradiction to B_t having the core-property. Hence, we know that $\sum_{t \in T_i} B_t$ is a lower bound on every deviation cost. Finally, note that for every directly connected terminal $t \in T_i$ player i pays B_t. A terminal t becomes indirectly connected only if it is unconnected and tight to a facility f by the time f is definitely closed. f becomes definitely closed only if there is another previously opened facility f' at distance $2B_t$ from f. Hence, there is an edge $c(t, f') \leq 3B_t$ by the metric inequality. So in the end player i pays at most $3B_t$ when connecting an indirectly connected terminal to the closest opened facility. This establishes the bound on the stability ratio.   □

In terms of lower bounds there is no polynomial time algorithm with an approximation ratio of 1.463 until $NP \subset DTIME(n^{O(\log \log n)})$ [9]. The following theorem shows that in the game of Fig. 1(a) the cost of any feasible solution cannot be distributed to get an approximate Nash equilibrium with stability ratio $\alpha \leq 1.097$.

Theorem 5. *There is a metric UFL game in which for every (α, β)-approximate Nash equilibrium $\alpha > 1.097$.*

Relation to cooperative games. In the cooperative game each terminal is a single player. The foremost stability concept is the core – the set of cost allocations assigning any coalition of players at most the cost of the optimum solution for this coalition only. A Nash equilibrium in our game guarantees this property only for the coalitions represented by our players. On the other hand the investments of a player now alter the cost of optimal solutions for other players. This feature makes overcovering the central problem that needs to be resolved to provide cheap solutions with low incentives to deviate. For deriving cheap approximately budget balanced core solutions the method of dual fitting can be applied, which scales the assigned payments of players to dual feasibility. The scaling factor then yields a factor for competitiveness, the notion in cooperative games analog to the stability ratio. In our non-cooperative framework the same simple scaling unfortunately does not work. In particular, for recently proposed greedy methods with better approximation ratios the factor for the approximation ratio does not translate.

Lemma 2. *The payments computed by recent greedy algorithms [12, 16] yield a stability ratio of $\Omega(k)$.*

Algorithm 1. Primal-dual algorithm for (3,3)-approximate Nash equilibria

---

In the beginning all terminals are unconnected, all budgets B_t are 0, and all facilities closed. Raise budgets of *unconnected* terminals at the same rate until one of the following events occurs. We denote the current budget of unconnected terminals by B. We call a terminal t *tight* with facility f if $B_t \geq c(t, f)$.

1. An unconnected terminal t goes tight with an opened facility f.
   In this case set t *connected* to f and assign player i_t to pay $p_{i_t}^c(t, f) = c(t, f)$.
2. For a facility f not yet definitely closed the sum of the budgets of unconnected and indirectly connected terminals t pays for opening and connection costs:
   $\sum_t \max(B_t - c(t, f), 0) = c(f)$. Then stop raising the budgets of the unconnected tight terminals. Also,
   (a) if there are opened facility f' and terminal t' with $c(t', f) + c(t', f') \leq 2B$, set f *definitely closed* and all unconnected terminals t tight with f *indirectly connected*.
   (b) Otherwise open f and set all terminals *directly connected* to f, which are tight with f and not yet directly connected to some other facility. For each such terminal assign player i_t to pay $p_{i_t}^c(t, f) = c(t, f)$ and $p_{i_t}^o(f) = B_t - c(t, f)$.

In the end connect all indirectly connected terminals to the closest opened facility and assign the corresponding players to pay for the connection cost.

2.2 Extensions

Connection-Restricted Facility Location Games. We extend the game from UFL to connection-restricted facility location (CRFL) problems as considered in [8]. Instead of the constraints $y_f - x_{tf} \geq 0$ there is for each facility f a set of feasible subsets of terminals that can be connected simultaneously to f. This formulation allows for instance capacity, quota, or incompatibility constraints and thus encompasses several well-known generalizations of the problem. For these games some of the previous results can be extended to hold.

Theorem 6. *For CRFL games, in which a partially conic relaxation of the underlying CRFL problem has integrality gap 1, the price of stability is 1.*

Theorem 7. *For CRFL games with singleton players the price of stability is 1.*

Single source connection games. By appropriately changing opening and connection costs most of the previous results translate in some reduced form to the SSC game with any number of terminals per player. As the previous algorithms in [2, 10] construct approximate Nash equilibria purchasing S^*, we explicitly examine a lower bound for this case.

Corollary 1. *There is a SSC game, in which for every (α, β)-approximate Nash equilibrium $\alpha > 1.0719$. For approximate Nash equilibria with $\beta = 1$ purchasing S^* the bound increases to $\alpha > 1.1835$.*

Corollary 2. *In the SSC game the price of stability is at least $k - 2$.*

Corollary 3. *It is NP-hard to decide whether a SSC game has a Nash equilibrium.*

3   Covering Games

Covering games and their equilibria are defined similarly to the facility location case. An integer covering problem (ICP) is given as

$$\text{Min} \quad \sum_{f=1}^{n} c(f)x_f$$

$$\text{subject to} \quad \sum_{f=1}^{n} a(t,f)x_f \geq b(t) \quad \text{for all } t = 1, \ldots, m \tag{2}$$

$$x_f \in \mathbb{N} \qquad\qquad \text{for all } f = 1, \ldots, n.$$

All constants are assumed to have non-negative (rational) entries $a(t,f), b(t), c(f) \geq 0$ for all $t = 1, \ldots, m$ and $f = 1, \ldots, n$. Associated with each of the k non-cooperative players is a subset of the constraints C_i, which she strives to satisfy. Integral units of a resource f have cost $c(f)$. They must be bought to be available for constraint satisfaction. Each player i chooses as a strategy a *payment function* $p_i : \{1, \ldots, n\} \rightarrow \mathbb{R}_+^n$, which specifies her non-negative contribution to each resource f. Then an integral number of x_f units of resource f are considered *bought* if x_f is the largest integer such that $\sum_i p_i(f) \geq c(f)x_f$. A bought unit can be used by all players for constraint satisfaction – no matter whether they contribute or not. We assume that if player i offers some amount $p_i(f)$ to resource f, and x_f units are bought in total, then her contribution to each unit is $p_i(f)/x_f$. Each player strives to minimize her cost, but insists on satisfying her constraints. We can translate definitions of exact and approximate Nash equilibria in this game directly from the UFL game. In addition, observations similar to the ones made in Sect. 2 can be used to simplify a game. Hence, in the following we will assume w.l.o.g. that the constraint sets C_i of the players form a partition of the constraints of the ICP. Note that in a Nash equilibrium no player contributes to an unbought unit, so the equality $\sum_i p_i(f) = c(f)x_f$ holds.

In the covering game prices of anarchy and stability behave similarly as in the metric UFL game. Using the results for vertex cover games in [3] and similar observations for the price of anarchy as in Sect. 2.1, we can see that the price of anarchy in the covering game is exactly k and the price of stability is at least $k - 1$. Furthermore, even for vertex cover games it is NP-hard to decide, whether a covering game has a Nash equilibrium. Hence, we again focus on classes of games, for which cheap Nash equilibria exist.

Theorem 8. *If the integrality gap of the ICP is 1, the price of stability is 1 and an optimal Nash equilibrium can be found in polynomial time.*

Theorem 9. *If for each player $|C_i| = 1$, the price of stability is 1.*

For set cover games there is an efficient algorithm to compute cheap approximate Nash equilibria. $\mathcal{F}$ denotes the maximum frequency of any element in the sets.

Theorem 10. *There is a primal-dual algorithm to compute a $(\mathcal{F}, \mathcal{F})$-approximate Nash equilibrium for set cover games.*

Acknowledgement. Part of this work was done during a visit at Dortmund University. I am grateful to Piotr Krysta and Patrick Briest for enlightening discussions on the topic.

References

1. E. Anshelevich, A. Dasgupta, J. Kleinberg, T. Roughgarden, É. Tardos, and T. Wexler. The price of stability for network design with fair cost allocation. In *Proc 45th FOCS*, pages 295–304, 2004.
2. E. Anshelevich, A. Dasgupta, É. Tardos, and T. Wexler. Near-optimal network design with selfish agents. In *Proc 35th STOC*, pages 511–520, 2003.
3. J. Cardinal and M. Hoefer. Selfish serive installation in networks. In *Proc 2nd Workshop Internet & Network Economics (WINE)*, 2006.
4. X. Deng, T. Ibaraki, and H. Nagamochi. Combinatorial optimization games. In *Proc 8th SODA*, pages 720–729, 1997.
5. N. Devanur, N. Garg, R. Khandekar, V. Pandit, A. Saberi, and V. Vazirani. Price of anarchy, locality gap, and a network service provider game. In *Proc 1st Workshop Internet & Network Economics (WINE)*, pages 1046–1055, 2005.
6. N. Devanur, M. Mihail, and V. Vazirani. Strategyproof cost-sharing mechanisms for set cover and facility location problems. In *Proc 4th EC*, pages 108–114, 2003.
7. H. Eiselt, G. Laporte, and J.-F. Thisse. Competitive location models: A framework and bibliography. *Transport. Sci.*, 27:44–54, 1993.
8. M. Goemans and M. Skutella. Cooperative facility location games. In *Proc 11th SODA*, pages 76–85, 2000.
9. S. Guha and S. Khuller. Greedy strikes back: Improved facility location algorithms. *J. Algorithms*, 31:228–248, 1999.
10. M. Hoefer. Non-cooperative tree creation. In *Proc 31st MFCS*, pages 517–527, 2006.
11. N. Immorlica, M. Mahdian, and V. Mirrokni. Limitations of cross-monotonic cost sharing schemes. In *Proc 16th SODA*, pages 602–611, 2005.
12. K. Jain, M. Mahdian, E. Markakis, A. Saberi, and V. Vazirani. Greedy facility location algorithms analyzed using dual fitting with factor-revealing LP. *J. ACM*, 50(6):795–824, 2003.
13. K. Jain and V. Vazirani. Applications of approximation algorithms to cooperative games. In *Proc 33rd STOC*, pages 364–372, 2001.
14. E. Koutsoupias and C. Papadimitriou. Worst-case equilibria. In *Proc 16th STACS*, pages 404–413, 1999.
15. X. Li, Z. Sun, and W. Wang. Cost sharing and strategyproof mechanisms for set cover games. In *Proc 22nd STACS*, pages 218–230, 2005.
16. M. Mahdian, Y. Ye, and J. Zhang. Improved approximation algorithms for metric facility location problems. In *Proc. 5th APPROX*, pages 229–242, 2002.
17. R. Mettu and G. Plaxton. The online median problem. *SIAM J. Comp*, 32(3):816–832, 2003.
18. T. Miller, T. Friesz, and R. Tobin. *Equilibrium Facility Location in Networks*. Springer Verlag, 1996.
19. M. Pál and É. Tardos. Group strategyproof mechanisms via primal-dual algorithms. In *Proc 44th FOCS*, pages 584–593, 2003.
20. Z. Sun, X. Li, W. Wang, and X. Chu. Mechanism design for set cover games when elements are agents. In *Proc 1st Intl Conf Algorithmic Applications in Management (AAIM)*, 2005.
21. A. Vetta. Nash equilibria in competitive societies with application to facility location, traffic routing and auctions. In *Proc 43rd FOCS*, page 416, 2002.

Optimal Algorithms for the Path/Tree-Shaped Facility Location Problems in Trees

Binay Bhattacharya[1,*], Yuzhuang Hu[1], Qiaosheng Shi[1], and Arie Tamir[2]

[1] School of Computing Science, Simon Fraser University, Burnaby B.C.,
Canada. V5A 1S6
{binay, yhu1, qshi1}@cs.sfu.ca
[2] School of Mathematical Sciences, Tel Aviv University, Ramat Aviv,
Tel Aviv 69978, Israel
atamir@post.tau.ac.il

Abstract. In this paper we consider the problem of locating a path-shaped or tree-shaped (extensive) facility in trees under the condition that existing facilities are already located. We introduce a parametric-pruning method to solve the conditional extensive weighted 1-center location problems in trees in linear time. This improves the recent results of $O(n \log n)$ by Tamir et al. [16].

1 Introduction

In a typical facility location problem, a set of demand points are embedded in some metric space and the goal is to locate a specified number of facilities in this space, such that the quality of service provided by these facilities is optimized. In general, the quality of service is measured by some objective function. There are many different objective functions of possible interests, among which the mostly studied ones are the minimization of average service distance and the minimization of maximum distance. The corresponding problems are referred to as *the median problem* and *the center problem* in the literature [5].

Usually a facility is represented by a point in the metric space [1,2,3,4,5,8,9, 10]. However, in recent years there has been a growing interest in studying the location of connected structures (referred to as *extensive* facilities) [6,7,11,13,14,15,16,17,18]. These studies were motivated by concrete decision problems related to routing and network design [16]. Also, in many practical situations there may exist some facilities and the problem is to find locations for a specified number of new facilities, which is referred to as *conditional location problem* by Minieka [12].

Our study in this paper is restricted to extensive facility location problems in a tree network where the objective function is to minimize the maximum distance. In specific terms our problem is to locate a path/subtree-shaped facility, whose length is no more than a predefined nonnegative value, in the underlying tree network such that the maximum weighted distance from demand points to the facility

---

* Research was partially supported by MITACS and NSERC.

T. Asano (Ed.): ISAAC 2006, LNCS 4288, pp. 379–388, 2006.
© Springer-Verlag Berlin Heidelberg 2006

is minimized. In the case when there are no existing facilities, Hedetmiemi et al. [7] proposed optimal algorithms for locating a path-shaped facility without length constraint that take advantage of a canonical recursive representation of a tree. Wang [17] proposed an optimal parallel algorithm for locating a path-shaped facility of a specified length in a weighted tree network. For locating a tree-shaped facility, Shioura and Shigeno [14] showed that these problems in tree networks are related to the bottleneck knapsack problems, and presented linear-time algorithms by using this relationship. When the existing facilities are taken into considerations, Mesa [11] provided an $O(n \log n)$ time algorithm (n is the number of vertices in the tree) for the conditional path-shaped center problem in the pure topological case of a tree, where vertices and edges in the tree are unweighted.

All the results mentioned above [7, 11, 14, 17] considered the tree network to be edge-weighted but not vertex-weighted. In recent papers [15, 16], Tamir et al. proposed $O(n \log n)$-time algorithms to solve the conditional problems in tree networks where the vertices and the edges are weighted. The basic technique used in their algorithms is parametric search. In this paper, we introduce a method (called *parametric pruning*) to optimally solve the weighted 1-center problem in a tree (that is, locate a point facility to minimize the maximum weighted distance from a demand point to it). This method is more general than Megiddo's method [9]. It is an integral part of the proposed optimal algorithms for locating a single path/tree-shaped center facility of a specified length in a tree network with/without existing facilities. These results improve the recent $O(n \log n)$ results of Tamir et al. [16].

The paper is organized as follows. Section 2 introduces notations and presents formal definitions of our problems. In Sect. 3, the main ideas of our algorithms are presented. The linear-time algorithms for the conditional path-shaped and tree-shaped center problems in tree networks are provided in sections 4 and 5 respectively. Section 6 gives a brief summary and shows that our approach can be extended to optimally solve the problem for more general service cost functions.

2   Notations and Problem Formulation

Let $T = (V, E, w, l)$ be an undirected tree network with the vertex set V, and the edge set E where each vertex $v \in V$ is associated with a positive weight $w(v)$ and each edge $e \in E$ is associated with a positive length $l(e)$. Let $A(T)$ denote the continuum set of points on the edges of T. For a subtree network T' of T, let $V(T'), E(T')$ and $A(T')$ denote the vertex set, the edge set and the continuum set of points on the edges of T', respectively. $P(u, v)$, $u, v \in A(T)$ denotes the *simple path* (also the shortest path) in T from u to v, whose length is denoted by $d(u, v)$. The length of a subtree network T' is the total length of edges in T'.

Let $\delta_{T'}(v)$ be the degree of vertex v in the subtree T'. A vertex v is called an anchor vertex of T' if $v \in V(T'), \delta_{T'}(v) = 1$ and $\delta_{T'}(u) = \delta_T(u), u \in V(T') \setminus \{v\}$. A subtree network is called *discrete* if all its leaf points are the vertices of T, and is called *continuous* otherwise. Let S represent the set of existing facilities, which by itself can be a subtree network or even a forest network [16].

The Weighted Path/Tree-Shaped Center Problem. Let C_1 (resp. C_2) be the set of all the path networks (resp. subtree networks) in T whose lengths are at most a predefined nonnegative value L, and let D_1 (resp. D_2) be the set of all the discrete path networks (resp. subtree networks) in T whose lengths are at most L. The goal is to establish one facility Y^*, either a path network in C_1/D_1 or a subtree network in C_2/D_2, such that

$$Cost(Y^* \cup S, T) = \min_{Y \in C_1(\text{ or } D_1, C_2, D_2)} Cost(Y \cup S, T),$$

where

$$Cost(Y \cup S, T) = \max_{v \in V} w(v) \cdot d(Y \cup S, v) \text{ with } d(Y \cup S, v) = \min_{x \in A(Y) \cup S} d(x, v)$$

is the *service cost function* of a new extensive facility Y and a set S of existing vertex/extensive facilities for all the demand points in T. When the facility Y is selected from C_1 or C_2, we call the model *continuous*, and if Y is chosen from D_1 or D_2, the model is called *discrete*. An extensive facility is *valid* if it is in C_1 (or $D_1/C_2/D_2$) for the continuous path-shaped (or discrete path-shaped/continuous tree-shaped/discrete tree-shaped) center problem. If S is empty we refer to the model as the *unconditional* model, and otherwise the model is *conditional*. We also consider the path-shaped center problem without the length constraint.

Since a problem in the unconditional model is a special case of the conditional one, it suffices to develop algorithms for problems in the conditional model only.

3   Main Idea of Our Algorithms

Given an extensive facility Y and a set S of existing facilities, a vertex v is called a *dominating vertex* of $Y \cup S$ if $w(v) \cdot d(Y \cup S, v) = Cost(Y \cup S, T)$. Observe that, in the center problem, an optimal facility $Y^* \cup S$ always has an equal or smaller service cost to the dominating vertices of a valid facility $Y \cup S$. In other words, given such dominating information of a valid facility, we are able to determine the relative location of an optimal facility with respect to the location of this valid facility. In Path Lemma and Tree Lemma discussed later, we will see how this idea works in solving our problems.

For the weighted 1-center problem, Megiddo [9] designed a 'prune-and-search' algorithm, which is carried out in two phases. The first phase is to locate a subtree network T', containing an optimal 1-center, that is anchored to a centroid vertex o of T. It is easy to see that the optimal 1-center provides services to all the clients in $T \setminus T'$ through the vertex o. Therefore, the topology of the subtree network $T \setminus T'$ is not important. For each vertex in $T \setminus T'$, we only need to keep its distance information to o. Since $|V(T \setminus T')| \geq n/2$, we call the subtree $T \setminus T'$ a *big component*. The second phase answers the following *key question*: determine whether there is an optimal 1-center in T' within distance t to o. An appropriate value of t is determined in the following way. We arbitrarily pair the vertices in $T \setminus T'$. Let $(u_1, u_1'), (u_2, u_2'), \ldots, (u_l, u_l')$ be the pairs where $w(u_i) \geq w(u_i')$. For every such pair $(u_i, u_i'), 1 \le i \le l$ let $t_i = [w(u_i')d(u_i', o) - w(u_i)d(u_i, o)]/(w(u_i) - w(u_i'))$, that is, u_i and u_i' have the same weighted distance to a point with the distance t_i to o. t is taken to be the median of these values. Once the answer to the key question is known, approximately $1/4$ of the vertices in $T \setminus T'$ cannot be dominating vertices of an optimal solution in T', and therefore can be discarded. The algorithm performs $O(\log n)$ such iterations. Each iteration takes linear time, linear in the size of the current tree. Therefore, the 1-center problem can be solved in linear time.

In this paper, we present a *parametric pruning* method for the weighted 1-center problem in T. For every pair $(u_i, u_i'), 1 \le i \le l$, $w(u_i) \ge w(u_i')$, let $c_i = w(u_i) \cdot (d(u_i, o) + t_i) = w(u_i') \cdot (d(u_i', o) + t_i)$ (t_i is described above). c_i is called the *switch service cost* of this pair. If the optimal service cost is larger (resp. smaller) than c_i then u_i (resp. u_i') is a dominating candidate. Let $R(u_i)$ and $R(u_i')$ denote the dominating regions of u_i, u_i', respectively. That is, $R(u_i) = (c_i, \infty)$ and $R(u_i') = [0, c_i]$. Let c be the median of these switch service costs. We can find either $c^* > c$ or $c^* \le c$ after solving the following decision problem: does there exist a point $p \in A(T)$ such that $Cost(p, T) \le c$? Here c^* denotes the optimal service cost. As we know, the decision problem in a tree network can be solved in linear time [10]. Therefore, the pruning of the vertices in $T \setminus T'$ can also be performed using this parametric-pruning method.

The advantage of parametric pruning over Megiddo's technique is that it is more general and is applicable to the problems considered in this paper. In Megiddo's method, we need to locate one big component served by a 1-center from outside so that a fraction of the vertices from the big component can be identified for pruning. Here the center facility serves the demand points in the big component through the vertex o only. However, the parametric-pruning method still works even if $O(n)$ disjoint components are found which are served from the outside by the facilities. We will see the details later in our algorithms. Basically, our main idea is to 'prune' the vertices that do not determine the optimal service cost (i.e. vertices are not dominating), and to 'shrink' the facility if some path network or subtree network is known to be a part of an optimal facility.

3.1   Locating Non-dominating Vertices in an Optimal Solution

For our problems, more ideas are needed to make the parametric pruning to work. In the conditional model we cannot afford to keep the information of the existing facilities in S at each pruning iteration as S could be $O(n)$. However, it is not difficult to design a linear-time step to find the distance $d(S, v)$ for each vertex $v \in V$ [16]. After this step, it is safe to discard the vertices of S in the subsequent steps. The following important lemma is established in [16].

Lemma 1. *[16] Given a point p and a nonnegative value c, the facility Y of smallest length with $p \in A(Y)$ and $Cost(Y \cup S, T) \le c$ can be computed in linear time.*

From Lemma 1, it is not hard to see that the feasibility test can be solved in linear time [16]. The feasibility test can be formally described as follows. Given a real number c, determine whether there exists a subtree/path Y of length not exceeding L such that $C(Y \cup S, T) \leq c$.

Number of Switch Service Costs for a Pair of Nodes. Suppose that facility Y serves a pair of vertices (u, v) through o, see Fig. 1.

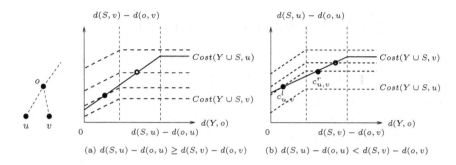

(a) $d(S, u) - d(o, u) \geq d(S, v) - d(o, v)$ (b) $d(S, u) - d(o, u) < d(S, v) - d(o, v)$

Fig. 1. The number of switch service costs for (u, v) in the conditional model

In the unconditional model, it is trivial to see that there exists at most one switch service cost. However, there may be more than one switch service cost in the conditional model. In the following we show that in the conditional model at most two switch service costs exist for a given pair of vertices (u, v). If $d(o, v) \geq d(S, v)$ then v will always be served by some existing facility. Without loss of generality, assume that $d(o, u) < d(S, u), d(o, v) < d(S, v)$, and $w(u) \geq w(v)$. Fig. 1 shows the service cost functions $Cost(Y \cup S, u), Cost(Y \cup S, v)$ with the change of $d(Y, o)$. We use the dashed lines to represent all possibilities. If $d(S, u) - d(o, u) \geq d(S, v) - d(o, v)$, see Fig. 1(a), there is at most one switch service cost. In the case when $d(S, u) - d(o, u) < d(S, v) - d(o, v)$, see Fig. 1(b), it is possible to have two switch service costs, but no more than two. When there are two switch service costs for (u, v), we call the switch service cost with smaller value *left switch service cost* and call the other one *right switch service cost*.

In fact, it is not hard to see that there are a constant number of intersection points between a pair of nondecreasing piecewise linear functions with a constant number of breakpoints. That is, there are a constant number of dominating regions for each vertex. In [18], Zemel provided a 'prune-and-search' method for the case where there are a constant number of dominating regions for each vertex. Here we show an algorithm to locate non-dominating vertices in our problems.

For those pairs having only one switch service cost, we can locate the non-dominating vertices easily by checking the feasibility of the median switch service cost. In this case there are half of such pairs in which one vertex is identified as a non-dominating vertex. Suppose that $(u_1, v_1), (u_2, v_2), \ldots, (u_k, v_k)$ are the pairs of vertices with two switch service costs, where $w(u_i) \geq w(v_i), 1 \leq i \leq k$. Let c_i^l (resp. c_i^r) be the left (resp. right) switch service cost of $(u_i, v_i), i = 1, \ldots, k$. Clearly, $R(u_i) = (c_i^l, c_i^r]$ and $R(u_i') = [0, c_i^l] \cup (c_i^r, \infty)$. Select one value c^l (resp. c^r)

such that one third of left (resp. right) switch service costs $c_i^l > c^l$ (resp. $c_i^r \leq c^r$) and the remaining ones are no more than (resp. larger than) it. We call c^l (resp. c^r) called the *left switch value* (resp. *right switch value*). After solving the decision problems with c^l and c^r, we can find at least $\lfloor \frac{k}{3} \rfloor$ non-dominating vertices for an optimal facility.

4   Weighted Path-Shaped Center Problems

In this section we apply the ideas introduced in the previous section to design a linear-time algorithm that solves the path-shaped center location problem.

Lemma 2 (Path Lemma). *Given a point q in T, we can find in linear time either the optimal service cost c^*, one subtree network anchored to q containing an optimal path facility, or two subtree networks anchored to q containing an optimal path facility and q is on it.*

Proof. Let $T_1, \ldots, T_m$ be the subtree networks anchored to q such that $Cost(\{q\} \cup S, T_1) \geq Cost(\{q\} \cup S, T_2)$ and $Cost(\{q\} \cup S, T_2) \geq Cost(\{q\} \cup S, T_i), i = 3, \ldots, m$. Let F be the set of dominating vertices for $\{q\} \cup S$.

- If some dominating vertices are in $T \setminus \{T_1 \cup T_2\}$, $c^* = Cost(\{q\} \cup S, T_1)$ and q is an optimal path facility.
- When $Cost(\{q\} \cup S, T_1) = Cost(\{q\} \cup S, T_2) > Cost(\{q\} \cup S, T_3)$, an optimal facility lies in $T_1 \cup T_2$, and q lies on it.
- Otherwise, T_1 contains all the dominating vertices. Let $c = Cost(\{q\} \cup S, T_2)$. If c is infeasible, an optimal path facility lies in T_1, otherwise, $c^* \leq c$. Note that q must be on an optimal path facility if $c^* < c$. By Lemma 1, we can find whether or not there is a valid path facility containing q with service cost no more than c. If there exists such facility then an optimal path facility lies in $T_1 \cup T_2$ and q lies on it. If not, $c^* = c$. □

One of the following cases occurs when the Path Lemma is applied to a centroid vertex o of T. Let $T_1, T_2, \ldots, T_m$ be the subtree networks anchored to o, as described in the proof of Path Lemma.

- *Case 1*: An optimal path facility lies in a subtree network T_1 anchored to o.
- *Case 2*: There is an optimal path facility lying in $T_1 \cup T_2$ anchored to o, and o lies on it.

In Case 1, vertices in subtree $T \setminus T_1$ are served by the new facility through o and $|V(T \setminus T_1)| \geq n/2$. Our goal is to prune a fraction of the vertices in the big component $T \setminus T_1$. Randomly pair the vertices in $T \setminus T_1$, and compute the switch service costs for each pair. By the method described in Sect. 3.1, at least $\frac{n}{4} \times \frac{1}{3}$ non-dominating vertices in $T \setminus T_1$ can be found and be discarded in linear time.

In Case 2, o is the closest point in an optimal path facility to any vertex in $T \setminus (T_1 \cup T_2)$. We discard all the vertices in $T \setminus (T_1 \cup T_2)$ except one vertex v with $w(v) \times d(\{o\} \cup S, v) = \max_{u \in V(T \setminus T_1 \cup T_2)} w(u) \cdot d(\{o\} \cup S, u)$. If $|V(T) \setminus$

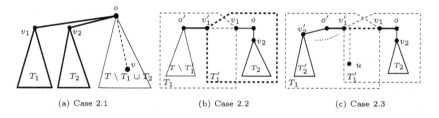

(a) Case 2.1   (b) Case 2.2   (c) Case 2.3

Fig. 2. Case 2: an optimal path facility lies in T_1, T_2 and o is on it

$V(T_1 \cup T_2)| \geq n/3$ (*Case 2.1*, see Fig. 2(a)), at least $n/3 - 1$ vertices in T can be removed. So assume that $|V(T_1) \cup V(T_2)| > 2n/3$ and $|V(T_1)| > n/3$. Let o' be a centroid vertex of T_1. We get the following cases after applying Path Lemma on o':

- *Case 2.2*: The path facility lies in one subtree network T_1', anchored to o' and contains o. In this case, the path facility serves the vertices in $T \setminus T_1'$ via o' and $|V(T_1 \setminus T_1')| > n/6$. Using the same method as the one used in Case 1, we can discard at least $\frac{n}{12} \times \frac{1}{3}$ vertices in $T_1 \setminus T_1'$.
- *Case 2.3*: The path facility lies in two subtrees T_1' and T_2' anchored to o'. Observe that $P(o, o')$ is a part of new facility in an optimal solution. It implies that the closest facility point of every vertex in $V(T_1 \setminus T_2')$ in an optimal solution is determined. Contract $P(o, o')$ into one point o and discard all the other vertices in $T_1 \setminus T_2'$ except the vertex u with the maximum service cost. Since $|V(T_1 \setminus T_2')| \geq \frac{n}{6}$, at least $\frac{n}{6} - 1$ vertices are removed from T.

Denote by T' the new tree thus computed. Since the size of T' is at most $\frac{35}{36}n$, the process terminates within $O(\log n)$ iterations. Since each iteration takes linear time, linear in the size of the current underlying tree, the total cost is therefore linear in n.

It is not hard to see that the algorithm works for the discrete case as well. Summing up, we have the following theorem.

Theorem 1. *The unconditional/conditional discrete/continuous weighted path-shaped center problem with/without length constraint in tree networks can be solved in linear time.*

5   Weighted Tree-Shaped Center Problems

In this section we present a linear-time algorithm to locate a tree-shaped center in a weighted tree. The following lemma shows a property for the tree-shaped facility location problem, which is similar in spirit to Path Lemma. Its proof is also very similar.

Lemma 3 (Tree Lemma). *Given a vertex u in T, we can find in linear time either the optimal service cost c^*, one subtree network anchored to u containing an optimal facility, or u lies in an optimal facility.*

After the application of the Tree Lemma to a centroid vertex o of T, there are two cases to consider. In Case 1 where one subtree network anchored to o is found to contain an optimal facility, we can discard $\frac{n}{12}$ vertices in linear time by a procedure similar to the one described for Case 1 in the path-shaped center problem.

In Case 2, o lies in an optimal facility. In this case, we focus on pruning the vertices of T of degree no more than two. For a leaf vertex u, let $p(u)$ denote the vertex adjacent to it in T and let e_u be the edge incident to it. If the optimal service cost is smaller than $w(u) \cdot d(\{p(u)\} \cup S, u)$, then a part of the edge e_u must lie in an optimal facility, and u is served by the new facility. Therefore, u is a dominating vertex of an optimal facility (if we want to minimize the length of the new facility without increasing the optimal service cost). Lemma 4 provides a process to prune such dominating leaf vertices. Otherwise, the new facility lies outside the edge e_u in an optimal solution.

Basically, we are interested in pairing the vertices (u, v), $\delta(u), \delta(v) \leq 2$, satisfying one of the following relations:

- (Type-I) u and v have a 'ancestor-descendent' relation, i.e. v is on the path $P(u, o)$ or u is on $P(v, o)$. In this case we can also show that there are at most two switch service costs no matter where the new facility lies. This is due to the fact that u, v and o all lie on a path and the new facility contains o.
- (Type-II) u and v are 'sibling' leaf vertices with $p(u) = p(v)$. If the new facility doesn't lie on e_u and e_v, surely one of them can be pruned.

In the following, we first describe a simple algorithm to determine disjoint Type-I pairs of vertices, and show that the output size depends on the number of leaf vertices in the tree. We then present the algorithm for the case where there are lots of leaf vertices.

We remove all the vertices of degree more than two. For each connected path, output the disjoint pairs of vertices (leave one if the number of vertices in one path is odd). Let m be the number of leaf vertices in T. Then the number of vertices not in any pair cannot be more than $3m$. In the case where the number of leaf vertices is no more than $\frac{n}{5}$, we can use the proposed parametric pruning in Sect. 3.1 to prune the tree. Since there are at least $\frac{n}{5}$ output (Type-I) pairs of vertices, at least $\frac{1}{3} \times \frac{n}{5}$ vertices can be discarded after solving the corresponding decision problems.

We now consider the problem where the number of leaf vertices in T is more than $\frac{n}{5}$. For each leaf vertex v, compute $c_v = w(v) \cdot d(\{p(v)\} \cup S, v)$. If there are lots of leaf vertices such that their neighboring vertices are of degree two, say at least one quarter of them, then we have at least $\frac{1}{4} \times \frac{n}{5}$ Type-I pairs of vertices (each pair consists of a leaf vertex and its neighbor vertex) and the algorithm described above can be applied here. Otherwise, there are at least $\frac{3}{4} \times \frac{n}{5}$ leaf vertices whose neighbors have degrees greater than two. We arbitrarily pair such sibling leaf vertices. It is not hard to see that there are at least $\frac{1}{4} \times \frac{n}{5}$ such disjoint Type-II pairs. We leave one leaf vertex if it has odd number of siblings.

For each leaf vertices pair (u, v), let $c_{(u,v)} = \max\{c_u, c_v\}$. c is chosen to be the median of these values.

- $c^* > c$. Then, for each pair (u, v) with $c_{(u,v)} \leq c$ it is not necessary for e_u and e_v to intersect the new facility in an optimal solution. Therefore, the algorithm used for Case 1 can be adapted again to prune vertices (at least $\frac{n}{5} \times \frac{1}{8}$ vertices).
- $c^* \leq c$. For each pair (u, v) with $c_{(u,v)} > c$, assume that $c_u = c_{(u,v)}$, without any loss of generality. Then u is a dominating vertex, and it is served by the new facility. Hence, there are at least $\frac{1}{8} \times \frac{n}{5}$ such dominating vertices. The following lemma provides the base to prune these dominating vertices. But first we need to shorten the edge e_u if $d(S, u) < l(e_u)$. In this case we update $L = L - l(e_u) + d(S, u))$ and set $l(e_u) = d(S, u)$.

Lemma 4. *Given two dominating leaf vertices u, v for an optimal facility with its service cost no more than $\min\{w(u) \cdot l(e_u), w(v) \cdot l(e_v)\}$, where $d(S, u) \geq l(e_u)$ and $d(S, v) \geq l(e_v)$, the optimal service cost in the new tree, constructed by deleting u and e_u, and updating $l(e_v) = l(e_u) + l(e_v)$ and $w(v) = \frac{w(u) \cdot w(v)}{w(u) + w(v)}$, is equal to the optimal service cost in original tree.*

Proof. It is not hard to see that $p(u)$ and $p(v)$ must lie in the optimal facility and u, v are served by it (not by some existing facility). Let d_1, d_2 be the distance from u, v to the optimal facility respectively. Clearly, $w(u) \cdot d_1 = w(v) \cdot d_2$ is equal to the optimal service cost. In the new tree, only vertices u and v and edges e_u and e_v are changed. Since $(d_1 + d_2) \cdot \frac{w(u) \cdot w(v)}{w(u) + w(v)} = w(u) \cdot d_1 = w(v) \cdot d_2$, the length of the part of the new facility on the new edge e_v is equal to the length of the part of the new facility on e_u and e_v (that is $l(e_u) + l(e_v) - d_1 - d_2$), for any optimal service cost. Therefore, the optimal service cost in the new tree is equal to the original optimal service cost. □

Putting everything together, we establish Theorem 2. Adapting the algorithm for the discrete case is straightforward.

Theorem 2. *The unconditional/conditional discrete/continuous weighted tree-shaped center problem with length constraint in trees can be solved in linear time.*

6   Conclusion and Future Work

In this paper, we propose optimal algorithms for the extensive facility location problems in tree networks, where the new facility (center) is either path-shaped or tree-shaped. The main technique is to 'prune' the nondominating vertices and to 'shrink' the facility if some path or subtree is known to be a part of an optimal facility. These results improve the recent $O(n \log n)$ results of Tamir et al. [16].

For the case where the service cost $f_i(x)$ of a client (vertex) v_i is a nondecreasing, piecewise linear function (with a fixed number of breakpoints) of the service distance to the facility (in the 'conditional' case $f_i(x)$ has only one breakpoint), all the ideas presented in this paper can be extended to achieve an optimal algorithm for locating the facility (either a point, a path, or a subtree). Actually, our method works even when the piecewise linearity assumption is relaxed to piecewise polynomiality (e.g. quadratic, or cubic) of fixed degree.

References

1. B. Ben-Moshe, B. Bhattacharya, Q. Shi, "An optimal algorithm for the continuous/discrete weighted 2-center problem in trees", in *LATIN*, Chile, 2006.
2. B. Bhattacharya, Q. Shi, "Optimal algorithms for weighted p-center problem in trees, any fixed p", manuscript, 2006.
3. G.N. Frederickson, "Parametric search and locating supply centers in trees", In Workshop on Algorithms and Data Structures (WADS) (1991), 299-319.
4. G.N. Frederickson, D.B. Johnson, "Finding k-th paths and p-centers by generating and searching good data structures", J. of Alg. 4 (1983) 61-80.
5. S.L. Hakimi, "Optimum location of switching centers and the absolute centers and medians of a graph", Oper. Res. 12 (1964) 450-459.
6. S.L. Hakimi, E.F. Schmeichel, M. Labbe, "On locating path or tree shaped facilities on networks", Networks 23 (1993) 543-555.
7. S.M. Hedetniemi, E.J. Cockaine, S.T. Hedetniemi, "Linear algorithms for finding the Jordan center and path center of a tree" Transport. Sci. 15 (1981) 98-114.
8. M. Jeger, O. Kariv, "Algorithms for finding p-centers on a weighted tree (for relatively small p)", Networks, 15 (1985) 381-389.
9. N. Megiddo, "Linear-time algorithms for linear programming in R^3 and related problems", SIAM J. Comput. 12 (1983) 759-776.
10. N. Megiddo, A. Tamir, E. Zemel, and R. Chandrasekaran, "An $O(n \log^2 n)$ algorithm for the kth longest path in a tree with applications to location problems", SIAM J. Comput. 10 (1981) 328-337.
11. J.A. Mesa, "The conditional path center problem in tree graphs", unpublished paper presented to EWGLA8 held in Lambrecht (Germany), 1995.
12. E. Minieka, "Conditional centers and medians on a graph", Networks 10 (1980) 265-272.
13. E. Minieka, "The optimal location of a path or tree in a tree network", Networks 15 (1985) 309-321.
14. A. Shioura, M. Shigeno, "The tree center problems and the relationship with the bottleneck knapsack problems", Networks 29 (1997) 107-110.
15. A. Tamir, J. Puerto, D. Pérez-Brito, "The centdian subtree on tree networks", Disc. Appl. Math. 118 (2002) 263-278.
16. A. Tamir, J. Puerto, J.A. Mesa, A.M. Rodriguez-Chia, "Conditional location of path and tree shaped facilities on trees", J. of Alg., 56 (2005) 50-75.
17. B.F. Wang, "Efficient parallel algorithms for optimally locating a path and a tree of a specified length in a weighted tree network", J. of Alg., 34 (2000) 90-108.
18. E. Zemel, "An $O(n)$ algorithm for the linear multiple choice knapsack problem and related problems", Information Processing Letters, 18 (1984) 123-128.

Multiobjective Optimization: Improved FPTAS for Shortest Paths and Non-linear Objectives with Applications*

George Tsaggouris and Christos Zaroliagis

Computer Technology Institute, Patras University Campus, 26500 Patras, Greece;
and Dept of Computer Eng & Informatics, University of Patras, 26500 Patras, Greece
{tsaggour, zaro}@ceid.upatras.gr

Abstract. We provide an improved FPTAS for multiobjective shortest paths, a fundamental (NP-hard) problem in multiobjective optimization, along with a new generic method for obtaining FPTAS to *any* multiobjective optimization problem with *non-linear* objectives. We show how these results can be used to obtain better approximate solutions to three related problems that have important applications in QoS routing and in traffic optimization.

1 Introduction

Multiobjective shortest paths (MOSP) is a core problem in the area of multiobjective optimization [3,4] with numerous applications. Informally, the problem consists in finding a set of paths that captures not a single optimum but the trade-off among $d > 1$ objective functions in a digraph whose edges are associated with d-dimensional attribute (cost) vectors. In general, an instance of a multiobjective optimization problem is associated with a set of feasible solutions Q and a d-vector function $\mathbf{f} = [f_1, \ldots, f_d]^T$ (d is typically a constant) associating each feasible solution $q \in Q$ with a d-vector $\mathbf{f}(q)$ (w.l.o.g. we assume that all objectives f_i, $1 \leq i \leq d$, are to be minimized). In a multiobjective optimization problem, we are interested not in finding a single optimal solution, but in computing the trade-off among the different objective functions, called the *Pareto set or curve* $\mathcal{P}$, which is the set of all feasible solutions in Q whose vector of the various objectives is *not* dominated by any other solution (a solution p *dominates* another solution q iff $f_i(p) \leq f_i(q)$, $\forall 1 \leq i \leq d$). Multiobjective optimization problems are usually NP-hard (as indeed is the case for MOSP). This is due to the fact that the Pareto curve is typically exponential in size (even in the case of two objectives). On the other hand, even if a decision maker is armed with the entire Pareto curve, s/he is left with the problem of which is the "best" solution for the application at hand. Consequently, three natural approaches to solve multiobjective optimization problems are to: (i) study approximate versions of the Pareto curve; (ii) optimize one objective while bounding the rest (*constrained approach*);

---

* This work was partially supported by the FET Unit of EC (IST priority – 6th FP), under contracts no. FP6-021235-2 (ARRIVAL) and no. IST-2002-001907 (DELIS).

T. Asano (Ed.): ISAAC 2006, LNCS 4288, pp. 389–398, 2006.

and (iii) proceed in a normative way and choose the "best" solution by introducing a utility (typically non-linear) function on the objectives (*normalization approach*). In this paper, we investigate all of them for MOSP.

Multiobjective Shortest Paths. Despite so much research in multiobjective optimization [3,4], only recently a systematic study of the complexity issues regarding the construction of approximate Pareto curves has been initiated [11,14]. Informally, an $(1 + \varepsilon)$-*Pareto curve* $\mathcal{P}_\varepsilon$ is a subset of feasible solutions such that for any Pareto optimal solution, there exists a solution in $\mathcal{P}_\varepsilon$ that is no more than $(1 + \varepsilon)$ away in all objectives. Papadimitriou and Yannakakis show in a seminal work [11] that for any multiobjective optimization problem there exists a $(1 + \varepsilon)$-Pareto curve $\mathcal{P}_\varepsilon$ of (polynomial) size $|\mathcal{P}_\varepsilon| = O((4B/\varepsilon)^{d-1})$, where B is the number of bits required to represent the values in the objective functions (bounded by some polynomial in the size of the input); $\mathcal{P}_\varepsilon$ can be constructed by $O((4B/\varepsilon)^d)$ calls to a GAP routine that solves (in time polynomial in the size of the input and $1/\varepsilon$) the following problem: given a vector of values $\mathbf{a}$, either compute a solution that dominates $\mathbf{a}$, or report that there is no solution better than $\mathbf{a}$ by at least a factor of $1 + \varepsilon$ in all objectives.

For the case of MOSP (and some other problems with linear objectives), it is shown in [11] how a GAP routine can be constructed (based on a pseudopolynomial algorithm for computing exact paths), and consequently a FPTAS is provided. Note that FPTAS for MOSP were already known in the case of two objectives [8], as well as in the case of multiple objectives in directed acyclic graphs (DAGs) [15]. In particular, the 2-objective case has been extensively studied [4], while for $d > 2$ very little has been achieved; actually, the results in [11,15] are the only and currently best FPTAS known (see Table 1). Let C^{max} denote the ratio of the maximum to the minimum edge weight (in any dimension), and let n (resp. m) be the number of nodes (resp. edges) in a digraph.

Our first contribution in this work (Section 3) is a new and remarkably simple FPTAS for constructing a set of approximate Pareto curves (one for every node) for the *single-source* version of the MOSP problem in *any* digraph. For any $d > 1$, our algorithm runs in time $O(nm(\frac{n \log(nC^{max})}{\varepsilon})^{d-1})$ for general digraphs, and in $O(m(\frac{n \log(nC^{max})}{\varepsilon})^{d-1})$ for DAGs. Table 1 summarizes the comparison of our results with the best previous ones. Our results improve significantly upon previous approaches for general digraphs [11,14] and DAGs [14,15], for all $d > 2$. For $d = 2$, our running times depend on ε^{-1}; those in [14] are based on repeated applications of a stronger variant of the GAP routine, like a FPTAS for the restricted shortest path (RSP) problem (see e.g., [9], and thus depend on ε^{-2}. Hence, our algorithm gives always better running times for DAGs, while for general digraphs we improve the dependence on $1/\varepsilon$.

Non-linear Objectives. Our second contribution in this work concerns two fundamental problems in multiobjective optimization: (i) Construct a FPTAS for the normalized version of a multiobjective optimization problem when the utility function is *non-linear*. (ii) Construct a FPTAS for a multiobjective optimization problem with *non-linear* objectives.

Table 1. Comparison of new and previous results for MOSP. T_{GAP} denotes the time of a GAP routine, which is polynomial in the input and $1/\varepsilon$ (but exponential in d).

		Best previous	This work
General digraphs	$d = 2$	$O\left(nm\frac{1}{\varepsilon}\log\left(nC^{max}\right)\left(\log\log n + \frac{1}{\varepsilon}\right)\right)$ [14]	$O\left(n^2 m\frac{1}{\varepsilon}\log\left(nC^{max}\right)\right)$
	$d > 2$	$O((\log(nC^{max})/\varepsilon)^d \cdot T_{GAP})$ [11]	$O\left(nm\left(\frac{n\log(nC^{max})}{\varepsilon}\right)^{d-1}\right)$
DAGs	$d = 2$	$O\left(nm\frac{1}{\varepsilon}\log n \log\left(nC^{max}\right)\right)$ [15] $O\left(nm\frac{1}{\varepsilon^2}\log\left(nC^{max}\right)\right)$ [14]	$O\left(nm\frac{1}{\varepsilon}\log\left(nC^{max}\right)\right)$
	$d > 2$	$O\left(nm(\frac{n\log(nC^{max})}{\varepsilon})^{d-1}\log^{d-2}(\frac{n}{\varepsilon})\right)$ [15]	$O\left(m\left(\frac{n\log(nC^{max})}{\varepsilon}\right)^{d-1}\right)$

An algorithm for the first problem was given in [12] (earlier version of this work) for $d \geq 2$ objectives and polynomial utility function, and independently in [1] for $d = 2$ objectives and quasi-polynomially bounded utility function. Let $T(1/\varepsilon, m')$ denote the time to generate an $(1+\varepsilon)$-Pareto curve for an instance of a multiobjective optimization problem of size m'. The algorithm in [1] provides a FPTAS with time complexity $T(\Lambda_1/\varepsilon^2, m')$, where Λ_1 is polylogarithmic on the maximum cost in any dimension.

We show in Section 4 that we can construct a FPTAS for the normalized version of *any* multiobjective optimization problem with $d \geq 2$ objectives and quasi-polynomially bounded utility function in time $T(\Lambda_2/\varepsilon, m')$, where $\Lambda_2 < \Lambda_1$ is polylogarithmic on the maximum cost in any dimension. Our results are based on a *novel* and simple analysis, and improve upon those in [1] both w.r.t. the running time (better dependence on $1/\varepsilon$ and $\Lambda_2 < \Lambda_1$) and the number of objectives – as well as upon those in [12] w.r.t. the class of utility functions.

The only generic method known for addressing the second problem is that in [11], which assumes the existence of a GAP routine. Such routines for the case of non-linear objectives are not known. The GAP routines given in [11] concern problems with linear objectives only.

We show in Section 4 that a FPTAS for *any* multiobjective optimization problem $\mathcal{M}'$ with quasi-polynomially bounded non-linear objective functions can be constructed from a FPTAS for a much simpler version $\mathcal{M}$ of the problem. $\mathcal{M}$ has the same feasible solution set with $\mathcal{M}'$ and objectives the *identity functions* on the attributes of the non-linear objective functions of $\mathcal{M}'$. In other words, our result suggests that restricting the study of approximate Pareto curves to identity (on the attributes) objectives suffices for treating the non-linear case. Our approach constitutes the first generic method for obtaining FPTAS for any multiobjective optimization problem with quasi-polynomial non-linear objectives.

Applications. The following problems play a key role in several domains.

Multiple Constrained (Optimal) Paths. One of the key issues in networking [10] is how to determine paths that satisfy QoS constraints, a problem known as QoS routing or constraint-based routing. The two most fundamental problems in QoS routing are the *multiple constrained optimal path* (MCOP) and the *multiple constrained path* (MCP) problems (see e.g., [7,10]). In MCOP, we are given a

d-vector of costs $\mathbf{c}$ on the edges and a $(d-1)$-vector $\mathbf{b}$ of QoS-bounds. The objective is to find an s-t path p that minimizes $c_d(p) = \sum_{e \in p} c_d(e)$, and obeys the QoS-bounds, i.e., $c_i(p) = \sum_{e \in p} c_i(e) \leq b_i, \forall 1 \leq i \leq d-1$. MCOP is NP-hard, even when $d = 2$ in which case it is known as the restricted shortest path problem and admits a FPTAS (see e.g., [9]). In MCP, the objective is to find an s-t path p that simply obeys a d-vector $\mathbf{b}$ of QoS-bounds, i.e., $c_i(p) = \sum_{e \in p} c_i(e) \leq b_i$, $\forall 1 \leq i \leq d$. MCP is NP-complete. For both problems, the case of $d = 2$ objectives has been extensively studied and there are also very efficient FPTAS known (see e.g., [9]). For $d > 2$, apart from the generic approach in [11], only heuristic methods and pseudopolynomial time algorithms are known [10]. We are able to show how (quality guaranteed) approximate schemes to both MCOP and MCP can be constructed that have the same complexity with MOSP, thus improving upon all previous approaches for any $d > 2$.

Non-Additive Shortest Paths. In this problem (NASP), we are given a digraph whose edges are associated with d-dimensional cost vectors and the task is to find a path that minimizes a certain d-attribute non-linear utility function. NASP is a fundamental problem in several domains [5,6], the most prominent of which is finding traffic equilibria [5]. NASP is an NP-hard problem. By virtue of the results in [1,12], there exists a FPTAS for $d = 2$ and quasi-polynomial utility function [1], and a FPTAS for any $d \geq 2$ and polynomial utility function [12].

In Section 5, we show how our FPTAS for MOSP, along with our generic framework for dealing with non-linear objectives, can be used to obtain a FPTAS for NASP for *any* $d > 1$ and a larger than quasi-polynomially bounded family of utility functions. Our results improve considerably upon those in [1,12] w.r.t. time (dependence on $1/\varepsilon$), number of objectives, and class of utility functions.

2   Preliminaries

Recall that an instance of a multiobjective optimization problem is associated with a set of feasible solutions Q and a d-vector function $\mathbf{f} = [f_1, \ldots, f_d]^T$ associating each feasible solution $q \in Q$ with a d-vector $\mathbf{f}(q)$. The *Pareto set or curve* $\mathcal{P}$ of Q is defined as the set of all undominated elements of Q. Given a vector of approximation ratios $\boldsymbol{\rho} = [\rho_1, \ldots, \rho_d]^T$ ($\rho_i \geq 1$, $1 \leq i \leq d$), a solution $p \in Q$ $\boldsymbol{\rho}$-*covers* a solution $q \in Q$ iff it is as good in each objective i by at least a factor ρ_i, i.e., $f_i(p) \leq \rho_i \cdot f_i(q)$, $1 \leq i \leq d$. A set $\Pi \subseteq Q$ is a $\boldsymbol{\rho}$-*cover* of Q iff for all $q \in Q$, there exists $p \in \Pi$ such that p $\boldsymbol{\rho}$-covers q (note that a $\boldsymbol{\rho}$-cover may contain dominated solutions). A $\boldsymbol{\rho}$-cover is also called $\boldsymbol{\rho}$-*Pareto set*. If all entries of $\boldsymbol{\rho}$ are equal to ρ, we also use the terms ρ-cover and ρ-Pareto set.

A *fully polynomial time approximation scheme* (FPTAS) for computing the Pareto set of an instance of a multiobjective optimization problem is a family of algorithms that, for any fixed constant $\varepsilon > 0$, contains an algorithm that always outputs an $(1 + \varepsilon)$-Pareto set and runs in time polynomial in the size of the input and $1/\varepsilon$. W.l.o.g. we make the customary assumption that $\varepsilon \leq 1$, yielding $\ln(1 + \varepsilon) = \Theta(\varepsilon)$, which will be used throughout the paper.

If $\mathbf{a} = [a_1, a_2, \cdots, a_d]^T$ is a d-dimensional vector and λ a scalar, then we denote by $\mathbf{a}^\lambda = [a_1^\lambda, a_2^\lambda, \cdots, a_d^\lambda]^T$. A vector with all its elements equal to zero is denoted by $\mathbf{0}$.

3   Single-Source Multiobjective Shortest Paths

In the multiobjective shortest path problem, we are given a digraph $G = (V, E)$ and a d-dimensional function vector $\mathbf{c} : E \rightarrow [\mathbb{R}^+]^d$ associating each edge e with a cost vector $\mathbf{c}(e)$. We extend the cost function vector to handle paths by extending the domain to the powerset of E, thus considering the function $\mathbf{c} : 2^E \rightarrow [\mathbb{R}^+]^d$, where the cost vector of a path p is the sum of the cost vectors of its edges, i.e., $\mathbf{c}(p) = \sum_{e \in p} \mathbf{c}(e)$. Given two nodes v and w, let $P(v, w)$ denote the set of all v-w paths in G. In the *multiobjective shortest path* problem, we are asked to compute the Pareto set of $P(v, w)$ w.r.t. $\mathbf{c}$. In the *single-source multiobjective shortest path* (SSMOSP) problem, we are given a node s and the task is to compute the Pareto sets of $P(s, v)$ w.r.t. $\mathbf{c}$, $\forall v \in V$.

Given a vector $\boldsymbol{\varepsilon} = [\varepsilon_1, \varepsilon_2, \cdots, \varepsilon_{d-1}]^T$ of error parameters ($\varepsilon_i > 0$, $1 \leq i \leq d - 1$) and a source node s, we present below an algorithm that computes, for each node v, a $\boldsymbol{\rho}$-cover of $P(s, v)$, where $\boldsymbol{\rho} = [1 + \varepsilon_1, 1 + \varepsilon_2, \cdots, 1 + \varepsilon_{d-1}, 1]^T$. Note that we can be *exact* in one dimension (here w.l.o.g. the d-th one), without any impact on the running time. In the following, let $c_i^{min} \equiv \min_{e \in E} c_i(e)$, $c_i^{max} \equiv \max_{e \in E} c_i(e)$, and $C_i = \frac{c_i^{max}}{c_i^{min}}$, for all $1 \leq i \leq d$. Let also $P^i(v, w)$ denote the set of all v-w paths in G with no more than i edges; clearly, $P^{n-1}(v, w) \equiv P(v, w)$.

3.1   The SSMOSP Algorithm

Our algorithm resembles the classical (label correcting) Bellman-Ford method. Previous attempts to straightforwardly apply such an approach [2,3,4] had a very poor (exponential) performance, since all undominated solutions (exponentially large sets of labels) have to be maintained. The key idea of our method is that we can implement the label sets as arrays of polynomial size by relaxing the requirements for strict Pareto optimality to that of $\boldsymbol{\rho}$-covering.

We represent a path $p = (e_1, e_2, \cdots, e_{k-1}, e_k)$ by a label that is a tuple $(\mathbf{c}(p), \text{pred}(p), \text{lastedge}(p))$, where $\mathbf{c}(p) = \sum_{e \in p} \mathbf{c}(e)$ is the d-dimensional cost vector of the path, $\text{pred}(p) = q$ is a pointer to the label of the subpath $q = (e_1, e_2, \cdots, e_{k-1})$ of p, and $\text{lastedge}(p) = e_k$ points to the last edge of p. An empty label is represented by $(\mathbf{0}, null, null)$, while a single edge path has a *null* pred pointer. This representation allows us to retrieve the entire path, without implicitly storing its edges, by following the pred pointers. Let $\mathbf{r} = [r_1, \ldots, r_{d-1}, 1]$ be a vector of approximation ratios. The algorithm proceeds in rounds. In each round i and for each node v the algorithm computes a set of labels Π_v^i, which is an $\mathbf{r}^i$-cover of $P^i(s, v)$. We implement these sets of labels using $(d - 1)$-dimensional arrays $\Pi_v^i[0..\lfloor \log_{r_1}(nC_1) \rfloor, 0..\lfloor \log_{r_2}(nC_2) \rfloor, \cdots, 0..\lfloor \log_{r_{d-1}}(nC_{d-1}) \rfloor]$, and index these arrays using $(d-1)$-vectors. This is done by defining a function $\mathbf{pos} : 2^E \rightarrow [\mathbb{N}_0]^{d-1}$. For a path p, $\mathbf{pos}(p) = [\lfloor \log_{r_1} \frac{c_1(p)}{c_1^{min}} \rfloor, \lfloor \log_{r_2} \frac{c_2(p)}{c_2^{min}} \rfloor, \cdots, \lfloor \log_{r_{d-1}} \frac{c_{d-1}(p)}{c_{d-1}^{min}} \rfloor]^T$ gives us the position in Π_v^i corresponding to p. The definition of $\mathbf{pos}$ along with the fact that for any path p we have $c_i(p) \leq (n - 1)c_i^{max}$, $\forall 1 \leq i \leq d$, justifies the size of the arrays.

Initially, $\Pi_v^0 = \emptyset$, for all $v \in V - \{s\}$, and Π_s^0 contains only the trivial empty path. For each round $i \geq 1$ and for each node v the algorithm computes Π_v^i as follows. Initially, we set Π_v^i equal to Π_v^{i-1}. We then examine the incoming edges of v, one by one, and perform an *Extend-&-Merge* operation for each edge examined. An *Extend-&-Merge* operation takes as input an edge $e = (u, v)$ and the sets Π_u^{i-1} and Π_v^i. It extends all labels $p \in \Pi_u^{i-1}$ by e, and merges the resulting set of s-v paths with Π_v^i. Since each extension results in a new label (path) $q = (\mathbf{c}(p) + \mathbf{c}(e), \boldsymbol{p}, e)$ whose $\mathbf{pos}(q)$ leads to an array position which may not be empty, the algorithm maintains in each array position the (at most one) path that covers all other paths with the same $\mathbf{pos}(\cdot)$ value, which turns out to be the path with the smallest c_d cost. This keeps the size of the sets polynomially bounded. In particular, q is inserted in the position
$$\mathbf{pos}(q) = [\lfloor \log_{r_1} \tfrac{c_1(q)}{c_1^{min}} \rfloor, \lfloor \log_{r_2} \tfrac{c_2(q)}{c_2^{min}} \rfloor, \cdots, \lfloor \log_{r_{d-1}} \tfrac{c_{d-1}(q)}{c_{d-1}^{min}} \rfloor]^T$$ of Π_v^i, unless this position is already filled in with a label q' for which $c_d(q') \leq c_d(q)$.

The next two lemmas establish the algorithm's correctness and complexity.

Lemma 1. *For all $v \in V$ and for all $i \geq 0$, after the i-th round Π_v^i $\mathbf{r}^i$-covers $P^i(s, v)$.*

Proof. It suffices to prove that for all $p \in P^i(s, v)$, there exists $q \in \Pi_v^i$ such that $c_\ell(q) \leq r_\ell^i c_\ell(p), \forall 1 \leq \ell \leq d$. We prove this by induction.

For the basis of the induction ($i = 1$) consider a single edge path $p \equiv (e) \in P^1(s, v)$. At each round all incoming edges of v are examined and an *Extend-&-Merge* operation is executed for each edge. After the first round and due to the **if** condition of the *Extend-&-Merge* operation, position $\mathbf{pos}(p)$ of Π_v^1 contains a path q for which: (i) $\mathbf{pos}(q) = \mathbf{pos}(p)$; and (ii) $c_d(q) \leq c_d(p)$. From (i) it is clear that for all $1 \leq \ell \leq d - 1$, we have $\lfloor \log_{r_\ell} \tfrac{c_\ell(q)}{c_\ell} \rfloor = \lfloor \log_{r_\ell} \tfrac{c_\ell(p)}{c_\ell} \rfloor$, and therefore $\log_{r_\ell} \tfrac{c_\ell(q)}{c_\ell^{min}} - 1 \leq \log_{r_\ell} \tfrac{c_\ell(p)}{c_\ell^{min}}$. This, along with (ii) and the fact that $r_d = 1$, implies that $c_\ell(q) \leq r_\ell c_\ell(p), \forall 1 \leq \ell \leq d$.

For the induction step consider a path $p \equiv (e_1, e_2, \ldots, e_k = (u, v)) \in P^i(s, v)$, for some $k \leq i$. The subpath $p' \equiv (e_1, e_2, \ldots, e_{k-1})$ of p has at most $i - 1$ edges and applying the induction hypothesis we get that there exists a path $q' \in \Pi_u^{i-1}$ such that $c_\ell(q') \leq r_\ell^{i-1} c_\ell(p'), \ 1 \leq \ell \leq d$. Let now $\bar{q}$ be the concatenation of q' with edge e_k. Then, we have:
$$c_\ell(\bar{q}) \leq r_\ell^{i-1} c_\ell(p), \ 1 \leq \ell \leq d \tag{1}$$

It is clear by our algorithm that during the *Extend-&-Merge* operation for edge e_k in the i-th round $\bar{q}$ was examined. Moreover, at the end of the i-th round and due to the **if** condition of the *Extend-&-Merge* operation, position $\mathbf{pos}(\bar{q})$ of Π_v^i contains a path q for which: (iii) $\mathbf{pos}(q) = \mathbf{pos}(\bar{q})$; and (iv) $c_d(q) \leq c_d(\bar{q})$. From (iii) it is clear that $\lfloor \log_{r_\ell} c_\ell(q) \rfloor = \lfloor \log_{r_\ell} c_\ell(\bar{q}) \rfloor, \ \forall \ 1 \leq \ell \leq d - 1$, and therefore $\log_{r_\ell} c_\ell(q) - 1 \leq \log_{r_\ell} c_\ell(\bar{q}), \ \forall \ 1 \leq \ell \leq d - 1$, which implies that
$$c_\ell(q) \leq r_\ell c_\ell(\bar{q}), \ 1 \leq \ell \leq d - 1. \tag{2}$$

Since $r_d = 1$, combining now (iv) and (2) with (1), we get that $c_\ell(q) \leq r_\ell^i c_\ell(p)$, $\forall \ 1 \leq \ell \leq d$. $\square$

Lemma 2. *Algorithm SSMOSP computes, for all $v \in V$, an $\mathbf{r}^{n-1}$-cover of $P(s,v)$ in total time $O(nm \prod_{j=1}^{d-1}(\lfloor \log_{r_j}(nC_j) \rfloor + 1))$.*

Proof. From Lemma 1, it is clear that, for any $v \in V$, Π_v^{n-1} is an $\mathbf{r}^{n-1}$-cover of $P^{n-1}(s,v) \equiv P(s,v)$, since any path has at most $n-1$ edges. The algorithm terminates after $n-1$ rounds. In each round it examines all of the m edges and performs an *Extend-&-Merge* operation. The time of this operation is proportional to the size of the arrays used, which equals $\prod_{j=1}^{d-1}(\lfloor \log_{r_j}(nC_j) \rfloor + 1)$ and therefore the total time complexity is $O(nm \prod_{j=1}^{d-1}(\lfloor \log_{r_j}(nC_j) \rfloor + 1))$. $\qquad\square$

Applying Lemma 2 with $\mathbf{r} = [(1+\varepsilon_1)^{\frac{1}{n-1}}, (1+\varepsilon_2)^{\frac{1}{n-1}}, \cdots, (1+\varepsilon_{d-1})^{\frac{1}{n-1}}, 1]$, and taking into account that $\ln(1+\delta) = \Theta(\delta)$ for small δ, yields our main result.

Theorem 1. *Given a vector $\boldsymbol{\varepsilon} = [\varepsilon_1, \varepsilon_2, \cdots, \varepsilon_{d-1}]^T$ of error parameters and a source node s, there exists an algorithm that computes, for all $v \in V$, a $\boldsymbol{\rho}$-cover of $P(s,v)$ (set of all s-v paths), where $\boldsymbol{\rho} = [1+\varepsilon_1, 1+\varepsilon_2, \cdots, 1+\varepsilon_{d-1}, 1]^T$, in total time $O(n^d m \prod_{j=1}^{d-1}(\frac{1}{\varepsilon_j} \log(nC_j)))$.*

Let $C^{max} = \max_{1 \le j \le d-1} C_j$. In the special case, where $\varepsilon_i = \varepsilon$, $\forall 1 \le i \le d-1$, we have the following result.

Corollary 1. *For any error parameter $\varepsilon > 0$, there exists a FPTAS for the single-source multiobjective shortest path problem with d objectives on a digraph G that computes $(1+\varepsilon)$-Pareto sets (one for each node of G) in total time $O(nm(\frac{n \log(nC^{max})}{\varepsilon})^{d-1})$.*

Further improvements can be obtained in the case of DAGs; see [13].

4 Non-linear Objectives

In this section, we present two generic methods to construct a FPTAS for the normalized version of any multiobjective optimization problem with a non-linear utility function, as well as a FPTAS for any multiobjective optimization problem with non-linear objectives, for a quite general family of non-linear functions. The only precondition is the existence of a FPTAS for a much simpler version of the problems.

Let $\mathcal{M}$ be (an instance of) a multiobjective optimization problem with set of feasible solutions Q and vector of objective functions $\mathbf{c} = [c_1, \ldots, c_d]^T$, associating each feasible solution $q \in Q$ with a d-vector of attributes $\mathbf{c}(q)$; i.e., the i-th objective is the identity function of the i-th attribute.

Let $\mathcal{N}$ be the normalized version of $\mathcal{M}$ w.r.t. a non-decreasing, non-linear utility function $\mathcal{U} : [\mathbb{R}^+]^d \to \mathbb{R}$; i.e., the objective of $\mathcal{N}$ is $\min_{q \in Q} \mathcal{U}(\mathbf{c}(q))$. We will show that a FPTAS for $\mathcal{M}$ can provide a FPTAS for $\mathcal{N}$. To obtain such a FPTAS, we consider a quite general family of non-linear functions $\mathcal{U}(\mathbf{x})$.

A multiattribute function $\mathcal{U}(\mathbf{x})$ is called *quasi-polynomially bounded* (see e.g., [1]) if there exist some constants γ and δ such that $\frac{\frac{\partial \mathcal{U}}{\partial x_i}(\mathbf{x})}{\mathcal{U}(\mathbf{x})} \le \gamma \frac{1}{x_i} \prod_{k=1}^{d} \ln^\delta x_k$,

$1 \leq i \leq d$. For instance, the function $\mathcal{U}([x_1, x_2]^T) = x_1^{polylog(x_1)} + x_2^{polylog(x_2)}$ is quasi-polynomially bounded, while the function $\mathcal{U}([x_1, x_2]^T) = 2^{x_1^\mu} + 2^{x_2^\mu}$, for some $\mu > 0$, is not. Note also that this class includes all non-decreasing polynomials.

Let $\mathcal{C}_i = \max_{q \in Q} c_i(q)$ be the maximum cost in the i-th dimension, and let $\log \mathcal{C}_i$ be polynomial to the input size (as indeed is the case for MOSP and other problems, like the multiobjective versions of spanning tree, perfect matching, knapsack, etc). We can prove the following.

Theorem 2. *Let the objective function $\mathcal{U}$ of $\mathcal{N}$ be quasi-polynomially bounded. If there exists a FPTAS for $\mathcal{M}$ with time complexity $T(1/\varepsilon, m')$, then there exists a FPTAS for $\mathcal{N}$ with complexity $T(\Lambda/\varepsilon, m')$, where m' is the input size of $\mathcal{M}$ and $\Lambda = \gamma d \prod_{i=1}^{d} \ln^\delta \mathcal{C}_i$.*

Proof. We construct an $(1 + \varepsilon')$-Pareto set Π for $\mathcal{M}$, where ε' will be chosen later. Pick $q = \operatorname{argmin}_{p \in \Pi}(\mathcal{U}(\mathbf{c}(p)))$. Let p^* denote the optimal solution with cost vector $\mathbf{c}^* = \mathbf{c}(p^*)$. By the definition of Π, we know that there exists some $p' \in \Pi$ such that $c_i(p') \leq \min\{(1 + \varepsilon')c_i^*, \mathcal{C}_i\}$. By the choice of q we have that $\mathcal{U}(\mathbf{c}(q)) \leq \mathcal{U}(\mathbf{c}(p'))$, thus it suffices to bound $\frac{\mathcal{U}(\mathbf{c}(p'))}{\mathcal{U}(\mathbf{c}(p^*))}$.

Let $\mathbf{c}'$ be the vector whose elements are given by $c_i' = \min\{(1 + \varepsilon')c_i^*, \mathcal{C}_i\}$, $\forall 1 \leq i \leq d$. Since $\mathcal{U}(\cdot)$ is non-decreasing, $\frac{\mathcal{U}(\mathbf{c}(p'))}{\mathcal{U}(\mathbf{c}(p^*))} \leq \frac{\mathcal{U}(\mathbf{c}')}{\mathcal{U}(\mathbf{c}^*)} = \exp[\ln \mathcal{U}(\mathbf{c}') - \ln \mathcal{U}(\mathbf{c}^*)]$. We write the exponent as a telescopic sum $\ln \mathcal{U}(\mathbf{c}') - \ln \mathcal{U}(\mathbf{c}^*) = \sum_{k=1}^{d} [F_k(c_k') - F_k(c_k^*)]$, where $F_k(x) = \ln \mathcal{U}([c_1', \ldots, c_{k-1}', x, c_{k+1}^*, \ldots, c_d^*]^T)$. On each term k of the sum, we apply the well-known Mean Value Theorem[1] for $F_k(x)$ on the interval (c_k^*, c_k'). Hence, $\forall 1 \leq k \leq d$, there exists some ζ_k with $c_k^* < \zeta_k < c_k'$ such that $F_k(c_k') - F_k(c_k^*) = F_k'(\zeta_k)(c_k' - c_k^*) \leq \frac{\frac{\partial \mathcal{U}}{\partial x_k}(\mathbf{c}^{[k]})}{\mathcal{U}(\mathbf{c}^{[k]})} \varepsilon' c_k^*$, where $\mathbf{c}^{[k]}$ are vectors with $c_i^{[k]} = \begin{cases} c_i' & \text{if } 1 \leq i < k \\ \zeta_k & \text{if } i = k \\ c_i^* & \text{if } k < i \leq d \end{cases}$. Consequently, $\frac{\mathcal{U}(\mathbf{c}')}{\mathcal{U}(\mathbf{c}^*)} \leq \exp\left[\varepsilon' \sum_{k=1}^{d} \left[\frac{\frac{\partial \mathcal{U}}{\partial x_k}(\mathbf{c}^{[k]})}{\mathcal{U}(\mathbf{c}^{[k]})} c_k^*\right]\right]$.

Observe now that the term $\sum_{k=1}^{d} \left[\frac{\frac{\partial \mathcal{U}}{\partial x_k}(\mathbf{c}^{[k]})}{\mathcal{U}(\mathbf{c}^{[k]})} c_k^*\right]$ is bounded by $\Lambda = \gamma d \prod_{i=1}^{d} \ln^\delta \mathcal{C}_i$. Hence, choosing $\varepsilon' = \frac{\ln(1+\varepsilon)}{\Lambda}$, yields an $1 + \varepsilon$ approximation in time $T(\Lambda/\varepsilon, m')$. □

The above result improves upon that of [1] both w.r.t. d (number of objectives) and time; the time in [1] ($d = 2$) is $T(\Lambda'/\varepsilon^2, m')$, where $\Lambda' = \gamma 2^{\delta+4} \prod_{i=1}^{2} \ln^{\delta+1} \mathcal{C}_i$.

Now, let $\mathcal{M}'$ be a multiobjective optimization problem, defined on the same with $\mathcal{M}$ set of feasible solutions Q, but having a vector of objective functions $\mathbf{U} = [U_1, \ldots, U_h]^T$ associating each $q \in Q$ with an h-vector $\mathbf{U}(q)$. These objective functions are defined as $U_i(q) = \mathcal{U}_i(\mathbf{c}(q))$, $1 \leq i \leq h$, where $\mathcal{U}_i : [\mathbb{R}^+]^d \to \mathbb{R}$ are non-linear, non-decreasing, quasi-polynomially bounded functions. By working similarly to Theorem 2, we can show the following (details in [13]).

---

[1] **Mean Value Theorem:** Let $f(x)$ be differentiable on (a, b) and continuous on $[a, b]$. Then, there is at least one point $c \in (a, b)$ such that $f'(c) = (f(b) - f(a))/(b - a)$.

Theorem 3. *Let the objective functions of $\mathcal{M}'$ be quasi-polynomially bounded. If there exists a FPTAS for $\mathcal{M}$ with time complexity $T(1/\varepsilon, m')$, then there exists a FPTAS for $\mathcal{M}'$ with complexity $T(\Lambda/\varepsilon, m')$, where m' is the input size of $\mathcal{M}$ and $\Lambda = \gamma d \prod_{i=1}^{d} \ln^{\delta} C_i$.*

5 Applications

Multiple Constrained (Optimal) Paths. Let $\rho = [1 + \varepsilon_1, 1 + \varepsilon_2, \cdots, 1 + \varepsilon_{d-1}, 1]^T$ and let Π be a ρ-cover Π of $P(s,t)$, constructed using the SSMOSP algorithm as implied by Theorem 1. For MCOP, choose $p' = \mathrm{argmin}_{p \in \Pi} \{c_d(p); c_i(p) \leq (1+\varepsilon_i)b_i, \forall 1 \leq i \leq d-1\}$. This provides a so-called *acceptable* solution in the sense of [7] by slightly relaxing the QoS-bounds; that is, the path p' is at least as good as the MCOP-optimum and is nearly feasible, violating each QoS-bound b_i, $1 \leq i \leq d-1$, by at most an $1+\varepsilon_i$ factor. For MCP, choose a path $p' \in \Pi$ that obeys the QoS-bounds, or answer that there is no path p in $P(s,t)$ for which $c_i(p) \leq b_i/(1+\varepsilon_i)$, $\forall 1 \leq i < d$. In the latter case, if a feasible solution for MCP exists, then (by the definition of Π) we can find a solution in Π that is nearly feasible (i.e., it violates each QoS-bound b_i, $1 \leq i \leq d-1$, by at most an $1+\varepsilon_i$ factor). By Theorem 1, the required time for both cases is $O(n^d m \prod_{j=1}^{d-1}(\frac{1}{\varepsilon_j} \log(nC_j))$, which can be reduced to $O(n^d m \prod_{j=1}^{d-1}(\frac{1}{\varepsilon_j} \log(\min\{nC_j, b_j/c_j^{min}\}))$ by observing that it is safe to discard any path p for which $c_j(p) > (1 + \varepsilon_j)b_j$ for some $1 \leq j \leq d - 1$ (thus reducing the size of the Π_v^i arrays).

Non-Additive Shortest Paths. In this problem (NASP) we are given a digraph $G = (V, E)$ and a d-dimensional function vector $\mathbf{c} : E \rightarrow [\mathbb{R}^+]^d$ associating each edge e with a vector of attributes $\mathbf{c}(e)$ and a path p with a vector of attributes $\mathbf{c}(p) = \sum_{e \in p} \mathbf{c}(e)$. We are also given a d-attribute non-decreasing and *non-linear* utility function $\mathcal{U} : [\mathbb{R}^+]^d \rightarrow \mathbb{R}$. The objective is to find a path p^*, from a specific source node s to a destination t, that minimizes the objective function, i.e., $p^* = \mathrm{argmin}_{p \in P(s,t)} \mathcal{U}(\mathbf{c}(p))$. (It is easy to see that in the case where $\mathcal{U}$ is linear, NASP reduces to the classical single-objective shortest path problem.) For the general case of non-linear $\mathcal{U}$, it is not difficult to see that NASP is NP-hard.

Theorem 2 suggests that our FPTAS for MOSP yields an (improved w.r.t. [1,12]) FPTAS for NASP for the case of quasi-polynomially bounded functions. We show that we can do better by taking advantage of the fact that our FPTAS for MOSP is *exact* in one dimension (w.l.o.g. the d-th). This allows us to provide a FPTAS for an even more general (than quasi-polynomial) family of functions. Specifically, we consider d-attribute functions for which there exist some constants γ and δ such that $\frac{\frac{\partial \mathcal{U}}{\partial x_i}(\mathbf{x})}{\mathcal{U}(\mathbf{x})} \leq \gamma \frac{1}{x_i} \prod_{k=1}^{d} \ln^{\delta} x_k$, $1 \leq i \leq d - 1$. The fact that we do not require that this condition holds for the d-th attribute allows $\mathcal{U}$ to be even *exponential* on x_d; e.g., $\mathcal{U}([x_1, x_2]^T) = x_1^{polylog(x_1)} + 2^{x_2^\mu}$, for any $\mu > 0$. Note that this does not contradict the inapproximability result in [1], which applies to functions of the form $\mathcal{U}([x_1, x_2]^T) = 2^{x_1^\mu} + 2^{x_2^\mu}$, for $\mu > 0$.

Our result makes the gap between NASP approximability and inapproximability even tighter. Let $\mathcal{C}_i$ denote the maximum path cost in the i-th dimension, i.e., $\mathcal{C}_i = (n-1)\max_{e \in E} c_i(e)$. We can show the following (see [13]).

Theorem 4. *Let $\mathcal{U}$ be a non-decreasing function for which $\frac{\frac{\partial \mathcal{U}}{\partial x_i}(\mathbf{x})}{\mathcal{U}(\mathbf{x})} \leq \gamma \frac{1}{x_i} \prod_{k=1}^{d}$ $\ln^\delta x_k$, $1 \leq i \leq d-1$. Then, for any $\varepsilon > 0$, there exists an algorithm that computes in time $O(n^d m (\frac{\log(nC^{max})\Lambda}{\varepsilon})^{d-1})$ an $(1+\varepsilon)$-approximation to the NASP optimum w.r.t. $\mathcal{U}(\mathbf{x})$, where $\Lambda = \gamma(d-1)\prod_{i=1}^{d}\ln^\delta \mathcal{C}_i$.*

References

1. H. Ackermann, A. Newman, H. Röglin, and B. Vöcking, "Decision Making Based on Approximate and Smoothed Pareto Curves", in *Algorithms and Computation – ISAAC 2005*, LNCS Vol. 3827 (Springer 2006), pp. 675-684; full version as Tech. Report AIB-2005-23, RWTH Aachen, December 2005.
2. H. Corley and I. Moon, "Shortest Paths in Networks with Vector Weights", *Journal of Optimization Theory and Applications*, 46:1(1985), pp. 79-86.
3. M. Ehrgott, *Multicriteria Optimization*, Springer, 2000.
4. M. Ehrgott and X. Gandibleux (Eds), *Multiple Criteria Optimization – state of the art annotated bibliographic surveys*, Kluwer Academic Publishers, Boston, 2002.
5. S. Gabriel and D. Bernstein, "The Traffic Equilibrium Problem with Nonadditive Path Costs", *Transportation Science* 31:4(1997), pp. 337-348.
6. S. Gabriel and D. Bernstein, "Nonadditive Shortest Paths: Subproblems in Multi-Agent Competitive Network Models", *Computational & Mathematical Organization Theory* 6(2000), pp. 29-45.
7. A. Goel, K. G. Ramakrishnan, D. Kataria, and D. Logothetis, "Efficient Computation of Delay-Sensitive Routes from One Source to All Destinations", in *Proc. IEEE Conf. Comput. Commun. – INFOCOM 2001*.
8. P. Hansen, "Bicriterion Path Problems", *Proc. 3rd Conf. Multiple Criteria Decision Making – Theory and Applications*, LNEMS Vol. 117 (Springer, 1979), pp. 109-127.
9. D.H. Lorenz and D. Raz, "A simple efficient approximation scheme for the restricted shortest path problem", *Operations Res. Lett.*, 28 (2001) pp.213-219.
10. P. Van Mieghem, F.A. Kuipers, T. Korkmaz, M. Krunz, M. Curado, E. Monteiro, X. Masip-Bruin, J. Sole-Pareta, and S. Sanchez-Lopez, "Quality of Service Routing", Chapter 3 in *Quality of Future Internet Services*, LNCS Vol. 2856 (Springer-Verlag, 2003), pp. 80-117.
11. C. Papadimitriou and M. Yannakakis, "On the Approximability of Trade-offs and Optimal Access of Web Sources", in *Proc. 41st Symp. on Foundations of Computer Science – FOCS 2000*, pp. 86-92.
12. G. Tsaggouris and C. Zaroliagis, "Improved FPTAS for Multiobjective Shortest Paths with Applications", CTI Techn. Report TR-2005/07/03, July 2005.
13. G. Tsaggouris and C. Zaroliagis, "Multiobjective Optimization: Improved FPTAS for Shortest Paths and Non-linear Objectives with Applications", CTI Techn. Report TR-2006/03/01, March 2006.
14. S. Vassilvitskii and M. Yannakakis, "Efficiently Computing Succinct Trade-off Curves", in *Automata, Languages, and Programming – ICALP 2004*, LNCS Vol. 3142 (Springer, 2004), pp. 1201-1213.
15. A. Warburton, "Approximation of Pareto Optima in Multiple-Objective Shortest Path Problems", *Operations Research* 35(1987), pp. 70-79.

Algorithms for Computing Variants of the Longest Common Subsequence Problem

Extended Abstract

M. Sohel Rahman*,** and Costas S. Iliopoulos***

Algorithm Design Group
Department of Computer Science, King's College London,
Strand, London WC2R 2LS, England
{sohel, csi}@dcs.kcl.ac.uk
http://www.dcs.kcl.ac.uk/adg

Abstract. The *longest common subsequence*(LCS) problem is one of
the classical and well-studied problems in computer science. The computation of the LCS is a frequent task in DNA sequence analysis, and has
applications to genetics and molecular biology. In this paper we define
new variants, introducing the notion of gap-constraints in LCS problem
and present efficient algorithms to solve them.

1 Introduction

The *longest common subsequence*(LCS) problem is one of the classical and well-
studied problems in computer science which has extensive applications in diverse
areas ranging from spelling error corrections to molecular biology [7, 2]. This
paper introduces the notion of gap-constraints in LCS. Our versions of LCS, on
one hand, offers the possibility to handle gap-constraints between the consecutive
matches among the sequences. On the other hand, they provide us with the
tool to handle motif finding problems where not all positions of the motif is
important [6]. Before going into details we need to present some preliminary
concepts.

Suppose we are given two strings $X[1..n] = X[1] \ X[2] \ldots X[n]$ and $Y[1..n] = Y[1] \ Y[2] \ldots Y[n]$. A subsequence $S[1..r] = S[1] \ S[2] \ldots S[r]$ of X is obtained
by deleting $n - r$ symbols from X. A common subsequence of two strings X
and Y, denoted $cs(X, Y)$, is a subsequence common to both X and Y. The
longest common subsequence of X and Y, denoted $lcs(X, Y)$ or $LCS(X, Y)$, is
a common subsequence of maximum length. We denote the length of $lcs(X, Y)$
by $r(X, Y)$.

Problem "LCS". *Given 2 strings X and Y, we want to find out the Longest
Common Subsequence of X and Y.*

---

* Supported by the Commonwealth Scholarship Commission in the UK under the
Commonwealth Scholarship and Fellowship Plan (CSFP).
** On Leave from Department of CSE, BUET, Dhaka-1000, Bangladesh.
*** Supported by EPSRC and Royal Society grants.

T. Asano (Ed.): ISAAC 2006, LNCS 4288, pp. 399–408, 2006.
© Springer-Verlag Berlin Heidelberg 2006

In what follows we assume that the two given strings are of equal length. But our results can be easily extended to handle two strings of different length.

Definition. (Correspondence Sequence): *Given a string $X[1..n]$ and a subsequence $S[1..r]$ of X, we define a correspondence sequence (not necessarily unique) of X and S, $C(X, S) = C[1]\ C[2]\ \ldots C[r]$ to be a strictly increasing sequence of integers taken from $[1, n]$ such that $S[i] = X[C[i]]$ for all $1 \leq i \leq r$.*

Definition. (Fixed Gapped Correspondence Sequence): *A correspondence sequence of a string X of length n and one of its subsequences S of length r is said to be a Fixed Gapped Correspondence Sequence with respect to a given integer K if and only if we have $C[i] - C[i-1] \leq K+1$ for all $2 \leq i \leq r$. We sometimes use $C_{FG(K)}$ to denote a Fixed Gapped Correspondence Sequence with respect to K.*

Definition. (Elastic Gapped Correspondence Sequence): *A correspondence sequence of a string X of length n and one of its subsequences S of length r is said to be a Elastic Gapped Correspondence Sequence with respect to given integers K_1 and K_2, $K_2 > K_1$, if and only if we have $K_1 < C[i] - C[i-1] \leq K_2+1$ for all $2 \leq i \leq r$.*

Definition. (Fixed and Elastic Gapped Common Subsequence): *Suppose we are given two strings $X[1..n]$ and $Y[1..n]$ and an integer K. A common subsequence $S[1..r]$ of X and Y is a Fixed Gapped Common Subsequence, if there exists Fixed Gapped Correspondence Sequences $C_{FG(K)}(X, S)$ and $C_{FG(K)}(Y, S)$. For a Fixed Gapped Common Subsequence to be **rigid** we must also have, for all $2 \leq i \leq r$, $C_{FG(K)}(X, S)[i] - C_{FG(K)}(X, S)[i-1] = C_{FG(K)}(Y, S)[i] - C_{FG(K)}(Y, S)[i-1]$. Elastic Gapped Common Subsequences (both non-rigid and rigid) can be defined analogously.*

The problems we handle in this paper are defined in Figure 1. In this paper we use the following notions. We say a pair $(i, j), 1 \leq i, j \leq n$ defines a match, if $X[i] = Y[j]$. The set of all matches, M, is defined as follows: $M = \{(i, j) \mid X[i] = Y[j], 1 \leq i, j \leq n\}$. We define $|M| = \mathcal{R}$.

PROBLEM	INPUT	OUTPUT (of maximum length)
FIG (Fixed Gap)	X, Y and K	Fixed Gapped Common Subsequence
ELAG (Elastic Gap)	X, Y, K_1 and K_2	Elastic Gapped Common Subsequence
RIFIG (Rigid Fixed Gap)	X, Y and K	Rigid Fixed Gapped Common Subsequence
RELAG (Rigid Elastic Gap)	X, Y, K_1 and K_2	Rigid Elastic Gapped Common Subsequence

Fig. 1. Problems handled in this paper

2   An Algorithm for FIG

In the traditional dynamic programming technique to solve LCS [11], the idea is to determine the longest common subsequences for all possible prefix combinations of the input strings. The recurrence relation for extending $r(X[1..i], Y[1..j])$, is as follows [11]:

$$T[i,j] = \begin{cases} 0 & \text{if } i = 0 \text{ or } j = 0, \\ T[i-1, j-1] + 1 & \text{if } X[i] = Y[j], \\ max(T[i-1, j], T[i, j-1]) & \text{if } X[i] \neq Y[j]. \end{cases} \tag{1}$$

Here we have used the tabular notion $T[i, j]$ to denote $r(X[1..i], Y[1..j])$. After the table has been filled, $r(X, Y)$ can be found in $T[n, n]$ and $lcs(X, Y)$ can be found by backtracking from $T[n, n]$. Unfortunately, the attempt to generalize this algorithm in a straightforward way doesn't give us an algorithm to solve our problems. Note that, in FIG, due to the gap constraint, a continuing common sequence may have to stop at an arbitrary $T[i, j]$ because the next match is not within the gap constraint. In order to cope with this situation what we do is as follows. For each tabular entry $T[i, j], (i, j) \in M$ we calculate and store two values namely $T_{local}[i, j]$ and $T_{global}[i, j]$. For all other $(i, j), T_{local}[i, j]$ is irrelevant and, hence, is undefined. The recurrence relations are defined below:

$$T_{local}[i,j] = \begin{cases} \text{Undefined} & \text{if } (i, j) \notin M, \\ \max\limits_{\substack{i-1-K \leq \ell_i < i \\ j-1-K \leq \ell_j < j \\ (\ell_i, \ell_j) \in M}} (T_{local}[\ell_i, \ell_j]) + 1 & \text{if } (i, j) \in M. \end{cases} \tag{2}$$

Remark 1. The max operation in Equation 2 returns 0, when there is no $(\ell_i, \ell_j) \in M, i - 1 - K \leq \ell_i < i, j - 1 - K \leq \ell_j < j$.

$$T_{global}[i,j] = \begin{cases} 0 & \text{if } i = 0 \text{ or } j = 0, \\ \max(T_{global}[i-1, j], T_{global}[i, j-1]) & \text{if } (i, j) \notin M, \\ \max(T_{global}[i-1, j], T_{global}[i, j-1], T_{local}[i, j]) & \text{if } (i, j) \in M. \end{cases} \tag{3}$$

It is easy to see that $T_{global}[i, j]$ is used to store the information of the LCS so far, i.e. the *'global'* LCS, on other hand $T_{local}[i, j]$ tracks any *'local'* LCS in growth. As soon as a local LCS becomes the global one the value of the corresponding T_{global} changes. What will be the running time of this algorithm? Since, for each $(i, j) \in M$, we have to check a $(K+1)^2$ area to find the maximum of T_{local} in that area, the total running time is $O(n^2 + \mathcal{R}(K+1)^2)$. The space requirement is $\theta(n^2)$. Also note that by keeping appropriate pointer information or by examining the calculation of the two recurrences we can easily construct $lcs(X, Y)$ as can be done in the case of the traditional solution of LCS problem.

Theorem 1. *Problem FIG can be solved in $O(n^2 + \mathcal{R}(K+1)^2)$ time using $\theta(n^2)$ space.* $\qquad\square$

Remark 2. Unfortunately, this strategy, if used for the LCS problem, would lead to an inefficient algorithm with $O(n^2 + \sum_{(i,j) \in M}(i-1)(j-1))$ worst case running time.

As in the case of the traditional solution of LCS problem the complete tabular information is required for our algorithm, basically, to provide the solution for the subproblems if required. If that is not required, we can get rid of T_{global} altogether by keeping a variable to keep track of the current global LCS. As soon as a local LCS becomes a global one we just change this variable. This would give us a running time of $O(\mathcal{R}(K+1)^2)$ provided we have a preprocessing step to construct the set M in sorted order according to their position they would be considered in the algorithm. This preprocessing step is as follows. We construct for each symbol $a \in \Sigma$ two separate lists, $L_X[a]$ and $L_Y[a]$. For each $a \in \Sigma$, $L_X[a]$ ($L_Y[a]$) stores, in sorted order, the positions of a in X (Y), if any. We now use an elegant data structure invented by Emde Boas [10] that allows us to maintain a sorted list of integers in the range $[1..n]$ in $O(\log \log n)$ time per insertion and deletion. In addition to that it can return $next(i)$ (successor element of i in the list) and $prev(i)$ (predecessor element of i in the list) in constant time. We construct an Emde Boas data structure E_P where we insert each pair $(i,j), i \in L_X[a], j \in L_Y[a], a \in \Sigma$. In this case the order of the elements in E_P is maintained according to the value $(i*(n-1)+j)$. Note that these values are within the range $[1..n^2]$ and hence the cost is $O(\log \log n^2) = O(\log \log n)$ per insertion and deletion. It is easy to verify that, using E_P, we can get all the pairs in the correct order to process them in row by row manner. We now analyze the running time of this preprocessing step. The $2 * |\Sigma|$ lists can be constructed in $O(n)$ by simply scanning X and Y in turn. Since there are in total $\mathcal{R}$ elements in M, the construction of E_P requires $O(R \log \log n)$ time. The space requirement is $O(\mathcal{R})$. We note, however, that for the complete algorithm, we still need the $\theta(n^2)$ space to guaranty a linear time search for the highest T_{local} in the $(K+1)^2$ area.

Theorem 2. *Given a preprocessing time of $O(\mathcal{R} \log \log n)$, Problem FIG can be solved in $O(\mathcal{R}(K+1)^2)$ time.* □

3   An Improved Algorithm for FIG

In this section we try to improve the running time of Algorithm presented in Section 2. Ideally, we would like to reduce the quadratic term $(K+1)^2$ to linear. Note that we can easily improve the running time of the algorithm to LCS problem reported in Remark 2 using the following interesting facts.

Fact 1. *Suppose $(i,j) \in M$. Then for all $(i',j), i' > i$ $((i,j'), j' > j)$, we must have $T[i',j] \geq T[i,j]$ $(T[i,j'] \geq T[i,j])$, where T is the table filled up by the traditional dynamic programming algorithm using Equation 1.* □

Fact 2. *The calculation of a $T[i,j], (i,j) \in M, 1 \leq i,j \leq n$ is independent of any $T[\ell,q], (\ell,q) \in M, \ell = i, 1 \leq q \leq n$.* □

The idea is to avoid checking the $(i-1)(j-1)$ entries and check only $(j-1)$ (or $(i-1)$) entries instead. We maintain an array H of length n where, for $T[i,j]$ we have, $H[\ell] = \max_{1 < k < i, (i,\ell) \in M}(T[k,\ell]), 1 \leq \ell \leq n$. The 'max' operation, here, returns 0 if there exists no $(i,\ell) \in M$ within the range. Given the updated array H, we can easily perform the task by checking only the $(j-1)$ entries of H. And Fact 1 makes it easy to maintain the array H on the fly as we proceed as follows. As usual, we proceed in a row by row manner. We use another array S, of length

n, as a temporary storage. When we find an $(i,j) \in M$, after calculating $T[i,j]$ we store $S[j] = T[i,j]$. We continue to store in this way as long as we are in the same row. As soon as we find an $(i',j) \in M, i' > i$, i.e. we start processing a new row, we update H with new values from S.

The correctness of the above procedure[1] follows from Fact 1 and 2. But this idea doesn't work for FIG because Fact 1 doesn't hold when we consider FIG. This is because due to the gap constraint a new local LCS may start which would surely have lesser T-value than another previous local LCS. We, however, use the similar idea to improve our previous running time. But we need to do something more than just maintaining an array. In the rest of this section we present a novel technique to present the improved algorithm. The basic idea depends on the following fact which is, basically, an extension of Fact 2.

Fact 3. *The calculation of a $T_{local}[i,j], (i,j) \in M, 1 \leq i,j \leq n$ is independent of any $T_{local}[\ell,q], (\ell,q) \in M, (\ell = i$ or $\ell < i - K - 1), 1 \leq q \leq n$.* □

We maintain n Emde Boas data structures $E_i, 1 \leq i \leq n$, one for each column. We also need to maintain one *insert list, $\mathcal{I}$* and one *delete list, $\mathcal{D}$*. Recall that we proceed in a row by row manner. Suppose we are starting to process row $i+K+2$ i.e. we are considering the *'first'* match in this row, namely, $(i + K + 2, j) \in M$. So we need to calculate $T_{local}[i+K+2, j]$ and $T_{global}[i+K+2, j]$. At this instant the delete list $\mathcal{D}$ contains all (i, ℓ) such that $1 \leq \ell \leq n, (i, \ell) \in M$ and the insert list $\mathcal{I}$ contains all $(i + K + 1, \ell)$ such that $1 \leq \ell \leq n, (i + K + 1, \ell) \in M$. In other words, when we consider the first match in row $(i + K + 2)$, $\mathcal{D}$ contains all the matches in row i and $\mathcal{I}$ contains all the matches in row $i + K + 1$. For each $(m, n) \in \mathcal{D}$ we delete (m, n) from E_n and for each $(m, n) \in \mathcal{I}$ we insert (m, n) in E_n. Note that the sorted order in E_n is maintained according to the T_{local}-value. We then calculate $T_{local}[i + K + 2, j]$ for all $1 \leq j \leq n$ such that $(i + K + 2, j) \in M$. It should be clear that we can calculate $T_{local}[i + K + 2, j]$ as follows in $O(K)$ time:

$$T_{local}[i + K + 1, j] = \max_{j-K-1 \leq \ell \leq j-1} (value(\max(E_\ell))) \tag{4}$$

Note that $value(\max(E_\ell)) = T_{local}[m, n]$ when $\max(E_\ell) = (m, n)$. The running time of $O(K)$ follows from the fact that we can find the maximum of each E_ℓ in constant time. And the correctness follows from Fact 3.

What should be the running time of this algorithm? It is clear that we spend $O(n^2 + \mathcal{R}K)$ time in computing the LCS for FIG. But this improved time comes at the cost of maintaining n Emde Boas data structure $E_i, 1 \leq i \leq n$. It is easy to verify that the total time to maintain $E_i, 1 \leq i \leq n$ is $O(\mathcal{R} \log \log n)$ because we never insert nor delete more than $\mathcal{R}$ elements in total from/to $E_i, 1 \leq i \leq n$. Note that the values to be inserted is always within the range $[1..n]$ since no subsequence can be of length greater than n.

Theorem 3. *Problem FIG can be solved in $O(n^2 + \mathcal{R}K + \mathcal{R} \log \log n)$ time using $\theta(n^2)$ space.* □

---

[1] Although we still don't achieve a good running time for Problem LCS in the worst case.

If the solutions for the subproblems are not required, then we need only compute $\mathcal{T}_{local}[i, j]$, $(i, j) \in M$. We, however, would need to use a variable to finally report $r(X, Y)$ and use appropriate pointers to construct $lcs(X, Y)$. Therefore, we get the following theorem (The corresponding algorithm is formally stated in the form of Algorithm 1).

Theorem 4. *Problem FIG can be solved in $O(\mathcal{R}K + \mathcal{R}\log\log n)$ time using $\theta(R)$ space.* $\qquad\square$

Algorithm 1.

1: Compute the set M using the preprocessing step suggested in Section 2. Let $M_i = (i, j) \in M, 1 \leq j \leq n$.
2: **for** $i = j$ to n **do**
3:    $E_j = \epsilon$ {Initialize the n Emde Boas structure one for each column}
4: **end for**
5: globalLCS.Instance $= \epsilon$
6: globalLCS.Value $= \epsilon$
7: **for** $i = 1$ to n **do**
8:    Insert $(i - 1, j) \in M_{i-1}$ in $E_j, 1 \leq j \leq n$ {If $i - 1 \leq 0$ then insert nothing}
9:    Delete $(i - K - 2, j) \in M_{i-k-2}$ in $E_j, 1 \leq j \leq n$ {If $i - K - 2 \leq 0$ then delete nothing}
10:    **for** each $(i, j) \in M_i$ **do**
11:       $maxresult = \max_{(j-K-1) \leq \ell \leq (j-1)} (\max(E_\ell))$
12:       $\mathcal{T}.Value[i, j] = maxresult.Value + 1$
13:       $\mathcal{T}.Prev[i, j] = maxresult.Instance$
14:       **if** globalLCS.Value $< \mathcal{T}.Value[i, j]$ **then**
15:          globalLCS.Value $= \mathcal{T}.Value[i, j]$
16:          globalLCS.Instance $= (i, j)$
17:       **end if**
18:    **end for**
19: **end for**
20: **return** globalLCS

---

4   A K-Independent Algorithm for FIG

In this section we try to devise an algorithm for FIG that is independent of K. As we shall see later that this would give us an efficient algorithm for Problem LCS as well. We make use of a classical problem in computer science, namely, Range Maxima Query Problem.

Problem "RMAX" (Range Maxima Query Problem). *Given a sequence $A = a_1 a_2 \ldots a_n$, a Range Maxima (minima) Query specifies an interval $I = (i_s, i_e), 1 \leq i_s \leq i_e \leq n$ and the goal is to find the index ℓ with maximum (minimum) value a_ℓ for $\ell \in I$.*

Theorem 5. *([4, 3]). The RMAX problem can be solved in $O(n)$ preprocessing time and $O(1)$ time per query.* $\qquad\square$

With Theorem 5 in our hand, we can modify Algorithm 1 as follows. We want to implement Step 11 in constant time so that we can avoid the dependency on K completely. Before we start processing a particular row, just after Step 9, we create an array of length n, $A = \max(E_j), 1 \leq j \leq n$. Now we simply replace the Step 11 with an *appropriate* Range Maxima Query. It is easy to see that, due to Fact 3, this will work correctly. Since we have a constant time implementation for Step 11, we now can escape the dependency on K. However there is a preprocessing time of $O(n)$ in case any E_j gets updated. But since this preprocessing is needed once per row (due to Fact 3), the computational effort added is $O(n^2)$ in total.

Theorem 6. *Problem FIG can be solved in $O(n^2 + \mathcal{R} \log \log n)$ time using $\theta(max(\mathcal{R}, n))$ space.* $\square$

Finally, it is easy to see that this algorithm can be easily used to solve LCS problem, virtually, without any modification. We, however, can do better using Fact 1 and the subsequent discussion in Section 3. We can get rid of the Emde Boas structures altogether and use a simple array (array H in Section 3) instead. So we get the following theorem.

Theorem 7. *Problem LCS can be solved in $O(n^2 + \mathcal{R} \log \log n)$ time using $\theta(max(\mathcal{R}, n))$ space.* $\square$

We can shave off the $\log \log n$ term from the running time of Theorem 7 by not computing M as a preprocessing step. In this case, however, we need to process each entry of $T[i, j], 1 \leq i \leq n, 1 \leq i \leq n$, instead of processing only each $(i, j) \in \mathcal{M}$.

Theorem 8. *Problem LCS can be solved in $O(n^2 + \mathcal{R})$ time using $\theta(n^2)$ space.* $\square$

5   Algorithm for Elastic Gapped LCS

In this section we modify the algorithms in Section 2, 3, and 4 to solve ELAG. Recall that, in ELAG, we have two parameters, namely K_1 and K_2. Note also that, FIG is a special case of ELAG when $K_1 = 0$ and $K_2 = K$. The obvious modification to Equation 2 in Section 2 to solve ELAG is as follows:

$$\mathcal{T}_{local}[i, j] = \begin{cases} \text{Undefined} & \text{if } (i, j) \notin M, \\ \max_{\substack{i-1-K_2 \leq \ell_i < i-K_1 \\ j-1-K_2 \leq \ell_j < j-K_1 \\ (\ell_i, \ell_j) \in M}} (\mathcal{T}_{local}[\ell_i, \ell_j]) + 1 & \text{if } (i, j) \in M. \end{cases} \quad (5)$$

Theorem 9. *Problem ELAG can be solved in $O(n^2 + \mathcal{R}(K + 1)^2)$ time using $\theta(n^2)$ space where $K = K_2 - K_1$.* $\square$

For Algorithm 1, described in Section 3, the only modification that is needed to solve ELAG, is in Step 11. The modified statement is as follows:

$$maxresult = \max_{(j-K_2-1) \leq \ell \leq (j-K_1)} (\max(E_\ell))$$

Theorem 10. *Problem ELAG can be solved in $O(\mathcal{R}K + \mathcal{R} \log \log n)$ time, using $\theta(max(\mathcal{R}, n))$ space, where $K = K_2 - K_1$.* $\square$

The following result holds if we need the solutions to the subproblems.

Theorem 11. *Problem ELAG can be solved in $O(n^2 + \mathcal{R}K + \mathcal{R}\log\log n)$ time, using $\theta(n^2)$ space, where $K = K_2 - K_1$.* □

Finally, it should be clear that in the algorithm in Section 4, virtually, there is no modification at all except for that we have to adjust the Range Maxima Query to incorporate the elastic gap constraint.

Theorem 12. *Problem ELAG can be solved in $O(n^2 + \mathcal{R}\log\log n)$ time, using $\theta(max(\mathcal{R}, n))$ space.* □

6   Algorithms for Rigid Gapped LCS

This section is dedicated to solve Problem RIFIG and Problem RELAG. RIFIG, by nature, is a bit more restricted because, in addition to the K-gap constraint, the consecutive characters in the common subsequence must have the same distance between them (rigidness) both in X and Y. Interestingly enough, this restriction makes this problem rather easier to solve. And in fact we will see that we can modify the algorithm in Section 2 easily to solve RIFIG and this slight modification would even improve the running time of the algorithm. The key idea lies in the fact that to calculate a $\mathcal{T}_{local}[i,j]$ we just need to check the $K + 1$ diagonal entries before it. This is true because of the required *rigidness*. The modified version of Equation 2 to handle RIFIG is given below.

$$\mathcal{T}_{local}[i,j] = \begin{cases} \text{Undefined} & \text{if } (i,j) \notin M, \\ \max_{\substack{(\ell_i,\ell_j) \in \{(i-1,j-1),(i-2,j-2)\ldots(i-1-K,j-1-K)\} \\ (\ell_i,\ell_j) \in M}} (\mathcal{T}_{local}[\ell_i,\ell_j]) + 1 & \text{if } (i,j) \in M. \end{cases}$$

(6)

Also we can easily modify Equation 6 to solve RELAG. So we get the following theorems.

Theorem 13. *Problem RIFIG can be solved in $O(n^2 + \mathcal{R}K)$ time using $\theta(n^2)$ space.* □

Theorem 14. *Problem RELAG can be solved in $O(n^2 + \mathcal{R}(K_2 - K_1))$ time using $\theta(n^2)$ space.* □

In the rest of this section we will try to achieve better solutions for RIFIG. We first introduce a variant of RIFIG where the gap constraint is withdrawn. In other words we can say that in this variant we have $K = n$.

Problem "RLCS" (Rigid LCS Problem). *Given two strings X and Y, each of length n, we want to find out a Rigid Common Subsequence of the maximum length[2]. A Rigid Common Subsequence of X and Y is a subsequence $S[1..r] = S[1]\, S[2]\, ...S[r]$ of both X and Y such that $C(X,S)[i] - C(X,S)[i-1] = C(Y,S)[i] - C(Y,S)[i-1]$ for all $2 \leq i \leq r$.*

---

[2] The generalized version of RLCS was introduced and proved to be Max-SNP hard in [6].

It is easy to see that using Equation 6 we can easily solve Problem RLCS by assuming $K = n$. But this will not give us a very good running time at all. On the other hand, it turns out that, we can achieve a better running time by appropriate modification to Equation 1. To solve RLCS, however, for each tabular entry $T[i, j]$ we calculate and store two values namely $T_{local}[i, j]$, $T_{global}[i, j]$. The recurrence relations are defined below:

$$T_{local}[i, j] = \begin{cases} 0 & \text{if } i = 0 \text{ or } j = 0, \\ T_{local}[i - 1, j - 1] + 1 & \text{if } X[i] = Y[j], \\ T_{local}[i - 1, j - 1] & \text{if } X[i] \neq Y[j]. \end{cases} \quad (7)$$

$$T_{global}[i, j] = \begin{cases} 0 & \text{if } i = 0 \text{ or } j = 0, \\ \max(T_{global}[i - 1, j], T_{global}[i, j - 1]) & \text{if } (i, j) \notin M, \\ \max(T_{global}[i - 1, j], T_{global}[i, j - 1], T_{local}[i, j] & \text{if } (i, j) \in M. \end{cases} \quad (8)$$

It is easy to see that Equation 7 preserves the rigidness of the subsequence. Note that, Equation 8 is required to keep track of the global solution.

Theorem 15. *Problem RLCS can be solved in $O(n^2)$ time using $\theta(n^2)$ space.* □

Inspired by the idea of above solution to RLCS in the rest of this section we try to devise an algorithm to solve RIFIG in $O(n^2)$ time. The idea is to some how propagate the *constraint information* up through the diagonal entries as soon as we find a match and whenever a match is found check this information. What we plan to do is as follows. For the calculation of T_{local} we apply K-modulo arithmetic. The actual length of LCS would be $\lceil T_{local}[n, n]/K \rceil$.

$$T_{local}[i, j] = \begin{cases} 0 & \text{if } i = 0 \text{ or } j = 0, \\ 0 & \text{if } X[i] \neq Y[j] \text{ and} \\ & T_{local}[i - 1, j - 1] \bmod K = 1, \\ \lceil T_{local}[i - 1, j - 1]/K \rceil * K + K & \text{if } X[i] = Y[j], \\ T_{local}[i - 1, j - 1] - 1 & \text{if } X[i] \neq Y[j] \text{ and} \\ & T_{local}[i - 1, j - 1] > 0, \\ 0 & \text{if } X[i] \neq Y[j] \text{ and} \\ & T_{local}[i - 1, j - 1] = 0. \end{cases} \quad (9)$$

$$T_{global}[i, j] = \begin{cases} T_{global}[i, j] & \text{if } i = 1 \text{ or } j = 1, \\ \max(T_{global}[i - 1, j], T_{global}[i, j - 1]) & \text{if } (i, j) \notin M, \\ \max(T_{global}[i - 1, j], T_{global}[i, j - 1], T_{local}[i, j] & \text{if } (i, j) \in M. \end{cases} \quad (10)$$

Theorem 16. *Problem RIFIG can be solved in $O(n^2)$ time using $\theta(n^2)$ space.* □

7   Conclusion

In this paper we have introduced new variants of LCS problem and presented efficient algorithms to solve them. Our algorithms can be used to solve some other interesting problems, specially in bioinformatics, as well. We can tackle the degenerate strings in biological applications, for e.g., using a clever technique invented by Lee et al. [5] (the details are omitted for space constraints). Moreover we can solve the Longest Common Substring problem for degenerate strings, a very common problem in molecular biology simply by putting $K = 1$

in the Problem FIG. Also, our algorithms should be useful in extracting long multiple repeats in DNA sequences as follows. One approach to solve this problem efficiently is to apply 'lossless' filters [1, 9] where filters apply a necessary condition that sequences must meet to be part of repeats. One way to improve the computational time and the sensitivity of the filters is to compute Longest Common Subsequences between the ordered sequences of exact k-mers used in the filtering technique. However in the case of the filter, the LCS that needs to be computed has bounded span [8] which, again, can be obtained by applying the gap-constraints in LCS.

References

1. ED'NIMBUS. *http://igm.univ-mlv.fr/ peterlon/officiel/ednimbus/.*
2. S. F. Altschul, W. Gish, W. Miller, E. W. Meyers, and D. J. Lipman. Basic local alignment search tool. *Journal of Molecular Biology*, 215(3):403–410, 1990.
3. M. A. Bender and M. Farach-Colton. The lca problem revisited. In *Latin American Theoretical INformatics (LATIN)*, pages 88–94, 2000.
4. H. Gabow, J. Bentley, and R. Tarjan. Scaling and related techniques for geometry problems. In *Symposium on the Theory of Computing (STOC)*, pages 135–143, 1984.
5. I. Lee, A. Apostolico, C. S. Iliopoulos, and K. Park. Finding approximate occurrence of a pattern that contains gaps. In *Australasian Workshop on Combinatorial Algorithms (AWOCA)*, pages 89–100, 2003.
6. B. Ma and K. Zhang. On the longest common rigid subsequence problem. In *CPM*, pages 11–20, 2005.
7. W. Pearson and D. Lipman. Improved tools for biological sequence comparison. *Proceedings of National Academy of Science, USA*, 85:2444–2448, 1988.
8. P. Peterlongo. Private communication.
9. P. Peterlongo, N. Pisanti, F. Boyer, and M.-F. Sagot. Lossless filter for finding long multiple approximate repetitions using a new data structure, the bi-factor array. In *SPIRE*, pages 179–190, 2005.
10. P. van Emde Boas. Preserving order in a forest in less than logarithmic time and linear space. *Information Processing Letters*, 6:80–82, 1977.
11. R. A. Wagner and M. J. Fischer. The string-to-string correction problem. *J. ACM*, 21(1):168–173, 1974.

Constructing Labeling Schemes Through Universal Matrices

Amos Korman[1,*], David Peleg[2,**], and Yoav Rodeh[3]

[1] Information Systems Group, Faculty of IE&M, The Technion, Israel
`pandit@tx.technion.ac.il`
[2] Dept. of Computer Science, Weizmann Institute, Israel
`david.peleg@weizmann.ac.il`
[3] Dept. of Computer Science, Tel Hai Academic College, Israel
`yoavr@telhai.ac.il`

Abstract. Let f be a function on pairs of vertices. An f-*labeling scheme* for a family of graphs $\mathcal{F}$ labels the vertices of all graphs in $\mathcal{F}$ such that for every graph $G \in \mathcal{F}$ and every two vertices $u, v \in G$, $f(u,v)$ can be inferred by merely inspecting the labels of u and v. The *size* of a labeling scheme is the maximum number of bits used in a label of any vertex in any graph in $\mathcal{F}$. This paper illustrates that the notion of universal matrices can be used to efficiently construct f-labeling schemes.

Let $\mathcal{F}(n)$ be a family of connected graphs of size at most n and let $\mathcal{C}(\mathcal{F}, n)$ denote the collection of graphs of size at most n, such that each graph in $\mathcal{C}(\mathcal{F}, n)$ is composed of a disjoint union of some graphs in $\mathcal{F}(n)$. We first investigate methods for translating f-labeling schemes for $\mathcal{F}(n)$ to f-labeling schemes for $\mathcal{C}(\mathcal{F}, n)$. In particular, we show that in many cases, given an f-labeling scheme of size $g(n)$ for a graph family $\mathcal{F}(n)$, one can construct an f-labeling scheme of size $g(n) + \log \log n + O(1)$ for $\mathcal{C}(\mathcal{F}, n)$. We also show that in several cases, the above mentioned extra additive term of $\log \log n + O(1)$ is necessary. In addition, we show that the family of n-node graphs which are unions of disjoint circles enjoys an adjacency labeling scheme of size $\log n + O(1)$. This illustrates a non-trivial example showing that the above mentioned extra additive term is sometimes not necessary.

We then turn to investigate distance labeling schemes on the class of circles of at most n vertices and show an upper bound of $1.5 \log n + O(1)$ and a lower bound of $4/3 \log n - O(1)$ for the size of any such labeling scheme.

Keywords: Labeling schemes, Universal graphs, Universal matrices.

1 Introduction

Motivation and related work. In the fields of communication networks and distributed computing, network representation schemes have been studied exten-

* Supported in part at the Technion by an Aly Kaufman fellowship.
** Supported in part by grants from the Israel Science Foundation and the Israel Ministry of Science and Art.

T. Asano (Ed.): ISAAC 2006, LNCS 4288, pp. 409–418, 2006.
© Springer-Verlag Berlin Heidelberg 2006

sively. This paper studies a type of representation based on assigning *informative labels* to the vertices of the network. In most traditional network representations, the names or identifiers given to the vertices betray nothing about the network's structure. In contrast, the labeling schemes studied here involve using more informative and localized labels for the network vertices. The idea is to associate with each vertex a label selected in a such way, that will allow us to infer information about any two vertices *directly* from their labels, without using *any* additional information sources. Hence in essence, this method bases the entire representation on the set of labels alone.

Obviously, without restricting the label size one can encode any desired information, including in particular, the entire graph structure. Our focus is thus on informative labeling schemes using *short* labels. Labeling schemes were previously developed for different graph families and for a variety information types, including adjacency [16, 5], distance [27, 21, 13, 12, 10, 17, 30, 7, 1], tree routing [9, 31], flow and vertex connectivity [20, 15], tree ancestry [4, 3, 17, 2, 18, 19], nearest common ancestor in trees [28, 2] and various other tree functions. See [11] for a survey on (static) labeling schemes. The dynamic version was studied in [23, 22, 8, 14].

The *size* of a graph is the number of vertices in it. The *size* of a labeling scheme for a family $\mathcal{F}$ of graphs is defined as the maximum number of bits assigned in a label of any vertex in any graph in $\mathcal{F}$. The following is an example (see [16]) of an adjacency labeling scheme on the family of n-node forests.

Example: Given an n-node forest, first assign each vertex v a disjoint identifier $id(v)$ in the range $1, 2, \cdots, n$. Then assign each non-root vertex v, the label $L(v) = \langle id(v), id(p(v)) \rangle$, where $p(v)$ is v's parent in the corresponding tree, and assign each root r the label $L(r) = \langle id(r) \rangle$. Note that two nodes in a forest are neighbors iff one is the parent of the other. Therefore, given the labels of two nodes, one can easily determine whether these nodes are neighbors or not. Clearly, the size of this labeling scheme is $2 \log n$. ∎

As shown in [16], the notion of adjacency labeling schemes is strongly related to the notion of vertex induced *universal graphs*. Given a graph family $\mathcal{F}$, a graph $\mathcal{U}$ is $\mathcal{F}$-induced universal if every graph in $\mathcal{F}$ is a vertex induced subgraph of $\mathcal{U}$. In the early 60's, induced universal graphs were studied in [29] for infinite graph families. Induced universal graphs for the family of all n-node graphs were studied in [26], for trees, forests, bounded arboricity graphs and planar graphs in [6, 16, 4], for hereditary graphs in [24] and for several families of bipartite graphs in [25].

As proved in [16], a graph family $\mathcal{F}$ (of n-node graphs) has an adjacency labeling scheme of size $g(n)$ iff there exists an $\mathcal{F}$-induced universal graph of size $2^{g(n)}$. Therefore, Example 1 implies the existence of an n^2-node induced universal graph for the family of n-node forests. This bound was further improved in [5] to $2^{\log n + O(\log^* n)}$ by constructing an adjacency labeling scheme for forests with label size $\log n + O(\log^* n)$.

An extension of the notion of an $\mathcal{F}$-induced universal graph was given in [10]. An $\mathcal{F}$-*universal distance matrix* is a square matrix $\mathcal{U}$ containing the distance matrix of every graph in $\mathcal{F}$ as an induced sub-matrix. It was shown in [10] that a graph family $\mathcal{F}$ has a distance labeling scheme of size $g(n)$ iff there exists an $\mathcal{F}$-universal distance matrix with dimension $2^{g(n)}$. We note that to the best of our knowledge, despite the above mentioned relation between labeling schemes and universal distance matrices, no attempt has been made so far to construct labeling schemes based on this relation.

Our results. We first notice that the notion of a *universal distance matrix* can be generalized into a *universal f-matrix* for any type of function f on pairs of vertices. This paper investigates this notion of universal f-matrices for various functions and graph families and uses it to explicitly construct upper bounds and lower bounds on the sizes of the corresponding f-labeling schemes. To the best of our knowledge, this is the first attempt to explicitly construct labeling schemes based on such notions.

Let $\mathcal{F}(n)$ be a family of connected graphs of size at most n and let $\mathcal{C}(\mathcal{F}, n)$ denote the collection of graphs of size at most n, such that each graph in $\mathcal{C}(\mathcal{F}, n)$ is composed of a disjoint union of some graphs in $\mathcal{F}(n)$. We first investigate methods for translating f-labeling schemes for $\mathcal{F}(n)$ to f-labeling schemes for $\mathcal{C}(\mathcal{F}, n)$. In particular, using the notion of universal f-matrices we show that in many cases, given an f-labeling scheme of size $g(n)$ for a graph family $\mathcal{F}(n)$, one can construct an f-labeling scheme of size $g(n) + \log \log n + O(1)$ for $\mathcal{C}(\mathcal{F}, n)$. We also show that in several cases, the above mentioned extra additive term of $\log \log n + O(1)$ is necessary. In addition, using the notion of universal induced graphs, we show that the family of n-node graphs which are unions of disjoint circles enjoys an adjacency labeling scheme of size $\log n + O(1)$. This illustrates a non-trivial example showing that the above mentioned extra additive term is sometimes not necessary.

We then turn to investigate distance labeling schemes on the class of circles of size at most n. Using the notion of universal distance matrices we construct a distance labeling scheme for this family of size $1.5 \log n + O(1)$ and then show a lower bound of $\frac{4}{3} \log n - O(1)$ for the size of any such scheme.

Throughout, all additive constant terms are small, so no attempt was made to optimize them.

2   Preliminaries

Let f be a function on pairs of vertices. An *f-labeling scheme* $\pi = \langle \mathcal{L}_\pi, \mathcal{D}_\pi \rangle$ for a graph family $\mathcal{F}$ is composed of the following components:

1. A *labeler* algorithm $\mathcal{L}_\pi$ that given a graph in $\mathcal{F}$, assigns labels to its vertices.
2. A polynomial time *decoder* algorithm $\mathcal{D}_\pi$ that given the labels $L(u)$ and $L(v)$ of two vertices u and v in some graph in $\mathcal{F}$, outputs $f(u,v)$.

The *size* of a labeling scheme $\pi = \langle \mathcal{L}_\pi, \mathcal{D}_\pi \rangle$ for a graph family $\mathcal{F}$ is the maximum number of bits in a label assigned by $\mathcal{L}_\pi$ to any vertex in any graph in $\mathcal{F}$.

We mainly consider the following functions on pairs of vertices u and v in a graph G.

The *adjacency* (respectively, *connectivity*) function: $f(u,v)=1$ if u and v are adjacent (resp., connected) in G and 0 otherwise.

The *distance* function: $f(u,v) = d_G(u,v)$, the (unweighted) distance between u and v in G; we may sometimes omit the subscript G when it is clear from the context.

The *flow* function: $f(u,v)$ is the maximum flow possible between u and v in a weighted graph G.

The *size* of a graph G, denoted $|G|$, is the number of vertices in it. Let $\mathcal{F}^{paths}(n)$ (respectively, $\mathcal{F}^{circles}(n)$) denote the family of paths (resp. circles) of size at most n. Given a family $\mathcal{F}(n)$ of connected graphs of size at most n, let $\mathcal{C}(\mathcal{F},n)$ denote the collection of graphs of size at most n, such that each graph in $\mathcal{C}(\mathcal{F},n)$ is composed of a disjoint union of some graphs in $\mathcal{F}(n)$.

A graph G with vertex set $\{v_1, v_2, \ldots, v_n\}$ is a *vertex induced subgraph* of a graph $\mathcal{U}$ with vertex set $\{u_1, u_2, \ldots, u_k\}$ if there exist indices $1 \le s_1, s_2, \ldots, s_n \le k$ such that for every $i, j \in \{1, 2, \ldots, n\}$, v_i and v_j are neighbors in G iff u_{s_i} and u_{s_j} are neighbors in $\mathcal{U}$. Given a graph family $\mathcal{F}$, a graph $\mathcal{U}$ is $\mathcal{F}$-induced universal if every graph in $\mathcal{F}$ is a vertex induced subgraph of $\mathcal{U}$.

Proposition 1. [16] *A graph family $\mathcal{F}$ has an adjacency labeling scheme with label size g iff there exists an $\mathcal{F}$-induced universal graph with 2^g nodes.*

The *dimension* of a square matrix M, denoted $\dim(M)$, is the number of rows in it. An $n \times n$ square matrix $B = (b_{i,j})_{1 \le i,j \le n}$ is an *induced sub-matrix* of a $k \times k$ square matrix $A = (a_{i,j})_{1 \le i,j \le k}$ if there exists a sequence $(s_1, s_2, \ldots, s_n)$ of distinct indices $1 \le s_\ell \le k$ such that $b_{i,j} = a_{s_i, s_j}$ for every $i, j \in \{1, 2, \ldots, n\}$. As defined in [10], given a graph family $\mathcal{F}$, an $\mathcal{F}$-*universal distance matrix* is a square matrix M containing the distance matrix of every graph in $\mathcal{F}$ as an induced sub-matrix.

Proposition 2. [10] *If a graph family $\mathcal{F}$ enjoys a distance labeling scheme with label size g, then there exists an $\mathcal{F}$-universal distance matrix of dimension $2^{g+O(1)}$. Conversely, if there exists an $\mathcal{F}$-universal distance matrix of dimension 2^g then $\mathcal{F}$ enjoys a distance labeling scheme of size $g + O(1)$.*

Let G be an n-node graph and let $u_1, u_2, \ldots, u_n$ denote its vertices. Given a function f on pairs of vertices, the f-*matrix of G* is an n-dimensional square matrix B such that $B_{i,j} = f(u_i, u_j)$ for every $i, j \in \{1, 2, \ldots, n\}$. We first notice that the notion of a universal distance matrix can be extended into a *universal f matrix* for any type of function f on pairs of vertices. Formally, given a graph family $\mathcal{F}$, an $\mathcal{F}$-*universal f-matrix* is a square matrix M containing the f-matrix of every graph in $\mathcal{F}$ as an induced sub-matrix of M. Going along the same steps as the proof of Proposition 2 in [10], we obtain the following proposition.

Proposition 3. *If a graph family $\mathcal{F}$ enjoys an f-labeling scheme with label size g, then there exists an $\mathcal{F}$-universal f-matrix of dimension $2^{g+O(1)}$. Conversely, if there exists an $\mathcal{F}$-universal f-matrix of dimension 2^g then $\mathcal{F}$ enjoys an f-labeling scheme of size $g + O(1)$.*

3 Transforming f-Labeling Schemes for Connected Graphs to Non-connected Graphs

3.1 The General Transformation

Let $\mathcal{F}(n)$ be a family of connected graphs, each of size at most n. In this section we show that in many cases one can transform an f-labeling scheme for $\mathcal{F}(n)$ to an f-labeling scheme for $\mathcal{C}(\mathcal{F}, n)$ with a size increase of $\log \log n + O(1)$. We then show that this additive term is necessary in some cases but not always.

Let f be a function on pairs of vertices, with the property that there exists some value $\sigma \in [0, \infty]$ such that $f(u, v) = \sigma$ for every two non-connected vertices u and v. For example, f can be the distance function (with $\sigma = \infty$), the flow function (with $\sigma = 0$) or the adjacency function (with $\sigma = 0$). First note that there exists a straightforward translation allowing us to transform a given f-labeling scheme $\pi = \langle \mathcal{L}, \mathcal{D} \rangle$ for a graph family $\mathcal{F}(n)$ of size $g(n)$ into an f-labeling scheme $\pi' = \langle \mathcal{L}', \mathcal{D}' \rangle$ for $\mathcal{C}(\mathcal{F}, n)$ of size $g(n) + \log n + O(1)$ as follows. Given a graph $H \in \mathcal{F}(n)$ and a vertex $v \in H$, let $L_H(v)$ be the label given to v by the labeler algorithm $\mathcal{L}$ applied on H. Given a graph $G \in \mathcal{C}(\mathcal{F}, n)$, let $G_1, G_2, \ldots, G_k$ be its connected components (which belong to $\mathcal{F}(n)$). For $1 \leq i \leq k$, given a vertex $v \in G_i$, the label $\mathcal{L}'(v)$ to be assigned to v by the labeler algorithm $\mathcal{L}'$ is composed of the concatenation of two sublabels, $M_1'(v)$ and $M_2'(v)$. The first sublabel, $M_1'(v)$, consists of precisely $\lceil \log n \rceil$ bits which are used to encode i (padded by 0's to the left as necessary). The second sublabel is $M_2'(v) = L_{G_i}(v)$. Since the first sublabel consists of precisely $\lceil \log n \rceil$ bits, given a label $\mathcal{L}'(v)$ one can easily distinguish the two sublabels $\mathcal{L}_1'(v)$ and $\mathcal{L}_2'(v)$. Given two labels $\mathcal{L}'(v)$ and $\mathcal{L}'(u)$ of two vertices v and u in G, the decoder $\mathcal{D}'$ outputs $\mathcal{D}(\mathcal{L}_2'(v), \mathcal{L}_2'(u))$ if the two labels agree in their first sublabels, and σ otherwise. Clearly π' is an f-labeling scheme of size $g(n) + \log n + 1$ for $\mathcal{C}(\mathcal{F}, n)$.

Lemma 1. *Let $g(n)$ be an increasing function satisfying (1) $g(n) \geq \log n$ and (2) for every constant $c \geq 1$, $2^{g(c \cdot n)} \geq c \cdot 2^{g(n)}$[1]. If, for every $m \leq n$, there exists an f-labeling scheme $\pi(m)$ of size $g(m)$ for $\mathcal{F}(m)$ then there exists an f-labeling scheme of size $g(n) + \log \log n + O(1)$ for $\mathcal{C}(\mathcal{F}, n)$.*

Proof. For every $1 \leq i \leq n$, let $m_i = \lfloor n/i \rfloor$ and let M_i be the $\mathcal{F}(m_i)$-universal f-matrix obtained from the f-labeling scheme $\pi(m_i)$. Let M be the matrix

$$\mathbf{M} = \begin{pmatrix} M_1 & \sigma & \sigma & \cdots \\ \sigma & M_2 & \sigma & \cdots \\ \sigma & \sigma & M_3 & \cdots \\ \vdots & \vdots & \vdots & \ddots \end{pmatrix}$$

where all entries except for the diagonal of M_i's contain the value σ. Since the size of $\pi(m_i)$ is at most $g(n/i)$, we obtain that $\dim(M_i) \leq 2^{g(n/i)+O(1)}$. By our assumption on $g(n)$, we get that $\dim(M_i) \leq 2^{g(n)}/i + O(1)$, hence $\dim(M) \leq$

---

[1] These requirements are satisfied by all functions of the form $\alpha \log^\beta n$, where $\alpha, \beta \geq 1$.

$O(n) + 2^{g(n)} \sum_{i=1}^{n} 1/i \leq O(n) + 2^{g(n)} \cdot \log n$. Since we assume $g(n) \geq \log n$, we obtain that $\dim(M) = O(2^{g(n)} \cdot \log n)$.

The lemma follows once we show that M is an $\mathcal{C}(\mathcal{F}, n)$-universal distance matrix. Let $G = \bigsqcup_{i=1}^{k} G_i$ be a graph in $\mathcal{C}(\mathcal{F}, n)$ such that for every $1 \leq i < k$, $|G_i| \geq |G_{i+1}|$. If follows that for every $1 \leq i \leq k$, $|G_i| \leq m_i$. For each $1 \leq i \leq k$, using the $\mathcal{F}(m_i)$-universal f-matrix M_i, we map the vertices of G_i to the corresponding indices of M_i in M. ∎

3.2   Labeling Schemes for Path Collections

The following easy to prove claim shows that the requirements from $g(n)$ in the previous lemma are necessary.

Claim. The size of a connectivity labeling scheme on $\mathcal{F}^{paths}(n)$ is $O(1)$ and the size of a connectivity labeling scheme on $\mathcal{C}(\mathcal{F}^{paths}, n)$ is $\Omega(\log n)$.

Note that one can easily show that the size of a distance labeling scheme for $\mathcal{F}^{paths}(n)$ is $\log n + \Theta(1)$. Therefore, the following lemma shows, in particular, that the extra additive term of $\log \log n + O(1)$ mentioned in Lemma 1 is sometimes necessary.

Lemma 2. *Any distance labeling schemes for $\mathcal{C}(\mathcal{F}^{paths}, n)$ must have size at least $\log n + \log \log n - O(1)$.*

Proof. It is enough to show that any $\mathcal{C}(\mathcal{F}^{paths}, n)$-universal distance matrix must have dimension $\Omega(n \log n)$. For simplicity of presentation, we assume that $n = m!$ for some m. The general case follows using similar arguments. Let M be any $\mathcal{C}(\mathcal{F}^{paths}, n)$-universal distance matrix and let $k = \dim(M)$, i.e., M is a square $k \times k$ matrix.

For every $1 \leq i \leq n$, let G_i be the graph consisting of a disjoint union of i paths of size n/i each. We now define inductively n sets of integers, $X_1, \ldots, X_n$, with the following properties.

1. $X_i \subset \{1, 2, \ldots, k\}$,
2. $X_i \subset X_{i+1}$,
3. $|X_n| = \Omega(n \log n)$,
4. for every $1 \leq i \leq n$, the set X_i can be partitioned into i disjoint subsets $Q_1, Q_2, \ldots, Q_i$, such that the following property is satisfied.
   Partition property: for every $1 \leq j \leq i$ and every two integers $x, y \in Q_j$, $M_{x,y} \neq \infty$.

The sets $X_1, \ldots, X_n$ are defined inductively as follows. Enumerate the vertices in G_1 from 1 to n, i.e., let the set of vertices in G_1 be $a_1, a_2, \ldots, a_n$. Let X_1 be a set of integers $\{s_1, s_2, \ldots, s_n\}$ satisfying that for every two vertices a_h and a_j in G_1, $f(a_h, a_j) = M_{s_h, s_j}$. The required properties trivially hold. Now assume that the sets X_j are already defined for each $j \leq i$. Then X_{i+1} is constructed as follows. Let $Q_1, Q_2, \ldots, Q_i$ be the subsets of X_i satisfying the partition property and let $P_1, P_2, \ldots, P_{i+1}$ be the $n/(i+1)$-node paths of G_{i+1}. Let $a_1, a_2, \ldots, a_n$

denote the set of vertices in G_{i+1} and let $(s_1, s_2, \ldots, s_n)$ be a sequence of integers such that for every $1 \leq a_h, a_j \leq n$, $f(a_h, a_j) = M_{s_h, s_j}$. For every $1 \leq j \leq i+1$, let $\varphi(P_j)$ be the set $\{s_k \mid a_k \in P_j\}$. By the partition property, for every $1 \leq h \leq i$, there exists at most one $1 \leq j \leq i+1$ such that $Q_h \cap \varphi(P_j) \neq \emptyset$. Therefore, there exists some j such that $Q_h \cap \varphi(P_j) = \emptyset$ for every $1 \leq h \leq i$. Let $X_{i+1} = X_i \cup \varphi(P_j)$. Clearly, the partition property is satisfied for X_{i+1}. Moreover, $|X_{i+1}| = |X_i| + n/(i+1)$, and therefore $|X_n| \approx n(1 + \frac{1}{2} + \ldots + \frac{1}{n}) = \Omega(n \log n)$. Since $|X_n| \leq k$, the lemma follows. ∎

It can be shown that $\mathcal{C}(\mathcal{F}^{paths}, n)$ enjoys an adjacency labeling scheme of size $\log n + O(1)$. It follows that the extra additive term of $\log \log n + O(1)$ (Lemma 1) is not necessary in this case. In the following subsection we show another example for the fact that this additive term is not necessary, using the notion of universal induced graphs.

3.3   An Adjacency Labeling Scheme for $\mathcal{C}(\mathcal{F}^{Circles}, n)$

In this subsection we consider adjacency labeling schemes for $\mathcal{C}(\mathcal{F}^{circles}, n)$. Clearly, any adjacency labeling scheme for $\mathcal{F}^{circles}(n)$ must have size at least $\log n$. We now describe an adjacency labeling scheme for $\mathcal{C}(\mathcal{F}^{circles}, n)$ of size $\log n + O(1)$.

Let us first note that the following straightforward labeling scheme for the class $\mathcal{C}(\mathcal{F}^{circles}, n)$ uses labels of size $3 \log n$. Given a graph $G \in \mathcal{C}(\mathcal{F}^{circles}, n)$, let $C_1, C_2, \ldots$ be the circles of G. For each i, enumerate the vertices of C_i clockwise and label each node $u \in C_i$ by $\mathcal{L}(u) = \langle i, |C_i|, n(u) \rangle$, where $n(u)$ is the number given to u in the above mentioned enumeration. Given the labels $\mathcal{L}(u)$ and $\mathcal{L}(v)$ of two vertices in G, the decoder can easily identify whether u and v belong to the same circle or not. If u and v do not belong to the same circle, then the decoder outputs 0. Otherwise, the adjacency between u and v in their common circle C_i can easily be determined using $|C_i|, n(u)$ and $n(v)$.

An *even* (respectively, *odd*) circle is a circle of even (resp., odd) size. Let $\mathcal{F}^{e-circles}$ be the graph family containing all m-node circles, where $m \geq 8$ is an even number. Let us first describe an $\mathcal{C}(\mathcal{F}^{e-circles}, n)$-universal graph. Let $\mathcal{U}^{e-circles}$ be a $3 \times n$-grid graph (see the bottom graph in Figure 2). Consider some graph $G \in \mathcal{C}(\mathcal{F}^{e-circles}, n)$ and let $C_1, C_2, \ldots$ be its collection of disjoint circles. Map each circle C_i into $\mathcal{U}^{e-circles}$ leaving a gap of one column between any two consecutive mapped circles. An example of such a mapping is depicted in Figure 1.

It follows that $\mathcal{U}^{e-circles}$ is an $\mathcal{C}(\mathcal{F}^{e-circles}, n)$-universal graph of size $O(n)$. In the full paper we describe how to extend $\mathcal{U}^{e-circles}$ to obtain an $\mathcal{C}(\mathcal{F}^{circles}, n)$-universal graph of size $O(n)$. The following lemma follows.

Lemma 3. *There exists an adjacency labeling scheme of size $\log n + O(1)$ for* $\mathcal{C}(\mathcal{F}^{circles}, n)$.

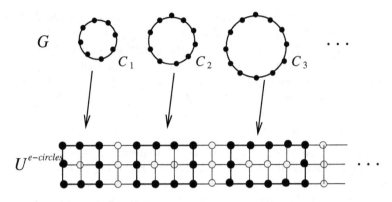

Fig. 1. A mapping of a graph $G \in \mathcal{C}(\mathcal{F}^{e-circles}, n)$ into the universal graph $\mathcal{U}^{e-circles}$

4   A Distance Labeling Scheme for $\mathcal{F}^{Circles}(n)$

In this section we construct a distance labeling scheme for $\mathcal{F}^{circles}(n)$ of size $1.5 \log n + O(1)$ and establish a lower bound of $4/3 \log n - O(1)$ for the size of any such labeling scheme. Due to lack of space, this extended abstract contains only the lower bound proof.

4.1   A Size Lower Bound on Distance Labeling Schemes for $\mathcal{F}^{Circles}(n)$

Lemma 4. *Any distance labeling scheme for $\mathcal{F}^{circles}(n)$ must have size at least $4/3 \log n - O(1)$.*

Proof. For simplicity of presentation, assume n is divisible by 12. The general case follows using similar arguments. Let $\pi = \langle \mathcal{L}, \mathcal{D} \rangle$ be a distance labeling scheme for $\mathcal{F}^{circles}(n)$ and denote the set of labels assigned to the vertices of graphs in $\mathcal{F}^{circles}(n)$ by $X = \{\mathcal{L}(v) \mid v \in C, \ C \in \mathcal{F}^{circles}(n)\}$.

For $m = 1, 2, \ldots, n/12$, let $c_m = n/2 + 6m$ (note that c_m is divisible by 6). For every $1 \le m \le n/12$, let C_m be the circle with c_m nodes. For any such circle C_m, let $I_m^1, I_m^2, \ldots, I_m^6$ be six vertex disjoint arcs of C_m, each of size $m/6$, ordered clockwise. Figure 2 shows this division of C_{12} into 6 disjoint arcs. Define

$$\Psi_m = \{\langle \mathcal{L}(a), \mathcal{L}(b), \mathcal{L}(c) \rangle \mid a \in I_m^1, b \in I_m^3, c \in I_m^5\}.$$

It is easy to show that for every two vertices $v, u \in C_m$, $\mathcal{L}(v) \ne \mathcal{L}(u)$, therefore Ψ_m contains $(c_m/6)^3 \ge (n/12)^3$ elements. Note that given any circle C_m, if $a \in I_m^1, b \in I_m^3$ and $c \in I_m^5$ then $d(a, b) + d(b, c) + d(c, a) = m$, so necessarily $\mathcal{D}(\mathcal{L}(a), \mathcal{L}(b)) + \mathcal{D}(\mathcal{L}(b), \mathcal{L}(c)) + \mathcal{D}(\mathcal{L}(c), \mathcal{L}(a)) = m$. We therefore obtain the following claim.

Claim. For every $1 \le m < m' \le n/12$, $\Psi_m \cap \Psi_{m'} = \emptyset$.

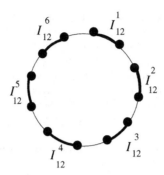

Fig. 2. The division of C_{12} into 6 disjoint arcs $I_{12}^1, I_{12}^2, \ldots, I_{12}^6$

Let $\Psi = \bigcup_{m=1}^{n/12} \Psi_m$. By the above claim, Ψ contains $\Omega(n^4)$ distinct elements. Since $\Psi \subset X \times X \times X$, we obtain that X contains $\Omega(n^{4/3})$ elements. Therefore, there exists a label in X encoded using $\log(\Omega(n^{4/3})) = \frac{4}{3}\log n - O(1)$ bits. The lemma follows. ∎

References

[1] S. Alstrup, P. Bille and T. Rauhe. Labeling schemes for small distances in trees. In *Proc. 14th ACM-SIAM Symp. on Discrete Algorithms*, Jan. 2003.

[2] S. Alstrup, C. Gavoille, H. Kaplan and T. Rauhe. Nearest Common Ancestors: A Survey and a new Distributed Algorithm. *Theory of Computing Systems* **37**, (2004), 441–456.

[3] S. Abiteboul, H. Kaplan and T. Milo. Compact labeling schemes for ancestor queries. In *Proc. 12th ACM-SIAM Symp. on Discrete Algorithms*, Jan. 2001.

[4] S. Alstrup and T. Rauhe. Improved Labeling Scheme for Ancestor Queries. In *Proc. 19th ACM-SIAM Symp. on Discrete Algorithms*, Jan. 2002.

[5] S. Alstrup and T. Rauhe. Small induced-universal graphs and compact implicit graph representations. In *Proc. 43rd IEEE Symp. on Foundations of Computer Science*, Nov. 2002.

[6] F.R.K. Chung. Universal graphs and induced-universal graphs. *J. of Graph Theory* **14(4)**, (1990), 443–454.

[7] E. Cohen, E. Halperin, H. Kaplan and U. Zwick. Reachability and Distance Queries via 2-hop Labels. In *Proc. 13th ACM-SIAM Symp. on Discrete Algorithms*, Jan. 2002.

[8] E. Cohen, H. Kaplan and T. Milo. Labeling dynamic XML trees. In *Proc. 21st ACM Symp. on Principles of Database Systems*, June 2002.

[9] P. Fraigniaud and C. Gavoille. Routing in trees. In *Proc. 28th Int. Colloq. on Automata, Languages & Prog.*, 757–772, July 2001.

[10] C. Gavoille and C. Paul. Split decomposition and distance labelling: an optimal scheme for distance hereditary graphs. In *Proc. European Conf. on Combinatorics, Graph Theory and Applications*, Sept. 2001.

[11] C. Gavoille and D. Peleg. Compact and Localized Distributed Data Structures. *J. of Distributed Computing* **16**, (2003), 111–120.

[12] C. Gavoille, M. Katz, N.A. Katz, C. Paul and D. Peleg. Approximate Distance Labeling Schemes. In *9th European Symp. on Algorithms*, Aug. 2001, 476–488.

[13] C. Gavoille, D. Peleg, S. Pérennes and R. Raz. Distance labeling in graphs. In *Proc. 12th ACM-SIAM Symp. on Discrete Algorithms*, pages 210–219, Jan. 2001.

[14] A. Korman. General Compact Labeling Schemes for Dynamic Trees. In *Proc. 19th Symp. on Distributed Computing*, Sep. 2005.

[15] A. Korman and S. Kutten. Distributed Verification of Minimum Spanning Trees. In *25th Annual ACM SIGACT-SIGOPS Symp. on Principles of Distributed Computing*, July 2006.

[16] S. Kannan, M. Naor, and S. Rudich. Implicit representation of graphs. In *SIAM J. on Descrete Math* **5**, (1992), 596–603.

[17] H. Kaplan and T. Milo. Short and simple labels for small distances and other functions. In *Workshop on Algorithms and Data Structures*, Aug. 2001.

[18] H. Kaplan and T. Milo. Parent and ancestor queries using a compact index. In *Proc. 20th ACM Symp. on Principles of Database Systems*, May 2001.

[19] H. Kaplan, T. Milo and R. Shabo. A Comparison of Labeling Schemes for Ancestor Queries. In *Proc. 19th ACM-SIAM Symp. on Discrete Algorithms*, Jan. 2002.

[20] M. Katz, N.A. Katz, A. Korman and D. Peleg. Labeling schemes for flow and connectivity. *SIAM Journal on Computing* **34** (2004),23–40.

[21] M. Katz, N.A. Katz, and D. Peleg. Distance labeling schemes for well-separated graph classes. In *Proc. 17th Symp. on Theoretical Aspects of Computer Science*, 516–528, February 2000.

[22] A. Korman and D. Peleg. Labeling Schemes for Weighted Dynamic Trees. In *Proc. 30th Int. Colloq. on Automata, Languages & Prog.*, July 2003.

[23] A. Korman, D. Peleg and Y. Rodeh. Labeling schemes for dynamic tree networks. *Theory of Computing Systems* **37**, (2004), 49–75.

[24] V. V. Lozin. On minimal universal graphs for hereditary classes. *J. Discrete Math. Appl.*,**7(3)**, (1997), 295–304.

[25] V. V. Lozin and G. Rudolf. Minimal universal bipartite graphs. *To appear in Ars Combinatoria*.

[26] J. W. Moon. On minimal n-universal graphs. In *Proc. Galasgow Math. Soc.* **7**, 32–33, 1965.

[27] D. Peleg. Proximity-preserving labeling schemes and their applications. In *Proc. 25th Int. Workshop on Graph-Theoretic Concepts in Computer Science*, pages 30–41, June 1999.

[28] D. Peleg. Informative labeling schemes for graphs. In *Proc. 25th Symp. on Mathematical Foundations of Computer Science*, 579–588, Aug. 2000.

[29] R. Rado. Universal graphs and universal functions. *Acta. Arith.*, (1964), 331–340.

[30] M. Thorup. Compact oracles for reachability and approximate distances in planar digraphs. *J. of the ACM* **51**, (2004), 993–1024.

[31] M. Thorup and U. Zwick. Compact routing schemes. In *Proc. 13th ACM Symp. on Parallel Algorithms and Architecture*, pages 1–10, Hersonissos, Crete, Greece, July 2001.

Making Arbitrary Graphs Transitively Orientable: Minimal Comparability Completions*

Pinar Heggernes, Federico Mancini, and Charis Papadopoulos

Department of Informatics, University of Bergen, N-5020 Bergen, Norway
{pinar, federico, charis}@ii.uib.no

Abstract. A transitive orientation of an undirected graph is an assignment of directions to its edges so that these directed edges represent a transitive relation between the vertices of the graph. Not every graph has a transitive orientation, but every graph can be turned into a graph that has a transitive orientation, by adding edges. We study the problem of adding an inclusion minimal set of edges to an arbitrary graph so that the resulting graph is transitively orientable. We show that this problem can be solved in polynomial time, and we give a surprisingly simple algorithm for it.

1 Introduction

A transitive orientation of an undirected graph is an assignment of a direction to each of the edges, such that the edges represent a binary transitive relation on the vertices. An undirected graph is a *comparability graph* if there is a transitive orientation of its edges, and hence comparability graphs are also called *transitively orientable graphs*. This is a wide and well known graph class studied by many authors, and and it has applications in areas like archeology, psychology, and political sciences [1,12]. Comparability graphs are perfect, and they can be recognized in polynomial time. Many interesting optimization problems that are NP-hard on arbitrary graphs, like coloring and maximum (weighted) clique, are polynomially solvable on comparability graphs [1]. Hence, computing a comparability supergraph of an arbitrary graph, and solving a generally NP-hard problem in polynomial time on this supergraph, is a way of obtaining approximation algorithms for several hard problems. For graphs coming from the application areas mentioned above, there may be missing edges due to lacking data so that the graph fails to be comparability, in which case one is again interested in computing a comparability supergraph. A comparability graph obtained by adding edges to an arbitrary graph is called a *comparability completion* of the input graph. Unfortunately, computing a comparability completion with the minimum number of added edges (called a *minimum* completion) is an NP-hard problem [2].

---

* This work is supported by the Research Council of Norway through grant 166429/V30.

T. Asano (Ed.): ISAAC 2006, LNCS 4288, pp. 419–428, 2006.
© Springer-Verlag Berlin Heidelberg 2006

A *minimal* comparability completion H of G is a comparability completion of G such that no proper subgraph of H is a comparability completion of G. Although the number of added edges in a minimal comparability completion may be far from minimum, computing a few different minimal comparability completions, and choosing the one with the smallest number of edges is a possible approach to finding a comparability completion close to minimum. Furthermore, the set of minimal comparability completions of a graph contains the set of minimum comparability completions. Therefore, the study of minimal comparability completions is a first step in the search for minimum comparability completions, possibly through methods like exact exponential time algorithms or parameterized algorithms. In this paper, we give the first polynomial time algorithm for computing minimal comparability completions of arbitrary graphs, and hence we show that this problem is solvable in polynomial time, as opposed to computing minimum comparability completions.

The study of minimal completions of arbitrary graphs into a given graph class started with a polynomial-time algorithm for minimal chordal completions in 1976 [13], before it was known that minimum chordal completions are NP-hard to compute [15]. Since then the NP-hardness of minimum completions has been established for several graph classes (summarized in [10]). Recently, several new results, some of which have been presented at recent years' SODA and ESA conferences, have been published on completion problems, leading to faster algorithms for minimal chordal completions [6,8,9], and polynomial-time algorithms for minimal completions into split, interval, and proper-interval graphs [3,5,11]. The complexity of computing minimal comparability completions has been open until now.

There are simple examples to show that a minimal comparability completion cannot be obtained by starting from an arbitrary comparability completion, and removing unnecessary edges one by one (as opposed to minimal completions into chordal and split graphs). To overcome this difficulty, we use a vertex incremental approach in our algorithm. A vertex incremental algorithm has also proved useful for minimal completions into interval graphs [5], and therefore we find it worthwhile to give a more general result here, describing classes of graphs into which minimal completions of arbitrary graphs can be computed with such a vertex incremental approach. Notice, however, that the algorithm for each step is completely different for, and dependent on, each graph class, and polynomial time computability is not guaranteed by the vertex incremental approach.

2 Notation and Background

We consider undirected finite graphs with no loops or multiple edges. For a graph G, we denote its vertex and edge set by $V(G)$ and $E(G)$, respectively, with $n = |V(G)|$ and $m = |E(G)|$. For a vertex subset $S \subseteq V(G)$, the subgraph of G induced by S is denoted by $G[S]$. Moreover, we denote by $G - S$ the graph $G[V(G) - S]$ and by $G - v$ the graph $G[V(G) - \{v\}]$.

The *neighborhood* $N_G(x)$ of a vertex x of the graph G is the set of all the vertices of G which are adjacent to x. The *closed neighborhood* of x is defined as $N_G[x] = N_G(x) \cup \{x\}$. If $S \subseteq V(G)$, then the neighbors of S, denoted by $N_G(S)$, are given by $\left(\bigcup_{x \in S} N_G(x) \right) - S$. For a vertex x of G, the set $N_G(N_G(x)) - \{x\}$ is denoted by $N_G^2(x)$. For a pair of vertices x, y of a graph G we call xy a *non-edge* of G if $xy \notin E(G)$. A vertex x of G is *universal* if $N_G[x] = V(G)$.

Given a new vertex $x \notin V(G)$ and a set of vertices N_x of G, we denote by G_x the graph obtained by adding x to G and making x adjacent to each vertex in N_x, i.e., $V(G_x) = V(G) \cup \{x\}$ and $E(G_x) = E(G) \cup \{xv \mid v \in N_x\}$; thus $N_{G_x}(x) = N_x$. For a vertex $x \notin V(G)$, we denote by $G + x$ the graph obtained by adding an edge between x and *every* vertex of $V(G)$, thus x is universal in $G + x$.

All the results presented here are new except those that contain references. Due to limited space, the proofs of most of the results of this extended abstract will be omitted. The proofs can be found in the full version of the paper [4].

2.1 Comparability Graphs

A *digraph* is a directed graph, and an *arc* is a directed edge. While we denote an undirected edge between vertices a and b equivalent by ab or ba, we denote an arc from a to b by (a, b), and an arc in the opposite direction by (b, a). A directed acyclic graph (*dag*) is *transitive* if, whenever (a, b) and (b, c) are arcs of the dag, (a, c) is also an arc. An undirected graph is a *comparability* graph if directions can be assigned to its edges so that the resulting digraph is a transitive dag, in which case this assignment is called a *transitive orientation*.

We consider an undirected graph G to be a symmetric digraph, that is, if $xy \in E(G)$ then (x, y) and (y, x) are arcs of G. Two arcs (a, b) and (b, c) of an undirected graph G are called *incompatible* if ac is not an edge of G. We say, then, that (a, b) is incompatible with (b, c) and vice versa, or that $((a, b), (b, c))$ is an incompatible pair. The *incompatibility graph* B_G of an undirected graph G is defined as follows: In B_G there is one vertex for each arc of G, and therefore we will (somewhat abusively) denote a vertex of B_G that corresponds to arc (a, b) of G by (a, b). For each edge ab of G, there are two adjacent vertices (a, b) and (b, a) in B_G. In addition, there is an edge between two vertices (a, b) and (b, c) of B_G if and only if arcs (a, b) and (b, c) are incompatible in G. We will refer to the edges of B_G of this latter type as *incompatibilities*. Since we consider an undirected graph to be a symmetric digraph, if $(a, b)(b, c)$ is an edge (incompatibility) of B_G then $(c, b)(b, a)$ is also an edge (incompatibility) of B_G. An example of a graph G and its incompatibility graph B_G is given in Figure 1.

A graph is *bipartite* if its vertex set can be partitioned into two independent sets. Bipartite graphs are exactly the class of graphs that do not contain cycles of odd length. The incompatibility graph will be our main tool to compute minimal comparability completions, and the following result from Kratsch et al. [7] is central to our algorithm.

Theorem 1 ([7]). *An undirected graph G is a comparability graph if and only if its incompatibility graph B_G is bipartite.*

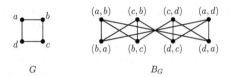

Fig. 1. A graph G and its incompatibility graph B_G

2.2   A Vertex Incremental Approach for Minimal Completions

A comparability graph can be obtained from any graph G by adding edges, and the resulting graph is called a *comparability completion* of G. An edge that is added to G to obtain a comparability completion H is called a *fill edge*. A comparability completion $H = (V, E \cup F)$ of $G = (V, E)$, with $E \cap F = \emptyset$, is *minimal* if $(V, E \cup F')$ fails to be a comparability graph for every $F' \subset F$. We will now show that minimal comparability completions can be obtained vertex incrementally.In fact, we give a more general result here, describing graph classes into which minimal completions of arbitrary graphs can be computed by a vertex incremental approach. A graph class Π is called *hereditary* if all induced subgraphs of graphs in Π also belong to Π.

Property 1. We will say that a graph class Π has the *universal vertex property* if, for every graph $G \in \Pi$ and a vertex $x \notin V(G)$, $G + x \in \Pi$.

Lemma 1. *Let H be a minimal Π completion of an arbitrary graph G, and let G_x be a graph obtained from G by adding a new vertex x adjacent to some vertices of G. If Π is hereditary and has the universal vertex property, then there is a minimal Π completion H' of G_x such that $H' - x = H$.*

An important consequence of Lemma 1 is that for a hereditary graph class Π with the universal vertex property, a minimal Π completion of any input graph G can be computed by introducing the vertices of G in an arbitrary order $x_1, x_2, \ldots, x_n$. Given a minimal Π completion H_i of $G_i = G[x_1, \ldots, x_i]$, we compute a minimal Π completion of $G_{i+1} = G[x_1, \ldots, x_i, x_{i+1}]$ by actually computing a minimal Π completion of the graph $H_{x_{i+1}} = (\{x_1, \ldots, x_{i+1}\}, E(H_i) \cup \{x_{i+1}v \mid v \in N_{G_{i+1}}(x_{i+1})\})$. In this completion, we add *only* fill edges incident to x_{i+1}. Meanwhile, notice that this minimal completion is not necessarily easy to obtain, and some major challenges might need to be overcome, depending on the graph class Π.

Observation 1. *The class of comparability graphs is hereditary and satisfies the universal vertex property.*

The real challenge is how to do the computations of each vertex incremental step. This is exactly the problem that we solve in the rest of this paper. Thus for the rest of the paper, due to Lemma 1 and Observation 1, we consider as input a comparability graph G and a new vertex $x \notin V(G)$ together with a list of vertices N_x in G. Our aim is to compute a minimal comparability completion of

$G_x = (V(G) \cup \{x\}, E(G) \cup \{xv \mid v \in N_x\})$. We do this by finding an appropriate set of fill edges F_x incident to x such that we obtain a comparability graph by adding F_x to G_x, and no proper subset F_x yields a comparability graph when added to G_x.

3   An Algorithm for Minimal Comparability Completion of G_x

In this section, we give an algorithm that computes a minimal comparability completion H of G_x, for a given comparability graph G and a new vertex $x \notin V(G)$ together with a neighborhood N_x in G. Our main tool will be the incompatibility graph B_G of G, which we know is bipartite by Theorem 1. We will proceed to update B_G with the aim of obtaining the incompatibility graph B_{G_x} of G_x. We will keep this partial incompatibility graph a bipartite graph at each step. If G_x is not a comparability graph, we will have to add fill edges to G_x to be able to achieve this goal.

Let $E_x = \{xv \mid v \in N_x\}$ (thus $G_x = (V \cup \{x\}, E \cup E_x)$). Our first step in obtaining B_{G_x} from B_G is to add vertices corresponding to edges of E_x and the edges and incompatibilities between these. We will make a separate graph B_x to represent the incompatibilities among the edges of E_x. Let B_x be the graph that has two adjacent vertices (x, v) and (v, x) for each $xv \in E_x$, and that has all incompatibilities that are implied by non-edges of G_x between vertices of N_x. To be more precise, if $\mathcal{E} = \{(x, v) \mid xv \in E_x\} \cup \{(v, x) \mid xv \in E_x\}$, and $B_{G_x[N_x \cup \{x\}]}$ is the incompatibility graph of $G_x[N_x \cup \{x\}]$, then B_x is the subgraph of $B_{G_x[N_x \cup \{x\}]}$ induced by $\mathcal{E}$. An example is given in Figure 2. Observe that the graph $G_x[N_x \cup \{x\}]$ is a comparability graph, since $G[N_x]$ is comparability by the hereditary property, and x is a universal vertex in $G_x[N_x \cup \{x\}]$. Following the above arguments, B_x is a bipartite graph by Theorem 1.

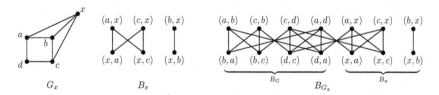

Fig. 2. An example that shows G_x, B_x, and B_{G_x}, for the graph G given in Figure 1

For our purposes, we also need to define the set of incompatibilities of B_G implied by a given non-edge xv of G. We call this set $C_G(xv)$, and define it as follows for each non-edge xv of G.

$$C_G(xv) = \{(x, w)(w, v) \mid w \in N_G(x) \cap N_G(v)\} \cup \{(v, w)(w, x) \mid w \in N_G(x) \cap N_G(v)\}$$

Observe that $C_G(e_1) \cap C_G(e_2) = \emptyset$ for any pair of non-edges e_1 and e_2 of G, and $\bigcup_{e \notin E(G)} C_G(e)$ is exactly the set of all incompatibilities in B_G.

Lemma 2. *By adding the set of edges* $C_{G_x}(xv)$ *for each* $v \in N^2_{G_x}(x)$ *into the graph* $B_G \cup B_x$, *we obtain the incompatibility graph* B_{G_x} *of* G_x.

Assume that we want to compute the incompatibility graph B_{G_x} of G_x. We start with the partial incompatibility graph $B_G \cup B_x$, which is bipartite by the above arguments. By Lemma 2, to get B_{G_x} it is sufficient to scan all non-edges of G_x between x and $N^2_{G_x}(x)$ one by one, and add the incompatibilities that are implied by each non-edge into the partial incompatibility graph. If G_x is a comparability graph, then by Theorem 1, the partial incompatibility graph will stay bipartite at each step, since we never delete edges from it. By the same argument, if G_x is not a comparability graph, then at some step, when we add the incompatibilities implied by a non-edge, we will get an odd cycle in the partial incompatibility graph. For computing a minimal comparability completion H of G_x, we augment this approach as follows: If adding the incompatibilities implied by non-edge xv results in a non-bipartite partial incompatibility graph, then we do not add these incompatibilities, and instead, we decide that xv should become a fill edge of H.

At start, we let $L = \{xv \mid v \in N^2_{G_x}(x)\}$, $B = B_G \cup B_x$, and $H = G_x$. For each non-edge $xv \in L$, we check whether or not non-edge xv should become a fill edge of the intermediate graph H, using the information given by $C_H(xv)$ and B. If $B \cup C_H(xv)$ is a bipartite graph, then we update $B = B \cup C_H(xv)$ and decide that xv *will never become a fill edge*. In the opposite case, we add *fill edge* xv to H, and update B as follows.

1. Add the two adjacent vertices (x, v) and (v, x) in B.
2. For each new incompatible pair $((z, x), (x, v))$ or $((v, x), (x, z))$ in H, add the corresponding edge (incompatibility) to B connecting the vertices of the pair. (We will show that this can never introduce odd cycles in the incompatibility graph.)
3. For each new incompatible pair $((x, v), (v, u))$ or $((u, v), (v, x))$ in H, add the corresponding edge (incompatibility) to B connecting the vertices of the pair *only if* xu *is a non-edge that has already been processed and decided to stay a non-edge (marked)*. If not, either $xu \in L$ or we add it to L.

The second case takes care of new incompatibilities among the edges incident to x, and the last case takes care of all other new incompatibilities. In the last case, when we encounter new incompatibilities that are implied by a non-edge e which we have not yet processed, we do not add these incompatibilities to B at this step, and we wait until we come to the step which processes e. The reason for this is the following: If we add these incompatibilities now, and later decide that e should become a fill edge, then we have to delete these incompatibilities from B. This causes problems regarding minimality, because deleting "old" incompatibilities can make some previously added fill edges become redundant, and thus we might have to examine each initial non-edge several times. When we do not add the incompatibilities before they are needed, we never have to delete anything from B, and B can only grow at each step. This way, the intermediate graph B will at all steps be a supergraph of $B_G \cup B_x$ and a subgraph of B_H.

This is the clue to the simplicity of our algorithm, which makes it sufficient to examine each non-edge incident to x *once*.

The non-edges that are removed from L are marked, which means that they will stay non-edges. This marking is necessary since new non-edges enter L during the algorithm, and we need to test for every incompatibility we discover, whether it is already implied by a marked non-edge so that we can add it at this step, or we should wait.

Algorithm: Minimal_Comparability_Completion (MCC)

Input: A comparability graph G, B_G, and G_x for a vertex $x \notin V(G)$
Output: A minimal comparability completion H of G_x, and $B = B_H$

1 $B = B_G \cup B_x$; $L = \{xv \mid v \in N^2_{G_x}(x)\}$; $H = G_x$;
2 Unmark all non-edges of H incident to x;
3 **while** $L \neq \emptyset$ **do**
4   Choose a non-edge $xv \in L$;
5   **if** $B \cup C_H(xv)$ is a bipartite graph **then**
6         $B = B \cup C_H(xv)$;
7   **else**
8   Add fill edge xv to H;
9   Add vertices (x, v) and (v, x) and an edge between them to B;
10   **forall** $z \in N_H(x)$ and $z \notin N_H[v]$ **do**
11   Add edges $(v, x)(x, z)$ and $(z, x)(x, v)$ to B;
12   **forall** $u \in N_H(v)$ and $u \notin N_H[x]$ **do**
13   **if** xu is marked **then**
14   Add edges $(x, v)(v, u)$ and $(u, v)(v, x)$ to B;
15   **else if** $xu \notin L$ **then**
16   Add xu to L;
17   Mark xv and remove it from L;

---

4   Correctness of Algorithm MCC

Although our algorithm is surprisingly simple due to the fact that each non-edge is examined once, its proof of correctness is quite involved, and requires a series of observations and lemmas, some of which with long proofs. Let us define a *step* of the algorithm to be one iteration of the while–loop given between lines 3–17. For the proof of correctness, we will sometimes need to distinguish between the graph H at the start of a step and the updated graph H at the end of a step, to consider the changes made at one step. Throughout the rest of the paper, let H_I be the graph H at the start of step I, and let H_{I+1} be the graph obtained at the end of this step, and define B_I and B_{I+1} analogously.

Observation 2. *Let I be the step of the algorithm that processes the non-edge $xv \in L$. Then B_I contains no edge belonging to $C_{H_I}(xv)$.*

Lemma 3. *At the end of each step of the algorithm, B_I is a subgraph of the incompatibility graph B_{H_I} of H_I.*

We have thus proved that B_I is at all times a partial incompatibility graph of the intermediate graph H_I. At the end of the algorithm, since all non-edges that can cause incompatibilities are scanned, and all such incompatibilities are added, we will argue that B_I is indeed the correct incompatibility graph of H_I. What remains to prove is that B_I is a bipartite graph at all steps. This is obvious if xv is not added as a fill edge at the step that processes xv, but it has to be shown in the case xv is added as a fill edge. First we introduce the notion of conflicts.

Definition 1. *At each step of the algorithm, a non-edge xv of the intermediate graph H_I is called a* conflict *if $B \cup C_{H_I}(xv)$ is not a bipartite graph.*

Lemma 4. *Let I be the step of the algorithm that processes non-edge $xv \in L$. If xv is a conflict then H_I is not a comparability graph.*

Now we start the series of results necessary to prove that at each step B_I is a bipartite graph. We will prove this by induction on the number of steps. For each step I, we will assume that B_I is bipartite, and show that this implies that B_{I+1} is bipartite. Since $B_1 = B_x \cup B_G$ is bipartite, the result will follow.

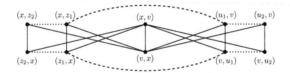

Fig. 3. Adding the fill edge xv in B

Let z_1, z_2 and u_1, u_2 be vertices of H_I which fulfill the conditions of the first for–loop and the second for–loop of Algorithm MCC, respectively. With the following result we establish the situations that occur in B_I whenever an odd cycle appears in B_{I+1} (see also Figure 3).

Observation 3. *Assume that B_I is bipartite. If xv is conflict at step I, then B_{I+1} is not bipartite only if there is a path on even number of vertices in B_I between the following pair of vertices: (i) $((x, z_1), (x, z_2))$ or (ii) $((v, u_1), (v, u_2))$ or (iii) $((x, z_1), (u_1, v))$.*

Our goal is to show that these cases cannot happen in B_I, and therefore B_{I+1} remains a bipartite graph. We prove each case by showing that if such a path exists then there is an odd cycle in B_I which is a contradiction to our assumption that B_I is a bipartite graph. For a complete proof of the case analysis we refer to [4].

Lemma 5. *At each step of the algorithm B_I is a bipartite graph.*

Theorem 2. *The graph H returned by Algorithm MCC is a minimal comparability completion of G_x.*

Proof. First we show that H is a comparability completion of G_x. During the algorithm, every time a new incompatible pair is created, the corresponding incompatibility is added to B_I unless it is implied by a non-edge of L. Incompatibilities implied by members of L that remain non-edges are added one by one until L is empty. At the end of the algorithm, the graph B contains all incompatibilities implied by the non-edges in H, since $L = \emptyset$. Thus B is the correct incompatibility graph of H, i.e., $B = B_H$. Since B_H is bipartite graph by Lemma 5, the resulting graph H is a comparability graph by Theorem 1.

Now we want to prove that H is minimal, that is, if any subset of the fill edges is removed the remaining graph is not comparability. Recall that at any step of the algorithm we do not remove any edges from the graph B_I (see also Lemma 3). Assume for the sake of contradiction that there is a subset F of the fill edges such that $H' = H - F$ is a comparability graph. First note that $B_{H'}$ is obtained from B_H by removing the vertices (x, u) and (u, x), and then adding the set $C_{H'}(xu)$, for every $xu \in F$. Let I be the earliest step in which Algorithm MCC adds a fill edge $xv \in F$. Thus no non-edge of G_x belonging to F has been processed before step I, and H_I is a subgraph of H'. Furthermore, B_I does not contain any edge belonging to $\bigcup_{xu \in F} C_{H'}(xu)$, and B_I does not contain any pair of vertices (x, u) and (u, x), for $xu \in F$. Thus B_I is a subgraph of $B_{H'}$. Now, observe that for each $xu \in F$, $C_{H_I}(xu) \subseteq C_{H'}(xu)$, since $N_{H_I}(x) \subseteq N_{H'}(x)$. In particular, $C_{H_I}(xv) \subseteq C_{H'}(xv)$. Since xv is a non-edge of H', all edges of $C_{H'}(xv)$ are present in $B_{H'}$. Therefore $B_I \cup C_{H_I}(xv)$ is a subgraph of $B_{H'}$. In Algorithm MCC, at step I, we know that $B_I \cup C_{H_I}(xv)$ contains an odd cycle, otherwise xv would not be a fill edge. Since it is not possible to remove an odd cycle by adding edges or vertices, this means that there is an odd cycle in $B_{H'}$. This gives the desired contradiction, because by Theorem 1 H' cannot be a comparability graph as assumed.

5   Time Complexity and Concluding Remarks

We point out that given an incompatible pair $((a, b)(b, c))$ of G there is an $O(n + m)$ time algorithm deciding whether its incompatibility graph has an odd cycle [7]. However, it is not straightforward to use this result for checking whether the graph $B_I \cup C_{H_I}(xv)$ of Algorithm MCC is bipartite in $O(n + m)$ time, since at each step of the algorithm, B_I is merely a subgraph of B_{H_I}, and B_I is not necessarily equal to B_{H_I} before the last step. The following result follows from Lemma 1 and Algorithm MCC.

Theorem 3. *There is an algorithm for computing a minimal comparability completion of an arbitrary graph G in $O(n^3 m)$ time.*

In this paper, we have shown that minimal comparability completions of arbitrary graphs can be computed in polynomial time. Our focus has been on the polynomial time complexity, and we believe that the running time of the given algorithm can be improved. Comparability graphs can be recognized in time $O(n^{2.38})$ [14], and even the straight forward $O(n^3 m)$ time complexity of our algorithm for computing minimal comparability completions is thus comparable to the time complexity of recognizing comparability graphs. As a comparison, both chordal and interval graphs can be recognized in linear time; the best known time for minimal chordal completions is $O(n^{2.38})$ [6], and for minimal interval completions is $O(n^5)$ [5].

References

1. M. C. Golumbic. *Algorithmic Graph Theory and Perfect Graphs.* Academic Press, 1980.
2. S. L. Hakimi, E. F. Schmeichel, and N. E. Young. Orienting graphs to optimize reachability. *Information Processing Letters*, 63(5):229–235, 1997.
3. P. Heggernes and F. Mancini. Minimal split completions of graphs. In *LATIN 2006: Theoretical Informatics*, pages 592–604. Springer Verlag, 2006. LNCS 3887.
4. P. Heggernes, F. Mancini, and C. Papadopoulos. Minimal comparabiltiy completions. Reports in Informatics 317, University of Bergen, 2006. http://www.ii.uib.no/publikasjoner/texrap/pdf/2006-317.pdf
5. P. Heggernes, K. Suchan, I. Todinca, and Y. Villanger. Minimal interval completions. In *Algorithms - ESA 2005*, pages 403 – 414. Springer Verlag, 2005. LNCS 3669.
6. P. Heggernes, J. A. Telle, and Y. Villanger. Computing minimal triangulations in time $O(n^\alpha \log n) = o(n^{2.376})$. In *Proceedings of SODA 2005 - 16th Annual ACM-SIAM Symposium on Discrete Algorithms*, pages 907–916, 2005.
7. D. Kratsch, R. M. McConnell, K. Mehlhorn, and J. P. Spinrad. Certifying algorithms for recognizing interval graphs and permutation graphs. *SIAM J. Comput.*, 2006. To appear.
8. D. Kratsch and J. P. Spinrad. Between $O(nm)$ and $O(n^\alpha)$. In *Proceedings of SODA 2003 - 14th Annual ACM-SIAM Symposium on Discrete Algorithms*, pages 709–716, 2003.
9. D. Kratsch and J. P. Spinrad. Minimal fill in $o(n^3)$ time. *Discrete Math.*, 306(3):366–371, 2006.
10. A. Natanzon, R. Shamir, and R. Sharan. Complexity classification of some edge modification problems. *Disc. Appl. Math.*, 113:109–128, 2001.
11. I. Rapaport, K. Suchan, and I. Todinca. Minimal proper interval completions. In *Proceedings of WG 2006 - 32nd International Workshop on Graph-Theoretic Concepts in Computer Science*, 2006. To appear.
12. F. S. Roberts. *Graph Theory and Its Application to Problems of Society.* Society for Industrial and Applied Mathematics, Philadelphia, PA, 1978.
13. D. Rose, R.E. Tarjan, and G. Lueker. Algorithmic aspects of vertex elimination on graphs. *SIAM J. Comput.*, 5:266 – 283, 1976.
14. J. Spinrad. On comparability and permutation graphs. *SIAM J. Comput.*, 14:658 – 670, 1985.
15. M. Yannakakis. Computing the minimum fill-in is NP-complete. *SIAM J. Alg. Disc. Meth.*, 2:77–79, 1981.

Analyzing Disturbed Diffusion on Networks*

Henning Meyerhenke and Thomas Sauerwald

Universität Paderborn
Fakultät für Elektrotechnik, Informatik und Mathematik
Fürstenallee 11, D-33102 Paderborn, Germany
{henningm, sauerwal}@upb.de

Abstract. This work provides the first detailed investigation of the disturbed diffusion scheme FOS/C introduced in [17] as a type of diffusion distance measure within a graph partitioning framework related to Lloyd's k-means algorithm [14]. After outlining connections to distance measures proposed in machine learning, we show that FOS/C can be related to random walks despite its disturbance. Its convergence properties regarding load distribution and edge flow characterization are examined on two different graph classes, namely torus graphs and distance-transitive graphs (including hypercubes), representatives of which are frequently used as interconnection networks.

Keywords: Disturbed diffusion, Diffusion distance, Random walks.

1 Introduction

Diffusive processes can be used to model a large variety of important transport phenomena arising in such diverse areas as heat flow, particle motion, and the spread of diseases. In computer science one has studied diffusion in graphs as one of the major tools for balancing the load in parallel computations [5], because it requires only local communication between neighboring processors. Equally important, the migrating flow computed by diffusion is $\| \cdot \|_2$-optimal [6].

Recently, disturbed diffusion schemes have been developed as part of a graph partitioning heuristic [17, 20]. Applied within a learning framework optimizing the shape of the partitions, disturbed diffusion is responsible for identifying densely connected regions in the graph. As partitions are placed such that their centers are located within these dense regions, this heuristic yields partitions with few boundary nodes. This is desirable particularly in scientific computing, where the boundary nodes model the communication within parallel numerical solvers. The disturbed diffusion scheme and the algorithm employing it are described in more detail in Sections 2.1 and 2.2, respectively.

---

* This work is partially supported by German Science Foundation (DFG) Research Training Group GK-693 of the Paderborn Institute for Scientific Computation (PaSCo) and by Integrated Project IST-15964 "Algorithmic Principles for Building Efficient Overlay Computers" (AEOLUS) of the European Union.

T. Asano (Ed.): ISAAC 2006, LNCS 4288, pp. 429–438, 2006.

While the connection between diffusion and random walks on graphs is well-known (see, e.g., [15]), the relation of disturbed diffusion the way considered here to random walks has not been explored yet. In Section 3 we thus show that random walk analysis can be applied despite the disturbance in the diffusion scheme. Before that, we draw connections to machine learning applications that employ distance measures based on diffusion and random walks in Section 2.3.

Using random walk theory, we analyze the load distribution and the edge flow induced by FOS/C in its convergence state on torus graphs in Section 4. Although random walks on infinite and finite tori have been investigated before (cf. Pólya's problem [7] and [8]), our monotonicity result on the torus provides an important theoretical property, to the best of our knowledge previously unknown. It is of high relevance for the graph partitioning heuristic, since the torus corresponds to structured grids that stem from the discretization of numerical simulation domains with cyclic boundary conditions. A simple characterization of the convergence flow by shortest paths is shown not to hold on the two-dimensional torus in general. Interestingly, this characterization is true on distance-transitive graphs, as shown in Section 5, which is supplemented by a more rigorous result for the hypercube, a very important representative of distance-transitive graphs.

These insights provide a better understanding of FOS/C, its properties within the graph partitioning framework, and its connection to similar distance measures in machine learning. They are expected to improve the partitioning heuristic in theory and practice and its applicability for graph clustering and related applications.

2   Disturbed Diffusion, Graph Partitioning, and Diffusion Distances

2.1   Disturbed Diffusion: FOS/C

Diffusion is one method for iteratively balancing the load on vertices of a graph by performing load exchanges only between neighboring vertices [26]. The idea of the FOS/C algorithm (C for constant drain) is to modify the first order diffusion scheme FOS [5] by letting some of the load on each vertex drain away after each diffusion step. The total drain of the graph is then sent back to some specified source vertex s, before the next iteration begins. For this algorithm the underlying graph has to be connected, undirected, and loop-free. Additionally, we assume it to be unweighted and simple throughout this paper.

Definition 1. *(comp. [17]) Given a graph $G = (V, E)$ with n nodes, a specified source vertex s, and constants $0 < \alpha \le (\deg(G) + 1)^{-1}$ and $\delta > 0$. Let the initial load vector $w^{(0)}$ and the disturbing drain vector d be defined as follows:*

$$w_v^{(0)} = \begin{cases} n & v = s \\ 0 & \text{otherwise} \end{cases} \qquad d_v = \begin{cases} \delta(n-1) & v = s \\ -\delta & \text{otherwise} \end{cases}$$

The FOS/C diffusion scheme performs the following operations in each iteration:

$$f_{e=(u,v)}^{(t)} = \alpha(w_u^{(t)} - w_v^{(t)}), \qquad w_v^{(t+1)} = w_v^{(t)} + d_v + \sum_{e=(*,v)} f_e^{(t)}.$$

This can be written in matrix-vector notation as $w^{(t+1)} = \mathbf{M}w^{(t)} + d$, where $\mathbf{M} = \mathbf{I} - \alpha\mathbf{L}$ is the stochastic diffusion matrix of G (and $\mathbf{L}$ its Laplacian [10]). It is shown in [17] that FOS/C converges for any d that preserves the total load amount in every iteration, i.e., $d \perp (1, \ldots, 1)^T$. Moreover, in this case the convergence load vector w can be computed by first solving the linear system $\mathbf{L}w = d$ and then normalizing w such that the total load is n again. The entries of this vector can then be interpreted as the diffusion distances between s and the other nodes.

Comparing this notation to [6] and [11], it is clear that the convergence state of FOS/C is equivalent to the following flow problem: Find the $\|\cdot\|_2$-minimal flow from the producing source s sending the respective load amount δ to all other vertices in the graph, which act as δ-consuming sinks. One therefore knows that s always has the highest load.

Remark 1. Using Lemma 4 of [6], it follows that $f = \mathbf{A}^T w$ (with $\mathbf{A}$ being the incidence matrix of G [10, p. 58]) is the $\|\cdot\|_2$-minimal flow via the edges of G induced by the flow problem equivalent to FOS/C. Hence: $f_{e=(u,v)} = w_u - w_v$.

Furthermore, it holds for every path between any $u, v \in V$ that the sum of the load differences $\sum_{i=0}^{l-1}(w_{v_i} - w_{v_{i+1}})$ on the path edges $e_i = (v_i, v_{i+1})$ is equal to $w_u - w_v$ (with $u = v_0$ and $v = v_l$).

The following proposition states a basic monotonicity result that holds on any graph, whereas stricter results will be proven in the forthcoming sections.

Proposition 1. *Let the graph $G = (V, E)$ and the load vector w be given. Then for each vertex $v \in V$ there is a path $(v = v_0, v_1, \ldots, v_l = s)$ with $(v_i, v_{i+1}) \in E$ such that $w_{v_i} < w_{v_{i+1}}, 0 \le i < l$.*

2.2   FOS/C for Graph Partitioning

The graph partitioning framework in which FOS/C is applied transfers Lloyd's algorithm [14] well-known from k-means-type cluster analysis and least square quantization to graphs. Starting with k (the number of partitions) randomly chosen center vertices, all remaining nodes are assigned to the closest center based on FOS/C. This means that we solve one FOS/C diffusion problem per partition (its center acts as source s) and assign each vertex to the partition which sends the highest load in the convergence state. After that, each partition computes its new center (based on a similar FOS/C problem again) for the next iteration. This can be repeated until a stable state, where the movement of all centers is small enough, is reached. For a detailed discussion of this iterative algorithm called Bubble-FOS/C the reader is referred to [17].

2.3   Diffusion Distances in Graphs

One can view FOS/C as a means to determine the distance from each vertex to the different center vertices within Bubble-FOS/C (hence, ordinary FOS is not applicable, because it converges to a completely balanced load situation), where this distance reflects how well-connected the two vertices are (comp. [19] and [7, p. 99f.]). Thus, it is able to identify dense regions of the graph. A similar idea is pursued by other works that make use of distance measures based on random walks and diffusion. They have mostly been developed for machine learning, namely, clustering of point sets and graphs [18, 19, 23, 25, 27], image segmentation [16], and dimensionality reduction [4]. However, their approaches rely on very expensive matrix operations, amongst others computation of matrix powers [23, 25], eigenvectors of a kernel matrix [4, 16, 18], or the pseudoinverse of the graph's Laplacian [19, 27]. This mostly aims at providing a distance between every pair of nodes.

   Yet, this is not necessary for Lloyd's algorithm, because distance computations are relative to the current centers and the determination of the new partition centers can also be replaced by a slightly modified FOS/C operation, as mentioned above. The sparse linear system $\mathbf{L}w = d$, where w can be seen as the result of the pseudoinverse's impact on the drain vector, can be solved with $\mathcal{O}(n^{3/2})$ and $\mathcal{O}(n^{4/3})$ operations for typical $2D$ and $3D$ finite-element graphs, respectively, using the conjugate gradient algorithm [21]. This can even be enhanced by (algebraic) multigrid methods [24], which have linear time complexity when implemented with care. Note that only a constant number of calls to FOS/C are sufficient in practice. Thus, this approach is faster (unless distances between every pair of nodes are necessary in a different setting) than the related methods, which all require at least $\mathcal{O}(n^2)$ operations in the general case.

3   Relating FOS/C to Random Walks

In order to examine the relationship between disturbed diffusion and random walks, we expand the original definition of FOS/C and obtain

$$w^{(t+1)} = \mathbf{M}^{t+1}w^{(0)} + (\mathbf{I} + \mathbf{M}^1 + \ldots + \mathbf{M}^t)d.$$

Note that the doubly stochastic diffusion matrix $\mathbf{M}$ of G in the classical FOS diffusion scheme can be viewed as the transition matrix of a random walk [15] on $V(G)$, i.e., $\mathbf{M}_{u,v}$ denotes the probability for a random walker located in node u to move to node v in the next timestep. Despite its disturbance, a similar connection holds for FOS/C, since its load differences in the convergence state (a.k.a. *stationary distribution* in random walk theory) can be expressed as scaled differences of hitting times, as shown below. In the following let $X_u^{(t)}$ be the random variable representing the node visited in timestep t by a random walker starting in u in timestep 0.

Definition 2. *Let the balanced distribution vector be* $\pi = (\frac{1}{n}, \ldots, \frac{1}{n})^T$ *and let* τ_u *be defined as* $\tau_u := \min\{t \geq 0 : X_u^{(t)} = s\}$ *for any* $u \in V$. *Then, the hitting time* H *is defined as* $H[u, s] := \mathbb{E}[\tau_u]$.

Theorem 1. *In the convergence state it holds for two nodes* $u, v \in V$ *not necessarily distinct from* s

$$w_u - w_v = \lim_{t \to \infty} n\delta \left(\sum_{i=0}^{t} \mathbf{M}_{u,s}^i - \sum_{i=0}^{t} \mathbf{M}_{v,s}^i \right) = \delta(H[v, s] - H[u, s]).$$

Proof. We denote the component corresponding to node u in a vector w by $[w]_u$ and assume that the nodes are ordered in such a way that the source node is the first one. Then some rearranging of the FOS/C iteration scheme yields

$$[w^{(t+1)}]_u = [\mathbf{M}^{t+1} w^{(0)}]_u + [(\mathbf{I} + \mathbf{M}^1 + \ldots + \mathbf{M}^t) \cdot (\delta(n-1), -\delta, \ldots, -\delta)^T]_u$$

$$= [\mathbf{M}^{t+1} w^{(0)}]_u + \sum\nolimits_{i=0}^{t} (\delta(|V|-1)) \mathbf{M}_{u,s}^i + \sum\nolimits_{i=0}^{t} \sum\nolimits_{v \in V, v \neq s} (-\delta) \mathbf{M}_{u,v}^i$$

$$= [\mathbf{M}^{t+1} w^{(0)}]_u + n\delta \sum\nolimits_{i=0}^{t} \mathbf{M}_{u,s}^i - (t+1)\delta.$$

As $\mathbf{M}^{t+1} w^{(0)}$ converges towards the balanced load distribution [5], we only have to consider $\lim_{t \to \infty} \sum_{i=0}^{t} (\mathbf{M}_{u,s}^i - \mathbf{M}_{v,s}^i)$. By a result of [12, p. 79] it holds that $H[u, s] = (-\sum_{k=1}^{\infty} \mathbf{M}_{u,s}^t + \sum_{k=1}^{\infty} (1/n) + Z_{s,s}) \cdot n$, where Z is the so-called fundamental matrix. Now, subtracting and dividing by n yields the desired result. $\qed$

4 FOS/C on the Torus

In this section we analyze two properties of FOS/C on torus graphs in the convergence state, namely, its edge flow and the corresponding load distribution.

Definition 3. *The* k-*dimensional torus* $T[d_1, \ldots, d_k] = (V, E)$ *is defined as:*

$$V = \{(u_1, \ldots, u_k) \mid 0 \leq u_\nu \leq d_\nu - 1 \text{ for } 1 \leq \nu \leq k\} \text{ and}$$
$$E = \{\{(u_1, \ldots, u_k), (v_1, \ldots, v_k)\} \mid \exists 1 \leq \mu \leq k$$
$$\text{with } v_\mu = (u_\mu + 1) \bmod d_\mu \text{ and } u_\nu = v_\nu \text{ for } \nu \neq \mu\}.$$

Torus graphs are very important in theory [13] and practice [22], e.g., because they have bounded degree, are regular and vertex-transitive[1], and correspond to the structure of numerical simulation problems that decompose their domain by structured grids with cyclic boundary conditions. Note that the load distribution on a torus and a grid graph are equal if their d_i are all odd and s is located at the center of the graphs, because then there is no flow via the wraparound edges of the torus.

---

[1] A graph $G = (V, E)$ is vertex-transitive if for any two distinct vertices of V there is an automorphism mapping one to the other.

Since the number of shortest paths from a source s to another vertex u does not depend on its distance to s alone, the following flow distribution among the shortest paths is not optimal on the torus in general. As we will see later, this optimality holds for graphs that are distance-transitive, an even stronger symmetry property than vertex-transitivity.

Definition 4. *Consider the flow problem where s sends a load amount of δ to every other vertex of G, which acts as a δ-consuming sink. If the flow is distributed such that for all $v \in V \backslash \{s\}$ the same flow amount is routed on every (not necessarily edge-disjoint) shortest path from s to v, we call this the* uniform flow distribution.

Proposition 2. *The uniform flow distribution on the 2D torus yields the $\| \cdot \|_2$-minimal flow for $d_1 = d_2 \in \{2, 3, 5\}$, but not for odd $d_1 = d_2 \geq 7$.*

Intuitively, the reason is that near the diagonal there are more shortest paths than on an axis and thus, by rerouting some of the uniform flow towards the diagonal, the costs can be reduced.

In the remainder of this section we exploit the simple structure and symmetries of the torus to show monotonicity w.r.t. the FOS/C convergence load distribution. Since we are only interested in the convergence state, we will set $\alpha = (\deg(G) + 1)^{-1}$, so that all entries of the diffusion matrix $\mathbf{M}$ are either 0 or α. This is a usual choice for transition matrices in random walk theory.

Now consider an arbitrary k-dimensional torus $T[d_1, \ldots, d_k]$. Each vertex u can be uniquely represented as a k-dimensional vector $u = (u_1, \ldots, u_k), \forall i \in 1, \ldots, k : 0 \leq u_i < d_i$. Since any torus is vertex-transitive, we assume w.l.o.g. that the source node is the zero-vector. Denote by $\mathbf{e}_i = (0, \ldots, 0, 1, 0, \ldots, 0)$ the unit-vector containing exactly one 1, namely in the i-th component. Note that all edges correspond to the addition (or subtraction) of some $\mathbf{e}_i$, where we always assume that the i-th component is meant to be modulo d_i. It is also easy to see that the distance between two nodes (vectors) is given by $\mathrm{dist}(u, v) = \sum_{i=1}^{k} \min\{|u_i - v_i|, d_i - |u_i - v_i|\}$.

Let u, v, s be pairwise distinct nodes such that $\mathrm{dist}(u, s) = \mathrm{dist}(v, s) - 1$ and u and v are adjacent, i.e., there exists a shortest path from s to v via u. Assume w.l.o.g. that u and v are adjacent along the j-th dimension: $v = u + \mathbf{e}_j$, so that

$$\forall i \in \{1, \ldots, k\}, i \neq j : \mathrm{dist}(v, s) - \mathrm{dist}(v, s \pm \mathbf{e}_i) = \mathrm{dist}(u, s) - \mathrm{dist}(u, s \pm \mathbf{e}_i),$$

implying the existence of a shortest path from $s \pm \mathbf{e}_i$ to v via $u \; \forall i \neq j$.

For vertex-transitive graphs G, all $\varphi \in \mathrm{Aut}(G)$, and all timesteps t we have $\mathbf{M}_{u,v}^t = \mathbf{M}_{\varphi(u), \varphi(v)}^t$ [2, p. 151]. Using this and the automorphisms of the next lemma, we prove the following theorem, which may be of independent interest for random walks in general.

Lemma 1. *The following functions are automorphisms for all $i \in \{1, \ldots, k\}$: $\psi_i : u \mapsto u + \mathbf{e}_i$, $\varphi_i : u \mapsto u + (d_i - 2u_i)\mathbf{e}_i$, and $\sigma_i : u \mapsto u + (d_i - 1 - 2u_i)\mathbf{e}_i$.*

Theorem 2. *Let* $T[d_1, \ldots, d_k] = (V, E)$, k *arbitrary, be a torus graph. For* $\alpha = (\deg(G) + 1)^{-1}$ *and all adjacent nodes* $u, v \in V$ *distinct from* s *with* $\operatorname{dist}(u, s) = \operatorname{dist}(v, s) - 1$ *it holds*

$$\forall t \in \mathbb{N}_0 : \mathbf{M}_{u,s}^t \geq \mathbf{M}_{v,s}^t.$$

Proof. We will prove the statement by induction on the number of timesteps t. Obviously, the claim is true for $t = 0$. By the Chapman-Kolmogorov equation, see e.g. [9], we have

$$\mathbf{M}_{u,s}^t = \frac{1}{\Delta + 1} \left(\mathbf{M}_{u,s}^{t-1} + \sum\nolimits_{i \in \{1, \ldots, k\}} \mathbf{M}_{u,s+\mathbf{e}_i}^{t-1} + \sum\nolimits_{i \in \{1, \ldots, k\}} \mathbf{M}_{u,s-\mathbf{e}_i}^{t-1} \right). \quad (1)$$

Obviously, the same equation holds also for $M_{v,s}^t$. Our strategy is now to find for any summand in $M_{v,s}^t$ a proper summand in $M_{u,s}^t$ which is not smaller by using the induction hypothesis for $t - 1$. Of course, if this is done bijectively, we have shown that $M_{u,s}^t \geq M_{v,s}^t$. To proceed, we divide this proof into two cases.

1. Case $u_j = 0$: By Lemma 1 we have

$$\mathbf{M}_{u,s}^{t-1} = \mathbf{M}_{\psi_j(u), \psi_j(s)}^{t-1} = \mathbf{M}_{v,s+\mathbf{e}_j}^{t-1}.$$

$$\mathbf{M}_{u,s+\mathbf{e}_j}^{t-1} = \mathbf{M}_{\varphi_j(u), \varphi_j(s+\mathbf{e}_j)}^{t-1} = \mathbf{M}_{u,s-\mathbf{e}_j}^{t-1} = \mathbf{M}_{\psi_j(u), \psi_j(s-\mathbf{e}_j)}^{t-1} = \mathbf{M}_{v,s}^{t-1}.$$

   To show $\mathbf{M}_{u,s-\mathbf{e}_j}^{t-1} \geq \mathbf{M}_{v,s-\mathbf{e}_j}^{t-1}$, we have to distinguish the following cases:

   (a) Ignoring the trivial case $d_j = 2$, we now consider the case where $d_j = 3$:

   $$\mathbf{M}_{u,s-\mathbf{e}_j}^{t-1} = \mathbf{M}_{u,s+\mathbf{e}_j}^{t-1} \overset{\psi_j^{-1}}{=} \mathbf{M}_{u-2\mathbf{e}_j, s-\mathbf{e}_j}^{t-1} = \mathbf{M}_{v,s-\mathbf{e}_j}^{t-1}.$$

   (b) $d_j \geq 4$: Then, $\operatorname{dist}(v, s - \mathbf{e}_j) = \operatorname{dist}(v, s) + 1$, implying the existence of a shortest path from v to $s - \mathbf{e}_j$ via u. Due to vertex-transitivity there exists an automorphism which maps $s - \mathbf{e}_j$ onto s and we can apply the induction hypothesis to conclude $\mathbf{M}_{u,s-\mathbf{e}_j}^{t-1} \geq \mathbf{M}_{v,s-\mathbf{e}_j}^{t-1}$.

   Recall that for all $i \in \{1, \ldots, k\}, i \neq j$, there exists a shortest path from v to $s \pm \mathbf{e}_i$ via u, so that we can again conclude inductively that $\mathbf{M}_{u,s\pm\mathbf{e}_i}^{t-1} \geq \mathbf{M}_{v,s\pm\mathbf{e}_i}^{t-1}$. With Equation (1) and its analogon for v the claim $\mathbf{M}_{u,s}^t \geq \mathbf{M}_{v,s}^t$ follows.

2. Case $u_j \neq 0$: One distinguishes two subcases by the parity of u_j and uses similar methods as before to prove this case. It is therefore omitted due to space constraints. □

Note that one can show with a modified three-dimensional hypercube as a counterexample that this monotonicity does not hold for all vertex-transitive graphs in all timesteps. Furthermore, the general result $\mathbf{M}_{u,u}^{2t} \geq \mathbf{M}_{u,v}^{2t}$ for random walks without loops on vertex-transitive graphs can be found in [2, p. 150], which is improved significantly by our last theorem on torus graphs. As one can prove by induction, on the torus the source vertex is the unique node with the highest load in all timesteps due to the choice of α and the back-flow of the drain. Thus, by combining Theorems 1 and 2, one can derive the following corollary for any pair of vertices.

Corollary 1. *On any torus graph $T = (V, E)$ it holds for all $u, v \in V$: $\forall t < \text{dist}(u, s) : w_u^{(t)} = w_v^{(t)}$, $\forall t \in \{\text{dist}(u, s), \ldots, \infty\} : w_u^{(t)} > w_v^{(t)}$.*

Using this monotonicity and the symmetry properties of the torus, it is easy (but rather technical) to show that Bubble-FOS/C produces connected partitions on this graph class, which is desirable in some applications.

5 FOS/C on Distance-Transitive Graphs

We have seen that the convergence flow does not equal the uniform flow distribution on the torus, despite its symmetry. Yet, in this section we show that this equality holds if the symmetry is extended to distance-transitivity.

Definition 5. *[3, p. 118] A graph $G = (V, E)$ is distance-transitive if, for all vertices $u, v, x, y \in V$ such that $\text{dist}(u, v) = \text{dist}(x, y)$, there exists an automorphism φ for which $\varphi(u) = x$ and $\varphi(v) = y$.*

One important subclass of distance-transitive graphs are Hamming graphs, which occur frequently in coding theory [1, p. 46]. A very well-known representative is the hypercube network [13]. It is not difficult to show that distance-transitive graphs $G = (V, E)$ have a level structure w.r.t. to an arbitrary $s \in V$, where level i consists of the vertex set $L_i := \{v \in V \mid \text{dist}(v, s) = i\}$ and Λ denotes the number of such levels. For the k-dimensional hypercube $Q(k)$, for instance, we have $\Lambda = k + 1$.

Now, the results of this section can be derived by means of this level structure and the aforementioned equivalence of FOS/C to a $\| \cdot \|_2$-minimal flow problem.

Proposition 3. *Let G be a distance-transitive graph. Then, $w_u^{(t)} = w_v^{(t)}$ holds for all vertices u, v with the same graph distance to s and all timesteps $t \geq 0$.*

We know by Proposition 1 that for each vertex $v \in V \backslash \{s\}$ of an arbitrary graph there exists a path from v to s such that by traversing it the load amount increases. Now we can show that for distance-transitive graphs this property holds on *every shortest* path.

Theorem 3. *If G is distance-transitive, then for all $u, v \in V$ with $\text{dist}(u, s) < \text{dist}(v, s)$ it holds that $w_u > w_v$.*

Note that, although the order induced by the FOS/C diffusion distance corresponds to the one induced by the ordinary graph distance, the load differences across levels reflect their connectivity (see also Theorem 5). We now state the following characterization of the convergence flow.

Theorem 4. *The uniform flow distribution of Definition 4 yields the $\| \cdot \|_2$-minimal FOS/C convergence flow on every distance-transitive graph.*

As this is not true for general tori, the following implication is not an equivalence.

Proposition 4. *If on a graph $G = (V, E)$ the uniform flow distribution is $\| \cdot \|_2$-minimal, then for $(u, v) \in E$ and $\mathrm{dist}(u, s) < \mathrm{dist}(v, s)$ it holds that $w_u > w_v$.*

Due to the explicitly known structure of the hypercube we obtain:

Theorem 5. *For the k-dimensional hypercube $Q(k) = (V, E)$ the result of Theorem 3 holds in all timesteps $t \geq 0$. Also, the FOS/C convergence flow f_e on an edge $e = (u, v) \in E$ (u in level i, v in level $i{+}1$, $0 \leq i < \Lambda$) is*
$$w_u - w_v = f_e = \frac{\delta}{\binom{k}{i}(k-i)} \cdot \sum_{l=i+1}^{k} \binom{k}{l}.$$

6   Conclusions

We have shown that the disturbed diffusion scheme FOS/C can be related to random walks despite its disturbance, since its load differences in the convergence state correspond to scaled differences of hitting times. Exploiting this correspondence, we have shown that load diffuses monotonically decreasing from a source vertex into the graph on torus and distance-transitive graphs. Furthermore, while the uniform flow division among shortest paths does not yield the $\| \cdot \|_2$-minimal flow on the torus in general, it does so on distance-transitive graphs. For the hypercube, one of its highly relevant representatives, the convergence flow has been stated explicitly.

Future work includes the extension of the results to further graph classes and simple characterizations of the convergence flow as in the case of distance-transitive graphs. Naturally, different disturbed diffusion schemes and drain concepts and therefore different distance measures could be examined as well. Moreover, while connectedness of partitions can be observed in experiments and verified easily for torus and distance-transitive graphs with the results of this paper, a rigorous proof for general graphs remains an object of further investigation, likewise a convergence proof for Bubble-FOS/C on general graphs. All this aims at further improvements to the heuristic in theory and practice for graph partitioning and its extension to graph clustering.

References

1. J. Adámek. *Foundations of Coding*. J. Wiley & Sons, 1991.
2. N. Alon and J. H. Spencer. *The Probabilistic Method*. J. Wiley & Sons, 2nd edition, 2000.
3. N. Biggs. *Algebraic Graph Theory*. Cambridge University Press, 1993.
4. R. R. Coifman, S. Lafon, A. B. Lee, M. Maggioni, B. Nadler, F. Warner, and S. W. Zucker. Geometric diffusions as a tool for harmonic analysis and structure definition of data. Parts I and II. *Proc. Natl. Academy of Sciences*, 102(21):7426–7437, 2005.
5. G. Cybenko. Dynamic load balancing for distributed memory multiprocessors. *Parallel and Distributed Computing*, 7:279–301, 1989.
6. R. Diekmann, A. Frommer, and B. Monien. Efficient schemes for nearest neighbor load balancing. *Parallel Computing*, 25(7):789–812, 1999.

7. P. G. Doyle and J. L. Snell. *Random Walks and Electric Networks*. Math. Assoc. of America, 1984.
8. R. B. Ellis. Discrete green's functions for products of regular graphs. In *AMS National Conference, invited talk, special session on Graph Theory*, 2001.
9. G. R. Grimmett and D. R. Stirzaker. *Probability and Random Processes*. Oxford University Press, second edition, 1992.
10. J. L. Gross and J. Yellen (eds.). *Handbook of Graph Theory*. CRC Press, 2004.
11. Y. F. Hu and R. F. Blake. An improved diffusion algorithm for dynamic load balancing. *Parallel Computing*, 25(4):417–444, 1999.
12. J. G. Kemeny and J. L. Snell. *Finite Markov Chains*. Springer-Verlag, 1976.
13. F. T. Leighton. *Introduction to Parallel Algorithms and Architectures: Arrays, Trees, Hypercubes*. Morgan Kaufmann Publishers, 1992.
14. Stuart P. Lloyd. Least squares quantization in PCM. *IEEE Transactions on Information Theory*, 28(2):129–136, 1982.
15. L. Lovász. Random walks on graphs: A survey. *Combinatorics, Paul Erdös is Eighty*, 2:1–46, 1993.
16. M. Meila and J. Shi. A random walks view of spectral segmentation. In *Eighth International Workshop on Artificial Intelligence and Statistics (AISTATS)*, 2001.
17. H. Meyerhenke, B. Monien, and S. Schamberger. Accelerating shape optimizing load balancing for parallel FEM simulations by algebraic multigrid. In *Proc. 20th IEEE Intl. Parallel and Distributed Processing Symp. (IPDPS'06)*, page 57 (CD). IEEE, 2006.
18. B. Nadler, S. Lafon, R. R. Coifman, and I. G. Kevrekidis. Diffusion maps, spectral clustering and eigenfunctions of fokker-planck operators. In *NIPS*, 2005.
19. M. Saerens, P. Dupont, F. Fouss, and L. Yen. The principal components analysis of a graph, and its relationships to spectral clustering. In *ECML 2004, European Conference on Machine Learning*, pages 371–383, 2004.
20. S. Schamberger. A shape optimizing load distribution heuristic for parallel adaptive FEM computations. In *Parallel Computing Technologies, PACT'05*, number 2763 in LNCS, pages 263–277, 2005.
21. J. R. Shewchuk. An introduction to the conjugate gradient method without the agonizing pain. Technical Report CMU-CS-94-125, Carnegie Mellon University, 1994.
22. The BlueGene/L Team. An overview of the BlueGene/L supercomputer. In *Proc. ACM/IEEE Conf. on Supercomputing*, pages 1–22, 2002.
23. N. Tishby and N. Slonim. Data clustering by markovian relaxation and the information bottleneck method. In *NIPS*, pages 640–646, 2000.
24. U. Trottenberg, C. W. Oosterlee, and A. Schüller. *Multigrid*. Academic Press, 2000.
25. S. van Dongen. *Graph Clustering by Flow Simulation*. PhD thesis, Univ. of Utrecht, 2000.
26. C. Xu and F. C. M. Lau. *Load Balancing in Parallel Computers*. Kluwer, 1997.
27. L. Yen, D. Vanvyve, F. Wouters, F. Fouss, M. Verleysen, and M. Saerens. Clustering using a random-walk based distance measure. In *ESANN 2005, European Symposium on Artificial Neural Networks*, 2005.

Exact Algorithms for Finding the Minimum Independent Dominating Set in Graphs

Chunmei Liu[1] and Yinglei Song[2]

[1] Dept. of Systems and Computer Science, Howard University, Washington, DC
20059, USA
chunmei@scs.howard.edu
[2] Dept. of Computer Science, University of Georgia, Athens, GA 30602, USA
song@cs.uga.edu

Abstract. In this paper, we consider the MINIMUM INDEPENDENT DOM-INATING SET problem and develop exact exponential algorithms that break the trivial $O(2^{|V|})$ bound. A simple $O^*(\sqrt{3}^{|V|})$ time algorithm is developed to solve this problem on general graphs. For sparse graphs, e.g. graphs with degree bounded by 3 and 4, we show that a few new branching techniques can be applied to these graphs and the resulting algorithms have time complexities $O^*(2^{0.465|V|})$ and $O^*(2^{0.620|V|})$, respectively. All our algorithms only need polynomial space.

1 Introduction

A *dominating set* in a graph $G = (V, E)$ is a vertex subset $D \subseteq V$ such that each vertex $v \in V$ is either in D or connected to a vertex in D. Similarly, an *independent dominating set* is a dominating set I in G such that each pair of the vertices in I are not connected. The goal of the MINIMUM INDEPENDENT DOMINATING SET problem is to find the independent dominating set of the minimum size in a given graph. This problem is of importance and has many practical applications in data communication and networks [13]. It has been shown to be NP-hard [8]. In [10], Halldórsson shows that this problem is unlikely to be approximated within $|V|^{1-\epsilon}$, where ϵ is any positive number less than 1. For graphs of bounded degree, Kann [11] has shown that this problem is APX-hard, which suggests it is NP-hard to approximate the minimum independent dominating set in such a graph within some constant ratio in polynomial time.

Due to these inapproximability results for the minimum independent dominating set problem, exact solutions that need time $O(2^{\alpha|V|})$, where α is a positive number less than 1.0, is more desirable in practice. In particular, a slight reduction in α may significantly reduce the amount of computation time when the graph is of large size. However, such algorithms are still not available for the minimum independent dominating set problem.

Recently, with the growth of interests in developing exact exponential time algorithms for NP-hard optimization problems, the upper bound time complexities for many NP-hard problems have been significantly improved. For example,

T. Asano (Ed.): ISAAC 2006, LNCS 4288, pp. 439–448, 2006.

a sophisticated algorithm [16] that needs time $O(2^{|V|/4})$ has been developed to find the maximum independent set in a general graph $G = (V, E)$. A $O(2^{0.114|E|})$ time algorithm [1] has been available for computing the maximum independent set in a sparse graph. A few other efficient algorithms [2,4] have also been developed for this problem.

The trivial $O(2^{|V|})$ bound for the MINIMUM DOMINATING SET problem, which needs to find the dominating set of minimum size in a graph, was broken very recently. Fomin et al. [7] used a deep graph theoretical result proposed by Reed [15] and developed an algorithm that can find the minimum dominating set in time $O(2^{0.955|V|})$. Randerath and Schiermeyer [14] showed that matching based techniques are useful for reducing the size of the search space and as a result, an elegant algorithm of time complexity $O^*(2^{0.919|V|})$ can be developed. Grandoni [9] considered a reduction from the MINIMUM DOMINATING SET problem to the MINIMUM SET COVER problem and was able to find the minimum dominating set in time $O^*(2^{0.919|V|})$ with polynomial space. Using a dynamic programming technique that may require exponential space, the time complexity of this algorithm can be further improved to $O^*(2^{0.850|V|})$. Recent work [6] on new measures for analyzing backtracking algorithms showed that the worst time complexities of the Grandoni's algorithms are in fact at most $O(2^{0.610|V|})$ and $O(2^{0.598|V|})$, respectively.

On the other hand, recent work has shown that many NP-hard problems can be solved with improved time complexity on sparse graphs. For example, in [1], exact algorithms of time complexity $O(2^{0.171|V|})$ and $O(2^{0.228|V|})$ are designed to compute the maximum independent sets in graphs of degree bounded by 3 and 4, respectively. In [12], Kneis et al. proved a generic result on the tree width of cubic graphs, which leads to an $O^*(2^{0.5|V|})$ time algorithm for the MINIMUM DOMINATING SET problem. Fomin and Hoie [5] showed that the path width of a cubic graph is bounded by $(\frac{1}{6} + \epsilon)|V|$, where ϵ is any positive number and the minimum dominating set in a graph with degree bounded by 3 can be found in time $O^*(2^{0.265|V|})$. Fomin and Hois's result on the pathwidth of cubic graph can also be used to compute the maximum independent set in graphs with degree bounded by 3 in time $O^*(2^{0.167|V|})$. However, tree decomposition or path decomposition based dynamic programming requires maintaining a dynamic programming table in each tree node and this algorithm thus may need exponential space to store the intermediate results needed in the dynamic programming.

In this paper, we develop exact algorithms to solve the MINIMUM INDEPENDENT DOMINATING SET problem in both general graphs and sparse graphs. Based on a maximal matching, we show that a simple algorithm can solve this problem in $O^*(\sqrt{3}^{|V|})$ time and polynomial space. For sparse graphs, we use the number of edges in a graph as a new measure for analyzing its computational complexity and develop new branching techniques based on the fundamental topological units in these graphs. We show that these branching techniques can lead to algorithms that can find a minimum independent dominating set on graphs of degree bounded by 3 and 4 in time $O^*(2^{0.465|V|})$ and $O^*(2^{0.620|V|})$ respectively with polynomial space. Here, $O^*(.)$ implies the existence of an additional polynomial factor in the corresponding time complexity result.

2   Preliminaries

The graphs in this paper are undirected graphs without loops. For a given graph $G = (V, E)$ and a vertex $v \in V$, $N(v)$ is the set of vertices that are connected to v in G and $|N(v)|$ is the *degree* of v. $N[v]$ is $\{v\} \cup N(v)$. The *degree* of a graph is the maximum degree of all its vertices. To simplify the notation, we use $G - v$ to represent the graph obtained by removing v and all the edges incident on v from G. For a vertex subset U, we use $G - U$ to denote the graph obtained from G by removing all vertices in U and the edges incident on them from G. For a subset $U \subseteq V$, $N(U)$ denotes the set of all vertices that are connected to at least one vertex in U; $G[U]$ is the subgraph induced on U in G. We use $d(G)$ to denote the size of the minimum independent dominating set in graph G. A *path* in a graph is a sequence of vertices $v_1, v_2, \cdots, v_l$ such that there is a graph edge between v_i and v_{i+1} $(1 \leq i < l)$. We use $P = (v_1, v_2, \cdots, v_l)$ to represent a path P of length l. A *cycle* of length l is a sequence of vertices $v_1, v_2, \cdots, v_l$ such that there is a graph edge between v_i and $v_{(i+1) \bmod l}$. For simplicity of notations, we focus on the exponential part of time complexity results and ignore all their polynomial factors.

3   Algorithms

3.1   On General Graphs

A *matching* in a graph $G = (V, E)$ is a set of edges $M \subseteq E$ such that no two edges in M are incident on the same vertex. A *maximal matching* in G is a matching M such that any edge set $M' \subseteq V$ is not a matching if $M \subset M'$. The *size* of a matching is the number of edges in M.

Theorem 1. For a given graph $G = (V, E)$, there exists an algorithm that can compute the minimum independent dominating set in G in time $O^*(\sqrt{3}^{|V|})$ and polynomial space.

Proof. We can obtain a maximal matching M in graph G in polynomial time. We then consider the vertex set $V_M = \{v \mid \exists e \in M, \text{ such that } e \text{ incident on } v\}$. It is not difficult to see that $V - V_M$ form an independent set I in G. We now consider a minimum independent dominating set D in G. We must have $D \cap I = I - N(D \cap V_M)$, since if it is not the case, there must exist a vertex $v \in I$ such that v is not connected to any vertex in D. This is contradictory to the fact that D is also a dominating set. Based on this fact, we can enumerate subsets in V_M that can be $D \cap V_M$. For each of the enumerated subset S, we can obtain $D_S = S \cup (I - N(S))$. We now only need to return the set D_S that is a dominating set and has the minimum number of vertices. The number of such subsets S is bounded by $\sqrt{3}^{|V|}$ and the time complexity of the algorithm is bounded by $O^*(\sqrt{3}^{|V|})$. Enumeration of all such subsets only need polynomial space.

3.2   On Graphs of Degree Bounded by 3

We now consider finding the minimum independent dominating set problem on graphs of degree bounded by 3.

Definition 1. In a given graph $G = (V, E)$, a path $P = (h, v_1, v_2, \cdots, v_l)$ $(l \geq 1)$ is a *sword* if vertices $v_1, v_2, \cdots, v_{l-1})$ are all of degree 2, v_l is of degree 1, and h is of degree at least 3. h is the *head* of the sword while v_l is the *tail*; l is the *length* of the sword.

Definition 2. In a given graph $G = (V, E)$, a path $P = (h, v_1, v_2, \cdots, v_l, t)$ $(l \geq 1)$ is a *bridge* if vertices $v_1, v_2, \cdots, v_l$ are all of degree 2 and both h and t are of degree at least 3. h and t are the *head* and *tail* of the bridge; l is the *length* of the bridge.

Definition 3. In a given graph $G = (V, E)$, a cycle $C = (b, v_1, v_2, \cdots, v_l)$ $(l \geq 2)$ is a *circle* if vertices $v_1, v_2, \cdots, v_l$ are of degree 2 and h is of degree at least 3. b and l are the *base* and the *length* of the circle, respectively.

Lemma 1. For a given graph $G = (V, E)$ and a given vertex $v \in V$, we have $d(G) = \min_{u \in N[v]}\{d(G - N[u])\}+1.$

Proof. This is an obvious fact, since v must be dominated by a minimum dominating set D, i.e., $\exists u \in N[v]$, $u \in D$. In addition, since D is independent, if $u \in D$, $N(u) \notin D$. We thus have $d(G) = \min_{u \in N[v]}\{d(G - N[u])\}+1.$

Lemma 1 describes a *simple branching* operation based on which we can recursively solve the MINIMUM INDEPENDENT DOMINATING SET problem by solving a few subproblems of smaller size. We use the number of edges E in a graph as the measure to evaluate the computation time of the branching.

Lemma 2. Given a graph $G = (V, E)$ of degree bounded by 3 and a sword (h, v_1). If the number of vertices in each connected component is at least 10, the simple branching on h can be simplified by $d(G) = \min\{d(G - N[h]), d(G - N[v_1])\}$. This branching contributes a factor of at most $2^{0.310}$ to the overall time complexity.

Proof. Since v_1 is of degree 1, a minimum dominating set in G must include one of h and v_1. If v_1 is included in the dominating set, h is dominated by v_1. Thus we do not need to consider other neighbors of h. This leads to a simplified branching operation and we have $d(G) = \min\{d(G - N[h]), d(G - N[v_1])\}$.

Now, the removal of $N[v_1]$ reduces at least 3 edges from G and the removal of $N[h]$ reduces at least 4 edges, because a connected component contains at least 10 vertices. Assume the computation time is $T(E)$, the recursion for this branching is:

$$T(E) \leq T(E - 3) + T(E - 4) \tag{1}$$

The equation $x^4 - x - 1 = 0$ has a root in interval $(1, 2)$ and its value is less than $2^{0.310}$. This number bounds the factor this branching can contribute to the overall time complexity.

Lemma 3. Given a graph $G = (V, E)$ of degree bounded by 3 and free of sword of length 1, if G contains a circle $(b, v_1, v_2, \cdots, v_l)$ where $l \geq 2$, the simple branching on the base of the circle contributes a factor not larger than $2^{0.310}$ to the overall time complexity.

Proof. We assume the minimum independent dominating set in G is D. We analyze the simple case. When $l = 2$, the simple branching can be simplified as $d(G) = \min \{d(G - b - v_1 - v_2), d(G - N[b])\}$. This is due to the fact that if $b \notin D$, one of the v_1, v_2 must be in D and we thus do not need to consider the third neighbor of b in this branching. $v_1 \in D$ and $v_2 \in D$ are the same and we thus do not need to distinguish between them. The recursion for time complexity $T(E)$ is $T(E) \leq 2T(E - 4)$, which shows that it contributes a factor less than $2^{0.310}$.

Now, we need to consider the cases where $l > 2$. In particular, we need to utilize the symmetry in the circle to avoid recomputation. It is not difficult to see that graphs $G - N[v_1]$ and $G - N[v_l]$ differs only in the connected component induced by the rest of the vertices in the circle. This component is only a path and its minimum independent dominating set can be easily computed in polynomial time. We thus only need to solve the problem on $G - N[v_1]$ and obtain the solution for $G - N[v_l]$ from that on $G - N[v_1]$ with some additional operations that take polynomial time. Based on this observation and the fact that G does not contain sword of length 1, the recursion relation is:

$$T(E) \leq T(E - l - 3) + 2T(E - l - 2) \qquad (2)$$

and the branching contributes a factor less than $2^{0.310}$.

Lemma 4. Given a graph $G = (V, E)$ of degree bounded by 3 and free of circles and swords of length 1, if G contains a sword $(h, v_1, v_2, \cdots, v_l)$ where $l \geq 4$, the simple branching operation on h contributes a factor of at most $2^{0.310}$ to the overall time complexity.

Proof. Figure 1(a) shows the branching on such a sword. Since G does not contain circles and sword of length 1, it is not difficult to check that the removal of $N[v]$ reduces the number of edges by at least $l + 4$ since the path in the sword is now disconnected from the rest part of the graph and its minimum independent dominating set can be computed in polynomial time. Similarly, the removal of v_1 reduces it by at least $l + 2$ and for each of the two neighbors of h, the number of edges is reduced by at least $l + 3$. We thus have the following recursion relation:

$$T(E) \leq T(E-l-4)+T(E-l-2)+2T(E-l-3) \leq T(E-8)+T(E-6)+2T(E-7) \qquad (3)$$

We consider the equation $x^8 - x^2 - 2x - 1 = 0$ and it has a root in the interval $(1, 2)$ and its value is less than $2^{0.310}$.

Lemma 5. Given a graph $G = (V, E)$ of degree bounded by 3 and free of swords of length 1, if each of the connected component of G contains at least 10 vertices and it also contains a bridge (h, v_1, t), a simple branching on v_1 contributes a factor less than $2^{0.310}$ to the overall time complexity.

Proof. We consider two different cases, i.e., h is connected to t or not connected to t. The correctness of this branching is obvious.

If h is connected to t, as shown in Figure 1(c), the removal of $N[v_1]$ reduces the number of edges in G by at least 5. In other two cases where h and l are included in the minimum dominating set respectively, at least 6 edges can be removed from G since the connected component G contains at least 10 vertices. We thus have $T(E) \leq T(E - 5) + 2T(E - 6)$.

If h is not connected to t, as shown in Figure 1(d), a similar analysis can show that the recursion is $T(E) \leq 3T(E - 6)$. Both cases show that this branching contributes a factor less than $2^{0.310}$ to the overall time complexity.

Lemma 6. Given a graph $G = (V, E)$ of degree bounded by 3 and free of swords of length 1, if each of the connected component of G contains at least 10 vertices and it also contains a bridge (h, v_1, v_2, t), we consider the following branching operations:

1. Neither v_1 nor v_2 is in the minimum independent dominating set.
2. One of v_1 and v_2 is in the minimum dominating set.

These branching operations contribute a factor less than $2^{0.310}$ to the overall time complexity.

Proof. We also need to analyze two possible cases, where h is either connected t or not connected to it.

In the case where h is connected to t, as shown in Figure 1(e), if none of v_1 and v_2 are in the minimum independent dominating set, both h and t must be in it. Therefore, at least 7 edges can be removed from G. In the second case where one of v_1 and v_2 is in the minimum independent dominating set, we can reduce the number of edges by at least 5. The recursion thus is $T(E) \leq T(E-7)+2T(E-5)$.

If h is not connected to t, as shown in Figure 1(f), a similar analysis can show that the recursion is $T(E) \leq T(E - 9) + 2T(E - 5)$. Combining the two cases together, we can see that this branching contributes a factor less than $2^{0.310}$ to the time complexity.

Lemma 7. Given a graph $G = (V, E)$ of degree bounded by 3 and free of swords of length 1, if each of the connected component of G contains at least 10 vertices and it also contains a bridge $(h, v_1, v_2, \cdots, v_l, t)$ where $l \geq 3$, we consider the following branching operations:

1. v_{l-1} is in the minimum independent dominating set;
2. v_{l-2} is in the minimum independent dominating set, v_l is not;
3. v_l is in the minimum independent dominating set, v_{l-2} is not;
4. both v_l and v_{l-2} are included in the minimum independent dominating set.

Proof. The correctness of the recursion is obvious. Note that if v_l is not in the minimum independent dominating set, t must be in it. Similar analysis holds for v_{l-2}. If $l = 3$, we can consider the two cases where h and t are connected and not connected. If h and t are connected, as shown in Figure 1(g), based on the

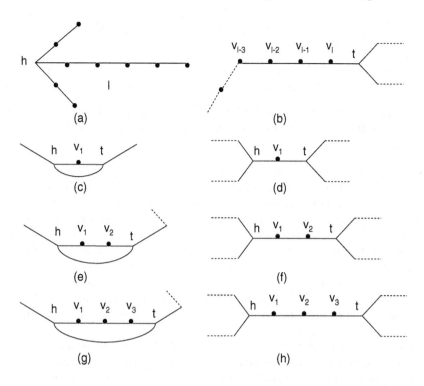

Fig. 1. A few different cases where the algorithm branches on bridges and swords. (a) a sword of length l; (b) a bridge of length larger than 3; (c)(d)(e)(f)(g)(h) describe the branching on bridges of length 1, 2 and 3.

above observation that t must be in the minimum dominating set when v_l is not, we have the recursion:

$$T(E) \leq T(E - 4) + T(E - 9) + T(E - 8) + T(E - 7) \tag{4}$$

If h and t are not connected, as shown in Figure 1(h), the recursion is:

$$T(E) \leq T(E - 4) + 2T(E - 10) + T(E - 8) \tag{5}$$

In the case where $l > 3$, as shown in Figure 1(b), we can get the following recursion:

$$T(E) \leq T(E - 4) + T(E - 9) + T(E - 8) + T(E - 7) \tag{6}$$

Considering all three possible cases, we are able to see that the branching factor is less than $2^{0.310}$.

Lemma 8. Given a graph $G = (V, E)$ of degree bounded by 3 and free of circles, bridges and swords of length 1, if G contains a sword $(h, v_1, \cdots, v_l)$ where $l = 2$ or 3 and at least one neighbor of h is of degree 3, the simple branching on h contributes a factor less than $2^{0.310}$ to the overall time complexity.

Proof. The proof is based on similar case analysis and proof details will appear in a full version of the paper.

Putting all the results we have obtained in the above lemmata together, we can obtain the following Theorem.

Theorem 2. For a given graph $G = (V, E)$ of degree bounded by 3, we can compute a minimum independent dominating set in G in time $O^*(2^{0.465|V|})$ and polynomial space.

Proof. We sketch the algorithm as follows. This algorithm only considers applying the branching rules on connected components that have more than 10 vertices. For a connected component that has less than 10 vertices, we can find a minimum independent dominating set by exhaustive enumeration. Now, if a large component contains a sword of length 1, we can branch on this sword as described in Lemma 2 until no sword of length 1 exists in the graph. We then branch on the circles in G as described in Lemma 3 until the graph is free of circles. Next, we branch on the bridges in the graph as described in Lemmata 5, 6, 7 until the graph is free of bridges. We then consider branching on swords of length at least 4 as described in lemma 4 until the graph does not contain such swords.

Now we consider a vertex that is of degree 3. If all its neighbors are of degree 2, we know that this vertex is the head of three swords; this connected component is a tree and we can use a dynamic programming algorithm to find its minimum independent dominating set in polynomial time. If 1 or 2 of its neighbors are of degree 2, we can branch on this vertex as described in lemma 8 until no such vertices exist. Last, if all of its three neighbors are of degree 3, a simple branching on this vertex will lead to the following recursion:

$$T(E) \le 4T(E-7) \tag{7}$$

since including any vertex in $N[v]$ removes at least 7 edges from G. This branching is again applied until no such v exists in G. It also contributes a branching factor less than $2^{0.310}$ to the overall time complexity. After all these branchings are applied, the graph now only contains paths and cycles that are vertex disjoint and we can use a dynamic programming algorithm to compute its minimum independent dominating set in polynomial time. Putting all these together, the above algorithm only needs time $O^*(2^{0.310|E|})$. Because $|E| \le 3|V|/2$, the time complexity of this algorithm is at most $O^*(2^{0.465|V|})$. This algorithm only needs polynomial space for branchings.

3.3   On Graphs of Degree Bounded by 4

In this section, we show that on graphs of degree bounded by 4, a similar algorithm can find the minimum independent dominating set in time $O^*(2^{0.620|V|})$. In fact, Lemmata 5, 6, and 7 can also be used to branch on the bridges in such a graph since the number of terms on the branching recursions does not increase and the number of edges reduced by branching is even more or at least the same. For the same reason, the branching in Lemma 2 can also be used to remove all the swords of length 1 from the graph.

Lemma 9. For a graph $G = (V, E)$ of degree bounded by 4 and containing a circle $(b, v_1, v_2, \cdots, v_l)$, if G is free of sword of length 1, the simple branching on base b contributes a factor less than $2^{0.310}$ to the overall time complexity.

Proof. The proof is similar to that of Lemma 3, details are shown in the full version of the paper.

Lemma 10. For a graph $G = (V, E)$ of degree bounded by 4, if G is free of circles and bridges and v is a vertex of degree 4, if at least one of the neighbors of v is of degree 3, a simple branching on v contributes a factor less than $2^{0.310}$ to the overall computation time.

Proof. The proof is similar to that of previous Lemmata and will be shown in the full version of the paper.

Theorem 3. For a graph $G = (V, E)$ of degree bounded by 4, we are able to compute a minimum independent dominating set in G in time $O^*(2^{0.620|V|})$ and polynomial space.

Proof. We can develop an algorithm similar to the one we have described in the proof of Theorem 2. Proof details will be shown in a full version of the paper.

4   Conclusions

In this paper, we develop exact exponential algorithms to solve the MINIMUM INDEPENDENT DOMINATING SET problem in both general and sparse graphs. The measure we have used to analyze the time complexities of our algorithms on sparse graphs is the number of edges in the graph. The recent work of Eppstein [3] and Fomin *et al.* [4,6], has shown that an improved measure for analyzing branching algorithms can sometimes lead to improved analysis results. It is thus possible that the measure we have used to evaluate the time complexities of our algorithms is not the optimal one and the actual time complexities of them are possibly better than we have obtained.

Acknowledgment

We thank the anonymous reviewers for their comments and suggestions on an earlier version of the paper.

References

1. R. Beigel, "Finding maximum independent sets in sparse and general graphs", *Proceedings of SODA 1999*, 856-857, 1999.
2. J. Chen, I. A. Kanj, and W. Jia, "Vertex Cover: Further Observations and Further Improvements", *Journal of Algorithms*, 41: 280-301, 2001.
3. D. Eppstein, "Quasiconvex analysis of backtracking algorithms", *Proceedings of SODA 2004*, 788-797, 2004.
4. F. V. Fomin, F. Grandoni, and D. Kratsch, "Measure and Conquer: A Simple $O(2^{0.288}n)$ Independent Set Algorithm", *Proceedings of SODA 2006*, to appear.
5. F. V. Fomin and K. Hoie, "Pathwidth of cubic graphs and exact algorithms", *Information Processing Letters*, 97(5): 191-196, 2006.
6. F. V. Fomin, F. Grandoni, and D. Krastch, "Measure and conquer: domination - a case study", *Proceedings of ICALP 2005*, 191-203, 2005.
7. F. V. Fomin, D. Kratsch, and G. J. Woeginger, "Exact(exponential) algorithms for the dominating set problem", *Proceedings of WG 2004*, 245-256, 2004.
8. M. R. Garey and D. S. Johnson, "Computers and intractability, A guide to the theory of NP-completeness", *Freeman and Company*, New York, NY, 1979.
9. F. Grandoni, "A note on the complexity of minimum dominating set", *Journal of Discrete Algorithms*, in press.
10. M. M. Halldórsson, "Approximating the minimum maximal independence number", *Information Processing Letter*, 46: 169-172, 1993.
11. V. Kann, "On the approximability of NP-complete optimization problems", *Ph.D. Thesis, Department of Numerical Analysis and Computing Science, Royal Institute of Technology*, Stockholm, 1992.
12. J. Kneis, D. Mölle, S. Richter, and P. Rossmanith, "Algorithms based on the treewidth of sparse graphs", *Proceedings of WG 2005*, 385-396, 2005.
13. F. Kuhn, T. Nieberg, T. Moscibroda, and R. Wattenhofer, "Local approximation schemes for ad hoc and sensor networks", *2005 Workshop on Discrete Algorithms and Methods for Mobile Computing and Communications*, 97-103, 2005.
14. B. Randerath, I. Schiermeyer, "Exact algorithms for MINIMUM DOMINATING SET", *Technical Report zaik-469, Zentrum für Angewandte Informatik Köln*, Germany, 2004.
15. B. Reed, "Paths, stars and the number three", *Combinatorial Probabilistic Computing*, 5: 277-295, 1996.
16. J. M. Robson, " Finding a maximum independent set in time $O(2^{n/4})$, *Technical Report 1251-01, LABRI, Université Bordeaux I*, 2001.

On Isomorphism and Canonization of Tournaments and Hypertournaments

V. Arvind, Bireswar Das, and Partha Mukhopadhyay

Institute of Mathematical Sciences
C.I.T Campus, Chennai 600 113, India
{arvind, bireswar, partham}@imsc.res.in

Abstract. We give a polynomial-time oracle algorithm for Tournament Canonization that accesses oracles for Tournament Isomorphism and Rigid-Tournament Canonization. Extending the Babai-Luks Tournament Canonization algorithm, we give an $n^{O(k+\log n)}$ algorithm for canonization and isomorphism testing of k-hypertournaments, where n is the number of vertices and k is the size of hyperedges.

1 Introduction

Computing canonical forms for graphs (and other finite structures) is a fundamental problem. Graph canonization, in particular, is well-studied for its close connection to Graph Isomorphism GRAPH-ISO. Let $\mathcal{G}$ be a class of graphs on n vertices closed under isomorphism. Then $f : \mathcal{G} \longrightarrow \mathcal{G}$ is a *canonizing function* for $\mathcal{G}$ if for all $X, X' \in \mathcal{G}$: $f(X) \cong X$ and $f(X) = f(X')$ if and only if $X_1 \cong X_2$. I.e., f assigns a *canonical form* to each isomorphism class of graphs. E.g. we could define $f(X)$ as the lexicographically least graph isomorphic to X. This particular canonizing function is computable in FP^{NP} by prefix search, but it is NP-hard [BL83, Lu93]. Whether there is *some* canonizing function for graphs that is polynomial-time computable is a long-standing open question. No better bound than FP^{NP} is known for computing any canonizing function. Clearly GRAPH-ISO is polynomial-time reducible to graph canonization. However, it is an intriguing open question if the converse reduction holds.

The seminal paper of Babai and Luks [BL83] takes a group-theoretic approach to canonization. As we use their approach we explain the group-theoretic setting. We first recall definitions and some basic facts about permutation groups [Wi64, Lu93]. A permutation group G is a subgroup of $\text{Sym}V$, where $\text{Sym}V$ is the group of all permutations on an n-element set V. We write $H \leq G$ when H is a subgroup of G. The image of $v \in V$ under $g \in G$ is denoted v^g. We apply permutations from left to right so that $v^{gh} = (v^g)^h$. The set $v^G = \{v^g \mid g \in G\}$ is the *G-orbit* of v, and G is *transitive* on V if $v^G = V$ for $v \in V$. The group $\langle S \rangle$ *generated* by a set $S \subseteq \text{Sym}V$ is the smallest subgroup of $\text{Sym}V$ containing S.

Let $G \leq \text{Sym}V$ be transitive on V. A subset $B \subseteq V$ is a *G-block* if either $B^g = B$ or $B^g \cap B = \emptyset$, for each $g \in G$. For any transitive group G, the set V and the singleton sets $\{u\}$, $u \in V$ are the *trivial* blocks. A transitive group G is *primitive* if it does not have any nontrivial blocks, otherwise it is *imprimitive*. If B is a G-block, then B^g is

T. Asano (Ed.): ISAAC 2006, LNCS 4288, pp. 449–459, 2006.
© Springer-Verlag Berlin Heidelberg 2006

also a G-block, for every $g \in G$. The collection of blocks $\{B^g : g \in G\}$ is a partition of V, called the B *block system*. Notice that G acts transitively on the B block system (since every $g \in G$ naturally maps blocks to blocks). Call $B \subset V$ a *maximal block* if there is no other block B' such that $B \subset B' \subset V$. Then, $\{B^g : g \in G\}$ is a *maximal block system*. An important fact is that G acts *primitively* on a maximal block system. For $\triangle \subseteq V$, the *set stabilizer* for $\triangle$, $G_\triangle = \{g \in G \mid \triangle^g = \triangle\}$.

Let $G \leq S_n$. For $X_1, X_2 \in \mathcal{G}$, we say X_1 is G-isomorphic to X_2, denoted by $X_1 \cong_G X_2$ if $X_2 = X_1^g$ for some $g \in G$. Suppose $\mathcal{G}$ is any class of graphs closed under G-isomorphisms. Assume w.l.o.g. that the vertex set of any n-vertex graph in $\mathcal{G}$ is $[n]$. Call $\mathrm{CF}_G : \mathcal{G} \to \mathcal{G}$ a *canonizing function* w.r.t. to G, if $\mathrm{CF}_G(X) \cong_G X$, for $X \in \mathcal{G}$, and $X_1 \cong_G X_2$ iff $\mathrm{CF}_G(X_1) = \mathrm{CF}_G(X_2)$, for $X_1, X_2 \in \mathcal{G}$. As $\mathcal{G}$ is closed under G-isomorphisms, $\mathcal{G}^\sigma$ is closed under $\sigma^{-1}G\sigma$-isomorphisms, for $\sigma \in S_n$. Thus, for a coset $G\sigma$ of G we can define $\mathrm{CF}_{G\sigma}(X) = \mathrm{CF}_{\sigma^{-1}G\sigma}(X^\sigma)$. Finally, the *canonical labeling coset* $\mathrm{CL}(X, G\sigma)$ is defined as $\{\tau \in G\sigma \mid X^\tau = \mathrm{CF}_{G\sigma}(X)\}$. Notice that $\mathrm{CL}(X, G\sigma) = (G \cap \mathrm{Aut}(X))\pi$ for any $\pi \in \mathrm{CL}(X, G\sigma)$.

Babai and Luks [BL83] give a canonizing algorithm that exploits the group structure of G. The algorithm is recursive and works by a divide-and-conquer guided by the orbits and blocks of G. In the following theorem, we state the main result of their paper.

Theorem 1 (Babai-Luks Theorem). *Suppose $G \leq S_n$ is a permutation group in the class Γ_d (i.e. all nonabelian composition factors of G are subgroups of S_d), then a canonical labeling coset of an n-vertex graph $X = (V, E)$ w.r.t. a coset $G\sigma$ of $\mathrm{Sym}V$, can be found in time $n^{O(d)}$.*

Theorem 1 crucially uses the fact that primitive subgroups of S_n in Γ_d are of size at most $n^{O(d)}$ (e.g. see [LS99]). This result yields an $n^{O(\log n)}$ algorithm for Tournament Canonization, T-CANON, and Tournament Isomorphism TOUR-ISO [BL83]. The algorithm exploits the fact that automorphism groups of tournaments are solvable and hence in Γ_d for $d = 1$.

Remark 1 (Generalized Babai-Luks theorem). We give a useful restatement of Theorem 1 applicable to more general structures like edge-colored tournaments or hyper-tournaments. Suppose Γ is a class of finite groups such that for each n and for any primitive subgroup $G \leq S_n$ in Γ we have $|G| \leq n^c$, where $c > 0$ is a fixed constant. Let $\mathcal{R}$ be the class of finite relational structures of any signature (For example, we can take $\mathcal{R}$ to be the class of all k-hypertournaments or the class of vertex-colored graphs). Consider the *canonization problem* for a structure X from $\mathcal{R}$ under permutation action of a group $G \in \Gamma$. Because of the size bound on primitive permutation groups in Γ, the Babai-Luks algorithm (Theorem 1) will compute a G-canonical form and the canonical labeling coset $(G \cap \mathrm{Aut}(X))\sigma$ for X in polynomial time. The proof of this restatement is identical to that of Theorem 1.

In this paper we study the complexity of canonization and isomorphism of tournaments as well as hypertournaments. A central motivation for our study is the question whether T-CANON is polynomial-time reducible to TOUR-ISO. While we are not able to settle this question, we prove an interesting weaker result: T-CANON has a polynomial-time oracle algorithm with oracle access to TOUR-ISO and an oracle for canonizing *rigid tournaments.* Rigid tournaments have no nontrivial automorphism. The other main result is an $n^{O(k+\log n)}$ algorithm for canonization and isomorphism of k-hypertournaments which builds on [BL83] and uses quite different properties of the automorphism groups of hypertournaments. In this extended abstract we omit some proof details.

2   Gadget Construction for Tournaments

We recall the definition of tournaments.

Definition 1 (tournament). *A directed graph* $T = (V, A)$ *is a* tournament *if for each pair of distinct vertices* $u, v \in V$, *exactly one of* (u, v) *or* (v, u) *is in* A.

In this section, we explain some polynomial-time reductions concerning TOUR-ISO that are useful for our algorithm presented in Theorem 6. A key technique here is "fixing" nodes in a tournament. A node v in a graph X is a *fixpoint* if $v^{\pi} = v$ for every $\pi \in$ Aut(X). By the *fixing* of v in X we mean a construction that modifies X to another graph X' using a gadget so that v is forced to be fixed in X'. We now show a gadget construction for fixing several nodes in a tournament so that the resulting graph is again a tournament. We use it to show that the color-tournament isomorphism problem is polynomial-time many-one reducible to TOUR-ISO. As a consequence, we derive some facts related to tournament isomorphism and automorphism (Theorem 2), useful for canonization. Let $u_1, u_2, \cdots, u_l$ be the nodes of a tournament T that we want to fix. The gadget we use is shown in Figure 1. Call the resulting tournament T'.

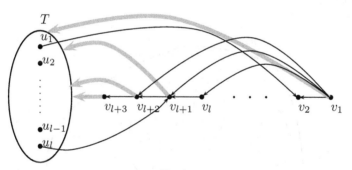

Fig. 1.

Here, $v_1, v_2, \cdots, v_{l+3}$ are $l + 3$ news vertices used in the gadget. Notice that v_1 is the unique vertex that beats all other vertices of T'. For $2 \leq j \leq l + 1$, v_j beats v_k for $k > j$, and beats all the vertices of T except u_{j-1}. Vertex v_{l+2} beats v_{l+3}, and both v_{l+2} and v_{l+3} beat all vertices of T. The thick gray edge between v_1 and T indicates that v_1 beats all the vertices of T. All thick gray edges have similar meaning.

Lemma 1. *Any automorphism of* T' *fixes* $\{u_1, u_2, \cdots, u_l\}$.

Proof. Notice that $v_1, v_2, v_3, \cdots, v_l$ are the unique vertices of in-degree $0, 2, 3, \cdots, l$, respectively. Hence they are fixed by any automorphism of T'. Also, v_{l+1} and v_{l+2} are the only vertices of in-degree $l + 1$. But, the directed edge (v_{l+1}, v_{l+2}) forces the fixing of these two vertices by all automorphisms. As v_{i+1} has a unique incoming edge from u_i, $1 \leq i \leq l$, each of $u_1, u_2, \cdots, u_l$ is fixed by all automorphisms of T'.   ∎

Search and decision for GRAPH-ISO are known to be polynomial-time equivalent to computing a generator set for the automorphism group $\mathrm{Aut}(X)$ of a graph X. We show similar results for tournaments. In fact, we give a general approach to proving this equivalence for any class of graphs and apply it to tournaments.

For a class of graphs $\mathcal{G}$, let GRAPH-ISO$_\mathcal{G}$ denote the decision problem: GRAPH-ISO$_\mathcal{G}$ $= \{\langle X_1, X_2 \rangle \in \mathcal{G} \times \mathcal{G} \mid X_1, X_2 \text{ are isomorphic}\}$. Two vertex-colored graphs[1] $X_1, X_2 \in \mathcal{G}$ are said to be isomorphic if there is a color preserving graph isomorphism between them. Let C-GRAPH-ISO$_\mathcal{G}$ be the corresponding decision problem. The graph automorphism problem is: GA$_\mathcal{G} = \{X \in \mathcal{G} \mid X \text{ has a nontrivial automorphism}\}$. For $X \in \mathcal{G}$, let AUT$_\mathcal{G}$ be the problem of computing a generating set for the automorphism group of X. The following theorem is easy to prove using standard techniques from [KST93].

Theorem 2. *Let $\mathcal{G}$ be any class of graphs. If* C-GRAPH-ISO$_\mathcal{G}$ *is polynomial-time many-one reducible to* GRAPH-ISO$_\mathcal{G}$ *then*

1. GA$_\mathcal{G}$ *is polynomial-time Turing reducible to* GRAPH-ISO$_\mathcal{G}$.
2. *Search version of* GRAPH-ISO$_\mathcal{G}$ *is polynomial-time Turing reducible to decision version of* GRAPH-ISO$_\mathcal{G}$.
3. AUT$_\mathcal{G}$ *is polynomial-time Turing reducible to* GRAPH-ISO$_\mathcal{G}$.

We now show C-TOUR-ISO $\leq_m^P$ TOUR-ISO, implying that tournaments satisfy the conditions of Theorem 2.

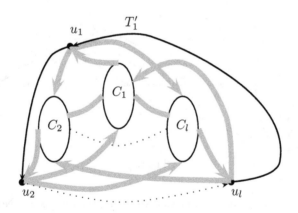

Fig. 2.

Theorem 3. *Color tournament isomorphism problem is polynomial time many-one reducible to tournament isomorphism problem.*

---

[1] In this paper, vertex and edge colorings are simply labels without any constraints like proper vertex/edge colorings etc.

Proof. Let T_1, T_2 be tournaments with vertices colored using l distinct colors $\{c_i\}_{i=1}^l$. Let C_i denotes the set of vertices colored with c_i. Our reduction transforms T_1 and T_2 into uncolored tournaments T_1', T_2'. We show the construction for T_1 (Fig 2). Construction for T_2 is likewise.

In T_1', u_i beats the vertices in each color class C_j with $j \neq i$, and u_i is beaten by all vertices in color class C_i. Using the gadget of Lemma 1, we fix $u_1, u_2, \cdots, u_l$ in T_1'. Call the resulting tournament T_1''. Likewise, T_2'' is obtained from T_2 by first constructing T_2' by introducing new vertices $v_1, v_2, \cdots, v_l$ and then fixing them in T_2'. v_i has the same edge relation with all the color classes, as u_i has. Now it is easy to see that $T_1 \cong T_2$ if and only if $T_1'' \cong T_2''$. ∎

3   Canonical Labeling of Tournaments

We recall an important fact about tournaments and the Babai-Luks result on tournament canonization [BL83].

Fact 4. *The automorphism group of a tournament has an odd number of elements and, therefore, is a solvable group (follows from [FT63]).*

Theorem 5. [BL83, Theorem 4.1] *There is an $n^{O(\log n)}$ algorithm for T-CANON, the Tournament Canonization problem.*

We sketch the proof of Theorem 5 as we need the main ideas. Let $T = (V, A)$ be a tournament with $|V| = n$. If T is not *regular*, partition V as $V = \cup_{i=1}^k V_i$, where V_i are the vertices of out-degree i in T. Let T_i be the subtournament induced by V_i. The algorithm recursively computes $\mathrm{CL}(T_i, \mathrm{Sym}V_i) = H_i\rho_i$, for all i, where $H_i = \mathrm{Aut}(T_i)$. Then, we set $\mathrm{CL}(T, \mathrm{Sym}V) = \mathrm{CL}(T, H_1\rho_1 \times H_2\rho_2 \times \cdots \times H_k\rho_k) = \mathrm{CL}(T, H\rho)$, where $H = H_1 \times H_2 \times \cdots \times H_k$ and $\rho = (\rho_1, \rho_2, \cdots, \rho_k)$). As each H_i is a solvable group, H is also solvable. By Theorem 1 $\mathrm{CL}(T, H\rho)$ can be computed in polynomial time. Thus, if $t(n)$ is the overall running time bound then for this case it satisfies the recurrence relation: $t(n) = \sum_{i=1}^k t(n_i) + n^{O(1)}$, where $n_i = |V_i|$.

Now, suppose T is a *regular* tournament. Fix $v \in V$. Put $V' = V \setminus \{v\}$ and let T' be the subtournament induced by V'. Consider the partition $V' = V_1' \cup V_2'$, where V_1' is the set of $(n-1)/2$ vertices in T that beat v and V_2' is the set of $(n-1)/2$ vertices beaten by v. Let the subtournaments induced by V_1' and V_2' be T_1' and T_2', respectively. The algorithm now recursively computes $\mathrm{CL}(T_i', \mathrm{Sym}V_i') = H_i\rho_i$ for $i = 1, 2$. Compute $\mathrm{CL}(T', \mathrm{Sym}V') = \mathrm{CL}(T, H_1\rho_1 \times H_2\rho_2)$. Repeat this process for every $v \in V$. This yields n cosets in all. We compute $\mathrm{CL}(T, \mathrm{Sym}V)$ as the union of those cosets that give rise to the lex-least canonical labeling among the n. Clearly, this union will itself paste into a coset $\mathrm{Aut}(T)\sigma$ of $\mathrm{Aut}(T)$ [BL83]. For this case, $t(n)$ satisfies the recurrence relation $t(n) = n(t(\frac{n-1}{2}) + n^{O(1)})$. Solving the two recurrence relations for $t(n)$ yields the running time bound $n^{O(\log n)}$.

We turn to the problem of this section: can T-CANON be polynomial-time reduced to TOUR-ISO? We make some progress on the problem by giving a polynomial-time oracle algorithm for T-CANON that accesses oracle TOUR-ISO with an additional oracle for canonizing *rigid* tournaments. Thus, canonizing rigid tournaments seems to be the bottleneck in reducing T-CANON to TOUR-ISO. Let RT-CANON denote the functional oracle for computing the canonical form of a rigid tournament. Since rigid tournaments have trivial automorphism groups, the canonical form trivially gives the canonical labeling coset as well.

Theorem 6. *There is a polynomial-time oracle algorithm for* T-CANON *that accesses oracles for* TOUR-ISO *and* RT-CANON.

Proof. Let $T = (V, A)$ be the input tournament. The function T-CANON(T) computing the canonical labeling coset CL$(T, \mathrm{Sym}V)$ of T has the following recursive description:

T-CANON(T):

1. **Orbit computing:** With oracle queries to TOUR-ISO and using the vertex fixing technique of Theorem 2 we can compute the partition of V into Aut(T)-orbits in polynomial time.
2. **If orbits are singletons:** This happens precisely when T is a rigid tournament. In this case we query the RT-CANON oracle to obtain a canonical form for T.
3. **Single orbit:** If V has only one orbit w.r.t. Aut(T) then the tournament is *vertex-transitive*. As T is vertex-transitive it follows that T is regular. Now, we can individualize (fix) *any vertex* v of T, and find canonical labeling coset with respect to $\mathrm{Sym}V'$, where $V' = V \setminus \{v\}$, because each such v will give rise to the same canonical form.[2] Then v defines the partition $V' = V_1 \cup V_2$, where V_1 is the set of all vertices that beat v and V_2 is the set of all vertices beaten by v. As T is regular, $|V_1| = |V_2| = (n-1)/2$. Suppose the tournaments induced by V_1 and V_2 are T_1 and T_2 respectively. Recursively compute $H_1\rho_1 := $ T-CANON(T_1) and $H_2\rho_2 := $ T-CANON(T_2). Let T_v be the tournament induced by V'. Applying Theorem 1 we compute T-CANON$(T_v) = $ CL$(T_v, H_1\rho_1 \times H_2\rho_2)$ in polynomial time as H_1 and H_2 are solvable, being automorphism groups of tournaments. This gives the canonical ordering for T_v. Placing v as the overall first vertex gives the canonical ordering for T. Let T' denote the resulting tournament (which is the canonical form for T). Finally, the canonical labeling coset is easy to compute from T and T' with queries to TOUR-ISO by applying Theorem 2.
4. **Nonrigid with more than one orbit:** This is the general case when there is more than one orbit and T is not rigid. Let $O_1, O_2, \cdots, O_\ell$ be the orbits of T.
   - [**Case (a)**] Let T_i be the tournament induced by O_i, for $1 \leq i \leq \ell$. We first consider a case that yields an easy recursive step. Suppose not all T_i are isomorphic to each other (which we can easily find with queries to TOUR-ISO). Then we partition the O_j's into k collections $S_1, S_2, \cdots, S_k$, where S_i contains all orbits O_j such that the corresponding T_j's are all isomorphic. Now, for $1 \leq j \leq k$, let $\hat{T}_j$ be the tournament induced by the union of the orbits in S_j. Recursively compute $H_j\rho_j := $ T-CANON$(\hat{T}_j)$ for all j. Then, we set T-CANON(T) as CL$(T, H\rho)$ — where $H = H_1 \times H_2 \times \cdots \times H_k$ and $\rho = (\rho_1, \cdots, \rho_k)$ — which can be computed in polynomial time using the Babai-Luks algorithm of Theorem 1 as each H_j, being the automorphism group of a tournament, is solvable.

---

[2] This is not true in the case of a tournament that is regular but not vertex transitive. Recall from proofsketch of Theorem 5 that n recursive calls are made in the regular case.

– [**Case (b)**] We are now in the case when the tournaments T_i induced by O_i are all isomorphic, for $1 \leq i \leq \ell$. Since T_i are induced by orbits they are all regular tournaments. That forces $|O_i|$ to be odd. Furthermore, all O_i are of same size since T_i are all isomorphic. Thus, $|O_i| = t$ is an odd positive integer. Now, from the ℓ orbits of T, we will construct a tournament $\mathcal{T}$ with ℓ vertices. The vertices v_i of $\mathcal{T}$ represent the orbits O_i. We still have to define the edges of $\mathcal{T}$. To that end, let X_{ij} denote the directed bipartite graph between O_i and O_j. As O_i and O_j are orbits of $\mathrm{Aut}(T)$ and $|O_i| = |O_j|$, the directed bipartite graph X_{ij} has the following property: there is a positive integer α such that, in the graph X_{ij}, the indegree of each vertex in O_i is α and the outdegree of each vertex in O_j is α. Since $|O_i|$ is odd, $|O_i| - \alpha \neq \alpha$. The edges of $\mathcal{T}$ can now be defined as follows: for $1 \leq i \neq j \leq \ell$, (v_i, v_j) is an edge in $\mathcal{T}$ if $|O_i| - \alpha > \alpha$, otherwise (v_j, v_i) is an edge in $\mathcal{T}$. Notice that $\mathcal{T}$ does not carry all the information about the original tournament T since the X_{ij} have been replaced by single directed edges. We now color the edges (v_i, v_j) of $\mathcal{T}$ by the isomorphism type of X_{ij}. To that end, we first recursively compute $H_i\rho_i := \text{T-CANON}(T_i)$ for $1 \leq i \leq \ell$. Then compute $\mathrm{CL}(X_{ij}, H_i\rho_i \times H_j\rho_j)$ using Theorem 1 for each pair $1 \leq i \neq j \leq \ell$ which actually canonizes each X_{ij}. Now, we color the edges of $\mathcal{T}$ so that two edges (v_i, v_j) and (v_s, v_t) of $\mathcal{T}$ get the same color iff the corresponding directed bipartite graphs X_{ij} and X_{st} are isomorphic (i.e. have same canonical forms as computed above). Let $\mathcal{T}'$ denote this edge-colored tournament obtained from $\mathcal{T}$. Now, recursively we compute, $\mathrm{Aut}(\mathcal{T})\rho = \text{T-CANON}(\mathcal{T})$. Then, as $\mathrm{Aut}(\mathcal{T})$ is solvable, using the Babai-Luks algorithm (Theorem 1), we can compute $H'\rho' = \mathrm{CL}(\mathcal{T}', \mathrm{Aut}(\mathcal{T})\rho)$ in polynomial time. From ρ' we obtain the canonical labeling of $\mathcal{T}'$. This canonical labeling effectively gives a canonical ordering of the orbits O_i of T, since vertices of $\mathcal{T}'$ represent orbits of T. Within each orbit O_i, we already have the canonical labeling given by the recursive calls $\text{T-CANON}(T_i)$. This defines the entire canonical labeling for T. Let T' denote the resulting tournament, i.e., the canonical form obtained from T by the canonical labeling. Now, by Theorem 2, the canonical labeling coset is easily computable from T and T' with oracle queries to TOUR-ISO.

Claim. Step 3 correctly computes the canonical labeling coset of a vertex transitive tournament.

Proof of Claim. It suffices to argue that the computed canonical form is independent of the choice of v. The proof is by an induction on n. Let u and v be two vertices. Let T_u and T_v be the tournaments induced by $V \setminus \{u\}$ and $V \setminus \{v\}$ respectively. Furthermore, let T_{u1}, T_{u2} be the tournaments induced by in-neighbors and out-neighbors of u. Similarly, define T_{v1}, T_{v2} for v. Vertex transitivity of T implies that $T_u \cong T_v$, $T_{u1} \cong T_{v1}$, and $T_{u2} \cong T_{v2}$. Consequently, by induction we have $\text{T-CANON}(T_u) = \text{T-CANON}(T_v)$. It easily follows that we get the same canonical form for T by choosing either u or v as first vertex.

Claim. In Step 4 Case(b), $H'\rho'$ is the canonical labeling coset for $\mathcal{T}'$.

Proof of Claim. This follows directly from the following property of the canonical labeling coset (Theorem 1 and Remark 1). Suppose $\mathcal{T} = (\hat{V}, A)$. Recall that $\mathcal{T}'$ is obtained from $\mathcal{T}$ by coloring its edges in some manner. Then we have $\mathrm{CL}(\mathcal{T}', \mathrm{Sym} V) =$

$\mathrm{CL}(\mathcal{T}', \mathrm{CL}(\mathcal{T}, \mathrm{Sym}\hat{V}))$. As $\mathrm{CL}(\mathcal{T}, \mathrm{Sym}\hat{V}) = \mathrm{Aut}(\mathcal{T})\rho$, the claim follows. The correctness of the algorithm follows from the above claims. We now analyze the running time. Let $T(n)$ bound the running time. In Step 1, we compute the orbits in polynomial time with queries to the TOUR-ISO oracle. If the tournament is rigid then we canonize it with a single query to RT-CANON. The remaining steps involve recursive calls. The recurrence relation for $T(n)$ in Step 3 is $T(n) = 2T((n-1)/2) + n^{O(1)}$, and in Step 4a it is given by $T(n) = \ell T(n/\ell) + n^{O(1)}$ for $\ell > 1$ because we need to compute the canonical labeling coset for ℓ tournaments induced by n/l-sized orbits. For Step 4 Case (b), the recurrence is $T(n) = \sum_{i=1}^{k} T(n_i) + n^{O(1)}$. It follows by induction that $T(n) = n^{O(1)}$.   ∎

Remark 2. It seems unlikely that a similar reduction can be carried out for general graphs. This is because our reduction heavily uses the fact that the automorphism group of tournaments are solvable and hence in Γ_d, enabling us to use Theorem 1. In case of general graphs, it is unlikely that in the intermediate stages of recursion we will have groups in Γ_d (or even Γ: see Remark 1) to canonize with.

4   Hypertournament Isomorphism and Canonization

Hypertournaments are a generalization of tournaments and are well-studied by graph theorists over the years (see e.g. [GY97]). We recall the definition.

Definition 2 (Hypertournament). *A k-hypertournament T on n vertices is a pair (V, A) where V is a set of n vertices and A is a set of k-tuples of vertices called* arcs *so that for each subset $S \in \binom{V}{k}$, A contains exactly one of the $(k!$ many) k-tuples whose entries belong to S.*

It is easy to show that Hypergraph Isomorphism (HGI) is polynomial-time many-one equivalent to GRAPH-ISO. Thus, complexity-theoretic upper bounds for GRAPH-ISO like NP∩coAM and SPP apply to HGI. However, consider an instance of HGI: (X_1, X_2), with n vertices and m hyperedges each. The reduction to GRAPH-ISO maps it to a pair of graphs (Y_1, Y_2) with vertex sets of size $m + n$. The best known isomorphism testing algorithm due to Luks and Zemlyachenko (see [BL83]) which has running time $c^{\sqrt{n \lg n}}$ will take time $c^{\sqrt{(m+n) \lg(m+n)}}$ when combined with the above reduction and applied to HGI. In [Lu99] a different, dynamic-programming based algorithm with running time $2^{O(n)}$ was developed.

Motivated by the above we study the analogous question for hypertournaments in this section. We consider k-Hypertournament Isomorphism (HYPER-TOUR-ISO$_k$) and give an $n^{O(k+\log n)}$ algorithm for the problem for k-hypertournaments, for each k. In fact, we actually give an $n^{O(k+\log n)}$ algorithm for the corresponding canonization problem. We first establish some observations about hypertournaments. We are interested in automorphisms of k-hypertournaments.

Lemma 2. *For $k \geq 2$, the automorphism group $\mathrm{Aut}(T)$ of a k-hypertournament T has the following property: for any prime factor p of k it holds that p does not divide the size of $\mathrm{Aut}(T)$.*

Proof. Let $T = (V, A)$. For $k = 2$, T is a usual tournament and in this case it is a well-known fact that $\mathrm{Aut}(T)$ has odd cardinality. Suppose $k > 2$ and p is any prime factor of k. Suppose p divides $\mathrm{Aut}(T)$. Let $\pi \in \mathrm{Aut}(T)$ be an order p element. Since $\pi \in \mathrm{Sym}(V)$, we can write it as a product of disjoint p-cycles, $\pi = C_1 C_2 \cdots C_\ell$, where the remaining $n - p\ell$ elements of V are fixed by π. Let $k/p = t$. If $k \leq p\ell$ then let $S = \cup_{i=1}^{t} C_i$. Notice that π maps S to S. Now, suppose $e \in A$ is the unique hyperedge defined by S. Then $e^\pi \neq e$, since π *reorders* the sequence defining hyperedge e. Thus, e^π is not a hyperedge of T, contradicting $\pi \in \mathrm{Aut}(T)$. If $k > p\ell$, choose S' as any subset of size $k - p\ell$ of the $n - p\ell$ points fixed by π, and let $S = S' \cup C_1 \cup \cdots \cup C_\ell$. Again, let $e \in A$ be the hyperedge defined by S. Then e^π is not a hyperedge of T, since π will reorder the sequence defining e. Again, this contradicts $\pi \in \mathrm{Aut}(T)$.   ■

Recall that a *section* of a group G is a quotient group of some subgroup of G. An easy corollary of the above lemma is the following.

Corollary 1. *For $k \geq 2$, the automorphism group $\mathrm{Aut}(T)$ of a k-hypertournament T does not have the alternating group A_k as section.*

Proof. For $k = 2$ (i.e. when T is a usual tournament) it is well known that $\mathrm{Aut}(T)$ has odd cardinality. We consider the case $k > 2$. For, suppose A_k is a section of $\mathrm{Aut}(T)$. Then $|A_k|$ divides $|\mathrm{Aut}(T)|$. As $|A_k| = (k!)/2$, $k!$ divides $2|\mathrm{Aut}(T)|$ which implies k divides $|\mathrm{Aut}(T)|$. Thus, any prime factor of k divides $|\mathrm{Aut}(T)|$, contradicting Lemma 2.   ■

Some notation: we denote by $\mathcal{C}_k$ the class of finite groups G such that A_k is *not* a section of G. Corollary 1 implies $\mathrm{Aut}(T) \in \mathcal{C}_k$ for any k-hypertournament T. This property is crucial for our canonization algorithm. First recall a celebrated result about *primitive permutation groups* not containing A_k as a section.

Theorem 7. *[BCP82, LS99] Let k be a positive integer and $G \leq S_n$ be a primitive group in $\mathcal{C}_k$ (i.e. G does not involve the alternating group A_k as a section), then $|G|$ is bounded by $n^{O(k)}$.*

At this point, in order to put Theorem 7 in perspective, we recall the discussion in Remark 1 explaining a general form of Theorem 1 applicable to hypertournaments.

Let $T = (V, A)$ be a k-hypertournament with n vertices. We define the i-*degree* of a vertex $v \in V$ as the number d_{vi} of hyperedges in which v occurs at the ith position, $1 \leq i \leq k$. Thus, to each $v \in V$ we can associate its *degree vector* $(d_{v1}, d_{v2}, \ldots, d_{vk})$. We say T is a *regular* k-hypertournament if all $v \in V$ have the same degree vector $(d_1, d_2, \ldots, d_k)$. It is easy to see that $d_i = \frac{1}{n}\binom{n}{k}$ for each i and $n\sum_{i=1}^{k} d_i = k\binom{n}{k}$. Moreover, each $v \in V$ occurs in exactly $\binom{n-1}{k-1}$ hyperedges. We are now ready to prove the main result of this section.

Theorem 8. *Canonization of k-hypertournaments can be done in $n^{O(k+\log n)}$ time. As a consequence, there is an $n^{O(k+\log n)}$ time isomorphism testing algorithm for k-hypertournaments.*

Proof. The canonization algorithm is recursive and we give a high level description of its phases. Let $T = (V, A)$ be the input k-hypertournament.

Phase 0 (**usual tournament case**): If $k = 2$ then we can invoke the Babai-Luks canonizing algorithm that runs in time $n^{O(\log n)}$.

Phase 1 **Vertex partitioning by degree vectors**: If T is not regular, partition V as, $V = V_1 \cup V_2 \cup \cdots \cup V_m$, where V_i ($1 \le i \le m$) is the set of all vertices having the same degree vector, where the degree vectors are sorted in lexicographic order. For $1 \le i \le m$, let T_i be the k-hypertournament induced by V_i. We recursively compute $\mathrm{CL}(T_i, \mathrm{Sym}V_i)$ for all i. Let $\mathrm{CL}(T_i, \mathrm{Sym}V_i) = H_i\rho_i$ where $H_i = \mathrm{Aut}(T_i)$ and $\rho_i \in \mathrm{CL}(T_i, \mathrm{Sym}V_i)$. Let $H = H_1 \times H_2 \times \cdots \times H_k$ and $\rho = (\rho_1, \rho_2, \cdots, \rho_k)$. Then $\mathrm{CL}(T, \mathrm{Sym}V) = \mathrm{CL}(T, H\rho)$. Notice that H is in $\mathcal{C}_k$ since each H_i is in $\mathcal{C}_k$. Thus, by Theorem 1 and Remark 1, $\mathrm{CL}(T, H\rho)$ can be computed in time $n^{O(k)}$. Repeated application of this phase eventually reduces the original hypertournament into an *ordered set* of regular k-hypertournaments, and it suffices to canonize each regular k-hypertournament in this list. In the next phase we explain the canonization of regular k-hypertournaments.

Phase 2 **Regular k-hypertournament phase**: If $k = 2$ then we invoke Phase 0. So, $k > 2$ and $T = (V, A)$ is a regular k-hypertournament. We will make n recursive calls, trying each of the n vertices $v \in V$ as the first vertex in the canonical ordering. Among these, we will pick the lexicographically least ordering. We now describe one of these recursive calls after placing v as the first vertex. Using v, we will decompose T into a $(k-1)$-hypertournament T' on $n-1$ vertices and a k-hypertournament T'' on $n-1$ vertices. Let A' denote the set of the $\binom{n-1}{k-1}$ many $(k-1)$-sequences obtained by taking each of the $\binom{n-1}{k-1}$ hyperedges of T containing v and dropping v from the sequence. Let $V' = V \setminus \{v\}$. Notice that $T' = (V', A')$ is a $(k-1)$-hypertournament. Let A'' denote all hyperedges of T not containing v. Then $T'' = (V', A'')$ is a k-hypertournament. We recursively canonize T'. Let $\mathrm{CL}(T', \mathrm{Sym}V') = G\rho$. By Corollary 1, $G \in \mathcal{C}_k$. Thus, invoking Theorem 1 (its general form explained in Remark 1) we can now directly canonize T'' w.r.t. the coset $G\rho$ in time $n^{O(k)}$. Suppose that algorithm returns the coset $\mathrm{CL}(T'', G\rho) = H_v\tau_v$. From the collection $\{H_v\tau_v\}_{v \in V}$ we can find the collection that gives the lexleast ordering. If several of these cosets give the least ordering then they will paste together into a single coset $\mathrm{Aut}(T)\tau$, which is $\mathrm{CL}(T, \mathrm{Sym}V)$.

For the overall algorithm, an easy inductive proof shows that $\mathrm{CL}(T, \mathrm{Sym}V)$ computes a canonizing coset for the input k-hypertournament T. Next we analyze the running time of the algorithm. Let $t(n, k)$ denote the running time taken by the algorithm for n-vertex k-hypertournaments. Clearly, for Phase 1 and $k > 2$ we have $t(n, k) = \sum_{i=1}^{m} t(|V_i|, k) + n^{O(1)}$, and for Phase 2 and $k > 2$ we have $t(n, k) = n(t(n-1, k-1) + n^{O(k)})$. When $k = 2$, $t(n, 2) = n^{c \log n}$ by Theorem 5. An easy induction yields $t(n, k) = n^{O(k + \log n)}$. ∎

References

[BCP82]   L. BABAI, P.J. CAMERON, AND P.P. PÁLFY. On the order of primitive groups with restricted nonabelian composition factors. *Journal of Algebra*, 79:161–168, 1982.

[BL83]   L. BABAI AND E.M. LUKS. Canonical labeling of graphs. *Proceedings of the Fifteenth Annual ACM Symposium on Theory of Computing*, pages 171–183, 1983.

[FT63]   W. FEIT AND J. THOMPSON, Solvability of groups of odd order, *Pacific Journal of Mathematics*, 13, 775-1029, 1963.

[GY97]   G. GUTIN AND A. YEO, Hamiltonian Paths and Cycles in Hypertournaments, *Journal of Graph Theory*, 25(4):277-286, 1997.

[KST93]   J. KÖBLER, U. SCHÖNING AND J. TORÁN, *The Graph Isomorphism Problem: Its Structural Complexity*, Birkhäuser, Boston, 1993.

[LS99]   M. LIEBECK AND A. SHALEV, Simple groups, permutation groups and probability, *Journal of Amer. Math. Soc.* 12, 497-520, 1999.

[Lu93]   E.M. LUKS, Permutation groups and polynomial time computations. *DIMACS Series in Discrete Mathematics and Theoretical Computer Science*, 11:139–175, 1993.

[Lu99]   E.M. LUKS, Hypergraph isomorphism and structural equivalence of boolean functions. *Proc. 31st ACM Symposium on Theory of Computing*, 652–658. ACM Press, 1999.

[Wi64]   H. WIELANDT, Finite Permutation Groups, *Acad.Press, New York*, 1964.

Efficient Algorithms for the Sum Selection Problem and K Maximum Sums Problem*

Tien-Ching Lin** and D.T. Lee**

Department of Computer Science and Information Engineering,
National Taiwan University, Taipei, Taiwan
{kero, dtlee}@iis.sinica.edu.tw

Abstract. Given a sequence of n real numbers $A = a_1, a_2, \ldots, a_n$ and a positive integer k, the SUM SELECTION PROBLEM is to find the segment $A(i,j) = a_i, a_{i+1}, \ldots, a_j$ such that the rank of the sum $s(i,j) = \sum_{t=i}^{j} a_t$ is k over all $\frac{n(n-1)}{2}$ segments. We present a deterministic algorithm for this problem that runs in $O(n \log n)$ time. The previously best known randomized algorithm for this problem runs in expected $O(n \log n)$ time. Applying this algorithm we can obtain a deterministic algorithm for the K MAXIMUM SUMS PROBLEM, i.e., the problem of enumerating the k largest sum segments, that runs in $O(n \log n + k)$ time. The previously best known randomized and deterministic algorithms for the K MAXIMUM SUMS PROBLEM run respectively in expected $O(n \log n + k)$ and $O(n \log^2 n + k)$ time in the worst case.

Keywords: k maximum sums problem, sum selection problem, maximum sum problem, maximum sum subarray problem.

1 Introduction

Given a sequence of n real numbers $A = a_1, a_2, \ldots, a_n$, the MAXIMUM SUM PROBLEM is to find the segment $A(i,j) = a_i, a_{i+1}, \ldots, a_j$ whose sum $s(i,j) = \sum_{t=i}^{j} a_t$ is the maximum among all possible $1 \leq i \leq j \leq n$. This problem was first introduced by Bentley [6,7] and can be easily solved in $O(n)$ time [7,13].

Given an $m \times n$ matrix of real numbers (assuming that $m \leq n$), the MAXIMUM SUM SUBARRAY PROBLEM is is to find the submatrix, the sum of whose entries is the maximum among all $O(m^2 n^2)$ submatries. The problem can be solved in $O(m^2 n)$ time [7,13,18]. Tamaki and Tokuyama [19] gave the first sub-cubic time algorithm for this problem and Takaoka [20] later gave a simplified algorithm achieving sub-cubic time as well. Many parallel algorithms under different parallel models of computation were also obtained [3,16,17,18].

---

* Research supported in part by the National Science Council under the Grants No. NSC-94-2213-E-001-004, NSC-95-2221-E-001-016-MY3, and NSC 94-2752-E-002-005-PAE, and by the Taiwan Information Security Center (TWISC), National Science Council under the Grant No. NSC94-3114-P-001-001-Y.
** Also with Institute of Information Science, Academia Sinica, Nankang, Taipei 115, Taiwan.

T. Asano (Ed.): ISAAC 2006, LNCS 4288, pp. 460–473, 2006.

The MAXIMUM SUM PROBLEM can find many applications in pattern recognition, image processing and data mining [1,12]. A natural generalization of the above MAXIMUM SUM PROBLEM is the K MAXIMUM SUMS PROBLEM which is to find the k segments such that their sums are the k largest over all $\frac{n(n-1)}{2}$ segments. Bae and Takaoka [4] presented an $O(kn)$ time algorithm for this problem. Bengtsson and Chen [5] gave an $O(\min\{k + n\log^2 n, nk^{\frac{1}{2}}\})$ time algorithm, or $O(n\log^2 n + k)$ time in the worst case. Cheng et al. [8] recently gave an $O(n + k\log(min\{n, k\}))$ time algorithm for this problem which is superior to Bengtsson and Chen's when k is $o(n\log n)$, but it runs in $O(n^2\log n)$ time in the worst case. Lin and Lee [14] recently gave an expected $O(n\log n + k)$ time randomized algorithm based on a randomized algorithm which finds in expected $O(n\log n)$ time the segment whose sum is the k-th smallest, for any given positive integer $1 \leq k \leq \frac{n(n-1)}{2}$. The latter problem is referred to as the SUM SELECTION PROBLEM. In this paper we will give a deterministic $O(n\log n + k)$ time algorithm for the K MAXIMUM SUMS PROBLEM based on a deterministic $O(n\log n)$ time algorithm for the SUM SELECTION PROBLEM as well.

The rest of the paper is organized as follows. Section 2 give a deterministic algorithm for the SUM SELECTION PROBLEM. Section 3 gives a deterministic algorithm for the K MAXIMUM SUMS PROBLEM. Section 4 gives some conclusion.

2 Algorithm for the Sum Selection Problem

We define the *rank* $r(x, P)$ of an element x in a set $P \subseteq \mathbf{R}$ of real numbers to be the number of elements in P no greater than x, i.e. $r(x, P) = |\{y|y \in P, y \leq x\}|$. Given a sequence A of real numbers $a_1, a_2, \ldots, a_n$, and a positive integer $1 \leq k \leq \frac{n(n-1)}{2}$, the SUM SELECTION PROBLEM is to find the segment $A(i^*, j^*)$ over all $\frac{n(n-1)}{2}$ segments such that the rank of the sum $s(i^*, j^*) = \sum_{t=i^*}^{j^*} a_t$ in the set of possible subsequence sums is k. That is, we would like to find $s^* = s(i^*, j^*)$ for some $i^* < j^*$ such that $r(s^*, P) = k$ where $P = \{s(i, j) \mid s(i, j) = \sum_{t=i}^{j} a_t, 1 \leq i \leq j \leq n\}$.

We will transform the SUM SELECTION PROBLEM into a problem of arrangements of lines in computational geometry in $O(n)$ time as follows. We first define the set $S = \{s_0, s_1, \ldots, s_n\}$ according to the prefix sums of the sequence A, where $s_i = \sum_{t=1}^{i} a_t, i = 1, 2, \ldots, n$ and $s_0 = 0$. We then define two sets of lines $H = \{h_i \mid h_i : y = -s_i, i = 0, 1, \ldots, n\}$ and $V = \{v_i \mid v_i : y = x - s_i, i = 0, 1, \ldots, n\}$ in the plane respectively. For any two lines $h_i \in H$ and $v_j \in V$ with $i < j$, they intersect at the point $p_{ij} = (x_{ij}, y_{ij})$ with abscissa $x_{ij} = s_j - s_i$. It means that the abscissa of the intersection point of any two lines $h_i \in H$ and $v_j \in V$ with $i < j$ is equal to the sum $s(i + 1, j)$ of the segment $A(i + 1, j)$. We say that an intersection point of two lines $h_i \in H$ and $v_j \in V$ is *feasible* if $i < j$. Note that there are totally n^2 intersection points in the arrangements of lines $\mathcal{A}(H \cup V)$ and it contains $\frac{n(n-1)}{2}$ feasible intersection points and $\frac{n(n+1)}{2}$ non-feasible intersection points. An example of the arrangements of lines $\mathcal{A}(H \cup V)$ is shown in Figure 1. Let $X_f = \{x_{ij} \mid p_{ij} = (x_{ij}, y_{ij})$ is a feasible intersection point of

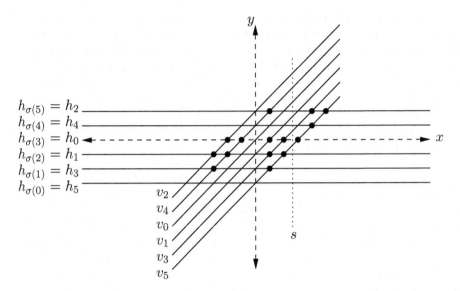

Fig. 1. Given $A = a_1, a_2, a_3, a_4, a_5 = 1, -3, 4, -3, 4$, we have $S = \{s_0, s_1, s_2, s_3, s_4, s_5, s_6\} = \{0, 1, -2, 2, -1, 3\}$, $H = \{h_i \mid h_i : y = -s_i, i = 0, 1, \dots, 6\}$ and $V = \{v_i \mid v_i : y = x - s_i, i = 0, 1, \dots, 6\}$ respectively. The intersection points shown in dark solid dots are the feasible intersection points and others are the non-feasible intersection points.

$\mathcal{A}(H \cup V)\}$. The SUM SELECTION PROBLEM now is equivalent to the following problem.

> Given a set of lines $L = H \cup V = \{h_0, v_0, h_1, v_1, \dots, h_n, v_n\}$ in $\mathbf{R}^2$, where $h_i : y = -s_i$ and $v_j : y = x - s_j$, find the feasible intersection point $p_{i^*j^*} = (x_{i^*j^*}, y_{i^*j^*})$ such that $r(x_{i^*j^*}, X_f) = k$.

Given a set of n lines in the plane and an integer k, $1 \le k \le \frac{n(n-1)}{2}$, the well-known dual problem of the SLOPE SELECTION PROBLEM[1] in computational geometry is to find the intersection point whose x-coordinate is the k-th smallest among all intersection points of these n lines. Cole et al. [10] develop an approximate counting scheme combining the AKS sorting network and parametric search to obtain an optimal $O(n \log n)$ algorithm for this problem. Brönnimann and Chazelle [9] modify their approximate counting scheme combining ε-net to obtain another optimal algorithm for this problem. The SUM SELECTION PROBLEM can be viewed as a variant of the SLOPE SELECTION PROBLEM. Since we don't know how many non-feasible intersection points of $\mathcal{A}(L)$ are to the left of the k-th feasible intersection point, and thus we don't know the actual rank of

---

[1] Given a set of n points in the plane and an integer k, $1 \le k \le \frac{n(n-1)}{2}$, the slope selection problem is to select the pair of points determining the line with the k-th smallest slope.

the k-th feasible intersection point in the set of all intersection points of $\mathcal{A}(L)$. The actual rank of the k-th feasible intersection point may lie between k and $k + \frac{n(n+1)}{2}$ in the set of all intersection points of $\mathcal{A}(L)$. Therefore, we can not solve the SUM SELECTION PROBLEM by fixing some specific rank and applying the slope selection algorithms [9,10] directly. We will give a deterministic algorithm for this problem that runs in $O(n \log n)$ time based on the ingenious parametric search technique of Megiddo [15], AKS sorting network [2] and a new approximate counting scheme. This new approximate counting scheme is a generalization of the approximate counting schemes developed by Cole et al. [10] and Brönnimann and Chazelle [9].

Given a vertical line $x = s$, the number of intersection points of $\mathcal{A}(L)$ on it or to its left is denoted $\mathcal{I}(L, s)$ and the number of feasible intersection points of $\mathcal{A}(L)$ on it or to its left is denoted $\mathcal{I}_f(L, s)$. The vertical order of the lines of L at $x = s$ defines a *permutation* $\pi(s)$ of L at s with $\pi(-\infty)$ being the identity permutation. An example of $\pi(s) = (h_5, h_3, h_1, v_5, h_0, v_3, h_4, v_1, h_2, v_0, v_4, v_2)$ is shown in Figure 1. An *inversion* of a permutation $(p_1, p_2, \ldots, p_n)$ of $\{1, 2, \ldots, n\}$ is a pair of indices $i < j$ with $p_i > p_j$. It is easy to see that the number of inversions, denoted by $I(\pi(s))$, of a permutation $\pi(s)$ is exactly $\mathcal{I}(L, s)$. We define the number of feasible inversions, denoted by $I_f(\pi(s))$, of $\pi(s)$ to be $\mathcal{I}_f(L, s)$. Therefore, the SUM SELECTION PROBLEM is also equivalent to finding some s^* such that $I_f(\pi(s^*)) = k$.

The problem for finding s^* can be viewed as an unusual sorting problem attempting to sort the set of lines L at $x = s^*$ without knowing the value of s^*, i.e. to sort $h_0(s^*), v_0(s^*), h_1(s^*), v_1(s^*), \ldots, h_n(s^*), v_n(s^*)$ in vertical order without knowing the value of s^*. We know that this sort may be achieved in $O(n \log n)$ comparisons. In particular, the questions of the forms "$h_i(s^*) \leq h_j(s^*)$" and "$v_i(s^*) \leq v_j(s^*)$" can be solved in $O(n \log n)$ time by any usual optimal sorting algorithm, since the ordering of h_i's, which is identical to that of v_j's, is independent of s^*. However, the question, q_{ij} of the form "$h_i(s^*) \leq v_j(s^*)$", can be answered by a counting subroutine that given any vertical line $x = s$ it can quickly compute $\mathcal{I}_f(L, s)$, the number of feasible intersection points of $\mathcal{A}(L)$ that lie on it or to its left. There is a simple way to perform this task in $O(n \log n)$ time by Lemma 1 with $s_\ell = -\infty$ and $s_r = s$. Even though we don't know s^*, we can answer the question q_{ij} by finding the x_{ij}, the x-coordinate of intersection point of h_i and v_j in constant time and call the counting subroutine at $x = x_{ij}$. If the return of the subroutine is less than or equal to k, we get $h_i(s^*) \leq v_j(s^*)$. Otherwise we get $h_i(s^*) > v_j(s^*)$. After solving the unusual sorting problem we can obtain the permutation $\pi(s^*)$ without knowing the value of s^*. Then, we can obtain $s^* = \max\{x_{\pi(s^*)[i]\pi(s^*)[i+1]}\}$.

Lemma 1. *([14], Lemma 2) Given a sequence A of n real numbers $a_1, a_2, \ldots, a_n$ and two real numbers s_ℓ, s_r with $s_\ell \leq s_r$, it takes $O(n)$ space and $O(n \log n)$ time to count the total number of segments $A(i, j)$, $1 \leq i \leq j \leq n$, among all $\frac{n(n-1)}{2}$ segments such that their sums $s(i, j)$ satisfy $s_\ell \leq s(i, j) \leq s_r$.*

How can we solve the unusual sorting problem? We will use the parametric search approach running a sequential simulation of a generic parallel sorting algorithm, which attempts to sort the lines along the line $x = s^*$, where s^* is the x-coordinate of the desired k-th leftmost feasible intersection point, without knowing the value of s^*. A naive algorithm is to use a parallel sorting algorithm of depth $O(\log n)$ and $O(n)$ processors developed by Ajtai, Komlós, and Szemerédi [2], and at each parallel step we may perform $\frac{n}{2}$ comparisons between pairs of lines. Since each comparison can be solved in $O(n \log n)$ time and $O(n)$ space following Lemma 1, it takes $O(n^2 \log n)$ time at each parallel step, and $O(n^2 \log^2 n)$ time overall.

However, we can improve it by the following slightly complicated algorithm. That is, we compute the median x_m of the x-coordinates of all the intersection points of these $\frac{n}{2}$ pairs of lines in each parallel step, and call the counting subroutine at x_m, which can answer half of the questions in $O(n \log n)$ time. For the $\frac{n}{4}$ unresolved questions at the same step, we again find the median among the $\frac{n}{4}$ x-coordinates and call the counting subroutine at the median, which can answer half of these $\frac{n}{4}$ unresolved questions in $O(n \log n)$ time. Repeating the above *binary search* process $O(\log n)$ times we can answer all $\frac{n}{2}$ comparisons in $O(n \log^2 n)$ time in each parallel step. We thus obtain an algorithm that runs in $O(n \log^3 n)$ time.

We can further improve $O(n \log^3 n)$ to $O(n \log^2 n)$ by using a technique due to Cole [11] as follows. Instead of invoking $O(\log n)$ counting subroutine calls at each parallel step to resolve all comparisons at this step, we call the counting subroutine only a constant number of times. This of course does not resolve all comparisons of this parallel step, but it does resolve a large fraction of them. All the unresolved comparisons at this step will be deferred to the next parallel step. Suppose that each of the unresolved comparisons can affect only a constant number of comparisons executed at the next parallel step. Each parallel step is now a mixture of many parallel steps. Cole shows that if it is implemented carefully by assigning an appropriate time-dependent weight to each unresolved comparison and choosing the weighted median at each step of the binary search, the number of the parallel steps of the algorithm increases only by an additive $O(\log n)$ steps. Since each of these steps uses only a constant number of counting subroutine calls, the whole running time improves to $O(n \log^2 n)$.

The final step to improve the sum selection algorithm from $O(n \log^2 n)$ to $O(n \log n)$ is to develop an approximate counting scheme. Note that the expensive counting subroutine, Lemma 1, can be used not only to find $\mathcal{I}_f(L, s)$ for each point s given by the sorting network but also to determine the relative ordering of s and s^* in $O(n \log n)$ time. Instead of invoking the expensive counting subroutines $O(\log n)$ times, we shall develop an approximate counting scheme, that counts the number of inversions of desired permutations only approximately, with an error that gets smaller and smaller as we get closer to the desired s^*. The idea of the approximate counting scheme is to use an approximation algorithm in $O(n)$ time for each point s chosen by the sorting network. If the error for the approximation algorithm is small enough, then we can decide the relative ordering of s and s^* directly. Otherwise, we will refine the approximation until we can decide the relative ordering of s and s^*. It turns out that an amortized

$O(n \log n)$ extra time is sufficient to refine approximations throughout the entire course of the algorithm.

We first define an m-block *left-compatible* (resp. *right-compatible*) permutation $\pi_l(s)$ (resp. $\pi_r(s)$) of permutation $\pi(s)$ such that it satisfies $I(\pi_l(s)) \leq I(\pi(s)) \leq I(\pi_l(s)) + mn$ (resp. $I(\pi_r(s)) - mn \leq I(\pi(s)) \leq I(\pi_r(s))$). Let $(\sigma(0), \sigma(1), \ldots, \sigma(n))$ denote the permutation of $\{0, 1, \ldots, n\}$ such that $h_{\sigma(0)}$, $h_{\sigma(1)}, \ldots, h_{\sigma(n)}$ are in the ascending vertical order, i.e. $-s_{\sigma(0)} \leq -s_{\sigma(1)} \leq \ldots \leq -s_{\sigma(n)}$. Let $G_0, G_1, \ldots, G_{\frac{n}{m}}$ be an m-*block* of H for some fixed size m, where group $G_i = \{h_{\sigma(i \cdot m)}, h_{\sigma(i \cdot m+1)}, \ldots, h_{\sigma(i \cdot m+m-1)}\}$. For any i, j, we say that v_i is greater than group G_j at s, denoted by $v_i(s) \succ G_j$, if $v_i(s) = x - s_i > h_{\sigma(j \cdot m+m-1)}(s) = -s_{\sigma(j \cdot m+m-1)}$ where $h_{\sigma(j \cdot m+m-1)}$ is the largest element in group G_j, and we say that v_i is in group G_j at s, denoted by $v_i(s) \sqsubset G_j$, if $h_{\sigma((j-1) \cdot m+m-1)}(s) < v_i(s) \leq h_{\sigma(j \cdot m+m-1)}(s)$.

We define a permutation $\pi_l(s)$ (resp. $\pi_r(s)$) as an m-block *left-compatible* (resp. *right-compatible*) permutation of $\pi(s)$ as follows: Given $h_{\sigma(0)}, h_{\sigma(1)}, \ldots,$ $h_{\sigma(n)}$ sorted in ascending σ-order, $\pi_l(s)$ and $\pi_r(s)$ will be obtained by inserting $v_{\sigma(i)}$, $i = 0, 1, \ldots, n$, one by one in between $h_{\sigma(q_i)}$ and $h_{\sigma(q_i+1)}$ for some q_i such that $v_{\sigma(i)}$'s, $i = 0, 1, \ldots, n$, are also in ascending σ-order. For each $v_{\sigma(i)} \in V$, if $v_{\sigma(i)}(s) \sqsubset G_j$ for some j then we insert $v_{\sigma(i)}$ in between $h_{\sigma((j-1) \cdot m+m-1)}$ and $h_{\sigma(j \cdot m)}$, where $h_{\sigma((j-1) \cdot m+m-1)}$ is the largest element in group G_{j-1} and $h_{\sigma(j \cdot m)}$ is the smallest element in group G_j. (resp. For each $v_{\sigma(i)} \in V$, if $v_{\sigma(i)}(s) \sqsubset G_j$ for some j then we insert $v_{\sigma(i)}$ in between $h_{\sigma(j \cdot m+m-1)}$ and $h_{\sigma((j+1) \cdot m)}$, where $h_{\sigma(j \cdot m+m-1)}$ is the largest element in group G_j and $h_{\sigma((j+1) \cdot m)}$ is the smallest element in group G_{j+1}.) For example, the 2-block left-compatible permutation $\pi_l(s) = (h_5, h_3, v_5, h_1, h_0, v_3, v_1, h_4, h_2, v_0, v_4, v_2)$ and right-compatible permutation $\pi_r(s) = (h_5, h_3, h_1, h_0, v_5, h_4, h_2, v_3, v_1, v_0, v_4, v_2)$ in Figure 1. Therefore, we have

$$I(\pi_l(s)) \leq I(\pi(s)) \leq I(\pi_l(s)) + mn, \; I_f(\pi_l(s)) \leq I_f(\pi(s)) \leq I_f(\pi_l(s)) + mn.$$

$$I(\pi_r(s)) - mn \leq I(\pi(s)) \leq I(\pi_r(s)), \; I_f(\pi_r(s)) - mn \leq I_f(\pi(s)) \leq I_f(\pi_r(s)).$$

Thus, we see that maintaining left-compatible (right-compatible) permutation with $\pi(s)$ gives a good approximation on the number of inversions of the permutation: the smaller the block size m, the finer the approximation.

We now give an $O(n \log n)$ algorithm for the SUM SELECTION PROBLEM as follows. We will first sketch the algorithm and then explain and analyze it in detail in subsequent paragraphs. We assume, for simplicity, that $n = 2^g$ for some integer g and the fractions in this algorithm are integers taken by floor or ceiling functions. We define $sign(s)$ to be 1 if s is a positive real number, 0 if s is zero and -1 if s is a negative real number. The algorithm maintains an interval (s_l, s_r) containing s^*, an m_l-block left-compatible permutation $\pi_l(s_l)$ at s_l and an m_r-block right-compatible permutation $\pi_r(s_r)$ at s_r such that they satisfy invariant conditions $(I1)$ and $(I2)$. An example of an interval (s_l, s_r) containing s^* is shown in Figure 2.

($I1$) $I_f(\pi_l(s_l)) + m_l n \leq I_f(\pi(s^*)) \leq I_f(\pi_r(s_r)) - m_r n.$

($I2$) $I_f(\pi_r(s_r)) - 2m_r n \leq I_f(\pi(s^*)) \leq I_f(\pi_l(s_l)) + 2m_l n.$

($I1$) means that s^* lies within the interval (s_l, s_r). Since $I_f(\pi(s_l)) \leq I_f(\pi_l(s_l)) + m_l n \leq I_f(\pi(s^*)) \leq I_f(\pi_r(s_r)) - m_r n \leq I_f(\pi(s_r))$, we have $s_l \leq s^* \leq s_r$. ($I2$) means that the left-compatible and right-compatible permutations are no finer than needed.

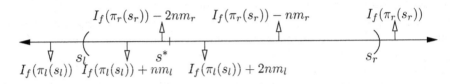

Fig. 2. The sum selection algorithm maintains an interval (s_l, s_r) containing s^* satisfying invariant conditions ($I1$) and ($I2$)

If $k < n$, then we can solve the sum selection problem by using the algorithm due to Cheng et al. [8]. Let us assume $k \geq n$ in the following. To initialize the algorithm we set $m_l = \frac{k}{n}$, $s_l = -\infty$, $\pi_l(s_l) = (v_{\sigma(0)}, v_{\sigma(1)}, \ldots, v_{\sigma(n)}, h_{\sigma(0)}, h_{\sigma(1)}, \ldots, h_{\sigma(n)})$, $I_f(\pi_l(s_l)) = 0$, $m_r = \frac{(n-1)}{4} - \frac{k}{2n}$, $s_r = \infty$, $\pi_r(s_r) = (h_{\sigma(0)}, h_{\sigma(1)}, \ldots, h_{\sigma(n)}, v_{\sigma(0)}, v_{\sigma(1)}, \ldots, v_{\sigma(n)})$, $I_f(\pi_r(s_r)) = \frac{n(n-1)}{2}$ and $I_f(\pi(s^*)) = k$. It is easy to check that this initial condition satisfies ($I1$) and ($I2$).

After coming in a new point s from AKS network combining Cole's technique, we will decide an interval, called *winning interval*, which contains s^* between (s_l, s) and (s, s_r) and maintain invariant conditions ($I1$) and ($I2$) for the winning interval. In order to decide the winning interval and maintain ($I1$) and ($I2$), we need the following four subroutines, each costing $O(n)$ time. The left reblocking subroutine allows us to construct an m_l-block left-compatible permutation $\pi_l(s)$ at s when $I_f(\pi_l(s_l)) + 2m_l n \geq I_f(\pi_l(s))$ holds. We will show later that if $I_f(\pi_l(s_l)) + 2m_l n < I_f(\pi_l(s))$ then (s, s_r) can not be the winning interval so we don't need to construct $\pi_l(s)$. The right reblocking subroutine allows us to construct an m_r-block right-compatible permutation $\pi_r(s)$ at s when $I_f(\pi_r(s)) \geq I_f(\pi_r(s_r)) - 2m_r n$ holds. We will also show later that if $I_f(\pi_r(s)) < I_f(\pi_r(s_r)) - 2m_r n$ then (s_l, s) can not be the winning interval so we don't need to construct $\pi_r(s)$. The left halving subroutine is to construct a $\frac{m_l}{2}$-block left-compatible permutation $\pi_l(s)$ at s. The right halving subroutine is to construct a $\frac{m_r}{2}$-block right-compatible $\pi_r(s)$ at s.

After coming in a new point s, we first do left reblocking m_l and right reblocking m_r at s to construct $\pi_l(s)$ and $\pi_r(s)$ respectively. It divides into three cases: For the case 1: If $I_f(\pi_l(s_l)) + 2m_l n < I_f(\pi_l(s))$ then (s_l, s) will be the winning interval. But if the winning interval (s_l, s) doesn't satisfy ($I1$) and ($I2$), then we will do the right halving $\frac{m_r}{2^1}$, $\frac{m_r}{2^2}$, ... until $\frac{m_r}{2^t}$ such that ($I1$) and ($I2$) hold for (s_l, s). For the case 2: If $I_f(\pi_r(s)) < I_f(\pi_r(s_r)) - 2m_r n$ then (s, s_r) will be the winning interval. But if the winning interval (s, s_r) doesn't satisfy $(I1)$ and $(I2)$, then we will do the left halving $\frac{m_l}{2^1}, \frac{m_l}{2^2}, \ldots$ until $\frac{m_l}{2^t}$ such that $(I1)$ and $(I2)$ hold for (s, s_r). For the case 3: If $I_f(\pi_l(s_l)) + 2m_l n \geq I_f(\pi_l(s))$ and $I_f(\pi_r(s)) \geq I_f(\pi_r(s_r)) - 2m_r n$ then we can not decide the winning interval yet. We will do the left halving and the right halving interleavingly $\frac{m_l}{2^1}, \frac{m_r}{2^1}, \ldots$ until $\frac{m_l}{2^t}$ $(\frac{m_r}{2^t})$ such that $(I1)$ and $(I2)$ hold, then (s, s_r) $((s_l, s))$ will be the winning interval and it satisfies $(I1)$ and $(I2)$ automatically.

After deciding the winning interval, we can decide the relative order of s and s^*. Therefore, we can answer the comparison question at s and the relevant s_i's of which s was the weighted median such that $sign(s - s_i) = sign(s^* - s)$. Then another new point will come in and repeat above procedure again. The algorithm will continue above procedure to make approximations until $m_l < 10$ and $m_r < 10$. When $m_l < 10$ and $m_r < 10$, we know that the winning interval (s_l, s_r) will contain s^* and $O(n)$ feasible intersection points. Let k' be the total number of feasible intersection points in $(-\infty, s_l]$ which can be obtained by the counting subroutine in Lemma 1. Then, we can enumerate all feasible intersection points in the winning interval (s_l, s_r) in $O(n \log n + n) = O(n \log n)$ time by the enumerating subroutine in Lemma 2, and select from those feasible intersection points the $(k - k')$-th feasible intersection point with sum s^* by using any standard selection algorithm in $O(n)$ time. If after the algorithm ends we have either $m_l \geq 10$ or $m_r \geq 10$, at this moment we have solved the unusual sorting problem to obtain $\pi(s^*)$ without knowing the value of s^*. Therefore, we can obtain $s^* = \max\{x_{\pi(s^*)[i]\pi(s^*)[i+1]}\}$.

Lemma 2. (*[14], Lemma 1) Given a sequence A of n real numbers $a_1, a_2, \ldots, a_n$ and two real numbers s_ℓ, s_r with $s_\ell \leq s_r$, it costs $O(n)$ space and $O(n \log n + h)$ time, where h is the output size, to find all segments $A(i, j)$, $1 \leq i \leq j \leq n$, among all $\frac{n(n-1)}{2}$ segments such that their sums $s(i, j)$ satisfy $s_\ell \leq s(i, j) \leq s_r$.*

We now develop the left reblocking, right reblocking, left halving and right halving subroutines and then explain the algorithm and analyze its complexity in detail. We develop left reblocking subroutine as an example since the right reblocking subroutine can be done similarly. The left reblocking subroutine will either construct an m_l-block left-compatible permutation $\pi_l(s)$ when $I_f(\pi_l(s)) - I_f(\pi_l(s_l)) \leq 2m_l n$ holds or output "fail" otherwise. Given an m_l-block left-compatible permutation $\pi_l(s_l)$, the left reblocking subroutine is to find the m_l-block left-compatible permutation $\pi_l(s)$ at s for some $s > s_l$ only if $I_f(\pi_l(s)) - I_f(\pi_l(s_l)) \leq 2m_l n$. At the beginning of the subroutine we just know $s > s_l$, but we don't know whether $I_f(\pi_l(s)) - I_f(\pi_l(s_l))$ is greater than $2m_l n$ or not. But once we found that $I_f(\pi_l(s)) - I_f(\pi_l(s_l)) > 2m_l n$ during running the left reblocking subroutine, we will halt the subroutine immediately and output "fail". Assume that we have had an m_l-block $G_1, G_2, \ldots, G_{\frac{n}{m_l}}$ of H, an m_l-block left-compatible permutation $\pi_l(s_l)$ and $I_f(\pi_l(s_l))$ and maintained an array $d_l[i] = j$ at s_l such that $v_i(s_l) \sqsubset G_j$ for each i, we want to find an m_l-block left-compatible permutation $\pi_l(s)$ and $I_f(\pi_l(s))$ and maintain an array $d_l[i] = j$ at s such that $v_i(s) \sqsubset G_j$ for each i. Let us process the lines of L one by one according the order $v_0, h_0, v_1, h_1, \ldots, v_n, h_n$ to construct $\pi_l(s)$ at s. Initially we set $I_f(\pi_l(s))$ to be $I_f(\pi_l(s_l))$ and current size $c[j] = 0$ for each group G_j, where $c[j]$ denotes the total number of lines in G_j processed so far. While processing v_i we will do the following steps until $v_i(s) \sqsubset G_{d_l[i]}$. If $v_i(s) \succ G_{d_l[i]}$, then $I_f(\pi_l(s))$ is increased by $c[d_l[i]]$ and $d_l[i]$ is increased by 1. While processing h_i, if it is in group G_j then the current size $c[j]$ in group G_j is increased by 1. A detailed description of the left reblocking subroutine is shown in the pseudo code. The whole procedure can be done in $O(n)$ time if $I_f(\pi_l(s)) - I_f(\pi_l(s_l)) \leq 2m_l n$. This can be easily seen by the fact that the total processing time is proportional to the number of times each v_i steps up the groups. But doing so increases rank m_l, and we know that there are at most $2m_l n$ rank between $I(\pi_l(s))$ and $I(\pi_l(s_l))$. Therefore, going up the groups cannot happen more than $O(n)$ times. And once we have found that $I_f(\pi_l(s)) - I_f(\pi_l(s_l)) > 2m_l n$, we will halt the subroutine immediately and output "fail". Therefore, it also costs $O(n)$ time if $I_f(\pi_l(s)) - I_f(\pi_l(s_l)) > 2m_l n$. Thus we have Lemma 3.

Lemma 3. (*Reblocking*) *Given an m_l-block left-compatible permutation $\pi_l(s_l)$ with approximation rank $I_f(\pi_l(s_l))$, we can compute in $O(n)$ time an m_l-block left-compatible permutation $\pi_l(s)$ with approximation rank $I_f(\pi_l(s))$ for any $s > s_l$ when $I_f(\pi_l(s)) - I_f(\pi_l(s_l)) \leq 2m_l n$ holds.*

Subroutine LeftReblocking($s, s_l, m_l, d_l[\cdot]$).
Input: An m_l-block left-compatible permutation $\pi_l(s_l)$
Output: An m_l-block left-compatible permutation $\pi_l(s)$

1. **for** $i = 0$ **to** n **do** $t[i] \leftarrow d_l[i]$;
2. $I_f(\pi_l(s)) \leftarrow I_f(\pi_l(s_l))$; $g \leftarrow 0$;
3. **for** $i = 0$ **to** $\frac{n}{m_l}$ **do** $c[i] \leftarrow 0$;
4. **for** $i = 0$ **to** n **do**
5.   **while** $v_i(s) \succ G_{t[i]}$
6. $I_f(\pi_l(s)) \leftarrow I_f(\pi_l(s)) + c[t[i]]$; $g = g + m_l$; $t[i] \leftarrow t[i] + 1$;
7.   **if** $g > 2m_l n$ **then return** fail;
8.   **if** $h_i(s)$ is in group G_j **then** $c[j] \leftarrow c[j] + 1$;
9. **for** $i = 0$ **to** $\frac{n}{m_l}$ **do** insert $h_{\sigma(i \cdot m_l)}, h_{\sigma(i \cdot m_l + 1)}, \ldots, h_{\sigma(i \cdot m_l + m_l - 1)}$ one by one into list $B[i]$;
10. **for** $i = 0$ **to** n **do** insert $v_{\sigma(i)}$ into list $B[t[i] - 1]$;
11. Concatenate the lists $B[0], B[1], \ldots, B[\frac{n}{m_l}]$ to obtain $\pi_l(s)$;
12. **for** $i = 0$ **to** n **do** $d_l[i] \leftarrow t[i]$;
13. **return** $\pi_l(s)$;

We develop the left halving subroutine as an example as follows. Given an m_l-block left-compatible permutation $\pi_l(s)$, the left halving subroutine is to find an $\frac{m_l}{2}$-block left-compatible permutation $\pi_l(s)$. Assume that we have had an m_l-block $G_1, G_2, \ldots, G_{\frac{n}{m_l}}$ of H, an m_l-block left-compatible permutation $\pi_l(s)$ and $I_f(\pi_l(s))$ and maintained an array $d_l[i] = j$ at s such that $v_i(s) \sqsubset G_j$ for each i, we want to find a $\frac{m_l}{2}$-block left-compatible permutation $\pi_l(s)$ and $I_f(\pi_l(s))$ and maintain the array $d_l[i]$ at s for each i.

Let $G'_1, G'_2, \ldots, G'_{\frac{2n}{m_l}}$ be a $\frac{m_l}{2}$-block of H. Let us process the lines of L one by one according the order $v_0, h_0, v_1, h_1, \ldots, v_n, h_n$ to construct an $\frac{m_l}{2}$-block left-compatible permutation $\pi_l(s)$ at s. It is easy to see that v_i is either in the group $G'_{2d_l[i]}$ or $G'_{2d_l[i]+1}$. Initially we set $d_l[i]$ to be $2d_l[i]$ for each i, and current size $c[j] = 0$ for each group G'_j, where $c[j]$ denotes the total number of lines in G'_j processed so far. While processing v_i, if $v_i(s) \succ G'_{d_l[i]}$, then $I_f(\pi_l(s))$ is increased by $c[d_l[i]]$ and $d_l[i]$ is increased by 1. While processing h_i, if it is in group G'_j then the current size $c[j]$ of group G'_j is increased by 1. A detailed description of the left halving subroutine is shown in the pseudo code. The whole procedure can be done in $O(n)$ time since each v_i steps up at most one group. Thus we have Lemma 4.

Lemma 4. *(Halving) Given an m_l-block left-compatible permutation $\pi_l(s)$ with approximation rank $I_f(\pi_l(s))$ for some s, we can compute in $O(n)$ time a $\frac{m_l}{2}$-block left-compatible permutation $\pi_l(s)$ with approximation rank $I_f(\pi_l(s))$.*

Subroutine LeftHalving(s, m_l, $d_l[\cdot]$).
Input: An m_l-block left-compatible permutation $\pi_l(s)$
Output: A $\frac{m_l}{2}$-block left-compatible permutation $\pi_l(s)$

1. $m'_l \leftarrow \frac{m_l}{2}$;
2. **for** $i = 0$ **to** n **do** $d_l[i] \leftarrow 2d_l[i]$;
3. **for** $i = 0$ **to** $\frac{n}{m_l}$ **do** $c[i] \leftarrow 0$;
4. **for** $i = 0$ **to** n
5.   **if** $v_i(s) \succ G'_{d_l[i]}$ **then** $I_f(\pi_l(s)) \leftarrow I_f(\pi_l(s)) + c[d_l[i]]$; $d_l[i] \leftarrow d_l[i] + 1$;
6.   **if** h_i is in group G'_j **then** $c[j] \leftarrow c[j] + 1$;
7. **for** $i = 0$ **to** $\frac{n}{m_l}$ **do** insert $h_{\sigma(i \cdot m'_l)}, h_{\sigma(i \cdot m'_l + 1)}, \ldots, h_{\sigma(i \cdot m'_l + m'_l - 1)}$ one by one into list $B[i]$;
8. **for** $i = 0$ **to** n **do** insert $v_{\sigma(i)}$ into list $B[d_l[i] - 1]$;
9. Concatenate the lists $B[0], B[1], \ldots, B[\frac{n}{m_l}]$ to obtain $\pi_l(s)$;
10. **return** $\pi_l(s)$;

We now explain the algorithm and analyze its complexity. After coming in a new point s from sorting network, we first do left reblocking m_l and right reblocking m_r at s. If left blocking fails, we have $I_f(\pi(s)) > I_f(\pi_l(s)) > I_f(\pi_l(s_l)) + 2m_l n > I_f(s^*)$. It implies $s > s^*$. Therefore, we can decide (s_l, s) to be the winning interval. But m_r may not be small enough such that (s_l, s) satisfies $(I1)$ and $(I2)$. If so, we do right halving at s until both $(I1)$ and $(I2)$ hold. Similarly, if right blocking fails, we have $I_f(\pi(s)) < I_f(\pi_r(s)) < I_f(\pi_r(s_r)) - 2m_r n < I_f(s^*)$. It implies $s < s^*$. Therefore, we can decide (s, s_r) to be the winning interval. But m_l may not be small enough such that (s, s_r) satisfies $(I1)$ and $(I2)$. If so, we do left halving at s until both $(I1)$ and $(I2)$ hold. If both left blocking and right blocking don't fail then we have $I_f(\pi_l(s_l)) + 2m_l n \geq I_f(\pi_l(s))$ and $I_f(\pi_r(s)) \geq I_f(\pi_r(s_r)) - 2m_r n$. It means that both m_l and m_r are not fine enough to decide the winning interval, so we can not decide the winning interval yet. We will do left halving and right halving at s interleavingly $\frac{m_l}{2^1}$, $\frac{m_r}{2^1}$, ... until $\frac{m_l}{2^t}$ ($\frac{m_r}{2^t}$) such that both ($I1$) and ($I2$) hold, then (s, s_r) $((s_l, s))$ will be the winning interval and it will satisfy ($I1$) and ($I2$) automatically. After deciding the winning interval, we can decide the relative order of s and s^*. Therefore, we can answer the comparison question at s and the relevant s_i's of which s was the weighted median such that $sign(s - s_i) = sign(s^* - s)$. Then another new point will come in and repeat above procedure again.

Since our sorting network is an AKS sorting network combining Cole's technique, the algorithm will invoke $O(1)$ left blocking and right blocking subroutines at each parallel step to resolve all comparisons at this step, each costing $O(n)$, it totally costs $O(n)$ time at each step. The sorting network has depth $O(\log n)$, each parallel step requires $O(n)$, so the algorithm totally costs $O(n \log n)$ time to do left blocking and right blocking. But during the execution of the algorithm the approximation sometimes is not small enough to distinguish the relative ordering of s and s^*, we will refine the approximation until we can decide relative ordering of s and s^*. The algorithm will at most invoke $O(\log n)$ left halving and right halving subroutines, each costing $O(n)$. It turns out that an amortized $O(n \log n)$ extra time will be done to refine approximations throughout the entire course of the algorithm. The correctness of this algorithm follows from the above discussion. Thus, we conclude with the following theorem.

Theorem 1. *The* SUM SELECTION PROBLEM *can be solved in $O(n)$ space and $O(n \log n)$ time.*

The complete pseudo code of the algorithm follows.

Algorithm Sum Selection Problem.
Input: A set of lines $L = H \cup V = \{h_0, v_0, h_1, v_1, \ldots, h_n, v_n\}$ in $\mathbf{R}^2$, where $h_i : y = -s_i$ and $v_j : y = x - s_j$.
Output: The feasible intersection pt. $p_{i^*j^*} = (x_{i^*j^*}, y_{i^*j^*})$ s.t. $r(x_{i^*j^*}, X_f) = k$.

1. $m_l \leftarrow \frac{k}{n}$; $s_l \leftarrow -\infty$; $\pi_l(s_l) \leftarrow (v_{\sigma(0)}, v_{\sigma(1)}, \ldots, v_{\sigma(n)}, h_{\sigma(0)}, h_{\sigma(1)}, \ldots, h_{\sigma(n)})$;
2. $m_r \leftarrow \frac{(n-1)}{4} - \frac{k}{2n}$; $s_r \leftarrow \infty$; $\pi_r(s_r) \leftarrow (h_{\sigma(0)}, h_{\sigma(1)}, \ldots, h_{\sigma(n)}, v_{\sigma(0)}, v_{\sigma(1)}, \ldots, v_{\sigma(n)})$;
3. **for** $i = 0$ **to** n **do** $d_l[i] \leftarrow 0$; $d_r[i] \leftarrow 0$;
4. **while** $m_l > 10$ **or** $m_r > 10$
5.   get next s from AKS network
6.   **if** s is not in (s_l, s_r)
7.   **then** resolve s and the relevant s_i's such that $sign(s - s_i) = sign(s^* - s)$;
8.   **else**
9.   **for** $i = 0$ **to** n **do** $t_l[i] \leftarrow d_l[i]$; $t_r[i] \leftarrow d_r[i]$;
10. $m_l' \leftarrow m_l$; $m_r' \leftarrow m_r$;
11. $\pi_l(s) \leftarrow$ LeftBlocking$(s, s_l, m_l', t_l[\cdot])$;
12. $\pi_r(s) \leftarrow$ RightBlocking$(s, s_r, m_r', t_r[\cdot])$;
13.   **if** LeftBlocking subroutine outputs "fail"
14.   **then**

15.   **if** (s_l, s) doesn't satisfy $(I1)$ and $(I2)$

16.   **then do** $\pi_r(s) \leftarrow$ RightHalving$(s, m_r', t_r[\cdot])$; $m_r' \leftarrow \frac{m_r'}{2}$; **until** $(I1), (I2)$ hold

17.   **if** RightBlocking subroutine outputs "fail"

18.   **then**

19.   **if** (s, s_r) doesn't satisfy $(I1)$ and $(I2)$

20.   **then do** $\pi_l(s) \leftarrow$ LeftHalving$(s, m_l', t_l[\cdot])$; $m_l' \leftarrow \frac{m_l'}{2}$; **until** $(I1)$, $(I2)$ hold

21.   **if** LeftBlocking **and** RightBlocking subroutines don't output "fail"

22.   **then do**

23. $\pi_l(s) \leftarrow$ LeftHalving$(s, m_l', t_l[\cdot])$; $m_l' \leftarrow \frac{m_l'}{2}$;

24. $\pi_r(s) \leftarrow$ RightHalving$(s, m_r', t_r[\cdot])$; $m_r' \leftarrow \frac{m_r'}{2}$;

25.   **until** (s_l, s) satisfies $(I1)$ and $(I2)$ **or** (s, s_r) satisfies $(I1)$ and $(I2)$

26.   **if** (s, s_r) satisfies $(I1)$ and $(I2)$

27.   **then** $s_l \leftarrow s$; $m_l \leftarrow m_l'$; $d_l[\cdot] \leftarrow t_l[\cdot]$; resolve s and the relevant s_i's such that $sign(s - s_i) = sign(s^* - s)$

28.   **else** $s_r \leftarrow s$; $m_r \leftarrow m_r'$; $d_r[\cdot] \leftarrow t_r[\cdot]$; resolve s and the relevant s_i's such that $sign(s - s_i) = sign(s^* - s)$

29.   **if** $m_l \leq 10$ **and** $m_r \leq 10$

30.   **then**

31. $k' \leftarrow$ total number of feasible points in $(-\infty, s_l]$ by Lemma 1

32. $S \leftarrow$ the set of all feasible points in (s_l, s_r) by Lemma 2

33.   **return** $s^* \leftarrow (k - k')$-th element in S by any optimal selection alg.

34. **return** $s^* \leftarrow \max\{x_{\pi(s^*)[i]\pi(s^*)[i+1]}\}$

3 Algorithm for k Maximum Sums Problem

After obtaining the algorithm for the SUM SELECTION PROBLEM, we can use it to obtain the algorithm for K MAXIMUM SUMS PROBLEM directly. We have the following result.

Theorem 2. *The* K MAXIMUM SUMS PROBLEM *can be solved in* $O(n)$ *space and* $O(n \log n + k)$ *time.*

Proof. Let $\ell = \frac{n(n-1)}{2} - k + 1$ and $r = \frac{n(n-1)}{2}$. We can run the algorithm of the SUM SELECTION PROBLEM to obtain the ℓ-th smallest segment s_ℓ and r-th smallest segment s_r respectively in $O(n \log n)$ time and then we can enumerate them by the enumerating subroutine Lemma 2 in the interval $[s_\ell, s_r]$ in $O(n \log n + k)$ time.

4 Conclusion

In the paper we have presented an algorithm for the SUM SELECTION PROBLEM that runs in $O(n \log n)$ time. We then use it to give a more efficient algorithm for the K MAXIMUM SUMS PROBLEM that runs in $O(n \log n + k)$ time. It is better than the previously best known result for the problem, but whether or not one can prove a $\Omega(n \log n)$ lower bound for the SUM SELECTION PROBLEM is of great interest.

References

1. Agrawal, R., Imielinski, T. Swami, A. Data mining using two-dimensional optimized association rules: scheme, algorithms, and visualization. *Proceedings of the 1993 ACM SIGMOD international conference on management of data*, 207-216, 1993.
2. M. Ajtai, J. Komlós, E. Szemerédi. An $O(n \log n)$ sorting networks. *Combinatorica*, 3:1–19, 1983.
3. Alk, S., Guenther, G. Application of broadcasting with selective reduction to the maximal sum subsegment problem. *International journal of high speed computating*, 3:107-119, 1991.
4. Bae, S. E., Takaoka, T. Algorithms for the problem of k maximum sums and a VLSI algorithm for the k maximum subarrays problem. *2004 International Symposium on Parallel Architectures, Algorithms and Networks*, 247-253, 2004.
5. Bengtsson, F., Chen, J. Efficient Algorithms for K Maximum Sums. *Algorithms and Computation, 15th International Symposium, ISAAC 2004*, 137-148.
6. Bentley, J. Programming perals: algorithm design techniques. *Commun. ACM*, 27, 9:865-873, 1984.
7. Bentley, J. Programming perals: algorithm design techniques. *Commun. ACM*, 27, 11:1087-1092, 1984.
8. Chih-Huai Cheng, Kuan-Yu Chen, Wen-Chin Tien, and Kun-Mao Chao Improved Algorithms for the k Maximum-Sums Problems. *Algorithms and Computation, 16th International Symposium, ISAAC 2005*.
9. H. Brönnimann, B. Chazelle. Optimal slope selection via cuttings. *Computational Geometry*, 10:23–29, 1998.
10. R. Cole, J. S. Salowe, W. L. Steiger, and E. Szemeredi. An optimal-time algorithm for slope selection. *SIAM Journal on Computing*, 18(4):792–810, 1989.
11. R. Cole. Slowing down sorign networks to obtain faster sorting algorithm. *Journal of the Association for Computing Machinery*, Vol. 34, No. 1:200–208, 1987.
12. Fukuda, T., Morimoto, Y., Morishita, S. Tokuyama, T. Mining association rules between sets of items in large databases. *Proceedings of the 1996 ACM SIGMOD international conference on management of data*, 13-23, 1996.
13. Gries, D. A note on the standard strategy for developing loop invariants and loops. *Science of Computer Programming*, 2:207-214, 1982.
14. Tien-Ching Lin, D. T. Lee Randomized algorithm for the Sum Selection Problem. *Algorithms and Computation, 16th International Symposium, ISAAC 2005*, 515-523.
15. N. Megiddo. Applying parallel computation algorithms in the design of serial algorithm. *Journal of the Association for Computing Machinery*, Vol. 30, No. 4:852–865, 1983.
16. Perumalla, K., Deo, N. Parallel algorithms for maximum subsequence and maximum subarray. *Parallel Processing Letters*, 5:367-373, 1995.
17. Qiu, K., Alk, S. Parallel maximum sum algorithms on interconnection networks. *Technical Report No. 99-431, Jodrey School of Computer Science, Acadia University, Canada*, 1999.

18. Smith, D. Applications of a strategy for designing divide-and-conquer algorithms. *Science of Computer Programming*, 8:213-229, 1987.
19. Tamaki, H., Tokuyama, T. Algorithms for the maximum subarray problem based on matrix multiplication. *Proceedings of the ninth annual ACM-SIAM symposium on Discrete algorithms*, 446-452, 1998.
20. Takaoka, T. Efficient algorithms for the maximum dubarray problem by fistance matrix multiplication. *Proceedings of the 2002 australian theory symposium*, 189-198, 2002.

Deterministic Random Walks
on the Two-Dimensional Grid

Benjamin Doerr and Tobias Friedrich

Max-Planck-Institut für Informatik, Saarbrücken, Germany

Abstract. Deterministic and randomized balancing schemes are used to distribute workload evenly in networks. In this paper, we compare two very general ones: The random walk and the (deterministic) Propp machine. Roughly speaking, we show that on the two-dimensional grid, the Propp machine always has the same number of tokens on a node as does the random walk in expectation, apart from an additive error of less than eight. This constant is independent of the total number of tokens and the runtime of the two processes. However, we also show that it makes a difference whether the Propp machine serves the neighbors in a circular or non-circular order.

1 Introduction

Given an arbitrary graph, a *random walk* is a path which begins at a given starting point and chooses the next node from the set of its current neighbors uniformly at random. Random walks have been used to model a wide variety of processes in economics (share prices), physics (Brownian motion of molecules in gases and liquids), medicine (cascades of neurons firing in the brain), and mathematics (estimations and modeling of gambling). In computer science, they are the heart of many randomized algorithms.

Jim Propp suggested the following quasirandom analogue to the random walk. The study of quasirandom approaches is motivated by the experience that in many applications they proved to be superior to random ones. Propp's *rotor-router model*, which we prefer to call *Propp machine*, is a simple deterministic process. Each vertex is equipped with a "rotor" which points to one of its neighbors. Instead of leaving a vertex in a random direction, the Propp walk follows the direction of the rotor. Afterwards, the rotor is updated to point to a new neighbor. For this, it follows a fixed cyclic permutation of the adjacent vertices.

The rotors ensure a very high degree of fairness. If a Propp walk visits some vertex v exactly k times, then for each neighbor w of v it does the passage $v \to w$ either $\lfloor k/\deg(v) \rfloor$ or $\lceil k/\deg(v) \rceil$ times. While in the random walk all these numbers are $k/\deg(v)$ in expectation, with high probability they deviate from that by $\Theta(\sqrt{k/\deg(v)})$. Therefore, in this respect the Propp machine is a better simulation for the "expected random walk" than the random walk itself.

The Propp machine found considerable attention recently [2, 3, 5, 6]. In this paper, we compare the two models with respect to their balancing behavior,

T. Asano (Ed.): ISAAC 2006, LNCS 4288, pp. 474–483, 2006.

another well-known application of random walks in computer science. In this setting, each vertex holds a number of chips. These chips simultaneously and in a synchronized manner perform a walk (random or Propp, depending on the model). Clearly, we expect both model to reduce imbalances between the occupation of vertices in each time step. As an idealized balancing scheme, we also regard the *linear machine*. Here, we allow fractional chips, and in each time step each vertex sends out exactly the same number (possibly non-integral) of chips to each neighbor. Clearly, the linear machine describes what the random walk does *in expectation*.

A well-studied problem is *balancing the workload* in a massively parallel computer or distributed network [9]. Each node/processor initially has a collection of tasks and the object is to balance the number of tasks at each node by moving them to neighboring nodes. Hence, processors are modeled as vertices of an undirected connected graph and links between them as edges. Jobs are modeled as unit-size tokens. There are two models prevalent in the literature: the dimension exchange and the diffusion paradigm. In both models each node's decisions are based only on local knowledge.

The intuition behind *diffusion* is that the number of tokens a processor sends to its neighbor is proportional to the differential between the two processors and some constant depending on the connecting edge [4]. A standard choice for these constants is uniform diffusion, in which each processor simply averages the loads of its neighbors at each step. This is usually modeled by allowing fractional tasks and ignoring the roundings to whole tasks at each local balancing step. However, the resulting deviations can become quite large, as was shown in [8]. To quantify the deviation between the fractional problem and the (real-world) integer problem is an important step in understanding diffusive load-balancing.

Dimension exchange refers to (periodic) balancing circuits [7]. This model is particularly suited to single-port architectures, where processors can only communicate with one of its d neighbors at a time. It decomposes the network in a sequence $M_1, \ldots, M_d$ of perfect matchings and adds a balancer at each edge. The balancer is a simple toggling device pointing to one of the incident nodes. Its purpose is to balance the flow of chips along the wires. Each balancing round consists of d steps, one for each matching. In step k, two nodes i, j which are matched in M_k balance their loads x_i, x_j as closely as possible, i.e., their loads become $\lfloor \frac{x_i + x_j}{2} \rfloor$ and $\lceil \frac{x_i + x_j}{2} \rceil$ with the excess chip following the balancer of the edge $\{i, j\}$.

Both models raise the question how well such quasirandom approaches simulate the random one, in particular, how close they get to the idealized linear approach (the random approach "in expectation").

In order to stick not too closely to a particular workload balancing approach for a particular distributed network, we analyze what we think is a sufficiently general but simple model. We compare the classical random walk (respectively the linear machine describing its expectation) with the Propp machine. Hence we do not have weights attached to the edges. As underlying graph we choose the two-dimensional infinite grid. Clearly, infinity is not a realistic assumption in a computer network setting. However, since in finite time chips can walk only a finite distance, the behavior we detect also occurs on finite grids. Hence the infinity assumption is rather used to get rid of some extra technicalities. Also, we assume that our results can be extended to settings with weights attached to the edges, as well as to other graph topologies or a setting where chips may also stay on a vertex. We would like to stress that the focus of our research is more fundamental than oriented to direct applicability. We do feel, however, that the same questions for a particular balancing setting are highly relevant from the view-point of application. Also, we are convinced that our methods can be applied there in an analogous manner.

To measure the difference between the two models, we estimate the maximal difference of the number of chips at any position and time in the Propp model compared to the number in the linear model if both models start with the same initial configuration. Apart from a technicality, which we defer to Section 2, Cooper and Spencer [2] proved for all grids $\mathbb{Z}^d$ that this discrepancy can be bounded by a constant c_d, which only depends on the dimension. In particular, c_d is independent of the initial configuration, the runtime, and the cyclic permutation of the cardinal directions (*rotor sequence*). For the graph being the infinite path, Cooper et al. [3] showed that this constant is $c_1 \approx 2.3$. They further proved that the discrepancy at the origin is maximized if each position sends exactly one odd chip at a certain time in the direction of the origin.

In this paper, we continue this work and rigorously analyze the Propp machine on the two-dimensional grid $\mathbb{Z}^2$. In comparison to the one-dimensional case, a number of new effects appear. In particular, the order in which the four neighbors are served now makes a difference. We prove $c_2 \approx 7.8$ for circular (i.e., clockwise or counterclockwise) rotor sequences and $c_2 \approx 7.3$ for all other rotor sequences. This is the first paper which regards the influence of these rotor sequences. We also precisely characterize the respective worst-case configurations. In particular, we prove that the worst-case is already reached when each position sends at most nine odd chips at at most three different times.

2   Preliminaries

To simplify the calculations, we rotate the grid by $45°$ and consider neighbors in directions $\mathrm{DIR} := \{ \nearrow, \searrow, \swarrow, \nwarrow \}$. Note that by this, we only allow chips on positions $\mathbf{x} = \binom{x_1}{x_2}$ with $x_1 \equiv x_2 \pmod 2$. Since this model is isomorphic to the standard two-dimensional grid model with neighbors $\{\uparrow, \rightarrow, \downarrow, \leftarrow\}$, our result also holds for the standard model.

First, we fix some notation to describe chips on the grid. For $\mathbf{x} \in \mathbb{Z}^2$ and $t \in \mathbb{N}_0$, $\mathbf{x} \sim t$ denotes that $x_1 \equiv x_2 \equiv t \pmod 2$ and $\mathbf{x} \sim \mathbf{y}$ denotes that $x_1 \equiv x_2 \equiv y_1 \equiv y_2 \pmod 2$. A position $\mathbf{x}$ is called "even" or "odd" if $\mathbf{x} \sim 0$ or $\mathbf{x} \sim 1$, respectively. A configuration is called "even" or "odd" if all chips are at an even or all at an odd positions, respectively.

As pointed out in the introduction, there is one limitation without which neither the results of [2, 3] nor our results hold. Note that since $\mathbb{Z}^2$ is a bipartite graph, chips that start on *even* positions never mix with those starting on *odd* positions. It looks like we are playing two games at once. However, this is not true, because chips of different parity may affect each other through the rotors. Within each game the number of chips send in the four directions is not balanced at each position. One can cleverly arrange piles of off-parity chips to reorient rotors and steer them away from random walk simulation. We therefore require the starting configuration to have chips only on *one* parity. Without loss of generality, we consider only even starting configurations.

A random walk on $\mathbb{Z}^2$ can be described nicely by its probability density. By $H(\mathbf{x}, t)$ we denote the probability that a chip from location $\mathbf{x}$ arrives at the origin after t random steps (at time t) in a simple random walk. On a grid as defined above, this is

$$H(\mathbf{x}, t) = 4^{-t} \binom{t}{(t+x_1)/2} \binom{t}{(t+x_2)/2}$$

for $\mathbf{x} \sim t$ and $\|\mathbf{x}\|_\infty \leq t$, and $H(\mathbf{x}, t) = 0$ otherwise.

The order, in which the four neighbors in directions DIR are served, has a significant impact on the discrepancy between Propp and linear machine. We use the same rotor sequence for all positions and describe it by a cyclic function NEXT: DIR → DIR. Implicitly in the following notations, we fix the rotor sequence as well as the starting configuration. That is, the number of chips on vertices and rotor directions. Let $f(\mathbf{x}, t)$ denote the number of chips and ARR$(\mathbf{x}, t)$ the direction of the arrow at position $\mathbf{x}$ after t steps of the Propp machine. Note that with this we can determine the resulting arrow after one Propp step via ARR$(\mathbf{x}, t + 1) = \text{NEXT}^{f(\mathbf{x}, t)}(\text{ARR}(\mathbf{x}, t))$.

Let $E(\mathbf{x}, t)$ denote the expected number of chips at location $\mathbf{x}$ after a random walk of all chips for t steps. In the proofs, we also need the following mixed notation. By $E(\mathbf{x}, t_1, t_2)$ we denote the expected number of chips at location x after first performing t_1 Propp and then $t_2 - t_1$ random walk steps.

3 Parity-Forcing Theorem

For a deterministic process like the Propp machine, it is obvious that the initial configuration (that is, the position of each chip and the direction of each rotor), determines all subsequent configurations. The following theorem shows a partial converse, namely that (roughly speaking) we may prescribe the number of chips modulo 4 on all vertices at all times and still find an initial configuration leading to such a game. An analogous result for the one-dimensional Propp machine has been shown in [3].

Theorem 1 (Parity-forcing Theorem). *For any fixed rotor sequence, any initial position of the arrows and any $\pi: \mathbb{Z}^2 \times \mathbb{N}_0 \to \{0, 1, 2, 3\}$ with $\pi(\mathbf{x}, t) = 0$ for all $\mathbf{x} \not\sim t$, there is an initial even configuration $f(\mathbf{x}, 0)$, $\mathbf{x} \in \mathbb{Z}^2$ that results in a game with $f(\mathbf{x}, t) \equiv \pi(\mathbf{x}, t) \pmod{4}$ for all $\mathbf{x}$ and t.*

The proof is based on the observation that a pile of 4^t chips splits evenly t times. The details of this proof (and all proofs coming) are deferred to the full version of this paper.

4   The Basic Method

In this section we derive the main equations to compare Propp and linear machine based on the number of chips on a single vertex. We are interested in bounding the discrepancies $f(\mathbf{x},t) - E(\mathbf{x},t)$ for all vertices $\mathbf{x}$ and all times t. Since we aim at bounds independent of the starting configuration, it suffices to regard the vertex $\mathbf{x} = \mathbf{0}$. With

$$E(\mathbf{0},0,t) = E(\mathbf{0},t),$$
$$E(\mathbf{0},t,t) = f(\mathbf{0},t),$$

we get

$$f(\mathbf{0},t) - E(\mathbf{0},t) = \sum_{s=0}^{t-1} (E(\mathbf{0},s+1,t) - E(\mathbf{0},s,t)). \tag{1}$$

By $\mathrm{REM}_s^{(j)}$ we denote the set of positions that are occupied by k chips with $k \equiv j \pmod 4$ at time s. Note that if a position contains four chips, then these four chips behave identically on the Propp and linear machine. With this, we obtain

$$
\begin{aligned}
E(\mathbf{0}, s+1, t) &- E(\mathbf{0}, s, t) \tag{2}\\
&= \sum_{\mathbf{x}\in\mathrm{REM}_s^{(1)}} \big(H(\mathbf{x}+\mathrm{ARR}(\mathbf{x},s), t-s-1) - H(\mathbf{x}, t-s)\big)\\
&+ \sum_{\mathbf{x}\in\mathrm{REM}_s^{(2)}} \big(H(\mathbf{x}+\mathrm{ARR}(\mathbf{x},s), t-s-1)\\
&\qquad +H(\mathbf{x}+\mathrm{NEXT}(\mathrm{ARR}(\mathbf{x},s)), t-s-1)\\
&\qquad -2H(\mathbf{x}, t-s))\\
&+ \sum_{\mathbf{x}\in\mathrm{REM}_s^{(3)}} \big(H(\mathbf{x}+\mathrm{ARR}(\mathbf{x},s), t-s-1)\\
&\qquad +H(\mathbf{x}+\mathrm{NEXT}(\mathrm{ARR}(\mathbf{x},s)), t-s-1)\\
&\qquad +H(\mathbf{x}+\mathrm{NEXT}^2(\mathrm{ARR}(\mathbf{x},s)), t-s-1)\\
&\qquad -3H(\mathbf{x}, t-s)).
\end{aligned}
$$

We now regard single chips and define $s_i(\mathbf{x}) := \min\{u \ge 0 \mid i < \sum_{t=0}^{u} f(\mathbf{x},t)\}$ for all $i \in \mathbb{N}_0$, i.e., at time $s_i(\mathbf{x})$ the location $\mathbf{x}$ is occupied by its i-th chip (counting from 0). With

$$\mathrm{INF}(\mathbf{x}, \mathbf{A}, t) := H(\mathbf{x}+\mathbf{A}, t-1) - H(\mathbf{x}, t)$$

for $\mathbf{x} \sim t$ and $\mathrm{INF}(\mathbf{x},\mathbf{A},t) = 0$ otherwise, we denote the influence of position $\mathbf{x}$ with the arrow pointing to $\mathbf{A}$ at time t to the discrepancy between Propp and linear machine. A simple calculation yields

$$\mathrm{INF}(\mathbf{x}, \mathbf{A}, t) = \big((A_1 x_1 \cdot A_2 x_2)t^{-2} - (A_1 x_1 + A_2 x_2)t^{-1}\big) H(\mathbf{x}, t).$$

With these notations, Equations (1) and (2) give

$$f(\mathbf{0}, T) - E(\mathbf{0}, T) \;=\; \sum_{\mathbf{x} \in \mathbb{Z}^2} \sum_{i \geq 0} \text{INF}(\mathbf{x}, \text{NEXT}^i(\text{ARR}(\mathbf{x}, 0)), T - s_i(\mathbf{x})). \qquad (3)$$

This is the main equation, which will be examined in the remainder. It shows that the discrepancy is just the sum of the contributions

$$\text{CON}(\mathbf{x}) := \sum_{i \geq 0} \text{INF}(\mathbf{x}, \text{NEXT}^i(\text{ARR}(\mathbf{x}, 0)), T - s_i(\mathbf{x}))$$

at all positions $\mathbf{x}$. This is a very important observation since this allows us to examine each position $\mathbf{x}$ separately.

5 The Modes of $\text{INF}(\mathbf{x}, \mathbf{A}, t)$

In the previous section, we expressed the discrepancy as a sum of certain influences $\text{INF}(\mathbf{x}, \mathbf{A}, t)$. We now analyze $\text{INF}(\mathbf{x}, \mathbf{A}, t)$. Let $X \subseteq \mathbb{R}$. We call a mapping $f \colon X \to \mathbb{R}$ *unimodal*, if there is an $m \in X$ such that $f|_{x \leq m}$ as well as $f|_{x \geq m}$ are monotone. We call a mapping $f \colon X \to \mathbb{R}$ *bimodal*, if there are $m_1, m_2 \in X$ such that $f|_{x \leq m_1}$, $f|_{m_1 \leq x \leq m_2}$, and $f|_{m_2 \leq x}$ are monotone. We call a mapping $f \colon X \to \mathbb{R}$ *strictly bimodal*, if it is bimodal, but not unimodal.

Unimodal functions are popular in optimization and probability theory. The probability $H(\mathbf{x}, t)$ that a chip from the origin arrives at location $\mathbf{x}$ at time t in a simple random walk is unimodal in t. For our purposes it is important that $\text{INF}(\mathbf{x}, \mathbf{A}, t)$ is bimodal. To prove this, we need Descartes' Rule of Signs, which can be found in [1]. With this, we are now well equipped to characterize $\text{INF}(\mathbf{x}, \mathbf{A}, t)$ precisely.

Lemma 2. *For all $\mathbf{x}$ and $\mathbf{A}$, $\text{INF}(\mathbf{x}, \mathbf{A}, t)$ is bimodal in t. It is strictly bimodal in t if and only if (i) $\|\mathbf{x}\|_\infty > 6$ and (ii) $-A_1 x_1 > A_2 x_2 > (-A_1 x_1 + 1)/2$ or $-A_2 x_2 > A_1 x_1 > (-A_2 x_2 + 1)/2$.*

We will also need the extrema of certain sums of INF's. The following lemma shows that

$$\text{INF2}(\mathbf{x}, \mathbf{A}^{(1)}, \mathbf{A}^{(2)}, t) := \text{INF}(\mathbf{x}, \mathbf{A}^{(1)}, t) + \text{INF}(\mathbf{x}, \mathbf{A}^{(2)}, t)$$

has even nicer properties than INF itself. In particular, $\text{INF2}(\mathbf{x}, \mathbf{A}^{(1)}, \mathbf{A}^{(2)}, t)$ is never strictly bimodal in t for $\mathbf{A}^{(1)} \neq \mathbf{A}^{(2)}$.

Lemma 3. *For all $\mathbf{x}$ and $\mathbf{A}^{(1)} \neq \mathbf{A}^{(2)}$, $\text{INF2}(\mathbf{x}, \mathbf{A}^{(1)}, \mathbf{A}^{(2)}, t)$ is unimodal in t.*

6 Worst-Case Behavior

In Section 4 we derived that for a fixed initial configuration, the single vertex discrepancy is the sum of the contributions

$$\text{CON}(\mathbf{x}) := \sum_{i \geq 0} \text{INF}(\mathbf{x}, A^{(i)}, t_i)$$

of all positions $\mathbf{x}$ with $A^{(i)} := \text{NEXT}^i(\text{ARR}(\mathbf{x}, 0))$ and $t_i := T - s_i(\mathbf{x})$. In this section we determine for each rotor sequence and each position $\mathbf{x}$ the maximum of $\text{CON}(\mathbf{x})$ for all initial configurations. We denote this by $\text{MAXCON}(\mathbf{x})$. Hence the sum of all $\text{MAXCON}(\mathbf{x})$ gives an upper bound for the single vertex discrepancy. Due to the parity-forcing Theorem 1, there is also an initial configuration which sends from all locations $\mathbf{x}$ (apart from multiples of four) exactly the number of chips from $\mathbf{x}$ as the configurations with $\text{CON}(\mathbf{x}) = \text{MAXCON}(\mathbf{x})$. Thus, the upper bound is tight:

$$f(\mathbf{0}, T) - E(\mathbf{0}, T) = \sum_{\mathbf{x} \in \mathbb{Z}^2} \text{MAXCON}(\mathbf{x}). \tag{4}$$

We now fix a rotor sequence and a position $\mathbf{x}$ and examine $\text{MAXCON}(\mathbf{x})$. Lemmas 2 and 3 prove that $\text{INF}(\mathbf{x}, \mathbf{A}, t)$ and $\text{INF}(\mathbf{x}, \mathbf{A}^{(1)}, t) + \text{INF}(\mathbf{x}, \mathbf{A}^{(2)}, t)$ are bimodal in t. We observe that sending a chip in each direction at the same time does not change $\text{CON}(\mathbf{x})$. For all t we have

$$\sum_{\mathbf{A} \in \{\nearrow, \searrow, \swarrow, \nwarrow\}} \text{INF}(\mathbf{x}, \mathbf{A}, t) = 0. \tag{5}$$

This shows that all $\sum_i \text{INF}(\mathbf{x}, \mathbf{A}^{(i)}, t)$ are bimodal in t. Since $\text{INF}(\mathbf{x}, \mathbf{A}, 0) = \lim_{t \to \infty} \text{INF}(\mathbf{x}, \mathbf{A}, t) = 0$ for all $\mathbf{A}$, each $\sum_i \text{INF}(\mathbf{x}, \mathbf{A}^{(i)}, t)$ has at most two extremal times. The set of all extremal times of all $\sum_i \text{INF}(\mathbf{x}, \mathbf{A}^{(i)}, t)$ can be defined as follows.

$$\text{EX}(\mathbf{x}) := \bigcup_{\substack{\mathbf{A}^{(1)}, \mathbf{A}^{(2)} \in \{\nearrow, \searrow, \swarrow, \nwarrow\} \\ \max_t \text{INF2}(\mathbf{x}, \mathbf{A}^{(1)}, \mathbf{A}^{(2)}, t) > 0}} \operatorname*{argmax}_t \text{INF2}(\mathbf{x}, \mathbf{A}^{(1)}, \mathbf{A}^{(2)}, t) \ \cup$$

$$\bigcup_{\substack{\mathbf{A}^{(1)}, \mathbf{A}^{(2)} \in \{\nearrow, \searrow, \swarrow, \nwarrow\} \\ \min_t \text{INF2}(\mathbf{x}, \mathbf{A}^{(1)}, \mathbf{A}^{(2)}, t) < 0}} \operatorname*{argmin}_t \text{INF2}(\mathbf{x}, \mathbf{A}^{(1)}, \mathbf{A}^{(2)}, t)$$

with $\operatorname{argmax}_t f(t) := \max\{s \mid f(s) = \max_t f(t)\}$ and $\operatorname{argmin}_t f(t) := \max\{s \mid f(s) = \min_t f(t)\}$. Notice that

$$\text{INF}(\mathbf{x}, \mathbf{A}, t) = \tfrac{1}{2}\text{INF2}(\mathbf{x}, \mathbf{A}, \mathbf{A}, t),$$

$$\sum_{i=0}^{1} \text{INF}(\mathbf{x}, \mathbf{A}^{(i)}, t) = \text{INF2}(\mathbf{x}, \mathbf{A}^{(0)}, \mathbf{A}^{(1)}, t), \text{ and}$$

$$\sum_{i=0}^{2} \text{INF}(\mathbf{x}, \mathbf{A}^{(i)}, t) = -\tfrac{1}{2}\text{INF2}(\mathbf{x}, \mathbf{A}^{(3)}, \mathbf{A}^{(3)}, t).$$

Lemmas 2 and 3 imply $|\text{EX}(\mathbf{x})| \leq 7$. By calculating the roots of the polynomials $p(\mathbf{x}, \mathbf{A}, t)$ and $p(\mathbf{x}, \mathbf{A}^{(1)}, t) + p(\mathbf{x}, \mathbf{A}^{(2)}, t)$ given in Lemma 2, it is easy to determine $\text{EX}(\mathbf{x})$. In Lemma 4 below we will show that there are configurations with $\text{CON}(\mathbf{x}) = \text{MAXCON}(\mathbf{x})$ where chips are only send from $\mathbf{x}$ at times $\text{EX}(\mathbf{x})$. This proof is based on the following blocking argument.

A *phase* (of $\mathbf{x}$) denotes a maximal period of time, in which all sums of $\text{INF}(\mathbf{x}, \mathbf{A}, t)$'s are monotonic. That is, the upper and lower limit of a phase is either an extremal time $t \in \text{EX}(\mathbf{x})$, or 0, or T. Note that we can assume $T > \max \text{EX}(\mathbf{x})$ and that the monotonicity is uniquely determined for all $\mathbf{A}$ in each phase.

A *block* denotes four consecutive chips, at (possibly different) times $t_j, \dots, t_{j+3}$ send from $\mathbf{x}$ within one phase such that $\sum_{i=j}^{j+k} \text{INF}(\mathbf{x}, \mathbf{A}^{(i)}, t)$ is monotonically increasing in t for all $k \in \{0, 1, 2\}$. By Equation (5), this is equivalent to $\sum_{i=j+k}^{j+3} \text{INF}(\mathbf{x}, \mathbf{A}^{(i)}, t)$ being monotonically decreasing in t for all $k \in \{1, 2, 3\}$.

A *block prefix* denotes $0 \le \ell \le 3$ consecutive chips at times $t_j, \dots, t_{j+\ell-1}$ send from $\mathbf{x}$ within one phase with $\sum_{i=j}^{j+k} \text{INF}(\mathbf{x}, \mathbf{A}^{(i)}, t)$ monotonically increasing in t for all $0 \le k < \ell$. A *block suffix* denotes $0 \le \ell \le 3$ consecutive chips at times $t_j, \dots, t_{j+\ell-1}$ send from $\mathbf{x}$ within one phase with $\sum_{i=j+k}^{j+\ell-1} \text{INF}(\mathbf{x}, \mathbf{A}^{(i)}, t)$ monotonically decreasing in t for all $0 \le k < \ell$.

To describe chips in a phase, we use $\rightarrow$ and $\leftarrow$ to denote chips send in arrow direction $\mathbf{A}^{(i)}$ whose $\text{INF}(\mathbf{x}, \mathbf{A}^{(i)}, t)$ is increasing or decreasing in t, respectively. With this notation, there are four types of blocks: $\rightarrow\leftarrow\leftarrow\leftarrow$, $\rightarrow\rightarrow\leftarrow\leftarrow$, $\rightarrow\rightarrow\rightarrow\leftarrow$, and $\rightarrow\leftarrow\rightarrow\leftarrow$. There are four important properties of blocks, which are shown easily:

- For all blocks, there is a common time t with $t_j \le t \le t_{j+3}$ such that $\sum_{i=j}^{j+3} \text{INF}(\mathbf{x}, \mathbf{A}^i, t_i) \le \sum_{i=j}^{j+3} \text{INF}(\mathbf{x}, \mathbf{A}^i, t) = 0$. Hence, removing a block does not decrease $\text{CON}(\mathbf{x})$.
- For all block suffixes and prefixes, there is a common time t with $t_j \le t \le t_{j+\ell-1}$ such that $\sum_{i=j}^{j+\ell-1} \text{INF}(\mathbf{x}, \mathbf{A}^i, t_i) \le \sum_{i=j}^{j+\ell-1} \text{INF}(\mathbf{x}, \mathbf{A}^i, t)$. Hence, sending the ℓ chips of a block suffix or prefix at a common time t instead at times $t_j, \dots, t_{j+\ell-1}$ does not decrease $\text{CON}(\mathbf{x})$.
- In each phase, the block type is uniquely determined by the monotonicity of INF and INF2.
- Any sequence of chips sent within one phase, can be partitioned in a block suffix, zero or more blocks, and a block prefix.

Lemma 4. *There is an initial configuration with* $\text{CON}(\mathbf{x}) = \text{MAXCON}(\mathbf{x})$ *such that there are only chips send from $\mathbf{x}$ at times* $\text{EX}(\mathbf{x})$.

Lemma 4 shows that configurations with $\text{CON}(\mathbf{x}) = \text{MAXCON}(\mathbf{x})$ that send a minimal number of chips, only send chips from $\mathbf{x}$ at times $\text{EX}(\mathbf{x})$. With $\text{MAXCON}_t(\mathbf{x})$ denoting the contribution at time $t \in \text{EX}(\mathbf{x})$, we obtain

$$\text{MAXCON}(\mathbf{x}) = \sum_{t \in \text{EX}(\mathbf{x})} \text{MAXCON}_t(\mathbf{x}).$$

The following lemma proves that $\text{MAXCON}_t(\mathbf{x})$ is uniquely determined.

Lemma 5. $\text{MAXCON}_t(\mathbf{x})$ *is uniquely determined by the monotonicity of* INF *and* INF2 *in the two adjacent phases.*

For all $\mathbf{x}$ we can now characterize exactly the configuration with $\text{CON}(\mathbf{x}) = \text{MAXCON}(\mathbf{x})$ that sends the least number of chips. Note that by Equation (4),

$\sum_{\mathbf{x}} \text{MAXCON}(\mathbf{x})$ is not only an upper bound for the single vertex discrepancy $f(\mathbf{0}, T) - E(\mathbf{0}, T)$, but also a lower bound. With the help of a computer, it is now easy to sum up over a large number of positions $\mathbf{x}$ and to calculate

$$\sum_{\|x\|_\infty \leq 800} \text{MAXCON}(\mathbf{x}) \approx \begin{cases} 7.831 & \text{for circular rotor sequences} \\ 7.285 & \text{for other rotor sequences.} \end{cases}$$

Notice that these constants are just lower bounds for the single vertex discrepancy. To prove that they are upper bounds as well, we have to bound $E := \sum_{\|x\|_\infty > 800} \text{CON}(\mathbf{x})$. Equation (3) yields

$$E \leq \sum_{\|x\|_\infty > 800} \left(\sum_{i \geq 0} -A_1^{(i)}(\mathbf{x}) \frac{x_1 H(\mathbf{x}, t - s_i(\mathbf{x}))}{t - s_i(\mathbf{x})} + \right.$$
$$\sum_{i \geq 0} -A_2^{(i)}(\mathbf{x}) \frac{x_2 H(\mathbf{x}, t - s_i(\mathbf{x}))}{t - s_i(\mathbf{x})} +$$
$$\left. \sum_{i \geq 0} A_1^{(i)}(\mathbf{x}) A_2^{(i)}(\mathbf{x}) \frac{x_1 x_2 H(\mathbf{x}, t - s_i(\mathbf{x}))}{(t - s_i(\mathbf{x}))^2} \right) \tag{6}$$

for all times $t \in \mathbb{N}_0$. Note that, independent of the chosen rotor sequence, each of the sequences $(A_1^{(i)}(\mathbf{x}))_{i \geq 0}$, $(A_2^{(i)}(\mathbf{x}))_{i \geq 0}$, and $(A_1^{(i)}(\mathbf{x}) A_2^{(i)}(\mathbf{x}))_{i \geq 0}$ are either strictly or in groups of two alternating. To bound the alternating sums of Equation (6), we need the following elementary fact.

Lemma 6. *Let $t_0, \ldots, t_n \in X \subseteq \mathbb{R}$ such that $t_0 \leq \ldots \leq t_n$. Let $f : X \to \mathbb{R}$ be non-negative and unimodal. If $A^{(i)}$ is either strictly alternating or alternating in groups of two, then*

$$\left| \sum_{i=0}^n A^{(i)} f(t_i) \right| \leq 2 \max_{x \in X} f(x).$$

The following lemma shows that (in contrast to INF) the three summands of Equation (6) are indeed always unimodal.

Lemma 7. *$H(\mathbf{x}, t)/t$ and $H(\mathbf{x}, t)/t^2$ are unimodal functions in t. Denote their global maxima with $t_{\max}(\mathbf{x})$ and $t'_{\max}(\mathbf{x})$, respectively. Then, $(x_1^2 + x_2^2)/4 - 2 \leq t_{\max}(\mathbf{x}) \leq (x_1^2 + x_2^2)/4 + 1$ and $(x_1^2 + x_2^2)/6 - 1 \leq t'_{\max}(\mathbf{x}) \leq (x_1^2 + x_2^2)/6 + 2$.*

By bounding the infinite sums with definite integrals, and applying Lemmas 6 and 7 we get $E < 0.15$, which finally proves

$$f(\mathbf{0}, T) - E(\mathbf{0}, T) \approx \begin{cases} 7.8 & \text{for circular rotor sequences} \\ 7.3 & \text{for other rotor sequences.} \end{cases}$$

References

[1] G. E. Collins and A. G. Akritas. Polynomial real root isolation using descarte's rule of signs. In *SYMSAC '76: Proceedings of the third ACM symposium on Symbolic and algebraic computation*, pp. 272–275, New York, NY, USA, 1976. ACM Press.

[2] J. Cooper and J. Spencer. Simulating a random walk with constant error. *Combinatorics, Probability and Computing*. (Also available at arXiv:math.CO/0402323).

[3] J. Cooper, B. Doerr, J. Spencer, and G. Tardos. Deterministic random walks. In *ANALCO'06: Proceedings of the Workshop on Analytic Algorithmics and Combinatorics*, pp. 185–197, Philadelphia, PA, 2006. SIAM.

[4] G. Cybenko. Dynamic load balancing for distributed memory multiprocessors. *J. Parallel Distrib. Comput.*, 7(2):279–301, 1989.

[5] M. Kleber. Goldbug Variations. *The Mathematical Intelligencer*, 27(1), 2005.

[6] L. Levine and Y. Peres. The rotor-router shape is spherical. *The Mathematical Intelligencer*, 27(3):9–11, 2005.

[7] Y. Rabani, A. Sinclair, and R. Wanka. Local divergence of markov chains and the analysis of iterative load-balancing schemes. In *FOCS'98: Proceedings of the 39th Annual Symposium on Foundations of Computer Science*, pp. 694–705. IEEE Computer Society, 1998.

[8] R. Subramanian and I. D. Scherson. An analysis of diffusive load-balancing. In *SPAA*, pp. 220–225, New York, NY, USA, 1994. ACM Press.

[9] C.-Z. Xu, B. Monien, R. Lüling, and F. C. M. Lau. An analytical comparison of nearest neighbor algorithms for load balancing in parallel computers. In *IPPS*, pp. 472–479. IEEE Computer Society, 1995.

Improving Time and Space Complexity for Compressed Pattern Matching

Shirou Maruyama[1], Hiromitsu Miyagawa[1], and Hiroshi Sakamoto[2]

[1] Graduate School of Computer Science and Systems Engineering
{s_maruyama, miyagawa}@donald.ai.kyutech.ac.jp
[2] Faculty of Computer Science and Systems Engineering
Kyushu Institute of Technology, Kawazu 680-4, Iizuka 820-8502, Japan
hiroshi@ai.kyutech.ac.jp

Abstract. The compressed pattern matching problem is to find all occurrences of a given pattern in a compressed text. In this paper an efficient grammar-based compression algorithm is presented for the compressed pattern matching. The algorithm achieves the worst-case approximation ratio $O(g_* \log g_* \log n)$ for the optimum grammar size g_* with an input text of length n. This upper bound improves the complexity of the compressed pattern matching problem to $O(g_* \log g_* \log m + \frac{n}{m} + m^2 + r)$ time and $O(g_* \log g_* \log m + m^2)$ space for any pattern shorter than m and the number r of pattern occurrences.

1 Introduction

In this paper we propose a dictionary-based compression algorithm and using the obtained upper bound of the compression ratio, we improve the time and space complexity for the *compressed pattern matching problem*. We give the related studies on the problem for our motivation.

The compressed pattern matching problem is to find all the occurrences of a pattern in a compressed text without decoding. This problem was first presented by Amir and Benson [1], and many algorithms were proposed by individual compression methods. Farach and Thorup's [5] algorithm on LZ77 compression achieved $O(n \log^2 \frac{N}{n} + m)$ time, where n is the compressed text size, N is the original text length, and m is the pattern length. Amir et al. [2] proposed an algorithm on LZW compression runs in $O(n + m^2)$ time and its experimental results [8] reported that it is approximately twice faster than Agrep [26] searching the original text. Their algorithm directly simulates the Knuth-Morris-Pratt automaton [6] on the compressed text, and the method was extended to the *multiple* pattern matching [8] in $O(n + m^2 + r)$ time, where m is the total length of the patterns and r is the number of pattern occurrences. Navarro and Raffinot [16] developed a technique for pattern matching which abstracts both LZ77 and LZ78 and its implementation for LZW is in $O(\frac{nm}{w} + m + r)$ time for the machine word length w.

Kida et al. [7] proposed a general framework for the compressed pattern matching called the *collage system* to represent a string by a pair of dictionary

T. Asano (Ed.): ISAAC 2006, LNCS 4288, pp. 484–493, 2006.
© Springer-Verlag Berlin Heidelberg 2006

D and token sequence S of in D. This system is unifying almost dictionary-based compression methods such as Lempel-Ziv family (LZ77, LZSS, LZ78, LZW). These dictionary-based compression methods are represented by a class of context-free grammar (CFG), called *grammar-based compression* by Kieffer et al. [27,10,9]. Their idea is to build a small CFG G that produces the original text uniquely. Then, we can consider that G encodes the text. Since the class of grammar-based compression corresponds to a subclass of collage system, other dictionary-based compression methods [12,17] are also encoded by the system. Kida et al. [7] showed that the compressed pattern matching problem can be solved in $O(|D| + |S| + m^2 + r)$ time and in $O(|D| + m^2)$ space, where $|D|$ is the size of dictionary and $|S|$ is the length of the token sequence.

The pattern matching algorithm on collage system is fast in practice [7]. However the time/space complexity is dependent on the size of the dictionary D and the token sequence S produced by the compression method, which maybe $O(n)$ in worst case. For example, it is known that the size of dictionary by LZW is almost $\Omega(n)$ for input size n [14]. This difficulty is implicitly contained in all results mentioned in the above, and this motivate us to ensure the efficiency of our compressed pattern matching algorithm by improving the upper bound of the complexity.

In order to improve the bounds, the most fundamental problem is to find a smallest context-free grammar that generates the given text uniquely. The factor $\frac{g}{g_*}$ is called the *approximation ratio* of an algorithm for the compressed grammar size g and the optimum grammar size g_*. Lehman and Shelat [14] showed that this problem is APX-hard, i.e., it is hard to approximate this problem within a constant factor (see [3]). Their result also suggests that $o(\log n/\log \log n)$ ratio is computationally hard due to the relationship between the grammar-based compression and the classical *semi-numerical problem* [11] consider to be hard.

The first $O(\log n)$-approximation algorithm was developed by Charikar et al. [4]. Their algorithm achieves the ratio $O(\log \frac{n}{g_*})$, where g_* is the size of a minimum deterministic CFG for an input. Independently, Rytter [19] presented other $O(\log \frac{n}{g_*})$-approximation algorithm that employs a suffix tree and the LZ-factorization technique for strings. Sakamoto [20] also proposed a simple linear-time algorithm based on Re-pair [12] and achieving ratio $O(\log n)$; Now this ratio has been improved to $O(\log \frac{n}{g_*})$.

The ratio $O(\log \frac{n}{g_*})$ achieved by these new algorithms is sufficiently small. However, all these algorithms require $\Omega(n)$ space, and it prevents us to apply the algorithms to huge texts, which is crucial to obtain a good compression ratio in practice. On the other hand, Sakamoto et al. [21] proposed a space-saving compression algorithm which guarantees at least $O(\log^2 n)$ approximation ratio within sub-linear space using a partial order on alphabetical symbols. Since the technique requires no real special data structure, such as suffix tree or occurrence frequency table, the space complexity is nearly equal to the total number of created nonterminal symbols, each of which corresponds to a production rule in Chomsky normal form.

Our compression method is also based on the alphabetical order only. We introduce three criteria to replace *digrams* in a string. The compression algorithm achieves $O(\log g_* \log n)$ approximation ratio and $O(g_* \log g_*)$ space as well as the derivation tree generated by the grammar is *balanced*, i.e., any token symbol encodes a sufficiently long substring, which is not obtained in [21] and enable us the efficient pattern matching on the compressed text. Using these results, we improve the results by Kida et al. [7] to $O(g_* \log g_* \log m + \frac{n}{m} + m^2 + r)$ time and $O(g_* \log g_* \log m + m^2)$ space.

The remaining part of this paper is organized as follows. In Section 2, we prepare the definitions related to the compressed pattern matching and the grammar-based compression. The compression algorithm is presented in Section 3 and we analyze the approximation ratio and estimate the space efficiency. In the final section, we summarize this study.

2   Notions and Definitions

In this section we give the notations and definitions for strings, grammar-based compression, and compressed pattern matchings.

2.1   Strings

We assume a finite *alphabet* Σ for the symbols forming input strings throughout this paper. The set of all strings over Σ is denoted by Σ^*, and Σ^i denotes the set of all strings of length just i. The length of a string $w \in \Sigma^*$ is denoted by $|w|$, and also for a set S, the notion $|S|$ refers to the cardinality of S.

Strings x and z are said to be a *prefix* and *suffix* of the string $w = xyz$, respectively. They are said to be *proper* if it is not w itself. Also x, y, z are called *substrings* of w. The ith symbol of w is denoted by $w[i]$. For an interval $[i, j]$ with $1 \le i \le j \le |w|$, the substring of w from $w[i]$ to $w[j]$ is denoted by $w[i, j]$. The integer i is called an *occurrence* of the substring $w[i, j]$ in w.

A *repetition* is a string x^k for a symbol x and an integer $k \ge 2$. A repetition $w[i, j] = x^k$ is *maximal* if $w[i - 1], w[j + 1] \ne x$. It is simply referred by x^+ if the length is unnecessary. Substrings $w[i, j]$ and $w[i', j']$ with $i < i'$ are *overlapping* if $i' \le j < j'$. A string of length two is called a *pair*.

2.2   Grammar-Based Compression

A *context-free grammar* (*CFG*) is a quadruple $G = (\Sigma, N, D, s)$ of disjoint finite alphabets Σ and N, a finite set (a dictionary) $D \subseteq N \times (N \cup \Sigma)^*$ of *production rules*, and the *start symbol* $s \in N$. Symbols in N are called *nonterminals*. A production rule $a \to b_1 \cdots b_k$ in D *derives* $\beta \in (\Sigma \cup N)^*$ *from* $\alpha \in (\Sigma \cup N)^*$ by replacing an occurrence of $a \in N$ in α with $b_1 \cdots b_k$. In this paper, we assume that any CFG is *deterministic*, that is, for each nonterminal $a \in N$, exactly one production rule from a is in D. Thus, the language $L(G)$ defined by G is a singleton set. We say a CFG G *derives* $w \in \Sigma^*$ if $L(G) = \{w\}$. The *size* *of* G is the total length of strings in the right hand sides of all production rules, and is denoted by $|G|$. The aim of grammar-based compression is formalized as a combinatorial optimization problem, as follows:

Problem 1 GRAMMAR-BASED COMPRESSION
INSTANCE: *A string* $w \in \Sigma^*$.
SOLUTION: *A deterministic CFG* G *that derives* w.
MEASURE: *The size of* G.

From now on, we assume that every deterministic CFG is in Chomsky normal form, i.e., the size of strings in the right-hand side of production rules is two, and we use $|N|$ for the size of a CFG. Note that for any CFG G there is an equivalent CFG G' in Chomsky normal form whose size is no more than $2 \cdot |G|$.

It is known that there is an important relation between a deterministic CFG and the following factorization. The *LZ-factorization* $LZ(w)$ *of* w is the decomposition of w into $f_1 \cdots f_k$, where $f_1 = w[1]$, and for each $1 < \ell \leq k$, f_ℓ is the longest prefix of the suffix $w[|f_1 \cdots f_{\ell-1}| + 1, |w|]$ that appears in $f_1 \cdots f_{\ell-1}$. Each f_ℓ is called a *factor*. The size $|LZ(w)|$ of $LZ(w)$ is the number of its factors. The following result is used in the analysis of the approximation ratio of our algorithm.

Theorem 1 ([19]). For any string w and its deterministic CFG G, the inequality $|LZ(w)| \leq |G|$ holds.

2.3   Compressed Pattern Matching

Let an input text w be represented by a deterministic CFG $G = (\Sigma, N, D, s)$. The grammar is denoted by (S, D) for short, where S is the string $X_1 \cdots X_k$ such that $s \rightarrow X_1 \cdots X_k \in D$ for the start symbol s. Each symbol X_i is called a *token* and S is called the *token sequence*. The expression $X.u$ denotes the string in Σ^* obtained by decoding the token X. As is mention in the above, any other production is of the form $X \rightarrow YZ$ for some $X \in N$ and $Y, Z \in \Sigma \cup N$. For such CFGs, the compressed pattern matching problem is defined as follows.

Problem 2 COMPRESSED PATTERN MATCHING
INSTANCE: *A CFG* (S, D) *for a text* $w \in \Sigma^*$ *and a pattern* $\pi \in \Sigma^*$.
SOLUTION: *The positions of all occurrences of* π *in* w.

For this problem, Kida et al. [7] developed the *collage system* unifying framework of the compressed pattern matching: For a compressed text (S, D) and a pattern π, the pattern matching algorithm outputs the required positions of π by simulating Knuth-Morris-Pratt (KMP) automaton [6] for π without decoding (S, D). They showed that the KMP automaton for π to the text $S.u$ can be simulated in linear time using $O(|D| + |\pi|^2)$ time preprocessing for D and π.

Theorem 2 ([7]). If any production is restricted to the form $X \rightarrow YZ$ for $X \in N$ and $X, Y \in \Sigma \cup N$, the problem of compressed pattern matching can be solved in $O(|D| + |S| + m^2 + r)$ time and $O(|D| + m^2)$ space.

In their result, the size $|D|$ and $|S|$ are dependent on the compression scheme. For example, LZW compression is contained in the framework of collage system but the lower bound of $|D|$ for LZW is $\Omega(\frac{n^{\frac{2}{3}}}{\log n})$, which is almost equal to the input size (See [14] for other bounds).

In this paper we present an algorithm which approximates the optimum compression in nearly log-scale and we improve the size $O(|D|)$ to $O(g_* \log g_* \log n)$ and the size $O(|S|)$ to $O(\frac{n}{m})$ in Theorem 2. We then obtain our results in the upper bounds for the time and space complexity for the compressed pattern matching problem.

3   Compression Algorithm

In this section we introduce a compression algorithm for the grammar-based compression problem and analyze its performance.

3.1   Key Idea

The task of the algorithm is only to replace a pair XY occurring in a current string by a new symbol Z and generate a production $Z \to XY$ to D, where for all occurrences of XY, different two productions are not generated. However, not all occurrences of XY are replaced by Z. Thus, the key idea of the compression algorithm is the decision rule for which pair should be replaced. Consequently, the aim of this algorithm is to minimize the number of different nonterminals generated. Here we explain the three decision rules for the replaced pairs.

The first rule (maximal repetition): If a current string contains a maximal repetition $w[i, j] = X^k$ ($k \geq 2$), then $w[i, i+1]$ and $w[j-1, j]$ are replaced by a same symbol, and $w[i+2, j-2]$ is replaced recursively, where the middle symbol of w is remained in case that $|w|$ is odd. For example, for the maximal repetition a^5, its an occurrence is transformed to AaA by an appropriate nonterminal A and the production $A \to aa$ is generated.

The second rule (minimal pair): We assume that all symbols in $\Sigma \cup N$ have a fixed partial order, that is, any symbol is represented by an integer. If a current string contains a substring $A_i A_j A_k$ such that $j < i, k$, then the occurrence of A_j is called *minimal*. The second decision rule is to replace all such pairs $A_j A_k$ in $A_i A_j A_k$ by an appropriate nonterminal.

In advance of the third decision rule, we explain the notion of the lowest common ancestor defined on an index tree.

Definition 1. Let d be a positive integer and $k = \lceil \log_2 d \rceil$. The *index tree* T_d is the rooted, ordered complete binary tree whose leaves are labeled with $1, \ldots, 2^k$ from the left. The *height* of an internal node refers to the number of edges in a path from the node to a descendant leaf. Then, the height of the lowest common ancestor of leaves i, j is denoted by $lca(i, j)_d$, or denoted by $lca(i, j)$ for short.

The third rule (maximal pair): For a fixed order of alphabet, let a current string contain a substring $A_{i_1} A_{i_2} A_{i_3} A_{i_4}$ such that the integers are increasing or decreasing order. If $lca(i_2, i_3) > lca(i_1, i_2), lca(i_3, i_4)$, then the occurrence of the middle pair $A_{i_2} A_{i_3}$ is called *maximal*. The third decision rule is to replace all such pairs by an appropriate nonterminal.

We have explained the decision rule to replaced pairs in the current string. However, we can not apply the three types of rules simultaneously since there maybe a case that a pair $w[i, i+1]$ satisfies one of the first/second rule and the overlapping pair $w[i+1, i+2]$ satisfies the other. For example, the substring $a_2 a_1 a_3 a_3 a_3$ contains such overlapping pairs. So we apply the first, second, and third rules in this order to hold the uniqueness of the replacement. Indeed, any cases are not overlapping by this priority.

3.2   Compression Algorithm

Outline of algorithm: For a current token string S(initially the input), this compression algorithm categorizes all pairs in S to the class of repetitions, maximal pairs, minimal pairs, and the others. These classes also define the priority for the replacement in this order, where the left occurrence has the higher priority in a same class. According to the priority, the algorithm replaces all such occurrences by appropriate nonterminals and updates the current token string to the resulting string. The algorithm repeats the above process until there is no expectation for compression, i.e., all pairs in the current token string are mutually different. Then the final token string and dictionary are returned.

Theorem 3. The time complexity of the compression algorithm is $O(|w|)$.

1: **Input:** a string w; ($S = w$, $D = \emptyset$)
2: **Output:** a deterministic CFG (S, D) for w;
3: **for each** maximal repetition in S **do**
4:      replace the pairs in the repetition;   /*first rule replacement*/
5: $i = 1$;
6: **while**($i < |S|$) **do**   /*second, third rule replacement*/
7:      **if** $S[i, i+1]$ is minimal or maximal pair **then**
8:          replace it, $i = i + 2$, update D;
9:      **else if** $S[i+1, i+2]$ is minimal or maximal pair **then**
10:         replace it, $i = i + 3$, update D;
11:     **else** replace $S[i, i+1]$, $i = i + 2$, update D;
12: update S to the replaced string;
13: **goto** line 3 until all pairs in S are different;
14: output (S, D);

Fig. 1. The compression algorithm

Proof. Since the algorithm replaces at least one of three pairs $w[i, i+1]$, $w[i+1, i+2]$, or $w[i+2, i+3]$, the loop is repeated at most $O(\log |w|)$ times. Moreover, for each iteration of the loop from line 3 to 13, the length of a token string becomes at least $\frac{2}{3}$ to the previous one. For each step, we can verify whether an occurrence of a pair is repetition, maximal, or minimal in $O(1)$ time[1]. Thus, the compression algorithm runs in linear time.   □

3.3   Performance Analysis

The approximation ratio of the compression algorithm is the upper bound of the factor $\frac{g}{g_*}$ for the output grammar size g and the optimum solution size g_*. Next we show that $\frac{g}{g_*}$ is smaller than $O(\log^2 n)$.

Definition 2. Let w be a string and $w[i, j] = \alpha$ be an occurrence of a substring α in w. We call $w[i, j]$ a *boundary occurrence* if $w[i-1] \neq w[i]$ and $w[j] \neq w[j+1]$.

Definition 3. Let w be a current token sequence. Let $R(i, j)$ be the set of occurrences of pairs in $w[i, j]$ such that they are replaced by the compression algorithm at the first execution of line 3–13 for w.

Lemma 1. Let $w[i_1, j_1] = w[i_2, j_2]$ be any occurrences of a substring α in w. There exists an integer $k \leq \log n$ such that $R(i_1 + k, j_1 - k) = R(i_2 + k, j_2 - k)$.

Proof. We first show the case that $w[i_1, j_1]$ and $w[i_2, j_2]$ are not boundary. Then, there is an interval $[\ell, r]$ containing $[i_1, j_1]$ such that $w[\ell, r] = a^+ \beta b^+$, where $\alpha = x\beta y$ for some repetition x of a and y of b. We can also assume $w[\ell', r'] = a^+ \beta b^+$ for some $[\ell', r']$. Since the replacement of a pair $w[i, i+1]$ depends on the three pairs in $w[i-1, i+2]$ only, the replacements for the two β in $w[i_1, j_1]$ and $w[i_2, j_2]$ completely synchronize. Thus, the difference between $R(i_1, j_1)$ and $R(i_2, j_2)$ is bounded for the replacements of x and y. A disagreement for x may happen only if exactly one of the length of a^+ in $w[\ell, r]$ and $w[\ell', r']$ is odd. In this case, exactly one of $w[i_1, i_1+1]$ and $w[i_2, i_2+1]$ is replaced, so $R(i_1+k, j_1-k) = R(i_2+k, j_2-k)$ holds for at least $k = 2$.

We assume that $w[i_1, j_1]$ and $w[i_2, j_2]$ are boundary string. Let Σ be the current alphabet containing all symbols generated. Here we consider the index tree $T_{|\Sigma|}$ such that any $lca(i, j)$ is defined for the leaves i, j. If a string $\alpha = a_{\ell_1} a_{\ell_2} \cdots a_{\ell_m}$ contains no repetition and minimal/maximal pair, then the sequence $\ell_1, \ell_2, \ldots \ell_m$ is monotonic, i.e., $\ell_1 < \ell_2 < \ldots < \ell_m$ or $\ell_1 > \ell_2 > \ldots > \ell_m$ and $lca(\ell_1, \ell_2), lca(\ell_2, \ell_3), \ldots, lca(\ell_{m-1}, \ell_m)$ is monotonic. The length of such string α is bounded by $\log |\Sigma|$. Thus, a prefix of $w[i_1, j_1]$ longer than $\log |\Sigma|$ contains at least one of repetition, minimal, or maximal pair and the pair also appears in $w[i_2, j_2]$ at the same position from the left, and their short suffixes have one of repetition, minimal, or maximal pair at the same position. Thus, the replacements of $w[i_1, j_1]$ and $w[i_2, j_2]$ completely synchronize between such the leftmost and rightmost pairs. The case that one of $w[i_1, j_1]$ and $w[i_2, j_2]$ is boundary

---

[1] We can get the *lca* of any two leaves i and j of complete binary trees by an **xor** operation between binary numbers in $O(1)$ time under our RAM model [6].

and the other is not is similarly proved. Hence, $R(i_1+k, j_1-k) = R(i_2+k, j_2-k)$ for $k = \log |\Sigma| \leq \log n$.   □

Lemma 2. The worst-case approximation ratio $\frac{g}{g_*}$ is $O(\log g_* \log n)$, where g is the output grammar size and g_* is the minimum grammar size.

Proof. We estimate the number of different nonterminals produced by the compression algorithm at a single loop. Let $w_1 \cdots w_m$ be the *LZ*-factorization of the input string w. We denote by $\#(w)$ the number of different nonterminals produced at a single loop. From the definition of *LZ*-factorization, any factor w_i occurs in the prefix $w_1 \cdots w_{i-1}$, or $|w_i| = 1$. By lemma 1, any factor w_i and its leftmost occurrence are compressed into almost the same strings $\alpha\beta\gamma$ and $\alpha'\beta\gamma'$, such that $|\alpha\gamma|, |\alpha'\gamma'| = O(\log|\Sigma|)$. Thus, we can estimate $\#(w) = \#(w_1 \cdots w_{m-1}) + O(\log|\Sigma|) = O(m \log|\Sigma|) = O(g_* \log|\Sigma|)$. This estimation can be applied to the occurrences of β until $|\beta| < \log|\Sigma| = O(\log n)$. Hence, $O(g_* \log n)$ is the maximum number of different nonterminals produced at a single loop. Since the depth of loop is at most $O(\log n)$, the size of the finial dictionary $O(g_* \log^2 n)$ is derived. This bound is easily improved to $O(g_* \log g_* \log n)$ since $\#(w)$ converges to $O(g_* \log g_*)$ for any constant alphabet Σ.   □

Next we show that the derivation tree of (D, S) is "*balanced*", i.e., any token X_i in S encodes a sufficiently long substring in the text.

Lemma 3. Let (S_k, D_k) be the grammar generated by the compression algorithm at the kth loop for an input text w. Then, for any substring $w[i, j]$, there is k $(k \leq 3\log(j - i + 1))$ and a substring XYZ of S_k whose length is at most three such that it encodes $w[i, j]$, that is, $(X.u)(Y.u)(Z.u)$ contains $w[i, j]$.

Proof. In the kth loop of the compression algorithm, at least one of three pairs in $w[i, i + 3]$ is replaced by an appropriate symbol. From the fact, we show this lemma by induction with the length ℓ of $w[i, j]$ $(\ell = j - i + 1)$. Clearly, the lemma is true for $\ell = 1$, so we assume the induction hypothesis on any substring of length at most ℓ. Let $w[i, j]$ be any substring of length $\ell + 1$. We split the substring $w[i, j] = w[i, j']w[j' + 1, j]$ for $j' = \lfloor \frac{\ell+1}{2} \rfloor$. Since the hypothesis is true for $w[i, j']$ and $w[j' + 1, j]$, a substring $X_1X_2X_3Y_1Y_2Y_3$ encodes $w[i, j]$, $X_1X_2X_3$ encodes $w[i, j']$, and $Y_1Y_2Y_3$ encodes $w[j' + 1, j]$. The string $X_1X_2X_3Y_1Y_2Y_3$ becomes to be at least length three within the next three loops. Thus, $w[i, j]$ is encoded by a consecutive three tokens and the depth k of the loop is at most $3\log \ell' + 3$ for $\ell = 2\ell' + 1$. Hence, we conclude the induction is true by $k \leq 3\log \ell' + 3 < 3\log(2\ell' + 1) = 3\log \ell < 3\log(\ell + 1)$.   □

Lemma 3 ensures that the length of the token sequences generated by the algorithm is sufficiently short according to the depth of loop and any token encodes sufficiently long substring. By lemma 2 and 3, we can obtain the main result of this paper.

Theorem 4. The problem of compressed pattern matching can be solved in $O(g_* \log g_* \log m + \frac{n}{m} + m^2 + r)$ time and $O(g_* \log g_* \log m + m^2)$ space for any pattern of length at most m, any text of length n, the optimum compression size g_*, and the number r of pattern occurrences.

Proof. For the time complexity, the bound $O(g_* \log g_* \log m)$ is derived from Lemma 2, which is the size of the grammar (S_k, D_k) for depth $k = 3 \log m$. On the other hand, the bound $O(\frac{n}{m})$ is derived from Lemma 3 since the length of the substring encoded by a token in S_k is greater than $2^{\log m} = m$. Also the preprocessing can be computed within the dictionary size. Thus, we obtain the time and space complexity. □

4   Conclusion

We introduced a compression algorithm based on a partial order of alphabets and showed that the algorithm ensures the compression ratio $O(g_* \log g_* \log n)$ within $O(g_* \log g_*)$ memory space for the length n of input and the optimum compression size g_*. Moreover our dictionary-based compression generates the balanced derivation tree for input string. We applied this property to the compressed pattern matching and obtain the improved time and space complexity $O(g_* \log g_* \log m + \frac{n}{m} + m^2 + r)$ and $O(g_* \log g_* \log m + m^2)$, respectively, where m is the pattern length and r is the number of occurrences of the pattern.

References

1. A. Amir, G. Benson, Efficient two-dimensional compressed matching, in: Proc. Data Compression Conference, p.279, 1992.
2. A. Amir, G. Benson, M. Farach, Let sleeping files lie: pattern matching in Z-compressed files, J. Comput. System Sci. 52:299–307, 1996.
3. G. Ausiello, P. Crescenzi, G. Gambosi, V. Kann, A. Marchetti-Spaccamela, M. Protasi, Complexity and Approximation: Combinatorial Optimization Problems and Their Approximability Properties, Springer, 1999.
4. M. Charikar, E. Lehman, D. Liu, R. Panigrahy, M. Prabhakaran, A. Rasala, A. Sahai, A. Shelat, Approximating the Smallest Grammar: Kolmogorov Complexity in Natural Models, in: Proc. 29th Ann. Sympo. on Theory of Computing, 792-801, 2002.
5. M. Farach, M. Thorup, String-matching in Lempel-Ziv compressed strings, in: 27th ACM STOC, pp. 703–713, 1995.
6. D. Gusfield, Algorithms on Strings, Trees, and Sequences, Computer Science and Computational Biology, Cambridge University Press, 1997.
7. T. Kida, Y. Shibata, M. Takeda, A. Shinohara, S. Arikawa, Collage System: a Unifying Framework for Compressed Pattern Matching, Theoret. Comput. Sci. 298:253–272.
8. T. Kida, M. Takeda, A. Shinohara, M. Miyazaki, S. Arikawa, Multiple pattern matching in LZW compressed text, J. Discrete Algorithms 1(1):133–158, 2000.
9. J. C. Kieffer, E.-H. Yang, Grammar-Based Codes: a New Class of Universal Lossless Source Codes, IEEE Trans. on Inform. Theory, 46(3):737–754, 2000.
10. J. C. Kieffer, E.-H. Yang, G. Nelson, P. Cosman, Universal Lossless Compression via Multilevel Pattern Matching, IEEE Trans. Inform. Theory, IT-46(4), 1227–1245, 2000.

11. D. Knuth, Seminumerical Algorithms, Addison-Wesley, 441-462, 1981.
12. N. J. Larsson, A. Moffat, Offline Dictionary-Based Compression, Proceedings of the IEEE, 88(11):1722-1732, 2000.
13. E. Lehman, Approximation Algorithms for Grammar-Based Compression, PhD thesis, MIT, 2002.
14. E. Lehman, A. Shelat, Approximation Algorithms for Grammar-Based Compression, in: Proc. 20th Ann. ACM-SIAM Sympo. on Discrete Algorithms, 205-212, 2002.
15. M. Lothaire, Combinatorics on Words, volume 17 of Encyclopedia of Mathematics and Its Applications, Addison-Wesley, 1983.
16. G. Navarro, M. Raffinot, A general practical approach to pattern matching over Ziv-Lempel compressed text, in: Proc. 10th Ann. Symp. on Combinatorial Pattern Matching, LNCS 1645, pp. 14–36, 1999.
17. C. Nevill-Manning, I. Witten, Compression and Explanation Using Hierarchical Grammars, Computer Journal, 40(2/3):103–116, 1997.
18. C. Nevill-Manning, I. Witten, Identifying hierarchical structure in sequences: a linear-time algorithm, J. Artificial Intelligence Research, 7:67–82, 1997.
19. W. Rytter, Application of Lempel-Ziv Factorization to the Approximation of Grammar-Based Compression, in: Proc. 13th Ann. Sympo. Combinatorial Pattern Matching, 20-31, 2002.
20. H. Sakamoto, A Fully Linear-Time Approximation Algorithm for Grammar-Based Compression, Journal of Discrete Algorithms, 3:416-430, 2005.
21. H. Sakamoto, T. Kida, S. Shimozono, A Space-Saving Linear-Time Algorithm for Grammar-Based Compression, in: Proc. 11th International Symposium on String Processing and Information Retrieval, pp.218-229, 2004.
22. D. Salomon, Data compression: the complete reference, Springer, second edition, 1998.
23. J. Storer, T. Szymanski, Data compression via textual substitution, J. Assoc. Comput. Mach., 29(4):928–951, 1982.
24. J. A. Storer, T. G. Szymanski, The Macro Model for Data Compression, in: Proc. 10th Ann. Sympo. on Theory of Computing, pp. 30–39, 1978.
25. T. A. Welch, A Technique for High Performance Data Compression, IEEE Comput., 17:8-19, 1984.
26. S. Wu, U. Manber, Agrep–a fast approximate pattern-matching tool, in: Usenix Winter 1992 Technical Conference, pp. 153–162, 1992.
27. E.-H. Yang, J. C. Kieffer, Efficient Universal Lossless Data Compression Algorithms Based on a Greedy Sequential Grammar Transform–Part One: without Context Models, IEEE Trans. on Inform. Theory, 46(3):755-777, 2000.
28. J. Ziv, A. Lempel, A Universal Algorithm for Sequential Data Compression, IEEE Trans. on Inform. Theory, IT-23(3):337-349, 1977.
29. J. Ziv, A. Lempel, Compression of Individual Sequences via Variable-Rate Coding, IEEE Trans. on Inform. Theory, 24(5):530-536, 1978.

Improved Multi-unit Auction Clearing Algorithms with Interval (Multiple-Choice) Knapsack Problems

Yunhong Zhou

HP Labs, 1501 Page Mill Rd, Palo Alto, CA 94304, USA
yunhong.zhou@hp.com

Abstract. We study the *interval knapsack problem* (I-KP), and the *interval multiple-choice knapsack problem* (I-MCKP), as generalizations of the classic 0/1 knapsack problem (KP) and the multiple-choice knapsack problem (MCKP), respectively. Compared to singleton items in KP and MCKP, each item i in I-KP and I-MCKP is represented by a $([a_i, b_i], p_i)$ pair, where integer interval $[a_i, b_i]$ specifies the possible range of units, and p_i is the unit-price. Our main results are a FPTAS for I-KP with time $O(n \log n + n/\epsilon^2)$ and a FPTAS for I-MCKP with time $O(nm/\epsilon)$, and pseudo-polynomial-time algorithms for both I-KP and I-MCKP with time $O(nM)$ and space $O(n + M)$. Here n, m, and M denote number of items, number of item sets, and knapsack capacity respectively. We also present a 2-approximation of I-KP and a 3-approximation of I-MCKP both in linear time.

We apply I-KP and I-MCKP to the single-good multi-unit sealed-bid auction clearing problem where M identical units of a single good are auctioned. We focus on two bidding models, among them the interval model allows each bid to specify an interval range of units, and XOR-interval model allows a bidder to specify a set of *mutually exclusive* interval bids. The interval and XOR-interval bidding models correspond to I-KP and I-MCKP respectively, thus are solved accordingly. We also show how to compute VCG payments to all the bidders with an overhead of $O(\log n)$ factor. Our results for XOR-interval bidding model imply *improved algorithms* for the *piecewise constant* bidding model studied by Kothari et al. [18], improving their algorithms by a factor of $\Omega(n)$.

1 Introduction

Combinatorial auctions have been proposed as expressive, economically efficient mechanisms for resource distributions and procurement auctions [16,22,23,26]. The winner determination problem in combinatorial auctions, unfortunately, is NP-hard and inapproximable in general [24]. Consequently, there is enormous interest in finding the right level of generality at which to address this problem.

There have been increasing activities of procurement auctions conducted by large firms, inspired and facilitated by the emergence of electronic commerce. For various reasons, firms tend to conduct separate auctions for different types

T. Asano (Ed.): ISAAC 2006, LNCS 4288, pp. 494–506, 2006.
© Springer-Verlag Berlin Heidelberg 2006

of goods, thus eliminating the main source of complexity in combinatorial auctions. There are also a flurry of research activities [6,1] for procurement auctions. But they either assume divisible goods, or use mixed integer programming or heuristics to solve the auction clearing problem, while none of these work has an *algorithmic* focus. And traditional single-good multi-unit auctions use either singleton bids or XOR of singleton bids, resulting in either inflexibility or inefficiency.

In this paper we focus on single-good multi-unit combinatorial auctions with increasingly expressive bidding models, and where *(weakly) tractable* algorithms are possible. Specifically, we focus on two bidding models, interval bids and XOR-interval bids (the former extends the traditional singleton bids and the later extends the *piecewise constant* bids), and formalize the corresponding auction clearing problem as new variants of knapsack problems. The connection between winner determination problems in sealed-bid combinatorial auctions and generalized knapsack problems (especially the equivalence of the Multi-good Multi-unit Combinatorial Auction and the Multidimensional Knapsack Problem) was observed recently [9,15,17]. This connection tremendously expedites the progress of auction clearing research as there is abundance of research literature on variants of knapsack problems. Our work is benefited from this connection, meanwhile advances the state of art in knapsack research as we introduce new variants of knapsack problems and design near-optimal algorithms to solve them.

1.1   Problem Statement

We consider the one-round sealed-bid single-good multi-unit auction problem (SMAP) where there is only one single good and M identical units of it. The auctioneer wants to either sell at most M units of the good (forward auction) with maximum revenue, or acquire at least M units of the good (reverse auction) with minimum cost. We focus on the forward auction version throughout the paper, since maximization and minimization are dual to each other, and most of our results for the forward auction version can be translated to the reverse auction version (sometimes requiring small tricks for the translation.). [1] There are n bidders, each submitting a sealed bid, and the auctioneer decides winning bidders, the number of units allocated to them as well as the associated unit-prices. We consider the following four bidding models:

1. **Point Bid:** a pair (x, p) where x is the number of units and p is the unit-price.
2. **Interval Bid:** a tuple $([x_l, x_u], p)$ where interval $[x_l, x_u]$ gives the range of units and p is the unit-price.
3. **XOR-Point Bid:** a collection of point bids, with at most one point bid taken.
4. **XOR-Interval Bid:** a collection of interval bids, with at most one interval bid taken.

---

[1] It is even possible to handle exchanges where agents are both buyers and sellers.

Point bids correspond to items in the classic 0/1 knapsack problem (KP) where point bid (x, p) corresponds to an item with weight x and profit xp. XOR bids encode mutually exclusive bids. For example, a buyer submits a bid $(x_1, p_1) \otimes (x_2, p_2)$, with the intention of buying either x_1 units at unit-price p_1 or x_2 units at unit-price p_2, but not both. SMAP with XOR-point bids corresponds to the *multiple-choice knapsack problem* (MCKP), which is defined as follows: Given m sets of items, where item j in set i has weight w_i^j and profit p_i^j, and a knapsack of capacity M, select a subset of items, at most one from each item set, to maximize their total profit while their total weight is at most M.

The interval bidding model generalizes the atomic point bidding model to an interval range of units. The XOR-interval bidding model is the most expressive model in this paper and it generalizes the piecewise constant bidding model [18,19]. In many procurement auction settings, in order to improve their total profit, suppliers offer *volume discount*, where the unit-price decreases as the number of units increases. The XOR-interval bid is one way of offering volume discount, and it is an extension to base bidding models.

1.2   Our Contributions

We define the *interval knapsack problem* (I-KP), and the *interval multiple-choice knapsack problem* (I-MCKP), as new generalizations of the classic 0/1 knapsack problem (KP) and the multiple-choice knapsack problem (MCKP), respectively. Our main results are a FPTAS for I-KP with time $O(n \log n + n/\epsilon^2)$ and a FPTAS for I-MCKP with time $O(nm/\epsilon)$, and pseudo-polynomial-time algorithms for both I-KP and I-MCKP with time $O(nM)$ and space $O(n + M)$. Here n, m, and M denote number of items, number of item sets, and knapsack capacity respectively. We also present a 2-approximation of I-KP and a 3-approximation of I-MCKP both in $O(n)$ time. Most of our algorithms for both I-KP and I-MCKP match the corresponding time bounds for the best algorithms of KP and MCKP. Since KP and MCKP are well studied problems in Operations Research and Theoretical Computer Science, our results for I-KP and I-MCKP will be hard to beat.

We apply I-KP and I-MCKP to the single-good multi-unit auction clearing problem (SMAP) where M identical units of a single good are auctioned. We focus on two bidding models, among them the interval model allows each bid to specify an interval range of units, and XOR-interval model allows a bidder to specify a set of *mutually exclusive* interval bids. The interval and XOR-interval bidding models correspond to I-KP and I-MCKP respectively, thus are solved accordingly. We also show how to compute VCG payments to all the bidders with only an overhead of $O(\log n)$ for various bidding models, while the straightforward approach takes an overhead of factor $O(n)$.

Our results for XOR-interval bidding model imply *improved algorithms* for the piecewise constant bidding model studied by Kothari et al. [18]. Specifically, Kothari et al. developed a FPTAS for this auction model with time $O(n^3/\epsilon)$ while our results for XOR-interval model implies a FPTAS for this problem with time $O(n^2/\epsilon)$. They also developed an ϵ-approximate and ϵ-efficient VCG payment scheme with time $O((n^3/\epsilon)\alpha \log(\alpha n/\epsilon))$ where α is a constant related to "market no-monopoly". We improve their algorithm to $O((n^2/\epsilon)\log n)$ time. We also point out that a constant-factor approximation for I-MCKP can be computed in $O(n)$ time, while they computed a 2-approximation in $O(n^2)$ time. Our algorithms are substantially simpler than previous approaches, as we identify one technical lemma for vector merging and another technical lemma for VCG computations, and get rid of unnecessary steps.

1.3   Related Work

Single-good Multi-unit Auction. The single-good multi-unit auction clearing problem with *piecewise linear* supply/demand curves were studied by [25,4]. Sandholm and Suri [25] also studied the point bid model and XOR-point bid model. However the relationship between the XOR-point bidding model and MCKP is not stated, and their FPTAS for the XOR-bidding model is actually based on a FPTAS for KP, thus incorrect.

The single-good multi-unit auction clearing problem with *piecewise constant* bidding curves were studied by Kothari et al. [18]. Kothari et al. proposed an algorithm for the auction clearing problem with time $O(n^3/\epsilon)$. They also proposed another algorithm to compute an ϵ-approximate and ϵ-efficient VCG payments to all bidders with time $O((n^3/\epsilon)\alpha \log(\alpha n/\epsilon))$ where α is a constant related to "market no-monopoly". The ϵ-approximate VCG mechanism is only *approximately truthful*, as a possible deviation of a single bidder might be high. Lehmann et al. [21] gave a sufficient condition of truthful mechanisms for single-minded combinatorial auctions, which requires the corresponding approximation algorithm to be *monotone* in an appropriate sense. Briest et al. [2] designed monotone approximation algorithms for multi-unit auctions, and their monotone FPTAS for single-commodity multi-unit auctions (only point bids are considered) runs in time $O((n^3/\epsilon)\log(n/(1-\epsilon)))$.

The single-good multi-unit auction clearing problem with *piecewise constant* bidding curves was also studied by Kothari, Suri and Zhou [19]. Their work differs from ours as they considered the special case of *uniform-price* auction clearing and reduced their problem to the *interval subset-sum problem*. Here we allow winning bids to take different clearing prices, thus improving the revenue of the auctioneer with the expense of *discriminatory pricing*.

Knapsack Problems. Variants of knapsack problems were studied intensively in Operations Research and Theoretical Computer Science. For a comprehensive exposition of this topic, see Kellerer et al. [15]. Here we review only literatures related to our work. The earliest FPTAS for KP (and also one of the first FPTAS in general) was given by Ibarra and Kim [11]. The best results were obtained by Kellerer and Pferschy [13,14], where a FPTAS with time $O(n\min\{\log n, \log(1/\epsilon)\} + 1/\epsilon^2 \log(1/\epsilon)\min\{n, 1/\epsilon \log(1/\epsilon)\})$ and space $O(n + 1/\epsilon^2)$ is given.

For MCKP, Dyer [5] and Zemel [27] independently developed linear time algorithms to compute the optimal solution of its linear relaxation. The maximum of the linear relaxation of MCKP and the split item is at least $1/2$ of the optimal solution for MCKP, thus a 2-approximation of MCKP can be computed in $O(n)$ time. Gens and Levner [8] gave a $(5/4)$-approximation of MCKP in $O(n \log m)$ time. The first FPTAS for MCKP was given by Chandra et al. [3] and the best FPTAS was given by Lawler [20] with running time $O(nm/\epsilon)$.

The rest of the paper is organized as follows. We define I-KP and I-MCKP formally in Section 2. We present a crucial technical lemma in Section 2.1, which is used subsequently for merging a vector with an interval vector in linear time. In Sections 3 and 4, we present both exact and approximation algorithms for I-KP and I-MCKP respectively. Section 5 summarizes single-good multi-unit auction clearing algorithms for various bidding models. We describe algorithms for VCG computations in Section 6 and conclude in Section 7.

2   Definitions and Preliminaries

In this section, we define the *interval knapsack problem* (I-KP) and the *interval multiple-choice knapsack problem* (I-MCKP) formally. The interval knapsack problem is a variant (generalization) of the classic 0/1 knapsack problem. Instead of singleton items in KP, I-KP associates with each item a unit-price and an *interval* range of units. It also generalizes the classic *integer knapsack problem*, either bounded or unbounded. Formally, I-KP is defined as follows:

Instance: Given a set of items S, each represented as an interval $[a_i, b_i]$ paired with a unit-price p_i, for $i = 1, \ldots, n$, and a capacity bound M.

Objective: Find a subset $\{x_i \mid i \in S'\}$ and $S' \subseteq S$, such that $x_i \in [a_i, b_i]$, $\forall i \in S'$, $\sum_{i \in S'} x_i \leq M$, and $\sum_{i \in S'} x_i p_i$ is maximized.

Next we define the *interval multiple-choice knapsack problem* (I-MCKP), a variant (generalization) of the classic *multiple-choice knapsack problem* (MCKP). Formally, I-MCKP is defined as follows:

Instance: Given a set of item sets $S_1, \ldots, S_m$, where item set S_i contains n_i items and each item $s_i^j \in S_i$ is represented by a tuple $([a_i^j, b_i^j], p_i^j)$, for $i = 1, \ldots, m$, and a capacity bound M.

Objective: Find a subset $\{x_i^j \mid s_i^j \in S_i, i \in I\}$, such that $x_i^j \in [a_i^j, b_i^j]$, $\forall i \in I$, $\sum_{i \in I, j} x_i^j \leq M$, there are at most one $x_i^j \neq 0$ for each $i \in I$, and $\sum_{i \in I, j} x_i^j p_i^j$ is maximized.

Here $n = \sum_{1 \leq i \leq m} n_i$ is the total number of items in all the item sets.

Without loss of generality, we assume that $b_i \leq M$ for all i in I-KP, and $b_i^j \leq M$ for all (i, j) pair in I-MCKP. For the special case of I-KP where $a_i = b_i$ for all i, I-KP degenerates to the classic knapsack problem where item i has weight a_i and profit $a_i p_i$. For the special case of I-MCKP where $a_i^j = b_i^j$ for all (i, j) pairs, each item corresponds to a singleton element with weight a_i^j and profit $a_i^j p_i^j$, thus I-MCKP degenerates to MCKP. And if each item set contains exactly one item, I-MCKP degenerates into I-KP. Since KP is NP-hard [12], both I-MCKP and

I-KP are NP-hard. It is easy to verify that both are in NP, thus both I-MCKP and I-KP are NP-complete. Fortunately, they are *weakly* NP-complete, as we will show in subsequent sections that both accept pseudo-polynomial-time exact algorithms and fully-polynomial-time approximation schemes (FPTAS).

2.1   A Linear Time Merging Subroutine

Next we describe a crucial technical lemma which is used as a subroutine to merge a vector with an interval item to obtain a new vector. Let $A = (A_1, \ldots, A_d)$ be a vector with length d, and $[a_r, b_r]$ be an integer interval. We want to compute a new integer vector $C = (C_1, \ldots, C_d)$ where $C_k = \min\{A_k, C'_k\}$, and

$$C'_k = \min\left\{A_\ell + x_r \mid 1 \leq \ell \leq k, a_r \leq x_r \leq b_r, x_r \geq (k - \ell)c_r\right\}, \quad \forall\, k = 1, \ldots, d.$$

Here c_r is a positive constant, not necessarily integer. Intuitively, vector C is the result of merging vector A with an interval item $([a_r, b_r], p_r)$ where the unit-price p_r is scaled down to a fractional value $1/c_r$. If $C_k = A_k$, no element is taken from the interval item. If $C_k = C'_k$, an element $x_r \in [a_r, b_r]$ is taken. Furthermore, we want to make sure that $\ell + x_r/c_r \geq k$, i.e., the combined solution has its scaled value at least k. The simple approach takes time $O(d)$ to compute one single value C'_k thus $O(d^2)$ time for the whole vector C'. The following crucial technical lemma shows that vectors C and C' can be computed in linear time:

Lemma 1. *We can compute vectors C and C' in time $O(d)$.*

The problem of merging a vector with an interval item is actually a special case of the so-called *vector merge problem*. Kellerer and Pferschy [14] showed how to solve the vector merge problem in $O(d \log d)$ time. By taking advantage of interval items, together with an advanced data structure (deque with heap order) [10,7], we are able to improve their bound and get an asymptotically optimal result. Not surprisingly, the same result applies to the corresponding maximization version.

Lemma 2. *Let $A = (A_1, \ldots, A_d)$ be a vector and $([a_r, b_r], p_r)$ an interval item. Let $C = (C_1, \ldots, C_d)$ where $C_k = \max\{A_k, C'_k\}$ and*

$$C'_k = \max\left\{A_\ell + x_r p_r \mid 1 \leq \ell \leq k, a_r \leq x_r \leq b_r, x_r + \ell \leq k\right\}, \quad \forall\quad k = 1, \ldots, d.$$

We can compute vectors C and C' in time $O(d)$.

3   The Interval Knapsack Problem

In this section we design both a pseudo-polynomial-time exact algorithm and a FPTAS for I-KP. We first develop a pseudo-polynomial-time exact algorithm using DP-by-unit.

Let $T(i, w)$ denote the maximum value for all solutions selected from the first i interval items with the total number of units bounded by w, for $1 \leq i \leq n$,

$1 \le w \le M$. Let $T(i)$ denote the vector with length M where the w-th position stores $T(i, w)$ for all w. It is easy to compute $T(1)$ since $T(1, w) = 0$ if $w \in (0, a_1)$, $T(1, w) = p_1 w$ if $w \in [a_1, b_1]$, and $T(1, w) = p_1 b_1$ if $w \in (b_1, M]$. Next we show how to compute $T(i)$ based on $T(i-1)$ and $([a_i, b_i], p_i)$:

$$T(i, w) = \max \{T(i-1, w), \max \{T(i-1, w - x_i) + x_i p_i \mid a_i \le x_i \le b_i\}\}.$$

By Lemma 2, $T(i)$ can be computed in time $O(M)$ based on $T(i-1)$, for each i. So that $T(n)$ can be computed in time $O(nM)$ and the optimal solution value is given by $T(n, M)$. The space complexity is $O(nM)$, however it can be reduced to $O(n + M)$ using standard storage reduction technique in Dynamic Programming [15]. In summary, we have:

Theorem 1. *I-KP is weakly NP-complete and pseudo-polynomial-time solvable. We can compute the exact solution of I-KP in time $O(nM)$ and space $O(n+M)$.*

3.1   FPTAS for I-KP

Next we design a FPTAS for I-KP based on dynamic programming. There are two ways to build the DP table, either by unit or by value. The DP-by-unit approach described above is relatively simple and incurs low overhead, but it is hard to convert into an approximation scheme. So we design a FPTAS based on DP-by-value. We first give a 2-approximation of I-KP in linear time, which is subsequently used for the FPTAS.

Lemma 3. *We can compute a 2-approximation of I-KP in $O(n)$ time.*

Next we describe a FPTAS for I-KP. Let V_0 be twice the value of the 2-approximation solution, then $V^* \le V_0 \le 2V^*$ where V^* is the optimal solution value. We use V_0 for a value rounding procedure in order to get the approximation scheme. Since a solution consists of at most n elements, a naive approach is to use $(\epsilon V_0/n)$ as the rounding factor, which results in a time complexity $\Omega(n^2)$. To get a $\tilde{O}(n)$ running time ignoring other factors, we divide items into two classes: *small* items and *large* items. Small items are those $([a_i, b_i], p_i)$ where $a_i p_i \le \epsilon V_0$, large items are all others. For simplicity, assume that there are n_S small items, n_L large items, $n = n_S + n_L$ and all items with $i \le n_L$ are large items. We consider large items first, and use the following *value-rounding procedure* for each item with value v:

$$v' = f(v) \equiv \lfloor \frac{v}{\epsilon^2 V_0} \rfloor \cdot (\epsilon^2 V_0).$$

Notice that we only need to consider values $v \le V_0$, thus the value-rounding procedure essentially reduces the number of distinct values to at most $1/\epsilon$ of them: $j \cdot (\epsilon^2 V_0)$ for $1 \le j \le 1/\epsilon^2$. Let $L(i, k)$ denote the minimum weight over all solutions with profit at least $k \cdot \epsilon^2 V_0$ and elements selected from the first i rounded item sets, and $L(i)$ denote the vector with length $1/\epsilon^2$ where the k-th cell stores $L(i, k)$, for $k = 1, \ldots, 1/\epsilon^2$. The following recursive formula is used to compute $L(i)$: $L(i, k) = \min\{L(i - 1, k), L'(i, k)\}$ where

$$L'(i, k) \equiv \min\left\{L(i - 1, \ell) + x_i \mid x_i p_i \geq (k - \ell)\epsilon^2 V_0, x_i \in [a_i, b_i]\right\}$$
$$= \min\left\{L(i - 1, \ell) + x_i \mid x_i \geq (k - \ell)c_i, x_i \in [a_i, b_i]\right\},$$

for $k = 1, \ldots, 1/\epsilon^2$, and $c_i = \epsilon^2 V_0/p_i$. Let $A_k = L(i - 1, k)$, $C_k = L(i, k)$, for $k = 1, \ldots, 1/\epsilon^2$, then Lemma 1 shows that $L(i)$ can be computed from $L(i - 1)$ in $O(1/\epsilon^2)$ time, for all $1 \leq i \leq n_L$.

After vector $L(n_L)$ is computed, $L(n_L, k)$ is the minimum weight among large item solutions with value at least $k \cdot \epsilon^2 V_0$, for $k = 1, \ldots, 1/\epsilon^2$. For each fixed $v = k \cdot \epsilon^2 V_0$ and the corresponding large item solution, we use a greedy algorithm to pack the small items to the end. This requires sorting all small items with decreasing unit-prices and it takes time $O(n \log n)$. After the sorting, we can simply pack small items with decreasing unit-prices to the end of the solution corresponding to value v, and stop immediately before the first time when the total number of units exceeds M. Let S_v denote the solution consisting of large items with total value at least v and small items packed in the end. Then it takes $O(n)$ time to obtain S_v for each v. Actually the greedy packing only needs to walk through the small item list once: (1) Start with $k = k_0$ corresponding to the largest $v = k \cdot \epsilon^2 V_0$ such that $L(n_L, v) \leq M$; (2) Walk through the small items list from the beginning and stop at position i_{k_0} to obtain S_{v_0}; (3) Set $k := k - 1$ and walk through the small item list from its current position to the right until obtaining S_v for the current $v = k \cdot \epsilon^2 V_0$; (4) Continue until $k = 0$ or reaching the end of the small item list.

The greedy packing procedure takes time linear of the vector length, thus the total processing time for merging large items and small items is $O(n \log n + 1/\epsilon^2)$. Once we obtain S_v for each $v = k \cdot \epsilon^2 V_0$ and $k = 1, \ldots, 1/\epsilon^2$, we select among all S_v the one with the maximum total value. We can recover all the elements of the solution using standard backtracking techniques. In summary, we have:

Theorem 2. *We can compute a* $(1+\epsilon)$*-approximation of I-KP in time* $O(n \log n + n/\epsilon^2)$.

4   The Interval Multiple-Choice Knapsack Problem

We first describe a pseudo-polynomial-time algorithm for I-MCKP which is similar to the algorithm described in Section 3 for I-KP. Let $T(i)$ denote the vector with length M where the w-th position $T(i, w)$ stores the maximum value for all solutions selected from the first i item sets with the total number of units bounded by w, for $1 \leq i \leq m$, $1 \leq w \leq M$. Let $T^j(i)$ denote the vector with length M where $T^j(i, w)$ denotes the maximum value over all solutions with elements selected from the first $i - 1$ item sets, together with an element from $([a_i^j, b_i^j], p_i^j)$, where $1 \leq j \leq n_i$. It is easy to compute $T(1)$: $T(1, w) = \min\{T^j(1, w) \mid 1 \leq j \leq n_1\}$ where $T^j(1, w)$ is determined easily from

$([a_1^j, b_1^j], p_1^j)$ for $1 \leq j \leq n_1$. Next we show how to compute $T(i)$ from $T(i-1)$. It is easy to know that

$$T^j(i,w) = \max\left\{T(i-1, w - x_i^j) + p_i^j x_i^j \mid x_i^j \in [a_i^j, b_i^j], x_i^j \leq w\right\}.$$

By Lemma 2, $T^j(i)$ can be computed in time $O(M)$ for each $1 \leq j \leq n_i$. Once we have computed $T^j(i)$ for all j, then

$$T(i,w) = \max\{T(i-1, w), \max\{T^j(i,w) \mid 1 \leq j \leq n_i\}\}, \quad \forall\, w = 1, \ldots, M.$$

So that it takes time $O(n_i M)$ to compute $T(i)$, for $i = 1, \ldots, m$. Since $\sum_{1 \leq i \leq m} n_i = n$, it takes in total $O(nM)$ time to compute vector $T(i)$, for $i = 1, \ldots, m$. The optimal solution is given by $T(n, M)$. In summary, we have:

Theorem 3. *I-MCKP is weakly NP-complete and pseudo-polynomial-time solvable. We can compute the exact solution of I-MCKP in time $O(nM)$ and space $O(n + M)$.*

4.1   FPTAS for I-MCKP

Next we design a FPTAS for I-MCKP based on DP-by-value. We first describe constant-factor approximations of I-MCKP, which are subsequently used for the FPTAS of I-MCKP. Our constant-factor approximations of I-MCKP are summarized as follows.

Lemma 4. *We can compute a 3-approximation of I-MCKP in $O(n)$ time, and another $(9/4)$-approximation in $O(n \log m)$ time.*

Comment: Kothari et al. [18] gave a 2-approximation with $O(n^2)$ time for the so-called general knapsack problem, which is actually a special case of I-MCKP. We observe that *a minor modification of their algorithm gave a worst-case 2-approximation with $O(n \log n)$ time.* Instead of running Greedy(ℓ, j) for each tuple $([u_\ell^j, u_\ell^{j+1}), p_\ell^j)$, we can simply run *one* Greedy and take care of the case where the next element will be taken from the interval $[u_\ell^j, u_\ell^{j+1})$. It remains an interesting problem to compute a 2-approximation of I-MCKP in linear time, or a $(5/4)$-approximation in $O(n \log n)$ time.

We are now ready to describe a FPTAS to I-MCKP. Let V_0 be three times the value of the 3-approximation solution, then $V^* \leq V_0 \leq 3V^*$ where V^* is the optimal solution value. We still use a value rounding procedure to get an approximation scheme, using $(\epsilon V_0/m)$ as the rounding factor. Formally, given v, its rounded value is:
$$v' = g(v) \equiv \lfloor \frac{vm}{\epsilon V_0} \rfloor \cdot \frac{\epsilon V_0}{m}.$$

Let $L(i, k)$ denote the minimum weight over all solutions with profit at least $k \cdot \epsilon V_0/m$, and elements selected from the first i rounded item sets, for $k = 1, \ldots, m/\epsilon$. Let $L(i)$ denote the vector with length m/ϵ where the k-th position stores $L(i, k)$. The following formula is used to compute $L(i)$ based on $L(i-1)$: $L(i, k)$ is the minimum of $L(i-1, k)$ and $\min\{L^j(i, k) \mid 1 \leq j \leq n_i\}$ where

$$L^j(i, k) \equiv \min\left\{L(i-1, \ell) + x_i^j \mid x_i^j p_i^j \geq (k - \ell)\epsilon V_0/m, x_i^j \in [a_i^j, b_i^j]\right\}.$$

Let $A_k = L(i-1,k)$, $C_k^j = L^j(i,k)$, $c_r = \epsilon V_0/(mp_i^j)$, for $k = 1,\ldots,m/\epsilon$, $j = 1,\ldots,n_i$. By Lemma 1, we can compute $C^j = L^j(i)$ for each j in $O(m/\epsilon)$ time. Once $L^j(i)$ is obtained for each $j = 1,\ldots,n_i$, then $L(i)$ can be computed easily as the minimum over these vectors and $L(i-1)$. In total, it takes $O(n_i m/\epsilon)$ to compute $L(i)$ based on $L(i-1)$. Since $\sum_i n_i = n$, thus in total it takes time $O(nm/\epsilon)$ to compute all vectors $L(i)$ for $1 \leq i \leq m$. The optimal solution value is given by walking through the last vector $L(m)$ and finding the largest $v = k \cdot \epsilon V_0/m$ such that $L(m,k) \leq M$. We can recover all the elements of the solution using standard backtracking techniques. In summary, we have:

Theorem 4. *We can compute a $(1 + \epsilon)$-approximation of I-MCKP in time $O(nm/\epsilon)$.*

5   Applications to Multi-unit Auction Clearing

In this section we apply algorithms for I-KP and I-MCKP to solve the single-good multi-unit auction clearing problem (SMAP). For SMAP with point bids, it is equivalent to the classic $0/1$ knapsack problem. By Kellerer and Pferschy [14], we have the following result:

Theorem 5. *For SMAP with point bids, we can compute the exact solution in time $O(nM)$ and space $O(n + M)$, and a $(1 + \epsilon)$-approximation in time $O(n \min\{\log n, \log(1/\epsilon)\} + 1/\epsilon^2 \log(1/\epsilon) \min\{n, 1/\epsilon \log(1/\epsilon)\})$ and space $O(n + 1/\epsilon^2)$.*

For SMAP with interval bids, it is reduced to I-KP studied in Section 3. By Theorems 1 and 2, we have the following result:

Theorem 6. *For SMAP with interval bids, we can compute the exact solution in time $O(nM)$ and a $(1 + \epsilon)$-approximation in time $O(n \log n + n/\epsilon^2)$.*

For SMAP with XOR-point bids, it corresponds to exactly MCKP. For SMAP with XOR-interval bids, it corresponds to I-MCKP. By Theorems 3 and 4, we have the following combined results:

Theorem 7. *For SMAP with XOR-point bids or XOR-interval bids, we can compute the optimal solution in time $O(nM)$ and a $(1 + \epsilon)$-approximation in time $O(nm/\epsilon)$.*

As said before, the XOR-interval bidding model covers the piecewise constant bidding model as a special case. Therefore we have the following corollary, which improves the corresponding algorithm of Kothari et al. [18] by a factor of n.

Corollary 1. *Given a buyer with M units of a single good, and n suppliers with piecewise constant bidding curves where each curve has $O(1)$ pieces, we can compute an exact solution with time $O(nM)$ and a $(1 + \epsilon)$-approximation with time $O(n^2/\epsilon)$.*

6   Improved Algorithms for VCG Computations

In this section we consider how to compute VCG payments to all bidders under various bidding models. Vickey-Clark-Grove (VCG) mechanism maximizes the expected payoff to the auctioneer and it is strategyproof for all bidders. For any sealed-bid combinatorial auction, we can embed a VCG payment scheme to it: bidder i is given a discount of $V(I) - V(I \setminus \{i\})$ to its payment, $\forall\, i \in I$. Here I denotes the set of all bidders, $V(S)$ denotes the maximum revenue for bids from S, for any $S \subseteq I$. While a straightforward approach requires solving $n + 1$ winner determination problems with n bidders, here we show that an overhead of factor $O(\log n)$ is sufficient for our bidding models.

We start with the interval bidding model with n interval items. For each $1 \le r \le n, 1 \le w \le M$, let $T(I \setminus \{r\}, w)$ denote the maximum value for solutions consisting of elements excluding the r-th item, and total units bounded by w. We can compute $T(I \setminus \{r\}, w)$ for all r, w values in $O(n^2 M)$ time using DP. In the following lemma, we show how to improve the running time to $O(nM \log n)$.

Lemma 5. *We can compute $T(I \setminus \{r\}, w)$ for all $1 \le r \le n$, $1 \le w \le M$ with time $O(nM \log n)$.*

Next we state results for VCG payment computations of SMAP with different bidding models. All our results are based on variations of Lemma 5.

Theorem 8. *For SMAP with point bids, we can compute the VCG payments to all the bidders in time $O(nM \log n)$, and an ϵ-approximate VCG payment in time $O(T(n) \log n)$ where $T(n)$ is the running time of KP.*

Theorem 9. *For SMAP with interval bids, we can compute the VCG payments to all bidders in time $O(nM \log n)$, and an ϵ-approximate VCG payment in time $O((n/\epsilon^2) \log n)$.*

Theorem 10. *For SMAP with XOR-point bids or XOR-interval bids, we can compute the VCG payments to all bidders in time $O(nM \log m)$ and an ϵ-approximate VCG payment in time $O((nm/\epsilon) \log m)$.*

Kothari et al. [18] designed an algorithm to compute an ϵ-approximate VCG payment to all bidders with time $O((n^3/\epsilon)\alpha \log(\alpha n/\epsilon))$ under the piecewise constant bidding model. Here $\alpha = \max_i V(I)/V(I \setminus \{i\})$ is a constant related to "market no-monopoly". Since the XOR-interval bidding model covers the piecewise constant bidding model as a special case, we obtain the following corollary, which gives a factor $\Omega(n)$ improvement over their algorithm.

Corollary 2. *Given a buyer with M units of a single good, and n suppliers with piecewise constant bidding curves where each curve has $O(1)$ pieces, we can compute an ϵ-approximate VCG payment in time $O((n^2/\epsilon) \log n)$.*

7   Conclusion

In this paper, we have studied the single-good multi-unit auction problem with interval bids and XOR-interval bids, and formalize the auction clearing problem as new variants of knapsack problems. We have designed both exact and approximate algorithms to solve these knapsack problems, as well as better algorithms for the auction clearing problem. Our results are built upon a crucial technical lemma, which is used to merge a vector with an interval item in linear time. This technical lemma essentially allows us to treat an interval item the same as a singleton item, and we are optimistic that similar techniques can be applied to other variants of knapsack problems to obtain better bounds.

Acknowledgement. We thank Anshul Kothari for introducing us to this problem and working on it initially, Bob Tarjan for pointing us to the deque with heap order data structure, and Terence Kelly for comments.

References

1. J. K. A. Davenport and H. Lee. Computational aspects of clearing continuous call double auctions with assignment constraints and indivisible demand. *Electronic Commerce Research*, 1(3):221–238, 2001.
2. P. Briest, P. Krysta, and B. Vcking. Approximation techniques for utilitarian mechanism design. In *Proc. STOC*, pages 39–48, 2005.
3. A. K. Chandra, D. S. Hirschberg, and C. Wong. Approximate algorithms for some generalized knapsack problems. *Theoretical Computer Science*, 3:293–304, 1976.
4. V. D. Dang and N. R. Jennings. Optimal clearing algorithms for multi-unit single-item and multi-unit combinatorial auctions with demand/supply function bidding. In *Proc. 5th ICEC*, pages 25–30, 2003.
5. M. E. Dyer. An $O(n)$ algorithm for the multiple-choice knapsack linear program. *Mathematical Programming*, 29:57–63, 1984.
6. M. Eso, S. Ghosh, J. Kalagnanam, and L. Ladanyi. Bid evaluation in procurement auctions with piecewise linear supply curves. *J. Heuristics*, 11(2):147–173, 2005.
7. H. Gajewska and R. E. Tarjan. Deques with heap order. *Information Processing Letters*, 22(4):197–200, 1986.
8. G. V. Gens and E. V. Levner. Approximation algorithms for certain universal problems in scheduling theory. *Soviet J. Comput. System Sci.*, 6:31–36, 1978.
9. R. C. Holte. Combinatorial auctions, knapsack problems, and hill-climbing search. In *Proc. Canadian Conf. on AI, LNCS 2056*, pages 57–66, 2001.
10. R. Hood and R. Melville. Real-time queue operations in pure LISP. *Information Processing Letters*, 13:50–54, 1981.
11. O. H. Ibarra and C. E. Kim. Fast approximation algorithms for the knapsack and sum of subset problems. *Journal of the ACM*, 22:463–468, 1975.
12. R. M. Karp. Reducibility among combinatorial problems. In *Complexity of Computer Computations*, pages 85–103. Plenum Press, New York, 1972.
13. H. Kellerer and U. Pferschy. A new fully polynomial time approximation scheme for the knapsack problem. *J. of Comb. Opt.*, 3(1):59–71, 1999.
14. H. Kellerer and U. Pferschy. Improved dynamic programming in connection with an FPTAS for the knapsack problem. *J. of Comb. Opt.*, 8(1):5–11, 2004.

15. H. Kellerer, U. Pferschy, and D. Pisinger. *Knapsack Problems*. Springer, 2004.
16. F. Kelly and R. Steinberg. A combinatorial auction with multiple winners for universal services. *Management Science*, 46:586–596, 2000.
17. T. P. Kelly. Generalized knapsack solvers for multi-unit combinatorial auctions. In *Workshop on Agent Mediated E-Commerce, LNAI 3435*, 2004.
18. A. Kothari, D. C. Parkes, and S. Suri. Approximately-strategyproof and tractable multi-unit auctions. *Decision Support Systems*, 39:105–121, 2005.
19. A. Kothari, S. Suri, and Y. Zhou. Interval subset-sum and uniform-price auction clearing. In *Proc. COCOON, LNCS 3595*, pages 608–620, 2005.
20. E. L. Lawler. Fast approximation algorithms for knapsack problems. *Mathematics of Operations Research*, 4:339–356, 1979.
21. D. Lehmann, L. I. O'Callaghan, and Y. Shoham. Truth revelation in approximately efficient combinatorial auctions. In *Proc. ACM EC*, pages 96–102, 1999.
22. N. Nisan and A. Ronen. Algorithmic mechanism design. *Games and Economic Behavior*, 35:166–196, 2001.
23. M. H. Rothkopf, A. Pekec, and R. M. Harstad. Computationally manageable combinatorial auctions. *Management Science*, 44(8):1131–1147, 1998.
24. T. Sandholm. Algorithm for optimal winner determination in combinatorial auctions. *Artificial Intelligence*, 135(1-2):1–54, 2002.
25. T. Sandholm and S. Suri. Market clearability. In *IJCAI*, pages 1145–1151, 2001.
26. W. E. Walsh, M. P. Wellman, and F. Ygge. Combinatorial auctions for supply chain formation. In *Proc. ACM EC*, pages 260–269, 2000.
27. E. Zemel. An $O(n)$ algorithm for the linear multiple choice knapsack problem and related problems. *Information Processing Letters*, 18:123–128, 1984.

A Simple Message Passing Algorithm for Graph Partitioning Problems

Mikael Onsjö and Osamu Watanabe

[1] Dept. of Computer Sci. and Eng., Chalmers Univ. of Technology, Sweden
[2] Dept. of Math. and Comput. Sci., Tokyo Inst. of Technology, Japan

Abstract. Motivated by the *belief propagation*, we propose a simple and deterministic message passing algorithm for the Graph Bisection problem and related problems. The running time of the main algorithm is linear w.r.t. the number of vertices and edges. For evaluating its average-case correctness, planted solution models are used. For the Graph Bisection problem under the standard planted solution model with probability parameters p and r, we prove that our algorithm yields a planted solution with probability $> 1 - \delta$ if $p - r = \Omega(n^{-1/2} \log(n/\delta))$.

1 Introduction

We begin by introducing problems discussed in this paper. A *Graph Bisection problem* is to find an equal size partition of a given undirected graph with the smallest number of crossing edges. Throughout this paper, we consider undirected graphs with no loop nor multiple edge, and assume that the number of vertices is even. We use $2n$ and m to denote the number of vertices and edges respectively. For a graph $G = (V, E)$, an *equal size partition* is a pair of disjoint subsets V_+ and V_- of V such that $V = V_+ \cup V_-$ and $|V_+| = |V_-|$. The Graph Bisection problem is to find such a partition V_+ and V_- minimizing $|(V_+ \times V_-) \cap E|$, i.e., the number of edges between them. In the case where the optimal solution is not unique, we only require to compute one of them. The same requirement is assumed for the other problems.

We consider another graph partitioning problem, the Most Likely Partition problem. Intuitively, the problem is to find, for a given graph $G = (V, E)$, a partition that is most likely under the condition that G is observed. For defining the problem precisely, we need to specify a certain random graph model, i.e., a way to generate a graph randomly. For any n, consider a set V of vertices; we let $V = \{v_1, ..., v_{2n}\}$. For generating a graph, we first generate a partition of V. This is done by simply dividing V into two equal size sets V_+ and V_- uniformly at random. Define a vector $\boldsymbol{a} = (a_1, ..., a_{2n}) \in \{+1, -1\}^{2n}$ so that $a_i = +1$ if $v_i \in V_+$ and $a_i = -1$ if $v_i \in V_-$; this $\boldsymbol{a}$ is called an *assignment* for the partition (V_+, V_-). Then for a priori determined parameters p and r, we generate undirected edges as follows: for any vertices $v_i, v_j \in V$, put an edge (v_i, v_j) to E with probability p if $a_i = a_j$, and put an edge (v_i, v_j) to E with probability r if $a_i \neq a_j$. This is the way of generating a graph randomly. Note that, for any size parameter n, and parameters p and r, this model defines a probability distribution on graphs of size $2n$. For any graph $G = (V, E)$, consider any partition (V_+, V_-) of V, and

T. Asano (Ed.): ISAAC 2006, LNCS 4288, pp. 507–516, 2006.

let $\boldsymbol{a}$ be its assignment. Then for given parameters p and r, the following is the probability that G is generated from (V_+, V_-) in the way specified above. (Below by $\overline{E}$ we denote the set of ordered pairs of V not in E.)

$$
\Pr[G \mid (V_+, V_-)] = \prod_{(v_i, v_j) \in E} p^{[a_i = a_j]} r^{[a_i \neq a_j]} \cdot \prod_{(v_i, v_j) \in \overline{E}} (1 - p)^{[a_i = a_j]} (1 - r)^{[a_i \neq a_j]}
$$

(1)

where $[\cdots]$ takes 1 if $\cdots$ holds and 0 otherwise. We call this probability the *likelihood* of (V_+, V_-). Note that the likelihood of (V_+, V_-) for observed G (that is, $\Pr[(V_+, V_-)|G]$) should be computed as $\Pr[G|(V_+, V_-)] \cdot \Pr[(V_+, V_-)] / \Pr[G]$. But both $\Pr[G]$ and $\Pr[(V_+, V_-)]$ are the same for all possible partitions, we use this probability $\Pr[(V_+, V_-)|G]$ for determining the most likely partition.

Now our second graph partitioning problem — Most Likely Partition (MLP) problem — is defined as follows: For a given graph $G = (V, E)$ and parameters p and r, the problem is to find a partition V_+ and V_- of V with the max. likelihood w.r.t. p and r. We also consider a problem where parameters p and r are not given, which requires to compute also these parameters besides a partition. In this case, parameters p and r to be computed are those maximize $\Pr[(V_+, V_-)|G]$ with most likely partition (V_+, V_-) w.r.t. p and r. This harder version is called a *parameterless version*. The Most Likely Partition problem is considered as a basic problem for various clustering problems; see, e.g., [CK01, DLP03] for the background of the problem.

1.1   Planted Solution Models: Our Average-Case Scenario

There are some NP-hard problems, for which we can show some algorithm that solves the problem correctly/efficiently *on average* under a reasonable average-case scenario. For discussing average-case performance of algorithms, the choice of an average-case scenario, that is, the choice of a probability model for determining an input distribution is important. The notion of "planted solution" has been used for defining reasonable probability models. Here we follow this approach and consider the standard planted solution model for our graph partitioning problems.

Jerrum and Sorkin [JS98] studied a planted solution model for the Graph Bisection problem, which has been used as a standard model. Here we use this model for our two graph partitioning problems. This model specifies a way to generate graph from a planted solution, which is almost the same as the one used for defining the most likely partition. We first fix probability parameters p and r, $0 < r < p \le 1$. Then for a given size parameter n, and a given equal size partition V_+^* and V_-^* of $V = \{v_1, ..., v_{2n}\}$, generate undirected edges in E as follows (here let $\boldsymbol{a}^*$ denote the assignment for (V_+^*, V_-^*)): for any vertices $v_i, v_j \in V$, put an edge (v_i, v_j) to E with probability p if $a_i^* = a_j^*$, and put an edge (v_i, v_j) to E with probability r if $a_i^* \neq a_j^*$. Since $r < p$, we have *on average* more edges among vertices in V_+^* (resp., V_-^*) than between V_+^* and V_-^*. Hence, we can expect that the partition (V_+^*, V_-^*) achieves the smallest number of cut edges, that is, it is optimal for the Graph Bisection problem. Thus, the partition (V_+^*, V_-^*) is called a *planted solution*.

The above intuition can be formally justified for our two graph partitioning problems. It has been shown [Betal87] that if $p - r = \Omega(n^{-1/2})$, then with high probability, a planted solution is the unique optimal solution of the Graph Bisection problem. We can show a similar property for the MLP problem. That is, it can be shown [Ons05] that if $p - r = \Omega(n^{-1/2})$, then a planted solution is, with high probability, the unique solution of the MLP problem for the generated instance. Thus, under the above planted solution model, both of our graph partitioning problems ask for the same solution for a wide range of parameters p and r.

1.2   Main Results

We propose a simple deterministic algorithm for our two graph partitioning problems. Since these two problems ask for the same answer (under the planted solution model with reasonable parameters), we explain the algorithm for the MLP problem.

First consider the case that the probability parameters p and r are given as input. Figure 1 (of the next section) states a general message passing algorithm for the problem. The algorithm aims to compute, for each vertex v_i, the belief b_i for its assignment a_i; the sign of b_i determines whether $a_i = +1$ or $a_i = -1$, and its absolute value reflects the strength of the belief. (Without losing generality, we may assume that $a_1 = +1$; hence, b_1 is set $+\infty$ at every round.) Roughly speaking, at each round, it updates (*in parallel*) the current belief b_i, which is propagated to its neighbor vertices in the next round. In general, this computation is repeated until all beliefs get stabilized. But for our theoretical analysis, we consider the algorithm that terminates in two rounds (i.e., MAXSTEP = 2) and outputs an assignment based on the obtained beliefs. We call this algorithm GraphPart2. (The two round restriction is necessary for our current analysis; roughly speaking, during the first two rounds, all edges are touched only once, and we can make use of the independence of edge existence for analyzing the variance of beliefs. Also some more minor changes are made in GraphPart2 for simplifying our analysis; see the next section for the details.) Though the general GraphPart requires some floating number computation, only simple counting is enough for executing this simplified GraphPart2, and it is easy to see that GraphPart2 runs in time $\mathcal{O}(n + m)$. For the correctness of the algorithm, we prove the following theorem.

Theorem 1. *For any n, and p and r, $0 < r < p < 1$, consider the execution of* GraphPart2 *on randomly generated graphs under the planted solution model. Then with some constant ϵ_1 we have*

$$\Pr[\text{ the algorithm yields the planted solution }] \geq 1 - 2n \cdot e^{-\epsilon_1 n \cdot \frac{(p-r)^4}{p^2}},$$

This accuracy can be stated in terms of the bound for $p - r$, which is more useful for comparing with the other algorithms. For simplicity, we consider the case such that $cp < r < p$ for some constant $c > 0$. Then the condition of the theorem can be restated as $p - r \geq c_1(n \log(n/\delta))^{-1/2}$ for some constant $c_1 > 0$. That is, if $p - r = \Omega(\sqrt{\log n/n})$, then the algorithm gives a correct answer with high probability.

We can also prove the same theorem even if the size of each class is not the same. That is, for any fixed pair of n_1 and n_2, we have the same accuracy bound for $n = \min(n_1, n_2)$, under the planted solution model for the partition of size n_1 and n_2. Note that the algorithm does not need to know n_1 or n_2. This property may be useful for multiclass partitioning; we may first separate vertices of class 1 from the others, and then separate those of class 2, and so on.

We next consider the prameterless version, in particular, for the Graph Bisection problem. Note here that we may assume that a planted partition is of equal size; hence, we can estimate $p + r$ by counting the number of edges in a given graph, and we can expect that this estimated value can be quite accurate. Then within some appropriate range, we search for approximations $\widetilde{p}$ and $\widetilde{r}$ of p and r by binary search. This needs to run the algorithm GraphPart2 for $\mathcal{O}(\log n)$ times (if $p - r = \Omega(n^{-1/c})$ for some $c > 1$). Notice that with the same strategy, we can also solve the Graph Bisection problem. The accuracy of this strategy is guaranteed by the following theorem.

Theorem 2. *For any n, and p and r, $0 < r < p < 1$, consider the execution of* GraphPart2 *on randomly generated graphs under the planted solution model, but here execute it with parameters $\widetilde{p}$ and $\widetilde{r}$ such that $\widetilde{p} + \widetilde{r} = p + r$ and $p - r \leq \widetilde{p} - \widetilde{r} < (5/4)(p - r)$. Even in this case with some constant ϵ_2 we have*

$$\Pr[\text{ the algorithm yields the planted solution }] \geq 1 - 2n \cdot e^{-\epsilon_2 n \cdot \frac{(p-r)^4}{p^2}},$$

The proofs of the above two theorems are not so hard. It is easy to estimate the expected belief computed at each vertex. On the other hand, we can make use of independency of assigning edges because all pairs of vertices are touched only once during the execution. Thus, by a standard argument, we can show that the actual values stay, with high probability, within a reasonable range from the expected beliefs.

1.3 Related Work

The Graph Bisection problem has been studied by many researchers, and a good number of *theoretically guaranteed* algorithms have been already proposed. Here we mention some of them related to our algorithm; see, e.g., [CK01, McS99, Coj05]. Boppana is one of those who gave polynomial-time algorithms in 80's, and his spectral algorithm [Bop87] has been the best until quite recently; see, e.g., [Coj05] for the details. On the other hand, for using his algorithm, one needs to solve some convex optimization problem, which is not so easy. More recently, McSherry [McS99] gave a more general spectral algorithm, which performs almost as well as Boppana's and which can be implemented in quasi linear time by some randomized computation. Among those having accuracy bounds close to Boppana's algorithm, the one proposed by Condon and Karp [CK01] achieves linear time. But note that theirs is also a randomized algorithm.

Compared with these known algorithms, we may claim that our algorithm is very simple and deterministic. Unfortunately, the range of parameters for which high accuracy is guaranteed by Theorem 2 is much weaker than the one for the algorithms by Boppana, though it is close to the one for the algorithm of Condon and Karp. Nevertheless, we think that our algorithm is worth studying theoretically. Firstly, since it is very simple, we may be able to clarify the reason why previously proposed randomized algorithms work. For example, instead of computing beliefs numerically, if we compute them by some appropriate random simulation, we may have a randomized algorithm that is conceptually similar to the one by Condon and Karp. Secondly, the algorithm works for the case with very unbalanced partition size, which has not been shown for the other algorithms. Thirdly, though our analysis is only for the case MAXSTEP = 2 (due to the technical reason), computer experiments show that the algorithm performs much better when it is allowed to update beliefs more than twice. It is an important open problem to justify this performance theoretically.

Belief Propagation

The algorithm `GraphPart` is derived from Pearl's belief propagation [Pea88] with some modification. Roughly speaking, the belief propagation is a way to compute a marginal probability of the state of each node in a given Bayesian network. We use this technique for the MLP problem. For any input G, p, and r for the MLP problem, we can define a Bayesian network on which a belief propagation algorithm (in short, the BP algorithm) is expected to compute $P(i) = \Pr[v_i \in V_+|G]$, where the probability is defined under our random model for defining the most likely partition. Intuitively, a *belief* (that v_i belongs to V_+) is the approximation of $P(i)$. The BP algorithm computes beliefs in rounds; at each round, it updates beliefs and we would like to have correct $P(i)$'s at some round. In fact, it is shown that the BP algorithm converges in finite rounds and yields the correct probabilities if a given Bayesian network is a tree; although such a convergence cannot be guaranteed in general Bayesian networks, it is often the case that the BP algorithm converges and gives quite accurate values even for Bayesian networks with cycles. Now suppose that the BP algorithm computes $P(i)$ correctly at some round, then a natural solution for our partition problem is to compute V_+ (resp., V_-) as a set of vertices v_i with $P(i) > 0.5$ (resp., $P(i) < 0.5$), which we may expect to give a partition with the max. likelihood. Our algorithm is derived from this *BP-based partition algorithm*. see, [OW05] for the derivation and related issues.

2   Algorithm and Its Analysis

We explain the algorithm `GraphPart` of Figure 1 and its analysis. As explained in Introduction, the algorithm updates beliefs for each vertex $v_i \in V_+$ each round. An updated value of b_i is computed by summing up the beliefs of *all* vertices v_j, multiplied by eigher $h_+ > 0$ (if an edge (v_i, v_j) exists) and by $-h_- <$

procedure GraphPart (G, p, r);
begin
  set all b_i to 0;
  repeat MAXSTEP times **do** {
    $b_1 \leftarrow +\infty$;
    for each $v_i \in V$ **do in parallel** {
      $b_i \leftarrow \sum_{v_j \in N_i} h_+ \cdot \mathrm{Th}_+(b_j)$
      $- \sum_{v_j \notin N_i} h_- \cdot \mathrm{Th}_-(b_j)$;
    }
    if all b_i's get stabilized **then break**;
  }
  output $(+1, \mathrm{sg}(b_2), ..., \mathrm{sg}(b_{2n}))$;
end-procedure

parameters & functions

$$c_- = \frac{1-p}{1-r}, \ c_+ = \frac{p}{r},$$

$$h_- = \left| \frac{c_- - 1}{c_- + 1} \right|, \ h_+ = \left| \frac{c_+ - 1}{c_+ + 1} \right|,$$

$$\mathrm{th}_- = \left| \frac{\ln c_-}{h_-} \right|, \ \mathrm{th}_+ = \left| \frac{\ln c_+}{h_+} \right|,$$

$$\mathrm{Th}_+(z) = \mathrm{sg}(z) \min(|z|, \mathrm{th}_+),$$

$$\mathrm{Th}_-(z) = \mathrm{sg}(z) \min(|z|, \mathrm{th}_-),$$

$\mathrm{sg}(z) =$ the sign of z, and

$N_i =$ the set of v_i's neighbors.

Fig. 1. Computation of pseudo beliefs for the MLP problem

0 (otherwise). This is intuitively reasonable because one can expect that two vertices v_i and v_j are in the same class (resp., in the different classes); if an edge exists (resp., does not exist) between them. The algorithm uses threshold functions $\mathrm{Th}_+(z)$ and $\mathrm{Th}_-(z)$ so that too large (or too small) beliefs are not sent to the other vertices. The algorithm terminates (before the time bound) if b_i gets stabilized for every i, i.e., either the change of b_i becomes small, or $|b_i|$ exceeds the threshold value $\max(\mathrm{Th}_+, \mathrm{Th}_-)$.

The theoretical analysis stated in Introduction is for the case that the algorithm is terminated in two rounds, i.e., MAXSTEP = 2. For simplifying the analysis, we further consider the following modifications: (i) use some small $\theta < \min(\mathrm{th}_+, \mathrm{th}_-)$ for the initial value of b_1 (for avoiding the thresholding), and (ii) set $b_1 = 0$ before the second round (for ignoring the effect from v_1 in the second round). Precisely speaking, this is the algorithm GraphPart2 investigated in our theorems.

We give the outline of the proof of Theorem 1. (Due to the space limit, we omit the proof of each lemma, which can be found in [OW05].) Below let $G = (V, E)$ be a random graph of size $|V| = 2n$ generated from the planted solution $V_+^* = \{v_1, ..., v_n\}$ and $V_-^* = \{v_{n+1}, ..., v_{2n}\}$ with parameters p and r, $0 < r < p$. Let a^* be the assignment of vertices in the planted solution.

Let us introduce some notations. Define $\alpha = p + r$ and $\beta = p - r$. Since the vertex v_1 is treated separately, we omit v_1 from our discussion, and by, e.g., "$v_i \in V_+^*$" we always mean v_i from $\{v_2, ..., v_n\}$. We introduce random variables (where the randomness is due to the random graph G). For any $v_i, v_j \in V$, let $E_{i,j}$ be a random variable taking a value in $\{0, 1\}$ that indicates whether there exists an edge between vertices v_i and v_j in G; hence, $E_{i,j} = 1$ with prob. p if $a_i^* = a_j^*$, and otherwise $E_{i,j} = 1$ with prob. r. Let P_1 and P_0 be the set of vertices in V_+^* that respectively does/does not have an edge with v_1; sets Q_1 and Q_0 are defined similarly for V_-^*.

Now consider any $v_i \in V$. we use b_i to denote the final pseudo belief computed by the algorithm. On the other hand, the value of the variable b_i after the first round is denoted as b'_i. It is easy to see that $b'_j = h_+ \cdot \theta$ if $v_j \in P_1 \cup Q_1$ and $b'_j = -h_- \cdot \theta$ if $v_j \in P_0 \cup Q_0$, and these values are used to compute the final value b_i. (Note that both $h_+ \cdot \theta$ and $h_- \cdot \theta$ are less than $\min(\text{th}_+, \text{th}_-)$.) Then we have the following lemma.

Lemma 1. $\mathrm{E}[b_i] = a_i^* \cdot \varphi(n, \alpha, \beta)$, where $\varphi(n, \alpha, \beta) \overset{\text{def}}{=} \dfrac{2n\theta\beta^4}{\alpha^2(2 - \alpha)^2}$.

This lemma shows that b_i gives *on average* the correct classification. Thus, it now suffices to show a condition that b_i is close to their expectations so that $b_i > 0$ and the algorithm outputs the correct assignment for v_i.

Let us consider the case where $\alpha \le 1$ and $v_i \in V_+^*$, and we discuss a condition that b_i becomes positive, i.e., the algorithm yields a correct assignment for v_i. Arguments for the other cases are similar and omitted. Here we introduce some random variables taking values in $[0, 1)$ and consider the following situation.

$$
\begin{aligned}
Y^+ &= \mathrm{E}[Y^+] + \delta_+ n = pn + \delta_+ n, \\
Y^- &= \mathrm{E}[Y^-] + \delta_- n = rn + \delta_- n, \\
X_i^{+,1} &= (pn + \delta_+ n)(p - \gamma_+) = p^2 n + (p\delta_+ - p\gamma_+ - \delta_+\gamma_+)n, \\
X_i^{-,1} &= (rn + \delta_- n)(r - \gamma_-) = r^2 n + (r\delta_- - r\gamma_- - \delta_-\gamma_-)n, \\
X_i^{+,0} &= ((1 - p)n - \delta_+ n)(p + \gamma'_+) = (1 - p)pn + ((1 - p)\gamma'_+ - p\delta_+ - \delta_+\gamma'_+)n, \\
X_i^{-,0} &= ((1 - r)n - \delta_- n)(r + \gamma'_-) = (1 - r)rn + ((1 - r)\gamma'_- - r\delta_- - \delta_-\gamma'_-)n.
\end{aligned}
\tag{2}
$$

We may also consider the other situations such as the case where $Y^+ = \mathrm{E}[Y^+] - \delta_+ n$; but it is easy to check that the above choice makes b_i the smallest. By using these variables, we express below a sufficient condition that b_i is close to its expectation and $b_i > 0$.

Lemma 2. *Assume that the following bounds hold for the estimators defined by (2).*

$$
\delta_+ < \min\left(p, \frac{\beta^2}{8p\alpha}\right), \quad \delta_- < \min\left(p, \frac{\beta}{8}, \frac{\beta^2}{8r\alpha}\right),
$$

$$
\gamma_+ < \frac{\beta^2}{8p(2 - \alpha)}, \qquad \gamma_- < \frac{\beta^2}{8p(2 - \alpha)}, \qquad \text{and} \qquad \gamma < \frac{\beta^2}{8\alpha(2 - \alpha)}.
$$

Then we have $b_i > \mathrm{E}[b_i] - \varphi(n, \alpha, \beta) = 0$.

Next we show a bound for the probability that the above condition holds.

Lemma 3. *There exists some constant ϵ_1 such that for any p, r, $0 < r < p < 1$, we have*

$$
\Pr[\text{ all bounds of Lemma 2 hold}] \ge 1 - e^{-\epsilon_1 n \cdot \frac{\beta^4}{p^2}}.
\tag{3}
$$

Note that this bound is for the event that the algorithm yields the planted solution for one vertex. The bound of the theorem is obtained by considering the event that the algorithm answers the planted solution for all vertices.

2.1   Robustness of the Algorithm

First we show that the algorithm works as well even if a planted solution (V_+^*, V_-^*) is not of the equal size. The argument is almost the same. For example, we can show the following generalization of Lemma 1.

Lemma 4. $\mathrm{E}[b_i] = a_i^* \cdot \dfrac{(n_+ + n_-)\theta\beta^4}{\alpha^2(2-\alpha)^2}$, where $n_+ = |V_+^*|$ and $n_- = |V_-^*|$.

Then by a similar argument, we can prove the same theorem for the general case with $n = \min(n_+, n_-)$.

Next consider the situation that we use parameters $\widetilde{p}$ and $\widetilde{r}$ that are different from those used for generating instances. This situation occurs when we want to solve the MLP problem of the parameterless version. Precisely speaking, under the planted solution model, our goal is to obtain a planted solution and parameters p' and r' close enough to those used to generate the input graph from the planted solution. For this goal, we may consider the following algorithm: First by counting the number of edges, we compute the estimation $\widetilde{\alpha}$ of α $(= p + r)$, which should be very close to α. Then by using a guess $\widetilde{\beta}$ of β, run the algorithm GraphPart2 with guessed $\widetilde{p}$ and $\widetilde{r}$, where $\widetilde{p} = (\widetilde{\alpha} + \widetilde{\beta})/2$ and $\widetilde{r} = (\widetilde{\alpha} - \widetilde{\beta})/2$. The initial guess of $\widetilde{\beta}$ is the largest candidate, i.e., $\widetilde{\alpha}$, and repeat the algorithm by revising $\widetilde{\beta}$ with $(4/5)\widetilde{\beta}$ until any "consistent" equal size partition is obtained. The consistency of the partition can be tested by checking whether the same partition can be obtained by the algorithm with parameters p' and r' that are estimated by counting the number of edges respectively within and between two partitioned sets.

In this situation, the algorithm is executed by using parameters $\widetilde{p}$ and $\widetilde{r}$ that are different from those used for generating instances; but we may assume that $\widetilde{p} + \widetilde{r}$ $(= \widetilde{\alpha}) \approx \alpha$, and $\widetilde{p} - \widetilde{r}$ $(= \widetilde{\beta})$ satisfies $\beta \le \widetilde{\beta} < (5/4)\beta$. For simplicity, we assume that $\widetilde{\alpha} = \alpha$ and $\beta \le \widetilde{\beta} < (5/4)\beta$. Theorem 2 states that, even with such parameters $\widetilde{p}$ and $\widetilde{r}$, the algorithm still yields the planted solution with high probability.

Let us see the proof outline of Theorem 2. Consider any $v_i \in V_+^*$ (again the case $v_i \in V_-^*$ can be argued similarly); $\mathrm{E}[b_i]$ is now the expected value of b_i when the algorithm is executed with $\widetilde{p}$ and $\widetilde{r}$. Consider also the execution of the algorithm (with $\widetilde{p}$ and $\widetilde{r}$) on a random graph generated with these parameters $\widetilde{p}$ and $\widetilde{r}$, and let $\widetilde{\mathrm{E}}[b_i]$ denote the expected value of b_i in this execution. By Lemma 1, we have $\widetilde{\mathrm{E}}[b_i] = \varphi(n, \alpha, \widetilde{\beta})$. Then by essentially the same way as Lemma 2, we obtain the following lemma. (Below let $\Delta = \widetilde{p} - p$, and recall $\widetilde{\alpha} = \alpha$.)

Lemma 5. $\mathrm{E}[b_i] = \widetilde{\mathrm{E}}[b_i] - a_i^* \left(\dfrac{2n(4\widetilde{\beta}^2)}{\alpha^2(2-\alpha)^2} \right) (\widetilde{\beta}\Delta - \Delta^2) \ge \widetilde{\mathrm{E}}[b_i]/2.$

From our assumption $\widetilde{\beta} \ge \beta$, it is easy to see that the tolerance against deviation is stronger by using $\widetilde{\beta}$ for β. More specifically, the condition of Lemma 2 using $\widetilde{\beta}$ instead of β implies that $b_i > \mathrm{E}[b_i] - \varphi(n, \alpha, \widetilde{\beta})$. Hence, from some stronger but

procedure SimplePart (G, p, r);
begin
    set all b_i to 0; $b_1 \leftarrow +1$; $b_{n+1} \leftarrow -1$;
    repeat MAXSTEP times **do** {
        for each $v_i \in V$ **do in parallel** $b_i \leftarrow \sum\limits_{v_j \in N_i} b_j$;
        if all b_i's get stabilized **then break**;
    }
    output $(\mathrm{sg}(b_1), ..., \mathrm{sg}(b_{2n}))$;
end-procedure

Fig. 2. Yet simpler message passing algorithm the MLP problem

still similar condition, we would have $b_i > \mathrm{E}[b_i] - \varphi(n, \alpha, \tilde{\beta})/2 = \mathrm{E}[b_i] - \widetilde{\mathrm{E}}[b_i]/2$, where the last expression is greater than equal to 0 by the above lemma. Then Theorem 2 is proved by the same last argument for Theorem 1.

3   Some Remarks

Simpler Algorithm Works?
From the argument of the previous section, one may wonder that much simpler message passing algorithm also works. For example, we can consider an algorithm SimplePart stated in Figure 2. The idea is clear. Assume that $v_1 \in V_+^*$ and $v_{n+1} \in V_-^*$, and positive and negative beliefs are passed to their neighbor vertices, from which beliefs are computed at all vertices. Then these beliefs are sent again to neighbor vertices, which are again used to update beliefs. For this algorithm, we may stop the iteration if for every i, either the change of b_i becomes small, or $|b_i|$ exceeds some large number.

We confirmed by some computer experiment that this algorithm works when p and r are in a certain range. Also again for a two round version of SimplePart (i.e., the one with MAXSTEP $= 2$), we can prove a property similar to the one that we showed in Theorem 1. In fact, it is easy to check that $\mathrm{E}[b_i]$, the average belief at vertex $v_i \in V_+^*$ after two updating rounds, is $n(p-r)^2$ $(= n\beta^2)$, and we can argue that the deviation is small enough if p and r satisfy a similar condition.

Unfortunately, however, this simple algorithm fails badly if instances are generated from an unbalanced partition, a partition such that $|V_+^*| \neq |V_-^*|$. This is because this algorithm uses only the information of the existence of edges. On the other hand, the BP based algorithm makes use of also the information of the nonexistence of edges. This point is important for keeping our algorithm to work even for instances from an unbalanced partition.

Some Observations from Experiments
Some interesting observations are also obtained from computer experiments on the algorithm GraphPart. We executed the algorithm on graphs with $n = 6000$ vertices that are generated randomly under the planted solution model with various values of parameters p and r. We had the following observations: (1)

The algorithm shows much better performance if it is executed until the beliefs are stabilized. (2) The algorithm yields a most likely partition for a quite wide range of $p - r$. Even when $p - r$ is "small" and the planted solution is not the optimal, the algorithm gives a solution whose likelihood is better than the planted solution. (3) The number of rounds required until the stabilization is not so large; approximately 10 rounds for large $p - r$, and up to 50 rounds even for the above "small" $p-r$. (4) The algorithm GraphPart works as well even without the thresholding; it yields an answer very close to the planted solution with high probability. On the other hand, without the thresholding, the execution becomes less stable, and the chance of obtaining the planted solution exactly gets smaller.

References

[Bop87]   R.B. Boppana, Eigenvalues and graph bisection: an average-case analysis, in *Proc. Symposium on Foundations of Computer Science*, 280-285, 1987.

[Betal87] T. Bui, S. Chaudhuri, F. Leighton, and M. Spiser, Graph bisection algorithms with good average behavior, in *Combinatorica* 7, 171–191, 1987.

[Coj05]   A. Coja-Oghlan, A spectral heuristic for bisecting random graphs, in *Proc. SODA 2005*, 850–859, 2005. (The journal version will appear in *Random Structures and Algorithms*.)

[CK01]   A. Condon and R. Karp, Algorithms for graph partitioning on the planted partition model, *Random Str. and Algorithms* 18, 116–140, 2001.

[DLP03]   D. Dubhashi, L. Laura, and A. Panconesi, Analysis and experimental evaluation of a simple algorithm for collaborative filtering in planted partition models, in *Proc. FST TCS 2003*, 168–182, 2003.

[DF89]   M.E. Dyer and A.M. Frieze, The solution of some random NP-hard problems in polynomial expected time, *J. of Algorithms* 10, 451–489, 1989.

[GJ79]   M.R. Garey, D.S. Johnson, *Computers and Intractability*, Bell Telephone Laboratories, Incorporated, 1979.

[GJS76]   M. Garey, D. Johnson, and L. Stockmeyer, Some simplified NP-complete graph problems, in *Theoret. Comput. Sci.* 1, 237–267, 1976.

[JS98]   M. Jerrum and G. Sorkin, The Metropolis algorithm for graph bisection, *Discrete Appl. Math* 82(1-3), 155–175, 1998.

[HSS03]   R. Hahnloser, H. Seung, and J. Slotine, Permitted and forbidden sets in threshold-linear networks, in *Neural Computation* 15, 621–638, 2003.

[McS99]   F. McSherry, Spectral partition of random graphs, in *Proc. 40th IEEE Sympos. on Foundations of Computer Science* (FOCS'99), IEEE, 529–537, 1999.

[Ons05]   M. Onsjö, Master Thesis, 2005.

[OW05]   M. Onsjö and O. Watanabe, Simple algorithms for graph partition problems, Research Report C-212, Dept. of Math. and Comput. Sci., Tokyo Inst. of Tech, 2005.

[Pea88]   J. Pearl, *Probabilistic Reasoning in Intelligent Systems: Networks of Plausible Inference*, Morgan Kaufmann Publishers Inc., 1988.

Minimal Interval Completion
Through Graph Exploration

Karol Suchan[1,2] and Ioan Todinca[1]

[1] LIFO, Université d'Orléans, 45067 Orléans Cedex 2, France
[2] Department of Discrete Mathematics, Faculty of Applied Mathematics,
AGH - University of Science and Technology, Cracow, Poland
{Karol.Suchan, Ioan.Todinca}@univ-orleans.fr

Abstract. Given an arbitrary graph $G = (V, E)$ and an interval graph $H = (V, F)$ with $E \subseteq F$ we say that H is an *interval completion* of G. The graph H is called a *minimal interval completion* of G if, for any sandwich graph $H' = (V, F')$ with $E \subseteq F' \subset F$, H' is not an interval graph. In this paper we give a $\mathcal{O}(nm)$ time algorithm computing a minimal interval completion of an arbitrary graph. The output is an interval model of the completion.

1 Introduction

Various well-known graph parameters, like *treewidth, minimum fill-in, pathwidth* or *bandwidth* are defined in terms of graph embeddings. The general framework consists in taking an arbitrary graph $G = (V, E)$ and adding edges to G in order to obtain a graph $H = (V, E \cup E')$ belonging to a specified class $\mathcal{H}$. For example, if H is chordal then it is called a *triangulation* of G. The *treewidth* can be defined as $\min(\omega(H)) - 1$, where the minimum is taken over all triangulations of G (here $\omega(H)$ denotes the maximum size of a clique in H). If, instead of minimizing $\omega(H)$, we minimize $|E'|$, the number of added edges, we define the *minimum fill-in* of G. If $H = (V, E \cup E')$ is an interval graph, we say that H is an interval completion of G. The *pathwidth* of G can be defined as $\min\{\omega(H)) - 1 \mid H$ is an interval completion of $G\}$. The minimum number of edges that we need to add for obtaining an interval completion is called the *profile* of the graph.

For each of the parameters cited above, as well as for similar embedding problems into other type of graph classes, the problem of computing the parameter is NP-hard. Obviously, for all of them, the optimal solution can be found among the *minimal* embeddings. We say that $H = (V, E \cup E')$ is a *minimal triangulation* (*minimal interval completion*) if no proper subgraph of H is a triangulation (interval completion) of G.

Computing minimal triangulations is a standard technique used in heuristics for the treewidth or the minimum fill-in problem. The deep understanding of minimal triangulations lead to many theoretical and practical results for the treewidth and the minimum fill-in. We believe that, similarly, the study of other types of minimal completions might bring new powerfull tools for the corresponding problems.

T. Asano (Ed.): ISAAC 2006, LNCS 4288, pp. 517–526, 2006.

Related work. Much research has been devoted to the minimal triangulation problem. Tarjan and Leuker propose the first algorithm solving the problem in $O(nm)$ time. Several authors give different approaches for the same problem, with the same running time. Only recently this $O(nm)$ (in the worst case $O(n^3)$) time complexity has been improved, and the fastest algorithm is due to Heggernes, Telle and Villanger ([11], running in $\mathcal{O}(n^\alpha \log n)$ time where $\mathcal{O}(n^\alpha)$ is the time needed for the multiplication of two $n \times n$ matrices). The latter algorithm is the fastest up to now for the minimal triangulation problem.

A first polynomial algorithm solving the minimal interval completion problem is given in [10], using an incremental approach. Recent results relate to minimal completions into split and comparability graphs [8,9].

Our result. We study the minimal interval completion problem. Our main result is an $\mathcal{O}(nm)$ time algorithm computing a minimal interval completion of an arbitrary graph, faster and simpler than the result of [10]. The latter result is based on characterization of interval graph by existence of its clique path. Here, we use the characterization by a special ordering of the vertex set, called interval ordering [12]. Its role is similar to the simplicial elimination scheme for chordal graphs. We define a family of orderings such that the associated proper interval graph is a minimal interval completion. Eventually, we give an $\mathcal{O}(nm)$ time algorithm computing such an ordering. Our algorithm is based on a breadth-first search of the input graphs, using special tie-break rules. In particular, we use the LexBFS algorithm for tie-breaks. The ordering can be efficiently transformed into an interval model.

2   Definitions and Basic Results

Let $G = (V, E)$ be a finite, undirected and simple graph. Moreover we only consider connected graphs — in the non-connected case each connected component can be treated separately. Denote $n = |V|$, $m = |E|$. If $G' = (V', E')$ is a spanning subgraph of $G = (V, E)$ (i.e. $V' = V$ and $E \subseteq E'$) we write $G \subseteq G'$ (and $G \subset G'$ if $G \subseteq G', G \neq G'$). The *neighborhood* of a vertex v in G is $N_G(v) = \{u \mid \{u, v\} \in E\}$. Similarly, for a set $A \subseteq V$, $N_G(A) = \bigcup_{v \in A} N_G(v) \setminus A$. The *closed neighborhood* of A (of v) is $N_G[A] = A \cup N_G(A)$ $(N_G[v] = \{v\} \cup N_G(v))$. As usual, the subscript is sometimes omitted.

A graph G is an *interval* graph if continuous intervals can be assigned to each vertex of G such that two vertices are neighbors if and only if their intervals intersect. The family of intervals is called the *interval model* of the graph.

Theorem 1 ([5]). *A graph G is interval if and only if there is a path P whose vertex set is the set of all maximal cliques of G, such that the subgraph of P induced by the maximal cliques of G containing vertex v is connected, for each vertex v of G.*

Such a path will be called a *clique path* of G. Notice, that a clique path P gives an interval model of G, with an interval (subpath) of maximal cliques assigned to each vertex. For our purpose, we also use the caracterization of interval graphs in terms of vertex orderings (also called layouts).

Definition 1 (interval ordering [12]). *An* interval ordering *of the vertices of a graph* $H = (V, F)$ *is a linear ordering* $\sigma = (v_1, v_2, \ldots, v_n)$ *of* V *such that, for any* $1 \leq i < j \leq k \leq n$, *if* $\{v_i, v_k\} \in F$ *then also* $\{v_i, v_j\} \in F$.

Theorem 2 ([12]). *A graph* $H = (V, F)$ *is an interval graph if and only if there exists an interval ordering of its vertex set.*

Definition 2. *Let* $G = (V, E)$ *be an arbitrary graph and* $\sigma = (v_1, \ldots, v_n)$ *be an ordering of* V. *The graph* $G(\sigma) = (V, F)$ *is defined by*

$$F = \{\{v_i, v_k\} \mid \text{ there is } j \text{ such that } 1 \leq i < k \leq j \leq n \text{ and } \{v_i, v_j\} \in E\}.$$

The following Lemma is a direct consequence of Theorem 2.

Lemma 1. $G(\sigma)$ *is an interval graph.*

Remark 1. Let $\sigma = (v_1, v_2 \ldots, v_n)$ be an interval ordering of an interval graph H. An interval model of H can be obtained by associating to each vertex v_i the interval $[i, j]$, where $j \geq i$ is the largest index such that $\{v_i, v_j\} \in F$.

Conversely, given an interval model of the graph H, we obtain an interval ordering by ordering the vertices according to the left-end point of their intervals, from left to right. Ties can be broken arbitrarily. For technical reasons, in this article, we decide to use the right-ends as a tie-break, from left to right, too.

Given an interval model, a clique path can be obtained by traversing the model from left to right and, at each point p where an interval finishes, adding the clique of intervals intersecting p to the model if it is not included in the (maximal) clique added right before. If $H = G(\sigma)$, for some simple graph G, let $P(G, \sigma)$ denote the clique path obtained in that way.

Theorem 3. *Let* $G = (V, E)$ *be an arbitrary graph and* $H = (V, F)$ *be a minimal interval completion of* G. *Then there is an ordering* σ *such that* $H = G(\sigma)$.

Proof. By Theorem 2, there is an ordering σ of V such that $H = H(\sigma)$. As a straight consequence of Definition 2, $E(G(\sigma)) \subseteq E(H)$. By Lemma 1, $G(\sigma)$ is also an interval graph. Thus, by minimality of H, we deduce that $E(G(\sigma)) = E(H)$. $\square$

Definition 3. *An ordering* $\sigma = (v_1, \ldots, v_n)$ *is called* nice *if* $G(\sigma)$ *is a minimal interval completion of* G. *Any prefix* $(v_1, \ldots, v_k)$, $k \leq n$ *of a nice ordering is called a* nice *prefix.*

Our goal will be to find a nice ordering σ of an arbitrary graph G. This will be achieved through ordered partitions of the vertex set, which are to be refined into a linear ordering.

Definition 4. *A tuple of disjoint subsets of* V, $OP = (V_1, \ldots, V_k)$ *whose union is exactly* V *is called an* ordered partition *of* V. *A* refinement *of* OP *is an ordered partition* OP' *obtained by replacing each set* V_i *by an ordered partition of* V_i.

Definition 5. *Given an ordered partition* $OP = (V_1, \ldots, V_k)$, *any tuple* $OP' = (V_1, \ldots, V_j)$, *with* $0 \leq j \leq k$, *is called a* prefix *of* OP. *We use* $V(OP')$ *to denote* $\bigcup \{V_i \mid 1 \leq i \leq j\}$.

In the particular case where $OP = (V_1)$, we simply write V_1. Moreover if V_1 is formed by a single vertex x, we write x instead of $\{x\}$. Given two tuples $OP' = (V_1, \ldots, V_k)$, $OP'' = (V_{k+1}, \ldots, V_{k+l})$, their concatenation $OP = (V_1, \ldots, V_k, V_{k+1}, \ldots, V_{k+l})$ is denoted by $OP' \bullet OP''$.

Notice that an ordering $\sigma = (v_1, \ldots, v_n)$ of V is a special case of an ordered partition.

3   Nice Orderings and Nice Prefixes

3.1   Choosing a First Vertex

A *module* is a set of vertices M such that for any $x, y \in M$, $N(x) \backslash M = N(y) \backslash M$. A clique module is a module inducing a clique. An inclusion-maximal clique module will be simply called a *maximal clique-module*.

A *minimal separator* S is a set of vertices such that there exist two connected components of $G - S$ with vertex sets C and D satisfying $N(C) = N(D) = S$.

Lemma 2 (see e.g. [6]). *Let P be a clique path of an interval graph H. For any minimal separator S of H, there exist two maximal cliques of H, consecutive in P, whose intersection is S.*

Definition 6 ([1]). *A moplex is a maximal clique module M, such that $N(M)$ is a minimal separator of G. The vertices of a moplex are called* moplexian *vertices.*

The LexBFS (Lexicographic Breadth-First Search) algorithm, introduced by Rose, Leuker and Tarjan [14], is a famous linear-time algorithm that numbers the vertices of an arbitrary graph from n to 1. Initially designed to obtain a simplicial ordering for chordal graphs, we use it here to obtain the first vertex of a nice ordering. LexBFS is a particular breadth-first search algorithm.Each vertex x has a label $lab(x)$, which is a tuple of integers. During the algorithm, each vertex x also receives a number. The algorithm may start the exploration of the graph on any vertex.

Theorem 4 ([1]). *The algorithm LexBFS ends on a moplexian vertex.*

A vertex v numbered 1 by some execution of LexBFS is called a *LexBFS-terminal vertex*. A moplex M such that some execution of LexBFS terminates on a vertex of M is called a *LexBFS-terminal moplex.*

Lemma 3 ([1,2]). *Let M be a LexBFS-terminal moplex and $S = N_G(M)$. Denote by $C_1, C_2, \ldots, C_k$, with $C_k = M$, the connected components of $G - S$ in the order in which the LexBFS execution encounters them. Then the following equation are satisfied:*

$$N(C_1) \subseteq N(C_2) \subseteq \cdots \subseteq N(C_k). \tag{1}$$

$$\forall i, j, x, y : 1 \leq i < j \leq k, x \in N(C_i), y \in N[C_j] \setminus N(C_i) \\ \Rightarrow \{x, y\} \in E(G). \tag{2}$$

Lemma 4. *Consider a non-complete graph $G = (V, E)$. Let v be a vertex of a moplex M and $S = N_G(M)$. Let $C_1, C_2, \ldots, C_k$, with $C_k = M$, the connected components of $G - S$, satisfy Equations 1 and 2 of Lemma 3. Then there exists a minimal interval completion H of G such that $N_G(v) = N_H(v)$.*

For any such H, there exists a clique path P of H such that $M \cup S$ is one of its end cliques.

Proof. Let H' be the graph obtained from G by transforming $N_G[C_i]$ into a clique, from each $1 \leq i \leq q$. By Equation 1 (see Lemma 3), $(N_G[C_1], \ldots, N_G[C_k])$ is a clique path of H', in particular H' is an interval graph. Consequently H' contains some minimal interval completion H of G as required.

Now let H be any minimal interval completion of G such that $N_H(v) = N_G(v)$. We first show that S induces a clique in H. Let D be a component of $G - S$, different from M, such that $N_G(D) = S$. Note that S is a v, u-minimal separator of G, for some $u \in D$. Let T be a minimal v, u separator of H such that $T \subseteq N_H(v)$. Clearly T exists because u and v are non-adjacent in H. We claim that $S \subseteq T$. For each vertex $s \in S$, there is a u, v path of G contained in $D \cup \{v, s\}$. This path intersects $N_G(v)$ only in s, so also in the graph H the only possible intersection between T and the path is s. It follows that $s \in T$, so $S \subseteq T$. The minimal separator T induces a clique in H by Lemma 2. Hence S also induces a clique in H. Note that, by definition of a moplex, $M \cup S$ also induces a clique in H.

For each i, $1 \leq i \leq k$ let $H_i = H[N_G[C_i]]$. Let H'' be the graph with vertex set V and edge set $E(H_1) \cup E(H_2) \cup \cdots \cup E(H_k)$. Therefore $G \subseteq H'' \subseteq H$. We will construct a clique path P of H'', showing that H'' is an interval graph. By minimality of H, this implies that $H'' = H$. Moreover, the clique path P will have $M \cup S = N_G[M]$ as one of its end cliques.

Let $S_i = N_G(C_i)$. By Equation 2, the vertices of S_{i-1} are adjacent to all vertices of $C_i - S_{i-1}$ in the graph G, so also in H_i. Combined with the fact that $S_{i-1} \subseteq S$ induces a clique in H we have that S_{i-1} is contained in each maximal clique of H_i. We claim that for each i, $1 \leq i < k$, there exists a clique path of H_i such that S_i is contained in the rightmost clique of P_i. Indeed, the graph $H_i^+ = H[C_i \cup S \cup M]$ is an interval graph and $M \cup S$ is one of its maximal cliques. Take any clique path P_i^+ of H_i^+, we prove that $M \cup S$ is an end clique. By contradiction, let x (resp y) be a vertex appearing in the clique left (resp. right) to $S \cup M$, but not appearing in $S \cup M$. By the properties of a clique path, $S \cup M$ must separate x and y in H_i^+. This contradicts the fact that $x, y \in C_i$ and there exists an x, y-path in $G[C_i]$. So the only possibility is that $S \cup M$ is at an end of P_i^+. Since $S_i \subseteq S$ and every vertex of S has a neighbour in M, S_i is contained in the clique next to $S \cup M$ in P_i^+. The clique path P_i of H_i obtained by removing $S \cup M$ from P_i^+ has the required property. Eventually, by concatenating the clique paths $P_1, P_2, \ldots, P_k$, it is easy to check that we obtain a clique path P of H''. Indeed if a vertex x appears in the subpaths P_i and P_j with $i < j$, then $x \in N_G[C_i] \cap N_G[C_j] = S_i$ (see Equation 1). By Equation 2, x appears in every clique of P_k, for each $k, i < k \leq j$. Since H_k is the complete graph with vertex set $N_G[M] = S \cup M$, the clique path P has $S \cup M$ as rightmost clique.   □

Theorem 5. *Let G be a non-complete graph and v be a LexBFS-terminal moplexian vertex of G. For any minimal interval completion H of G such that $N_G(v) = N_H(v)$, there is an interval ordering of H starting with v.*

Proof. By Lemma 4, there exists a clique path of H such that the left-most clique is $M \cup S$, where $S = N(M)$. We can reverse this path so that $S \cup M$ becomes the leftmost clique of the clique path P. By construction, H has no fill edges incident to v, in particular the v only appears in the left-most clique of P. By Remark 1, there is an interval ordering of H starting with v.   □

3.2   A Family of Nice Orderings

Notation 1. *We denote by $\rho = (v_1, \ldots, v_k)$ a prefix, and $R = V \setminus V(\rho)$. Let Nxt be a non-empty subset of R, such that $\mathrm{Nxt} = N_G(v_i) \cap R$ for some $v_i \in V(\rho)$ and Nxt is inclusion-minimal for this property. We denote $R \setminus \mathrm{Nxt}$ by Rst.*

Lemma 5. *Let σ be a refinement of $\rho \bullet \mathrm{Nxt} \bullet \mathrm{Rst}$ and σ' be a refinement of $\rho \bullet R$ such that $G(\sigma') \subseteq G(\sigma)$. Then σ' also is a refinement of $\rho \bullet \mathrm{Nxt} \bullet \mathrm{Rst}$.*

Proof. Let $v_i \in V(\rho)$ such that $\mathrm{Nxt} = R \cap N_G(v_i)$. Suppose that σ' is not a refinement of $\rho \bullet \mathrm{Nxt} \bullet \mathrm{Rst}$, so there is some vertex $u \in \mathrm{Rst}$ and a vertex $w \in \mathrm{Nxt}$ such that u appears before w in σ'. Since u appears in σ after all vertices of Nxt, v_i and u are not adjacent in $G(\sigma)$. Now in σ', u appears after v_i and before w. Since $\mathrm{Nxt} \subseteq N_G(v_i)$, v_i and w are adjacent in G and therefore v_i and u are adjacent in $G(\sigma')$ – a contradiction.   □

Lemma 6. *Consider two vertex orderings σ and σ' of G that are refinements of $\rho \bullet \mathrm{Nxt} \bullet \sigma_{\mathrm{Rst}}$, where σ_{Rst} is an ordering of Rst. That is to say, σ and σ' differ only by a permutation of Nxt. Let u, v be two vertices adjacent in $G(\sigma')$ but non-adjacent in $G(\sigma)$. Then both $u, v \in \mathrm{Nxt}$.*

Proof. By construction of $G(\sigma)$ and $G(\sigma')$, at least one of the vertices u, v are in Nxt. By contradiction, suppose that the other is not in Nxt.

First we consider the case when $u \in V(\rho)$ and $v \in \mathrm{Nxt}$. Suppose that u has a neighbour $u' \in \mathrm{Rst}$. In both $G(\sigma)$ and $G(\sigma')$ all vertices of Nxt are adjacent to u as they appear after u and before u' in the corresponding ordering – a contradiction. So $N_G(u) \cap R \subseteq \mathrm{Nxt}$. By definition (minimality) of Nxt, either $\mathrm{Nxt} \subseteq N_G(u)$ or $\mathrm{Nxt} \cap N_G(u) = \emptyset$. Clearly, in the first case Nxt is contained in the neighborhood of u in both $G(\sigma)$ and $G(\sigma')$. In the second, for both $G(\sigma)$ and $G(\sigma')$ the vertex u has no neighbours in Nxt – a contradiction.

It remains to consider the situation when $u \in \text{Nxt}$ and $v \in \text{Rst}$. Since u and v are adjacent in $G(\sigma')$, there is a neighbour v' of u in G, appearing after v in σ'. But u, v, u' appear in the same order in σ, so u and v are adjacent in $G(\sigma)$ – a contradiction. $\qquad\square$

3.3   Nice Orderings: A Sufficient Condition

Notation 2. *Let σ be a vertex ordering of G and let ρ be a prefix of σ. We denote by T the set of vertices of Nxt having neighbours in Rst. G_{Nxt} denotes the graph obtained from $G[\text{Nxt}]$ by adding a dummy vertex d_1, adjacent to each vertex of T, and a dummy vertex d_2 adjacent only to d_1. The graph G_{Nxt}^+ is obtained from G_{Nxt} by completing T into a clique. Given a clique path P of H, let $P[\text{Nxt}]$ denote the clique path of $H[\text{Nxt}]$ obtained by restricting all the bags of P to their intersections with Nxt and then removing the redundant ones (leaving only unique maximal cliques of $H[\text{Nxt}]$).*

Theorem 6. *Let σ be a vertex ordering of G with the following properties:*

1. *σ starts with a LexBFS-terminal vertex v_1.*
2. *For any non-empty prefix $\rho = (v_1, \ldots, v_i)$*
   - *σ respects ρ, i.e. σ is a refinement of $\rho \bullet \text{Nxt} \bullet \text{Rst}$,*
   - *the next vertex in σ is a LexBFS-terminal vertex of G_{Nxt}^+ obtained by running LexBFS starting from d_2.*

Then σ is a nice ordering.

Proof. Suppose that $\sigma = (v_1, \ldots, v_n)$ is not nice and let σ' be an ordering such that $H' = G(\sigma')$ is a minimal interval completion of G strictly contained in $H = G(\sigma)$. Take σ' such that the maximal common prefix ρ of σ and σ' is the longest possible.

Claim. ρ is not empty.

The first vertex v_1 of σ is LexBFS-terminal. $N_G(v_1) = N_{G(\sigma)}(v_1)$, since σ respects the prefix (v_1) and thus the neighbours of v_1 in G appear right after v_1 in σ. So the Claim follows by Theorem 5.

Let v (resp. u) be the vertex right after ρ in σ (resp. in σ'). By Lemma 5, we have:

Claim (1). σ' is a refinement of $\rho \bullet \text{Nxt} \bullet \text{Rst}$, in particular $u \in \text{Nxt}$.

Claim (2). Let σ'' be any refinement of $\rho \bullet \text{Nxt} \bullet \text{Rst}$ and $H'' = G(\sigma'')$. Let $P'' = P(G, \sigma'')$ (see Remark 1). Then $H''[\text{Nxt}]$ is an interval completion of $G[\text{Nxt}]$, where the clique path $P''[\text{Nxt}]$ has the set $T = N_G(\text{Rst}) \cap \text{Nxt}$ contained in one of the end-cliques. In particular, T is a clique in H''.

Clearly, $P''[\text{Nxt}]$ is a clique path of $H''[\text{Nxt}]$. The last clique contains T, since the corresponding intervals in the model intersect the interval of a vertex in Rst.

Claim (3). $H'[\text{Nxt}]$ is an interval completion of $G[\text{Nxt}]$, minimal with respect to the property expressed in the previous claim.

Since σ' defines a minimal interval completion H' of G, σ' has to yield $H'[\text{Nxt}]$ minimal with this property. Suppose it is not minimal, and let $H'''[\text{Nxt}]$ be the corresponding completion strictly included in $H'[\text{Nxt}]$. Then we can take the corresponding clique path $P'''[\text{Nxt}]$ to create an interval order σ'''_{Nxt} of $H'''[\text{Nxt}]$ (see Remark 1). By Lemma 6, $\sigma''' = \rho \bullet \sigma'''_{\text{Nxt}} \bullet \sigma'_{\text{Rst}}$ yields $H''' = G(\sigma''')$ strictly contained in H'. A contradiction with minimality of H'.

Following Notation 2, let H'_{Nxt} be obtained from $H'[\text{Nxt}]$ by adding a dummy vertex d_1 adjacent to the vertices of T and a vertex d_2 adjacent to d_1.

Claim (4). H'_{Nxt} is a minimal interval completion of G^+_{Nxt}.

Let P'_{Nxt} denote the clique path of H'_{Nxt}, obtained from $P'[\text{Nxt}]$ by ading two bags $q_1 = T \cup \{d_1\}$ and $q_2 = \{d_1, d_2\}$ after the clique containing T (see Claim 2). It is a clique path indeed, so H'_{Nxt} is an interval completion of G^+_{Nxt}. Suppose it is not minimal. So there is a minimal one H''_{Nxt} strictly included in H'_{Nxt}. Notice that this graph has a clique path P''_{Nxt}, that also has $- - q_1 - -q_2$ at an end. Indeed, the moplex $M = d_2$ satisfies the conditions of Lemma 4 in H''_{Nxt}, so there is a clique path of H''_{Nxt} with $\{d_1, d_2\}$ as one of the end-cliques. Therefore $P''[\text{Nxt}]$, obtained by removing $- - q_1 - -q_2$ from P''_{Nxt}, is a clique path of $H''[\text{Nxt}]$ with T contained in one of the end-cliques. Which contradicts Claim 3, since $H''[\text{Nxt}]$ is a strict subgraph of $H'[\text{Nxt}]$.

Claim (5). There is an interval ordering of H'_{Nxt} starting with v.

Let us prove that $N_{H'_{\text{Nxt}}}(v) = N_{G^+_{\text{Nxt}}}(v)$. Indeed if $v \in T$, since v is the last vertex encountered by LexBFS launched on G^+_{Nxt} from d_2, we have $T = \text{Nxt}$. In this case the neighborhood of v in both graphs is $T \setminus \{v\} \cup \{d_1\}$, and the equality follows.

Now if $v \notin T$ then $N_G(v) \cap R \subset \text{Nxt}$. By second condition of the theorem, σ respects $\rho \bullet v$, so $N_G(v) \cap \text{Nxt}$ is put before $R \setminus N_G(v)$ in σ and $N_{H[\text{Nxt}]}(v) = N_{G[\text{Nxt}]}(v)$. Therefore $N_{H'_{\text{Nxt}}}(v) = N_{G^+_{\text{Nxt}}}(v)$, since

$$N_{G^+_{\text{Nxt}}}(v) \subseteq N_{H'_{\text{Nxt}}}(v) \subseteq N_{H_{\text{Nxt}}}(v) = N_{G_{\text{Nxt}}}(v) \subseteq N_{G^+_{\text{Nxt}}}(v).$$

The claim follows from Theorem 5 and Claim 4.

Claim (6). There is an ordering σ'', with $G(\sigma'') = G(\sigma')$, sharing a longer prefix with σ – a contradiction.

We restrict the ordering from the previous claim to Nxt and obtain σ''_{Nxt}. Let $\sigma'' = \rho \bullet \sigma''_{\text{Nxt}} \bullet \sigma'_{\text{Rst}}$. By Lemma 6, $G(\sigma'') = G(\sigma')$. So σ'' defines the same completion and shares a longer prefix. Which contradicts the choice of σ'.

This achieves the proof of our theorem. $\qquad\square$

Theorem 7. *There is an $\mathcal{O}(nm)$-time algorithm computing a minimal interval completion of an arbitrary graph.*

Proof. Algorithm MIC_Ordering of Figure 1 computes in $\mathcal{O}(nm)$ time a vertex ordering satisfying the conditions of Theorem 6. The full proof is given in [15]. Let us simply point out that unlike the Theorem 6, our algorithm computes the vertex v_i by launching LexBFX from d_2 on the graph G_{Nxt} and not G_{Nxt}^+. The reason is related to the running time. Indeed G_{Nxt} has $\mathcal{O}(n + m)$ edges, while if we compute G_{Nxt}^+, the number of edges of G_{Nxt}^+ might be up to $\Omega(n^2)$. Nevertheless we prove [15] that v_i is also a LexBFS-terminal vertex obtained by using G_{Nxt}^+ instead of G_{Nxt}.   □

Algorithm MIC_Ordering
Input: $G = (V, E)$ connected
Output: a nice ordering σ and the corresponding interval model
let v_1 be the last vertex encountered by **LexBFS**(G)
$\rho := (v_1)$
$\text{Nxt} = N_G(v_1); \text{Rst} := V \setminus N_G[v_1]$
$OP := v_1 \bullet \text{Nxt} \bullet \text{Rst}$
for $i := 2$ **to** n **do**
    let Nxt be the class appearing after ρ in OP
    let v_i be the last vertex encounterd by **LexBFS**
        launched on G_{Nxt} starting from d_2 (see Theorem 6)
    $\rho = \rho \bullet v_i$
    if $|\text{Nxt}| \geq 2$ **then**
        replace Nxt in OP by $v_i \bullet (\text{Nxt} \setminus \{v_i\})$
    let C be the last class of OP such that $N_G(v_i) \cap C \neq \emptyset$
    if $C \setminus N_G(v_i) \neq \emptyset$ **then**
        replace C in OP by $(C \cap N_G(v_i)) \bullet (C \setminus N_G(v_i))$
$\sigma := OP$
return IntervalModel(σ)

Fig. 1. Algorithm MIC_Ordering

4 Conclusion

We give in this paper an $\mathcal{O}(nm)$ time algorithm computing a minimal interval completion of an arbitrary input graph. The algorithm is based on the notion of nice orderings, which characterize a minimal interval completion, and on Theorem 6 which gives a sufficient condition for a nice ordering. We point out that there are nice orderings satisfying the conditions of Theorem 6, which cannot be produced by the algorithm. Such examples can be easily obtained when the input graph is a cycle. In particular an ordering produced by our algorithm is always a breadth-first search ordering, which is not required by the theorem.

There are two very natural directions for further research. One is to obtain a faster algorithm for the minimal interval completion problem. The second important question is to characterize all nice orderings. For the minimal triangulation problem, the perfect elimination orderings (which play the same role as the nice orderings here) have been completely characterized. In our case, we have examples of nice orderings that do not satisfy the conditions of Theorem 6.

References

1. A. BERRY, J. P. BORDAT, *Separability Generalizes Dirac's Theorem.* Discrete Applied Mathematics, 84(1-3): 43-53, 1998.
2. A. BERRY, J. P. BORDAT, *Local LexBFS Properties in an Arbitrary Graph.* Proceedings of Journes Informatiques Messines, 2000. http://www.isima.fr/berry/lexbfs.ps.
3. H. L. BODLAENDER, *A Linear-Time Algorithm for Finding Tree-Decompositions of Small Treewidth.* SIAM Journal on Computing, 25(6):1305-1317, 1996.
4. L. CAI, *Fixed-Parameter Tractability of Graph Modification Problems for Hereditary Properties.* Information Processing Letters, 58(4):171-176, 1996.
5. P. C. GILMORE AND A. J. HOFFMAN, *A characterization of comparability graphs and of interval graphs.* Canadian Journal of Mathematics, 16:539-548, 1964.
6. M. C. GOLUMBIC, *Algorithmic Graph Theory and Perfect Graphs.* Academic Press, 1980.
7. M. HABIB, C. PAUL, L. VIENNOT, *Partition Refinement Techniques: An Interesting Algorithmic Tool Kit.* International Journal of Foundations of Computer Science, 10(2): 147-170, 1999.
8. P. HEGGERNES, F. MANCINI, *Minimal Split Completions of Graphs.* Proceedings of LATIN 2006, Lecture Notes in Computer Science, 3887:592-604, 2006.
9. P. HEGGERNES, F. MANCINI, C. PAPADOPOULOS *Minimal Comparability Completions.* Tech. Report, University of Bergen, 2006, http://www.ii.uib.no/publikasjoner/texrap/pdf/2006-317.pdf
10. P. HEGGERNES, K. SUCHAN, I. TODINCA,Y. VILLANGER, *Minimal Interval Completions.* Proceedings of the 13th Annual European Symposium on Algorithms - ESA 2005, Lecture Notes in Computer Science, 3669:403-414, 2005.
11. P. HEGGERNES, J. A. TELLE, Y. VILLANGER, *Computing minimal triangulations in time $O(n^\alpha logn) = o(n^{2.376})$.* Proceedings of the 16th Annual ACM-SIAM Symposium on Discrete Algorithms - SODA 2005, SIAM, 907-916, 2005.
12. S. OLARIU, *An optimal greedy heuristic to color interval graphs.* Information Processing Letters, 37(1): 21–25, 1991.
13. I. RAPPAPORT, K. SUCHAN, I. TODINCA, *Minimal proper interval completions.* To appear in Proceedings of the 32nd Workshop on Graph-Theoretic Concepts in Computer Science (WG'06), 2006. http://www.univ-orleans.fr/SCIENCES/LIFO/prodsci/rapports/RR2006.htm.en.
14. D. ROSE, R.E. TARJAN, AND G. LUEKER, *Algorithmic aspects of vertex elimination on graphs.* SIAM J. Comput., 5:146–160, 1976.
15. K. SUCHAN, I. TODINCA, *Minimal interval vompletion through graph exploration.* Research Report RR 2006-08, Université d'Orléans. http.//www.univ-orleans.fr/SCIENCES/LIFO/prodsci/rapports/RR2006.htm.en.

Balanced Cut Approximation in Random Geometric Graphs*

Josep Diaz[1], Fabrizio Grandoni[2], and Alberto Marchetti Spaccamela[3]

[1] Departament de Llenguatges i Sistemes Informatics, Universitat Politecnica de Catalunya, Campus Nord - Ed. Omega, 240 Jordi Girona Salgado, 1-3 E-08034, Barcelona
diaz@lsi.upc.edu
[2] Dipartimento di Informatica, Università di Roma "La Sapienza", via Salaria 113, 00198 Roma, Italy
grandoni@di.uniroma1.it
[3] Dipartimento di Informatica e Sistemistica, Università di Roma "La Sapienza", via Salaria 113, 00198, Roma, Italy
alberto.marchetti@dis.uniroma1.it

Abstract. A random geometric graph $\mathcal{G}(n,r)$ is obtained by spreading n points uniformly at random in a unit square, and by associating a vertex to each point and an edge to each pair of points at Euclidian distance at most r. Such graphs are extensively used to model wireless ad-hoc networks, and in particular sensor networks. It is well known that, over a critical value of r, the graph is connected with high probability.

In this paper we study the robustness of the connectivity of random geometric graphs in the supercritical phase, under deletion of edges. In particular, we show that, for a sufficiently large r, any cut which separates two components of $\Theta(n)$ vertices each contains $\Omega(n^2 r^3)$ edges with high probability. We also present a simple algorithm that, again with high probability, computes one such cut of size $O(n^2 r^3)$. From these two results we derive a constant expected approximation algorithm for the β-balanced cut problem on random geometric graphs: find an edge cut of minimum size whose two sides contain at least βn vertices each.

Keywords: ad-hoc networks, sensor networks, random geometric graphs, balanced cut, approximation algorithms.

1 Introduction

Let us consider a wireless network of sensors on a terrain, where the sensors communicate by radio frequency, using an omnidirectional antenna. Each sensor

---

* Partially supported by EU Integrated Project AEOLUS (FET-15964); the first author was partially supported by *La distinció de la Generalitat de Catalunya, 2002*, the second author by project ALGO-NEXT of the Italian Ministry of University and Research and the third author by project FIRB RBIN047MH9, Italy-Israel of the Ministry of University and Research.

T. Asano (Ed.): ISAAC 2006, LNCS 4288, pp. 527–536, 2006.

broadcasts with the same power to the same distance. Two sensors can communicate if and only if they are within the transmission radius of each other. Sensor networks, and more in general ad-hoc wireless networks, are often modelled via random geometric graphs [1,5]. A *random geometric graph* $\mathcal{G}(n, r)$ [9] is a graph resulting from placing a set V of n vertices uniformly at random on the unit square $[0, 1]^2$, and connecting two vertices if and only if their Euclidean distance is at most the given radius r.

Random geometric graphs in general, and in particular their connectivity properties, have been intensively studied, both from the theoretical and from the empirical point of view. For the present paper, the most interesting result on random geometric graphs is the fact that, for $r = r(n) = \sqrt{(\ln n + c(n))/(\pi n)}$, for any $c(n)$ such that $c(n) \to \infty$ when $n \to \infty$, $\mathcal{G}(n, r)$ is connected whp [11,14,15]. (Throughout this paper, "whp" will abbreviate *with high probability*, that is with probability tending to 1 as n goes to ∞). Once the connectivity is achieved, it is natural to wonder how robust it is: how many edges one needs to remove in order to disconnect the graph? In most applications the disconnection of one vertex, or of a few vertices, does not affect significantly the behavior of the network. So we can reformulate the question above in the following more general way: given $\beta \in [0, 1/2]$, how many edges one needs to remove in order to isolate two components (not necessarily connected) of βn vertices each?

Our results. We can formalize the question above in the following way. A *cut* of a graph is a partition of its vertices into two subsets W and B, the *sides* of the cut. The *size* of cut (W, B) is the number of edges $\delta(W, B)$ between W and B. Given $\beta \in [0, 1/2]$, $\beta n \in \mathbb{N}$, a β-*balanced cut* is a cut where both sides contain at least βn vertices. The β-*balanced cut problem* is to compute a β-balanced cut of minimum size. Here we prove that, if $r = r(n) = \sqrt{R \ln n/n}$ for $R \geq R^*$, with $R^* > 0$ a sufficiently large constant, with high probability any β-balanced cut of $\mathcal{G}(n, r)$ has size $\Omega(\min\{\beta n R \log n, \sqrt{\beta n R^3 \log^3 n}\})$.

We also present a simple algorithm that with high probability computes a cut of size $O(\min\{\beta n R \log n, \sqrt{\beta n R^3 \log^3 n}\})$, thus matching the lower bound. The two mentioned results imply a probabilistic constant expected approximation algorithm for the β-balanced cut problem. We eventually show how to extend such result to a constant expected approximation algorithm.

We remark that the above results hold also if R is a function of n, and that the hidden constants in the O and Ω notations do not depend on n, R and β.

Related Work. Nothing is known on β-balanced cut approximation in random geometric graphs, for arbitrary values of β. For $\beta = 1/2$, the β-balanced cut problem is the well-know *minimum edge bisection problem*. Minimum edge bisection is a difficult problem which has received a lot of attention due to its numerous applications (see e.g. [10]). The problem is known to be NP-Hard for general graphs [8], and in such case there is a $O(\log^{1.5} n)$ approximation [6]. In the same paper, the authors prove that if the graph is planar, the approximation can be reduced to $O(\log n)$. If the input graph is dense, i.e. each vertex has degree $\Theta(n)$, there is a polynomial time approximation scheme (PTAS) for the minimum bisection problem [2]. In the case of random geometric graphs, it is known how to obtain a constant approximation whp for the special case $R = R(n) \to \infty$ for $n \to \infty$ [3]. Our approximation algorithm improves on the algorithm in [3] in several ways: (i) it holds for arbitrary values of β, including the case $\beta = o(1)$; (ii) it holds for constant values of R as well; (iii) the value of the approximation ratio is constant in expectation, not only with high probability. We remark that each of the mentioned improvements is achieved by introducing new, simple techniques (which do not trivially follow from [3,13]).

One of the first papers to introduce the general problem of the minimum β-balanced cut was [4]. In this paper, the authors also show that given an $\epsilon > 0$, it is NP-hard to approximate the minimum bisection within an additive term of $n^{2-\epsilon}$. The β-balanced cut problem admits a PTAS for $\beta \leq 1/3$, if every vertex has degree $\Theta(n)$ [2]. For planar graphs there is a 2-approximation for the β-balanced cut, if $\beta \leq 1/3$ [7]. However, it is still open whether bisection and β-balanced cut are NP-hard for planar graphs.

Preliminaries. Given a region Q of the unit square, $|Q|$ denotes the area of Q, and $\|Q\|$ the number of points falling in Q. Note that $\|Q\|$ is a Binomial random variable of parameters n and $|Q|$, for which the following standard Chernoff's Bounds hold [12]. Let $\mu = E[\|Q\|] = |Q|\, n$. Then:

$$Pr[\|Q\| < (1-\delta)\mu] \leq e^{-\delta^2 \mu/2} \qquad \text{for } \delta \in [0,1); \qquad (1)$$

$$Pr[\|Q\| > (1+\delta)\mu] \leq e^{-\delta^2 \mu/3} \qquad \text{for } \delta \in [0,1); \qquad (2)$$

$$Pr[\|Q\| > (1+\delta)\mu] \leq e^{-\delta^2 \mu/4} \qquad \text{for } \delta \in [1, 2e-1); \qquad (3)$$

$$Pr[\|Q\| > (1+\delta)\mu] \leq e^{-\delta\mu \ln 2} \qquad \text{for } \delta \geq 2e-1. \qquad (4)$$

From now on $r = r(n) = \sqrt{R \ln n / n}$. For the sake of simplicity, we will assume $R = o(n/\log n)$. For $R = \Omega(n/\log n)$, the problems considered here become trivial. In particular, for $R \geq 2n/\ln n$ the graph is a clique (deterministically).

2   A Lower Bound

In this section we show that, for any $\beta \in [0, 1/2]$, $\beta n \in \mathbb{N}$, and for $R \geq 240$, the size of any β-balanced cut is $\Omega(\min\{\beta n R \log n, \sqrt{\beta\, n R^3 \log^3 n}\})$ with high probability.

In order to prove the mentioned lower bound, we consider a partition of the unit square into $5n/(R \ln n)$ non-overlapping square *cells* of the same size. Each cell is *adjacent* to the cells to its right, left, top, and bottom. Observe that, since the side of each cell has length $\sqrt{R \ln n/(5n)}$, a vertex is adjacent to all the vertices in the same cell and in all the adjacent cells. This property is crucial in the analysis. The number of points $\|C\|$ in each cell C satisfies the following probabilistic bounds.

Lemma 1. *For any $R \geq 240$, each cell C of the partition above contains $\|C\|$ vertices of $\mathcal{G}(n,r)$, $R \ln n/10 \leq \|C\| \leq 3\, R \ln n/10$, with probability $1 - o(1/n^2)$.*

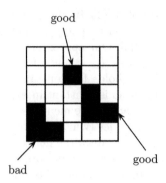

good

bad

good

Fig. 1. Possible configuration of black and white cells. There are 3 black clusters and 1 white cluster.

Proof. Consider any cell C. Observe that $E[\|C\|] = R \ln n/5$. By Chernoff's Bounds (1) and (3),

$$Pr\left(\|C\| \notin \left[\frac{R \ln n}{10}, \frac{3 R \ln n}{10}\right]\right) \le e^{-(1/2)^2 R \ln n/10} + e^{-(1/2)^2 R \ln n/20} = O(1/n^3).$$

The claim follows by applying the union bound to the $O(n/(R \ln n))$ cells. □

Let (W, B) be any given cut, with $|W| = \beta n$. Let us call the vertices in W *white*, and the vertices in B *black*. A cell is *white* if at least one half of its points are *white*, otherwise the cell is *black*. We define a *cluster* C to be a maximal connected component of cells of the same color, with respect to the adjacency between cells defined above. The *frontier* ∂C of C is the subset of its cells which either touch the border of the unit square, or are adjacent to a cell of different color. We call *good* the cells of ∂C which are adjacent to a cell of different color, and *bad* the other cells of ∂C. In particular, a cell is bad if it touches the border of the unit square and it is surrounded by cells of the same cluster (see Figure 1).

In order to prove the lower bound, we need the following two observations.

Lemma 2. *Given a cluster of k cells, its frontier contains at least $\sqrt{\pi k/4}$ cells.*

Proof. Suppose that the frontier contains $h < \sqrt{\pi k/4}$ cells. Thus the perimeter of the cluster has length at most $4hL$, where $L = \sqrt{R \ln n/(5n)}$ is the length of the side of one cell. Such perimeter can enclose an area of size at most $(4hL)^2/(4\pi)$ (case of a disk of radius $4hL/(2\pi)$), and thus at most $4h^2/\pi < k$ cells, which is a contradiction. □

Lemma 3. *Consider a cluster touching either 0, or 1, or 2 consecutive sides of the square. Then at least one third of the cells on its frontier are good.*

Proof. Consider any cluster C. Without loss of generality, let C be white. If C does not touch any side of the square, all the cells of ∂C are good. Thus the claim is trivially true.

Now suppose $\mathcal{C}$ touches one or two consecutive sides of the square, say the left side and possibly the top side. Let $\partial\mathcal{C}_{good}$ be the good cells of $\partial\mathcal{C}$, and $\partial\mathcal{C}_{bad} = \partial\mathcal{C} \setminus \partial\mathcal{C}_{good}$ the bad ones. Moreover, let $\partial\mathcal{C}^{out}$ be the cells of $\partial\mathcal{C}$ touching the border of the square, and $\partial\mathcal{C}^{in} = \partial\mathcal{C} \setminus \partial\mathcal{C}^{out}$. Note that $\partial\mathcal{C}^{in} \subseteq \partial\mathcal{C}_{good}$ since the cells in $\partial\mathcal{C}^{in}$ do not touch any side of the square.

At least one half $\partial\mathcal{C}'$ of the cells of $\partial\mathcal{C}^{out}$ touches one between the left and the top side of the square, say the left one. Consider any cell $C' \in \partial\mathcal{C}'$. If C' is bad, we can univocally associate to C' a good cell $C'' \in \partial\mathcal{C}^{in}$ in the following way. Consider the sequence of consecutive white cells at the right of C' (there must be at least one such cell, since C' is bad). We let C'' be the rightmost of such cells. As a consequence, the number of good cells is lower bounded by $|\partial\mathcal{C}'|$, and $|\partial\mathcal{C}_{good}| \geq |\partial\mathcal{C}'| \geq |\partial\mathcal{C}^{out}|/2$. Thus

$$|\partial\mathcal{C}| = |\partial\mathcal{C}^{in}| + |\partial\mathcal{C}^{out}| \leq |\partial\mathcal{C}_{good}| + |\partial\mathcal{C}^{out}| \leq 3|\partial\mathcal{C}_{good}|.$$

The claim follows.   □

Theorem 1. *With probability $1 - o(1/n^2)$, for any $\beta \in [0, 1/2]$, $\beta\, n \in \mathbb{N}$, and for any $R \geq 240$, the size of any β-balanced cut of $\mathcal{G}(n, r)$ is*

$$\Omega(\min\{\beta\, n\, R \log n, \sqrt{\beta\, n\, R^3 \log^3 n}\}).$$

Proof. By Lemma 1, with probability $1 - o(1/n^2)$ for each cell C,

$$\|C\| \in \left[\frac{R\ln n}{10}, \frac{3R\ln n}{10}\right]. \tag{5}$$

Thus it is sufficient to show that, given (5), the lower bound holds (deterministically) for any $\beta \in [0, 1/2]$ and for any cut (W, B) with $|W| = \beta\, n$.

We need some notation. By $\mathcal{W}$ and $\mathcal{B}$ we denote the set of white and black cells respectively. Moreover, $W_{black} \subseteq W$ ($B_{white} \subseteq B$) is the subset of white (black) vertices in black (white) cells.

Since each vertex is adjacent to all the other vertices in the same cell, each vertex $w \in W_{black}$ contained into a (black) cell C contributes with at least $\|C\|/2 \geq R\ln n/20$ edges to the edges of the cut. It follows that, if $|W_{black}| \geq |W|/2 = \beta\, n/2$, the size of the cut is at least

$$|W_{black}|\frac{R\ln n}{20} \geq \frac{\beta\, n\, R\ln n}{40} = \Omega(\beta\, n\, R \log n).$$

Analogously, if $|B_{white}| \geq |B|/2 = (1 - \beta)\, n/2$, then the size of the cut is at least

$$|B_{white}|\frac{R\ln n}{20} \geq \frac{(1 - \beta)\, n\, R\ln n}{40} = \Omega(\beta\, n\, R \log n).$$

Thus, let us assume $|W_{black}| < |W|/2$ and $|B_{white}| < |B|/2$. Note that, since all the vertices in adjacent cells are adjacent, each pair of adjacent (good) cells (C', C''), with $C' \in \mathcal{W}$ and $C'' \in \mathcal{B}$ contributes with at least

$$\frac{\|C'\|}{2}\frac{\|C''\|}{2} \geq \frac{R^2 \ln^2 n}{400} = \Omega(R^2 \log^2 n).$$

distinct edges to the total number of edges in the cut. Since there must be at least one such pair (C', C''), if $\beta = O(R \log n/n)$, trivially the size of the cut is $\Omega(R^2 \log^2 n) = \Omega(\beta n R \log n)$.

For $\beta = \Omega(R \log n/n)$ we need to bound the number of distinct pairs (C', C'') in a more sophisticated way. In particular, we will show that the number of good cells, either white or black, is $\Omega(\sqrt{\beta n/(R \log n)})$, from which it follows that the size of the cut is at least

$$\Omega(R^2 \log^2 n) \, \Omega(\sqrt{\beta n/(R \log n)}) = \Omega(\sqrt{\beta n R^3 \log^3 n}).$$

Observe that, from Equation (5) and from the assumption $|W_{black}| < |W|/2$ and $|B_{white}| < |B|/2$,

$$|W| \geq \frac{\beta n/2}{3R \ln n/10} = \frac{5\beta n}{3R \ln n} \quad \text{and} \quad |B| \geq \frac{(1-\beta)n/2}{3R \ln n/10} = \frac{5(1-\beta)n}{3R \ln n} \tag{6}$$

We distinguish three sub-cases, depending on the existence of white clusters with some properties.

(B.1) There is a white cluster C touching either 3 or 2 opposite sides of the square (but not 4). Without loss of generality, let the right side of the square be untouched. Consider all the cells of C which have no cell of the same cluster to their right. Note that such cells belong to the frontier ∂C of the cluster. Moreover, they are all good (they have a black cell to their right). The number of such cells is exactly $\sqrt{5n/(R \ln n)} = \Omega(\sqrt{\beta n/(R \log n)})$.

(B.2) Every white cluster touches 0, 1, or 2 consecutive sides of the square. Recall that the white cells are $|W| \geq 5\beta n/(3R \ln n)$ by (6). Let $C_1, C_2, \ldots, C_p$ be the p white clusters. It follows by Lemmas 2 and 3, that the total number of white good cells is at least

$$\sum_{i=1}^{p} \frac{1}{3} \sqrt{\frac{\pi |C_i|}{4}} \geq \frac{1}{3} \sqrt{\frac{\pi |W|}{4}} \geq \frac{1}{3} \sqrt{\frac{\pi 5 \beta n}{12 R \ln n}} = \Omega(\sqrt{\beta n/(R \log n)}).$$

(B.3) There is a white cluster touching the 4 sides of the square. It follows that each black cluster touches 0, 1, or 2 consecutive sides of the square. Thus, by basically the same argument as in case (B.2), the number of black good cells is at least

$$\frac{1}{3} \sqrt{\frac{\pi |B|}{4}} \geq \frac{1}{3} \sqrt{\frac{\pi 5 (1-\beta) n}{12 R \ln n}} = \Omega(\sqrt{\beta n/(R \log n)}).$$

$\square$

3   A Simple Cutting Algorithm

In this section we describe a simple algorithm `simpleCut` which, for a given input $\beta \in [0, 1/2]$, $\beta n \in \mathbb{N}$, computes a β-balanced cut. We will show that,

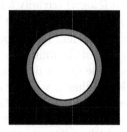

Fig. 2. The white disk D contains one side W of the cut, $\|W\| = \beta n$. The annulus A of D, of width $\sqrt{R \ln n/n}$, is drawn in gray.

for $R \geq 3/\pi$, the size of the cut computed is $O(\min\{\beta n R \log n, \sqrt{\beta n R^3 \log^3 n}\})$ with high probability. This, together with Theorem 1, implies that simpleCut is a probabilistic constant approximation algorithm for the β-balanced cut problem for $R \geq 240$. We later show how to convert such result into a constant expected approximation algorithm.

Algorithm 1. (simpleCut) *Take the βn vertices which are closest to $(1/2, 1/2)$ (breaking ties arbitrarily). Such vertices form one side W of the cut.*

Observe that simpleCut can be easily implemented in polynomial time.

In order to bound the size of the cut produced by simpleCut, we need the following simple probabilistic bound on the degree of the vertices.

Lemma 4. *For $R > 3/\pi$, the degree of each vertex of $\mathcal{G}(n, r)$ is upper bounded by $(3\pi R \ln n)$ with probability $1 - o(1/n^2)$.*

Proof. Consider the ball of radius $\sqrt{R \ln n/n}$ centered at vertex v, and let X_v be the number of vertices it contains. Clearly, the degree of v is $X_v - 1$. By denoting $\mu_v = E[X_v]$ we have

$$\pi R \ln n/4 \leq \mu_v \leq \pi R \ln n,$$

where the upper and lower bounds correspond to the case v is in the middle of the unit square and in one corner, respectively. By Chernoff's Bounds (2)-(4),

$$Pr(X_v > 3\pi R \ln n) \leq e^{-\ln 2(3\pi R \ln n/\mu_v - 1)\mu_v} \leq e^{-\ln 2(2\pi R \ln n)} = o(1/n^3).$$

Hence, from the union bound,

$$Pr(\exists v \in V : X_v > 3\pi R \ln n) \leq \sum_{v \in V} Pr(X_v > 3\pi R \ln n) = o(1/n^2).$$

$\square$

Theorem 2. *For any $\beta \in [0, 1/2]$, $\beta n \in \mathbb{N}$, and for $R > 3/\pi$, the size of the cut of $\mathcal{G}(n, r)$ computed by simpleCut is $O(\min\{\beta n R \log n, \sqrt{\beta n R^3 \log^3 n}\})$ with probability $1 - o(1/n^2)$.*

Proof. The upper bound $O(\beta n R \log n)$ trivially follows from Lemma 4. So, it is sufficient to show that for $\beta = \Omega(R \ln n/n)$, the size of the cut is $O(\sqrt{\beta n R^3 \log^3 n})$. In particular, $\beta \geq 8\pi R \ln n/n$ is sufficient for our purposes.

Recall that, for a given region Q of the unit square, $|Q|$ denotes the area of Q, and $\|Q\|$ the number of points inside Q. Let us denote by D the disk centered in $(1/2, 1/2)$, of minimum possible radius ρ, which contains all the vertices in W (see Figure 2). In the following we will assume $\|D\| = \beta n$, which happens with probability one by standard probabilistic techniques.

Let A denote the annulus of width $\sqrt{R \ln n/n}$ surrounding D. The edges of the cut are a subset of the edges incident to the vertices in A. Hence, from Lemma 4, it is sufficient to show that the number $\|A\|$ of vertices in A is $O(\sqrt{\beta n R \log n})$ with probability $1 - o(1/n^2)$.

Consider the disk D' centered in $(1/2, 1/2)$ of radius $\rho' = \sqrt{(3/2)\beta/\pi}$, and let A' be the annulus of width $\sqrt{R \ln n/n}$ surrounding D'. Since $\rho' \leq \sqrt{3/(4\pi)} < 1/2$, for n large enough, D' and A' are entirely contained in the unit square.

Observe that, given $\rho \leq \rho'$, the density of points in both A and A' is the same, that is $(n - \beta n)/(1 - |D|)$. The density is maximized when $\rho = \rho'$. Note that whp, $\rho \neq \rho'$, but, for our purposes of getting and upper bound to the size of $\|A'\|$, the argument below is valid. Thus, for any $c > 0$,

$$Pr[\|A\| > c \,|\, \rho \leq \rho'] \leq Pr[\|A'\| > c \,|\, \rho \leq \rho'] \leq Pr[\|A'\| > c \,|\, \rho = \rho'].$$

For $\rho = \rho'$,

$$\frac{n - \beta n}{1 - |D|} = \frac{n - \beta n}{1 - 3\beta/2} \quad \text{and} \quad |A'| = \pi \sqrt{\frac{R \ln n}{n}} \left(2\sqrt{\frac{3\beta}{2\pi}} + \sqrt{\frac{R \ln n}{n}} \right).$$

Therefore

$$\mu = E[\|A'\| \,|\, \rho = \rho'] = \frac{n - \beta n}{1 - 3\beta/2} \pi \sqrt{\frac{R \ln n}{n}} \left(2\sqrt{\frac{3\beta}{2\pi}} + \sqrt{\frac{R \ln n}{n}} \right).$$

In particular

$$\sqrt{108} \ln n \leq \sqrt{(3/2)\pi \beta R n \ln n} \leq \mu \leq 12\sqrt{(3/2)\pi \beta R n \ln n}.$$

It follows from Chernoff's Bound (3) that

$$Pr[\|A'\| > 2\mu \,|\, \rho = \rho'] \leq e^{-\mu/4} \leq e^{-\sqrt{108/16} \ln n} = o(1/n^2).$$

Moreover, being $E[\|D'\|] = (3/2)\beta n$, from Chernoff's Bound (1),

$$Pr[\rho > \rho'] = Pr[\|D'\| < \beta n] \leq e^{-(1/3)^2 (3/2)\beta n/2} \leq e^{-\beta n/12} = o(1/n^2).$$

Altogether

$$\begin{aligned}
Pr[\|A\| > 2\mu] &\leq Pr[\rho > \rho'] + Pr[\|A\| > 2\mu \,|\, \rho \leq \rho'] \, Pr[\rho \leq \rho'] \\
&\leq o(1/n^2) + Pr[\|A'\| > 2\mu \,|\, \rho = \rho'] \\
&= o(1/n^2).
\end{aligned}$$

It follows that $\|A\| \leq 2\mu = O(\sqrt{\beta \, n \, R \log n})$ with probability $1 - o(1/n^2)$.   □

Theorems 1 and 2 imply that simpleCut is a probabilistic constant approximation algorithm for the β-balanced cut problem. We next show how to extend this result to a constant expected approximation algorithm for the same problem. Consider the following algorithm zeroCut to compute a cut of size zero, if any. Compute the connected components of $\mathcal{G}(n, r)$. For any integer m, $\beta n \leq m \leq n/2$, check whether there is a subset of components whose total size is m. If yes, return such subset of components as one side of the partition. Note that for each of the $O(n)$ possible values of m, we have to solve an instance of the *subset sum* problem. Since the sum of the sizes of the connected components is n, it follows that dynamic programming allows to solve all such instances in total time $O(n^2)$ and space $O(n)$ [8]. Combining zeroCut and simpleCut, one obtains the desired constant expected approximation algorithm.

Algorithm 2. (refinedCut) *If* zeroCut *returns a solution, return it. Otherwise, return the solution computed by* simpleCut.

Theorem 3. *For any $\beta \in [0, 1/2]$, $\beta \, n \in \mathbb{N}$, and for any $R \geq 240$, refinedCut is a constant expected approximation algorithm for the β-balanced cut problem on $\mathcal{G}(n, r)$.*

Proof. Let z^H and z^* denote the size of the solution found by refinedCut and the size of the optimum cut, respectively. Let moreover $\mathcal{A}$ denote the event that

$$z^* \geq c \min\{\beta \, n \, R \, \log n, \, \sqrt{\beta \, n \, R^3 \log n^3}\}$$

and

$$z^H \leq C \min\{\beta \, n \, R \, \log n, \, \sqrt{\beta \, n \, R^3 \log n^3}\},$$

where the constants c and C are as in the proofs of Theorems 1 and 2. Note that $Pr[\mathcal{A}] = 1 - o(1/n^2)$. Given $\mathcal{A}$, the approximation ratio of refinedCut is at most $C/c = O(1)$. Given $\overline{\mathcal{A}}$, if the size of the optimum cut is zero, zeroCut computes the optimum solution and the approximation ratio is 1 by definition. Otherwise, any cut, and hence also the cut computed by simpleCut, is a $O(n^2)$ approximation. Altogether the expected approximation ratio is

$$E(z^H / z^*) = Pr[\mathcal{A}] \, O(1) + Pr[\overline{\mathcal{A}}] O(n^2) = O(1).$$

□

Remark 1. The threshold 240 can be reduced to a value arbitrarily close to 30 by adapting the constants in Lemma 1. However, this would increase the approximation ratio. If we only desire a probabilistic constant approximation, such threshold can be made arbitrarily close to 10, with the same drawback as above.

Acknowledgments. We thank an anonymous referee for the comments that have improved the presentation of the paper.

References

1. I. Akyildiz, W. Su, Y. Sankarasubramaniam, and E. Cayirci. Wireless sensor networks: a survey. *Computer Networks*, 38:393–422, 2002.
2. S. Arora, D. Karger, and M. Karpinski. Polynomial time approximation schemes for dense instances of NP-hard problems. In *ACM Symposium on the Theory of Computing (STOC)*, pages 284–293, 1995.
3. J. Díaz, M. Penrose, J. Petit, and M. Serna. Approximating layout problems on random geometric graphs. *Journal of Algorithms*, 39:78–116, 2001.
4. T.N. Bui, and C. Jones. Finding good approximate vertex and edge partitions is NP-hard. *Information Processing Letters*, 42:153–159, 1992.
5. J. Díaz, J. Petit, and M. Serna. Evaluation of basic protocols for optical smart dust networks. *IEEE Transactions on Mobile Networks*, 2:189–196, 2003.
6. U. Feige, and R. Krauthgamer. A polylogarithmic approximation of the minimum bisection. *SIAM Journal on Computing*, 31(3): 1090–1119, 2002.
7. G. Garg, H. Saran, and V. Vazirani. Finding separator cuts in planar graphs within twice the optimal. In *IEEE Symposium on Foundations of Computer Science (FOCS)*, pages 14–23, 1994.
8. M. Garey, and D. Johnson. *Computers and Intractability*. Freeman. N.Y., 1979.
9. E. Gilbert. Random plane networks. *Journal of the Society for Industrial and Applied Mathematics*, 9:533–543, 1961.
10. S.H. Gerez. *Algorithms for VLSI design automation*. Wiley, 2003.
11. A. Goel, S. Rai, and V. Krishnamachari. Sharp thresholds for monotone properties in random geometric graphs. In *ACM Symposium on Foundations of Computer Science (FOCS)*, pages 13–23, 2004.
12. R. Motwani and P. Raghavan. *Randomized Algorithms*. Cambridge University Press, 1995.
13. S. Muthukrishnan and G. Pandurangan. The Bin-covering technique for thresholding random geometric graph properties graphs. In *ACM-SIAM Symposium on Discrete Algorithms (SODA)*, pages 989–998, 2005.
14. M. Penrose. The longest edge of the random minimal spanning tree. *The Annals of Applied Probability*, 7(2):340–361, 1997.
15. M. Penrose. *Random Geometric Graphs*. Oxford Studies in Probability. Oxford U.P., 2003.

Improved Algorithms for the Minmax-Regret 1-Center Problem

Tzu-Chin Lin, Hung-I Yu, and Biing-Feng Wang

Department of Computer Science, National Tsing Hua University,
Hsinchu, Taiwan 30043, Republic of China
rems@cs.nthu.edu.tw, herbert@cs.nthu.edu.tw, bfwang@cs.nthu.edu.tw

Abstract. This paper studies the problem of finding the 1-center on a graph where vertex weights are uncertain and the uncertainty is characterized by given intervals. It is required to find a minmax-regret solution, which minimizes the worst-case loss in the objective function. Averbakh and Berman had an $O(mn^2\log n)$-time algorithm for the problem on a general graph. On a tree, the time complexity of their algorithm becomes $O(n^2)$. In this paper, we improve these two bounds to $O(mn\log n)$ and $O(n\log^2 n)$, respectively.

Keywords: location theory, minmax-regret optimization, centers.

1 Introduction

Over three decades, location problems on networks have received much attention from researchers in the fields of transportation and communication [10,11,12,13,18]. Traditionally, network location theory has been concerned with networks in which the vertex weights and edge lengths are known precisely. However, in practice, it is often impossible to make an accurate estimate of all these parameters [14,15]. Real-life data often involve a significant portion of uncertainty, and these parameters may change with time. Thus, location models involving uncertainty have attracted increasing research efforts in recent years [2,3,4,5,6,7,8,9,14,15,17,19,20,21,22].

Several ways for modeling network uncertainty have been defined and studied [14,17,19]. One of the most important models is the *minmax-regret approach*, introduced by Kouvelis [14]. In the model, uncertainty of network parameters is characterized by given intervals, and it is required to minimize the worst-case loss in the objective function that may occur because of the uncertain parameters. During the last ten years, many important location problems have been studied on the minmax-regret model. The 1-center problem was studied in [3,4,7], the p-center problem was studied in [3], and the 1-median problem was studied in [5,6,8,14].

The minmax-regret 1-center problem is the focus of this paper. For a general graph with uncertain edge lengths, the problem is strongly NP-hard [2]. For a general graph with uncertain vertex weights, Averbakh and Berman [3] gave an $O(mn^2\log n)$-time algorithm, where n is the number of vertices and m is the

T. Asano (Ed.): ISAAC 2006, LNCS 4288, pp. 537–546, 2006.

number of edges. For a tree with uncertain vertex weights, the time complexity of their algorithm becomes $O(n^2)$ [3,4]. For a tree with uncertainty in both vertex weights and edge lengths, Averbakh and Berman [4] presented an $O(n^6)$-time algorithm and Burkard and Dollani [7] had an $O(n^3\log n)$-time algorithm. For a tree with uncertain edge lengths, assuming uniform vertex weights, Averbakh and Berman [4] presented an $O(n^2\log n)$-time algorithm and Burkard and Dollani [7] had an $O(n\log n)$-time algorithm. In this paper, efficient algorithms are presented for the minmax-regret 1-center problem on a general graph and a tree with uncertain vertex weights. For general graphs, we improve the upper bound from $O(mn^2\log n)$ to $O(mn\log n)$. The bottleneck of the previous algorithm is the computation of the classical 1-centers of the input graph under n different weight assignments. The key idea of our improvement is to identify the similarity among these assignments and preprocess the input graph under this similarity, such that the classical 1-center for each assignment can be derived efficiently. For trees, we improve the upper bound from $O(n^2)$ to $O(n\log^2 n)$ using the same concept.

The remainder of this paper is organized as follows. In Section 2, notation and preliminary results are presented. In Sections 3 and 4, improved algorithms for the minmax-regret 1-center problem are proposed. Finally, in Section 5, we conclude this paper.

2 Notation and Preliminaries

Let $G = (V, E)$ be an undirected connected graph, where V is the vertex set and E is the edge set. Let $n = |V|$ and $m = |E|$. In this paper, G also denotes the set of all points of the graph. Thus, the notation $x \in G$ means that x is a point along any edge of G which may or may not be a vertex of G. Each edge $e \in E$ has a nonnegative length. For any two points $a, b \in G$, let $d(a, b)$ be the distance of the shortest path between a and b. Suppose that the matrix of shortest distances between vertices of G is given. Each vertex $v \in V$ is associated with two positive values w_v^- and w_v^+, where $w_v^- \leq w_v^+$. The *weight* of each vertex $v \in V$ can take any value randomly from the interval $[w_v^-, w_v^+]$. Let Σ be the Cartesian product of intervals $[w_v^-, w_v^+], v \in V$. Any element $S \in \Sigma$ is called a *scenario* and represents a feasible assignment of weights to the vertices of G. For any scenario $S \in \Sigma$ and any vertex $v \in V$, let w_v^S be the weight of v under the scenario S.

For any scenario $S \in \Sigma$, and a point $x \in G$, we define
$$F(S, x) = max_{v \in V}\{w_v^S \times d(v, x)\},$$
which is the maximum weighted distance from all the vertices to x according to S. Given a specific scenario $S \in \Sigma$, the *classical 1-center problem* is to find a point $x^* \in G$ that minimizes $F(S, x^*)$. The point x^* and the value of $F(S, x^*)$ are, respectively, called a *classical 1-center* and the *1-radius* of G under the scenario S. For any scenario $S \in \Sigma$, denote $F^*(S)$ as the 1-radius of G under the scenario S. For any point $x \in G$, the *regret* of x with respect to a scenario $S \in \Sigma$ is $max_{y \in G}\{F(S, x) - F(S, y)\}$ and the *maximum regret* of x is

$$Z(x) = max_{S \in \Sigma} max_{y \in G} \{F(S, x) - F(S, y)\}.$$

The *minmax-regret 1-center problem* is to find a point $x \in G$ minimizing $Z(x)$.

The *upper envelope* of a set H of functions is the function U defined as $U(x) = max_{f \in H}\{f(x)\}$. A function is *unimodal* if it increases to a maximum value and then decreases. Kariv and Hakimi [12] gave the following result.

Theorem 2.1 [12]. *The upper envelope of a set of n unimodal piecewise linear functions with at most two segments defined on the same interval is a piecewise linear function having $O(n)$ linear segments.*

In this paper, we assume that a piecewise linear function is represented by an array storing the sequence of its breakpoints such that $f(x)$ can be determined in $O(\log |f|)$ time for any given x and the upper envelope of f and g can be constructed in $O(|f| + |g|)$ time, where f and g are piecewise linear functions that have, respectively, $|f|$ and $|g|$ linear segments.

3   Minmax-Regret 1-Center on a General Graph

Averbakh and Berman [3] had an $O(mn^2 \log n)$-time algorithm for finding a minmax-regret 1-center of a graph $G = (V, E)$. Their algorithm is firstly described in Subsection 3.1. Then, our improved algorithm is presented in Subsection 3.2.

3.1   Averbakh and Berman's Algorithm

For each $i \in V$, let S_i be the scenario in which the weight of i is w_i^+ and the weight of any other vertex v is w_v^-. Averbakh and Berman solved the minmax-regret 1-center problem by an elegant transformation to the classical 1-center problem. Define an auxiliary graph G' as follows. Let $M = (max_{v \in V} w_v^+) \times \sum_{e \in E} l(e)$, where $l(e)$ is the length of e. The graph G' is obtained from G by adding for each $i \in V$ a vertex i' and an edge (i, i') with length $(M - F^*(S_i))/w_i^+$. Specific weights are assigned to the vertices of G'. For each $i \in V$, the weight of i is zero and the weight of i' is w_i^+. Averbakh and Berman gave the following important property for solving the minmax-regret 1-center problem.

Lemma 3.1 [3]. *Any classical 1-center of G' is a minmax-regret 1-center of G.*

Based upon Lemma 3.1, Averbakh and Berman solved the minmax-regret 1-center problem as follows. First, for each $i \in V$, the 1-radius $F^*(S_i)$ is computed. Kariv and Hakimi [12] had an $O(mn \log n)$-time algorithm for the classical 1-center problem. By applying their algorithm, this step is done in $O(mn^2 \log n)$ time. Next, the auxiliary graph G' is constructed, which requires $O(m + n)$ time. Finally, a solution is obtained in $O(mn \log n)$ time by applying Kariv and Hakimi's algorithm again to G'.

Theorem 3.2 [3]. *The minmax-regret 1-center problem on a general graph can be solved in $O(mn^2 \log n)$ time.*

3.2   The Improved Algorithm

The bottleneck of Averbakh and Berman's algorithm is the computation of $F^*(S_i)$ for every $i \in V$. In this subsection, we improve their upper bound by showing that the computation can be done in $O(mn\log n)$ time. For any scenario $S \in \Sigma$ and $e \in E$, the *local-radius* of G on e under the scenario S is $F_e^*(S) = min_{x \in e}\{F(S, x)\}$. In order to compute the 1-radius of G under a scenario, it is enough to compute the local-radius on each of the edges. Therefore, let us focus on the determination of a local-radius. Let e be an edge in G and l be its length. For ease of discussion, e is regarded as an interval $[0, l]$ on the real line so that any point on e corresponds to a real number $x \in [0, l]$. Let $S \in \Sigma$ be a scenario. For any vertex $v \in V$ and any point x on e, define $D(S, v, x) = w_v^S \times d(v, x)$, which is the weighted distance from v to x under the scenario S. Clearly, each $D(S, v, \cdot)$ is a unimodal piecewise linear function, which consists of at most two linear segments. For convenience, we define $F_e(S, x) = max_{v \in V}\{D(S, v, x)\}$ for $x \in [0, l]$. Note that $F_e(S, x) = F(S, x)$ for $x \in [0, l]$. Since $F_e(S, \cdot)$ is the upper envelope of n unimodal piecewise linear functions with at most two segments, by Theorem 2.1, the following is obtained.

Lemma 3.3 [12]. $F_e(S, \cdot)$ *is a continuous piecewise linear function having $O(n)$ breakpoints and can be computed in $O(n\log n)$ time.*

For any scenario $S \in \Sigma$, the minimum of $F_e(S, \cdot)$ is the local-radius $F_e^*(S)$. Therefore, by Lemma 3.3, $F_e^*(S_i)$ of all $i \in V$ can be computed in $O(n^2\log n)$ time. In the following, we show how to reduce the time complexity. Let S^- be the scenario in which the weight of every $v \in V$ is w_v^-. For any $i \in V$, the scenario S_i can be obtained from S^- by simply increasing the weight of i from w_i^- to w_i^+. Therefore, for any $i \in V$ and $x \in e$, we have $D(S_i, i, x) \geq D(S^-, i, x)$ and $D(S_i, v, x) = D(S^-, v, x)$ for all $v \neq i$. Consequently, the following relation between $F_e(S_i, \cdot)$ and $F_e(S^-, \cdot)$ is established.

Lemma 3.4. *For any $i \in V$ and $x \in e$, $F_e(S_i, x) = max\{F_e(S^-, x), D(S_i, i, x)\}$.*

According to Lemma 3.4, $F_e(S_i, \cdot)$ is the upper envelope of $F_e(S^-, \cdot)$ and $D(S_i, i, \cdot)$. Since $F_e(S^-, \cdot)$ contains $O(n)$ linear segments and $D(S_i, i, \cdot)$ contains at most two linear segments, it is easy to compute their upper envelope in $O(n)$ time. Therefore, after $F_e(S^-, \cdot)$ is determined, the time for computing each $F_e^*(S_i)$ can be reduced to $O(n)$. With some preprocessing on $F_e(S^-, \cdot)$, the time for computing each $F_e^*(S_i)$ can be further reduced to $O(\log n)$. The trick is not to construct the whole function $F_e(S_i, \cdot)$, but only to determine its minimum from $F_e(S^-, \cdot)$ and $D(S_i, i, \cdot)$. For convenience, we define the *radius-adjustment problem* as follows. Let R be a continuous piecewise linear function that has $O(n)$ breakpoints and is defined on an interval $I = [0, l]$. The *radius-adjustment problem* is to preprocess R so as to answer the following queries efficiently: given a unimodal piecewise linear function A with at most two segments defined on I, determine the minimum of $max\{R(x), A(x)\}$ over all $x \in I$. Later, we will present an efficient algorithm for the radius-adjustment problem. The presented algorithm requires $O(n)$ preprocessing time and $O(\log n)$ query time. Such a result immediately leads to a procedure to compute $F_e^*(S_i)$ for all $i \in V$ in $O(n\log n)$ time. Consequently, we have the following theorem.

Theorem 3.5. *The minmax-regret 1-center problem on a general graph can be solved in $O(mn\log n)$ time.*

In the remainder of this subsection, we complete the proof of Theorem 3.5 by presenting an efficient algorithm for the radius-adjustment problem. Let U be the upper envelope of A and R. Our problem is to compute the minimum of U. The query function A is piecewise linear with at most two linear segments. In the following, we first discuss the case that A contains only a linear segment with positive slope.

We define $\Phi(x) = min_{0 \le z \le x}\{R(z)\}$ for $0 \le x \le l$, which is called the *prefix-minimum function* of R. Clearly, Φ is a piecewise linear function having $O(n)$ breakpoints and each of its breakpoints is a point of R. Two vertical linear segments are associated with Φ. One is from $(0, \Phi(0))$ to $(0, \infty)$ and the other is from $(l, \Phi(l))$ to $(l, -\infty)$. These two linear segments are called, respectively, the *left* and *right boundaries* of Φ. Let Φ^* be the set of points on Φ and its boundaries. The function Φ is non-increasing. Thus, it is easy to see that there is a unique point at which A intersects Φ^*. With some efforts, the following can be obtained.

Lemma 3.6. *Let $r = (x_r, y_r)$ be the unique point at which A intersects Φ^*. If r lies on the right boundary of Φ, the minimum of U is $\Phi(l)$; otherwise, the minimum of U is y_r.*

According to Lemma 3.6, after Φ is computed, our problem becomes to find the unique point at which A intersects Φ^*. Since Φ is non-increasing and A is increasing, the finding can be done in $O(\log n)$ time by performing binary search on the linear segments of Φ. Therefore, we have the following.

Lemma 3.7. *If A is a linear segment with positive slope, by using the prefix-minimum function Φ, the minimum of U can be determined in $O(\log n)$ time.*

Next, consider the case that A is a linear segment with negative slope. Define the *suffix-minimum function* of R as $\Pi(x) = min_{x \le z \le l}\{R(z)\}$ for $0 \le x \le l$. Similarly, we have the following.

Lemma 3.8. *If A is a linear segment with negative slope, by using the suffix-minimum function Π, the minimum of U can be determined in $O(\log n)$ time.*

When A consists of two linear segments, by combining Lemmas 3.7 and 3.8, it is not difficult to compute the minimum of U in $O(\log n)$ time.

By scanning the breakpoints of R from left to right, Φ can be computed in $O(n)$ time. Similarly, Π can be computed in $O(n)$ time. We conclude this subsection with the following theorem.

Theorem 3.9. *With an $O(n)$-time preprocessing on R, each query of the radius-adjustment problem can be answered in $O(\log n)$ time.*

4 Minmax-Regret 1-Center on a Tree

In this section, we assume that the underlying graph G is a tree $T = (V, E)$. For a tree, the auxiliary graph defined in Subsection 3.1 is also a tree. Megiddo [18] had an $O(n)$-time algorithm for finding the classical 1-center of a tree. Thus, as indicated in [3,4], on a tree the time complexity of Averbakh and Berman's algorithm becomes $O(n^2)$. The bottleneck is still the computation of $F^*(S_i)$ for every $i \in V$. In this section, we improve this upper bound by showing that the computation can be done in $O(n\log^2 n)$ time. In Subsection 4.1, we preprocess T under the scenario S^- to construct a data structure. The constructed data structure is useful in evaluating the value of $F(S^-, x)$ for any point $x \in T$. Then, in Subsection 4.2, we describe the computation of $F^*(S_i)$ for each $i \in V$.

4.1   Preprocess

A *top tree* [1] is adopted to provide a hierarchical representation of the tree T. It is defined as follows. For any subtree X of T, we call a vertex in X having a neighbor in T outside X a *boundary vertex*. A *cluster* of T is a subtree having at most two boundary vertices. Two clusters A and B can be *merged* if they intersect in a single vertex and $A \cup B$ is still a cluster. There are five different cases of merging, which are illustrated in Figure 4.1. A *top tree* of T is a binary tree with the following properties [1].

1. Each node represents a cluster of T.
2. The leaves represent the edges of T.
3. Each internal node represents the cluster merged from the two clusters represented by its children.
4. The root represents T.
5. The height is $O(\log n)$.

A top tree of T defines a way to recursively decompose T into subtrees, until each of the subtrees contains only a single edge. It was shown in [1] that a top tree of T always exists and can be constructed in $O(n)$ time. For each node α of

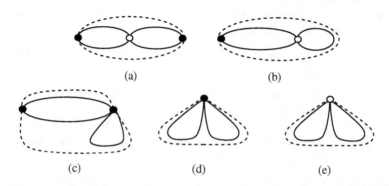

(a)   (b)

(c)   (d)   (e)

Fig. 4.1. Five cases of merging A and B into C. Black nodes are C's boundary vertices.

a top tree, let $C(\alpha)$ denote the cluster represented by α, $B(\alpha)$ denote the set of boundary vertices of $C(\alpha)$, and $V(\alpha)$ and $E(\alpha)$ denote, respectively, the vertex set and edge set of $C(\alpha)$.

We preprocess T under the scenario S^- to construct a data structure τ. The data structure τ is a top tree of T; in addition, each node α stores a function $U_{\alpha,b}$ for every $b \in B(\alpha)$, where

$$U_{\alpha,b}(t) = max_{v \in V(\alpha)}\{w_v^- \times (d(v,b) + t)\} \text{ for } t \geq 0.$$

Before presenting the computation of the functions $U_{\alpha,b}$, their usage is described as follows. For any subset $K \subset V$ and any point $x \in T$, define

$$F^-(K, x) = max_{v \in K}\{w_v^- \times d(v,x)\},$$

which is the largest weighted distance from any vertex $v \in K$ to x under the scenario S^-. For ease of description, for any node α of τ, we simply write $F^-(\alpha, x)$ in place of $F^-(V(\alpha), x)$. Then, we have the following.

Lemma 4.1. *Let α be a node of τ. For any given point $x \in (T\backslash C(\alpha)) \cup B(\alpha)$, $F^-(\alpha, x) = U_{\alpha,b}(d(b,x))$, where b is the boundary vertex nearer to x in $B(\alpha)$.*

For convenience, we call each $U_{\alpha,b}$ a *dominating function* of α. Since $U_{\alpha,b}$ is the upper envelope of $|V(\alpha)|$ linear functions, by Theorem 2.1, it is a piecewise linear function having $O(|V(\alpha)|)$ breakpoints. According to Lemma 4.1, with the dominating functions of a node α of τ, the largest weighted distance from any vertex in $C(\alpha)$ to any point x outside $C(\alpha)$ can be determined in $O(\log |V(\alpha)|)$ time.

We now proceed to discuss the computation of the dominating functions. We do the computation for each node α of τ, layer by layer from the bottom up. Each leaf of τ represents an edge of T. Thus, if α is a leaf, the computation of its dominating functions takes $O(1)$ time. Next, consider the case that α is an internal node. Let b be a boundary vertex of $C(\alpha)$. Let α_1 and α_2 be the children of α, and let c be the intersection vertex of $C(\alpha_1)$ and $C(\alpha_2)$. Let $U_1(t) = max_{v \in V(\alpha_1)}\{w_v^- \times (d(v,b) + t)\}$ and $U_2(t) = max_{v \in V(\alpha_2)}\{w_v^- \times (d(v,b) + t)\}$. Since $V(\alpha) = V(\alpha_1) \cup V(\alpha_2)$, $U_{\alpha,b}(t) = max\{U_1(t), U_2(t)\}$. Therefore, $U_{\alpha,b}$ is the upper envelope of U_1 and U_2. Consider the function U_1. According to the definition of a top tree, $B(\alpha) \subseteq B(\alpha_1) \cup B(\alpha_2)$ and $\{c\} = B(\alpha_1) \cap B(\alpha_2)$. If $b \in B(\alpha_1)$, $U_1(t) = U_{\alpha_1,b}(t)$; otherwise, we have $d(v,b) = d(v,c) + d(c,b)$ for each $v \in C(\alpha_1)$ and thus $U_1(t) = U_{\alpha_1,c}(d(c,b) + t)$. Therefore, we obtain U_1 from a dominating function of α_1 in $O(|V(\alpha_1)|)$ time. Similarly, we obtain U_2 from a dominating function of α_2 in $O(|V(\alpha_2)|)$ time. With U_1 and U_2, we then construct $U_{\alpha,b}$ in $O(|V(\alpha_1)| + |V(\alpha_2)|) = O(|V(\alpha)|)$ time. Since α has at most two dominating functions, the computation for α requires $O(|V(\alpha)|)$ time.

The time complexity for computing all the dominating functions is analyzed as follows. For any node α, the computation time is $O(|V(\alpha)|) = O(|E(\alpha)|)$. No two clusters of the same layer of τ share a common edge. Thus, for any layer of τ, $\sum_\alpha |E(\alpha)| = O(n)$. Therefore, the computation for all nodes in a layer requires $O(n)$ time. Since there are $O(\log n)$ layers, we have the following.

Lemma 4.2. *The data structure τ can be constructed in $O(n\log n)$ time.*

4.2   An Improved Algorithm

In this subsection, by presenting an efficient algorithm to compute $F^*(S_i)$ for every $i \in V$, we show that the minmax-regret 1-center problem on T can be solved in $O(n\log^2 n)$ time. We begin with several important properties of this problem. For any $i \in V$ and $x \in T$, define $D_i(x) = w_i^+ \times d(i,x)$. Trivially, Lemma 3.4 can be extended to the following.

Lemma 4.3. *For any $i \in V$ and $x \in T$, $F(S_i, x) = max\{F(S^-, x), D_i(x)\}$.*

Lemma 4.4 [18]. *For any scenario $S \in \Sigma$, the function $F(S, \cdot)$ is convex on every simple path in T.*

Let c^- be the classical 1-center of T under the scenario S^-. For each $i \in V$, let c_i be the classical 1-center of T under the scenario S_i. For any two points $x, y \in T$, let $P(x, y)$ be the unique path from x to y. The following lemma suggests a possible range for searching each c_i. Due to page limit, the proof is omitted.

Lemma 4.5. *For each $i \in V$, c_i is a point on $P(c^-, i)$.*

Let $i \in V$ be a vertex and x be a point on $P(c^-, i)$. By Lemma 4.4, the value of $F(S^-, x)$ is non-decreasing along the path $P(c^-, i)$. On the contrary, the value of $D_i(x)$ is decreasing. Thus, by Lemma 4.3, c_i is the point $x \in P(c^-, i)$ at which $F(S^-, \cdot)$ and D_i intersect. Moreover, $F(S^-, x) < D_i(x)$ for any $x \in P(c^-, c_i)\backslash\{c_i\}$ and $F(S^-, x) > D_i(x)$ for any $x \in P(c_i, i)\backslash\{c_i\}$. We have the following lemma, which plays a key role in our algorithm.

Lemma 4.6. *Let $i \in V$ be a vertex. For any point $x \in P(c^-, i)$, $c_i = x$ if $F(S^-, x) = D_i(x)$; otherwise, $c_i \in P(c^-, x)$ if $F(S^-, x) > D_i(x)$, and $c_i \in P(x, i)$ if $F(S^-, x) < D_i(x)$.*

According to Lemma 4.6, if the values $F(S^-, v)$ of all $v \in V$ are available, by examining the vertices on $P(c^-, i)$ in a binary-search manner, the range for searching c_i can be further restricted to an edge in $O(\log n)$ time. We can compute the values $F(S^-, v)$ by using the divide-and-conquer strategy on the top tree τ. Due to page limit, the details are omitted here.

Lemma 4.7. *$F(S^-, v)$ can be computed in $O(n\log n)$ time for all $v \in V$.*

With the values $F(S^-, v)$, in $O(n\log n)$ time, we compute the edge e_i containing c_i for every $i \in V$ as follows. First, we compute c^- and then orient T into a rooted tree with root c^-. (In case c^- is not a vertex, a dummy vertex is introduced.) Then, we perform a depth-first traversal on T, during which we maintain the path from the root c^- to the current vertex in an array. And, for each $i \in V$, when i becomes the current vertex, we perform a binary search to find the edge e_i.

Next, we show how to determine the exact position of c_i and the value $F^*(S_i)$ for a fixed $i \in V$. If an edge $e \in T$ is removed from T, two subtrees are induced. We denote the vertex set of the subtree containing c^- by $Y(e)$ and the vertex set of the other subtree by $Z(e)$. With some efforts, the following can be proved.

Lemma 4.8. *Let $e \in E$ be an edge. For any point $x \in e$, $F(S^-, x) = F^-(Y(e), x)$.*

By Lemmas 4.6 and 4.8, c_i is the unique point $x^* \in e_i$ at which $F^-(Y(e_i), \cdot)$ and D_i intersect. Moreover, $F^*(S_i) = D_i(x^*)$. Therefore, the problem remained is to determine the unique point x^* on e_i at which $F^-(Y(e_i), \cdot)$ and D_i intersect. The construction of the function $F^-(Y(e_i), \cdot)$ is a costly computation. To avoid it, we replace it with $O(\log n)$ dominating functions. The replacement is based upon the following lemma. The proof is omitted due to page limit.

Lemma 4.9. *For any $e \in E$, we can determine a set Q of $O(\log n)$ dominating functions in τ such that $F^-(Y(e), x) = max_{U_{\alpha,b} \in Q}\{U_{\alpha,b}(d(b, x))\}$ for any $x \in e$ in $O(\log n)$ time.*

Now, we proceed to show how to determine the unique point x^* on e_i at which $F^-(Y(e_i), \cdot)$ and D_i intersect. According to Lemma 4.9, we compute a set Q of dominating functions in τ such that $F^-(Y(e_i), x) = max_{U_{\alpha,b} \in Q}\{U_{\alpha,b}(d(b, x))\}$ for any $x \in e_i$. Let $e_i = (y, z)$, where $y \in Y(e)$ and $z \in Z(e)$. For ease of discussion, in the following, e_i is regarded as an interval $[0, d(y, z)]$ on the real line, where y and z correspond, respectively, to 0 and $d(y, z)$. For convenience, for any $U_{\alpha,b} \in Q$, we say that D_i intersects $U_{\alpha,b}$ at a number $x \in e_i$ if $D_i(x) = U_{\alpha,b}(d(b, x))$. We compute I as the set of numbers $x \in e_i$ at which D_i intersects the functions in Q. For each $U_{\alpha,b} \in Q$, since D_i is decreasing and $U_{\alpha,b}$ is increasing, D_i and $U_{\alpha,b}$ have at most one intersection point, which can be found in $O(\log n)$ time by binary search. Thus, the computation of I takes $O(\log^2 n)$ time. Then, in $O(\log n)$ time, we compute x^* as the smallest number in I. The correctness is ensured by the following lemma. We omit the proof due to page limit.

Lemma 4.10. *Let Q be a set of increasing functions defined on the same interval and U be the upper envelope of the functions in Q. Let f be a decreasing function that intersects U. The smallest number at which f intersects a function in Q is the unique number at which f intersects U.*

As mentioned, c_i is x^* and $F^*(S_i) = D_i(x^*)$. Therefore, we have the following.

Lemma 4.11. *We can compute c_i and $F^*(S_i)$ for all $i \in V$ in $O(n\log^2 n)$ time.*

Theorem 4.12. *The minmax-regret 1-center problem on a tree can be solved in $O(n\log^2 n)$ time.*

5 Concluding Remarks

During the last decade, minmax-regret optimization problems have attracted significant research efforts. For many location problems, however, there are still large gaps between the time complexities of the solutions to their classical versions and those to their minmax-regret versions. For example, the classical 1-center problem on a tree can be solved in $O(n)$ time, while the current upper bound for its minmax-regret version on a tree with uncertainty in both vertex weights and edge lengths is $O(n^3\log n)$ [7]. It would be a great challenge to bridge the gaps.

References

1. Alstrup, S., Lauridsen, P. W., Sommerlund, P., and Thorup, M.: Finding cores of limited length. Technical Report. The IT University of Copenhagen (2001)
2. Averbakh, I.: On the complexity of a class of robust location problems. Working Paper. Western Washington University. Bellingham, WA. (1997)
3. Averbakh, I., Berman, O.: Minimax regret p-center location on a network with demand uncertainty. *Location Science* **5** (1997) 247–254
4. Averbakh, I., Berman, O.: Algorithms for the robust 1-center problem on a tree. *European Journal of Operational Research* **123** (2000) 292–302
5. Averbakh, I., Berman, O.: Minmax regret median location on a network under uncertainty. *Informs Journal on Computing* **12** (2000) 104–110
6. Averbakh, I., Berman, O.: An improved algorithm for the minmax regret median problem on a tree. *Networks* **41** (2003) 97–103
7. Burkard, R. E., Dollani, H.: A note on the robust 1-center problem on trees. *Annals of Operations Research* **110** (2002) 69–82
8. Chen, B. T., Lin, C. S.: Minmax-regret robust 1-median location on a tree. *Networks* **31** (1998) 93–103
9. Drezner, Z.: Sensitivity analysis of the optimal location of a facility. *Naval Research Logistics Quarterly* **33** (1980) 209–224
10. Goldman, A. J.: Optimal center location in simple networks. *Transportation Science* **5** (1971) 212–221
11. Hakimi, S. L.: Optimal locations of switching centers and the absolute centers and medians of a graph. *Operations Research* **12** (1964) 450–459
12. Kariv, O., Hakimi, S. L.: An algorithmic approach to network location problems. I: The p-centers. *SIAM Journal on Applied Mathematics* **37** (1979) 513–538
13. Kariv, O., Hakimi, S. L.: An algorithmic approach to network location problems. II: The p-medians. *SIAM Journal on Applied Mathematics* **37** (1979) 539–560
14. Kouvelis, P., Vairaktarakis, G., Yu, G.: Robust 1-median location on a tree in the presence of demand and transportation cost uncertainty. Working Paper 93/94-3-4. Department of Management Science and Information Systems, Graduate School of Business, The University of Texas at Austin (1994)
15. Kouvelis, P., Yu, G.: Robust discrete optimization and its applications. Kluwer Academic Publishers, Dordrecht (1997)
16. Ku, S. C., Lu, C. J., Wang, B. F., Lin, T. C.: Efficient algorithms for two generalized 2-median problems on trees. in *Proceedings of the 12th International Symposium on Algorithms and Computation* (2001) 768–778
17. Labbe, M., Thisse, J.-F., Wendell, R.: Sensitivity analysis in minisum facility location problems. *Operations Research* **38** (1991) 961–969
18. Megiddo, N.: Linear-time algorithms for linear-programming in R^3 and related problems. *SIAM Journal on Computing* **12** (1983) 759–776
19. Mirchandani, P. B., Odoni, A. R.: Location of medians on stochastic networks. *Transportation Science* **13** (1979) 85–97
20. Mirchandani, P. B., Oudjit, A., Wong, R. T.: Multidimensional extensions and a nested dual approach for the M-median problem. *European Journal of Operational Research* **21** (1985) 121–137
21. Oudjit, A.: Median locations on deterministic and probabilistic multidimensional networks. PhD Dissertation. Rennselaer Polytechnic Institute, Troy (1981)
22. Weaver, J. R., Church, R. L.: Computational procedures of location problems on stochastic networks. *Transportation Science* **17** (1983) 168–180

On Approximating the Maximum Simple Sharing Problem*

Danny Z. Chen[1],**, Rudolf Fleischer[2],***, Jian Li[2],
Zhiyi Xie[2], and Hong Zhu[2],†

[1] Department of Computer Science and Engineering, University of
Notre Dame, Notre Dame, IN 46556, USA
dchen@cse.nd.edu
[2] Department of Computer Science and Engineering, Shanghai Key Laboratory of
Intelligent Information Processing, Fudan University, Shanghai, China
{rudolf, lijian83, xie_zhiyi, hzhu}@fudan.edu.cn

Abstract. In the *maximum simple sharing problem (MSS)*, we want to
compute a set of node-disjoint simple paths in an undirected bipartite
graph covering as many nodes as possible of one layer of the graph, with
the constraint that all paths have both endpoints in the other layer. This
is a variation of the *maximum sharing problem (MS)* that finds important
applications in the design of molecular quantum-dot cellular automata
(QCA) circuits and physical synthesis in VLSI. It also generalizes the
maximum weight node-disjoint path cover problem. We show that MSS
is NP-complete, present a polynomial-time $\frac{5}{3}$-approximation algorithm,
and show that it cannot be approximated with a factor better than $\frac{740}{739}$
unless $P = NP$.

1 Introduction

Let $G = (U, V; E)$ be an undirected bipartite graph with *upper nodes* U and
lower nodes V. An upper node $u \in U$ forms a *sharing* with two distinct lower

---

* This work was supported in part by a grant from the Shanghai Key Laboratory of
Intelligent Information Processing, Fudan University, Shanghai, China. The order
of authors follows the international standard of alphabetic order of the last name.
In China, where first-authorship is a particularly important aspect of a publication,
the order of authors should be Zhiyi Xie, Jian Li, Hong Zhu, Danny Z. Chen, and
Rudolf Fleischer.
** The research of this author was supported in part by the US National Science Foun-
dation under Grant CCF-0515203. This work was partially done while the author
was visiting the Shanghai Key Laboratory of Intelligent Information Processing at
Fudan University, China.
*** The work described in this paper was partially supported by a grant from the
National Natural Science Fund China (grant no. 60573025).
† The work described in this paper was partially supported by a grant from the
National Natural Science Fund China (grants #60496321 and #60573025) and the
Shanghai Science and Technology Development Fund (grant #03JC14014).

T. Asano (Ed.): ISAAC 2006, LNCS 4288, pp. 547–556, 2006.
© Springer-Verlag Berlin Heidelberg 2006

Fig. 1. The MSS problem: A non-optimal maximal solution (left), and an optimal solution (right)

nodes $v_1, v_2 \in V$ if (u, v_1) and (u, v_2) are both edges in E. In the *maximum simple sharing problem (MSS)*, we want to cover the maximum number of upper nodes by sharings such that the edges of the sharings form a set of node-disjoint simple paths in G, where every path has both endpoints in V. See Figure 1 for an example.

MSS is a variant of the *maximum sharing problem (MS)* where a node in U may be involved in multiple sharings (i.e., the paths formed by the sharings may be non-simple and may overlap in the nodes of U). In a companion paper [10] we give a 1.5-approximation for MS. Unfortunately, the techniques used for that result do not carry over to the restricted variant MSS.

MS can help us to solve the *node-duplication based crossing elimination problem (NDCE)* [1,4]. In a two-layered (bipartite) graph, we want to duplicate as few nodes as possible in the first layer such that afterwards all connections between the two layers can be realized crossing-free (after a suitable permutation of the nodes in both layers). Intuitively, each sharing in an MS solution tells us how to avoid a node duplication. MSS is equivalent to *restricted NDCE* where duplicated nodes can only have a single neighbor in V. Variants of NDCE play a key role in the design of molecular quantum-dot cellular automata (QCA) circuits [1,12,15] and physical synthesis [3] which is similar to the key role played by the well studied *crossing minimization problem* [9,11,14] in the design of traditional VLSI circuit layouts.

MSS also generalizes the NP-hard *maximum weight node-disjoint path cover problem (PC)* where we want to find in an undirected graph a set of node-disjoint paths maximizing the number (or total weight) of the edges used by the paths. It is easy to see that PC is equivalent to MSS when all nodes in U have degree two (V and U correspond to the nodes and edges of the PC instance, respectively). PC is equivalent to the $(1, 2)$-TSP problem in the following sense. An approximation ratio of γ for one problem yields an approximation ratio of $\frac{1}{2-\gamma}$ for the other [13] (note that we adapted their formula for the approximation ratio to our different definition of approximation ratio). Since $(1, 2)$-TSP can be approximated with a factor of $\frac{8}{7}$ [2], PC, and thus the case of MSS where all nodes in U have degree two, can be approximated with a factor of $\frac{7}{6}$. On the other hand, it is NP-hard to approximate $(1, 2)$-TSP better than with a factor of $\frac{741}{740}$ [7]. Thus, MSS cannot be approximated with a factor better than $\frac{740}{739}$ unless $P = NP$.

An MSS solution is *maximal* if it cannot be enlarged by extending any path from its endpoints without destroying the current solution; otherwise, it is *extendible*. Note that we can enlarge any given solution to a maximal solution in polynomial time. The greedy algorithm that always chooses an unused node in V and arbitrarily extends a maximal node-disjoint path in both endpoints obviously constructs a maximal solution. Since a sharing touches exactly three nodes, each sharing in a maximal solution can only block three sharings in an optimal solution. Thus, any maximal solution is a 3-approximation for MSS.

Our main contribution is to show that MSS can be approximated with a factor of $\frac{5}{3}$. Our algorithm is based on a relaxation of the path constraint to allow solutions to contain node-disjoint simple cycles as well as paths, still maximizing the number of nodes of U covered by the paths and cycles. We call this relaxed version the *cyclic maximum simple sharing problem (CMSS)*. While MSS is NP-hard, CMSS can be solved optimally in polynomial time as a *maximum weight perfect matching problem (MWPM)* [5]. A similar phenomenon occurs in the 2-matching relaxation of TSP [8] which can be computed in polynomial time [6]. The difference is that a 2-matching is a pure cycle cover, whereas CMSS computes a mixture of cycles and paths. Since each cycle in a CMSS solution contains at least two sharings, we can get a 2-approximate MSS solution by removing one sharing from each cycle (thus breaking the cycle into a path). To obtain an approximation factor of $\frac{5}{3}$, we must carefully construct a set of node-disjoint simple paths from an optimal CMSS solution.

We assume in this paper that every node $v \in V$ has degree at least two. In MS we can get rid of degree-one nodes by adding a parallel edge to the one edge connecting the node, giving rise to the so-called q-MS problem in [10]. This approach does not work for MSS. Instead, we must duplicate the upper node adjacent to a degree-one lower node (we can w.l.o.g. assume that there is only one degree-one neighbor) together with all its adjacent edges. Then, maximizing the sharings is still equivalent to minimizing the node duplications.

The rest of this paper is organized as follows. First, we show in Section 2 that MSS is equivalent to restricted NDCE. In Section 3 we show how to solve CMSS in polynomial time. In Section 4 we then show how to transform an optimal CMSS solution into a $\frac{5}{3}$-approximation for MSS.

2   MSS and Restricted NDCE

The input to NDCE is a bipartite graph $G = (U, V; E)$. We want to duplicate as few nodes of U as possible to achieve a crossing-free drawing of G with the nodes U (and their copies) drawn (in some suitable order) along a line (the upper layer) and the nodes V drawn along another parallel line (the lower layer). In restricted NDCE the copies of nodes in U can only have a single neighbor in V. The following theorem shows that minimizing the number of duplications is equivalent to maximizing the number of simple sharings.

Theorem 1. *Given a bipartite graph $G = (U, V; E)$ in which every node has degree at least two, there is a solution of the MSS problem containing m simple sharings if and only if we can duplicate $|E| - |U| - m$ nodes of U to eliminate all wire crossings.*

Proof. Denote the layout of the circuit without wire crossings by $G' = (U', V'; E')$. U' consists of $|U|$ original nodes and the newly duplicated nodes. One way to achieve crossing-free wires is to duplicate $|E| - |U|$ nodes, i.e., every edge of E has a distinct endpoint in U'. To reduce the number of node duplications we observe that the original nodes (not the duplicated nodes) in U' can connect to more than one node, thus reducing the number of duplications.

Consider a permutation of the nodes in V', and let v_i and v_{i+1} be two consecutive nodes. It is easy to see that v_i and v_{i+1} can have at most one common neighbor in U'; otherwise there would be some wire crossings. It can also be seen that in G' the degree of a node in U' cannot be bigger than two; otherwise edge crossings cannot be avoided because we cannot duplicate nodes in V.

In G, if there are m simple sharings, then we can arrange m pairs of nodes in V consecutively so that each pair of nodes shares a common neighbor in U, thus reducing the duplication number by m. The other direction can be proved by a similar argument.   □

3   The Cyclic Maximum Simple Sharing Problem (CMSS)

The *cyclic maximum simple sharing problem (CMSS)* is defined as follows. Given a bipartite graph $G = (U, V; E)$, find a set $\mathcal{C}$ of node-disjoint simple cycles and simple paths in G such that every path begins at a node of V and ends at another node of V, maximizing the number of nodes in U covered by $\mathcal{C}$ (i.e., maximizing the number of sharings). Since any MSS solution is also a CMSS solution, the optimal objective function value of MSS is upper-bounded by the optimal CMSS value.

We now show how to solve CMSS by reducing it to the *maximum weight perfect matching problem (MWPM)* on undirected graphs which can be solved optimally in polynomial time [5]. Given a bipartite graph $G = (U, V; E)$, we construct an undirected graph H as follows. We want to represent every node and every edge of G by a pair of adjacent nodes in H. If a node or an edge in G is not used by any sharing, then the corresponding paired nodes in H are matched by their connecting edge. Otherwise, they are matched by other edges.

Figure 2 shows an example of the construction. For each node $v \in U \cup V$, we add to H two nodes $v^{(1)}$ and $v^{(2)}$ connected by an edge of weight zero. Similarly, for each edge $e \in E$, we add to H two nodes $e^{(1)}$ and $e^{(2)}$ connected by an edge of weight zero. In addition, for each edge $e = (u, v)$, with $u \in U$ and $v \in V$, we add the four edges $(u^{(1)}, e^{(1)})$, $(u^{(2)}, e^{(1)})$, $(v^{(1)}, e^{(2)})$, and $(v^{(2)}, e^{(2)})$, where the first edge has weight one and the other three edges have weight zero. Finally, for any two nodes v_1 and v_2 in V, we add an edge $(v_1^{(2)}, v_2^{(2)})$ of weight zero to H. It is easy to see that we can construct H in time $O(|E| + |V|^2)$.

Theorem 2. *G has a CMSS solution with k sharings if and only if H has a perfect matching of weight k.*

Proof. Figure 2 illustrates the proof. We prove the "only if" direction first. Given a set $\mathcal{C}$ of node-disjoint simple paths and cycles in G with k sharings, we construct a perfect matching M of weight k in H, as follows. We treat $\mathcal{C}$ as a subgraph

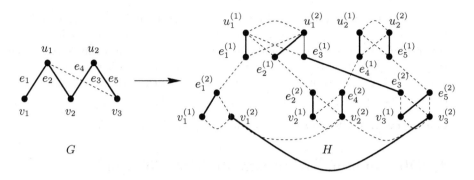

Fig. 2. Illustrating the proof of Theorem 2: The equivalence between CMSS in G and MWPM in H

of G. For each node v of G not covered by $\mathcal{C}$ we add the edge $(v^{(1)}, v^{(2)})$ to M. Similarly, we add the edge $(e^{(1)}, e^{(2)})$ to M for each edge $e \in E$ not in $\mathcal{C}$.

We now classify the nodes of $\mathcal{C}$ into three types: (i) upper nodes, (ii) lower nodes of degree two in $\mathcal{C}$, and (iii) lower nodes of degree one in $\mathcal{C}$. All nodes of type (i) have degree two in $\mathcal{C}$, and the number of nodes of type (iii) is even.

If $u \in U$ is of type (i), let e_1 and e_2 be the two edges adjacent to u in $\mathcal{C}$. We add the two edges $(u^{(1)}, e_1^{(1)})$ and $(u^{(2)}, e_2^{(1)})$ to M, increasing the weight of M by one. If $v \in V$ is of type (ii), let f_1 and f_2 be the two edges adjacent to v in $\mathcal{C}$. We add the two edges $(v^{(1)}, f_1^{(2)})$ and $(v^{(2)}, f_2^{(2)})$ to M. If $w \in V$ is of type (iii), let g be the edge adjacent to w in $\mathcal{C}$. We add the edge $(w^{(1)}, g^{(2)})$ to M.

All these operations are possible because a node or an edge in G can appear at most once in $\mathcal{C}$. Now, all nodes in H are matched except those of the form $w^{(2)}$ corresponding to a lower node $w \in V$ of type (iii). Because there is an even number of such nodes and they are all pairwise connected, we can arbitrarily match them. Now we have a perfect matching M in H. The weight of M is k, the number of nodes of type (i).

Next, we prove the "if" direction. Given a perfect matching M of weight k in H, we construct a CMSS solution in G, as follows. We call a node $v \in U \cup V$ *used* if the corresponding edge in H, $(v^{(1)}, v^{(2)})$, does not belong to M. Similarly, we call an edge $e \in E$ *used edge* if the corresponding edge in H, $(e^{(1)}, e^{(2)})$, does not belong to M.

Let $e = (u, v)$ be a used edge, where $u \in U$ and $v \in V$. Since $(e^{(1)}, e^{(2)})$ is not in M, $e^{(1)}$ must be matched either with $u^{(1)}$ or $u^{(2)}$, and $e^{(2)}$ must be matched either with $v^{(1)}$ or $v^{(2)}$. Thus, both u and v are used nodes. On the other hand, let $u \in U$ be a used upper node. Since $(u^{(1)}, u^{(2)})$ is not in M, $u^{(1)}$ must be matched with some node $e_1^{(1)}$ and $u^{(2)}$ must be matched with some node $e_2^{(1)}$, where e_1 and e_2 are two edges in E adjacent to u. Thus, both e_1 and e_2 are used edges and there are no other edges of E corresponding to used edges adjacent to u in G. Only the used upper nodes contribute one to the weight of M. Similarly, each used lower node $v \in V$ must be adjacent to one or two used edges in E. In summary, every used edge connects two used nodes, every used upper node is adjacent to exactly two used edges, and every used lower node is adjacent to either one or two used edges. Thus, all used nodes and used edges form a subgraph C of G consisting of node-disjoint simple cycles and simple paths such that every path begins at a node of V and ends at another node of V. The number of used nodes in U (i.e., the number of sharings contained in C) equals the weight k of M.   □

Corollary 3. *CMSS can be solved optimally in polynomial time.*   □

4   Obtaining a $\frac{5}{3}$-Approximate MSS Solution

Given a bipartite graph $G = (U, V; E)$, let S denote a (not necessarily optimal) CMSS solution, i.e., S is a subgraph of G. We first classify the lower nodes V into three types: (i) *white nodes*, which are not covered by S, (ii) *gray nodes*, which have degree one in S (i.e., the endpoints of the paths in S), and (iii) *black nodes*, which have degree two in S (i.e., the lower nodes lying in the interior of a path or on a cycle in S). Nodes on a cycle in S are also called *cycle nodes*. Cycle nodes are always black. Note that the color of a lower node depends on the subgraph S and may vary while the subgraph S changes.

Let C be a cycle in S. An edge not belonging to C but connected to an upper node in C is a *short tail* of C. A *long tail* is a chain of two edges not belonging to C starting at a lower node of C with the middle (upper) node not in S (i.e., a long tail is a sharing). We often do not distinguish between short and long tails, just calling them *tails*. A tail of C has the color of its endpoint not in C. Note that the edges of a tail never belong to S. See Figure 3 for an example.

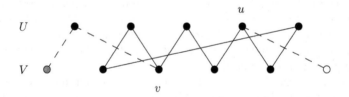

Fig. 3. A gray long tail at v and a white short tail at u

Lemma 4. *Let C be a cycle in S.*

(a) *If C has a white tail, we can break C into a path with the same number of sharings.*
(b) *If C has a gray tail ending at an endpoint of a path D in S, then we can break C by merging it with D into a single path with the same number of sharings.*
(c) *If C has a tail ending at a node of another cycle D in S, then we can merge C and D into a single path at the cost of losing one sharing.*

Proof. See Fig. 4.   □

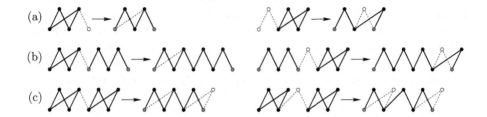

Fig. 4. Three ways of breaking a cycle, proving Lemma 4

We now present our approximation algorithm. For the input graph $G = (U, V; E)$, let $OPT^\star$ denote an optimal CMSS solution. We apply the following algorithm to $OPT^\star$.

Cycle-breaking Algorithm

Step 0:   Let $S = OPT^\star$.

Step 1:   Repeatedly pick a cycle C in S with a white or gray tail and break C into a path as in Lemma 4 (a) or (b), until no such cycle exists. Then go to Step 2.

Step 2:   Pick a pair of cycles C and D such that C has a tail ending at a cycle node in D, merge C and D into a path P as in Lemma 4 (c), and then go to Step 3. Go to Step 4 if no such pair exists.

Step 3:   Let v_1 and v_2 denote the two (gray) endpoints of P. Perform the following two substeps:

    Step 3(a):   Repeatedly pick a cycle C that has a tail ending at v_1 or v_2, and break C as in Lemma 4 (b). Then go to Step 3(b).

    Step 3(b):   For the two endpoints v_1 and v_2 of P obtained in Step 3(a), if there is an upper node $u \in U$ not in S and both (v_1, u) and (v_2, u) are edges in E, then add the sharing (v_1, u, v_2) to S, closing a cycle. Then go to Step 2.

Step 4:   Break all remaining cycles in S by arbitrarily removing one sharing from each cycle.

The output from the algorithm above is our approximate MSS solution. It is easy to see that the algorithm takes only polynomial time. Note that Steps 2 and 3 are iterated at most $\lfloor \frac{|U|}{2} \rfloor$ times since each iteration merges at least two cycles into either one path or one cycle. Moreover, if at the end of an iteration of Steps 2 and 3 a path is generated, then it will stay as a path from that point on.

Let $SOL^\star$ denote the CMSS solution before we begin with Step 4. Furthermore, let $|OPT^\star|$ and $|SOL^\star|$ denote the number of sharings contained in $OPT^\star$ and $SOL^\star$, respectively.

Lemma 5. *Let p be the number of paths that are generated and added to $SOL^\star$ in Steps 2 and 3. Then $|OPT^\star| = |SOL^\star| + p$.*

Proof. Let $s(S)$ denote the number of sharings in S during the algorithm. By Lemma 4 (a) and (b), we do not lose any sharings in Step 1. Thus, $s(S) = |OPT^\star|$ at the end of Step 1. For each iteration of Steps 2 and 3, by Lemma 4 (b) and (c), we lose one sharing in Step 2, but we do not lose any sharing in Step 3(a). Therefore, if the algorithm does not find a suitable upper node in Step 3(b), then one path is generated that will stay as a path from that point on and we lose one sharing. On the other hand, if the algorithm forms a cycle in Step 3(b), then no path is generated and no sharing is lost. Hence, the number of paths generated is equal to the number of sharings lost in Steps 2 and 3, which is exactly p. Therefore, $|OPT^\star| = |SOL^\star| + p$. □

Lemma 6. *Let (v_1, u, v_2) be a sharing in G such that the upper node u is not covered by a path in $SOL^\star$.*

(a) Then at least one of v_1 and v_2 is black in $SOL^\star$.
(b) If v_1 is a cycle node in $SOL^\star$, then v_2 is also black.
(c) If v_1 and v_2 are both cycle nodes, then they belong to the same cycle in $SOL^\star$.

Proof. Part (c) follows immediately from the termination codition of Step 2. The other two parts we prove by induction on the number of iterations of Steps 2 and 3. Note that $|S| = |OPT^\star|$ at the end of Step 1. The termination condition of Step 1 implies part (b) at the end of Step 1, and part (a) if u is covered by a cycle in S at that time. If u is not covered by S and both v_1 and v_2 were not black after Step 1, we could add the sharing (v_1, u, v_2) to S and get a CMSS solution better than $OPT^\star$, which is impossible.

In Steps 2 and 3 we do not destroy a path, and we do not create new black nodes or cycle nodes. However, it may happen that we break a cycle into a path, thus turning two black nodes into gray nodes (the endpoints of the path). If a tail ends at one of these gray nodes, Step 3(a) does not terminate. This implies that part (b) holds after each iteration of Steps 2 and 3.

To prove part (a), assume there is a sharing (v_1, u, v_2) with u not covered by a path at the end of an iteration of Steps 2 and 3. Since part (a) holds at the beginning of the iteration, at least one of the two lower nodes must have been black at that time. Since only cycle nodes can change their color during an iteration, it was a cycle node. By part (b), the other node must also have been black before the iteration. Since both nodes became gray, they are the two endpoints of the path created in the iteration. But then we would add the sharing (v_1, u, v_2) to S in Step 3(b), making both nodes black again. □

Now we have all the ingredients to prove our main result. Let SOL denote the final MSS solution obtained by the "Cycle-breaking Algorithm" on G.

Theorem 7. *SOL is a $\frac{5}{3}$-approximate MSS solution.*

Proof. We partition the paths in SOL into three sets: (i) SOL_1, the paths that exist right after Step 1, (ii) SOL_2, the paths created in Steps 2 and 3, and (iii) SOL_4, the paths created in Step 4. We denote the number of sharings in SOL_i

by s_i, for $i = 1, 2, 4$, and we denote the number of paths in SOL_i by p_i. Each path in SOL_1 or SOL_2 is a path in $SOL^\star$, and each path in SOL_4 corresponds to a cycle in $SOL^\star$ with the same set of lower nodes.

Let OPT denote an optimal MSS solution. We partition the sharings in OPT into three disjoint subsets.   1) The set OPT_{1+2} of all sharings whose upper nodes are contained in the paths in SOL_1 or SOL_2.   2) The set OPT_4 of all sharings whose upper nodes are not contained in any paths in SOL_1 or SOL_2, and whose two lower nodes are contained in some paths in SOL_4.   3) The set OPT_{other} of all other sharings.

For each sharing (v_1, u, v_2) in OPT_4, in $SOL^\star$ both v_1 and v_2 are cycle nodes and u is not contained in any path. Thus, by Lemma 6 (c), v_1 and v_2 are in the same cycle in $SOL^\star$, i.e., in the same path in SOL_4. Let $s(P)$ denote the number of sharings in a simple path P. For each path P in SOL_4 there are at most $s(P)$ sharings that are in OPT_4. Summing over all paths in SOL_4, we obtain $|OPT_4| \leq s_4$.

Similarly, for each sharing (v_1, u, v_2) in OPT_{other}, u is not in any path in $SOL^\star$. Thus, if v_1 (or v_2) is a cycle node in $SOL^\star$, then by Lemma 6 (b), v_2 (or v_1) is a black node in a path in $SOL^\star$. On the other hand, if neither v_1 nor v_2 is a cycle node in $SOL^\star$, then by Lemma 6 (a) at least one of v_1 and v_2 is a black node on a path in $SOL^\star$. Hence, in either case, the sharing (v_1, u, v_2) has at least one black node on a path in $SOL^\star$, i.e., on a path in SOL_1 or SOL_2. Since a lower node can appear in at most two sharings in OPT, we have $|OPT_{\text{other}}| \leq 2 \cdot \#(\text{black nodes in a path in } SOL_1 \text{ or } SOL_2) = 2(s_1 - p_1 + s_2 - p_2)$.

Since the number of sharings in OPT_{1+2} cannot exceed the number of sharings in SOL_1 and SOL_2, we have $|OPT_{1+2}| \leq s_1 + s_2$. Altogether, we have $|OPT| = |OPT_{1+2}| + |OPT_4| + |OPT_{\text{other}}| \leq s_1 + s_2 + s_3 + 2(s_1 - p_1 + s_2 - p_2) = |SOL| + 2(s_1 - p_1 + s_2 - p_2)$.

Note that we lost p_4 sharings in Step 4 of the algorithm. Moreover, by Lemma 5, $|OPT^\star| = |SOL^\star| + p_2$. Thus, $|OPT| \leq |OPT^\star| = |SOL| + p_2 + p_4$ and therefore $|OPT| + 2 \cdot |OPT| \leq |SOL| + 2(s_1 - p_1 + s_2 - p_2) + 2(|SOL| + p_2 + p_4) = 3 \cdot |SOL| + 2(s_1 + s_2 + p_4) - 2p_1$. Since $p_4 \leq s_4$ and $p_1 \geq 0$, this implies $3 \cdot |OPT| \leq 3 \cdot |SOL| + 2(s_1 + s_2 + s_4) = 5 \cdot |SOL|$.   □

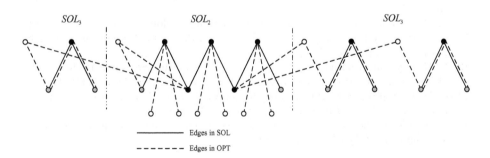

Fig. 5. An example showing that our approximation ratio $\frac{5}{3}$ is tight

The $\frac{5}{3}$ approximation ratio of our algorithm is tight, as shown by the example in Fig. 5.

References

1. D. A. Antonelli, D. Z. Chen, T. J. Dysart, X. S. Hu, A. B. Khang, P. M. Kogge, R. C. Murphy, and M. T. Niemier. Quantum-dot cellular automata (QCA) circuit partitioning: problem modeling and solutions. *Proc. 41st ACM/IEEE Design Automation Conference (DAC)*, pp. 363–368, 2004.
2. P. Berman and M. Karpinski. $\frac{8}{7}$-approximation algorithm for $(1,2)$-TSP. *Proc. 17th Annual ACM-SIAM Symp. on Discrete Algorithms (SODA'06)*, pp. 641–648, 2006.
3. A. Cao and C.-K. Koh. Non-crossing OBDDs for mapping to regular circuit structures. *Proc. IEEE International Conference on Computer Design*, pp. 338–343, 2003.
4. A. Chaudhary, D. Z. Chen, X. S. Hu, M. T. Niemier, R. Ravinchandran, and K. M. Whitton. Eliminating wire crossings for molecular quantum-dot cellular automata implementation. *Proc. IEEE/ACM International Conference on Computer-Aided Design*, pp. 565–571, 2005.
5. W. Cook and A. Rohe. Computing minimum-weight perfect matchings. *INFORMS J. on Computing*, 11(2):138–148, 1999.
6. J. Edmonds. Maximum matching and a polyhedron with 0,1-nodes. *J. Res. Nat. Bur. Stand. B*, 69:125–130, 1965.
7. L. Engebretsen and M. Karpinski. TSP with bounded metrics. *Journal of Computer and System Sciences*, 72(4):509–546, 2006.
8. S. R. Kosaraju, J. K. Park, and C. Stein. Long tours and short superstrings. *Proc. 35th Annual Symp. on Foundations of Computer Science (FOCS'94)*, pp. 166–177, 1994.
9. T. Lengauer. *Combinatorial Algorithms for Integrated Circuit Layout*. Wiley, New York, 1990.
10. J. Li, A. Chaudhary, D. Z. Chen, R. Fleischer, X. S. Hu, M. T. Niemier, Z. Xie, and H. Zhu. Approximating the Maximum Sharing Problem. Submitted for publication, 2006.
11. M. Marek-Sadowska and M. Sarrafzadeh. The crossing distribution problem, *IEEE Trans. on Computer-Aided Design of Integrated Circuits and Systems*, 14(4):423–433, 1995.
12. M. T. Niemier and P. M. Kogge. Exploring and exploiting wire-level pipelining in emerging technologies, *Proc. 28th Annual International Symp. on Computer Architecture*, pp. 166–177, 2001.
13. C. H. Papadimitriou and M. Yannakakis. The Traveling Salesman Problem with distances one and two, *Mathematics of Operations Research*, 18(1):1–11, 1993.
14. C. D. Thompson. Area-time complexity for VLSI, *Proc. 11th Annual ACM Symp. on Theory of Computing (STOC'79)*, pp. 81–88, 1979.
15. P. D. Tougaw and C. S. Lent. Logical devices implemented using quantum cellular automata, *J. of App. Phys.*, 75:1818, 1994.

Approximation Scheme for Lowest Outdegree Orientation and Graph Density Measures

Łukasz Kowalik[1,2,*]

[1] Institute of Informatics, Warsaw University, Warsaw, Poland
[2] Max-Planck-Institute für Informatik, Saarbrücken, Germany
kowalik@mimuw.edu.pl

Abstract. We deal with the problem of finding such an orientation of a given graph that the largest number of edges leaving a vertex (called the outdegree of the orientation) is small.

For any $\varepsilon \in (0,1)$ we show an $\tilde{O}(|E(G)|/\varepsilon)$ time algorithm[1] which finds an orientation of an input graph G with outdegree at most $\lceil (1 + \varepsilon)d^* \rceil$, where d^* is the maximum density of a subgraph of G. It is known that the optimal value of orientation outdegree is $\lceil d^* \rceil$.

Our algorithm has applications in constructing labeling schemes, introduced by Kannan *et al.* in [18] and in approximating such graph density measures as arboricity, pseudoarboricity and maximum density. Our results improve over the previous, 2-approximation algorithms by Aichholzer *et al.* [1] (for orientation / pseudoarboricity), by Arikati *et al.* [3] (for arboricity) and by Charikar [5] (for maximum density).

1 Introduction

In this paper we deal with approximating lowest outdegree orientation, pseudoarboricity, arboricity and maximum density. Let us define these notions as they are not so widely used.

Let $G = (V, E)$ be a graph. An *orientation* of G is a digraph $\vec{G} = (V, \vec{E})$ that is obtained from G by replacing every undirected edge uv by an arc, i.e., (u, v) or (v, u). The *outdegree* of an orientation is the largest of its vertices' outdegrees. In this paper we focus on the problem of finding for a given graph its orientation with minimum outdegree. We will call it a *lowest outdegree orientation*. This problem is closely related to computing pseudoarboricity and maximum density of graphs.

Density of a graph $G = (V, E)$, denoted by $d(G)$, is defined as $d(G) = |E|/|V|$, i.e., it is half of its average degree. In the *Densest Subgraph Problem*, given graph G one has to find its subgraph G^* such that any nonempty subgraph H of G satisfies $d(H) \leq d(G^*)$. The number $d(G^*)$ will be called *maximum density* of graph G, and we will denote it by $d^*(G)$. As it was shown by Charikar [5], the linear program for Densest Subgraph Problem is dual to the relaxation of the integer program for finding the lowest outdegree orientation. Moreover, it follows from a theorem by Frank and Gyárfás [11] that $\lceil d^*(G) \rceil$ equals the outdegree of the lowest outdegree orientation of G.

---

* Supported in part by KBN grant 4T11C04425.
[1] The $\tilde{O}(\cdot)$ notation ignores logarithmic factors.

T. Asano (Ed.): ISAAC 2006, LNCS 4288, pp. 557–566, 2006.
© Springer-Verlag Berlin Heidelberg 2006

A *pseudotree* is a connected graph containing at most one cycle. *Pseudoforest* is a union of vertex disjoint pseudotrees. *Pseudoarboricity* of graph G, denoted as $P(G)$, is the smallest number of pseudoforests needed to cover all edges of G. As it was noticed by Picard and Queyranne [22], $P(G) = \lceil d^*(G) \rceil$, which combined with the theorem of Frank and Gyárfás implies that pseudoarboricity equals the outdegree of the lowest outdegree orientation. This can be also easily proved directly (see Section 2).

Arboricity of graph G, denoted as $\mathrm{arb}(G)$, is the smallest number of forests needed to cover all edges of G. A classic theorem by Nash-Williams [21] says that arboricity is equal to $\max_J \lceil |E(J)|/(|V(J)| - 1) \rceil$ where J is any subgraph of G with $|V(J)| \geq 2$ vertices and $|E(J)|$ edges. Using this fact it is easy to show (see [22]) that $P(G) \leq \mathrm{arb}(G) \leq P(G) + 1$.

Applications. Arboricity is the most often used measure of graph sparsity. Complexity of many graph algorithms depends heavily on the arboricity of the input graph – see e.g. [6,4,8].

Kannan et al. [18] noticed that for any n-vertex graph of arboricity k one can label its vertices using at most $(k + 1) \log n$ bits for each label in such a way that adjacency of any pair of vertices can be verified using merely their labels. They call it a $(k + 1)$-*labeling scheme*. It is achieved as follows: (1) number the vertices from 1 to n, (2) find a partition of the graph into k forests, (3) in each tree in each forest choose a root, and (4) assign each vertex a label containing its number and the numbers of its parents in the at most k trees it belongs to. Then to test adjacency of vertices u and v it suffices to check whether u is the parent of v in some tree or vice versa.

Chrobak and Eppstein [7] observed that in order to get the labeling schemes one can use orientations instead the partition into forests. Then each vertex v stores in its label the numbers of endpoints of the arcs leaving v. As for any graph G, $P(G) \leq \mathrm{arb}(G)$ this is a little bit more efficient approach. Then the problem of building a $(P(G) + 1)$-labeling scheme reduces to the problem of finding a lowest degree orientation.

It should be noted that low outdegree orientations are used as a handy tool in many algorithms, see e.g., [2,19,9].

Related Work. Throughout the whole paper m and n denote the number of edges and vertices of the input graph, respectively.

The problem of computing pseudoarboricity and the related decomposition was first raised by Picard and Queyranne [22] who applied network flows and obtained $\mathcal{O}(nm \log^3 n)$ algorithm by using the maximum flow algorithm by Galil and Naamad [14]. It was improved by Gallo, Grigoriadis and Tarjan [15] to $\mathcal{O}(nm \log(n^2/m))$ by using parametric maximum flow. Next Gabow and Westermann [13] applied their matroid partitioning algorithm for the k-pseudoforest problem – the problem of finding in a given graph k edge-disjoint forests containing as many edges as possible. Their algorithm works in $\mathcal{O}(\min\{(kn')^{3/2}, k(n')^{5/3}\})$ time where $n' = \min\{n, m/k\}$. As pseudoarboricity is at least m/n it gives an $\mathcal{O}(m \min\{m^{1/2}, n^{2/3}\})$-time algorithm which verifies whether a given graph has pseudoarboricity at most k and if the answer is yes computes the relevant pseudoforest partition. Using binary search pseudoarboricity p can be computed in $\mathcal{O}(m \min\{m^{1/2}, n^{2/3}\} \log p)$ time. However, if the pseudoforest partition is not needed, they show also that this value can be found in $\mathcal{O}(m \min\{(m \log n)^{1/2}, (n \log n)^{2/3}\})$ time. For the related problem of finding arboricity and the relevant forest partition they describe an $\mathcal{O}(m^{3/2}\sqrt{\log n})$ algorithm.

Finally, Aichholzer et al. [1] claimed (without giving details) that one can solve the equivalent lowest outdegree orientation problem in $\mathcal{O}(m^{3/2} \log p)$ time by using Dinic's algorithm.

Since the problem seems to be closely related with network flows and matroid partitions it can be hard to get a near-linear algorithm for it. Hence one can consider approximation algorithms. Arikati et al. [3] showed a simple linear-time 2-approximation algorithm for computing arboricity and the corresponding partition into forests. Independently, Aichholzer et al. [1] showed a 2-approximation algorithm for the problem of finding lowest outdegree orientation (and hence also pseudoarboricity). In fact, these two algorithms are the same. Both can be viewed as finding for a given graph G its acyclic orientation of outdegree at most $2P(G)$.

Recently, Gabow [12] considered a related problem of orienting as many edges as possible subject to upper bounds on the indegree and outdegree of each vertex. He proves that the problem is MAXSNP-hard and shows a $3/4$-approximation algorithm.

For the densest subgraph problem the state of art is very similar to computing pseudoarboricity. A paper of Goldberg [16] contains a reduction to a network flow problem, which combined with the algorithm by Goldberg and Rao [17] gives an algorithm with time complexity $\tilde{\mathcal{O}}(m \min\{n^{2/3}, m^{1/2}\})$. On the other hand, Charikar [5] showed a simple linear-time 2-approximation algorithm.

Our Results. We show an algorithm which, given a number $\varepsilon > 0$ and a graph with maximum density d^*, finds a d-orientation so that $d \leq \lceil(1 + \varepsilon)d^*\rceil$. In other words, it is an approximation scheme with additional additive error (caused by rounding up) bounded by 1. For $0 < \varepsilon < 1$ the algorithm works in $\mathcal{O}(m \log n \max\{\log d^*, 1\}\varepsilon^{-1})$ time.

As $P(G) \leq \mathrm{arb}(G) \leq P(G) + 1$ and $P(G) = \lceil d^*(G) \rceil$ it is not surprising that our algorithm can be also used for efficient approximating arboricity and maximum density – for both these problems we get an approximation scheme with an additional small additive error (2 for arboricity, 1 for maximum density). In Section 2 we also note that finding a partition of edges of a graph into d pseudoforests is equivalent to the problem of finding an orientation with outdegree d. Thus our algorithms apply also to the pseudoforest partition problem.

In the particular case of *sparse graphs*, i.e., graphs of bounded arboricity, the running time of our algorithm is $\mathcal{O}((n \log n)/\varepsilon)$, as then $d^* = \mathcal{O}(1)$ and $m = \mathcal{O}(n)$. It is worth noting that for sparse graphs our algorithm can be used to efficiently find an orientation with outdegree $\lceil d^* + \delta \rceil$, for $\delta > 0$. Alternatively, we can use it for approximating arboricity (additive error $1 + \lceil \delta \rceil$) and maximum density (additive error $1 + \delta$). This can be done in $\mathcal{O}(m \log n \max\{\log d^*, 1\} \max\{\frac{d^*}{\delta}, 1\})$ time, which is $\mathcal{O}(n \log n \max\{\delta^{-1}, 1\})$ for sparse graphs. In particular, for sparse graphs this gives $\mathcal{O}(n \log n)$ time approximation algorithms with additive error 1 (for lowest outdegree orientation / pseudoarboricity), or 2 (for arboricity and maximum density).

The idea of our approximation algorithms is very simple. We start Dinic's maximum flow algorithm in some network which depends on the input graph and some parameter d. We stop it when augmenting paths grow too long. If d is greater, but not very close to the maximum density d^* we show that the augmenting paths will never grow too long and then we obtain a d-orientation. Otherwise we know d is too close to d^* — closer than we need. In order to find the smallest value of d such that the augmenting paths are always short we use binary search.

2   Preliminaries

We say that $\overrightarrow{G}$ is a d-orientation when vertex outdegrees in $\overrightarrow{G}$ do not exceed d. We assume that the reader is familiar with basic concepts concerning network flow algorithms. For details see e.g. [20]. Let us recall here only some basic notions.

Let $G = (V, E)$ be a directed graph with two special vertices s (called source) and t (called sink). Each arc of G is assigned a number called *capacity*. More precisely, *capacity* is a function $c : V^2 \to \mathbb{R}_{\geq 0}$ such that for $(v, w) \notin E$, $c(v, w) = 0$. Graph G with the capacity function c is called a *network*. *Flow* in a network G is any function $f : V^2 \to \mathbb{R}$ such that for any $u, v \in V$ (1) $f(u, v) \leq c(u, v)$, (2) $f(u, v) = -f(v, u)$, (3) if $v \neq s, t$, $\sum_{x \in V} f(v, x) = 0$. The *value* of flow f, denoted by $|f|$, is the value of $\sum_{x \in V} f(s, x)$. A *maximum flow* is a flow with largest possible value. For network G and flow f the *residual capacity* is a function $c_f : V^2 \to \mathbb{R}_{\geq 0}$ such that $c_f(u, v) = c(u, v) - f(u, v)$. The graph with vertex set V containing edge (u, v) if and only if $c_f(u, v) > 0$ is denoted as G_f. Graph G_f with c_f as capacity function is called a *residual network*. An *augmenting path* is any path from s to t in the residual network. Edge (u, v) of graph G is called *augmented* when $f(u, v) = c(u, v)$.

Below we show an important relation between partitions into pseudoforests and orientations.

Proposition 1. *The problems of finding p-orientation and partition into p pseudoforests are equivalent, i.e., from a given p-orientation of some graph one can find a partition of edges of this graph into p pseudoforests and vice versa. Both conversions take time linear in the number of edges.*

Proof. Every pseudotree has a 1-orientation, as it suffices to remove any edge of the cycle (if there is one), choose one of its ends as the root, orient all edges toward the root and finally add the removed edge oriented from the root to the other endpoint. Thus given a decomposition of a graph into p pseudoforests we can find its p-orientation in linear time.

Conversely, consider a connected graph G with a 1-orientation. We will show that G is a pseudotree. G has at least $|V(G)| - 1$ edges since it is connected, and at most $|V(G)|$ edges since it has 1-orientation. If it has $|V(G)| - 1$ then it is a tree. If it has $|V(G)|$ edges it contains a cycle. After removing any edge of this cycle we get a connected graph G' with $|V(G')| - 1$ edges. Hence G' is a tree which implies that G has precisely one cycle. It follows that a graph with a 1-orientation is a pseudoforest.

Now, given a p-orientation, for each vertex we remove any of the edges leaving it. Then we obtain a $(p-1)$-orientation and the removed edges form a 1-orientation, which is a pseudoforest when we forget about edge orientations. After repeating this step p times we obtain the desired decomposition into p pseudoforests. The whole process also takes linear time.   □

The above proposition implies that finding pseudoarboricity and the corresponding partition of edges into pseudoforests is equivalent to finding the lowest degree orientation.

3   Reduction to a Flow Problem

Here we present a reduction of finding a d-orientation of a given graph (if it exists) to finding a maximum flow in some network. Other reductions are used in [22,1].

Let $G = (V, E)$ be a graph and let d be a positive integer. Let $\overrightarrow{G}$ be an arbitrary orientation of G (we will call it *the initial orientation*). We build a network $\tilde{G}^d = (\tilde{V}, \tilde{E})$ with capacity function c^d as follows. Set $\tilde{V}$ contains all vertices from V and two new vertices s (source) and t (sink). Set $\tilde{E}$ contains all edges from $E(\overrightarrow{G})$, each with capacity 1, an edge (s, v) with capacity outdeg$(v) - d$ for each vertex v with outdegree in $\overrightarrow{G}$ greater than d and an edge (v, t) with capacity $d -$ outdeg(v) for each vertex v with outdegree in $\overrightarrow{G}$ smaller than d. Let $\tilde{G}^d_f$ denote the residual network for flow f. Note the following proposition.

Proposition 2. *For any integral flow f in network $\tilde{G}^d$, the subgraph of the residual network $\tilde{G}^d_f$ induced by set V is an orientation of graph G.*   □

Let us denote the subgraph described above by $\overrightarrow{G}_f$. From now on we assume that flows in $\tilde{G}^d$ are integral.

Lemma 1. *Let f be any flow in network $\tilde{G}^d$. There is an edge (s, v) in $\tilde{G}^d_f$ if and only if* outdeg$_{\overrightarrow{G}_f}(v) > d$. *Also, there is an edge (v, t) in $\tilde{G}^d_f$ if and only if* outdeg$_{\overrightarrow{G}_f}(v) < d$.

Proof. Let $\overrightarrow{G}$ be the initial orientation of G and let v be an arbitrary vertex in V. Clearly, outdeg$_{\overrightarrow{G}_f}(v) =$ outdeg$_{\overrightarrow{G}}(v) + \sum_{w \in V} f(w, v)$. As f is a flow, $0 = \sum_{w \in \tilde{V}} f(w, v) = \sum_{w \in V} f(w, v) + f(s, v) + f(t, v)$. Hence,

$$\text{outdeg}_{\overrightarrow{G}_f}(v) = \text{outdeg}_{\overrightarrow{G}}(v) - f(s, v) + f(v, t). \tag{1}$$

Now, if $c(s, v) = c(v, t) = 0$ then $f(s, v) = f(v, t) = 0$ and we see that both (s, v) and (v, t) are not in $\tilde{G}^d_f$ and outdeg$_{\overrightarrow{G}_f}(v) = d$.

If $c(s, v) > 0$ then $c(v, t) = 0$ and further $f(v, t) = 0$. Hence, by (1), we have $c(s, v) - f(s, v) =$ outdeg$_{\overrightarrow{G}}(v) - d - f(s, v) =$ outdeg$_{\overrightarrow{G}_f}(v) - d$. Then $c(s, v) - f(s, v) > 0$ if and only if outdeg$_{\overrightarrow{G}_f}(v) - d > 0$, which is equivalent to the first part of the lemma. Also, since $c(v, t) = 0$ and $c(t, v) = 0$ we get $(v, t) \notin E(\tilde{G}^d_f)$. Moreover, $0 \le c(s, v) - f(s, v) =$ outdeg$_{\overrightarrow{G}_f}(v) - d$ which implies that the second part of the lemma also holds in this case. The case $c(s, v) < 0$ can be verified analogously.   □

Corollary 1. *There is an augmenting path $sv_1v_2 \ldots v_k t$ in the residual network $\tilde{G}^d_f$ iff there is a path $v_1v_2 \ldots v_k$ in $\overrightarrow{G}_f$ such that* outdeg$_{\overrightarrow{G}_f}(v_1) > d$, outdeg$_{\overrightarrow{G}_f}(v_k) < d$.

Let $c^d(s, V - s)$ denote the capacity of the $(\{s\}, V \setminus \{s\})$ cut, i.e., $c^d(s, V - s) = \sum_{v \in V} c^d(s, v)$.

Theorem 1. *Let G be a graph and let f be a maximum flow in network $\tilde{G}^d$. There exists a d-orientation of G if and only if $|f| = c^d(s, V - s)$. Moreover, when $|f| = c^d(s, V - s)$ then $\overrightarrow{G}_f$ is a d-orientation of G.*

Proof. Assume that there is a d-orientation of G and the maximum flow f in $\tilde{G}^d$ is smaller than $c^d(s, V - s)$. Then at least one edge leaving s, say (s, v) is not augmented. Then from Lemma 1 outdeg$_{\overrightarrow{G}_f}(v) > d$. Let $W \subseteq V$ denote the set of vertices reachable from v in $\overrightarrow{G}_f$. Since there is a d-orientation of G, graph $G[W]$ contains at most $d|W|$ edges. If W contained no vertex of outdegree smaller than d then graph $G[W]$ would contain more than $d|W|$ edges, which would be a contradiction. Hence W contains a vertex w such that outdeg$_{\overrightarrow{G}_f}(w) < d$ and by Corollary 1 there is an augmenting path which contradicts the maximality of flow f.

Conversely, if f is a flow in $\tilde{G}^d$ of value $c^d(s, V - s)$ there are no edges leaving s in the residual network and Lemma 1 implies that outdegrees in $\overrightarrow{G}_f$ do not exceed d. $\square$

In order to analyze our approximation algorithm we will use the following lemma. The lemma and its proof is analogous to Lemma 2 in [4] (however, our Lemma 2 implies Lemma 2 in [4] and not vice-versa, hence we include the proof for completeness).

Lemma 2. *Let $\overrightarrow{G}$ be a d-orientation of some n-vertex graph G of maximum density d^* and let $d > d^*$. Then for any vertex v the distance in $\overrightarrow{G}$ to a vertex with outdegree smaller than d does not exceed $\log_{d/d^*} n$.*

Proof. Let v be an arbitrary vertex and let k be the distance from v to a vertex with outdegree smaller than d. For every $i = 0, \ldots, k$ let V_i be the set of vertices at distance at most i from v. We will show by induction that for each $i = 0, \ldots, k$, $|V_i| \geq (\frac{d}{d^*})^i$. We see that this inequality holds for $i = 0$. For the induction step assume that $i < k$. Let E_{i+1} be the set of edges with both ends in V_{i+1}. We see that exactly $d|V_i|$ edges leave V_i. Since all these edges belong to E_{i+1} it gives us $|E_{i+1}| \geq d|V_i|$. As $\frac{|E_{i+1}|}{|V_{i+1}|} \leq d^*$ we get $|V_{i+1}| \geq \frac{d}{d^*}|V_i|$. After applying the induction hypothesis we get the desired inequality. Then since $|V_k| \leq n$ we get $(\frac{d}{d^*})^k \leq n$ which ends the proof. $\square$

As an immediate consequence of Corollary 1 and Lemma 2 we get the following corollary.

Corollary 2. *Let G be an n-vertex graph of maximum density d^* and let for some integer $d > d^*$, $\tilde{G}^d$ be the corresponding network with some flow f. If $\tilde{G}^d$ contains an augmenting path then it contains an augmenting path of length at most $2 + \log_{d/d^*} n$.*

4   Approximation Algorithm

Let us now briefly recall Dinic's algorithm. Details can be found in many textbooks, e.g. [20]. Dinic's algorithm begins with the empty flow f. It consists of a sequence of *phases*. In the beginning of each phase it builds a *layered network*, i.e., a subgraph of the residual network containing only edges of shortest paths from source s to sink t.

The goal of each phase is to find a *blocking flow* in the layered network, i.e., such flow that each s, t-path in the layered network contains an augmented edge. In the end of each phase the blocking flow is added to flow f.

Dinic's algorithm finds the blocking flow by finding a number of augmenting paths, each time sending maximal amount of flow through the path. To find such path it uses the following method. Start from the empty path. Let v be the end of the path p found so far. If $v = t$ an augmenting path is found. If there is an edge leaving v add it to path p (this step is called *advance*). Otherwise remove the last edge of path p from both the layered network and p (this step is called *retreat*).

It is known that Dinic's algorithm finds a blocking flow in unit capacity networks in linear time (see e.g. [10]). It is not surprising that it is similarly fast in network $\tilde{G}^d$, which is "almost unit capacity". For completeness, below we give a proof.

Proposition 3. *For any graph G with m edges the Dinic's algorithm finds a blocking flow in network $\tilde{G}^d$ in $\mathcal{O}(m)$ time.*

Proof. The number of advance steps after which the sink t is reached is equal to the number of augmenting paths found, which is bounded by the value of maximum flow, which in turn is bounded by m. The total number of other advance steps is bounded by the sum of relevant edge capacities, i.e., $\sum_v c^d(s, v) + \sum_{v,w \in V} c^d(v, w) \leq \sum_v \text{outdeg}(v) + \sum_{v,w \in V} c^d(v, w) = 2m$. The number of retreat steps is bounded by the number of edges. We see that the total number of advance and retreat steps is at most $4m$. □

Let us also recall another crucial property of Dinic's algorithm (see e.g. [20]):

Proposition 4. *After each phase of Dinic's algorithm the length of the shortest augmenting path increases.*

Now let us describe our main result, algorithm $\text{ORIENT}(\varepsilon)$ which finds orientation of a given graph with outdegree close to optimal.

Algorithm 4.1 $\text{TEST}(k,d)$

1: Build $\tilde{G}^d$
2: **while** $\text{dist}_{\tilde{G}^d_f}(s, t) \leq k$ **do**
3:     Run another phase of Dinic's algorithm
4: **if** $|f| = c^d(s, V - s)$ **then return** $\overrightarrow{G}_f$ **else return** "FAIL"

We will use a subroutine $\text{TEST}(k,d)$. It builds network $\tilde{G}^d$, and runs the Dinic's algorithm until it finishes (i.e. when there is no augmenting path) or the augmenting paths become longer than k. If the resulting flow has value $c^d(s, V - s)$, it returns an orientation $\overrightarrow{G}_f$. Otherwise it returns "FAIL" message. As an immediate consequence of propositions 3 and 4 we get the following proposition.

Proposition 5. *Algorithm* TEST(k,d) *works in* $\mathcal{O}(km)$ *time.*   □

Lemma 3. *Let G be a graph with maximum density d^* and let $d \geq \lceil (1 + \varepsilon)d^* \rceil$ for some $\varepsilon > 0$. Then* TEST($2 + \log_{1+\varepsilon} n$, d) *returns a d-orientation of G.*

Proof. As $\varepsilon > 0$ it follows that $d > d^*$ and by Corollary 2 if there is an augmenting path, there is an augmenting path of length at most $2 + \log_{d/d^*} n$, which is not greater than $2 + \log_{1+\varepsilon} n$. Hence the **while** loop is stopped when there is no augmenting path, i.e., $\mathrm{dist}_{\vec{G}_f^d}(s, t) = \infty$, which implies that a maximum flow f is found. As $d \geq \lceil (1 + \varepsilon)d^* \rceil \geq \lceil d^* \rceil = P(G)$, there exists a d-orientation of G, so by Theorem 1, $|f| = c^d(s, V - s)$ and $\vec{G}_f$ is a d-orientation. It establishes the proof.   □

Algorithm 4.2 ORIENT(ε)

1: $d_1 \leftarrow 0; d_2 \leftarrow 1$
2: **while** TEST $(2 + \log_{1+\varepsilon} n, d_2) = $ "FAIL" **do**
3:    $d_1 \leftarrow d_2; d_2 \leftarrow 2d_2$
4: **while** $d_1 < d_2$ **do**
5:    $d' = \lceil \frac{d_1 + d_2}{2} \rceil$
6:    **if** TEST $(2 + \log_{1+\varepsilon} n, d') = $ "FAIL" **then** $d_1 \leftarrow d'$ **else** $d_2 \leftarrow d'$
7: **return** the orientation returned by the last call of TEST

Algorithm ORIENT(ε) uses binary search to find an integer d such that TEST($2 + \log_{1+\varepsilon} n$, $d - 1$) returns "FAIL" message, while TEST($2 + \log_{1+\varepsilon} n$, d) does not. (Note that it may happen that $d < \lceil (1 + \varepsilon)d^* \rceil$). It returns the d-orientation returned by the relevant call of TEST. Now we state the main result of the paper.

Theorem 2. *Let G be any graph of maximum density d^*. For any $\varepsilon > 0$ algorithm* ORIENT(ε) *finds a d-orientation of G such that $d \leq \lceil (1 + \varepsilon)d^* \rceil$. Its time complexity is* $\mathcal{O}(m \log n \max\{\log d^*, 1\} \max\{\varepsilon^{-1}, 1\})$.

Proof. Correctness of the algorithm is an immediate consequence of Lemma 3. By Proposition 5 each call of TEST($2 + \log_{1+\varepsilon} n, d$) subroutine takes $\mathcal{O}(m \log_{1+\varepsilon} n) = \mathcal{O}(m(\log n)(\log(1 + \varepsilon))^{-1})$ time. By Taylor expansion, for $\varepsilon < 1$, $\ln(1 + \varepsilon) = \varepsilon + \mathcal{O}(\varepsilon^2)$. Hence each call of TEST routine in algorithm ORIENT takes time bounded by $\mathcal{O}(m(\log n) \max\{\varepsilon^{-1}, 1\})$. Theorem 2 implies that ORIENT(ε) makes at most $\mathcal{O}(\lceil \log \lceil (1 + \varepsilon)d^* \rceil \rceil) = \mathcal{O}(\max\{\log d^*, 1\})$ calls of subroutine TEST. Hence we get the claimed time bound.   □

4.1 Approximating Graph Density Measures

Using our algorithm one can approximate efficiently graph density measures. The details are given in the following theorem.

Theorem 3. *Let G be any graph of maximum density d^*. For any $\varepsilon > 0$ there are algorithms with time complexity $\mathcal{O}(m \log n \max\{\log d^*, 1\} \max\{\varepsilon^{-1}, 1\})$ for the following problems:*

(i) *(pseudoarboricity approximation) Finding a partition of G into $\tilde{d}$ pseudoforests so that $\tilde{d} \leq \lceil (1 + \varepsilon)d^* \rceil \leq (1 + \varepsilon)P(G) + 1$.*

(ii) *(arboricity approximation) finding a number $\tilde{a}$ such that there exists a partition of G into $\tilde{a}$ forests so that $\tilde{a} \leq (1 + \varepsilon)\mathrm{arb}(G) + 2$.*

(iii) *(densest subgraph approximation) finding a number $\tilde{d}^*$ such that G contains a subgraph of density at least $\tilde{d}^*$ so that $\tilde{d}^* \geq (1 - \varepsilon)d^* - 1$.*

Proof. Part (i) follows immediately from Proposition 1 and Theorem 2.

To construct the algorithm described in (ii) it suffices to find the number $\tilde{d}$ using part (i) and report $\tilde{a} = \tilde{d} + 1$. Since $P(G) \leq \mathrm{arb}(G)$, the claimed bound follows. The relevant partition into forests exists because $\tilde{a} \geq P(G) + 1 \geq \mathrm{arb}(G)$.

Similarly, for part (iii) we also apply algorithm from part (i), but using different value of ε, namely using $\varepsilon' = \varepsilon/(1 - \varepsilon)$. Since $\max\{(\varepsilon')^{-1}, 1\} = \max\{\varepsilon^{-1} - 1, 1\} \leq \max\{\varepsilon^{-1}, 1\}$, the algorithm works in the claimed time. Then we report $\tilde{d}^* = (\tilde{d} - 1)/(1 + \varepsilon')$. Since $\tilde{d} \leq \lceil (1 + \varepsilon')d^* \rceil$, we see that $\tilde{d}^* \leq d^*$, hence there is a subgraph of density at least $\tilde{d}^*$. Finally, because $\tilde{d} \geq P(G) \geq d^*$, we get $\tilde{d}^* \geq \frac{d^*}{1+\varepsilon'} - \frac{1}{1+\varepsilon'} = (1 - \varepsilon)d^* - (1 - \varepsilon) > (1 - \varepsilon)d^* - 1$.   □

4.2 Approximation with Additive Error

Now we observe that for sparse graphs our algorithm can be used to efficiently find an orientation with outdegree $\lceil d^* + \delta \rceil$, for $\delta > 0$. To this end one finds a d'-orientation, $d^* < d' < \frac{3}{2}d^*$ using algorithm ORIENT($\frac{3}{2}$). If $\delta \geq d'$ the algorithm stops and returns the d'-orientation found. Otherwise it calls algorithm ORIENT($\frac{\delta}{d'}$). Clearly the second call returns an orientation with outdegree $\lceil (1 + \frac{\delta}{d'})d^* \rceil \leq \lceil (1 + \frac{\delta}{d^*})d^* \rceil = \lceil d^* + \delta \rceil$. Time complexity is $\mathcal{O}(m \log n \max\{\log d^*, 1\} \max\{\frac{d^*}{\delta}, 1\})$, which for sparse graphs can be rewritten as $\mathcal{O}(n \log n \max\{\delta^{-1}, 1\})$. Similarly as in Theorem 3 we obtain also algorithms with the same time complexity for approximating pseudoarboricity (additive error $\lceil \delta \rceil$), arboricity (additive error $1 + \lceil \delta \rceil$), and maximum density (additive error $1 + \delta$).

5 Further Research

We showed how to efficiently approximate *the values* of arboricity and maximum density. It is very natural to ask for near-linear algorithms for finding the relevant decomposition into forests and the relevant dense subgraph. In the context of the first problem it is particularly interesting whether there is a fast algorithm which transforms a decomposition of a graph into d pseudoforests to a decomposition into $d + 1$ forests (or, if this is infeasible, then into $\alpha \cdot d$ forests, for some $\alpha < 2$).

References

1. O. Aichholzer, F. Aurenhammer, and G. Rote. Optimal graph orientation with storage applications. SFB-Report F003-51, SFB 'Optimierung und Kontrolle', TU Graz, Austria, 1995.
2. N. Alon, R. Yuster, and U. Zwick. Color-coding. *J. ACM*, 42(4):844–856, 1995.

3. S. R. Arikati, A. Maheshwari, and C. D. Zaroliagis. Efficient computation of implicit representations of sparse graphs. *Discrete Appl. Math.*, 78(1-3):1–16, 1997.
4. G. S. Brodal and R. Fagerberg. Dynamic representations of sparse graphs. In *Proc. 6th Int. Workshop on Algorithms and Data Structures (WADS'99)*, volume 1663 of *LNCS*, pages 342–351, 1999.
5. M. Charikar. Greedy approximation algorithms for finding dense components in a graph. In *Proc. 13th Int. Workshop on Approximation Algorithms for Combinatorial Optimization (APPROX'00)*, volume 1913 of *LNCS*, pages 84–95, 2000.
6. N. Chiba and T. Nishizeki. Arboricity and subgraph listing algorithms. *SIAM J. Comput.*, 14(1):210–223, 1985.
7. M. Chrobak and D. Eppstein. Planar orientations with low out-degree and compaction of adjacency matrices. *Theoretical Computer Science*, 86(2):243–266, 1991.
8. D. Eppstein. Arboricity and bipartite subgraph listing algorithms. *Inf. Process. Lett.*, 51(4):207–211, 1994.
9. D. Eppstein. All maximal independent sets and dynamic dominance for sparse graphs. In *Proc. 16th Annual ACM-SIAM Symposium on Discrete Algorithms (SODA'05)*, pages 451–459, 2005.
10. S. Even and R. E. Tarjan. Network flow and testing graph connectivity. *SIAM J. Comput.*, 4(4):507–518, 1975.
11. A. Frank and A. Gyárfás. How to orient the edges of a graph? In *Combinatorics Volume I (Proc. of the Fifth Hungarian Colloquium on Combinatorics, Keszthely, 1976, A. Hajnal, V. T. Sós, eds.)*, pages 353–364, Amsterdam, 1976. North-Holland.
12. H. Gabow. Upper degree-constrained partial orientations. In *Proc. 17th Annual ACM-SIAM Symposium on Discrete Algorithms (SODA'06)*, 2006.
13. H. Gabow and H. Westermann. Forests, frames, and games: algorithms for matroid sums and applications. In *Proc. of the 20th Annual ACM Symposium on Theory of Computing (STOC '88)*, pages 407–421, New York, NY, USA, 1988. ACM Press.
14. Z. Galil and A. Naamad. An $O(EV \log^2 V)$ algorithm for the maximal flow problem. *J. Comput. System Sci.*, 21:203–217, 1980.
15. G. Gallo, M. D. Grigoriadis, and R. E. Tarjan. A fast parametric maximum flow algorithm and applications. *SIAM J. Comput.*, 18(1):30–55, 1989.
16. A. V. Goldberg. Finding a maximum density subgraph. Technical Report UCB/CSD-84-171, EECS Department, University of California, Berkeley, 1984.
17. A. V. Goldberg and S. Rao. Beyond the flow decomposition barrier. In *Proc. of the 38th Annual Symposium on Foundations of Computer Science (FOCS '97)*, page 2, Washington, DC, USA, 1997. IEEE Computer Society.
18. S. Kannan, M. Naor, and S. Rudich. Implicit representation of graphs. In *Proc. of the 20th Annual ACM Symposium on Theory of Computing (STOC '88)*, pages 334–343, New York, NY, USA, 1988. ACM Press.
19. Ł. Kowalik and M. Kurowski. Shortest path queries in planar graphs in constant time. In *Proc. 35th Symposium on Theory of Computing (STOC'03)*, pages 143–148. ACM, June 2003.
20. D. C. Kozen. *The design and analysis of algorithms*. Springer-Verlag New York, Inc., New York, NY, USA, 1992.
21. C. S. J. A. Nash-Williams. Decomposition of finite graphs into forests. *Journal of the London Mathematical Society*, 39:12, 1964.
22. J.-C. Picard and M. Queyranne. A network flow solution to some nonlinear 0-1 programming problems with application to graph theory. *Networks*, 12:141–159, 1982.

Improved Approximation Algorithms for Maximum Resource Bin Packing and Lazy Bin Covering Problems*

Mingen Lin, Yang Yang, and Jinhui Xu

Department of Computer Science and Engineering
University at Buffalo, the State University of New York
Buffalo, NY 14260, USA
{mlin6, yyang6, jinhui}@cse.buffalo.edu

Abstract. In this paper, we study two variants of the bin packing /covering problems called *Maximum Resource Bin Packing (MRBP)* and *Lazy Bin Covering (LBC)* problems, and present new approximation algorithms for each of them. For the offline MRBP problem, the previous best known approximation ratio is $\frac{6}{5} = 1.2$, achieved by the classical First-Fit-Increasing (FFI) algorithm [1]. In this paper, we give a new FFI-type algorithm with an approximation ratio of $\frac{80}{71} \approx 1.12676$. For the offline LBC problem, it has been shown in [2] that the classical First-Fit-Decreasing (FFD) algorithm achieves an approximation ratio of $\frac{71}{60} \approx 1.18333$. In this paper, we present a new FFD-type algorithm with an approximation ratio of $\frac{17}{15} \approx 1.13333$. Both algorithms are simple, run in near linear time (i.e., $O(n \log n)$), and therefore are practical.

1 Introduction

Bin packing is a fundamental problem in combinatorial optimization and finds numerous applications in various areas. The problem and many of its variants have been extensively studied in the past and a number of important results have been obtained [3,4,5,6,7,8,9,10,11]. In its most basic form, the bin packing problem seeks to pack a sequence of items of size between zero and one into a minimum number of unit-sized bins. Recently Boyar *et al.* studied an interesting variant of the classical bin packing problem, called *Maximum Resource Bin Packing (MRBP)* [1], which considers the bin packing problem from a reverse perspective and maximizes the total number of used unit-sized bins. In its offline version, the MRBP problem maintains an order of the packed bins such that the following constraint is satisfied.

Constraint 1. *No item in a latter bin fits into any earlier bin.*

MRBP has applications in real world. For details, we refer the readers to [1] for an interesting example. In [1], the authors showed that no algorithm for this

---

* The research of this work was supported in part by an NSF CARRER Award CCF-0546509.

T. Asano (Ed.): ISAAC 2006, LNCS 4288, pp. 567–577, 2006.

problem has an approximation ratio worse than $\frac{17}{10}$ and the First-Fit-Decreasing (FFD) algorithm has the worst possible approximation ratio. They proved that the First-Fit-Increasing (FFI) algorithm has a better ratio of $\frac{6}{5}$. In Section 2, we introduce a new FFI-type algorithm called *Modified FFI (MFFI)* for the offline MRBP, with an approximation ratio of $\frac{80}{71}$.

As the "dual" of the bin packing problem, bin covering problem has also been extensively studied, and its common objective is to pack items of size between zero and one into a maximum number of unit-sized bins so that the level (i.e. the total size of all contained items) of each bin is no less than one [12,13]. Motivated by the MRBP problem, in [2] the authors considered the bin covering problem from the reverse perspective of minimizing the number of covered bins and studied the *Lazy Bin Covering (LBC)* problem. In its offline version, LBC requires that all bins are covered, except for at most one bin and every covered bin satisfies the following constraint.

Constraint 2. *Removing any item from a covered bin should make it uncovered.*

The non-redundancy constraint (i.e., Constraint 2) implies that the level of each bin is strictly smaller than one plus the size of its smallest item. It has been shown in [2] that FFD has an approximation ratio of $\frac{71}{60}$, and there exists an asymptotic polynomial time approximation scheme (APTAS) (i.e., asymptotic $(1 + \epsilon)$-approximation) for the (offline) LBC problem. However, similar to the APTAS in [10] for the classical bin packing problem, the time complexity of their APTAS for the LBC problem is exponential in $\frac{1}{\epsilon}$, which makes the algorithm impractical. In Section 3, we give a new FFD-type algorithm called *Modified FFD (MFFD)* with an approximation ratio of $\frac{17}{15}$.

Both of our algorithms run in $O(n \log n)$ time for a sequence of n items. Moreover, our algorithms are simple and can be easily implemented for applications. Another interesting feature of our algorithms is the judiciary combination of computer programs and analytical analysis. In the design of both algorithms, we have relied on computer programs to enumerate all possible packing patterns and used the generated patterns to guide the design of the algorithms.

In the rest of this paper, we use the following notation for our approximation algorithms. An (asymptotic) approximation algorithm ALG is a c-approximation algorithm for $c \geq 1$, if there is a constant b such that for all possible input sequences L, $ALG(L) \leq cOPT(L) + b$ for minimization problems (or $OPT(L) \leq cALG(L) + b$ for maximization problems). The infimum of all such c is called the approximation ratio of the algorithm, R_{ALG}.

Due to space limit, many proofs are omitted or shortened. Details are left for the full paper.

2   MFFI for the Offline MRBP

In this section, we study the offline MRBP and present a Modified First-Fit-Increasing (MFFI) algorithm with an approximation ratio of $\frac{80}{71}$. We start with some definitions and notations.

Let $l(B)$ denote the level of a bin B and $\min(B)$ denote the smallest item in B. Let L be a given input sequence, and intervals $I_k = (\frac{1}{k+1}, \frac{1}{k}], 1 \le k \le 8, I_9 = (0, \frac{1}{9}]$. We define a weighting function $w : L \to \Re$ as follows:

$$w(a_i) = \begin{cases} a_i, & \text{for } a_i \in I_9, \\ \frac{1}{k}, & \text{for } a_i \in I_k, 1 \le k \le 8. \end{cases}$$

It is clear from the weighting function that the weight of each item is no less than its size. Let W be the total weight of all items in a given input sequence.

As mentioned in Section 1, the classical FFI achieves a $\frac{6}{5}$-approximation [1]. To obtain a better approximation, we consider an optimal packing OPT. First, observe those bins containing at least one item in $(0, \frac{1}{9}]$. By Constraint 1, there is at most one such bin whose level is less than or equal to $\frac{8}{9}$. Since the weight of any item is no less than its size, we know that there is at most one such bin whose total weight of its items is less than $\frac{71}{80} < \frac{8}{9}$. Second, if a bin contains one item in $(\frac{1}{2}, 1]$, the total weight of its items is at least one.

To better characterize a packing, we define a pattern P [1] to be a multiset of numbers (or elements) in $\{\frac{1}{2}, \frac{1}{3}, \frac{1}{4}, \frac{1}{5}, \frac{1}{6}, \frac{1}{7}, \frac{1}{8}\}$ whose sum is at most 1, and $w(P)$ to be the sum of all elements in P. Denote $\gamma(P)$ as the number of elements in pattern P. The type of a pattern is the inverse of its smallest element. A pattern P of type j is a maximal pattern if adding another copy of $\frac{1}{j}$ to P will result in a non-pattern (i.e., $P \cup \{\frac{1}{j}\}$ no longer forms a pattern). The pattern of a packed bin is the multiset formed by the weights of its items. By Constraint 1, it is not difficult to see that for any feasible packing, there is at most one bin whose pattern is of type j, for $j = 2, 3, \cdots, 8$, but not maximal. Let $\psi = \{P_i | P_i \text{ is maximal, and } w(P_i) < \frac{71}{80}\}$. Then, there are in total 15 maximal patterns (enumerated by a computer program) in ψ, and we list them below in a non-decreasing order of their weights and types (If two patterns have the same weight, then order them by their types).

$P_1 = (\frac{1}{2}, \frac{1}{3})$ $P_6 = (\frac{1}{4}, \frac{1}{4}, \frac{1}{5}, \frac{1}{6})$ $P_{11} = (\frac{1}{5}, \frac{1}{5}, \frac{1}{6}, \frac{1}{6}, \frac{1}{7})$

$P_2 = (\frac{1}{3}, \frac{1}{4}, \frac{1}{4})$ $P_7 = (\frac{1}{4}, \frac{1}{5}, \frac{1}{6}, \frac{1}{6})$ $P_{12} = (\frac{1}{3}, \frac{1}{5}, \frac{1}{5}, \frac{1}{7})$

$P_3 = (\frac{1}{4}, \frac{1}{5}, \frac{1}{5}, \frac{1}{5})$ $P_8 = (\frac{1}{5}, \frac{1}{5}, \frac{1}{6})$ $P_{13} = (\frac{1}{4}, \frac{1}{5}, \frac{1}{7}, \frac{1}{7}, \frac{1}{7})$

$P_4 = (\frac{1}{3}, \frac{1}{3}, \frac{1}{5})$ $P_9 = (\frac{1}{4}, \frac{1}{6}, \frac{1}{6}, \frac{1}{7}, \frac{1}{7})$ $P_{14} = (\frac{1}{6}, \frac{1}{7}, \frac{1}{7}, \frac{1}{7}, \frac{1}{7}, \frac{1}{7})$

$P_5 = (\frac{1}{5}, \frac{1}{6}, \frac{1}{6}, \frac{1}{6}, \frac{1}{6})$ $P_{10} = (\frac{1}{3}, \frac{1}{4}, \frac{1}{7}, \frac{1}{7})$ $P_{15} = (\frac{1}{5}, \frac{1}{5}, \frac{1}{5}, \frac{1}{7}, \frac{1}{7})$

From the above patterns, we have the following observation.

Observation 1. *For any two patterns P_i and P_j with $i < j$, $w(P_i) \le w(P_j)$ and the smallest number in P_i is no less than that in P_j.*

Let OPT_i denote the number of bins in OPT whose pattern is P_i and OPT_0 denote the number of all other bins in OPT. By the above arguments, we have $OPT = \sum_{i=0}^{15} OPT_i$ and

---

[1] In [1], a similar definition of patterns has been used to analyze FFI algorithm.

$$W \geq \frac{71}{80}(OPT_0 - 8) + \sum_{i=1}^{15} w(P_i)OPT_i = \frac{71}{80}OPT - \frac{71}{10} - \sum_{i=1}^{15}\left(\frac{71}{80} - w(P_i)\right)OPT_i(1)$$

Now consider a packing generated by our to-be-designed algorithm MFFI. Let A be the total number of bins in the packing, A_i be the number of bins of pattern P_i (for $1 \leq i \leq 15$), and A_0 be the number of remaining bins. Then, $A = \sum_{i=0}^{15} A_i$. Further, we assume that A_i, $0 \leq i \leq 15$, and W satisfies the following inequality.

$$W \leq \sum_{i=1}^{15} w(P_i)A_i + A_0 + C = A - \sum_{i=1}^{15}(1 - w(P_i))A_i + C, \text{ for some constant } C. (2)$$

In order for MFFI to achieve an asymptotic $\frac{71}{80}$-approximation (i.e., $\frac{71}{80}OPT \leq A + D$ for some constant D), it is sufficient to ensure a) Inequality (2) is satisfied; and b) $\sum_{i=1}^{15}(\frac{71}{80} - w(P_i))OPT_i \leq \sum_{i=1}^{15}(1 - w(P_i))A_i + E$, for some constant E. To satisfy the two conditions, we have the following main steps for our MFFI algorithm.

Algorithm 1. MFFI

1: Greedily pack bins of pattern P_i ($P_i = P_1, P_2, \cdots, P_{15}$) until no bin of pattern P_i can be packed.
2: Pack the remaining items separately by using FFI algorithm.
3: Merge the two packings.

For each pattern P_i, let $\alpha(P_i) = 1 - w(P_i)$ and $\beta(P_i) = \frac{71}{80} - w(P_i)$. Since $w(P_1) \leq w(P_2) \leq \cdots \leq w(P_{15})$, we have $\alpha(P_1) \geq \alpha(P_2) \geq \cdots \geq \alpha(P_{15})$ and $\beta(P_1) \geq \beta(P_2) \geq \cdots \geq \beta(P_{15})$. Let $\rho(MFFI) = \sum_{i=1}^{15}(\alpha(P_i)A_i)$ be the mffi-gain and $\rho(OPT) = \sum_{i=1}^{15}(\beta(P_i)OPT_i)$ be the opt-gain. As mentioned before, we need to show that $\rho(OPT) \leq \rho(MFFI) + E$ for some constant E.

Next we discuss the details of each step in the above algorithm.

Step 1 includes fifteen phases. In each phase we have a partial output bin list X and a remaining (i.e. unpacked) item set R. Initially $X = \emptyset$ and R is the whole input input sequence L. Starting from phase one, in each phase i we keep packing bins of pattern P_i. More specifically, for each element $e_j \in P_i$, we select the largest item of weight e_j from R, and form a pattern-P_i bin once we found an item for each element in P_i. Note that since the size of each item is no larger than its weight, the level of each generated pattern-P_i bin is no larger than $w(P_i)$, which is less than $\frac{71}{80}$. All generated bins are placed in a reverse order (i.e., a newly produced bin becomes the first bin in X). Phase i ends when one of the following two things happens: a) For some $e_j \in P_i$, there exists no item of weight e_j in R, and therefore can no longer form a pattern-P_i bin. b) Even though the selected items form a bin B of pattern P_i, the total size of the items is not large enough so that the smallest item in the first bin in X (which is also the smallest item in all the packed bins in X) fits into B, thus violating Constraint 1. In either case, the selected items are put back to R and the algorithm moves to the next phase (i.e., phase $i+1$). At the end of Step 1, if there is only one bin of pattern P_i for some $1 \leq i \leq 15$ in X, we remove it from X. We call such a removal as X-refinement. Obviously, the total mffi-gain of all removed bins by the X-refinement procedure is at most $\sum_{i=1}^{15} \alpha(P_i)$.

To estimate the quality of the packing obtained by MFFI, we consider the following scenario. Whenever we successfully pack the selected items from the remaining set R into a bin, we remove the bins in OPT that contain at least one selected item (if they have not been removed yet), and call the remaining packing of OPT as the updated OPT. In addition, at the end of phase i, if there is one bin of pattern P_i in the updated OPT, remove it. We call it additional removal. We have the following lemmas.

Lemma 1. *In Phase i, if there are at least two bins of pattern P_i in the updated OPT, MFFI always successfully packs one bin of pattern P_i.*

The overall opt-gain of the bins in OPT removed by the additional removals is at most $\sum_{i=1}^{15} \beta(P_i)$. Lemma 2 bounds the opt-gain of the removed bins in the updated OPT for each successfully packed bin of pattern P_i in phase i.

Lemma 2. *The total opt-gain of the removed bins in the updated OPT is at most $\gamma(P_i)\beta(P_i)$ for each successfully packed bin of pattern P_i in phase i for $1 \leq i \leq 15$.*

Lemma 3. $\rho(MFFI) + \sum_{i=1}^{15}(\alpha(P_i) + \beta(P_i)) \geq \rho(OPT)$.

Proof. By Lemma 1 and the additional removals, in Step 1 of MFFI when phase i ends, there is no bin of pattern P_i ($1 \leq i \leq 15$) in the updated OPT. By Lemma 2, the additional removals, and the X-refinement, we have $\rho(OPT) \leq \sum_{i=1}^{15}(\gamma(P_i)\beta(P_i)(A_i+1)) + \sum_{i=1}^{15} \beta(P_i)$. From definition, we have $\rho(MFFI) = \sum_{i=1}^{15}(\alpha(P_i)A_i)$. To show $\rho(MFFI) + \sum_{i=1}^{15}(\alpha(P_i) + \beta(P_i)) \geq \rho(OPT)$, it is sufficient to show $\alpha(P_i) \geq \beta(P_i)\gamma(P_i)$ for each $1 \leq i \leq 15$. This is clearly true by examining each row of Table 1. □

Lemma 4. *Any bin B in X of type k has a remaining capacity less than $\frac{1}{k}$.*

In Step 2 of MFFI, we pack the remaining items in R by the FFI algorithm without using any bins in X. FFI handles the items in a non-decreasing order of their sizes, and places each of them in the first bin in which it fits. For MRBP, FFI behaves in the exactly the same way as Next-Fit-Increasing (NFI). Once a new bin is opened, no previous opened bin will be used any more. Let the output bin list be Y. For each $k = 9, 8, \cdots, 1$, remove the largest indexed bin containing items in I_k from Y. Let the set of removed bins be T and the resulting packing be Y'. Obviously, $Y' \geq Y - 9$. We know that each removed bin has a total weight less than two since each removed bin has a total size of at most one and the maximum ratio of the weight of any item over its size is less than two. It is clear that all bins in Y' only contain items from one interval I_k. This implies that for any bin in Y', the total weight of its items is at most one. This is clear if the items are all in I_9 whose weights are equal to their sizes. For other items, there are exactly k items in I_k with the sum of weights equal to one. Thus we have

$$W \leq Y + 9 + \sum_{i=1}^{15} w(P_i)A_i, \tag{3}$$

where W is the total sum weight of all items.

Lemma 5. *Any bin in Y' whose pattern is of type k for $2 \leq k \leq 8$ has a remaining capacity less than $\frac{1}{k+1}$.*

In Step 3 of MFFI, we first merge packing X with packing Y' by inserting the bins in X into Y'. Let the resulting packing be Z. Then we pack the items in T into Z by using FFI. The merging procedure of X and Y' consists of several phases. At each phase, we only deal with all the patterns of the same type k ($k = 7, 6, 5, 4, 3$). Let $k' = \min\{h | h \geq k$ and there exists a bin whose pattern is of type h in Y' }. If there exists a such k', we insert all the bins in X whose patterns are of type k into Y', right behind the last bin of type k' in Y' while preserving their orders in X. Otherwise we insert all these bins into the top of Y' while preserving their orders in X. By Lemma 5, every bin appeared before the inserted bins in Y' has a remaining capacity less than $\frac{1}{k+1}$, and the smallest item of the inserted bins is larger than $\frac{1}{k+1}$. Thus there is no violation of Constraint 1 between the inserted bins and the bins before them in Y'. Since the smallest item of the bins after the inserted bins in Y' is larger than $\frac{1}{k}$ and by Lemma

Table 1. The values of $w, \alpha, \beta, \gamma, \beta\gamma$ for $P_1, P_2, \cdots, P_{15}$

	w	α	β	γ	$\beta\gamma$
P_1	$\frac{5}{6}$	$\frac{1}{6} \approx 0.16667$	$\frac{13}{240}$	2	$\frac{13}{120} \approx 0.10833$
P_2	$\frac{5}{6}$	$\frac{1}{6} \approx 0.16667$	$\frac{13}{240}$	3	$\frac{13}{80} = 0.1625$
P_3	$\frac{17}{20}$	$\frac{3}{20} = 0.15$	$\frac{3}{80}$	4	$\frac{3}{20} = 0.15$
P_4	$\frac{13}{15}$	$\frac{2}{15} \approx 0.13333$	$\frac{1}{48}$	3	$\frac{1}{16} = 0.0625$
P_5	$\frac{13}{15}$	$\frac{2}{15} \approx 0.13333$	$\frac{1}{48}$	5	$\frac{5}{48} \approx 0.10417$
P_6	$\frac{13}{15}$	$\frac{2}{15} \approx 0.13333$	$\frac{1}{48}$	4	$\frac{1}{12} \approx 0.08333$
P_7	$\frac{13}{15}$	$\frac{2}{15} \approx 0.13333$	$\frac{1}{48}$	4	$\frac{1}{12} \approx 0.08333$
P_8	$\frac{13}{15}$	$\frac{2}{15} \approx 0.13333$	$\frac{1}{48}$	3	$\frac{1}{16} = 0.0625$
P_9	$\frac{73}{84}$	$\frac{11}{84} \approx 0.13095$	$\frac{31}{1680}$	5	$\frac{31}{336} \approx 0.09226$
P_{10}	$\frac{73}{84}$	$\frac{11}{84} \approx 0.13095$	$\frac{31}{1680}$	4	$\frac{31}{420} \approx 0.07381$
P_{11}	$\frac{92}{105}$	$\frac{13}{105} \approx 0.12381$	$\frac{19}{1680}$	5	$\frac{19}{336} \approx 0.05655$
P_{12}	$\frac{92}{105}$	$\frac{13}{105} \approx 0.12381$	$\frac{19}{1680}$	4	$\frac{19}{420} \approx 0.04524$
P_{13}	$\frac{123}{140}$	$\frac{17}{140} \approx 0.12143$	$\frac{1}{112}$	5	$\frac{5}{112} \approx 0.04464$
P_{14}	$\frac{37}{42}$	$\frac{5}{42} \approx 0.11905$	$\frac{1}{1680}$	6	$\frac{11}{280} \approx 0.03929$
P_{15}	$\frac{31}{35}$	$\frac{4}{35} \approx 0.11429$	$\frac{1}{560}$	5	$\frac{1}{112} \approx 0.00893$

4 each of the inserted bins has a remaining capacity less than $\frac{1}{k}$, there is no violation of Constraint 1 between the bins in Y' after the inserted bins and the inserted bins. Therefore packing Z is valid packing. After placing the items in T into Z by using FFI, we have the final packing MFFI. It is clear that $MFFI \geq X + Y' = X + Y - 9$.

Theorem 1. *For the Offline Maximum Resource Bin Packing Problem, the approximation ratio of MFFI is at most $\frac{80}{71}$.*

Proof. First we have $MFFI \geq X + Y - 9$. By (1), (3) and Lemma 3, we are able to show $MFFI \geq \frac{71}{80}OPT - \frac{251}{10} - \sum_{i=1}^{15}(\alpha(P_i) + \beta(P_i))$. □

Theorem 2. *For the Offline Maximum Resource Bin Packing Problem, the approximation ratio of MFFI is $\frac{80}{71}$.*

3 MFFD for the Offline LBC

In this section, we study the offline lazy bin covering problem and present a Modified First-Fit-Decreasing (MFFD) algorithm with an approximation ratio of $\frac{17}{15}$.

Let $l(B)$ denote the level of a bin B. Let L be a given input sequence of items and let intervals $I_k = [\frac{1}{k}, \frac{1}{k-1}), 1 \leq k \leq 8, I_9 = (0, \frac{1}{8})$. Here we assume $\frac{1}{0} = +\infty$. We define a weighting function $w : L \to \Re$ as follows:

$$w(a_i) = \begin{cases} a_i, & \text{for } a_i \in I_9, \\ \frac{1}{k}, & \text{for } a_i \in I_k, 1 \leq k \leq 8 \end{cases}$$

It is clear that the weight of each item is at most its size. Let $w(B) = \sum_{a_i \in B} w(a_i)$ denote the weight of B. By the problem description, we have the following fact:

Fact 1. *For any bin B, $w(B) < 1 + \min\{w(a_i) \mid a_i \in B\}$.*

Let W be the total weight of all items in a given input sequence. Consider an optimal packing OPT. First of all, consider the bins containing at least one item in I_8 or I_9. By Fact 1, we have that the weight of each such bin is less than $1 + \frac{1}{8} < \frac{17}{15}$. Secondly, consider the bins containing at least one item in I_1, we know that by the non-redundancy constraint, each of these bins contains exactly one item of size one. So the weight of each such bin is one, which is less than $\frac{17}{15}$.

We define a pattern P to be a multiset of numbers (or elements) in $\{\frac{1}{2}, \frac{1}{3}, \frac{1}{4}, \frac{1}{5}, \frac{1}{6}, \frac{1}{7}\}$ whose sum is less than one plus the smallest number in P, and $w(P)$ to be the sum of all elements in P. We denote $\gamma(P)$ to be the number of elements in P. The pattern of a bin is the multiset formed by the weights of its items. Below are nine patterns (enumerated by a computer program) with weight larger than $\frac{17}{15}$, listed in a non-increasing order of their weights.

$$P_1 = (\tfrac{1}{3}, \tfrac{1}{4}, \tfrac{1}{5}, \tfrac{1}{5}, \tfrac{1}{5}) \qquad P_4 = (\tfrac{1}{4}, \tfrac{1}{5}, \tfrac{1}{5}, \tfrac{1}{6}, \tfrac{1}{6}, \tfrac{1}{6}) \qquad P_7 = (\tfrac{1}{2}, \tfrac{1}{4}, \tfrac{1}{5}, \tfrac{1}{5})$$
$$P_2 = (\tfrac{1}{3}, \tfrac{1}{3}, \tfrac{1}{4}, \tfrac{1}{4}) \qquad P_5 = (\tfrac{1}{4}, \tfrac{1}{4}, \tfrac{1}{4}, \tfrac{1}{5}, \tfrac{1}{5}) \qquad P_8 = (\tfrac{1}{5}, \tfrac{1}{5}, \tfrac{1}{6}, \tfrac{1}{7}, \tfrac{1}{7}, \tfrac{1}{7}, \tfrac{1}{7})$$
$$P_3 = (\tfrac{1}{2}, \tfrac{1}{3}, \tfrac{1}{3}) \qquad P_6 = (\tfrac{1}{3}, \tfrac{1}{4}, \tfrac{1}{5}, \tfrac{1}{5}, \tfrac{1}{6}) \qquad P_9 = (\tfrac{1}{4}, \tfrac{1}{5}, \tfrac{1}{5}, \tfrac{1}{5}, \tfrac{1}{7}, \tfrac{1}{7})$$

Let OPT_i denote the number of bins in OPT whose pattern is P_i, $1 \le i \le 9$, and OPT_0 denote the number of all other bins in OPT. We have $OPT = \sum_{i=0}^{9} OPT_i$, and by the above arguments we know

$$W \le \frac{17}{15}OPT_0 + \sum_{i=1}^{9} w(P_i)OPT_i = \frac{17}{15}OPT + \sum_{i=1}^{15}\left(w(P_i) - \frac{17}{15}\right)OPT_i. \qquad (4)$$

To achieve a better approximation than the one in [2], let A be the number of bins in the packing produced by our to-be-designed MFFD algorithm. Let A_i be the number of bins whose patterns are P_i in A, $1 \le i \le 9$, and A_0 be the number of all other bins in A. Then we have $A = \sum_{i=0}^{9} A_i$. Further, we assume A_i, $0 \le i \le 9$, and W satisfies the following inequality

$$W \ge \sum_{i=1}^{9} w(P_i)A_i + A_0 + C = A + \sum_{i=1}^{9}(w(P_i) - 1)A_i + C, \text{ for some constant } C. \ (5)$$

From (4) and (5), we know that in order to make MFFD have an approximation ratio of $\frac{17}{15}$ (or more precisely, $\frac{17}{15}OPT \ge A + D$, for some constant D), it is sufficient to ensure: a) Inequality (5) is satisfied; and b) $\sum_{i=1}^{9}(w(P_i) - \frac{17}{15})OPT_i \le \sum_{i=1}^{9}(w(P_i) - 1)A_i + E$ for some constant E. To satisfy these two conditions, we have our MFFD algorithm include the following main steps.

Algorithm 2. MFFD

1: Greedily pack bins of pattern P_i ($P_i = P_2, \cdots, P_9$) until no bin of pattern P_i can be packed.

2: Pack the remaining items separately by using FFD algorithm.

3: Merge the two packings.

It is interesting to point out that even though the MFFD algorithm shares some similar ideas with the MFFI algorithm, it has also one major difference. To achieve better approximation ratio, in Step 1 of the MFFD algorithm, it starts with P_2 for the greedy packing, instead of P_1. The reason will be clear later on.

For each pattern P_i, let $\alpha(P_i) = w(P_i) - 1$ and $\beta(P_i) = w(P_i) - \frac{17}{15}$. Since $w(P_1) \ge w(P_2) \ge \cdots \ge w(P_9)$, we have $\alpha(P_1) \ge \alpha(P_2) \ge \cdots \ge \alpha(P_9)$ and $\beta(P_1) \ge \beta(P_2) \ge \cdots \ge \beta(P_9)$. Let $\rho(MFFD) = \sum_{i=1}^{9}(\alpha(P_i)A_i)$ be the mffd-gain and $\rho(OPT) = \sum_{i=1}^{9}(\beta(P_i)OPT_i)$ be the opt-gain. As mentioned before, one of the objectives is to show is that $\rho(OPT) \le \rho(MFFD) + E$, for some constant E.

Next we discuss the details of each step in MFFD.

Step 1 consists of eight phases. In each phase we have a partial output bin list X and a remaining (i.e. unpacked) item set R. Initially $X = \emptyset$ and R is the whole set of the input items. Starting with phase one, in each phase i we only consider a particular pattern P_{i+1}. More specifically, for each element $e_j \in P_{i+1}$, we select the smallest item of weight e_j from R, and try to form a bin of pattern P_{i+1}. Since the size of each item is no smaller than its weight, we know that the level of each packed bin in phase i is no smaller than $w(P_{i+1})$, which is greater than $\frac{17}{15}$. Therefore each bin generated in Step 1 is a covered bin. Phase i ends when one of the following two events occurs: a) For some $e_j \in P_{i+1}$, there exists no item of weight e_j in R and therefore the selected items can not form a bin of pattern P_{i+1}; b) The selected items indeed form a bin B of pattern P_{i+1}, but the total size of the items is greater than the sum of one and the smallest selected item, thus violating the non-redundancy constraint. In either case, the attempting packing is rolled back (i.e., the selected items are put back to R), and the algorithm enters phase $i + 1$.

To measure the quality of the packing obtained by MFFD, we consider the following scenario. Whenever we successfully pack the selected items from R into a bin B, we remove the bins in OPT that contain at least one selected item in B if they have not been removed yet. The remaining packing of OPT is called updated OPT. In addition, at the end of phase 1 (note that phase one deals with pattern P_2), if there is one bin of pattern P_1 in the updated OPT, remove it. We call this as the additional removal. We have the following lemmas.

Lemma 6. *If there are at least two bins of pattern P_1 in the updated OPT, MFFD always packs one bin of pattern P_2 in phase one. Moreover, $A_2 \geq \lfloor \frac{OPT_1}{2} \rfloor$.*

Lemma 7. *If there is at least one bin of pattern P_i in the updated OPT, MFFD always successfully packs one bin of pattern P_i in phase $i - 1$ for $2 \leq i \leq 9$.*

Lemma 8. *The total opt-gain of the removed bins in the updated OPT is at most $\gamma(P_i)\beta(P_i)$ for each successfully packed bin of pattern P_i in phase $i - 1$ for $3 \leq i \leq 9$.*

Lemma 9. *At the end of phase one, the overall opt-gain of the removed bins from OPT is at most $\frac{1}{6}A_2 + \frac{1}{20}$.*

Lemma 10. $\rho(MFFD) + \frac{1}{20} \geq \rho(OPT)$.

Proof. When Step 1 completes, by Lemma 6, Lemma 7, and the additional removal there is no bin of pattern P_i in the updated OPT. By Lemma 8 and Lemma 9, we have $\rho(OPT) \leq \sum_{i=3}^{9}(\gamma(P_i)\beta(P_i)A_i) + \frac{1}{6}A_2 + \frac{1}{20}$. From the definition, $\rho(MFFD) = \sum_{i=1}^{9}(\alpha(P_i)A_i)$. To show $\rho(MFFD) + \frac{1}{20} \geq \rho(OPT)$, it is sufficient to have $\alpha(P_i) \geq \beta(P_i)\gamma(P_i)$ for each $3 \leq i \leq 9$ and $\alpha(P_2) \geq \frac{1}{6}$. This is ensured by the corresponding values in Table 2. □

Table 2. The values of $w, \alpha, \beta, \gamma, \beta\gamma$ for $P_1, P_2, \cdots, P_9$

	w	α	β	γ	$\beta\gamma$
P_1	$\frac{71}{60}$	$\frac{11}{60} \approx 0.18333$	$\frac{1}{20}$	5	$\frac{1}{4} = 0.25$
P_2	$\frac{7}{6}$	$\frac{1}{6} \approx 0.16667$	$\frac{1}{30}$	4	$\frac{2}{15} \approx 0.13333$
P_3	$\frac{7}{6}$	$\frac{1}{6} \approx 0.16667$	$\frac{1}{30}$	3	$\frac{1}{10} = 0.1$
P_4	$\frac{23}{20}$	$\frac{3}{20} = 0.15$	$\frac{1}{60}$	6	$\frac{1}{10} = 0.1$
P_5	$\frac{23}{20}$	$\frac{3}{20} = 0.15$	$\frac{1}{60}$	5	$\frac{1}{12} \approx 0.08333$
P_6	$\frac{23}{20}$	$\frac{3}{20} = 0.15$	$\frac{1}{60}$	5	$\frac{1}{12} \approx 0.08333$
P_7	$\frac{23}{20}$	$\frac{3}{20} = 0.15$	$\frac{1}{60}$	4	$\frac{1}{15} \approx 0.066667$
P_8	$\frac{239}{210}$	$\frac{29}{210} \approx 0.13810$	$\frac{1}{210}$	7	$\frac{1}{30} = 0.03333$
P_9	$\frac{159}{140}$	$\frac{19}{140} \approx 0.13571$	$\frac{1}{420}$	6	$\frac{1}{70} \approx 0.01429$

Note that in Table 2, we have $\alpha(P_1) < \beta(P_1)\gamma(P_1)$. Therefore MFFD will not achieve the approximation ratio of $\frac{17}{15}$ if it starts with pattern P_1 (instead of P_2) for the greedy packing in Step 1.

In Step 2, we pack the remaining items in R by FFD without using any bins in X. FFD handles the items in a non-increasing order of their sizes, and places each of them in the first bin in which it fits. Note that for LBC, FFD behaves in the exactly same way as Next-Fit-Decreasing (NFD). Once a new bin is opened, no previous opened bin will be used any more. We denote the output bin list as Y. Consider a bin B in Y which is not the last bin and contains only items from a single interval I_j for some j. If $1 \leq j \leq 8$, B will contain exactly j items and $w(B) = 1$. If $j = 9, w(B) = l(B) \geq 1$. Thus $w(B) < 1$ only if B contains items from more than one interval or B is the last bin in Y. Let C be the set of bins in Y with weight less than one. Obviously, $C \leq 8$ and we have

$$Y - 8 + \sum_{i=2}^{9} A_i < W. \tag{6}$$

In Step 3, we merge all the bins in X with all the bins in Y (i.e., append the list of bins in Y at the end of the list of bins in X; denote $MFFD = X|Y$).

Theorem 3. *For the offline Lazy Bin Covering Problem, the approximation ratio of $MFFD$ is at most $\frac{17}{15}$.*

Proof. We have $MFFD = X + Y$ and by (4), (6) and Lemma 10, we are able to show $MFFD < \frac{17}{15}OPT + \frac{161}{20}$. □

Theorem 4. *For the offline Lazy Bin Covering Problem, the approximation ratio of $MFFD$ is $\frac{17}{15}$.*

References

1. Boyar, J., Epstein, L., Favrholdt, L.M., Kohrt, J.S., Larsen, K.S., Pedersen, M.M., Wøhlk, S.: The maximum resource bin packing problem. In: FCT. (2005) 397–408
2. Lin, M., Yang, Y., Xu, J.: On lazy bin covering and packing problems. In: CO-COON. (2006)
3. Garey, M.R., Graham, R.L., Johnson, D.S.: Resource constrained scheduling as generalized bin packing. J. Comb. Theory, Ser. A **21** (1976) 257–298
4. Csirik, J.: The parametric behavior of the first-fit decreasing bin packing algorithm. J. Algorithms **15** (1993) 1–28
5. Csirik, J., Johnson, D.S.: Bounded space on-line bin packing: Best is better than first. Algorithmica **31** (2001) 115–138
6. Johnson, D.S., Garey, M.R.: A 71/60 theorem for bin packing. J. Complexity **1** (1985) 65–106
7. Galambos, G., Woeginger, G.: Repacking helps in bounded space on-line bin-packing. Computing **49** (1993) 329–338
8. Woeginger, G.J.: Improved space for bounded-space, on-line bin-packing. SIAM J. Discrete Math. **6** (1993) 575–581
9. Shachnai, H., Tamir, T.: On two class-constrained versions of the multiple knapsack problem. Algorithmica **29** (2001) 442–467
10. Friesen, D.K., Langston, M.A.: Analysis of a compound bin packing algorithm. SIAM J. Discrete Math. **4** (1991) 61–79
11. Bar-Noy, A., Ladner, R.E., Tamir, T.: Windows scheduling as a restricted version of bin packing. In: SODA '04. (2004) 224–233
12. Csirik, J., Kenyon, C., Johnson, D.S.: Better approximation algorithms for bin covering. In: SODA. (2001) 557–566
13. Assmann, S.F., Johnson, D.S., Kleitman, D.J., Leung, J.Y.T.: On a dual version of the one-dimensional bin packing problem. J. Algorithms **5** (1984) 502–525

Partitioning the Nodes of a Graph to Minimize the Sum of Subgraph Radii[*]

Guido Proietti[1,2] and Peter Widmayer[3]

[1] Dipartimento di Informatica, Università di L'Aquila, 67010 L'Aquila, Italy
[2] Istituto di Analisi dei Sistemi ed Informatica, CNR, 00185 Roma, Italy
[3] Institut für Theoretische Informatik, ETH, 8092 Zürich, Switzerland
`proietti@di.univaq.it, widmayer@inf.ethz.ch`

Abstract. Let $G = (V, E)$ denote a weighted graph of n nodes and m edges, and let $G[V']$ denote the subgraph of G induced by a subset of nodes $V' \subseteq V$. The *radius* of $G[V']$ is the maximum length of a shortest path in $G[V']$ emanating from its *center* (i.e., a node of $G[V']$ of minimum eccentricity). In this paper, we focus on the problem of partitioning the nodes of G into exactly p non-empty subsets, so as to minimize the *sum* of the induced subgraph radii. We show that this problem – which is of significance in facility location applications – is NP-hard when p is part of the input, but for a fixed constant $p > 2$ it can be solved in $O(n^{2p}/p!)$ time. Moreover, for the notable case $p = 2$, we present an efficient $O(mn^2 + n^3 \log n)$ time algorithm.

Keywords: Graph partition, Facility location problems, Clustering problems, Graph radius, NP-hardness.

1 Introduction

Locating facilities on a network requires to identify a set of distinguished points in the network, so as to ideally minimize the global effort needed from a set of customers to benefit from the service provided by the facilities. The basic graph-theoretic definition of a facility location problem, also known as a *graph location problem*, is the following: Given an undirected graph $G = (V, E)$ of n nodes and m edges, with positive (rational) weights on the edges inducing a symmetric *distance function* $\delta(\cdot, \cdot)$ which associates with each pair of nodes in G the length (i.e., total weight) of a shortest path between them, and given a positive integer $p \leq n$, find a subset of nodes $X \subseteq V$ of size p such that some distance criteria $\phi(X, G)$ is minimized. Once that a subset of the nodes of the graph are identified as being facilities, each demand node remains associated with one facility. Up to now, the literature was mainly concentrated on two general types of distance criteria for graph location problems, always under the

---

[*] Work partially supported by the Research Project GRID.IT, funded by the Italian Ministry of Education, University and Research, by the European Union under COST 295 (DYNAMO), and by the Swiss SBF under grant no. C05.0047. Part of this work has been developed while the first author was visiting ETH Zürich.

T. Asano (Ed.): ISAAC 2006, LNCS 4288, pp. 578–587, 2006.
© Springer-Verlag Berlin Heidelberg 2006

condition that each demand node is associated with its nearest facility: (1) a *min-sum* criterion, in which the facilities have to be located so as to minimize the total sum of all the distances to the associated nodes; (2) a *min-max* criterion, in which the facilities have to be located so as to minimize the maximum of their radii, where the *radius* of a facility is the distance from a farthest node associated to it. Correspondingly, under the problem definition given above, a set of nodes of cardinality p which minimizes the former (resp., the latter) criterion is known as a *p-median* (resp., a *p-center*) of G.

1.1   Related Work

The p-median and the p-center problem are central to the field of location theory (for a survey, see [10]). We briefly recall some results known for the p-center problem. Not surprisingly, the problem is NP-hard [9], and the fastest exact algorithm (exponential in p) is based on an exhaustive search which considers all the combinations of p-tuples of nodes in G, and, for each of these tuples, associates each remaining node with the closest node in the tuple. Since the association step can be performed in linear time, this amounts to an $O(n^{p+1}/(p-1)!)$ time algorithm. In particular, for the case $p = 2$, this implies the existence of an $O(n^3)$ time algorithm. For the metric case, a 2-approximation algorithm [7,8] is known, and this ratio is tight [12], while for the Euclidean case, where centers can lie everywhere and so the exhaustive search cannot take place, there exists a polynomial time approximation scheme (PTAS) which is exponential in p [2].

In spite of its relevance, the minimization of the maximum radius induces a negative side effect, the *dissection effect* [5]. Recall that in clustering, we partition a set of objects into a collection of disjoint subsets (i.e., the *clusters*) so that a certain measure defined over these clusters is minimized. Thus, a clustering that minimizes the maximum radius tends to remove peripheral objects from large clusters (so as to have fairly equal cluster radii), despite of their natural similarity with the elements in the original cluster. This dissection effect can be avoided by partitioning a set of objects according to the minimum *sum* of the cluster radii/diameters. For the diameter case, the problem remains hard, even hard to approximate [6]. Conversely, when we minimize the sum of radii, which is of interest here, the complexity of the problem is still open. On the positive side, for metric spaces there exists an algorithm that computes an $O(1)$-approximate solution with a constant factor blow-up in the number of clusters [4]. On the other hand, in a geometric setting, the problem can be formulated, in its widest generality, as that of covering a given set of n points in a d-dimensional Euclidean space by means of a set of at most p circles of radii r_i, whose center positions can be constrained in several different ways. The objective function is to minimize the sum of r_i^α over all these circles, where α is a constant that can be 1 or larger, depending on the boundary conditions. For a summary of results in this general framework, see [1]. However, for the case $\alpha = 1$, which is close in spirit to our setting, the complexity of the problem is still unknown. In Euclidean spaces of constant dimension, two PTASs exist, for both the case in which the the center positions are restricted to a given set of feasible locations but the number of circles is left unspecified [11], and for the (more general) case in which the number of circles is bounded and centering a circle at any given point has a variable non-negative cost [3].

1.2   Our Results

The dissection effect is easy to be reported also in a facility location framework, in which distances between elements are actually lengths of shortest paths. The traditional idea of minimizing the maximum distance to a facility is *customer-centric*, meaning that the objective is to minimize the maximum effort that a customer has to make *to benefit* from a facility. Many practical situations, however, are *facility-centric*, in the sense that the set-up cost of installing a facility depends upon the maximum effort that a facility has to make *to serve* a customer. Consider, for instance, a delivery service that needs to maintain in each location a vehicle whose properties (such as speed or fuel efficiency) and (as a consequence) whose cost depend on the distance the vehicle needs to travel: A small and cheap car may serve a small radius well, but for a larger delivery radius a more powerful and expensive car may be useful. The corresponding optimization problem (which locations, and which car for each location, so that the total cost for all cars is smallest) can be modelled in abstract terms by what we call the *p-radius problem*, which asks for *partitioning* the nodes of G into exactly p non-empty subsets, so as to minimize the *sum of the radii* of the induced subgraphs. In this paper, we therefore aim to solve this problem, and we provide: (1) an NP-hardness proof for the general case; (2) for the case $p = 2$, an algorithm requiring $O(mn^2 + n^3 \log n)$ time and $O(n^3)$ space; (3) for the case $p > 2$, an algorithm requiring $O(n^{2p}/p!)$ time and $O(n^{p+1}/(p+1)!)$ space.

The paper is organized as follows: in Section 2, we show that the p-radius problem is NP-hard, while in Section 3 we show, for the case $p = 2$, the general idea behind our algorithm; finally, in Section 4 we outline the extension of our algorithm to the case $p > 2$.

2   The p-Radius Problem is NP-Hard

Let $G = (V, E)$ denote a graph of n nodes and m edges with positive (rational) weights. For any two given nodes a, b in G, we denote by $\delta_G(a, b)$ the *distance* (i.e., the length of a shortest path) in G between a and b. Let $G[V']$ be the subgraph of G induced by a subset of nodes $V' \subseteq V$. Recall that the *center* of $G[V']$ is a node for which the maximum distance in $G[V']$ to any node in V' is minimum (notice that such a node is not necessarily unique). This distance is called the *radius* of $G[V']$, and is denoted by $r(G[V'])$. Given a graph G and an integer value p, the *p-radius problem* asks for a partition of the nodes of G into exactly p disjoint non-empty subsets, say $V_1, \ldots, V_p$, in such a way that $\sum_{k=1}^{p} r(G[V_k])$ is minimum. We start by proving the following:

Theorem 1. *The p-radius problem is* NP-*hard.*

Proof. We show the NP-hardness by reduction from the NP-complete SATISFI-ABILITY problem. In SATISFIABILITY, we are given a set $X = \{x_0, \ldots, x_{n-1}\}$ of n variables, a set $\mathcal{C} = \{c_1, \ldots, c_m\}$ of m clauses over X, and we want to find a truth assignment $\tau : X \to \{0, 1\}$ satisfying $\mathcal{C}$. For a given instance I of SATISFI-ABILITY, we build an instance I' for the p-radius problem as follows. Each clause c_j defines a vertex v_{c_j}, and each variable x_i defines $n + 2$ vertices: two vertices v_{x_i} and $\bar{v}_{x_i}$ (for the non-negated and the negated variable), and n vertices y_i^k, for $k = 1, \ldots, n$. We have an edge (v_{x_i}, v_{c_j}) iff x_i appears in c_j in non-negated form, and an edge $(\bar{v}_{x_i}, v_{c_j})$ iff x_i appears in negated form in c_j. Moreover, each of v_{x_i} and $\bar{v}_{x_i}$ is connected with an edge to each of the $y_i^k, k = 1, \ldots, n$, and v_{x_i} and $\bar{v}_{x_i}$ are connected with an edge as well. The weights of all the edges incident to v_{x_i} or $\bar{v}_{x_i}$ are 2^i. Notice that this is still a polynomially long representation in the logarithmic cost measure, with at most n bits for an edge weight. For the p-radius problem we require $p = n$ subsets. We show that the SATISFIABILITY instance I has a positive answer iff there exists a solution for I' having sum of radii at most $b = 2^n - 1$. One direction is immediate. Given a satisfying assign-ment τ, we define the following n centers. For $i = 0, \ldots, n - 1$, if $\tau(x_i) = 1$, then we pick v_{x_i} as a center, otherwise (i.e., $\tau(x_i) = 0$) we pick $\bar{v}_{x_i}$. The subset V_i contains $v_{x_i}, \bar{v}_{x_i}$, the y_i^k for $k = 1, \ldots, n$, and all v_{c_j} reachable from the center with a single edge (if v_{c_j} is reachable from different centers with one edge, we put v_{c_j} in one of the corresponding subsets arbitrarily). By construction, each V_i has radius 2^i, the totality of all chosen centers covers all vertices, and the value of the solution is exactly b. Concerning the other direction, suppose that there is a solution for I' having value at most b. We show that such a solution can be transformed in polynomial time into a satisfying assignment for I. For each i, we define the set $U_i = \{v_{x_i}, \bar{v}_{x_i}, y_i^1, \ldots, y_i^n\}$. Notice that for each i, since $|U_i| > n$, there must exist two vertices in U_i belonging to the same subset in the solution of I'. Let us call such a subset V_i. We prove the following two facts.

Fact 1. *Given a solution of I', there exists another solution of I' having the same cost and with no subset V' consisting of a single vertex of U_i, for each i.*

Proof. If there is such a singleton subset, it is always possible to move the vertex of V' in V_i without increasing the radius of the subgraph induced by V_i. □

Fact 2. *Any subset V' with $|V'| > 1$ containing a vertex $x \in U_i$ induces a subgraph having radius at least 2^i.*

Proof. The distance between x and any other vertex in V' is at least 2^i. □

Now, we prove the following.

Lemma 1. *For every i, it holds: (i) $U_i \subseteq V_i$, and (ii) $V_i \cap V_j = \emptyset, \forall j > i$.*

Proof. The proof is by induction on i, starting at $i = n - 1$ and ending at $i = 0$. For the base case $i = n - 1$, (ii) is trivial, and (i) follows from Fact 1 and 2, because otherwise there would be two subsets with associated radius at least 2^{n-1}, together exceeding b. Now assume that the claim is true for $j = i, \ldots, n-1$, and let us prove it for $i - 1$. For the sake of contradiction, suppose that (i) does not hold. Then, there are two subsets V' and V'' both containing vertices in U_{i-1}. If both V' and V'' do not contain vertices in $U_j, j \geq i$, then the overall sum of the induced subgraph radii is at least $2^{i-1} + 2^{i-1} + \sum_{j=i}^{n-1} 2^j > b$. Otherwise, w.l.o.g. assume that $V' = V_k$ for some $k \geq i$, while V'' is different from every $V_j, j \geq i$. Because there is no edge between any two vertices for different variables, the radius associated with V_k must be at least $2^k + 2^{i-1}$, and the sum of the induced subgraph radii is strictly greater than b. Similarly, if both V' and V'' contain some vertices in V_j for some $j \geq i$, the value of the solution exceeds b. In order to prove (ii), notice that if $V_{i-1} \cap V_j \neq \emptyset$ for some $j \geq i$, then, since we have just proven that $U_{i-1} \subseteq V_{i-1}$ and $U_j \subseteq V_j$, then necessarily it must be $V_{i-1} = V_j$. But then, the radius associated with V_j is at least $2^j + 2 \cdot 2^{i-1}$, which yields to a solution having value greater than b.   □

From the above lemma, it is easy to see that the center associated with each V_i must be either v_{x_i} or $\bar{v}_{x_i}$. Now, for each i, we set

$$\tau(x_i) = \begin{cases} 1 & \text{if } v_{x_i} \text{ is the center associated with } V_i; \\ 0 & \text{otherwise.} \end{cases}$$

Since, if V_i contains a vertex v_{c_j}, then v_{c_j} must be adjacent to the center associated with V_i (otherwise by applying induction as we did above, we can prove that the solution will exceed the bound b), it follows that τ is a truth assignment satisfying C.   □

3   An Efficient Solution of the 2-Radius Problem

In this section, we will concentrate on the 2-radius problem. At a very high level, our approach works as follows. Basically, we consider all the possible pairs of nodes in G. In fact, each of these pairs is a feasible pair of subgraph centers. Thus, when we fix one of these pairs, we compute an optimal bipartition of V, namely a distribution of the nodes into two subsets in such a way that the sum of the radii with respect to the fixed centers is minimum. This is exactly the crucial step of our algorithm. Finally, out of all these feasible pairs of centers, we select a pair minimizing the sum of the radii. We start by observing that for the 2-radius problem, an approach based on the association of each node with its closest center (which works for the p-median and the p-center problem) does not work, as suggested from the example in Figure 1: If we associate node v with the closest root, namely b, then the sum of radii will be equal to 3, while for the bipartition on the right side of the picture, this sum is $2 + \epsilon$. This suggests that it might help to keep track of the distances from a root which avoid the use of the other root. More generally, we will have to determine paths that avoid the nodes on the other side of the bipartition, while at the same time finding that bipartition. It is this mutual dependence of the path lengths on the bipartition, and of the bipartition on the path lengths, that distinguishes the p-radius problem from other location or clustering problems (such as the p-center

Fig. 1. On the right side, an optimal bipartition for nodes a, b

or the p-median) and that makes it difficult to solve. In the following, for a node $v \in V$, we denote by $G - v$ the graph G deprived of the node v and all the incident edges. For a node a of graph G, let $S_G(a)$ denote a *shortest-paths tree* (SPT) of G rooted at a. We start by computing the following set of SPTs

$$\mathcal{S} = \{S_{G-a}(b) : a, b \in V, a \neq b\}.$$

This takes $O(mn^2 + n^3 \log n)$ time. Afterwards, we sort in $O(n^3 \log n)$ time the $n(n-1)/2$ obtained distance arrays (one for each SPT in $\mathcal{S}$). Each array is sorted in increasing lexicographic order of the distance value and the identifier of the associated node. In total, each node $a \in V$ is associated with $n - 1$ arrays for shortest path trees rooted in a and avoiding one specific other node. Let $Q^a_{-b}[1..n-1]$ denote the sorted distance array associated with $S_{G-b}(a)$. Let each of its entries contain the following three fields: (i) $Q^a_{-b}[k].\texttt{dist}$, the distance from a of the kth closest node in $G - b$; (ii) $Q^a_{-b}[k].\texttt{node}$, the identifier of this node; and (iii) $Q^a_{-b}[k].\texttt{pred}$, the identifier of the predecessor in $S_{G-b}(a)$ of this node. Each array can be found in $O(n)$ time, once we have computed $\mathcal{S}$.

We now give the following:

Definition 1. *Let a, b, i, j be distinct nodes of G, and let*

$$V_a(i) = \{v \in V : \delta_{G-b}(a, v) \leq \delta_{G-b}(a, i)\};$$
$$V_b(j) = \{v \in V : \delta_{G-a}(b, v) \leq \delta_{G-a}(b, j)\}. \tag{1}$$

Then, $\langle i, j \rangle$ is a 2-periphery for a, b if and only if $V_a(i) \cup V_b(j) = V$.

Recall that the radius of a rooted tree T, say $r(T)$, is the length of a maximum-length root-leaf path in T. We now give the following:

Definition 2. *Let $a, b \in V$. Then, a legal tree pair (LTP) for a, b is a pair of trees $\{T_a, T_b\}$ together spanning G and rooted at a, b, respectively, and each having a radius associated with a root-leaf path which is a shortest path in $G - b$ and $G - a$, respectively.*

We can now state the following proposition, whose proof is only sketched due to lack of space:

Proposition 1. *Let $\langle i, j \rangle$ be a 2-periphery for $a, b \in V$. Then, there exists a LTP for a, b associated with $\langle i, j \rangle$, say $\{T_a, T_b\}$, whose total radius is at most $\delta_{G-b}(a, i) + \delta_{G-a}(b, j)$, and which can be computed in $O(n)$ time, once Q^a_{-b} and Q^b_{-a} are given.*

Proof sketch. We create the two trees T_a and T_b by visiting concurrently $S_{G-b}(a)$ and $S_{G-a}(b)$ in a *closest-first-search* (CFS) order: The kth node to be examined is exactly the kth closest node to any of the roots a, b. This is done until the neighborhood of either a or b w.r.t the maximum allowed distances induced by $\langle i, j \rangle$ is exhausted. Afterwards, we focus on the root whose neighborhood has not yet been exhausted, say a, and we go ahead in adding the unreached node to T_a by maintaining the invariant that a farthest node from a in T_a is on a shortest path in $G - b$. This might require the detachment of some subtrees from T_b, more precisely all the subtrees of T_b which are rooted at the nodes of T_b belonging to the path used to reach a new node (for an illustration of this step, see Figure 2). Concerning the time complexity, it is not hard to see that by exploiting the information contained in Q^a_{-b} and Q^b_{-a}, all the computations require $O(n)$ time, from which the claim follows. □

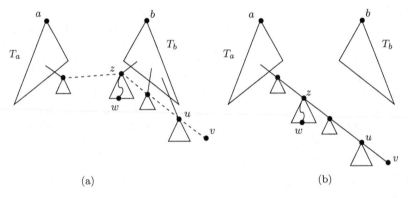

(a)   (b)

Fig. 2. In (a), a node v is visited starting from a, but its predecessor belongs to T_b; in (b), the two trees after the detachment of the associated path (triangles denote subtrees)

We can now prove the following:

Proposition 2. *Let $a, b \in V$. Then, there exists a 2-periphery $\langle i^*, j^* \rangle$ for a, b such that the associated LTP induces an optimal (in the sense of minimizing the sum of radii) bipartition of G, once a and b are fixed to be the centers of the induced subgraphs.*

Proof. Let r_a and r_b be the radii of an optimal bipartition of G when a and b are fixed to be the centers. W.l.o.g., let us assume that $r_a \geq r_b$. We now exhibit a LTP of a, b such that its total radius is $r_a + r_b$. To this aim, we select the nodes i and j in V such that

$$i \in \arg\max \big\{ \delta_{G-b}(a, v) : v \in V \text{ and } \delta_{G-b}(a, v) \leq r_a \big\};$$
$$j \in \arg\max \big\{ \delta_{G-a}(b, v) : v \in V \text{ and } \delta_{G-a}(b, v) \leq r_b \big\}.$$

We now show that $\langle i, j \rangle$ is a 2-periphery for a, b, namely we prove that for any $v \in V$, we have that either $\delta_{G-b}(a, v) \leq \delta_{G-b}(a, i)$, or $\delta_{G-a}(b, v) \leq \delta_{G-a}(b, j)$. Indeed, if there exists a node $\hat{v} \in V$ such that $\delta_{G-b}(a, \hat{v}) > \delta_{G-b}(a, i)$ and $\delta_{G-a}(b, \hat{v}) > \delta_{G-a}(b, j)$, then by definition of i and j it must be $\delta_{G-b}(a, \hat{v}) > r_a$ and $\delta_{G-a}(b, \hat{v}) > r_b$. But then it would also be $\delta_{T_a}(a, \hat{v}) \geq \delta_{G-b}(a, \hat{v}) > r_a$ and $\delta_{T_b}(b, \hat{v}) \geq \delta_{G-a}(b, \hat{v}) > r_b$, a contradiction. This means that $\langle i, j \rangle$ is a 2-periphery for a, b, and therefore we can build an associated LTP by following the algorithm suggested in Proposition 1. Clearly, since the radii of the two resulting trees cannot be larger than $\delta_{G-b}(a, i) \leq r_a$ and $\delta_{G-a}(b, j) \leq r_b$, respectively, it follows that the total radius of the LPT cannot be larger than $r_a + r_b$. More precisely, from the assumed minimality of $r_a + r_b$, it must be exactly equal to this value. □

To select an optimal LTP for a and b, we can therefore check out all the 2-peripheries associated with them. This can be done efficiently as stated in the following:

Proposition 3. *Let $a, b \in V$. Then, an optimal LTP of a, b can be computed in $O(n)$ time, once Q^a_{-b} and Q^b_{-a} are given.*

Proof. It is easy to see that when we fix the first node i of a 2-periphery, then there exists a set of corresponding nodes which can be associated with i in order to generate a 2-periphery. Among all these nodes, let us consider the subset of nodes which are closer to b in $G - a$, and let j_i be any such node. Then, $\langle i, j_i \rangle$ is a 2-periphery for a, b, and from the minimality of j_i it follows that for any $j \in V$ such that $\delta_{G-a}(b, j) \geq \delta_{G-a}(b, j_i)$, we have that $\langle i, j \rangle$ is a 2-periphery, while for any $j \in V$ such that $\delta_{G-a}(b, j) < \delta_{G-a}(b, j_i)$, we have that $\langle i, j \rangle$ is not a 2-periphery. We name $\langle i, j_i \rangle$ a *minimal 2-periphery* (M2P) *for i*. Then, from the analysis of the algorithm proposed in Proposition 1, it follows that an optimal LTP is associated with a M2P $\langle i^*, j^* \rangle$ such that $\delta_{G-b}(a, i^*) + \delta_{G-a}(b, j^*)$ is minimum. Therefore, this value can be selected within the set $\mathcal{P}(a, b)$ of all the possible M2Ps for the nodes a and b. The set $\mathcal{P}(a, b)$ can be found in $O(n)$ time as follows. Initially, we set $\mathcal{P}(a, b) := \{\langle a, j_a \rangle\}$, where $j_a = Q^b_{-a}[n - 1]$.node. Next, we consider $i = Q^a_{-b}[2]$.node (and, implicitly, all the nodes having in $S_{G-b}(a)$ the same distance from a as i); let $V_a(i)$ be this set of nodes, as defined in (1). It is not hard to see that if we maintain for each $v \in V$ a set of pointers leading to its images in all the distance arrays, then we can mark in $O(|V_a(i)|)$ time the set of nodes in Q^b_{-a} corresponding to $V_a(i)$. Thus, we search in reverse order the array entries $Q^b_{-a}[j]$.node, until we encounter the first node which is not marked as belonging to $V_a(i)$; then, we set j_i exactly equal to this node, and for an arbitrary $v \in V_a(i)$, we add $\langle v, j_i \rangle$ to $\mathcal{P}(a, b)$. At the next step, we consider the immediately farthest node in $G - b$ from a, and we continue to browse in reverse order the array entries $Q^b_{-a}[j]$.node from where we stopped at the previous step. We go ahead in this way until we complete the retrieval of all the M2Ps for the endnode a (i.e., we exhaust Q^a_{-b}). Then, we pass to compute the M2Ps for the endnode b (some of which are already available from a by symmetry), by inverting the roles of Q^a_{-b} and Q^b_{-a}. Finally, we select in $\mathcal{P}(a, b)$ the M2P $\langle i^*, j^* \rangle$. It is not hard to see that all the computations require $O(n)$ time, once Q^a_{-b} and Q^b_{-a} are given. Finally, we compute in $O(n)$ time the LTP associated with $\langle i^*, j^* \rangle$.   □

From the above proposition, it follows that in $O(n^3)$ time we compute all the optimal LTPs of G, and finally we select an optimal bipartition of G as that induced by an optimal LTP of minimum total radius. Then, we can claim the following:

Theorem 2. *Given a positively real-weighted graph $G = (V, E)$ of n nodes and m edges, the 2-radius problem can be solved in $O(mn^2 + n^3 \log n)$ time and $O(n^3)$ space.*

4   A Solution of the p-Radius Problem

In this section, we will concentrate on the general case $p > 2$. Due to lack of space, we will only sketch our technique, which is indeed an extension of the case $p = 2$. Basically, we consider all the possible p-tuples of nodes in G as feasible centers. Thus, when we fix one of these sets of centers, we compute an optimal p-partition of V, namely a distribution of the nodes into p subsets in such a way that the sum of the radii with respect to the fixed centers is minimum. Finally, out of all these feasible sets of centers, we select a p-tuple minimizing the sum of the radii. As for the case $p = 2$, the p-partition is performed as follows: assume that for a given set of centers $v_1, \ldots, v_p \in V$, the optimal associated radii are $r_1, \ldots, r_p$, respectively, and let $(V_1, \ldots, V_p)$ be the corresponding optimal p-partition. Then, this partition can be obtained by incrementally building p spanning trees of $G[V_i]$ rooted at v_i and having radius $r_i, i = 1, \ldots, p$, in a way analogous to the case $p = 2$. These particular spanning trees are associated with an *optimal p-periphery*, selected by checking all the possible p-peripheries associated with the current set of centers. Once again, an optimal p-periphery can be found by making use of the distance arrays associated with the following set of SPTs

$$\mathcal{S} = \{S_{G-\{v_1, \ldots, v_{p-1}\}}(v_p) : \{v_1, \ldots, v_p\} \in C(V, p)\},$$

where $C(V, p)$ is the set of combinations in p-tuples of all the nodes in V. Thus, we have $O(n^p/(p+1)!)$ SPTs, and the space needed for storing them is $O(n^{p+1}/(p+1)!)$. The computation of all the SPTs, along with the construction of the sorted distance arrays, takes $O(mn^p + n^{p+1} \log n)$ time. Then, the selection of an optimal p-periphery is done as follows: (1) first, consider exhaustively all the possible p-tuples of indexes of the arrays associated with the centers under consideration; since each array has $n - p + 1$ entries, this amounts to a total of $(n - p + 1)^p = O(n^p)$ possible p-tuples; (2) then, for each of these tuples, verify by scanning the distance arrays in $O(pn)$ time if it identifies a p-periphery, and if so, compute in $O(p)$ time the associated radius measure. Then, an optimal p-periphery can be computed in $O(pn^p)$ time, and since we have $O(n^p/(p+1)!)$ feasible p-tuples of centers, we can finally state the following:

Theorem 3. *Given a positively real-weighted graph $G = (V, E)$ of n nodes and m edges, the p-radius problem for $p > 2$ can be solved in $O(n^{2p}/p!)$ time and $O(n^{p+1}/(p+1)!)$ space.*

Acknowledgements. The authors would like to thank Davide Bilò, Jörg Derungs, and Luciano Gualà for inspiring discussions, and a referee for helpful comments.

References

1. H. Alt, E.M. Arkin, H. Brönnimann, J. Erickson, S.P. Fekete, C. Knauer, J. Lenchner, J.S.B. Mitchell, and K. Whittlesey, Minimum-cost coverage of point sets by disks, *Proc. 22nd ACM Symp. on Computat. Geometry (SoCG'06)*, 449–458, 2006.
2. P.K. Agarwal and C.M. Procopiuc, Exact and approximation algorithms for clustering, *Algorithmica*, 33(2):201–226, 2002.
3. V. Bilò, I. Caragiannis, C. Kaklamanis, and P. Kanellopoulos, Geometric clustering to minimize the sum of cluster sizes, *Proc. 13th Europ. Symp. on Algorithms (ESA'05)*, Vol. 3669 of LNCS, Springer-Verlag, 460–471, 2005.
4. M. Charikar and R. Panigrahy, Clustering to minimize the sum of cluster diameters, *J. of Computer and Systems Sciences*, 68(2):417–441, 2004.
5. R.M. Cormack, A review of classification, *J. of the Royal Statistical Society*, 134:321–367, 1971.
6. S.R. Doddi, M.V. Marathe, S.S. Ravi, D.S. Taylor, and P. Widmayer, Approximation algorithms for clustering to minimize the sum of diameters, *Nordic Journal of Computing*, 7(3):185–203, 2000.
7. T.F. Gonzalez, Clustering to minimize the maximum intercluster distance, *Theor. Comp. Science*, 38(23)293–306, 1985.
8. D.S. Hochbaum and D.B. Shmoys, A best possible heuristic for the k-center problem, *Mathematics of Operations Research*, 10:180–184, 1985.
9. O. Kariv and S.L. Hakimi, An algorithmic approach to network location problems. I: The p-centers, *SIAM J. Applied Mathematics*, 37(3):519–538, 1979.
10. M. Labbe, D. Peeters, and J.F. Thisse, Location on networks, in *Handbooks in Operations Research and Management Science: Network Routing*, M. Ball, T. Magnanti, and R. L. Francis Eds., Elsevier, Amsterdam, 1995.
11. N. Lev-Tov and D. Peleg, Polynomial time approximation schemes for base station coverage with minimum total radii, *Computer Networks*, 47(4):489–501, 2005.
12. J. Plesník, On the computational complexity of centers locating in a graph, *Aplikace Matematiky*, 25(6):445–452, 1980.

Efficient Prüfer-Like Coding and Counting Labelled Hypertrees

Saswata Shannigrahi[1] and Sudebkumar Prasant Pal[2]

[1] School of Technology and Computer Science, Tata Institute of Fundamental
Research, Homi Bhabha Road, Mumbai-400005, India
saswata@tcs.tifr.res.in
[2] Department of Computer Science and Engineering, and Centre for Theoretical
Studies, Indian Institute of Technology Kharagpur, 721302, India
spp@cse.iitkgp.ernet.in

Abstract. We show that r-uniform hypertrees can be encoded in linear
time using as little as $n - 2$ integers in the range $[1, n]$. The decoding
algorithm also runs in linear time. For general hypertrees, we require
codes of length $n + e - 2$, where e is the number of hyperedges. We
show that there are at most $\frac{n^{(n-2)} - f(n,r)}{(r-1)^{(r-2)*\frac{n-1}{r-1}}}$ distinct labeled r-uniform
hypertrees, where $f(n, r)$ is a lower bound on the number of trees with
vertex degrees exceeding $(r-1) + \frac{n-1}{r-1} - 2$. We suggest a counting scheme
for determining $f(n, r)$.

Keywords: hypertree, Prüfer code, coding, counting, r-uniform.

1 Introduction

Hypergraphs have been studied extensively in the combinatorics literature [1,5].
A *hypergraph* represents an arbitrary set of subsets of its vertex set, where each
subset is called a *hyperedge*. *Graphs* may be viewed as special hypergraphs where
each hyperedge has exactly two vertices. From the combinatorial viewpoint, a
natural and interesting connection can be made between *multipartite quantum
entanglement* and hyperedges; representing n nodes in a network as vertices, an
m-partite maximally entangled quantum state $\frac{(|0^m\rangle + |1^m\rangle)}{\sqrt{2}}$ $(m \leq n)$ betweem m
nodes can be represented as a hyperedge with m vertices. Such a state is called an
m-CAT state, or simply a CAT or GHZ state [2]. We can represent an ensemble
or collection of such states over the n-node network using a hypergraph made of
several such hyperedges, where each hyperedge represents a multipartite entangled (CAT state) state [10,11,8]. Given two distinct *multipartite entanglement
configurations* $H1$ and $H2$, it is always possible to construct $H1$ from $H2$, or vice
verse, if sufficient quantum communication is permitted between the n nodes of
the network. Such transformations between entanglement configurations is not
always possible with only *local operations* within nodes, and *classical communication* between nodes. Such a restricted scenario is termed $LOCC$ (Local Operations and Classical Communication). The notion of $LOCC$ *comparability* between

T. Asano (Ed.): ISAAC 2006, LNCS 4288, pp. 588–597, 2006.
© Springer-Verlag Berlin Heidelberg 2006

Table 1. Encoding the hypertree in Figure 1, where layers in the encoding procedure are numbered in increasing order starting from the outermost layer. The code for the hypertree is the Prüfer code of the tree in the final step.

i	h_i	$V(S_i)$	p_i	Prüfer code of T_i	Remaining lowest layer hyperedges
1	a	$\{1,2,3\}$	4	$< 2, 3 >$	$\{d, e, f, h\}$
2	d	$\{8,9,10\}$	5	$< 9, 9 >$	$\{e, f, h\}$
3	e	$\{23, 24, 25\}$	22	$< 25, 23 >$	$\{f, h\}$
4	f	$\{11, 12, 13\}$	7	$< 12, 12 >$	$\{h\}$
5	h	$\{17, 18, 19\}$	16	$< 17, 17 >$	$\{c, g\}$
6	c	$\{20, 21, 22(T_3)\}$	6	$< 20, 20, 22,$ $25, 23 >$	$\{g\}$
7	g	$\{14, 15, 16(T_5)\}$	$7(T_4)$	$< 12, 12, 7, 14,$ $15, 16, 17, 17 >$	$\{b\}$
8	b	$\{4(T_1), 5(T_2), 6(T_6)\}$	$7(T_7)$	$< 2, 3, 4, 9, 9, 5,$ $12, 12, 7, 17, 17, 16$ $15, 14, 7, 4, 5,$ $6, 20, 20, 22, 23, 25 >$	$\{\}$

two distinct entanglement configurations H and H' is studied in [10,11,8]; the problem is to determine whether LOCC transformations can generate an entanglement configuration H' from another entanglement configuration H, written as $H' <_{LOCC} H$. If neither of $H' <_{LOCC} H$ or $H <_{LOCC} H'$ hold, then we say that H and H' are *LOCC incomparable*. It is known that any two multipartite entanglement configurations represented by distinct labelled r-uniform hypertrees are mutually *LOCC incomparable* [10,11]. The derivation of this result was done using a non-trivial property of r-uniform hypertrees as established in [10,11]. This property may be interpreted as follows: two labeled r-uniform hypertrees defined on the same set of vertices are identical if and only of their *vertex pairing relations* are identical [8]. The *vertex pairing* relation $R(H)$ of a hypergraph is the set of *all possible pairs* of vertices (u, v), where u and v share a hyperedge in H.

In this paper we give a constructive proof of the unique reconstruction of r-uniform hypertrees from their succinct codes, in contrast to the non-constructive characterizations of [10,11,8] as mentioned above. In particular, we settle the open question in [8], by showing that r-uniform hypertrees can be encoded efficiently in sequences of $n - 2$ integers, and decoded efficiently from such succinct codes in section 2. Following Prüfer codes for trees, we call our codes Prüfer-like codes for hypertrees; our codes are of smaller length compared to the trivial bound of $\frac{n-1}{1-1/r}$ integers required to represent all hyperedges with labelled vertex numbers in each hyperedge. We are not aware of better bounds in the literature. We also show in section 3 that on the average, our codes for random r-uniform trees can be compressed significantly due to the redundancy in the codes developed in this paper. This follows from our upper bound on the number of distinct

labelled r-uniform hypertrees on n vertices as developed in section 3. Earlier,
Renyi and Renyi [9] developed Prüfer-like codes for k-trees and also devised
some counting techniques.

Let S be a set of n vertices, and $F = \{E_1, E_2, \cdots, E_m\}$, where $E_i \subseteq S; i = 1, 2, \cdots, m$. Then, the set system $H = (S, F)$ is called a *hypergraph* with *hyperedges* E_i, $1 \leq i \leq m$. We define *connectedness* for hypergraphs as follows. A
sequence of j hyperedges E_1, E_2, ..., E_j in a hypergraph $H = (S, F)$ is called
a *hyperpath* (path) from a vertex $a \in S$ to a vertex $b \in S$ if E_i and E_{i+1} have
a common vertex v_i in S, for all $1 \leq i \leq j - 1$, $a \in E_1$, and $b \in E_j$, where the
vertices v_i are distinct. A connected hypergraph $H = (S, F)$ is a *hypertree* if it
contains no cycles. In other words, no pair of vertices from S has two distinct
hyperpaths connecting them. An r-uniform hypertree is a hypertree where there
are exactly r vertices in every hyperedge. Here r is a fixed integer greater than
1. Note that any two hyperedges in a hypertree can share at most one vertex;
sharing two vertices would introduce cycles. Any vertex shared between two hyperedges in a hypertree is called a *pivot*. Any hyperedge having a single pivot is
called a *peripheral hyperedge*. In ordinary graphs, a vertex belonging to a single
edge is called a *pendant* vertex. This concept is extended to the case of hypergraphs. A vertex of a hypergraph $H = (S, F)$ belonging to exactly one hyperedge
from the set F is called a *pendant* vertex in H. Note that each hyperedge in any
(multi-hyperedge) connected hypergraph must have at least one non-pendant
vertex. Throughout this paper, we deal with vertex labelled hypertrees.

2   Efficient Coding of r-Uniform Hypertrees

We know thet the number of hyperedges $e = (n - 1)/(r - 1)$, for any n-vertex
r-uniform hypertree as shown in [10,11]. So, a labelled r-uniform hypertree can
be encoded using $e * r = \frac{n-1}{1-\frac{1}{r}}$ integers in the range $[1, n]$. We show that such

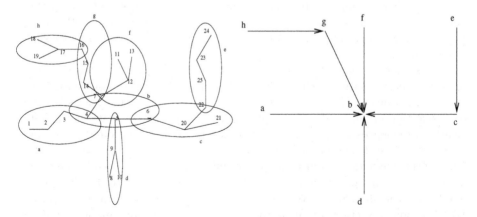

Fig. 1. A 4-uniform hypertree with 25 vertices and its (i) encoding tree (left), and (ii)
hypergraph partial order (right)

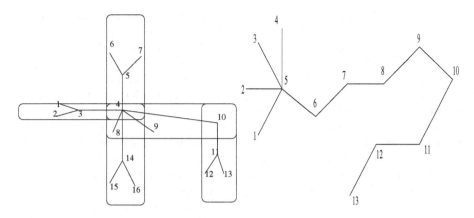

Fig. 2. (i) A 4-uniform hypertree generating an encoding tree with maximum degree (left), and (ii) a tree that is not an encoding tree of any 4-uniform hypertree

Table 2. Decoding the 4-uniform hypertree of Figure 1. (All steps are not shown). The chosen separators are $\{4, 5, 7, 16, 22, 6, 7, 7\}$. The reconstructed edges are $\{a, d, f, h, e, c, g, b\}$.

i	$V(R_i)$	V Vertices to be deleted from T_i	$sep(R_i, T_i)$	h_i
1	$\{1, 2, 3, 4\}$	$\{1, 2, 3\}$	4	$a = \{1, 2, 3, 4\}$
2	$\{5, 8, 9, 10\}$	$\{8, 9, 10\}$	5	$d = \{5, 8, 9, 10\}$
8	$\{4, 5, 6, 7\}$	$\{4, 5, 6\}$	7	$b = \{4, 5, 6, 7\}$

hypertrees can be encoded using as little as $n - 2$ integers in the range $[1, n]$. It is a well-known fact that any n-vertex labelled tree has a unique encoding in the form of a sequence of $n - 2$ integers, popularly known as Prüfer coding (see [4,7]). Note that a tree is actually a 2-uniform hypertree. We start with the following useful property, omitting the proof in this version (see [12]).

Lemma 1. *Any r-uniform hypertree has at least one peripheral hyperedge.*

We require the notion of *elimination* of a peripheral hyperedge from the hypertree. *Elimination* of a peripheral hyperedge is the removal of the hyperedge and all the terminal vertices in the hyperedge.

Observation 1. *The elimination of a peripheral hyperedge from an r-uniform hypertree results in a r-uniform hypertree with $n - r + 1$ vertices.*

Let $HT(S, F)$ be a hypertree defined on vertex set S with the set E of hyperedges. We define the *hyperedge partial order* $(hpo(HT))$ on the hyperedges of the hypertree $HT(S, F)$ by defining a DAG with vertex set F where each vertex represents a hyperedge of HT. For each peripheral hyperedge $a \in F$, introduce a directed edge (a, b) if $b \in F$ is not a peripheral hyperedge and a and b share a vertex in

S; delete all such peripheral hyperedges a from the hypertree HT. These deleted hyperedges form a *layer* of the DAG. Repeat this process on the resulting hypertree until all hyperedges are deleted, layer by layer. See Figure 1(ii). Note that this is a unique DAG for a given hypertree since the peripheral hyperedges in each layer are uniquely determined. The approach we adopt is to first compute an undirected graph underlying the hyperedge partial order, and then traverse the this graph in order to identify and process hyperedges, layer by layer.

Consider the undirected graph $G(F, E)$ defined on the set of hyperedges F as the vertex set; the edge set E is made of unordered pairs of hyperedges from F that share a vertex of the given hypertree $HT(S, F)$. We start with a vertex in $G(F, E)$ (as root), corresponding to an arbitrary peripheral hyperedge in the given hypertree $HT(S, F)$, and perform a depth first search on $G(F, E)$ to build a DFS tree. At each vertex $v \in F$, we maintain the $depth(v)$ and $height(v)$, where $depth(v)$ is the usual DFS depth of the node v and $height(v)$ is the maximum over $height(u)$, u being the children of v in the DFS. The minimum over $depth(v)$ and $height(v)$ is the layer number of the hyperedge corresponding to the vertex v. Clearly, these layer numbers can therefore be computed in linear time, thereby yielding the hyperedge partial order $hpo(HT)$.

The encoding procedure: Let $H_1(= HT)$, H_2, ..., H_{e+1} be a sequence of r-uniform hypertrees, where H_{i+1} is generated by the deletion of a peripheral hyperedge h_i from H_i, $1 \leq i \leq e$. [By Lemma 1, such a hyperedge h_i exists in H_i. By Observation 1 we know that H_i, $1 \leq i \leq e$, are r-uniform.] Here HT is the given hypertree, e is the number of its hyperedges, and H_{e+1} is the empty hypertree. We ensure that peripheral hyperedges h_i are deleted layer by layer; as a peripheral hyperedge h_i is deleted from H_i, any new peripheral hyperedge introduced in H_{i+1} can be determined by keeping track of hyperedges sharing the common pivot p_i with h_i. This whole procedure is performed in linear time since the layers have already been computed in linear time in the hyperedge partial order as explained above. Also, note that the last (innermost) layer will have either a single hyperedge or a set of hyperedges, all sharing a single vertex. For the latter case, we say that the single common vertex is the pivot for each hyperedge in the last layer. For the case of a single hyperedge in the last layer, the pivot is defined as the vertex which was a (common) pivot of the maximum number of hyperedges of the penultimate layer (ties are broken arbitrarily).

Let p_i be the pivot of h_i in H_i, and S_i be any spanning tree of the pendant vertices of h_i in H_i. Now, let v be any vertex of S_i. For each v, if v is the pivot p_j of h_j in H_j, for some $j < i$, then assume that the *partial encoding subtree* T_j is already attached to S_i at $v = p_j$. Construct a (partial encoding) subtree tree T_i connecting r vertices including (i) the attached partial encoding subtrees T_j rooted at pivot vertices $p_j \in S_i$, $j \leq i$, (ii) the remaining (non-pivot) vertices of S_i, and (iii) the pivot vertex p_i of h_i; some partial encoding tree T_j might already have been created and attached to this vertex $p_i(= p_j)$ (see Table 1, the cases of T_4 and T_7). Make p_i the root of T_i by connecting p_i to an arbitrary vertex w of S_i. This completes the construction of the *partial encoding subtree* T_i. [In order to complete the inductive definition of $T_i's$, we define T_1 as S_1 connected to the pivot p_1 of h_1 in $H_1 = HT$, connected to S_1 at an arbitrary vertex v of S_1.] We call the final spanning tree T_e the *encoding tree* $T(HT)$ of the hypertree HT and use the Prüfer code $E(HT)$ of T_e as an encoding of HT. Observe that there can be several codes for HT because the spanning trees S_i can be built in many different ways, and the vertex w of S_i chosen to be attached to pivot p_i can also be chosen in many ways. Also, note that all these steps can be performed in linear time; in particular, the Prüfer coding of $T(HT)$ can be done in linear time by the algorithm of Chen and Wang [3].

Theorem 1. *The encoding algorithm generates the code $E(HT)$ of the given r-uniform hypertree HT by generating the tree T_e and its Prüfer code in linear time.*

The decoding procedure: Now consider the problem of reconstructing a labelled r-uniform hypertree from its encoding $E(HT)$. Here $E(HT)$ is the Prüfer code of the encoding tree $T(HT)$ of HT. It is possible to reconstruct the encoding tree $T(HT)$ of HT from the Prüfer code $E(HT)$ in $O(n)$ time [6]. We need a few definitions. Let $V(R)$ denote the set of vertices of a tree R. Let R be any proper subtree of a tree T. We call a vertex $sep(R, T)$ of R, a *separator of R in T*, if the deletion of $sep(R, T)$ from T gives two or more subtrees, at least one of which is exclusively made of one or more vertices of $T \setminus R$. Let R_1 be such a subtree of $T(HT)$ with r vertices and a single separator vertex $sep(R_1, T(HT))$ whose degree in R_1 is 1 (in Table 2, R_1 has vertex set $\{1, 2, 3, 4\}$). We say that such a subtree R_1 satisfies the *reconstruction criterion*. Note that the set of vertices of each peripheral hyperedge of HT satisfies the reconstruction criterion. [This follows from the construction of the encoding tree T_e of HT in the encoding algorithm; in the construction of each T_i, the pivot p_i is the separator $sep(T_i, T(HT))$, with degree 1 in T_i.] We now argue that no subtree of r vertices can satisfy the reconstruction criterion if it does not constitute a peripheral hyperedge in HT. Suppose the subtree chosen does not constitute a hyperedge of HT. Then it must have two vertices such that no hyperedge in HT has both these vertices. A subtree including such two vertices must have more than one separator if it has r vertices, or must have at least $2r - 1$ vertices if there is only one separator. So $V(R_1)$ must constitute a hyperedge of HT; we now show that $V(R_1)$ is, in particular, also a peripheral hyperedge of HT.

Lemma 2. *The vertices of R_1 form a peripheral hyperedge in HT with pivot $sep(R_1, T(HT))$.*

Proof: For the sake of contradiction, we assume that the set $V(R_1)$ does not constitute a peripheral hyperedge in HT. So, there are 2 or more pivots in $V(R_1)$, a contradiction because the constitution of R_1 permits only one pivot, which is the same as $sep(R_1, T(HT))$. □

The reconstruction of HT from $E(HT)$ is done in e steps; in the process hyperedges h_1, h_2, ..., h_e of HT are identified. We claim that all the identified hyperedges are exactly the hyperedges of HT. Following Lemma 2, we observe that $V(S_1)$ constitutes a peripheral hyperedge in HT; so we call this hyperedge h_1. For the remaining hyperedges we have the following lemma whose proof is similar to that of Lemma 2. Suppose we have already identified i hyperedges of HT, h_1, h_2, ..., h_i, $i \leq e$, following the hyperedge partial order, layer by layer. The computation of h_i from T_i has already been done transforming T_i to T_{i+1} by the deletion of $r - 1$ vertices of h_i. [The r-th vertex of h_i which is retained in T_{i+1} is the separator vertex $sep(R_i, T_i)$.] Let R_{i+1} be a subtree of T_{i+1} satisfying the reconstruction criterion. More precisely, R_{i+1} has r vertices and a single separator vertex $sep(R_{i+1}, T_{i+1})$, whose degree in R_{i+1} is 1. By the definition of the hyperedge partial order and the way we designed the encoding procedure, it is also clear that R_{i+1} constitutes the hyperedge h_{i+1} belonging to the proper layer of $hpo(HT)$. All we need to do now is to establish the two *hereditary properties* by induction that (i) T_{i+1} is a (valid) encoding tree of $HT\backslash\{h_1, h_2, ..., h_i\}$. [Here, $HT\{h_1, h_2, ..., h_i\}$ denotes the remaining hypertree after the deletion of the sequence $h_1, \cdots, h_i$ of hyperedges from HT], and (ii) $hpo(HT \backslash \{h_1, h_1, ..., h_{i-1}\})$ is identical to h_i appended with $hpo(HT \backslash \{h_1, h_1, ..., h_i\})$. [We omit details of induction in this version.] Following arguments similar to those given for R_1 and the Lemma 2, we have the following claim about the induction step.

Lemma 3. $V(R_{i+1})$ *form a peripheral hyperedge in* $HT \backslash \{h_1, h_2, ..., h_i\}$ *with pivot* $sep(R_{i+1}, T_{i+1})$.

The reconstruction with hyperedges h_i of HT proceeds obeying the hyperedge partial order, in batches, layer by layer. The entire reconstruction is possible in linear time (see [12]); we omit the analysis in this version. We can now claim the following result.

Theorem 2. *Given the Prüfer code $E(HT)$ of the encoding tree $T(HT)$ of an r-uniform hypertree HT, it is possible to reconstruct HT from $E(HT)$ in linear time.*

Proof: The sets R_i and separators $sep(R_i, T_i)$ are identified by the decoding algorithm, such that the sets R_i satisfy the reconstruction criterion in the remaining tree (encoding tree) T_i. By Lemmas 2 and 3, these identified hyperedges are exactly those of HT.   □

3   Counting r-Uniform Hypertrees

Our encoding algorithm for labelled r-uniform hypertrees produces a code of $n - 2$ integers, each of which is an integer in the range $[1, n]$. This code is not unique however. In the following lemma we derive the exact number of distinct codes possible for any given input.

Lemma 4. *The encoding algorithm generates one of $(r - 1)^{(r-2)*\frac{n-1}{r-1}}$ distinct labelled encoding trees for any r-uniform hypertree.*

Proof: In each step of our encoding algorithm, we select one peripheral hyperedge and delete its pendant vertices. There are $r - 1$ pendant vertices in a peripheral hyperedge. These $r - 1$ pendant vertices can be connected to form one of $(r-1)^{(r-3)}$ different labelled (partial encoding) subtrees. The root of each of these labelled trees can be selected in $r - 1$ ways. So, from each peripheral hyperedge, we can get $(r-1)^{(r-2)}$ different rooted subtrees. After we delete these $r - 1$ vertices, we consider the second hyperedge in the deletion sequence. We can again create one of $(r - 1)^{(r-2)}$ different rooted labelled (partial encoding) subtrees corresponding to the different subtrees with the r vertices of this second hyperedge. Since a total of $\frac{n-1}{r-1}$ hyperedges are processed, the total number of distinct labelled encoding trees of the original hypertree is $(r - 1)^{(r-2)*\frac{n-1}{r-1}}$. The order in which hyperedges are deleted does not matter in the above count, because hyperedges are deleted layer by layer. □

We know by Cayley's formula that the number of distinct labelled trees on n vertices is n^{n-2}. The above lemma gives the exact number of possible distinct labelled encoding trees that our encoding algorithm can generate for any given r-uniform hypertree. All these labelled encoding trees, when decoded, return the unique original encoded hypertree. So, based on our coding scheme, we get an upper bound on the number of distinct r-uniform hypertrees with n vertices as $\frac{n^{n-2}}{(r-1)^{(r-2)*\frac{n-1}{r-1}}}$.

Improving the upper bound: This upper bound can be improved if we observe that all possible labelled trees on n vertices are not encoding trees of a r-uniform hypertree. See Figure 2(ii) for an example of a labelled tree which is not a valid encoding tree for any 4-uniform hypertree. We need better estimates of the number of potential encoding trees for r-uniform hypertrees. One way is to use necessary conditions as follows.

Lemma 5. *Given an r-uniform hypertree HT, let $T(HT)$ be encoding tree for HT. The degree of each vertex in $T(HT)$ is less than or equal to $(r-1)+(\frac{n-1}{r-1})-2$.*

Proof: Observe that in the last step of the encoding procedure, the last hyperedge f processed can give an encoding tree in which one vertex v of this hyperedge is connected to the $r - 1$ remaining vertices. We show that there could have been up to $\frac{n-1}{r-1} - 2$ hyperedges in the very previous stage or layer of hyperedges (call this set K) with vertex v as the pivot, leading to exactly $\frac{n-1}{r-1} - 2$ vertices connected to v in the encoding tree. The hyperedge f must survive the stage in which the hyperedges in K are processed so that vertex v in f is connected to $r - 1$ vertices of f as already mentioned. This will give the maximum possible degree of $(r - 1) + (\frac{n-1}{r-1}) - 2$; the deficiency of 2 is due to the fact that at least one peripheral hyperedge connected to a vertex $u \neq v$ of f must also be eliminated in the layer of the hyperedges in the set K, thereby limiting the size of K to $\frac{n-1}{r-1} - 2$. See illustration in Figure 2(i). In this figure, $r = 4$, $n = 16$, the number of hyperedges $e = 5$, $v = 4$ and $u = 10$. □

We wish to obtain a lower bound $f(n,r)$ on the number of trees with maximum vertex degree exceeding $d = (r - 1) + (\frac{n-1}{r-1}) - 2$. We consider trees where one vertex v, has degree $d + i$, where $1 \leq i \leq n - d - 1$. This vertex v is connected to $d + i$ non-empty subtrees with a total of $n - 1$ vertices. We have to therefore partition the $n-1$ labelled vertices into $d+i$ unlabelled non-empty subtrees, and generate all possible labelled subtrees in each partition. So, we write $f(n, r) =$

$\Sigma_{i=1}^{n-d-1} g(n, r, i)$, where $g(n, r, i)$ is the number of trees with at least one node of degree $d + i$. Let (i) $p_j, 1 \le j \le d + i$ be integers such that $p_j > 0$, (ii) $p_j \le p_{j+1}, 1 \le j \le d + i - 1$, and (iii) $\Sigma_{j=1}^{d+i} p_j = n - 1$. We can write $g(n, r, i)$ as the sum of $h(n, r, i, j)$ over all partitions $p_1, p_2, \ldots, p_{d+i}$ of $n - 1$, satisfying conditions (i), (ii) and (iii), where $h(n, r, i, j)$ is the number of possibilities with the jth tree has p_j vertices. The choice of these p_j vertices from the total of $n - 1$ vertices can be done in after p_k vertices are selected for subtrees $k = 1, 2, \ldots, j - 1$. Then, these p_j labelled vertices can be built in to a subtree in $p_j^{p_j - 2}$ ways and connected by any of the p_j vertices, to the vertex v with degree $d + i$, resulting in a total of $p_j^{p_j - 1}$ possibilities. However, parts that have the same cardinality may be permuted arbitrarily without any effect, so we need to divide by $count(p_j)!$ only once, where $count(p_j)$ is the number of times p_j appears in the partition. So, $h(n, r, i, j) = \dfrac{\Pi_{j=1}^{d+i} [^{(n-1-\Sigma_{k=1}^{j-1} p_k)} C_{p_j}] p_j^{p_j - 1}}{\Pi_{j=1, (p_j \ne p_{j+1})}^{d+i} count(p_j)!}$. Here, $^n C_r$ is the number of combinations without repetitions of n objects taken r at a time. Thus we define the scheme for determining the lower bound for $f(n, r)$.

Using our encoding algorithm in Section 2, any r-uniform hypertree with n vertices can be encoded as a sequence of $n - 2$ integers, where each integer is in the range [1,n]. If we randomly choose such a sequence of $n - 2$ integers, the sequence may or may not represent a r-uniform hypertree. Using our decoding algorithm of Section 2, the generating party can determine whether the randomly generated sequence indeed encodes an r-uniform hypertree. Since all sequences can be generated with equal probability, the probability p that a code for a specific r-uniform hypertree is generated exceeds $\dfrac{(r-1)^{(r-2)*\frac{n-1}{r-1}}}{n^{n-2}}$ (see Lemma 4). So the entropy $\log \frac{1}{p} \le (n - 2) \log n - \dfrac{(r-2)*(n-1)*\log(r-1)}{(r-1)}$. This upper bound on the entropy is strictly less than $(n - 2) \log n$ for $r > 2$.

4   Conclusion

We can extend our results to general hypertrees where each hyperedge has at least 2 vertices. The hyperedge partial order and the encoding tree can be computed by applying the same encoding procedure as in Section 2. The additional cost to be paid is the increase in the length of the code; we need to append a list of pivots. Since the number of pivots is bounded by the number e of hyperedges, we need at most $(n-2)+e$ integers to encode a general hypertree. For decoding, we can apply the same decoding procedure of Section 2, with slight modification. Instead of checking whether a subtree of size $r - 1$ has been identified for R_i, as required in the definition of the separator $sep(R_i, T_i)$, we now need to check whether we have reached a pivot, where the pivot is a separator of degree 1 in the subtree T_i. The rest of the decoding algorithm remains unchanged. Proving better bounds for the length of Prüfer-like codes for general hypertrees is an interesting open question.

Distinct hypertrees generated using some combinatorial enumerative scheme may be used to compute distinct codes for hypertrees using our encoding algorithm. The generation of such codes requires time proportional to the length of the code. Such a scheme may be used for allocating unique IDs or PINs to users where each group of users can be associated with a distinct hypertree whereas users within a group can be allocated distinct codes of the same hypertree associated with the group.

Acknowledgements

The authors would like to sincerely thank Subir Ghosh, Arnab Basu and Prajakta Nimbhorkar for their valuable comments and suggestions. The first author also thanks Center for Theoretical Studies, IIT Kharagpur for inviting him for a visit during which the rudiments of this problem were conceptualized. The second author expresses his thanks to the School of Technology and Computer Science, TIFR, for sponsoring a summer visit to the School during which this problem was solved and the present manuscript took its final form.

References

1. C. Berge, *Hypergraphs*, Elsevier Science Publishers, N.H. (1989).
2. D. Bouwmeester, A. Ekert and A. Zeilinger (Eds.), *The Physics of Quantum Information*, Springer (2000).
3. H. -C. Chen and Y. -L. Wang, An efficient algorithm for generating Prüfer codes from labelled trees, Theory of Computer Systems 33: 97-105 (2000).
4. N. Deo, *Graph Theory: With Applications to Engineering and Computer Science*, Prentice Hall (1974).
5. R. L. Grahmam, M. Grotschel, and L. Lovasz (Eds), *Handbook of Combinatorics*, vol-1, Elsevier Science Publishers , N.H. (1995).
6. R. Greenlaw, M. M. Halldorsson and R. Petreschi, On computing Prüfer codes and their corresponding trees optimally, *Proceedings Journees de l'Informatique Messine, Graph Algorithms*, 2000.
7. L. Lovász, J. Pelikán and K. Vesztergombi, *Discrete Mathematics: Elementary and Beyond*, Springer, 2003.
8. S. P. Pal, S. Kumar and R. Srikanth, *Multipartite entanglement configurations: Combinatorial offshoots into (hyper)graph theory and their ramifications*, to appear in the Proceedings of the workshop on *Quantum Computing: Back Action*, IIT Kanpur, March 2006.
9. C. Rényi and A. Rényi, *The Prüfer code for k-trees*, Combinatorial Theory and its Applications III, pp. 945-971, (Eds.) P. Erdös, A. Rényi and Vera T. Sós, North-Holland Publsihing Comapny, 1970.
10. S. K. Singh, *Combinatorial Approaches in Quantum Information Theory*, M.Sc. Thesis, Dept. of Mathematics, IIT Kharagpur, India, eprint quant-ph/0405089 (2004).
11. S. K. Singh, S. P. Pal, S. Kumar and R. Srikanth, A combinatorial approach for studying local operations and classical communication transformations of multipartite states, J. Math. Phys. 46, 122105 (2005); eprint quant-ph/0406135 v3.
12. Saswata Shannigrahi and Sudebkumar Prasant Pal, Efficient Prüfer-like coding and counting labelled hypertrees (manuscript).

Intuitive Algorithms
and t-Vertex Cover[*]

Joachim Kneis, Daniel Mölle, Stefan Richter, and Peter Rossmanith

Department of Computer Science, RWTH Aachen University, Fed. Rep. of Germany
{kneis, moelle, richter, rossmani}@cs.rwth-aachen.de

Abstract. Many interesting results that improve on the exponential running times of exact algorithms for NP-hard problems have been obtained in recent years. One example that has attracted quite some attention of late is t-VERTEX COVER, the problem of finding k nodes that cover at least t edges in a graph. Following the first proof of fixed-parameter tractability, several algorithms for this problem have been presented in rapid succession. We improve on the best known runtime bound, designing and analyzing an intuitive randomized algorithm that takes no more than $O(2.0911^t n^4)$ steps. In fact, we observe and encourage a renewed vigor towards the design of intuitive algorithms within the community. That is, we make a plea to prefer simple, comprehendable, easy-to-implement and easy-to-verify algorithms at the expense of a more involved analysis over more complicated algorithms that are specifically tailored to ease the analysis.

1 Introduction

Under the assumption P≠NP, which is widely believed to hold true, thousands of well-known problems — many of which have important practical applications — cannot be solved in polynomial time. This troublesome fact has been addressed by several concepts such as approximation, heuristics, randomization and parameterization.

In spite of the unquestionable success of approximation methods, examples abound of problems where this approach is not viable, either because approximation is not possible or because an exact result is required. The obvious practical need has inspired renewed effort in the field of exact algorithms, and the exponential runtime bounds for many problems have been improved in recent years. Some examples for such improvement have been achieved with new algorithms for 3-SATISFIABILITY [17], INDEPENDENT SET [4,26], DOMINATING SET [14], and MAX-CUT [18,28,30].

Still, it seems that our community is lacking powerful analytical methods. In practice, many hard problems can be solved extremely fast using heuristics — for which it is usually impossible to prove non-trivial runtime bounds. On the other hand, the analysis of exact algorithms for these problems may only result in prohibitively large bounds.

---

[*] Supported by the DFG under grant RO 927/7-1.

T. Asano (Ed.): ISAAC 2006, LNCS 4288, pp. 598–607, 2006.
© Springer-Verlag Berlin Heidelberg 2006

There are two possible reasons for this effect: Firstly, it may be the case that no worst-case instances do occur in practical applications. There has been a lot of research that aims to identify the complexity inherent in a problem instance, such as Spielman and Teng's smoothed analysis approach [29] or the whole field of parameterized complexity [12]. A lot of notable results have been obtained using the latter approach, which allows for closer inspection of NP-hard problems. The underlying idea of this paradigm is to identify a measure describing the hardness of a problem instance that, if bounded, eases the solution. For example, the famous Steiner tree problem — given an edge-weighted graph with n nodes k of which are marked as terminals, find a cheapest tree that spans the terminals — is NP-hard, but since there are algorithms with running times of the form $O(c^k n^{O(1)})$ where $2 < c \leq 3$, it can be tackled even for very large n provided that k is reasonably small [13,20].

Secondly, it may be that we have hold of a rather efficient exact algorithm but have not been able to prove that it is as fast as we think it is. Indeed, there are many exact algorithms that perform well on every tested instance, and surprisingly better than suggested by the runtime bounds obtained through a theoretical analysis.

In many cases, the only known way of improving on the bounds for exact algorithms lies in introducing manifold case distinctions into the algorithm. Tailoring the analytical tools to each of the cases then leads to a better bound, but in fact, this often decreases the actual practical performance. Moreover, it makes the algorithms less understandable, more error-prone, harder to maintain, and often nearly impossible to verify.

In contrast to such tendencies, the paper at hand can be seen to be a plea for *intuitive* algorithms. We aim at obtaining algorithms that are very easy to comprehend and that can be turned into short and concise high-level programs. This, of course, is most likely to happen at the expense of a more involved analysis — a price we are willing to pay. Depending on the actual problem, new methods and techniques may be required. Intuitive algorithms are by no means a new idea. In recent years, there have been quite some surprising results that could be obtained with very simple algorithms, such as Schöning's algorithm for k-SAT [27]. As another illustrative instance, new bounds on the treewidth of sparse graphs have led to intuitive algorithms that come close to or even beat the hitherto best known bounds for problems like MAX-CUT or MAX-2-SAT [18].

A fruitful concept toward the construction of intuitive algorithms, which also combines very well with parameterized complexity, lies in randomization. A very elegant and famous example is the color-coding technique by Alon et al. [2], which yields an $O(5.44^k poly(n))$ algorithm with exponentially small error probability for LONGEST PATH (i.e., finding a path of length k in a graph) and similar results for several graph packing problems. Whereas it is possible to derandomize color-coding algorithms, the resulting running times are impractically large. Even in the face of some recent results [25] that call into question the complexity-theoretical power of randomization, probabilistic algorithms often outperform the best known deterministic method while being much more simple and intuitive at the same time.

2 Previous and New Results

In this paper, we design and analyze a simple algorithm for the VERTEX COVER variant t-VERTEX COVER. VERTEX COVER, the problem of checking for a set of at most k nodes that cover all the edges of a graph $G = (V, E)$, arguably constitutes the most intensely studied problem in parameterized complexity. Its importance is reflected by a long history of improved runtime bounds [3,21,10,22,9] culminating in the algorithm by Chen, Kanj, and Xia with running time $O(1.2738^k + kn)$ [11]. Moreover, many generalizations have been introduced and investigated, for instance CONNECTED VERTEX COVER, TREE COVER, and CAPACITATED VERTEX COVER [15,19].

Given a graph $G = (V, E)$ and numbers $k, t \in \mathbf{N}$, the problem t-VERTEX COVER asks for a set of at most k nodes that cover at least t edges. There are several positive results regarding the approximability of t-VERTEX COVER [23,6,16]. In this setting, the minimum number of nodes to cover t edges is approximated. As in the case of VERTEX COVER itself, an approximation ratio of two can be achieved. The parameterized complexity, however, of this problem has long remained unclear.

Recently, Markus Bläser [5] has shown that the problem is in fact fixed-parameter tractable when parameterized in t: Using color-coding, it is possible to obtain a randomized $O(5.44^t poly(n))$ algorithm. Just as in the case of LONGEST PATH, the running time of the derandomized version is impractically large. Obviously, fixed-parameter tractability also holds when parameterizing in both k and t.

Cai, Chan and Chan have applied the new random separation method [8] to several problems [7], including t-VERTEX COVER. The resulting randomized algorithm has a running time of $O(4^t poly(n))$, but compared to Bläser's algorithm, it allows for a much better derandomization. This can be achieved using (ϵ, k)-independent sets [1], yielding a deterministic version solving t-VERTEX COVER in $O(16^t poly(n))$ steps. This is the best deterministic bound currently known. Raman and Saurabh[24] have furtherwise come up with an $O(t^k)$ algorithm for t-VERTEX COVER this year, a result that is only relevant for small values of k. On the other hand, note that t-VERTEX COVER is known to be W[1]-hard when only k is chosen as a parameter [15, Theorem 11].

In the next section we present an intuitive randomized algorithm that solves t-VERTEX COVER in $O(2.0911^t n \cdot (n + m)k)$ steps with exponentially small error probability. As a kind of by-product of the somewhat involved analysis, a problem kernel of size $O(t^3)$ can be constructed efficiently.

Given natural numbers k and t, we say that a t-vertex cover C is *optimal*, under the implicit constraint that $|C| \leq k$, if it covers a maximal number of edges. We also adhere to the following standard notation for neighborhoods of nodes and induced subgraphs.

Definition 1. *Let $G = (V, E)$ be a graph and $V' \subseteq V$. We define:*

- $N(v) := \{\, w \in V \mid \{v, w\} \in E \,\}$,
- $N[v] := N(v) \cup \{v\}$,
- $V'[a, b] := \{\, v \in V' \mid a \leq \deg_G(v) \leq b \,\}$, *and*
- $G[V']$ *is the subgraph of G induced by V'.*

3 A Randomized Algorithm for t-Vertex Cover

In what follows, we present a randomized algorithm (see Table 1) that takes $O((n + m)k)$ steps where $n = |V|$ and $m = |E|$. Its error probability is bounded by $1 - 2.091^{-t}$. In effect, this algorithm just chooses a vertex v of maximum degree d. If $d < 3$, the solution can be easily found in linear time[1]. Otherwise, it selects either v itself (with probability $1/2^d$) or a randomly choosen neighbor of v (with probability $1 - 1/2^d$) for the vertex cover. The selected node u is removed from the graph, and the algorithm searches a $(t - deg(u))$-vertex cover for the remaining graph.

Table 1. The randomized algorithm (TVC)

$\mathrm{TVC}(G, k, t)$:
if $t \leq 0$ **then return** *true* **fi**;
if $E = \emptyset$ **or** $k \leq 0$ **then return** *false* **fi**;
choose $v \in V$ of maximum degree;
$d := \deg(v)$;
if $d < 3$ **then** find the solution in linear time **fi**;

randomly choose $u \begin{cases} \in N(v) \text{ with probability } 1 - 2^{-d} \\ = v \text{ with probability } 2^{-d}; \end{cases}$

if $\mathrm{TVC}(G[V - \{u\}], k - 1, t - \deg(u))$ **then return** *true* **fi**;
return *false*

While it is quite intuitive that either a node of maximum degree or some of its neighbors should be part of an optimum solution, it is not so obvious that behaving according to this particular distribution yields a favorable success probability. When aiming at a randomized algorithm with a running time of about 2^t, however, we have to pick this maximum-degree node with probability at least 2^{-d}: Consider the case that iterating this choice is the only way to obtain the optimum t-vertex cover $\{v_1, \ldots, v_k\}$, and that this solution covers exactly t edges (as exemplified by k isolated t/k-stars). Then the overall success probability is at least $2^{-d_1} \cdot \ldots \cdot 2^{-d_k} = 2^{-t}$, where $d_i := \deg(v_i)$. Therefore, it seems reasonable to use exactly this probability in our algorithm.

The following key observation formalizes the fact that in some optimal solution, many edges will be covered by either v or some neighbors of v all having the same degree.

---

[1] In this case, the graph only consists of paths and cycles.

Lemma 1. *Let $G = (V, E)$ be a graph and $k, t \in \mathbf{N}$. Let v be a node of maximum degree in G and $d = \deg(v) \geq 3$, $S = N[v]$. Assume that there is no optimal t-vertex cover C with either $v \in C$ or $N(v) \subseteq C$. Then there is an i, $3 \leq i \leq d$, and an optimal t-vertex cover C' such that*

$$\left| C' \cap S[i, i] \right| \geq \frac{d}{2^i}.$$

Proof. Let first C be an optimal t-vertex cover with $v \notin C$. If $C \cap N(v)$ contains a node u of degree two, then $(C \cup \{v\}) \setminus \{u\}$ is an optimal t-vertex cover whenever $N(v) \not\subseteq C$. We may thus assume that an optimal t-vertex cover not containing v never contains a neighbor u of v with $\deg(u) \leq 2$.

Now let C' be an optimal t-vertex cover with

$$\left| C' \cap S[i, i] \right| < \frac{d}{2^i}$$

for $3 \leq i \leq d$. Obviously, the nodes from $C' \cap S[i, i]$ cannot cover $di/2^i$ or more edges. Hence, all the nodes from $S[3, d]$ cannot cover $\sum_{i=3}^{d} i/2^i$ or more edges. Note that

$$\sum_{i=3}^{\infty} \frac{i}{2^i} = \sum_{i=3}^{\infty} \sum_{j=1}^{i} \frac{1}{2^i} = \sum_{j=3}^{\infty} \sum_{i=j}^{\infty} \frac{1}{2^i} + 2 \sum_{i=3}^{\infty} \frac{1}{2^i} = \sum_{j=3}^{\infty} \frac{1}{2^{j-1}} + \frac{1}{2} = 1.$$

If $C' \cap S \neq \emptyset$, then C' cannot be optimal, since replacing a node from $C' \cap S$ with v yields a better t-vertex cover—a contradiction to the optimality of C'. Otherwise, if $C' \cap S = \emptyset$, replacing an arbitrary node in C' with v yields another optimal t-vertex cover, because $d = \deg(v)$ is maximal. This contradicts the assumption in the statement of the lemma. $\qquad\square$

Now we can estimate the success probability for a single call of the algorithm. Notice that for a Monte-Carlo algorithm like this, the notion of success probability replaces the distinct notions of correctness and running time in the deterministic case.

Lemma 2. *On a yes-instance, Algorithm TVC returns true with probability at least $2^{-t}(7/8)^{t/3}$. It always returns false on a no-instance.*

Proof. The second statement is easily observed. We show the first statement by induction on t: Let (G, t, k) be a yes-instance. Except for some special cases in which the algorithm behaves deterministically, it considers a node v with maximum degree $d \geq 3$. If $t \leq d$, the algorithm returns *true* in any case.

According to Lemma 1, either a maximum degree node v or one of its neighbors belongs to an optimal solution. In the first case, the algorithm guesses correctly with probability 2^{-d}. By induction, the overall success probability in this case is at least

$$2^{-d} \cdot 2^{-(t-d)} (7/8)^{(t-d)/3} \geq 2^{-t} (7/8)^{t/3}.$$

Otherwise, v is not part of any optimal solution. There are two subcases to investigate:

Assume first that there is a degree-two neighbor u of v in some optimal solution C. This implies $N(v) \subseteq C$, because otherwise $C \cup \{v\} \setminus \{u\}$ is an optimal solution as well: While retaining the same cardinality as C, it covers the same edges except for $\{u, x\}$ with $x \neq v$, which is replaced by $\{v, u'\}$ for some $u' \in N(v) \setminus C$. The algorithm chooses some node from $N(v)$ with probability $1 - 2^{-d} \geq 7/8$. Using the induction hypothesis, we obtain a lower bound to the overall success probability of

$$(7/8) \cdot 2^{-(t-1)}(7/8)^{(t-1)/3} \geq 2^{-t}(7/8)^{t/3}.$$

In the remaining subcase, Lemma 1 guarantees that for some i, $3 \leq i \leq d$, there are at least $d/2^i$ nodes of degree i in $N(v)$ that belong to some optimal solution. The probability that the algorithm chooses one of these nodes is at least $(1 - 2^{-d})2^{-i} \geq (7/8)2^{-i}$. By induction, the overall success probability is at least

$$(7/8)2^{-i} \cdot 2^{-(t-i)}(7/8)^{(t-i)/3} \geq 2^{-t}(7/8)^{t/3}. \qquad \square$$

We can now state the following central theorem which establishes our main result.

Theorem 1. t-VERTEX COVER *can be solved with failure probability* e^{-n} *in* $O(2.0911^t n \cdot (n + m)k)$ *steps.*

Proof. Algorithm TVC only takes time $O((n + m)k)$, because there are at most k recursive calls, each of which takes time $O(n+m)$. By Lemma 2, a single call of Algorithm TVC fails to find an existing t-vertex cover with probability at most $1 - 2^{-t}(7/8)^{t/3}$. After $n2^t(8/7)^{t/3}$ repetitions the failure probability drops to

$$\left(1 - 2^{-t}(7/8)^{t/3}\right)^{n2^t(8/7)^{t/3}} < e^{-n}. \qquad \square$$

4   A Problem Kernel for t-Vertex Cover

Kernelization constitutes an important tool in the field of parameterized complexity. If each instance of a problem with parameter k can be reduced to an instance of size $f(k)$ in polynomial time for some fixed function f, we say that the problem allows for a *problem kernel* of that size.

Given the ideas from the previous section, a problem kernel of size $O(t^3)$ for t-VERTEX COVER can be developed in a straightforward fashion. Let us first state the following simple proposition, which will be used several times.

Proposition 1. *Let* $G = (V, E)$ *be a graph with maximum degree* d. *Then* G *contains an independent set of size at least* $|V|/(d + 1)$.

To ease the proof of the kernelization theorem, we begin with a lemma regarding properties of graphs containing only small independent sets of certain degrees.

Lemma 3. *Let* $G = (V, E)$ *be a graph and* $c, k, t \in \mathbf{N}$, $c \leq t/k - 1$. *If* G *contains at most* k *independent nodes of degree between* c *and* $t/k - 1$, *then* $|V[c, t/k - 1]| \leq t$.

Proof. Assume there are $t+1$ nodes of degree between c and $t/k-1$. Proposition 1 then states that $k + 1$ of these nodes constitute an independent set.

Let us now prove the main result of this section: A small problem kernel for t-VERTEX COVER can be constructed efficiently. The proof is based on two crucial ideas. Firstly, if there are many nodes of relatively high degree, the input cannot be a *no*-instance. Secondly, if there are many nodes of relatively low degree, many of them can be dropped due to an equivalence argument.

Theorem 2. *Any instance of* t-VERTEX COVER *can be reduced to a kernel of size at most* $2t^3 + 4t^2 + 3t$ *in linear time.*

Proof. The problem is trivial for $k \geq t$. Given a connected graph $G = (V, E)$ and $k, t \in \mathbf{N}$, $1 \leq k < t$, we proceed as follows. Note that all the operations employed take linear time.

There are three cases in which we may replace G by a trivial *yes*-instance. Firstly, if G contains a node v of degree t or more, then v alone forms a t-vertex cover. Secondly, if $|V[t/k + k, t - 1]| \geq k$, then any k nodes from $V[t/k + k, t - 1]$ constitute a t-vertex cover: Since their degrees sum up to $t + k^2$ and at most $\binom{k}{2}$ edges can be incident to two of them, these nodes cover at least t edges. Thirdly, if $|V[t/k, t/k + k - 1]| \geq t + k^2$, then Proposition 1 guarantees that $V[t/k, t/k + k - 1]$ contains an independent set of size at least k. It is easy to see that this set automatically forms a t-vertex cover.

Otherwise, $|V_1| < t + k^2$ and $|V_2| < k$ for $V_1 = V[t/k, t/k + k - 1]$ and $V_2 = V[t/k+k, t-1]$. In the following kernelization, we will consider $N[N[V_1 \cup V_2]]$ rather than just $N[V_1 \cup V_2]$ in order to preserve the degrees of nodes in the neighborhood of $V_1 \cup V_2$. The above inequalities imply $|N[V_1]| < (t+k^2)(t/k+k)$ and $|N[V_2]| < kt$, and consequently $|N[V_1 \cup V_2]| < t^2/k + 3tk + k^3$ as well as

$$|N[N[V_1 \cup V_2]]| < (t^2/k + 3tk + k^3)t/k = t^3/k^2 + 3t^2 + tk^2.$$

Define $G' = (V', E')$ to be the graph obtained from G by removing $N[N[V_1 \cup V_2]]$ as well as all isolated nodes. If $|V'[1, t/k - 1]| \leq t$, then G is small enough, and we may stop immediately. Otherwise, let $c \in \mathbf{N}$ be the largest number such that $|V'[c, t/k-1]| > t$. In this case, Lemma 3 implies that $V'[c, t/k-1]$ contains an independent set I of more than k nodes. By choice of c, the size j of a largest independent set from $V'[c + 1, t/k - 1]$ is bounded by k.

Construct a node set V_3 by combining $V'[c + 1, t/k - 1]$ and a set A of $k - j$ additional nodes from I. We claim that the subgraph G_K of G induced by

$N[N[V_1 \cup V_2]] \cup N[V_3]$ constitutes a problem kernel. Since $|V_3| \leq t + k$ and each $v \in V_3$ has degree at most t/k in G, the number of nodes in G_K is bounded by

$$(t + k)(t/k + 1) + t^3/k^2 + 3t^2 + tk^2 = t^2/k + 2t + k + t^3/k^2 + 3t^2 + tk^2.$$

It remains to show that G_K indeed has a t-vertex cover of size k iff G does. Clearly, if G_K admits such a cover C, then C also consitutes a t-vertex cover of size at most k for G. On the other hand, if C is such a cover for G, the set

$$C_K = \{ v \in C \mid \deg(v) > c \} \cup \{ v \in C \cap V_3 \mid \deg(v) = c \} \cup A$$

constitutes a t-vertex cover of size at most k for G_K:

By construction, there are no edges between A and $C_K \setminus A$ in G or G_K. Moreover, A is an independent set in G and G_K. Let $B = C \setminus C_K$. Since A contains $|B|$ independent nodes of degree c and $\deg(v) \leq c$ for each $v \in B$, replacing B with A does not decrease the number of covered edges.   □

Note that only $O(t^2)$ of the nodes in the kernel have high degrees. In fact, the construction could easily be modified to output only $O(t^2)$ nodes of degree greater than one. An $f(n)$-time algorithm for t-VERTEX COVER that processes nodes of degree one in an adequate way would thus only take $f(t^2)$ steps on the problem kernel. More precisely, it only makes sense to choose a node of degree one if the maximum degree is one as well. In the case of our algorithm, preprocessing the input by a reduction to its problem kernel yields the following runtime bound:

Theorem 3. *We can solve t-VERTEX COVER with failure probability at most p in $O(2.0911^t t^4 \ln(1/p) + n + m)$ steps.*

Proof. First construct a problem kernel G' with t^3 nodes in linear time. Note that there are at most t^4 edges as each node has degree less than t. Then call Algorithm TVC $\ln(1/p)2^t(8/7)^t$ times on G'.   □

5   Concluding Remarks

We argued for the concept of *intuitive algorithms* using the example of t-VERTEX COVER. The resulting randomized algorithm is rather simple and has a running time of $O(2.0911^t poly(n))$, improving vastly over previous bounds. It remains an open question whether our algorithm can be derandomized without a large computational overhead. Whereas rather sophisticated derandomization methods have been developed in earlier scholarship, none of them seems to be applicable in this case. This poses the question as to whether randomization should be considered more than merely a design tool. In fact, we believe that the use of probabilistic methods can be justified from the practical viewpoint of actual performance as well.

References

1. N. Alon, O. Goldreich, J. Håstad, and R. Peralta. Simple constructions of almost k-wise independent random variables. *Journal of Random structures and Algorithms*, 3(3):289–304, 1992.
2. N. Alon, R. Yuster, and U. Zwick. Color-coding. *Journal of the ACM*, 42(4): 844–856, 1995.
3. R. Balasubramanian, M. R. Fellows, and V. Raman. An improved fixed parameter algorithm for vertex cover. *Information Processing Letters*, 65(3):163–168, 1998.
4. R. Beigel. Finding maximum independent sets in sparse and general graphs. In *Proc. of 10th SODA*, pages 856–857, 1999.
5. M. Blser. Computing small partial coverings. *Information Processing Letters*, 85:327–331, 2003.
6. N. H. Bshouty and L. Burroughs. Massaging a linear programming solution to give a 2-approximation for a generalization of the vertex cover problem. In M. Morvan, C. Meinel, and D. Krob, editors, *Proc. of 15th STACS*, number 1373 in LNCS, pages 298–308. Springer, 1998.
7. L. Cai, S. M. Chan, and S. O. Chan. Random separation: A new method for solving fixed-cardinality optimization problems. In *Proc. of 2nd IWPEC*, number 4169 in LNCS. Springer, 2006. To appear.
8. L. Cai, S. M. Chan, and S. O. Chan. Random separation: A new method for solving fixed-parameter problems. Manuscript, 2006.
9. L. Sunil Chandran and F. Grandoni. Refined memorization for vertex cover. *Information Processing Letters*, 93:125–131, 2005.
10. J. Chen, I. A. Kanj, and W. Jia. Vertex cover: Further observations and further improvements. *Journal of Algorithms*, 41:280–301, 2001.
11. J. Chen, I. A. Kanj, and G. Xia. Simplicity is beauty: Improved upper bounds for vertex cover. Technical Report TR05-008, School of CTI, DePaul University, 2005.
12. R. G. Downey and M. R. Fellows. *Parameterized Complexity*. Springer-Verlag, 1999.
13. S. E. Dreyfus and R. A. Wagner. The Steiner problem in graphs. *Networks*, 1: 195–207, 1972.
14. F. V. Fomin, F. Grandoni, and D. Kratsch. Measure and conquer: Domination – A case study. In *Proc. of 32d ICALP*, LNCS. Springer, 2005.
15. J. Guo, R. Niedermeier, and S. Wernicke. Parameterized complexity of generalized vertex cover problems. In *Proc. of 9th WADS*, number 3608 in LNCS, pages 36–48, Waterloo, Canada, 2005. Springer.
16. D. S. Hochbaum. The t-vertex cover problem: Extending the half integrality framework with budget constraints. In K. Jansen and D. S. Hochbaum, editors, *Proc. of 1st APPROX*, number 1444 in LNCS, pages 111–122. Springer, 1998.
17. K. Iwama and S. Tamaki. Improved upper bounds for 3-SAT. In *Proc. of 15th SODA*, pages 328–328, 2004.
18. J. Kneis, D. Mölle, S. Richter, and P. Rossmanith. Algorithms based on the treewidth of sparse graphs. In *Proc. of 31st WG*, number 3787 in LNCS, pages 385–396. Springer, 2005.
19. D. Mölle, S. Richter, and P. Rossmanith. Enumerate and expand: Improved algorithms for connected vertex cover and tree cover. In *Proc. of 1st CSR*, number 3967 in LNCS, pages 270–280. Springer, 2006.
20. D. Mölle, S. Richter, and P. Rossmanith. A faster algorithm for the Steiner tree problem. In *Proc. of 23rd STACS*, number 3884 in LNCS, pages 561–570. Springer, 2006.

21. R. Niedermeier and P. Rossmanith. Upper bounds for Vertex Cover further improved. In *Proc. of 16th STACS*, number 1563 in LNCS, pages 561–570. Springer, 1999.
22. R. Niedermeier and P. Rossmanith. On efficient fixed parameter algorithms for Weighted Vertex Cover. *Journal of Algorithms*, 47:63–77, 2003.
23. E. Petrank. The hardness of approximation: Gap location. *Computational Complexity*, 4:133–157, 1994.
24. V. Raman and S. Saurabh. Triangles, 4-cycles and parameterized (in-) tractability. In *Proc. of 10th SWAT*, number 4059 in LNCS. Springer, 2006. To appear.
25. O. Reingold. Undirected ST-connectivity in Log-Space. In *Proc. of 37th STOC*, pages 376–385, 2005.
26. J. M. Robson. Algorithms for maximum independent sets. *Journal of Algorithms*, 7:425–440, 1986.
27. U. Schöning. A probabilistic algorithm for k-SAT and constraint satisfaction problems. In *Proc. of 40th FOCS*, pages 410–414, 1999.
28. A. Scott and G. B. Sorkin. Faster algorithms for Max-CUT and Max-CSP, with polynomial expected time for sparse instances. In *Proc. of 7th RANDOM*, number 2764 in LNCS, pages 382–395. Springer, 2003.
29. D. A. Spielman and S.-H. Teng. Smoothed analysis: Why the simplex algorithm usually takes polynomial time. In *Proc. of 33d STOC*, pages 296–305, 2001.
30. R. Williams. A new algorithm for optimal constraint satisfaction and its implications. In *Proc. of 31st ICALP*, number 3142 in LNCS, pages 1227–1237. Springer, 2004.

Politician's Firefighting

Allan E. Scott[1,*], Ulrike Stege[1,**], and Norbert Zeh[2,***]

[1] Department of Computer Science, University of Victoria, Victoria, Canada
{aescott, stege}@cs.uvic.ca
[2] Faculty of Computer Science, Dalhousie University, Halifax, Canada
nzeh@cs.dal.ca

Abstract. Firefighting is a combinatorial optimization problem on graphs that models the problem of determining the optimal strategy to contain a fire and save as much from the fire as possible. We introduce and study a new version of firefighting, *Politician's Firefighting*, which exhibits more locality than the classical one-firefighter version. We prove that this locality allows us to develop an $O(bn)$-time algorithm on trees, where b is the number of nodes initially on fire. We further prove that Politician's Firefighting is NP-hard on planar graphs of degree at most 5. We present an $O(m + k^{2.5}4^k)$-time algorithm for this problem on general graphs, where k is the number of nodes that burn using the optimal strategy, thereby proving that it is fixed-parameter tractable. We present experimental results that show that our algorithm's search-tree size is in practice much smaller than the worst-case bound of 4^k.

1 Introduction

Firefighting can be thought of as a puzzle game where the player's goal is to save nodes in a graph from an advancing fire. Given a graph $G = (V, E)$ and a set $B_0 \subset V$ of initially burning nodes, the game proceeds in rounds, numbered 0 through r. In each round, first the fire advances and then the player places firefighters on one or more nodes that are neither burning nor occupied (by a firefighter). Once a node is burning or occupied, it stays that way for the rest of the game. Round 0 is special; all nodes in B_0 are set on fire. In subsequent rounds, the fire spreads from each burning node to every adjacent unoccupied node. The game ends when the fire can no longer spread, that is, when all neighbours of burning nodes are burning or occupied. Viewed as an optimization problem, the player's goal is to find a firefighter-placement strategy that minimizes the number of burning nodes at the end of the game. The problem can be seen as a

---

* Research supported by a University of Victoria fellowship and a grant of the Natural Sciences and Engineering Research Council of Canada, grantholder Ulrike Stege.
** Research supported by the Natural Sciences and Engineering Research Council of Canada.
*** Research supported by the Natural Sciences and Engineering Research Council of Canada and the Canadian Foundation for Innovation.

T. Asano (Ed.): ISAAC 2006, LNCS 4288, pp. 608–617, 2006.
© Springer-Verlag Berlin Heidelberg 2006

model of the spread of forest fires or diseases in social networks, which motivated the initial study of this problem [3, 4, 7].

The classical version of firefighting, introduced in [4], allows the player to place *one* firefighter on an arbitrary node in each round, so long as the node is not yet occupied or burning. Finding an optimal strategy under these conditions is NP-hard even on trees of degree three and with $|B_0| = 1$, provided that the node in B_0 has degree three [1]. A simple greedy algorithm produces an optimal strategy on binary trees. On arbitrary trees, the greedy strategy that always protects the heaviest threatened subtree produces a 2-approximation w.r.t. the number of saved nodes [5]. In [8], algorithms for containing the fire on 2- and 3-dimensional grids are studied. Apart from the results cited here, we are not aware of any algorithmic results for this problem. What makes this version of firefighting hard is the complete freedom where to place the firefighter, that is, the non-local nature of the problem.

In this paper we study *Politician's Firefighting*, a more localized version of the problem: In each round, we are allowed to deploy as many firefighters as there are burning nodes (politicians allocate resources according to how dire the situation is). However, if a node x "generates" a firefighter by being on fire, this firefighter can only be placed on an unoccupied, non-burning neighbour of x. In other words, we can use only as many firefighters as there are burning nodes with unoccupied non-burning neighbours. It seems more realistic to allow more than one firefighter to be placed in each round because typically not just one fire brigade fights a forest fire. The constraints imposed on where firefighters may be placed reflect the political reality that politicians and local inhabitants would prefer to see their fire brigade protect them or their neighbours, rather than somebody miles away. This constraint can also be seen as a logistic one since fire trucks travel at a finite speed. Another motivation for the locality is that, when using vaccination to contain the spread of a disease, one usually vaccinates persons interacting with infected persons before using a much wider "radius" of vaccination, particularly if vaccine is expensive or hard to obtain.

We prove in Section 2 that this problem can be solved in $O(bn)$ time on trees, where $b = |B_0|$. In Section 3, we show that Politician's Firefighting is NP-hard even on planar graphs of degree at most 5. In Section 4, we present an $O(m + k^{2.5}4^k)$-time algorithm for general graphs, which shows that the problem is fixed-parameter tractable when parameterized by k, the number of nodes that is allowed to burn. The worst-case bound on the size of the search tree in our algorithm is tight. However, experimental results discussed in Section 4 indicate that, in practice, the search-tree size is much smaller.

2   Trees

We start by arguing that the locality of Politician's Firefighting helps to solve it in polynomial time on trees. We choose an arbitrary node in B_0 as the root of the tree. For every node v, let T_v be the subtree rooted in v, let p_v be v's parent, and let C_v be the set of v's children. Our strategy is to consider all possible cases how a node v may be set on fire or saved and, based on their analysis, develop

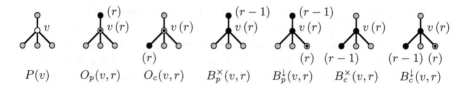

Fig. 1. The different states of a node v and the possible choices how v can attain this state. Burning nodes are black, occupied nodes are white with a black dot inside, and unoccupied non-burning nodes are white. The state of a gray node is unspecified. Labels in parentheses show when the nodes attain the shown state.

a recurrence for the number of nodes that burn in T_v in each case. This allows us to use dynamic programming to determine the number of nodes that burn using the optimal strategy (The corresponding strategy is then easily found). A straightforward evaluation of these recurrences leads to a running time of $\Theta(n^4)$ in the worst case. We then discuss how to reduce the running time to $O(bn)$.

2.1   The Basic Algorithm

Consider a node v. At the end of the game, v is in one of three states: *burning*, *occupied* by a firefighter, or *protected*; the latter means that there is no firefighter on v, but there is a firefighter on every path between v and a burning node. A burning node v either belongs to B_0 or is set on fire by one of its neighbours in some round r. We distinguish whether it is p_v or a child of v that sets v on fire. We further distinguish whether or not we choose to place v's firefighter on a child of v in round $r+1$. Similarly, an occupied node has a firefighter placed on it in some round r. This firefighter is available because p_v or a child of v catches fire in round r. We use the following notation to denote the number of nodes that burn in T_v in each of the resulting cases (see Figure 1 for an illustration): $P(v)$ if v is protected; $O_p(v,r)$ and $O_c(v,r)$ if v is occupied in round r; $B_c^\times(v,r)$, $B_c^\downarrow(v,r)$, $B_p^\times(v,r)$, and $B_p^\downarrow(v,r)$ if v is set on fire in round r. Subscripts p and c denote the subcases when v is set on fire by its parent or a child, respectively, or when the parent's or a child's firefighter is placed on v. Superscripts indicate whether we place v's firefighter on one of v's children ($\downarrow$) or not ($\times$). In addition, we use the following notation: $B_p(v,r) = \min(B_p^\downarrow(v,r), B_p^\times(v,r))$, $B_c(v,r) = \min(B_c^\downarrow(v,r), B_c^\times(v,r))$, $O_c^*(v,r) = \min_{1 \le r' \le r} O_c(v,r')$, $B_c^*(v,r) = \min_{r \le r' \le n} B_c(v,r')$, and $L(v,r) = \min(B_c(v,r-1), B_c(v,r), B_c(v,r+1), O_c^*(v,r), B_p(v,r+1))$.

Since $|B_0| = b$, every node v in T can be occupied or set on fire only in b different rounds, corresponding to the lengths of the paths from v to the nodes in B_0. If node v cannot be set on fire or occupied at time r, we define its corresponding $B_\cdot(v,r)$ or $O_\cdot(v,r)$ value to be $+\infty$.

Next we derive a recurrence for $B_c^\downarrow(v,r)$. Due to lack of space, we only state the recurrences for the other cases; they are easily obtained using similar, but simpler, analyses. For technical reasons, we treat every node $v \in B_0$ as being set on fire in round 0 by both an imaginary child and an imaginary parent. Thus, for $v \in B_0$ and $r > 0$, we have $B_c^\downarrow(v,r) = +\infty$; for $r = 0$, we have

$B_c^{\downarrow}(v,0) = 1 + \min_{w \in C_v}\left(O_p(w,0) + \sum_{w' \in C_v \setminus \{w\}} L(w',0)\right)$ because, after being set on fire, node v chooses one child w to occupy in round 0; any other child $w' \in C_v \setminus \{w\}$ is then either set on fire by v in round 1, set on fire by a child in round 0 or 1, or occupied using one of its children's firefighters in round 0.

For $v \notin B_0$, node v is set on fire by one of its children, say w_1, and again node v chooses a child $w_2 \in C_v \setminus \{w_1\}$ to occupy using v's firefighter; any other node $w' \in C_v \setminus \{w_1, w_2\}$ is set on fire by v in round $r+1$, set on fire by one of its own children in round $r-1$, r, or $r+1$, or occupied by one of its children's firefighters no later than round r. This leads to the following recurrence:

$$B_c^{\downarrow}(v,r) = 1 + \min_{\substack{w_1,w_2 \in C_v \\ w_1 \neq w_2}}\left(B_c(w_1, r-1) + O_p(w_2, r) + \sum_{w' \in C_v \setminus \{w_1, w_2\}} L(w', r)\right)$$

Using similar analyses, we obtain for $v \in B_0$: $O_c(v,r) = O_p(v,r) = P(v) = +\infty$. For $r > 0$, we have $B_c^{\times}(v,r) = B_p^{\downarrow}(v,r) = B_p^{\times}(v,r) = +\infty$. For $r = 0$, we have $B_p^{\downarrow}(v,0) = B_c^{\downarrow}(v,0)$ and $B_c^{\times}(v,0) = B_p^{\times}(v,0) = 1 + \sum_{w \in C_v} L(w,0)$. For $v \notin B_0$, we obtain

$$B_c^{\times}(v,r) = 1 + \min_{w \in C_v}\left(B_c(w, r-1) + \sum_{w' \in C_v \setminus \{w\}} L(w', r)\right)$$

$$B_p^{\downarrow}(v,r) = 1 + \min_{w \in C_v}\left(O_p(w, r) + \sum_{w' \in C_v \setminus \{w\}} L(w', r)\right)$$

$$B_p^{\times}(v,r) = 1 + \sum_{w \in C_v} L(w, r)$$

$$O_c(v,r) = \min_{w \in C_v}\left(B_c^{\times}(w, r) + \sum_{w' \in C_v \setminus \{w\}} \min(P(w'), B_c^*(w', r), O_c^*(w', n))\right)$$

$$O_p(v,r) = \sum_{w \in C_v} \min(P(w), B_c^*(w, r), O_c^*(w, n))$$

$$P(v) = \sum_{w \in C_v} \min(O_c^*(w, n), P(w))$$

Each of these recurrences for a given node v depends only on values of the recurrences on children of v. Hence, they can be computed bottom-up. Since the game ends after at most n rounds, we must consider up to n different time values. The most expensive recurrence to evaluate is $B_c^{\downarrow}(v,r)$, where we must consider all pairs of children (w_1, w_2) of v. The number of these pairs, summed over all nodes in T, is $\Theta(n^2)$ in the worst case. For each pair, we spend linear time to evaluate the expression inside the outer parentheses, leading to a $\Theta(n^3)$ bound per round. Summing over all n rounds gives a running time of $\Theta(n^4)$.

2.2 A Faster Algorithm

To reduce the running time to $O(n^2)$, we need to evaluate every recurrence for a given pair (v,r) in $O(1 + |C_v|)$ time. This is easy for $P(v)$, $O_p(v,r)$, and $B_p^{\times}(v,r)$.

Next we discuss in detail how to achieve this bound for evaluating $B_p^{\downarrow}(v, r)$; the same ideas also speed up the computation of the other recurrences $O_c(v, r)$, $B_c^{\downarrow}(v, r)$, and $B_c^{\times}(v, r)$. If we precompute the sum $L^*(v, r) = \sum_{w \in C_v} L(w, r)$, which takes $O(|C_v|)$ time, we can rewrite the recurrence for $B_c^{\downarrow}(v, r)$ as

$$B_c^{\downarrow}(v, r) = 1 +$$
$$\min_{\substack{w_1, w_2 \in C_v \\ w_1 \neq w_2}} \left(L^*(v, r) + B_c(w_1, r - 1) - L(w_1, r) + O_p(w_2, r) - L(w_2, r) \right),$$

which can be evaluated in $O(1 + |C_v|^2)$ time. Looking more closely at the rewritten form of $B_c^{\downarrow}(v, r)$, we observe that $B_c^{\downarrow}(v, r)$ is minimized if $B'(w_1, r) = B_c(w_1, r-1) - L(w_1, r)$ and $B''(w_2, r) = O_p(w_2, r) - L(w_2, r)$ are minimized, except that w_1 and w_2 cannot be the same node. Thus, if w_1' and w_1'' are the two children of v that minimize $B'(w, r)$ and w_2' and w_2'' are the two children of v that minimize $B''(w, r)$, we have three cases: Assume w.l.o.g. that $B'(w_1', r) \leq B'(w_1'', r)$ and $B''(w_2', r) \leq B''(w_2'', r)$. If $w_1' \neq w_2'$, let $w_1 = w_1'$ and $w_2 = w_2'$. If $w_1' = w_2'$ and $B(w_1'', r) - B(w_1', r) \leq B(w_2'', r) - B(w_2', r)$, then let $w_1 = w_1''$ and $w_2 = w_2'$; otherwise, let $w_1 = w_1'$ and $w_2 = w_2''$.

Nodes w_1', w_1'', w_2', and w_2'' can be found in $O(|C_v|)$ time. Once this is done, $B_c^{\downarrow}(v, r)$ can be evaluated in constant time because one of the three combinations $(w_1', w_2'), (w_1', w_2''), (w_1'', w_2')$ minimizes $B_c^{\downarrow}(v, r)$. Hence, each recurrence can be evaluated in $O(1 + |C_v|)$ time per pair (v, r), and the total cost of evaluating the recurrences over all nodes is $O(n)$ per time value r. Since we have to consider only $1 \leq r \leq n$, the total running time is $O(n^2)$.

To reduce the running time to $O(bn)$, we observe that every node can be set on fire or occupied by a neighbour at only b different times, determined by the distances from v to the nodes in B_0. Thus, we must evaluate each recurrence for only b different time values for each node in T; we define every value that is not computed explicitly to be $+\infty$. This reduces the running time to $O(bn)$.

Theorem 1. *Politician's Firefighting can be solved in $O(bn)$ time on a tree with n nodes of which b are initially on fire.*

3   NP-Hardness on Planar Graphs

Theorem 2. *Politician's Firefighting is NP-hard, even on planar graphs with vertices of degree at most five and only one node initially on fire.*

We prove NP-hardness of Politician's Firefighting by reduction from Planar Vertex Cover [2]. In particular, given a planar graph G, we construct another planar graph G' with nodes[1] of degree at most 5, and a set $B_0 = \{\rho\}$, where ρ is an almost arbitrary node of G', such that G has a vertex cover of size k if and only if G' has a firefighting strategy that burns only "a few" nodes.

---

[1] To avoid confusion, we refer to the vertices of G as "vertices" and to the vertices of G' as "nodes".

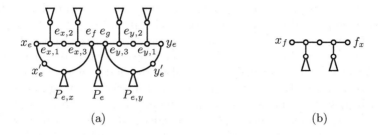

Fig. 2. (a) The edge widget $\mathcal{E}_e$ for the edge $e = xy$. (b) A connector.

The construction of G' has the following intuition: First we replace the vertices and edges of G with subgraphs called *vertex widgets* and *edge widgets*. A vertex widget is built such that if any one of its nodes burn, n^3 nodes burn in the widget. An edge widget is built such that letting one or both of two special *incineration nodes* burn sacrifices n^5 nodes, unless we let at least one of the vertex widgets corresponding to the endpoints of the edge burn as well. We complete the construction by superimposing two additional graph structures on the vertex and edge widgets. The first one allows the fire to spread from ρ to all incineration nodes, and is built so that we cannot prevent the spread unless we sacrifice n^5 nodes elsewhere. Thus, if the size of a minimum vertex cover of G is k, at least n^5 nodes in G' will burn unless we let k vertex widgets in G' burn, which means that roughly kn^3 nodes burn. The second graph structure allows the fire to spread to the k vertex widgets corresponding to vertices in a vertex cover, without using nodes in edge widgets.

We use *penalizers* to ensure that letting certain nodes in G' burn causes many more nodes to burn. These are complete ternary trees whose leaves are at depth d, for some $d > 0$. When the root of a penalizer P catches fire, the optimal strategy burns a complete binary subtree of P of height d. Thus, we have

Lemma 1. *If the root of a penalizer P of height d catches fire, the optimal strategy burns $2^{d+1} - 1$ nodes in P.*

We call a penalizer *small* if $2^{d+1} - 1 = n^3$, and *big* if $2^{d+1} - 1 = n^5$. Both are of polynomial size: small penalizers have size $O(n^{3\log 3})$, big penalizers have size $O(n^{5\log 3})$. We say a node v of G' is *adjacent to a penalizer* P if v is adjacent to the root of P and no other node in P is adjacent to a node in $G' - P$.

Next we define the different widgets that comprise G' and discuss how they are connected. For the construction, we assume that we are given a planar embedding of G. From the construction, it will be obvious that G' is planar and that every node in G' has degree at most five.

Vertex widgets. Let x be a vertex of G, and let $e_1, f_1, \ldots, e_d, f_d$ be the edges and faces incident to x, in clockwise order. The vertex widget $\mathcal{V}_x$ consists of a simple cycle $(x_{e_1}, x_{f_1}, \ldots, x_{e_d}, x_{f_d})$. Each node of this cycle is adjacent to a small penalizer, except x_{e_1}, which is adjacent to two small penalizers.

Note that, once a single node of the cycle burns we have two choices: let the fire spread around the cycle and protect one penalizer per node, or protect a cycle-neighbour of a burning node. In the former case, we let the second penalizer attached to x_{e_1} burn, incurring a penalty of n^3 burning nodes. In the latter case, the penalizers attached to the node whose cycle-neighbour we protect burn. This incurs a penalty of at least n^3 burning nodes.

Edge widgets. Let e be an edge with endpoints x and y and incident faces f and g. In G', edge e is represented by an edge widget $\mathcal{E}_e$, shown in Figure 2a. The endpoints x_e and y_e are shared between the edge widget $\mathcal{E}_e$ and the vertex widgets $\mathcal{V}_x$ and $\mathcal{V}_y$; that is, the endpoints of $\mathcal{E}_e$ are the same nodes as the nodes with the same names in $\mathcal{V}_x$ and $\mathcal{V}_y$. All penalizers in the edge widget are big. We argue later that we have to let both e_f and e_g burn. We call e_f and e_g *incineration nodes*, as we cannot protect all three penalizers threatened by these two nodes unless we let at least one of the nodes x'_e and y'_e burn, which can be achieved only by letting $\mathcal{V}_x$ or $\mathcal{V}_y$ (or both) burn.

Face widgets. G' contains one face widget $\mathcal{F}_f$ per face f of G. Similar to a vertex widget, the face widget for a face f with incident vertices and edges $x_1, e_1, \ldots, x_d, e_d$, in this order clockwise around f, consists of a cycle $(f_{x_1}, f_{e_1}, \ldots, f_{x_d}, f_{e_d})$, each of whose nodes has an attached penalizer; but this time the penalizers are big. Once one node in $\mathcal{F}_f$ catches fire, the only way we can prevent n^5 nodes from burning is to let the fire spread around $\mathcal{F}_f$, while protecting the roots of all penalizers in $\mathcal{F}_f$.

The last two widgets build two additional graph structures within G'. The first allows us to cheaply set fire to vertex widgets corresponding to vertices in a vertex cover of G. The second ensures that every node e_f or e_g in an edge widget burns eventually, forcing us to set fire to at least one vertex widget incident to each edge widget in order to save all penalizers in the edge widget.

Channels. A channel is a path of length $42n$. Each of its internal nodes has a big penalizer attached to it. The first endpoint of the channel belongs to a face widget; the second endpoint belongs to an edge widget. More precisely, for every face f and every edge e on its boundary, there is a channel in G' whose endpoints are f_e and e_f. Note that this implies that, once one of the endpoints of the channel burns, we have to protect its big penalizer inside the face or edge widget. This sets fire to its neighbour inside the channel. In order to prevent n^5 nodes from burning in the channel, we now have to let the fire spread along the channel path and, for every node on the path, protect its adjacent penalizer.

Connectors. A connector is used to let the fire spread cheaply between face and vertex widgets. For every face f and every vertex x on its boundary, there is a connector with endpoints x_f and f_x, which belong to $\mathcal{V}_x$ and $\mathcal{F}_f$; see Figure 2b. For each connector, if f_x burns, we have two choices: Either we let the fire spread along the connector, thereby forcing all cycle nodes in $\mathcal{V}_x$ to burn, or we let the fire spread to the middle node and then stop the fire by placing this node's firefighter on the neighbour of x_f in the connector.

To finish the construction of the firefighting instance, we choose an arbitrary non-penalizer node ρ in a face widget and define $B_0 = \{\rho\}$. The following lemma proves that connectors and face widgets allow us to cheaply and quickly set fire to the appropriate vertex widgets in G'. Lemma 3 then uses this fact to prove that G has a small vertex cover if and only if G' has a firefighting strategy that lets few nodes burn.

Lemma 2. *Let $V' = \{x_1, \ldots, x_k\}$ be a vertex cover of G. Then G' contains a connected subgraph containing only nodes from face widgets, connector widgets, and vertex widgets $C_{x_1}, \ldots, C_{x_k}$. This graph includes ρ and has diameter at most $42n - 84$.*

Lemma 3. *Graph G has a vertex cover of size k if and only if G' has a strategy that burns at most $kn^3 + 252n^2 - 432n - 144$ nodes.*

Proof sketch. We prove the "only if" part; the "if" part can be proved using similar arguments. Let $V' = \{x_1, \ldots, x_k\}$ be a vertex cover of size k, and let H be a subgraph of G' as in Lemma 2. Then, due to the diameter bound of H, we can make sure that every node in H burns by time $42n - 84$. Moreover, we can protect all penalizers incident to nodes in H, except one small penalizer per vertex widget. Thus, we let kn^3 nodes in penalizers burn.

Now observe that an incineration node in an edge widget $\mathcal{E}_e$ for an edge $e = xy$ catches fire no earlier than round $42n$, unless we let the fire spread to this node from x_e or y_e. In summary, for every edge widget $\mathcal{E}_e$, at least one of x_e and y_e catches fire at least 84 rounds before e_f or e_g can be set on fire through a channel. We sketch here what happens if only x_e burns, that is, $y \notin V'$. The other two cases are similar.

In this case, we let the fire spread from x_e to x'_e and $e_{x,1}$, using x_e's firefighter to protect its incident penalizer inside $\mathcal{V}_x$. In the next time step, we use the firefighters of x'_e and $e_{x,1}$ to protect the adjacent penalizers, letting $e_{x,2}$ burn. Next we let the fire spread from $e_{x,2}$ to $e_{x,3}$ and then on to e_f, e_g, $e_{y,3}$, and $e_{y,2}$. Each of these nodes has only one unprotected adjacent penalizer by the time it catches fire because x'_e has already protected $P_{e,x}$ and then e_f protects P_e before e_g catches fire. Thus, each node can use its firefighter to protect the one penalizer it threatens. Finally, we use $e_{y,2}$'s firefighter to protect $e_{y,1}$, thereby preventing the fire from spreading into $\mathcal{V}_y$.

To obtain the claimed bound on the total number of nodes that burn, we count the total number of non-penalizer nodes in vertex widgets, face widgets, channels, and connectors, and add the kn^3 nodes that burn in small penalizers in the k vertex widgets $\mathcal{V}_{x_1}, \ldots, \mathcal{V}_{x_k}$. $\qquad\qquad\square$

4   Fixed-Parameter Tractability on General Graphs

We present a bounded search-tree algorithm that solves Politician's Firefighting in $O(m + k^{2.5}4^k)$ time. Rather than deciding an entire round at once, our algorithm is based on the idea of choosing a single *threatened node* v (i.e., a non-burning node adjacent to a burning node), and branching recursively on two cases: place a firefighter on v, or let v burn. However, there are two problems that must be addressed for the algorithm to work.

The first problem arises because we decouple the recursion from the rounds. Specifically, we have to track the set of nodes threatened from the beginning of the round since we place fires during the round rather than at the beginning of the next round. Otherwise, new nodes would become threatened during the round as we place fires, which would spread fires indiscriminately.

The other problem is that this approach creates illegal firefighter placements, since the branching step does not associate firefighters with fires. To overcome this, before adding a node v to our set F of nodes to be occupied by firefighters, we check the size of a maximum matching between the nodes in $F \cup \{v\}$ and the nodes in B. If there is a matching that includes every vertex in $F \cup v$, then every firefighter can be matched to a unique fire, so putting a firefighter on v does not create an illegal placement. If we delete the edges between two burning nodes or two firefighters, the subgraph induced by $B \cup F \cup v$ is bipartite with at most $2k$ vertices and k^2 edges. A maximum matching in a bipartite graph can be computed in $O(\sqrt{n}m)$ time [6], or in this case in $O(k^{2.5})$ time.

Algorithm politiciansFirefighting(V, E, B, F, T, k)
    if $k < 0$ **then return** false
    if T is empty **then**
        $T \leftarrow \{v \in V \setminus (F \cup B) : v$ is adjacent to a node in $B\}$
        if T is still empty **then return** true
        if $|T| -$ max_match$(B, T, E) > k$ **then**
            return false (more than k fires will spawn this round)
    Choose any $v \in T$
    if max_match$(B, F \cup \{v\}, E) = |F| + 1$ and
        politiciansFirefighting($V, E, B, F \cup \{v\}, T \setminus \{v\}, k$) **then return** true
    return politiciansFirefighting($V, E, B \cup \{v\}, F, T \setminus \{v\}, k - 1$)

Algorithm politiciansFirefighting runs in time $O(m + k^{2.5}4^k)$, which can be verified as follows. The height of the search tree is bounded by $2k$: Overall we cannot let more than k nodes burn; furthermore we cannot place more than k firefighters because every fire gives us one firefighter to place. This results in a search-tree size of at most $2^{2k} = 4^k$ nodes.[2] The time per node is dominated by the cost of procedure max_match, which computes a maximum matching for a given bipartite graph and takes $O(k^{2.5})$ time.

We have implemented our algorithm to measure the average search-tree size in practice. Our experimental results indicate that, although the size of our search tree is 4^k in the worst-case, in practice the running times are much better.

For five different densities, we tested 1000 random (connected) graphs from $G_{n,p}$ where n is the number of nodes and p is the probability of any given edge

---

[2] This worst-case search-tree size for our algorithm is indeed tight up to a polynomial factor. If the number of threatened nodes is exactly twice the number of burning neighbours, the number of legal firefighter placements that must be generated is $\geq \frac{2^{2b}}{2b}$, which is greater than x^{2b} for any $x < 2$ and sufficiently large b.

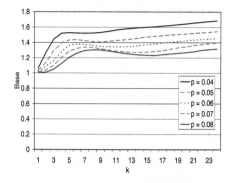 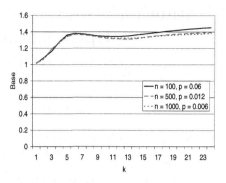

Fig. 3. Left: search-tree size for $n = 100$, $k \leq 25$. Base refers to the x in our x^k search tree size. Right: search-tree sizes for $n = 100, 500, 1000$, $k \leq 25$.

occuring. As shown in Figure 3 (left), running time decreases as graph density increases. This is likely due to nodes in the graph burning more quickly, causing the algorithm to reach a no-answer sooner. Therefore, we concentrated on sparse graphs. We also checked several larger test cases, but as Figure 3 (right) shows, the number of search-tree nodes actually decreases slightly as n increases and relative density is maintained.

References

1. S. Finbow, A. King, G. MacGillivray, and R. Rizzi. The firefighter problem for graphs of maximum degree three. In *Proceedings of the European Conference ond Combinatorics, Graph Theory and Applications*, 2003.
2. M. Garey, D. Johnson, and L. Stockmeyer. Some simplified NP-complete problems. In *Proceedings of the 6th ACM Symposium on the Theory of Computing*, pages 47–63, 1974.
3. S. G. Hartke. *Graph-Theoretic Models of Spread and Competition*. PhD thesis, Rutgers University, 2004.
4. B. Hartnell. Firefighter! an application of domination, 1995. Presentation at the *24th Manitoba Conference on Combinatioral Mathematics and Computing*.
5. B. Hartnell and Q. Li. Firefighting on trees: How bad is the greedy algorithm? *Congressus Numerantium*, 145:187–192, 2000.
6. J. E. Hopcroft and R. M. Karp. A $n^{5/2}$ algorithm for maximum matchings in bipartite graphs. *SIAM Journal on Computing*, 2:225–231, 1973.
7. F. Roberts *Challenges for discrete mathematics and theoretical computer science in the defense against bioterrorism*, pages 1–34. SIAM Frontiers in Applied Mathematics. 2003.
8. P. Wang and S. Moeller. Fire control in graphs. *Journal of Combinatorial Mathematics and Combinatorial Computing*, 41:19–34, 2002.

Runtime Analysis of a Simple Ant Colony Optimization Algorithm

Extended Abstract

Frank Neumann[1] and Carsten Witt[2],[*]

[1] Institut für Informatik, CAU Kiel, 24098 Kiel, Germany
`fne@informatik.uni-kiel.de`
[2] FB Informatik, LS 2, Univ. Dortmund, 44221 Dortmund, Germany
`carsten.witt@cs.uni-dortmund.de`

Abstract. Ant Colony Optimization (ACO) has become quite popular in recent years. In contrast to many successful applications, the theoretical foundation of this randomized search heuristic is rather weak. Building up such a theory is demanded to understand how these heuristics work as well as to come up with better algorithms for certain problems. Up to now, only convergence results have been achieved showing that optimal solutions can be obtained in finite time. We present the first runtime analysis of an ACO algorithm, which transfers many rigorous results on the runtime of a simple evolutionary algorithm to our algorithm. Moreover, we examine the choice of the evaporation factor, a crucial parameter in ACO algorithms, in detail for a toy problem. By deriving new lower bounds on the tails of sums of independent Poisson trials, we determine the effect of the evaporation factor almost completely and prove a phase transition from exponential to polynomial runtime.

1 Introduction

The analysis of randomized search heuristics with respect to their runtime is a growing research area where many results have been obtained in recent years. This class of heuristics contains well-known approaches such as Randomized Local Search (RLS), the Metropolis Algorithm (MA), Simulated Annealing (SA), and Evolutionary Algorithms (EAs). Such heuristics are often applied to problems whose structure is not known or if there are not enough resources such as time, money, or knowledge to obtain good specific algorithms. It is widely acknowledged that a solid theoretical foundation for such heuristics is needed.

Some general results on the runtime of RLS can be found in Papadimitriou, Schäffer and Yannakakis (1990). The graph bisection problem has been subject to analysis of MA (Jerrum and Sorkin, 1998), where MA can be seen as SA with a fixed temperature. For a long time, it was an open question whether there is a natural example where SA outperforms MA for all fixed temperatures. This question has recently been answered positively by Wegener (2005) for instances of the minimum spanning tree problem.

---

[*] This work was supported by the Deutsche Forschungsgemeinschaft (DFG) as a part of the Collaborative Research Center "Computational Intelligence" (SFB 531).

T. Asano (Ed.): ISAAC 2006, LNCS 4288, pp. 618–627, 2006.
© Springer-Verlag Berlin Heidelberg 2006

In this paper, we focus on another kind of randomized search heuristics, namely Ant Colony Optimization (ACO). Like EAs, these heuristics imitate optimization processes from nature, in this case the search of an ant colony for a common source of food. Solving problems by ACO techniques has become quite popular in recent years. Developed by Dorigo, Maniezzo and Colorni (1991), they have shown to be a powerful heuristic approach to solve combinatorial optimization problems such as the TSP (see Dorigo and Stützle, 2004, for an overview on numerous applications). From a theoretical point of view, there are no results that provide estimates of the runtime of ACO algorithms. Despite interesting theoretical investigations of models and dynamics of ACO algorithms (Merkle and Middendorf, 2002; Dorigo and Blum, 2005), convergence results are so far the only results related to their runtimes. Dorigo and Blum (2005) explicitly formulate the open problem to determine the runtime of ACO algorithms on simple problems in a similar fashion to what has been done for EAs.

We solve this problem, starting the analysis of ACO algorithms with respect to their expected runtimes and success probability after a specific number of steps. RLS, SA, MA, and simple EAs search more or less locally, and runtime bounds are often obtained by considering the neighborhood structure of the considered problem. Considering ACO algorithms, this is different as search points are obtained by random walks of ants on a so-called construction graph. The traversal of an ant on this graph is determined by values on the edges which are called pheromone values. Larger pheromone values correspond to a higher probability of traversing a certain edge, where the choice of an edge usually fixes a parameter in the current search space. The pheromone values are updated if a good solution has been constructed in this random walk. This update depends on the traversal of the ant and a so-called evaporation factor ρ.

The choice of ρ seems to be a crucial parameter in an ACO algorithm. Using a large value of ρ, the last accepted solution changes the pheromone values by a large amount such that there is a large probability of producing this solution in the next step. In contrast to this, the use of a small evaporation factor leads to a small effect of the last accepted solution such that an improvement may be hard to find in the next step. We show that a simple ACO algorithm behaves for very large values of ρ (namely $\rho \geq 1/3$) as the simplest EA called (1+1) EA. This algorithm has been studied extensively with respect to its runtime on pseudo-boolean functions $f \colon \{0,1\}^n \to \mathbb{R}$ (see, e. g., Droste, Jansen and Wegener, 2002) as well as on combinatorial optimization problems. The list of problems where runtime bounds have been obtained include some of the best-known polynomially solvable problems such as maximum matchings (Giel and Wegener, 2003) and minimum spanning trees (Neumann and Wegener, 2004). It should be clear that we cannot expect such general heuristics to outperform the best-known algorithms for these mentioned problems. The main aim of such analyses is to get an understanding how these heuristics work. In the case of NP-hard problems, one is usually interested in good approximations of optimal solutions. Witt (2005) has presented a worst-case and average-case analysis of the (1+1) EA for the partition problem, which is one of the first results on NP-hard problems. All these results immediately transfer to our ACO algorithm with very large ρ.

After these general results, we consider the effect of the evaporation factor ρ on the runtime of our ACO algorithm in detail. As proposed in the open problem stated by Dorigo and Blum (2005), we examine the simplest non-trivial pseudo-boolean function called ONEMAX and analyze for the first time for which choices of ρ the runtime with high probability is upper bounded by a polynomial and for which choices it is exponential. We observe a phase transition from exponential to small polynomial runtime when ρ crosses the threshold value $1/n$. Larger values of ρ imply that the expected function value of a new solution is determined by the function value of the best seen solution. Then an improvement will be achieved after an expected polynomial number of steps. In the case of smaller ρ, an improvement does not increase the expected function value sufficiently. Here exponential lower bounds are obtained by showing that there is a large gap between the expected value and the best-so-far function value. Both the proof of the upper and the lower runtime bound contain new analytical tools to lower bound the tail of a sum of independent trials with different success probabilities. The new tools may be of independent interest in other probabilistic analyses.

In Section 2, we introduce the simple ACO algorithm which we will consider. We investigate its relation to the (1+1) EA in Section 3 and transfer the results on this EA to our algorithm. In Section 4, we investigate the choice of the evaporation factor ρ for the function ONEMAX in great detail and finish with some conclusions. Several proofs have been omitted in this extended abstract. A technical report with full proofs is available (Neumann and Witt, 2006).

2   The Algorithm

Gutjahr (2003) has considered a graph-based ant system and investigated under which conditions such an algorithm converges to an optimal solution. We consider a simple graph-based ant system metaheuristic that has been inspired by this algorithm. Such a heuristic produces solutions by random walks on a construction graph. Let $C = (V, E)$ be the construction graph with a designated start vertex s and pheromone values τ on the edges. Starting at s, an ant traverses the construction graph depending on the pheromone value using Algorithm 1. Assuming that the ant is at vertex v, the ant moves to a successor w of v, where w is chosen proportionally to the pheromone values of all non-visited successors of v. The process is iterated until a situation is reached where all successors of the current vertex v have been visited.

Algorithm 1 (Construct(C, τ))
1.) $v := s$, mark v as visited.
2.) While there is a successor of v in C that has not been visited:
    a.) Let N_v be the set of non-visited successors of v and $T := \sum_{(v,w)|w \in N_v} \tau_{(v,w)}$.
    b.) Choose one successor w of v where the probability of selection of any fixed $u \in N_v$ is $\tau_{(v,u)}/T$.
    c.) Mark w as visited, set $v := w$ and go to 2.).
3.) Return the solution x and the path $P(x)$ constructed by this procedure.

Based on this construction procedure, solutions of our simple ACO algorithm (see Algorithm 2) called 1-ANT are constructed. In the initialization step, each edge gets a pheromone value of $1/|E|$ such that the pheromone values sum up to 1. After that, an initial solution x^* is produced by a random walk on the construction graph and the pheromone values are updated with respect to this walk. In each iteration, a new solution x is constructed and the pheromone values are updated if this solution is not inferior to the currently best solution x^*. We formulate our algorithm for maximization problems although it can be easily adapted to minimization.

Algorithm 2 (1-ANT)

1.) Set $\tau_{(u,v)} = 1/|E|$ for all $(u,v) \in E$.
2.) Compute x (and $P(x)$) using Construct(C, τ).
3.) Update$(\tau, P(x))$ and set $x^* := x$.
4.) Compute x (and $P(x)$) using Construct(C, τ).
5.) If $f(x) \geq f(x^*)$, Update$(\tau, P(x))$ and set $x^* := x$.
6.) Go to 4.).

For theoretical investigations, it is common to have no termination condition in such an algorithm. One is interested in the random optimization time which equals the number of constructed solutions until the algorithm has produced an optimal search point. Usually, we try to bound the expected value of this time.

We take a general view and consider optimization for pseudo-boolean goal functions $f \colon \{0,1\}^n \to \mathbb{R}$ for $n \geq 3$ using the canonical construction graph in our setting, $C_{\text{bool}} = (V, E)$ (see Figure 1) with $s = v_0$. In the literature, this graph is also known as *Chain* (Gutjahr, 2006). Optimizing bitstrings of length n, the graph has $3n+1$ vertices and $4n$ edges. The decision whether a bit x_i, $1 \leq i \leq n$, is set to 1 is made at node $v_{3(i-1)}$. In case that the edge $(v_{3(i-1)}, v_{3(i-1)+1})$ is chosen, x_i is set to 1 in the constructed solution. Otherwise $x_i = 0$ holds. After this decision has been made, there is only one single edge which can be traversed in the next step. In case that $(v_{3(i-1)}, v_{3(i-1)+1})$ has been chosen, the next edge is $(v_{3(i-1)+1}, v_{3i})$, and otherwise the edge $(v_{3(i-1)+2}, v_{3i})$ will be traversed. Hence, these edges have no influence on the constructed solution and we can assume $\tau_{(v_{3(i-1)}, v_{3(i-1)+1})} = \tau_{(v_{3(i-1)+1}, v_{3i})}$ and $\tau_{(v_{3(i-1)}, v_{3(i-1)+2})} = \tau_{(v_{3(i-1)+2}, v_{3i})}$ for $1 \leq i \leq n$. We call the edges $(v_{3(i-1)}, v_{3(i-1)+1})$ and $(v_{3(i-1)+1}, v_{3i})$ 1-*edges* and the other edges 0-*edges*. The edges $(v_{3(i-1)}, v_{3(i-1)+1})$ and $(v_{3(i-1)}, v_{3(i-1)+2})$ as well as $(v_{3(i-1)+1}, v_{3i})$ and $(v_{3(i-1)+2}, v_{3i})$ are called *complementary* to each other.

The pheromone values are chosen such that at each time $\sum_{(u,v) \in E} \tau_{(u,v)} = 1$ holds. In addition, it seems to be useful to have bounds on the pheromone values (see, e.g., Dorigo and Blum, 2005) to ensure that each search point has a positive probability of being chosen in the next step. We restrict each $\tau_{(u,v)}$ to the interval $\left[\frac{1}{2n^2}, \frac{n-1}{2n^2}\right]$ and ensure $\sum_{(u,\cdot) \in E} \tau_{(u,\cdot)} = \frac{1}{2n}$ for $u = v_{3i}$, $0 \leq i \leq n-1$, and $\sum_{(\cdot,v)} \tau_{(\cdot,v)} = \frac{1}{2n}$ for $v = v_{3i}$, $1 \leq i \leq n$. This can be achieved by normalizing the pheromone values after an update and replacing the current value by $\frac{1}{2n^2}$ if $\tau_{(u,v)} < \frac{1}{2n^2}$ and by $\frac{n-1}{2n^2}$ if $\tau_{(u,v)} > \frac{n-1}{2n^2}$ holds. Depending on whether edge (u, v)

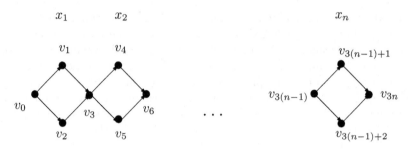

Fig. 1. Construction graph for pseudo-boolean optimization

is contained in the path $P(x)$ of the accepted solution x, the pheromone values are updated to τ' in the procedure Update$(\tau, P(x))$ as follows:

$$\tau'_{(u,v)} = \min\left\{ \frac{(1-\rho) \cdot \tau_{(u,v)} + \rho}{1 - \rho + 2n\rho}, \frac{n-1}{2n^2} \right\} \quad \text{if } (u,v) \in P(x)$$

and

$$\tau'_{(u,v)} = \max\left\{ \frac{(1-\rho) \cdot \tau_{(u,v)}}{1 - \rho + 2n\rho}, \frac{1}{2n^2} \right\} \quad \text{if } (u,v) \notin P(x).$$

Due to the bounds on the pheromone values, the probability of fixing x_i as in an optimal solution is at least $1/n$. Hence, the 1-ANT finds an optimum for each pseudo-boolean function f regardless of ρ in expected time at most n^n.

3   1-ANT and (1+1) EA

We consider the relation between the 1-ANT and a simple evolutionary algorithm called (1+1) EA, which has extensively been studied with respect to its runtime distribution. The (1+1) EA starts with a solution x^* that is chosen uniformly at random and produces in each iteration a new solution x from a currently best solution x^* by flipping each bit of x^* with probability $1/n$. Hence, the probability of producing a certain solution x with Hamming distance $H(x, x^*)$ to x^* is $(1/n)^{H(x,x^*)} \cdot (1 - 1/n)^{n-H(x,x^*)}$.

Algorithm 3 ((1+1) EA)
1.) Choose $x^* \in \{0,1\}^n$ uniformly at random.
2.) Construct x by flipping each bit of x^* independently with probability $1/n$.
3.) Replace x^* by x if $f(x) \geq f(x^*)$.
4.) Go to 2.).

In the following, we consider the 1-ANT with values of ρ at least $\frac{n-2}{3n-2}$, which is for large n approximately $1/3$. In this case, the pheromone values attain their upper and lower bounds $\frac{n-1}{2n^2}$ respectively $\frac{1}{2n^2}$. Theorem 1 shows that the 1-ANT behaves as the (1+1) EA on each function for the mentioned choice of ρ. This also means that the 1-ANT has the same expected optimization time as the (1+1) EA on each function.

Theorem 1. *Choosing $\rho \geq (n-2)/(3n-2)$, the 1-ANT has the same runtime distribution as the (1+1) EA on each function.*

4   1-ANT on OneMax

In the following, we inspect the choice of ρ in detail for a simple pseudo-boolean function called ONEMAX defined by $\text{ONEMAX}(x) = \sum_{i=1}^{n} x_i$. This is the simplest non-trivial function that can be considered and analyses of ACO algorithms for such simple functions are explicity demanded by Dorigo and Blum (2005). Note that due to results on the (1+1) EA by Droste, Jansen and Wegener (2002), the expected optimization time of the 1-ANT is $O(n \log n)$ on each linear function if $\rho \geq (n-2)/(3n-2)$ holds.

We prepare ourselves by considering the effects of pheromone updates for a solution x^* in greater detail. Let $\tau(e)$ and $\tau'(e)$ be the pheromone values on edge e before resp. after the update. If $e \in P(x^*)$, $\tau'(e) \geq \tau(e)$ and $\tau'(e) \leq \tau(e)$ otherwise. The amount by which the pheromone value is increased on a 1-edge equals the amount the pheromone value is decreased on the complementary 0-edge. However, the change of a pheromone value depends on the previous value on the edge. In the following lemma, we bound the relative change of pheromone values. We call an edge *saturated* iff its pheromone value is either $\frac{1}{2n^2}$ or $\frac{n-1}{2n^2}$.

Lemma 1. *Let e_1 and e_2 be two edges of C_{bool} and let τ_1 resp. τ_2 be their current pheromone values in the 1-ANT. Let τ'_1 resp. τ'_2 be their updated pheromone values for the next accepted solution x. If $e_1, e_2 \in P(x^*)$ and none of the edges is saturated before or after the update, then $|(\tau'_1 - \tau_1) - (\tau'_2 - \tau_2)| \leq \rho|\tau_1 - \tau_2|$.*

In the following, we will figure out which values of ρ lead to efficient runtimes of the 1-ANT and which do not. Intuitively, $1/n$ is a threshold value for ρ since the denominator $1 - \rho + 2n\rho$ of the normalization factor diverges for $\rho = \omega(1/n)$ and is $1 - \rho - o(1)$ for $\rho = o(1/n)$. We will make precise that there is a phase transition in the behavior of the 1-ANT on ONEMAX when ρ is asymptotically smaller resp. larger than $1/n$.

4.1   Exponential Lower Bounds

Choosing $\rho = 0$, the pheromone value on each edge is $1/(4n)$ at each time step. This implies that the expected optimization time of the 1-ANT on ONEMAX is 2^n as each solution is chosen uniformly at random from $\{0,1\}^n$. In the following, we show that the optimization time with overwhelming proability still is exponential if ρ is convergent to 0 only polynomially fast.

Assume that the currently best solution x^* has value k. Then the following lemma gives a lower bound on the probability of overshooting k by a certain amount in the next accepted step.

Lemma 2. *Let $X_1, \ldots, X_n \in \{0, 1\}$ be independent Poisson trials with success probabilities p_i, $1 \leq i \leq n$. Let $X := X_1 + \cdots + X_n$, $\mu := E(X) = p_1 + \cdots + p_n$ and $\sigma := \sqrt{\mathrm{Var}(X)}$. For any $0 \leq k \leq n - \sigma$, let $\gamma_k = \max\{2, (k - \mu)/\sigma\}$. If $\sigma \to \infty$ then $\mathrm{Prob}(X \geq k + \sigma/\gamma_k \mid X \geq k) = \Omega(1)$.*

Using this lemma, we are able to prove an exponential lower bound on the runtime of the 1-ANT on ONEMAX. In order to show that the success probability in an exponential number of steps is still exponentially small, we assume that $\rho = O(n^{-1-\epsilon})$ for some constant $\epsilon > 0$.

Theorem 2. *Let $\rho = O(n^{-1-\epsilon})$ for some constant $\epsilon > 0$. Then the optimization time of the 1-ANT on ONEMAX is $2^{\Omega(n^{\epsilon/3})}$ with probability $1 - 2^{-\Omega(n^{\epsilon/3})}$.*

Proof. The main idea is to keep track of the so-called 1-*potential*, defined as the sum of pheromone values on 1-edges. Note that the 1-potential multiplied by n equals the expected ONEMAX-value of the next constructed solution x. If the 1-potential is bounded above by $1/2 + O(1/\sqrt{n})$, Chernoff bounds yield that the probability of $\mathrm{ONEMAX}(x) \geq n/2 + n^{1/2+\epsilon/3}$ is bounded above by $2^{-\Omega(n^{\epsilon/3})}$. We will show that with overwhelming probability, the 1-potential is bounded as suggested as long as the ONEMAX-value of the so far best solution is bounded above by $n/2 + n^{1/2+\epsilon/3}$.

Starting with initialization, we consider a phase of length $s := \lfloor 2^{cn^{\epsilon/3}} \rfloor$ for some constant c to be chosen later and show that the success probability in the phase is $2^{-\Omega(n^{\epsilon/3})}$. A main task is to bound the number of successful steps of the phase, i. e., of steps where the new solution is accepted and a pheromone update occurs. In a success with ONEMAX-value $n/2 + i$, $n + 2i$ pheromone values on 1-edges are increased and $n - 2i$ are decreased. Suppose all pheromone values are $1/(4n) \pm o(1/n)$ in the phase. Then Lemma 1 yields that the 1-potential is changed by at most $4i(1 \pm o(1))\rho$ due to the considered success. Hence, if the best solution always had ONEMAX-value at most $n/2 + n^{1/2+\epsilon/3}$, the total change of the 1-potential due to at most $O(n^{2\epsilon/3})$ successes would be at most

$$O(n^{2\epsilon/3}) \cdot 4n^{1/2+\epsilon/3} \cdot (1 \pm o(1))\rho = O(n^{1/2+\epsilon}) \cdot O(1/n^{1+\epsilon}) = O(1/n^{1/2})$$

by our assumption on ρ. This would prove the theorem since the initial 1-potential is $1/2$.

Under the assumption on the pheromone values, we want to show that with probability $1 - 2^{-\Omega(n^{\epsilon/3})}$, at most $c'n^{2\epsilon/3}$ successes occur in the phase, where c' is an appropriate constant. We already know that then the probability of a success with value at least $n/2 + n^{1/2+\epsilon/3}$ is $2^{-\Omega(n^{\epsilon/3})}$ in each step of the phase. If c is chosen small enough, this probability is $2^{-\Omega(n^{\epsilon/3})}$ for the whole phase. Moreover, the initial value is at least $n/2 - n^{1/2+\epsilon/3}$ with probability $1 - 2^{-\Omega(n^{\epsilon/3})}$.

Let the so far best value be k. We apply Lemma 2 with respect to the expected ONEMAX-value μ of the next constructed solution. Note that $k - \mu = O(n^{1/2+\epsilon/3})$ holds at each time step we consider. Moreover, $p_i = 1/2 \pm o(1)$ is assumed to hold for all bits, implying $\sigma = \Theta(n^{1/2})$. Hence, with probability $\Omega(1)$ the next success leads to a value at least $k + \Omega(n^{1/2-\epsilon/3})$. Using Chernoff bounds, with probability $1 - 2^{-\Omega(n^{\epsilon/3})}$, $c'n^{2\epsilon/3}$ successes increase the ONEMAX-value by at least $c''n^{1/2+\epsilon/3}$, where c'' is an appropriate constant.

We still have to show the statement on the pheromone values. This is not too difficult for our choice of ρ if the number of successes is bounded by $O(n^{2\epsilon/3})$. Then the total change of pheromone on any fixed edge is bounded above by

$$\rho \cdot O(n^{2\epsilon/3}) = O(n^{-1-\epsilon}) \cdot O(n^{2\epsilon/3}) = o(1/n)$$

with probability $1 - 2^{-\Omega(n^{\epsilon/3})}$. Since the number of edges is bounded by $4n$, this holds also for all edges together. Since the sum of all failure probabilities is $2^{-\Omega(n^{\epsilon/3})}$, this completes the proof. □

4.2 Polynomial Upper Bounds

In the following, we consider for which values of ρ the optimization time of the 1-ANT on ONEMAX with high probability is still polynomial. We will show that the function value of the last accepted solution determines the expected value of the next solution almost exactly if $\rho = \Omega(n^{-1+\epsilon})$, $\epsilon > 0$ an arbitrary constant. To determine the expected time to reach an improvement, we give a lower bound on the probability of overshooting the expected value by at least a small amount.

Lemma 3. *Let $X_1, \ldots, X_k \in \{0,1\}$, $k \leq n$, be independent Poisson trials with success probabilities $p_i \in [1/n, 1-1/n]$, $1 \leq i \leq k$. Let $X := X_1 + \cdots + X_k$ and $\mu := E(X) = p_1 + \cdots + p_k$. If $\mu \leq k - 1$ then $\text{Prob}(X \geq \mu + 1/2) = \Omega(1/n)$.*

The proof of Lemma 3 makes use of ideas by Hoeffding (1956). It is shown that $\text{Prob}(X \geq \mu + 1/2)$ is minimized if the p_i take on at most three different values, namely $1/n$, $1 - 1/n$ and some third value a, $1/n \neq a \neq 1 - 1/n$. This property allows us to prove the lemma by distinguishing a small number of cases.

Finally, we prove that the runtime switches from exponential to polynomial when ρ crosses the threshold $1/n$.

Theorem 3. *Choosing $\rho = \Omega(n^{-1+\epsilon})$, $\epsilon > 0$ a constant, the optimization time of the 1-ANT on ONEMAX is $O(n^2)$ with probability $1 - 2^{-\Omega(n^{\epsilon/2})}$.*

Proof. We assume $\rho \leq 1/2$ since the result follows from Theorem 1 otherwise. In contrast to previous definitions, an edge is called saturated if its pheromone value is $\frac{n-1}{2n^2}$ and called unsaturated otherwise. Let x^* be a newly accepted solution and denote by $\mathcal{S}$ the set of saturated 1-edges and by $\mathcal{U}$ the set of unsaturated 1-edges *after* the pheromone update. Let $k = \text{ONEMAX}(x^*)$ and decompose k according to $k = k_s + k_u$, where k_s denotes the number of ones in x^* whose corresponding 1-edges belong to $\mathcal{S}$ and k_u to the number of ones in x^* whose 1-edges belong to $\mathcal{U}$. The probability that the edges of $\mathcal{S}$ contribute at least k_s to the next (not necessarily accepted) solution x is at least $(1 - 1/n)^{k_s} = \Omega(1)$.

Consider the 1-potential (i. e., the sum of pheromone values) P^* of all edges of $\mathcal{U}$ *before* x^* updates the pheromone values. Let $\mu^* = P^*n$ be the expected

ONEMAX-value w.r.t. these edges before the update. Depending on P^* and k_u, we compute $P(\rho)$, their 1-potential *after* the update:

$$P(\rho) = \frac{(1-\rho)P^* + 2k_u\rho}{(1-\rho) + 2n\rho}.$$

We denote by $\mu = P(\rho) \cdot n$ the expected ONEMAX-value w.r.t. the edges of $\mathcal{U}$ *after* the update has occured. Under certain assumptions, we will prove that with probability $1 - 2^{-\Omega(n^\epsilon)}$, $\mu + 1/2 > k_u$. Since k_u is an integer, Lemma 3 shows that the probability of producing in the next solution x at least $\lceil \mu + 1/2 \rceil \geq k_u + 1$ ones by the $\mathcal{U}$-edges is at least $\Omega(1/n)$. Consider the difference

$$\mu - k_u \geq \frac{(1-\rho)P^* + 2k_u\rho}{(1-\rho) + 2n\rho} \cdot n - k_u = \frac{(\mu^* - k_u)(1-\rho)}{(1-\rho) + 2n\rho}.$$

We exploit that $\rho \leq 1/2$, implying $1 - \rho \geq 0$. Hence, if $\mu^* - k_u \geq 0$ then $\mu \geq k_u > k_u - 1/2$ anyway. Assuming $\mu^* - k_u < 0$, we can lower bound the (negative) last fraction by $(\mu^* - k_u)/(2n\rho)$. Hence, if we can prove that $k_u - \mu^* < n\rho$, we obtain $\mu > k_u - 1/2$ as desired. We will bound the probability of a large deviation $k_u - \mu^*$ keeping track of the variance of the random number of ones on the $\mathcal{U}$-edges. Let v^* be the variance *before* the pheromone values have been updated with respect to x^* and denote by v the variance *after* the update. If $v^* \leq (n\rho)^{3/2}$, then a Chernoff-Hoeffding-type bound (Theorem 3.44 in Scheideler, 2000) yields

$$\text{Prob}(k_u - \mu^* \geq n\rho) \leq e^{-\frac{(n\rho)^2}{2v^*(1+n\rho/(3v^*))}} = 2^{-\Omega(\sqrt{n\rho})} = 2^{-\Omega(n^{\epsilon/2})}.$$

However, we cannot show that $v^* \leq (n\rho)^{3/2}$ is likely for all points of time. Therefore, we will prove $v \geq v^*/(4n\rho)$ for any time step. This will show that v is large enough to compensate a large $k_u - \mu^*$ in the following step (constructing x).

Suppose $v^* > (n\rho)^{3/2}$. Then $v^* \geq \sqrt{v^* n\rho}$, and by the above bound,

$$\text{Prob}(k_u - \mu^* \geq \sqrt{v^* n\rho}) \leq e^{-\frac{(\sqrt{v^* n\rho})^2}{2v^* + 2\sqrt{v^* n\rho}/3}} \leq e^{-\frac{v^* n\rho}{2v^* + 2v^*/3}} = 2^{-\Omega(n^\epsilon)}.$$

Hence, with probability $1 - 2^{-\Omega(n^\epsilon)}$, $(k_u - \mu^*)/(2n\rho) \leq \sqrt{v^*/(2n\rho)}$, implying $\mu \geq k_u - \sqrt{v^*/(2n\rho)}$. Due to the assumptions $v^* > (n\rho)^{3/2}$, $v \geq v^*/(4n\rho)$ and $n\rho = \Omega(n^\epsilon)$, it follows that $v \to \infty$. Hence, we can apply Lindeberg's generalization of the Central Limit Theorem for the number of ones on $\mathcal{U}$. The probability of producing at least $k_u + 1$ ones on these edges is bounded below by the probability of producing at least $1 + \mu + \sqrt{v^*/(2n\rho)}$ ones on these edges. By the Central Limit Theorem, this has probability $\Omega(1)$ since $\sqrt{v} \geq \sqrt{v^*/(2n\rho)}$.

We still have to show that $v \geq v^*/(4n\rho)$. It is sufficient to show a statement on the success probability for each edge (u, v) of the construction graph. Consider the expression $\tau'_{(u,v)} \geq \frac{(1-\rho)\tau_{(u,v)}}{1-\rho+2n\rho}$. The last fraction is at least $\frac{\tau_{(u,v)}}{4n\rho}$ since $\rho \leq 1/2$.

The $\mathcal{S}$-edges contribute with probability $\Omega(1)$ at least k_s to the next solution, and (if no failure of probability $2^{-\Omega(n^{\epsilon/2})}$ occurs) with probability $\Omega(1/n)$, the $\mathcal{U}$-edges contribute at least $k_u + 1$. At most $n - 1$ improvements suffice, and, by

Chernoff bounds, cn^2 steps contain at least $n - 1$ improvements with probability $1 - 2^{-\Omega(n)}$ for an appropriate constant c. Since $\rho \leq 1/2$, $\epsilon \leq 1$ must hold. Hence, the sum of all failure probabilities in $O(n^2)$ steps is $2^{-\Omega(n^{\epsilon/2})}$. $\quad\square$

5 Conclusions

For the first time, bounds on the runtime of a simple ACO algorithm have been obtained. Choosing a large evaporation factor, it behaves like the (1+1) EA and all results on this algorithm transfer directly to our ACO algorithm. In addition, we have inspected the effect of the evaporation factor in detail for the function ONEMAX and figured out the border between a polynomial and an exponential optimization time almost completely. Thereby, we have developed new techniques for the analysis of randomized search heuristics.

References

Dorigo, M. and Blum, C. (2005). Ant colony optimization theory: A survey. *Theor. Comput. Sci.*, **344**, 243–278.

Dorigo, M., Maniezzo, V., and Colorni, A. (1991). The ant system: An autocatalytic optimizing process. Tech. Rep. 91-016 Revised, Politecnico di Milano.

Dorigo, M. and Stützle, T. (2004). *Ant Colony Optimization*. MIT Press.

Droste, S., Jansen, T., and Wegener, I. (2002). On the analysis of the (1+1) evolutionary algorithm. *Theor. Comput. Sci.*, **276**, 51–81.

Giel, O. and Wegener, I. (2003). Evolutionary algorithms and the maximum matching problem. In *Proc. of STACS '03*, vol. 2607 of *LNCS*, 415–426.

Gutjahr, W. J. (2003). A generalized convergence result for the graph-based ant system metaheuristic. *Probab. Eng. Inform. Sc.*, **17**, 545–569.

Gutjahr, W. J. (2006). On the finite-time dynamics of ant colony optimization. *Methodology and Computing in Applied Probability*. To appear.

Hoeffding, W. (1956). On the distribution of the number of successes in independent trials. *Ann. Math. Stat.*, **27**, 713–721.

Jerrum, M. and Sorkin, G. B. (1998). The Metropolis algorithm for graph bisection. *Discrete Appl. Math.*, **82**(1–3), 155–175.

Merkle, D. and Middendorf, M. (2002). Modelling the dynamics of ant colony optimization algorithms. *Evolutionary Computation*, **10**(3), 235–262.

Neumann, F. and Wegener, I. (2004). Randomized local search, evolutionary algorithms, and the minimum spanning tree problem. In *Proc. of GECCO 04*, vol. 3102 of *LNCS*, 713–724.

Neumann, F. and Witt, C. (2006). Runtime analysis of a simple ant colony optimization algorithm. Tech. Rep. TR06-084, Electr. Colloq. on Comput. Compl. (ECCC).

Papadimitriou, C. H., Schäffer, A. A., and Yannakakis, M. (1990). On the complexity of local search. In *Proc. of STOC '90*, 438–445. ACM Press.

Scheideler, C. (2000). *Probabilistic Methods for Coordination Problems*. HNI-Verlagsschriftenreihe 78, University of Paderborn. Habilitation Thesis, available at http://www14.in.tum.de/personen/scheideler/index.html.en.

Wegener, I. (2005). Simulated annealing beats metropolis in combinatorial optimization. In *Proc. of ICALP '05*, vol. 3580 of *LNCS*, 589–601.

Witt, C. (2005). Worst-case and average-case approximations by simple randomized search heuristics. In *Proc. of STACS '05*, vol. 3404 of *LNCS*, 44–56.

Lower Bounds on the Deterministic and Quantum Communication Complexities of Hamming-Distance Problems[*]

Andris Ambainis[1], William Gasarch[2], Aravind Srinivasan[2], and Andrey Utis[3]

[1] University of Waterloo, Dept. of Combinatorics and Optimization and Instuitute for Quantum Computing, University of Waterloo, 200 University Avenue West, Waterloo, ON, Canada N2L 3G1
ambainis@uwaterloo.ca
[2] Department of Computer Science and University of Maryland Institute for Advanced Computer Studies, University of Maryland at College Park, College Park, MD 20742, USA
{gasarch, srin}@cs.umd.edu
[3] Department of Computer Science, University of Maryland at College Park, College Park, MD 20742, USA
utis@cs.umd.edu

Abstract. Alice and Bob want to know if two strings of length n are almost equal. That is, do they differ on *at most* a bits? Let $0 \leq a \leq n - 1$. We show that any deterministic protocol, as well as any error-free quantum protocol (C^* version), for this problem requires at least $n - 2$ bits of communication. We show the same bounds for the problem of determining if two strings differ in *exactly* a bits. We also prove a lower bound of $n/2 - 1$ for error-free Q^* quantum protocols. Our results are obtained by employing basic tools from combinatorics and calculus to lower-bound the ranks of the appropriate matrices.

1 Introduction

Given $x, y \in \{0, 1\}^n$ one way to measure how much they differ is the Hamming distance.

Definition 1. *If $x, y \in \{0, 1\}^n$ then $\mathrm{HAM}(x, y)$ is the number of bits on which x and y differ.*

If Alice has x and Bob has y then how many bits do they need to communicate such that they both know $\mathrm{HAM}(x, y)$? The trivial algorithm is to have Alice send x (which takes n bits) and have Bob send $\mathrm{HAM}(x, y)$ (which takes $\lceil \lg(n + 1) \rceil$ bits) back to Alice. This takes $n + \lceil \lg(n + 1) \rceil$ bits. Pang and El Gamal [15] showed that this is essentially optimal. In particular they showed that HAM requires at least $n + \lg(n + 1 - \sqrt{n})$ bits to be communicated. (See [1, 3, 12, 14] for more on the communication complexity of HAM. See [5] for how Alice and Bob can approximate HAM without giving away too much information.)

---

[*] The research of the first author was supported in part by IQC University Professorship and CIAR, that of the second author in part by NSF grant CCR-01-05413, and that of the third author in part by NSF grant CCR-0208005 and NSF ITR Award CNS-0426683.

T. Asano (Ed.): ISAAC 2006, LNCS 4288, pp. 628–637, 2006.
© Springer-Verlag Berlin Heidelberg 2006

What if Alice and Bob just want to know if $\mathrm{HAM}(x, y) \leq a$?

Definition 2. *Let $n \in \mathbb{N}$. Let a be such that $0 \leq a \leq n - 1$. $HAM_n^{(a)} : \{0,1\}^n \times \{0,1\}^n \to \{0,1\}$ is the function $HAM_n^{(a)}(x, y) = 1$ if $\mathrm{HAM}(x, y) \leq a$, and is 0 otherwise.*

The problem $HAM_n^{(a)}$ has been studied by Yao [18] and Gavinsky et al [6]. Yao showed that there is an $O(a^2)$ public coin simultaneous protocol for $HAM_n^{(a)}$ which yields (by Newman [13], see also [10]) an $O(a^2 + \log n)$ private coin protocol and also an $O(2^{a^2} \log n)$ quantum simultaneous message protocol with bounded error [18]. Gavinsky et al. give an $O(a \log n)$ public coin simultaneous protocol, which yields an $O(a \log n)$ private coin protocol; recently, Huang et al. have presented an improved $O(a \log a)$ public coin simultaneous protocol [7]. See [8] for lower bounds. All of the protocols mentioned have a small probability of error. How much communication is needed for this problem if we demand no error? There is, of course, the trivial $(n + 1)$-bit protocol. Is there a better one?

In this paper we show the following; in the list of results below, the "c" (in the "$c\sqrt{n}$" terms) is some positive absolute constant.

1. For any $0 \leq a \leq n - 1$, $HAM_n^{(a)}$ requires at least $n - 2$ bits in the deterministic model.
2. For $a \leq c\sqrt{n}$, $HAM_n^{(a)}$ requires at least n bits in the deterministic model.
3. For any $0 \leq a \leq n - 1$, $HAM_n^{(a)}$ requires at least $n - 2$ bits in the quantum model with Alice and Bob share an infinite number of EPR pairs, using a classical channel, and always obtain the correct answer.
4. For $a \leq c\sqrt{n}$, $HAM_n^{(a)}$ requires at least n bits in the quantum model in item 3.
5. For any $0 \leq a \leq n - 1$, $HAM_n^{(a)}$ requires at least $\frac{n}{2} - 1$ bits in the quantum model with Alice and Bob share an infinite number of EPR pairs, using a quantum channel, and always obtain the correct answer.
6. For $a \leq c\sqrt{n}$, $HAM_n^{(a)}$ requires at least $n/2$ bits in the quantum model in item 5.

Note that if $a = n$ then $(\forall x, y)[HAM_n^{(a)}(x, y) = 1$, hence we do not include that case.

What if Alice and Bob need to determine if $\mathrm{HAM}(x, y) = a$ or not?

Definition 3. *Let $n \in \mathbb{N}$. Let a be such that $0 \leq a \leq n$. $HAM_n^{(=a)} : \{0,1\}^n \times \{0,1\}^n \to \{0,1\}$ is the function $HAM_n^{(=a)}(x, y) = 1$ if $\mathrm{HAM}(x, y) = a$, and is 0 otherwise.*

We show the exact same results for $HAM_n^{(=a)}$ as we do for $HAM_n^{(a)}$. There is one minor difference: for $HAM_n^{(a)}$ the $a = n$ case had complexity 0 since all pairs of strings differ on at most n bits; however, for $HAM_n^{(=a)}$ the $a = n$ case has complexity $n + 1$ as it is equivalent to equality.

All our results use the known "log rank" lower bounds on classical and quantum communication complexity: Lemmas 1 and 2. Our approach is to lower-bound the ranks of the appropriate matrices, and then to invoke these known lower bounds. It has been pointed out to us by anonymous referees of this paper that our results may follow from known results [9] on the zeroes of the Krawtchouk polynomials. While these results employ analysis and a number of other theorems, our method is elementary (just requires generating functions and basic combinatorics), and is self-contained. Also, to the best of our understanding, our results are new for the case where n is odd and $a = (n-1)/2$.

2 Definitions, Notations, and Useful Lemmas

We give brief definitions of both classical and quantum communication complexity. See [10] for more details on classical, and [4] for more details on quantum.

Definition 4. *Let f be any function from $\{0,1\}^n \times \{0,1\}^n$ to $\{0,1\}$.*

1. *A protocol for computing $f(x,y)$, where Alice has x and Bob has y, is defined in the usual way (formally using decision trees). At the end of the protocol both Alice and Bob know $f(x,y)$.*
2. *$D(f)$ is the number of bits transmitted in the optimal deterministic protocol for f.*
3. *$Q^*(f)$ is the number of bits transmitted in the optimal quantum protocol where we allow Alice and Bob to share an infinite number of EPR pairs and communicate over a quantum channel.*
4. *$C^*(f)$ is the number of bits transmitted in the optimal quantum protocol where we allow Alice and Bob to share an infinite number of EPR pairs and communicate over a classical channel.*
5. *M_f is the $2^n \times 2^n$ matrix where the rows and columns are indexed by $\{0,1\}^n$ and the (x,y)-entry is $f(x,y)$.*

Let lg denote the logarithm to the base two. Also, as usual, if $x < y$, then $\binom{x}{y}$ is taken to be zero. The following theorem is due to Mehlhorn and Schmidt [11]; see also [10]:

Lemma 1. *If $f : \{0,1\}^n \times \{0,1\}^n \to \{0,1\}$ then $D(f) \geq \lg(\mathrm{rank}(M_f))$.*

Buhrman and de Wolf [2] proved a similar theorem for quantum communication complexity:

Lemma 2. *If $f : \{0,1\}^n \times \{0,1\}^n \to \{0,1\}$ then the following hold: $Q^*(f) \geq \frac{1}{2}\lg(\mathrm{rank}(M_f))$, and $C^*(f) \geq \lg(\mathrm{rank}(M_f))$.*

3 The Complexity $HAM_n^{(a)}$ for $a \leq O(\sqrt{n})$

We start by presenting results for general a, and then specialize to the case $a \leq c\sqrt{n}$.

Definition 5. *Let M_a be $M_{HAM_n^{(a)}}$, the $2^n \times 2^n$ matrix representing $HAM_n^{(a)}$.*

Lemma 3. *M_a has 2^n orthogonal eigenvectors.*

Proof. This follows from M_a being symmetric.   □

We know that M_a has 2^n real eigenvalues; we will bound the multiplicity of 0 as an eigenvalue of M_a. This leads to a lower bound on $D(HAM_n^{(a)})$ by Lemma 1.

Definition 6. *Let $z \in \{0,1\}^n$.*

1. *$v_z \in R^{2^n}$ is defined by, for all $x \in \{0,1\}^n$, $v_z(x) = (-1)^{\sum_i x_i z_i}$. The entries $v_z(x)$ of v_z are ordered in the natural way: in the same order as the order of the index x in the rows (and columns) of M_a.*
2. *We show that v_z is an eigenvector of M_a. Once that is done we let $eig(z)$ be the eigenvalue of M_a associated with v_z.*

Lemma 4.

1. *The vectors $\{v_z : z \in \{0,1\}^n\}$ are orthogonal.*
2. *For all $z \in \{0,1\}^n$, v_z is an eigenvector of M_a.*
3. *If z has exactly m 1's in it, then*

$$eig(z) = \sum_{j=0}^{a} \sum_{k=\max\{0,j+m-n\}}^{\min\{j,m\}} \binom{m}{k}\binom{n-m}{j-k}(-1)^k.$$

Proof. (**Sketch**) The first assertion (orthogonality) follows by simple counting. We omit the proofs of the other two assertions due to the lack of space. Similar ideas are used in [16], but while estimates suffice in the context of [16], we need exact results.   □

Definition 7. *Let*

$$F(a,n,m) = \sum_{j=0}^{a} \sum_{k=\max\{0,j+m-n\}}^{\min\{j,m\}} \binom{m}{k}\binom{n-m}{j-k}(-1)^k.$$

The following lemma will be used in this section to obtain a lower bound when $a = O(\sqrt{n})$, and in Section 5 to obtain a lower bound for general a.

Lemma 5. *$D(HAM_n^{(a)})$ and $C^*(HAM_n^{(a)})$ are both lower-bounded by the quantity $\lg \sum_{m:F(a,n,m)\neq 0} \binom{n}{m}$. Also, $Q^*(HAM_n^{(a)}) \geq \frac{1}{2} \cdot \lg \sum_{m:F(a,n,m)\neq 0} \binom{n}{m}$.*

Proof. By Lemma 4, the eigenvector v_z has a nonzero eigenvalue if v_z has m 1's and $eig(z) \neq 0$. The rank of M_a is the number of nonzero eigenvalues that correspond to linearly independent eigenvectors. This is $\sum_{m:F(a,n,m)\neq 0} \binom{n}{m}$. The lemma follows from Lemmas 1 and 2.   □

Lemma 6. *The number of values of m for which $F(a,n,m) = 0$ is $\leq a$.*

Proof. View the double summation $F(a,n,m)$ as a polynomial in m. The jth summand has degree $k + (j-k) = j$. Since $j \leq a$ the entire sum can be written as a polynomial in m of degree a. This has at most a roots.   □

Theorem 1. *There is a constant $c > 0$ such that if $a \leq c\sqrt{n}$ then: $D(HAM_n^{(a)}) \geq n$, $Q^*(HAM_n^{(a)}) \geq n/2$, and $C^*(HAM_n^{(a)}) \geq n$.*

Proof. By Lemma 5, $D(f), C^*(f) \geq \lg(\sum_{m:F(a,n,m)\neq 0} \binom{n}{m})$, and $Q^*(f)$ is at least half of this latter quantity (i.e., half of the "log-sum"). Note that

$$2^n = \sum_{m:F(a,n,m)\neq 0} \binom{n}{m} + \sum_{m:F(a,n,m)=0} \binom{n}{m}.$$

By Lemma 6 $|\{m : F(a, n, m) = 0\}| \leq a$. Hence,

$$\sum_{m:F(a,n,m)=0} \binom{n}{m} \leq |\{m : F(a, n, m) = 0\}| \cdot \max_{0 \leq m \leq n} \binom{n}{m} \leq a\binom{n}{n/2} \leq \frac{a2^n}{\sqrt{n}}.$$

So, if $a \leq \frac{1}{4}\sqrt{n}$, then

$$\sum_{m:F(a,n,m)\neq 0} \binom{n}{m} \geq 2^n - \frac{a2^n}{\sqrt{n}} \geq 2^n - 2^{n-2}.$$

Hence,

$$\lg\left(\sum_{m:F(a,n,m)\neq 0} \binom{n}{m}\right) \geq \lg(2^n - 2^{n-2}); \quad \text{i.e.,} \quad \left\lceil\lg\left(\sum_{m:F(a,n,m)\neq 0} \binom{n}{m}\right)\right\rceil \geq n.$$

$\square$

4   The Complexity of $HAM_n^{(=a)}$ for $a \leq O(\sqrt{n})$

We again start by deducing results for general a, and then specialize to the case where $a \leq c\sqrt{n}$.

Definition 8. *Let $M_{=a}$ be $M_{HAM_n^{(=a)}}$, the $2^n \times 2^n$ matrix representing $HAM_n^{(=a)}$.*

The vectors v_z are the same ones defined in Definition 6. We show that v_z is an eigenvector of M. Once that is done we let $eig(z)$ be the eigenvalue of M associated to z. The lemmas needed, and the final theorem, are very similar (in fact easier) to those in the prior section. Hence we just state the needed lemmas and final theorem.

Lemma 7.

1. *For all $z \in \{0, 1\}^n$ v_z is an eigenvector of $M_{=a}$.*
2. *If z has exactly m 1's in it then*

$$eig(z) = \sum_{k=\max\{0,a+m-n\}}^{\min\{a,m\}} \binom{m}{k}\binom{n-m}{a-k}(-1)^k.$$

Definition 9.

$$f(a, n, m) = \sum_{k=\max\{0,a+m-n\}}^{\min\{a,m\}} \binom{m}{k}\binom{n-m}{a-k}(-1)^k.$$

Using our convention "if $x < y$, then $\binom{x}{y} \equiv 0$", we can also write

$$f(a, n, m) = \sum_{k=0}^{a} \binom{m}{k} \binom{n-m}{a-k} (-1)^k.$$

The following lemma will be used in this section to obtain a lower bound when $a = O(\sqrt{n})$, and in Section 5 to obtain a lower bound for general a.

Lemma 8. $D(HAM_n^{(=a)}) \geq \lg \sum_{m: f(a,n,m) \neq 0} \binom{n}{m}$; also, $Q^*(HAM_n^{(=a)})$ is at least $\frac{1}{2} \cdot \lg \sum_{m: f(a,n,m) \neq 0} \binom{n}{m}$, and $C^*(HAM_n^{(=a)}) \geq \lg \sum_{m: f(a,n,m) \neq 0} \binom{n}{m}$.

Lemma 9. The number of values of m for which $f(a, n, m) = 0$ is $\leq a$.

Theorem 2. There is a constant $c > 0$ such that if $a \leq c\sqrt{n}$ then the following hold: $D(HAM_n^{(=a)}) \geq n$, $Q^*(HAM_n^{(=a)}) \geq n/2$, and $C^*(HAM_n^{(=a)}) \geq n$.

5 The Complexity of $HAM_n^{(a)}$ and $HAM_n^{(=a)}$ for General a

We now consider the case of general a. As above, we will show that $F(a, m, n)$ and $f(a, m, n)$ are nonzero for many values of m. This will imply that the matrices M_a and $M_{=a}$ have high rank, hence $HAM_n^{(a)}$ and $HAM_n^{(=a)}$ have high communication complexity. We will use general generating-function methods to derive facts about these sums. A good source on generating functions is [17].

One of our main results will be Lemma 11, which states that if $0 \leq a \leq m < n$, then "$f(a, m, n) = 0$" implies "$f(a, m+1, n) \neq 0$". The idea behind our proof of Lemma 11 will be the following: we will show a relationship between the sum $f(a, m, n)$ and a certain new sum $h(a, m, n)$. Then we will derive generating functions for f and h, and translate this relationship into a relation between their generating functions. Finally, we will show that this relation cannot hold under the assumption that $f(a, m, n) = f(a, m + 1, n) = 0$, thus reaching a contradiction. Some auxiliary results needed for this are now developed in Section 5.1.

5.1 Auxiliary Notation and Results

Define $[x^b]g(x)$ to be the coefficient of x^b in the power series expansion of $g(x)$ around $x_0 = 0$. Also let $t^{(i)}(x)$ denote the i'th derivative of $t(x)$.

We will make use of the following lemma, which follows by an easy induction on i:

Lemma 10. Let $t(x)$ be an infinitely differentiable function. Let $T_1(x) = (x-1)t(x)$, and $T_2(x) = (x+1)t(x)$. Then for any $i \geq 1$: $T_1^{(i)}(x) = (x-1)t^{(i)} + i \cdot t^{(i-1)}(x)$, and $T_2^{(i)}(x) = (x+1)t^{(i)} + i \cdot t^{(i-1)}(x)$.

For the rest of Section 5.1, the integers a, m, n are arbitrary subject to the constraint $0 \leq a \leq m \leq n$, unless specified otherwise.

Definition 10. *Let* $h(a, m, n) = \sum_{i=0}^{a} \binom{m}{i}\binom{n-m}{a-i}\frac{(-1)^i}{m-i+1}$. *Also define the function*
$g(x) = \frac{x^{m+1}-(x-1)^{m+1}}{m+1} \cdot (x+1)^{n-m}$.

We will show an interesting connection between h and f.

Proposition 1. *Suppose* $f(a, m, n) = 0$. *Then* $f(a, m+1, n) = 0$ *iff* $h(a, m, n) = 0$.

Proof.

$$
\begin{aligned}
f(a, m+1, n) &= \sum_{i=0}^{a} \binom{m+1}{i}\binom{n-m-1}{a-i}(-1)^i \\
&= \frac{m+1}{n-m}\sum_{i=0}^{a}\binom{m}{i}\binom{n-m}{a-i}(-1)^i \cdot \frac{n-m-a+i}{m-i+1} \\
&= \frac{m+1}{n-m}\left((n+1-a)\sum_{i=0}^{a}\binom{m}{i}\binom{n-m}{a-i}\frac{(-1)^i}{m-i+1}\right) - \sum_{i=0}^{a}\binom{m}{i}\binom{n-m}{a-i}(-1)^i\right) \\
&= \frac{m+1}{n-m}\left((n+1-a)h(a, m, n) - f(a, m, n)\right)
\end{aligned}
$$

Thus, if $f(a, m, n) = 0$, then $f(a, m+1, n) = 0$ iff $h(a, m, n) = 0$.   □

We next show a connection between $g(x)$ and h.

Proposition 2. $h(a, m, n) = (-1)^m \cdot [x^a]g(x)$.

Next, define an auxiliary function $\phi(u, v, w)$ as the w'th derivative of the function $(x+1)^u(x-1)^v$ evaluated at $x = 0$. We now relate ϕ and h.

Proposition 3. $h(a, m, n) = 0$ *iff* $\phi(n-m, m+1, a) = 0$.

The proof of Propositions 2 and 3 are omitted due to the lack of space. Now we can relate the zeroes of f with those of ϕ:

Proposition 4. $f(a, m, n) = 0$ *iff* $\phi(n-m, m, a) = 0$.

Proof.

$$
\begin{aligned}
(x-1)^m(x+1)^{n-m} &= \sum_{i=0}^{m}\binom{m}{i}x^i(-1)^{m-i} \cdot \sum_{j=0}^{n-m}\binom{n-m}{j}x^j \\
&= (-1)^m \sum_{i=0}^{m}\binom{m}{i}x^i(-1)^i \cdot \sum_{j=0}^{n-m}\binom{n-m}{j}x^j \\
&= (-1)^m \sum_{b=0}^{n}\sum_{k=0}^{b}\binom{m}{k}\binom{n-m}{b-k}(-1)^k x^b \\
&= (-1)^m \sum_{b=0}^{n} f(b, m, n) \cdot x^b.
\end{aligned}
$$

So $f(a, m, n) = \frac{(-1)^m}{a!} \cdot \phi(n-m, m, a)$, and the proposition follows.   □

Proposition 5. *Suppose* $m < n$ *and* $\phi(n-m, m, a) = 0$. *Then* $\phi(n-m-1, m+1, a) = 0$ *iff* $\phi(n-m, m+1, a) = 0$.

Proof. This proposition follows from Propositions 1, 3, and 4.   □

We are now able to prove a recursive relation between values of ϕ:

Proposition 6. *If* $k > 0$, $a > 0$, *and* $\phi(k, m, a) = \phi(k, m, a-1) = 0$, *then* $\phi(k-1, m, a) = \phi(k-1, m, a-1) = 0$.

Proof. Suppose $\phi(k, m, a) = \phi(k, m, a - 1) = 0$. By Lemma 10,

$$\phi(k, m + 1, a) = -\phi(k, m, a) + a \cdot \phi(k, m, a - 1) = 0. \tag{5.1}$$

By Proposition 5, since $\phi(k, m, a) = 0$, we know that

$$\phi(k - 1, m + 1, a) = 0 \text{ iff } \phi(k, m + 1, a) = 0.$$

Now, (5.1) yields $\phi(k - 1, m + 1, a) = 0$. Applying Lemma 10 again, we obtain:

$$0 = \phi(k - 1, m + 1, a) = -\phi(k - 1, m, a) + a \cdot \phi(k - 1, m, a - 1);$$
$$0 = \phi(k, m, a) = \phi(k - 1, m, a) + a \cdot \phi(k - 1, m, a - 1)$$

Solving these equations, we get $\phi(k - 1, m, a) = \phi(k - 1, m, a - 1) = 0$.   □

5.2   The Main Results

We are now ready to prove our main lemma.

Lemma 11. *Let $0 \le a \le m < n$. If $f(a, m, n) = 0$, then $f(a, m + 1, n) \neq 0$.*

Proof. The lemma holds trivially for $a = 0$, since both $f(a, m, n)$ and $f(a, m + 1, n)$ are nonzero if $a = 0$. So suppose $a \ge 1$. Suppose $f(a, m, n) = f(a, m + 1, n) = 0$. Then by Propositions 4 and 5, we know that

$$\phi(n - m, m, a) = \phi(n - m - 1, m + 1, a) = \phi(n - m, m + 1, a) = 0.$$

By Lemma 10, $\phi(n - m, m + 1, a) = -\phi(n - m, m, a) + a \cdot \phi(n - m, m, a - 1)$, i.e., $\phi(n - m, m, a - 1) = 0$. Hence $\phi(n - m, m, a - 1) = \phi(n - m, m, a) = 0$. Now, an iterative application of Proposition 6 eventually yields $\phi(0, m, a) = \phi(0, m, a-1) = 0$. By definition, $\phi(0, m, a)$ is the a'th derivative of

$$(x - 1)^m = \sum_{i=0}^{m} \binom{m}{i} x^i (-1)^{m-i}$$

evaluated at $x = 0$. But $m \ge a$, so this is clearly not zero. Thus we have reached a contradiction, and Lemma 11 is proved.   □

Theorem 3. *For large enough n and all $0 \le a \le n$: $D(HAM_n^{(=a)}) \ge n - 2$, $Q^*(HAM_n^{(=a)}) \ge \frac{n}{2} - 1$, and $C^*(HAM_n^{(=a)}) \ge n - 2$.*

Proof. By Lemma 8,

$$D(f), C^*(f) \ge \lg(\sum_{m:f(a,m,n) \neq 0} \binom{n}{m})$$

and

$$Q^*(f) \ge \frac{1}{2} \lg(\sum_{m:f(a,m,n) \neq 0} \binom{n}{m}).$$

First suppose $a \leq n/2$. We have

$$\sum_{m:f(a,m,n)\neq 0} \binom{n}{m} \geq \sum_{m\geq n/2:f(a,m,n)\neq 0} \binom{n}{m}. \qquad (5.2)$$

Let us lower-bound the r.h.s. of (5.2). First of all, since the r.h.s. of (5.2) works in the regime where $m \geq n/2 \geq a$, Lemma 11 shows that no two consecutive values of m in this range satisfy the condition "$f(a, m, n) = 0$". Also, for $m \geq n/2$, $\binom{n}{m}$ is a non-increasing function of m. Thus, if we imagine an adversary whose task is to keep the r.h.s. of (5.2) as small as possible, the adversary's best strategy, in our regime where $m \geq n/2$, is to make $f(a, m, n) = 0$ exactly when $m \in S$, where

$$S \doteq \{\lceil n/2 \rceil, \lceil n/2 \rceil + 2, \lceil n/2 \rceil + 4, \ldots\}. \qquad (5.3)$$

Now,

$$2^{n-1} \leq \sum_{m\geq n/2} \binom{n}{m} \leq 2^{n-1} + O(2^n/\sqrt{n}). \qquad (5.4)$$

(We need the second inequality to handle the case where n is even.) Also, recall that an $(1 - o(1))$ fraction of the sum $\sum_{m\geq n/2} \binom{n}{m}$ is obtained from the range $n/2 \leq m \leq n/2 + \sqrt{n \log n}$, for instance. In this range, the values of $\binom{n}{m}$ for any two consecutive values of m are within $(1 + o(1))$ of each other. In conjunction with (5.4), this shows that

$$\sum_{m\geq n/2:f(a,m,n)\neq 0} \binom{n}{m} \geq \sum_{m\geq n/2:m\notin S} \binom{n}{m} \geq (1/2 - o(1))2^{n-1}.$$

Thus,

$$\left\lceil \lg\left(\sum_{m\geq n/2:f(a,m,n)\neq 0} \binom{n}{m} \right) \right\rceil \geq n - 2,$$

completing the proof for the case where $a \leq n/2$.

Now we apply symmetry to the case $a > n/2$: note that Alice can reduce the problem with parameter a to the problem with parameter $n - a$, simply by complementing each bit of her input x. Thus, the same communication complexity results hold for the case $a > n/2$. $\qquad\square$

Lemma 12. *Let* $0 \leq a < m < n$. *If* $F(a, m, n) = 0$, *then* $F(a, m + 1, n) \neq 0$.

Proof. We have $f(j, m, n) = (-1)^m [x^j]((x - 1)^m (x + 1)^{n-m})$. By definition,

$$\begin{aligned}
F(a, m, n) &= \sum_{j=0}^{a} f(j, m, n) \\
&= (-1)^m \sum_{j=0}^{a} [x^j]((x - 1)^m (x + 1)^{n-m}) \\
&= (-1)^m [x^a]((x - 1)^m (x + 1)^{n-m} \cdot \sum_{j=0}^{\infty} x^j) \\
&= (-1)^m [x^a]((x - 1)^m (x + 1)^{n-m} \cdot \frac{1}{1-x}) \\
&= (-1)^{m-1} [x^a]((x - 1)^{m-1} (x + 1)^{n-m}) = f(a, m - 1, n - 1).
\end{aligned}$$

So $F(a, m, n) = F(a, m + 1, n) = 0$ iff $f(a, m - 1, n - 1) = f(a, m, n - 1) = 0$. But the latter is impossible by Lemma 11, thus the lemma is proved. $\qquad\square$

By a proof mostly similar to that of Theorem 3, we get

Theorem 4. *For large enough n and all $0 \leq a \leq n - 1$: $D(HAM_n^{(a)}) \geq n - 2$, $Q^*(HAM_n^{(a)}) \geq \frac{n}{2} - 1$, and $C^*(HAM_n^{(a)}) \geq n - 2$.*

Acknowledgments. We thank Jaikumar Radhakrishnan and the anonymous referees for their helpful comments.

References

1. K. Abdel-Ghaffar and A. E. Ababdi. An optimal strategy for comparing file copies. *IEEE Transactions on Parallel and Distributed Systems*, 5:87–93, 1994.
2. H. Buhrman and R. de Wolf. Communication complexity lower bounds by polynomials. In *Proc. of the 16th IEEE Conf on Complexity Theory*. IEEE Computer Society Press, 2001.
3. G. Cormode, M. Paterson, S. Sahinalp, and U. Vishkin. Communication complexity of document exchange. In *Proc. of the 11th ACM Symp. on Discrete Algorithms*, pages 197–206, 2000.
4. R. de Wolf. Quantum communication and complexity. *Theoretical Comput. Sci.*, 12: 337–353, 2002.
5. J. Feigenbaum, Y. Ishai, T. Malkin, K. Nissim, M. Strauss, and R. Wright. Secure multiparty computation of approximations. In *Proc. of the 28th ICALP (LNCS 2076)*, volume 2076 of *Lecture Notes in Computer Science*, pages 927–938, Berlin, 2001. Springer-Verlag.
6. D. Gavinsky, J. Kempe, and R. de Wolf. Quantum communication cannot simulate a public coin, 2004. arxiv.org/abs/quant-ph/0411051.
7. W. Huang, Y. Shi, S. Zhang, and Y. Zhu. The communication complexity of the Hamming distance problem. arxiv.org/abs/quant-ph/0509181.
8. H. Klauck. Lower Bounds for Quantum Communication Complexity. In *Proc. IEEE Symposium on Foundations of Computer Science*, pages 288–297, 2001.
9. I. Krasikov and S. Litsyn. On integral zeros of Krawtchouk polynomials. *J. Comb. Theory Ser. A*, 74:71–99, 1996.
10. E. Kushilevitz and N. Nisan. *Communication Complexity*. Cambridge University Press, 1997.
11. K. Mehlhorn and E. Schmidt. Las Vegas is better than determinism for VLSI and distributed systems. In *Proc. of the 14th ACM Symp. on Theory of Computing*, pages 330–337, 1982.
12. J. Metzner. Efficient replicated remote file comparison. *IEEE Transactions on Computers*, 40:651–659, 1991.
13. I. Newman. Private vs. common random bits in communication complexity. *Inf. Process. Lett.*, 39:67–71, 1991.
14. A. Orlitsky. Interactive communication: balanced distributions, correlated files, and average-case complexity. In *Proc. of the 32st IEEE Symp. on Found. of Comp. Sci.*, pages 228–238, 1991.
15. K. Pang and A. E. Gamal. Communication complexity of computing the Hamming distance. *SIAM Journal of Computing*, 15, 1986.
16. R. Raz. Fourier analysis for probabilistic communication complexity. *Journal of Computational Complexity*, 5:205–221, 1995.
17. H. Wilf. *Generatingfunctionology*. Academic Press, 1994.
18. A. Yao. On the power of quantum fingerprinting. In *Proc. of the 35th ACM Symp. on Theory of Computing*, pages 77–81, 2003.

Resources Required for Preparing Graph States

Peter Høyer[1,*], Mehdi Mhalla[2,**], and Simon Perdrix[2,***]

[1] Dept. of Comp. Sci., University of Calgary, Canada
hoyer@cpsc.ucalgary.ca
[2] Leibniz Laboratory, Grenoble, France
{mehdi.mhalla, simon.perdrix}@imag.fr

Abstract. Graph states have become a key class of states within quantum computation. They form a basis for universal quantum computation, capture key properties of entanglement, are related to quantum error correction, establish links to graph theory, violate Bell inequalities, and have elegant and short graph-theoretical descriptions. We give here a rigorous analysis of the resources required for producing graph states. Using a novel graph-contraction procedure, we show that any graph state can be prepared by a linear-size constant-depth quantum circuit, and we establish trade-offs between depth and width. We show that any minimal-width quantum circuit requires gates that acts on several qubits, regardless of the depth. We relate the complexity of preparing graph states to a new graph-theoretical concept, the local minimum degree, and show that it captures basic properties of graph states.

Keywords: Quantum Computing. Algorithms. Foundations of computing.

1   Introduction

What are the minimal resources required for universal quantum computation? This single question is one of the most fundamental questions related to building quantum computers, and it is one of the most studied questions within quantum computing. In 2000, in seminal work, Raussendorf and Briegel [15] proposed a new model for quantum computations. They show that if certain initial quantum states, called graph states, are provided, then the mere ability to perform one-qubit measurements suffices for quantum computations.

Graph states have been studied extensively within the last five years. The recent survey [11] by Hein *et al.* provides an excellent introduction to the area. These efforts have established several fundamental results on the universality of quantum computations based on graph states, physical implementations of graph states, the entanglement embodied by graph states, and have proved links to basic concepts within graph theory. In this paper, we study computational

* Work supported by Canada's CIAR, MITACS, and NSERC, France's FFCR, and the US ARO.
** Work conducted in parts while at the University at Calgary.
*** Work conducted in parts while visiting the University at Calgary.

aspects of graph states. We study in particular the question of characterizing the resources required for producing graph states and we establish stronger links to graph theory.

We first and foremost prove that any graph state can be prepared in constant time. That is, given a classical description of a graph $G = (V, E)$, we can produce the corresponding graph state $|G\rangle$ by a constant-depth quantum circuit that has size linear in the input size $|V| + |E|$ and that consists only of one-qubit operations and control-not operations. This implies that all two-qubit operations ever required by any quantum algorithm can be conducted at the outset of the algorithm in parallel, after which all operations act only on one qubit. We also show that our circuit is robust against various alterations. If we for instance do not wish to conduct all two-qubit operations at the outset, they can be postponed, and if for instance we want to limit the number of qubits used, i.e., the size of the Hilbert space acted upon, we can trade width for depth without compromising the overall linear upper bound on the size of the circuit.

The ability to efficiently procedure arbitrary graph states has several advantages: it reduces the number of qubits involved in the computation, sometimes even quadratically, and hence decreases the possibilities of errors, it replaces two-qubit quantum operations by simple and reliable classical computations, and it allows tailoring the preparation to specific quantum algorithms such as Shor's factoring algorithm [1].

We then introduce a new graph-theoretical measure, the local minimum degree, denoted δ_{loc}, and show that it is intimately linked to the complexity of preparing graph states. For instance, we use it to prove that any measurement-based quantum circuit for preparing graph states requires either ancilla qubits or multi-qubit measurements that act on at least $\delta_{\text{loc}} + 1$ qubits (or both). We also establish that the local minimum degree is related to the entanglement in graph states, and we give a family of graphs for which the local minimum degree is large. Such families may be suitable for cryptographic purposes, though likely difficult to create in practice [10]. Other graph-theoretical measures related to graph states have recently and independently been considered in [12, 16, 18].

2   Graph States and Signed Graph States

A *graph state* on n qubits is a state that is a superposition over all basis states,

$$|G\rangle = \frac{1}{\sqrt{2^n}} \sum_{x \in \{0,1\}^n} (-1)^{q_\Gamma(x)} |x\rangle, \tag{1}$$

where Γ is the adjacency matrix of a graph $G = (V, E)$ on $n = |V|$ vertices and $q_\Gamma(x) = \sum_{i<j:(i,j)\in E} x_i x_j$. The quadractic form q_Γ satisfies that $q_\Gamma(x) = x^T \Gamma^{\text{upper}} x$ where Γ^{upper} is the upper-triangle of Γ obtained by setting entries $\Gamma_{i,j}$ with $i \geq j$ to zero, and where T denotes taking transpose.

For technical reasons, it is sometimes convenient to associate signs to graph states. Given any graph state $|G\rangle$ and any subset $S \subseteq V$ the *signed graph state* $|G; S\rangle$ is the state

$$|G; S\rangle = \mathsf{Z}_S|G\rangle, \tag{2}$$

where $\mathsf{Z}_S = \bigotimes_{v \in S} \mathsf{Z}_v$ where Z_v denotes that the local operator $\mathsf{Z} = |0\rangle\langle 0| - |1\rangle\langle 1|$ acts on qubit v. We sometimes omit the signs when they are not essential for the discussion. We use similar abbreviated notation, $\mathsf{X}_S = \bigotimes_{v \in S} \mathsf{X}_v$ and $\mathsf{Y}_S = \bigotimes_{v \in S} \mathsf{Y}_v$, for the two other pauli-operators $\mathsf{X} = |1\rangle\langle 0| + |0\rangle\langle 1|$ and $\mathsf{Y} = \sqrt{-1}|1\rangle\langle 0| - \sqrt{-1}|0\rangle\langle 1|$.

Proposition 1. *For all graphs $G = (V, E)$ and all non-empty subsets $S \subseteq V$,*

$$\langle G; S|G\rangle = 0,$$

and hence $\langle G; S|G; S'\rangle = 0$ for all distinct subsets $S, S' \subseteq V$, and the $2^{|V|}$ states $\{|G; S\rangle\}_{S \subseteq V}$ form an orthonormal basis.

3   Preparation of Graph States

There is a simple algorithm for preparing the graph state $|G\rangle$ corresponding to any graph $G = (V, E)$ on $n = |V|$ vertices with $m = |E|$ edges. We first prepare n qubits in a superposition of all 2^n basis states,

$$|\Psi_0\rangle = \frac{1}{\sqrt{2^n}} \sum_{x \in \{0,1\}^n} |x\rangle,$$

by applying for instance the Hadamard operator $\mathsf{H} = \frac{1}{\sqrt{2}}\big(|0\rangle\langle 0| + |0\rangle\langle 1| + |1\rangle\langle 0| - |1\rangle\langle 1|\big)$ on each of the n qubits in the initial state $|0\rangle$. Each of the qubits corresponds to a vertex of G. We then modify the phases of the basis states by applying a sequence of m two-qubit operations, producing the graph state $|G\rangle$,

$$|G\rangle = \prod_{(u,v) \in E} \Delta_u(\mathsf{Z}_v)|\Psi_0\rangle. \tag{3}$$

For each edge $(u, v) \in E$, we apply the controlled phase change operator defined by

$$\Delta_u(\mathsf{Z}_v) = |00\rangle\langle 00| + |01\rangle\langle 01| + |10\rangle\langle 10| - |11\rangle\langle 11|$$

on the two qubits labelled by the endpoints u, v of the edge. These m two-qubit operations are diagonal and thus commute, allowing us to apply them in any order of our choosing. Summarizing, we can prepare any graph state $|G\rangle$ using n single-qubit operations and m two-qubit $\Delta(\mathsf{Z})$ operations. Considering this simple algorithm a quantum circuit, we can prepare a graph state by a circuit on n qubits of size $n + m$ and depth $m + 1$ using only single-qubit and two-qubit operations.

The depth of the circuit may be improved by parallelizing the two-qubit operations by choosing an edge coloring of G [1]. An *edge coloring* using χ' colors is a mapping $c : E \to \{1, 2, \ldots, \chi'\}$ satisfying that if two distinct edges e and

e' share one endpoint, then they are assigned different colors. Any graph has an edge coloring using at most $\Delta(G)+1$ colors, where $\Delta(G)$ is the maximum degree of the vertices in G, and we can find such an edge coloring in time polynomial in n and m, for instance by Vizing's (classical) algorithm [19]. This implies that we can arrange the m two-qubit operations in our circuit such that it has depth at most $\Delta(G) + 2$.

Proposition 2 ([1]). *Any graph state $|G\rangle$ can be prepared by a quantum circuit consisting of single-qubit and two-qubit operations of size $O(n + m)$ and depth $O(\Delta(G))$ acting on n qubits, where $\Delta(G)$ is the maximum degree of any vertex in G.*

The above proposition implies that graphs of bounded degree can be prepared by constant depth quantum circuits. In particular, two-dimensional cluster states, which is the common name for graph states arising from two-dimensional grid graphs, can be prepared by constant depth quantum circuits. We now extend this result and prove that arbitrary graphs can be prepared in constant depth.

Theorem 1 (Constant depth graph state preparation). *For any graph G, we can prepare some signed graph state $|G; S\rangle$ by a constant depth quantum circuit consisting of single-qubit and two-qubit operations of size $O(n+m)$ acting on $n + O(m)$ qubits.*

A key idea in our proof is to embed G as an induced minor of a larger graph of bounded degree, and then utilize that taking induced minors can be obtained by Pauli measurements.

We consider four types of substructures of a graph $G = (V, E)$. A *deletion of a vertex* $v \in V$ in G is the graph $G \setminus v$ obtained from G by deleting v and all edges incident to v. A *deletion of an edge* $e \in E$ in G is the graph $G \setminus e$ obtained from G by simply deleting the edge e. A *contraction of an edge* $(u, v) \in E$ in G is the graph $G/(u, v)$ obtained from G by introducing edges between v and each vertex in $N_G(u) \setminus N_G(v)$, and then deleting u and all its incident edges. A graph G is an *induced subgraph* of G' if G is isomorphic to a graph that can be obtained from G' by vertex deletions. It is a *subgraph* of G' if we can obtain a graph isomorphic to G by edge deletions. It is a *minor* of G' if it is isomorphic to a graph that can be obtained from G' by edge deletions and edge contractions, and it is an *induced minor* of G' if it is isomorphic to a graph that can be obtained from G' by vertex deletions and edge contractions. Any induced subgraph is also a subgraph, and any induced minor is also a minor.

The next two lemmas provide our main technical tools for generating the graph state $|G\rangle$.

Lemma 1 (Vertex deletion). *Let $v \in V$ be any vertex. Conducting a Z measurement of the qubit v maps $|G\rangle$ to $|G \setminus v\rangle$ and thus corresponds to deleting the vertex v.*

Proof. We first note that for any $S \subseteq V$,

$$|G; S\rangle = \frac{1}{\sqrt{2}}\big(|G \setminus v; S\rangle|0\rangle_v + |G \setminus v; S \oplus N_G(v)\rangle|1\rangle_v\big)$$

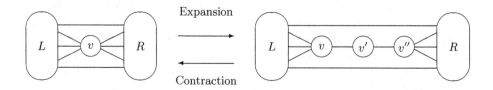

Fig. 1. Embedding G as an induced minor of a bounded degree graph $\tilde{G}$ by repeated applications of graph expansions. Conducting X measurements of qubits v' and v'' contracts the two edges (v, v') and (v', v'').

which follows from Eqs. 2 and 3. If qubit v is Z-measured (that is, measured in the basis of the eigenvectors of the Pauli operator Z), the resulting state is thus either $|G \setminus v; S\rangle$ or $|G \setminus v; S \oplus N_G(v)\rangle$ (where $\oplus$ denotes the symmetric difference), and hence is equal to $|G \setminus v\rangle$ up to (known) signs.   □

Lemma 2 (Edge contraction). *Let $v, v', v'' \in V$ so that $N_G(v') = \{v, v''\}$ and $N_G(v) \cap N_G(v'') = \{v'\}$. Conducting an X measurement of each of the two qubits v' and v'' maps $|G\rangle$ to $|G/(v, v')/(v', v'')\rangle$ and thus corresponds to contracting the two edges (v, v') and (v', v'').*

Proof. Let $v, v', v'' \in V$ be as in the lemma. For any subset $S \subseteq V$,

$$|G; S\rangle = \frac{Z_S}{2} \sum_{k,l \in \{0,1\}} \left(Z_{v''}^k Z_{N_G(v) \setminus \{v'\}}^l |G/vv'/v'v''\rangle \right) \left(|0\rangle + Z_{v'}^l |1\rangle \right)_{v'} \left(|0\rangle + Z_{v''}^k |1\rangle \right)_{v''}$$

and hence, if qubits v' and v'' are X-measured, the resulting state is given by $|G/(v, v')/(v', v'')\rangle$ up to (known) signs.   □

Proof of Theorem 1. We first embed G as an induced minor of a larger graph $\tilde{G}$ of bounded degree by repeatedly expanding any vertex v of degree $d \geq 4$ into three vertices v, v', v'' of strictly smaller degrees, as illustrated in Figure 1. Formally, we partition the neighborhood $N_G(v)$ of v in two sets L and R of sizes $\lceil \frac{d}{2} \rceil$ and $\lfloor \frac{d}{2} \rfloor$, respectively. We then set $\hat{V} = V \cup \{v', v''\}$ and $\hat{E} = E \setminus (\{v\} \times R) \cup \{(v, v'), (v', v'')\} \cup (\{v''\} \times R)$. We set $\hat{G} = (\hat{V}, \hat{E})$, and recursively expand any vertex v of $\hat{G}$ till all vertices have degree at most three. The thus obtained graph $\tilde{G}$ will have $O(m)$ vertices and maximum degree three.

We first prepare the graph state $|\tilde{G}\rangle$ by a constant-depth circuit on $O(m)$ qubits by applying Proposition 2, and then apply Lemma 2 to contract $|\tilde{G}\rangle$ to $|G\rangle$ by applying X measurements on all vertices introduced during the expansion of G to $\tilde{G}$. The X measurements commute and can thus all be conducted in parallel by a circuit of depth one.   □

Proposition 2 and Theorem 1 give two linear-size circuits for preparing any graph state, up to signs. The former has small width and large depth, the latter large width and small depth. We can trade width for depth in the above construction, without compromising the overall size of the circuit, by stopping the expansion once all vertices have degree at most T.

Theorem 2 (Small depth graph state preparation). *For any graph G, we can prepare some signed graph state $|G; S\rangle$ by a quantum circuit consisting of single-qubit and two-qubit operations of size $O(n + m)$ and depth $O(T)$ acting on $n + O(m/T)$ qubits, for any integer T.*

In the above theorems, we can replace the unitary operations by projective measurements. Any single-qubit unitary operation can be simulated using two single-qubit projective measurements, one two-qubit projective measurement, and one ancilla qubit [14]. Similarly, a two-qubit control-not operation can be simulated using two single-qubit projective measurements, two two-qubit projective measurements, and one ancilla qubit [14]. In Theorem 2, we may thus replace the unitary operations by projective measurements. We re-use the ancilla qubits in each of the $\Theta(T)$ layers of the circuit so that the the total number of additional ancilla qubits required by the simulation is only $O(m/T)$.

Theorem 3 (Measurement-based preparation). *For any graph G, we can prepare some signed graph state $|G; S\rangle$ by a quantum circuit consisting of single-qubit and two-qubit projective measurements of size $O(n + m)$ and depth $O(T)$ acting on $n + O(m/T)$ qubits, for any integer T.*

In the standard circuit for graph preparation, in which each of the $O(m)$ layers of the circuit consists of exactly one gate, we may similarly replace the unitary operations by projective measurements, yielding that any graph state can be prepared using $n + 1$ qubits by single-qubit and two-qubit projective measurements. We require only one ancilla qubit for iteratively simulating each of the $O(m)$ two-qubit operations.

Proposition 3 (Measurement-based preparation using one ancilla qubit). *For any graph G, we can prepare some signed graph state $|G; S\rangle$ by a quantum circuit consisting of single-qubit and two-qubit projective measurements of size $O(n + m)$ and depth $O(m)$ acting on $n + 1$ qubits.*

It would be interesting to extend Theorem 1 and its corollaries to incorporate noise and errors, for instance as discussed for cluster states in [6].

4 Circuits and Local Complementation

The circuit constructions for preparation of graph states given above are based on the concept of graph minors. To improve the constructions further, we require the additional concept of local complementation in graphs. A local complementation is a graph operation that maps one graph into another. Kotzig first showed that the class of circle graphs are closed under local complementation (see [8, 4]) and it was then used by Bouchet [5] to give a characterization of circle graphs.

A *local complementation* of a graph G at a vertex $v \in V$ is the operation which replaces the subgraph of G induced by the neighborhood $N_G(v)$ of v by its complement. We denote the thus obtained graph by $G \star v$. Local complementation defines an equivalence relation. We say two graphs G_1 and G_2 are *locally equivalent* if one can be obtained from the other by a sequence of local complementations, and write $G_1 \approx_{\text{loc}} G_2$.

A *pivoting* on a edge $(u, v) \in E$ is the operation that maps G to $G \star u \star v \star u$. This operation is well-defined as $G \star u \star v \star u = G \star v \star u \star v$, and we denote it by $G \wedge (u, v)$. Following Oum [13], we say a graph H is a *vertex minor* of G if H can be obtained from G by vertex deletions and local complementations. A graph H is a *pivot minor* of G if H can be obtained from G by vertex deletions and pivotings.

The most important property of local complementation for this paper, is that it can be implemented by local quantum operations. Let G and G' be two locally equivalent graphs, $G \approx_{\text{loc}} G'$. Then there exists a tensor product $U = \bigotimes_{v \in V} U^{(v)}$ of n single-qubit unitary operations $U^{(v)}$ such that $|G'\rangle = U|G\rangle$. This implies that if $\mathcal{C}$ is a circuit that maps $|\Psi_0\rangle$ to $|G'\rangle$, then $U\mathcal{C}$ is a circuit that maps $|\Psi_0\rangle$ to $|G\rangle$. Thus, any two locally equivalent graphs can be implemented by circuits of the same depths, up to an additive constant of one.

Let $\delta(G) = \min\{\deg_G(v) : v \in V\}$ denote the minimum degree of any vertex in G, where $\deg_G(v)$ denotes the degree of v in G. Let $\delta_{\text{loc}}(G) = \min\{\delta(G') : G' \approx_{\text{loc}} G\}$ denote the minimum degree achievable by local complementations. We refer to δ_{loc} as the *local minimum degree of* G. Similarly, let $m_{\text{loc}}(G) = \min\{|E'| : (V, E') \approx_{\text{loc}} G\}$ denote the minimum total number of edges achievable by local complementations. Unfortunately, there is no known polynomial-time algorithm for computing either of the two quantities $\delta_{\text{loc}}(G)$ and $m_{\text{loc}}(G)$, given a graph G as input. The thus far best result in this direction is a result of Bouchet [2] stating that the problem of deciding if two graphs are locally equivalent is polynomial-time computable. Van den Nest [17] gives in his Ph.D. thesis a short description of Bouchet's algorithm.

The quantity m_{loc} is related to the size of any quantum circuit preparing a graph state. Suppose we could find a polynomial-time algorithm that given any graph, outputs a locally equivalent graph of minimum total degree. Then in Theorems 1, 2 and 3, we could replace m by m_{loc}, and still have polynomial-time constructable quantum circuits. However, currently no such result is in sight.

We now show that δ_{loc} is related to the usage of ancilla qubits in the circuits for preparing graph states. To prove this, we first give three equivalent definitions of δ_{loc}, the first graph theoretical, the second combinatorial, and the third algebraic. We require the following notation and concepts.

For any subset $X \subseteq V$ of vertices, let $\text{Odd}_G(X) = \{u \in V \setminus X \mid N_G(u) \cap X = 1 \mod 2\}$ denote the set of vertices that is adjacent to an odd number of vertices in X in G. Similarly, let $\text{Even}_G(X) = \{u \in V \setminus X \mid N_G(u) \cap X = 0 \mod 2\}$. We say that the vertices in $\text{Odd}_G(X)$ are *odd neighbors* of X in G, and that the vertices in $\text{Even}_G(X)$ are *even neighbors* of X in G.

The cut-matrix of a subset $X \subseteq V$ of vertices is the submatrix $\Gamma_G(X, V \setminus X)$ indexed by $X \times (V \setminus X)$ of the adjacency matrix Γ_G of G. The cut-rank $\mathrm{Cutrk}(X)$ of X is the rank of its cut-matrix, where we define the rank over GF(2). The cut-rank of X is invariant under local complementation [3], though the null-space of $\Gamma_G(X, V \setminus X)$ may change under local complementation. It was used by Bouchet [2] and others under the name "connectivity function", and coined the cut-rank by Oum [13]. We say that a set of vertices $L \subseteq V$ is *local* if $L = X \cup \mathrm{Odd}_G(X)$ for some subset $X \subseteq L$. Note that a local set L does not have full cut-rank, and that $\{v\} \cup N_G(v)$ is local for any vertex $v \in V$.

Lemma 3. *Any local set L is invariant under local complementation. Moreover, for all $y \in L$, there exists a graph G' locally equivalent to G such that $\{y\} \cup \mathrm{Odd}_{G'}(\{y\}) \subseteq L$.*

Proof. Suppose that $L = X \cup \mathrm{Odd}_G(X)$. We consider how the three-way partition $V = X \cup \mathrm{Odd}_G(X) \cup \mathrm{Even}_G(X)$ changes under local complementation at a vertex $v \in V$. Let $G' = G \star v$. Then the three-way partition changes as follows.

	X'	$\mathrm{Odd}_{G'}(X')$	$\mathrm{Even}_{G'}(X')$
$v \in \mathrm{Even}_G(X)$	X	$\mathrm{Odd}_G(X)$	$\mathrm{Even}_G(X)$
$v \in \mathrm{Odd}_G(X)$	$X \cup \{v\}$	$\mathrm{Odd}_G(X) \setminus \{v\}$	$\mathrm{Even}_G(X)$
$v \in X$ and $\|N_X(v)\|$ is odd	$X \setminus \{v\}$	$\mathrm{Odd}_G(X) \cup \{v\}$	$\mathrm{Even}_G(X)$
$v \in X$ and $\|N_X(v)\|$ is even	X	$\mathrm{Odd}_G(X)$	$\mathrm{Even}_G(X)$

The forth and last column implies that any local set L is invariant under local complementation, and thus only the internal structure of L changes. By the second row, we can move vertex y into X, if $y \in \mathrm{Odd}_G(X)$. By the third row, we can move vertices out of X as long as there exists a vertex in X having an odd number of neighbors in X. If all vertices in X have an even number of neighbors in X, and if any vertex in X has a neighbor z in $\mathrm{Even}_G(X)$, then a local complementation at z creates at least two vertices in X having an odd number of neighbors in X. One of these must be a vertex different from y. Thus, by a sequence of local complementations, we can map G to some graph G' in which there are no edges between X and $\mathrm{Even}_{G'}(X)$, and in which $y \in X$. Hence $N_{G'}(y) \subseteq L$ and thus $\{y\} \cup \mathrm{Odd}_{G'}(\{y\}) \subseteq L$. □

Corollary 1. *Let $x \in V$ be any vertex of degree $d = \delta_{\mathrm{loc}}(G)$. Then for all neighbors $y \in N_G(x)$, there exists a graph G' locally equivalent to G for which $N_{G'}(y) = (N_G(x) \cup \{x\}) \setminus \{y\}$. In particular, each neighbor of x can be locally reduced to having degree d.*

Theorem 4 (Characterization of local minimum degree). *For any graph G, the local minimum degree $\delta_{\mathrm{loc}}(G)$ is equal to*

1. $\min \left\{ \delta(G') \mid G' \approx_{\mathrm{loc}} G \right\}$.
2. $\min \left\{ |L| : L \text{ is nonempty and local} \right\} - 1$.
3. $\min \left\{ |X| \mid \mathrm{Cutrk}(X) < |X| \right\} - 1$.

Proof. We first show that the quantity in (1) is an upper bound on the quantity in (2). Let $y \in V$ be a vertex of degree $\delta_{\mathrm{loc}}(G)$ in $G' \approx_{\mathrm{loc}} G$. Then $\{y\} \cup N_{G'}(y)$ is local. Similarly, we show that the quantity in (2) is an upper bound on the quantity in (3). Let $L = X \cup \mathrm{Odd}_G(X)$ be local. Then $\chi_X \Gamma_G[L, V \setminus L] = 0$, where χ_X is the indicator function of X in L, and thus L does not have full cutrank. Finally, we show that the quantity in (3) is an upper bound on the quantity in (1). Let $X \subset V$ be a set that does not have full cutrank. Let $Y \subseteq V \setminus X$ be such that $\chi_Y \Gamma[X, V \setminus X] = 0$. Then $\mathrm{Odd}_G(Y) \subset X$. By Lemma 3, for all $y \in Y$, there is a graph G' locally equivalent to G such that $\deg_{G'}(y) \leq |X| - 1$. □

By Theorem 4, for any fixed integer d, there exists a polynomial-time algorithm for deciding if $\delta_{\mathrm{loc}} > d$. If d is part of the input, no polynomial-time algorithm is known. Though plausible, it is not known whether the concept of cut-rank is helpful in computing δ_{loc} in polynomial time. One result in this direction is by Oum [13], who gives a polynomial-time algorithm that given any two disjoint non-empty sets of vertices $A, B \subset V$ as input, computes the value $\min\{\mathrm{Cutrk}(Z) \mid X \subseteq Z \subseteq V \setminus B\}$ by greedily searching for blocking sequences as introduced by Geelen [9].

We now use the above characterization to show that if no ancilla qubits are available for preparing a graph state, then joint projective measurements on at least $\delta_{\mathrm{loc}}(G) + 1$ qubits are required. As a consequence, for all graphs for which $\delta_{\mathrm{loc}} > 1$, there does not exist a measurement-based preparation using only single-qubit and two-qubit projective measurements without the use of ancilla qubits.

Theorem 5 (Lower bound on measurement-based preparation). *Let G be any graph. Any preparation of $|G\rangle$ by a quantum circuit acting on n qubits and consisting of projective measurements requires measurements acting on $\delta_{\mathrm{loc}}(G) + 1$ qubits.*

Proof. By contradiction. Assume the last measurement of the preparation acts on at most $\delta_{\mathrm{loc}}(G)$ qubits X and that it produces the signed graph state $|G; S\rangle$. Let W be the observable describing this measurement. Then $\mathsf{W}|G; S\rangle = |G; S\rangle$, and thus $\langle G; S|\mathsf{W}|G; S\rangle = 1$.

Let $U \subseteq X$ be any subset of the measured vertices X. Since $|X| \leq \delta_{\mathrm{loc}}(G)$, subset X has full cutrank by Theorem 4, and thus there exists a subset $Y \subseteq V \setminus X$ such that $\chi_U = \Gamma[X, V \setminus X]\chi_Y$. The operator X_Y acts only on qubits not in X, and thus commutes with W. In addition, $\mathsf{Z}_{\mathrm{Odd}(Y)}\mathsf{X}_Y|G; S\rangle = \pm|G; S\rangle$, and hence

$$1 = \langle G; S|\mathsf{W}|G; S\rangle = \langle G; S|\mathsf{X}_Y \mathsf{W}\mathsf{X}_Y|G; S\rangle = \langle G; S|\mathsf{Z}_{\mathrm{Odd}(Y)}\mathsf{W}\mathsf{Z}_{\mathrm{Odd}(Y)}|G; S\rangle$$
$$= \langle G; S|\mathsf{Z}_U \mathsf{W}\mathsf{Z}_U|G; S\rangle = \langle G; S \oplus U|\mathsf{W}|G; S \oplus U\rangle.$$

It follows that W acts trivially on $|G; S \oplus U\rangle$ for all subsets $U \subseteq X$. Since these $2^{|X|}$ states are pairwise orthogonal, W is the identity, which is a contradiction. □

It is natural to consider recursive methods for preparing a graph state G, for instance by partitioning the vertex set V into parts which then are considered individually. The next lemma states that deleting any one vertex or edge may decrease the local degree by at most one.

Lemma 4. *For any graph $G = (V, E)$, any vertex $u \in V$, and any edge $e = (v, w) \in E$, $\delta_{\mathrm{loc}}(G \setminus v) \geq \delta_{\mathrm{loc}}(G) - 1$ and $\delta_{\mathrm{loc}}(G \setminus e) \geq \delta_{\mathrm{loc}}(G) - 1$.*

Proof. Let $X \subseteq V \setminus \{u\}$ be any set of vertices satisfying that $\mathrm{Cutrk}_{G \setminus u}(X) < |X|$. Then $\mathrm{Cutrk}_G(X \cup \{u\}) \leq \mathrm{Cutrk}_{G \setminus u}(X) + 1 < |X \cup \{u\}|$. Now, let $X \subseteq V$ be any set of vertices satisfying that $\mathrm{Cutrk}_{G \setminus e}(X) < |X|$, and consider an edge $e = (v, w) \in E$. Firstly, if $v, w \in X$ or $v, w \notin X$, then $\mathrm{Cutrk}_G(X) = \mathrm{Cutrk}_{G \setminus e}(X)$. Secondly, if $v \in X$ and $w \notin X$, then $\mathrm{Cutrk}_G(X \cup \{w\}) = \mathrm{Cutrk}_{G \setminus e}(X \cup \{w\}) \leq \mathrm{Cutrk}_{G \setminus e}(X) + 1 < |X| + 1 = |X \cup \{w\}|$. □

Suppose we are given an oracle $\mathcal{O}_\delta$ that given any graph G returns $\delta_{\mathrm{loc}}(G)$. Then there exists a deterministic algorithm that given any graph G outputs a graph G' locally equivalent to G with $\delta(G') = \delta_{\mathrm{loc}}(G)$. The algorithm runs in time polynomial in n and uses at most a linear number of oracle queries. We omit the proof.

Theorem 6. *The following two computational problems are polynomially equivalent: (1) computing $\delta_{\mathrm{loc}}(G)$ and (2) finding a graph G' with $G' \equiv G$ and $\delta(G') = \delta_{\mathrm{loc}}(G)$.*

5   Bi-separability and δ_{loc}

An n-qubit state ρ is *bi-separable* if there exists a partition A, B of V such that ρ can be written on the form $\rho = \sum_{i=1}^{k} \alpha_i \rho_i^A \otimes \rho_i^B$ for a finite set of density operators ρ_i^A, ρ_i^B and non-negative weights α_i, where each ρ_i^A acts on A only and each ρ_i^B acts on B only. In this section, we refer to bi-separability simply as separability.

Theorem 7. *For any graph state $|G\rangle$, there exists a subset $U \subseteq V$ of vertices of size $\delta_{\mathrm{loc}}(G)$ such that the reduced density operator $\rho = \mathrm{Tr}_U(|G\rangle\langle G|)$ is separable.*

Proof. Let G' be a graph that is locally equivalent to G and contains a vertex v of degree $d = \delta_{\mathrm{loc}}(G)$. We show that the state ρ obtained from $|G'\rangle\langle G'|$ by tracing out the d qubits corresponding to the neighbors $N_{G'}(v)$ of v is separable. We do this by giving a procedure for preparing ρ that preserves separability.

Let $H = G' \setminus N_{G'}(v)$ be the subgraph of G' obtained by removing the neighbors $N_{G'}(v)$ of v. The subgraph H consists of (at least) two components, $\{v\}$ and the rest. The graph state $|H\rangle$ is thus separable and can be written on the form $|v\rangle|H \setminus \{v\}\rangle$.

Consider we first prepare the separable pure state $|H\rangle$. Now for each neighbor $w \in N_{G'}(v)$ in turn, we randomly flip an unbiased coin. If the outcome of the coinflip is head, we apply a (single-qubit) σ_z operation on each of the qubits of $|H\rangle$ corresponding to the neighbors $N_{G'}(w) \cap H$ of w. If the outcome of the coinflip is tail, we do not alter the state. This maps the state $|v\rangle|H \setminus \{v\}\rangle$ to some other pure state $|v\rangle|H'\rangle$ that is also separable with respect to the same partitioning of vertices. We take sum of the density operators $|v\rangle|H'\rangle\langle H'|\langle v|$ over all 2^d possible outcomes, yielding a density operator ρ' that is separable

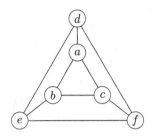

Fig. 2. The prism graph P on six vertices. Tracing out vertices b and e from $|G\rangle$ creates a separable mixed state.

with respect to the same partitioning. The density operator ρ' is the same as the density operator ρ obtained by tracing out the neighbors $N_{G'}(v)$ of v in G', and thus ρ is separable. $\qquad\square$

The upper bound of $\delta_{\mathrm{loc}}(G)$ on separability in Theorem 7 is tight for some graphs, but not all. An example for which the bound is not tight is the prism graph P on six vertices, illustrated in Figure 2. On the one hand, any local set in P has size at least 4, and hence its local minimum degree is (at least) 3 by Theorem 4. On the other hand, if we trace out qubits b and e in $|P\rangle$, the remaining state is separable across the cut $(\{a,d\},\{c,f\})$. One way of seeing this, is to first delete the edge (b,e), do a local complementation on b and then e, deleting b and e, and noticing that the remaining graph consists of two components, $\{a,d\}$ and $\{c,f\}$. It would be interesting to explore the relationship between δ_{loc} and separability further, and also to consider k-separability and full separability [7].

We can also show that there exists a natural family of graphs for which δ_{loc} is polynomial in the input graph. It seems plausible that this family of graphs contains entanglement that is very robust against quantum operations acting on a sublinear number of qubits, and thus could be useful in for instance quantum cryptography and quantum communication complexity.

Theorem 8. *There exists a constant $c > 0$ and a family of graphs G for which $\delta_{\mathrm{loc}}(G) \in \Omega(|G|^c)$.*

Acknowledgements

We thank Jop Briet, David Feder, Michael Garrett, and Mark Tame for stimulating conversations.

References

1. P. Aliferis and D. W. Leung. Computation by measurements: A unifying picture. *Physical Review A*, 70:062314, 2004.
2. A. Bouchet. Diagraph decompositions and eulerian systems. *SIAM J. Algebraic Discrete Methods*, 8:323–337, 1987.

3. A. Bouchet. Connectivity of isotropic systems. In *Combinatorial Mathematics: Proc. of the Third International Conference*, volume 555 of *Ann. New York Acad. Sci.*, pages 81–93, 1989.
4. A. Bouchet. κ-transformations, local complementations and switching. In *Cycles and rays: Basic structures in finite and infinite graphs*, volume C 301 of *NATO Adv. Sci. Inst. Ser.*, pages 41–50. Kluwer Acad. Publ., Dordrecht, 1990.
5. A. Bouchet. Circle graph obstructions. *J. Comb. Theory Ser. B*, 60(1):107–144, 1994.
6. L. M. Duan and R. Raussendorf. Efficient quantum computation with probabilistic quantum gates. *Phys. Rev. Lett.*, 95:080503, 2005.
7. J. Eisert and D. Gross. *Lectures on Quantum Information*, chapter Multi-particle entanglement. Wiley-VCH, Berlin, 2006.
8. H. de Fraysseix. Local complementation and interlacement graphs. *Discrete Mathematics*, 33(1):29–35, 1981.
9. J. F. Geelen. *Matchings, Matroids and Unimodular Matrices*. PhD thesis, Univ. Waterloo, 1995.
10. K. Goyal, A. McCauley, and R. Raussendorf. Purification of large bi-colorable graph states, May 2006. quant-ph/0605228.
11. M. Hein, W. Dür, J. Eisert, R. Raussendorf, M. Van den Nest, and H. J. Briegel. Entanglement in graph states and its applications. In *Proc. of the Int. School of Physics "Enrico Fermi" on "Quantum Computers, Algorithms and Chaos"*, July 2005. quant-ph/0602096.
12. I. Markov and Y. Shi. Simulating quantum computation by contracting tensor networks. In *Ninth Workshop on Quantum Information Processing*, Jan. 2006. (No proceedings).
13. S.-i. Oum. Approximating rank-width and clique-width quickly. In D. Kratsch, editor, *Graph-Theoretic Concepts in Computer Science, WG 2005*, volume 3787 of *Lecture Notes in Computer Science*, pages 49–58. Springer, 2005.
14. S. Perdrix. State transfer instead of teleportation in measurement-based quantum computation. *International Journal of Quantum Information*, 3(1):219–224, 2005.
15. R. Raussendorf and H. J. Briegel. A one-way quantum computer. *Physical Review Letters*, 86:5188–5191, May 2001.
16. Y. Shi, L. M. Duan, and G. Vidal. Classical simulation of quantum many-body systems with a tree tensor network. (In completion), Feb. 2006.
17. M. Van den Nest. *Local equivalence of stabilizer states and codes*. PhD thesis, Faculty of Engineering, K. U. Leuven, Belgium, May 2005.
18. M. Van den Nest, A. Miyake, W. Dür, and H. J. Briegel. Universal resources for measurement–based quantum computation, Apr. 2006. quant-ph/0604010.
19. V. G. Vizing. On an estimate of the chromatic class of a p-graph. *Metody Diskret. Analiz.*, 3:25–30, 1964. In Russian.

Online Multi-path Routing in a Maze*

Stefan Rührup[1] and Christian Schindelhauer[2]

[1] Heinz Nixdorf Institute, University of Paderborn, Germany
`sr@uni-paderborn.de`
[2] Computer Networks and Telematics, University of Freiburg, Germany
`schindel@informatik.uni-freiburg.de`

Abstract. We consider the problem of route discovery in a mesh network with faulty nodes. The number and the positions of the faulty nodes are unknown. It is known that a flooding strategy like expanding ring search can route a message linear in the minimum number of steps d while it causes a traffic (i.e. the total number of messages) of $\mathcal{O}(d^2)$. For optimizing traffic a single-path strategy is optimal producing traffic $\mathcal{O}(d+p)$, where p is the number of nodes that are adjacent to faulty nodes. We present a deterministic multi-path online routing algorithm that delivers a message within $\mathcal{O}(d)$ time steps causing traffic $\mathcal{O}(d + p \log^2 d)$. This algorithm is asymptotically as fast as flooding and nearly traffic-optimal up to a polylogarithmic factor.

1 Introduction and Overview

Sending a message is the most fundamental feature of communication networks. We consider two-dimensional mesh networks, which can be found in parallel computers, in integrated circuits, FPGAs (Field Programmable Gate Arrays) and also some kinds of wireless sensor networks. In all these networks nodes may fail or may be unavailable. A node's failure can only be noticed by its neighbors. A straight-forward approach is to regularly test the neighbors of each node, to collect this data and to distribute a map of all failed and working nodes throughout the network. We investigate scenarios where this knowledge is not available when the message is on its way. Due to the lack of global information this routing problem states an online problem.

The basic problem is that the faulty nodes are barriers to the routing algorithm and that the algorithm does not know these barriers. There is no restriction on the size and the shape of the barriers, so even labyrinths are possible. In such situation a fast message delivery can only be guaranteed by flooding the complete network, which results in a tremendous increase of traffic, i.e. the number of node-to-node transmissions. If the algorithm uses a single-path strategy, then the additional effort necessary for searching a path to the destination increases the time.

---

* Partially supported by the DFG Sonderforschungsbereich 376 and by the EU within 6th Framework Programme under contract 001907 "Dynamically Evolving, Large Scale Information Systems" (DELIS).

We analyze algorithms with respect to the length of the shortest path d between source and target and with respect to the number of border nodes p, which are the nodes adjacent to faulty nodes. Regarding the time, no single-path online algorithm can beat the optimal offline algorithm and in worst case scenarios it has to investigate all the barriers, i.e. a traffic proportional to the number of border nodes p is inevitable. There are single-path algorithms that use only $\mathcal{O}(d+p)$ messages in total, but they need $\mathcal{O}(d+p)$ time steps. Time-optimal algorithms are parallel multi-path algorithms (e.g. expanding ring search) with time $\mathcal{O}(d)$ and traffic $\mathcal{O}(d^2)$ in the worst case.

We are interested in optimizing time and traffic at the same time. One might expect a trade-off situation between these measures. However, our research shows that there are algorithms that approximate the offline time bound and the optimal online traffic bound by a factor of $\mathcal{O}(\sqrt{d})$ [21] at the same time. The quotient comparing to the offline time bound is called the competitive time ratio, while the quotient comparing to the traffic bound of the optimal online algorithm is called the comparative traffic ratio. Subsequent work showed that both bounds could be improved below any polynomial bound to a term of $\tilde{O}(d^{\sqrt{\frac{\log\log d}{\log d}}})$ [22]. We call a bound on both ratios the combined comparative ratio $\mathcal{R}_c$ (Def. 5).

Strategy	Time	Traffic	$\mathcal{R}_c$
Exp. Ring Search [9,18]	$\mathcal{O}(d)$	$\mathcal{O}(d^2)$	$\mathcal{O}(d)$
Lucas' Algorithm [13]	$\mathcal{O}(d+p)$	$\mathcal{O}(d+p)$	$\mathcal{O}(d)$
Alternating Strategy [21]	$\mathcal{O}(d^{3/2})$	$\mathcal{O}(\min\{d^2, d^{3/2}+p\})$	$\mathcal{O}(\sqrt{d})$
Selective Flooding [22]	$d \cdot 2^{\mathcal{O}\left(\sqrt{\frac{\log d}{\log\log d}}\right)}$	$\mathcal{O}(d)+p\,d^{\mathcal{O}\left(\sqrt{\frac{\log\log d}{\log d}}\right)}$	$d^{\mathcal{O}\left(\sqrt{\frac{\log\log d}{\log d}}\right)}$
JITE (this paper)	$\mathcal{O}(d)$	$\mathcal{O}(d+p\log^2 d)$	$\mathcal{O}(\log^2 d)$
Online Lower Bound (cf. [3])	$\Omega(d)$	$\Omega(d+p)$	$\Omega(1)$

In this paper we achieve a break-through in this line of research showing a ratio of $\mathcal{O}(\log^2 d)$. More specifically we present a deterministic algorithm that delivers the message on a multi-path route within time $\mathcal{O}(d)$ and with traffic $\mathcal{O}(d+p\log^2 d)$. This shows, that one can route a message asymptotically as fast as flooding while increasing the traffic by a factor of only $\mathcal{O}(\log^2 d)$ compared to the traffic-optimal online algorithm.

This paper is organized as follows. We continue this section by presenting related research. In the following section we describe the basic definitions and techniques more formally. In Section 3, we present an overview of the algorithm and its components. In Section 4, we sketch the time and traffic analysis which concludes the paper.

1.1 Related Work

The problem studied in this paper has a strong relation to online search and navigation problems. These problems have been investigated in different research communities, which Angluin et al. [1] called "the online competitive analysis community" and the "theoretical robotics community". The fundamental goal in online searching is to find a point in an unknown environment. In theoretical robotics the scenarios contain obstacles of arbitrary shape and the performance of algorithms is expressed by comparing the distance traveled by the robot to the sum of the perimeters of the obstacles [15,1] (see also [2] for a survey and [14,19] for an overview of path-planning and maze traversal algorithms). The competitive analysis community has studied various kinds of scenarios with restrictions on the obstacles (e.g. quadratic or convex obstacles). The performance is expressed by the *competitive ratio*, which is the ratio of the distance traveled by the robot and the length of the shortest obstacle-free path [17,3].

Our model connects these two lines of research. Scenarios considered in online navigation with a lower bound on distance between s and t and with finite obstacle perimeters can be modeled by a faulty mesh network. We also investigate the problem of finding a path to a given point in an unknown environment, but here, the search can also be done in parallel. For robot navigation problems it is not clear how unbounded parallelism can be modeled in a reasonable way. Usually, navigation strategies are only considered for a constant number of robots. The model of a mesh network with faulty parts enables us to study the impact of parallelism on the time needed for finding the target. For the time analysis we use the competitive ratio as used by the competitive analysis community. Traffic is compared to the perimeters of the barriers which gives the *comparative traffic ratio*. This ratio expresses the amount of parallelism used by the algorithm.

Routing in faulty networks has also been considered as an offline problem. In the field of parallel computing the fault-tolerance of networks is studied, e.g. by Cole et al. [8]. The problem is to construct a routing scheme that emulates the original network. Zakrevski and Karpovski [26] investigate the routing problem for two-dimensional meshes. The model is similar to ours as they consider two-dimensional meshes under the store-and-forward model. Their algorithm needs an offline pre-routing stage, in which fault-free rectangular clusters are identified. Routing algorithms for two-dimensional meshes, that need no pre-routing stage are presented by Wu [24]. These algorithms use only local information, but the faulty regions in the mesh are assumed to be be rectangular blocks. In [25] Wu and Jiang present a distributed algorithm that constructs convex polygons from arbitrary fault regions by excluding nodes from the routing process. This is advantageous in the wormhole routing model, because it helps to reduce the number of virtual channels. We will not deal with virtual channels and deadlock-freedom as we consider the store-and-forward model.

Bose and Morin [6,5] study the online routing problem for triangulations and plane graphs with certain properties and present constant-competitive algorithms for routing in these graphs. In triangulations, where no local minima exist, routing can be done by a greedy strategy. Such strategies are also used for position-based routing. Position-based routing is a reactive routing used in wireless networks, where the nodes are equipped with a positioning system, such that a message can be forwarded in the direction of the target (see [16] for a survey). Due to the limited range of the radio transceivers, there are local minima and messages have to be routed around void regions (an analog to the fault regions in the mesh network). There are various single-path strategies, e.g. [11,7,12]. Position-based routing strategies have been mainly analyzed in a worst case setting, i.e. the void regions have been constructed such that the connections form a labyrinth. In this case the traffic-efficient single-path strategies produce as much traffic as flooding. In our analysis we take the perimeters of fault regions into account, so that we can express performance beyond the worst case point of view.

2 Basic Definitions and Techniques

A two-dimensional mesh network with faulty nodes is defined by a set of nodes $V \subseteq \mathbb{N} \times \mathbb{N}$ and a set of edges $E := \{(v, w) : v, w \in V \wedge |v_x - w_x| + |v_y - w_y| = 1\}$. There is no restriction on the size of the network, because time and traffic are analyzed with respect to the position of the given start node s and target node t. We assume a synchronized communication: Each message transmission to a neighboring node takes one *time step*. Furthermore, we assume the messages to be transported in a store-and-forward fashion and that the nodes do not fail while a message is being transported. However, there is no global knowledge about faulty nodes. Only adjacent nodes can determine whether a node is faulty.

Barriers, Borders and Traversals. The network contains *active* (functioning) and *faulty* nodes. Faulty nodes neither participate in communication nor can they store information. Faulty nodes which are orthogonally or diagonally neighboring form a *barrier*. A barrier consists only of faulty nodes and is not connected to or overlapping with other barriers. Active nodes adjacent to faulty nodes are called *border nodes*. All the nodes in the neighborhood (orthogonally or diagonally) of a barrier B form the *perimeter* of B. A path around a barrier in (counter-)clockwise order is called a *right-hand (left-hand) traversal path*, if every border node is visited and only nodes in the perimeter of B are used. The *perimeter size* $p(B)$ of a barrier B is the number of directed edges of the traversal path. The *total perimeter size* is $p := \sum_{i \in \mathbb{N}} p(B_i)$. The perimeter size is the number of steps required to send a message from a border node around the barrier and back to the origin, whereby each border node of the barrier is visited. It reflects the time consumption of finding a detour around the barrier.

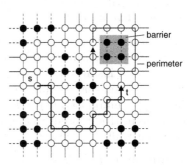

Fig. 1. Mesh network with faulty nodes (black), routing path and right-hand traversal path

The Competitive Time Ratio. Time is the number of steps needed by the algorithm to deliver a message and equivalent to the length of a path a message takes. Comparing the time of the algorithm with the optimal time leads to the competitive ratio, which is well known in the field of online algorithms [4].

Definition 1. *An algorithm A has a* competitive ratio *of c, if $\forall x \in \mathcal{I} : C_A(x) \leq c \cdot C_{\text{opt}}(x)$, where $\mathcal{I}$ is the set of all instances of the problem, $C_A(x)$ the cost of algorithm A on input x and $C_{\text{opt}}(x)$ the cost of an optimal offline algorithm.*

We compare the time of the algorithm with the length d of the shortest path to the target. Note, that the shortest path uses only non-faulty nodes.

Definition 2. *Let d be the length of the shortest barrier-free path between source and target. A routing algorithm has* competitive time ratio $\mathcal{R}_t := T/d$ *if the message delivery is performed in T steps.*

The Comparative Traffic Ratio. Traffic is the number of messages an algorithm needs. A comparison with the traffic of the best offline algorithm would be unfair, because no online algorithm can reach this bound. Therefore, we define a *comparative ratio* based on a *class* of instances of the problem, which is a modification of the definition given by Koutsoupias and Papadimitriou [10]:

Definition 3. *An algorithm A has a* comparative ratio *$f(P)$, if*

$$\forall p_1 \ldots p_n \in P : \max_{x \in \mathcal{I}_P} C_A(x) \leq f(P) \cdot \min_{B \in \mathcal{B}} \max_{x \in \mathcal{I}_P} C_B(x),$$

where $\mathcal{I}_P$ is the set of instances which can be described by the parameter set P, $C_A(x)$ the cost of algorithm A and $C_B(x)$ the cost of an algorithm B from the class of online algorithms $\mathcal{B}$.

With this definition we address the difficulty that is caused by a certain class of scenarios that can be described in terms of the two parameters d and p. For any such instance the online traffic bound is $\min_{B \in \mathcal{B}} \max_{x \in \mathcal{I}_{\{d,p\}}} C_B(x) = \Theta(d + p)$. Note, that for any scenario one can find an optimal offline algorithm: $\max_{x \in \mathcal{I}_{\{d,p\}}} \min_{B \in \mathcal{B}} C_B(x) = d$. This requires the modification of the comparative ratio in [10] in order to obtain a fair measure. So, we use the online lower bound for traffic to define the *comparative traffic ratio*.

Definition 4. *Let d be the length of the shortest barrier-free path between source and target and p the total perimeter size. A routing algorithm has* comparative traffic ratio $\mathcal{R}_{Tr} := M/(d + p)$ *if the algorithm needs altogether M messages.*

The combined comparative ratio addresses time efficiency *and* traffic efficiency:

Definition 5. *The* combined comparative ratio *is the maximum of the competitive time ratio and the comparative traffic ratio: $\mathcal{R}_c := \max\{\mathcal{R}_t, \mathcal{R}_{Tr}\}$*

Basic Strategies. Lucas' algorithm [13] is a simple single-path strategy that works as follows: (1.) Follow the straight line connecting source and target node. (2.) If a barrier is in the way, then traverse the barrier, remember all points where the straight line is crossed, and resume step 1 at that crossing point that is nearest to the target. This algorithm needs at most $d + \frac{3}{2}p$ steps, where d is the length of the shortest barrier-free path and p the total perimeter size. This is an optimal single-path strategy [14], which matches the asypmtotical lower bound for traffic.

Expanding ring search is a straightforward multi-path strategy [9,18], which is nothing else than to start flooding with a restricted search depth and repeat flooding while doubling the search depth until the destination is reached. This strategy is asymptotically time-optimal, but it causes a traffic of $\mathcal{O}(d^2)$, regardless of the presence of faulty nodes.

We modify the previous strategy as follows: The source starts flooding without a depth restriction, but with a delay of $\sigma > 1$ time steps for each hop. If the target is reached, a notification message is sent back to the source. Then the source starts flooding a second time, and this second wave, which is not slowed down, is sent out to stop the first wave. This *continuous ring search* needs time $\sigma \cdot d$ and causes a traffic of $\mathcal{O}\left(\left(\frac{\sigma+1}{\sigma-1}d\right)^2\right)$ (see [20,23] for a proof). The asymptotic performance is no improvement to expanding ring search, but an area is flooded at most two times, whereas expanding ring search visits some areas $\mathcal{O}(\log d)$ times. We use this advantage for our algorithm.

3   The JITE Algorithm

The Just-In-Time Exploration (JITE) algorithm consists of two parts: *Slow Search* and *Fast Exploration*. Slow Search is a modified breadth-first search (BFS) algorithm which uses (in contrast to flooding) a path system that is generated just-in-time by Fast Exploration. This path system is similar to a quadtree-style grid subdivision and consists of the borders of quadratic subnetworks, called *frames*, and perimeters of barriers. It is constructed while Slow Search is running and provides a traffic-efficient search for the target. Due to space limitations, we present the basic ideas in the following and refer to [20,23] for a detailed description.

The algorithm starts with a search area consisting of four connected frames with s lying on the common corner (see Fig. 2). These frames are examined by Fast Exploration[1]: Messages are sent on a *traversal path*[2] along the frame and—if the frame is intersected by a barrier—along the perimeter of the barrier (see Fig. 3). If the traversal needs too much time because of a detour caused by barriers, then the frame is subdivided and the traversal starts again. The recursive subdivision stops when a partition of a frame is *simple*, i.e. a traversal path contains only a bounded number of border nodes. This way the path system becomes denser in the proximity of barriers. Slow Search uses this path system and propagates a slowly proceeding *shoreline* (i.e. the leaves of the BFS tree) through the network. The path system is not constructed completely when Slow Search starts. The shoreline triggers the exploration of new frames in its proximity, so that the traffic-producing exploration is restricted to areas which are visited by the Slow Search.

---

[1] Here, exploration means that only frames are investigated not their interior.

[2] A traversal uses the well-known *right-hand rule*: By keeping the right hand always in touch of the wall, one will find the way out of the maze.

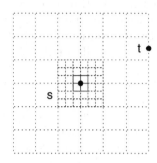

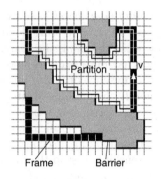

Fig. 2. Initial frames (solid) and extended search area

Fig. 3. Partition of a frame, defined by a right-hand traversal path

The search area is successively enlarged until t is reached[3]. For the expansion of the search area we use the idea of continuous ring search: When the target is found, the source is notified, which sends out messages to stop the search.

As the exploration takes some time, we have to slow down the shoreline, so that there is always enough time for the exploration. Furthermore, to achieve a constant slow down, the size of the neighboring frames has to be restricted, which is expressed by the *Subdivision Rule*: A simple partition of a $3g \times 3g$ frame is subdivided, if there is an orthogonally neighboring frame of size $g \times g$ or if there is diagonally neighboring frame of size $\frac{g}{3} \times \frac{g}{3}$. We use 3×3 subdivisions instead of 2×2 ones, because the Subdivision Rule would cause a domino effect in the initial subdivision, where a small barrier can trigger a cascade of subdivisions of neighboring frames.

The shoreline may enter a frame from different sides, so that a coordination mechanism is necessary in order to avoid that the same frame is explored several times. We solve this by a schedule of traversal messages that count border nodes and coordinate the concurrent exploration. This Frame Exploration Algorithm performs a constant number of traversals, where each traversal is done in $\mathcal{O}((1 + \frac{1}{\gamma(t)})g)$ time steps. A detailed description is given in [20,23].

4   Time and Traffic Analysis

Due to the space limitations we present only proof sketches for the time and traffic analysis. Complete proofs can be found in [20,23].

Time. The constant competitive time ratio can be achieved because Fast Exploration constructs a path system that contains a constant-factor approximation

---

[3] We assume, that t also lies on a frame, i.e. $||s - t||_\infty = 3^k$ for some $k \in \mathbb{N}$. If this is not the case, then we search for any node s' with $||s' - t||_\infty = 3^k$ with the same algorithm and restart the algorithm from s'. This increases time and traffic only by a constant factor.

of the shortest path tree. Slow Search performs a breadth-first search on this
path system which is delayed only by a constant factor. The allowed detours
do not affect the linear time behavior because of the following reason. A simple
partition in a $g \times g$-frame contains at most $g/\gamma(t)$ border nodes. $\gamma(t)$ is a function
of the time t when the exploration of a frame starts. We choose $\gamma(t) := \log(t)$, so
that the accepted detours decrease with the time. In the beginning, $g/\gamma(t)$ can
only be bound by $\mathcal{O}(g)$. Thus, summing up the allowed detours in all the frames
of a recursive subdivision (with frame side lengths ranging from 1 to $\log(d)$)
would result in logarithmic time overhead. But this holds only for a fraction of
$1/\log(d)$ of the frames. For most of the frames $g/\gamma(t)$ is bound by $\mathcal{O}(g/\log d)$.
With these observations we can prove the following theorem:

Theorem 1. *Let P be the shortest path connecting s and t with length d . The
algorithm finds a path P' connecting s and t of length $\mathcal{O}(d)$.*

The shoreline is slowed down by a constant factor σ and uses frames that provide
a constant factor approximation of the shortest path. The exploration of new
frames can be always performed in time because the Subdivision Rule guarantees
that neighboring frames differ only by a constant factor in the side length.

Corollary 1. *Let d be the length of the shortest path connecting s and t. The
algorithm finds a path connecting s and t in time $\mathcal{O}(d)$.*

Traffic. The traffic depends on the size of the path system, i.e. the number and
size of the frames that are explored and subdivided by the algorithm and that
constitute the search area. In the analysis we first consider a single frame and
sum up the traffic of all possible subdivisions. We distinguish between *barrier-
induced subdivisions* and *neighbor-induced subdivisions*. A barrier-induced subdivision occurs, if at least $g/\gamma(t)$ barrier nodes are inside the frame (whether
they are found or not). The other subdivisions are neighbor-induced and due to
the Subdivision Rule. In the traffic analysis, a barrier that causes a subdivision
"pays" for the (barrier-induced) subdivision of the current frame as well as for
the (neighbor-induced) subdivision of the neighboring frames. This extra cost
adds a constant factor to the traffic for exploring a single frame.

In each level of the subdivision there are at most $\frac{p}{g/\gamma(t)}$ frames that have to be
subdivided into smaller frames. For each of these frames the exploration causes
a traffic of $\mathcal{O}((1 + \frac{1}{\gamma(t)})g)$. The sum of this over $\log g$ frame sizes gives a total
traffic of $\mathcal{O}(g + p \log g \cdot \log t)$ for the recursive subdivision of a $g \times g$ frame. The
sum over all frames which are constructed in the search area yields the following
bound.

Theorem 2. *The algorithm produces traffic $\mathcal{O}(d + p \log^2 d)$.*

Corollary 2. *The JITE Algorithm has a constant competitive time ratio and
a comparative traffic ratio of $\mathcal{O}(\log^2 d)$. It has a combined comparative ratio of
$\mathcal{O}(\log^2 d)$.*

5   Conclusions and Open Problems

Conclusions. In this paper we present an algorithm for the routing problem in faulty meshes which can route a message asymptotically as fast as a the fastest algorithm and (up to a term of $\mathcal{O}(\log^2 d)$) with only as many messages as the number of faulty nodes obstructing the messages plus the minimal path length. This considerably improves the known factors of $\mathcal{O}(\sqrt{d})$ [21] and more previously of $\tilde{O}(d^{\sqrt{\frac{\log \log d}{\log d}}})$ [22].

This is achieved by the JITE Algorithm combining several techniques. First we use a continuously expanding search area and establish an adaptive grid of frames which becomes denser near barriers. On this grid a slowly proceeding shoreline simulates a flooding mechanism. This shoreline triggers the Just-In-Time Exploration (JITE) of new frames that are used by the shoreline. The artful combination of these techniques lead to an algorithm which needs time $\mathcal{O}(d)$ and traffic $\mathcal{O}(d + p \log^2 d)$ where d denotes the length of the shortest path and p denotes the number of border nodes being adjacent to faulty nodes.

Open problems. This gives rise to the open question whether these bounds are tight, whether there is a small trade-off between time and traffic. The routing time is delayed by a large constant factor. It seems achievable to decrease this factor without an asymptotic increase of the traffic. However, it is not clear how. Another chance of improvement could be the use of randomized algorithms, which for many other problems outperform deterministic online algorithms.

A straight-forward generalization of this problem are three-dimensional meshes with faulty nodes. The JITE Algorithm, however, in its straight-forward generalization causes a significant increase in traffic. So the question for efficient online routing in higher dimensions is wide open.

References

1. D. Angluin, J. Westbrook, and W. Zhu. Robot navigation with distance queries. *SIAM Journal on Computing*, 30(1):110–144, 2000.
2. P. Berman. On-line searching and navigation. In A. Fiat and G. J. Woeginger, editors, *Online Algorithms: The State of the Art*, pages 232–241. Springer, 1998.
3. A. Blum, P. Raghavan, and B. Schieber. Navigating in unfamiliar geometric terrain. *SIAM Journal on Computing*, 26:110–137, 1997.
4. A. Borodin and R. El-Yaniv. *Online Computation and Competitive Analysis*. Cambridge University Press, 1998.
5. P. Bose and P. Morin. Competitive online routing in geometric graphs. *Theoretical Computer Science*, 324(2-3):273–288, September 2004.
6. P. Bose and P. Morin. Online routing in triangulations. *SIAM Journal on Computing*, 33(4):937–951, May 2004.
7. P. Bose, P. Morin, I. Stojmenovic, and J. Urrutia. Routing with guaranteed delivery in ad hoc wireless networks. *Wireless Networks*, 7(6):609–616, 2001.
8. R. J. Cole, B. M. Maggs, and R. K. Sitaraman. Reconfiguring Arrays with Faults Part I: Worst-case Faults. *SIAM Journal on Computing*, 26(16):1581–1611, 1997.

9. D. B. Johnson and D. A. Maltz. Dynamic Source Routing in Ad Hoc Wireless Networks. In *Mobile Computing*, pages 152–181. Kluwer, 1996.
10. E. Koutsoupias and Ch. H. Papadimitriou. Beyond competitive analysis. *SIAM Journal on Computing*, 30(1):300–317, 2000.
11. E. Kranakis, H. Singh, and J. Urrutia. Compass routing on geometric networks. In *Proc. 11th Canadian Conference on Computational Geometry*, pages 51–54, 1999.
12. F. Kuhn, R. Wattenhofer, and A. Zollinger. Asymptotically optimal geometric mobile ad-hoc routing. In *Proc. of the 6th Int. Workshop on Discrete Algorithms and Methods for Mobile Computing and Communications*, pages 24–33, 2002.
13. C. Lucas. Comments on "dynamic path planning for a mobile automation with limited information on the environment". *IEEE Transactions on Automatic Control*, 33(5):511, May 1988.
14. V. J. Lumelsky. Algorithmic and complexity issues of robot motion in an uncertain environment. *J. Complex.*, 3(2):146–182, 1987.
15. V. J. Lumelsky and A. A. Stepanov. Path-planning strategies for a point mobile automaton moving amidst unknown obstacles of arbitrary shape. *Algorithmica*, 2:403–430, 1987.
16. M. Mauve, J. Widmer, and H. Hartenstein. A survey on position-based routing in mobile ad hoc networks. *IEEE Network Magazine*, 15(6):30–39, November 2001.
17. Ch. H. Papadimitriou and M. Yannakakis. Shortest paths without a map. In *Proc. of the 16th Int. Colloq. on Automata, Languages, and Programming (ICALP'89)*, pages 610–620, 1989.
18. C. E. Perkins, E. M. Belding-Royer, and S. Das. Ad hoc on-demand distance vector (AODV) routing. IETF RFC 3561, July 2003.
19. N. Rao, S. Kareti, W. Shi, and S. Iyenagar. Robot navigation in unknown terrains: Introductory survey of non-heuristic algorithms, 1993.
20. S. Rührup. *Position-based Routing Strategies*. PhD thesis, University of Paderborn, 2006.
21. S. Rührup and Ch. Schindelhauer. Competitive time and traffic analysis of position-based routing using a cell structure. In *Proc. of the 5th IEEE Int. Workshop on Algorithms for Wireless, Mobile, Ad Hoc and Sensor Networks (WMAN'05)*, page 248, 2005.
22. S. Rührup and Ch. Schindelhauer. Online routing in faulty mesh networks with sublinear comparative time and traffic ratio. In *Proc. of the 13th European Symposium on Algorithms (ESA'05), LNCS 3669*, pages 23–34. Springer, 2005.
23. S. Rührup and Ch. Schindelhauer. Improved bounds for online multi-path routing in faulty mesh networks. Technical Report TR-RSFB-06-078, University of Paderborn, 2006.
24. J. Wu. Fault-tolerant adaptive and minimal routing in mesh-connected multicomputers using extended safety levels. *IEEE Transactions on Parallel and Distributed Systems*, 11:149–159, February 2000.
25. J. Wu and Z. Jiang. Extended minimal routing in 2-d meshes with faulty blocks. In *Proc. of the 1st Intl. Workshop on Assurance in Distributed Systems and Applications*, pages 49–55, 2002.
26. L. Zakrevski and M. Karpovsky. Fault-tolerant message routing for multiprocessors. In *Parallel and Distributed Processing*, pages 714–730. Springer, 1998.

On the On-Line k-Truck Problem with Benefit Maximization

Weimin Ma[1,2] and Ke Wang[1]

[1] School of Economics and Management
Beijing University of Aeronautics and Astronautics, Beijing, 100083, P.R. China
mawm@buaa.edu.cn, wangke@sem.buaa.edu.cn
[2] School of Economics and Management, Xi'an Technological University
Xi'an, Shaanxi Province, 710032, P.R. China

Abstract. Based on some results of the on-line k-truck problem with cost minimization, a realistic model of the on-line k-truck problem with benefit maximization is proposed. In the model, the object of optimization is how to design on-line algorithms to maximize the benefit of all trucks' moving. In this paper, after the model's establishment, several on-line algorithms, e.g., Position Maintaining Strategy, Partial Greedy Algorithm, are employed to address the problem. The analyses concerning the competitive ratios of the algorithms are given in detail. Furthermore, the lower bound of competitive ratio is discussed.

1 Introduction

Over the past two decades, on-line problems and their competitive analysis have received considerable interest. Since the approach of competitive analysis was first applied to analyze on-line problem by Sleator and Tarjian [1], it has been investigated in a wide variety of areas such as competitive auctions [2, 3], scheduling problem for jobs [4] and foreign currency trading [5, 6].

The k-server problem [7], introduced by Manasse, McGeoch and Sleator [8], is a famous on-line problem which had been studied extensively. In the k-server problem we have k servers that reside and move in a metric space. When a request is made by a point, one of the servers must be moved to the point to satisfy this request immediately. An algorithm A which decides a server to satisfy the request at each step, is said to be *on-line* if its decisions are made without the information about future requests. The cost of all servers equals to the distance of their moving. Our goal is to minimize the total cost.

The k-truck problem is a generalization of the k-server problem. In the k-truck problem, each request contains two points, one of which is the point made the request and the other is the destination point. When a request occurs, an empty truck must be moved to serve it immediately without the information about future possible requests. Some results on the k-truck problem have been proposed[9,10]. All the previous researches on this problem aim at minimizing the distance or cost of all trucks' moving. However, we usually use the ability

T. Asano (Ed.): ISAAC 2006, LNCS 4288, pp. 660–669, 2006.
© Springer-Verlag Berlin Heidelberg 2006

of getting benefit to appraise a company rather than the cost in real life. Thus, we discuss the k-truck problem aiming at maximizing the benefit of all trucks in this paper.

The realistic background of the on-line k-truck problem with benefit maximization is described as follows. There are k trucks on a traffic net to supply service. When a service request that transporting goods from one point to anther point occurs, an empty truck must be moved to serve it immediately without the information about future requests. It is assumed that all the trucks are scheduled by a control center. When the truck is empty, the cost of running one unit distance is M. And the cost of trucks with goods is different from that without goods on the same distance. For simplicity, we assume that all trucks have same load weight and the cost of the truck with goods is θ times of that without goods on the same distance, and $\theta \geq 1$. The customer who releases the request pays N for every one unit distance between the origin point and the destination point. That is to say, the truck with goods gains $N - \theta \cdot M$ every one unit distance, and the truck would lose M if it runs without goods. And then how to maximize the benefit? If $\theta = 1$, the problem is also called on-line k-taxi problem with benefit maximization[11].

For simplicity, let η denote the ratio of the gains of truck with goods to the losses of truck without goods on the same distance, i.e. $\eta = \frac{N-\theta \cdot M}{M} > 0$. In the following discussion it is assumed that the losses of truck without goods equal to the distance of its moving, therefore the gains of truck with goods are η times of the distance of its moving.

The rest of paper is organized as follows: in section 2, we formulate the model of on-line k-truck problem with benefit maximization and some preliminary knowledge of on-line theory is also described in this part. Section 3 presents some results concerning the on-line algorithms for the problem. In section 4, the lower bound of competitive ratio for this problem is developed. Finally, in section 5, we conclude the paper and discuss the future research directions.

2   The Model

There are k trucks that reside and move on a weighted graph G (with n points)to supply service. For any points x and y in G, $d(x,y)$ denotes their shortest distance, and the distance is symmetric, i.e., for all x, y, $d(x,y) = d(y,x)$. A service request $r = (a,b)(a, b$ are different points in the given graph G), implies there are some goods on point a that must be moved to point b by truck (for simplicity, it is assumed that the weights of goods are same all the time). A service request sequence R consists of some service request in turn, namely $R = (r_1, r_2, \cdots, r_m)$ where $r_i = (a_i, b_i)$. It is assumed that $a_i \neq b_i$ in the following discussion, because if $a_i = b_i$, the request r_i does not exist in fact and this case is of no significance to discuss in this model. When the trucks move on the given graph G, the losses of truck without goods equal to the distance of its moving, and the gains of truck with goods are η times of the distance its moving. All discussions are based on an essential assumption: when a new service request occurs, k trucks are all free.

The on-line k-truck problem with benefit maximization is to decide which truck should be moved when a new service request occurs on the basis that we have no information about future possible requests.

In this model, for a known request sequence $R = (r_1, r_2, \cdots, r_m)$, let $B_{\text{OPT}}(R)$ be the total benefit after finishing it with optimal off-line algorithm. An optimal off-line algorithm knows the entire input sequence in advance and can process it optimally. For every new service request r_i, if algorithm ON can schedule without information of the sequence after r_i, we call ON an on-line algorithm. For on-line algorithm ON, if there are constants α and β for any possible R satisfying:

$$\alpha \cdot B_{\text{ON}}(R) \geq B_{\text{OPT}}(R) - \beta \qquad (1)$$

ON is called a competitive algorithm, where $B_{\text{ON}}(R)$ is the total benefit with algorithm ON to satisfy the sequence R. α is the competitive ratio.

3   Competitive Ratios

3.1   Position Maintaining Strategy (PMS for Short)

Position Maintaining Strategy [12]. *For the present request* $r_i = (a_i, b_i)$, *PMS schedules the trucks with the following three steps.*

(1) Firstly, schedules a truck to a_i using algorithm A which is for the k-server problem.

(2) Then moves the truck reaching a_i from a_i to b_i with goods to complete r_i.

(3) Finally, moves the truck at b_i back to a_i before the next request arrives.

The famous on-line k-server problem is presented as a special case of the on-line truck problem. Using PMS to address the k-truck problem, we could get the following theorem.

Theorem 1. *For a given graph G if there is a c-competitive algorithm for the k-server problem on G and $\eta > c + 1$ holds, then there is a $(\frac{\eta}{\eta - c - 1})$-competitive algorithm for the k-truck problem with benefit maximization on G, where η has the same meaning as defined above.*

To prove the theorem 1, we should present two lemmas at first. On a given graph G, for an request sequence $R = (r_1, r_2, \cdots, r_m)$, let $L_{\text{OPT}}(R)$ be the total losses of all trucks running without goods after finishing R with the optimal off-line algorithm. It equals to the distance of all trucks' moving without goods, i.e., we have $B_{\text{OPT}}(R) = \eta \cdot \sum_{i=1}^{m} d(a_i, b_i) - L_{\text{OPT}}(R)$. ($L_{\text{OPT}}(R)$ is different from $C_{\text{OPT}}(R)$ which denotes the total cost of all trucks after finishing R with the optimal off-line algorithm in the traditional k-truck problem [9], and $L_{\text{OPT}}(R) = C_{\text{OPT}}(R) - \theta \cdot \sum_{i=1}^{m} d(a_i, b_i)$ holds). Then we have the following lemma.

Lemma 1. *On a given graph G, for an request sequence $R = (r_1, r_2, \cdots, r_m)$, let $C_{\text{OPT}}(\sigma)$ denote the total cost after finishing σ with the optimal off-line algorithm in the k-server problem, where $\sigma = (a_1, a_2, \cdots, a_m)$. Then we have*

$$L_{\text{OPT}}(R) \geq C_{\text{OPT}}(\sigma) - \sum_{i=1}^{m} d(a_i, b_i) \tag{2}$$

Proof. Using reduction to absurdity. We assumed that the losses of truck without goods equal to the distance of its running, so $L_{\text{OPT}}(R) + \sum_{i=1}^{m} d(a_i, b_i)$ denotes the total distance of all trucks' moving after finishing R. If $L_{\text{OPT}}(R) + \sum_{i=1}^{m} d(a_i, b_i) < C_{\text{OPT}}(\sigma)$, then we could use the optimal algorithm for k-truck problem to schedule the servers in k-server problem, and the distance of servers' moving is less. Thus, the original algorithm for k-server problem is not optimal. Therefore the inequality $L_{\text{OPT}}(R) + \sum_{i=1}^{m} d(a_i, b_i) \geq C_{\text{OPT}}(\sigma)$ holds.   □

Lemma 2. *On a given graph G, for any request sequence $R = (r_1, r_2, \cdots, r_m)$,*

$$B_{\text{OPT}}(R) \leq \eta \cdot \sum_{i=1}^{m} d(a_i, b_i) \tag{3}$$

Proof. $\sum_{i=1}^{m} d(a_i, b_i)$ is the total distance that all trucks have to run with goods for finishing R. If there is no loss of empty truck's moving, the total benefit would be $\eta \cdot \sum_{i=1}^{m} d(a_i, b_i)$, i.e. it is the upper bound of benefit.   □

Then we begin to prove theorem 1.

Proof. Because $B_{\text{OPT}}(R) = \eta \cdot \sum_{i=1}^{m} d(a_i, b_i) - L_{\text{OPT}}(R)$, from lemma 1 we have

$$C_{\text{OPT}}(\sigma) \leq (\eta + 1) \cdot \sum_{i=1}^{m} d(a_i, b_i) - B_{\text{OPT}}(R) \tag{4}$$

As described above, for current service request $r_i = (a_i, b_i)$, the PMS first schedules a truck to a_i using the algorithm A which is a c-competitive algorithm for the k-server problem. Then the truck reaching a_i is moved from a_i to b_i with the goods to complete r_i and get back to a_i before the next request arrives. Let $B_{\text{PMS}}(R)$ denote the benefit of this algorithm, we have

$$B_{\text{PMS}}(R) = \eta \cdot \sum_{i=1}^{m} d(a_i, b_i) - C_A(\sigma) - \sum_{i=1}^{m} d(a_i, b_i) \tag{5}$$

Since A is a c-competitive algorithm, and from (4), we also get

$$C_A(\sigma) \leq c \cdot C_{\text{OPT}}(\sigma) \leq c \cdot (\eta + 1) \cdot \sum_{i=1}^{m} d(a_i, b_i) - c \cdot B_{\text{OPT}}(R) \tag{6}$$

Taking (6) into (5), we have

$$B_{\text{PMS}}(R) \geq c \cdot B_{\text{OPT}}(R) + (\eta - c \cdot \eta - c - 1) \sum_{i=1}^{m} d(a_i, b_i) \tag{7}$$

From (7) and lemma 2, we have

$$B_{\mathrm{PMS}}(R) \geq \frac{\eta - c - 1}{\eta} \cdot B_{\mathrm{OPT}}(R) \tag{8}$$

Because $\eta > c + 1$ holds as defined above,

$$\frac{\eta}{\eta - c - 1} \cdot B_{\mathrm{PMS}}(R) \geq B_{\mathrm{OPT}}(R) \tag{9}$$

Therefore theorem 1 has been proved. By the way, we could see that the benefit might be 0 when $\eta = c + 1$ holds. If $0 < \eta < c + 1$, the benefit may be negative (assuming that every request can not be rejected). When benefit is negative, form (8) we have $-B_{\mathrm{PMS}}(R) \leq \frac{c+1-\eta}{\eta} \cdot B_{\mathrm{OPT}}(R)$, which means the total losses of all trucks must be less than $\frac{c+1-\eta}{\eta}$ times of $B_{\mathrm{OPT}}(R)$. □

3.2   Partial Greedy Algorithm (PGA for Short)

For a given graph G with n points, let $d_{max} = \max d(x, y)$, $d_{min} = \min d(x, y)$, where x and y are different points in G, and let $\lambda = \frac{d_{max}}{d_{min}}$. When $2 \leq k \leq n-2$, we employ a Partial Greedy Algorithm [10] (which is so called because the Greedy Algorithm is used for some of the cases in this problem) as follows. It is assumed that there is at most one truck located at a point before the first request arrives. Otherwise, we can precondition the truck locations such that each point has at most one truck. Furthermore, the losses of this precondition is at most a constant $(k-1)d_{max}$, and it has no influence on the competitive radio. PGA will keep that no point has more than one truck after finishing every request in the whole game.

Partial Greedy Algorithm[10]. *For the present request $r_i = (a_i, b_i)$,*

(1) If there is a truck at a_i and no truck at b_i, then PGA moves the truck at a_i to b_i to complete the request.

(2) If there is no truck at a_i and there is a truck at b_i, then PGA moves the truck at b_i to a_i first, and then moves from a_i to b_i to complete the request.

(3) If there are trucks on both a_i and b_i, then PGA moves the truck at a_i to b_i to complete the request and at the same time moves the truck at b_i to a_i.

(4) If there is no truck at a_i and b_i, then PGA moves the truck which is closest to a_i (supposing that the truck is locate at c_i) to a_i and then moves to b_i to complete the request.

Theorem 2. *For a given graph G, if $2 \leq k \leq n-2$ and $\eta > \lambda$ holds, PGA is a $(\frac{\eta}{\eta - \lambda})$-competitive algorithm for the k-truck problem with benefit maximization on G.*

Proof. For case (1), the benefit of PGA finishing r_i is $\eta \cdot d(a_i, b_i)$. For case (2) and (3), the benefit is $(\eta - 1) \cdot d(a_i, b_i)$. For case (4), the benefit is $\eta \cdot d(a_i, b_i) - d(c_i, a_i)$. So for any case, the benefit of PGA for r_i can not be less than $\eta \cdot d(a_i, b_i) - d_{max}$.

$$B_{\mathrm{PGA}}(R) = \sum_{i=1}^{m} B_{\mathrm{PGA}}(r_i) - \beta \geq \sum_{i=1}^{m} [\eta \cdot d(a_i, b_i) - d_{max}] - \beta$$

$$= \sum_{i=1}^{m} \left[\eta - \frac{d_{max}}{d(a_i, b_i)}\right] \cdot d(a_i, b_i) - \beta \geq \sum_{i=1}^{m} \left[\eta - \frac{d_{max}}{d_{min}}\right] \cdot d(a_i, b_i) - \beta$$

$$= (\eta - \lambda) \cdot \sum_{i=1}^{m} d(a_i, b_i) - \beta \tag{10}$$

Where β is the losses of preconditioning the trucks such that each point has at most one truck. From (10) and lemma 2, we have

$$B_{\mathrm{PGA}}(R) \geq \left(1 - \frac{\lambda}{\eta}\right) \cdot B_{\mathrm{OPT}}(R) - \beta \tag{11}$$

When $\eta > \lambda$ holds as defined above,

$$\frac{\eta}{\eta - \lambda} \cdot B_{\mathrm{PGA}}(R) \geq B_{\mathrm{OPT}}(R) - \frac{\eta}{\eta - \lambda} \cdot \beta \tag{12}$$

The proof is completed. □

For a given graph G with n points, when $k = n - 1$, we can employ Partial Greedy Algorithm to design a similar algorithm B as follows. It is assumed that there is at most one truck located at each point before the first request arrives too. Otherwise, we can precondition the truck locations such that each point has at most one truck. $k = n - 1$ holds, so there is only one point with no truck. The algorithm B will keep that there is only one point with no truck after finishing every request in the whole game.

Algorithm B. *For the present request $r_i = (a_i, b_i)$,*

(1) If there is a truck at a_i and there is no truck at b_i, then B moves the truck at a_i to b_i to complete the request.

(2) If there is no truck at a_i, b_i must have a truck, then B moves the truck at b_i to a_i first, and then moves from a_i to b_i to complete the request.

(3) If there are trucks on both a_i and b_i, then B moves the truck at a_i to b_i to complete the request and at the same time moves the truck at b_i to a_i.

Theorem 3. *For a given graph G, if $k = n - 1$ and $\eta > 1$ holds, there is a $(\frac{\eta}{\eta-1})$-competitive algorithm for the k-truck problem with benefit maximization.*

Proof. The discussion is similar to the proof of theorem 2. For case (1), the benefit of B finishing r_i is $\eta \cdot d(a_i, b_i)$. For case (2) and (3), the benefit is $(\eta-1) \cdot d(a_i, b_i)$. So for any case, the benefit of B for r_i is not less than $(\eta-1) \cdot d(a_i, b_i)$.

$$B_{\mathrm{B}}(R) = \sum_{i=1}^{m} B_{\mathrm{B}}(r_i) - \beta \geq \sum_{i=1}^{m} (\eta - 1) \cdot d(a_i, b_i) - \beta$$

$$= (\eta - 1) \cdot \sum_{i=1}^{m} d(a_i, b_i) - \beta \tag{13}$$

Where β is the losses of preconditioning the trucks such that each point has at most one truck. From (13) and lemma 2, we have

$$B_B(R) \geq \left(1 - \frac{1}{\eta}\right) \cdot B_{OPT}(R) - \beta \tag{14}$$

When $\eta > 1$ holds, we get $\frac{\eta}{\eta-1} \cdot B_B(R) \geq B_{OPT}(R) - \frac{\eta}{\eta-1} \cdot \beta$ □

For the case $k = n$, we can design a similar algorithm B' as follows. It is also assumed that there is at most one truck located at a point before the first request arrives. That is to say every point just has one truck. Otherwise, we can precondition the truck locations such that each point has one truck. For the present request $r_i = (a_i, b_i)$, because there are trucks at both a_i and b_i, we can schedule the truck at a_i to b_i to complete the request and at the same time moves the truck at b_i to a_i. Then the benefit of B' finishing r_i is $(\eta - 1) \cdot d(a_i, b_i)$, and at present each point still has one truck.

Similar to the case $k = n - 1$, we can prove that the algorithm B' is a $(\frac{\eta}{\eta-1})$-competitive algorithm. So we have the following theorem.

Theorem 4. *For a given graph G, if $k = n$ and $\eta > 1$ holds, there is a $(\frac{\eta}{\eta-1})$-competitive algorithm for the k-truck problem with benefit maximization.*

4   A Lower Bound

In this section we will give a lower bound of competitive ratio for the on-line k-truck problem with benefit maximization. The approach we will take in was proposed in [8], where the lower bound and matching upper bound were given for the traditional k-server problem. In the paper [10], the approach was used to get a lower bound as $\frac{\theta \cdot k + k}{\theta \cdot k + 2}$ for the traditional k-truck problem. And then we will use it to investigate the k-truck problem with benefit maximization as follows.

Suppose we wish to compare an on-line algorithm with k trucks to an off-line one with $h \leq k$ trucks. Naturally, the competitive factor decreases when the on-line algorithm gets more trucks than the off-line algorithm. We have actually proven a slightly more general lower bound as $\frac{(\eta-2)\cdot k+2h-2}{(\eta-1)\cdot k}$ for a given graph G with $\eta \geq \max(2, \lambda)$. The constraint $\eta \geq \max(2, \lambda)$ is used to make the benefit positive in the following discussion.

Theorem 5. *On a constraint graph G with at least $k + 2$ points and $\eta \geq \max(2, \lambda)$, let C be an on-line algorithm for the k-truck problem with benefit maximization. Then, for any $2 \leq h \leq k$, there exists request sequences $R_1, R_2, R_3, \cdots$ such that: (1) For all i, R_i is an initial subsequence of R_{i+1}; (2)There is an off-line h-truck algorithm D (which may start with its trucks anywhere) such that for all i, $\frac{(\eta-2)\cdot k+2h-2}{(\eta-1)\cdot k} \cdot B_C(R_i) < B_D(R_i)$.*

Proof. Without loss of generality, assume C is an on-line algorithm and that the k trucks start out at different points. Let H (of size $k + 2$) be a subgraph of G, induced by the k initial positions of C's trucks and two other points. Define

$R = (r_1, r_2, \cdots, r_m)$, C's nemesis sequence on H, such that $R(i)$ and $R(i-1)$ are the two unique points in H not covered by C and a request $r_i = (R(i), R(i-1))$ occurs at time i, where all $i \geq 1$. $R_i = (r_1, r_2, \cdots, r_i)$ is the subsequence of R by the time i. At each step R requests the point just vacated by C, thus we have

$$
\begin{aligned}
B_C(R) = \sum_{i=1}^{m} B_C(r_i) &= \sum_{i=1}^{m} [\eta \cdot d(R(i), R(i-1)) - d(R(i+1), R(i))] \\
&= (\eta - 1) \cdot \sum_{i=1}^{m-1} d(R(i+1), R(i)) + \eta \cdot d(R(1), R(0)) \\
&\quad - d(R(m+1), R(m))
\end{aligned}
\tag{15}
$$

Let S be any h-element subset of H containing $R(1)$ but not $R(0)$. We can define an off-line h-truck algorithm $D(S)$ as follows: the trucks finally occupy the points in set S. To process a request $r_i = (R(i), R(i-1))$, the following rule is applied: if S contains $R(i)$, move the truck at $R(i)$ to $R(i-1)$ with goods to complete the request, and update S to reflect this change. Otherwise move the truck at $R(i-2)$ to $R(i)$ without goods and then to $R(i-1)$ with goods to complete the request, and update S to reflect this change.

It is easy to see that for all $i > 1$, the set S contains $R(i-2)$ and does not contain $R(i-1)$ when step i begins. The following observation is the key to the rest of the proof: if we run the above algorithm starting with distinct equal-sized sets S and T, then S and T never become equal, for the reason described in the following paragraph.

Suppose that S and T differ before r_i is processed. We shall show that the versions of S and T created by processing r_i, as described above, also differ. If both S and T contain $R(i)$, they both move the truck at $R(i)$ to $R(i-1)$, on which there is exactly not any truck. The other points have no changes, so S and T are still different and both S and T contain $R(i-1)$. If exactly one of S or T contains $R(i)$, then after the request exactly one of them contains $R(i-2)$, so they still differ. If neither of them contains $R(i)$, then both change by dropping $R(i-2)$ and adding $R(i-1)$, so the symmetric difference of S and T remains the same (non-empty).

Let us consider simultaneously running an ensemble of algorithms $D(S)$, starting from each h-element subset S of H containing $R(1)$ but not $R(0)$. There are $\binom{k}{h-1}$ such sets. Since no two sets ever become equal, the number of sets remains constant. After processing $R(i)$, the collection of subsets consists of all the h element subsets of H which contain $R(i-1)$.

By our choice of starting configuration, step 1 benefits $\eta \cdot d(R(1), R(0))$. At step i (for $i \geq 2$), each of these algorithms either moves the truck at $R(i)$ to $R(i-1)$ (if S contains $R(i)$), with benefit $\eta \cdot d(R(i), R(i-1))$, or moves the truck at $R(i-2)$ to $R(i)$ and then to $R(i-1)$ (if S does not contain $R(i)$), with benefit $\eta \cdot d(R(i), R(i-1)) - d(R(i-2), R(i))$. Of the $\binom{k}{h-1}$ algorithms being run, $\binom{k-1}{h-1}$ of them (the ones which not contain $R(i)$) get the benefit of $\eta \cdot d(R(i), R(i-1)) - d(R(i-2), R(i))$. The remaining $\binom{k-1}{h-2}$ of algorithms get the benefit of $\eta \cdot d(R(i), R(i-1))$. Therefore, for step i, the total benefit of running all of the algorithms is

$$\binom{k}{h-1} \cdot \eta \cdot d(R(i), R(i-1)) - \binom{k-1}{h-1} \cdot d(R(i-2), R(i)) \qquad (16)$$

The total benefit of running all of the algorithms to finish R (i.e., there is m steps in all) is

$$\sum_{i=1}^{m} \binom{k}{h-1} \cdot \eta \cdot d(R(i), R(i-1)) - \sum_{i=2}^{m} \binom{k-1}{h-1} \cdot d(R(i-2), R(i)) \qquad (17)$$

Thus the expected benefit of one of these algorithms chosen at random is

$$B_{\mathrm{EXP}}(R) = \eta \cdot \sum_{i=1}^{m} d(R(i), R(i-1)) - \frac{\binom{k-1}{h-1}}{\binom{k}{h-1}} \cdot \sum_{i=2}^{m} d(R(i-2), R(i))$$

$$= \eta \cdot \sum_{i=1}^{m} d(R(i), R(i-1)) - \frac{k-h+1}{k} \cdot \sum_{i=2}^{m} d(R(i-2), R(i)) \qquad (18)$$

Since the weights of edges on graph G satisfy the triangle inequality (i.e., $d(R(i-2), R(i)) \le d(R(i-2), R(i-1)) + d(R(i-1), R(i))$, for $i \ge 2$), we have

$$\sum_{i=2}^{m} d(R(i-2), R(i)) \le 2 \sum_{i=1}^{m-1} d(R(i), R(i+1))$$

$$+ d(R(0), R(1)) - d(R(m-1), R(m)) \qquad (19)$$

Taking (19) into (18), we get

$$B_{\mathrm{EXP}}(R) \ge \frac{(\eta - 2) \cdot k + 2h - 2}{k} \cdot \sum_{i=1}^{m-1} d(R(i), R(i+1))$$

$$+ \frac{(\eta - 1) \cdot k + h - 1}{k} \cdot d(R(0), R(1)) + \frac{k-h+1}{k} \cdot d(R(m-1), d(m)) \quad (20)$$

Comparing (20) with (15),we can see that the two summations of the $B_{\mathrm{EXP}}(R)$ and $B_C(R)$ are identical, except that both of the benefits include some extra terms, which are bounded as a constant. When $\eta \ge \max(2, \lambda)$ holds, both $B_{\mathrm{EXP}}(R)$ and $B_C(R)$ are positive. Therefore, after some mathematical manipulation (e.g., let $m \to \infty$), we obtain

$$\frac{B_{\mathrm{EXP}}(R)}{B_C(R)} \ge \frac{(\eta - 2) \cdot k + 2h - 2}{(\eta - 1) \cdot k} \qquad (21)$$

Finally, there must be some initial sets whose performance is often no worse than the average of the benefit. Let S be one of these sets, and $D(S)$ be the algorithm starting from set S. Let R_i be an initial subsequence of R, for which $D(S)$ does no worse than average. $\qquad \square$

Corollary 1. *For any k-truck problem with benefit maximization on a constraint graph G with $\eta \ge \max(2, \lambda)$, there is no c-competitive algorithm for $c < \frac{\eta \cdot k - 2}{(\eta - 1) \cdot k}$.*

5   Conclusion

In this paper, a realistic model of the on-line k-truck problem with benefit maximization is proposed. And then several on-line algorithms are designed to address the problem, the analyses concerning the competitive ratios of the algorithms are given in detail. The lower bound of competitive ratio is also discussed in this paper. However, there are still many theoretical problems that need to be studied further. For example, we assumed that all trucks have same load weight in the problem. If the load weight varies from one request to anther, how would the results of competitive ratios be? Although we have get a lower bound of competitive ratio for the k-truck problem with benefit maximization, the optimal lower bound of the competitive ratio for it is still open. Furthermore, whether there are some better on-line algorithms than PMS or PGA needs further investigation.

Acknowledgements. The work was partly supported by the National Natural Science Foundation of China (70401006, 70501015, and 70521001).

References

1. D. D. Sleator, R. E. Tarjan, Amortized efficiency of list update and paging rules, *Communication of the ACM*, 28: 202-208, 1985.
2. A. V. Goldberg, and J. D. Hartline, Competitiveness via Consensus, *In 14th Annual ACM-SIAM Symposium on Discrete Algorithms (SODA '03)*, 2003.
3. J. D. Hartline, and A. V. Goldberg, Envy-Free Auctions for Digital Goods , In *Proc. 4th ACM Conf. on Electronic Commerce*, 2003.
4. Y. Bartal, F.Y.L. Chin, M. Chrobak, S.P.Y. Fung, W. Jawor, R. Lavi, J. Sgall, and T. Tichy, Online competitive algorithms for maximizing weighted throughput of unit jobs, In Proc. *21st Annual Symposium on Theoretical Aspects of Computer Science (STACS'04)*, LNCS 2996, Springer 2004, 187-198.
5. R. El-Yaniv, A. Fiat, R. M. Karp, et al. Competitive analysis of financial games. *Proc. 33rd Annual Symposium on Foundations of Computer Science*, 1992. 327-333.
6. R. El-Yaniv, A. Fiat, R. M. Karp, et al. Optimal search and one-way trading online algorithms. *Algorithmica*, 30: 101-139, 2001.
7. E. Koutsoupias and C. Papadimitriou. On the k-server conjecture *Journal of ACM*. 42(5):971-983, 1995.
8. M. S. Manasse, L. A. McGeoch, and D. D. Sleator, Competitive algorithms for server problems, *Journal of Algorithms*,1990(11),208-230.
9. W. M. Ma, Y. F. Xu, and K. L. Wang, On-line k-truck problem and its competitive algorithm, *Journal of Global Optimization*, 21 (1): 15-25, 2001.
10. W. M. Ma, Y. F. Xu, J. You, J. Liu and K. L. Wang, On the k-truck scheduling problem, *International Journal of Foundations of Computer Science*, 15 (1): 127-141, 2004.
11. W. M. Ma, K. Wang. On-line taxi problem on the benefit-cost graphs. *Proceedings of the Fifth International Conference on Machine Learning and Cybernetics (ICMLC 2006)*, 2006, 900-905.
12. Y. F. Xu, K. L. Wang, and B. Zhu, On the k-taxi problem, *Information*, Vol.2, No.4, 1999.

Energy-Efficient Broadcast Scheduling for Speed-Controlled Transmission Channels

Patrick Briest[1],[*] and Christian Gunia[2],[**]

[1] Dortmund University, Dept. of Computer Science, Dortmund, Germany
`patrick.briest@cs.uni-dortmund.de`
[2] Freiburg University, Dept. of Computer Science, Freiburg, Germany
`gunia@informatik.uni-freiburg.de`

Abstract. We consider the problem of computing broadcast schedules for a speed-controlled channel minimizing overall energy consumption. Each request defines a strict deadline and we assume that sending at some speed s for t time units consumes energy $t \cdot s^\alpha$. For the case that the server holds a single message and the speed of a broadcast needs to be fixed when it is started, we present an $\mathcal{O}(2^\alpha)$-competitive deterministic online algorithm and prove that this is asymptotically best possible even allowing randomization. For the multi-message case we prove that an extension of our algorithm is $(4c-1)^\alpha$-competitive if the lengths of requests vary by at most a factor of c. Allowing the speed of running broadcasts to be changed, we give lower bounds that are still exponential in α.

1 Introduction

Classical objectives in online broadcasting usually abstract away from the precise hardware architecture of the underlying computing machinery. They mostly aim at producing solutions that ensure a high degree of convenience for the serviced clients, but do not take into account the cost of actually realizing the solution. While this approach is quite reasonable in many traditional scenarios, recent years have brought about an increasing number of applications in which these issues become non-negligible.

The most important factor determining the *cost* of running a broadcasting algorithm in practical applications is the algorithm's energy consumption. In fact, energy efficiency has become a premier objective in many different areas of computer science and recent advances have already led to changes in the structure of processors, graphic cards and other parts of computer hardware. Reduced energy consumption offers new application areas for computer systems. Multi-agent systems consisting of dozens of small, self-sustaining units support engineers and researchers in an increasing number of applications that range from production process planning to geographical observations [6]. Multi-agent systems depend on a reliable communication link between them that is typically provided by

---

[*] Supported by DFG grant Kr 2332/1-2 within Emmy Noether program.
[**] Supported by DFG research training program No 1103 'Embedded Microsystems'.

T. Asano (Ed.): ISAAC 2006, LNCS 4288, pp. 670–679, 2006.

means of a wireless connection. Due to characteristics of their operation areas they are likely to be small and, consequently, carry a limited power supply. To use this as efficiently as possible specialized hardware like, e.g., low-power CPUs are utilized.

However, the energy consumed by the wireless connection is also far from being negligible. As wireless communication is implicitly done via broadcasts we propose to exploit this fact. We focus on a single agent that acts as a data server and adapt the situation introduced by Yao et al. in their seminal work [13]: we consider requests that have individual release times and deadlines and allow that multiple requests for the same piece of data can be answered by a single broadcast. While doing this results in a smaller number of broadcasts needed to answer all requests, it also reduces the time slot left for the broadcast at hand and, thus, requires higher transmission speed.

In previous works (e. g., [4] and [9]) the transmission power is merely used to adjust the transmission range in order to construct a topology that supports broadcasts but minimizes the energy consumption. We propose a completely different usage of the transmission power and use it to adjust the maximal transmission speed of the broadcast channel. As observed, e.g., for the 802.11a-WLAN technology, the signal-to-noise ratio needed to send a transmission increases with the transmission speed [12] and, thus, higher speed results in increased energy consumption. Looking at it from the optimistic point of view, the server can reduce its energy consumption by simply keeping transmission speed low. We follow the standard model and assume that at speed s the power consumption is s^α per time unit, where $\alpha \geq 2$ is constant.

1.1 Related Work

Extensive research on various versions of online broadcasting has been going on for several years. The most popular problem variation aims at flowtime minimization, i.e., minimizing the time span between the arrival of a request and the time by which it is answered [1, 5]. Other objectives include minimization of the number of unsatisfied requests [8] and different QoS-measures allowing messages to be split up into an arbitrary number of smaller objects [11]. In [7] flowtime minimization is combined with an additional constraint defining the maximum energy consumption allowed for servicing the requests. Dynamic adjustment of the transmission energy has been used for interference reduction in wireless networks by Burkhart et al. [3] and by Moscibroda et al. [10].

A related problem that has received a lot of attention also from the energy perspective is job scheduling with deadlines [13]. Here, a sequence of jobs, each with release time, deadline and a certain workload need to be scheduled on a single speed-controlled processor, such that all jobs are finished in time and the overall energy consumption is minimized. Yao et al. [13] present a polynomial time offline algorithm and propose different online strategies. Bansal et al. [2] present an $\mathcal{O}(e^\alpha)$-competitive online algorithm and show that this is asymptotically best possible.

1.2   Preliminaries

As the base problem of this paper we consider a server that is confronted with a sequence $R = (r_1, r_2, \ldots)$ of requests for the same piece of information. This piece of information has a transfer volume of one, i.e., it can be broadcasted completely in $1/s$ time units at speed s. Request $r_j = (t_j, d_j)$ is posed at its release time t_j and has to be answered until its deadline d_j, i.e., the server's message has to be broadcasted completely between times t_j and d_j at least once. A broadcast performed at speed s for t time units consumes $t \cdot s^\alpha$ energy units for $\alpha \geq 2$. Therefore broadcasting the message completely in time t at fixed speed $1/t$ consumes $(1/t)^{\alpha-1}$ energy units. Due to the convexity of the energy function, it is not difficult to see that this is the minimal amount of energy needed to deliver the whole information within t time units. For the first part of this paper this will also be the only allowed type of broadcast, i.e., we will assume that the transmission speed for each broadcast needs to be fixed the moment it is started and cannot be changed while the broadcast is running. We consider the problem of finding a feasible schedule of broadcasts (i.e., answering all requests within their deadlines) that minimizes the overall energy consumption.

We will also consider two extensions of the problem defined above. First, we will investigate the case in which the server holds a larger number $k \in \mathbb{N}$ of messages. Every request $r_j = (t_j, d_j, m_j)$ then asks for a single message m_j to be delivered before its deadline. We then turn to the variation in which the speed of a running broadcast can be adapted by the algorithm. As before, sending at speed s for t time units causes an energy consumption of $t \cdot s^\alpha$.

Finally, let us introduce some notation that will be used throughout the rest of the paper. Given a sequence of requests R, we let $B = (b_1, b_2, \ldots)$ and $B^* = (b_1^*, b_2^*, \ldots)$ denote the corresponding sequences of broadcasts sent by an an online strategy or the optimal offline strategy. We assume that B and B^*, as well as R, are sorted chronologically by their starting and release times, respectively. Sometimes it will be convenient to associate a broadcast $b_i = (s_i, f_i)$ with the interval $[s_i, f_i]$ defined by its starting and finishing times. For requests r_j as well as for broadcasts b_i we let $|r_j| = d_j - t_j$ and $|b_i| = f_i - s_i$ refer to their lengths.

1.3   Contributions

To the authors' best knowledge, this is the first analysis directly addressed to the minimization of energy consumption for broadcasting by speed scaling. We start by considering the restricted version of the problem in which the server holds only a single message and transmission speed cannot be changed while a broadcast is being performed. We first point out that an optimal schedule can be found in polynomial time in the offline setting. We then present an easy to implement online algorithm and prove that it achieves competitive ratio $\mathcal{O}(2^\alpha)$ in general and is $\mathcal{O}((3/2)^\alpha)$-competitive for requests of identical length. These results are found in Sections 2.1 and 2.2.

It turns out that our algorithm's competitive ratio is best possible, as we proceed by showing a matching lower bound of $\omega((2 - \varepsilon)^\alpha)$ for any $\varepsilon > 0$ that holds even for randomized online algorithms. The lower bound is based on what could be called a *growing gap* argument. Assuming that a given online algorithm achieves a better competitive ratio we construct a series of requests, such that in each newly added request there is a time interval of monotonically increasing relative length that the algorithm cannot use for the broadcast answering it. Details are found in Section 2.3.

We continue by investigating the multiple-message scenario, in which the server holds any larger number $k \in \mathbb{N}$ of different messages. We present an extension of our online algorithm and show that it is $(4c - 1)^\alpha$-competitive if request lengths vary by at most a factor of c. Finally, we take a look at the effect of allowing the algorithm to adapt the speed of running broadcasts. We prove a lower bound of $\omega((1.38 - \varepsilon)^\alpha)$ for arbitrary $\varepsilon > 0$ on the competitive ratio of any online algorithm in the general case and a lower bound of $\Omega(1.09^\alpha)$ for requests of identical length. For results on these extensions see Section 3.

2   Single-Message Broadcasting

We consider the situation in which our server holds a single message and transmission speed for each broadcast needs to be fixed the moment it is started. Section 2.1 deals with the offline setting. We then present our main result by deriving an $\mathcal{O}(2^\alpha)$-competitive online algorithm in Section 2.2 and proving a matching lower bound in Section 2.3.

2.1   The Offline Setting

Consider a given sequence of requests R and assume that we are given an optimal broadcast schedule B^*. It is not difficult to argue that w.l.o.g. B^* consists of blocks of equally distributed consecutive broadcasts. Using this observation one can apply a dynamic programming approach to compute optimal schedules efficiently. The proof is left for the long version of this paper.

Theorem 1. *In the offline setting the single-message broadcasting problem can be solved in polynomial time.*

2.2   An Online Algorithm

Algorithm ONLINE-SM below proceeds as follows. If a request $r = (t, d)$ arrives at time t while the channel is idle, we start a broadcast that uses the full length $d - t$ of the request. If the channel is busy and the currently running broadcast is scheduled to finish at time τ, we abort and start a new broadcast if at least half of the interval $[t, d]$ defined by r lies before τ. In our implementation τ denotes the end of the currently running broadcast, ρ refers to the earliest deadline of any request that needs to be answered by a broadcast starting after τ.

```
1   τ ← +∞, ρ ← +∞                              14  if a broadcast finishes and ρ < +∞ then
2   if a request r = (t, d) arrives then         15  |   τ ← ρ
3   |   if channel is idle then                  16  |   ρ ← +∞
4   |   |   τ ← d                                 17  |   start broadcast at speed (τ − t)⁻¹
5   |   |   start broadcast at speed (τ − t)⁻¹
6   |   if channel is busy then
7   |   |   if τ − t ≥ d − τ then
8   |   |   |   abort current broadcast
9   |   |   |   τ ← min{τ, d}
10  |   |   |   ρ ← +∞
11  |   |   |   start broadcast at speed (τ − t)⁻¹
12  |   |   else
13  |   |   |   ρ ← min{ρ, d}
```

Algorithm 1: Online-Sm

Theorem 2. *Let E_{ON} denote the energy consumption of algorithm* Online-Sm *on any sequence of requests, E_{OPT} the value of an optimal offline solution on the same sequence. It holds that*

$$E_{ON} \leq \left(\frac{\alpha}{\alpha - 1}\right)^2 2^{\alpha} \cdot E_{OPT}.$$

Before presenting the proof of Theorem 2 we point out that a better competitive ratio is obtained if we require all requests to have identical length.

Theorem 3. *Let E_{ON} denote the energy consumption of algorithm* Online-Sm *on any sequence of requests of identical length, E_{OPT} the value of an optimal offline solution on the same sequence. It holds that*

$$E_{ON} \leq \frac{2\alpha}{\alpha - 1} \left(\frac{3}{2}\right)^{\alpha} \cdot E_{OPT}.$$

We proceed by proving Theorem 2. It is straightforward to show that algorithm Online-Sm outputs a feasible solution, i.e., every request is indeed answered by one of the algorithm's broadcasts. The formal proof is omitted due to space limitations. In order to prove the competitive ratio claimed above we first need to bound the cost incurred by our algorithm due to aborted broadcasts. The following lemma states that this cost becomes negligible for larger values of α. The proofs of Lemma 1 and Theorem 3 are omitted.

Lemma 1. *Let E_{ab} denote the energy consumed by algorithm* Online-Sm *due to aborted broadcasts, E_{co} the energy used by completed broadcasts. Then $E_{ab} \leq (1/(\alpha - 1))E_{co}$ for any given sequence of requests.*

Proof of Theorem 2: For the remainder of the proof it will be important to know to which request to assign the cost of a single (completed) broadcast. Given broadcast b sent by algorithm Online-Sm we say that b *is linked to* request r, if b is started due to r in lines 5, 11 or 17. It is straightforward to observe that $|b| \geq |r|/2$, whenever b is linked to r.

Consider the optimal broadcast schedule B^* and let us fix a single broadcast $b^* \in B^*$ sent by the offline strategy. By R we refer to the set of requests answered by b^*. Clearly, for each request $r \in R$ it holds that $|r| \geq |b^*|$.

We are going to bound the energy E consumed by algorithm ONLINE-SM for sending (completed) broadcasts linked to requests in R. To this end, let $b^* = (s, f)$ and $E = E_1 + E_2$, where E_1 and E_2 denote energy consumption before and after time s, respectively.

Let us first consider E_1. We know that every request in R has a deadline no earlier than f, as otherwise it could not be answered by b^*. Now assume that algorithm ONLINE-SM is sending a broadcast b linked to some request $r \in R$ at time $u < s$. It follows that $|b| \geq (1/2)|r| \geq (1/2)(s - u + |b^*|)$ and, thus, the algorithm is sending at speed $|b|^{-1} \leq 2(s - u + |b^*|)^{-1}$ at time u. We can then write that

$$E_1 \leq 2^{\alpha} \int_0^s (s - u + |b^*|)^{-\alpha} du = \frac{2^{\alpha}}{\alpha - 1} u^{-\alpha+1} \Big|_{s+|b^*|}^{|b^*|} \leq \frac{2^{\alpha}}{\alpha - 1} |b^*|^{-\alpha+1}.$$

We next consider E_2. As observed before, broadcasts linked to requests in R have length at least $|r|/2 \geq |b^*|/2$ and, thus, their energy consumption never exceeds $(|b^*|/2)^{-\alpha+1}$. It is also straightforward to argue that algorithm ONLINE-SM starts at most one such broadcast after time s, as this will clearly satisfy all previously unanswered requests in R. It follows that energy consumption after time s can only be due to the last broadcast started before s and one additional broadcast started afterwards. Hence, we may write that $E_2 \leq 2(|b^*|/2)^{-\alpha+1} = 2^{\alpha}|b^*|^{-\alpha+1}$.

On the other hand, the energy consumed by the offline strategy answering requests from R is $E^* = |b^*|^{-\alpha+1}$ and we finally obtain

$$E = E_1 + E_2 \leq \left(\frac{2^{\alpha}}{\alpha - 1} + 2^{\alpha} \right) |b^*|^{-\alpha+1} = \frac{\alpha}{\alpha - 1} 2^{\alpha} \cdot E^*.$$

Summing over all $b^* \in B^*$ and taking into account energy consumption due to aborted broadcasts yields the theorem. □

2.3   A Lower Bound

We proceed by showing that algorithm ONLINE-SM is asymptotically best possible. We start with a matching lower bound for deterministic algorithms in Theorem 4. Theorem 5 states an extension to randomized algorithms.

Theorem 4. *The competitive ratio of every deterministic online algorithm for single-message broadcasting is $\omega((2 - \varepsilon)^{\alpha})$ for any $\varepsilon > 0$.*

Here is the high-level idea of the proof: Assume that some online algorithm with competitive ratio $\mathcal{O}((2-\varepsilon)^{\alpha})$, $\varepsilon > 0$, is given. We construct a sequence of requests of exponentially decreasing lengths. In each step of the construction, there will be a time interval contained within the very last request, which the algorithm cannot use for sending a broadcast answering it. We will refer to this time interval as a *gap*. The key ingredient of the proof is Lemma 2, which states that the relative length of the gap (i.e., compared to the length of the last request) is strictly increasing, forcing the algorithm to violate its competitive ratio after a number of requests.

Proof of Theorem 4: Towards a contradiction assume that deterministic online algorithm A has competitive ratio $\mathcal{O}((2 - \varepsilon)^{\alpha})$ for some constant $\varepsilon > 0$. We denote our sequence of requests as $r_0, r_1, \ldots$ and let $R_j = (r_0, \ldots, r_j)$. Since A is deterministic, we can construct the input sequence in a step by step manner, i.e., we can define request r_{j+1} depending on the algorithm's observed behavior on the sequence R_j of previous requests.

Before we give a detailed description of our construction, we need to define the notion of a gap more formally. We have to consider two different types of gaps. We say that r_j has a gap of relative length δ at *its beginning*, if a broadcast that has been started before r_j is posed is not aborted and finishes $\delta \cdot |r_j|$ time units after r_j's release time. Thus, the broadcast answering r_j must be started at least $\delta \cdot |r_j|$ time units after the request is actually posed. On the other hand, we say that r_j has a gap of relative length δ at *its end*, if there exists a request r_i, $i < j$, that has a deadline $\delta \cdot |r_j|$ time units before the deadline of r_j and needs to be answered by the same broadcast as r_j. In this situation, the broadcast answering r_j clearly needs to finish $\delta \cdot |r_j|$ time units before its deadline, as otherwise r_i were left unanswered. Gap positions are depicted in Figure 2.3.

Let now $E_{OPT}(R_j)$ denote the cost of an optimal offline solution on R_j and assume for the moment that $E_{OPT} = \mathcal{O}(|r_j|^{-\alpha+1})$, i.e., assume that the overall cost is dominated by the cost of answering the last request. Having this it is clear that algorithm A's broadcast answering r_j must have length at least $(1/2+\varepsilon')|r_j|$ for some appropriately chosen $\varepsilon' > 0$ in order to guarantee its competitive ratio. Let now $r_j = (t_j, d_j)$ and $m_j = (t_j + d_j)/2$ refer to the middle of the interval defined by r_j. We set $r_0 = (0, 1)$ and say that r_0 has a gap of relative length 0 at its beginning. For the definition of r_{j+1} we distinguish two cases. If r_j has a gap of relative length δ at its beginning, we set $r_{j+1} = (m_j + (\delta + \varepsilon')|r_j|/2, d_j + (\delta + \varepsilon')|r_j|/2)$. If a gap of the same length is at the end of r_j, we set $r_{j+1} = (m_j - (\delta - \varepsilon')|r_j|/2, d_j - (\delta - \varepsilon')|r_j|/2)$. Intuitively, r_{j+1} has length 2^{-j-1} spanning the second half of request r_j and is shifted according to the gap's position in r_j.

Assuming that the gap in request r_j has relative length at most $1/2 - \varepsilon'$ it is not difficult to check that in the above construction the release time of request r_{j+1} always lies after the starting time of the broadcast answering r_j. Thus, the gap position in r_j is known by the time we need to decide on r_{j+1} and we can iterate our construction while the gap's relative length is at most $1/2 - \varepsilon'$.

By Lemma 2 the relative length increases by at least ε' in each iteration. Hence, it exceeds $1/2 - \varepsilon'$ at some point and there must exist a request r_n that algorithm A answers by a broadcast of length $(1/2 + \varepsilon' - \gamma')|r_n|$ for some

$\gamma' > 0$. Thus, the energy consumption of algorithm A on R_n is $E_A(R_n) = \Omega((2 - \varepsilon + \gamma)^{(\alpha-1)}|r_n|^{-\alpha+1})$ for some $\gamma > 0$.

It remains to bound $E_{OPT}(R_n)$ from above. The optimal schedule uses the full length of r_n for its last broadcast. For any other request r_j we distinguish two cases. If r_j contains some r_k, $k > j$, then it is answered by the broadcast answering r_k. Otherwise, all requests r_k, $k > j$, are posed after m_j and r_j can be answered by a broadcast of length $m_j - t_j = |r_j|/2$. Now remember that $|r_j| = 2^{-j}$ and we obtain that

$$E_{OPT}(R_n) \leq \sum_{j=0}^{n-1} (|r_j|/2)^{-\alpha+1} + |r_n|^{-\alpha+1} = \mathcal{O}(|r_n|^{-\alpha+1}),$$

which contradicts the assumption that A is $\mathcal{O}((2 - \varepsilon)^\alpha)$-competitive.   □

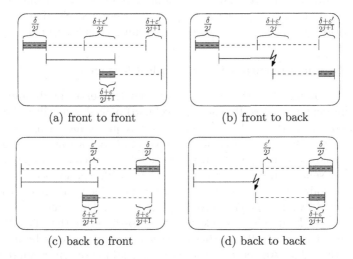

(a) front to front   (b) front to back

(c) back to front   (d) back to back

Fig. 1. Construction for a gap of relative length δ with respect to its position in the j-th request. Intervals corresponding to requests and broadcasts are depicted as dashed and solid lines, respectively. The flash symbol denotes abortion of a running broadcast. Boxes mark gaps as described in Theorem 4.

Lemma 2. *Given online algorithm A as in the proof of Theorem 4, the relative length of the gap increases by at least ε' with each newly added request.*

Figure 2.3 gives some idea of the proof of Lemma 2, which is rather technical and is omitted here due to space limitations.

The next theorem extends our lower bound to randomized online algorithms. We briefly sketch the key idea. As the length of the broadcast answering the last request dominates the cost of the solution, we know that the expected length of the broadcast answering r_j must not fall below $(1/2 + \varepsilon')|r_j|$. Assuming that each request is added as before, we obtain that the expected length of the gap increases in every step. However, as an adversary we do not know the random coin flips of the algorithm and, thus, do not know the position of the gap. We solve this problem by shifting requests in either of the two possible directions randomly. The key observation is that for constant $\varepsilon' > 0$ we need only a constant number of consecutive successful steps to reach a sufficient gap size and, thus, the randomized online algorithm fails to achieve its expected competitive ratio with sufficiently high probability.

Theorem 5. *The expected competitive ratio of every (randomized) online algorithm for single-message broadcasting is $\omega((2 - \varepsilon)^\alpha)$ for any $\varepsilon > 0$.*

3   Extensions

Finally, we will briefly sketch a number of results for some natural extensions of our problem. Section 3.1 presents a competitive online algorithm for the case that more than a single message needs to be broadcasted. Section 3.2 discusses the implications of allowing the speed of a running broadcast being changed.

3.1   An Online Algorithm for Multiple Messages

We sketch an online algorithm for the case that the server holds some number $k \in \mathbb{N}$ of different messages. Request $r = (t, d, m)$ now also specifies the message m that needs to be received. We assume that requests have lengths that vary between ℓ and $c\ell$ for some positive constants c and ℓ. Algorithm ONLINE-MM is a simple extension of our algorithm from Section 2.2. Instead of single broadcasts ONLINE-MM sends sequences of broadcasts (of different messages) of overall length $\ell/2$. Again we let τ refer to the earliest deadline of a request that needs to be answered by the following broadcast sequence. This time we set τ to the deadline of a newly arriving request if it overlaps by more than $\ell/2$. We collect requests until time $\tau - \ell/2$ and then answer collected requests by a broadcast sequence. After that, we set τ to the earliest deadline of any request that has not been answered.

It can be shown that algorithm ONLINE-MM has a constant competitive ratio that depends only on c, i.e., the factor by which request lengths may vary. We specifically note that this competitive ratio is independent from the number k of messages held by the server.

Theorem 6. *Let E_{MM} denote the energy consumption of algorithm ONLINE-MM on any sequence of requests with lengths varying between ℓ and $c\ell$ for some fixed $c \geq 1$, E_{OPT} the value of an optimal offline solution on the same sequence. It holds that $E_{MM} \leq (4c - 1)^\alpha \cdot E_{OPT}$.*

3.2   More Flexible Speed-Adjustment

We briefly address another extension to our problem. It is conceivable to allow the server to adapt the speed of a running broadcast without enforcing a restart.

While our algorithmic results can obviously still be applied, the lower bound given in Section 2.3 does not hold in this situation. However, the next theorem states that the competitive ratio of any algorithm must be exponential in α, even if we allow speed adaptation.

Theorem 7. *The competitive ratio of any (randomized) online algorithm for single-message broadcasting capable of speed adaptation is $\omega((\gamma - \varepsilon)^\alpha)$ for every $\varepsilon > 0$, where $\gamma = (5 + 5\sqrt{5})/(5 + 3\sqrt{5}) > 1.38$. If request lengths are identical the competitive ratio is at least $\Omega(1.09^\alpha)$.*

References

1. N. Bansal, D. Coppersmith, and M. Sviridenko. Improved Approximation Algorithms for Broadcast Scheduling. In *Proceedings of the Symposium on Discrete Algorithms (SODA)*, 2006.
2. N. Bansal, T. Kimbrel, and K. Pruhs. Dynamic Speed Scaling to Manage Energy and Temperature. In *Proceedings of the Symposium on Foundations of Computer Science (FOCS)*, 2004.
3. M. Burkhart, P. von Rickenbach, R. Wattenhofer, and A. Zollinger. Does Topology Control Reduce Interference? In *Proceedings of the Symposium on Mobile Ad Hoc Networking and Computing (MobiHoc)*, 2004.
4. A. Clementi, P. Crescenzi, P. Penna, G. Rossi, and P. Vocca. On the Complexity of Computing Minimum Energy Consumption Broadcast subgraphs. In *Proceedings of the Symposium on Theoretical Aspects of Computer Science (STACS)*, 2001.
5. J. Edmonds and K. Pruhs. Multicast Pull Scheduling: When Fairness is Fine. *Algorithmica*, 36, 3:315–330, 2003.
6. G.-Y. Gao and S.-K. Wang. A Multi-Agent System Architecture for Geographic Information Gathering. *Journal of Zhejiang University SCIENCE*, 5(11):1367–1373, 2004.
7. C. Gunia. On Broadcast Scheduling with Limited Energy. In *Proceedings of the Conference on Algorithms and Complexity (CIAC)*, 2006.
8. B. Kalyanasundaram, K. Pruhs, and M. Velauthapillai. Scheduling Broadcasts in Wireless Networks. In *Proceedings of the European Symposium on Algorithms (ESA)*, 2000.
9. X-Y. Li, W-Z. Song, and W. Wang. A Unified Energy-Efficient Topology for Unicast and Broadcast. In *Proceedings of the International Conference on Mobile Computing and Networking (MobiCom)*, 2005.
10. T. Moscibroda and R. Wattenhofer. Minimizing Interference in Ad Hoc and Sensor Networks. In *Proceedings of the Joint Workshop on Foundations of Mobile Computing*, 2005.
11. K. Pruhs and P. Uthaisombut. A Comparison of Multicast Pull Models. In *Proceedings of the European Symposium on Algorithms (ESA)*, 2002.
12. D. Qiao, S. Choi, and K.-G. Shin. Goodput Analysis and Link Adaptation for IEEE 802.11a Wireless LANs. *Transactions on Mobile Computing*, 1(4):278–292, 2002.
13. F. Yao, A. Demers, and S. Shenker. A Scheduling Model for Reduced CPU Energy. In *Proceedings of the Symposium on Foundations of Computer Science (FOCS)*, 1995.

Online Packet Admission and Oblivious Routing in Sensor Networks

Mohamed Aly[1],[*] and John Augustine[2],[**]

[1] Dept. of Computer Science, University of Pittsburgh,
210 South Bouquet Street,
Pittsburgh, PA 15260
maly@cs.pitt.edu
[2] Donald Bren School of Information and Computer Sciences,
University of California at Irvine,
444 Computer Science Bldg,
Irvine, CA 92697
jea@ics.uci.edu

Abstract. The concept of Oblivious Routing for general undirected networks was introduced by Räcke [12] when he showed that there exists an oblivious routing algorithm with polylogarithmic competitive ratio (w.r.t. edge congestion) for any undirected graph. In a following result, Räcke and Rosén [13] presented admission control algorithms achieving a polylogarithmic fraction (in the size of the network) of the optimal number of accepted messages. Both these results assume that the network incurs a cost only after it is accepted and the message is routed. Admission control and routing algorithms for sensor networks under energy constraints, however, need to account for the energy spent in checking for feasible routes prior to the acceptance of a message and hence, it is unclear if these algorithms achieve polylogarithmic bounds under this condition. In this paper, we address this problem and prove that such algorithms do not exist when messages are generated by an adversary. Furthermore, we show that an oblivious routing algorithm cannot have a polylogarithmic competitive ratio without a packet-admission protocol. We present a deterministic $O(\rho \log n)$-competitive algorithm for tree networks where the capacity of any node is in $[k, \rho k]$. For grids, we present an $O(\log n)$-competitive algorithm when nodes have capacities in $\Theta(\log n)$ under the assumption that each message is drawn uniformly at random from all pairs of nodes in the grid.

1   Introduction

In his seminal work, Räcke [12] proved the existence of oblivious routing algorithms (with respect to edge congestion) for general undirected graphs, which

---

[*] Funded in part by NSF grants ANI-0325353, CCF-0448196, CCF-0514058, and IIS-0534531.

[**] Supported in part by NSF grant CCF-0514082.

was a breakthrough in routing with local information. The paper defined the concept of *oblivious routing* where the routing path for a message between a source s and a target t is independent of all other messages and only depends on s, t and some random input. Räcke presented the idea of decomposing a general graph into a tree and routing demands on the underlying tree decomposition. Many researchers have employed Räcke's technique in other network models such as the work of Hajiaghayi, Kim, Leighton and Räcke [8] on directed graphs.

In addition, Räcke's technique has also been used to provide algorithms for objective functions other than edge congestion. For instance, Räcke and Rosén [13], present the first known distributed online admission control algorithms for general network with polylogarithmic competitive ratios in the size of the network with respect to the number of accepted messages. Based on the use of Räcke's tree decomposition, the algorithms consisted of path-establishment and message admission procedures that can operate in a hop-by-hop manner through the network. For a given request, a sequence of request-messages is sent along a path (determined in a distributed hop-by-hop fashion) between the source and destination. The message is accepted in case the loads of all the edges along such path have enough capacity to route the message. In case a message is accepted, it is routed on the same path.

Routing issues in sensor networks has gained considerable attention in the recent years [3, 9, 15, 11, 10, 6]. Unlike general networks, a sensor network is typically composed of a large number of sensors with *limited* energy, processing, communication, and storage capabilities [1]. This node-centric constraint motivates us to consider routing algorithms in node capacitated graph models. Hajiaghayi *et al.* [7] study routing in node capacitated graphs where they minimize the node congestion. Admission control algorithms in the sensor network context for node capacitated graphs have not been studied so far to the best of our knowledge. Can the known algorithms for admission control provide us a guarantee of polylogarithmic competitive ratio in this scenario? While this is a promising direction, the answer is unclear because the messages sent to check whether a path can be established consume energy just like the input messages. Obviously, this is not an issue in the analysis of Räcke's technique in general networks. Additionally, a message that is dropped mid-way before reaching its destination consumes energy at all the sensor nodes that it visited before being dropped. Based on these observations, an oblivious routing algorithm for sensor networks should not accept messages if it can somehow anticipate that an intermediate node on the message path is likely to drop such message.

In this paper, we study the problem of *Online Packet-Admission and Oblivious Routing* in the context of sensor networks. We define the problem as follows. A sequence of messages (packets) is received in an online manner. Each message arises in a sensor node, referred to as the *source*, and it should be sent to another sensor node, namely the *destination*. The sender first determines whether to send the message or drop it. Then, the sender decides which is the best neighbor to forward the message to. The sender makes both decisions based on its local energy status, the destination location and information that it has received via control messages whose energy consumption is duly accounted. The path of the message to its destination is determined in a distributed manner. Any sensor node can locally decide to drop the message according to its local information. The packet-admission and oblivious routing algorithm aims at maximizing the number of successfully sent messages.

2   Our Results

We analyze online distributed packet-admission and oblivious routing algorithms for sensor networks. Our primary goal is to design algorithms that achieve a logarithmic competitive ratio in terms of throughput (number of successfully sent messages) when compared to the optimal offline algorithm, OPT. Our results are twofold. In the first part of the paper, we derive lower bounds on distributed oblivious routing algorithms. A major result that we prove is that any distributed packet-admission and oblivious routing algorithm cannot achieve polylogarithmic competitiveness when given a set of adversarial messages. This result forms a strong constraint on the input of any routing protocol in sensor networks, and in capacitated graphs in general. Furthermore, we consider a distributed oblivious routing algorithm that *always sends* packets without applying any packet-admission protocol. The messages however can be dropped when it reaches a node that has used up all its energy, i.e., a *dead node*. For such algorithms, we prove the following theorem.

Theorem 2.1. *Given a balanced binary tree $T(V,E)$ and a set of demands D, an always-send distributed oblivious routing algorithm A_{as} cannot maintain polylogarithmic competitiveness in either of the following cases:*

- *D is a set of adversarial demands, or follows a general distribution that is unknown to all sensor nodes.*
- *an adversary sets the tree node capacities (internal nodes or leaf nodes).*

This theorem shows that any distributed oblivious routing algorithm needs a concrete packet-admission protocol in order to achieve polylogarithmic competitiveness with respect to throughput in the context of sensor networks. Since we prove these lower bound results for balanced binary trees, they extend to general trees, as well as general undirected graphs.

We complement the lower bounds in the second part by providing two $O(\log n)$-competitive algorithms, in terms of throughput, for undirected tree and grid networks, respectively. These topologies arise naturally in the study of sensor networks lending credence to the our choice. For example, the work on data aggregation uses trees in collecting data from sensor nodes, even when sensor networks have general graph topologies (see [3, 9, 15, 11, 10, 6]). We also have several examples in the sensor networks literature that study grid topologies (see [4, 5, 14]). Our first algorithm achieves an $O(\rho \log n)$ competitive ratio for any sequence of input messages, assuming the capacity of any node is in $[k, \rho k]$ where k is in $\Omega(\log n)$, for a sensor network that can be modeled as a tree. Our second algorithm is $O(\log n)$-competitive assuming the capacity of any node is $k \geq 104 \log^2 n + 6 \log n$ and is a uniform value for all nodes.

Our work leads to many interesting open problems in terms of deriving tighter lower and upper bounds for distributed oblivious routing algorithms for sensor networks, as well as, considering the oblivious routing problem for more general types of graphs with energy-capacitated nodes.

3 Preliminaries

We represent the sensor network as a graph $G(V, E)$, where $v \in V$ is a sensor node with energy capacity $cap(v)$ units, $|V| = n$, and $|E| = m$. An edge $u = (i, j) \in E$ iff there exists a wireless link between sensors i and j. The two particular types of graphs that we consider in this paper are trees, which we represent by $T(V, E)$ in Section 5, and grid networks. A *message* is an ordered pair (s, t) that is generated at s. We are presented with a sequence of calls D, also known as the message demand or simply *demand*, in an online manner, i.e., each message (s, t) is only known to s when it is generated and the sequence of calls is not known in advance. Our distributed routing algorithm has to either accept or reject each message (s, t) as it is presented and if accepted, the message should be routed to its destination t. In addition to rejection, the message may also be dropped enroute to its destination. A node spends 1 energy unit in sending a message while receiving is free. We present distributed algorithms that execute without any central decision maker. All decisions are made within a a sensor node. We allow control messages to be passed along by the distributed algorithm as long as the messages are of size at most $O(\log n)$. Each control message, like the calls in the demand, that hops from node a to node b consumes 1 energy unit at node a. Each node is oblivious of the energy levels of the other nodes and can only use the information about itself and that which is explicitly passed on to it via control messages. Let $OPT(G, D)$ be the maximum number of messages that can be successfully routed by an optimal algorithm that knows D in advance and has full knowledge of the energy levels of every node in G at all times. Our algorithms are $O(\log n)$, i.e., the number of messages they route successfully when presented with the demand D is asymptotically within a logarithmic factor of $OPT(G, D)$.

4 Lower Bounds

In this section, we present lower bounds on the competitive ratio of distributed oblivious routing algorithms in tree networks. In other words, we are studying the conditions preventing an oblivious routing algorithm from having a poly-logarithmic competitive ratio. We consider a balanced binary tree $T(V, E)$ such that $|V| = 4n - 1$, i.e., it is a tree with n leaves in each half. A node is assumed to consume 1 unit in receiving a message. We assume, without loss of generality, that all messages are from leaf nodes of the left subtree to the leaf nodes of right subtree through the root node. In some cases, we assume that the node capacities follow the nesting property whereby, a non-leaf node has capacity exactly equal to the sum of the capacities of its children. Note that when a tree follows the nesting property, we only need to set the capacities of the leaves in order to define the capacities of all the nodes in the tree. Using such settings, we prove theorem 2.1 using the next four lemmas.

4.1   Adversarial Demands

We first assume that the set of input demands D is formed by an adversary that knows the capacities of all the tree nodes at any time during the network operation. Further, T is assumed to follow the nesting property, and each leaf node has a capacity of e energy units, assuming $(e \ll n)$. For this setting, we claim the following lemma.

Lemma 4.1. *For the above tree setting, there exists at least one set of adversarial demands D that makes any deterministic distributed routing algorithm A at least $\Omega(n)$ competitive.*

Proof. We denote leaf nodes of the left and right subtrees by l_i, r_j $1 \leq i, j \leq n$, respectively. Each left subtree leaf node is assumed to have an n-bit vector numbered from 1 to n such that each bit represents the status of one of the n right subtree leaf nodes. Node l_i sends messages to node r_j only if the jth bit in l_i's vector is set to 1. Assuming any arbitrary A, A does not change the bit vector of any node l_i except when l_i either sends (or drops) a message, or receives a control bit. Given any A, we assume the presence of an adversary that knows the bit vectors of every node l_i, $1 \leq i \leq n$, at any point in time. The adversary is assumed to know the mechanism A uses to change the bit vector of any node l_i. The adversary is supposed to form an input sequence of $D = (M(l_i, r_j) : 1 \leq i, j \leq n)$ messages in a way that minimizes the throughput of A when compared to that of OPT. Also, the adversary continuously inputs messages till totally depleting the energy of all tree nodes.

The main idea of the adversarial strategy consists of making A send a message that OPT would not have sent (as the message destination is dead). Before starting its algorithm, the adversary checks the bit vectors of all l_i's and stops if any bit in any of these vectors is set to zero. In such case, A has an infinite competitive ratio compared to OPT. If this is not the case, the adversary applies the algorithm explained below.

The algorithm operates in n rounds. At each round, the adversary selects a destination r_j, which is a non-dead right subtree leaf node. Then, the adversary iterates on the alive left subtree leaf nodes, one by one, and for every node l_i, it keeps inserting $M(l_i, r_j)$ messages in D until node l_i switches its jth bit to zero. The iteration ends when all left subtree leaf nodes are either dead or have their j^{th} bit set to zero. At any iteration $0 \leq k < n$, the destination r_j, selected by the adversary at the start of k, will be dead after it receives the first e successful messages sent by A. The rest of the messages sent to r_j until the end of iteration k will be considered as *falsely accepted messages* by A.

Now, we analyze the above strategy. At every iteration, OPT achieves exactly e successful messages. Similarly, at each iteration i where $0 \leq i < e$, A achieves at most e successful messages. However, the adversary is able to force A to loose at least n energy units (i.e., at least one energy unit per sensor) at each of the first e iterations by making each alive node at least send one message (or receive one control bit). After e iterations, the number of accepted messages by OPT is e^2 and it has $n \cdot e - e^2$ energy units left ($n - e$ left subtree leaf nodes each having e energy units). On the other hand, A has successfully sent at most e^2 messages and has all left subtree leaf nodes dead. Thus, after n rounds, OPT's throughput is $n \cdot e$, while that of A is at most e^2, i.e. A is $(\frac{n}{e})$-competitive, which ends the proof. $\square$

4.2   Demands Drawn from an Unknown Distribution

Using the same tree setting as 5.1, we now drop the assumption of adversarial demands and assume that the messages in D are drawn from a distribution, but the sensors are oblivious to the nature of the distribution. We further assume that n is a power of 2, and leaf nodes have equal capacity belonging to $\log n$. We now claim the following lower bound. (proof omitted)

Lemma 4.2. *For the above tree setting, there exists a distribution unknown to the nodes $v \in V$ from which we draw messages in D that makes an always-send oblivious routing algorithm A_{as} at least $\Omega(n)$-competitive, assuming D is drawn from a distribution that is unknown for all nodes $v \in V$.*

4.3   Adversarial Leaf Node Capacities

We now consider the case where the adversary can set the energy capacities of the nodes in the tree. We restrict ourselves to the trees that follow the nesting property. Therefore, the adversary sets the capacities of the leaves and the rest of the capacities are determined by the nesting property. We now present the following lower bound (proof omitted)

Lemma 4.3. *There exists at least one adversarial setting for leaf node capacities that makes an always-send oblivious routing algorithm A_{as} $\Omega(n)$ competitive, assuming messages in D are drawn from a uniform distribution.*

Theorem 2.1 directly follows from the previous three lemmas. A direct implication from the theorem is that any distributed oblivious routing algorithm A_o should have a packet-admission protocol in order to achieve a polylogarithmic competitive ratio. In the following two sections, we complement these lower bound results by presenting oblivious routing algorithms that are logarithmic for undirected tree and grid networks (under uniform random demands assumption).

5   Oblivious Routing in Tree Networks

In this section, we present a deterministic packet admission and oblivious routing algorithm for a balanced binary tree $T(V, E)$ of height $\log n$. We later show

how this can be extended to any tree network. All sensors $v \in V$ have energy capacity in $[k, \rho k]$, where k, where $k \in \Omega(\log n)$ and $\rho > 1$. Our algorithm is $O(\rho \log n)$-competitive against an omniscient adversary. Given a message $M(s, t)$, the distributed algorithm is assumed to know the level of the lowest common ancestor (LCA) of s and t.

Algorithm. We classify each message (s, t) according to the level of its lowest common ancestor (LCA). Each node uses at most k units of energy. It divides its quota of energy k into $\log n$ shares with each containing energy $\lfloor k/\log n \rfloor$. We denote the shares s_i, where $1 \leq i \leq \log n$. Each message has a unique route and LCA because our network is a balanced binary tree. When a message with an LCA at level i reaches (or starts at) a node, it allows it to pass if and only if share $s_i > 0$ and then the node decrements s_i. The message is dropped if $s_i = 0$.

We now analyze our algorithm.

Lemma 5.1. *Consider any node $v \in V$. For any fixed sequence $\mathcal{I}$ of input messages, if $CS_{\mathcal{I}}(v)$ and $OPT_{\mathcal{I}}(v)$ be the number of messages with LCA v that are successfully routed by CS and OPT, respectively, then*

$$\frac{OPT_{\mathcal{I}}(v)}{CS_{\mathcal{I}}(v)} \leq \rho \log n \tag{1}$$

Proof. Let i be the level of node v. CS allows the first s_i messages whose LCA is v and rejects subsequent messages with LCA v. Hence, the ratio between CS and OPT is maintained trivially. OPT can pass at most ρk messages whose LCA is v because the capacity of v is in $[k, \rho k]$. Since, $s_i = \lfloor k/\log n \rfloor$, $\rho k \leq \rho s_i (1 + \log n)$, thus, Equation 1 holds. □

Lemma 5.2. *CS is $O(\rho \log n)$-competitive for balanced binary trees.*

Proof. For a fixed input sequence $\mathcal{I}$, let $CS_{\mathcal{I}}$ and $OPT_{\mathcal{I}}$ be the number of messages successfully routed by CS and OPT respectively throughout the tree T. We know that

$$CS_{\mathcal{I}} = \sum_{v \in V} CS_{\mathcal{I}}(v)$$

$$OPT_{\mathcal{I}} = \sum_{v \in V} OPT_{\mathcal{I}}(v)$$

Therefore,

$$\frac{OPT_{\mathcal{I}}}{CS_{\mathcal{I}}} = \frac{\sum_{v \in V} OPT_{\mathcal{I}}(v)}{\sum_{v \in V} CS_{\mathcal{I}}(v)} = \frac{\rho \log n \sum_{v \in V} \frac{OPT_{\mathcal{I}}(v)}{\rho \log n}}{\sum_{v \in V} CS_{\mathcal{I}}(v)}$$

$$\geq \frac{\rho \log n \sum_{v \in V} CS_{\mathcal{I}}(v)}{\sum_{v \in V} CS_{\mathcal{I}}(v)} = \rho \log n.$$

□

We can easily extend this algorithm to work in arbitrary tree networks using a well-known tree partitioning technique [2]. We know that we can easily find a node in a tree that divides the tree into two parts such that neither part has more than $\frac{2n}{3}$ nodes. We find such a node and call it the pivot node at level one. This spawns two subtrees and we recursively perform this operation to spawn subtrees at subsequent levels along with their corresponding pivot nodes. It is easy to see that we generate $\ell = O(\log n)$ levels. Similar to the case of balanced binary trees, we designate a fraction $\frac{1}{\ell}$ of the energy available at each node as a quota for messages that are entirely within subtrees at each level, say level i, but span multiple subtrees at higher numbered levels. Consider the jth tree at level i. The set of messages at level i and pertaining to that tree is denoted by Ψ_i^j. Like before, the pivot node at level i and tree j sends out a broadcast when the quota is reached. Therefore, we successfully transmit $\max(|\Psi_i^j|, \frac{1}{\ell}OPT(\Psi_i^j)$ messages, where $OPT(\Psi_i^j)$ is the number of messages (at the very best) that OPT can transmit when provided with Ψ_i^j as its input . It is easy to see that at least a fraction $\Omega(\frac{1}{\log n})$ messages in Ψ_i^j that OPT can transmit will be transmitted if that quota is observed. Summed over all trees that are spawned, we get the required competitive ratio for arbitrary trees.

Theorem 5.3. *CS is $O(\rho \log n)$-competitive for arbitrary tree networks.*

6   Oblivious Routing in Grids

In this section, we consider the sensors to form a 3-dimensional $n \times n \times n$ grid of vertices represented by the set V. For now, we assume that $n = 2^\ell - 1$. We will show how this requirement can be eliminated. Each node $v \in V$ in the grid can be addressed by its co-ordinates (x_v, y_v, z_v). Each node has $k = 6c\log^2 n + 6\log n$ units of energy and an unit of energy is required to transmit a message to a neighboring node. We assume that messages in our demand sequence are generated independent of each other and the source and destination are chosen uniformly at random. We present a call control algorithm and show that it is $O(\log n)$-competitive with high probability

Consider a message $M(s, t)$ where s and t are the source and destination nodes, respectively. We denote the positions of s and t by (x_s, y_s, z_s) and (x_t, y_t, z_t). Consider the sequence $(1, 2, \ldots, n)$ that can be used to represent a single line of nodes along one of the three axes. We decompose it into pivot levels as follows: node $\frac{n+1}{2}$ is at pivot level 1. Nodes $\frac{n+1}{4}$ and $\frac{(n+1)3}{4}$ are in pivot level 2. In general, $\frac{(n+1)i}{2^\ell}$ is at pivot level ℓ, where $\ell \le \log(n+1)$, $i < 2^\ell$ and i and ℓ are positive integers, if and only if i and 2^ℓ are mutually prime. We define the x-pivot (resp., y-pivot and z-pivot) for a message M to be the node with the smallest pivot level value inclusively between x_s (resp., y_s and z_s) and x_t (resp., y_t and z_t). We have boolean variables $L_x(\ell)$ and $R_x(\ell)$ stored in each node s that are used to decide whether a message should be accepted. $L_x(\ell)$ (resp. $R_x(\ell)$) in node s is set to true initially if there is a pivot level ℓ to the left (resp. right) of x_s and false otherwise, thereby allowing messages originating at s with a destination t

such that $x_s > x_t$, i.e., traveling left (resp. $x_s < x_t$, i.e., traveling right), and pivoted at level ℓ along the x-axis. We define the pivot plane $P_x(i, \ell)$ (resp. $P_y(i, \ell)$ and $P_z(i, \ell)$) to be the set of nodes with x value (resp. y and z values) equal to $(n+1)\frac{i}{2^\ell}$. The pivot plane $P_x(i, \ell)$ has a reach associated with it and is defined as the set of nodes whose x values are in $((n+1)\frac{i-1}{2^\ell}, (n+1)\frac{i+1}{2^\ell})$. Each node p in $P_x(i, \ell)$ is allocated a quota $2c\log n$ of messages that can pivot at p. When the quota is reached, node p broadcasts a message to all nodes in its reach stating that plane $P_x(i, \ell)$ has reached its quota. Each node v that receives such a broadcast message in turn switches its $L_x(\ell)$ (if $x_v > (n+1)\frac{i}{2^\ell}$) or $R_x(\ell)$ (if $x_v < (n+1)\frac{i}{2^\ell}$) to false. Since there are $\log(n+1)$ pivot levels and each node has two variables that can be switched off, there are only $2\log(n+1)$ such messages that pass through each node. When a node $p \in P_x(i, \ell)$ sends out such a broadcast message, we say that $P_x(i, \ell)$ is saturated. I.e., the algorithm no longer admits messages that pivot at $P_x(i, \ell)$.

The algorithm works as follows: when presented with a message $M(s, t)$ such that its pivot levels along the three axes are ℓ_x, ℓ_y and ℓ_z, we accept this message if and only if the three boolean variable corresponding to this message along the three axes stored at node s are set to true. Notice that we need to use either the 'L' or the 'R' according the direction M travels along each axis. For instance, if $x_s < x_t$, then we need to use R_x. The message is routed from s to t along these nodes: $s \rightarrow (x_t, y_s, z_s) \rightarrow (x_t, y_t, z_s) \rightarrow t$. It is easy to see that all messages that are accepted can be routed. It remains to show that this algorithm successfully transmits at least a fraction $\frac{1}{O(\log n)}$ of the number of messages that an optimal algorithm that knows all the messages in advance can transmit.

Consider an arbitrary pivot plane P. Let Ψ be the set of all messages that pivot about P. The probability that $|P| < cn^2\log n$ is at most $2n^{4-\frac{c}{4}}$ and it is obtained by viewing each node in the plane as a bin and the messages going through that node are the balls in that bin. If every node allows less than $2c\log n$ messages to pivot on it, then no message is discarded on account of P. In other words, P is a non-saturated plane. However, if a node reaches $2c\log n$, i.e., its corresponding plane is saturated, then we know with probability at most $2n^{4-\frac{c}{4}}$ that $|P| < cn^2\log n$. The maximum number of messages that can pivot at this plane is $6cn^2\log^2 n + 6\log n$. However, if we consider the saturated planes, we also know that $cn^2\log n$ messages successfully pivoted at each of these planes with probability at least $1 - 2n^{3(4-\frac{c}{4})}$. Therefore, we maintain a competitive ratio of $O(\log n)$ with high probability when $c \geq \frac{52}{3}$.

Notice that the key point of this result is that we partition each axis into $O(\log n)$ levels thereby allowing us to classify messages based on its pivot level. Since we trivially know that a similar partition with $O(\log n)$ levels can be obtained for all positive values of n, our algorithm works for unrestricted values of n.

Theorem 6.1. *Our distributed and oblivious level based algorithm is $O(\log n)$-competitive for $n \times n \times n$ grid when the messages in our demand sequence are generated uniformly at random.*

Acknowledgments. We thank Kirk Pruhs, Sandy Irani, Harald Räcke, and MohammadTaghi Hajiaghayi for several insightful comments.

References

[1] I.F. Akyildiz, Su Weilian, Y. Sankarasubramaniam, and E. Cayirci. A survey on sensor networks. *IEEE Communications Magazine*, 40:102– 114, 2002.

[2] Baruch Awerbuch, Yair Bartal, Amos Fiat, and Adi Rosén. Competitive non-preemptive call control. In *Proc. of the ACM Symp. on Discrete Algorithms (SODA)*, 1994.

[3] Philippe Bonnet, Johannes Gehrke, and Praveen Seshadri. Towards sensor database systems. In *MDM '01: Proceedings of the Second International Conference on Mobile Data Management*, pages 3–14, London, UK, 2001. Springer-Verlag.

[4] Krishnendu Chakrabarty, S. Sitharama Iyengar, Hairong Qi, and Eungchun Cho. Grid coverage for surveillance and target location in distributed sensor networks. *IEEE Transactions on Computers*, 2002.

[5] Santpal S. Dhillon, Krishnendu Chakrabarty, and S. S. Iyengar. Sensor placement for grid coverage under imprecise detections. In *Proc. International Conference on Information Fusion*, 2002.

[6] Himanshu Gupta, Vishnu Navda, Samir R. Das, and Vishal Chowdhary. Efficient gathering of correlated data in sensor networks. In *Proc. of MobiHoc*, 2005.

[7] Mohammad Taghi Hajiaghayi, Robert D. Kleinberg, Tom Leighton, and Harald Raecke. Oblivious routing on node-capacitated and directed graphs. In *Proc. of the ACM Symp. on Discrete Algorithms (SODA)*, 2005.

[8] MohammadTaghi Hajiaghayi, Jeong Han Kim, Tom Leighton, and Harald Räcke. Oblivious routing in directed graphs with random demands. In *Proc. of the ACM Symp. on Theory of Computing (STOC)*, 2005.

[9] Samuel Madden, Michael J. Franklin, Joseph M. Hellerstein, and Wei Hong. Tag: a tiny aggregation service for ad-hoc sensor networks. volume 36, pages 131–146, New York, NY, USA, 2002. ACM Press.

[10] Seung-Jong Park, Ramanuja Vedantham, Raghupathy Sivakumar, and Ian F. Akyildiz. A scalable approach for reliable downstream data delivery in wireless sensor networks. In *Proc. of MobiHoc*, 2004.

[11] Tri Pham, Eun Jik Kim, and W. Melody Moh. On data aggregation quality and energy efficiency of wireless sensor network protocols. In *Proc. of BROADNETS*, 2004.

[12] Harald Räcke. Minimizing congestion in general networks. In *Proc. of the ACM Symp. on Foundations of Computer Science (FOCS)*, 2002.

[13] Harald Räcke and Adi Rosén. Distributed online call control on general networks. In *SODA '05: Proceedings of the sixteenth annual ACM-SIAM symposium on Discrete algorithms*, pages 791–800, Philadelphia, PA, USA, 2005. Society for Industrial and Applied Mathematics.

[14] R. Stoleru and J. Stankovic. Probability grid: A location estimation scheme for wireless sensor networks. In *Proceedings of the IEEE Communications Society Conference on Sensor and Ad Hoc Communications and Networks*, 2004.

[15] Yong Yao and Johannes Gehrke. Query processing for sensor networks. In *Proceedings of the First Biennial Conference on Innovative Data Systems Research (CIDR 2003)*, 2003.

Field Splitting Problems in Intensity-Modulated Radiation Therapy*

Danny Z. Chen and Chao Wang**

Department of Computer Science and Engineering
University of Notre Dame
Notre Dame, IN 46556, USA
{chen, cwang1}@cse.nd.edu

Abstract. Intensity-modulated radiation therapy (IMRT) is a modern cancer treatment technique that delivers prescribed radiation dose distributions, called *intensity maps* (IMs), to target tumors via the help of a device called the *multileaf collimator* (MLC). Due to the *maximum leaf spread constraint* of the MLCs, IMs whose widths exceed a given threshold cannot be delivered as a whole, and thus must be split into multiple subfields. Field splitting problems in IMRT normally aim to minimize the total beam-on time (i.e., the total time when a patient is exposed to actual radiation during the delivery) of the resulting subfields. In this paper, we present efficient polynomial time algorithms for two general field splitting problems with guaranteed output optimality. Our algorithms are based on interesting observations and analysis, as well as new techniques and modelings. We formulate the first field splitting problem as a special integer linear programming (ILP) problem that can be solved *optimally* by linear programming due to its geometry; from an optimal integer solution for the ILP, we compute an optimal field splitting by solving a set of shortest path problems on graphs. We tackle the second field splitting problem by using a novel *path-sweeping* technique on IMs.

1 Introduction

In this paper, we study a few geometric partition problems, called *field splitting*, which arise in *intensity-modulated radiation therapy* (IMRT). IMRT is a modern cancer treatment technique that aims to deliver highly conformal prescribed radiation dose distributions, called *intensity maps* (IMs), to target tumors while sparing the surrounding normal tissues and critical structures. An IM is specified by a set of nonnegative integers on a uniform 2-D grid (see Figure 1(b)). The

---

* This research was supported in part by the Faculty Research Program of the University of Notre Dame, the National Science Foundation under Grant CCF-0515203, and NIH NIBIB Grant R01-EB004640-01A2.
** Corresponding author. The research of this author was supported in part by two Fellowships in 2004-2006 from the Center for Applied Mathematics of the University of Notre Dame.

T. Asano (Ed.): ISAAC 2006, LNCS 4288, pp. 690–700, 2006.

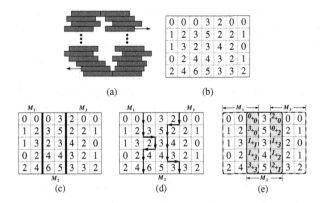

(a)   (b)

(c)   (d)   (e)

Fig. 1. (a) An MLC. (b) An IM. (c)-(e) Examples of splitting an IM into three subfields, M_1, M_2, and M_3, using vertical lines, y-monotone paths, and with overlapping, respectively. The dark cells in (e) show the overlapping regions of the subfields; the dose in each dark cell is divided into two parts and allocated to two adjacent subfields.

value in each grid cell indicates the intensity level of prescribed radiation at the body region corresponding to that IM cell. The delivery is done by a set of cylindrical radiation beams orthogonal to the IM grid.

One of the most advanced tools for IM delivery is the *multileaf collimator* (MLC) [15,16]. An MLC consists of many pairs of tungsten alloy leaves of the same rectangular shape and size (see Figure 1(a)). The opposite leaves of each pair are aligned to each other, and can move left or right to form a y-monotone rectilinear polygonal region. The cross-section of a cylindrical radiation beam is shaped by such a region. All IM cells exposed under the radiation beam receive a uniform radiation dose proportional to the exposure time. The mechanical design of the MLCs restricts what kinds of beam-shaping regions are allowed [15]. A common constraint is called the **maximum leaf spread**: No leaf can move away from the vertical center line of the MLC by more than a threshold distance.This means that an MLC cannot enclose an IM of a too large width.

One key criterion used to measure the quality of an IMRT treatment is the **beam-on time**, which is the total time while a patient is exposed to actual radiation irradiation. Minimizing the beam-on time without compromising the prescribed dose distributions is an effective way to enhance the efficiency of the radiation machine, referred to as the *monitor-unit (MU) efficiency* in medical literature, and to reduce the patient's risk under irradiation [2]. The beam-on time also constitutes a significant portion of the total treatment time [1,15,16]. Thus, minimizing the beam-on time lowers the treatment cost of each patient and increases the patient throughput of the hospitals.

For example, on one of the most popular MLC systems called Varian, the maximum leaf spread constraint limits the maximum allowed field width to about 15 *cm*. Hence, this necessitates a large-width IM field to be split into two or more adjacent subfields, each of which can be delivered separately by the MLC subject to the maximum leaf spread constraint [8,17]. But, such IM splitting may result in a prolonged beam-on time and thus affect the treatment quality. The **field splitting problem**, roughly speaking, is to split an IM of a large width into several subfields whose widths are all no bigger than a threshold value, such that the total beam-on time of these subfields is minimized.

Geometrically, we distinguish three versions of the field splitting problem based on how an IM is split (see Figures 1(c)-1(e)): (1) splitting using vertical lines; (2) splitting using y-monotone paths; (3) splitting with overlapping. Note that in versions (1) and (2), an IM cell belongs to exactly one subfield; but in version (3), a cell can belong to two adjacent subfields, with a nonnegative value in both subfields, and in the resulting sequence of subfields, each subfield is allowed to overlap only with the subfield immediately before or after it.

Algorithm research for IMRT problems has been active in areas such as medical physics [3,14,19], operations research [1,2,7], computational geometry [4,5,6], and computer algorithms [9,10,11]. Engel [7] showed that for an IM M of size $m \times n$, when n is no larger than the maximum allowed field width, the **minimum beam-on time (MBT)** of M is captured by the following formula

$$MBT(M) = \max_{i=1}^{m}\{M_{i,1} + \sum_{j=2}^{n} \max\{0, M_{i,j} - M_{i,j-1}\}\} \qquad (1)$$

Engel also described a class of algorithms achieving this minimum value. Geometrically, if we view a row of an IM as representing a directed x-monotone rectilinear curve f, called the *dose profile* (see Figure 2(a)), then the MBT of that IM row is actually the total sum of lengths of all upward edges on f. Formula (1) may be explained as follows: Each IM row is handled by one pair of MLC leaves; for an upward edge e, say, of length l, on the dose profile of an IM row, the tip of the left MLC leaf for this row must stay at the x-coordinate of e for at least l time units (assuming one unit of dose is delivered in one unit of time) while the beam is on, so that the dose difference occured here is l.

A few known field splitting algorithms aim to minimize the *total MBT*, i.e., the sum of the MBTs of the resulting subfields. Kamath *et al.* [11] first gave an $O(mn^2)$ time algorithm to split a size $m \times n$ IM using vertical lines into at most 3 subfields (i.e., $n \leq 3w$ for their algorithm, where w is the maximum allowed field width). Wu [18] formulated the problem of splitting an IM of an arbitrary width into $k \geq 3$ subfields using vertical lines as a k-link shortest path problem, giving an $O(mnw)$ time algorithm. Recently, Kamath *et al.* studied the important problem of splitting an IM into at most three *overlapping* subfields, minimizing the total MBT of the resulting subfields [9][1]; a greedy algorithm is proposed that produces optimal solutions in $O(mn)$ time when the overlapping regions of the subfields are fixed. (In fact, by considering all possible overlapping regions, we can extend their algorithm to solving the field splitting with overlapping problem in $O(mn\Delta^{d-2})$ time, where d is the number of resulting subfields

---

[1] The paper [9] was among the ten John R. Cameron Young Investigator Competition finalists of the *47th Annual Meeting of the American Association of Physicists in Medicine* (AAPM'2005) over more than 1100 submissions.

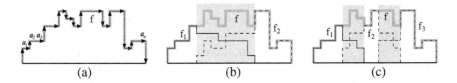

Fig. 2. (a) The dose profile of an IM of one row. The MBT (minimum beam-on time) of the IM is equal to the sum of the lengths of all upward edges on the curve. (b) Splitting the IM in (a) into two subfields with overlapping. (c) Splitting the IM in (a) into three subfields with overlapping.

and $\Delta = w\lceil n/w\rceil - n + 1$.) To our best knowledge, no algorithms are known so far for computing an optimal field splitting of IMs using y-monotone paths.

In this paper, we give efficient algorithms for the following problems.

(1) **Field splitting with overlapping (FSO) problem**: Given an IM M of size $m \times n$ and a maximum field width $w > 0$, split M into $d = \lceil \frac{n}{w}\rceil$ *overlapping* subfields $M_1, M_2, \ldots, M_d$, each with a width $\leq w$, such that the total MBT of these d subfields is minimized (e.g., Figures 2(b)-2(c)). Here, d is the minimum number of subfields required to deliver M subject to the maximum field width w. We assume that d is a constant, i.e., $w = O(n)$, which is clinically practical (under the current clinical settings, $d \leq 3$). Our FSO algorithm takes $O(mn + m\Delta^{d-2})$ time, where $\Delta = w\lceil n/w\rceil - n + 1$. This improves the time bound of Kamath *et al.*'s algorithm [9] by a factor of $\min\{\Delta^{d-2}, n\}$.

(2) **Field splitting using y-monotone paths (FSMP) problem**: Given an IM M of size $m \times n$ and a maximum field width $w > 0$, split M using y-monotone paths into $d = \lceil \frac{n}{w}\rceil$ subfields $M_1, M_2, \ldots, M_d$, each with a width $\leq w$, such that the total MBT is minimized. Again, we assume $d = O(1)$. Our FSMP algorithms take $O(mn + m\Delta \log m)$ time for the case with $d = 2$, and $O(mn + m^{d-2}\Delta^{d-1}\log(m\Delta))$ time with $d \geq 3$.

The FSO problem seems quite non-trivial. Our key ideas are to reduce it to simpler and simpler cases of the problem, as summarized below.

(1) At the lowest level, we consider a very basic case of the FSO problem, called the *basic row splitting (BRS) problem*, which seeks to optimally split an IM of size $1 \times l$ into two overlapping subfields with fixed sizes and positions. Geometrically, the BRS problem partitions the region enclosed between an x-monotone rectilinear dose curve f and the x-axis into two regions, each enclosed by an x-monotone rectilinear curve f_i ($i = 1, 2$) and the x-axis, such that (i) $f_1(x) + f_2(x) = f(x)$ for all $x \in \mathbb{R}$ (we assume $g(x) = 0$ for any x outsides the domain interval of a function g), and (ii) the sum of lengths of all upward edges on f_1 and f_2 is minimized (see Figure 2(b)). Although this basic case can be solved by Kamath *et al.*'s algorithm [9], we take a very different approach in order to exploit some interesting geometric structures. Our approach formulates this problem as a shortest path problem on a direct acyclic graph (DAG) G of a pseudo-polynomial size. By utilizing the properties of this DAG, we obtain an optimal $O(l)$ time BRS algorithm without explicitly constructing and searching

G. Further, we prove that all the *MBT tuples* (i.e., the i-th element of an MBT tuple is the MBT of the i-th subfield) induced by the optimal solutions of the BRS problem form a set of lattice points on a line segment in $\mathbb{R}^2$. Moreover, for any lattice point on this line segment, we can obtain a corresponding optimal BRS solution in $O(l)$ time, by finding a *restricted* shortest path in G.

(2) At the second level, we study the *general row splitting (GRS) problem*: Given a d-tuple $\tau \in \mathbb{Z}_+^d$, split a size $1 \times n$ IM into d overlapping subfields of fixed sizes and positions, such that the resulting MBT tuple $t \leq \tau$, i.e., $t_k \leq \tau_k$ for each $k = 1, 2, \ldots, d$. Observe that there are $d-1$ overlapping regions, each involving a BRS problem instance (see Figure 2(c)). Using the geometric properties revealed by our BRS algorithm, we prove that the GRS problem has a solution if and only if $\tau \in P$, where P is a convex polytope in $\mathbb{R}^d$. Also, for any lattice point $\tau \in P$, the GRS problem is solvable in $O(n)$ time.

(3) At the third level, we solve the *field splitting with fixed overlapping (FSFO) problem* on a size $m \times n$ IM. Basically, this is the FSO problem subject to the constraint that the sizes and positions of the d sought subfields are all fixed. This problem is closely related to the GRS problem, and can be transformed to an integer linear programming (ILP) problem for which we seek an optimal lattice point in the common intersection of m convex polytopes in $\mathbb{R}^d$. Interestingly, the constraint matrix of this ILP problem is totally unimodular, and thus the ILP can be solved optimally by linear programming (LP). Further, the dual of this LP turns out to be a shortest path problem on a DAG of $O(d)$ vertices, which implies that the original ILP is solvable in $O(1)$ time (assume $d = O(1)$). To transform an optimal ILP solution (i.e., a d-D lattice point) back to an optimal FSFO solution, we apply our GRS algorithm $O(m)$ times. Hence, we solve the FSFO problem in totally $O(mn)$ time.

(4) At the top level, we reduce the original FSO problem to a set of $O(\Delta^{d-2})$ FSFO problem instances, where $\Delta = w\lceil n/w \rceil - n + 1$. We show that under our ILP formulation of the FSFO problem, with an $O(mn)$ time preprocess, each FSFO instance is solvable in only $O(m)$ time. (Note that in Kamath *et al.*'s algorithm [9], it takes $O(mn)$ time to solve each FSFO instance due to their greedy approach.) This gives an $O(mn + m\Delta^{d-2})$ time FSO algorithm.

For the FSMP problem on a size $m \times n$ IM, the approaches for splitting using vertical lines [11,18] do not seem to work, since there are $O((\Delta + 1)^m)$ candidate paths for each of the $d - 1$ sought y-monotone splitting paths. Our key observation is that only $m\Delta + 1$ candidates for the leftmost y-monotone path used in the splitting need to be considered. We then show that all these $m\Delta + 1$ candidate paths can be enumerated efficiently by using a heap data structure and an interesting new method called *MBT-sweeping*. Further, we show how to exploit the geometric properties of the FSMP problem to speed up the computation of the total MBT of the induced subfields.

Due to the space constraints, we omit the proofs of lemmas and theorems and leave our detailed discussion on the FSMP problem to the full paper.

2   Field Splitting with Overlapping (FSO)

2.1   Notation and Definitions

We define the function $C(x) = \sum_{j=2}^{l} \max\{0, x_j - x_{j-1}\}$, for $x = (x_1, x_2, \dots, x_l) \in \mathbb{R}^l$. It is easy to show that $x_1 + C(x)$ is the MBT of an IM with only one row x, when no splitting is applied.

We say that intervals $[\mu_1, \nu_1], [\mu_2, \nu_2], \dots, [\mu_d, \nu_d]$ $(d \geq 2)$ form an *interweaving list* if $\mu_1 < \mu_2 \leq \nu_1 < \mu_3 \leq \nu_2 < \mu_4 \leq \cdots < \mu_{k+1} \leq \nu_k < \cdots < \mu_d \leq \nu_{d-1} < \nu_d$. For a subfield S restricted to start from column $\mu + 1$ and end at column ν, we call $[\mu, \nu]$ the *bounding interval* of S. We say that subfields $S_1, S_2, \dots, S_d$ form a *chain* if their corresponding bounding intervals form an interweaving list (i.e., subfields S_k and S_{k+1} are allowed to overlap, $k = 1, 2, \dots, d-1$). Further, $t = (t_1, t_2, \dots, t_d) \in \mathbb{Z}_+^d$ is called the *MBT tuple* of a chain of d subfields $S_1, S_2, \dots, S_d$ if t_k is the MBT of S_k $(k = 1, 2, \dots, d)$. For $t = (t_1, t_2, \dots, t_d)$ and $\tau = (\tau_1, \tau_2, \dots, \tau_d)$, we say $t \leq \tau$ if $t_k \leq \tau_k$ for every $k = 1, 2, \dots, d$.

2.2   The Basic Case: The Basic Row Splitting (BRS) Problems

Precisely, the **basic row splitting (BRS) problem** is: Given a vector (row) $a = (a_1, a_2, \dots, a_l) \in \mathbb{Z}_+^l$ $(l \geq 3)$, seek $\min_{x,y}\{C(x) + C(y)\}$, subject to: (1) $x = (x_1, x_2, \dots, x_l), y = (y_1, y_2, \dots, y_l) \in \mathbb{Z}_+^l$, (2) $x + y = a$, and (3) $x_1 = 0, x_l = a_l, y_1 = a_1, y_l = 0$. Observe that for any feasible solution (x, y), the total MBT of x and y is $C(x) + C(y) + a_1$. Hence the BRS problem is to partition $(a_1, a_2, \dots, a_l)$ into $(0, x_2, x_3, \dots, x_{l-1}, a_l)$ and $(a_1, y_2, y_3, \dots, y_{l-1}, 0)$ such that their total MBT is minimized. Observe that the range of x (resp., y) is actually restricted to interval $[2, l]$ (resp., $[1, l-1]$). Note that the more general case, in which we seek an optimal partition of $(a_1, a_2, \dots, a_l)$ into x and y such that the range of x (resp., y) is restricted to the interval $[p, l]$ (resp., $[1, q]$), where p, q are fixed and $1 < p < q < l$, can be easily reduced to the above version of the BRS problem.

Denote by $BRS(a)$ the BRS problem on a vector a. The BRS problem can be solved by the greedy algorithm in [9]. However, we are more interested in the useful structures of *all* optimal solutions for the BRS problem. As it turns out, the set of all optimal solutions can be mapped to a set of lattice points on a line segment in a plane. To show this, we need to study the following **restricted BRS problem**: Given a vector (row) $a \in \mathbb{Z}_+^l$ and an integer $t \geq 0$, find (x, y) such that (x, y) is an optimal solution of $BRS(a)$ and $C(x) = t$.

Our main idea is to model the BRS (or restricted BRS) problem on $a \in \mathbb{Z}_+^l$ as a shortest path (SP) (or restricted SP) problem on a DAG of a pseudo-polynomial size. By exploiting a set of interesting properties of this DAG, we are able to solve both the SP and restricted SP problems in an optimal $O(l)$ time without explicitly constructing and searching the graph. In the meantime, we will establish the geometric structures of the set of all optimal solutions for any BRS problem. We summarize our results (proofs left to the full paper) below.

Theorem 1. *Given a vector* $a = (a_1, a_2, \ldots, a_l) \in \mathbb{Z}_+^l$ *($l \geq 3$), BRS(a) has an the optimal objective function value* $\rho(a) = \max\{a_l, \sum_{j=2}^l \max\{0, a_j - a_{j-1}\}\}$. *Moreover, we can compute in* $O(l)$ *time an optimal solution* (x^*, y^*) *for* $BRS(a)$.

Theorem 2. *Given a vector* $a = (a_1, a_2, \ldots, a_l) \in \mathbb{Z}_+^l$ *($l \geq 3$), and an integer* t, *the restricted BRS problem on* a *and* t *has a solution if and only if* $t \in [a_l, \rho(a)]$, *where* $\rho(a) = \max\{a_l, \sum_{j=2}^l \max\{0, a_j - a_{j-1}\}\}$. *Moreover, for any integer* $t \in [a_l, \rho(a)]$, *we can solve the restricted BRS problem in* $O(l)$ *time.*

Theorem 2 implies that the MBT tuples induced by all optimal solutions of $BRS(a)$ form a set of lattice points on a line segment $\{(t, \rho(a) - t) \mid a_l \leq t \leq \rho(a)\}$ in $\mathbb{R}^2$. The corollary below thus follows.

Corollary 1. *Given a vector* $a \in \mathbb{Z}_+^l$ *($l \geq 3$), an integer* t, *and a feasible solution* (x, y) *of* $BRS(a)$, *we can compute in* $O(l)$ *time an optimal solution* $(\bar{x}, \bar{y})$ *of* $BRS(a)$, *such that* $(C(\bar{x}), C(\bar{y})) \leq (C(x), C(y))$.

2.3   The General Row Splitting (GRS) Problem

Precisely, the **general row splitting (GRS) problem** is: Given a vector (row) $\alpha \in \mathbb{Z}_+^n$, a d-tuple $\tau \in \mathbb{Z}_+^d$ ($d \geq 2$), and an interweaving interval list IL: $I_k = [\mu_k, \nu_k]$, $k = 1, 2, \ldots, d$, with $\mu_1 = 0$ and $\nu_d = n$, split α into a chain of d subfields $S_1, S_2, \ldots, S_d$ such that the bounding interval of S_k is I_k ($k = 1, 2, \ldots, d$) and the MBT tuple t of the resulting subfield chain satisfies $t \leq \tau$.

We seek to answer two questions: (1) When does the GRS problem have a solution? (2) If the GRS problem has a solution, how to find such a solution?

Note that the resulting d subfields have $d - 1$ overlapping regions, each of which involves an instance of the basic row splitting problem. Define $a^{(k)} = (\alpha_{\mu_{k+1}}, \alpha_{\mu_{k+1}+1}, \ldots, \alpha_{\nu_k+1}) \in \mathbb{Z}_+^{l_k}$ ($l_k \triangleq \nu_k - \mu_{k+1} + 2$) for $k = 1, 2, \ldots, d-1$, and $c_k = \sum_{i=\nu_{k-1}+2}^{\mu_{k+1}} \max\{0, \alpha_i - \alpha_{i-1}\}$ for $k = 1, 2, \ldots, d$ (we assume $\nu_0 = -1, \alpha_0 = 0$, and $\mu_{d+1} = n$). It is easy to show that the GRS problem is equivalent to the following problem (denoted by **GRS'**): Find $x^{(k)}, y^{(k)} \in \mathbb{Z}_+^{l_k}$, $k = 1, 2, \ldots, d-1$, subject to: (1) $x^{(k)} + y^{(k)} = a^{(k)}$, for $k = 1, 2, \ldots, d-1$, (2) $x_1^{(k)} = 0, x_{l_k}^{(k)} = a_{l_k}^{(k)}$, $y_1^{(k)} = a_1^{(k)}$, and $y_{l_k}^{(k)} = 0$, for $k = 1, 2, \ldots, d-1$, and (3) $C(y^{(1)}) + c_1 \leq \tau_1$, $C(x^{(d-1)}) + c_d \leq \tau_d$, and $C(x^{(k-1)}) + C(y^{(k)}) + c_k \leq \tau_k$, for $1 < k < d$.

Constraints (1) and (2) mean that for each $k = 1, 2, \ldots, d-1$, $(x^{(k)}, y^{(k)})$ is a basic row splitting solution for $a^{(k)}$, occurring in the k-th overlapping region. Since $C(x^{(k-1)}) + C(y^{(k)}) + c_k$ is the MBT of the k-th subfield, constraint (3) means that the MBT tuple t of the resulting subfields satisfies $t \leq \tau$.

Denote by $GRS(\alpha, \tau, IL)$ (or $GRS'(\alpha, \tau, IL)$) the GRS (or GRS') problem on the instance α, τ, and IL. We say that a solution $(x^{(1)}, y^{(1)}, \ldots, x^{(d-1)}, y^{(d-1)})$ of $GRS'(\alpha, \tau, IL)$ is *primal* if for every $k \in \{1, 2, \ldots, d-1\}$, $(x^{(k)}, y^{(k)})$ is an optimal solution of $BRS(a^{(k)})$. By Corollary 1, we can prove the next lemma.

Lemma 1. *For any* α, τ, *and* IL, $GRS'(\alpha, \tau, IL)$ *has a solution only if it has a primal solution.*

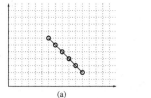 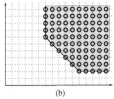

(a) (b)

Fig. 3. (a) Illustrating the convex polytope $Q \subset \mathbb{R}^2$ when $d = 2$ (Q is the line segment in (a)). The lattice points in Q are marked as circles, and form the set $Q \cap \mathbb{Z}^2$. (b) Illustrating the Minkowski sum $(Q \cap \mathbb{Z}^2) \oplus \mathbb{Z}_+^2$, which consists of all the points marked as circles. The darkened region is the convex polytope $P \subset \mathbb{R}^2$ with $P \cap \mathbb{Z}^2 = (Q \cap \mathbb{Z}^2) \oplus \mathbb{Z}_+^2$.

Define $\lambda_k = \sum_{q=1}^{k-1} \rho(a^{(q)}) + \sum_{q=1}^{k} c_q$ (for $k = 1, 2, \ldots, d$), and convex polytope

$$Q = \left\{ (t_1, t_2, \ldots, t_d) \in \mathbb{R}^d \;\middle|\; \begin{array}{l} \sum_{q=1}^{d} t_q = \lambda_d, \\ \lambda_k \leq \sum_{q=1}^{k} t_q \leq \lambda_k + \rho(a^{(k)}) - a_{l_k}^{(k)}, 1 \leq k \leq d-1 \end{array} \right\}.$$

Q is closely related to the GRS' problem, as stated in the next lemma.

Lemma 2. $GRS'(\alpha, \tau, IL)$ has a primal solution if and only if $\tau \in (Q \cap \mathbb{Z}^d) \oplus \mathbb{Z}_+^d$, where $\oplus$ is the Minkowski sum defined as $A \oplus B = \{a + b \mid a \in A, b \in B\}$ on sets A and B (see Figure 3). Moreover, given any $\tau \in (Q \cap \mathbb{Z}^d) \oplus \mathbb{Z}_+^d$, a primal solution of $GRS'(\alpha, \tau, IL)$ can be computed in $O(n)$ time.

Geometrically, Lemma 2 states that $GRS'(\alpha, \tau, IL)$ has a primal solution if and only if the given point (vector) $\tau \in \mathbb{Z}^d$ "dominates" some lattice point in the polytope $Q \subset \mathbb{R}^d$, or equivalently, $\tau \in (Q \cap \mathbb{Z}^d) \oplus \mathbb{Z}_+^d$ (see Figure 3). The set $(Q \cap \mathbb{Z}^d) \oplus \mathbb{Z}_+^d$, in fact, consists of all lattice points in another convex polytope $P \subset \mathbb{R}^d$ according to the following lemma.

Lemma 3. Let $\eta = (\eta_1, \eta_2, \ldots, \eta_d)$ and $\xi = (\xi_1, \xi_2, \ldots, \xi_d)$ be two vectors in $\mathbb{Z}^d$ with $\eta \leq \xi$, and let $Z = \left\{ (t_1, t_2, \ldots, t_d) \in \mathbb{Z}^d \;\middle|\; \eta_k \leq \sum_{q=1}^{k} t_q \leq \xi_k, \forall k : 1 \leq k \leq d \right\}$, and $W = \left\{ (t_1, t_2, \ldots, t_d) \in \mathbb{Z}^d \;\middle|\; \sum_{q=k}^{k'} t_q \geq \eta_{k'} - \xi_{k-1}, \forall (k, k') : 1 \leq k \leq k' \leq d \right\}$ (for convenience, $\xi_0 \triangleq 0$). Then $Z \oplus \mathbb{Z}_+^d = W$. Moreover, given any $w \in W$, we can find a $z \in Z$ and $\delta \in \mathbb{Z}_+^d$ in $O(d)$ time, such that $w = z + \delta$.

Note that $\rho(a^{(k)}) = \max\{a_{l_k}^{(k)}, \sum_{j=2}^{l_k} \max\{0, a_j^{(k)} - a_{j-1}^{(k)}\}\} = \max\{\alpha_{\nu_k+1}, \sum_{j=\mu_k+1+1}^{\nu_k+1} \max\{0, \alpha_j - \alpha_{j-1}\}\}$. After an $O(n)$ time preprocess on α, given any IL, we can compute all c_k's, $\rho(a^{(k)})$'s, and λ_k's in $O(d)$ time. Hence the polytope Q as well as the set $(Q \cap \mathbb{Z}^d) \oplus \mathbb{Z}_+^d$, by Lemma 3, can be computed in $O(d)$ time. Combining this with Lemmas 1 and 2, we have the result below.

Theorem 3. $GRS(\alpha, \tau, IL)$ has a solution if and only if $\tau \in P \cap \mathbb{Z}^d$, where $P = \{(t_1, t_2, \ldots, t_d) \in \mathbb{R}^d \mid \sum_{q=k}^{k'} t_q \geq b_{k,k'}, \forall (k, k') : 1 \leq k \leq k' \leq d\}$, and $b_{k,k'}$'s are integers depending only on α and IL. Moreover, given any $\tau \in P \cap \mathbb{Z}^d$, we can solve $GRS(\alpha, \tau, IL)$ in $O(n)$ time.

2.4 The Field Splitting with Fixed Overlapping (FSFO) Problem

In this section, we study a special case of the field splitting with overlapping problem, i.e., the sizes and positions of the subfields are all fixed. Precisely, the **field splitting with fixed overlapping (FSFO) problem** is: Given an IM M of size $m \times n$ and an interweaving list IL of d intervals $I_1, I_2, \ldots, I_d$, split M into a chain of d subfields $M_1, M_2, \ldots, M_d$, such that I_k is the bounding interval of M_k ($k = 1, 2, \ldots, d$) and the total MBT of the d resulting subfields is minimized.

Recall that the MBT of a subfield M_k is the maximum MBT over each of the m rows of M_k (see Formula (1)). For any feasible solution of the FSFO problem, let τ be the corresponding MBT tuple. Then for any $i \in \{1, 2, \ldots, m\}$, $GRS(\alpha_i, \tau, IL)$ must have a solution, where α_i denotes the i-th row of M. By Theorem 3, we have $\tau \in P_i \cap \mathbb{Z}^d$, where P_i is a convex polytope of the form of $\left\{ (t_1, t_2, \ldots, t_d) \in \mathbb{R}^d \left| \sum_{q=k}^{k'} t_q \geq b_{k,k'}^{(i)}, \forall (k, k') : 1 \leq k \leq k' \leq d \right. \right\}$. Hence, the FSFO problem can be transformed to the following integer linear programming (ILP) problem: seek $\min \sum_{k=1}^{d} \tau_k$, subject to $(\tau_1, \tau_2, \ldots, \tau_d) \in P \cap \mathbb{Z}^d$, where $P = \bigcap_{i=1}^{m} P_i = \left\{ (t_1, t_2, \ldots, t_d) \in \mathbb{R}^d \left| \sum_{q=k}^{k'} t_q \geq \max_{i=1}^{m} b_{k,k'}^{(i)}, \forall (k, k') : 1 \leq k \leq k' \leq d \right. \right\}$.

Note that the constraint matrix is a (0,1) interval matrix, and is thus totally unimodular [12]. Hence, this ILP can be solved as a linear programming (LP) problem. We introduce a variable π_0, and define $\pi_k = \pi_0 + \sum_{q=1}^{k} \tau_q$, for $k = 1, 2, \ldots, d$. Then the above LP is equivalent to seeking $\max \pi_0 - \pi_d$, subject to $\pi_k - \pi_{k'} \leq w_{k,k'}, \forall 0 \leq k < k' \leq d$, where $w_{k,k'} = -\max_{i=1}^{m} b_{k+1,k'}^{(i)}$. As shown in [13], the dual LP of this kind of LP is an s-t shortest path (SP) problem on a graph G, where G has $d+1$ vertices $v_0 (= s), v_1, \ldots, v_d (= t)$, and for every pair (k, k') with $k < k'$, there is an edge from v_k to v'_k with a weight $w_{k,k'}$. Clearly, G is a DAG with $O(d)$ vertices and $O(d^2)$ edges. Therefore, this SP problem, and further, the original ILP can both be solved in $O(d^2) = O(1)$ time, assuming $d = O(1)$. Moreover, with an optimal MBT tuple $\tau \in \mathbb{Z}_+^d$, by Theorem 3 we can obtain the corresponding optimal splitting of M in $O(mn)$ time. Thus, the FSFO problem can be solved optimally in $O(mn)$ time.

2.5 Our Field Splitting with Overlapping (FSO) Algorithm

We now study the FSO problem on a size $m \times n$ IM M. Our observation is that we can always assume that the size of each subfield is $m \times w$, where w is the maximum field width. This is because we can introduce columns filled with 0's to the subfield without increasing its total MBT. Also, note that among the d sought subfields, the leftmost and rightmost ones are fixed. Based on these observations, it is sufficient to consider only $O(\Delta^{d-2})$ possible subfield chains, where $\Delta = n - w\lfloor n/w \rfloor + 1$. Therefore, the FSO problem can be solved by solving $O(\Delta^{d-2})$ FSFO problems, to find an optimal MBT tuple τ. Further, we can show that with an $O(mn)$ time preprocess on M, an optimal MBT tuple for each FSFO instance can be computed in $O(m)$ time. Thus, the FSO problem is solvable in $O(mn + m\Delta^{d-2})$ time for $d = O(1)$.

3 Implementation and Experiments

We implemented our new field splitting algorithms FSMP and FSO using C on Linux, and experimented with them on 58 IMs of various sizes for 11 clinical cases obtained from the Dept. of Radiation Oncology, Univ. of Maryland.Our FSMP and FSO programs run in only a few seconds (mostly under one second). We conducted comparisons with a most popular commercial treatment planning software CORVUS 5.0, the field splitting algorithm in [11,18] (denoted by FSSL), which splits along vertical lines, and the algorithm by Kamath *et al.* [9] (denoted by FSO_K), which splits with overlapping. Specifically, we examined the total beam-on time of the output subfields of these splitting approaches. The widths of the tested IMs range from 15 to 31, and the maximum allowed subfield width is 14. The maximum intensity level of each IM is normalized to 100.

For each IM, algorithms FSO_K and FSO produce exactly the same result in terms of the beam-on time, due to the fact that both these two algorithms obtain optimal splittings. (Theoretically, our FSO algorithm considerably outperforms FSO_K in terms of the running time.) For each IM, the total beam-on times of the four methods, CORVUS 5.0, FSSL, FSMP, and FSO, are always in decreasing order. For all 58 tested IMs, the sums of the total beam-on times of these four methods are 21838, 19511, 18159, and 17221, respectively. In terms of the beam-on time, our FSMP algorithm showed an improvement of 17% over CORVUS 5.0, and 7% over the FSSL algorithm [11,18]. Our FSO algorithm gave the best splitting results, and on average reduced the total beam-on time by 21% over CORVUS 5.0, and 12% over the FSSL algorithm.

References

1. R.K. Ahuja and H.W. Hamacher. A Network Flow Algorithm to Minimize Beam-on Time for Unconstrained Multileaf Collimator Problems in Cancer Radiation Therapy. *Networks*, 45:36–41, 2005.
2. N. Boland, H.W. Hamacher, and F. Lenzen. Minimizing Beam-on Time in Cancer Radiation Treatment Using Multileaf Collimators. *Networks*, 43(4):226–240, 2004.
3. T.R. Bortfeld, A.L. Boyer, W. Schlegel, D.L. Kahler, and T.L. Waldron. Realization and Verification of Three-Dimensional Conformal Radiotherapy with Modulated Fields. *Int. J. Radiat. Oncol. Biol. Phys.*, 30:899–908, 1994.
4. D.Z. Chen, X.S. Hu, S. Luan, S.A. Naqvi, C. Wang, and C. Yu. Generalized Geometric Approaches for Leaf Sequencing Problems in Radiation Therapy. *Int. Journal of Computational Geometry and Applications*, 16(2-3):175–204, 2006.
5. D.Z. Chen, X.S. Hu, S. Luan, C. Wang, and X. Wu. Geometric Algorithms for Static Leaf Sequencing Problems in Radiation Therapy. In *Proc. of 19th ACM Symposium on Computational Geometry*, pages 88–97, 2003.
6. D.Z. Chen, X.S. Hu, S. Luan, C. Wang, and X. Wu. Mountain Reduction, Block Matching, and Applications in Intensity-Modulated Radiation Therapy. In *Proc. of 21th ACM Symposium on Computational Geometry*, pages 35–44, 2005.
7. K. Engel. A New Algorithm for Optimal Multileaf Collimator Field Segmentation. *Discrete Applied Mathematics*, 152(1-3):35–51, 2005.

8. L. Hong, A. Kaled, C. Chui, T. Losasso, M. Hunt, S. Spirou, J. Yang, H. Amols, C. Ling, Z. Fuks, and S. Leibel. IMRT of Large Fields: Whole-Abdomen Irradiation. *Int. J. Radiat. Oncol. Biol. Phys.*, 54:278–289, 2002.

9. S. Kamath, S. Sahni, J. Li, J. Palta, and S. Ranka. A Generalized Field Splitting Algorithm for Optimal IMRT Delivery Efficiency. *The 47th Annual Meeting and Technical Exhibition of the American Association of Physicists in Medicine (AAPM)*, 2005. Also, *Med. Phys.*, 32(6):1890, 2005.

10. S. Kamath, S. Sahni, J. Palta, and S. Ranka. Algorithms for Optimal Sequencing of Dynamic Multileaf Collimators. *Phys. Med. Biol.*, 49(1):33–54, 2004.

11. S. Kamath, S. Sahni, S. Ranka, J. Li, and J. Palta. Optimal Field Splitting for Large Intensity-Modulated Fields. *Med. Phys.*, 31(12):3314–3323, 2004.

12. G.L. Nemhauser and L.A. Wolsey. *Integer and Combinatorial Optimization*. John Wiley, 1988.

13. C. H. Papadimitriou and K. Steiglitz. *Combinatorial Optimization: Algorithms and Complexity*. Prentice-Hall, New Jersey, 1982.

14. R.A.C. Siochi. Minimizing Static Intensity Modulation Delivery Time Using an Intensity Solid Paradigm. *Int J. Radiation Oncology Biol. Phys.*, 43(3):671–680, 1999.

15. S. Webb. *The Physics of Three-Dimensional Radiation Therapy*. Bristol, Institute of Physics Publishing, 1993.

16. S. Webb. *The Physics of Conformal Radiotherapy — Advances in Technology*. Bristol, Institute of Physics Publishing, 1997.

17. Q. Wu, M. Arnfield, S. Tong, Y. Wu, and R. Mohan. Dynamic Splitting of Large Intensity-Modulated Fields. *Phys. Med. Biol.*, 45:1731–1740, 2000.

18. X. Wu. Efficient Algorithms for Intensity Map Splitting Problems in Radiation Therapy. In *Lecture Notes in Computer Science, Springer-Verlag, Proc. 11th Annual International Computing and Combinatorics Conference*, volume 3595, pages 504–513, 2005.

19. P. Xia and L.J. Verhey. MLC Leaf Sequencing Algorithm for Intensity Modulated Beams with Multiple Static Segments. *Med. Phys.*, 25:1424–1434, 1998.

Shape Rectangularization Problems in Intensity-Modulated Radiation Therapy[*]

Danny Z. Chen[1], Xiaobo S. Hu[1], Shuang Luan[2,**],
Ewa Misiołek[3], and Chao Wang[1,***]

[1] Department of Computer Science and Engineering
University of Notre Dame
Notre Dame, IN 46556, USA
{chen, shu, cwang1}@cse.nd.edu
[2] Department of Computer Science
University of New Mexico
Albuquerque, NM 87131-0001, USA
sluan@cs.unm.edu
[3] Mathematics Department
Saint Mary's College
Notre Dame, IN 46556, USA
misiolek@saintmarys.edu

Abstract. In this paper, we present a theoretical study of several geometric shape approximation problems, called *shape rectangularization (SR)*, which arise in *intensity-modulated radiation therapy* (IMRT). Given a piecewise linear function f such that $f(x) \geq 0$ for any $x \in \mathbb{R}$, the SR problems seek an optimal set of *constant window functions* to approximate f under a certain error criterion, such that the sum of the resulting constant window functions equals (or well approximates) f. A constant window function $W(\cdot)$ is defined on an interval I such that $W(x)$ is a fixed value $h > 0$ for any $x \in I$ and is 0 otherwise. Geometrically, a constant window function can be viewed as a rectangle (or a *block*). The SR problems find applications in micro-MLC scheduling and dose calculation of the IMRT treatment planning process, and are closely related to some well studied geometric problems. The SR problems are NP-hard, and thus we aim to develop theoretically efficient and provably good quality approximation SR algorithms. Our main results include a polynomial time $(\frac{3}{2} + \epsilon)$-approximation algorithm for a general key SR problem and an efficient dynamic programming algorithm for an important SR case that has been studied in medical literature. Our key ideas include the following. (1) We show that a crucial subproblem of the key SR problem can be reduced to the multicommodity

[*] This research was supported in part by the Faculty Research Program of the University of Notre Dame, the National Science Foundation under Grants CCR-9988468 and CCF-0515203, and NIH NIBIB Grant R01-EB004640-01A2.
[**] The research of this author was supported in part by a faculty start-up fund from the Department of Computer Science, University of New Mexico.
[***] Corresponding author. The research of this author was supported in part by two Fellowships in 2004-2006 from the Center for Applied Mathematics of the University of Notre Dame.

T. Asano (Ed.): ISAAC 2006, LNCS 4288, pp. 701–711, 2006.

demand flow (MDF) problem on a path graph (which has a known $(2+\epsilon)$-approximation algorithm); further, by extending the result of the known $(2+\epsilon)$-approximation MDF algorithm, we develop a polynomial time $(\frac{3}{2}+\epsilon)$-approximation algorithm for our first target SR problem. (2) We show that the second target SR problem can be formulated as a k-MST problem on a certain geometric graph G; based on a set of interesting geometric observations and a non-trivial dynamic programming scheme, we are able to compute an optimal k-MST in G efficiently.

1   Introduction

In this paper, we present a theoretical study of several geometric shape approximation problems, called *shape rectangularization (SR)*, which arise in *intensity-modulated radiation therapy* (IMRT). IMRT is a modern cancer treatment technique that aims to deliver highly conformal prescribed radiation dose distributions, called *intensity maps* (IMs), to target tumors while sparing the surrounding normal tissues and critical structures. An IM is specified by a set of nonnegative integers on a uniform 2-D grid (see Figure 1(a)). The value in each grid cell indicates the intensity level of the prescribed dose at the body region corresponding to that IM cell. The delivery is done by a set of cylindrical radiation beams orthogonal to the IM grid.

Currently, an IM is delivered using the *multileaf collimator* (MLC) [23, 24]. An MLC consists of many pairs of metal leaves of the same rectangular shape and size (see Figure 1(b)). These metal leaves can move left or right to form a y-monotone rectilinear polygonal beam-shaping region. The cross-section of a cylindrical radiation beam is shaped by such a region. Note that each IM row is handled by one MLC leaf pair. A special type of MLC, called *micro-MLC*, was introduced in recent years. Compared with an ordinary MLC, the micro-MLC has a much smaller leaf width (3-5 *mm*) and hence yields a higher accuracy [5, 6]. Currently, micro-MLC is often used in dynamic conformal arc therapy [15], during which its leaves are continuously moving and dynamically adjusting the beam shape while the radiation beam source traverses along a path (an arc in 3-D). Reducing the number of arcs used in each treatment is a key to reducing the total treatment time of the dynamic conformal arc therapy [17]. During our visits to and research at University of Maryland School of Medicine in 2004-2005, we observed that the scheduling of micro-MLC treatments can be modeled by the following shape rectangularization problem.

Let B be a set of *blocks* in which each block b_i is defined by an interval I_i and an integer height $h_i > 0$ (b_i can be viewed as a rectangle). Let C_B denote the rectilinear upper boundary curve of the area in $\mathbb{R}^2$ resulted by "stacking up" all blocks of B on the x-axis (some parts of the blocks may "fall" to the lower levels; see Figure 1(c)), i.e., for any $x \in \mathbb{R}$, $C_B(x) = \sum_{x \in I_i, b_i \in B} h_i$ (the sum of heights of all blocks in B whose intervals contain x). Precisely, the **shape rectangularization (SR) problem** is: Given an x-monotone rectilinear curve f with integral vertices such that f is on or above the x-axis, find a block set B

such that $C_B = f$ and $|B|$ is minimized. In micro-MLC scheduling applications, f is a dose profile, and each block specifies an arc for the treatment on f.

We are also interested in the **generalized shape rectangularization (GSR) problem**: Given an x-monotone polygonal curve f, find a block set B such that the rectilinear curve C_B well approximates f. There are two versions of the GSR problem: (a) Given an integer $k > 0$, find a block set B such that $|B| \leq k$ and the error between f and C_B is minimized; (b) given an error bound $\mathcal{E}$, find a block set B such that the error between f and C_B is $\leq \mathcal{E}$ and $|B|$ is minimized. These two versions are dual problems to each other; further, it is not hard to see that an efficient solution for (a) leads to an efficient solution for (b). We denote these two versions as **GSR(a)** and **GSR(b)**. Clearly, with an error bound $\mathcal{E} = 0$, the GSR(b) problem on a rectilinear curve f becomes the SR problem.

The GSR problem is important. On one hand, it is a theoretical extension of the SR problem; on the other hand, it has applications to dose calculation in IMRT treatment planning. During the IMRT treatment planning process, the shapes, sizes, and relative positions of a tumor volume and other surrounding tissues are determined by 3-D image data, and an "ideal" dose distribution is computed. Without loss of generality (WLOG), let the z-axis be the beam orientation. Then this "ideal" dose distribution is a function defined on the xy-plane (geometrically, it is a 3-D functional surface above the xy-plane), which is usually not deliverable by the MLC. Thus, the "ideal" distribution must be simplified or converted to a discrete IM, i.e., an IM approximates the "ideal" distribution under certain criteria. Since an IM is delivered using many MLC leaf pairs, with one leaf pair for each IM row, it is sufficient (as shown in [7, 11]) to consider using one IM row to approximate each "strip" of the "ideal" distribution's 3-D surface, which is exactly what the GSR problem models.

Due to the radiation treatment applications [7, 11, 24], we consider several criteria for measuring the error $\mathcal{E}(C_B, f)$ between the input curve f and C_B: (1) symmetric difference [7]: $\mathcal{E}(C_B, f) = \int |C_B(x) - f(x)| dx$; (2) mean square [11]: $\mathcal{E}(C_B, f) = \int (C_B(x) - f(x))^2 dx$; (3) "no overdose": $\mathcal{E}(C_B, f)$ is defined as in (1) but $C_B(x) \leq f(x)$ is required for every x (for treating tumors that are very close to some vital organs [7]); (4) "no underdose": $\mathcal{E}(C_B, f)$ is as in (1) but $C_B(x) \geq f(x)$ is required for every x (for maximizing the cure of a tumor that is relatively far away from any vital organs [7]).

Both the SR and GSR problems are mathematically fundamental. Note that a block b defined by an interval I and a height h corresponds to a *constant window function* $W_b(x)$ whose value is h for $x \in I$ and 0 otherwise. Thus, the SR (or GSR) problem finds the fewest number of constant window functions such that their sum equals (or well approximates) a given functional curve f.

Therefore, although the SR and GSR problems are derived from medical applications, the problems themselves are of an important theoretical nature. In fact, theoretical studies of IMRT problems have so far been lacking in both the medical and computer science literature. In this paper, we focus on solving these two problems from a theory point of view and present efficient algorithms.

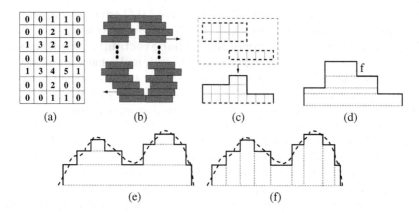

Fig. 1. (a) An IM. (b) An MLC. (c) Illustrating the SR problem: "Stacking up" two blocks to form a given (solid) rectilinear curve. (d) A horizontal trapezoidal decomposition of f naturally yields a solution for the SR problem. (e) The FBT problem: The dashed curve is for an "ideal" dose distribution "strip", and the solid rectilinear curve is C_B defined by a set B of 5 blocks. (f) The OQ problem: It uses 9 blocks to approximate the dashed curve in (e).

The SR problem is NP-hard; we can give a proof very similar to that in [10], by reducing to it the knapsack problem [13]. Thus we seek to develop efficient and good quality approximation SR algorithms. It is not hard to show that a horizontal trapezoidal decomposition of the input curve f yields a 2-approximation solution for the SR problem (e.g., see Figure 1(d)). To our best knowledge, we are not aware of any efficient SR algorithms that can achieve an approximation ratio less than 2.

The GSR problem, as a generalization of the SR problem, is clearly NP-hard. The following special case of the GSR(a) problem is interesting: The blocks in the output set B are required to satisfy the *inclusion-exclusion constraint* [7, 11], meaning that for any two distinct blocks $b_i, b_j \in B$, either their intervals do not intersect each other (except possibly at an endpoint), or one interval, say I_i for b_i, fully contains the other interval I_j. Note that blocks satisfying the inclusion-exclusion constraint can be stacked up such that no parts of the blocks fall to the lower levels (see Figures 1(d)-1(e)). Intuitively, such blocks can be viewed as forming a *forest of towers of blocks* (see Figure 1(e)). Thus, we call this case the **forest of block towers (FBT) problem**. The FBT problem has been studied by medical researchers [7, 11] as a version for solving the GSR(a) problem. In fact, we can show that an optimal solution for the FBT problem immediately yields a 2-approximation for the GSR(a) problem. This is because for any solution B of the GSR(a) problem on f, a horizontal trapezoidal decomposition of the curve C_B immediately gives a solution B' for the FBT problem on f, with $|B'| \leq 2|B|$.

Some previous work has been done on the FBT problem in the medical field [7, 11, 24]. An approach based on Newton's method and calculus techniques was used in [11]; but, it works well only when the input curve f has few "peaks", and must handle exponentially many cases as the number of "peaks" of f increases. Such methods did not give any optimality guarantee. The FBT problem is also closely related to some geometric problems. One such problem is called *optimal quantization (OQ)* [8, 25], arising in coding and information theory: Given a 2-D x-monotone curve f and an integer $k > 0$, find a set B of k blocks such that no two distinct blocks overlap in their intervals (except possibly at an endpoint) and the curve C_B approximates f with the minimum error (e.g., see Figure 1(f)). Clearly, the OQ problem is a special case of the FBT problem. In fact, in the worst case, an optimal block set B' for the OQ problem can have a size almost twice the size of an optimal block set B for the FBT problem (see Figures 1(e)-1(f)). Observe that the output of the OQ problem can be represented as a k-node list. Thus, the OQ problem can be modeled as a k-link shortest path problem on a DAG [25]; further, the paths in the DAG for this problem satisfy the Monge property [1, 20], and hence can be computed very efficiently [1, 2, 22, 25]. In contrast, the FBT output can be represented as a k-node forest; thus, the FBT problem can be modeled as a k-**MST problem** [3, 4, 12, 14, 18, 19, 21, 26] on a geometric graph. Although the k-MST problem is NP-hard both on general graphs and in general geometric settings [12, 21, 26], we are able to solve our FBT problem optimally and efficiently.

In this paper, we present a polynomial time $(\frac{3}{2} + \epsilon)$-approximation algorithm for the SR problem. We exploit a set of interesting geometric observations and show that the SR problem is closely related to the **multicommodity demand flow (MDF) problem** [9]: Given a capacitated tree network $T = (V, E, c)$, with an integer capacity $c_e > 0$ on each edge e, and a set of demands, each of which is defined as a flow from a vertex u to another vertex v in T and has an associated integer demand value $d_{u,v} > 0$ and a real valued profit $w_{u,v} > 0$, find a subset S of the demands that can be simultaneously routed without violating any edge capacity of T such that the total profit $w(S)$ is maximized. We prove that for any constant $\mu \geq 2$, if a μ-approximate solution for the MDF problem defined on a *path* is available, then we can find a $(2 - \frac{1}{\mu})$-approximate solution for the SR problem. Chekuri *et al.* [9] gave a polynomial time $(2 + \epsilon)$-approximation algorithm for the MDF problem on a path when the maximum demand d_{max} is $\leq$ the minimum capacity c_{min}. We extend Chekuri *et al.*'s result [9] and give a $(2 + \epsilon)$-approximation algorithm for the MDF problem on a path when $d_{max} \leq \lambda \cdot c_{min}$, where $\lambda > 0$ is any constant. This leads to a $(\frac{3}{2} + \epsilon)$-approximation algorithm for the SR problem when $M_f \leq \lambda \cdot m_f$, where m_f (or M_f) is the global positive minimum (or maximum) of the input curve f. Note that $M_f \leq \lambda \cdot m_f$ is satisfied by all IMs used in current clinical treatments.

For the FBT problem, we present a unified algorithmic approach producing optimal FBT solutions under each of the above four error criteria. Our main idea is as follows. We first show a set of geometric observations, which imply that only a finite set of rectangles needs to be considered as candidates for the sought blocks, and a graph G on such rectangles can be built. We then use a dynamic programming scheme to compute an optimal k-MST in G. Interestingly, in this dynamic programming scheme, computing a k-link shortest path is a repeatedly used key subroutine. Although it is not clear whether the Monge property [1, 20] holds for our particular k-link shortest path problem, this problem is solvable in a much faster manner than a straightforward k-link shortest path algorithm on general graphs [16]. Our FBT algorithm thus obtained runs in $O(k^2 M^3 P^3 + n)$ time, where n (resp., P) is the number of vertices (resp., "peaks") of f (clearly, $P < n$), and M is the value of f's global maximum. In fact, our FBT algorithm is quite practical since k, M, and P are very small for real medical data. Our approach can also be extended to the generalized FBT problem in which the heights of the blocks need not be integers, resulting in an ϵ-approximation algorithm for the generalized FBT problem.

Due to the space constraints, we omit most of the proofs and leave our detailed discussion of the FBT problem to the full paper.

2 Our Shape Rectangularization (SR) Algorithm

In this section, we present our polynomial time $(\frac{3}{2} + \epsilon)$-approximation SR algorithm. We first show that for the SR problem, it is sufficient to consider only a special type of block sets, called *canonical block sets*. Then we show that an optimal block set, if canonical, is isomorphic to a weighted forest. This observation inspires us to study an interesting combinatorial optimization problem, called the *primary block set* (PBS) problem, which is related to the SR problem in that for any $\mu \geq 2$, a μ-approximation PBS algorithm immediately implies a $(2 - \frac{1}{\mu})$-approximation SR algorithm. We further show that the PBS problem can be reformulated, in polynomial time, as a multicommodity demand flow (MDF) problem on a path. By extending Chekuri *et al.*'s algorithm [9], we get a $(2 + \epsilon)$-approximation algorithm for this MDF problem, which leads to our final $(\frac{3}{2} + \epsilon)$-approximation SR algorithm.

2.1 Notation and Preliminaries

For a rectilinear x-monotone curve f that is on or above the x-axis and starts and ends at the x-axis, denote by $A(f)$ the area enclosed between f and the x-axis. We classify the vertical edges of f into two sets: The *left edges* (the interior of $A(f)$ is to the right of each such edge) and *right edges* (defined symmetrically), denoted by $E^L(f)$ and $E^R(f)$, respectively. For any vertical edge e, denote by $x(e)$ the x-coordinate of e. Define the *L-end set* of f, denoted by $left(f)$, as $left(f) \triangleq \{x(e) \mid e \in E^L(f)\}$. The *R-end set* $right(f)$ is defined symmetrically.

For a block b defined by an interval $I = [\alpha, \beta]$ and a height h, we encode it as a 3-tuple (α, β, h); we say that α (or β) is the *left* (or *right*) *end* of block b. For a block set S, define the *L-end set* of S, denoted by $left(S)$, as $left(S) \triangleq \{\alpha \mid (\alpha, \beta, h) \in S\}$. The *R-end set* $right(S)$ is defined symmetrically.

Let f be a rectilinear x-monotone curve, and B be a block set that *builds* f, i.e., $C_B = f$. Clearly, $left(f) \subseteq left(B)$ and $right(f) \subseteq right(B)$. Note that a horizontal trapezoidal decomposition of $A(f)$ naturally induces a block set of a size $\leq |left(f)| + |right(f)| - 1$ that builds f. Thus, the size of an optimal block set (i.e., of the minimum size) that builds f is always $\leq |left(f)| + |right(f)| - 1$.

2.2 Optimality of Canonical Block Sets

Let f be a rectilinear x-monotone curve, and B be a block set that builds f. We say that B is *L-canonical* (or *R-canonical*) if $left(B) = left(f)$ (or $right(B) = right(f)$). Since $left(f) \subseteq left(B)$ always holds, B is L-canonical if and only if $left(B) \subseteq left(f)$. Symmetrically, B is R-canonical if and only if $right(B) \subseteq right(f)$. A block set is said to be *canonical* if it is both L-canonical and R-canonical. Geometrically, for any block in an L-canonical (or R-canonical) block set, the left (or right) endpoint of its corresponding interval is an L-end (or R-end) of f. The next lemma shows that to solve the SR problem on f, it is sufficient to consider only the canonical block sets that build f.

Lemma 1. *Let f be a rectilinear x-monotone curve. For any block set B that builds f, there exists a canonical block set B^* that builds f with $|B^*| \leq |B|$.*

Lemma 1 implies the existence of an optimal block set for the SR problem that is also canonical. We call such a block set a *canonical optimal block set*. In the next subsection, we show that a canonical optimal block set has an interesting geometric structure: It is isomorphic to a forest of weighted trees.

2.3 Geometric Structures of a Canonical Optimal Block Set

Let B be a canonical optimal block set that builds the curve f. Note that for any block $b = (\alpha, \beta, h) \in B$, $\alpha \in left(B) = left(f)$ and $\beta \in right(B) = right(f)$. Construct a weighted bipartite graph, called the *LR-graph* of B, as follows: $\mathcal{G} = (left(f), right(f), E)$, where $E = \{(u,v) \in left(f) \times right(f) \mid (u,v,h) \in B\}$. For any $(u,v) \in E$, since B is optimal, clearly there exists a unique h such that $(u,v,h) \in B$. Thus for an edge $e = (u,v) \in E$, we can define its *weight* $w(e) = h$.

It is clear that an edge in $\mathcal{G}$ has a one-to-one correspondence with a block in B. For any subgraph $\mathcal{H}$ of $\mathcal{G}$, the set of all edges in $\mathcal{H}$ naturally *induces* a block set $B(\mathcal{H}) \subseteq B$. Let $\mathcal{C}_1, \mathcal{C}_2, \ldots, \mathcal{C}_K$ ($K \geq 1$) be the connected components of $\mathcal{G}$. Denote by B_i the induced block set of $\mathcal{C}_i$, $i = 1, 2, \ldots, K$.

Clearly, $B = \bigcup_{i=1}^{K} B_i$ and $f = C_B = \sum_{i=1}^{K} C_{B_i}$. For any $i \in \{1, 2, \ldots, K\}$, since B is an optimal block set that builds f, B_i must be an optimal block set that builds C_{B_i}. Hence $|B_i| = n(C_{B_i}) \leq |left(C_{B_i})| + |right(C_{B_i})| - 1$. Observe that $left(C_{B_i}) \subseteq left(B_i)$ and $right(C_{B_i}) \subseteq right(B_i)$ (since B_i is a block set that builds C_{B_i}). Thus we have $|B_i| \leq |left(B_i)| + |right(B_i)| - 1$. Since $\mathcal{C}_i$ is a connected component of $\mathcal{G}$, we also have $|E(\mathcal{C}_i)| \geq |V(\mathcal{C}_i)| - 1$. Note that $|E(\mathcal{C}_i)| = |B_i|$ and $|V(\mathcal{C}_i)| = |left(B_i)| + |right(B_i)|$; thus $|B_i| \geq |left(B_i)| + |right(B_i)| - 1$. Hence $|B_i| = |left(B_i)| + |right(B_i)| - 1$, or equivalently, $|E(\mathcal{C}_i)| = |V(\mathcal{C}_i)| - 1$. Therefore, $\mathcal{C}_i$ is a tree, and we have the following lemma.

Lemma 2. *For any canonical optimal block set B that builds the input curve f, the LR-graph of B is a forest. In other words, B can be partitioned into $B_1, B_2, \ldots, B_K$, such that (1) $\{left(B_i)\}_{i=1}^{K}$ and $\{right(B_i)\}_{i=1}^{K}$ form a partition of $left(B)$ and $right(B)$, respectively, and (2) $|B_i| = |left(B_i)| + |right(B_i)| - 1$ holds for $i = 1, 2, \ldots, K$, where K is the number of connected components of $\mathcal{G}$.*

2.4   The Primary Block Set (PBS) Problem

Let $len(e)$ denote the length of an edge (or line segment) e. Given an x-monotone rectilinear curve f, we define a bipartite graph $\mathcal{G}_f = (E^L(f), E^R(f), A_f)$, where $A_f = \{(e^L, e^R) \in E^L(f) \times E^R(f) \mid x(e^L) < x(e^R) \text{ and } len(e^L) = len(e^R)\}$. In this subsection, we refer to edges in $\mathcal{G}_f$ as *arcs* to avoid possible confusion. For each arc $a = (e^L, e^R)$ in $\mathcal{G}_f$, we associate with it a block $(x(e^L), x(e^R), len(e^L))$, denoted by $blk(a)$. For a matching M in $\mathcal{G}_f$, define its *induced block set* as $B(M) = \{blk(a) \mid a \in M\}$. A block set B' is said to be *primary* if (i) $C_{B'} \leq f$, i.e., $C_{B'}(x) \leq f(x)$ for any $x \in \mathbb{R}$, and (ii) B' is induced by some matching M in $\mathcal{G}_f$, i.e., $B' = B(M)$. The **primary block set (PBS) problem** seeks a largest size primary block set for f. Its close relation to the SR problem is shown below.

Lemma 3. *A polynomial time μ-approximation PBS algorithm implies a polynomial time SR algorithm with an approximation ratio of $(2 - \frac{1}{\mu})$ if $\mu \geq 2$ and of no more than $\frac{3}{2}$ if $\mu < 2$.*

Proof. (Sketch) Given a μ-approximation PBS algorithm, our SR algorithm works on a curve f as follows. We first apply the μ-approximation PBS algorithm to f to obtain a primary block set B' with $|B'| \geq |\overline{B}|/\mu$, where $\overline{B}$ is an optimal PBS solution for f. Since B' is primary, $C_{B'} \leq f$. Consider two possible cases. Case I: $C_{B'} = f$. In this case, $|B'| = |left(f)| = |right(f)|$. We simply output B', which is clearly an optimal SR solution for f. Case II: $C_{B'} \neq f$. In this case, we perform a horizontal trapezoidal decomposition of $f - C_{B'}$ to obtain a block set B'', and output $B \triangleq B' \cup B''$. To show the correctness of this SR algorithm, it suffices to show that for Case II, the output block set B is the sought approximate solution. In fact, we can prove that (details are left to the full paper) $|B|$ has an upper bound $|left(f)| + |right(f)| - |\overline{B}|/\mu$, and the size of a canonical optimal block set, say B^*, for the SR problem on f, has a lower bound $\frac{2}{3}(|left(f)| + |right(f)| - |\overline{B}|/2)$. Since $|left(f)| \leq |\overline{B}|$ and $|right(f)| \leq |\overline{B}|$, we have $|B|/|B^*| \leq 2 - \frac{1}{\mu}$ when $\mu \geq 2$, and $|B|/|B^*| \leq \frac{3}{2}$ when $\mu < 2$.

Henceforth, we focus on finding a polynomial time $(2 + \epsilon)$-approximation PBS algorithm, which leads to a $(\frac{3}{2} + \epsilon)$-approximation SR algorithm by Lemma 3.

2.5   Reformulation of the Primary Block Set (PBS) Problem

This section gives interesting geometric observations on the PBS problem, which allow us to reformulate the PBS problem as a multicommodity demand flow (MDF) problem on a special path graph.

   Clearly, the PBS problem can be viewed as a *constrained* matching problem on the bipartite graph $\mathcal{G}_f$. Note that $\mathcal{G}_f$ may have $O(|E^L(f)| \cdot |E^R(f)|)$ arcs in the worst case. Our first observation is that, for this constrained matching problem, we can ignore a significant portion of the arcs of $\mathcal{G}_f$ without sacrificing the optimality of the solution. As to be shown later, for each vertex v of $\mathcal{G}_f$, we need to keep only at most one arc that is incident to v. These special arcs are computed as follows. We partition all left and right edges of f into groups

$E_1, E_2, \ldots, E_s$ ($s \geq 1$) based on their lengths. For the left and right edges in the same group, say E_i, we treat the left (resp., right) edges as the left (resp., right) parentheses, and perform a left-to-right scan along f for E_i; when encountering a right parenthesis, we pair it with the rightmost scanned yet unpaired left parenthesis, if any. The resulting parenthesis pairs naturally induce a set of arcs in $\mathcal{G}_f$, called *critical arcs*. The *critical arc set*, denoted by M^{cr}, contains all critical arcs obtained from all the edge groups $E_1, E_2, \ldots, E_s$. Clearly, M^{cr} is a matching in $\mathcal{G}_f$. The next lemma shows that M^{cr} is crucial to the PBS problem.

Lemma 4. *For any primary block set B for f, we can find a primary block set $\overline{B}$ for f such that $|\overline{B}| = |B|$ and $\overline{B}$ is induced by a matching $\overline{M} \subseteq M^{cr}$.*

Recall that the PBS problem seeks a primary block set of the maximum size for f. By Lemma 4, it is sufficient to consider only those primary block sets that are subsets of B^{cr}, where B^{cr} is the block set induced by the critical arc set M^{cr}. In other words, we seek a largest size subset B of B^{cr} subject to $C_B \leq f$. Hence, the PBS problem can be modeled as a multicommodity demand flow (MDF) problem [9] on a special path graph G_P. The vertex set of the path graph G_P consists of all left and right ends of f, sorted by the x-coordinates. There is an edge (x', x'') in G_P between every two consecutive ends x' and x'' of f in the sorted list, and the edge (x', x'') has an integer *flow capacity* equal to the height value of f between x' and x''. Each block $b = (\alpha, \beta, h)$ in B^{cr} has a corresponding source-sink pair (α, β) in G_P with a *demand flow* of a value h. The objective is to maximize the *number* of demand flows that can be simultaneously routed entirely in G_P without violating its edge capacity constraints.

2.6   Our Algorithm for the MDF Problem on a Path

The MDF problem on a path graph was studied by Chekuri *et al.* [9], who gave a polynomial time $(2+\epsilon)$-approximation algorithm for the case when the maximum demand value d_{max} in the path graph is $\leq$ the minimum edge capacity c_{min}. We extend their result and present a $(2 + \epsilon)$-approximation MDF algorithm for the case with $d_{max} \leq \lambda \cdot c_{min}$, where λ is any positive constant.

We borrow some definitions and notation from [9]. Let $d(F)$ be the value of a demand flow F. The *bottleneck capacity* $b(F)$ of a demand F is the smallest edge capacity on the path of F. For a given $\delta > 0$, we say that a demand F is δ-*large* if $d(F) \geq \delta \cdot b(F)$; else, F is δ-*small*. The next lemma is a generalization of a similar lemma in [9], and is crucial to our extended algorithm.

Lemma 5. *When $d_{max} \leq \lambda \cdot c_{min}$ for a constant $\lambda > 0$, in any feasible solution of the MDF problem on a path graph G_P, the number of δ-large demands crossing any edge in G_P is at most $2\lfloor \lambda/\delta^2 \rfloor$.*

Following the same outline developed in [9], we further have the next lemma.

Lemma 6. *For the MDF problem on a path graph G_P with $d_{max} \leq \lambda \cdot c_{min}$, there is a $(2 + \epsilon)$-approximation algorithm which runs in $O(nm^{O(\lambda/\epsilon^4)})$ time, where n is the number of edges in G_P, m is the number of demands, and λ is any positive constant.*

We conclude our discussion on the SR problem with the following result.

Theorem 1. *Let f be an x-monotone rectilinear curve of n vertices. If $M_f \leq \lambda \cdot m_f$ holds for a constant $\lambda > 0$, where M_f (or m_f) is the global positive maximum (or minimum) of f, then we can compute a $(\frac{3}{2} + \epsilon)$-approximate solution for the SR problem on f in $O(n^{O(\lambda/\epsilon^4)})$ time.*

References

1. A. Aggarwal, M.M. Klawe, S. Moran, P. Shor, and R. Wilber. Geometric Applications of a Matrix-Searching Algorithm. *Algorithmica*, 2:195–208, 1987.
2. A. Aggarwal, B. Schieber, and T. Tokuyama. Finding a Minimum-weight k-link Path in Graphs with the Concave Monge Property and Applications. In *Proc. 9th Annual ACM Symp. on Computational Geometry*, pages 189–197, 1993.
3. E.M. Arkin, J.S.B. Mitchell, and G. Narasimhan. Resource-Constrained Geometric Network Optimization. In *Proc. 14th ACM Symp. on Computational Geometry*, pages 307–316, 1998.
4. S. Arora. Polynomial-Time Approximation Schemes for Euclidean TSP and Other Geometric Problems. *J. of the ACM*, 45(5):753–782, 1998.
5. S.H. Benedict, R.M. Cardinale, Q. Wu, R.D. Zwicker, W.C. Broaddus, and R. Mohan. Intensity-Modulated Stereotactic Radiosurgery Using Dynamic Micro-Multileaf Collimation. *Int. J. Radiation Oncology Biol. Phys.*, 50(3):751–758, 2001.
6. D.E. Boccuzzi, S. Kim, J. Pryor, A. Berenstein, J.A. Shih, S.T. Chiu-Tsao, and L.B. Harrison. A Dosimetric Comparison of Stereotactic Radiosurgery Using Static Beams with a Micro-Multileaf Collimator versus Arcs for Treatment of Arteriovenous Malformations. In *Proc. of the 41st Annual American Society for Therapeutic Radiology and Oncology (ASTRO) Meeting*, page 413, 2000.
7. T.R. Bortfeld, D.L. Kahler, T.J. Waldron, and A.L. Boyer. X-ray Field Compensation with Multileaf Collimators. *Int. J. Radiat. Oncol. Biol. Phys.*, 28:723–730, 1994.
8. J.D. Bruce. *Optimal Quantization*. PhD thesis, MIT, May 1964.
9. C. Chekuri, M. Mydlarz, and F.B. Shepherd. Multicommodity Demand Flow in a Tree and Packing Integer Problem. In *Proc. of 30th International Colloquium on Automata, Languages and Programming*, pages 410–425, 2003.
10. D.Z. Chen, X.S. Hu, S. Luan, S.A. Naqvi, C. Wang, and C. Yu. Generalized Geometric Approaches for Leaf Sequencing Problems in Radiation Therapy. *Int. Journal of Computational Geometry and Applications*, 16(2-3):175–204, 2006.
11. P.M. Evans, V.N. Hansen, and W. Swindell. The Optimum Intensities for Multiple Static Collimator Field Compensation. *Med. Phys.*, 24(7):1147–1156, 1997.
12. M. Fischetti, H.W. Hamacher, K. Jørnsten, and F. Maffioli. Weighted k-Cardinality Trees: Complexity and Polyhedral Structure. *Networks*, 24:11–21, 1994.
13. M.R. Garey and D.S. Johnson. *Computers and Intractability: A Guide to the Theory of NP-Completeness*. Freeman, San Francisco, CA, 1979.
14. N. Garg. A 3-Approximation for the Minimum Tree Spanning k Vertices. In *Proc. 37th Annual IEEE Symp. on Foundations of Comp. Sci.*, pages 302–309, 1996.
15. G. Grebe, M. Pfaender, M. Roll, and L. Luedemann. Dynamic Arc Radiosurgery and Radiotherapy: Commissioning and Verification of Dose Distributions. *Int. J. Radiation Oncology Biol. Phys.*, 49(5):1451–1460, 2001.

16. E. Lawler. *Combinatorial Optimization: Networks and Matroids*. Holt, Rinehart and Winston, 1976.
17. L. Ma. Personal communication. Department of Radiation Oncology, University of Maryland School of Medicine, July 2004.
18. J.S.B. Mitchell. Guillotine Subdivisions Approximate Polygonal Subdivisions: A Simple New Method for the Geometric k-MST Problem. In *Proc. 7th Annual ACM-SIAM Symp. on Discrete Algorithms*, pages 402–408, 1996.
19. J.S.B. Mitchell. Guillotine Subdivisions Approximate Polygonal Subdivisions: Part II – A Simple Polynomial-Time Approximation Scheme for Geometric TSP, k-MST, and Related Problems. *SIAM J. Comput.*, 28(4):1298–1309, 1999.
20. G. Monge. Déblai et Remblai. In *Mémories de l'Académie des Sciences*, Paris, 1781.
21. R. Ravi, R. Sundaram, M.V. Marathe, D.J. Rosenkrantz, and S.S. Ravi. Spanning Trees Short and Small. In *Proc. 5th Annual ACM-SIAM Symp. on Discrete Algorithms*, pages 546–555, 1994.
22. B. Schieber. Computing a Minimum Weight k-link Path in Graphs with the Concave Monge Property. *Journal of Algorithms*, 29(2):204–222, 1998.
23. S. Webb. *The Physics of Three-Dimensional Radiation Therapy*. Bristol, Institute of Physics Publishing, 1993.
24. S. Webb. *The Physics of Conformal Radiotherapy — Advances in Technology*. Bristol, Institute of Physics Publishing, 1997.
25. X. Wu. Optimal Quantization by Matrix Searching. *Journal of Algorithms*, 12:663–673, 1991.
26. A. Zelikovsky and D. Lozevanu. Minimal and Bounded Trees. In *Tezele Cong. XVIII Acad. Romano-Americane*, pages 25–26. Kishinev, 1993.

A New Approximation Algorithm for Multidimensional Rectangle Tiling

Katarzyna Paluch*

Institute of Computer Science, University of Wroclaw, Poland
Max-Planck-Institute für Informatik, Saarbrücken, Germany
abraka@ii.uni.wroc.pl

Abstract. We consider the following tiling problem: Given a d-dimensional array A of size n in each dimension, containing non-negative numbers and a positive integer p, partition the array A into at most p disjoint rectangular subarrays called *rectangles* so as to minimise the maximum weight of any rectangle. The weight of a subarray is the sum of its elements.

In the paper we give a $\frac{d+2}{2}$-approximation algorithm that is tight with regard to the only known and used lower bound so far.

1 Introduction

In some applications including databases, load balancing and video compression we would like to partition data into sets of roughly the same weight. We consider the following tiling problem: Given a d-dimensional array A of size n in each dimension, containing non-negative numbers and a positive integer p, partition the array A into at most p disjoint rectangular subarrays called *rectangles* so as to minimise the maximum weight of any rectangle. The weight of a subarray is the sum of its elements.

The problem, restricted to two dimensions, was introduced by Khanna et al in [4], where it is shown that a 5/4-approximation for this problem is NP-hard. Successive approximation algorithms were constructed for this problem, beginning with the one having factor 5/2 by Khanna et al([4], through factors 7/3 ([6],[11]), 9/4 ([7]), 11/5 ([1]) and ending with the one having factor 17/8([10]).

The multidimensional version was first considered by Smith and Suri in [12], where they give an algorithm with approximation ratio $\frac{d+3}{2}$, that runs in time $O(n^d + p \log n^d)$ and the constant is of the order of $d!$. Next, Sharp in [11] gave a $(d^2 + 2d - 1)/(2d - 1)$-approximation algorithm that runs in time $O(n^d + 2^d p \log n^d)$.

In this paper we give a $\frac{d+2}{2}$-approximation algorithm that runs in time $O(n^d + 2^d p \log n^d)$. Additionally, this algorithm is tight with regard to the only known and used lower bound so far.

* Partially done while at University of Dortmund and supported by Emmy Noether DFG grant KR 2332/1-2.

T. Asano (Ed.): ISAAC 2006, LNCS 4288, pp. 712–721, 2006.

The general approach has a similar spirit as that in [10]. We also classify the arrays and subarrays into types. In the multidimensional case, however, there are many kinds of subarrays with a short type (having length 2) that are difficult to partition (whereas in a twodimensional case there is only one kind of such subarrays). As previously, we also have to consider arbitrarily large subarrays i.e. having arbitrarily long type. Fortunately subarrays that are difficult to partition display a regular structure that can be handled by appropriate linear programs. Curiously, linear programs describing large difficult subarrays disintegrate into small linear programs that can be treated independently and in this respect they are much easier to analyze than linear programs describing large difficult subarrays in a twodimensional version, where they cannot be decomposed into small linear programs.

Organization of the paper. In Section 2 we give some basic notions and notation. In Section 3 we introduce the notion of a simple subarray and show the way in which we use linear programs. In Section 4 we define the classification into types of arrays and subarrays and show which subarrays having short type are difficult to partition. In Section 5 we give the algorithm and prove its correctness. Also in that section Lemma 7 explains why large linear programs disintegrate into smaller ones.

2 Preliminaries

Let M denote the value of the element(s) of maximal weight in A and $w(S)$ the weight of a subarray S.

In any partition of A, clearly, at least one rectangle must have weight greater or equal $W = \max\{\frac{w(A)}{p}, M\}$. Thus W is a simple lower bound for the maximum rectangle weight in an optimal solution.

For convenience sake we can rescale the array A by dividing all its elements by W and thus assume that we deal only with arrays of elements from the interval $[0, 1]$ and that the lower bound on the optimal solution is equal to 1.

To represent subarrays we will use the notation $[a_1, b_1] \times [a_2, b_2] \times \ldots \times [a_d, b_d]$. Individual elements will be represented by $(a_1, a_2, \ldots, a_d)$.

Definition 1. *We say that the array or subarray is α-partitioned if it is partitioned into rectangles having weight not greater than α. If we additionally require that the number of tiles used does not exceed $\lfloor w(A) \rfloor$ ($\lceil w(A) \rceil$, resp.) then we say that the array is* well α-partitioned *(nearly well α-partitioned, resp.).*

From [7] we have

Fact 1. *If we partition the input array A into a number of disjoint subarrays $A_1, \ldots, A_l, A_{l+1}$ and well α-partition each $A_i (1 \leq i \leq l)$ and nearly well α-partition A_{l+1}, then we will get the solution within α of the optimal one.*

Definition 2. *A slice in dimension f of an array or a subarray is its subarray consisting of the elements having the same index in dimension f.*

From now on, we will assume that $\alpha = \frac{d+2}{2}$.

3   Simple Subarrays, Their Complexity and Difficulty

The key role in the analysis of the possible partitions is played by simple subarrays, into which we will appropriately decompose A.

Definition 3. *Let $\beta \leq \alpha$.*
A subarray S is called β-simple (or simple if we know which β we mean) if in every dimension there exists one slice that separates two subarrays, each having weight less than β (i.e. S is a disjoint sum of two subarrays having weight less than β and the slice). The element that is the common part of all the separating slices of a simple subarray is called its center.

Suppose we have a simple subarray S $[a_1, b_1] \times [a_2, b_2] \times \ldots \times [a_d, b_d]$ that has a center $(c_1, c_2, \ldots, c_d)$. β is arbitrary.

In each of the dimensions S can have *complexity* $0, 1$ or 2. In dimension i it has complexity 0 iff $a_i = c_i = b_i$, it has complexity 1 iff $a_i = c_i < b_i$ or $a_i < c_i = b_i$ and it has complexity 2 iff $a_i < c_i < b_i$. The overall complexity of S is the sum of its complexities in all dimensions.

One of the interpretations of the complexity is reflected in the following fact.

Fact 2. *If we have a simple subarray S that has complexity p and cut off one rectangle that contains only its center, then the rest of S can be covered by p rectangles and no fewer.*

Proof. We can do it as follows. In the first step: if $a_1 < c_1$ we cut off a rectangle $[a_1, c_1 - 1] \times [a_2, b_2] \times \ldots \times [a_d, b_d]$ and also if $b_1 > c_1$ a rectangle $[c_1 + 1, b_1] \times [a_2, b_2] \times \ldots \times [a_d, b_d]$.
In the ith step $(i \leq d)$: if $a_i < b_i$ - a rectangle $[c_1, c_1] \times [c_2, c_2] \times \ldots \times [c_{i-1}, c_{i-1}] \times [a_i, c_i - 1] \times \ldots [a_d, b_d]$ and if $a_i > b_i$ a rectangle $[c_1, c_1] \times [c_2, c_2] \times \ldots \times [c_{i-1}, c_{i-1}] \times [c_i + 1, b_i] \times \ldots [a_d, b_d]$.

We will use vectors $v = (v_1, v_2, \ldots, v_d)$ such that $v_i\{-1, 1, 0, 2\}$ to point certain subarrays of S. Namely, we will say that v *cuts off* a subarray $[a_1', b_1'] \times [a_2', b_2'] \times \ldots \times [a_d', b_d']$, where $[a_i', b_i'] = [c_i, c_i]$ if $v_i = 0$,
$[a_i', b_i'] = [c_i, b_i]$ if $v_i = 1$,
$[a_i', b_i'] = [a_i, c_i]$ if $v_i = -1$,
$[a_i', b_i'] = [a_i, b_i]$ if $v_i = 2$.

Of course, it does not make much sense to put $v_i = -1$ if $a_i = c_i$ or $v_i = 1$ when $c_i = b_i$. Therefore we will say that v is *valid* if in every dimension i it holds: if $v_i = -1$, then $a_i < c_i$, if $v_i = 1$, then $c_i < b_i$ and if $v_i = 2$, then $a_i < c_i < b_i$.
Let $|v| = \sum |v_i|$.
A subarray is going to be described as *m-difficult* if it cannot be α-partitioned into m rectangles.

Lemma 1. *A simple subarray S that has complexity p is m-difficult iff every vector v such that $|v| = p - m + 1$ cuts off a subarray having weight greater than α.*

Proof. If some vector v such that $|v| = p - m + 1$ cuts off a subarray having weight at most α, then we can α-partition S by taking one rectangle cut off by v and covering the rest of S by $m - 1$ rectangles of weight less than β in the way similar as that in the proof of Fact 2. We are able to do so, because in each dimension i a separating slice placed in c_i separates two subarrays having weight less than β.

In the other direction. Suppose we have some α-partition of S into m rectangles. We can show that one of them must contain a rectangle that is cut off by a valid vector v such that $|v| \geq p - m + 1$.

We can calculate the minimal weight of simple subarrays that are m-difficult using linear programming.

Let us explain it by an example. Suppose we have a 3-difficult simple 5×5 array A, whose center is $(3, 3)$. Then the distribution of weight on this array can be described by the following array

$$
\begin{array}{ccc}
x_1 & x_2 & x_3 \\
x_4 & s & x_5 \\
x_6 & x_7 & x_8
\end{array}
$$

where s denotes the weight of the center, $x_2 = a_{1,3} + a_{2,3}, x_7 = a_{4,3} + a_{5,3}$ and $x_1 = a_{1,1} + a_{1,2} + a_{2,1} + a_{2,2}$ and the remaining variables denote analogous sums of elements.

The complexity of array A is 4 and by Lemma 1 we know that each vector v, such that $|v| = 2$ cuts off a rectangle of weight greater than α. Hence, the lower bound on the minimal weight of A is the solution to the following linear program.

minimize $s + \sum_{i=1}^{8} x_i$

subject to
$$
\begin{aligned}
s & \leq 1 \\
s + x_1 + x_2 + x_4 & \geq \alpha \\
s + x_2 + x_3 + x_5 & \geq \alpha \\
s + x_5 + x_7 + x_8 & \geq \alpha \\
s + x_4 + x_6 + x_7 & \geq \alpha \\
s + x_2 + x_7 & \geq \alpha \\
s + x_4 + x_5 & \geq \alpha
\end{aligned}
$$

Let us notice that in estimating the lower bound of this subarray we can use only five variables s, x_2, x_4, x_5 and x_7 and without loss of generality assume that variables x_1, x_3, x_6 and x_8 are equal to 0. It is so because if the array, whose weight is distributed as above is 3-difficult, then so is the array, whose weight is distributed as follows, because all of the inequalities from the above linear program are satisfied as well.

$$
\begin{array}{ccc}
0 & x_2 + \frac{x_1}{2} + \frac{x_3}{2} & 0 \\[2mm]
x_4 + \frac{x_1}{2} + \frac{x_6}{2} & s & x_5 + \frac{x_3}{2} + \frac{x_8}{2} \\[2mm]
0 & x_7 + \frac{x_6}{2} + \frac{x_8}{2} & 0
\end{array}
$$

We can generalize this reasoning and state the following lemma.

Lemma 2. *The minimal weight of a simple m-difficult subarray S of complexity p is greater than the solution of the following linear program*

$$\text{minimize} \quad s + \sum_{i=1}^{p} x_i$$

$$\begin{aligned}\text{subject to} \quad & s \leq 1 \\ & s + x_{i_1} + x_{i_2} + \ldots + x_{i_{p-m+1}} \geq \alpha \\ & \text{for all } 1 \leq i_1 < i_2 < \ldots < i_{p-m+1} \leq p\end{aligned}$$

and thus it is greater than $1 + p\frac{\alpha-1}{p-m+1}$.

Proof. Similarly as above we can assume that the weight of the whole subarray is concentrated in p elements and the center (i.e. only those elements are non-zero) and by Lemma 1 the weight of every rectangle cut off by a vector v such that $|v| = p - m + 1$ is greater than α. One can easily check that the sum of the variables in the above linear program is minimal when all the inequalities are satisfied with equality and thus when $s = 1$ and $x_1 = x_2 = \ldots = x_p = \frac{\alpha-1}{p-m+1}$.

We will say that an element $(e_1, e_2, \ldots, e_d)$ is a *satellite* of the center if there exists exactly one i such that $e_i \neq c_i$. Using this terminology we can say that the weight of a simple m-difficult subarray is minimal when its weight is concentrated in its center and the p satellites of the center.

4   Blocks

We look at the array A as a sequence of its slices in dimension one, which will be called *sheets*. We will distinguish two classes of them: those with weight at least 1 ($>$-*sheets*) and those with weight less than 1 ($<$-*sheets*). In [6] we have analogous notions of $<$- and $>$-columns which actually are slices in a twodimensional array. There we have:

Lemma 3 ([6]).

1. *A $>$-column can be well 2-partitioned.*
2. *A subarray consisting solely of $<$-columns having weight at least 1 can be well 2-partitioned.*

This remains true in the case of $<$-sheets and α-partitioning. Therefore we can notice that a group of adjacent $<$-sheets can be treated like a single $>$-column if its overall weight is at least 1 ($<$-sheets are elements of a $>$-column) and thus can be not only α-partitioned but even well 2-partitioned and otherwise like a single $<$-sheet (its elements are the sums of the appropriate elements of $<$-sheets). Without loss of generality we will assume that every array consists of alternating $<$- and $>$-sheets and begins and ends with a $<$-sheet. It will sometimes be achieved by inserting artificial sheets of weight 0.

Further we are going to identify more classes of sheets.

Definition 4. *For every natural number m, a sheet of type m, also referred to as an m-sheet, denotes a $>$-sheet having weight from the interval $[m, m+1)$.*

The type of an array or a subarray is going to be described by the types of its sheets given in the order of their occurrence. A sheet of type m will be represented by a natural number m and a $<$-sheet by a symbol $\diamond$. The type of a subarray or an array will be given in the form $(\diamond)m_1 \diamond m_2 \ldots \diamond m_n(\diamond)$. In the algorithm the array A will be processed from the leftmost sheet to the rightmost one, however not completely in the sheet-wise manner but in the block-wise one.

A *block* is a subarray of type $\diamond m$. Among blocks of type $\diamond m$ we will distinguish two subclasses:

- $\llcorner$ m - simple subarrays that can be α-partitioned into m rectangles
- $\sqcup m$ - simple subarrays that cannot be α-partitioned into m rectangles

Now we will show some applications of Lemma 2.

Lemma 4. *Every simple $>$-sheet can be well α-partitioned.*

Proof. The complexity of a simple $>$-sheet is at most $2d - 2$. Suppose its weight falls in the interval $[m, m+1)$, which means that if we want to well α-partition it we can use at most m rectangles. From Lemma 2 we know that a simple $>$-sheet that cannot be α-partitioned into m rectangles has weight greater than $1 + (2d-2)\frac{\frac{d}{2}}{2d-m-1}$. One can easily check that $1 + (2d-2)\frac{\frac{d}{2}}{2d-m-1} < m+1$ holds only for $m \in (d-1, d)$. Therefore a simple $>$-sheet that cannot be α-partitioned into m rectangles has weight greater than $m + 1$.

As every $>$-sheet S can be easily decomposed into simple $>$-sheets, we can well α-partition S by well α-partitioning each simple $>$-sheet separately.

Corollary 1. *Every $>$-sheet can be well α-partitioned.*

This means that if the array consisted solely of $>$-sheets we would have a very easy $\frac{d+2}{2}$-approximation for d-dimensional arrays. Also, blocks of type $\llcorner$ m can be from the definition well α-partitioned (because they contain an m-sheet that has weight from the interval $[m, m+1)$ and thus for well α-partitioning we are allowed to use m rectangles). Let us now examine which blocks of type $\sqcup m$ can be well α-partitioned and which cannot.

The complexity of a block of type $\sqcup m$ is at most $2d-1$. Therefore by Lemma 2, its weight is greater than $1 + (2d-1)\frac{\frac{d}{2}}{2d-m}$. Next we solve the following inequality

$$1 + (2d-1)\frac{\frac{d}{2}}{2d-m} \geq m+1$$

and get that it is true for $m \in [d - \sqrt{d/2}, d + \sqrt{d/2}]$. Thus, for the above m, blocks of type $\sqcup m$ can be well α-partitioned, because in this case we can use $m+1$ rectangles and from Lemma 4 an m-sheet can be well partitioned (we use at most m rectangles for it) and we use 1 rectangle for a $<$-sheet. If m falls outside the above interval, then blocks of type m indeed cannot be well α-partitioned. This means that if we want to get an α-approximation we cannot restrict ourselves to subarrays consisting of only 2 sheets but are forced to examine larger subarrays.

5   Algorithm

We will process the array in the block-wise manner starting from the leftmost. As we have seen in the previous section, some blocks of type $\sqcup m$ cannot be well α-partitioned. Therefore, if we encounter such a block, we will extend it to the neighbouring one and see whether they together can be well α-partitioned. If it still turns out impossible, we will extend it further until the subarray finally can be well α-partitioned or we have reached the end of the array. In the other case we will be allowed to nearly well α-partition the subarray.

If a subarray S consists only of simple blocks, then its type T has the form $\diamond_1 m_1 \diamond_2 m_2 \ldots \diamond_k m_k$, where each $\diamond_i$ denotes either $\_$ or $\sqcup$. To types of this form we will ascribe a natural number $N(T)$:

$$N(T) = \sum_{i=1}^{k} m_i + \sum_{i=1}^{k} [\diamond_i = \sqcup] - 1.$$

Thus, if $T = \sqcup 4$, then $N(T) = 4$ and if $T = \sqcup 4 \sqcup 3$, then $N(T) = 8$.

A subarray that is not simple is *complex*. In the algorithm we will check whether a given block is 1-simple.

We say that two elements $(e_1, e_2, \ldots, e_d)$ and $(e'_1, e'_2, \ldots, e'_d)$ *coincide* if there exists at least one i such that $e_i = e'_i$.

$S := \emptyset$
while extension of S to the next block possible
    extend S to the next block
    if S is a block of type $\_ m$, well α-partition it, else
    if S ends with a complex block, well α-partition S as shown in Lemma 14 else
    if S of type T has weight at least $N(T) + 1$, well α-partition S as shown in Lemma 5, else
    if S contains a subarray of type m that can be α-partitioned
        into $m - 1$ rectangles or a subarray of type $m\sqcup$ that can be
        α-partitioned into m rectangles, then well α-partition S as shown in Lemma 12, else
    if the centers of some two neighbouring blocks coincide, well
        α-partition S as shown in Lemma 13
nearly-well-α-partition S together with the last $<$-sheet.

In the algorithm if S ends with a complex block, then it is relegated for well α-partitioning to Lemma 14. Therefore, if S has to be extended, then we know that it consists solely of simple blocks.

Lemma 5. *If subarray S of type T consists only of simple blocks and its weight is at least $N(T) + 1$, then we can well α-partition it.*

Proof. We can do it by α-partitioning each block of type $\sqcup m$ into $m+1$ rectangles (m rectangles for an m-sheet and one for a sheet denoted by $\sqcup$) and by α-partitioning each block of type $\_ m$ into m rectangles. This way we will use $N(T) + 1$ rectangles.

5.1   Minimal Weight of S

This whole subsection is to prove

Lemma 6. *In the algorithm the subarray S of type T that has to be extended has weight greater than $N(T) + \frac{1}{2}$*

From Lemma 2 we have

Fact 3. *The minimal weight of a block of type $\sqcup m$ is greater than $m + \frac{3}{2}$.*

In the following facts and lemmas we will assume that S' is a subarray of S that has to be extended by the algorithm.

Fact 4. *The weight of a subarray S' of type $\sqcup m \sqcup$ is greater than $1 + 2d\frac{\alpha-1}{2d-m} \geq m + 1$.*

Proof. The minimal weight of S' is equal to the minimal weight of an $(m + 1)$-difficult subarray having complexity $2d$ and thus by Lemma 2 it is greater than $1 + 2d\frac{\alpha-1}{2d-m}$ and the inequality $1 + 2d\frac{d/2}{2d-m} \geq m + 1$ implies $d^2 - 2dm + m^2 \geq 0$.

Next we prove a technical lemma that will prove very useful and will mean that often a large linear program describing the weight of a difficult subarray can be decomposed into smaller linear programs.

Lemma 7. *If we have a block of type $\sqcup m$ and it contains one element e having value b that does not coincide with the center and $m \geq 3$, then the minimal weight of such a block is greater than $b + 1 + (2d - 1)\frac{\alpha-1}{2d-m}$. In other words, it is greater than the minimal weight of a block of type $\sqcup m$ plus b.*

Proof. The linear program from Lemma 2 has the same solution as the following linear program

minimize $s + \sum_{i=1}^{p} x_i$

subject to $s \leq 1$

$s + x_{i \bmod p} + x_{i+1 \bmod p} + x_{i+2 \bmod p} + \cdots + x_{i+p-m \bmod p} \geq \alpha$
for $1 \leq i \leq p$

It means that if the vectors to which the above p inequalities correspond cut off rectangles that do not contain element e, then the lemma is proved. Since e does not coincide with the center, for each i, either $e_i < c_i$ or $e_i > c_i$, which means that a vector v cuts off a rectangle that contains e iff for each i it holds: $v_i = 2$ or ($v_i = 1$ and $e_i > c_i$) or ($v_i = -1$ and $e_i < c_i$). This in turn means that there exist d satellites of the center denoted by some variables $x_{i_1}, \ldots, x_{i_d}$ such that vector v cuts off a rectangle that contains an element e iff the rectangle cut off by this v contains all these d satellites. Therefore if no inequality in the linear program contains all the variables that represent those concrete d satellites, we are done. If we have a block of type $\sqcup m$ and $m \geq 3$, then its complexity is $p = 2d - 1$ and $p - m + 1 = 2d - m \leq 2d - 3$, which means that no inequality in the above linear program contains all the variables $x_1, x_3, x_5, \ldots, x_{2d-1}$. Therefore we can rename the variables x_i so that a vector v cuts off a rectangle that contains e iff it cuts off a rectangle that contains satellites denoted by $x_1, x_3, x_5, \ldots, x_{2d-1}$.

Lemma 8. *The weight of a subarray S' of type $\sqcup m_1 \sqcup m_2$, such that $m_2 \geq 3$ is greater than the minimal weight of $\sqcup m_1 \sqcup$ plus the minimal weight of $\sqcup m_2$ and thus it is greater than $m_1 + m_2 + \frac{3}{2}$.*

Proof. We use 2 variables for the center and $2d + 2d - 1$ variables that represent the satellites of the centers. One satellite of the center of $\sqcup m_1 \sqcup$ is contained in the subarray $\sqcup m_2$, however it does not coincide with the center of $\sqcup m_2$ and one satellite of the center of $\sqcup m_2$ falls in the subarray $\sqcup m_1 \sqcup$ and also it does not coincide with the center of $\sqcup m_1 \sqcup$. Thus by Lemma 7 the linear programs connected with these two subarrays can be considered separately and thus the minimal weight of S is greater than the sum of the solutions of these linear programs.

Similarly we can prove

Lemma 9. *For $m_1, m_2 \geq 3$, the weight of a subarray S' of type $\_ m_1 \sqcup m_2$ is greater than $m_1 + m_2 + 1$.*

From Lemmas 8 and 9 we get

Lemma 10. *If the subarray S' has type $T = \diamond_1 m_1 \diamond_2 m_2 \ldots \diamond_k m_k$ and each $m_i \geq 3$, then the weight of S' is greater than $N(T) + \frac{1}{2}$.*

To finish the proof of Lemma 7, we need to show it is also true if S contains m-sheets such that $m \leq 2$. It is done by enumerating the inequalities that the linear program contains and proving dual programming that the sum of the variables is minimized when all of them are satisfied with equality.

The immediate corollary of Lemma 6 is

Corollary 2. *In the algorithm each S that we encounter that is not ended with a complex subarray has weight at least $N(T)$.*

Proof. If S is a simple block, then its weight is at least $N(T)$. If S consists of more blocks, then it has the form $S'B$ and S' was relegated for an extension by the algorithm. Thus by Lemma 6 the weight of S' is greater than $N(T') + \frac{1}{2}$ and the weight of a block of type $\_ m$ is at least m and of a block of type $\sqcup m$ greater than $m + \frac{1}{2}$.

This means that if we want to well α-partition S, we can use $N(T)$ rectangles.

Lemma 11. *A subarray of type $m\_$ that is contained in S can be α-partitioned into m rectangles.*

Proof. If it cannot be α-partitioned into m rectangles, then it is m-difficult and then the weight of S is at least $N(T) + 1$.

Lemma 12. *If in S there exists a subarray of type m that can be α-partitioned into $m - 1$ rectangles or a subarray of type $m\sqcup$ that can be α-partitioned into m rectangles, then S can be partitioned into $N(T)$ rectangles.*

Lemma 13. *If in S the centers of two neighbouring blocks coincide, then S can be well α-partitioned.*

Lemma 14. *If S ends with a complex block, then it can be well α-partitioned.*

The running time of the algorithm is mostly spent in searching for separating slices in simple subarrays and it can be estimated similarly as the time spent by procedure Heavy-Search, Heavy-Cut and Medium-Tile in [11].

6 The Algorithm Is Tight

Suppose that we have an array A that is 1-simple and has complexity $2d$ and it has only $2d + 1$ non-zero elements and these are: the center, that has weight 1 and the $2d$ satellites that have weight $\frac{1}{2}$. Then the overall weight of A is equal to $1 + d$. Let $p = 1 + d$. Thus the lower bound $W = 1$. We can easily see that one rectangle in the partition must contain the center and at least d satellites, therefore its weight is at least $1 + \frac{d}{2}$.

References

1. Berman, P., DasGupta, B., Muthukrishnan, S., Ramaswami, S.: Efficient Approximation Algorithms for Tiling and Packing Problems with Rectangles. J. Algorithms **41(2)** (2001) 443-470
2. Grigni, M., Manne, F.: On the complexity of generalized block distribution. Proc. 3rd intern. workshop on parallel algorithms for irregularly structured problems (IRREGULAR'96), Springer, 1996, LNCS 1117, 319-326
3. Han, Y., Narahari, B., Choi, H.A.: Mapping a chain task to chained processors. Information Processing Letters **44** (1992) 141-148
4. Khanna, S., Muthukrishnan, S., Paterson, M.: On approximating rectangle tiling and packing. Proc. 19th SODA (1998) 384–393
5. Khanna, S., Muthukrishnan, S., Skiena, S.: Efficient array partitioning. Proc. 24th ICALP (1997) 616-626
6. Loryś, K., Paluch, K.: Rectangle Tiling. Proc. 3rd APPROX, Springer, 2000, LNCS 1923, 206-213
7. Loryś, K., Paluch, K.: A new approximation algorithm for RTILE problem. Theor. Comput. Sci. **2-3(303)** (2003) 517-537
8. Martin, G., Muthukrishnan, S., Packwood, R., Rhee, I.: Fast algorithms for variable size block matching motion estimation with minimal error. IEEE Trans. Circuits Syst. Video Techn. 10(1) (2000) 42-50
9. Muthukrishnan, S., Poosala, V., Suel, T.: On rectangular partitioning in two dimensions: algorithms, complexity, and applications. Proc. 7th ICDT (1999) 236-256
10. Paluch, K.: A 2(1/8)-Approximation Algorithm for Rectangle Tiling. ICALP (2004) 1054-1065
11. Sharp, J.: Tiling Multi-dimensional Arrays. Proc. 12th FCT, Springer, 1999, LNCS 1684, 500-511
12. Smith, A., Suri, S.: Rectangular Tiling in Multidimensional Arrays. J.Algorithms **37(2)** (2000) 451-467

Tessellation of Quadratic Elements

Scott E. Dillard[1,3], Vijay Natarajan[1,3], Gunther H. Weber[1,3],
Valerio Pascucci[2,3], and Bernd Hamann[1,3]

[1] Department of Computer Science, University of California, Davis
[2] Lawrence Livermore National Laboratory
[3] Institute for Data Analysis and Visualization

Abstract. Topology-based methods have been successfully used for the analysis and visualization of piecewise-linear functions defined on triangle meshes. This paper describes a mechanism for extending these methods to piecewise-quadratic functions defined on triangulations of surfaces. Each triangular patch is tessellated into monotone regions, so that existing algorithms for computing topological representations of piecewise-linear functions may be applied directly to piecewise-quadratic functions. In particular, the tessellation is used for computing the Reeb graph, which provides a succinct representation of level sets of the function.

1 Introduction

Scalar functions often represent physical quantities like temperature, pressure, etc. Scientists interested in understanding the local and global behavior of these functions study their level sets. A *level set* of a function f consists of all points $f^{-1}(c)$ where the function value is equal to a constant c. Various methods have been developed for the purpose of analyzing the topology of level sets of a scalar function. These methods primarily apply to piecewise linear functions. We discuss an extension of these methods to bivariate piecewise quadratic functions defined over a triangulated surface.

A *contour* is a single connected component of a level set. Level sets of a smooth bivariate function are simple curves. The *Reeb graph* of f is obtained by contracting each contour to a single point [1], see Fig. 1. The connectivity of level sets changes at *critical points* of a function. For smooth functions, the critical points occur where the gradient becomes zero. Critical points of f are the *nodes* of the Reeb graph, connected by *arcs* that represent a family of topologically equivalent contours.

1.1 Related Work

Methods to extract contours from bivariate quadratic functions have been explored in the context of geometric modeling applications [2]. A *priori* determination of the topology of contours has been studied also in the computer graphics and visualization [3,4]. Much research has focused on functions defined by bilinear and trilinear interpolation of discrete data given on a rectilinear

T. Asano (Ed.): ISAAC 2006, LNCS 4288, pp. 722–731, 2006.

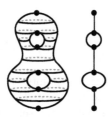

Fig. 1. Reeb graph of a height function defined over a double torus. Critical points of the surface become nodes of the Reeb graph.

grid. Work in this area has led to efficient algorithms for computing the contour tree, a special Reeb graph that has no cycles [5,6]. More general algorithms have been developed for computing Reeb graphs and contour trees for piecewise-linear functions [7,8,9].

Topological methods were first used in computer graphics and scientific visualization as a user interface element, to describe high-level topological properties of a dataset [10]. They are also used to selectively explore large scientific datasets by identifying important function values related to topological changes in a function [11,12], and for selective visualization [13]. Reeb graphs have also been used as the basis for searching large databases of shapes [14], and for computing surface parametrizations of three-dimensional models [15].

1.2   Results

Given a triangulated surface and a piecewise-quadratic function defined on it, we tessellate the surface into monotone regions. A graph representing these monotone regions is a valid input for existing algorithms that compute Reeb graphs and contour trees for piecewise-linear functions. The essential property we capture in this tessellation is that the Reeb graph of the function restricted to a single tile of the tessellation is a straight line. In other words, every contour contained in that tile intersects the triangle boundary at least once and at most twice. We tessellate each triangular patch by identifying critical points of the function within the triangle and connecting them by arcs to guarantee the required property.

2   Background

We consider bivariate, piecewise-quadratic functions defined over triangular meshes. Bivariate quadratics are functions $f : \mathbb{R}^2 \to \mathbb{R}$ of the form

$$f(x, y) = Ax^2 + Bxy + Cy^2 + Dx + Ey + F.$$

A critical point of f is a point $\mathbf{x}$ where $\nabla f(\mathbf{x}) = \mathbf{0}$. The partial derivatives are given by the following linear expressions

$$\frac{\partial f}{\partial x} = 2Ax + By + D \quad \text{and} \quad \frac{\partial f}{\partial y} = 2Cy + Bx + E.$$

The location of a critical point $(\hat{x}, \hat{y})$ is given by

$$\hat{x} = \frac{-2CD + BE}{4AC - B^2} \quad \text{and} \quad \hat{y} = \frac{-2AE + BD}{4AC - B^2}.$$

Bivariate quadratics and their critical points can be classified based on their contours. The contours are conic sections. Let $H = 4AC - B^2$ be the determinant of the Hessian matrix of f. We partition the set of bivariate quadratic functions into three classes:

1. $H > 0$: Contours are ellipses; The critical point is maximum or minimum.
2. $H < 0$: Contours are hyperbolas; The critical point is a saddle.
3. $H = 0$: Contours are parabolas, pairs of parallel lines, or single lines; No critical point exists.

We refer to members of these classes as *elliptic*, *hyperbolic*, and *parabolic*, respectively. We further classify the critical points of elliptic functions using the second-derivative test. The critical point is a minimum when $A > 0$, and a maximum when $A < 0$. When $A = 0$, the sign of C discriminates between maxima and minima.

2.1   Line Restrictions

Let $\ell(t) = \begin{pmatrix} x_0 + tx_d \\ y_0 + ty_d \end{pmatrix}$ be a parametrically defined line passing through $(x_0, y_0)^T$ in direction $(x_d, y_d)^T$. Now restrict the domain of the bivariate quadratic $f(x, y)$ to only those points on $\ell(t)$. We then have a univariate quadratic $r(t) = \alpha t^2 + \beta t + \gamma$, where

$$\alpha = Ax_d^2 + Bx_d y_d + Cy_d^2,$$
$$\beta = 2Ax_0 x_d + B(y_0 x_d + x_0 y_d) + 2Cy_0 y_d + Dx_d + Ey_d, \text{ and}$$
$$\gamma = Ax_0^2 + Bx_0 y_0 + Cy_0^2 + Dx_0 + Ey_0 + F.$$

We call $r(t)$ a *line restriction* of f. If $\alpha \neq 0$, r is a parabola with one critical point at $\hat{t} = -\beta/2\alpha$; if $\alpha = 0$, r is linear. We refer to a critical point of this univariate quadratic function as a *line-critical point*, while we refer to critical points of the bivariate function as *face-critical points*. The line restrictions have several useful properties:

1. There is at most one line-critical point on a line restriction, since the function along the line is either quadratic or linear.
2. If a line intersects any contour twice, it must contain a line-critical point between these intersections: Assume that the function value on the contour is zero. The sign of the line-restriction changes each time it crosses the contour. Applying the mean value theorem to the derivative of the line-restriction, there must be a critical point between two zero crossings.
3. A line-critical point is located at the point where the line is tangent to a contour, following from the previous property.

4. Any line restriction passing through a face-critical point has a line-critical point which is coincident with the face-critical point: The gradient at the face critical point is zero. Thus, all directional derivatives are zero, and, in particular, the derivative of the line restriction is zero.

2.2   Contours and Critical Points

A critical point is a point where the number of contours or the connectivity of existing contours changes. When the gradient is not defined, we may classify critical points based on the behavior of the function in a local neighborhood [16]. Figure 2 shows this classification. Consider a plane of constant height passing through the graph surface $(x, y, f(x, y))$ of f. The intersection of the surface and the plane is a set of contours, each homeomorphic to either a closed loop or a line segment. When this plane passes a minimum, a new contour is created. When the surface passes a maximum, an existing contour is destroyed. When the surface passes a saddle, two types of events occur: (a) Two segments may merge into a new segment or a segment may split into two. (b) The endpoints of a segment may connect with each other to form a loop or a loop may split into a segment.

Fig. 2. Interior minimum, maximum, saddle and regular point, and boundary minimum, maximum, saddle and regular point. Shaded areas are regions with function value less than the indicated point.

We consider critical points of a function restricted to a triangular patch. A face criticality of an elliptic function creates or destroys a loop when the sweep plane passes it. A face criticality of a hyperbolic function interchanges the connectivity of two segments. A line criticality can create or destroy a segment, merge two segments into a new segment or split a segment in two, transform a segment into a loop or a loop into a segment. However, a line-critical point cannot create or destroy loops. We determine whether a line criticality is an extremum or a saddle by examining the directional derivative perpendicular to the edge, at the critical point. Vertices, when not located exactly at a hyperbolic saddle point, can only create or destroy segments within a triangular patch. (See Sect. 4 for a proof.)

3   Tessellation

We are given a scalar-valued bivariate function defined by quadratic triangular patches. We aim to tessellate the patches into monotone regions, for the purpose of analyzing the topology of level sets of one patch, and, by extension, of a piecewise-defined function composed of many patches. The tessellation will decompose each triangular patch into subpatches, so that each subpatch has a Reeb graph which is a straight line. More specifically, every level set within each subpatch is a single connected component and is homeomorphic to a line segment. We achieve this by ensuring that each subpatch contains exactly one maximum and one minimum, and no saddles. Since we are interested in the topology of the subpatches but not their geometry, we only compute a combinatorial graph structure which captures the connectivity of contours. Some of the arcs of this graph originate from the patch, such as boundary edges, and thus their geometry can be inferred. The embeddings of remaining arcs are not computed since they are not required to construct the Reeb graph.

The construction of the tessellation proceeds by using a case analysis. For each patch, we count the number of line-critical points ($L = 0, 1, 2, 3$) and face-critical points ($F = 0, 1$). Each pair $\langle F, L \rangle$ is handled as an individual case to determine the appropriate tessellation. We first describe the composition of the tessellation. The nodes of a tessellation graph include: (1) all three vertices of the triangle, (2) all line-critical points that exist on the triangle boundary and are not coincident with a triangle vertex, and (3) the face-critical point, assuming that it lies within the triangle and is not coincident with the boundary. We are given (1) as input, but (2) and (3) must be computed in a pre-processing step. Numerical problems can arise. No root finding is needed to compute (2) and (3), so exact arithmetic may be used. However, if speed and consistency are to be favored over accuracy then we only need to ensure that the tessellation graphs of every patch agree on their boundary edges. The existence and location of (3) does not effect the tessellation boundary, so no consistency checks are needed for computing these points. The computation of (2) must agree between two triangles sharing an edge. To ensure this, we do not treat edges as line restrictions of bivariate quadratics, but rather as univariate quadratics defined by data prescribed for the edge. The arcs are included into the tessellation based on the following rules:

- If a line-critical point exists on an edge, we connect that node to both triangle vertices on that edge. Otherwise, if an edge has no line-critical point, then we connect its vertices by an arc.
- If a face criticality does not exist then
  - If there is only one line criticality, we connect it to the opposite vertex, as in Fig. 3 $\langle 0, 1 \rangle$.
  - If there are two line-critical points, we connect each one to the other. The tessellation is not yet a triangulation. There are two possible arcs that can be added to triangulate the quadrilateral shown in Fig. 3 $\langle 0, 2 \rangle$. One, but not both, may be incident on a boundary saddle. We include this arc into the tessellation.
  - If there are three line criticalities, we connect each one to the other two, as shown in Fig. 3 $\langle 0, 3 \rangle$.
- Otherwise, if a face-critical point exists in the triangle, connect it by an arc to every other node, as shown in Fig. 3 $\langle 1, 1 \rangle$, $\langle 1, 2 \rangle$ and $\langle 1, 3 \rangle$.

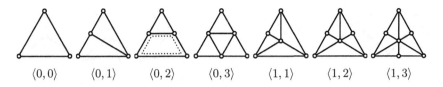

Fig. 3. Tessellation cases $\langle F, L \rangle$, where F is the number of face-critical points and L is the number of line-critical points

4   Case Analysis

The tessellation constructed using these rules satisfies the monotonicity property. We first state and prove some useful results about configurations of triangle vertices, line-critical points and face-critical points. We assume that all triangles in the triangulation are non-degenerate, i.e., every angle is between 0 and π and every triangle has non-zero, finite area. In the proofs below, we make significant use of two properties of the Reeb graph of the function defined over a triangular patch. The first property is that the Reeb graph does not contain any cycles because the domain is a topological disk [8]. The second property is that the number of extrema in the graph is twice the number of saddles.

Lemma 1. *If a triangle vertex is a boundary saddle of the function restricted to the triangle, then it lies at the intersection of the two hyperbolic asymptotes.*

Proof. We prove this by contradiction. Assume that there exists a vertex v that is a boundary saddle, but not the hyperbolic saddle point. As we sweep the level sets "downward in function value" and pass v, two contours merge or a single contour splits into two. Assume, without loss of generality, that a contour splits into two, as shown in Fig. 4. Above the value of v, the contour is contained entirely in the triangle. Below the value of v, the contour passes outside the triangle and then back in; the triangle cuts the contour into two segments. (These segments may be joined elsewhere, but locally they are distinct.) Consider the segment of the contour approaching v from the right. If this segment is to remain strictly inside the triangle until the sweep arrives at v, its tangent direction is constrained by the triangle edge to the right of v. Similarly, the tangent direction of the contour segment approaching from the left is constrained by the triangle edge to the left of v. All contours except for hyperbolic asymptotes are smooth, and so the tangent on the right of v must agree with the tangent on the left. The edges must be parallel, and therefore the triangle must be degenerate, which violates our assumption.

Lemma 2. *If a vertex of the triangular patch is a boundary saddle, then the triangle has zero face-critical points and one line-critical point.*

Proof. Considering Lemma 1, if a vertex v is a boundary saddle then v is a face criticality of a hyperbolic function. Therefore, the face criticality does not lie in

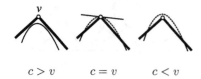

Fig. 4. A smooth contour c cannot both touch the triangle boundary at v and lie completely in the interior

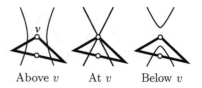

Fig. 5. A triangle containing a boundary saddle at a vertex v must contain a line-critical point on the opposite edge

the interior. The edges incident on v cannot have line criticalities because v is necessarily the line-critical point of all lines that pass through it. The edge opposite v intersects the asymptotes twice and therefore must have a line criticality, see Fig. 5.

Lemma 3. *If a triangular patch contains exactly one line criticality and no face criticality, then that line criticality is reachable by a monotone path from any other point in the triangle.*

Proof. Assume that none of the vertices is a boundary saddle. The line criticality may be an extremum or a saddle. If it is an extremum, the triangle does not contain any boundary saddle. The Reeb graph of the triangular patch is a straight line, and every contour in the patch is homeomorphic to a line segment. We can reach any point from the extremum by walking along contours that monotonically sweep the patch. If the line criticality is a saddle, then it is the only point at which the connectivity of contours in the patch change, because no other saddles exist. The Reeb graph has one internal node and three leaves. Starting from the saddle, we can walk to any other point in the patch by choosing the appropriate contour component to follow as it is swept away from the saddle. Assume that one vertex is a boundary saddle, as shown in Fig. 5. If the line-critical point is also a boundary saddle, the Reeb graph has two saddles, which implies at least four extrema. This is impossible because there are only two vertices left. Therefore, the line criticality must be an extremum. Assume without loss of generality that it is a maximum. The two vertices on that edge must be minima, so the Reeb graph has one maximum, one saddle and two minima. A monotone path exists from the maximum to all points in the triangle.

Lemma 4. *If a triangular patch contains zero face-critical points and two line-critical points, the two line-criticalities are connected by a monotone path.*

Proof. Lemma 2 implies that there are no vertex saddles. Let e_0 and e_1 be the line criticalities. If both e_0 and e_1 are saddles, then they are connected by a monotone path because they are the only two interior nodes in the Reeb graph of the triangular patch. If both are extrema, then one must be a maximum and the other a minimum and the Reeb graph is a straight line. If e_0 is a boundary saddle and e_1 is an extremum, then there is a monotone path from that boundary saddle to every point in the patch.

We now use the above lemmas to prove that in all cases our tessellations consist of monotone subpatches.

Case $\langle 0, 0 \rangle$: No line-critical point exists on any edge, and no face-critical point exists in the triangle. The function is monotone along the edges. There is exactly one maximum and one minimum, which occur at vertices. All contours in the triangle are homeomorphic to a line segment.

Case $\langle 0, 1 \rangle$: One line-critical point e exists on an edge, and no face-critical point exists in the triangle. Let v_0, v_1 and v_2 be the triangle vertices, where v_0 is the vertex opposite e. We split the triangle in two, creating patches e, v_1, v_0 and e, v_2, v_0, each containing zero face- and line criticalities. Considering Lemma 3, the arc e, v_0 is guaranteed to have a monotone embedding.

Case $\langle 0, 2 \rangle$: Two line-critical points exist, e_0 and e_1, and no face-critical point exists. We first subdivide the triangle along the monotone arc between e_0 and e_1. Considering Lemma 4, we know this arc exists. This arc splits the patch into a triangular subpatch and a quadrilateral subpatch. The triangular subpatch belongs to Case $\langle 0, 0 \rangle$. Both e_0 and e_1 may not be boundary saddles because this implies the Reeb graph of the triangular patch has two saddles and two extrema, an impossible configuration. Let e_0 be a saddle and e_1 an extremum. If we triangulate the quadrilateral by adding an arc which does not terminate at e_0, then e_0 will still be a saddle of its triangular subpatch, which violates our desired monotonicity property. To prevent this, we always triangulate the quadrilateral by adding an arc which has e_0 as an endpoint. If e_0 and e_1 are both extrema, the Reeb graph is a straight line because no other saddles exist.

Case $\langle 0, 3 \rangle$: Monotone arcs exist between all three pairs of line criticalities. Please refer to the extended version of this paper for proof [17].

Case $\langle 1, 0 \rangle$: This case is impossible. Face-critical points occur in elliptic and hyperbolic functions only. Consider a triangular patch containing an elliptic minimum. Tracking the topology of level sets during a sweep in the increasing function direction, we note that a loop, contained entirely inside the triangle, is created at the minimum. This loop cannot be destroyed without first being converted into a segment by a line criticality. Therefore, the triangle boundary should contain at least one line criticality. A similar argument holds when the triangular patch contains an elliptic maximum. Let us consider a triangular patch containing a hyperbolic face-critical point. The level set at this critical point consists of a pair of intersecting asymptotes. These asymptotes intersect the triangle boundary at four unique points. Since there are only three edges, at least one edge intersects the asymptotes twice. This triangle edge contains a line-critical point between the two intersection points.

Cases $\langle 1, 1 \rangle$, $\langle 1, 2 \rangle$, $\langle 1, 3 \rangle$: We tessellate the patch by connecting the face criticality to all vertices and line criticalities on the boundary. These new arcs are not only monotone, but they are straight lines as well, because any line-restriction containing the face criticality as an end-point is necessarily monotone. All possible critical points appear as nodes in the tessellation, all new subpatches are triangular and contain zero face and line criticalities.

5   Application to Reeb Graphs

We show how these tessellations may be used to compute a Reeb graph. Reeb graph algorithms, such as the algorithm of Cole-McClaughin *et al.* [8], operate by tracking contours of a level set during a plane sweep of the range of function values. Since all arcs of our tessellation are monotone, any contour intersects an arc at most once. Contours of function values that are not equal to any node intersect the boundary of every triangular patch twice, or not at all. We can follow a contour by starting at one arc that it intersects, and moving to another arc of the same triangle that intersects the contour.

When the domain of a function is planar, such as 2D grey-valued images or terrains/height fields, the Reeb graph contains no cycles and is called a contour tree. Efficient contour tree algorithms proceed in two distinct steps [5,7]. First, two separate trees, called the *join tree* and *split tree*, are constructed using a variety of methods. The join tree tracks topological changes in the subdomain lying above a level set, and the split tree tracks topological changes in the subdomain lying below a level set. In the second step, the join and split trees are merged to yield the contour tree. Our tessellation graph is a valid input for the join and split tree construction algorithms. Applying any of these algorithms to the tessellation graph of a piecewise-quadratic function yields the correct contour tree for that function. Carr *et al.* [7] and Chiang *et al.* [9] utilize the following properties to ensure the correctness of their algorithms [6]:

1. All critical points of the piecewise function appear as nodes in the graph.
2. For any value h, a path above (below) h exists in the graph if and only if a path above (below) h exists in the domain.

We explicitly include every potential critical point of f in the tessellation, so that the first property is satisfied. We assert that the monotonicity property of our tessellation ensures that properties 2 and 3 are also satisfied.

6   Conclusions and Future Work

We have described a tessellation scheme for piecewise-quadratic functions on surfaces. Our tessellation allows existing Reeb graph and contour tree construction algorithms to be applied to a new class of inputs, thereby extending the applications of these topological structures. We intend to develop a similar tessellation scheme for trivariate quadratic functions defined over tetrahedral meshes. We hope that this work will contribute to the development of robust and consistent topological methods for analysis of functions specified as higher-order interpolants or approximation functions over 2D and 3D domains.

References

1. Reeb, G.: Sur les points singuliers d'une forme de pfaff completement integrable ou d'une fonction numrique. Comptes Rendus Acad. Sciences **222** (1946) 847–849
2. Worsey, A., Farin, G.: Contouring a bivariate quadratic polynomial over a triangle. Comput. Aided Geom. Des. **7**(1-4) (1990) 337–351

3. Nielson, G., Hamann, B.: The asymptotic decider: resolving the ambiguity in marching cubes. In: Proc. Visualization '91, IEEE Computer Society Press (1991) 83–91

4. Nielson, G.: On marching cubes. IEEE Trans. Visualization and Comp. Graph. **9**(3) (2003) 283–297

5. Pascucci, V., Cole-McLaughlin, K.: Efficient computation of the topology of level sets. In: Proc. Visualization '02, IEEE Computer Society Press (2002) 187–194

6. Carr, H.: Topological Manipulation of Isosurfaces. PhD thesis, University of British Columbia (2004)

7. Carr, H., Snoeyink, J., Axen, U.: Computing contour trees in all dimensions. Comput. Geom. Theory Appl. **24**(2) (2003) 75–94

8. Cole-McLaughlin, K., Edelsbrunner, H., Harer, J., Natarajan, V., Pascucci, V.: Loops in reeb graphs of 2-manifolds. In: Proc. of the 19th annual symposium on computational geometry, ACM Press (2003) 344–350

9. Chiang, Y.J., Lenz, T., Lu, X., Rote, G.: Simple and optimal output-sensitive construction of contour trees using monotone paths. Comput. Geom. Theory Appl. **30**(2) (2005) 165–195

10. Bajaj, C.L., Pascucci, V., Schikore, D.R.: The contour spectrum. In: Proc. Visualization '97, IEEE Computer Society Press (1997) 167–174.

11. Takahashi, S., Nielson, G.M., Takeshima, Y., Fujishiro, I.: Topological volume skeletonization using adaptive tetrahedralization. In: Proc. Geometric Modeling and Processing 2004, IEEE Computer Society Press (2004) 227–236

12. Weber, G.H., Scheuermann, G., Hagen, H., Hamann, B.: Exploring scalar fields using critical isovalues. In: Proc. Visualization '02, IEEE Computer Society Press (2002) 171–178

13. Carr, H., Snoeyink, J., van de Panne, M.: Simplifying flexible isosurfaces using local geometric measures. In: Proc. Visualization '04, IEEE Computer Society Press (2004) 497–504

14. Hilaga, M., Shinagawa, Y., Kohmura, T., Kunii, T.L.: Topology matching for fully automatic similarity estimation of 3d shapes. In: Proce. 28th annual conference on computer graphics and interactive techniques, ACM Press (2001) 203–212

15. Zhang, E., Mischaikow, K., Turk, G.: Feature-based surface parameterization and texture mapping. ACM Trans. Graph. **24**(1) (2005) 1–27

16. Milnor, J.W.: Morse Theory. Princeton University Press, Princeton, New Jersey (1963)

17. Dillard, S.E.: Tessellation of quadratic elements. Technical report, University of California, Davis, Department of Computer Science (2006)

Effective Elections for Anonymous Mobile Agents

Shantanu Das[1], Paola Flocchini[1], Amiya Nayak[1], and Nicola Santoro[2]

[1] School of Information Technology and Engineering, University of Ottawa, Canada
{shantdas, flocchin, anayak}@site.uottawa.ca
[2] School of Computer Science, Carleton University, Canada
santoro@scs.carleton.ca

Abstract. We present distributed protocols for electing a leader among k mobile agents that are dispersed among the n nodes of a graph. While previous solutions for the agent election problem were restricted to specific topologies or under specific conditions, the protocols presented in this paper face the problem in the most general case, i.e. for an arbitrary topology where the nodes of the graph may not be distinctly labelled and the agents might be all identical (and thus indistinguishable from each other). In such cases, the agent election problem is often difficult, and sometimes impossible to solve using deterministic means. We have designed protocols for solving the problem that—unlike previous solutions—are *effective*, meaning that they always succeed in electing a leader under any given setting if at all it is possible, and otherwise detect the fact that election is impossible in that setting. We present several election protocols, all effective. Starting with the straightforward solution, that requires an exponential amount of edge-traversals by the agents, we describe significantly more efficient algorithms; in the latter the total number of edge-traversals made by the agents is always polynomial, their difference is in the amount of bits of storage they required at the nodes.

1 Introduction

1.1 The Framework

We consider the problem of leader election in distributed networked environments that support autonomous mobile agents. More specifically, there are k identical mobile agents dispersed among the n nodes of a network (or simply an undirected graph) and the agents can autonomously move from node to neighboring node throughout the network. Communication between agents is done using public whiteboards available at the nodes. An agent can communicate with another by leaving a written message at some node, which can be read by any agent visiting that node. The objective is to elect one of the agents as the leader.

We are interested in systems where both the networks and the agents are *anonymous*. The reason for this interest is because these systems provide the

T. Asano (Ed.): ISAAC 2006, LNCS 4288, pp. 732–743, 2006.
© Springer-Verlag Berlin Heidelberg 2006

(computationally) weakest environments; thus, they provide insights on the nature and amount of computational power necessary for the solvability of tasks. Furthermore, any solution protocol so designed would work without relaying upon (and thus not requiring) the availability of name servers in the system.

First observe that, in these systems, the election problem is not always solvable using deterministic algorithms. Thus, an important line of research is the investigation of under what conditions election is indeed possible, if at all, in such systems. A connected line of research is to design efficient protocols for electing a leader under specific conditions. A third related area of research is the one of designing *effective* solution protocols; that is, protocols that in each setting within finite time will elect a leader, if election is at all possible in that setting, otherwise report that the problem is not solvable in that setting. The focus of this paper is on the latter area.

The problem of leader election among mobile agents communicating by whiteboards, has been earlier studied by Barrière *et al.* [5] and Das *et al.* [9]. Both these papers solve the problem under the constraint that n (the size of the graph) and k (the number of agents) are co-prime to each-other. These solutions are therefore, not *effective* according to our definition, since there are many settings where election is possible even when n and k are not co-prime. In this paper, we aim to improve upon these solutions by designing *effective* protocols for leader election in the mobile agent model. We are also concerned about the efficiency of the proposed solutions. The main cost measure of an algorithm in this model is the number of moves performed by the agents during the execution; another measure is the minimum size of the whiteboards required for the execution of the algorithm.

1.2   Our Results

In this paper we first extend the characterization of the conditions for election in anonymous message-passing systems as given by Yamashita and Kameda[17], to the mobile agent model. Based on this characterization, we describe a straightforward technique for *effective* election (protocol *Compare-View* described in Section 2.2), showing that effective protocols are indeed possible.

We then present a more efficient yet effective solution (algorithm *Agent-Elect* described in Section 3), that achieves leader election using only $O(m \cdot k)$ agent moves for k agents in a graph with m edges. In Section 4, we give two improvements to this algorithm which bring down its memory usage at the cost of a slight increase in the total agent moves. Table 1 below shows the comparison between the various algorithms based on the two cost measures. (Here Δ indicates the maximum degree of the graph.) In comparison with these algorithms, the non-*effective* solutions presented in [5] and in [9] have cost of $O(k \cdot n)$ and $O(k \cdot m)$ edge-traversals respectively and require $O(\log n)$ bits of node memory.

1.3   Related Work

In the message-passing distributed computing model, the characterization of conditions for computability in general and election in particular has been object

Table 1. Comparison of the cost (moves vs storage) for the proposed election algorithms

Algorithm	Cost		Assumption
	Edge-traversals	Node-Storage	
Agent-Elect	$O(k\ m)$	$O(m \log n)$	one of n
Agent-Elect-2	$O(k\ m^2)$	$O(\log n)$	or k
Agent-Elect-3	$O(k\ n\ m^2)$	$O(\log \Delta)$	is known
Compare-View	$O(k\ \Delta^{2n})$	$O(1)$	n is known

of extensive investigations, starting from the pioneering work of Dana Angluin [2]. A complete characterization has been eventually provided by Yamashita and Kameda [17], and later refined by [8,15]. The present work is based on the concept of the *view* of a node, introduced in [17]. Norris [15] improved on some of these results while Boldi *et al.* [8] gave a similar characterization for directed graphs using the notion of *fibration*. Many other authors have focussed their investigations on the issue of computability in anonymous networks having specific topologies, noticeably rings [3], hypercubes [14], and tori [6]. For a recent survey see [13].

The leader election problem is also related to the problem of spanning tree construction in a graph. Many distributed algorithms have been proposed for minimum spanning tree construction in labelled graphs (i.e. where nodes are labelled with distinct identifiers), notably the one proposed by Gallager, Humblet and Spira [11]. For anonymous networks, Sakamoto[16] gave an algorithm that builds a spanning forest of the graph under a variety of initial conditions. Korach, Kutten and Moran [12] showed that the complexity of solving leader election in any arbitrary graph depends on how efficiently the graph can be traversed.

The traversal or exploration of anonymous graphs have also been studied extensively, using different models for marking the nodes (e.g. [7,10]). Another related problem in the mobile agent setting—that of gathering the agents (called the *Rendezvous* problem)—has been studied mainly for agents having distinct labels [1].

The agent election problem has been specifically studied by Barrière *et al.* [5] and Das *et al.* [9] but, as mentioned earlier, these solutions are not *effective*. Barrière *et al.* [4] have studied the agent election problem for networks where nodes are anonymous and edge/agent labels are distinct but incomparable.

2   Solving the Agent Election Problem

The Model: The network is modelled as an undirected graph $G(V, E)$, where the ordering or numbering on the nodes in V is unknown to the agents, i.e. the nodes are anonymous. At each node of the graph, the edges incident to it are locally labelled, so that an agent arriving at a node can distinguish among them.

The edge labelling of the graph G is given by $\lambda = \{\lambda_v : v \in V\}$, where for each vertex v of degree d, $\lambda_v : \{e(v, u) : u \in V \text{ and } e \in E\} \to \{1, 2, ...d\}$ is a bijection specifying the local labelling at v.

Each agent is initially located in a distinct node of the graph, called its home-base. For simplicity, we assume that no two agents have the same homebase. Those nodes which are homebases are initially marked. Thus, the initial place-ment of agents in the graph G is denoted by the bi-coloring $b : V \to \{0, 1\}$ on the set of vertices, where those vertices that are colored 1(or, black) are the homebases of the agents. The agents communicate by reading and writing in-formation on public whiteboards locally available at the nodes of the network. Access to the whiteboard is restricted by fair mutual exclusion. The agents are identical (i.e. they do not have distinct names or labels which can be used to distinguish among them) and each agent executes the same protocol. They are *asynchronous*, in the sense that every action an agent performs (computing, moving, etc.) takes a finite but otherwise unpredictable amount of time.

We define the leader election problem in mobile agent systems as follows:

The Problem: The agent election problem for k agents located in the network (G, λ, b) is said to have been *solved* when exactly one of the k agents reaches the final state 'LEADER', and all other agents reach the final state 'FOLLOWER'.

Solvable Instance: A given instance (G, λ, b) of the AEP problem is said to be *solvable*, if there exists a deterministic (distributed) algorithm $\mathcal{A}$ such that every execution of the algorithm $\mathcal{A}$ on that instance, solves the problem within some finite time.

Effective Algorithm: A deterministic agent election algorithm $\mathcal{A}$ is said to be *effective* if every execution of $\mathcal{A}$ on every instance of the problem detects whether the instance is solvable, and terminates in finite time, succeeding in solving the problem if and only if the given instance is solvable.

2.1   Characterization: Conditions for Solvability

In this paper we shall assume that the agents have prior knowledge of the value of at least one of the parameters n or k (which is a necessary condition as shown in [5]). Yamashita and Kameda [17] have determined the necessary and sufficient conditions for solution to the leader election problem (assuming that n is known) in the classical message passing network model. In that model, they introduced the concept of the *view* of a node v in a graph G, which is simply the infinite rooted tree with edge labels, that contains all (infinite) walks starting at v. In our model, we extend the concept of view to bi-colored views, where the vertices[1] in the view are colored black or white depending on whether or not they represent the homebase of some agent. The view of an agent is taken to be the bi-colored view of its homebase.

---

[1] A note on terminology: We use the word 'vertex' to refer to vertices in the view T, whereas the vertices of the original graph G are referred to as 'nodes'. Multiple vertices in the view may correspond to a single node in the graph G.

Definition 1. *The bi-colored view $T_v(G, \lambda, b)$ of node v, in the network (G, λ, b), is an infinite edge-labelled rooted tree T, whose root represents the node v and for each neighboring node u_i of v, there is a vertex x_i in T (with same color as u_i) and an edge from the root to x_i with the same labels as the edge from v to u_i in G. The subtree of T rooted at x_i is again the bi-colored view $T_{u_i}(G, \lambda, b)$ of the node u_i.*

The view from node u, truncated to a depth of h is denoted by T_u^h. An interesting observation is that the view T_u^h from node u, contains the view T_v^{h-d} for all such nodes v that are a distance of $d < h$ from node u. Thus, the view up to depth $2n$ from any node contains as sub-views, the views up to depth n, of all other nodes.

Property 1 ([15,17]). The views $T_u(G, \lambda, b)$ and $T_v(G, \lambda, b)$ of any two nodes u and v in a graph G of size n, are equal if and only if the truncated view up to depth $n - 1$, of these two nodes are equal, i.e. $T_u(G, \lambda, b) = T_v(G, \lambda, b)$ if and only if $T_u^{n-1}(G, \lambda, b) = T_v^{n-1}(G, \lambda, b)$.

The following result is known about the solvability of leader election in the message-passing network model.

Property 2 ([17]). The Leader Election Problem in a message-passing network represented by graph G having edge labelling λ, is solvable if and only if the view of a node is unique.

In [4], it was proved that there exists a simple transformation for converting a mobile agent based algorithm to distributed message passing algorithm, provided that the homebases are marked (i.e. the graph is bi-colored). This implies the following result:

Theorem 1. *The Agent Election Problem is solvable for a graph G with edge labelling λ, and bi-coloring b, if and only if the bi-colored view of an agent is unique.*

2.2   An Effective Election Protocol

When the agents know the value of n, we can use the following protocol for electing a leader (based on the approach of [17]). Each agent can traverse the graph to compute its view up to a depth of $2n - 1$ and thus obtain the views (up to depth n-1) of all other agents. Since there exists a total order on views (of the same depth), it is possible to order the agents based on their views and thus solve the leader election problem. The following algorithm implements such a strategy:

ALGORITHM *Compare-View*

Set h to $2n - 1$;
Call Construct-View(h) to get the bi-colored view T_u^h of the homebase u;
Set i to 1; $A[i] \leftarrow T_u^{n-1}$;
For each vertex v in T_u^h, which is at level less than n,

If $T_v^{n-1} \neq A[j]$ for any $1 \leq j \leq i$.
    then $i \leftarrow i + 1$; $A[i] \leftarrow T_v^{n-1}$;
If $i \neq n$, then terminate with failure;
$A_b \leftarrow \{T_x^{n-1} \in A : x \text{ is a black node}\}$; Sort the array A_b based on a fixed total order;
If $A_b[1] = T_u^{n-1}$, then become LEADER;
Else become FOLLOWER;

PROCEDURE *Construct-View*(h)

(To construct the view up to level h, for current node u)
Add the current node u to T_u^h,
If $h = 0$, return T_u^h;
Else,
    For i=1 to degree(u),
        traverse link i to reach node v_i;
        add the traversed edge and its labels to T_u^h;
        compute $T_{v_i}^{h-1} = $ Construct-View$(h-1)$, and add it to T_u^h;
    Return T_u^h;

Theorem 2. *The algorithm* Compare-View *is an* effective *election algorithm.*

Theorem 3. *The total number of agent moves in an execution of algorithm* Compare-View *is* $O(k \cdot \Delta^{2n})$ *for a graph of size n with maximum degree Δ, and k agents. The amount of node memory required by the algorithm is constant.*

3 Polynomial Solutions to the Agent Election Problem

3.1 Partial-Views

In the algorithm *Compare-View* from the previous section, each agent computes its view by traversing the complete graph which—in the absence of any topological information—takes an exponential number of moves. We want to reduce the number of the moves made by an agent to solve the problem. The approach we use for achieving that is to restrict the movements of each agent to a limited region around its homebase. Each agent would traverse only a subgraph of the original graph G and then exchange information with the other agents to obtain knowledge about the other parts of the graph. In other words, we partition the graph G into disjoint subgraphs, each *owned* by a single agent and called the agent's territory. Each agent executes the algorithm EXPLORE given below, to obtain its territory.

Algorithm *EXPLORE*:
1. Set *Path* to empty ;
    Mark the homebase as explored and include it in the Territory $\mathcal{T}$.

2. While there is another unexplored edge e at the current node u,
        mark link $\lambda_u(e)$ as a 'T'-edge and then traverse e to reach node v;
        If v is already marked (or v is a homebase),
            return back to u and re-mark the link $\lambda_u(e)$ as a 'NT'-edge;
        Otherwise
            mark v as explored and mark $\lambda_v(e)$ as a 'T'-edge;
            Add link $\lambda_v(e)$ to *Path*;
            Add edge e and node v to the territory $\mathcal{T}$;

3. When there are no more unexplored edges at the current node,
    If *Path* is not empty then,
        remove the last link from *Path*, traverse that link and repeat Step 2;
    Otherwise, Stop and return T;

After the agents execute algorithm EXPLORE, each agent has its own territory T, which forms a tree consisting of the nodes marked by the agent and the 'T'-edges joining them. It can be easily shown that the territories of the agents are disjoint from each-other and together they form a spanning forest of the graph, containing exactly k trees each rooted at the homebase of some agent.

Lemma 1. *The total number of edge traversals made by the agents in executing algorithm EXPLORE, is at most 4.m, irrespective of the number of agents.*

During algorithm EXPLORE, each agent A can label the nodes in its territory T_A by marking them with numeric identifiers, i.e. numbering them (in the order that they are visited). Thus, within the territory of an agent, each node could be uniquely identified. However, the nodes on two different trees may have the same label. So, once an agent traverses an 'NT'-edge to reach another marked node, it is generally not possible for the agent to determine if that node belongs to its own territory or that of some other agent. This fact complicates the design of our solution protocol.

Based on the territory of an agent, we define the *Partial-View* of an agent A having territory T_A, as the finite rooted tree, such that: (i) The root corresponds to the homebase v_0 of agent A. (ii) For every other node v_i in T_A, there is a vertex x_i in PV_A. (iii) For each edge (v_i, v_j) in T_A, there is a edge (x_i, x_j) in PV_A. (iv) For each outgoing edge $e = (v_i, u_i)$ such that $v_i \in T_A$ but $e \notin T_A$, PV_A contains an extra vertex y_i (called an external vertex) and an edge $\hat{e} = (x_i, y_i)$ that joins x_i to it. (v) Each edge in PV_A is marked with two labels, which are same as those of the corresponding edge in G. (vi) Each vertex in PV_A is colored black or white depending on whether the corresponding node in G is a homebase or not. (vii) Each vertex is also labelled with the numeric identifier assigned to the corresponding node of G.

During the execution of algorithm EXPLORE, each agent can build its Partial-View. We have the following important result:

Lemma 2. *If the agent election problem is solvable for an instance (G, λ, b) then, after the execution of algorithm EXPLORE on (G, λ, b), there would be at least two agents having distinct Partial-Views.*

The above result implies that we can use the Partial-Views to distinguish between some of the agents. We fix a particular encoding for the partial-views so that we can compare among the PV's of the agents. We show below how such comparisons can be used for an effective agent election algorithm.

3.2   Algorithm *Agent-Elect*

The following algorithm proceeds in phases, where in each phase, agents compete with each other to become the leader. An agent can be in one of the following states: *Active, Passive, Leader, Follower* or *Fail.* Each agent starts the algorithm in *active* state and knows the value of k at start[2]. The (encoded) Partial View of an agent A in phase i is denoted PV_{iA} and Agent-Count(A) denotes the number of homebases in the current territory of agent A.

```
ALGORITHM Agent-Elect
Execute EXPLORE to construct the territory T_A;
PV_0A ← COMPUTE-PV(T_A) ;
For phase i = 1 to k {
    If (AgentCount(A) = k ) {
        State ← Leader ;
        SEND-ALL("Success"); Exit();
    }
    SEND-ALL(PV_iA, i);
    S ← RECEIVE-PV (i);
    State ← COMPARE-PV(PV_iA, S);
    If (State = Passive) {
        SEND-MERGE(i);
        SEND-ALL("Defeated", i);
        Return to homebase and execute WAIT();
    }Else {
        RECEIVE-MERGE(i);
        execute UPDATE-PV() and continue;
    }
}
SEND-ALL("Failure"); Exit();
```

- **Procedure** SEND-ALL(M): During this procedure, agent A simply writes the message M on the whiteboard of each node in its territory.
- **Procedure** RECEIVE-PV(i): During this procedure, agent A visits each vertex u in its territory and traverses each NT-edge $e = (u, v)$ incident at u. On reaching the node v at the other end of the edge e, agent A waits till it finds the pair (PV_{iX}, i) written at node v (where PV_{iX} is an encoded Partial-View). Each Partial-View PV_{iX} that is read is added to the set S, which is returned at the end.
- **Procedure** COMPARE-PV(PV_{iA}, S): During this procedure, agent A compares its Partial-View PV_{iA} with those in the set S. If it finds any $PV_{iX} > PV_{iA}$, agent A changes its State to *Passive*. Else, for every PV_{iY} that is less than PV_{iA}, agent A stores the corresponding node v (where it was found) to the *Defeated-List*. The procedure returns the current state of the agent (Active or Passive).
- **Procedure** SEND-MERGE(i): During this procedure, the agent A returns to the node v where it found some Partial-View PV_{iX} that is greater than its own Partial-View PV_{iA}. On reaching node v, it writes (MERGE,i,$\lambda_v(e)$) on the whiteboard of node v, where e is the NT-edge joining v to T_A.
- **Procedure** RECEIVE-MERGE(i): During this procedure, agent A visits every node v in the *Defeated-List* and waits till it finds ("Defeated",i) on the whiteboard at node v. Finally agent A visits every node u in its territory and if it finds (MERGE,i,l) written at u, then the edge e having $\lambda_u(e) = l$

---

[2] With a simple modification the same algorithm can be executed if the value of n is known instead of k.

is re-marked as a T-edge (from both sides). In this case, we say that the territories at the two ends of edge e are merged.

– **Procedure** UPDATE-PV(): During this procedure an active agent updates its Partial-View and its territory as follows. For every edge e that it re-marked as T-edge during this phase, it finds the corresponding edge in its Partial-View and replaces the external node v incident on this edge with the Partial-View PV_{iX} that it read at node v. The new Partial-View obtained at the end of this procedure is called $PV_{(i+1)A}$ and the internal nodes in this partial view represents the new territory of agent A.

– **Procedure** WAIT(): This procedure is executed by a Passive agent A. The agent A simply waits at its homebase until it finds either "Success" or "Failure" written on the whiteboard. If it finds "Success", then State is changed to *Follower*; otherwise the State is changed to *Fail*, and the agent terminates the algorithm.

We have the following results showing the correctness of the above algorithm.

Definition 2. *(a) Γ_i denotes the set of agents that reach phase i of the algorithm in active state. (b) We say that the algorithm reaches phase i if at least one agent reaches phase i in active state.*

Lemma 3. *(a) During any phase i of the algorithm, the territories of the agents $\in \Gamma_i$ form a spanning forest of the graph G where each territory is a connected component of G. (b)For any phase i, if $|\Gamma_i| \geq 2$, and none of the agents in $Gamma_i$ become passive during phase i, then the AEP problem is unsolvable for the given instance (G, λ, b). (c)If at least one agent becomes passive in phase i then at least one agent reaches phase $(i+1)$ in active state.*

Theorem 4. *Given any instance (G, λ, b) of the AEP, algorithm AGENT-ELECT succeeds in electing a unique leader agent whenever (G, λ, b) is a solvable instance, and otherwise terminates with failure notification.*

Theorem 5. *The algorithm AGENT-ELECT requires $O(m \cdot k)$ agent moves, in total, for any given instance (G, λ, b) where m is the number of edges in G and k is the number of agents.*

Theorem 6. *The algorithm AGENT-ELECT requires $O(m \log n)$ memory at the nodes of the graph.*

We have the following lower bound on the cost of an effective election algorithm for mobile agents.

Lemma 4. *Any deterministic algorithm for effective leader election among k anonymous agents, dispersed in an arbitrary anonymous graph $G(V, E)$ with $|V| = n$ nodes and $|E| = m$ edges, would require $\Omega(k \cdot n)$ edge traversals, irrespective of the amount of memory available at the nodes.*

Thus from Theorem 5 and Lemma 4, we can say that the algorithm AGENT-ELECT is almost optimal in terms of agent moves, at least for sparse graphs (where $m \simeq n$).

4   Reducing the Size of the Whiteboards

Even though the number of agent moves used by algorithm AGENT-ELECT is quite efficient in terms of the number of agent moves; the memory requirements for the algorithm is much larger than that of algorithm *Compare-View* which uses constant memory. We would like to reduce the amount of memory required, without making an exponential number of agent moves. In the following we propose some modifications to our algorithm, in order to reduce the amount of memory required at the whiteboards of the nodes. As a result the number of agent moves made by the algorithm is slightly increased, even though it is still polynomial in the size of the graph.

4.1   Algorithm Agent-Elect-2

We propose the algorithm *Agent-Elect-2* which is a modified version of the previous algorithm (AGENT-ELECT) and works as follows:

Algorithm *Agent-Elect-2*
Execute EXPLORE-2 to construct the territory T_A;
$PV_{0A} \leftarrow$ COMPUTE-PV(T_A) ;
For phase $i = 1$ to k {
    If (AgentCount(A) = k) {
        State $\leftarrow$ Leader ;
        SEND-ALL("Success"); Exit();
    }
    SEND-ALL("Begin", i);
    $S \leftarrow$ RECEIVE-PV-2 (i);
    State $\leftarrow$ COMPARE-PV(PV_{iA}, S);
    If (State = Passive) {
        SEND-MERGE(i);
        SEND-ALL("Defeated", i);
        Return to homebase and execute WAIT();
    }Else {
        RECEIVE-MERGE-2(i);
        execute UPDATE-PV() and continue;
    }
}
SEND-ALL("Failure"); Exit();

- The procedure EXPLORE-2() is similar to procedure EXPLORE with the only difference that instead of marking the edges as T-edge and NT-edge, at each node v a parent-link would be stored which would point to the parent of node v in the territory tree.
- The procedure RECEIVE-PV-2() is different from RECEIVE-PV() in the following way. When an agent A executing RECEIVE-PV-2() reaches an external node v to read the Partial-View, it traverses the territory T_B that contains v (using the Parent-links), and computes the Partial-View PV_{iB}.
- The procedure RECEIVE-MERGE-2() is again similar to RECEIVE-MERGE() with the following changes. Whenever an agent A has to merge its territory T_A with the territory T_B at other end of an external edge (u,v), agent A sets the Parent-link at node v to point to node u and then updates (i.e. reverses) the Parent-Link for each edge on the path from v to the root of T_B.

Lemma 5. *The output of procedure RECEIVE-PV-2 is identical to the output of procedure RECEIVE-PV.*

The only difference between the two algorithms is that in the modified algorithm, an agent computes the Partial-View of its neighboring agents, instead of reading it from the whiteboard, as in the original algorithm AGENT-ELECT. Thus, the correctness of the algorithm *Agent-Elect-2* follows from the correctness of AGENT-ELECT, due to the above lemma.

Theorem 7. *The algorithm* Agent-Elect-2 *performs* $O(k \cdot m^2)$ *agent moves in total and requires* $O(\log n)$ *bits of node memory.*

4.2   Algorithm Agent-Elect-3

In order to further reduce the memory requirement of our algorithm, we can make the following modifications to algorithm *Agent-Elect-2*:

1. The procedure EXPLORE-2 is replaced by procedure EXPLORE-3, with the change that EXPLORE-3 would not explicitly write the labels of the nodes on the whiteboard of the nodes.
2. The procedure RECEIVE-PV-2() would be replaced by procedure RECEIVE-PV-3() where an agent A that is computing the Partial-View of a neighbor B and needs to read the label of a node x *external* to T_B, computes the label[3] by traversing the tree containing x.
3. During the execution of the algorithm, whenever an agent A has to write the phase number i on the whiteboard, it will write $(i \mod 3)$ instead.

This new algorithm is called *Agent-Elect-3*.

Theorem 8. *The algorithm* Agent-Elect-3 *performs* $O(k \cdot n \cdot m^2)$ *agent moves in total and requires* $O(\log \Delta)$ *bits of node memory.*

Acknowledgements

The authors would like to thank Masafumi Yamashita for the many helpful discussions.

References

1. S. Alpern and S. Gal. *The Theory of Search Games and Rendezvous.* Kluwer, 2003.
2. D. Angluin. Local and global properties in networks of processors. In *Proc. 12th ACM Symp. on Theory of Computing* (STOC '80), 82–93, 1980.
3. H. Attiya, M. Snir, and M.K. Warmuth. Computing on an anonymous ring. *Journal of ACM*, 35(4), 845–875, 1988.

---

[3] The label assigned to a node v belonging to tree $\mathcal{T}$, is uniquely determined as the rank of the path to v from the root of $\mathcal{T}$, when compared with the paths to other nodes belonging to the same tree $\mathcal{T}$.

4. L. Barrière, P. Flocchini, P. Fraigniaud, and N. Santoro. Can we elect if we cannot compare? In *Proc. 15th ACM Symp. on Parallel Algorithms and Architectures* (SPAA '03), 200–209, 2003.

5. L. Barrière, P. Flocchini, P. Fraigniaud, and N. Santoro. Election and rendezvous in fully anonymous networks with sense of direction. *Theory of Computing Systems*, 2006 (to appear). Preliminary version in *Proc. 10th Coll. on Structural Information and Communication Complexity* (SIROCCO '03), 17–32, 2003.

6. P.W. Beame and H.L. Bodlaender. Distributed computing on transitive grids: The torus. In *Proc. Symp. Theor. Aspects of Computer Science* (STACS '89), 294–303, 1989.

7. M. Bender, A. Fernandez, D. Ron, A. Sahai, and S. Vadhan. The power of a pebble: Exploring and mapping directed graphs. In *Proc. 30th ACM Symp. on Theory of Computing* (STOC '98), 269–287, 1998.

8. P. Boldi, S. Shammah, S. Vigna, B. Codenotti, P. Gemmell and J. Simon. Symmetry breaking in anonymous networks: Characterizations. In *Proc. 4th Israel Symp. on Theory of Computing and Systems*, 16–26, 1996.

9. S. Das, P. Flocchini, S. Kutten, A. Nayak, and N. Santoro. Map construction of unknown graphs by multiple agents. *Theo. Comp. Sci.* (Submitted). Preliminary version in *Proc. 12th Coll. on Structural Information and Communication Complexity* (SIROCCO '05), 99–114, 2005.

10. P. Fraigniaud and D. Ilcinkas. Digraph exploration with little memory. In *Proc. 21st Symp. on Theoretical Aspects of Computer Science* (STACS '04), 246–257, 2004.

11. R.G. Gallager, P.A. Humblet, and P.M. Spira. A distributed algorithm for minimum-weight spanning trees. *ACM Transactions on Programming Languages and Systems*, 5(1), 66–77, 1983.

12. E. Korach, S. Kutten, S. Moran. A modular technique for the design of efficient distributed leader finding algorithms. *ACM Transactions on Programming Languages and Systems*, 12(1), 84–101, 1990.

13. E. Kranakis. Symmetry and computability in anonymous networks: A brief survey. In *Proc. 3rd Int. Conf. on Structural Information and Communication Complexity* (SIROCCO'97), 1–16, 1997.

14. E. Kranakis and D. Krizanc. Distributed computing on anonymous hypercube networks. *Journal of Algorithms*, 23(1), 32–50, 1997.

15. N. Norris. Universal covers of graphs: Isomorphism to depth $n - 1$ implies isomorphism to all depths. *Discrete Applied Mathematics*, 56(1), 61–74, 1995.

16. N. Sakamoto. Comparison of initial conditions for distributed algorithms on anonymous networks. In *Proc. 18th ACM Symposium on Principles of Distributed Computing* (PODC '99), 173–179, 1999.

17. M. Yamashita and T. Kameda. Computing on anonymous networks: Parts I and II. *IEEE Trans. on Parallel and Distributed Systems*, 7(1), 69–96, 1996.

Gathering Asynchronous Oblivious Mobile Robots in a Ring

Ralf Klasing[1], Euripides Markou[2,*], and Andrzej Pelc[3,**]

[1] LaBRI - Université Bordeaux 1 - CNRS, 351 cours de la Libération,
33405 Talence cedex, France
klasing@labri.fr
[2] School of Computational Engineering & Science, McMaster University, 1280 Main
Street West, Hamilton, Ontario L8S4K1, Canada
emarkou@di.uoa.gr
[3] Département d'informatique, Université du Québec en Outaouais, Gatineau,
Québec J8X 3X7, Canada
pelc@uqo.ca

Abstract. We consider the problem of gathering identical, memoryless, mobile robots in one node of an anonymous unoriented ring. Robots start from different nodes of the ring. They operate in Look-Compute-Move cycles and have to end up in the same node. In one cycle, a robot takes a snapshot of the current configuration (Look), makes a decision to stay idle or to move to one of its adjacent nodes (Compute), and in the latter case makes an instantaneous move to this neighbor (Move). Cycles are performed asynchronously for each robot. For an odd number of robots we prove that gathering is feasible if and only if the initial configuration is not periodic, and we provide a gathering algorithm for any such configuration. For an even number of robots we decide feasibility of gathering except for one type of symmetric initial configurations, and provide gathering algorithms for initial configurations proved to be gatherable.

Keywords: asynchronous, mobile robot, gathering, ring.

1 Introduction

Mobile entities (robots), initially situated at different locations, have to gather at the same location (not determined in advance) and remain in it. This problem of distributed self-organization of mobile entities is known in the literature as the gathering problem. The main difficulty of gathering is that robots have to break symmetry by agreeing on a common meeting location. This difficulty is aggravated when (as in our scenario) robots cannot communicate directly but have to make decisions about their moves only by observing the environment.

* Research partly supported by the European Research Training Network COMB-STRU HPRN-CT-2002-00278 and MITACS.
** Research partly supported by NSERC discovery grant, by the Research Chair in Distributed Computing at the Université du Québec en Outaouais, and by a visiting fellowship from LaBRI.

T. Asano (Ed.): ISAAC 2006, LNCS 4288, pp. 744–753, 2006.

We study the gathering problem in a scenario which, while very simple to describe, makes the symmetry breaking component particularly hard. Consider an unoriented anonymous ring of stations (nodes). Neither nodes nor links of the ring have any labels. Initially, some nodes of the ring are occupied by robots and there is at most one robot in each node. The goal is to gather all robots in one node of the ring and stop. Robots operate in Look-Compute-Move cycles. In one cycle, a robot takes a snapshot of the current configuration (Look), then, based on the perceived configuration, makes a decision to stay idle or to move to one of its adjacent nodes (Compute), and in the latter case makes an instantaneous move to this neighbor (Move). Cycles are performed asynchronously for each robot. This means that the time between Look, Compute, and Move operations is finite but unbounded, and is decided by the adversary for each robot. The only constraint is that moves are instantaneous, and hence any robot performing a Look operation sees all other robots at nodes of the ring and not on edges, while performing a move. However a robot R may perform a Look operation at some time t, perceiving robots at some nodes, then Compute a target neighbor at some time $t' > t$, and Move to this neighbor at some later time $t'' > t'$ in which some robots are in different nodes from those previously perceived by R because in the meantime they performed their Move operations. Hence robots may move based on significantly outdated perceptions, which adds to the difficulty of achieving the goal of gathering. It should be stressed that robots are memoryless (oblivious), i.e. they do not have any memory of past observations. Thus the target node (which is either the current position of the robot or one of its neighbors) is decided by the robot during a Compute operation solely on the basis of the location of other robots perceived in the previous Look operation. Robots are anonymous and execute the same deterministic algorithm. They cannot leave any marks at visited nodes, nor send any messages to other robots.

This very weak scenario, similar to that considered in [1, 3, 5, 6, 10, 13, 14], is justified by the fact that robots may be very small, cheap and mass-produced devices. Adding distinct labels, memory, or communication capabilities makes production of such devices more difficult, and increases their size and price, which is not desirable. Thus it is interesting to consider such a scenario from the point of view of applications. On the theoretical side, this weak scenario increases the difficulty of gathering by making the problem of symmetry breaking particularly hard, and thus provides an interesting setting to study this latter issue in a distributed environment.

It should be noted that the gathering problem under the scenario described above is related to the well-known leader election problem (cf. e.g. [12]) but is harder than it for the following reason. If robots in the initial configuration cannot elect a leader among nodes (this happens for all periodic configurations and for some symmetric configurations) then gathering is impossible (see Section 3). However, even if leader election is possible in the initial configuration, this does not necessarily guarantee feasibility of gathering. Indeed, while the node elected as a leader is a natural candidate for the place to gather, it is not clear how to preserve the same target node during the gathering process, due to its asynchrony. (Recall that nodes do not have labels, and configurations perceived by robots during their Look operation change during the gathering process, thus robots may not "recognize" the previously elected node later on.)

An important and well studied capability in the literature on robot gathering is the *multiplicity detection* [10, 14]. This is the ability of the robots to perceive, during the Look operation, if there is one or more robots in a given location. In our case, we prove that without this capability, gathering of more than one robot is always impossible. Thus we assume the capability of multiplicity detection in our further considerations. It should be stressed that, during a Look operation, a robot can only tell if at some node there are no robots, there is one robot, or there are more than one robots: a robot does not see a difference between a node occupied by a or b robots, for distinct $a, b > 1$.

Related work. The problem of gathering mobile robots in one location has been extensively studied in the literature. Many variations of this task have been considered. Robots move either in a graph, cf. e.g., [2, 7, 8, 9, 11], or in the plane [1, 3, 4, 5, 6, 10, 13, 14, 15], they are labeled [7, 8, 11], or anonymous [1, 3, 4, 5, 6, 10, 13, 14, 15], gathering algorithms are probabilistic (cf. [2] and the literature cited there), or deterministic [1, 3, 4, 5, 6, 7, 9, 10, 11, 13, 14, 15]. Deterministic algorithms for gathering robots in a ring (which is a task closest to our current setting) have been studied e.g., in [7, 8, 9, 11]. In [7, 8, 11] symmetry was broken by assuming that robots have distinct labels, and in [9] it was broken by using tokens.

To the best of our knowledge, the very weak assumption of anonymous identical robots that cannot send any messages and communicate with the environment only by observing it, was used to study deterministic gathering only in the case of robots moving freely in the plane [1, 3, 4, 5, 6, 10, 13, 14, 15]. The scenario was further precised in various ways. In [4] it was assumed that robots have memory, while in [1, 3, 5, 6, 10, 13, 14, 15] robots were oblivious, i.e., it was assumed that they do not have any memory of past observations. Oblivious robots operate in Look-Compute-Move cycles, similar to those described in our scenario. The differences are in the amount of synchrony assumed in the execution of the cycles. In [3, 15] cycles were executed synchronously in rounds by all active robots, and the adversary could only decide which robots are active in a given cycle. In [4, 5, 6, 10, 13, 14, 15] they were executed asynchronously: the adversary could interleave operations arbitrarily, stop robots during the move, and schedule Look operations of some robots while others were moving. It was proved in [10] that gathering is possible in the asynchronous model if robots have the same orientation of the plane, even with limited visibility. Without orientation, the gathering problem was positively solved in [5], assuming that robots have the capability of multiplicity detection. A complementary negative result concerning the asynchronous model was proved in [14]: without multiplicity detection, gathering robots that do not have orientation is impossible.

Our scenario is the most similar to the asynchronous model used in [10, 14]. The only difference is in the execution of Move operations. This has been adapted to the context of the ring of stations (nodes): moves of the robots are executed instantaneously from a node to its neighbor, and hence robots always see other robots at nodes. All possibilities of the adversary concerning interleaving operations performed by various robots are the same as in the model from [10, 14], and the characteristics of the robots (anonymity, obliviousness, multiplicity detection) are also the same.

Our results. For an odd number of robots we prove that gathering is feasible if and only if the initial configuration is not periodic, and we provide a gathering algorithm for any such configuration. For an even number of robots we decide feasibility of gathering except for one type of symmetric configurations, and provide gathering algorithms for initial configurations proved to be gatherable.

Due to space limitations, most of the proofs have been omitted and will appear in the full version of the paper.

2   Terminology and Preliminaries

We consider an n-node anonymous unoriented ring. Initially, some nodes of the ring are occupied by robots and there is at most one robot in each node. The number of robots is denoted by k. During the gathering process robots move, and at any time they occupy some nodes of the ring, forming a *configuration*. A configuration is denoted by a pair of sequences $((a_1, \ldots, a_r), (b_1, \ldots, b_s))$, where the integers a_i and b_j have the following meaning. Choose an arbitrary node occupied by at least one robot as node u_1 and consider consecutive nodes $u_1, u_2, u_3, \ldots, u_r$, occupied by at least one robot, starting from u_1 in the clockwise direction. (Clockwise direction is introduced only for the purpose of definition, robots do not have this notion, as the ring is not oriented.) Integer a_i, for $i < r$, denotes the distance in the ring between nodes u_i and u_{i+1}, and integer a_r denotes the distance between nodes u_r and u_1 (in the clockwise direction). Next, consider those nodes among $u_1, u_2, u_3, \ldots, u_r$ which are occupied by more than one robot. Such nodes are called *multiplicities*. Suppose that $u_{v_1}, \ldots, u_{v_s}$ are these consecutive nodes (ordered in clockwise direction). Integer b_i is defined as the distance in the clockwise direction between node u_1 and node u_{v_i}. It should be clear that different choices of node 1 give rise to different pairs of sequences. Respective sequences in these pairs are cyclic shifts of each other and correspond to the same positioning of robots. So formally a configuration should be defined as an equivalence class of a pair of sequences with respect to those shifts. To simplify notation we will use pairs of sequences instead of those classes, and for configurations without multiplicities we will drop the second sequence, simply using sequence $(a_1, \ldots, a_r)$.

Consider a configuration $C = (a_1, \ldots, a_r)$ without multiplicities. The *range* of the configuration C is the set $\{a_1, \ldots, a_r\}$. For any integer a_i in the range of C, the *weight* of a_i is the number of times this integer appears in the sequence $(a_1, \ldots, a_r)$. C is called *periodic* if the sequence $(a_1, \ldots, a_r)$ is a concatenation of at least two copies of a subsequence p. The configuration C can be also represented as the set Z of nodes occupied by the robots. C is called *symmetric* if there exists an axis of symmetry of the ring, such that the set Z is symmetric with respect to this axis. If the number of robots is odd and S is an axis of symmetry of the set Z then there is exactly one robot on the axis S. This robot is called *axial* for this axis. Two robots are called *neighboring*, if at least one of the two segments of the ring between them does not contain any robots. A segment of the ring between two neighboring robots is called *free* if there is no robot in this segment.

We now describe formally what a robot perceives during a Look operation. Fix a robot R in a configuration represented by a pair of sequences $((a_1, \ldots, a_r),$ $(b_1, \ldots, b_s))$, where this particular representation is taken with respect to the node occupied by R (i.e., this node is considered as node u_1). The *view* of robot R is the set of two pairs of sequences $\{((a_1, \ldots, a_r), (b_1, \ldots, b_s)), ((a_r, a_{r-1}, \ldots, a_1),$ $(n - b_s, \ldots, n - b_1))\}$ (if the node occupied by R is a multiplicity then we define the view of R as $\{((a_1, \ldots, a_r), (0, b_2, \ldots, b_s)), ((a_r, a_{r-1}, \ldots, a_1), (0, n - b_s, \ldots, n - b_2))\}$). This formalization captures the fact that the ring is unoriented and hence the robot R cannot distinguish between a configuration and its symmetric image, if R is itself on the axis of symmetry. This is conveyed by defining the view as the *set* of the two couple of sequences because the sets $\{((a_1, \ldots, a_r), (b_1, \ldots, b_s)), ((a_r, a_{r-1}, \ldots, a_1), (n - b_s, \ldots, n - b_1))\}$ and $\{((a_r, a_{r-1}, \ldots, a_1), (n - b_s, \ldots, n - b_1)), ((a_1, \ldots, a_r), (b_1, \ldots, b_s))\}$ are equal. As before, if there are no multiplicities, we will drop the second sequence in each case and write the view as the set of two sequences: $\{(a_1, \ldots, a_r), (a_r, a_{r-1}, \ldots, a_1)\}$. For example, in a 9-node ring with consecutive nodes $1, \ldots, 9$ and three robots occupying nodes 1,2,4, the view of robot R at node 1 is the set $\{(1, 2, 6), (6, 2, 1)\}$.

A configuration without multiplicities is called *rigid* if the views of all robots are distinct. We will use the following geometric facts.

Lemma 1. *1. A configuration without multiplicities is non-rigid, if and only if it is either periodic or symmetric.*
 2. If a configuration without multiplicities is non-rigid and non-periodic then it has exactly one axis of symmetry.

Consider a configuration without multiplicities that is non-rigid and non-periodic. Then it is symmetric. Let S be its unique axis of symmetry. If the number of robots is odd then exactly one robot is situated on S and S goes through the antipodal node if the size n of the ring is even, and through the (middle of the) antipodal edge if n is odd. If the number of robots is even then two cases are possible:

- *edge-edge symmetry* : S goes through (the middles of) two antipodal edges;
- *node-on-axis symmetry* : at least one node is on the axis of symmetry.

Note that the first case can occur only for an even number of robots in a ring of even size.

We now establish two basic impossibility results. Note that similar results have been proven for gathering robots on the plane. However, these results do not directly imply ours.

Proposition 1. *1. Gathering any 2 robots is impossible on any ring.*
 2. If multiplicity detection is not available then gathering any $k > 1$ robots is impossible on any ring.

Proposition 1 justifies the two assumptions made throughout this paper: the number k of robots is at least 3 and robots are capable of multiplicity detection.

All our algorithms describe the Compute part of the cycle of robots' activities. They are written from the point of view of a robot R that got a view in a Look operation and computes its next move on the basis of this view.

The rest of the paper is organized as follows. In Section 3 we first establish two impossibility results: gathering is not feasible for periodic and edge-edge symmetric configurations. We then describe a procedure to gather configurations containing exactly one multiplicity and finally we propose a gathering procedure for rigid configurations. In Section 4 we give the complete solution of the gathering problem for any odd number of robots. Section 5 concludes the paper with a discussion of gathering for an even number of robots and with open problems.

3   Gatherable Configurations

In this section we first show two impossibility results. The first one concerns any number of robots, while the second one concerns only the case of an even number of robots on a ring of even size.

Theorem 1. *Gathering is impossible for any periodic configuration.*

Theorem 2. *Gathering is impossible for any edge-edge symmetric configuration.*

We now show a gathering procedure for any configuration containing exactly one multiplicity, say at node v.

Procedure Single-Multiplicity-Gathering

if R is at the multiplicity **then** do not move
else
    if none of the segments between R and the multiplicity is free
    then do not move
    else move towards the multiplicity along the shortest of the free
        segments or along any of them in the case of equality.

The idea is to gather all robots at v, avoiding creating another multiplicity (which could potentially create a symmetry, making the gathering process harder or even impossible). Procedure Single-Multiplicity-Gathering achieves this goal by first moving the robots closest to v towards v, then moving there the second closest robots, and so on.

Lemma 2. *Procedure Single-Multiplicity-Gathering performs gathering of robots for any configuration with a single multiplicity.*

Now we describe a gathering procedure for any rigid configuration, regardless of the number of robots.

The main idea of the procedure is to elect a single robot and move it until it hits one of its neighboring robots, thus creating a single multiplicity, and then to apply Procedure Single-Multiplicity-Gathering. The elected robot must be such that during its walk the rigidity property is not lost. In order to achieve this goal, we perform the election as follows. First the robots elect a pair of neighboring robots at maximum distance (there may be several such pairs, whence the need for election). Then they choose among them the robot which has the other neighboring robot closer. Ties can be broken easily.

In order to elect a robot we need to linearly order all possible views. This can be done in many ways. One of them is to order lexicographically all finite sequences of integers and number them by consecutive natural numbers. Then a view becomes a set of two natural numbers. Treat these sets as ordered pairs of natural numbers in increasing order, order these pairs lexicographically, and assign them consecutive natural numbers in increasing order. We fix the resulting linear order of views and this numbering beforehand, adding it to the algorithm for all robots. We call this procedure *Rigid-Gathering*.

Lemma 3. *Procedure Rigid-Gathering performs gathering of robots for any rigid configuration without multiplicities.*

4   Gathering an Odd Number of Robots

In this section we present a gathering algorithm for any non-periodic configuration of an odd number of robots. Together with Theorem 1 this solves the gathering problem for an odd number of robots.

Algorithm Odd-Gathering

if the configuration is periodic **then** output: gathering impossible
else

       if the configuration has a single multiplicity
       then Single-Multiplicity-Gathering
       else

              if the configuration is rigid **then** Rigid-Gathering
              else

                     if R is axial **then** move (to any of the adjacent nodes)

The idea of the algorithm is the following. Consider any non-periodic configuration of an odd number of robots (recall that initially there are no multiplicities). If it is rigid then apply Procedure Rigid-Gathering. Otherwise it must be symmetric, by Lemma 1. There is a unique axial robot for its unique axis of symmetry. Move this robot to any adjacent node. We prove that three cases can occur. (1) The resulting situation has a multiplicity (the adjacent node was occupied by a robot): then apply Procedure Single-Multiplicity-Gathering. (2) The resulting configuration is rigid: then apply Procedure Rigid-Gathering. (3) Another axis of symmetry has been created (the previous one has been obviously destroyed by the move). In this case there is a unique axial robot for the unique axis of symmetry. Move this robot to any adjacent node. Again one of the three above cases can occur. We prove that after a finite number of such moves, only cases (1) or (2) can occur, and thus gathering is finally accomplished either by applying Procedure Single-Multiplicity-Gathering or by applying Procedure Rigid-Gathering. In the proof of the correctness of Algorithm Odd-Gathering we will use the following lemmas.

Lemma 4. *Let C be a symmetric configuration of an odd number of robots, without multiplicities. Let C' be the configuration resulting from C by moving the axial node to any of the adjacent nodes. Assume that C' does not have multiplicities. Then C' is not periodic.*

Let C be a symmetric non-periodic configuration of an odd number of robots, without multiplicities. The unique value of odd weight in the configuration C is called the *chief* of C. Let C' be the configuration resulting from C by moving the axial robot to any of the adjacent nodes. If C' does not have multiplicities and is symmetric then we will call it *special*. The subset of the range of a special configuration C' consisting of integers of the same parity as that of the chief is called the *white part* of the range, and its complement is called the *black part* of the range. We denote by $b(C')$ the total number of occurrences in C' of integers from the black part of its range.

Lemma 5. *Consider a sequence $(C_1, C_2, \ldots)$ of special configurations, such that C_{i+1} results from C_i by moving the axial robot to any of the adjacent nodes. Then for some $i \leq k$, we have $b(C_i) = 0$.*

Lemma 6. *Consider a special configuration C, with $b(C) = 0$. Let C' be the configuration resulting from C by moving the axial robot to any of the adjacent nodes. If C' does not have multiplicities then it is not symmetric.*

We are now ready to prove the correctness of Algorithm Odd-Gathering.

Theorem 3. *Algorithm Odd-Gathering performs gathering of any non-periodic configuration of an odd number of robots.*

Proof. Consider an initial non-periodic configuration C of an odd number of robots. By assumption it does not contain multiplicities. If it is rigid then we are done by Lemma 3. Otherwise, it must be symmetric by Lemma 1. Let A be its unique axial robot. Let C_1 be the configuration resulting from C by moving robot A to any of the adjacent nodes. If C_1 contains a multiplicity then we are done by Lemma 2. If C_1 is rigid then we are done by Lemma 3. Otherwise, C_1 is either periodic or symmetric, in view of Lemma 1. By Lemma 4, it cannot be periodic, hence it must be symmetric, and thus special. Consider the configuration C_2 resulting by moving the axial robot of C_1 to any of the adjacent nodes. Again C_2 either contains a multiplicity, or is rigid, or is special. In the first two cases we are done, and in the third case the axial robot is moved again. In this way we create a sequence $C_1, C_2, \ldots$ of special configurations. By Lemma 5, there is a configuration C_i in this sequence, with $b(C_i) = 0$. Let C' be the configuration resulting from C_i by moving the axial robot to any of the adjacent nodes. By Lemma 6, the configuration C' either has a multiplicity, or cannot be symmetric, and thus must be rigid. In the first case we are done by Lemma 2 and in the second case by Lemma 3.   □

Theorem 3 and Theorem 1 imply the following corollary.

Corollary 1. *For an odd number of robots, gathering is feasible if and only if the initial configuration is not periodic.*

5   Conclusion

We completely solved the gathering problem for any odd number of robots, by characterizing configurations possible to gather (these are exactly non-periodic configurations) and providing a gathering algorithm for all these configurations. Corollary 1 is equivalent to the following statement: for an odd number of robots, gathering is feasible if and only if in the initial configuration, robots can elect a node occupied by a robot.

For an even number of robots, we proved that gathering is impossible if either the number of robots is 2, or the configuration is periodic, or when it has an edge-edge symmetry. On the other hand, we provided a gathering algorithm for all rigid configurations. This leaves unsettled one type of configurations: symmetric non-periodic configurations of an even number of robots with a node-on-axis type of symmetry. These are symmetric non-periodic configurations in which at least one node is situated on the unique axis of symmetry. This (these) node(s) may or may not be occuppied by robots. In this case, the symmetry can be broken by initially electing one of the axial nodes. This node is a natural candidate for the place to gather. However, it is not clear how to preserve the same target node during the gathering process, due to its asynchrony. Unlike in our gathering algorithm for an odd number of robots, where only one robot moves until a multiplicity is created, in the case of the above symmetric configuration of an even number of robots, some robots would have to move together. This creates many possible outcomes of Look operations for other robots, in view of various possible behaviors of the adversary, which can interleave their actions. We note here that for an even number of robots there are cases where gathering is feasible even when robots cannot initially elect a node occupied by a robot (they will be included in the full version of the paper).

The complete solution of the gathering problem for an even number of robots remains a challenging open question left by our research. We conjecture that in the unique case left open (non-periodic configurations of an even number of robots with a node-on-axis symmetry), gathering is always feasible. In view of our results, this is equivalent to the following statement.

Conjecture: For an even number of more than 2 robots, gathering is feasible if and only if the initial configuration is not periodic and does not have an edge-edge symmetry.

The validity of this conjecture would imply that, for any number of more than 2 robots, gathering is feasible if and only if, in the initial configuration robots can elect a node (not necessarily occupied by a robot).

References

1. N. Agmon, D. Peleg: Fault-Tolerant Gathering Algorithms for Autonomous Mobile Robots. SIAM J. Comput. 36(1): 56-82 (2006).
2. S. Alpern, S. Gal: The Theory of Search Games and Rendezvous, Kluwer Academic Publishers, 2002.
3. H. Ando, Y. Oasa, I. Suzuki, M. Yamashita: Distributed Memoryless Point Convergence Algorithm for Mobile Robots with Limited Visibility. IEEE Trans. on Robotics and Automation 15(5): 818-828 (1999).
4. M. Cieliebak: Gathering Non-oblivious Mobile Robots. Proc. 6th Latin American Symposium on Theoretical Informatics (LATIN 2004): 577-588.
5. M. Cieliebak, P. Flocchini, G. Prencipe, N. Santoro: Solving the Robots Gathering Problem. Proc. 30th International Colloquium on Automata, Languages and Programming (ICALP 2003), LNCS 2719: 1181-1196.
6. R. Cohen, D. Peleg: Robot Convergence via Center-of-Gravity Algorithms. Proc. 11th International Colloquium on Structural Information and Communication Complexity (SIROCCO 2004), LNCS 3104: 79-88.
7. G. De Marco, L. Gargano, E. Kranakis, D. Krizanc, A. Pelc, U. Vaccaro: Asynchronous deterministic rendezvous in graphs. Proc. 30th Inter. Symp. on Mathematical Foundations of Computer Science, (MFCS 2005), LNCS 3618: 271-282.
8. A. Dessmark, P. Fraigniaud, D. Kowalski, A. Pelc: Deterministic rendezvous in graphs. Algorithmica, to appear.
9. P. Flocchini, E. Kranakis, D. Krizanc, N. Santoro, C. Sawchuk: Multiple Mobile Agent Rendezvous in a Ring. Proc. 6th Latin American Symposium on Theoretical Informatics (LATIN 2004): 599-608.
10. P. Flocchini, G. Prencipe, N. Santoro, P. Widmayer: Gathering of Asynchronous Robots with Limited Visibility. Theoretical Computer Science 337(1-3): 147-168 (2005).
11. D. Kowalski, A. Pelc: Polynomial deterministic rendezvous in arbitrary graphs. Proc. 15th Annual Symposium on Algorithms and Computation (ISAAC'2004), LNCS 3341: 644-656.
12. N. Lynch: Distributed Algorithms, Morgan Kaufman 1996.
13. G. Prencipe: CORDA: Distributed Coordination of a Set of Autonomous Mobile Robots. Proc. ERSADS 2001: 185-190.
14. G. Prencipe: On the Feasibility of Gathering by Autonomous Mobile Robots. Proc. 12th International Colloquium on Structural Information and Communication Complexity (SIROCCO 2005), LNCS 3499: 246-261.
15. I. Suzuki, M. Yamashita: Distributed Anonymous Mobile Robots: Formation of Geometric Patterns. SIAM J. Comput. 28(4): 1347-1363 (1999).
16. M. Yamashita, T. Kameda: Computing on Anonymous Networks: Parts I and II. IEEE Trans. Parallel Distrib. Syst. 7(1): 69-96 (1996).

Provably Secure Steganography and the Complexity of Sampling*

Christian Hundt[1], Maciej Liśkiewicz[1], and Ulrich Wölfel[2]

[1] Institut für Theoretische Informatik, Universität zu Lübeck, Germany
{chundt, liskiewi}@tcs.uni-luebeck.de
[2] Bundesamt für Sicherheit in der Informationstechnik, Bonn, Germany
ulrich.woelfel@bsi.bund.de

Abstract. Recent work on theoretical aspects of steganography resulted in the construction of oracle-based stegosystems. It has been shown that these can be made secure against the steganography equivalents of common cryptographic attacks. In this paper we use methods from complexity theory to investigate the efficiency of sampling from practically relevant types of channels. We show that there are channels that cannot be efficiently used in oracle-based stegosystems. By classifying channels based on their usability for stegosystems, we provide a means to select suitable channels for their practical implementation.

1 Introduction

Alice and Bob want to communicate via a public channel $\mathcal{C}$ which is carefully monitored by Eve. The aim of steganography is for Alice to secretly send a message to Bob via $\mathcal{C}$, such that Eve cannot distinguish between "typical" communication on the channel $\mathcal{C}$ and communication that contains hidden messages.

In this paper we investigate oracle-based stegosystems, i.e., systems that have access to an oracle which samples according to a given channel $\mathcal{C}$. Research in this area started with works by Hopper et al. [1], who were the first to study steganography from a complexity theoretic point of view. The stegosystem they analyse uses rejection-sampling, which was first informally described in [2], and the main result of [1] says that the stegosystem is secure under standard cryptographic assumptions. Subsequent papers extend these analyses to public-key steganography [3], higher embedding rates [4], different attack types [5] or different types of oracles [6]. The fundamental building block of all the systems is the rejection-sampling according to a given channel $\mathcal{C}$.

Motivated by these theoretical results, the question arises whether such stegosystems could be practically implemented. We will therefore analyse oracle-based stegosystems from a complexity theoretic point of view. For any communication channel $\mathcal{C}$ it is natural to assume the existence of an oracle $M_{\mathcal{C}}$ sampling according to the channel distribution. Examples of freely sampleable channels are images from digital cameras or text written in natural languages. In the case

---

* Supported by DFG research grant RE 672/5-1.

T. Asano (Ed.): ISAAC 2006, LNCS 4288, pp. 754–763, 2006.

of oracle-based stegosystems, particularly rejection-sampling steganography, a stronger construction is needed, namely an oracle $M_{\mathcal{C}_h}$ that draws from the channel $\mathcal{C}_h$ with a distribution conditioned on the history h of already drawn samples. It is important to note the big difference between both types of oracles and it is not trivial to construct $M_{\mathcal{C}_h}$ from a given $M_{\mathcal{C}}$ in "natural" situations. In this paper we analyse the efficiency of such constructions.

Our main result states that there exist channels $\mathcal{C}$ that can be efficiently sampled by an oracle $M_{\mathcal{C}}$, but for which no efficiently sampling oracle $M_{\mathcal{C}_h}$ can be constructed, unless some widely believed complexity theoretic assumptions, like P $\neq$ NP, are false. For our analyses we describe a scenario in which Alice and Bob communicate using a set of context free languages. We thus introduce a connection between formal languages and channel distributions in order to apply results from complexity theory to channels. Furthermore, we characterise those properties of a channel which either lead to the existence of an oracle $M_{\mathcal{C}_h}$ that can be sampled efficiently or cause the sampling from $\mathcal{C}_h$ to be intractable. This way we provide a novel approach for classifying a given channel according to the practical applicability of the corresponding oracle-based stegosystem.

2   Preliminaries and Basic Concepts

Let Σ be a finite alphabet of the message space, let Σ^n denote the set of all sequences over Σ of lengths n, $\Sigma^\star$ denote the set of all finite sequences over Σ, and finally let Σ^∞ be the set of infinite sequences over Σ. A *channel* $\mathcal{C}$ is a distribution on Σ^∞. For $x \in \Sigma^\star$ we denote by $|x|$ the length of x. For a channel $\mathcal{C}$ let $\mathcal{C}^n$ be the distribution of the initial blocks of length n, i.e. for every sequence s of length n we have $P_{\mathcal{C}^n}[s] = \sum_{y \in \Sigma^\infty} P_{\mathcal{C}}[sy]$, where $P_Q[x]$ denotes $\Pr[X = x]$ for a random variable X of the probability distribution Q. We base our similarity measure for probability distributions on the Kullback-Leibler distance.

Definition 1. *Let $\mathcal{P}$ and $\mathcal{Q}$ be probability distributions on the same probability space. The relative entropy or Kullback-Leibler distance between $\mathcal{P}$ and $\mathcal{Q}$ is defined by $D_{KL}(\mathcal{P}; \mathcal{Q}) = \sum_x P_{\mathcal{P}}[x] \log \frac{P_{\mathcal{P}}[x]}{P_{\mathcal{Q}}[x]}$, where by convention $0 \cdot \log 0/q = 0$ and $p \cdot \log p/0 = \infty$. We define $D(\mathcal{P}, \mathcal{Q}) = D_{KL}(\mathcal{P}; \mathcal{Q}) + D_{KL}(\mathcal{Q}; \mathcal{P})$ and say that $\mathcal{P}$ and $\mathcal{Q}$ are ϵ close if $D(\mathcal{P}, \mathcal{Q}) \leq \epsilon$.*

For a complexity point of view on oracle-based stegosystems it is helpful to relate the notions of channel and formal language.

Definition 2. *Let $\mathcal{C}$ be a channel and let $L \subseteq \Sigma^\star$ be a language. We say that $\mathcal{C}$ is L-consistent if for all $x \in \Sigma^\star$ and $n = 2|x| + 1$ the following properties hold: (1) if $x \in L$ then $P_{\mathcal{C}^n}[1^{|x|}0x] > 0$ and (2) if $x \notin L$ then $P_{\mathcal{C}^n}[1^{|x|}0x] = 0$.*

Similarly as in [1] we assume the existence of an oracle $M_{\mathcal{C}}$ that perfectly samples according to the distribution of the channel $\mathcal{C}$, which seems to be a reasonable assumption (for more detail see Section 3). For their realisation of secure steganography, Hopper et al. further assume the existence of a stronger oracle $M_{\mathcal{C}_h}$ for the conditional channel distribution $\mathcal{C}_h^b$. This oracle draws blocks of fixed length b conditioned on the history h of previously drawn blocks, so for every sequence s of length b we have $P_{\mathcal{C}_h^b}[s] = \Pr[hs$ is a prefix of $z \mid h$ is a prefix of $z]$ where z is drawn from $\mathcal{C}$. To allow for steganography in this channel, we will assume (similarly as in [1]) the minimum entropy constraint:

$$\forall h \text{ drawn from } \mathcal{C} : \ H_\infty(\mathcal{C}_h^b) > 1 \tag{1}$$

(we say, for short, that h is drawn from $\mathcal{C}$ if h is a prefix block of a string drawn from $\mathcal{C}$). A stegosystem is a pair of probabilistic algorithms (SE, SD). SE takes a key $K \in \{0,1\}^k$, a hiddentext $m \in \{0,1\}^\star$, a message history h, and an oracle $M_{\mathcal{C}_h}$ which samples blocks according to a channel distribution $\mathcal{C}_h^b$. $SE(K, m, h)$ returns a sequence of blocks[1] $c_1||c_2||...||c_l$ (the stegotext) from the support of $\mathcal{C}_h^{l \cdot b}$. SD takes a key K, a sequence of blocks $c_1||c_2||...||c_l$, a message history h, and an oracle $M_{\mathcal{C}_h}$, and returns a hiddentext m. There must be a polynomial q, with $q(k) > k$, such that SE and SD satisfy: $\forall m, |m| < q(k)$: $\Pr[SD(K, SE(K, m, h), h) = m] \geq 2/3$, where the randomisation is over any coin tosses of SE, SD, and $M_{\mathcal{C}_h}$.

Informally speaking, a stegosystem is secure against chosen-hiddentext attacks if no polynomial time adversary can tell whether Alice's message to Bob encodes any hiddentext at all, even one of the adversary's choice (for formal definition see [1]). The provably secure stegosystems presented in [1] use the oracle $M_{\mathcal{C}_h}$ in the following way:

Procedure RS^F:
Input: target bit x, iteration $count$, history h
$i := 0$
repeat $c := M_{\mathcal{C}_h}$; $i := i + 1$
until $F(c) = x$ or $i = count$
Output: c

where $F : \{0,1\}^b \rightarrow \{0,1\}$ is chosen from a pseudorandom function family $F_K(\cdot, \cdot)$ indexed by $k = |K|$ key bits which maps a d bit number and a b bit string to $\{0,1\}$. For the stegosystem HLA we assume that Alice and Bob initially share a secret key K and a synchronised d bit counter N. $Enc(m)$ and $Dec(m)$ denote encoding and decoding algorithms for error correcting codes.

Stegosystem HLA

Procedure SE:	**Procedure SD:**
Input: key K; message m, history h	**Input:** key K; stego-text c
Let $m = Enc(m)$	Parse c as $c_1||c_2|| \ldots ||c_l$
Parse m as $m_1||m_2|| \ldots ||m_l$	for $i = 1, \ldots, l$ do
for $i = 1, \ldots, l$ do	$\quad m_i := F_K(N, c_i)$
$\quad c_i := RS^{F_K(N, \cdot)}(m_i, 2, h)$	$\quad N := N + 1$
$\quad h := h||c_i$; $N := N + 1$	$m := m_1||m_2|| \ldots ||m_l$
Output: $c_1||c_2|| \ldots ||c_l$	**Output:** $Dec(m)$

---

[1] In this paper we use both notations $u||v$ and uv for string concatenation.

Theorem 1 (Hopper, Langford, and von Ahn [1]). *If $F_K(\cdot, \cdot)$ is pseudorandom then the stegosystem HLA is secure against chosen-hiddentext attacks.*

Note that the stegosystem HLA is efficient if the rejection sampling procedure works in polynomial time.

3   The Complexity of Sampling

In this section we give definitions for the efficient sampling of the unconditional and conditional channel distributions $\mathcal{C}$ and $\mathcal{C}_h^b$. We say that a randomised Turing machine $M_{\mathcal{C}}$ samples according to the channel distribution $\mathcal{C}$ if for every positive integer n, $M_{\mathcal{C}}$ starting with input 1^n outputs sequences of length n according to the distribution $\mathcal{C}$ conditioned on the length n, i.e., if for every $s = s_1 s_2 \ldots s_n \in \Sigma^n$ it is true $\Pr[M_{\mathcal{C}}(1^n) = s] = P_{\mathcal{C}^n}[s]$, where $M_{\mathcal{C}}(z)$ denotes a random variable defined as the output of the Turing machine M working on the input z.

Definition 3 (Efficient Sampling). *We say that $\mathcal{C}$ can be sampled in time T and space S if there exists a randomised Turing machine M sampling $\mathcal{C}$ simultaneously in time T and space S, i.e. if for all n every computation path of M on 1^n is no longer than $T(n)$ and it uses no more than $S(n)$ space. Denote the class of all such channels by $\mathrm{TiSp}(T, S)$ and, for short, let $\mathrm{TiSp}(\mathrm{pol}, S)$ be the sum of $\mathrm{TiSp}(p, S)$ over all polynomials p. We say that $\mathcal{C}$ can be sampled efficiently if $\mathcal{C} \in \mathrm{TiSp}(p, p)$ for some polynomial p.*

Let $\mathcal{C}$ be a channel distribution and let $M_{\mathcal{C}}$ be an oracle which samples histories according to $\mathcal{C}$, i.e., on input 1^n, $M_{\mathcal{C}}$ draws h from $\mathcal{C}^n$. We say that a randomised algorithm M with access to $M_{\mathcal{C}}$ samples according to $\mathcal{C}_h$ if for every history $h \in \Sigma^\star$ drawn from $\mathcal{C}$ and for every positive integer b, M starting with input (h, b) and perhaps querying $M_{\mathcal{C}}$ generates a sequence of b symbols $s = s_1 s_2 \ldots s_b$ such that for every s it is true $\Pr[M(h, b) = s] = P_{\mathcal{C}_h^b}[s]$.

Definition 4 (Efficient Sampling for Conditional Distributions). *We say that $\mathcal{C}_h^b$ can be efficiently sampled if there exists a randomised Turing machine $M_{\mathcal{C}_h}$ with access to $M_{\mathcal{C}}$, sampling $\mathcal{C}_h$ in worst case polynomial time, i.e., if there exists a polynomial p such that every computation path of $M_{\mathcal{C}_h}$ on (h, b) is no longer than $p(|h| + b)$. In this model we charge oracle queries with unit costs.*

The existence of an oracle $M_{\mathcal{C}_h}$ sampling efficiently according to the conditional channel distribution $\mathcal{C}_h^b$ implies that there is also an efficient oracle $M_{\mathcal{C}}$ which samples according to $\mathcal{C}$. One of the main results of our paper says that the opposite implication does not hold in general.

Theorem 2. *There exist channels $\mathcal{C}$ that can be efficiently sampled by an oracle $M_{\mathcal{C}}$, but for which it is impossible to construct an oracle $M_{\mathcal{C}_h}$ that efficiently samples the channel $\mathcal{C}_h^b$, unless $\mathrm{P} = \mathrm{NP}$.*

Thus, any oracle-based stegosystem for such channels, and particularly the stegosystem HLA, cannot be implemented efficiently, unless $\mathrm{P} = \mathrm{NP}$. In the next section we prove the theorem using a natural channel $\mathcal{C}$.

4   The Intractability of Oracle-Based Steganography

Imagine some natural communication channel C, e.g., an internet chat room which is monitored by Eve. Alice and Bob want to chat using provably secure stegosystems to embed hidden messages into an innocent looking conversation. It is a realistic assumption that the messages exchanged during the cover conversation are structured in a certain way and belong to some specific language. Let us assume that the chat room allows communication in a language L which is the intersection of a small set of context free languages. Note that a real world conversation would have to be more complex to convince Eve.

Let us assume further that Alice possesses an efficient conditional oracle M_{C_h} which samples conditionally according to the distribution of the channel C_h described by the chat room and L. To secretly transmit a message m to Bob, Alice iteratively samples M_{C_h} on input m as described in Section 2 to obtain an unsuspicious cover message which Bob can easily decode to m.

We will show that even with slightly structured languages L the efficiency of M_{C_h} is not guaranteed. In fact we will give an example of L being the intersection of only three simple context free languages such that M_{C_h} can sample efficiently only if the widely believed assumption of P $\neq$ NP fails. Consider the following *Intersected-CFL-Prefix* problem (ICFLP, for short). Let the parameters of the problem be context free grammars $\mathcal{G}_1, \ldots, \mathcal{G}_g$ with $\mathcal{G}_i = (\Sigma, \Sigma_i, \sigma_i^0, \Pi_i)$ over a finite terminal alphabet Σ, variables Σ_i, start variable σ_i^0, and productions Π_i. Then for a given string $x = x_1 \ldots x_m$ over the finite alphabet Σ, 1^n with $n > m$ decide whether there is a string y which contains x as a prefix such that $|y| = n$ and $y \in L = L(\mathcal{G}_1) \cap \ldots \cap L(\mathcal{G}_g)$.

Lemma 1. *There are context free grammars* $\mathcal{G}_1, \mathcal{G}_2, \mathcal{G}_3$ *such that ICFLP is NP-complete.*

Due to space limit we omit this and also some other proofs in the paper.

Now let C be a channel which is consistent with the language $L = L_1 \cap L_2 \cap L_3$, with $L_i = L(\mathcal{G}_i)$ for $\mathcal{G}_i$ satisfying Lemma 1. One can additionally assume that C fulfills the minimum entropy constraint (1) and that it can be sampled efficiently by a probabilistic Turing machine M_C. One can construct such M_C in a similar way as in the proof of Theorem 3 but because of space limit we skip the description here.

To prove Theorem 2 assume there exists an efficient sampler N for the channel C_h^b working in polynomial time p. We show that using N we can construct a deterministic algorithm A solving the ICFLP problem in polynomial time. Let $x \in \Sigma^m$ and 1^n with $n > m$ be a given input. Initially A generates the string $h_0 = 1^n 0x$ (w.l.o.g. we assume that x is encoded in the channel C just as x). Then simulating the sampler N, algorithm A iteratively computes $h_j = h_{j-1} || N(h_{j-1}, b)$ for $j = 1, 2, \ldots, \lceil (n-m)/b \rceil$ such that every random choice $r \in_R \{0, 1\}$ of N is replaced by an assignment $r := 0$ and for every j at most $p(|h_{j-1}| + b)$ steps of N are simulated. If N does not stop after $p(|h_{j-1}| + b)$ steps for some j then A rejects x and halts the computation. Otherwise, let $h_{\lceil (n-m)/b \rceil} = 1^k 0xyz$ with

$|xy| = n$. The input will be accepted if $xy \in L$ and rejected otherwise. This completes the proof of Theorem 2.

5 Channels with Hard Conditional Sampling

In the present section we will analyse how the gap between the complexity of computing $M_\mathcal{C}$ or $M_{\mathcal{C}_h}$ is caused. Simply speaking, it results from the algorithmic structure of the channel $\mathcal{C}$. If L is a language, then in certain cases it may be much easier to compute a random word from L than it is to complete a given one. This phenomenon is well known in formal language theory [8]. As a consequence, conditionally sampling a channel $\mathcal{C}$, which is consistent with L, may be harder than sampling it without a given history.

To show the following theorem we apply the theory of NP-completeness, in particular the NP-complete problem 3SAT, to state the existence of hard conditional channels $\mathcal{C}_h$ for a large number of tractable channels $\mathcal{C}$. We refer to Garey and Johnson [7] for a detailed introduction.

Theorem 3. *Let $S : \mathbb{R}^+ \to \mathbb{R}$ be an increasing function such that $\log x \leq S(x) \leq x$ for every $x \geq 1$ that is space constructible in polynomial time. Moreover, let $\hat{S}$ be the inverse function of S, i.e. $\hat{S}(S(x)) = x$ for all $x \in \mathbb{R}^+$. Then there exist channels $\mathcal{C} \in \mathrm{TiSp}(\mathrm{pol}, S)$ fulfilling the minimum entropy constraint (1) such that every conditional distribution $\tilde{\mathcal{C}}_h^b$, which is δ close to $\mathcal{C}_h^b$ for some constant $\delta \geq 0$, cannot be sampled efficiently, unless the 3SAT problem can be solved by a deterministic algorithm in time $T(m) = (\hat{S}(m))^{O(1)}$, where m is the number of variables of the input 3CNF formula.*

From the theorem follows immediately:

Corollary 1.

1. *There exists $\mathcal{C} \in \mathrm{TiSp}(\mathrm{pol}, \log^2 n)$ such that any $\tilde{\mathcal{C}}_h^b$ which is δ close to $\mathcal{C}_h^b$ for some $\delta \geq 0$ cannot be sampled efficiently unless the 3SAT problem can be solved by a deterministic algorithm in time $T(m) = 2^{O(\sqrt{m})}$, where m is the number of variables of input 3CNF formulas.*
2. *There exists $\mathcal{C} \in \mathrm{TiSp}(\mathrm{pol}, 2^{\sqrt{\log n}})$ such that any $\tilde{\mathcal{C}}_h^b$ which is δ close to $\mathcal{C}_h^b$ for some $\delta \geq 0$ cannot be sampled efficiently unless $\mathrm{NP} \subseteq \mathrm{DTime}(n^{O(\log n)})$.*
3. *For every $c > 0$, there exists $\mathcal{C} \in \mathrm{TiSp}(\mathrm{pol}, n^c)$ such that any $\tilde{\mathcal{C}}_h^b$ which is δ close to $\mathcal{C}_h^b$ for some $\delta \geq 0$ cannot be sampled efficiently unless $\mathrm{P} = \mathrm{NP}$.*

It is clear that the three implications are decreasingly likely and that even implication 1 is far away from what is possible today. The best exact algorithm for 3SAT, by Iwama and Tamaki [9], runs in time $O(1.324^m)$ with respect to the number m of variables.

Proof (of Theorem 3). We construct a channel $\mathcal{C}$ over Σ which encodes instances of the 3SAT problem. Moreover, we assume that there are some fixed efficiently computable encodings $\mathcal{F}_m$ and $\mathcal{E}_m$ over Σ for 3CNF formulas of m variables and respectively for assignments $(b_1, b_2, \ldots, b_m)$ such that they guarantee the minimum entropy constraint (1) for $\mathcal{C}$. To fulfill this constraint we assume that for every 3CNF formula φ over $\{x_1, \overline{x}_1, \ldots, x_m, \overline{x}_m\}$, $\mathcal{F}_m(\varphi)$ is a set of code words over Σ such that for every word vw in $\mathcal{F}_m(\varphi)$ with $|w| \geq b$, the cardinality of the set $\{v : |v| = b$ and uv is the prefix of some code word in $\mathcal{F}_m(\varphi)\}$ is at least four. Similarly, $\mathcal{E}_m(b_1, \ldots, b_m)$ is a set of code words of equal length, say $d_{\mathcal{E}_m}$, such that for every prefix u of some word in $\mathcal{E}_m(b_1, \ldots, b_m)$ with $|u| \leq d_{\mathcal{E}_m} - b$, the cardinality of $\{v : |v| = b$ and uv is a prefix of some code word in $\mathcal{E}_m(b_1, \ldots, b_m)\}$ is at least four. Additionally, let ξ be a string over Σ such that ξ does not occur as a substring in any code word of $\mathcal{F}_m(\varphi)$ and $\mathcal{E}_m(b_1, \ldots, b_m)$ for all m. Thus, we get that for any $u \in \mathcal{F}_m(\varphi)$ and $v \in \mathcal{E}_m(b_1, \ldots, b_m)$ one detects uniquely in the concatenation $u\xi v$ the boundary between these two code words. Using these encodings we will construct the channel $\mathcal{C}$ having the following properties:

(i) For every $w \in \Sigma^\infty$ with $\Pr_{\mathcal{C}}[w] > 0$ and $w = 1^k 0z$, $k \geq 0$, $m = \lceil S(k) \rceil$, there exists a partition $z = z_1 z_2 z_3 z_4$ such that $z_1 \in \mathcal{F}_m(\varphi)$ for some satisfiable 3CNF formula φ, $z_2 = \xi$, $z_3 \in \mathcal{E}_m(b_1, \ldots, b_m)$ for some satisfying assignment $b_1, \ldots, b_m$ for φ, and z_4 is an arbitrary infinite string over Σ.

(ii) For every satisfiable 3CNF formula φ over m variables and for every satisfying assignment $b_1, \ldots, b_m$ of φ, for all k with $m = \lceil S(k) \rceil$, and for every $z = z_1 z_2 z_3$ with $z_1 \in \mathcal{F}_m(\varphi)$, $z_2 = \xi$ and $z_3 \in \mathcal{E}_m(b_1, \ldots, b_m)$, we have $\Pr_{\mathcal{C}^\ell}[1^k 0z] > 0$, where $\ell = |1^k 0z|$.

We define the channel $\mathcal{C}$ by giving a description of the sampler M of $\mathcal{C}$. For any integer $n \geq 1$ and the input 1^n the machine M works as follows.

1. Choose a positive integer k with $1 \leq k \leq n$ and with the probability distribution $\Pr[k] = 1/2^k$ for $k < n$ and $\Pr[n] = 1/2^{n-1}$, return the string $1^k 0$ and set the current length of the output $\ell := k + 1$.

2. Compute $m = \lceil S(k) \rceil$.

3. Choose independently uniformly at random an assignment $b_i \in_R \{0, 1\}$ for $i = 1, 2, \ldots, m$ and store the vector.

4. Choose independently uniformly at random three literals $L_1, L_2, L_3 \in_R \{x_1, \overline{x}_1, x_2, \overline{x}_2, \ldots, x_m, \overline{x}_m\}$. Let $\psi := L_1 \vee L_2 \vee L_3$. If $\psi(b_1, \ldots, b_m) = 0$ then assign to ψ a tautology (e.g. let $\psi := x_1 \vee \overline{x}_1 \vee \overline{x}_1$).

5. Choose randomly an encoding word $u \in \mathcal{E}_m(\psi)$. If $\ell + |u| \geq n$ then return the prefix of u of length $n - \ell$ and exit. Otherwise, return u, set $\ell := \ell + |u|$, and choose randomly $r \in_R \{0, 1\}$. If $r = 0$ then go to 4 else go to the next step.

6. Choose randomly an encoding word for the assignment $v \in \mathcal{F}_m(b_1, \ldots, b_m)$. If $\ell + |\xi v| \geq n$ then return the prefix of ξv of the length $n - \ell$ and exit. Otherwise, return ξv and set $\ell := \ell + |\xi v|$.

7. For $i = \ell + 1$ to n do: choose randomly a symbol $\sigma \in_R \Sigma$ and return σ.

If S is an efficiently constructible function, then M works in space $S(n)$ and in polynomial time. Hence for $\mathcal{C}$ sampled by M we have $\mathcal{C} \in \mathrm{TiSp}(\mathrm{pol}, S)$. Moreover, $\mathcal{C}$ fulfills the minimum entropy constraint (1).

Now, assume that the conditional distribution $\tilde{\mathcal{C}}_h^b$ which is δ close to $\mathcal{C}_h^b$ can be sampled efficiently and let N be a randomised Turing machine which samples according to $\tilde{\mathcal{C}}_h^b$ in polynomial time p. We show that using N we can construct a deterministic algorithm A which for a given 3CNF formula φ over $\{x_1, \overline{x}_1, \ldots, x_m, \overline{x}_m\}$ decides in time $T(m) = (\hat{S}(m))^{O(1)}$ whether φ is satisfiable or not.

The algorithm A initially computes an integer k, with $\lceil S(k) \rceil = m$. This can be done in polynomial time with respect to k, since S is efficiently constructible. Then A encodes φ over Σ choosing an arbitrary code word $u \in \mathcal{F}_m(\varphi)$ and generates the string $h_0 = 1^k 0 u \xi$. Recall that $d_{\mathcal{E}_m}$ denotes the length of code words in $\mathcal{E}_m$. Simulating the conditional sampler N, algorithm A computes iteratively $h_j = h_{j-1} \| N(h_{j-1}, b)$ for $j = 1, 2, \ldots, \lceil d_{\mathcal{E}_m}/b \rceil$ in such a way that every random choice $r \in_R \{0, 1\}$ of N is replaced by an assignment $r := 0$ and for every j at most $p(|h_{j-1}| + b)$ steps of N are performed. If N does not stop after $p(|h_{j-1}| + b)$ steps for some j then A rejects φ and halts the computation. Otherwise, let $h_{\lceil d_{\mathcal{E}_m}/b \rceil} = 1^k 0 z_1 z_2 z_3 z_4$ with $z_1 = u$, $z_2 = \xi$, z_3 a string of the length $d_{\mathcal{E}_m}$, and z_4 an arbitrary suffix over Σ. The formula φ will be rejected if z_3 does not encode any assignment in $\mathcal{E}_m$. If $z_3 \in \mathcal{E}_m(b_1, \ldots, b_m)$ for some assignment $(b_1, \ldots, b_m)$ then accept φ if $\varphi(b_1, \ldots, b_m) = 1$ and reject otherwise. The correctness of A follows directly from the properties (i) and (ii) of the channel $\mathcal{C}$. It is also easy to check that A works in time $(\hat{S}(m))^{O(1)}$. $\qquad \square$

As we applied 3SAT in the above proof it is also possible to define encodings $\mathcal{F}^A$ and $\mathcal{E}^A$ for any NP-complete problem A such that $\mathcal{F}^A$ encodes instances of A and $\mathcal{E}^A$ witnesses. Furthermore one can easily assure that a channel which is consistent to the set of strings encoded by $\mathcal{F}^A$ and $\mathcal{E}^A$ fulfills the minimum entropy constraint (1). Consequently for any NP-complete problem there are corresponding channels $\mathcal{C}$ with intractable oracles $M_{\mathcal{C}_h}$.

Corollary 2. *Let A be an NP-complete problem. Then there are redundant encodings $\mathcal{F}^A$ and $\mathcal{E}^A$ over Σ for the instances of A and the witnesses and a channel $\mathcal{C}$ over Σ which is consistent to $\{1^m 0 z_1 z_2 z_3 z_4 \in \Sigma^\infty | z_1 \in \mathcal{F}_m^A(x)$ for some $x \in A$ and $|x| = m, z_2 = \xi, z_3 \in \mathcal{E}_m^A(x), z_4 \in \Sigma^\infty\}$ and which fulfills the minimum entropy constraint (1) such that the distribution $\mathcal{C}$ can be sampled efficiently and the conditional distribution $\mathcal{C}_h^b$ cannot be sampled efficiently unless $P = NP$.*

The proof of Corollary 2 is analogous to Theorem 3. By the above results, the existence of an efficiently sampleable channel $\mathcal{C}_h$ becomes unlikely whenever the channel $\mathcal{C}$ has a certain structural complexity. It is remarkable that even channels with $\log^2$-space oracles $M_{\mathcal{C}}$ may already have intractable oracles $M_{\mathcal{C}_h}$.

6   Feasible Conditional Sampling

Having characterised channels $\mathcal{C}$ with feasible oracles $M_{\mathcal{C}}$ but hard $M_{\mathcal{C}_h}$, we will now establish constraints on $\mathcal{C}$ to assure an efficient oracle $M_{\mathcal{C}_h}$. We follow two approaches, namely sampling in logarithmic space and context free languages.

Whereas it is likely that it is not possible to sample $\mathcal{C}_h$ efficiently in case $\mathcal{C} \in \mathrm{TiSp}(\mathrm{pol}, \omega(\log))$ by Theorem 3, it becomes possible if $\mathcal{C} \in \mathrm{TiSp}(\mathrm{pol}, \log)$.

In this case there is a probabilistic Turing machine N sampling according to a conditional distribution $\tilde{C}_h^b$ which can be arbitrarily close to C_h^b. The slight difference between the distribution $\tilde{C}_h^b$ generated by N and C_h^b does not result from the computational complexity of C, as in the case when $C \in \text{TiSp}(\text{pol}, \omega(\log))$, but from the insufficient power of N to generate randomness. Equipped with a more powerful random generator than coin flipping, N would meet C_h^b exactly.

Theorem 4. *For every channel $C \in \text{TiSp}(\text{pol}, \log)$ and for all $0 < \epsilon < 1$ there is a probabilistic polynomial time Turing machine N which samples according to the conditional distribution $\tilde{C}_h^b$ which is ϵ close to C_h^b.*

In the previous section we assumed that channels C encode words of certain languages, like for example 3SAT. We classified channels according to the complexity of the encoded languages and observed that especially if $C \in \text{TiSp}(\text{pol}, \log)$ then C_h^b becomes tractable. Now we restrict ourselves to channels which encode context free languages. This family of languages is decidable in polynomial time by the CYK algorithm.

Let L be a CFL and $\mathcal{G} = (\Sigma, \Sigma_N, \sigma_0, \Pi)$ a context free grammar for L. We will define the channel C by giving an efficient sampler M_L. W.l.o.g. we assume that $\mathcal{G}$ is in Greibach Normal Form (GNF) and every variable in Σ_N is generating. On input 1^n the probabilistic machine M_L works as follows. Machine M_L chooses a positive integer k with the probability distribution: $\Pr[k] = \frac{1}{2^k}$ if $1 \leq k < n$ and $\Pr[k] = \frac{1}{2^{n-1}}$ for $k = n$. Then M_L computes the $k \times |\Sigma_N|$ matrix A with $A[i,j]$ containing the set of productions π of the form $\sigma_j \vdash s$ with

1. $s = a, a \in \Sigma$, if $i = 1$,
2. $s = a\sigma_u, a \in \Sigma, \sigma_u \in \Sigma_N$, if $A[i-1, u] \neq \emptyset$ or
3. $s = a\sigma_u\sigma_v, a \in \Sigma, \sigma_u, \sigma_v \in \Sigma_N$, if there is $1 \leq i' \leq i - 2$ with $A[i', u] \neq \emptyset$ and $A[i - i' - 1, v] \neq \emptyset$.

Notice that A can be computed in polynomial time with respect to k. If the entry $A[k, 0] = \emptyset$ then L contains no string of length k and in that case M_L writes $0z$ to its output tape, where z denotes a random string in Σ^{n-1} and then M_L halts. If L contains strings of length k, M_L generates such a string randomly by the help of A. For that M_L holds a stack containing in each cell a pair of integers $(i, j), 1 \leq i \leq k, 0 \leq j < |\Sigma_N|$ where each stack element indicates that a substring of length i has to be deduced from σ_j. M_L initially sets $h = \lambda$, pushes $(k, 0)$ on the stack and starts working iteratively on the stack until it is empty. In each iteration M_L takes the top element (i, j) from the stack and generates a list B of all tuples $(a, x, y, u, v), a \in \Sigma, 1 \leq x, y < i - 1, 1 \leq u, v < |\Sigma_N|$ such that

1. there is a production π in $A[i, j]$ of the form $\sigma_j \vdash a\sigma_u\sigma_v$. Thereby $v = -1$ indicates that π is actually of the form $\sigma_j \vdash a\sigma_u$ and if additionally $u = -1$ then $\pi = \sigma_j \vdash a$ and
2. if $v \neq -1$ then $A[u, x] \neq \emptyset$, $A[v, y] \neq \emptyset$, and $x + y = i - 1$, if only $u \neq -1$ then $A[u, x] \neq \emptyset$ and $x = i - 1$ and if $u = v = -1$ then i must be one.

Then M_L randomly chooses one tuple (a, x, y, u, v) in B by tossing a polynomial number of coins, adds a to h, pushes (y, v) on top of the stack if $v \neq -1$, and subsequently pushes (x, u) if $u \neq -1$. When the stack is empty, machine M_L returns the n-symbol prefix of $1^k 0hz$, where z denotes a random string in Σ^∞.

Machine M_L works in polynomial time since A can be constructed efficiently, the iteration stops after k steps and each iteration step takes at most a polynomial number of coin tosses.

Theorem 5. *For every context free language L and the channel $\mathcal{C}$ which is described by M_L there is a probabilistic polynomial time Turing machine N which samples according to the conditional distribution $\tilde{\mathcal{C}}_h^b$ that is ε close to $\mathcal{C}_h^b$ for arbitrary $\varepsilon > 0$.*

7 Conclusions and Future Work

In this paper we analysed the complexity of conditional channels $\mathcal{C}_h$ as used in rejection-sampling based steganography. Our main question was how efficiently a sampling oracle $M_{\mathcal{C}_h}$ for $\mathcal{C}_h$ can be constructed from an oracle $M_{\mathcal{C}}$ that efficiently samples the unconditional channel $\mathcal{C}$. We showed that there are channels for which such an efficient construction is impossible and the ability to perform such a construction is dependent on the type of channel. However, there remain some more problems to be solved in future work. We do not yet have precise characterisations of the channels in which conditional sampling is hard and those in which it is feasible. Such a characterisation will be needed to further investigate the feasibility of practical oracle-based stegosystems.

References

1. Hopper, N.J., Langford, J., von Ahn, L.: Provably secure steganography. In: Advances in Cryptology - CRYPTO 2002. LNCS(2002) Vol. 2442, 77–92.
2. Anderson, R.J., Petitcolas, F.A.P.: On the limits of steganography. IEEE Journal of Selected Areas of Communications **16**(4) (1998) 474–481.
3. von Ahn, L., Hopper, N.J.: Public-key steganography. In: Advances in Cryptology - EUROCRYPT 2004. LNCS(2004) Vol. 3027, 323–341.
4. Le, T.V., Kurosawa, K.: Efficient public key steganography secure against adaptively chosen stegotext attacks. Technical Report 2003/244, IACR Archive (2003).
5. Backes, M., Cachin, C.: Public-key steganography with active attacks. In: Theory of Cryptography Conference. LNCS(2005) Vol. 3378, 210–226.
6. Lysyanskaya, A., Meyerovich, M.: Provably secure steganography with imperfect sampling. In: Public Key Cryptography. LNCS(2006) Vol. 3958, 123–139.
7. Garey, M.R., Johnson, D.S.: Computers and Intractability; A Guide to the Theory of NP-Completeness. W. H. Freeman & Co., New York (1979).
8. Hopcroft, J.E., Motwani, R., Ullman, J.D.: Introduction to Automata Theory, Languages, and Computation. Addison-Wesley, Reading MA (2000).
9. Iwama, K., Tamaki, S.: Improved upper bounds for 3-sat. In: Proc. ACM-SIAM Symposium on Discrete algorithms - ACM 2004, SIAM(2004), 328–328.

Author Index

Lecture Notes in Computer Science

For information about Vols. 1–4238

please contact your bookseller or Springer

Vol. 4278: R. Meersman, Z. Tari, P. Herrero (Eds.), On the Move to Meaningful Internet Systems 2006: OTM 2006 Workshops, Part II. XLV, 1004 pages. 2006.

Vol. 4277: R. Meersman, Z. Tari, P. Herrero (Eds.), On the Move to Meaningful Internet Systems 2006: OTM 2006 Workshops, Part I. XLV, 1009 pages. 2006.

Vol. 4276: R. Meersman, Z. Tari (Eds.), On the Move to Meaningful Internet Systems 2006: CoopIS, DOA, GADA, and ODBASE, Part II. XXXII, 752 pages. 2006.

Vol. 4275: R. Meersman, Z. Tari (Eds.), On the Move to Meaningful Internet Systems 2006: CoopIS, DOA, GADA, and ODBASE, Part I. XXXI, 1115 pages. 2006.

Vol. 4274: Q. Huo, B. Ma, E.-S. Chng, H. Li (Eds.), Chinese Spoken Language Processing. XXIV, 805 pages. 2006. (Sublibrary LNAI).

Vol. 4273: I. Cruz, S. Decker, D. Allemang, C. Preist, D. Schwabe, P. Mika, M. Uschold, L. Aroyo (Eds.), The Semantic Web - ISWC 2006. XXIV, 1001 pages. 2006.

Vol. 4272: P. Havinga, M. Lijding, N. Meratnia, M. Wegdam (Eds.), Smart Sensing and Context. XI, 267 pages. 2006.

Vol. 4271: F.V. Fomin (Ed.), Graph-Theoretic Concepts in Computer Science. XIII, 358 pages. 2006.

Vol. 4270: H. Zha, Z. Pan, H. Thwaites, A.C. Addison, M. Forte (Eds.), Interactive Technologies and Sociotechnical Systems. XVI, 547 pages. 2006.

Vol. 4269: R. State, S. van der Meer, D. O'Sullivan, T. Pfeifer (Eds.), Large Scale Management of Distributed Systems. XIII, 282 pages. 2006.

Vol. 4268: G. Parr, D. Malone, M. Ó Foghlú (Eds.), Autonomic Principles of IP Operations and Management. XIII, 237 pages. 2006.

Vol. 4267: A. Helmy, B. Jennings, L. Murphy, T. Pfeifer (Eds.), Autonomic Management of Mobile Multimedia Services. XIII, 257 pages. 2006.

Vol. 4266: H. Yoshiura, K. Sakurai, K. Rannenberg, Y. Murayama, S. Kawamura (Eds.), Advances in Information and Computer Security. XIII, 438 pages. 2006.

Vol. 4265: L. Todorovski, N. Lavrač, K.P. Jantke (Eds.), Discovery Science. XIV, 384 pages. 2006. (Sublibrary LNAI).

Vol. 4264: J.L. Balcázar, P.M. Long, F. Stephan (Eds.), Algorithmic Learning Theory. XIII, 393 pages. 2006. (Sublibrary LNAI).

Vol. 4263: A. Levi, E. Savaş, H. Yenigün, S. Balcısoy, Y. Saygın (Eds.), Computer and Information Sciences – ISCIS 2006. XXIII, 1084 pages. 2006.

Vol. 4262: K. K. Havelund, M. Núñez, G. Roşu, B. Wolff, (Eds.), Formal Approaches to Software Testing and Runtime Verification. VIII, 255 pages. 2006.

Vol. 4261: Y. Zhuang, S. Yang, Y. Rui, Q. He (Eds.), Advances in Multimedia Information Processing - PCM 2006. XXII, 1040 pages. 2006.

Vol. 4260: Z. Liu, J. He (Eds.), Formal Methods and Software Engineering. XII, 778 pages. 2006.

Vol. 4259: S. Greco, Y. Hata, S. Hirano, M. Inuiguchi, S. Miyamoto, H.S. Nguyen, R. Słowiński (Eds.), Rough Sets and Current Trends in Computing. XXII, 951 pages. 2006. (Sublibrary LNAI).

Vol. 4257: I. Richardson, P. Runeson, R. Messnarz (Eds.), Software Process Improvement. XI, 219 pages. 2006.

Vol. 4256: L. Feng, G. Wang, C. Zeng, R. Huang (Eds.), Web Information Systems – WISE 2006 Workshops. XIV, 320 pages. 2006.

Vol. 4255: K. Aberer, Z. Peng, E.A. Rundensteiner, Y. Zhang, X. Li (Eds.), Web Information Systems – WISE 2006. XIV, 563 pages. 2006.

Vol. 4254: T. Grust, H. Höpfner, A. Illarramendi, S. Jablonski, M. Mesiti, S. Müller, P.-L. Patranjan, K.-U. Sattler, M. Spiliopoulou, J. Wijsen (Eds.), Current Trends in Database Technology – EDBT 2006. XXXI, 932 pages. 2006.

Vol. 4253: B. Gabrys, R.J. Howlett, L.C. Jain (Eds.), Knowledge-Based Intelligent Information and Engineering Systems, Part III. XXXII, 1301 pages. 2006. (Sublibrary LNAI).

Vol. 4252: B. Gabrys, R.J. Howlett, L.C. Jain (Eds.), Knowledge-Based Intelligent Information and Engineering Systems, Part II. XXXIII, 1335 pages. 2006. (Sublibrary LNAI).

Vol. 4251: B. Gabrys, R.J. Howlett, L.C. Jain (Eds.), Knowledge-Based Intelligent Information and Engineering Systems, Part I. LXVI, 1297 pages. 2006. (Sublibrary LNAI).

Vol. 4250: H.J. van den Herik, S.-C. Hsu, T.-s. Hsu, H.H.L.M. Donkers (Eds.), Advances in Computer Games. XIV, 273 pages. 2006.

Vol. 4249: L. Goubin, M. Matsui (Eds.), Cryptographic Hardware and Embedded Systems - CHES 2006. XII, 462 pages. 2006.

Vol. 4248: S. Staab, V. Svátek (Eds.), Managing Knowledge in a World of Networks. XIV, 400 pages. 2006. (Sublibrary LNAI).

Vol. 4247: T.-D. Wang, X. Li, S.-H. Chen, X. Wang, H. Abbass, H. Iba, G. Chen, X. Yao (Eds.), Simulated Evolution and Learning. XXI, 940 pages. 2006.

Vol. 4246: M. Hermann, A. Voronkov (Eds.), Logic for Programming, Artificial Intelligence, and Reasoning. XIII, 588 pages. 2006. (Sublibrary LNAI).

Vol. 4245: A. Kuba, L.G. Nyúl, K. Palágyi (Eds.), Discrete Geometry for Computer Imagery. XIII, 688 pages. 2006.

Vol. 4244: S. Spaccapietra (Ed.), Journal on Data Semantics VII. XI, 267 pages. 2006.

Vol. 4243: T. Yakhno, E.J. Neuhold (Eds.), Advances in Information Systems. XIII, 420 pages. 2006.

Vol. 4242: A. Rashid, M. Aksit (Eds.), Transactions on Aspect-Oriented Software Development II. IX, 289 pages. 2006.

Vol. 4241: R.R. Beichel, M. Sonka (Eds.), Computer Vision Approaches to Medical Image Analysis. XI, 262 pages. 2006.

Vol. 4239: H.Y. Youn, M. Kim, H. Morikawa (Eds.), Ubiquitous Computing Systems. XVI, 548 pages. 2006.